2ND EDITION

Veterinary Assisting
FUNDAMENTALS & APPLICATIONS

2ND EDITION

Veterinary Assisting
FUNDAMENTALS & APPLICATIONS

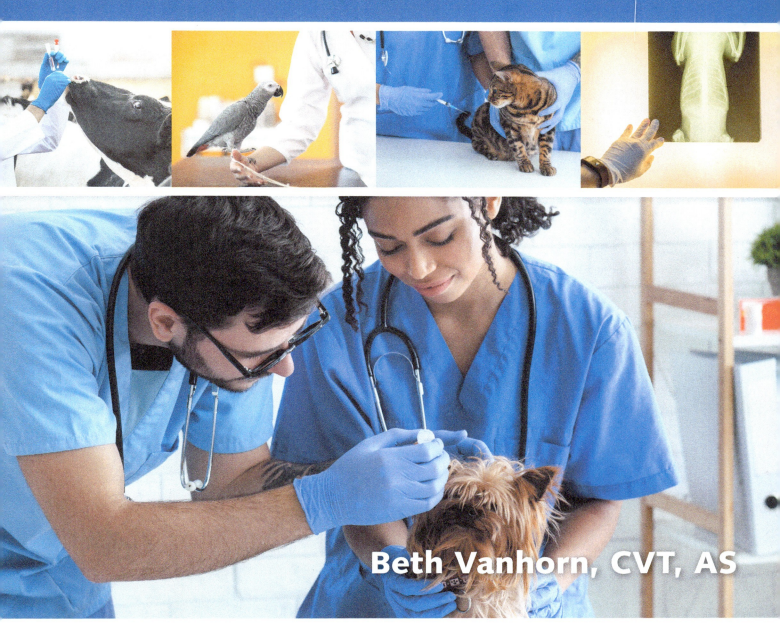

Beth Vanhorn, CVT, AS

Australia • Brazil • Canada • Mexico • Singapore • United Kingdom • United States

Veterinary Assisting Fundamentals and Applications, Second edition

Beth Vanhorn

SVP, Higher Education & Skills Product: Erin Joyner

VP, Higher Education & Skills Product: Michael Schenk

Product Director: Matthew Seeley

Product Manager: Lauren Whalen

Product Assistant: Dallas Wilkes

Learning Designer: Debbie Bordeaux

Senior Content Manager: Kara A. DiCaterino

Digital Delivery Lead: David O'Connor

Director, Marketing: Danaë April

Marketing Manager: Courtney Cozzy

IP Analyst: Ashley Maynard

IP Project Manager: Kelli Besse

Production Service: Lumina Datamatics, Inc.

Designer: Felicia Bennett

Cover Image Sources:

Jenoche/Shutterstock.com

iStock.com/LuckyBusiness

Yakov Oskanov/Shutterstock.com

Prostock-studio/ShutterStock.com

Prostock-studio/ShutterStock.com

Interior image Sources:

Jenoche/Shutterstock.com

iStock.com/LuckyBusiness

Yakov Oskanov/Shutterstock.com

Prostock-studio/ShutterStock.com

Prostock-studio/ShutterStock.com

© 2022, 2011 Cengage Learning, Inc. ALL RIGHTS RESERVED.

No part of this work covered by the copyright herein may be reproduced or distributed in any form or by any means, except as permitted by U.S. copyright law, without the prior written permission of the copyright owner.

> For product information and technology assistance, contact us at
> **Cengage Customer & Sales Support, 1-800-354-9706
> or support.cengage.com.**
>
> For permission to use material from this text or product,
> submit all requests online at **www.copyright.com.**

Library of Congress Control Number: 2020922393

ISBN: 978-1-305-49921-8

Cengage
200 Pier 4 Boulevard
Boston, MA 02210
USA

Cengage is a leading provider of customized learning solutions with employees residing in nearly 40 different countries and sales in more than 125 countries around the world. Find your local representative at **www.cengage.com.**

To learn more about Cengage platforms and services, register or access your online learning solution, or purchase materials for your course, visit **www.cengage.com.**

Notice to the Reader

Publisher does not warrant or guarantee any of the products described herein or perform any independent analysis in connection with any of the product information contained herein. Publisher does not assume, and expressly disclaims, any obligation to obtain and include information other than that provided to it by the manufacturer. The reader is expressly warned to consider and adopt all safety precautions that might be indicated by the activities described herein and to avoid all potential hazards. By following the instructions contained herein, the reader willingly assumes all risks in connection with such instructions. The publisher makes no representations or warranties of any kind, including but not limited to, the warranties of fitness for particular purpose or merchantability, nor are any such representations implied with respect to the material set forth herein, and the publisher takes no responsibility with respect to such material. The publisher shall not be liable for any special, consequential, or exemplary damages resulting, in whole or in part, from the readers' use of, or reliance upon, this material.

Printed in the United States of America
Print Number: 07   Print Year: 2022

*This book is dedicated to
all the veterinarians and staff members
who made working in and learning about
the veterinary industry worthwhile.*

Contents

Preface	xxiv
About the Authors	xxx
Acknowledgments	xxxi
Reviewers	xxxi

Section I ■ Practice Management and Client Relations

CHAPTER 1	**Veterinary Medical Terminology and Abbreviations**	**2**
	Objectives	2
	Introduction	2
	Veterinary Medical Terminology	2
	Learning to Dissect a Veterinary Term	2
	Putting It All Together	4
	Common Directional Terms	7
	Common Terms and Abbreviations Used in Veterinary Practice	8
	Spelling	11
	Summary	11
	Key Terms	12
	Review Questions	12
CHAPTER 2	**Medical Records**	**16**
	Objectives	16
	Introduction	16
	The Veterinary Medical Record	16
	Creating a Medical Record	17
	Invoicing	23
	Recording Information in the Medical Record	24
	Filing Medical Records	26
	Consent Forms and Certificates	26
	Medical Records as Legal Documents	28
	Computerized Veterinary Programs	30
	Summary	30
	Key Terms	30
	Review Questions	31

CHAPTER 3	**Scheduling Appointments and Computer Applications**	**35**
	Objectives	35
	Introduction	35
	The Veterinary Appointment Book	35
	Types of Appointment Schedules	36
	Types of Appointments	38
	Organization of the Appointment Book	39
	Scheduling Appointments	39
	Policies and Procedures	43
	Basic Computer Equipment	43
	Keyboarding Skills	45
	Summary	47
	Key Terms	47
	Review Questions	49
CHAPTER 4	**Veterinary Careers**	**54**
	Objectives	54
	Introduction	54
	The Veterinary Team	54
	Professional Dress and Appearance	58
	The Veterinary Facility	59
	Summary	63
	Key Terms	63
	Review Questions	65
CHAPTER 5	**Communication and Client Relations**	**68**
	Objectives	68
	Introduction	68
	The Communication Process	68
	Verbal Communication	68
	Nonverbal Communication	69
	Written Communication	70
	Appropriate Communication Skills	71
	Interacting with People	71
	Communicating with Difficult Clients	72
	Telephone Skills	73
	Grief and Communication	74
	Summary	74
	Key Terms	75
	Review Questions	76
CHAPTER 6	**Veterinary Ethics and Legal Issues**	**81**
	Objectives	81
	Introduction	81

	Veterinary Ethics	81
	Veterinary Laws and Veterinary Practice Acts	82
	Veterinary Common Law	83
	Federal Veterinary Laws	83
	Summary	85
	Key Terms	85
	Review Questions	86
CHAPTER 7	**Veterinary Safety and Aseptic Techniques**	**90**
	Objectives	90
	Introduction	90
	Safety in the Veterinary Facility	90
	OSHA Guidelines and Regulations	97
	Methods of Sanitation	101
	Veterinary Sanitation Chemicals and Cleaners	104
	Housekeeping and General Cleaning	107
	Exam Room Sanitation	110
	Aseptic Techniques	112
	Isolation Ward	113
	Summary	113
	Key Terms	114
	Review Questions	115

Section II ■ Veterinary Animal Production

CHAPTER 8	**Dog Breed Identification and Production Management**	**118**
	Objectives	118
	Introduction	118
	Veterinary Terminology	118
	Biology	119
	Breeds	119
	Breed Selection	119
	Nutrition	122
	Feeding Adult Dogs	123
	Behavior	124
	Basic Training	125
	Equipment and Housing Needs	127
	Restraint and Handling	127
	Grooming	133
	Basic Health Care and Maintenance	137
	Reproduction and Breeding	142
	Common Diseases	143
	Common Parasites and Prevention	144
	Common Canine Surgical Procedures	147

	Summary	148
	Key Terms	148
	Review Questions	152
CHAPTER 9	**Cat Breed Identification and Production Management**	**154**
	Objectives	154
	Introduction	154
	Veterinary Terminology	154
	Biology	155
	Breeds	157
	Breed Selection	157
	Nutrition	158
	Behavior	159
	Basic Training	160
	Equipment and Housing Needs	160
	Restraint and Handling	160
	Grooming	164
	Basic Health Preventative Care and Maintenance	165
	Vaccinations	166
	Reproduction and Breeding	166
	Common Diseases	167
	Common Parasites and Prevention	169
	Common Surgical Procedures	170
	Summary	171
	Key Terms	171
	Review Questions	173
CHAPTER 10	**Avian Breed Identification and Production Management**	**174**
	Objectives	174
	Introduction	174
	Veterinary Terminology	174
	Avian Biology	174
	Breeds	178
	Breed Selection	182
	Nutrition	182
	Behavior	183
	Equipment and Housing Needs	186
	Restraint and Handling	187
	Grooming	188
	Basic Health Care and Maintenance	189
	Vaccinations	190
	Fecal Exams	191
	Radiology	191
	Reproduction and Breeding	191
	Respiratory Emergencies	192

	Injuries and Fractures	192
	Common Diseases	192
	Common Parasites	193
	Common Surgical Procedures	193
	Summary	193
	Key Terms	193
	Review Questions	195
CHAPTER 11	**Pocket Pet Health and Production Management**	**197**
	Objectives	197
	Introduction	197
	Mice and Rats	197
	Hamsters	203
	Guinea Pigs	207
	Gerbils	211
	Ferrets	213
	Hedgehogs	217
	Chinchillas	220
	Summary	223
	Key Terms	223
	Review Questions	224
CHAPTER 12	**Rabbit Identification and Production Management**	**225**
	Objectives	225
	Introduction	225
	Veterinary Terminology	225
	Biology	226
	Breeds	226
	Breed Selection	227
	Nutrition	228
	Behavior	230
	Basic Training	230
	Equipment and Housing Needs	230
	Restraint and Handling	231
	Grooming	231
	Basic Health Care and Maintenance	232
	Vaccinations	233
	Reproduction and Breeding	233
	Common Diseases	234
	Common Parasites	235
	Common Surgical Procedures	235
	Summary	236
	Key Terms	236
	Review Questions	236

Contents　xi

CHAPTER 13	**Reptile and Amphibian Breed Identification and Production Management**	**238**
	Objectives	238
	Introduction	238
	Veterinary Terminology	238
	Biology of Reptiles	239
	Biology of Amphibians	242
	Reptile Breeds	243
	Reptile and Amphibian Breed Selection	246
	Nutrition	246
	Behavior	247
	Basic Training	248
	Equipment and Housing Needs	248
	Restraint and Handling	249
	Grooming	250
	Basic Health Care and Maintenance	251
	Vaccinations	251
	Reproduction and Breeding	251
	Common Diseases	253
	Common Parasites	253
	Common Surgical Procedures	253
	Summary	254
	Key Terms	254
	Review Questions	255
CHAPTER 14	**Ornamental Fish and Aquaculture Identification and Production Management**	**256**
	Objectives	256
	Introduction	256
	Veterinary Terminology	257
	Ornamental Fish Biology	257
	Ornamental Fish Breeds	258
	Breed Selection	261
	Nutrition	261
	Equipment Needs	261
	Water Sources and Quality	264
	Catching and Handling Ornamental Fish	265
	Basic Health Care and Maintenance	265
	Breeding and Reproduction in Fish	266
	Fish Disease and Health	266
	Aquaculture Industry	267
	Veterinary Terminology	268
	Aquatic Biology	268
	Aquaculture Production Systems	269
	Water Sources	271

	Water Quality	271
	Aquatic Reproduction and Breeding	272
	Breeds or Species	274
	Nutrition	275
	Summary	276
	Key Terms	276
	Review Questions	280
CHAPTER 15	**Zoo, Exotic, and Wildlife Identification and Production Management**	**282**
	Objectives	282
	Introduction	282
	The Exotic Animal	283
	Purposes and Uses of Exotics	283
	Exotic Animal Businesses	286
	Exhibition (Zoo) and Performance Animals	286
	Exotic Animal Care and Management	288
	The Importance of Wildlife	290
	Control of Wildlife	292
	The Role of Veterinary Medicine in Wildlife Management	292
	Classes of Wildlife	292
	Wildlife Health and Management Practices	294
	Capture and Restraint of Wildlife	295
	Wildlife Rehabilitation	295
	Summary	296
	Key Terms	297
	Review Questions	298
CHAPTER 16	**Laboratory and Animal Research Production Management**	**300**
	Objectives	300
	Introduction	300
	Animal Research	300
	Types of Research	301
	Animal Models and Laboratory Medicine	302
	Species of Animals Commonly Used in Research	303
	Research Regulations and Guidelines	304
	Lab Animal Management and Health Practices	305
	Lab Animal Careers	306
	Summary	307
	Key Terms	307
	Review Questions	308
CHAPTER 17	**Beef and Dairy Cattle Breed Identification and Production Management**	**310**
	Objectives	310
	Introduction	310
	Veterinary Terminology	310
	Biology	311

	Breeds	312
	Breed Selection	313
	Nutrition	315
	Behavior	316
	Basic Training	317
	Equipment and Housing Needs	317
	Restraint and Handling	319
	Grooming	321
	Basic Health Care and Maintenance	321
	Vaccinations	322
	Common Diseases	323
	Bovine Respiratory Disease Complex (BRDC)	323
	Common Parasites	325
	Common Surgical Procedures	325
	Reproduction and Breeding Beef Cattle	325
	Reproduction and Breeding of Dairy Cattle	326
	Summary	327
	Key Terms	328
	Review Questions	331
CHAPTER 18	**Equine and Draft Animal Breed Identification and Production Management**	**333**
	Objectives	333
	Introduction	333
	Veterinary Terminology	333
	Biology	335
	Breeds of Horses	337
	Nutrition	343
	Behavior	344
	Basic Training	345
	Equipment and Housing Needs	345
	Restraint and Handling	346
	Grooming	350
	Basic Health Care and Maintenance	351
	Vaccinations	354
	Reproduction and Breeding	355
	Common Diseases	358
	Common Parasites	360
	Summary	360
	Key Terms	361
	Review Questions	364
CHAPTER 19	**Swine Breed Identification and Production Management**	**365**
	Objectives	365
	Introduction	365
	Veterinary Terminology	365

	Biology	366
	Breeds	366
	Breed Selection	367
	Nutrition	368
	Behavior	369
	Basic Training	370
	Equipment and Housing Needs	370
	Restraint and Handling	371
	Grooming	372
	Basic Health Care and Maintenance	372
	Vaccinations	374
	Reproduction and Breeding	375
	Common Diseases	375
	Common Parasites	376
	Swine Production Industry	377
	Swine Production Methods and Systems	377
	Summary	378
	Key Terms	378
	Review Questions	379
CHAPTER 20	**Sheep Breed Identification and Production Management**	**381**
	Objectives	381
	Introduction	381
	Veterinary Terminology	381
	Biology	382
	Breeds	382
	Breed Selection	382
	Nutrition	383
	Behavior	384
	Equipment and Housing Needs	385
	Restraint and Handling	385
	Grooming	385
	Basic Health Care and Maintenance	385
	Vaccinations	386
	Breeding and Reproduction	386
	Common Diseases	387
	Sheep Production	389
	Summary	390
	Key Terms	390
	Review Questions	392
CHAPTER 21	**Goat Breed Identification and Production Management**	**393**
	Objectives	393
	Introduction	393

	Veterinary Terminology	393
	Biology	393
	Breeds	395
	Breed Selection	396
	Goat Nutrition	396
	Behavior	396
	Basic Training	396
	Equipment and Housing Needs	397
	Restraint and Handling	397
	Grooming	397
	Basic Health Care and Maintenance	398
	Vaccination	398
	Reproduction and Breeding	398
	Common Diseases	398
	Common Parasites	399
	Summary	399
	Key Terms	399
	Review Questions	400
CHAPTER 22	**Poultry Breed Identification and Production Management**	**401**
	Objectives	401
	Introduction	401
	Veterinary Terminology	401
	Biology	402
	Poultry Species and Classes	403
	Selection of Species or Breed	407
	Nutrition	407
	Behavior	407
	Basic Training	407
	Equipment and Housing Needs	407
	Restraint and Handling	408
	Grooming	409
	Basic Health Care and Maintenance	409
	Vaccinations	410
	Reproduction and Breeding	410
	Common Diseases	411
	Common Parasites	412
	Poultry Production	412
	Poultry Production Systems	412
	Summary	415
	Key Terms	416
	Review Questions	418

Section III — General Anatomy/Physiology and Disease Processes

CHAPTER 23	**The Structure of Living Things**	**422**
	Objectives	422
	Introduction	422
	Cells	422
	Clinical Concepts	423
	Cell Functions	425
	Clinical Concepts	425
	Genetics	428
	Clinical Concepts	429
	Tissues	429
	Epithelial Tissue	430
	Connective Tissue	432
	Muscle Tissue	432
	Clinical Concepts	433
	Clinical Concepts	433
	Organs	433
	Disease and Injury	434
	Summary	434
	Key Terms	434
	Review Questions	436
CHAPTER 24	**The Musculoskeletal System**	**438**
	Objectives	438
	Introduction	438
	Bones	438
	Joints	440
	Muscles and Connective Tissues	440
	Axial Skeleton	441
	Common Musculoskeletal Diseases and Conditions	444
	Muscular System	449
	Summary	449
	Key Terms	450
	Review Questions	453
CHAPTER 25	**The Digestive System**	**455**
	Objectives	455
	Introduction	455
	Teeth	455
	The Mouth	456
	Stomach Systems	456
	Clinical Concepts and Critical Thinking	461

	Common Digestive Conditions	465
	Summary	471
	Key Terms	472
	Review Questions	475
CHAPTER 26	**Circulatory System**	**477**
	Objectives	477
	Introduction	477
	Blood	477
	The Heart	480
	Blood Flow	480
	Veins	482
	Heart Sounds	482
	Electrocardiography	483
	Common Diseases and Conditions of the Circulatory System	484
	Summary	486
	Key Terms	486
	Review Questions	489
CHAPTER 27	**The Respiratory System**	**490**
	Objectives	490
	Introduction	490
	Structures of the Respiratory System	491
	Respiration	492
	Diseases and Conditions	494
	Summary	496
	Key Terms	497
	Review Questions	498
CHAPTER 28	**The Endocrine System**	**500**
	Objectives	500
	Introduction	500
	Structures of the Endocrine System	500
	Hormones and Estrus	504
	Diseases and Conditions	505
	Dog Breeds Prone to Hypothyroidism	505
	Summary	507
	Key Terms	507
	Review Questions	509
CHAPTER 29	**The Renal System**	**511**
	Objectives	511
	Introduction	511
	Structures of the Renal System	511

Common Diseases and Conditions	513
Summary	516
Key Terms	516
Review Questions	517

CHAPTER 30 The Reproductive System — 519

Objectives	519
Introduction	519
The Female Reproductive System	519
The Male Reproductive System	520
Fertilization	521
Estrus	522
Establishing Pregnancy	524
Gestation and Incubation	524
The Labor Process	527
Semen Collection	528
Artificial Insemination	528
Neutering	529
Reproductive Problems	529
Mammary Gland Problems	530
Summary	531
Key Terms	531
Review Questions	534

CHAPTER 31 The Immune System — 535

Objectives	535
Introduction	535
Antibodies	535
Innate Versus Adaptive Immunity	536
Vaccines	536
Vaccine Storage and Use	537
Infection	537
Spleen	538
Allergies	538
ELISA Testing	538
Neoplasia (Cancer)	539
Summary	540
Key Terms	541
Review Questions	542

CHAPTER 32 The Nervous System — 543

Objectives	543
Introduction	543
Neurons	543

	Structures of the Nervous System	543
	Coordinated Function of the CNS and PNS	545
	Receptors	545
	Nerve Testing	547
	Common Diseases and Conditions	547
	Summary	549
	Key Terms	550
	Review Questions	551
CHAPTER 33	**The Sensory System**	**553**
	Objectives	553
	Introduction	553
	The Eye	553
	Reflex Testing	555
	Ear	556
	Hearing	557
	Equilibrium	558
	Smell	558
	Taste	559
	Summary	559
	Key Terms	559
	Review Questions	561
CHAPTER 34	**Animal Nutrition**	**562**
	Objectives	562
	Introduction	562
	Nutritional Needs of Animals	562
	Nutrients	563
	Animal Nutrition and Concentrations	565
	Palatability	566
	Types of Diets	566
	Feeding Animals	567
	Ideal Weight	568
	Weight Management	568
	Food Analysis	570
	AAFCO	570
	Feeding Schedules	571
	Summary	572
	Key Terms	572
	Review Questions	574
CHAPTER 35	**Microbiology and Parasitology as Disease Processes**	**577**
	Objectives	577
	Introduction	577
	Microbiology	577
	Types of Microorganisms	578

	Parasitology	579
	Flukes	584
	Bots and Pinworms	584
	Anthelmintics	591
	Zoonosis	592
	Maintaining Good Animal Health	593
	Treating Disease	594
	Summary	594
	Key Terms	595
	Review Questions	597

Section IV ■ Clinical Procedures

CHAPTER 36	**Basic Veterinary Restraint and Handling Procedures**	**600**
	Objectives	600
	Introduction	600
	Instinctive Behaviors	600
	Learned Behaviors	601
	Interpreting an Animal's Body Language and Behavior	602
	Restraint Considerations	603
	Animal Safety	604
	Restraint Equipment	604
	Planning the Restraint Procedure	612
	Restraint Knots	612
	Restraint Positions	618
	Restraint for Blood Collection	623
	Restraint Procedures and Techniques	628
	Summary	654
	Key Terms	654
	Review Questions	655
CHAPTER 37	**Physical Examination Procedures and Patient History**	**657**
	Objectives	657
	Introduction	657
	The Patient History	657
	The Physical Examination	658
	Eye Exam	670
	Ear Exam	671
	Dental Examination	671
	Auscultation	675
	Recording the Physical Examination	679
	Summary	680
	Key Terms	680
	Review Questions	682

CHAPTER 38	**Veterinary Assistant Procedures**	**684**
	Objectives	684
	Introduction	684
	The Hospitalized Patient	685
	Pain Assessment	689
	General Nursing Care	689
	Evaluating Emergency Situations	689
	Common Hospital Procedures	690
	Dental Care	694
	Administering Injections	701
	Fluid Administration	706
	Socialization and Exercise of Patients	718
	Understanding Euthanasia	720
	Grooming Skills in the Veterinary Facility	723
	Summary	741
	Key Terms	741
	Review Questions	744
CHAPTER 39	**Laboratory Procedures**	**746**
	Objectives	746
	Introduction	746
	Veterinary Laboratory Equipment	747
	Hematology Sample Collection	755
	Recording Laboratory Results	755
	Blood Chemistry Procedures	755
	The Fecal Sample	765
	The Urine Sample	769
	Microscopic Exam	772
	Elements in Urinary Sediment	774
	Cytology and Histology	777
	Summary	779
	Key Terms	780
	Review Questions	782
CHAPTER 40	**Radiology and Imaging Procedures**	**784**
	Objectives	784
	Introduction	784
	Radiology Generation	784
	Radiation Safety	785
	Radiology Terminology	787
	Radiographic Detail	789
	Veterinary Animal Restraint	790
	Radiology Log	791
	Patient Measurement	791

	Technique Chart	792
	Machine Setting Procedures	793
	Milliamperage	794
	Radiographic Film	794
	Film Identification	794
	Developing Film	795
	Filing Film	800
	Darkroom Care and Maintenance	801
	Radiograph Artifacts and Errors	802
	Ultrasound Diagnostics	802
	Endoscopy	803
	CT and MRI Procedures	804
	Summary	805
	Key Terms	805
	Review Questions	807
CHAPTER 41	**Pharmacy Procedures**	**809**
	Objectives	809
	Introduction	809
	Medication Names	809
	Medication Forms	810
	The VCPR and Prescriptions	810
	Reading a Prescription	811
	Labeling a Prescription	811
	Dispensing Medications	813
	Metric System	813
	Types of Medications	816
	Educating Clients on Medication Use	817
	Medication Storage	824
	Pharmacokinetics	824
	Types of Medications	825
	Preanesthetics and Anesthetics	829
	Anthelmintics and Antiparasitics	830
	Summary	831
	Key Terms	831
	Review Questions	833
CHAPTER 42	**Surgical Assisting Procedures**	**834**
	Objectives	834
	Introduction	835
	Surgical Asepsis	835
	Surgical Logbook	835
	Anesthesia Logbook	835
	Surgical Suite Maintenance	837

Preanesthetic Patient Care	838
Fluid Therapy	839
Anesthesia Induction	840
Intubation Procedure	840
Patient Monitors	843
Surgical Preparation	844
Surgical Pack Preparation	847
Gowning the Surgeon and Assistants	855
Patient Surgical Positioning	856
Anesthesia Planes	857
The Anesthesia Machine	857
Breathing Systems	859
Postanesthetic Care	860
Postoperative Patient Care	860
Sterilization Techniques	862
Surgical Instruments	863
Summary	870
Key Terms	870
Review Questions	874

Glossary	875
Index	921

Preface

The popularity of and nationwide demand for veterinarians and trained staff have caused the industry to reach out to high school career and technical schools, seeking to give students experience with and an understanding of the veterinary field. Thus, Ms. Vanhorn began creating a high school–based veterinary assistant program to give students experience with and a basic introduction to the world of veterinary medicine. While developing the program, Vanhorn found that the majority of veterinary-related educational materials and textbooks were written at a college-to-post secondary level. In an effort to allow secondary students to understand and learn about veterinary assisting as a career opportunity, this textbook was created for veterinary assistant students and as a resource for practicing veterinary assistants, veterinary receptionists, hospital managers, and technicians. The author believes that the basic layout of the units, as well as the clinical experiences, offers students a way to see veterinary medicine. This approach provides the learner with a preview of the veterinary industry and the education needed to achieve success in the profession. It also allows students to critically think and problem-solve realistic scenarios. This textbook has all types of veterinary careers under one cover, including small, large, zoo, exotic, and production animals.

The target market for this textbook is veterinary assistants; however, any veterinary professional can benefit from the information in this textbook Employment of veterinary assistants and other veterinary professionals is projected to grow 16% by 2029, much more quickly than the average for all occupations. High turnover should result in good job opportunities. There are increasingly more medications and treatments for pets today, which means more pet owners will be coming into veterinary clinics seeking treatment for their animals. This increase in business will continue to drive up the demand for more professionally trained veterinary assistants. With spending levels on pets reaching an all-time high, more veterinary facilities are looking to hire more veterinary assistants.

Organization of This Text

The organization of this text is based on the National Association of Veterinary Technicians in America Approved Veterinary Assistant (NAVTA AVA) Essential Skills for Assistant Training curriculum. This includes business management skills, communication skills, safety procedures, hospital procedures, pharmacy skills, surgical assisting, nursing care, laboratory procedures, and radiology skills. The text includes topics related to small animals, large animals, production animals, and exotic and zoo animal species. The chapters are structured as follows:

Section I: Practice Management and Client Relations
Section II: Veterinary Animal Production
Section III: General Anatomy/Physiology and Disease Processes
Section IV: Clinical Procedures

Each section is broken down into significant chapters that address essential skills and curricula for veterinary assistants, veterinary technicians, veterinary receptionists, kennel attendants, groomers, animal health herd managers, practice managers, and agricultural educators.

Features

In recognition of the contributions veterinary assistants make to the practice team and to the health of pets, the National Association of Veterinary Technicians in America (NAVTA) has created the Approved Veterinary Assistant (AVA) designation. This designation is currently used across the United States. This edition includes updated chapters that focus on changes in veterinary medicine, both in medical standards and technology. Special features found consistently throughout all chapters include chapter objectives, competency skills, terminology tips, making the connection, key terms, review questions, and clinical situations.

New to This Edition

This edition has been condensed to decrease repetitive information by restructuring chapters to include a consistent outline based on chapter material and animal species. Each chapter has been placed in an order related to veterinary assistant teaching curricula in order of learning importance from a student and educator perspective. For example, Section I includes terminology, communication, ethics and legality, and safety procedures to better allow students to understand the terminology and veterinary medical language so that they can complete medical records efficiently and professionally, while understanding the ethics and laws involved in the veterinary medical field. Prior to learning about hands-on skills and advanced veterinary technical procedures, students will learn how to effectively communicate with veterinary team members and clients, as well as the safety and aseptic techniques involved in providing safe and sterile procedures. Section II is broken down by animal species, including breed identification and production management. Section III discusses anatomy and physiology concepts, animal nutrition, and parasitology and disease processes. Section IV then builds on the previous section's layout to discuss hands-on clinical skills and clinical procedures.

- Additional review questions have been added to chapters, including clinical situations and critical thinking concepts.
- Additional artwork has been added to most chapters to further engage students and enhance student comprehension.

Veterinary Assisting Fundamentals and Applications

Contents

Section I: Practice Management and Client Relations

Chapter 1: Veterinary Medical Terminology and Abbreviations

- Additional suffixes, root words, and prefixes
- Additional common abbreviations and veterinary terms

Chapter 2: Medical Records

- Additional invoicing and fee concepts
- Additional focus on computerized medical records and paperless facilities

Chapter 3: Scheduling Appointments and Computer Applications

- Combination of previous Chapters 3, 4, and 5 to include computer applications
- Additional inventory management concepts

Chapter 4: Veterinary Careers

- New chapter content about expanded veterinary career field opportunities
- Additional career options for groomers and hospital administrators
- Additional specialty certifications
- Additional veterinary facility types and specialty options
- Additional artwork to meet industry standards

Chapter 5: Communication and Client Relations

- Additional resources to help with the grief process

Chapter 6: Veterinary Ethics and Legal Issues

- Expanded information on federal agencies
- Additional resources on medical waste materials
- Additional resources on controlled substances and drug schedules

Chapter 7: Veterinary Safety and Aseptic Techniques

- Chapter moved to Section I to be in better alignment with curriculum
- Combination of previous Chapters 41 and 42
- Expanded discussion on personal protective equipment (PPE)
- Additional resources related to changes to Safety Data Sheets (SDS)
- Additional concepts on labeling secondary containers, radiation safety, anesthetic safety, and OSHA regulations

Section II: Veterinary Animal Production

Chapter 8: Dog Breed Identification and Production Management

- Additional updates to American Kennel Club (AKC)–approved breeds
- Combination of associated species behavior information from previous Chapter 39
- Expanded information on aggression

- Additional resources related to use of E-collar for restraint purposes
- Updated and expanded vaccination schedule recommendations, including core and noncore vaccines
- Updated flea and tick, heartworm, and parasite preventative care recommendations

Chapter 9: Cat Breed Identification and Production Management

- Additional updates to Cat Fanciers Association (CFA)–approved breeds
- Additional restraint information related to removing a cat from a carrier
- Updated and expanded vaccination schedule recommendations, including core and noncore vaccines
- Updated flea and tick, heartworm, and parasite preventative care recommendations

Chapter 10: Avian Breed Identification and Production Management

- Additional artwork for student engagement and understanding
- Additional information on fecal exams and radiology concepts
- Additional information related to respiratory emergencies, injuries, and fractures

Chapter 11: Pocket Pet Identification and Production Management

- Additional artwork for student engagement and understanding
- Additional information on barbering behavior

Chapter 12: Rabbit Breed Identification and Production Management

- Additional updates on recognized rabbit breeds
- Additional information on pododermatitis and malocclusion concepts

Chapter 13: Reptile and Amphibian Identification and Production Management

Chapter 14: Ornamental Fish and Aquaculture Identification and Production Management

- Combination of aquaculture information from previous Chapter 25

Chapter 15: Zoo, Exotic, and Wildlife Identification and Production Management

- Combination of zoo and exotic animal information from previous Chapter 16

Chapter 16: Laboratory and Animal Research Identification and Production Management

Chapter 17: Beef and Dairy Cattle Breed Identification and Production Management

- Additional updates on milk production by state
- Combination of associated species behavior information from previous Chapter 39

Chapter 18: Equine and Draft Breed Identification and Production Management

- Combination of associated species behavior information from previous Chapter 39
- Combination of draft animal information from previous Chapter 20
- Additional artwork for student engagement and understanding
- Additional information on chewing, cribbing, and tetanus

Chapter 19: Swine Breed Identification and Production Management

- Combination of associated species behavior information from previous Chapter 39
- Additional artwork for student engagement and understanding

Chapter 20: Sheep Breed Identification and Production Management

- Combination of associated species behavior information from previous Chapter 39

Chapter 21: Goat Breed Identification and Production Management

- Combination of associated species behavior information from previous Chapter 39

Chapter 22: Poultry Breed Identification and Production Management

Section III: General Anatomy/Physiology and Disease Processes

Chapter 23: The Structure of Living Things

- Expanded material on cell structure, cell function, and cell division
- Additional information on genetics and clinical concepts
- Expanded information on tissue types

Chapter 24: The Musculoskeletal System

- Expanded information on bone development and the muscular system
- Additional artwork and clinical concepts for student engagement and understanding

Chapter 25: The Digestive System

- Additional artwork and clinical concepts for student engagement and understanding

Chapter 26: The Circulatory System

- Expanded information on veins, heart sounds, and shock
- Additional artwork and clinical concepts for student engagement and understanding

Chapter 27: The Respiratory System

- Additional information on pneumothorax
- Additional artwork and clinical concepts for student engagement and understanding

Chapter 28: The Endocrine System

- Expanded information on male and female hormones
- Expanded information on hyperadrenocorticism and hypoadrenocorticism
- Additional artwork and clinical concepts for student engagement and understanding

Chapter 29: The Renal System

- Additional artwork and clinical concepts for student engagement and understanding

Chapter 30: The Reproductive System

- Expanded information on establishing pregnancy
- Additional artwork and clinical concepts for student engagement and understanding

Chapter 31: The Immune System

- Expanded information on antibodies and types of immunity
- Additional information on vaccine storage and use
- Additional information on the spleen
- Additional information on neoplasia
- Additional artwork and clinical concepts for student engagement and understanding

Chapter 32: The Nervous System

- Expanded information on neurons
- Additional artwork and clinical concepts for student engagement and understanding

Chapter 33: The Sensory System

- Additional information on conditions within the eye, including cataracts, nuclear sclerosis, and glaucoma
- Additional information on conditions within the ear, including otitis and aural hematoma
- Additional information on the senses of smell and taste
- Additional artwork and clinical concepts for student engagement and understanding

Chapter 34: Animal Nutrition

- Additional information on palatability
- Additional information on weight management
- Expanded information on food analysis
- Additional information on the Association of American Feed Control Officials (AFFCO)
- Additional artwork and clinical concepts for student engagement and understanding

Chapter 35: Microbiology and Parasitology as Disease Processes

- Expanded information on parasite hosts
- Additional information on flukes, bots, and pinworms
- Additional information on maggots
- Expanded information on anthelmintics
- Additional artwork and clinical concepts for student engagement and understanding

Section IV: Clinical Procedures

Chapter 36: Basic Veterinary Restraint and Handling Procedures

- Combination of behavior information from previous Chapter 39
- Expanded animal behavior information
- Expanded skill and hands-on concepts for clinical procedures
- Additional information on fear-free handling

Chapter 37: Physical Examination Procedures and Patient History

- Combination of physical exam information from previous Chapters 43 and 44
- Expanded information on weight assessment and blood pressure procedures
- Expanded information on ear and dental exams

Chapter 38: Veterinary Assistant Procedures

- Combination of hospital and grooming procedures information from previous Chapters 45, 46, and 47
- Additional information on pain assessment
- Expanded information on dental procedures and dental prophylaxis
- Expanded information on urinary catheter maintenance
- Expanded information on euthanasia procedures

Chapter 39: Laboratory Procedures

- Expanded information on serologic testing kits and blood chemistry procedures
- Additional information on hematology collection and handling of samples
- Expanded information on centrifugal and sediment fecal testing procedures
- Expanded information on the urinary microscopic exam and sediment exam
- Additional information on cytology and histology sampling
- Additional information on handling rabies samples

Chapter 40: Radiology and Imaging Procedures

- Additional information on the production and generation of radiation
- Expanded information on PPE and radiation safety
- Additional information on radiation detail and developing
- Expanded information on milliamperage and kilovoltage peak (kVp)
- Expanded information on digital radiology, ultrasound, and endoscopy techniques
- Additional information on computerized tomography or cat scan (CT) and magnetic resonance imaging (MRI) procedures

Chapter 41: Pharmacy Procedures

- Additional information on medication names and types of medications
- Expanded information on medication and the veterinary client patient relationship (VCPR) concept
- Additional information on the metric system
- Expanded information on medication identification
- Additional information on pharmacokinetics
- Additional information on anesthesia medications, anthelmintics, and antiparasitic medications

Chapter 42: Surgical Assisting Procedures

- Additional artwork and clinical concepts for student engagement and understanding
- Additional information on laser surgery
- Expanded information on the anesthesia machine
- Additional information on dental instruments

Supplements

To help you learn and teach from **Veterinary Assisting: Fundamentals & Applications**, a variety of additional materials have been prepared for you.

MindTap

MindTap™ is a fully online, highly personalized learning experience combining readings, activities, and assessments into a singular Learning Path. Instructors can personalize the Learning Path by customizing Cengage resources and adding their own content via apps that integrate into the MindTap framework seamlessly with Learning Management Systems.

Our MindTap product provides a Learning Path designed to meet critical needs while also allowing instructors to measure skills and outcomes with ease.

- Read: Students can prepare for class with interactive guided reading. MindTap comes standard with the ability to highlight and take notes within the chapter content. Concept Check quizzes are embedded in the reading to help students review the chapter's content.
- Study It: Flashcards, PowerPoint slides, and Review Questions help the students review the concepts and content, as well as key terms. Students can come to class better prepared to engage with the material.
- Apply It: NameIt Game, Critical Thinking Questions, and Final Exam for each chapter provide students the opportunity to check their knowledge of the chapter content. With an interactive game, analysis and discussion questions, and a final exam, students will test their mastery of the chapter topics. Feedback is provided along with links to the reading for in-context understanding.
- Certification Style Exam Practice: Questions written in the format of the Approved Veterinary Assistant Exam will help students prepare for certification.

Through the Progress app, instructors are provided data on class standings as well as specific activities and assignments that students may be finding difficult. Additionally, students are provided data about where they stand—both individually and compared to the highest performers in class.

To view a demo video and learn more about MindTap, please visit www.cengage.com/mindtap

Instructor Resources

This complementary site includes free materials the instructor can access to minimize preparation time and maximize student engagement.

Instructor Manual

The Instructor Manual provides the following for each chapter: Chapter Objectives, Key Terms, What's New in the Chapter, and a Chapter Outline.

Solution and Answer Guide

The Solution and Answer Guide provides answers to the Review Questions and the Clinical Situation questions.

PowerPoint® Lecture Review Slides

The PowerPoint slides consist of lecture outlines and select tables and figures used in the book. The slides are available for use by instructors for enhancing their lectures and by students as an aid to note taking. This edition's PowerPoint slides are all new and fully accessible.

Test Bank

Cognero is a flexible, online system that allows instructors to author, edit, and mange test bank content from Cengage; create multiple test versions in an instant; and deliver tests from the instructor's LMS, classroom, or wherever the instructor desires. The test bank contains more than 800 questions in multiple-choice and matching formats.

Competency Skills Checklists

Competency skills checklists are provided for the instructor to evaluate students on the demonstration and mastery of the skills presented.

About the Author

Beth Vanhorn has been a Certified Veterinary Technician (CVT) since 1995, working in clinical small animal and exotic practice upon graduating from Wilson College. Beth's background includes numerous years working in private and corporate veterinary practices in surgery and intensive care unit (ICU) settings and extensive development and continuing education in veterinary dentistry. Beth has served on the Pennsylvania State Board of Veterinary Medicine as a CVT member. Beth also developed the Dauphin County Technical School (DCTS) Veterinary Assistant Program in 2004 and then became one of the first high school–level institutions to be approved for creating a curriculum and submitting the program for NAVTA AVA program approval. Once approved, graduates of these programs are given the opportunity to take the national test. With successful completion of the exam, individuals will become an approved veterinary assistant (AVA) and be recognized for their accomplishment with the AVA designation. Beth also worked closely with several high schools and technical schools throughout the United States to help develop veterinary assistant programs that have also attained AVA status. She has also been invited over the years to speak at conferences on how to develop a veterinary assistant program within a high school or technical school environment, including the National Association of Agricultural Educator's, Association for Career and Technical Education, and the Pennsylvania Veterinary Medical Association. Her involvement with veterinary assistant program development allowed DCTS to win the Pennsylvania Veterinary Medical Association (PVMA) President's Award in 2007. Beth has been involved as an educator in the DCTS program's adult education program since its development in 2010. Beth currently works as a practice manager for Banfield, where she has served in the management career field since 2012. She also is a horse owner and has been involved with the American Quarter Horse Association most of her life, showing and breeding quarter horses. She has an additional textbook publication, *Veterinary Guide to Animal Breeds* (Wiley-Blackwell, 2017).

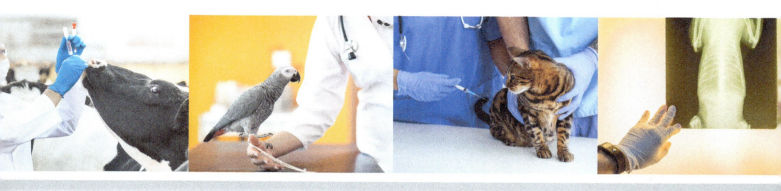

Acknowledgments

I would like to acknowledge and thank the reviewers who took the time to provide feedback on this textbook edition, as well as everyone at Cengage who has provided me with the tools, resources, and guidance to develop a second edition that follows industry standards. I would like to acknowledge the use of many diagrams and illustrations from Janet Romich's *An Illustrated Guide to Veterinary Medical Terminology*. I also would like to thank all my past students and veterinary associates who allowed the use of their time and skills to help me provide the photographs necessary to make this textbook project complete. Finally, I would like to thank all my past veterinary team members and doctors who guided me in my veterinary technology career field, and I appreciate the support of my family who allowed me the time to make this textbook a reality.

Reviewers

Jan Barnett, DVM
Program Director, Veterinary Technology Program
Tulsa Community College
Tulsa, OK

Katie Tyler, DVM
Veterinarian
Murray State College
Tishomingo, OK

Section I
Practice Management and Client Relations

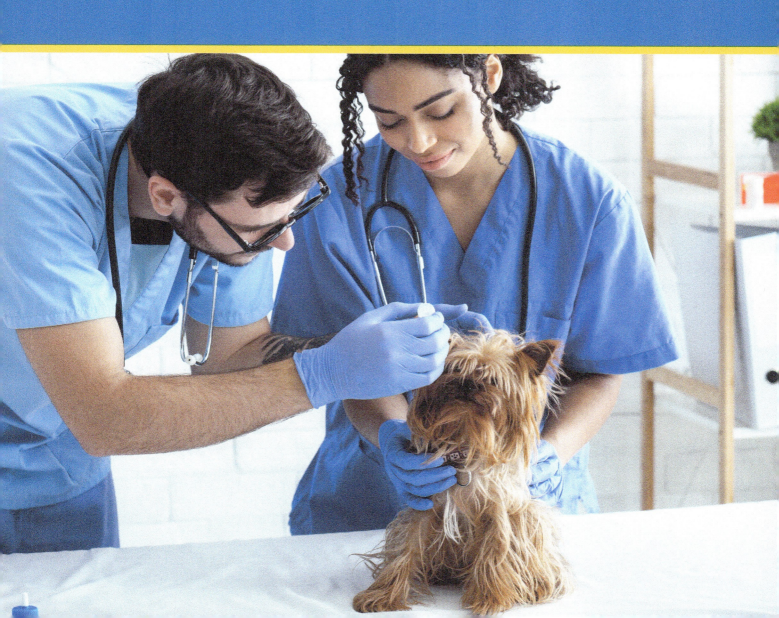

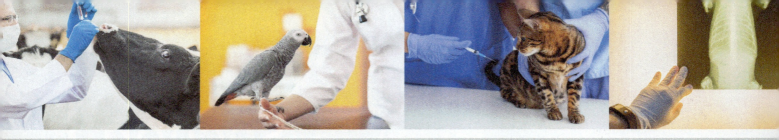

CHAPTER 1: Veterinary Medical Terminology and Abbreviations

Objectives

Upon completion of this chapter, the reader should be able to:

1.1 Analyze veterinary medical terms to determine the prefix, suffix, and root word
1.2 Identify the meaning of common prefixes
1.3 Identify the meaning of common suffixes
1.4 Identify the meaning of common root words
1.5 Define common veterinary medical terms related to direction, species, patient history, and pharmacy
1.6 Identify and use common veterinary medical abbreviations

Introduction

The field of veterinary medicine has adopted a common language used by all veterinary professionals working at all levels of the industry. A basic knowledge and foundation of veterinary medical terminology will help the veterinary assistant communicate with other veterinary professionals, understand treatments and procedures within the veterinary facility, and properly enter and input medical information in patient charts. This chapter will focus on the basics of how to pronounce a word, how to break down and interpret word parts, and how these words apply to animals and areas of veterinary medical practice. All veterinary facilities should have a good veterinary medical dictionary that is available for reference at all times.

Veterinary Medical Terminology

Learning new vocabulary terms is a part of learning any new language, and veterinary medical terminology is the language used in the field of veterinary medicine. Some of the terms may be familiar; however, many may be foreign. Veterinary terms are often based on Latin and Greek words. Learning how to break down a term to define the meaning is the key to learning this new language. Some words may seem long and complex, but working with each part of the word will create a structure for learning terminology. Veterinary assistants should also have a veterinary dictionary to verify the spelling and definitions of new words that they may encounter.

Learning to Dissect a Veterinary Term

Words are made of prefixes, root words, and suffixes. Learning each of these parts and being able to define their meaning will allow the **dissecting**, or separating, of medical terms in a logical way. The **prefix** is the part of the word at the beginning of a term and consists of one or more syllables. The prefix typically indicates time, number, location, or status. The **root word** is the part of the word that gives the term its essential meaning.

A term may consist of more than one root word. The **suffix** is the part of the word at the end of the term. The suffix typically indicates disease, condition, or procedure. Some words have a **combining vowel** that is attached to a root word and allows for certain terms to be pronounced more easily. The combining vowel is commonly the letter *o*. A **combining form** is a root word plus a combining vowel and often describes a body part. New words may be formed when combining forms are added to prefixes, root words, and suffixes. Learning to break down unfamiliar terms into recognizable word parts allows the veterinary assistant to increase and improve their medical vocabulary.

TERMINOLOGY TIP

Create flash cards to learn new terms. Use 3″ × 5″ or larger index cards and write the term, root word, prefix, or suffix on one side and a definition on the other side. Practice new terms 5–10 minutes each day (see Figure 1–1).

Common Root Words

A root word gives a term its essential meaning. Terms may have one or more root words. Some root words are formed with a combining vowel, usually the letter *o* that is placed at the end of the root word to allow the term to be more easily pronounced. Table 1–1 contains commonly used root words that veterinary assistants should be familiar with. Again, it is important to reference a veterinary dictionary when encountering new root words.

FIGURE 1–1 These veterinary professionals are using flashcards to practice terminology.

TABLE 1–1

Common Root Words

ROOT	MEANING
ABDOMIN/O-	abdomen
ARTHR/O-	joint
AUR/O-	ear
BRONCHI/O-	lungs and bronchial tube
CARDI/O-	heart
CEPHEL/O-	head
CHEM/O-	chemical
CHONDR/O-	cartilage
COL/O-	colon
COST/O-	rib
CUTANE/O-	skin
CYST/O-	urinary bladder
CYT/O-	cell
DENT/O-	teeth
DERM/O-/DERMAT/O-	skin
ELECTR/O-	electricity
ENTER/O-	intestines
GASTR/O-	stomach
GINGIV/O-	gums
HEM/O-	blood
HEPAT/O-	liver
HIST/O-	tissue
HYSTER/O-	uterus
LAPAR/O-	abdomen
LARYNG/O-	larynx
MAST-/MAMM-	mammary gland
MY/O-	muscle
NAS/O-	nose
NEPHR/O-	kidneys
NEUR/O-	nerve
OCUL/O-	eye
OPHTHALM/O-	eye
OR/O-	mouth
OSTE/O-	bone
OT/O-	ear
OVARI/O-	ovary
PNEUM/O-	lung
PULM/O-, PULMON/O-	lung
RADI/O-	radiation
RECT/O-	rectum
REN/O-	kidney
RHIN/O-	nose
SPLEN/O-	spleen

(*Continues*)

TABLE 1–1 (Continued)	
SPONDYL/O-	vertebra, spinal column
STETH/O-	chest
STOMAT/O-	mouth
THORAC/O-	chest
TRACHE/O-	trachea
URETHR/O-	urethra
URIN/O-, UR/O-	urine
UTER/O-	uterus
VEN/O-	vein

Common Prefixes

The meaning of a prefix never changes; however, changing the prefix of a root word will change the meaning of the word. Prefixes are added to the beginning of words to qualify as follows:

- Numbers: bi- = two, tri- = three, quadra- = four
- Measurements: hyper- = excessive, hypo- = less than normal
- Position and/or direction: sub- = under, supra- = above, peri- = around
- Negatives: an- = without, anti- = against
- Color: cyan/o = blue, erythr/o = red, jaund/o = yellow

Example

Cardi/o is the root word meaning "heart."
The suffix *–ia* means "condition."
The prefix *brady* means "abnormally slow"; therefore, *bradycardia* means the condition of an abnormally slow heartbeat.
The prefix *tachy* means "abnormally fast"; therefore, *tachycardia* means the condition of an abnormally fast heartbeat.

Many prefixes have opposites that can be used to learn their meanings.

Example

ab- = away from, as in *abduction*
ad- = toward, as in *adduction*
pre- = before, as in *preoperative*
post- = after, as in *postoperative*

Table 1–2 lists prefixes commonly used in veterinary medicine. When encountering an unfamiliar prefix, it is important to use a dictionary to determine the meaning. Using the dictionary to look up new vocabulary is a good habit to acquire.

Common Suffixes

Suffixes are located at the end of the word and commonly define surgical and medical procedures and conditions. Several suffixes have the same meaning and are called the "pertaining to" suffixes (see Table 1–3). To define these terms, look at the root word for the essential meaning. Table 1–4 lists suffixes commonly used in veterinary medicine. Locate the meaning of new suffixes in a dictionary.

Combining Forms

The combining form includes the combining vowel, which is usually the letter *o*, but may also be such vowels as *a, e, i,* and *u*. A combining vowel is placed between the prefix and suffix in certain words to allow the word to be more easily pronounced. Some words have a poor flow between the prefix, suffix, and root word combination; placing a vowel within the word allows the newly combined form to be easy to pronounce.

Putting It All Together

With a basic understanding of prefixes, suffixes, and root words, it becomes easy to break words down into parts to define new terminology. When dissecting a term, use a slash mark to isolate each part of the word. Then focus on the meaning of each dissected part of the term. Once the meaning of each word part is determined, place the meaning into a logically formed definition. Below are some examples of dissected veterinary terms.

Example

ARTHRITIS	inflammation of the joint
CARDIO/LOGY	the study of the heart
CARDIO/MEGALY	enlargement of the heart
CYSTO/CENTESIS	to puncture into the urinary bladder
CYTO/LOGY	the study of cells
HEMAT/URIA	blood in the urine
HEPATITIS	inflammation of the liver
GASTR/IC	pertaining to the stomach
LAPARO/TOMY	surgical incision into the abdomen

TABLE 1–2
Common Prefixes

PREFIX	MEANING	EXAMPLE	DEFINITION
A- or AN-	without; no	anemia	without or no blood cell production
AB-	away from	abduction	away from the center of the body
AD-	toward	adduction	toward the center of the body
ANTI-	against; to stop	anticoagulant	medicine used to stop bleeding
BRACHY-	short	brachycephalic	short head
BRADY-	slow	bradycardia	slow heart rhythm
CYANO-	blue	cyanotic	blue in color
DYS-	difficult; painful	dysuria	difficult or painful urination
ECTO-	outside	ectoskeleton	bones located outside the body
ENDO-	within, inside	endothermic	body temperature controlled within the body
EPI-	above	epidermis	top outer layer of skin
EX/EXO-	outside	exothermic	body temperature controlled outside of the body
ERYTHRO-	red	erythrocyte	red blood cell
HYPER-	above normal	hyperglycemia	high blood sugar
HYPO-	below normal	hypoglycemia	low blood sugar
INTER-	between	interdigital	between the toes
INTRA-	within	intramuscular	within the muscle
LEUKO-	white	leukocyte	white blood cell
MACRO-	large	macrobacteria	large-sized living organism
MICRO-	small	microscope	instrument used to visualize small items
MAL-	bad	malocclusion	poor tooth alignment
NEO-	new	neoplasia	new tissue formation or growth
OLIGO-	very little	oliguria	very little urine production
PERI-	around	perioperative	around or during the surgery
POLY-	many, excessive	polyuria	excessive urine production
POST-	after	postoperative	after the surgery
PRE-	before	preanesthetic	before anesthesia
PYO-	pus	pyometra	pus-filled uterus
SUB-	below	subcutaneous	below the skin
SUPER-	above	superimposed	above the surface
TACHY-	fast	tachycardia	fast heart rhythm

MAST/ECTOMY	the surgical removal of the mammary gland	RADIO/GRAPH	to record using radiation
ONYCH/ECTOMY	the surgical removal of the nail; declaw	RHINO/PLASTY	the surgical repair of the nose
OVARIO/HYSTER/ECTOMY Abbreviated OHE/OVH	the surgical removal of the ovaries and uterus	STOMAT/ITIS	inflammation of the mouth
		URINA/LYSIS	to break down urine

TABLE 1-3
Suffixes Meaning "Pertaining To"

SUFFIX	TERM	MEANING
-AC	cardiac	pertaining to the heart
-AL	renal	pertaining to the kidney
-AN	ovarian	pertaining to the ovary
-AR	lumbar	pertaining to the loin or lower back
-ARY	alimentary	pertaining to the gastrointestinal tract
-EAL	laryngeal	pertaining to the larynx
-IC	enteric	pertaining to the intestines
-INE	uterine	pertaining to the uterus
-OUS	cutaneous	pertaining to the skin
-TIC	nephrotic	pertaining to the kidneys

Remember, when defining, pronouncing, or spelling unfamiliar veterinary medical terms, it is best to consult a veterinary medical dictionary before recording the term in medical records or communicating it to others. Veterinary medical terminology is a foreign language for some that is necessary in the veterinary facility; over time and with consistent use it will become a comfortable second language.

TERMINOLOGY TIP

Practice makes perfect. Complete practice assignments to study and review veterinary terms. ■

TABLE 1-4
Common Suffixes

SUFFIX	MEANING	EXAMPLE	DEFINITION
-ALGIA	pain	arthralgia	pain in the joints
-CENTESIS	surgical puncture into	cystocentesis	surgical puncture with a needle into the urinary bladder
-CYTE	cell	erythrocyte	red blood cell
-ECTOMY	surgical removal of	ovariohysterectomy	surgical removal of the ovaries and uterus
-EMESIS	to vomit	antiemetic	to stop vomiting
-EMIA	blood	hypocalcemia	low blood calcium
-GRAM	a record of	electrocardiogram	a record of the electrical activity of the heart
-GRAPH	to record with an instrument	radiograph	a record made using radiation; X-ray
-GRAPHY	the act of recording using an instrument	radiography	the act of taking a picture using radiation
-IA	condition	leukopenia	condition causing decreased white blood cells
-IST	specialist	pathologist	specialist who studies diseases
-ITIS	inflammation	colitis	inflammation of the colon
-LOGY	the study of	histology	the study of tissues
-LYSIS	to break down	urinalysis	the breakdown of urine into parts
-MALACIA	abnormal softening	osteomalacia	the abnormal softening of bone
-MEGALY	enlargement of	cardiomegaly	enlargement of the heart
-OMA	tumor, mass	hematoma	localized collection of blood
-OPSY	to view or see	biopsy	the process of viewing living organisms
-OSIS	condition	osteoporosis	condition of bone loss
-PATHY	disease	cardiopathy	heart disease
-PENIA	lack of or deficiency	leukopenia	lack of white blood cells
-PEXY	to suture to	gastropexy	to suture to the stomach
-PHAG	to eat	coprophagic	to eat feces
-PLASIA	to grow or change	neoplasia	new growth (as in tissues)
-PLASTY	to surgically repair	rhinoplasty	surgical repair of the nose

-PNEA	breathing	dyspnea	difficulty breathing
-RRHAGE-	to burst	hemorrhage	the bursting of blood; bleeding
-RRHEA-	to flow	diarrhea	the flow of feces
-SCLEROSIS	abnormal hardening	arteriosclerosis	abnormal hardening of the arteries
-SCOPE	instrument used to view	microscope	instrument used to view small items
-SCOPY	the act of using an instrument for viewing	endoscopy	the act of using a scope to view the inside of the body
-STOMY	to create a new surgical opening	cystostomy	to create a new surgical opening in the urinary bladder
-THERAPY	treatment	chemotherapy	treatment of chemicals
-TOMY	to cut into surgically; to make an incision	cystotomy	to make an incision into the urinary bladder

Common Directional Terms

Directional terms relate to the body position and are useful in recording locations relating to the body; they are also useful in surgical or radiographic positioning. These terms relate to a specific area on the body of an animal and allow for better communication when referring to an animal's anatomy (see Table 1–5, Figure 1–2, and Figure 1–3).

TABLE 1–5

Common Directional Terms

TERM	MEANING
ANTERIOR	front of the body
ASPECT	area
CAUDAL	toward the tail
CRANIAL	toward the head
DEEP	far away from the surface
DISTAL	away from the center of the body
DORSAL	toward the back area
EXTERNAL	outer surface of the body
INTERNAL	deep inside the body
LATERAL	side of the body; toward the outside
MEDIAL	inside of an area; toward the inside
OBLIQUE	slanted or on an incline
PALMAR	the bottom of the front feet
PLANTAR	the bottom of the rear feet
POSTERIOR	rear end of the body
PROXIMAL	closer to the center of the body
RECUMBENCY	lying in position
RECUMBENT	lying
ROSTRAL	toward the nose
SUPERFICIAL	near the surface
TRANSVERSE	across an area dividing it into cranial and caudal sections
VENTRAL	toward the abdomen or belly area

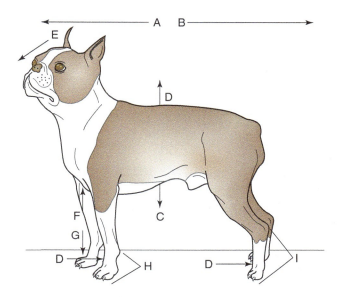

FIGURE 1–2 The arrows on this Boston Terrier represent the following directional terms: A = cranial, B = caudal, C = ventral, D = dorsal, E = rostral, F = proximal, G = distal, H = palmer, I = plantar.

FIGURE 1–3 Medial versus lateral. The lines on these cats represent the directional terms *medial* and *lateral*.

Common Terms and Abbreviations Used in Veterinary Practice

The veterinary field has developed common universal terms that are used in the industry to relate to patient history, species information, the physical exam, and pharmacy terms. Many of these terms also have common abbreviations used in medical recording. This section will outline commonly used terms and abbreviations used in recording medical records and treatment boards (see Table 1–6 through Table 1–15, and Figure 1–4).

TABLE 1–6
Species Terms

AMPHIBIAN	frog or toad
AVIAN	bird
BOVINE	cow
CANINE	dog
CAPRINE	goat
CAVY	guinea pig
EQUINE	horse
FELINE	cat
LAGOMORPH	rabbit
MURINE	rat or mouse
OVINE	sheep
PORCINE	pig or swine
POULTRY	chicken and turkey
PRIMATE	monkey and ape
REPTILE	snake and lizard
TERRAPIN	turtle

TABLE 1–7
Common Veterinary Abbreviations

C or cast	castrated
C-sect	C-section or caesarian section
d	day
d/c	discharge/discontinue
DLH	domestic longhair (cat)
DSH	domestic shorthair (cat)
d/o	drop off
DOA	dead on arrival
EX	exotic
F	female
K-9	dog or canine
M	male
Mo	month
NM	neutered male
o	owner
ohe or ovh	ovariohysterectomy, spay
p/u	pick up
rec	recommend
S or SF	spayed or spayed female
S/R	suture removal
sx	surgery
wk or w	week
y or yr	year

TABLE 1–8
Terms Related to Patient History

ANOREXIA	not eating or decreased appetite
BM	bowel movement
D	diarrhea
DYSURIA	difficulty or trouble with urination
Dz	disease
HBC	hit by car
HEMATURIA	blood in the urine
Hx	history
LETHARGIC	tired or inactive
PD	polydipsia (increased thirst)
PU	polyuria (increased urination)
U	urine
V	vomiting
V/D	vomiting and diarrhea

TABLE 1–9
Terms Related to Physical Examination

ACUTE	short term
AG	anal glands
BAR	bright, alert, responsive
BP	blood pressure
Bpm	beats per minute
Br/m	breaths per minute
CHRONIC	long term
CPR	cardiopulmonary resuscitation
CRT	capillary refill time
Dx	diagnosis
FeLV	feline leukemia virus

FIP	feline infectious peritonitis
FIV	feline immunodeficiency virus
HR	heart rate
L	left
LN	lymph node
mm	mucous membranes
N or (−)	negative
NR	nothing reported
NSF	no signs found/no significant findings
PE	physical exam
Px	prognosis
QAR	quiet, alert, responsive
R	right
rad	radiograph, X-ray
RR	respiratory rate
Rx	prescription
SOAP	Subjective–Objective–Assessment–Plan
TPR	temperature, pulse, respiration
Tx	treatment
URI	upper respiratory infection
UTI	urinary tract infection
WNL	within normal limits
Wt	weight
+	positive

TABLE 1–10
Laboratory Terms and Abbreviations

AI	artificial insemination
BUN	blood urea nitrogen
Bx	biopsy
CBC	complete blood count
CHEM	blood chemistry panel
C/S or C & S	culture and sensitivity
Cysto	cystocentesis
Fecal	fecal or stool sample
HW	heart worm
PCV	packed cell volume
RBC	red blood cell
T4	thyroid test
UA	urinalysis
WBC	white blood cell

TABLE 1–11
Pharmacy Terms and Abbreviations

BID	twice a day
Cap	capsule
cc	cubic centimeter
d	day
EOD	every other day
g	gram
gr	grain
h	hour
kg	kilogram
L	liter
mg	milligram
ml	milliliter
NPO	nothing by mouth
oz	ounces
PO	by mouth
prn	give/take as needed
q	every
qd	every day
QID	four times a day
qod	every other day
Rx	prescription
SID	once a day
Tab	tablet
tbsp	tablespoon
TID	three times a day
tsp	teaspoon
wk or w	week
/	per
# or lb	pound
#	number of tablets to dispense

TERMINOLOGY TIP
When reading a pharmacy prescription label, write out all the information on the label, including the abbreviations, and verify it is correct before typing it and labeling the container. ■

TABLE 1–12
Eyes and Ears

AD	right ear
AS	left ear
AU	both ears
OD	right eye
OS	left eye
OU	both eyes

TABLE 1-13
Routes of Medicinal Administration

Adm.	administer
ID	intradermal (within the layers of skin)
IM	intramuscular (into the muscle)
IN	intranasal (into the nasal cavity)
IO	intraosseous (into the bone)
IP	intraperitoneal (into the peritoneum or lining of the abdomen)
IT	intratracheal (into the trachea or windpipe)
IV	intravenous (into the vein)
PO	by mouth or orally
SQ	subcutaneous (under the skin)
SUB-Q	subcutaneous (under the skin)

TABLE 1-14
Association Abbreviations

AAHA	American Animal Hospital Association
AALAS	American Association of Laboratory Animal Science
AKC	American Kennel Club
AVMA	American Veterinary Medical Association
CFA	Cat Fanciers Association
NAVTA	National Association of Veterinary Technicians of America
OFA	Orthopedic Foundation of America

TABLE 1-15
Common Animal Terms

Bitch	intact female dog	Chick	young parrot; young chicken
Litter	group of newborn dogs	Cock	male parrot; male chicken
Puppy	young dog	Flock	group of birds; group of chickens, turkeys, or ducks
Stud dog	intact male dog	Hen	female parrot; female chicken; female turkey
Whelping	the labor process of dogs	Clutch	group of eggs
Kitten	young cat	Poult	young turkey; young chicken
Tom	intact male cat; male turkey	Capon	young castrated male chicken
Queen	intact female cat	Cockerel	immature male chicken
Queening	the labor process of cats	Pullet	immature female chicken
Buck	male rabbit; male goat; male deer	Rooster	male chicken
Doe	female rabbit; female goat; female deer	Drake	male duck
Kindling	labor process of rabbits and ferrets	Duck	female duck
Kit	young rabbit; young ferret	Duckling	young duck
Lapin	neutered male rabbit	Barrow	young castrated male pig
Gib	neutered male ferret	Farrowing	labor process of swine
Hob	intact male ferret	Gilt	young female pig that has not farrowed
Jill	intact female ferret	Piglet	young pig
Sprite	neutered female ferret	Stag	male pig castrated after maturity
Boar	male guinea pig; male pig	Colt	young male horse
Pup	young guinea pig; young mouse; young rat; young dog	Filly	young female horse
Sow	female guinea pig; female pig	Foal	young male or female horse
Dam	female rat; female mouse; term for a female parent that is breeding	Gelding	castrated male horse; castrated male llama
Sire	male rat; male mouse; term for a male parent that is breeding	Hand	the measurement of a horse equal to 4 inches

Herd	group of horses	Lambing	labor process of sheep
Horse	horse over 14.2 hands in height	Ram	intact male sheep
Mare	intact adult female horse	Wether	castrated male sheep; castrated male goat
Pony	horse under 14.2 hands in height	Freshening	labor process of dairy-producing animals
Stallion	intact adult male horse	Kid	young goat
Weanling	young horse under a year of age	Kidding	labor process of goats
Yearling	young horse between one and two years of age	Bull	intact male cow; intact male llama
Donkey	donkey crossed with a donkey	Cow	intact female cow; intact female llama
Hinny	cross of a male horse and female donkey	Cria	young llama
Jack	intact male donkey	Calf	young cow
Jenny	intact female donkey	Calving	labor process of cows
Mule	cross of a male donkey and female horse	Heifer	young female cow that has not been bred
Ewe	intact female sheep	Stag	mature castrated male cow
Lamb	young sheep	Steer	young castrated male cow

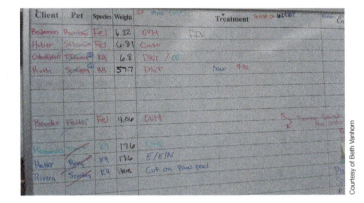

FIGURE 1–4 It is important to understand veterinary abbreviations, as shown on this treatment board, to ensure proper care and handling of the animals in the clinic.

Spelling

Spelling a word correctly is essential in terms of medical recording. A misspelled word is unprofessional, may give a word an entirely different meaning, and may delay proper care of a patient. Medical terms may consist of words that look alike but have entirely different definitions. Some words have similar pronunciation and sounds but are very different in meaning. A veterinary or medical dictionary can be helpful in determining the correct spelling and use of words.

 TERMINOLOGY TIP
Every staff member within the veterinary community should have a reliable veterinary dictionary available at all times. There are a variety of veterinary dictionaries and terminology books that can serve as a reference. ■

SUMMARY

Veterinary medical terminology and veterinary abbreviations are a vital part of the veterinary industry. Each member of the veterinary health care team must have a basic knowledge of terminology and abbreviations to complete his or her job. This may mean reading a hospital treatment board, a patient's medical record, a cage card, or directions on a medication label. For the veterinary assistant to perform his or her duties properly, it is essential to have a working knowledge and understanding of the veterinary language.

Key Terms

dissect   to separate into pieces, or break down into parts, to identify the meaning of a word

prefix   the word part at the beginning of a term

suffix   the word part at the end of a term

root word   the origin or main part of the word that gives the term its essential meaning

combining form   the root word plus a combining vowel; often describes a body part

combining vowel   a letter, usually a vowel, that is placed between the prefix and suffix and makes the word easier to pronounce

Review Questions

Matching

Match the prefix, suffix, or root word to its definition.

1. ___ -itis
2. ___ cardiac
3. ___ -ectomy
4. ___ hepatic
5. ___ hyper-
6. ___ post-
7. ___ hemo-
8. ___ therapy
9. ___ cysto-
10. ___ gastric
11. ___ laparo-
12. ___ -centesis
13. ___ poly-
14. ___ -megaly
15. ___ osteo-

a. pertaining to the stomach
b. blood
c. treatment
d. many, excessive
e. inflammation
f. above normal, increased, excessive
g. abdomen
h. pertaining to the heart
i. surgical removal of
j. pertaining to the liver
k. bone
l. urinary bladder
m. enlargement
n. to puncture into
o. after

Interpretation of Abbreviations

16. Write a description of the meaning of the following situation involving an animal:

    The veterinarian has given you Sassy Smith's Dx and Tx plan and you are to complete the directions and d/c Sassy. You read the following in the medical record: Dx: gastritis; NPO × 24 h, soft diet × 7 d and then return to normal diet; adm. 100 ml Lactated ringers SQ; d/c; recheck 3 d.o. to monitor v/d.

17. You read the following notation in a medical record as asked by the veterinarian:

    1/7/18   OHE sx.; NR; S/R 8-10 d; limit exercise and monitor sx site

    What do you tell the owner? Please write your complete response.

18. Interpret the following statement:

    Rx: 1 drop OU TID × 10 d

Problem Solving

Provide a solution for the following three pharmaceutical problems.

19. The veterinarian instructs you to call Mrs. Smith and have her change Sweetie's dose from (1) 100 mg capsule PO SID to (1) 100 mg capsule PO TID × 5 d. What will you tell Mrs. Smith?
20. You are asked to label the following prescription for Dr. Daniels: Place (3) drops AU BID × 10 d. What will you write on the label?
21. You read the medical record for Phoebe Jones. It instructs the veterinary assistant to "Give 3 ml of Amoxicillin EOD × 1 w prn. Keep medication refrigerated." What will you do for Phoebe's treatment?

Identification

Identify the common abbreviations:

22. F _____
23. SID _____
24. R _____
25. DSH _____
26. Wt _____
27. BAR _____
28. Hx _____
29. BM _____
30. TPR _____
31. HBC _____
32. L _____
33. NM _____
34. BID _____
35. V/D _____
36. q _____
37. TID _____
38. DLH _____
39. UA _____
40. Dx _____
41. S _____
42. IN _____
43. ml _____
44. AS _____
45. Tx _____

14 SECTION I Practice Management and Client Relations

Identify which animal species the following terms belong to:
46. Gelding _____
47. Lapin _____
48. Hen _____
49. Queen _____
50. Gilt _____
51. Doe _____
52. Capon _____
53. Cria _____
54. Heifer _____
55. Jill _____

Break down the following terms into prefix, root word and/or suffix. EXAMPLE: cysto/centesis
56. hepatomegaly
57. postoperative
58. renal
59. hypoglycemia
60. chemotherapy

Bonus Question

What is meant by the term *SOAP*?

Clinical Situation

Complete the following activity by providing the common abbreviation for the underlined medical terms.
    A six-year-old spayed female domestic long hair cat was seen at the veterinary clinic for a physical examination due to vomiting and diarrhea. The cat has a history of being within normal limits. A temperature, pulse, and respiration rate was evaluated with normal signs found. The cat's weight is recorded as 11 pounds. The vet evaluated a complete blood count and radiographs to rule out infection and disease. The diagnosis is found to be acute colitis. The treatment is prepared. The cat's owner is to give 2 tablets of Metronidazole by mouth once a day. The owner is to monitor the cat's bowel movements. A recheck visit was scheduled.

COMPETENCY SKILL 1

Learning Veterinary Medical Terminology and Abbreviations

Objective:

Have a basic knowledge and understanding of common veterinary terminology and abbreviations.

Materials:

- 3" × 5" or larger index cards
- Colored markers, pens, or pencils

Procedure:

1. Obtain a vocabulary terminology word list. Use the terms presented in the tables in this chapter.
2. Write a vocabulary term on one side of each index card.
3. Write the definition of the word on the opposite side of each index card.
4. Study each word term and then try reciting the definition.
5. Study each definition and then try reciting the word.

Practice daily by studying the terms and definitions.

Chapter 2: Medical Records

Objectives

Upon the completion of this chapter, the reader should be able to:

2.1 Identify the various forms included in the medical record

2.2 Properly create veterinary medical records

2.3 Demonstrate how to input information in the medical record

2.4 Properly locate, file, and refile medical records

2.5 Discuss the importance of the legal requirements of the veterinary medical record

2.6 Discuss the importance of consent forms

2.7 Identify the types of certificates provided by the veterinary facility

Introduction

The veterinary assistant is responsible for many office procedures and administrative duties, including the critically important creation, maintenance, and organization of medical records. The medical records furnish documentary evidence of the patient's illness, hospital care, and treatment and serve as a basis for review, study, and evaluation of the care and treatment rendered by the veterinarian.

The Veterinary Medical Record

The primary purpose of the veterinary **medical record** is to record detailed information for each veterinary **patient** (the animal seen by the facility; see Figure 2–1). This information should include patient and **client** (the animal's owner) information, patient history, medical and surgical records, progress notes, and laboratory information. The medical record serves as a diary of the animal's health. This is especially important in multiple-doctor facilities and when medical records are transferred from location to location. The veterinary medical record is owned by the veterinary facility and is the property of the facility that originally activated the record. It is a legal document that is private and confidential, even within the veterinary team. This is also true for any radiographs, ultrasound images, and other diagnostic tests that are produced by the facility.

This legal document allows for the **Veterinary–Client–Patient Relationship (VCPR)** to be established. The VCPR legally allows the veterinarian to treat the patient and dispense medications. The VCPR must be maintained on a yearly basis. It serves as documentation and protects the veterinarian and staff through written recordings. The VCPR relationship satisfies the following criteria: (1) the veterinarian has assumed responsibility for the patient in making judgment regarding the health of an animal and the need for veterinary treatment, and the owner has agreed to follow the instructions of the veterinarian; (2) the veterinarian has sufficient knowledge of the animal species to initiate a general, preliminary, or tentative diagnosis of the medical condition of the animal; (3) the veterinarian is competent with keeping and caring for the animal through examination and timely visits to the facility where the animal is housed; (4) the veterinarian is available for consultation in adverse reactions or failure of the deemed therapy; and (5) the veterinarian maintains medical records of the animal's treatments.

FIGURE 2–1 Accurate record keeping is an important duty of the veterinary assistant.

TABLE 2–1	
Owner and Patient Information	
OWNER INFORMATION	**PATIENT INFORMATION**
Name	Name
Address	Species
Phone	Breed
Work Phone	Color
Emergency Contact Numbers	Gender/Sex
Client ID or Chart Number	Age/Birth Date
	Vaccine History
	Allergies
	Surgical History
	Chief Complaint

The original medical record must remain in the facility for at least one to three years from any patient's last visit, depending on state law. Many facilities keep records for at least seven years, depending on the patient's status. This may be done in a storage area that allows easy access to the files. The original record must not be removed from the facility, but copies may be made and sent with the owner or placed in the mail for continued animal health treatments. A facility may not withhold the release of the veterinary medical record contents.

Creating a Medical Record

Each veterinary medical record must contain certain information and paperwork. Owner information and patient information should be obtained and recorded (see Table 2–1). This information may change and should be maintained and updated on a regular basis. Paper medical records often are kept in file folders that are 8 ½" × 11". The owner and patient information should be typed on a label and placed on the folder in the location established by the veterinary facility.

The medical file should include the following sections or forms:

- The client and patient information sheet should be at the beginning for easy retrieval (see Figure 2–2).
- The master problem list details a patient's history and previous medical problems, vaccines, or surgeries (see Figure 2–3).
- **Progress notes** allow for chronological log entries to be recorded each time a patient is seen and treatment is completed (see Figure 2–4).
- Laboratory reports should include veterinary test results.
- Radiology reports detail X-rays, including proper identification of hospital name, date, name of client, name of patient, and positional marker.
- Pharmacy records note all medications prescribed. If the medications prescribed are controlled substances, a controlled substance log must also be maintained.
- Surgical and anesthesia reports identify surgical procedures.
- **Consent forms** include any form signed by the client for surgery and clinical procedures (see Figure 2–5).

When assembling the medical record, it may also be helpful to make a **cage card** for the animal. The cage card is used to identify and locate each patient within the facility (see Figure 2–6).

Each veterinary facility will have a preference for medical forms that are used and the sequence in which they will be placed in the medical record. It is important that each medical file be kept in the same format, with the information in the same location, for ease of use.

Once the veterinarian recommends a specific treatment, an estimate sheet will be prepared for the client. This outlines the recommended procedures or treatments and all associated costs. This will be reviewed with the client. The client will be required to sign authorization or consent forms before the

CLIENT/PATIENT INFORMATION FORM

DATE _____   CASE NUMBER _____

Please provide the following information for our records: **PLEASE PRINT!**

OWNER INFORMATION

Owner's Name	Social Security Number

Street Address

City/State	Zip Code	Country

Telephone (Include Area Code)	Home	Business

Driver's License Number	Place of Employment	How Long Employed?

ANIMAL INFORMATION

Animal Species (Dog, Cat, Other)	Breed

Animal's Name	Sex	Has the animal been altered? ☐ YES ☐ NO

Color	Birth Date	THE UNDERSIGNED OWNER OR AGENT CERTIFIES THAT THE HEREIN DESCRIBED ANIMAL HAS A MAXIMUM VALUE OF APPROXIMATELY $._____

REFERRAL INFORMATION

Were you referred by a veterinarian?

☐ YES ☐ NO   If so, complete the following information.

Veterinarian's Name	Phone

Street Address

City/State	Zip Code

You will be advised of estimated cost and anticipated procedures. Please feel free to discuss the proposed treatment and any costs with the veterinarian. A minimum deposit of 50 percent of the initial estimated charges will be required for hospitalization of the patient.

STATEMENT OF OWNERSHIP AND CONSENT: I am the owner of the above–described animal or have authorization from the owner to consent to its treatment.

   I hereby authorize the performance of professionally accepted diagnostic, therapeutic, anesthetic, and surgical procedures necessary for its treatment.
   I accept financial responsibility for these services.
   I have read the above consent and understand why these procedures may be necessary. I have also been told of the possible complications and alternatives to the anticipated procedures.

PAYMENT CHOICE: ☐ Cash   ☐ Check   ☐ Bank Card

SIGNATURE (Owner/Agent) _____   DATE _____

FIGURE 2–2 Sample client/patient information form.

CITY ANIMAL HOSPITAL
Master Problem List

OWNER INFORMATION

Owner Name ☐ Mr. ☐ Miss ☐ Ms. ☐ Mrs.   Patient/Pet's Name

Address   City/State/Zip

Home Phone   Business Phone

PATIENT/PET INFORMATION

Chart # _____
Patient _____   Species _____   Breed _____
Color _____   Sex ☐ F ☐ M ☐ N   Birth Date _____
Vax History _____   Weight _____

IMMUNIZATION/PREVENTATIVE RECORD

DATE									
RABIES									
DA2PL									
FVR-CP									
FELV									
FECAL									

PROBLEM LIST

PROBLEM	DATE ENTERED	DATE RESOLVED
1.		
2.		
3.		
4.		
5.		
6.		
7.		
8.		
9.		
10.		
11.		
12.		
13.		

FIGURE 2–3 Sample master problem list.

veterinarian can begin to provide the recommended procedures or treatments. The authorization form serves as a contract between the client and the veterinarian.

After treatments or procedures are completed, the client will receive a discharge sheet. This contains instructions for the client on care of the animal while recuperating, such as how to give prescribed medications or care for surgical incision sites.

The invoice should be placed on the top of the chart so the veterinarian and all staff members have easy access to a working list of itemized charges. The invoice

	PROGRESS NOTES				
PROGRESS NOTES					
Client Name:			Address:		
			Phone:		
Pet Name:	Species:	Breed:	Color:	Gender:	
				Age:	

Date/Initials		
	S	
	O	
	A	
	P	

FIGURE 2–4 Sample progress note.

can easily be removed during the patient's discharge and totaled up for the client (see Figure 2–7).

Invoicing and Fees

It is important for every member of the veterinary health care team to have a basic knowledge of the invoicing and billing procedures used in the practice. Generally, the veterinary receptionist collects client payments and handles invoicing and receipts. However, some circumstances may occur when other members of the team may be in the situation of handling these tasks. This requires training in handling money, basic mathematical skills, and communication skills.

The **invoice** is commonly called the bill and outlines the fees that the client must pay (see Figure 2–8). The receipt states the bill has been paid and what type of payment was made. An invoice and receipt may be the same item that notes the bill has been paid. The fee structure in clinics is based on the type of services and procedures that are provided. The fee structure lists the prices or costs of the service. Some facilities will use a **travel sheet** that lists codes according to the procedure

```
                    CITY HOSPITAL
                    ANY STREET
                   ANY TOWN, XX 00000

                    CONSENT FORM

Owner's Name: _____    Animal's Name: _____
Address: _____     Species: _____
         _____     Breed: _____
Case Number: _____     Sex: _____
```

I am the owner or agent for the owner of the above-described animal and have the authority to execute this consent.

I hereby consent to the hospitalization of the above-described animal and authorize the veterinarian and staff to administer any tests, medications, anesthesia, or surgical procedures that the veterinarian deems necessary for the health, safety, and well-being of the animal.

I specifically request the following procedure(s) or operation(s):

_____

I understand that during the course of the above-mentioned procedure(s) or operation(s), unforeseen conditions may be discovered that necessitate an extension of the above-mentioned procedure(s) or operation(s) or additional procedure(s) or operation(s). I hereby consent to and authorize the performance of such procedure(s) or operation(s) as are necessary according to the veterinarian's professional judgment.

I also authorize the use of appropriate anesthetics and other medications. I understand that the veterinary support personnel will be employed as necessary according to the veterinarian's professional judgment.

I have been advised as to the nature of the procedure(s) or operation(s) to be performed and the risks involved. I realize that results cannot be guaranteed.

I understand that all fees for professional services are due at the time of discharge.

I have read and understand this authorization and consent.

Additional Comments/Information:

_____
_____
_____

Date _____   Signature of Owner or Agent _____

                               Signature of Witness _____

FIGURE 2–5 Sample consent form.

or service (see Figure 2–9). The travel sheet is used by the veterinary team to circle or note the services that have been provided to the patient. The veterinary receptionist or other veterinary team member will then enter the fees into a computer or onto an invoice. This is how a client is charged for the services the veterinary hospital has provided. It is important for all members of the team to understand the costs involved with the services. Many procedures involve the use of equipment, supplies, staff assistance, and inventory fees. Customers may ask why

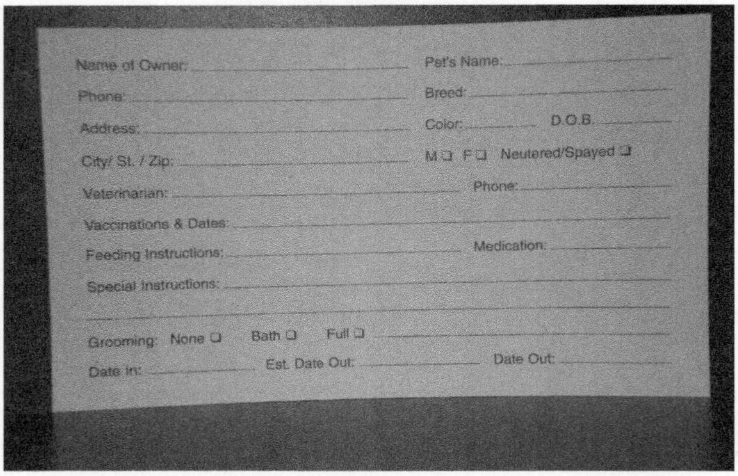

Name of Owner: Pet's Name:
Phone: Breed:
Address: Color: D.O.B.
City/ St. / Zip: M ❑ F ❑ Neutered/Spayed ❑
Veterinarian: Phone:
Vaccinations & Dates:
Feeding Instructions: Medication:
Special Instructions:
Grooming: None ❑ Bath ❑ Full ❑
Date In: Est. Date Out: Date Out:

Courtesy of Beth Vanhorn

FIGURE 2–6 A cage card helps identify and keep track of the animal while it is in the facility.

INSERT YOUR LOGO
INSERT YOU COMPANY NAME
INSERT YOUR ADDRESS
INSERT YOUR PHONE AND FAX
INSERT YOUR WEB ADDRESS

INVOICE

Bill To:	Send To:

Order Number	Invoice Date	Shipped Date	Shipped Via	Terms

Item	Description	Quantity	Price	Total

Subtotal	
Sales Tax	
Total	

FIGURE 2–7 An invoice will be completed and presented to the client upon discharge.

```
┌─────────────────────────────────────────────────────────────────────┐
│                         ANIMAL HOSPITAL                             │
│                          1823 Main St                               │
│                       Philadelphia, Pa 15555                        │
│                                                                     │
│  Client Name:            Patient Name:           Date:              │
│  Doctor:                 Breed:                  Age:     Gender:   │
│  DHLPPC Due:             FVRCP Due:              Rabies due:        │
│  ------------------------------------------------------------------ │
│  INVOICE                                                            │
│                                                                     │
│  Physical Exam/Office Visit      $35.00                             │
│  Ear Flush                       $14.00                             │
│  Betagen Otic Ear Drops          $12.00                             │
│  Amoxicillin Tablets             $15.00                             │
│  Nail Trim                       $ 8.00                             │
│                                                                     │
│  BALANCE DUE:                    $84.00                             │
│  ------------------------------------------------------------------ │
│  Paid—Check #:          Cash:                                       │
│  Thank you for your business!                                       │
└─────────────────────────────────────────────────────────────────────┘
```

FIGURE 2–8 A sample bill or invoice.

certain services cost the amount they do. The veterinary assistant must have a basic knowledge of these services to explain to the client all that is involved with the fees behind the procedures.

Payment plans should begin with the client receiving an estimate. The **estimate** details the approximate cost of procedures that are required by the patient. This will be helpful to the client to determine a type of payment plan that will be necessary at the time of discharge. Payment options can then be reviewed by the receptionist. Many clinics accept payments in cash, check, or credit card form. Some facilities also offer **credit applications**, which allow clients to obtain a line of credit specifically to pay the costs owed to the veterinary clinic. Occasionally, veterinarians may allow clients to set up a **specialized payment plan**. This usually includes a down payment with a set amount of money and smaller payments collected on set dates until the total balance is paid.

The invoice should itemize each fee so that the client understands how much each part of the procedure cost. Through an **account history** the customer's previous invoices can be reviewed at any time. This is also a way for the veterinarians to see what procedures were done and what charges incurred.

Invoicing

The veterinary assistant should be familiar with the invoicing procedures of the facility. This may be completed using a paper invoice or a computerized invoice. A written invoice is often kept with the medical record to note billable activities that are completed while the animal is in the facility. Computerized invoices are kept in the computer system and items are entered as they are completed, and the invoice is then printed out at the time the patient is discharged. Estimates are often initially provided to the client to make them aware of the procedures to be completed and the fees that will be involved (see Figure 2–10). This also provides a direct line of communication between the staff and client to prevent any negative situations and surprises at the time of checkout. Invoices should be detailed to outline what services were performed along with the fees involved to allow the client to see the value of the visit. Payments and outstanding balances are rarely allowed in any veterinary facility, but many hospitals offer third-party credit payment systems with approval of applications. One such provider is CareCredit, which offers credit exclusively for medical expenses, including veterinary fees. Credit decisions are often received within 10 minutes of submission. Any fees not paid

INVOICE SHEET

Owner Name: _____ Phone: _____
Address: _____ Date: _____
Patient: _____ Age: _____ Breed: _____

Office Call	OC Routine	OCM Multiple Pet	OCE Office Call extended	OCC Office Call Complimentary
Exam	PE Comprehensive PE	PEA PE Annual Visit	PER PE Recheck	PEC PE Complimentary
Treatment	HOSP Hospital care	# DAYS _____	IV Fluids	# DAYS _____
	SQ Fluids	# DAYS _____	CBC Complete Blood Count	CHEM Chemistry Panel
	FEC Fecal Sample	URIN Urine Sample	HWT Heartworm Test	FeLV Feline Leukemia test
Medication	HW Heartworm meds	FLEA Flea/tick meds	WORM Wormer medicine	TYPE _____ Qty _____
	MEDS:	NAME _____	QTY _____	
	MEDS:	NAME _____	QTY _____	
	MEDS:	NAME _____	QTY _____	
Diet	CD Hills c/d diet QTY _____	AD Hills a/d diet QTY _____	RD Hills r/d diet QTY _____	UD Hills u/d diet QTY _____
	FEED Daily hospital food	QTY _____		

FIGURE 2–9 A sample travel sheet.

FIGURE 2–10 Completing an invoice.

at the time that services are rendered will result in an outstanding debt collection, will likely affect a person's credit score, and may be reported to the credit bureau.

Recording Information in the Medical Record

Due to the legal requirements for veterinary medical records, there are several rules when recording on medical forms. First, *all* information must be recorded in black or blue ink—never in pencil or other ink colors. Second, all information should be accurate and legible. If a mistake is made, one should place a single line through the error, initial the error, and place the corrected statement after the entry. This shows that an error in writing occurred rather than suggesting information was

TABLE 2–2
Recording in Medical Records

DOS	DON'TS
Use blue or black ink	Use pencil
Write legibly and neat	Use red or light colors of ink
Place one single line through written errors	Use a correction fluid
Initial all mistakes	Erase mistakes
Check spelling and grammar	Cross or scratch out errors
Use proper veterinary medical terminology	Write illegibly
Record dates and times	Use common language
Identify the initials of the person treating the animal	Use bad grammar or spelling
Maintain one record per patient	Place multiple animals within one record without properly identifying the animals
Record all client communications	

changed. Never erase; instead, use a correction fluid or scratch out information! Table 2–2 lists the Dos and Don'ts of recording in medical records.

Record all communications and phone conversations held with clients and detail the conversation in the medical record following the date entry and the initials of the team member who was involved in the discussion. This allows others to properly identify and communicate with the patient's caregiver as necessary.

Each patient record should contain one medical record for that patient only. In the case of a veterinary facility that treats large animals and farm production systems, one chart may be used for the entire herd, as long as each animal treated is identified by a tag number or registration number to determine which animals have been examined and treated by the veterinarian. Laboratory research facilities should maintain their records in a similar fashion and may have regulations outlined by other veterinary laboratory animal associations.

When recording information in the medical record, some facilities use a format called **SOAP**. This stands for **Subjective–Objective–Assessment–Plan**. The **subjective information** is based on the animal's overall appearance and the health care team's descriptions of the animal. This may include the animal's attitude and the veterinary assistant's observations on the appearance of the pet. The **objective information** is measured facts about the patient that can be recorded. This includes vital signs, such as temperature, heart rate, and weight. The **assessment** of the patient is what the veterinarian determines to be the diagnosis or the patient's problem. The **plan** is the treatment or procedure that is to be given to the patient. The SOAP is commonly used in the progress notes and the physical examination section of the medical records (see Figure 2–11).

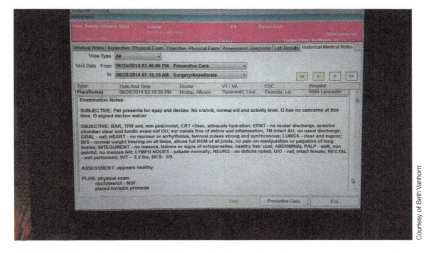

FIGURE 2–11 The SOAP outline of medical notes allows for proper medical recording and ease of patient care.

Filing Medical Records

The veterinary medical record must be kept in a file system, such as a paper file or a computer file. Many facilities will use a combination of both paper and computerized records. The paper-based filing system must meet the legal requirements outlined in the state's veterinary practice act. Paper forms and information are usually stored in a manila file that may have pockets or clasps to contain each form.

There are a variety of ways to file and store medical records in the veterinary office. The most common filing method is an **alphabetical filing system**. This system organizes files by letter, usually according to the client's last name. Each letter of the alphabet is assigned a color code; the first two to three letters of the owner's last name are attached to the outside of the medical file. It is easy to identify charts according to color and letter sequencing on a shelf or in a file cabinet (see Figure 2–12).

Another method used in veterinary facilities is the **numerical filing system**. The numerical filing system can be used in two ways: the client may be assigned a number, or each individual patient is assigned a number. Each number digit is assigned a **color code**. The color-code method assigns a color for each label so that when filed together, it is easy to note a misfiled chart by the color of the labels. The numbers are placed on labels and then placed on the file, aiming for ease of use and location of file sequencing (see Figure 2–13). Each file should also contain a label for the year the patient was last seen. This allows the staff to determine which patients are overdue for health exams. Most files are kept for seven years from the patient's last visit. This may be due to animal's

FIGURE 2–13 A numerical filing system uses groups of numbers that identify the patient.

passing away or clients moving or transferring to another hospital. When the medical record is no longer in use, it may be deleted or inactivated. After seven years it may be discarded and is often done so through document shredding to maintain confidentiality. Remember, some clients may move back to the area or decide to return to the clinic.

Consent Forms and Certificates

The veterinary consent form exists to identify for each client the procedures and prices that will be incurred while a patient is under the care of the veterinary facility. The client should read and sign the consent form to show agreement to the medical care necessary and understanding of any potential risks that may be involved (see Figure 2–14).

FIGURE 2–12 The use of alphabetical letters and color codes in a filing system makes filing easy and helps identify files that have not been filed correctly.

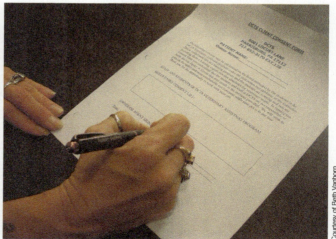

FIGURE 2–14 A consent form is signed by the client and outlines the patient services and procedures that will be completed.

The consent form also outlines the costs of the procedures and thus enters the client into a legal agreement and authorization for the veterinary facility to provide care to the patient. The owner should receive a copy of the consent form, and the original should be placed in the medical record. **Euthanasia release forms** must be signed by the owner of the animal. This form states that the owner agrees to have the pet humanely put to sleep and death will result. These forms often state that the pet was not knowingly exposed to rabies and no human or animal was known to have been bitten by the pet within the last 15 days (see Figure 2–15).

Patients that have undergone surgical alteration to remove the reproductive organs should be issued a **neuter certificate** (see Figure 2–16). This provides the owner with proof that the pet has been spayed or castrated and is no longer sexually intact or capable of reproducing. This certificate may allow pet owners to purchase licenses at a reduced fee or may be required if a pet has been adopted from a shelter.

A **rabies certificate** is issued after a pet has received its rabies vaccine (see Figure 2–17). Many animals are legally required to have rabies vaccines: owners must show proof if an animal bites or injures another person or pet. State laws for pet ownership vary from state to state and each pet owner should review their state's dog and animal laws. Rabies certificates present client and patient information as well as vaccination information that includes the manufacturer's name, vaccine expiration date, vaccine serial number, and rabies tag number. The veterinarian administering the vaccine must sign the certificate, or a stamp of the original signature is used.

A **health certificate** must be issued if an animal is being transported out of state or out of the country (see Figure 2–18). This requirement must include a physical examination from a licensed veterinarian that states the animal is healthy and free of disease and that the required vaccines are up to date. A health certificate usually includes vaccine records, a statement of health, and any required test results that may be necessary for travel or show purposes. A pet traveling within the United States may require an **interstate health certificate**; those traveling outside the United States may require an **international heath certificate**. These animal health documents are referred to as certificates of veterinary inspection (CVI). Veterinarians completing CVIs must be accredited by the United States Department of Agriculture (USDA). Different states and countries have various requirements for animals being imported or exported and animal owners and transporters should research and review all regulations. The most current regulations may be found on the USDA's website.

VETERINARY HOSPITAL
ANY STREET, ANY TOWN, XX 00000
NEUTER CERTIFICATE

Patient name: _____ Breed: _____

Gender: _____ Age: _____ Color: _____

Client Name: _____

Address: _____

This neuter certificate certifies that the above name animal has been surgically altered and is no longer capable of reproduction. The surgical procedure was performed by the veterinarian signing this certificate.

(Signature of Veterinarian) Date: _____

FIGURE 2-16 Sample neuter certificate.

EUTHANASIA RELEASE

I, the undersigned, hereby certify that I am the owner of the animal, and hereby give permission to Veterinary Animal Hospital full and complete authorization to humanely euthanize the stated animal in the appropriate manner the veterinarian deems fit. I hereby release the veterinarians and staff from any and all liability for the euthanasia procedure of said animal. I understand that the euthanasia procedure results in death. I do hereby certify that the stated animal has not bitten any person or animal in the last 15 days and to the best of my knowledge has not been exposed to rabies.

Signature of Owner:_____ Date: _____

FIGURE 2-15 Sample euthanasia release form.

28 SECTION I Practice Management and Client Relations

CITY HOSPITAL
ANY STREET
ANY TOWN, SS 00000

RABIES CERTIFICATE

Patient Information

Client Name: _____   Patient Name: _____

Address: _____   Species: _____

_____   Breed: _____

Telephone (H): _____   Sex: _____

(W): _____   License No.: _____

Vaccination Information

Vaccination Date: _____   Expiration Date: _____

Vaccine Serial No.: _____   Producer Killed: _____

Comments:
_____
_____
_____

_____
Signature of Veterinarian

FIGURE 2–17 Sample rabies certificate.

Medical Records as Legal Documents

Veterinarians and their staff shall protect the personal privacy of clients by maintaining **confidentiality** of all information of the client and patient records. Legally, all information within the veterinary medical record is private and not to be discussed unless the client gives approval for the information to be released. This may be necessary when veterinarians release records for clients who are moving to a new location, changing veterinary facilities, or in need of a referral to a veterinary specialist. When a record is released from the facility, it may only be done as a summary or copy of the record. Veterinary assistants should be familiar with using a copier. When requested by law or by the client, a veterinarian may not withhold the release of a copy of the medical record. The client should sign a waiver of confidentiality form to allow medical records to be released (see Figure 2–19). This legal confidentiality also pertains to all areas of social media, posting of pictures within the veterinary facility, and gossip-like discussions, such as statements made by clients.

Veterinary medical records serve as legal documents and as communication among members of the veterinary staff. The records provide documented evidence of an animal's illness, hospital care, treatment, and surgical history. The records must be maintained in a way that permits any veterinary staff member to read the notes and be able to proceed with the proper care and treatment of the animal. The records must be accurate and properly maintained should they ever be necessary to be reviewed in a legal case.

VETERINARY HOSPITAL
ANY STREET, ANY TOWN, XX 00000
CERTIFICATE OF VETERINARY INSPECTION

Date: _____

Client Name: _____ Patient Name: _____
Address: _____ Species/Breed: _____
Telephone: _____ Color: _____ Gender: _____ Age: _____
Vaccination: _____ Date: _____ Vaccination: _____ Date: _____
Vaccination: _____ Date: _____ Vaccination: _____ Date: _____
Vaccination: _____ Date: _____ Vaccination: _____ Date: _____
Test: _____ Result: _____ Date: _____
Test: _____ Result: _____ Date: _____
Test: _____ Result: _____ Date: _____

This health certificate certifies that the animal listed is in adequate health and has been vaccinated according to the above information. The animal has been examined by a licensed veterinarian with 30 days of issue of this health certificate.

(Signature of Veterinarian)

(License Number)

FIGURE 2–18 Sample CVI health certificate.

CONSENT TO DISCLOSURE OF MEDICAL RECORD
Waiver of Confidentiality
By Authorized Patient Representative

I, _____ (name of owner or agent) _____ , the _____ (owner or agent of owner) _____
of _____ (name of animal) _____ , a _____ (breed) _____ ,
do understand that the information contained in _____ (name of animal) _____ 's medical record
is confidential. However, I specifically give my consent for _____ (name of veterinary practice) _____ to
release the following information concerning _____ (name of animal) _____ to
_____ (name of party to whom information is being released) _____ .

The above-listed information is to be disclosed for the specific purpose of
_____
_____

It is further understood that the information released is for professional purposes only. This information may not be given in whole or part to any person other than that stated above.

_____
Signature of Authorized Representative

_____
Date

FIGURE 2–19 Sample consent of disclosure certificate.

Computerized Veterinary Programs

Many veterinary facilities are "going paperless" and keeping electronic medical records (see Figure 2–20). Computers are an excellent way to utilize efficiency practices and provide neat, professional-looking records. There are several types of veterinary software programs available from software manufacturers that are useful in many areas of the facility. Many programs allow for medical record keeping, inventory monitoring, laboratory reporting, scheduling of appointments, invoicing, payroll, printing of pharmacy labels, digital X-rays, and more. Many facilities now have computers located at the reception desk, in each exam room, the treatment area, the pharmacy, and lab areas.

Paperless medical records are often time-stamped when entries are made; the initials of the staff member completing the records may be added, too. Many programs prevent changes to a medical record after a 24-hour period for permanent documentation and to prevent records from being falsified. A paperless record is completed in the same manner as a written record; the same concepts of recording and maintaining records are followed. Computer systems and data should be backed up daily, with back-up disks kept in a fireproof location or back-up systems located outside of the facility.

FIGURE 2–20 Recording paperless medical records.

SUMMARY

All veterinary assistants are required to have a basic knowledge and understanding of veterinary medical record keeping. Medical records are used to summarize a patient's history. Medical records are legal documents that allow veterinarians and the veterinary health care team to properly care for each patient seen at the facility and as a means of communicating information to each other. A patient's medical records tell the history of the patient in a diary-like script each time an animal visits the facility. It is important that every member of the veterinary health care team learn to accurately record in the medical record. Medical records are a vital tool for every veterinarian and must be properly assembled, recorded, and maintained. Medical records are legal documents and valuable tools in the veterinary facility. The veterinarian and health care team rely on medical records to provide a history of each patient. This includes proper labeling, information, forms, and accurate notations. It is important to record everything and be accurate and neat.

Key Terms

account history a client's previous invoices, which can be viewed at any time to see past payments and charge

alphabetical filing system the organization of medical records by client or patient last name

assessment a veterinary diagnosis of problem

cage card the identification and location of patient while in care of facility

client the owner of the animal

color code a method used to file medical records for visual ease of use

confidentiality the maintenance of privacy of medical information

consent form   the permission given by owner to treat animal

credit application   paperwork completed by clients to allow them to obtain a line of credit to pay for veterinary services

estimate   the approximate cost of services

euthanasia release form   a signed permission from the owner to put a pet humanely to sleep, thus resulting in death

health certificate   documentation that an animal is in good health and can be transported out of state or out of the country

international health certificate   documentation of an animal's health and vaccine history, stating that it is approved for travel outside of its home country

interstate health certificate   documentation of an animal's health and vaccine history, stating that it is approved to travel within the United States; certificate of veterinary inspection (CVI)

invoice   a bill given to a client that lists the costs of services and procedures

medical record   written and recorded information on the care and treatment of the animal

neuter certificate   documentation of an animal's having been spayed or castrated and is no longer able to reproduce

numerical filing system   the organization of medical records by numbering and/or color-coding

objective information   measured facts (e.g., vital signs)

patient   the animal being treated

plan   treatments or procedures provided

progress notes   documentation chronologically detailing the patient's history, treatments, and outcomes

rabies certificate   documentation of the animal's having received a proper vaccination for rabies

SOAP   subjective information–objective information–assessment–plan; a method used for documentation in the medical record

specialized payment plan   a method approved by the veterinarian that allows a down payment with additional regular payments until the balance is paid

subjective information   observations of the animal's appearance and behavior made by the veterinary staff

travel sheet   paperwork list that codes, services, and procedures that allows the staff to track fees to charge the client

veterinary–client–patient relationship (VCPR)   a relationship established between the patient, client, and veterinary staff based on trust, expertise, and duty to care for animal

Review Questions

True or False

Read each statement and circle T if it is true or F if false. If a statement is false, rewrite or modify it to make it true.

1. The subjective information in a veterinary medical record includes factual information, such as the animal's respiratory rate and temperature.   T   F

2. When taking a patient's information over the phone, you may record the information in pencil, but use only blue or black ink when recording in the medical record.   T   F

3. Each medical file should have a format or sequence of where each form is placed in the file. T F
4. The SOAP format is used as a system for filing medical records. T F
5. Upon signing a consent form, the original should be given to the owner and a copy placed in the medical record. T F
6. Erase all errors in a medical chart or use correction fluid to cover mistakes and then correct them immediately. T F

Short Answer

The following items were recorded incorrectly in a medical chart. Describe the way to correct the mistakes.

7. Wt. 75.6 lbs—The weight should have been recorded as 57.6 lbs. How will you correct the mistake?
8. T-102°F—The temperature should have been recorded as 101°F. How will you correct the mistake?
9. Why is it standard practice to keep medical records for at least seven years after the patient's last visit?
10. What two systems are commonly used when filing medical records? Explain each one.
11. Discuss the importance of consent forms in the veterinary facility.
12. List the types of certificates that may be obtained by the veterinarian.
13. Discuss the difference between the: "patient" and the "client."
14. Define the abbreviation VCPR. Discuss the criteria of this term in the veterinary facility.
15. List five items that may be found in a medical record.

Clinical Situation

Mrs. Patel scheduled an examination for her dog, "Gretchen," with Dr. Camillion due to a strange swelling on its abdomen. The veterinary assistant greeted the client and escorted her to an exam room. The veterinary assistant quickly reviewed Gretchen's medical file and noted the size, location, and appearance of the swelling. When Dr. Camillion arrived to examine Gretchen, he reviewed the chart and asked Mrs. Patel some questions about the dog's recent health. The vet noticed in the chart that Gretchen had been seen in the clinic several years ago for a similar problem, which had been an abscess or buildup of fluid from a minor infection.

On examination of the dog, Dr. Camillion quickly determined that Gretchen had another abscess. He explained the situation to Mrs. Patel, who had forgotten about the previous problem. The vet drained the fluid and prescribed the same antibiotic used for the last infection. A recheck exam appointment was scheduled for 5 days.

1. How did the use of the medical records play out in this case?
2. What might have happened if the medical record had not been reviewed?
3. How did proper medical record filing help the veterinary assistant in this situation?

COMPETENCY SKILL 2

Medical Record Assembly

Objective:

To create a medical record for a new patient.

Preparation:

- Familiarize yourself with the proper procedures used in the facility to complete the medical record. Note the forms used and the sequence or order within the record they are to be placed.
- Select an adequate space to assemble the medical record.

Materials:

- Manila file folder
- Alphabetical or numerical labels
- Blank label
- Year labels
- Client/patient forms: patient history forms, progress notes, laboratory forms, surgical forms, radiology forms, miscellaneous forms, cage card
- Blue or black pen

Procedure:

1. Place the correct labels on the manila file folder according to the filing system used in the facility. This includes the owner and patient information label and filing labels.
2. Place the year label on the manila file folder.
3. Record the owner and patient information on each form using a blue or black ink pen.
4. Secure the forms and file them in proper sequence and location within the medical record.
5. Correctly file the medical record in the facility filing system.
6. Put all materials in their proper storage area.

COMPETENCY SKILL 3

Filing Medical Records

Objective:

To properly file records as per office protocol.

Preparation:

- Determine the type of filing system used by the facility.
- Locate completed files and the filing area.

Procedure—Alphabetical Method:

1. Locate the medical record file and the owner's last name.
2. The medical record is filed alphabetically by the owner's last name in the filing cabinet.

3. If there are multiple last names, the record is filed alphabetically according to the owner's first name.
4. If there are multiple files for the same name, the record is filed alphabetically according to the patient name.
5. Check each record filed to make sure no misfiling errors occur.

Procedure—Numeric Method:

1. Each client is assigned a number according to the facility filing system.
2. A separate recording system, either index cards listed alphabetically by name or a computer system, should record each client number.
3. Each record is then filed by number in the filing cabinet.

Procedure—Files for Appointments:

1. Pull the record from the shelf or cabinet according to the filing system method. Pull each file for the daily scheduled appointments.
2. Place the files in a holding area for the daily appointments. File each medical record in order of scheduled appointments.
3. After the appointment or patient discharge, place files in a bin for refiling.
4. Complete filing according to the facilities filing method. Recheck.
5. Before leaving the veterinary office area, make sure all files have been returned to the proper location.

CHAPTER 3: Scheduling Appointments and Computer Applications

Objectives

Upon completion of this chapter, the reader should be able to:

3.1 Properly schedule appointments in a veterinary clinic
3.2 Properly discuss the necessary client and patient information needed to set up an appointment
3.3 Identify the importance of hospital policies related to scheduling appointments
3.4 Implement hospital policies when scheduling appointments
3.5 Manage the appointment schedule efficiently and accurately
3.6 Identify the basic parts of the computer and computer equipment
3.7 Explain the importance of veterinary software programs
3.8 Discuss the uses of a computer system in a veterinary facility
3.9 Apply basic keyboarding skills within the veterinary clinic
3.10 Discuss inventory control and management procedures

Introduction

Appointment scheduling is an important part of the administrative duties in a veterinary medical facility. It is important to follow the facility's scheduling procedures to pace each veterinarian's day appropriately. This area of veterinary office procedures requires efficiency and knowledge of common medical problems so that the veterinary health care team is not overworked or stressed during office hours. Many veterinary offices use computers and veterinary software programs that maintain records, schedule appointments, record client and patient information, and handle invoicing and accounting for the veterinary facility. People have different levels of computer experience and knowledge. Many computerized facilities will have expectations of their employees to have a general understanding of computers and basic keyboarding skills. Therefore, it is important to develop an understanding of the computer system and software programs within the veterinary clinic.

The Veterinary Appointment Book

The appointment book is a valuable tool for recording notes and managing the daily workings of the veterinary facility (see Figure 3–1). Many styles and formats of appointment books or schedules can be used in a veterinary facility. The type of appointment book used is determined by the type of schedule that the hospital has established.

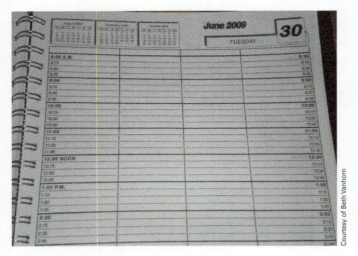

FIGURE 3-1 The appointment book.

Some facilities may use veterinary software that has a **computerized appointment book**. A computerized appointment book is scheduled and maintained on the veterinary computer system (see Figure 3-2). The procedures that accompany the software should be used when recording information in the computer. The computerized program works much like the traditional written appointment book schedule but allows a more organized scheduling system that saves all information for future use. Many veterinary practices are becoming completely computerized and moving away from using paper in facilities: this is known as "going paperless."

The appointment scheduling format should be developed to maximize the needs of the hospital and to minimize the wait time for clients. If appointments are running behind, it is important to notify all clients of the delay and allow them to continue to wait or reschedule for another date or time. The length of time of the appointment is critical in appropriately scheduling the patient. Certain visits will take longer than others; the facility should have an outline of necessary time frames for specific types of appointments. Facilities that employ credentialed veterinary technicians can utilize technician appointments scheduled to help allow the doctor to see sick patients. Some examples of technician appointments include nail trims, anal gland expressions, and blood sample re-testing procedures. This technician resource is an asset to every hospital and will maximize the facility's efficiency

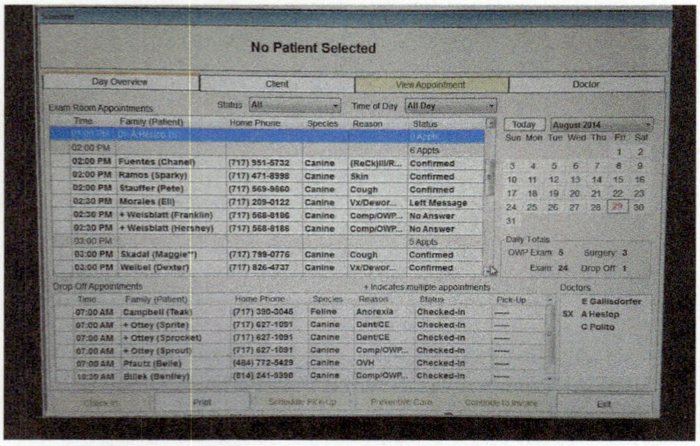

Types of Appointment Schedules

Veterinary facilities use three basic types of schedules: wave schedule, flow schedule, and fixed office hours.

Wave Schedule

The **wave schedule** is maintained by a veterinarian who sees a total number of patients within a set amount of time, typically an hour. Patients are usually seen in the order they arrive within the time frame. For example, a wave schedule may be set in the morning: the veterinarian would see four patients between 9 a.m. and 10 a.m., allowing 15 minutes for each patient. The wave schedule is helpful in situations where an appointment may not show up or is late.

Flow Schedule

The **flow schedule** is based on each patient being seen at a specific time, or **interval**, such as every 15 or 30 minutes. The most common interval is 15 minutes (see Figure 3-3); thus, appointments are scheduled for every 15 minutes in the appointment book. Some medical problems may require more or less time than the scheduled interval.

The flow schedule is the most common type of veterinary schedule. It allows for monitoring time and keeping the staff on schedule. However, if not managed appropriately, it is easy to get behind schedule or overbook patients.

Fixed Office Hours

A facility with **fixed office hours** may have available office hours throughout the day; patients are seen in the order in which they enter the facility as walk-in appointments. A disadvantage to this schedule type is a loss of control over the flow of patients.

FIGURE 3-2 Sample computerized appointment schedule.

CHAPTER 3 Scheduling Appointments and Computer Applications

8	00 / 15 / 30 / 45
9	00 / 15 / 30 / 45
10	00 / 15 / 30 / 45
11	00 / 15 / 30 / 45
12	00 / 15 / 30 / 45
1	00 / 15 / 30 / 45
2	00 / 15 / 30 / 45
3	00 / 15 / 30 / 45
4	00 / 15 / 30 / 45
5	00 / 15 / 30 / 45

FIGURE 3-3 Sample appointment book style.

Types of Appointments

It is important for employees who schedule appointments to be trained and knowledgeable in common medical conditions so they can determine approximately how long the veterinarian will need to be with the patient. Many facilities have guidelines of the amount of time needed for types of procedures. If an associate is unsure of the amount of time needed on the schedule, it is important they discuss this information with the veterinarian.

Routine Appointments

Routine appointments are visits scheduled in the appointment book. Examples include yearly physical exams, vaccinations, and sick patient visits. Each routine appointment may be spaced out according to how long the expected procedure or exam will take. Each veterinarian may need a different amount of time for a particular appointment.

Surgical Appointments

Surgical appointments are usually kept in a separate appointment book and schedule. Surgeries are commonly done at specific times in the day and set by the veterinarian who will be on surgical duty on a particular day. Surgeries involve anesthesia and a sterile or clean, specialized environment to perform specific procedures. A hospital may designate specific surgery days or times. The appointment schedule for surgical cases should include the veterinarian's name, the type of procedure, and the time of the surgery. Many veterinary hospitals have a set **drop-off time** for surgical patients, which is a specific time an animal arrives prior to a surgery or another procedure.

When scheduling a surgery appointment, the following information should be included on the schedule: owner name, phone number, patient name, breed or species, and surgical procedure. Veterinarians usually have pre-surgical instructions that are given to clients, and these instructions should be provided to the owner at the time the surgery appointment is set up. Common presurgical instructions that should be communicated to clients when they make the appointment include the following:

- Do not feed your pet 12 hours prior to surgery.
- Do not give your pet water 8 hours prior to surgery.
- Bring in your pet at the specified drop-off time.
- Sign the surgical release form.
- Sign presurgical blood work consent forms.
- Sign any estimates or form disclosing payment for services.
- Bring any medications currently prescribed.

Many hospitals will hand out or email these instructions to their clients.

Emergency and Walk-in Appointments

Many hospitals offer **emergency services** and **walk-in appointments**. An emergency is the need for an immediate appointment and may be a life or death situation. Common emergencies include being hit by a car, bloats, and severe bleeding from wounds. Emergencies should be clearly identified as needing immediate attention by the veterinarian. A walk-in client, on the other hand, does not have an appointment but comes to the hospital wishing to see the veterinarian. Even if the facility doesn't allow walk-in appointments, they still occasionally happen.

The best way to serve emergency situations and walk-ins is by reserving several slots in the daily appointment schedule for such cases. Clients will know that in an emergency, there is a good chance the clinic will see their animal. This flexibility will also keep the veterinary health care team on schedule and not delay services to other clients. If the emergency or walk-in slots are not used, the staff has some catch-up time. When emergencies or walk-ins call or arrive at the hospital, the veterinary technician or veterinarian should be informed immediately so the situation can be handled appropriately. Once the technician or veterinarian has control over the case, the patient's chart is pulled or assembled (for a new patient).

Drop-Off Appointments

Many veterinary facilities offer **drop-off appointments**. Clients may set a drop-off time for routine appointments, surgical procedures, or grooming services. This gives both the owner and the staff the flexibility to work on the patient throughout the day. There are set drop-off and pick-up times. Charts for these appointments are usually kept with surgical appointments and pulled when the patient arrives. Owners must complete consent forms for services rendered and provide contact numbers for the health care team if any questions come up.

House Calls

Many clinics offer **house calls** and **mobile services** that will be provided at the client's home. Mobile services usually include a vehicle that has equipment and a place within to perform procedures. The facility should have a policy on scheduling house calls. It may be necessary to take a message and call the client back with a date and time of a house call, as some facilities may have limited hours of offering this service. Once a date and time are scheduled, it should be noted in the appointment book. The staff will need to consider that the veterinarian will be off-premises and may need additional staff members to cover back at the hospital. The medical file for house calls and mobile services should be completed prior to the date and time of the appointment. It is a good idea to include extra forms that may be necessary. Also include the reason for the appointment so the veterinary health care team has adequate information regarding the equipment and supplies that will be needed for the appointment.

Organization of the Appointment Book

The appointment book must allow enough space to accommodate the hours that the office is open. It is also important to make sure there is enough room to record details about the appointments and list the scheduled time intervals (see Figure 3–4). Most facilities record the following information in the appointment book:

- client name
- phone number
- patient name
- breed or species
- brief note stating the reason for the visit

This critical information enables the veterinarian and the health care team to be prepared and efficient for each scheduled appointment. Some clinics or veterinarians may request other pieces of information in the appointment space. It is important to know the facility's expectations and rules for scheduling appointments.

		DAY DATE	MON 6/23
			00
Julie Simon, 555-0001; Persian; Amos; Eye Infection		8:	(15)
			30
			45

FIGURE 3–4 There should be enough space in the appointment book for the veterinary assistant to write in all the necessary information.

Scheduling Appointments

The top priority when scheduling appointments should be efficiency. Many facilities will have designated time frames for many common and routine procedures and appointments (see Figure 3–5). Information in the appointment book should always be recorded in pencil so that changes can easily be made should clients need to reschedule (see Figure 3–6). Appointments that need to be cancelled or rescheduled should be erased. It is not appropriate to cross out or use correction fluid to cover an entry in the appointment book. This can make the book illegible, which may confuse staff members and lead to errors. The appointment is always recorded in the book using the same method as the filing system. If the filing system method is alphabetical, then record the client name by last name first, then first name. If the filing system method is numeric, then record the client number and then name. When scheduling appointments, check to make sure the information is accurate and neat and all entries are complete. The more time spent managing the appointment book, the more punctual and less stressed the entire veterinary health care team will be.

Common Veterinary Procedures and Time Estimates	
• Yearly exam and vaccines	20 min
• New puppy or kitten exam	30 min
• Repeat puppy and kitten vaccines	15 min
• Recheck exam	10 min
• Suture removal	5 min
• Hospital discharge appointment	15 min
• Sick pet exam	30 min
• Consultation exam/second opinion	30 min
• Behavioral consult	45 min
• Bird, reptile, or exotic exam	45 min
• Reproductive exam	30 min
• Artificial insemination	30 min
• Chronic problem exams (ears, skin)	15 min
• Nail trims	10 min
• Pre-surgical blood-work collection	10 min
• X-rays or radiographs	30 min
• Heartworm check	10 min
• FeLV test	10 min
• Fecal check	10 min

FIGURE 3–5 The veterinary assistant should be knowledgeable about the times in the schedule that are allotted to routine types of appointments and procedures.

FIGURE 3-6 Enter information into the appointment book with a pencil.

FIGURE 3-7 The veterinary assistant should call the client in advance to confirm the appointment and note any changes in the appointment book.

Follow-up appointments should be scheduled before the client leaves the facility. Clients may well forget to call to schedule a next appointment.

Working with the Client

After an appointment has been scheduled, read the date and time back to the client and make sure all client and patient information is accurate. This will allow for any necessary changes to be updated in the charts. When possible, give **appointment cards** to remind clients of their upcoming dates. It is important to make sure new clients have directions to the facility so they do not get lost or arrive late. It is also a benefit to clients and the veterinary staff to confirm all appointments a day in advance. Note that the appointment has been confirmed in the schedule or medical chart (see Figure 3–7). Many hospitals have a policy on clients that do not show up for an appointment: this can serve as a history in the medical chart.

Many facilities with computerized systems now send appointment reminders and upcoming vaccine reminders through email or texting. Clients may also be able to access their pet's medical records online.

Cancellations

It is important to ask clients to call if they cannot keep their scheduled appointment. Many hospitals require cancellations 24 hours in advance. Make a note of any cancelled appointments or when owners don't show up for a scheduled visit.

Blocking Times

When scheduling the day, it is necessary to consider certain times during the day: when surgeries are being performed; the office is closed; employees have lunch, breaks, or days off; and emergencies are scheduled. Blocking off times for emergency appointments on the schedule will prevent the flow of scheduled appointments from being disrupted. This may be done by highlighting or drawing an "X" through areas that the veterinarian would prefer to use as catch-up time or for emergencies (see Figure 3–8). Computerized systems notes to be placed in the daily schedule to announce times when appointments should not be made. An example would be to block off one 30-minute window in the morning and another in the afternoon to account for emergencies that are brought into the clinic; this prevents the staff from getting behind schedule and keeping scheduled clients waiting.

Scheduling for Multiple Veterinarians

If the facility has multiple veterinarians on staff, doctors are scheduled according to their available appointment times. The appointment book should accommodate the number of veterinarians who work on any given day and who may work during the same office hours. Some veterinary facilities offer appointments with veterinary technicians; these facilities may need to keep an elaborate or customized appointment book. A color-coded system can denote multiple doctors' schedules and times when appointments should *not* be scheduled. Highlighters and colored pencils may be useful (see Figure 3–9).

Scheduling Mishaps

A carelessly created schedule will cause chaos, angry clients, and stress for the veterinary health care team and patients. No one likes to wait, and waiting time tends to lead to unhappy and dissatisfied clients. An unhappy client will not return to the facility. An employer will not be tolerant of poor scheduling, either. Figure 3–10 lists some of the dos and don'ts for scheduling appointments. It is important to note that certain days or times of the day may require

Monday, May 9

	00			
8	15		Out	
	30			
	45		Accountant	
	00			
9	15			
	30			
	45			
	00			
10	15			
	30			
	45			
	00			
11	15			
	30			
	45			
	00			
12	15		Lunch	
	30			
	45			

Tuesday, May 10

	00			
8	15		Out	
	30			
	45			
	00			
9	15		Dentist	
	30			
	45		Appt.	
	00			
10	15			
	30			
	45			
	00			
11	15			
	30			
	45			
	00			
12	15		Lunch	
	30			
	45			

FIGURE 3–8 Times that are not available for appointments should be blocked off in the appointment book.

Multiple Doctor Schedule

Dr. Jones	Dr. Smith
Date:	Date:
8:00	8:00
8:15	8:15
8:30	8:30
8:45	8:45
9:00	9:00
9:15	9:15
9:30	9:30
9:45	9:45
10:00	10:00
10:15	10:15
10:30	10:30
10:45	10:45
11:00	11:00
11:15	11:15
11:30	11:30

FIGURE 3–9 Sample of an appointment book for multiple veterinarians. (*Continues*)

Dr. Jones	Dr. Smith
11:45	11:45
12:00	12:00
12:15	12:15
12:30	12:30
12:45	12:45
1:00	1:00
1:15	1:15
1:30	1:30
1:45	1:45
2:00	2:00
2:15	2:15
2:30	2:30
2:45	2:45
3:00	3:00
3:15	3:15
3:30	3:30
3:45	3:45
4:00	4:00
4:15	4:15
4:30	4:30
4:45	4:45
5:00	5:00
5:15	5:15
5:30	5:30

FIGURE 3-9 (Continued)

Appointment Book Dos and Don'ts

DOS	DON'TS
• Use a pencil to schedule appointments. • Confirm all appointments. • Give reminder cards with the date and time of the appointment. • Give hospital directions to new clients. • Allow open slots for emergencies and walk-ins. • Make sure every appointment is accurate and complete. • Record cancellations and no-shows. • Keep the appointment schedule neat and easy to read. • Clearly identify multiple doctors' schedules. • Identify when there are NO office hours by drawing an X over the blocked-off times.	• Use pen to schedule appointments. • Assume clients remember everything they're told. • Overbook the veterinarian and staff. • Neglect to follow hospital scheduling policies. • Schedule house calls without consulting the veterinarian. • Cross out or use white out to make changes in the appointment book. • Forget to get a phone number for all appointments.

FIGURE 3-10 Appointment book dos and don'ts.

special scheduling: such as when staff members and veterinarians will not be available (for lunch time, surgery time, and treatment times). These issues may need to be addressed in the appointment schedule to avoid both conflicts and scheduling an appointment when a doctor is not available. Times when staff or veterinarians are not available should be noted in the appointment book. This may be best done by drawing an "X" through a time frame.

Policies and Procedures

The veterinary clinic should provide a detailed policy manual and procedures manual to each staff member so everyone understands the rules and regulations of the facility (see Figure 3–11). The **policy manual** explains the employee rules and covers the expectations for the staff—such as dress code, vacation time, personal time, and sick leave. Rules apply to all staff employees; some regulations may be based on the length of time that an individual has been employed by the facility. The **procedures manual** outlines the rules about the workings of the veterinary facility itself, such as client rules, scheduling methods, payment procedures, and client visitation times. This information is based on the rules of the clinic and is part of a training method for all new employees so that all staff members complete their duties and uphold their responsibilities in the same manner. These manuals will outline the employer's expectations of the employees and provide a basis of consistency on how issues at the practice are handled.

Basic Computer Equipment

Most computer systems in a veterinary facility use a simple setup with easy-to-use **hardware**. The hardware refers to the physical parts of the computer and its equipment: input devices, output devices, and the CPU. The primary input devices are the **keyboard**, which allows for typing; a **mouse**, which allows the user to move the **cursor** (or blinking line) on the computer screen; and the **scanner**, which allows documents and images to be copied and stored electronically. The primary output devices of the computer are the **monitor**, which is the screen; the **modem**, which allows Internet access; and the **printer**, which allows for printing forms and invoices from the computer system (see Figure 3–12). The **central processing unit (CPU)** is the brains of the computer. It carries out the instructions designed by **software** (a program or set of instructions the computer follows). It takes the data input into the system, configures it based on the instruction received from the software, and sends it to the output device. In addition to a general knowledge of operating the

Policy and Procedures Manuals

POLICY MANUAL ITEMS	PROCEDURES MANUAL ITEMS
Dress Code/Attire	Payment Policy
Personal Appearance	Appointment Policy
Vacation Time	Surgery Policy
Sick Time	Scheduling Policy
Personal Time	Opening/Closing Procedures
Attendance Policy	Admission/Discharge Procedures
Continuing Education	Emergency Procedures (fires, robbery, injuries, etc.)
Benefits (employee discounts, uniform allowance, reimbursements)	Answering Phones
Codes of Conduct	Invoicing and Billing
	Computer System Procedures
	Client Visitation Hours

FIGURE 3–11 Information found in the policy and procedures manuals.

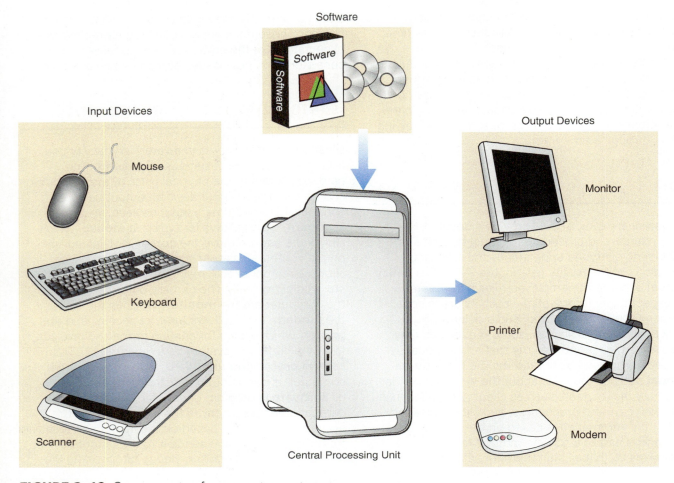

FIGURE 3-12 Components of a computer system.

computer system, the veterinary assistant should also know how to maintain the printer, such as how to load the paper or change the **toner**, which is either black or colored ink for printing.

Many clinics will purchase veterinary software that simplifies the tasks pertaining to office and hospital procedures (see Figure 3-13). These software programs include **word-processing programs** that allow typing of forms and notes (similar to using a typewriter). Common word-processing programs include Microsoft Word and WordPerfect. Other components to a veterinary software program include patient and client information; medical recording; surgical, laboratory, anesthesia, and radiology reports; scheduling appointments; invoicing and billing; inventory management; employee records and information; and consent forms. There are many veterinary software programs available. It is important to complete specific training on the program used in the clinic.

The veterinary facility should determine which type of veterinary software program meets the clinic needs. Many software programs require regular updates and

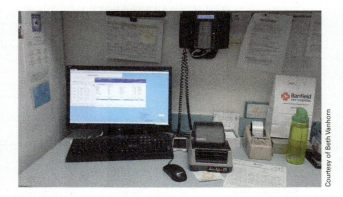

FIGURE 3-13 Example of a veterinary computer system.

new program additions. The software company should be reputable, available to help troubleshoot problems, and offer training opportunities within the clinic. Many computer programs require advanced computers and may require equipment upgrades after several years of use. The computer equipment must be suitable to meet the needs of the software and have enough memory,

RAM, and storage to meet the needs of a veterinary hospital. Security measures should be put in place to prevent a virus from entering the computer system.

Veterinary clinics also use such equipment as a copier machine, scanner and a fax machine. The **copier machine** is used to make paper copies of records and other documents (see Figure 3–14). Copiers must be maintained with toner and paper. They may need additional maintenance and service from the manufacturer. A **fax machine** is used to copy and send documents via a telephone line (see Figure 3–15). Fax machines copy paper documents and transmit them to other fax machines via a telephone line. A fax machine requires

FIGURE 3–16 Veterinary assistants may be required to operate a cash register or drawer in the veterinary facility.

toner and regular maintenance and service, too. Some veterinary offices may use a cash register to keep money sorted and contained (see Figure 3–16).

Keyboarding Skills

Keyboarding skills include a basic understanding and use of functions on the computer keyboard and overall typing skill. The veterinary assistant should know basic keyboarding techniques and habits and understand keyboard functions (see Figure 3–17). Keyboarding skills also come in handy when working with cash registers, typewriters, fax machines, and

FIGURE 3–14 Veterinary assistants will need to understand the use and maintenance of a copier.

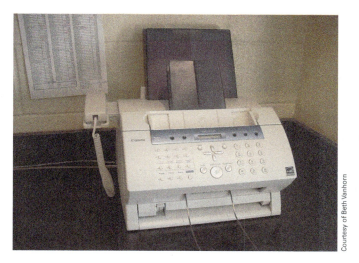

FIGURE 3–15 Veterinary assistants may often be required to fax information to other offices or to clients.

FIGURE 3–17 A basic understanding of keyboarding will assist a veterinary assistant in his or her ability to efficiently perform duties.

other electronic equipment with keyboards. It usually is not important how many **words per minute** the veterinary assistant can type, but rather that the typing and use of the keyboarding functions is accurate. The assistant should learn the positions of the keys and the functions that can be performed from the keyboard (see Figure 3–18). Programs and online tutorials allow for the practice and development of keyboarding skills. Excel is a popular program used to create **spreadsheets**. A spreadsheet is used to track large amounts of items in a single document; for veterinary facilities, this could be inventory supplies, clients, patients, and invoices. Common sites and resources for learning keyboarding and developing keyboarding skills include the following:

- Typing.com
- Mavis Beacon keyboarding software
- Typing Tutor
- UltraKey (www.ultrakeyonline.com)

Inventory Management and Control

Most veterinary facilities have systems to maintain and manage inventory supplies. One or more people may oversee **inventory management**. This involves knowing how much of each drug or supply is currently in the hospital. It also includes knowing the price, manufacturer, pharmaceutical company that handles the product, and expiration date. Routinely, each facility will do an **inventory** count. This is a physical count of all medications and supplies within the clinic to ensure that accurate records are kept and maintained. Expiration dates are checked and expired products must routinely be removed from the facility. A **want list** is helpful to determine when supplies of a product are running low and need to be reordered. Such a list is posted somewhere in the facility in a book, on a clipboard, or on a whiteboard for staff members to take note. Many facilities have a minimum amount of commonly used items that must be reordered before they are completely out. The hospital manager is typically in charge of ordering these items.

Many veterinary facilities use **computer inventory systems** to monitor the inventory and assist in purchasing items. The inventory of each product and supply item is entered into the computer system and relates to the invoicing system: any item sold or used is tied to the client invoice and removed from the inventory total. The hospital manager is typically in charge of maintaining the inventory and ordering supplies. Every staff member, including the veterinary assistant, plays a vital role in inventory management. Each person must record when items are running

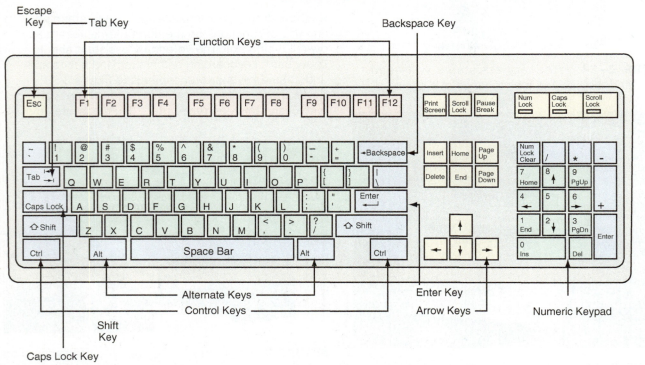

FIGURE 3–18 Standard keyboard.

low. They must also track inventory on **consumable products**, such as paper towels, gauze sponges, and cotton balls. These products are used daily, perhaps in treatment or surgical procedures, but are not itemized on invoices. Another important part of staff involvement in inventory control is unpacking order shipments. When orders arrive, the first step is to open the box and locate the **packing slip**. The packing slip itemizes each item that was ordered and has arrived in the package. It is important to make sure each item is accounted for and not damaged. The expiration dates of the new items should be noted in the computer and the new quantities ordered should also be placed in the computer inventory system. Items should be stored on shelves or in cabinets following the veterinary facility's procedures.

Some facilities may not be computerized or have a computer inventory system and instead use an **automated inventory system**. This is a physical inventory, meaning that items—office supplies, medical supplies, and pharmacy supplies—are counted by hand and numbers are logged into a notebook. This system must be controlled carefully to avoid running out of stock of supplies. Many facilities perform a yearly physical inventory to make certain all records are accurate.

SUMMARY

The written or computerized appointment book and the art of scheduling must be an important part of the clinic's policy manual and procedures manual. Each employee should know how to schedule appointments and understand the policies affecting each type of appointment. It is important to be efficient and manage the schedule closely. Having medical charts and equipment prepared ahead of time will make the veterinary facility's routines less stressful. Efficient and accurate scheduling will keep clients happy about the facility's services and keep them coming back. The veterinary health care team must be able to work together to create an efficient and safe environment. Computers are commonplace in today's veterinary profession. Employees should have a basic understanding of computers, equipment, and software programs. The accuracy of keyboarding skills and functions can be maximized by employee training and use of the veterinary software. The veterinary assistant should make sure all information entered is accurate and spelled properly. Training may be provided in the clinic or through use of a self-training program and printed manuals. Practice makes perfect!

Key Terms

appointment card a written reminder for clients that includes a date and time of the appointment scheduled for their pet

automated inventory system a system in which all supplies and products in the facility are counted by hand and a list is kept of how many items are in stock

central processing unit (CPU) the brain of the computer that carries out instructions asked by the software

computer inventory system a method of monitoring supplies and products within the facility in order to know when items need to be reordered

computerized appointment book a software program on the veterinary computer system used to schedule appointments and maintain medical information; allows future use of information in an organized manner

consumable products supplies used daily that are not included on the client invoice, such as paper towels, gauze sponges, and cotton balls

copier machine equipment used to make exact copies of paper or other documents

cursor a blinking line on the computer screen that determines where typing is occurring

drop-off appointment a set time for a client to bring in the patient for an exam, surgery, and other procedures according to the veterinarian's schedule

drop-off time a set time when an animal is expected at the veterinary facility for a later exam or surgery

emergency service a situation that must be seen to immediately and that may be a life or death situation

fax machine equipment used to copy and send documents via a telephone line

fixed office hours set times throughout the day when patients are seen in the order they enter the facility

flow schedule when appointments are scheduled based on each patient being seen at a specific time by the veterinarian during the day

hardware the physical parts of a computer and its equipment

house call when a veterinarian examines a patient in the owner's home

interval an amount of time set aside for a patient, such as 15 minutes

inventory a physical count of every medicine and supply item within the facility in order to keep accurate records

inventory management knowing the price, manufacturer, pharmaceutical company, amount, and expiration date of each item in the facility

keyboard buttons that have letters on them that allow typing and word formation on a computer

keyboarding the act of typing and development of skills used in computer functions and keys

mobile services a vehicle equipped with veterinary equipment, tools, and supplies that travels to a client's home to examine a patient

modem box-like equipment for a computer that allows access to the Internet

monitor the computer screen

mouse a movable instrument attached to a computer that allows the user to move the cursor

packing slip a paper enclosed in a package that details the items in the shipment

policy manual information provided to all staff members as to what the staff expectations are, such as the dress code, policies on vacation time, personal time, and sick leave

printer a machine that prints forms and invoices on a computer onto paper

procedures manual information provided to all staff members on the workings of the veterinary facility, such as client rules, scheduling methods, payment procedures, and client visitation times

routine appointment a patient visit scheduled during the office hours; examples are vaccinations, yearly exams, and rechecks

scanner equipment that allows documents and images to be copied and stored electronically

software a program or set of instructions that a computer follows

spreadsheet a program used to track large quantities of items in one file

surgical appointment an appointment scheduled for surgery that is usually outside of routine office hours

toner black or colored ink that allows printing with a printer, fax machine, or copier

walk-in appointment when a client without an appointment arrives at the hospital wishing to see the veterinarian

want list a list of items that are needed or low in stock within the facility that need to be reordered

wave schedule when a veterinarian sees a certain number of scheduled patients within a set amount of time during the day

word-processing program software that allows typing of a document

words per minute the number of words that can be typed within 60 seconds

CHAPTER 3 Scheduling Appointments and Computer Applications

Review Questions

Matching 1

Match the term to its definition.

1. appointment card
2. computerized appointment book
3. drop-off time
4. emergency services
5. fixed office hours
6. flow schedule
7. interval
8. mobile services
9. surgical appointment
10. wave schedule
11. procedures manual
12. routine appointment
13. personnel manual
14. drop-off appointment
15. house-call appointment
16. walk-in appointment

a. a set time when an animal is expected at the veterinary facility for a later exam or surgery

b. a situation that must be seen immediately and that could be a life or death situation

c. a written reminder for clients that includes a date and time of the appointment scheduled for their pet

d. information provided to all staff members as to what the staff expectations are, such as dress code, policies on vacation time, personal time, and sick leave

e. an appointment during office hours; examples include vaccinations, yearly exams, and rechecks

f. an appointment scheduled for surgery that is usually outside of routine office hours

g. when a veterinarian sees a certain number of scheduled patients within a set amount of time during the day

h. when patients are seen in the order they enter the facility during set times throughout the day

i. a software program on the veterinary computer system used to schedule appointments and maintain medical information; allows future use of information in an organized manner

j. when an animal is expected at the veterinary facility at a set time for a later exam or surgery

k. when appointments are scheduled based on each patient being seen by the veterinarian at a specific time

l. when the veterinarian examines a patient at the owner's home

m. a set amount of time that is required to see a patient, such as 15 minutes

n. a vehicle equipped with veterinary equipment, tools, and supplies that travels to a client's home to examine a patient

o. when a client without an appointment arrives at the hospital wishing to see the veterinarian

p. information provided to all staff members on the workings of the veterinary facility, such as client rules, scheduling methods, payment procedures, and client visitation times

SECTION I Practice Management and Client Relations

Matching 2

1. central processing unit
2. fax machine
3. hardware
4. keyboard
5. modem
6. monitor
7. mouse
8. scanner
9. software
10. toner
11. copier machine
12. printer
13. keyboarding
14. cursor
15. spreadsheet

a. box-like equipment that allows access to the Internet
b. equipment used to make exact copies of paper or other documents
c. movable instrument attached to a computer that allows the user to move the cursor
d. equipment that allows documents and images to be copied and stored electronically
e. the brain of the computer that carries out instructions asked by the software
f. the computer screen that looks similar to a TV screen
g. allows printing forms and invoices from the computer onto paper
h. program used to track large quantities of items in one file
i. act of typing and development of skills used in computer functions and keys
j. black or colored ink that allows printing in a printer, fax machine, or copier
k. a program or set of instructions the computer follows
l. buttons that have letters that allow typing and word formation on a computer
m. blinking line on the computer screen that determines where typing is occurring
n. equipment used to copy and send documents via a telephone line
o. the physical parts of the computer and its equipment

Short Answer

1. List the information you should obtain when setting up an appointment.
2. As a veterinary assistant, you work in a practice with multiple veterinarians. You need to set up the appointment book. List the factors you should consider and ways you can be accurate in scheduling.
3. What should be done to the schedule to ensure that no appointments are scheduled during lunch and when the office is closed?
4. What types of policies regarding appointment scheduling should you be aware of when you work at a veterinary facility?
5. Before scheduling which types of appointments should you consult the veterinarian?
6. List the components of a computer that are considered hardware.
7. What veterinary tasks can be accomplished using a veterinary software program?
8. Name some commonly used word-processing programs that may be used in a veterinary facility.
9. Medical records may be computerized.   T   F
10. Why is it important to develop good keyboarding skills?
11. Give three examples of routine appointment types.
12. What is the concept behind a drop-off appointment?
13. Why are word-processing programs used in veterinary facilities?
14. What are the differences between a copier and fax machine?
15. Explain how a want list is useful in the veterinary facility.

Clinical Situation 1

A veterinary assistant, Priya, is working at a small animal clinic, answering phones and handling the appointment schedule. Priya answers the phone: "Good afternoon, Green Valley Veterinary Hospital, this is Priya, how may I help you?"

"Yes, this is Mrs. Haines. I need to make an appointment for my dog, Timmy. He has not been feeling well."

"What has been the problem with Timmy, Mrs. Haines?"

"He has been vomiting for a few days and not eating. He has been very lazy and he has me concerned. Can Dr. Bloom see him today?"

"I think that Timmy should be seen as soon as possible. Can you bring him in at 2:00?"

"That is fine. Thank you so much."

"We will see you and Timmy today at 2:00," Priya says, placing Mrs. Haines's name and phone number in the appointment book. She writes Timmy's name down in the schedule and makes a note about his vomiting. When she pulls the medical record, she sees that Timmy is overdue for his yearly exam and vaccines. Priya notes this in the medical record and on the appointment schedule. She also notes that Timmy is a 10-year-old NM collie.

- Which form of scheduling does this clinic most likely use?
- How does placing the overdue vaccine and yearly exam information in the appointment book help the veterinary staff?
- What information can be determined by the age and breed of the animal when that is listed on the appointment schedule?

Clinical Situation 2

Kim is a veterinary assistant at the Smithville Vet Clinic. She works the morning shift and opens up the clinic. One of her daily responsibilities is to boot up the computers in the clinic and check the hospital email account to see if any clients have contacted the hospital. Kim notes that the computers are working slowly this morning; eventually, she is able to access the email account. Two clients have sent emails requesting medication refills. She notes the information and client names and writes a memo to Dr. Calvin, who is the veterinarian on duty this morning. One client has also sent an email about changing an appointment time to 9:00 this morning. "That is not enough time to contact the client who sent the email," she thought, as it was already 8 a.m. "If only she had emailed or called earlier, I would have been able to accommodate her." Unfortunately, the client has emailed too late to reschedule the appointment. Kim emails the client back with other available times and then logs off.

- How does Kim use the computer to aid her work?
- How could Kim have better served the client with the scheduling issue?
- What should Kim do about the computer problems she encountered?
- How does the use of a computer make the veterinary facility more efficient?

SECTION I Practice Management and Client Relations

 COMPETENCY SKILL 4

Appointment Scheduling

Objective:
To accurately and efficiently schedule appointments in the appointment book.

Preparation:
- Establish basic keyboarding skills (30–45 words per minute typed correctly is the average).
- Review basic computer terminology as well as components of hardware and software.
- Familiarize yourself with basic word processing.

Materials:
- Sharpened pencils
- Erasers
- Appointment book

Procedure:
1. Have the appointment book open to the current date, when appointments begin.
2. Block off unavailable times for appointments using a single line or X.
3. Determine the type of appointment as routine, surgical, emergency, drop-off, or house call.
4. Determine the amount of time necessary for the appointment according to the schedule intervals.
5. Ask the client for a preferred date or time for the appointment.
6. Check the availability of the date and time. Schedule the appointment based on the client and the clinic suitability.
7. Enter the data into the appointment book as follows:
   - Client's name (confirm)
   - List as alphabetical filing or numeric filing, according to the clinic medical record system
   - Patient's name, age, breed or species
   - Client's phone number (confirm)
   - Reason for appointment or chief complaint
   - Record the name of the veterinarian who will see the patient
8. If the client is present when the appointment is scheduled, fill out a reminder appointment card with the date and time and give it to the owner.
9. Call the client a day in advance to remind him or her of the date and time of the scheduled appointment.
10. Update and check for accurate information with each client and patient when he or she arrives for the appointment.

COMPETENCY SKILL 5

Computer Skills

Objective:

To perform basic computer operations and maintenance.

Preparation:

- Establish basic keyboarding skills (30 to 45 words per minute typed correctly is the average).
- Review basic computer terminology as well as components of hardware and software.
- Familiarize yourself with basic word processing.

Procedure:

1. Turn on the computer.
2. Learn the veterinary software program and how it is used in the veterinary facility. It is often helpful to train with someone who is competent and knows the following:
    - The responsibilities and duties that are computer based
    - The veterinary program components you are expected to use
    - How to navigate through the program
    - How to use the printers
3. Practice with the program and review your work for accuracy.
4. Double-check to make sure all information is accurate and properly recorded.

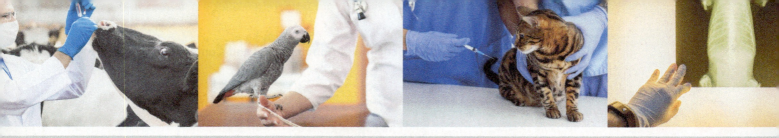

CHAPTER 4 Veterinary Careers

Objectives

Upon completion of this chapter, the reader should be able to:

4.1 Explain the roles and duties of each veterinary health care team member
4.2 Discuss the importance of veterinary dress and a professional appearance
4.3 Discuss the different types of veterinary practices
4.4 Discuss the veterinary foundations that set the standards of care in veterinary medicine
4.5 Explain the different locations within the veterinary facility

Introduction

Many veterinary facilities include staff associates that perform specific roles and responsibilities. Employees must be trained in their individual roles; many of the roles have specific career-related requirements. There are a variety of career options in today's world of veterinary medicine, and it is important to understand each role. The goal in veterinary medicine is to set high standards for the medical care provided to the patient and exceptional customer service to the client. Bear in mind that careers in the veterinary field are not limited to animal hospitals. Other opportunities include teaching, research, military, corporate, and government positions.

The Veterinary Team

The members of the veterinary health care team are involved in the facility's daily workings. These duties range from practice management to client communication to treating the patient. The team must work together in the most efficient manner to maintain and sometimes improve client satisfaction, associate engagement, and overall medical care.

The Veterinarian

The **veterinarian** is a doctor and holds the ultimate responsibility within the facility. Legally, the veterinarian is held responsible for the safety of the staff and the actions of each employee within the facility. Veterinarians are the only team members that may legally diagnose patients, prescribe medications, discuss prognosis of patients, and perform surgery (see Figure 4–1). A veterinarian's education includes four years of a pre-vet program that results in a **bachelor of science (BS)** degree in a science-related course of study. The student then enrolls in a veterinary medical college, which offers a four-year doctorate program. In the United States, veterinary schools must be accredited by the **American Veterinary Medical Association (AVMA)** for state licensure (see Appendix A for a list of accredited schools). The AVMA sets the standards of veterinary medicine and the quality of care that veterinarians provide to animal owners. The **American Animal Hospital Association (AAHA)** sets the standards for small companion animal hospitals. After students have completed a doctorate degree, they earn the title of

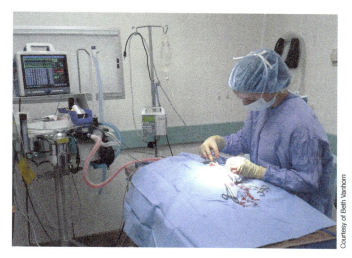

FIGURE 4–1 The veterinarian is the only staff member qualified to perform surgery.

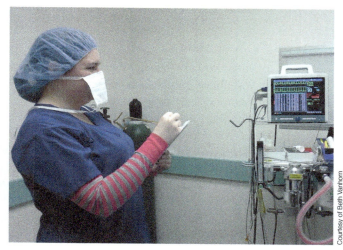

FIGURE 4–2 The veterinary technician performs various procedures within the veterinary facility.

doctor of veterinary medicine (DVM) or veterinary medical doctor (VMD). Upon graduation, veterinarians are required to take a **state board examination**. Once they have passed the exam, veterinarians are licensed to practice in the state or states in which they work. Veterinarians are required to update their license on a regular basis; they and licensed veterinarian technicians must also complete a number of hours of **continuing education (CE)**, or professional development, in a set amount of time. This education is meant to keep the staff up-to-date on changes in the industry and allow the facility to keep learning new information. It is important to note that all staff members should receive continuing education credits.

The Veterinary Technician and Veterinary Technologist

Veterinary technicians and **veterinary technologists** have specialized training in areas of veterinary health and surgical care of many types of animal species (see Figure 4–2). The specialized training is acquired at an AVMA-accredited institution (see Appendix B). The veterinary technician graduates with a two-year **associate degree** and the veterinary technologist graduates with a four-year bachelor degree. The associate degree results in an associate of science (AS) diploma. The bachelor degree results in a bachelor of science (BS) diploma.

The **National Association of Veterinary Technicians in America (NAVTA)** sets the standards of veterinary care for veterinary technicians and technologists. Graduates of the program take a state board examination as required by the state or states in which they intend to work. Passing this exam will result in state licensure. Depending on the state, the technicians will be recognized as a **certified veterinary technician (CVT)**, a **registered veterinary technician (RVT)**, or a **licensed veterinary technician (LVT)**. Technicians work under the indirect supervision of a veterinarian. The responsibilities of the veterinarian and veterinary technician are outlined by each state's **veterinary practice act**: a set of rules and regulations that govern veterinary medicine. There is a current proposal by NAVTA for a universal title for all credentialed veterinary technicians to be referred to as "veterinary nurses."

Veterinary technicians may gain advanced skills by specializing in a specific area of veterinary care (see Table 4–1). Technicians who choose to specialize must complete a required number of clinical and working hours related to skills within that field of study in a set amount of time. Each specialty typically requires a required amount of continuing education within the field of study in addition to the required license renewal CE hours.

The Veterinary Assistant

The **veterinary assistant** is a member of the team who is trained to help the veterinarian and veterinary technician in many duties in the facility (see Figure 4–3). Veterinary assistants are not licensed by the state, but may earn a national credential through NAVTA. The approved veterinary assistant (AVA) title is given to veterinary assistants who have completed a NAVTA-approved curriculum and passed the AVA exam. Veterinary assistants may also be trained on the job. Many state practice acts outline the regulations for veterinary assistants. Assistants work under the

TABLE 4–1
Veterinary Technician Specialties

Academy of Veterinary Emergency and Critical Care Technicians (AVECCT)
Academy of Veterinary Dental Technicians (AVDT)
Academy of Veterinary Technician in Anesthesia and Analgesia (AVTAA)
Academy of Veterinary Surgical Technicians (AVST)
Academy of Internal Medicine for Veterinary Technicians (AIMVT)
Academy of Veterinary Technicians in Clinical Practice (AVTCP)
Academy of Veterinary Zoological Medicine Technicians (AVZMT)
Academy of Veterinary Clinical Pathology Technicians (AVCPT)
Academy of Veterinary Behavior Technicians (AVBT)
Academy of Equine Veterinary Nursing Technicians (AEVNT)
Academy of Veterinary Nutrition Technicians (AVNT)
Academy of Veterinary Technicians in Diagnostic Imaging (AVTDI)
Academy of Laboratory Animal Veterinary Technicians and Nurses (ALAVTN)
Academy of Dermatology Veterinary Technicians (ADVT)
Academy of Veterinary Ophthalmic Technicians (AVOT)
Academy of Physical Rehabilitation Veterinary Technicians (APRVT)

FIGURE 4–3 The veterinary assistant is a critical part of the veterinary health care team.

FIGURE 4–4 The veterinary hospital manager runs the business aspects of the veterinary facility.

direct supervision of a veterinarian. Many veterinary assistants are cross-trained, which means they can work in every area of the veterinary hospital and complete many duties and responsibilities.

The Veterinary Hospital Manager

The **veterinary hospital manager** is in charge of the internal workings of the facility (see Figure 4–4) and must be trained in business management. The hospital or practice manager is responsible for scheduling staff, paying bills, maintaining inventory, and overseeing the day-to-day business within the hospital and usually interviews and hires employees. The manager maintains information in the **personnel manual** and the **procedures manual**. The personnel manual contains information about staff jobs, attendance, vacation and sick days, the dress code, and employee expectations and conduct. The procedures manual outlines the events of the facility, hospital rules, and staff and client rules. Some hospital managers may wish to become certified veterinary hospital managers. This requires taking educational courses in business administration and passing an exam to become a CVPM, among other requirements. The Veterinary Hospital Managers Association (VHMA) is a resourceful organization that sets the goal of supporting the manager to become successful and efficient in hospital management.

The Kennel Attendant

The **kennel attendant** is in charge of the kennels and animal wards within the facility (see Figure 4–5). Duties include cleaning cages, feeding and watering patients, monitoring and exercising patients, and cleaning and sanitizing the facility. Kennel attendants must be trained to handle animals and have a basic knowledge of animal care. They are responsible for noting any abnormal signs or changes in the patient's health and can identify emergencies that may occur while a patient is hospitalized.

The Office Manager

The **office manager** is typically responsible for overseeing and training the front desk receptionists. As part of their administrative role, the office manager must have skills in customer service and public relations. These responsibilities may be more limited if the hospital also has a practice manager. Office managers must be outgoing, courteous, and professional, as they are the face of the practice.

The Receptionist

The veterinary **receptionist** is responsible for maintaining the front office of the facility (see Figure 4–6). A receptionist and an administrative assistant have similar duties, which include scheduling appointments, answering telephones, invoicing and collecting payments, and admitting patients. The receptionist is vital to the daily flow of the hospital. The receptionist is usually the first person clients have contact with and must have excellent communication and office skills. The receptionist plays a significant role in the success of the hospital and must be outgoing, polite, and professional. This duty is not only one of the most important roles in the hospital but can also be quite challenging: one must be a good listener, a good speaker, able to show empathy, and capable of dealing with upset and difficult clients—all while serving as the client's first impression of the facility.

Hospital Administrators

A **hospital administrator** has ultimate responsibility of a facility's business and medical operations. This person may be a veterinarian, credentialed technician, or practice manager. The hospital administrator sets budgets, approves raises, creates organizational structure and training protocols, and plans events. It is important that this person have a general knowledge of veterinary medicine and of quality control and assurance.

Groomers

Groomers may be a part of the veterinary team that provides bathing and grooming services to the hospital's patients. This area of client services requires knowledge of grooming practices and restraint techniques, patience, and experience. Groomers must take great care in preventing injuries to the animals

FIGURE 4–5 The kennel attendant maintains the kennel area and the animals housed in the facility, including keeping the cages and kennel area clean.

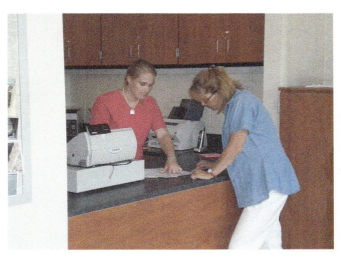

FIGURE 4–6 The receptionist meets and greets clients, answers phones, and fields questions in the veterinary facility.

they are handling. Groomers may enroll in professional grooming schools and courses; the National Dog Groomers Association is a good source of information. It is important that groomers are educated; many are often required to apprentice with certified groomers prior to employment. Some states may require licensing or certification within the grooming profession.

Professional Dress and Appearance

Professional dress and appearance will vary within veterinary facilities. The veterinary health care team—including veterinarians, veterinary technicians, and veterinary assistants—will typically wear professional "scrubs," including a top and pants (see Figure 4–7), or perhaps a lab coat. Scrubs and lab coats, which are easily washed and/or replaced, are worn to protect dress clothing from becoming contaminated, soiled, or torn. Some facilities have a set dress code or color to be worn by all staff members. The staff and receptionists of a kennel may have a similar dress code or be asked to wear special hospital shirts, such as polo shirts, and dress pants, such as khakis. All employees should wear closed-toe shoes that are nonslip, preferably sneaker-type shoes. This is for safety purposes as well as for ease of cleaning and controlling the spread of contaminants. Working in a veterinary facility may bring such hazards as animal bites, needles, and glass (causing a foot injury). In large-animal facilities, employees may wear heavier boot-type footwear for protection. Employees should have certain items at the ready (perhaps in their pockets), including a blue or black pen, a watch or timer, a calculator, a gauze roll, scissors, half-inch tape, a nylon leash, and a stethoscope. Long hair should be pulled back from the face. Jewelry should ideally not be worn or kept to a minimum. No long bracelets or necklaces or other dangling items, such as hoop earrings, should be worn, as they present a safety issue to the employee and to the animals. Fingernails should be kept short, neat, and clean and not extend beyond the end of the finger. Artificial nails are not to be worn by staff members handling animals due to an increased potential for bacterial contamination and injury to animals or other staff members. Tattoos and body piercings should be kept non-offensive; some facilities may have regulations against these items. The dress code of all hospital employees should be listed in the hospital manual; the expectations of the employer should be outlined as well. The image of the entire hospital is reflected in the image that a veterinary assistant presents. Remember, first impressions are lasting impressions. Clients will judge the staff based on their appearance.

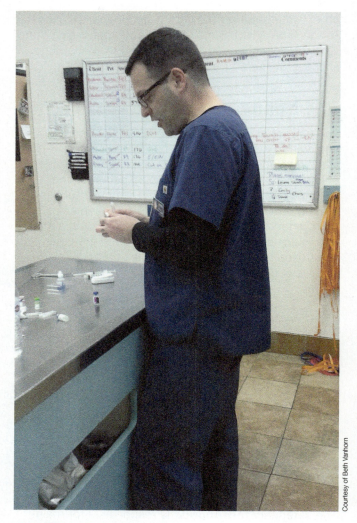

FIGURE 4–7 The veterinary assistant should dress in a professional manner and be well groomed.

Companion Animal Practice

The **companion animal practice** treats and maintains the health of small animals, such as dogs, cats, and exotic pets. This facility is often referred to as a "small animal hospital." Many small animal practices employ one to two licensed technicians and two to three veterinary assistants for each veterinarian in the facility. Additional staff often includes receptionists, kennel attendants, a practice manager, and possible groomers.

Large Animal Practice

The **large animal practice** treats livestock animals, such as horses, cattle, and pigs. This type of practice often has veterinarians traveling to farms and off-site locations to examine animals. Some large animal facilities may have clinical areas where exams and surgery are carried out. Most large animal facilities maintain a smaller staff, such as one to two technicians or assistants and one or two receptionists. The veterinarian is often assisted by the farm owner or a stable hand when off-site, resulting in fewer staff.

Mixed Practice

The **mixed-animal practice** treats both small and large animals. This type of facility may have on-site small animal clinical areas and treat large animals off-site (by traveling to farms) or see both large and small animals in the hospital setting. Depending on the facility, veterinarians may see all types of animals or focus specifically on either large or small animals.

Veterinary Referral Practice

The **veterinary referral practice** offers services by veterinarians who specialize in specific areas, such as cardiology or orthopedics. These facilities may vary in the number of veterinarians available and the type of specialty medicine that is practiced. Clients will seek a specialist consultation based on a referral from their regular veterinarian. This type of service is available in many emergency clinics, as well. Many general practice veterinarians do not have the knowledge, experience, or equipment for more advanced cases and may seek specialist referrals to better serve the patient.

Mobile Veterinary Services

The **mobile veterinary practice** offers services to small animal owners, such as dogs and cats, and employs veterinarians who travel to the homes of the client to provide exams and services. The mobile clinic is typically a converted van or a larger recreational vehicle that may have an exam area and house some limited veterinary equipment and supplies. Mobile veterinarians often provide vaccine services and routine examinations but may be limited in more diagnostic resources, such as blood machines and radiology equipment.

The Veterinary Facility

The veterinary hospital consists of multiple sections where various veterinary medical procedures and functions take place. The front entrance typically leads into the veterinary **waiting room** and **reception area** (see Figure 4–8). This is where clients sign in, are greeted by the front office staff, and wait to see the vet. Many small animal facilities have separate waiting area for dogs and cats to help reduce the stress of both clients and patients. Or, there may be separate check-in and checkout areas, with their own entrances and exits. The waiting area should be comfortable and clean and have client seating. There should be a restroom near the waiting area.

FIGURE 4–8 The waiting room and reception area.

The **exam room** is where patients receive routine exams. There may be one or more exam rooms, depending on the size of the facility and the number of veterinarians on staff (see Figure 4–9). Exam rooms should be clutter-free, with an exam table or area that allows enough room for several people and pets. The area should be easy to clean and mop between patients. A comfortable seating area should be available.

The **treatment area**, sometimes called the "prep area," is a place in the hospital where patients are treated or prepared for surgical procedures. This is typically an area for veterinary staff only (see Figure 4–10). It is often located in the back of the facility and has several treatment tables or wet tables where animals can be bathed or receive dental cleanings. The space should allow staff to work on animals from all angles.

Medications are prepared and stocked in the **pharmacy** (see Figure 4–11). This area is often located near the exam rooms and reception area for easy access on the part of pet owners. This area often has numerous shelves and drawers to hold medicine bottles and vials and should include an area where medications can be properly counted and packaged.

Patient tests and samples are prepared for sampling in the **laboratory**. Examples of laboratory procedures include fecal samples, urine samples, and blood tests (see Figure 4–12). The lab may occupy the same space as the pharmacy, if space is limited. Laboratory equipment must be placed far enough apart to keep it from overheating and to create enough space for it to be used correctly.

Radiology is a specific area that meets veterinary regulations for radiation and the use of special chemicals needed to develop X-rays (see Figure 4–13). This area must be in a lead-protected area and is often found at the back of a facility, away from clients, to reduce any chance of radiation exposure.

The **intensive care unit** (ICU) is where critical patients needing immediate care are treated and housed. This area has emergency equipment to care for injured or sick animals.

The **surgical suite** is another specialized area that houses the surgery table, anesthesia machine, and instruments and equipment used in surgical procedures (see Figure 4–14). This area must be clutter-free, be easily cleaned, and kept sterile.

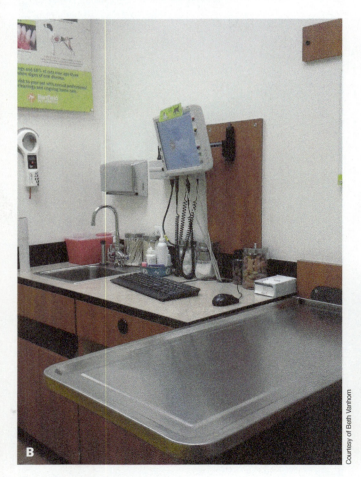

FIGURE 4–9 The exam room.

FIGURE 4–10 The treatment room.

CHAPTER 4 Veterinary Careers   61

FIGURE 4–11 The pharmacy area.

FIGURE 4–13 The radiology area.

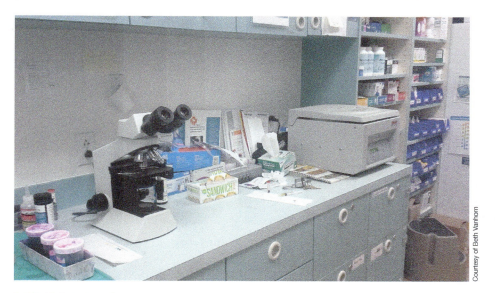

FIGURE 4–12 The laboratory area.

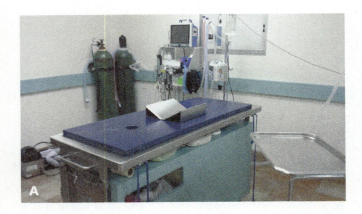

The **isolation ward** is a room that contains cages and equipment to treat animals that may be contagious. There must be a place to keep supplies within this area so the staff can treat animals without worrying about spreading contamination.

The **kennel ward** houses patients waiting for surgery or treatment; they may also board here (see Figure 4–15). This area often has cages of varying sizes and runs with access to the outside. Some facilities may have a fenced-in location to walk dogs during hospitalization.

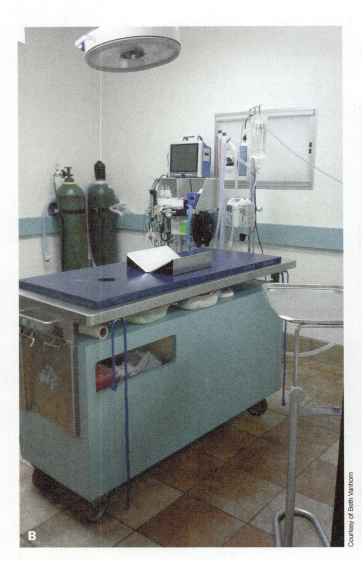

FIGURE 4–14 The surgical suite.

FIGURE 4–15 The kennel ward.

CHAPTER 4 Veterinary Careers 63

FIGURE 4–15 The kennel ward.

SUMMARY

The world of veterinary medicine has greatly evolved in the past 5 to 10 years, with many less-traditional career options. Being knowledgeable of the areas within the facility, the roles of each team member, and other career opportunities and specialties will better help students decide which avenue of veterinary medicine is best suited for them.

Key Terms

American Animal Hospital Association (AAHA) sets the standards of small animal and companion animal hospitals

American Veterinary Medical Association (AVMA) accredits college programs for veterinarian and veterinary technician state licensure and sets the standards of the veterinary medical industry

associate degree a two-year degree program

bachelor of science (BS) a four-year college degree program

certified veterinary technician (CVT) a veterinary technician who has passed a state licensing exam

companion animal practice a veterinary facility that treats and maintains the health of small animals, such as dogs, cats, and exotic pets

continuing education (CE) education provided for adults after they have left the formal education system, consisting typically of short or part-time courses in their career field

exam room an area where patients receive routine exams

hospital administrator a person with complete authority over the hospital business and medical aspects

intensive care unit (ICU) an area of the facility where critical patients needing immediate and constant care are treated and housed

isolation ward an area that contains cages and equipment to care for patients that are contagious and may spread disease

kennel attendant a person working in the kennel area and animal wards, providing food, exercise, and clean bedding

kennel ward an area where animals are housed in cages for surgery, boarding, and care

laboratory a location in the facility where tests and samples are prepared or completed for analysis

large animal practice a veterinary facility that treats livestock animals, such as horses, cattle, and pigs

licensed veterinary technician (LVT) a veterinary technician who has passed a state licensing exam

mixed-animal practice a veterinary facility that treats both small and large animals

mobile veterinary practice a veterinary facility that offers services to small animal owners, such as dogs and cats, and employs veterinarians who travel to the homes of the client to provide exams and services to pets

National Association of Veterinary Technicians in America (NAVTA) sets the standard of veterinary technology and the care of animals

office manager a person who oversees the training and management of the front office staff

personnel manual a handbook that describes staff job descriptions, attendance policy, dress code, and employer expectations

pharmacy the area where medications are prepared and stored for veterinary patients

procedures manual handbook that describes the rules and regulations of the facility, such as visiting hours and opening and closing procedures

radiology an area of the facility where X-rays are taken and developed; this area must meet specific regulations to guard against chemical and radiation exposure

reception area a location where clients arrive and check in with the receptionist

receptionist personnel greeting and communicating with patients as they come in and leave the veterinary facility

registered veterinary technician (RVT) a veterinary technician who has passed a state licensing exam

state board examination an exam that veterinarians and veterinary technicians must pass in order to earn state licensure to practice

surgical suite an area where surgery is performed and harmful anesthesia gases are used; the area must be kept sterile to prevent contamination

treatment area a place where animals are treated or prepared for surgery

veterinarian a doctor who holds a state license to practice veterinary medicine; the vet holds the ultimate responsibility within the facility

veterinary assistant   a staff member trained to assist veterinarians and technicians in various duties

veterinary hospital manager   a staff member in charge of the business workings of the entire hospital, including staff scheduling, paying bills, and payroll

Veterinary Practice Act   a set of rules and regulations that govern veterinary medicine in each state within the United States

veterinary referral practice   a veterinary facility that offers services by veterinarians who specialize in specific areas, such as cardiology or orthopedics

veterinary technician   a two-year associate degree in specialized training to assist veterinarians

veterinary technologist   a four-year bachelor degree in specialized training to assist veterinarians

waiting room   an area where clients and patients wait for their appointments

Review Questions

Matching

Match the term with the definition.

1. veterinary technologist
2. veterinary assistant
3. receptionist
4. intensive care unit (ICU)
5. referral practice
6. isolation ward
7. kennel attendant
8. laboratory
9. veterinarian
10. hospital administrator
11. radiology
12. veterinary technician
13. exam room
14. treatment area
15. waiting room

a. a facility that offers specialized services in specific areas
b. an animal doctor who has completed eight years of education
c. an area that contains cages and equipment to care for patients that are contagious
d. a staff member who helps both doctors and technicians to complete duties
e. holds a four-year bachelor degree and assists the doctor
f. area of the facility where X-rays are taken and developed; this area must meet certain regulations because of chemical and radiation exposure
g. a person working in the kennel area and animal wards, providing food, exercise, and clean bedding
h. a person who greets and communicates with clients as they enter the facility
i. an area of the facility where critical patients needing immediate and constant care are treated and housed
j. a person with complete authority over the management of a facility
k. a location in the facility where tests and samples are prepared or completed for analysis
l. an area where patients receive routine exams
m. an area where animals are treated or prepared for surgery
n. holds a two-year associate degree and assists the doctor
o. an area where surgery is performed and harmful anesthesia gases are used; the area must be kept sterile to prevent contamination
p. an area where clients and patients wait for their appointments
q. rules and regulations that govern each state

Short Answer

Please respond to the questions that follow:

1. List the essential members of the veterinary health care team and their duties and responsibilities within the hospital.
2. What association sets the standards of care for veterinary medicine?
3. What are some differences between the associate of science (AS) degree and a bachelor of science (BS) degree?
4. What associations are important to veterinary medicine?
5. What are the areas of a veterinary facility, and what is the importance of each one?
6. What items should each staff member receive when beginning a veterinary position?
7. What are other career options outside of being a veterinarian?
8. What are some veterinary technician specialties?
9. What is the importance of the Veterinary Practice Act?
10. Explain veterinary mobile services.

Clinical Situation

Kate, the hospital manager at All Creatures Animal Hospital, was having a discussion with a fellow veterinary assistant and licensed technician. She was reviewing several updates in their Veterinary Practice Act booklet.

Kate was concerned that several changes had occurred in the rules and regulations section. She wanted to make certain that each team member was aware of the changes and knew which duties they were permitted to perform and which could only be performed by licensed staff members. She didn't want to offend any team members by making them feel unimportant or underutilized based on their skills. She was concerned that team members may not agree and could create problems because of the changes.

- How might Kate handle the discussion in this situation?
- In what ways can Kate educate her team on the importance of following the Veterinary Practice Act?

COMPETENCY SKILL 6

Professional Appearance and Dress

Objective:

To conduct oneself in both manner and appearance befitting a professional.

Materials:

- Scrub outfit
- Nonslip, closed-toed shoes
- Gauze roll
- ½" to 1" adhesive tape roll
- Watch or timer
- Stethoscope

- Nylon leash
- Memo pad
- Scissors
- Pen
- Calculator

Procedure:

1. Review your personnel manual and procedures manual that outline the staff rules and dress code for the facility.
2. Adhere to the dress code, which may include the following:
   - A clean and durable scrub outfit, including a comfortable top and pants that fit well. The neckline should be modest, and pants should fit without falling down during bending and movement. Pockets are desired.
   - Nonslip, closed-toed shoes. White sneakers are preferable. Shoes should fit well and be comfortable and supportive to the feet.
   - The list of materials presented here should be carried in a pocket and readily available at all times.
   - Jewelry should not be worn or be kept to a minimum. Items may become damaged, cause patient or staff injury, or become contaminated. Loose items—earrings, necklaces, and bracelets—may cause injury to patients or the staff member.
   - Employees should pull their hair back from the face to allow safe patient handling and a clear view of the work area.
   - Fingernails should be kept trimmed and neat. Long and/or fake nails tend to harbor bacterial and infectious materials and present a higher risk of contamination to patients.
   - Conceal tattoos and piercings. Many facilities have rules that should be noted in the personnel manual.

CHAPTER 5: Communication and Client Relations

Objectives

Upon completion of this chapter, the reader should be able to:

5.1 Discuss the communication process and basic components of communication
5.2 Distinguish between appropriate and inappropriate communication and interactions with people
5.3 Demonstrate ways to improve speech and listening skills
5.4 Describe how to communicate with people on the phone
5.5 Explain how interpersonal communication skills apply to clients' perception
5.6 Discuss how to handle difficult clients and situations
5.7 Discuss strategies for helping people handle the grief process

Introduction

Veterinary assistants, as well as the entire health care team, rely on interpersonal communication skills when working closely with people. Interpersonal communication is any form of message and response between two or more people. It may be in the form of the written word, verbal exchanges, or body language cues. The people the veterinary assistant will regularly interact with include clients; staff members; veterinary professionals from other facilities; and businesspeople, such as a salesperson who sells items to the facility. It is important to understand that impressions are everything and occur within the first two to three minutes of the visit.

The Communication Process

Communication is a way of passing along information from one person to another or to a group of people. The **communication process** has five essential components: the sender, the message, the channel, the receiver, and the feedback (see Figure 5–1). The process begins with the **sender**, the person trying to relay an idea, or the "**message**." The message is sent by the sender to another person, the **receiver**. The receiver is the person intended to understand the message. The message is sent through a **channel**, which is the chosen route of communication, whether that is verbal, nonverbal, or written. Once this process of communication is complete, the receiver may return a message called **feedback**. This process is continued until the conversation is complete.

Verbal Communication

Verbal communication is spoken words used between two or more people to form an understanding. Verbal communication is the most common form of communication between people (see Figure 5–2). This type of communication is needed between the veterinary assistant and other staff members, as well as with clients.

CHAPTER 5 Communication and Client Relations

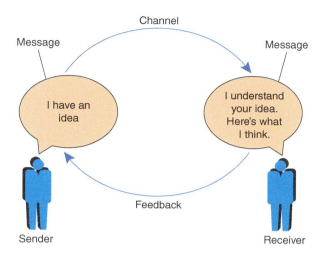

FIGURE 5–1 The communication process.

It is important to be able to speak well to other people. Verbal communication in medicine can now also mean recording notes on tape recorders to be written at a later date, allowing for more accurate notes. Greeting the client is often the first form of verbal communication. All verbal communication should be in a friendly and positive manner. It is important to speak to clients using common terms and avoiding the use of medical terminology. Clients will be interested in learning information needed to care for their pet but may not be able to understand certain words and care must be taken not to offend or confuse.

Nonverbal Communication

Nonverbal communication consists of interactions between people without the use of spoken words, such as through body language and physical expressions. Nonverbal communication is often used with staff members and may be helpful when dealing with upset, angry, or grieving clients. Examples of nonverbal communication include smiling, frowning, glaring, or shrugging the shoulders (see Figure 5–3). **Body language** consists of mannerisms and gestures that demonstrate how a person feels. It is easy to confuse the meaning of the message when only nonverbal communication is used. Incorporating positive body language can help clarify the message. Positive body language gestures include the following (see Figure 5–4):

- Hold arms at your side or gently folded in front of your body.
- Stand up straight, with a relaxed position.
- Face the person and maintain eye contact.
- Speak with a moderate voice and clear tone.
- Stay at least one arm's length from people to create a comfort zone.
- Dress professionally and wear a nametag.
- Smile and relax your face.
- Practice good hygiene.

FIGURE 5–2 Verbal communication plays a large role in the duties of the veterinary assistant.

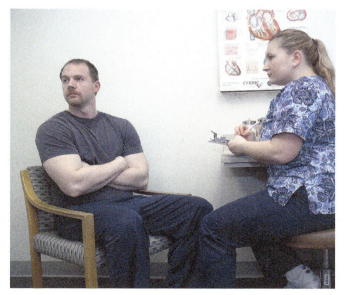

FIGURE 5–3 Body language can communicate more than spoken words.

FIGURE 5–4 Positive body language can enhance communication between the veterinary assistant and the client.

Written Communication

Members of a veterinary facility rely on **written communication** every day. Written communication may include writing letters, memos, notes for progress records, emails, and texts. Email and text messaging have become popular methods of communication. Written communication must be clear, accurate, and understandable. In the facility, it is important to have a memo pad and a pen or pencil located at every telephone so that messages (including details) will be recorded accurately. All telephone messages should be written up and given to the recipient immediately. It is important to write the date, time of call, caller's name and phone number, message, and the message taker's name on the memo. All information must be recorded legibly. Medical records and progress notes must contain accurate information to allow proper communication between staff members (see Figure 5–5). Written communication will also include client education materials. Clients may take information home after each visit. In fact, many visits will include an overwhelming amount of information. Clients will need to review and understand any materials relating to their pet. A discussion on health concerns, services provided, yearly exam findings, and future visits should

FIGURE 5–5 Excellent writing skills are an important part of the communication among the veterinary health care team.

be detailed and followed up with such take-home information. Surgery patients are often sent home with home care instructions (varying by procedure) that may include feeding instructions, medication needs, exercise instructions, and potential concerns to watch for. It is important to review this information with the client before the pet has been discharged to make sure the client has a clear understanding.

Appropriate Communication Skills

As people are communicating with each other, they should consider several qualities that will bring about a positive outcome, including courtesy, kindness, patience, tactfulness, sympathy, and empathy. **Courtesy** means putting someone else's needs and concerns before your own and includes such characteristics as sharing, giving, and cooperation. All people should be treated with respect and professionalism. **Kindness** means being helpful, understanding, and working in a friendly manner. Treat each client as you wish to be treated. **Patience** is defined as a calm demeanor in all situations and may need to be practiced. Some people and animals will require more patience than others. **Tactfulness** means doing and saying the appropriate things at the right time and is crucial for good relations with others. To be tactful, you must think before you speak; in this way you can avoid offending others. Having **sympathy** means showing support to someone in a difficult situation. **Empathy** is being able to understand someone else's feelings at a particular time. Table 5–1 provides examples of these traits in common situations.

TABLE 5–1

Examples of Appropriate Communication Skills

Courtesy	• Holding a door • Helping a person to his or her vehicle • Carrying a bag of dog food for someone • Walking someone's dog to an exam room
Empathy	• Telling someone you understand how they feel when an animal is being put to sleep
Kindness	• Asking if you can help someone • Making a phone call for someone • Drawing a diagram of how to do something
Patience	• Walking an elderly person to his or her car • Repeating information for a person who is hard of hearing • Offering children coloring books
Sympathy	• Saying you are sorry to a client who has lost a pet • Writing a sympathy note to a client • Sending flowers to a staff member who is sick
Tactfulness	• Speaking politely • Not swearing • Not yelling • Discussing a difficult situation professionally

Interacting with People

Interpersonal communication includes speaking, listening, and observing others. Clients may be pleasant or difficult based on their personality, the type of day they have had, or how their visit goes. Good interpersonal communication skills are necessary to deal with these situations.

Speaking

Speech communication requires practice and experience and can be a frustrating part of any career. Some clients begin conversations that will have no bearing on the reason they have come to the veterinary clinic. Or they may ask many questions, some of which may be of little importance. When interacting with people, show respect to everyone. Remember that people are different; therefore, treat each person as an individual. Speaking clearly with team members will keep everyone (including the animals) safe and make everyone's job a little easier. It may be necessary to interact with other veterinary health care team members in order to complete treatment or wrap up a progress report (see Figure 5–6). Respectful communication can include the following:

- Always greet people when you first come into contact.
- Treat everyone equally, with respect and honesty.
- Address people with professionalism: Dr. Jones, Ms. Smith, Professor Lee.
- Speak slowly and patiently when explaining information.
- Keep all areas of the facility neat and clean for the comfort of all who work in and enter the facility.
- Offer assistance to anyone who looks like they need help.
- Smile even when you don't feel like smiling.

TERMINOLOGY TIP

Students who are not certain whether a particular word or usage is correct should reference a veterinary medical dictionary or Merriam-Webster's dictionary. ■

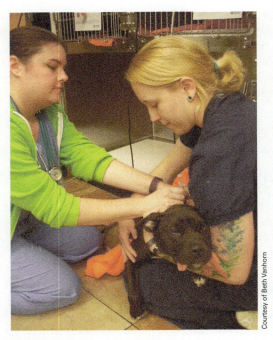

FIGURE 5–6 It is often necessary to talk and listen with other staff members to get help to complete an animal's care.

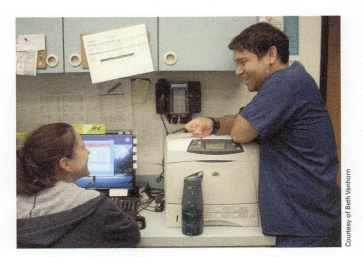

FIGURE 5–7 Stop and listen to what your coworker is saying to ensure your safety, the safety of those you are working with, and the animal's safety.

Listening

Another way of interacting with clients involves listening and reflection with clients. **Listening skills** are essential in the veterinary field. Listening involves training oneself to focus on what is being said. Good listeners hear exactly what is being said and think about the words to clearly understand the main point. Good listening skills also include watching a person's body language to determine his or her mood. Good listeners know when not to speak and allow others to finish their thoughts. You must take time to understand what is being said and think of an appropriate response. Remember that keeping silent gives you time to think about what has been communicated; it may be a good idea to let the speaker know that you are being quiet in order to think about what has been said! After reaching an understanding, it is worthwhile to repeat what you have understood to the speaker. This allows both the sender and receiver to have a common understanding and to verify that the information is accurate. In a veterinary situation, the veterinary assistant may need to communicate to another staff member how to properly restrain an animal or ask for help when necessary (see Figure 5–7).

Observation

Observation skills help us understand a person's speech or body language, including the gestures or expressions they use to communicate their feelings. Many people may not be aware that they are expressing nonverbal behaviors; others may use such expressions to get their point across. An employee in a veterinary practice may observe the following examples of client nonverbal communication:

- Smiling
- Heavy sighing
- Shaking the head
- Slouching
- Nervous tapping of the fingers
- Folding the arms
- Pacing
- Frowning
- Glaring

Communicating with Difficult Clients

The unfortunate occasion will occur when a veterinary assistant must handle an angry or difficult person. It is important to be able to read a person's body language and note when he or she is becoming upset. This may help avoid a difficult situation. If a difficult situation does arise, it may cause shock and frustration, but it is best to remain calm. There is no "right" way to deal with an angry person. It is best to begin by asking the person what the problem or situation is. Ask for an explanation of what happened. Many times, if an angry person has a chance to discuss or describe a problem, it relieves some of the tension. It will also give the assistant some time to collect his or her thoughts and come up with a plan of action. Consider these pointers when handling difficult people.

- Keep a positive and assertive attitude.
- Treat each client with courtesy and respect.
- Communicate invoice and procedure costs to clients before they pick up a pet.
- Explain all procedures and ask if there are any questions.
- Get client permission with signed consent forms at all times.
- Be helpful at all times.
- Stay calm and relaxed.
- Minimize interruptions.
- Do not argue or speak negatively.
- Do not place blame on anyone.
- Apologize for any misunderstanding that has occurred.
- Talk to the person in a private place, with a third-party witness.
- Remain objective and listen attentively.
- Assure the person the matter will be resolved.
- Find a neutral party to help resolve the matter.

Angry people may say things that they later regret. In this case, do not take personal offense. If a client becomes enraged, it may be best to alert a veterinarian or hospital manager about the situation. Treat every person who becomes difficult with tact and respect. These matters are not easy for anyone to handle but at times may be unavoidable. Consider a phone call from an upset client. It is important to listen to what a client has to say. Then, review the facts of what was said to make sure there is no miscommunication. If possible, direct the call to a manager. If a manager is not available, politely state that the person who will resolve the situation is not available but that you will have them call back as soon as possible. This may give the client time to calm down before a return call is made, allowing for an easier resolution.

Telephone Skills

Speaking on the telephone is usually the main line of communication in a veterinary facility (see Figure 5–8). Good or bad impressions can be made easily over the phone! Anyone who answers the phone should

FIGURE 5-8 Courtesy and a pleasant telephone manner will make a good impression on clients.

be comfortable and knowledgeable in telephone communication skills and follow certain rules:

- Answer the phone promptly—try to pick up by the third ring.
- When you speak into the phone, give the caller your name and the name of the facility.
- Be pleasant and polite.
- Give the caller your full attention.
- Speak clearly and loudly enough to be understood (some callers may be hard of hearing).
- Repeat important information back so that it is clearly understood.
- Thank the caller and end the conversation with a pleasant good-bye. Remember that a cheery "good morning" or "thank you for calling" can brighten someone's day.

If another call comes in while you are on the phone, politely ask the first caller if you may place him or her on hold. Wait for a response before you put the caller on hold: you need to make sure the first caller doesn't have an emergency or need immediate assistance. If you must put someone on hold for more than one minute, ask the caller for a name and phone number so you can call back promptly. Holding for one minute can feel ten! If you need someone to continue to hold or need to have someone else answer their question, ask the caller if you can call them back. Make sure to give them a time frame. If you do not, you may give the impression that the client and patient are not important.

Grief and Communication

Grief is the emotion that people experience after they have lost a loved one or a pet. This feeling is not reserved only for pet owners: members of the veterinary health care team may also grieve because they have become close to the client and patient through hospital visits. A person who is grieving may exhibit some of the following emotions. **Shock** is a loss of emotion that a person feels after the death of the pet—whether it was sudden or not. Shock may lead to **denial**, when a person cannot immediately accept the pet's death. **Bargaining** is the owner's attempt to resolve the pain that has resulted by any means possible; he or she may try to gain control over the situation and come up with "what if's" that could have altered the course of the death. That may lead to **acceptance**, when the owner accepts the pet's death. Acceptance may lead to **anger**—a natural emotion in the event of the trauma—directed at themselves, a family member, or a veterinary staff member. Some people experience other emotions after acceptance, including sorrow, depression, and guilt. **Sorrow** is a feeling sadness over the loss of the pet. This can be a positive emotion: it may allow a person to cry and talk about the pet. It is a large part of the healing process. **Depression** is a state of sadness where a person can't handle the normal functions of daily life. A person suffering from depression may require the help of a doctor or therapist to recover. **Guilt** is when a person feels that he or she is to blame for the death and should have been able to do something to save the pet. A veterinary assistant should be prepared to deal with these emotions and stages of the grief process by offering such helpful strategies as talking to a family member or friend, talking with other people who have lost pets, joining a pet loss support group, or seeking a professional grief counselor. Suggest to clients that they memorialize their pets through framing pictures, having the pets ashes returned as a keepsake, making a donation in a pet's name, or keeping a clay paw print.

The following table offers ways to help clients during the grieving process (see Table 5–2).

The relationship that develops between people and animals is known as the **human–animal bond**. Many people think of their pets as their children or members of their family. The company of an animal tends to make people happy. Because of this bond, it can be difficult for a pet owner to deal with an old or severely ill animal. A common procedure in the veterinary facility is **euthanasia**, which is the practice of humanely ending an animal's suffering and life. The term **humane** is used to describe what is considered acceptable by people in regards to an animal's physical, mental, and emotional well-being. Making the decision to euthanize a pet is a difficult choice for the owner and staff. The process should be explained to the client. It is very important to make clients feel as comfortable as possible during euthanasia.

TABLE 5–2

Ways to Help Clients Who Are Grieving

- Provide soft, high-quality facial tissues.
- Schedule euthanasia appointments during the slower parts of the day: clients won't have to wait and will have more privacy.
- Provide a quiet room that has comfortable seating and soft lighting.
- Provide a large, soft blanket for the pet to lie on.
- Collect fees for all services prior to the procedure so clients may leave immediately after.
- Provide handouts on the procedure and coloring books for children to help them understand the situation.
- Offer clients options for how the remains are handled: a private cremation, communal cremation, or burial.
- Send a sympathy card from the entire veterinary staff offering condolences for the family's loss.
- Help clients to their vehicle and make certain they are well enough to safely drive.

SUMMARY

Positive and effective interpersonal communications are vital to the smooth workings of a veterinary facility. The veterinary assistant must know how to get a message delivered clearly and be able to listen effectively. Proper communication skills, including speaking, listening, observing, and use of body language, are critical for the veterinary assistant to achieve. This is as important with the veterinary staff as it is with clients. The use of respect, courtesy, understanding, and caring coupled with the use of communication skills will make for a much more successful and enjoyable veterinary environment.

Key Terms

acceptance   when a person understands and accepts that a pet has passed away

anger   a strong feeling of displeasure felt when confronting the death of a pet

bargaining   stage of grief that allows the person an attempt to resolve the pet's death by any means possible

body language   the use of mannerisms and gestures that show how a person is feeling

channel   a chosen route of communication

communication process   the five essential steps to relaying information to others that includes the sender, the receiver, the message, the channel, and feedback

courtesy   placing one's needs and concerns before your own

denial   when a person will not accept a pet's death

depression   a state in which a person is so sad that he or she can't handle the normal functions of daily life

empathy   being able to understand another person's feelings

euthanasia   the process of humanely putting an animal to sleep

feedback   the return message sent by the receiver

grief   the sadness that people feel after the loss of a loved one or pet

guilt   when a person feels he or she is to blame for the pet's death and should have been able to save its life

human–animal bond   the close relationship developed between humans and their pets, where pets are often considered members of the family

humane   what is considered acceptable by people in regards to an animal's physical, mental, and emotional well-being

interpersonal communication   allows people to discuss and understand information with other people

kindness   being helpful and understanding

listening skills   the ability to hear and understand what someone has said

message   an idea passed along through a route of communication

nonverbal communication   communicating an idea without speaking

observation skills   the ability to watch and understand a person's body language and speech

patience   behaving calmly in all situations without any negative complaints

receiver   a person who gets a message in the route of communication

sender   a person relaying an idea or message in the route of communication

shock   a lack of feelings caused by the sudden death of an animal or loved one

sorrow   sadness over the loss of a pet that may result in crying and talking about the pet

speech communication   passing along information verbally

sympathy   showing support for others at a time of sadness

tactfulness   doing and saying the appropriate things at the correct time

verbal communication   the use of speech to pass along information in the communication process

written communication   writing, emailing, or texting to pass along information

SECTION I Practice Management and Client Relations

Review Questions

Matching

Match the term to its definition.

1. bargaining
2. body language
3. communication process
4. empathy
5. grief
6. humane
7. receiver
8. sender
9. sympathy
10. tactfulness

a. stage of grief that allows the person an attempt to resolve the pet's death by any means possible
b. what is considered acceptable by people in regards to an animal's physical, mental, and emotional well-being
c. trying to relay an idea or message
d. doing and saying the appropriate things at the correct time
e. the sadness that people feel after the loss of a loved one or pet
f. an idea passed along through a route of communication
g. showing support for another person at a time of sadness
h. the use of mannerisms and gestures that show how a person is feeling
i. a person who gets a message in the route of communication
j. being able to understand another person's feelings
k. the return message sent by the receiver
l. the five essential steps to relaying information to others that include the sender, the receiver, the message, the channel, and feedback

Please respond to the following questions:

1. Give three examples of channels of communication.
2. Which of the following responses is the best example of effective communication?
   a. "I am sorry to hear about Kitty being hit by a car, Mrs. Campbell. You must be taking it pretty hard. Cats get themselves into so much trouble. That is why I prefer dogs."
   b. "Of course there is no ashtray! Didn't you see the no smoking sign?"
   c. "I know you were in a hurry, Mrs. Smith, but payment is still expected when services are provided."
   d. "Hello, Mr. Tate! How are you and Fluffy today? Please have a seat and we will be with you shortly."
3. List seven ways that a person may display nonverbal communication.
4. The gestures and mannerisms that a person uses to communicate are called _____.
5. Explain how you would deal with a difficult client who was upset with his bill.
6. Re-write each statement to show proper and respectful verbal communication in the following telephone conversations:
   a. "Hello, please hold."
   b. "Can I help you?" (after three minutes being on hold)
   c. "What did you say your name was?"
   d. "See you later." (Laughs and hangs up)
   e. "Come at 6." (Sighs and hangs up)

Clinical Situation

A client is shouting at Emily, a veterinary assistant. The client has been waiting 35 minutes for the vet to see his sick dog. Emily is stunned by the client's behavior and responds, "It is not my fault. Stop screaming at me!"

  a. What did Emily do wrong?
  b. Write a script of how you would have responded to the client.

COMPETENCY SKILL 7

Interpersonal Communication Skills

Objective:

To effectively communicate with clients and coworkers.

Preparation:

- Determine what needs to be stated.
- Use references as needed, such as a dictionary or veterinary medical terminology text.
- If communication is written, use a memo form and dark pen.
- Use a computer with a word-processing program when possible.

Procedure:

Verbal Communication:

1. Listen carefully by focusing your attention on the speaker.
2. Maintain eye contact when listening.
3. Record notes as necessary for accuracy.
4. Speak in a tone that is audible and easy to understand.
5. Maintain eye contact when speaking.
6. Speak slowly.
7. Use correct grammar and be polite.

Written Communication:

1. Check all spelling, punctuation, and grammar for accuracy.
2. Proofread for meaning and understanding.
3. Use reference materials as necessary.
4. Be neat and legible and record in blue or black ink.
5. Place all written items in the proper location.

SECTION I Practice Management and Client Relations

COMPETENCY SKILL 8

Client Communication Skills

Objective:

To effectively communicate with clients.

Preparation:

- Determine the type of client communications used in the facility.
- Determine which types of forms are appropriate for the situation.

Procedure:

Written Communications:

1. Check all written information for accuracy and clarity.
2. Read all information to the client to ensure understanding.
3. Ask the client if he or she has any questions regarding the information.
4. Place all original paperwork in the client's file. Make sure the client receives a copy of the paperwork.

Verbal Communication:

1. Speak audibly, clearly, and slowly.
2. Maintain eye contact with the client.
3. Read the client's body language to ensure your understanding.
4. Ask the client if he or she has any questions regarding the information.
5. Note all verbal communication details within the patient's medical record.

COMPETENCY SKILL 9

Telephone Communication Skills

Objective:

To foster a clear and concise demeanor when communicating via telephone.

Preparation:

- Place a memo pad or paper and pen (that works!) next to every telephone.
- Have a good understanding and working knowledge of the telephone system.
- Have client and patient information forms near each telephone.
- Understand the proper scheduling technique of the facility.
- Understand how to properly schedule an appointment and maintain the appointment book.
- Have a price list available by each telephone.
- Keep a smile on your face to reflect a smile in your voice.
- Answer by the second to third ring.

Procedure:

1. Greet the caller, state the name of the facility, and state your own name: for example, "Good morning, this is Homestead Animal Hospital, Sue speaking."
2. Follow with, "How may I help you?"
3. Avoid putting a caller on hold, if possible. If you must, ask the caller if you may do so and wait for a response.
4. If the hold time is longer than one minute, ask the caller for a name and phone number to return the call promptly.
5. When answering a call that has been put on hold, tell the caller, "Sorry for the hold, this is Amanda, how may I help you?" and direct the call as necessary.
6. Record all telephone messages and give the memo to the recipient immediately. Write a date, time of call, caller's name and phone number, message, and your name on the memo. Be neat and accurate.
7. Direct all personal phone calls as needed. Ask the caller his or her name and the reason for calling so that you may direct the call appropriately.
8. Be polite and patient at all times. Speak slowly and clearly and in an audible tone.

COMPETENCY SKILL 10

Handling Difficult Clients

Objective:
To create a strategy for handling difficult client situations that may arise in the veterinary facility.

Preparation:
- Try to avoid creating a difficult situation.
- Treat each client with courtesy and respect.
- Use a warm, caring, and positive attitude at all times.
- Minimize interruptions during client communication.
- Minimize client waiting time.
- Communicate costs with estimates before the services are provided. Each client should be given a copy of an estimate.
- Obtain client permission with signed consent forms.
- Explain all procedure details and ask the client if he or she has any questions.
- Be helpful at all times.

Procedure:
1. Remain calm.
2. Ask the client to explain the problem. Let the client to talk before asking questions. Listen to the client's needs, problems, and perspectives.
3. Determine who can best resolve the situation.
4. Determine if the situation should be dealt with in private and with a third-party witness.
5. Do not be defensive or place blame on someone else.
6. Be understanding.
7. Ask the client what he or she would like done to help resolve the problem. Offer an alternative, as necessary.
8. If the client becomes disrespectful or verbally abusive, summon a veterinarian immediately.

COMPETENCY SKILL 11

Grief Counseling

Objective:

To create a strategy for handling the death of a client's animal and the client's grieving process.

Materials:

- Grief counseling phone numbers
- Handouts on euthanasia
- Coloring books on euthanasia for children

Procedure:

1. Limit discussions of grief if at all possible. Grief counseling should be done by professionals.
2. Offer grief counseling hotlines or phone numbers.
3. Give all clients handouts of the grief process and on euthanasia.
4. If the client wishes, children can be given a euthanasia coloring book that discusses the situation in a manner that helps them understand what is occurring.
5. A sympathy card should be prepared and signed by the veterinarian and staff members.

CHAPTER 6: Veterinary Ethics and Legal Issues

Objectives

Upon completion of this chapter, the reader should be able to:

6.1 Discuss the differences between ethical and unethical practices
6.2 Discuss veterinary responsibilities related to emergency care
6.3 Describe the Veterinary Practice Act
6.4 Explain the Rules and Regulations of the Veterinary Practice Act
6.5 Demonstrate professional responsibility in veterinary medicine
6.6 Describe the difference between common law, state law, and federal law

Introduction

Veterinarians and all veterinary staff members are required to uphold a standard of ethical conduct. There are laws in place in each state that put the needs of the patient first. It has become a responsibility to the profession to set a high standard of patient care and outline ethics and laws within the profession. Each state's veterinary practice act outlines these laws. Many practices go beyond the laws and determine rules and regulations for the facility.

Veterinary Ethics

Ethics are rules that govern proper conduct; they are often based on higher standards rather than those that are considered acceptable (and often minimal) requirements. Veterinary ethics dictate the **moral** conduct of veterinarians and their professional staff members. These morals indicate what a person believes as right and wrong. The AVMA has set standards in veterinary medical ethics. The most important responsibility of the veterinary staff is to put the patient's needs first and relieve any suffering and pain in animals. When these obligations are fulfilled, it is considered to be **ethical**. When these obligations are not completed, it is considered to be **unethical**. Examples of unethical practices in veterinary medicine include the following:

- Not following the laws pertaining to the Practice Act
- **Misrepresentation** as a veterinarian (not licensed but working as a vet)
- **Slander**—speaking negatively of other veterinary professionals
- Violating confidentiality
- Practicing below the standards of patient care
- Substance abuse problems
- Animal abuse or neglect
- Prescribing medicine without a VCPR

The veterinarian makes all final decisions based on the needs of patient care. These decisions are established through the veterinarian–client–patient relationship, or VCPR. Legally, the veterinarian must examine an animal before making any diagnosis or providing treatment, surgery, prescription medicines, or prognosis of diseases (see Figure 6–1). The VCPR relationship must be maintained within a year's time frame in order to provide the

FIGURE 6–1 The veterinarian is required to conduct an examination of the patient before administering treatments or rendering a diagnosis.

listed services. Each year a reexamination must take place to continue the legal responsibility to the client and patient. Once the veterinarian has begun this relationship, he or she is then responsible for continuing the standard of care for the patient. This includes emergency services as specified by the veterinary facility. The client may end the VCPR at any time. The veterinarian may end the VCPR only after an animal has received treatment.

Medical records are ethically considered **confidential**. This means that patient information is private and not to be shared by anyone except during the care of the patient by the staff members involved. Personal client information is not to be disclosed. Medical records are the property of the veterinary facility and are not owned by the client or veterinarian. Copies may be made and provided to the client. The client may also request copies of the record to be transferred, with written permission. The physical medical record shall legally remain in the veterinary facility at all times.

Veterinary Laws and Veterinary Practice Acts

Veterinary laws are based on each state's **Veterinary Practice Act**. The practice act is a legal document that outlines the rules and regulations of veterinary professionals. These laws will vary from state to state. It is important that all staff members obtain and understand their state's practice act and follow the rules. The practice act is regulated and overseen by **state boards of veterinary medicine**. Each state has veterinary professionals and public members, along with legal counsel that oversees the veterinary laws within the state. The board serves the public by overseeing complaints of veterinary professionals and taking disciplinary action as deemed necessary for the veterinary profession. The veterinary assistant should research his or her state Veterinary Practice Act to become familiar with the rules and regulations as well as the skills and tasks they are legally allowed to perform. This information can be obtained by visiting the American Association of Veterinary State Boards (AAVSB) website at http://www.aavsb.org/. The site includes licensing information and each state's laws according to the practice act. Each state's practice act states that all veterinary licensed professionals are required to keep abreast of changes in veterinary technology and medical information updates on a regular basis by attending continuing education (CE) seminars. Staff members receive updated education within their area of specialty and learn more about the profession (see Figure 6–2). It is considered ethical for all members of the profession, whether licensed or not, to stay up-to-date in their training and learn more about changes within the profession. Veterinary

FIGURE 6–2 Continuing education (CE) is a requirement in the field of veterinary medicine and may be completed by taking online courses.

staff members are expected to be monitored by the veterinarian through either direct or indirect supervision. Direct supervision requires that the supervisor be on the premises and within the same room in case there is a need for assistance. Indirect supervision requires the supervisor to be on the premises and near enough to be of assistance.

Veterinary Common Law

Common laws are based on legal precedents. **Statutory laws** are based on written government laws. Veterinary medicine is regulated through both types of laws. Every member of the veterinary industry must have knowledge of veterinary laws. It is not an acceptable excuse or defense for a veterinary assistant to simply say that he or she was not aware of the laws within the profession. Common law is based on a court and judge's decision. Statutory law is based on the legislature and government laws. Animals are considered private property, as is a car or house. Owners have rights based on damage, loss, or injury to their pets. This governing action also applies to the veterinary facility. This governing action is known as **liability** or a legal responsibility. This action may extend to the owner, as well as the pet, when on the premises of the facility. For example, if a client falls on a slippery floor the hospital is liable. **Malpractice** means that a professional is working below the standard of practice, knowingly or unknowingly. **Negligence** is failure to do what should be done. In these cases, the client may have a right to file legal actions .

Here are some ways to prevent malpractice and negligence in a veterinary facility:

- Document everything in the medical record.
- Have every client sign a consent form that lists the procedures that are to be carried out.
- Provide clients with estimates of the procedures that are to be carried out.
- Lock all windows, doors, and drug cabinets.
- Provide leashes and carriers for all small pets.
- Animal restraint should legally be performed by only trained staff members.
- Keep animals separate from each other.
- Use ID cards and neck bands on all hospitalized patients.
- Record all controlled substances in a carefully kept log that is maintained daily.
- Check all areas of the hospital for safety hazards on a daily basis.
- Post signs that warn clients of possible hazards.
- Suggest that any complaints be made to the specific state board of veterinary medicine.
- Follow all OSHA rules and regulations.

Three elements must be proven in cases of malpractice and negligence when a professional is subject to liability. The elements include: (1) the veterinarian agreed to treat the animal; (2) the veterinarian failed to provide the necessary treatment and legal obligation to the animal; and (3) neglectful injury to the animal. Negligence and malpractice also apply to veterinary technicians, assistants and any non-licensed employee of the facility. Veterinarians are responsible for all employees within a veterinary facility under the common law doctrine of respondent superior that states that veterinarians can be found negligent and guilty of malpractice for any injury or action by an associate that causes an injury or death to a patient. Licenses on veterinary technicians may also be suspended or revoked, as well.

Federal Veterinary Laws

Federal laws are passed by the U.S. Congress and upheld through the federal court system. They are broader and hold greater authority than state laws. Federal veterinary laws pertain to both people and pets. Federal laws affecting people pertain mostly to veterinary staff members. Laws regarding safety within the workplace are governed by the **Occupational Safety and Health Administration (OSHA)**. Any employer with one or more associates must be in compliance with OSHA regulations, which puts laws in place that provide for a safe and hazard-free workplace for all individuals. Guidelines on material safety data sheets, state rules and regulations, training information, fines, and reporting documents can be located on the official OSHA website at www.osha.gov. The **Americans with Disabilities Act (ADA)** prohibits discrimination against employees with disabilities. Another law is the **Fair Labor Standards Act (FLSA)**, which governs wages, overtime, and children and **child labor laws**. There are guidelines for hiring children at certain ages and for the duties that they may perform. Many employees' titles are **nonexempt**, meaning they are entitled to paid overtime when working more than 40 hours per week. **Equal Employment Opportunity (EEO)** laws provide protection from discrimination and provide a basis for any terminated employee to sue the employer for wrongful termination of employment. These laws state employers may not discriminate against race, color, gender, religion, national origin, sexual harassment, pregnancy or childbirth situations, age, or disabilities. Federal laws can pertain to animals, too. The **Animal Welfare Act** governs how animals are handled and cared for and takes a large interest in monitoring research facilities. Laws may also cover **animal rights**, stipulating that animals have feelings and emotions that entitle them to the same rights as people. Many of these rights relate to **animal abuse** laws. Humane societies, such as the American Society for the Prevention of Cruelty to Animals (ASPCA), and state dog wardens typically guard against the neglect and abuse of animals. (Practices such as dog fighting and cock fighting would be considered abuse.)

Medication

Veterinary professionals are also governed by federal agencies that monitor medications used within the industry. The **Food and Drug Administration (FDA)** sets manufacturing standards of food additives and medications used in animals. The FDA is responsible for protecting the public health by ensuring the safety, efficacy, and security of human and veterinary drugs, biological products, and medical devices. The **Drug Enforcement Agency (DEA)** is responsible for monitoring drugs and issuing veterinary DEA licenses to those professionals prescribing medicines that have the potential for addiction and abuse. Many of these drugs are referred to as **controlled substances**. Controlled substances must legally be kept locked at all times and are to be logged each time they are prescribed to a patient. These records must be kept updated on a daily basis.

Medical Waste

Medical waste disposal and management regulations vary from state to state and locally. Typical veterinary wastes include needles, syringes, vials containing live vaccines, vials containing controlled substances, animal remains exposed to infectious disease, zoonotic exposure pathogens, animal tissue, blood or other bodily fluids. Each state and county rules and regulations should be reviewed to determine proper disposal of all veterinary wastes.

Controlled Substances

Controlled substances, also called **scheduled drugs**, are medications that have the potential for abuse and addiction and must be kept in a locked cabinet by law. These medications are regulated by the Drug Enforcement Administration (DEA). Medications that are controlled are listed in schedules according to their potential for abuse, ranging from schedule I to schedule V, with V being the lowest potential and I being the highest potential for abuse or addiction (see Table 6–1).

Under DEA laws and regulations, veterinarians are required to obtain a DEA license to purchase and dispense any controlled substance. The license must be maintained yearly with the DEA, the registration number must be placed on all prescriptions, a **controlled substance log** must be maintained accurately at all times and be kept on file for two years, and all scheduled drugs must be locked at all times. The controlled substance log requires information that must be recorded and accurate and be available for inspection at any time. Table 6–2 outlines the required legal information that must be kept on every controlled substance prescribed by the veterinarian.

TABLE 6–1

Drug Schedules

- **Schedule I**: no medical use—high abuse (illegal substances, e.g., marijuana)
- **Schedule II**: accepted medical use—high abuse (e.g., morphine)
- **Schedule III**: accepted medical use—medium abuse (e.g., Phenobarbitol [seizures])
- **Schedule IV**: accepted medical use—low abuse (e.g., Tylenol with codeine)
- **Schedule V**: accepted medical use—very low abuse (e.g., liquid cough medications)

TABLE 6–2

Controlled Substance Log Required Information

Drug name	Drug strength
Date of use	Client name
Patient name	Amount used
Amount remaining	Initial of handler(s)

CHAPTER 6 Veterinary Ethics and Legal Issues

SUMMARY

Veterinary ethics and laws are important to everyone within the veterinary field. Knowledge of these legal responsibilities is essential to each employee, but the veterinarian has ultimate responsibility. Many general rules apply to safety and it is best to know the law and follow it. Each veterinary professional is expected to behave in a professional manner and pass along respect and a high standard of quality care to both the patient and the client according to their states practice act.

Key Terms

Americans with Disabilities Act (ADA) a law that prohibits discrimination against people with disabilities

animal abuse laws rules that protect against the neglect and abuse of animals; these laws are typically regulated by a humane society or dog warden

animal rights rules that govern how animals are handled and cared for, especially in research facilities

Animal Welfare Act rules that regulate how animals are handled and cared for, including within research facilities

child labor laws laws and guidelines that regulate and protect children who work and the conditions in which they work

common laws state regulations and rules based on legal precedents

confidential information that is private and not to be shared by anyone outside of the facility without permission

controlled substance drugs that have a potential for addition and abuse

controlled substance log a book used to keep track of when scheduled drugs are prescribed that must be kept on file for two years

Drug Enforcement Agency (DEA) the agency responsible for monitoring drugs that may be addictive and issuing DEA licenses to veterinarians

Equal Employment Opportunity (EEO) laws that prohibit discrimination against individuals for their race, color, gender, religion, age, sexual harassment, national origin, or pregnancy or childbirth situations and disabilities

ethical the act of doing what is right

ethics rules and regulations that govern proper conduct

Fair Labor Standards Act (FSLA) the agency that governs the age of children working and the duties they can perform

Food and Drug Administration (FDA) the agency responsible for protecting the public health by ensuring the safety, efficacy, and security of human and veterinary drugs, biological products, and medical devices

liability to have a legal responsibility toward something

malpractice the act of working below the standards of practice

misrepresentation working or acting as someone who you are not qualified to be

moral what a person believes to be right and wrong

negligence failure to do what is necessary or proper

nonexempt employee status that is entitled to paid overtime when working more than 40 hours per week

Occupational Safety and Health Administration (OSHA) agency that governs laws regarding safety in the workplace

scheduled drugs another name for controlled substances that have the potential for abuse

slander talking negatively about someone in an improper manner

state board of veterinary medicine a state agency of veterinary members and public members who take disciplinary action in the profession as deemed necessary and also monitor the rights of the public

statutory laws rules and regulations based on written government laws

unethical the act of knowingly doing something wrong or improper

Veterinary Practice Act a legal document that outlines the rules and regulations of veterinary professionals

Review Questions

Matching

Match the term to the proper definition.

1. common laws
2. confidential
3. controlled substance
4. direct supervision
5. ethics
6. indirect supervision
7. liability
8. malpractice
9. misrepresentation
10. moral
11. negligence
12. slander
13. statutory laws

a. the act of working below the standards of practice
b. failure to do what is necessary or proper
c. talking negatively about someone in an improper manner
d. drugs that have the potential for addition and abuse
e. state regulations and rules based on legal precedents
f. what a person believes is right and wrong
g. rules and regulations based on written government law
h. rules and regulations that govern proper conduct
i. working or acting as someone who you are not qualified to be
j. a legal responsibility toward something
k. information that is private and not to be shared by anyone outside of the facility without permission
l. individual who is working while a veterinarian is within the facility or area to assist as necessary
m. individual who is working while a veterinarian is not within the facility or property
n. individual who is working while a veterinarian is immediately within the same room if assistance is necessary

True/False

Read statements 1 through 10 and determine if they are TRUE or FALSE. If the statement is FALSE, correct it to be a true statement.

1. Veterinarians are required to provide emergency care to all patients.   T   F
2. Veterinary assistants should never allow clients to think they are the veterinarian.   T   F
3. The VCPR may be concluded at any time.   T   F
4. Only veterinarians should set a high standard of patient care.   T   F
5. It is acceptable to question the ability of a veterinarian.   T   F
6. Veterinarians are morally expected to care for each patient that needs medical care.   T   F
7. All veterinary health care members are required to take CE courses to update their training.   T   F
8. Common laws and state laws are the same.   T   F
9. The VCPR is legally required within a two-year time frame.   T   F
10. The FDA regulates controlled substances within the veterinary industry.   T   F

Short Answer

1. Describe five situations that would be considered unethical in a veterinary hospital.
2. List four areas in veterinary medicine that are governed by federal laws.
3. What government agencies regulate areas of veterinary ethics?
4. What is the difference between direct and indirect supervision?

Clinical Situation

Juan, a veterinary assistant at All Creatures Great and Small Veterinary Hospital, was working in the reception area when a client came through the door carrying a small dog that was bleeding.

"Please help my dog, she is hurt very badly."

Rushing into the waiting area, Juan said, "Please try to calm down and tell me what happened."

"I was walking Patches in the park this morning when she got away from me and ran into the street. A car hit her and I thought she was dead," the woman replied. "She needs to see the vet now!"

Juan looked at the dog and said, "She is bleeding very badly. Let me get you a towel so you can apply pressure to the wound. She needs X-rays and will probably need stitches. She also may be going into shock."

"Please get the vet to look at her."

"I'm sorry, but Dr. Hughes is not in yet and Dr. Carr is out on a farm call this morning. I will write down directions for you to an emergency clinic that is about 15 minutes away. I will call and let them know you are bringing in a dog that is in need of immediate attention. They are not the best doctors, but they should be able to help Patches."

"Please, can't you do anything to help Patches?"

"No, I'm sorry, there is nothing I can do."

"Please, she may die!"

"I can't help her. Your dog needs to see a veterinarian and the only one that is available is at the emergency clinic. I recommend that you apply pressure to the area that is bleeding and take her there immediately."

"I can't believe that you won't you help me!" she replied, walking to the door. "Fine, I'll take her somewhere that cares and will actually do something for her."

- What did Juan do that was ethical in this situation?
- What did Juan do that was unethical in this situation?
- What could Juan have done to better handle this situation?

COMPETENCY SKILL 12

Professional Veterinary Ethics

Objective:

To reach an understanding regarding what is appropriate and inappropriate practice in the veterinary facility.

Preparation:

- Know the rules and regulations of the veterinary facility.
- Know the difference between ethical and unethical practices.
- Use common sense and respect.

Procedure:

1. Reduce pain, suffering, disease, as the patient's needs come before all others.
2. Obey all laws.
3. Be fair and honest with everyone.
4. The veterinarian makes all decisions based on patient care.
5. Every patient must develop a VCPR within a year: this includes a physical exam in order to diagnose, prescribe medication, perform surgery, or supply a prognosis. This relationship must legally be updated every year.
6. The veterinarian is responsible for all aspects of patient care, including emergency services.
7. Medical records are confidential, even between staff members.

COMPETENCY SKILL 13

Professional Veterinary Laws

Objective:

To acquire an understanding of the legal obligations of the veterinary profession.

Preparation:

- Obtain a copy of the state's Veterinary Practice Act and rules and regulations.
- Know common and federal laws relating to the veterinary profession.
- Use common sense at all times.

Procedure:

1. Know all laws and obey them.
2. Always document everything within the medical record.
3. Have every client sign a consent form.
4. Give estimates to every client.
5. Keep all doors and windows closed and secure.
6. Provide leashes and carriers for all small pets.
7. Animal restraint should legally be performed by only trained staff members.
8. Keep animals separate from each other.

9. Use ID cards and neck bands on all hospitalized patients.
10. Record all controlled substances correctly in a log that is maintained daily.
11. Check all areas of the hospital for safety hazards on a daily basis.
12. Post signs that warn clients of possible hazards.
13. Report any misconduct to the state board of veterinary medicine.

Chapter 7: Veterinary Safety and Aseptic Techniques

Objectives

Upon completion of this chapter, the reader should be able to:

7.1 State the concerns of human and animal safety in the veterinary industry
7.2 Discuss the safety hazards within the veterinary industry
7.3 Discuss OSHA guidelines in veterinary medicine
7.4 Explain the use of SDS
7.5 Identify safety signs and equipment
7.6 Discuss safety plan guidelines in a veterinary facility
7.7 Explain the types of sanitation used in the veterinary industry
7.8 Discuss areas of importance in keeping clean in the veterinary facility
7.9 Describe aseptic techniques
7.10 Discuss the importance of the isolation ward

Introduction

Veterinary safety practices are important to the well-being of the veterinary staff, clients, and patients within the veterinary facility, thus, it is important that the veterinary assistant knows the hazard concerns within the veterinary industry. Safety issues within the veterinary facility include the handling and restraint of animals, OSHA rules and regulations, human medical emergencies, exposure to radiation and anesthetic gases, physical safety hazards, and sanitation concerns. This chapter discusses safety related to physical hazards and how to develop a safety plan within the facility. The care and cleanliness of the veterinary facility serve as a safety measure for staff, clients, and patients. Preventing the spread of disease is of utmost importance. Sanitation, disinfection, and aseptic techniques govern the standards of proper cleanliness in a veterinary care facility. All staff members should have proper knowledge of how to clean, disinfect, sanitize, and sterilize the areas of the veterinary facility. When clients enter the facility, the way it looks and smells will be their first impression, and first impressions are everything to the client. It is important that the veterinary assistant understand the hazard concerns within the veterinary industry.

Safety in the Veterinary Facility

The veterinary facility is required to have emergency plans prepared for multiple types of safety issues. All staff members are required to be properly trained in safety practices and to maintain a copy of any safety plan created in the veterinary facility. Training and yearly safety updates are required for all new and current employees.

Fire and Safety Plans

All veterinary facilities should develop a safety plan (with information about evacuation) that informs the staff members what to do during an emergency, such as a fire, severe weather, or other disaster situations. Emergency numbers for police, fire, and other emergency response teams should be posted near each telephone in the facility. A fire plan should outline evacuation routes for people and animals and list the responsibilities of each staff member during the emergency. All fire extinguishers should be checked regularly to make sure they are up-to-date and in working order (see Figure 7–1). A fire exit plan, clearly marked and easy to read, should be posted near every entrance and exit of the facility. Fire alarms should be checked regularly to make sure they work properly (see Figure 7–2). A meeting location that is at a safe distance from the facility should be established for the entire staff. Staff members should practice routine fire drills so they are knowledgeable of the plan and procedures.

Personal Protective Equipment

Personal protective equipment (PPE) is provided by the employer so that employees can be safe in their work environment. Exam gloves protect the hands from contaminants and should be used when handling any animal that may be potentially infectious or cause a person to become injured or contaminated in any way. Exam gloves should be changed before handling a different animal. Hands should be washed thoroughly, using warm water and antibacterial soap, after the handling of each animal. Barking dogs can be a threat to hearing, especially in indoor kennels and veterinary wards. Noise levels in dog wards can reach 110 m decibels, and exposure for excessive amounts of time can affect hearing. When working in these areas for extended periods, the use of personal hearing protectors rated to filter the noise by at least 20 decibels is advised.

Eye protection, such as visors, shields, or glasses, should be available to all personnel. Certain procedures within the veterinary facility—such as performing dental cleanings or mixing medications or solutions—may harm the eyes. Always wear safety glasses when required. Masks and respirators may be required when handling certain chemicals that emit noxious or toxic vapors.

FIGURE 7–1 All staff should know where the fire extinguishers are in the facility.

FIGURE 7–2 The facility should be equipped with fire alarms in proper working order.

COMPETENCY SKILL 14

Personal Protective Equipment

Objective:

To know what PPE is available and the proper uses and needs for it.

Guidelines:

- Wear exam gloves at all times when working with chemicals, animals, and hazardous materials.
- Wear goggles or safety glasses when mixing chemicals or other tasks that may risk eye injury.
- Wear face masks and shields when brushing teeth, working on dentals, using high-powered equipment, or when a risk of splashing items in the face is possible.
- Wear earplugs when working in noisy areas, such as kennels.
- Wear aprons or gowns when bathing animals or working in surgical assisting and isolation wards.

COMPETENCY SKILL 15

Glove Removal

Objective:

To properly remove gloves.

Procedure:

1. Firmly pinch the outer surface of one glove about 2 inches below the cuff (see Figure 7–3A).
2. Pull the grasped glove downward and pull the hand out of the glove. The removed glove will be inside out (see Figure 7–3B).
3. Holding on to the removed glove with the fingertips, gather it into the palm of the other gloved hand.
4. Firmly pinch the outer surface of the second glove about 2 inches below the cuff.
5. Pull the gloved hand upward removing the hand from the glove. The remaining glove will be inside out with the first glove inside it (see Figure 7–3C).
6. Discard both gloves in the medical waste container.
7. Wash hands with disinfectant soap immediately. Dry hands thoroughly. Apply hand sanitizer.

FIGURE 7–3 Proper removal of examination gloves. (*Continues*)

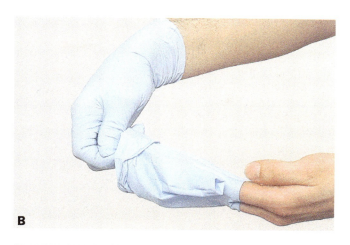
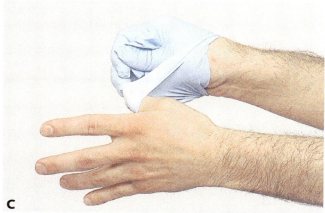

FIGURE 7–3 (Continued)

 COMPETENCY SKILL 16

Hand Hygiene

Objective:

To properly clean hands to prevent the spread of infection.

Guidelines:

- Wash hands after removing gloves.
- Wash hands after handling each animal.
- Wash hands using warm water and antibacterial soap, cleansing each surface of the hand for 5 minutes.

Instruments and Equipment

"**Sharps**" are sharp instruments and equipment that can injure a human or animal: a wound or cut from a contaminated sharp can transmit a contagious disease. Examples of sharps materials include glass, needles, and surgical blades. Veterinary staff members must be properly trained to deal with these safety hazards to prevent health risks. Any instrument capable of causing a puncture wound must be placed in a **sharps container** to prevent contamination and spread of disease (see Figure 7–4). To prevent accidents due to punctures and lacerations, always keep needles capped and surgical blades sheathed. If possible, do not try to recap needles; immediately place them in a sharps container. When necessary, needles can be re-capped using a one-handed method. Place the cap on a flat surface and use one hand to thread the needle into the cap. Then use both hands to firmly push the cap firmly into place. Dispose of the entire unit in the sharps container. Never open or place your hand into a sharps container.

Other safety hazards that commonly cause injuries to people and animals include heavy gates (may pinch fingers) and large animal restraint equipment (can crush a person).

FIGURE 7–4 Any object that has the potential to cause an injury such as a scratch, puncture wound, or cut should be placed in a sharps container.

 COMPETENCY SKILL 17

Biohazard and Sharps Disposal

Objective:

To identify the sharps container and properly dispose of hazardous materials into it.

Procedure:

1. Locate all biohazard containers (see Figure 7–5).
2. Place needles, blades, and any sharp materials that may have infectious waste in a sharps container.
3. Place all body fluids—including blood, urine, and feces—in medical waste bags.
4. Place all surgical drape material in medical waste bags.
5. Dispose of sharps containers and medical waste in proper location for incineration or medical waste pickup.

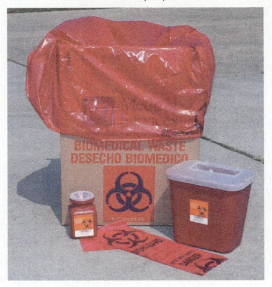

FIGURE 7–5 Biohazardous wastes should be placed in red bags, and sharps in sharps containers.

Physical Hazards

A **physical hazard** is a situation or agent with the potential to cause physical harm to a human or animal's body. Examples include animal bites, back injuries from lifting heavy objects, falls on wet surfaces, and radiation exposure.

Clean animal bites and scratches with warm water and soap and apply a bandage until the injury can be given medical attention (see Figure 7–6). Draw attention to wet floors by placing a sign in those areas where people are walking and could slip and fall (see Figure 7–7). When you are lifting heavy objects or animals, remember to keep your back straight. Do not bend over—rather, bend your hips and knees to squat down close to the object or animal. Keep it close to your body and straighten your legs to lift. When you are lifting more than 40 pounds, get someone to help you.

Chemical Hazards

A **chemical hazard** may cause injury to the skin, lungs, eyes, or other areas due to exposure. Examples include chemotherapy drugs, cleaning agents, insecticides, and anesthetic gases. The veterinary industry uses many hazardous chemicals. These chemicals may cause severe injury through direct contact or inhalation of vapors. Some vaccines pose a potential risk, as they may accidentally expose someone to disease. People may also be exposed to hazardous chemicals when they handle certain antibiotics. Always wear protective gloves and goggles and never mix chemicals unless you know it is safe to do so. When diluting solutions,

FIGURE 7–7 Cones or signs should be used to alert staff and clients of a wet floor or danger of slipping.

always start with the correct amount of water and then add the appropriate amount of concentrate. Never add the water to the concentrate, as the chemical may splash or produce an unexpected reaction.

Minor chemical spills may be cleaned up using paper towels or other absorbent materials that are then disposed of in the trash. More dangerous chemicals, such as formaldehyde or ethylene oxide, require special spill kit procedures. Always wear protective gloves and any suggested protective equipment as advised by the SDS information relating to the product.

 TERMINOLOGY TIP

Chemotherapy is the use of chemicals to treat disease. ■

Biological Hazards

A **biological hazard** poses a risk to humans and animals through the contamination of living organisms in body tissues and fluids. Biological hazards include blood, urine, feces, and live vaccines. **Medical waste** is a concern, as living tissue and fluids may have contagious and

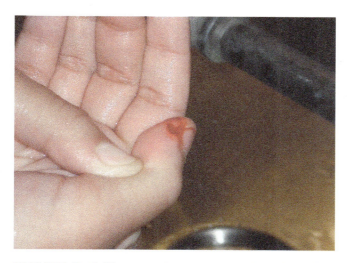

FIGURE 7–6 Thoroughly clean and administer first aid to any wound caused while on the job. Make sure facility protocols and procedures are followed in regard to care and reporting of the incident.

infectious organisms that can cause zoonotic disease. Medical waste may include surgical drapes, bandage materials, urine-soaked bedding, or saliva.

Zoonotic Hazards

A **zoonotic hazard** allows contagious organisms to spread to humans, causing infections by viral bacterial, fungal, and parasitic transmission. A **virus** causes diseases that are not treatable and can only run their course. Some may be fatal. Examples of a virus are the flu and rabies. **Bacteria** is a living organism that invades the body, causing illness. Some bacteria may spread through contamination of wounds; some may be released into the air and spread through the mucous membranes or inhaled. A **fungus** is a living organism that invades the external area of the body through direct and indirect contact. **Direct contact** is transmission of a disease through a direct source, such as saliva, a cage, or the ground. **Indirect contact** is transmission of a disease through an indirect source, such as air or water. These sources may be secondary to the host that is spreading the disease or contagion. The most common fungal infection and transmission from animals to humans is ringworm. The most effective way to prevent ringworm is by wearing proper protective gloves and practicing good personal hygiene—which means thoroughly washing hands after handling each patient. Parasitic sources are another route of disease transmission. **Parasites** may be **protozoans**, which are one-celled simple organisms; or **rickettsia**, which are more complex parasites that bite, such as fleas and ticks. Good sanitation practices help prevent and control disease transmission. Most common internal parasites are spread through contamination of eggs. Some parasite infections in humans can spread to organs in the body causing damage. The Companion Animal Parasite Council offers information on internal and external parasite protection and exposure considerations for veterinary professionals.

The handling of specimens, such as fecal samples and blood samples, should include the use of protective gloves and routine hand washing practices. Contamination of these types of substances can be cleaned with disinfectant solution and paper towels. Protective gloves should be worn when cleaning all contaminated areas. When treating an animal that may be infectious to people or other animals, protective gear including gloves, mask, gown, shoe covers, and eye protection should be worn. Any clothing that has been contaminated should be changed immediately and hands should be routinely washed with a disinfecting agent. Table 7–1 outlines some commonly encountered diseases in the veterinary facility and the sources of those diseases.

Rabies Suspects

Rabies is a serious and typically fatal virus that can affect humans and all warm-blooded animals. The rabies virus is spread through direct contact with infected animal saliva. This may be from the bite of an infected animal but may also transmit through contaminated saliva in open wounds or with contact of the mucous membranes, even from cleaning contaminated tables or food bowls. When handling an unvaccinated animal or one suspected of having the rabies virus, always wear protective gloves, a gown and protective eyewear. Care should be taken to ensure animals do not bite any people or personnel. Veterinary professionals may opt to be vaccinated with a safe and effective human prophylactic vaccine. Ask the hospital manager for information on vaccination and exposure, as many facilities and insurances offer discounts to high risk employees.

 MAKING THE CONNECTION
Review Chapter 35 on microbiology for additional content related to zoonosis. ■

TABLE 7–1

Examples of Diseases and Sources

DISEASE NAME	DISEASE TYPE
Brucellosis	Bacterial
Canine distemper	Viral
Ringworm	Fungal
Coccidia	Protozoan
Rabies	Viral
Lyme Disease	Bacterial
Rocky Mountain Spotted Fever	Rickettsial
Leptosporosis	Bacterial
Tetanus	Bacterial
Parvovirus	Viral
Hepatitis	Viral
Toxoplasmosis	Protozoan
Tick Paralysis	Rickettsial

OSHA Guidelines and Regulations

The **Occupational Safety and Health Act** (**OSHA**) of 1970 provides for safe and healthy work environments and conditions for all employees. OSHA guidelines were created by the U.S. Department of Labor and state that all employers are required to inform their employees of potential workplace hazards and risks so that everyone can work safely in their profession. OSHA regulates many safety aspects of business in all industries within the United States. Some states are regulated both by federal OSHA regulations and state-approved programs. Veterinary practices are regulated for the following areas so they:

- Ensure sanitation, cleanliness, and safety
- Provide clearly marked, adequate-sized, and unobstructed exits
- Ensure all compressed gas cylinders for anesthesia and oxygen purposes are properly marked and in working order
- Provide strict guidelines for the handling of hazardous chemicals
- Require protective devices for eyes, face, hands, feet, respiratory, and skin in situations posing any form of physical hazard
- Provide adequate toilet facilities and clean water
- Maintain operating fire extinguishers, sprinkler systems, and fire exit plans
- Require protective guards over all hazardous machinery
- Provide hazard-free electrical equipment
- Provide information to employees regarding any and all potential hazards, including SDS binders and OSHA accident reports

Safety Data Sheets (SDS)

The **right-to-know station** is the area where all OSHA binder information and the **Safety Data Sheets (SDS)** (formerly called MSDS sheets) should be kept allowing all employees access to information regarding any hazard within the facility. SDS are safety data sheets that provide information published by manufacturers of various products that have the potential to harm humans within the facility (see Figure 7–8). The information contained in the SDS is mostly the same as the MSDS, except the SDS sheets are now required to be presented in a consistent and user-friendly 16 section format. These may be chemicals or physical products. Parts of the SDS include the following:

- Product name
- Manufacturer information
- Hazard ingredients/identity information
- Physical/chemical characteristics
- Fire and explosion hazard data
- Reactivity data (permissible exposure limit [PEL]) in radiology
- Health hazard data
- Precautions for safe handling and use
- Control measures

The SDS must be in English or bilingual and stored in an area where all staff have access. Staff members must be trained on finding and using the SDS information. Each hazardous substance must have a file maintained with updated product information and is best kept filed alphabetically for ease of use. The use of a three-ring binder or computerized program is recommended and should be ready for inspections at all times. Employers must ensure that the SDS sheets are readily accessible to employees for all hazardous chemicals found in the workplace. Many employers offer a binder or a computerized system. Often a person is designated in charge of the SDS to keep the items updated.

The OSHA binder must include safety plans that contain an evacuation plan in the event of an emergency; locations of water, gas, and electric shutoffs; locations of fire extinguishers; and emergency telephone numbers for the police, fire department, and emergency service companies. All chemicals, disinfectants, and other products within the facility should be labeled with hazardous warning labels that identify which body systems they may be toxic to. Hazardous labels should identify the product name, the body systems or organs it affects, what PPE equipment is necessary when working with the item and include a picture that identifies it as hazardous (see Figure 7–9). The facility should have emergency and personal protection equipment, such as safety glasses or goggles, latex and vinyl gloves, earplugs, and radiation gear. A person should be designated to oversee the safety training within the practice of all new employees and design a yearly OSHA safety meeting that reviews all updated information regarding employee safety.

SAMPLE SDS

Product A
Safety Data Sheet
According To Federal Register / Vol. 77, No. 58 / Monday, March 26, 2012 / Rules And Regulations
Date of Issue: 02/23/2017   Version: 1.0

SECTION 1: IDENTIFICATION

1.1. Product Identifier
Product Form: Mixture
Product Name: Product A
Synonyms: Anionic Anti-Sludge Agent

1.2. Intended Use of the Product
Use of the Substance/Mixture: Anti-Sludge Agent. For professional use only

1.3. Name, Address, and Telephone of the Responsible Party
Company
Glendale Industries, Inc.
1234 Anywhere Way
Anytown, US 12345
1.888.362.2007

1.4. Emergency Telephone Number
Emergency Number : 1.888.362.2007
For Chemical Emergency, Spill, Leak, Fire, Exposure, or Accident, call GLENTREC– Day or Night

SECTION 2: HAZARDS IDENTIFICATION

2.1. Classification of the Substance or Mixture
GHS-US Classification
Flam. Liq. 2   H225
Skin Corr. 1B   H314
Eye Dam. 1   H318
STOT SE 3   H336
Full text of hazard classes and H-statements : see section 16

2.2. Label Elements
GHS-US Labeling
Hazard Pictograms (GHS-US) :

GHS02   GHS05   GHS07

Signal Word (GHS-US) : Danger
Hazard Statements (GHS-US) : H225 - Highly flammable liquid and vapor.
H314 - Causes severe skin burns and eye damage.
H318 - Causes serious eye damage.
H336 - May cause drowsiness or dizziness.
Precautionary Statements (GHS-US) : P210 - Keep away from heat, sparks, open flames, hot surfaces. - No smoking.
P240 - Ground/Bond container and receiving equipment.
P241 - Use explosion-proof electrical, ventilating, and lighting equipment.
P242 - Use only non-sparking tools.
P243 - Take precautionary measures against static discharge.
P260 - Do not breathe vapors, mist, spray.
P264 - Wash hands, forearms, and other exposed areas thoroughly after handling.
P271 - Use only outdoors or in a well-ventilated area.
P280 - Wear protective gloves, protective clothing, eye protection, face protection, respiratory protection.
P301+P330+P331 - If swallowed: rinse mouth. Do NOT induce vomiting.
P303+P361+P353 - If on skin (or hair): Take off immediately all contaminated clothing. Rinse skin with water/shower.
P304+P340 - If inhaled: Remove person to fresh air and keep at rest in a position comfortable for breathing.
P305+P351+P338 - If in eyes: Rinse cautiously with water for several minutes. Remove contact lenses, if present and easy to do. Continue rinsing.
P310 - Immediately call a poison center or doctor.

FIGURE 7–8 The first page of a Sample Safety Data Sheet. More information on Safety Data Sheets can be found at MSDSonline.com.

SAMPLE

CHEMICAL INVENTORY FORM

Date updated _____

Dental office _____

Chemical Name	Hazard Class				Physical State	Manufacturer	Comments
	(H)	(F)	(R)	(P)			

(H) Health	**(F) Fire Hazard**	**(R) Reactivity**	**(P) Protection**
0—Minimal	0—Will not burn	0—Stable	A—Goggles
1—Slight	1—Slight	1—Slight	B—Goggles/gloves
2—Moderate	2—Moderate	2—Moderate	C—Goggles/gloves/clothing
3—Serious	3—Serious	3—Serious	D—Goggles/gloves/clothing/mask
4—Extreme	4—Extreme	4—Extreme	E—Goggles/gloves/mask
			F—Gloves
			G—Face shield/gloves

FIGURE 7-9 Each SDS binder or computerized program should contain a list of the chemicals and the health dangers they may impose.

Employees are responsible for reading and understanding the facility's practice safety manual and regulations and knowing where all safety equipment is kept and how it is used. It is also the employees' responsibility to learn how to properly read the SDSs. Many practices require documents and accident reports to be completed and filed as soon as an injury occurs. Failure to report an incident may result in a violation of legal requirements and delay the processing of necessary insurance claims and benefits.

Table 7-2 lists safety equipment that should be in each veterinary facility.

Labeling Secondary Containers

All secondary containers in use in the facility must be labeled with an OSHA label. These labels indicate specific hazards and are divided into four color-coded sections. Blue indicates a health hazard, red indicates flammability, yellow indicates reactivity (radiation), and white indicates the need for personal protection.

TABLE 7-2

Safety Equipment in the Veterinary Facility

Fire extinguishers	Safety glasses/goggles
Exam gloves (latex and vinyl)	Wet floor signs
Warning labels	Biohazard signs
Radiation hazard signs	Lead apron, gloves, and thyroid shield
Ear plugs	Anesthetic gas scavenger system
Eye wash station	Sharps container
Medical waste bags	

Each color code has a box in which the hazard rating number is placed. Each hazard code must be properly identified and written on the container based on the SDS information for each product.

100   SECTION I   Practice Management and Client Relations

COMPETENCY SKILL 18

OSHA Veterinary Hospital Plan

Objective:

To become aware of a facility's safety plans and know what role to play when a plan needs to be activated.

Materials:

- OSHA plan
- Safety binder or policies and procedures manual
- SDS binder or computerized program

Procedure:

1. Review the veterinary OSHA plan.
2. Identify all hazards in the work area (physical, chemical, zoonotic, and biological).
3. Locate the SDS binder or computerized program.
4. Review the safety labeling on hazardous items.
5. Learn the emergency procedures for the hospital (accidents, hazardous materials spills, fire evacuation, gas releases).
6. Review emergency procedures for patient evacuation.
7. Locate the fire extinguishers.
8. Locate all PPE (personal protective equipment).
9. Know where to dispose of sharps containers and medical waste.

COMPETENCY SKILL 19

Lifting Heavy Objects

Objective:

To properly lift objects without injury.

Guidelines:

- Work with more than one person if the object is over 50 pounds.
- Each person should be in position to lift simultaneously.
- Bend your hips and knees to squat down close to the object or animal.
- Keep it close to your body and straighten your legs to lift.
- Keep your back straight.
- Move the object or animal to the desired location.

Radiation Hazards

Exposure to small amounts of radiation, such as routine radiographs, poses little threat to one's overall health. However, long-term exposure or exposure to large doses of radiation can cause health problems, such as skin damage, gastrointestinal disorders, bone marrow disorders, and poor organ function. It is important when using any radiographic equipment to never place any body part in the primary beam. The use of lead lined protective wear including apron, gloves and thyroid shield should always be worn. Everyone working within the radiology area is required to wear a **dosimetry badge** during all radiographic procedures. This badge measures any scatter radiation that may be received during the procedure. Reports are regularly provided to the facility with the measurements of exposure and alert if your exposure to radiation reaches a hazardous amount. The dosimetry badge number transfers with you to any hospital in which you are employed so that is can accurately track lifetime exposures.

Anesthetic Hazards

Long-term exposure to waste anesthetic gases is a health hazard and potentially causing spontaneous miscarriages, liver and kidney damage, and congenital abnormalities in young children. OSHA sets the guideline for safe exposure limits for inhaled anesthetic agents at 2 parts per million. Concerns with leaks from the anesthesia machine should be check daily before use and a scavenger system should be in place to capture excess gases and transport them to a safe exhaust port outside of the facility.

Compressed gas tanks enclosed in cylinders should be stored safely in a cool, dry location away from any heat sources. Tanks should be secured in an upright position by a chain or strap.

Methods of Sanitation

There are many methods of sanitation used in a veterinary facility. **Sanitation** is the process of keeping an area clean and neat. This includes appearance and odor control. The type of sanitation will depend on the location in the facility, the purpose of the area, and the type of chemical or product used for sanitation purposes. First, it is important to understand the levels of cleaning. **Cleaning** is the process of physically removing all visible signs of dirt and organic matter such as feces, blood, and hair. **Disinfecting** is the process of destroying most microorganisms on nonliving things by physical or chemical means (see Figure 7–10). **Sterilizing** is the process of destroying _all_ microorganisms and viruses on an object using chemicals and/or extreme heat or cold under pressure.

Physical Cleaning

Physical cleaning is the most common method of sanitary control within the veterinary facility. It involves the use of a chemical with a cleaning object, such as a mop, sponge, or washcloth and may include dusting, mopping, or cleaning up urine or feces within a cage. In physical cleaning, the hands remove dirt, debris, and organisms from all surfaces of the veterinary facility. This includes disinfecting areas to prevent the spread of disease (see Figure 7–11).

FIGURE 7–10 Disinfection of animal cages.

102 SECTION I Practice Management and Client Relations

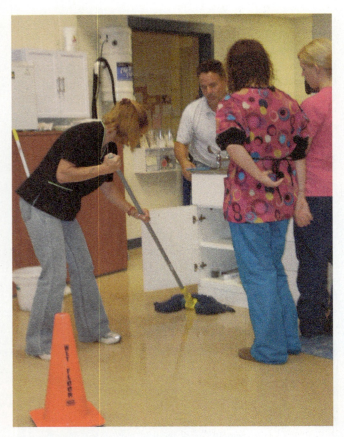

FIGURE 7–11 Mopping is one method used to physically clean the facility.

Disinfecting

Disinfecting is the destruction of most pathogens and organisms on inanimate objects, such as exam tables and cages. Most disinfectants are chemical microbial agents that kill microbes; others are bacterial agents that destroy bacterial and inhibit the growth of microbes. A variety of disinfectants are available for veterinary use and aid in destroying bacteria, microbes, spores, fungus, and viruses. Bleach is a common disinfectant and should never be mixed with any other chemical; it is often diluted in water. Many disinfecting agents create vapors that are harmful to patients and people.

Sterilization

Sterilization is the process of killing all living organisms on a surface, typically in exam rooms, treatment areas, and surgical suites to ensure that tables and instruments are free of disease. There are several classifications of sterilization, some being more effective than others.

Cold Sterilization

Cold sterilization is the process of soaking items in a disinfectant chemical until they are cleaned for reuse. A **cold tray** holds a chemical that acts as a sterilizing agent (see Figure 7–12). Items such as nail clippers, brushes, endotracheal tubes, and some surgical scissors may be prepared through cold sterilization techniques. These items are used repeatedly in the facility and require a simple and quick sanitation method. Table 7–3 lists common cold tray solutions. This technique is not used for sterile surgical procedures and not intended for surgical procedures.

Dry Heat or Incineration

Dry heat exposes an item to extreme heat with a flame or through incineration. An **incinerator** is used

FIGURE 7–12 A cold tray used for sterilization.

TABLE 7-3
Common Cold Tray Solutions

SOLUTION NAME	USE	CONTACT TIME	DILUTION
Chlorhexidine solution	Dilute with water for wiping, soaking, mopping, and spraying on external animate or inanimate objects.	5 minutes	1 ounce or 2 tablespoons per gallon of water
Nolvasan solution	Dilute with water for wiping, mopping, and spraying on or soaking inanimate objects.	10 minutes	1 ounce or 2 tablespoons per gallon of water
Cetylcide-G	Dilute with water for soaking inanimate objects.	40 minutes	60 ml per gallon water; 15 ml per quart water
Pink germicide	Dilute with water for soaking inanimate objects.	4 minutes	10 ml per quart water

to burn the remains of items that have the potential to spread disease. **Incineration** may be necessary with biological hazards, medical wastes, or animal carcasses that may be contagious or zoonotic. This method requires items to be burned to ashes to prevent the spread of disease.

Autoclave

Another common method for sterilizing items, especially surgical equipment, is to use an autoclave. An **autoclave** is a sealed chamber in which objects are exposed to heat and steam under pressure at extremely hot temperatures to kill all living organisms (see Figure 7–13).

FIGURE 7-13 An autoclave.

Gas Sterilization

Gas sterilization is most commonly performed with ethylene oxide. This is a flammable, explosive liquid that becomes an effective sterilization agent when mixed with carbon dioxide or Freon gas. Equipment that cannot withstand extreme temperatures and the pressure of steam sterilization can safely be sterilized with ethylene oxide. Such equipment includes endotracheal tubes, plastics, and items used in an endoscopy. All items sterilized by this method require 24 hours of air exposure prior to internal use. Items should be clean and dry prior to gas sterilization, as the mixture of ethylene oxide and moisture leaves a toxic residue.

Radiation and Ultrasound

Another method of sterilization may be through the use of radiation or ultrasound. **Radiation** uses ultraviolet or gamma rays to radiate and kill living organisms. **Ultrasound** passes high-frequency sound waves through a solution to create a vibration that scrubs an object to remove debris. This is commonly done with an ultrasonic cleaner machine in which items are soaked in a solution that vibrates to remove dirt and debris.

Filtration

Another method of sanitation within a facility is the use of filtration. **Filtration** removes particles from the air using a physical barrier and is commonly used in lab areas or research facilities. This method usually takes place in a room other than where the animals are housed or contained. This room is pressurized to prevent organisms from entering the facility. It is common for people entering a filtration area to change into sterile clothing and put on PPE equipment before they enter the area where the animals are housed.

Veterinary Sanitation Chemicals and Cleaners

A variety of chemicals are used for cleaning, disinfecting, and sterilization in a veterinary facility. Chemicals should always be handled with care and consideration: some are harmful to people, animals, or both. Certain chemicals have vapors that may be harmful if inhaled. Other chemicals may cause burns if contact is made with the skin or eyes. These are reasons to be careful when mixing or handling any chemical. It is important to always read the bottle label or SDS to determine the safety of the chemical. Always wear exam gloves when handling or using chemicals. When mixing or diluting chemicals, read all labels and directions beforehand. Some cleaning agents may be used in one location but not in another. A chemical solution that is used on the floor may be harmful to an animal and not be used to clean a cage or exam table.

Antiseptics are effective disinfecting agents that destroy microorganisms or inhibit their growth on living tissue. Disinfecting agents have a variety of properties that alter their uses and effectiveness. Some of these properties include the following:

- Spectrum of activity—what the agent will kill, such as viruses, bacteria, or fungus
- Concentration—its strength. Some agents must be diluted before use; some can be used at full strength
- Contact time—how long the disinfectant should sit before being cleaned from the surface
- Appropriate surface uses—the types of items the agent may be used to clean
- Inhibiting factors—uses that should be avoided when using the disinfectant
- Toxic effects—the hazardous effects the agent may have on humans or animals

TERMINOLOGY TIP
Spectrum refers to a wide variety of factors.

TERMINOLOGY TIP
Dilution the act of diluting, or lessening in strength by adding another component, such as water. Dilution requires adequate measuring skills.

Some common agents used in the veterinary facility are summarized in Table 7–4.

Hand Hygiene

Proper hygiene includes hand washing (see Figure 7–14). For animals and other staff members to be safe and sanitary, it is important that everyone in the veterinary health care field practice proper hand-washing techniques. The most common method of spreading disease is through direct hand contact.

Hand washing should take place:

- After visiting the restroom
- After coughing
- After sneezing
- After removing exam gloves
- After touching or handling each animal
- After handling money
- After contact with people (shaking hands)
- After contact with visibly sick people
- After touching phones
- After touching door handles
- After using a computer keyboard

A proper hand-washing procedure begins with an antibacterial soap and the use of hand sanitizer (see Figure 7–15). To decrease the possibility of contamination, staff members should avoid wearing jewelry, having artificial nails, and allowing nails to grow beyond the end of the finger. As frequent hand washing can dry out the skin, use a moisturizing lotion to keep hands soft and healthy. Signs should be placed throughout the facility to remind all staff members to wash their hands.

TABLE 7–4
Common Veterinary Cleaners and Disinfectants

CLEANER	TYPE OF AGENT	USE
Nolvasan solution (Chlorhexidine)	Disinfectant	Used on inanimate objects. Dilute with water at 1:40 dilution. Used in cold sterilization. Two-day activity. Inhibited by soaps.
Nolvasan scrub (Chlorhexidine)	Disinfectant and antiseptic	Used on humans and animals. Used for surgical preps. Sterile prep. Two-day activity. Inhibited by soaps.
Clorox or chlorine bleach	Cleaner disinfectant	Caution with odors and vapors. Toxic to skin. Treats all living organisms.
Pink germicide	Disinfectant	Used on inanimate objects. Dilute with water. Used in cold sterilization. Toxic to skin.
Instrument milk	Cleaner and disinfectant	Used on surgical equipment and instruments. Used in cold sterilization.
Roccal-D	Disinfectant	Dilute with water. Toxic to skin.
Citrus II	Cleaner and disinfectant	Dilute with water. Pleasant smelling. Used on inanimate objects.
Ethyl alcohol or isopropyl Alcohol	Disinfectant and antiseptic	Requires long-term contact to cause sterilization. Requires 70 percent concentration. Used on living and nonliving sites.
Iodine or Betadine	Disinfectant and antiseptic	Use with gloves due to staining effect on skin. Used on living and nonliving sites. 2 percent solution—1:10 dilution for skin contact—1:100 dilution for tissue lavage. Surgical scrub and solution. Inactive with alcohol or organic matter. Four- to 6-hour activity.
Quaternary ammonia	Disinfectant	Toxic to skin. Toxic vapors. Caution mixing with other chemicals. Inanimate objects.
Formaldehyde	Disinfectant	Preservative of tissues. Used less due to highly toxic affects on skin and toxic vapors. Potential for causing cancer.
Hydrogen peroxide	Antiseptic	Used to clean areas on patients, especially blood and skin wounds. Used at 3–20 percent concentration.

COMPETENCY SKILL 20

Dilution of a Substance or Disinfectant

Objective:
To properly work with disinfecting agents to achieve the best possible effects of cleanliness in the facility.

Materials:
- Disinfectant
- PPE
- Containers
- Dilutant

Procedure:
1. Assemble all disinfectants in one location.
2. Wear eye protection and gloves when handling chemicals.
3. Assemble secondary containers, such as buckets or bottles.
4. Dilute the product as indicated on the label. Use a syringe to measure the proper amount.
   - 1 teaspoon equals 5 mL
   - 1 tablespoon equals 15 mL
   - 1 ounce equals 30 mL
5. Add water to the container.
6. Place dilutants into the container without overfilling it.
7. Replace lids on containers, wipe from bottles any spillage, and place bottle in storage location.

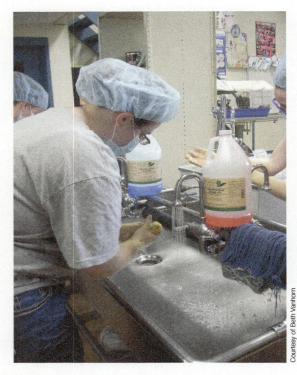

FIGURE 7–14 Proper hand washing is the key to preventing the spread of infection.

FIGURE 7–15 An antibacterial solution should initially be used when washing the hands.

COMPETENCY SKILL 21

Hand Washing

Objective:

To properly and effectively clean the hands to prevent cross contamination of oneself, patients, and clients.

Materials:

- Antibacterial soap
- Sink with warm water
- Paper towels
- Hand sanitizer
- Hand lotion

Procedure:

1. Gather paper towels.
2. Wet hands with warm water.
3. Place antibacterial soap into the palm of one hand. Use an amount the size of a nickel.
4. Use vigorous friction by rubbing hands together; lather up both hands to the wrist consistently for 15–30 seconds.
5. Cover all surfaces of the hand, including the back of the hands and between the fingers.
6. Rinse hands thoroughly, allowing the water to run downward off the skin.
7. Dry hands completely with paper towels.
8. Use a paper towel to turn off faucets and open doors.
9. Clean up area.

Housekeeping and General Cleaning

The most important aspect of general cleaning in a veterinary facility is to practice good hygiene and odor control. Cleanliness must be a priority to every staff member. If an area looks clean but smells dirty, then cleanliness has not been achieved. Many areas of the veterinary clinic need daily cleaning and routine disinfecting. Some areas will need to remain sterile for the safety of staff and patients. Areas of high traffic and animal contact require repeated methods of cleaning and sanitation. Cleaning protocols will vary from facility and area of the hospital. It is important that all staff members know proper disinfection and cleaning protocols for the entire hospital.

Many practices develop a daily and weekly cleaning routine. Daily cleaning is conducted on areas and objects that require strict cleansing (see Figure 7–16). Staff members should practice good hygiene and clean up after themselves, including the equipment and supplies that they have used. It is everyone's responsibility to help keep the facility clean on a daily basis. This career requires keeping an open mind of the duties that may be required as a veterinary professional. Many veterinary assistants are required to keep the

FIGURE 7–16 Animal enclosures must be cleaned daily.

TABLE 7-5
Veterinary Areas to Clean

Exam tables	Countertops	Sinks and faucets
Wash tubs	Desk tops	Floors
Walls	Light switches	Cages/kennels
Door handles	Scales	Drawer handles
Windows and glass	Cabinets and shelves	Restrooms
Instruments	Equipment	Computer keyboards
Isolation ward	Surgical suite	X-ray table
Telephones	Lab area	Air filters

facility clean, the animals clean, and equipment maintained. This may include animal bathing and grooming, cage and kennel disinfecting, and lawn care. As such, it is important to be familiar with the facility's cleaning and disinfecting methods and chemicals available for use.

It is important to note some rules for cleaning:

- Work from the top to the bottom when cleaning or disinfecting items.
- Work from the back of a space toward the door or entrance.
- Use towels, beds, or newspaper, as needed, to keep animals and cages clean.

All veterinary team members should practice the "clean as you go" attitude to keep the facility clean and safe. This means putting items back where they were originally located, replacing items after use, tidying up an area after a procedure, dusting and sweeping, and cleaning up messes as they happen. This practice helps minimize the spread of disease. Many factors can be considered in helping decrease the spread of disease to animals and humans, including the following:

- Limit area where animals eliminate (urinate/defecate).
- Routinely clean and change bedding in stalls and cages.
- Remove uneaten food.
- Change and clean water sources.
- Change clothing when contaminated.
- Treat animals in isolation last.
- Clean as you go.
- Use proper odor control.
- Clean rugs and chairs.
- Wipe down walls and doors.
- Wear gloves when handling animals.
- Keep a second set of clothes and scrubs to replace contaminated work clothes.
- Clean parking lots and walk areas.

Cleaning must be done properly to allow proper disinfection to follow. How to clean a surface depends on whether it is smooth or rough. Also, the wash ability of the surface must be considered: some items are best vacuumed or mopped. Fabrics may need laundering for proper cleaning. Hair is the most common issue within the veterinary clinic. It must be removed from all surfaces for proper disinfecting to take place. Hair may float around the clinic, causing contamination or the spread of disease. Sweeping, vacuuming, dusting, and mopping should be a routine part of the daily cleaning schedule. Windows and glass doors should be cleaned as well. Table 7–5 lists the various areas that must be cleaned in the veterinary facility.

COMPETENCY SKILL 22

Cage or Kennel Cleaning

Objective:

To properly clean animal enclosures to eliminate the risk or spread of infection.

Materials:

- Disinfectant
- Waste containers
- PPE
- Buckets
- Sponges

Procedure:

1. Remove all bedding, toys, or bowls.
2. Place dirty bedding into the laundry area, using the appropriate laundry detergent.
3. Disinfect all toys and bowls appropriately with soapy disinfectant and water; allow soaking.
4. Remove all feces, urine, or other materials with paper towels; discard in the garbage or medical waste.
5. Wash all sides of the cage or kennel walls—including the door, grates, and latches—with an appropriate disinfectant and warm water. Use a washcloth soaked in disinfectant.
6. Use a stiff brush for any items on the walls or other areas that are not easily removed with a washcloth.
7. Allow the disinfectant to stand on the cage items for the allotted contact time.
8. Rinse with warm water.
9. Dry all items.
10. Disinfect or launder all cleaning items.
11. Wipe all light switches, doors, and walls of the kennel area daily.
12. Replenish cleaning supplies as necessary.

COMPETENCY SKILL 23

Cleaning Surfaces and Supplies

Objective:

To understand how to properly maintain and clean all surfaces in the veterinary facility.

Materials:

- Disinfectants
- Buckets
- Sponges
- PPE

Procedure:

1. Evaluate the areas and surfaces that need to be cleaned.
2. Evaluate the cleaners and disinfectants that may be safely used on the areas of the facility.
3. Dust all areas to remove debris and hair.
4. Sweep or vacuum the floor to remove debris and hair.
5. Mop all floors and place wet floor signs for safety.
6. Dry all wet areas to prevent streaks from forming.

COMPETENCY SKILL 24

Laundry

Objective:
To properly launder and maintain cleanliness of the linens and clothing in the veterinary facility.

Materials:
- Detergent
- Washing machine
- Dryer

Procedure:
1. Sort laundry materials and determine the size of the laundry load.
2. Pre-treat all laundry in the washing machine by soaking.
3. Set the washing cycle for the appropriate load size and proper temperature, usually warm to hot water.
4. Add laundry detergent that disinfects bedding and other items.
5. Place items in the dryer when the washing cycle is complete.
6. Fold items properly and store them in a closed cabinet.

Exam Room Sanitation

Exam room sanitation is an important part of safety and disease control. It is important for not only the exam area to be sanitary but also the equipment, tools, and supplies should be clean and properly disinfected, and staff members should be using sanitary practices and proper hand-washing techniques.

The exam room should be cleaned after every patient and at the end of the day. This includes sweeping up hair and debris. It is best to clean from the top of the room to the bottom and from dry areas to wet. This will prevent dust, debris, and bacteria from spreading to cleaned areas. Areas of the exam room that contain the most dirt and debris should be cleaned first. Begin cleaning at a starting point and work clockwise around the room. This will allow every area to be cleaned. When using a spray disinfectant on tables and countertops, spray and wipe using an ample amount of disinfectant and allow enough contact time as noted on the label to provide proper sanitation. When using mixed chemicals with cloths or sponges, make sure the area is cleaned well and dried to prevent streak marks. Always clean using an up-and-down motion or side-to-side motion to ensure all surfaces have been covered. Chemicals should be safe to use around animals and produce little to no odor that may cause respiratory irritation.

The trash should be cleaned up on a regular basis to prevent the spread of disease and odors from forming. Spray and wipe out the empty trash cans and replace each can with a new liner. Place any used sharp items into the hazardous waste or sharps container. When the sharps containers are full, make sure they are closed securely and place them in a location for proper removal.

A cold tray should be located in each exam room to allow tools and equipment that may be safely submerged in disinfectant to be sanitized. Before tools are placed in cold trays, they should be wiped free of any blood, debris, or bodily waste materials using a disinfectant. The tool should then be placed in the cold tray and fully submerged. Soak time follows according to the disinfectant label. Most soak times are short, with an average of five minutes. After soaking, remove the tool from the cold tray, wash it with warm water, and dry well to prevent any rusting. The cold tray chemicals should be changed on a regular basis depending on how much they are used.

All areas of the exam room floor should be swept or vacuumed and cleared of debris prior to mopping. Use a bucket or large push bucket with a ringer for mopping so that the chemicals are properly diluted in water, decreasing the chance of cross-contamination in another area. A mop bucket should be used only in exam rooms and in no other areas of the facility to prevent the spread of disease. A *wet floor* sign should

COMPETENCY SKILL 25

Cleaning Exam Rooms and the Surgical Suite

Objective:

To properly maintain cleanliness and sterility in examination rooms and surgical suites.

Materials:

- Disinfectants
- Buckets
- Sponges
- PPE
- Waste containers

Procedure:

1. Gather all disposable materials and place them in a garbage or medical waste container.
2. Clean areas of hair and other debris.
3. Apply disinfectant solution to all areas of animal and human contact. Dry with paper towels.
4. Sweep or vacuum the floor.
5. Spot clean or mop the floor. Place a "wet floor" sign out for safety purposes.

COMPETENCY SKILL 26

Cleaning Surgical Instruments

Objective:

To properly clean and sterilize surgical instruments and prevent the spread of infection.

Materials:

- Disinfectants
- Sterilization tools
- PPE

Procedure:

1. Place all instruments in a cold tray solution.
2. Use a wire brush to scrub all debris from items and rinse with instrument milk or cleaner.
3. Rinse with distilled water.
4. Use ultrasonic cleaner if large amounts of debris. Place items in cleaner and soak in solution.
5. Follow recommendations of manufacturer.
6. Remove from solution and rinse with distilled water.
7. Dry instruments.
8. Use instrument milk to lubricate items by immersing for 30 seconds.
9. Lay on towel with all hinged instruments open to dry.

be placed near the area prior to mopping. The solution should be placed in the mop bucket and warm water added to the bucket in appropriate amounts. Place the mop into the cleaning solution and wring it out so that it is only damp. Begin at the farthest corner of the exam room and mop the floor area lengthwise along the base of the room. Then use figure-eight motions along the remaining area of the floor. Apply more disinfectant as the mop becomes dry and repeat as necessary. Work toward the door and allow the room to air-dry. When the bucket water becomes visibly soiled, change it immediately. When finished mopping, empty the mop bucket in a safe area. Wring out the mop and allow the mop to air dry. All mops should be rinsed thoroughly and laundered on a regular basis. Never use a dirty or odorous mop to clean.

Aseptic Techniques

The most important principle in the veterinary facility is maintaining aseptic technique. **Asepsis** is the absence of microorganisms. **Aseptic techniques** are the practice of keeping a sterile environment that is free of disease and contaminants. This is especially vital in the surgical suite. Aseptic technique governs how the facility is cleaned, how equipment and instruments are cleaned, and how surgical and medical procedures are performed. A break in aseptic technique leads to possible infection, disease, and potential patient death.

Four main sources of potential contamination in the surgical area include the surgical team personnel, surgical instruments and equipment, the surgical environment, and the patient. All personnel must be aware of what items are sterile and non-sterile. Sterile items should be kept together and separate from non-sterile items. Body movements and locations should be restricted in the surgical suite to reduce air currents and keep contaminants at a minimum. Only sterile items should touch patient tissues; if an item's sterility is in question, the item should be removed and considered contaminated.

When humans cause the spread of disease and contamination of an animal, this is known as a **nosocomial infection** or hospital acquired infection. A nosocomial infection may occur with unsterile surgical practices, contamination of a healthy animal due to unsafe sterile practices with hands or equipment not being cleansed or allowing contagious animals to be in contact with healthy animals.

COMPETENCY SKILL 27

Aseptic and Sterile Techniques

Objective:
To properly maintain asepsis in the veterinary facility.

Materials:
- Disinfectants
- PPE
- Appropriate waste containers

Procedure:
1. Treat every patient as if it were contagious.
2. Wear scrubs or a lab coat to protect clothing and the patients.
3. Wear a mask, goggles, foot covers, and hair covers, as necessary.
4. Disinfect everything a patient or human uses or is in contact with.
5. Place contagious animals in isolation.
6. Place all disposable items in the garbage or medical waste.
7. Disinfect all non-disposable items by soaking them in a cold tray or cleaning and sterilizing them.
8. Place used bedding in the laundry area for immediate attention.
9. Wash hands after removing exam gloves. Use an antibacterial soap and hand sanitizer.
10. Change scrubs when in contact with contagious animals or any body fluids.

Surgical Suite

Ideally, the surgical suite should be used for sterile surgical procedures only. The surgery room should be easy to clean and disinfect and be as closed off as possible. Closing the door will minimize traffic, increase cleanliness, reduce air circulation, and contain aseptic procedures. All items should be kept off the floor; any cabinets or shelves holding surgical materials should be closed to protect the packs and supplies from contaminants. The area should be free of clutter. Limited personnel should be allowed in the area to reduce airflow contamination. All surgical personnel should wear proper protective gear, including gowns, hair covers, masks, gloves, and shoe covers.

Isolation Ward

Hospital housing is arranged according to the type of cases and patient care needs. This may include healthy patients, outpatients, surgical patients, medical patients, and contagious patients. Contagious patients are kept away from all healthy patients in the **isolation ward**. This separate housing groups similar patients, making it safe for all animals and staff. Each ward should have its own medical supplies, cleaners, disinfectants, and equipment. The isolation ward should have items that are never removed from this location. The laundry and bedding should be disinfected separately from other laundered items. Signs should be kept on all cages and doors detailing the types of contagious diseases of the patients that are housed in the ward. All staff members who enter should wear exam gloves, sterile gowns, face masks, shoe covers, and hair covers (see Figure 7–17). All items should be discarded or appropriately disinfected after each use in the isolation ward.

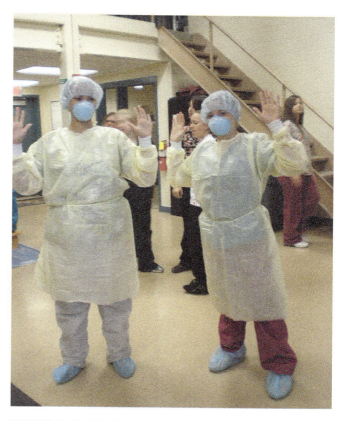

FIGURE 7–17 Proper PPE must be worn when working within an isolation ward.

SUMMARY

It is essential that all veterinary facility employees are properly trained in safety protection and have working knowledge of all safety plans. SDSs should be readily and easily accessible to all employees. Additional information and OSHA guidelines are available online at www.osha.gov, by calling 1-800-321-OSHA (6742), or at the U.S. Department of Labor, OSHA, 200 Constitution Avenue, N.W., Washington, DC 20210.

Sanitation involves cleaning and disinfecting the veterinary facility to control the direct spread of organisms on surfaces, in the air, and on other objects within the facility. All members of the veterinary staff must follow standard sanitation procedures. Failure to maintain a high standard of disinfection and sanitation will result in harm to the patients, loss of clients, a decrease in business, and the loss of employee jobs. Every staff member must have a high level of awareness for the potential transmission of diseases from patient to patient and patient to people. Everyone must be diligent in proper cleaning and sanitation techniques and follow protocols for aseptic procedures.

Key Terms

antiseptics effective disinfecting agents that destroy microorganisms or inhibit their growth on living tissue

asepsis the absence of disease and contamination

aseptic technique the practice of keeping a sterile environment

autoclave a sealed chamber in which objects are exposed to heat and steam under pressure at extremely hot temperatures to kill all living organisms

bacteria a living organism that invades the body, causing illness

biological hazard safety concern that poses a risk to humans and animals through contamination of living organisms through body tissues and fluids, such as blood or urine

chemical hazard safety concern over products that may cause injury or over vapors that can cause injuries to the eye, lungs, or skin

chemotherapy the use of chemical to treat disease

cleaning the process of physically removing all visible signs of dirt and organic matter

cold sterilization the process of soaking items in a disinfectant chemical until they are necessary for reuse but not considered truly sterile for surgery procedures

cold tray a container that holds a chemical that acts as a disinfecting agent

dilution the act of diluting, or lessening in strength by adding another component, such as water

direct contact transmission of a disease through a direct source, such as saliva or the ground

disinfecting the process of destroying most microorganisms on nonliving things

dosimetry badge measures the amount of scatter radiation received during a radiographic procedure

dry heat a method of sterilization that exposes an item to extreme heat with a flame or through incineration

filtration using a physical barrier to remove particles from the air; common in lab areas or research facilities

fungus a living organism that invades the external area of the body through direct or indirect contact

incineration the burning of infectious materials or animal carcasses

incinerator a device used to burn the remains of items

indirect contact the transmission of a disease through an indirect source, such as air or water

isolation ward separate housing that groups similar patients, making it safe for all animals and staff

medical waste the storage and trash for items contaminated by bodily fluids and living tissues that may be infected with contagious diseases

nosocomial infection when humans cause the spread of disease and contamination of an animal

Occupational Safety and Health Administration (OSHA) a federal agency that provides for safe and healthy work environments and conditions for all employees

parasites living organisms found on the internal or external area of the body and feed off the animal, causing disease

personal protective equipment (PPE) equipment worn for safety purposes and protection

physical cleaning the most common method of sanitary control within the veterinary facility and which involves a chemical with a cleaning object

physical hazard a situation or agent with the potential to cause physical harm to a human or animal's body

protozoans one-celled simple organisms or parasites

radiation the use of ultraviolet or gamma rays to radiate and kill living organisms

rickettsia more complex multi-cellular organism that are tick- or flea-borne

right-to-know station an area in the facility where the OSHA and SDS binders are kept

Safety Data Sheet (SDS) safety information published by the manufacturer of a product that has the potential to harm humans within the facility

sanitation the process of keeping an area sterile to prevent spread of disease

sharps sharp instruments and equipment that can injure a human or animal and cause a wound or cut that transmits a contagious disease due to contamination

sharps container an object that holds sharp items that may spread disease, such as needles, surgical blades, and glass

spectrum a wide variety of factors

sterilizing destroying all microorganisms

ultrasound passing high-frequency sound waves, which create a vibration and break up debris

virus something that causes a disease that is not treatable and must run it course

zoonotic hazards safety concerns that allow contagious organisms to be spread to humans, causing infections, viruses, bacterial, fungal, and parasitic transmission

Review Questions

1. What are the four types of hazards found in the veterinary industry?
2. What does OSHA stand for?
3. What is the importance of OSHA in the veterinary industry?
4. What are the four classes of zoonotic infection sources?
5. What is the difference between direct and indirect contact? Give examples of each.
6. What is an SDS?
7. What are the parts of the SDS?
8. How should needles be properly disposed?
9. What is a dosimetry badge?
10. How is the rabies virus spread?
11. What is the exposure limit for inhaled anesthetic agents?
12. What are the three levels of cleaning?
13. What are methods of sterilization?
14. What are the factors to consider when using disinfectants?
15. What are some chemicals and disinfectants used in the veterinary hospital?
16. What is aseptic technique?
17. What areas should be cleaned in a veterinary facility?
18. What is the importance of proper handwashing?
19. What is the proper way to mop an exam room floor?

Clinical Situation 1

Jessica, a new veterinary assistant at Glenmoor Vet Clinic, was checking the post-op surgery patients in the afternoon. Dr. Miles, the veterinarian who performed the daily surgeries, asked her to check each patient for any signs of pain or bleeding. Jessica carefully looked at each animal, opening and closing each cage, and observing each animal for signs of pain. She also looked at each animal's surgical incision. She did not wear gloves and did not wash her hands between each patient. She also decided to offer each animal small amounts of food and water. One small Beagle, named Daisy, appeared to be extremely sleepy. She gave her a new towel and patted her head. Another patient, a cat named Rascal, who had been declawed that day, was attempting to bite at his bandages. Jessica giggled and said, "Rascal, are you trying to chew your booties?" She wrapped the cat in a large towel, thinking this would distract him from the bandages. She quietly turned out the lights, tuned on a radio, and left the post-op ward.

- What did Jessica do correctly?
- What did Jessica do incorrectly?
- What should Jessica have done in the surgery ward?

Clinical Situation 2

Judy, a veterinary assistant, and Farah, a veterinary technician, have spent the whole morning cleaning the exam rooms and the surgery area. Judy has been disinfecting the exam rooms and equipment. Farah has been sterilizing the surgical suite and autoclaving instruments for the next day's surgery schedule.

"I am out of disinfectant in the exam rooms," says Judy. "What else can I use to disinfect the rooms?"

"Just get some Lysol out of the closet," says Farah. "That should work."

Judy locates the Lysol and uses it to clean the exam areas and equipment. She notices several other disinfectants in the closet but thinks the lemon-scented Lysol smells much better. When she is done, she takes the mop, bucket of water, and Lysol mixture over to where Farah is and begins mopping the surgery floor.

- Did Farah and Judy follow proper cleaning and disinfecting guidelines?
- How would have you handled cleaning the facility?
- Is Lysol a good cleaning agent? Explain.

Clinical Situation 3

Lawrence, a veterinary assistant at Woodbine Animal Clinic, is about to clean up the exam room to prepare it for the next patient. When he enters the room, Lawrence notes that there are several used needles and syringes, empty vaccine vials, dirty paper towels, and several exam tools—including an otoscope, thermometer, and a pair of bandage scissors—on the exam counter. The table is dirty and soiled with what appears to be urine. The floor has some areas of wetness and visible blood. Lawrence realizes he has a lot to clean up to prepare the room for the next patient.

- What protocol should Lawrence use?
- What should Lawrence do first to prepare the room?
- How should a client who is in the waiting area be handled if the room needs to be thoroughly cleaned?

Section II
Veterinary Animal Production

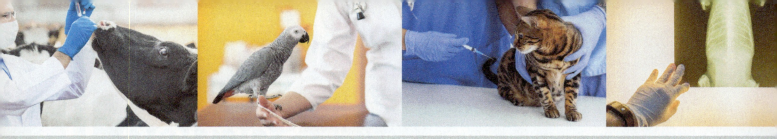

CHAPTER 8: Dog Breed Identification and Production Management

Objectives

Upon completion of this chapter, the reader should be able to:

8.1 Define common veterinary terms relating to the dog
8.2 Describe the biology and development of the dog
8.3 Identify common breeds of dogs recognized by the American Kennel Club (AKC)
8.4 Discuss proper dog selection methods
8.5 Discuss nutritional factors for dogs
8.6 Describe normal and abnormal dog behaviors
8.7 Explain basic training methods in dogs
8.8 Explain how to properly and safely restrain dogs for various procedures
8.9 Discuss proper grooming needs of dogs
8.10 Discuss basic health care and maintenance of dogs
8.11 Discuss basic vaccine health programs for dogs
8.12 Explain dog reproduction and breeding methods
8.13 Identify common dog diseases and prevention
8.14 Identify common internal and external parasites of dogs
8.15 Discuss common surgical procedures in dogs

Introduction

Dogs have become one of the most popular companion animals; they are also one of the most important species of animals in society by providing some type of service to people. This is one reason that dogs have become known as "man's best friend" (see Figure 8–1). Research has shown that owning a dog relieves stress, and this in turn results in a healthier immune system. Touching a dog and sharing time with a dog has a positive effect on people and their health. For this reason, veterinary medicine is largely dedicated to the health and well-being of the dog.

Veterinary Terminology

In veterinary medicine, it is essential to know species-specific terms and when to properly use them in discussions. Table 8–1 contains common canine terms that are used in the veterinary industry. The veterinary assistant should also be familiar with the external (outside) parts of the canine body structure. Figure 8–2 shows the location and terminology of each canine body part that the assistant should be familiar with.

CHAPTER 8 Dog Breed Identification and Production Management

FIGURE 8–1 Dogs are the second most popular companion animals.

TABLE 8–1
Canine-Specific Terminology

Canine (K-9)	Dog (derived from Latin)
Bitch	Female dog
Stud	Male dog of breeding age
Intact	Capable of breeding; still has reproductive organs
Neuter	Surgical removal of the reproductive organs
Spay	Surgical removal of the reproductive organs in the female
Castration	Surgical removal of the reproductive organs in the male
Litter	Group of young dogs from the same parents
Puppy	Dog that is less than 1 year old
Pack	Group of dogs in a household
Gestation	Length of pregnancy
Whelping	Labor process of a female dog

Biology

Dogs were bred from wolves and were domesticated over 10,000 years ago. **Domestication** means that the animal has become tame and is bred to be near humans for companionship. The canine can be described by its genus and species, ***Canis familiaris***. This identifies the dog as a relative of the wolf. Dogs are mammals that are known as **carnivores**, or meat eaters. They are able to regulate their body temperature internally in the same manner as humans and are thus called **endothermic** animals. Dogs have a simple stomach digestive system called **monogastric**. Dogs are built for specific uses and purposes and are categorized by size, weight, age, and coat type. They also have a unique **anatomy**, or body structure. The body structure has similar functions to the human body. The body functions are known as **physiology**. Dogs live for different lengths of time, known as their **life span**, based on size and health status. Smaller dogs typically live longer than larger dogs. Life spans vary by weight range, breed (some breeds may be at greater risk of genetic diseases), and how well the dogs are cared for by their owners.

Breeds

The American Kennel Club (AKC) recognizes over 160 breeds of dogs categorized into seven groups. Dog registries bring organization and facts to the diversity of dog breeds. Groups such as the AKC collect information on pedigrees, breeding records, and categories of each dog group to better improve each breed and maintain and promote responsible dog ownership and breeding methods. The seven groups are the sporting dogs, hounds, working dogs, terriers, toys, herding dogs, and non-sporting dogs (see Table 8–2). A miscellaneous group has been established to classify popular breeds that have not been admitted to one of the seven groups. **Hybrid** or **designer breeds** are also increasing in popularity. An example of a designer dog breed is the labradoodle, which is a mixture of a Labrador Retriever and a Poodle. This is a cross between two common **purebred** breeds. A purebred dog has parents that are registered and have known parentage (see Figure 8–3). Some dogs are **mixed breeds**, or a mixture of two or more breeds of dogs that have no known parentage (see Figure 8–4).

Breed Selection

It is common for clients to seek veterinary assistance and knowledge in selecting the proper dog for a household. When offering information, guide the client to think about factors that will be important to them. For instance, does the client want the responsibility of a puppy, or an adult dog? Puppies require a lot of time and care with housebreaking, training,

120 SECTION II Veterinary Animal Production

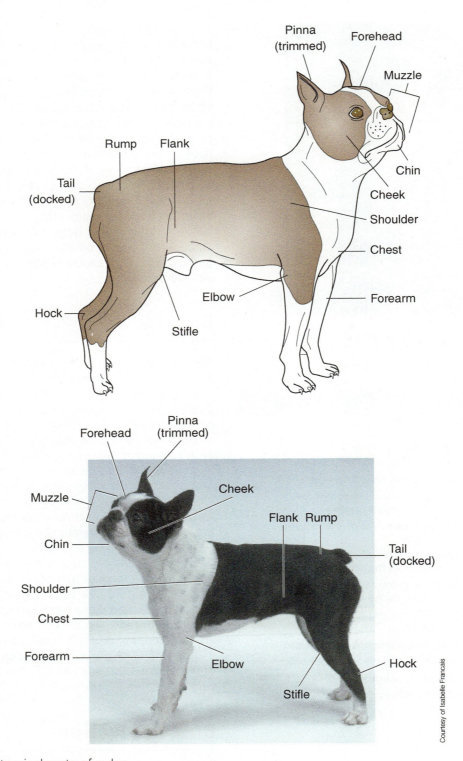

FIGURE 8–2 Anatomical parts of a dog.

and health care. Your discussion can include these other factors:

- Size of the dog as an adult
- Temperament and personality
- Indoor or outdoor dog
- Inside space
- Outside space: yard size and/or fencing
- Type of Shelter/Housing
- Short or long hair
- Grooming needs

TABLE 8–2
Dog Groups and Breeds

GROUP	NUMBER OF BREEDS	EXAMPLES OF BREEDS	CHARACTERISTICS
Sporting	30	Pointers, Retrievers, Setters, Spaniels	Bred for hunting and outdoor activities, active, need regular exercise, alert
Hound	32	Bassett, Beagle, Bloodhound, Greyhound	Bred for hunting, strong scent and sight abilities, good stamina, vocal
Terrier	32	Airedale, Bull, Fox	Bred to hunt vermin, feisty, energetic, not good with other animals, strong willed
Working	31	Boxer, Great Dane, Siberian Husky, Saint Bernard	Bred for jobs such as pulling sleds, guide dogs, police dogs, guard dogs; large, powerful, intelligent, quick learners
Herding	30	Collie, German Shepherd, Sheepdog, Corgi	Bred to control movement of other animals, intelligent, good for training exercises
Toy	21	Chihuahua, Pekingese, Pug, Shih Tzu	Small, do not require a lot of space, companion and family dogs
Non-sporting	20	Boston Terrier, Bulldog, Dalmatian, Poodle	Sturdy, varied sizes and shapes, personalities, and coat types
Miscellaneous	10	Barnet, Mudi	Dog breeds not yet placed in one of the seven groups recognized by the AKC
Hybrid		Puggle, Labradoodle, Schnoodle	Breed mixtures that are not yet recognized by AKC. Examples: Pug and Beagle (Puggle); Labrador and Poodle (Labradoodle); Schnauzer and Poodle (Schnoodle)

FIGURE 8–3 A purebred dog has characteristics and features unique to one specific breed.

- Health care needs
- Breed health problems
- Breeding or show purposes?
- Training needs
- Nutritional needs
- Restrictions in the neighborhood or community

The best advice to offer clients is that they research several breeds to determine the type or

FIGURE 8–4 A mixed-breed dog will exhibit characteristics and features of two or more different breeds.

types of dog that will best suit their lifestyle and family needs. Veterinary clinics may keep a list of reputable breeders in the area to provide to interested clients. Refer people to local humane shelters to seek information about adopting a dog. Keep reference materials that will allow future dog owners to become educated and responsible pet owners. Some websites that offer information on various dog breeds include the following:

- American Kennel Club—www.akc.org
- Dog Breed Info Center—dogbreedinfo.com
- Continental Kennel Club—continentalkennelclub.com

Nutrition

A dog's nutritional needs will change as it grows from a puppy to an adult. These changes are known as **life stages**. Each life stage requires different nutrients for the age, health, and lifestyle of the dog (see Table 8–3).

Feeding Puppies

Puppies should begin being **weaned** from their mother at about 3–5 weeks old, depending on the breed size. Weaning is the process of stopping nursing and beginning to eat a solid food diet. Some puppies may need to nurse longer depending on size and health conditions. Typically, small breed puppies begin weaning between 4 to 5 weeks old and large breed puppies between 3 to 4 weeks old. The goal is for a puppy to be completely weaned by 6 to 8 weeks old. During the weaning process, pups should be fed either canned food or dry puppy food that has been softened with warm water and made into a gruel or slurry. A **gruel** (or slurry mixture) is created by mixing equal parts of food to equal amounts of water: this results in a soup-like consistency that can gradually be made into an oatmeal-like consistency. Puppies should be feed three to four times daily during weaning. Once they have been weaned, the amount of water can be decreased until the puppy is eating a dry food diet (that is size appropriate). In the transition to dry food, the puppy should be fed several times a day and allowed food for 15 minutes each time. This will teach the puppy to

TABLE 8–3

Feeding Requirements for Dogs

LIFE STAGE	REQUIREMENTS
Puppy (orphaned)	• High-quality puppy formula, such as Esbilac or KMR, Feed 60 mL/day • Four daily feedings until 2 weeks old • Three daily feedings until weaned • High-protein, high-calcium, high-fat diet
Puppy weaning	• Wean at 3–5 weeks old • Begin on size-appropriate, high-quality dry puppy food • Add warm water to soften as necessary • High-protein, high-calcium diet • Large- or giant-breed dogs should be fed moderate to low protein
Puppy to 1 year	• High-quality puppy diet • Large-breed puppy diet • Moderate protein, calcium, and phosphorus diet
Adult (1 to 7 years old)—active	• Maintenance diet • Moderate protein, calcium, and phosphorus diet with low fat
Adult—overweight, inactive	• Reduced calorie diet • Healthy weight diet • Low-fat, low-calorie diet • Low-protein, moderate calcium and phosphorus diet
Senior (over 7 years old)	• Senior diet • Reduced-calorie diet • Low-fat, low-calorie diet • High-phosphorus, high-calcium diet with moderate protein

learn to eat at scheduled times. As the puppy grows, it should be fed twice a day. Sometimes breeders will deal with **orphaned** puppies due to the mother refusing to nurse, illness, or death. In this case, a specialized formula diet is necessary. Examples of commercial formulas include Esbilac and KMR (see Figure 8–5). A homemade formula can be prepared as well, but a specific, commercial, high-quality puppy formula is best for the health of the orphan. When preparing or opening a can of formula, always date it and then keep it refrigerated. It will need to be warmed for each use and should be used within 24 to 48 hours.

> **Homemade Puppy Formula**
> - Three egg yolks
> - One cup homogenized milk
> - One tablespoon corn oil
> - One dropper full of liquid vitamins such as Pet Tinic

When preparing a homemade formula, take care to maintain the purity of the ingredients. Do not use regular cow's milk, as it does not have the appropriate protein or lactose levels.

Formula may be fed with a pet nurser, an eye dropper, a syringe, or a rubber feeding tube (see Figure 8–6). A pet nurser is a bottle used to feed a puppy, similar to how a human baby is fed. It is important to weigh each puppy daily prior to feeding. This will allow the veterinary assistant to monitor weight gains. The typical formula rate for puppies is 60 mL per pound per day. The total amount should be divided into four feedings until 2 weeks of age and then decreased to three feedings until the puppy is weaned. Puppies typically gain 2 to 4 g/kg of their expected adult weight daily.

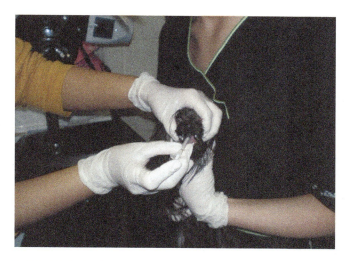

FIGURE 8–6 Syringe feeding of a puppy.

> **Calculating Puppy Formula Rates**
> - Formula Rate = 60 mL/pound/day
> - Feed puppies under 2 weeks of age four times daily
> - Feed puppies over 2 weeks of age, and until weaned, three times daily

Example

To calculate the amount per feeding for a 1.6-pound puppy that is 1 week old, use the following calculation:

Number of pounds x 60 mL = total amount feed in one day

1.6# x 60 mL = 96 mL

Total number of mL for day divided by number of feedings per day = amount to be fed per feeding

96 mL / 4 daily feedings = 24 mL per feeding.

It is important to remember that the term *pound* can be written with the number sign (#) to mean pounds of body weight.

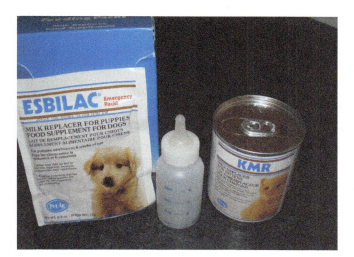

FIGURE 8–5 A commercial puppy formula can be bottle-fed to puppies that are orphaned before weaning age.

Feeding Adult Dogs

Adult dogs should ideally be fed a dry food diet to help prevent tartar buildup on teeth. The food size should be appropriate for the size of the dog. It is best to feed a high-quality pet food that is easily digestible. Examples of high-quality dog foods are Royal Canin, Hills, Eukanuba, Iams, Purina, Purina One, and Nestlé.

124 SECTION II Veterinary Animal Production

Adult dogs that get regular exercise and are on average between 1 and 7 years old should be fed a **maintenance diet**. This will allow the dog to maintain the same weight (see Figure 8–7). It is important to adjust the amount of food that a dog is given based on its body condition to ensure that it does not become obese or underweight. Dogs over 7 years old, especially medium to giant breeds, are considered senior pets and should be fed a **senior diet** (see Figure 8–8). These diets usually consist of fewer calories for less active dogs. They also include increased levels of protein, zinc, copper, and vitamins A, B, and E and decreased amounts of phosphorus to benefit kidney health. Overweight dogs may be placed on a **reduced-calorie diet**. These diets help dogs lose weight and meet the daily energy needs of an overweight or less-active dog. A dog should be fed for its ideal weight based on age, breed, and activity level. The pet-food label should list feeding requirements. Make sure to feed any overweight dogs for their ideal

FIGURE 8–8 Older dogs, particularly large breeds, should be fed a senior diet.

size rather than their actual weight. The label will state the necessary **ration**, or the amount of food that should be fed. Any changes in diet should be done gradually over a 7- to 10-day time frame; including more of the new food and less of the previous food each day.

 MAKING THE CONNECTION

For a general discussion of nutritional needs of animals, please refer to Chapter 34. ■

Behavior

Dogs may be happy, scared, or **aggressive** (ready to attack). The happy dog is alert, ears up and forward, mouth open and panting, tail up and wagging, with a relaxed stance and appearance. A scared dog has the tail tucked under its body, the body lowered to the ground, ears flat and back on the head, avoids eye contact, jumps when noises occur, raises a paw when approached, and shakes from nervousness. This type of dog may exhibit the fight-or-flight behavior response. Most scared dogs may want to get away from a threatening situation. However, some dogs will resort to the fight principle and become a fear biter during a threatening situation. An angry or aggressive dog may growl, show its teeth, have its hackles up over the back and shoulders, lower its head to the ground, stare with direct eye contact, raise and bristle the tail, and adopt a stiff stance. A dominant dog holds its body posture high, with a raised head and tail. A submissive dog will have a lowered head and neck as well as a lowered or tucked tail (see Figure 8–9).

FIGURE 8–7 Young active dogs should be fed a maintenance diet.

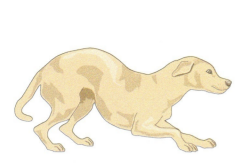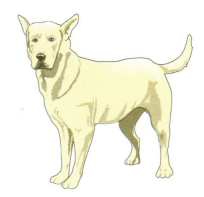

FIGURE 8–9 Dominance and submissive postures in dogs. The dominant dog has a raised head and tail. The submissive dog has a lowered body and tail.

Aggression can vary based on the dog's actions. **Dominance aggression** may be presented against a dog's pack family or human family. This type of aggression may be exhibited to lower-ranking dogs in the household or certain humans in the house over whom the dog is dominant. These situations may include being food possessive, toy aggressive, gender based, territorial, or cage aggressive. **Redirected aggression** is often a behavior that stimulates a response from another source, such as a dog outside of the yard or a car driving down the road. This type of aggression is the most dangerous and hardest to manage. When this type of aggressive dog feels threatened by an outside stimulus, its reaction is to turn on the closest pack member, who is often the dominant owner.

Dogs are naturally pack animals. They will bark to warn others of approaching danger or strangers. Barking is a natural behavior that must be taken into consideration when owning and housing a dog. Barking is a way of communication among dogs. Another common dog behavior is marking or urinating in areas to mark its territory. This creates a scent alerting other animals to where it lives. Marking may become a problem if a dog does this inside a home. Marking is decreased by neutering dogs at an early age.

It is important to know and use the following steps when approaching a dog in the hospital:

- Speak in a quiet, calm voice, using the dog's name
- Approach quietly and with confidence
- Lower yourself to the dog's level, staying at a safe distance
- Slowly offer the back of your hand to the dog to allow it to smell
- Allow enough space to safely handle and restrain the dog
- Work in a quiet area with decreased stress and noise

Basic Training

Veterinary assistants should discuss basic training information of puppies and dogs and offer referrals of local dog trainers or dog obedience classes. **Training** is the process of helping an animal understand and acquire specific desired habits. These habits should include housebreaking, obedience training, and socialization. **Housebreaking** includes teaching the dog to eliminate outside at specific times, typically when it wakes up, after eating, and when appearing to sniff out an area to eliminate around the house. A crate is often used to properly house train puppies and prevent destructive behaviors, such as chewing (this is called "crate training"). The crate should be considered to be a puppy's "bedroom" and be used as a positive reinforcement in training. To work, crates must be size appropriate. Small puppies kept in large crates will be able to use an area of the crate as their bathroom. Partitions may be used to make the area small to prevent this from occurring. Dogs do not like to eliminate in the same area they sleep and eat. It is important to remember that a puppy's bladder and bowels can only be controlled for a short amount of time. This may be as little as one hour or up to 4 to 5 hours at a time depending on age. During the house-training phase, the puppy must be taught where to eliminate. This includes taking the puppy out the same door or exit area every time and to the same location outside every time. When the puppy eliminates, give immediate positive reinforcement. This reward may be a small treat, verbal praise with excitement, or a pat. Accidents—urination or defecation—will happen, but physical punishment is not appropriate. You may interrupt the unwanted behavior by calling the puppy's name loudly; this may give you time to take the puppy outside. Punishing the puppy will only scare it and could even make it afraid of people.

Obedience includes teaching a dog to walk on a leash, sit, stay, and behave with "proper" manners. Obedience classes are useful in educating both the dog and the owner and often help socialize the dog. **Socialization** includes teaching a dog to act properly with other dogs and people. Some facilities will offer puppy socialization classes to discuss what protocols new puppy owners should address while their pets are young. It is not uncommon for clients to ask questions related to behavior and training issues for their dogs. Some common topics include excessive barking, marking territory, digging, and inappropriate elimination. Some pet owners may inquire about specialized training for their dogs, such as hunting or sport-related behaviors or tricks.

Inappropriate Elimination (House Soiling)

Many indoor companion animals have **inappropriate elimination** problems, also called "house soiling." These behaviors may be due to a physical problem, such as a urinary tract infection, and should first be ruled out by a veterinarian. When a behavior problem is established, it should be handled by an animal behavior specialist or a trained veterinary professional. Elimination problems in dogs may be the result of poor housebreaking techniques. A good place to start to remedy the problem behavior is to retrain the dogs in housebreaking guidelines (similar to training a puppy). It is highly recommended that all animals in the household be neutered, as this will decrease the chances of a house-soiling problem. Cats tend to have more elimination problems than dogs.

Barking

Barking is a mode of communication that seems to be more common in dogs than other canine species. Most dogs bark, howl, and whine to some degree. Sometimes this may be the result of human encouragement. Certain breeds, such as the Rottweiler or German Shepherd, have been bred to bark as part of their watchdog or herding duties. Dogs bark to alert or warn others; to defend a territory, to seek attention or play; to identify oneself to another dog; and in response to boredom, excitement, being startled, loneliness, anxiety, or being teased.

Excessive barking is a common behavior problem. It is important that a dog understand when barking is and is not appropriate. This is commonly achieved by training the dog to know when to stop barking. It is important to not reward a dog's barking behavior. Before it can be corrected, it is important to determine why the dog is barking in the first place. Dog trainers or animal behavior specialists can share training techniques for excessive barking. It is important that they be followed consistently.

Howling is used as a means of long-range communication in many different circumstances. Howls are more often associated with wolves, but dogs howl too. Wolves often howl to signify territorial boundaries; locate other pack members; coordinate activities, such as hunting; or attract other wolves for mating. Dogs may howl as a reaction to certain stimuli, such as sirens or other loud noises. Growling can occur during different activities. It is used to threaten, to warn, to defend, in aggression, and to show dominance. But growling is also used in play. You should be able to tell the difference by the dog's body posture. Growls in aggression are accompanied by a stare or snarl, and the growling dog often remains stationary. Play growls occur along with a happy tail and a play bow to signal willingness to play. These dogs are often moving and jumping about to entice play.

Digging

Dogs are known to cause destructive digging problems in yards (see Figure 8–10). Dogs may dig for a variety of reasons: wanting to escape the confines of the yard, trying to stay cool in the summer or warm in the winter, a predator instinct of wildlife, separation anxiety, boredom, curiosity (about unfamiliar objects in the soil), aggression directed at an object beyond the yard, and displaced aggression or digging out of frustration at not being able to get to an object. Some cases of digging can be resolved by using a screen or other distracting items to block the view of objects. Placing the animal in an area with limited access to soil or ground can limit digging. Providing toys and

FIGURE 8–10 Digging can be a problem behavior common in some dog breeds.

other items for play can redirect animals. In cases of separation anxiety, the dog may need medication for behavior modification.

Biting

Any animal has the potential to bite, especially when they are scared or threatened. This is a reflex inherited by animals in the wild and becomes a defensive reaction. Some animals may show signs of biting, such as growling, hissing, or bearing their teeth, while others may show no signs at all. Some dogs are naturally aggressive and bite out of anger. Others to show they want to be left alone. Yet others may bite as a reaction to pain or injury. To solve a biting problem, it is essential to determine why the animal is biting. Many biting issues can be resolved through training modification techniques.

Equipment and Housing Needs

Dogs will require some special facilities and equipment. All dogs should have a collar and leash, a crate or kennel area, a food and water bowl, and appropriate bedding (see Figure 8–11, A–D). Toys are optional but recommended to keep dogs from becoming bored.

Restraint and Handling

Animals have more developed senses than humans. They use these senses to perceive changes within the environment. Dogs, therefore, should be handled with care. Dogs see prolonged staring or direct eye contact as threatening. Quick movements from people can also cause a negative reaction. Large numbers

FIGURE 8–11 Essential equipment needed to care for a dog: (A) collar and leash; (B) food and water bowls; (C) crate and bedding; (D) toys.

of people can be intimidating to a dog. Restrainers should move slowly and calmly, talk quietly, and avoid eye contact with any nervous animal. Knowledge of dog behavior and body language is essential to a veterinary assistant for use in safe and adequate restraint skills. As a restrainer, the veterinary assistant can influence the behavior of the animals they are is working with. Each situation and patient should be monitored carefully to prevent injury to the handler, the patient, and others involved in working with the animal. Sometimes less restraint is more, as some dogs will do better with less use of physical restraint, but the veterinary assistant must monitor patients at all times for changes in behavior. There is **physical restraint**, **verbal restraint**, **chemical restraint**, or a combination of these methods. Physical restraint is when a person controls the position of an animal through the use of the body and/or equipment. Verbal restraint is the use of the voice to control an animal, such as the words "No," "Sit," or "Down." Chemical restraint is the use of sedatives or tranquilizers to calm the animal to allow procedures to be carried out.

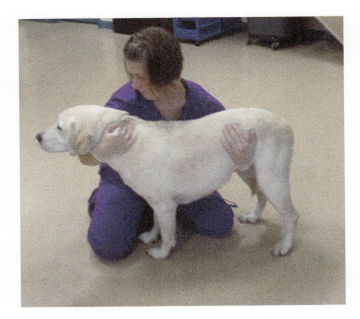

FIGURE 8–12 Standing restraint.

 TERMINOLOGY TIP
A **sedative** or **tranquilizer** is a type of medication used to calm an animal and prevent it from moving. It is important to note, however, that these medications may not relieve pain sensations. ∎

Physical Restraint

There are several common physical restraints that may be used for dogs. These include sitting and standing restraint techniques as well as placing the dog in a **recumbency** position. Standing restraint may be used on the floor or on a table. One arm is placed around the neck to control the head. This is called the *bear hug*. The other arm is placed under the stomach to prevent sitting and to control the body (see Figure 8–12). If the dog appears to become fearful or aggressive, the hand holding the neck can be moved to the muzzle to prevent biting.

Sitting restraint is done in the same manner, except the dog is sitting (see Figure 8–13). It is helpful to do this against a wall to prevent a large resistant dog from backing up. The restrainer can place the backside of their body against the wall and place the dog in a sitting position with the rear side between restrainer's legs. The head is then controlled by the restrainer's arms (see Figure 8–14). The same method can be applied for small dogs: in this case, the restrainer uses their own body as the wall.

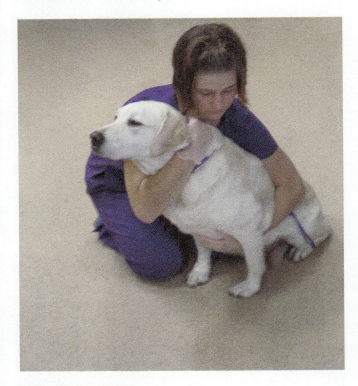

FIGURE 8–13 Sitting restraint.

Lateral recumbency, placing an animal on its side for restraint, is commonly used for X-ray positioning and other procedures. The dog should be restrained in a sitting position and carefully placed in a lying position on its chest. It is then carefully moved onto the necessary side. For small dogs, the restrainer must hold the legs that are on the downside and nearest

FIGURE 8–14 Sitting restraint of large dog with the use of a wall and the handler's body.

the floor or table to prevent the dog from standing. The restrainer uses his or her arms and upper body to gently hold the dog in position (see Figure 8–15). For large-sized dogs, this restraint will require two to three people. One person will restrain the front of the dog's body and one will restrain the rear (see Figure 8–16). It is important to continue the restraint of the legs on the downside of the dog, as this is what will keep the dog

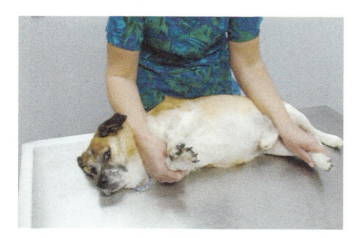

FIGURE 8–15 Lateral recumbency in a small dog.

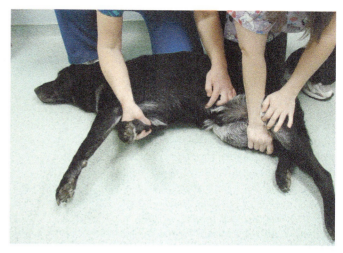

FIGURE 8–16 Lateral recumbency in a large dog.

from rising. Care must be taken so that the dog does not hit its head against the floor during restraint.

Sternal recumbency is a restraint position in which the dog is lying on its chest or sternum. This requires placing the dog in a sitting position and applying pressure to gently place it in a lying position on its chest (see Figure 8–17). For small dogs, the restrainer

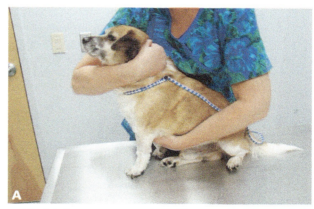

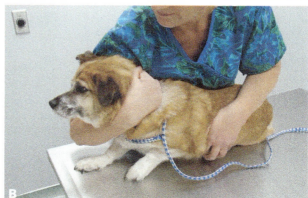

FIGURE 8–17 Sternal recumbency.

130 SECTION II Veterinary Animal Production

can use one arm to control the head and the other arm and body to gently hold the dog in position. For large dogs, one person can bear hug the head of the dog and gently apply body pressure to the shoulders and back, while another person restrains the rear of the dog.

Dorsal recumbency is when the dog is positioned on its back, commonly used in X-ray positioning. The dog is placed in lateral recumbency and then carefully rolled onto its back. Two or more people should restrain the head, middle of the body, and the rear of the body. The dog's front legs are pulled forward toward its head; its rear limbs are pulled backward toward the tail (see Figure 8–18).

For a **cephalic venipuncture**—blood collected from the cephalic vein located on the inside of the front legs—the dog is typically placed in sitting or sternal recumbency and a front leg is extended (see Figure 8–19). Some vets and technicians will ask the restrainer to hold off the vein by placing pressure around the elbow of the dog.

For a **jugular venipuncture**—blood collection from the jugular vein located on either side of the neck—the common restraints are sitting or sternal recumbency (see Figure 8–20). For smaller dogs, the front limbs can be extended over the edge of a table to allow more space to collect the blood sample.

For a **saphenous venipuncture**—blood collection from the inside of the thigh in the rear limbs—dogs are commonly placed in lateral recumbency; the restrainer may be asked to hold off the pressure around the upper thigh area of the rear limb (see Figure 8–21).

Lifting Techniques

Lifting dogs requires safety for both the restrainers and the animal. When lifting any sized dog, a restrainer should always keep his or her back straight and bend

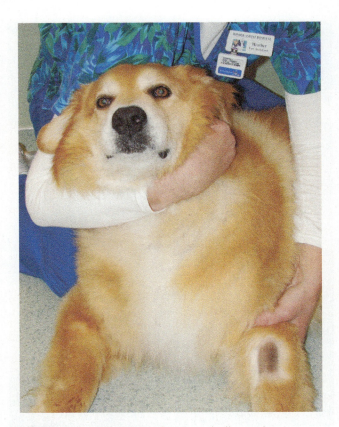

FIGURE 8–19 Restraint for a cephalic venipuncture.

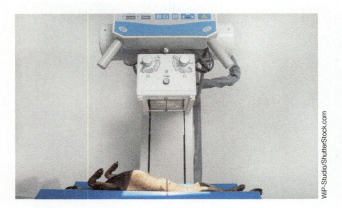

FIGURE 8–18 Dorsal recumbency.

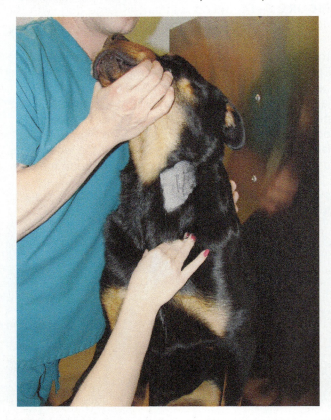

FIGURE 8–20 Restraint for a jugular venipuncture.

CHAPTER 8 Dog Breed Identification and Production Management 131

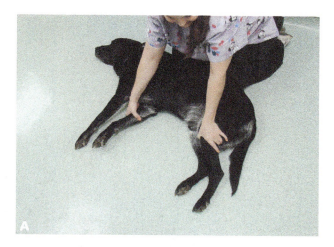

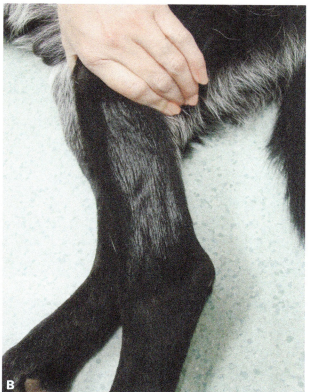

FIGURE 8–21 Restraint for a saphenous venipuncture.

Use and Application of a Muzzle

Dogs will sometimes need restraint equipment to help keep them from biting or to keep them calm. A common piece of equipment used to prevent biting is the **muzzle**. The muzzle may be made of commercial nylon, a plastic or wire basket, tape, a leash, or gauze material. A commercial muzzle is usually adjustable with a snap that locks in place (see Figure 8–22). They are come in various sizes to fit the size of the dog. The muzzle should provide a snug fit and only be used for short periods of time. The short side of the muzzle should be on top of the dog's nose and the wide side should be placed under the jaw. The snap goes behind the ear flaps and snaps securely in place (see Figure 8–23, A–B). The dog's legs should be properly restrained to prevent removing the muzzle. When applying the muzzle, note that for safety, the restrainer should stand at the side or the rear of the animal. Never work directly in front of the dog. A tape or gauze muzzle can be used if a commercial muzzle is not available. Make sure to use an adequate amount of material that will fit the dog. **Gauze** is thin material that may come on a roll and is made of a mesh-like cotton. Gauze material should be heavy to prevent tearing. Tape should be folded upon itself to prevent glue from attaching to the dog's hair or skin. Begin by making a large loop that is placed over the dog's nose and secured as in tying a shoelace. Pull tight and cross ends snugly under the jaw and tie in a knot under the jaw. Take each end on either side of the ears and tie in a bow behind the ear flaps (see Figure 8–24, A–C). When removing the muzzle, release the bow and gently pull off the muzzle (see Figure 8–25). If the dog is likely to try biting, untie or unsnap the muzzle and allow the dog to remove it. Dogs that have a flat face lacking

at the knees. This will help prevent any back injuries. Small dogs may be lifted by supporting the front and back ends, similar to standing restraint. Large dogs will require two or more people. One person should control the head and one or more will need to control the rear of the body. Lifting should occur with pressure on the abdomen and chest area. It is important for two or more people to always work on the same side of the dog. Teamwork involves communication to tell each other when to lift.

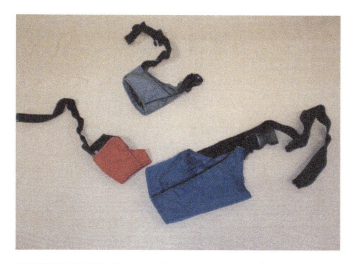

FIGURE 8–22 Various sizes of commercial muzzles.

132 SECTION II Veterinary Animal Production

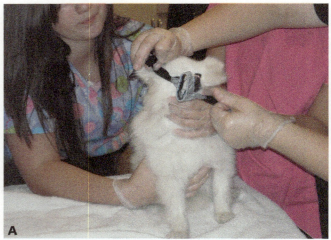

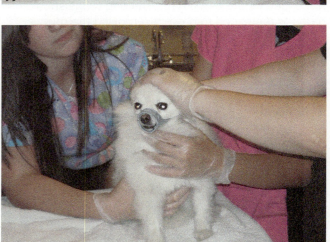

FIGURE 8–23 Use of a commercial muzzle: (A) applying a muzzle; (B) snapping the muzzle into place.

a long muzzle, such as pugs or Bulldogs, are called **brachycephalic**. These breeds are prone to difficulty breathing. Care must be taken when muzzling these breeds. A tape or gauze muzzle can be placed over the nose and one end looped under the material and pulled tight. Both ends are pulled between the eyes and over the head and tied securely in a bow around the nape of the neck. Caution should be used when removing any type of muzzle from a dog to ensure the dog does not bite during the removal process.

Use of a Rabies or Snare Pole

A restraint device used to remove aggressive dogs from a cage is called the **rabies pole** or **snare pole**. The pole is a long extension with a heavy loop on the end that is adjusted by sliding the pole handle (see Figure 8–26). It acts as a source to separate the

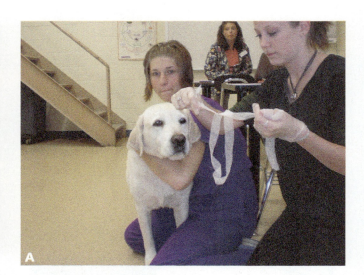

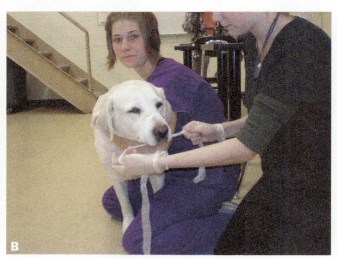

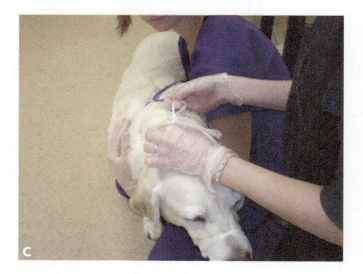

FIGURE 8–24 Use of gauze for a muzzle: (A) make a loop with the gauze; (B) secure the loop over the dog's nose and mouth; (C) tie the ends of the loop behind the dog's ears.

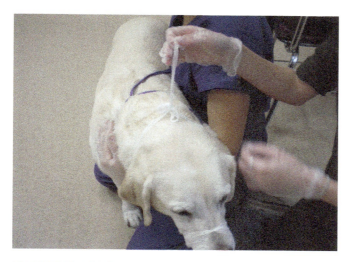

FIGURE 8–25 Release the muzzle by pulling the loose ends of the bowtie.

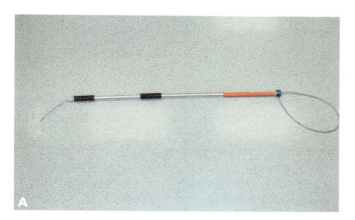

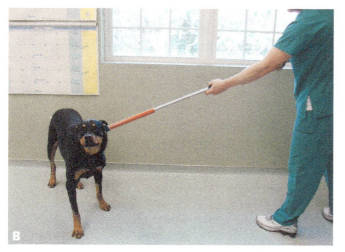

FIGURE 8–26 Restraint using a rabies pole.

restrainer from the dog in order to walk the dog or control the dog to be properly restrained. Once the loop is over the head, the dog should be guided forward with the restrainer behind the dog. Do not choke the animal by pulling it. Care must be used in this type of restraint to avoid injury to the dog. A quick release handle is used to prevent strangulation.

Use of an Elizabethan Collar (E-Collar)

An Elizabethan collar or E-collar may be used as a mobility limiting device to better restrain a dog. The cone shaped plastic collar is placed over the neck of the dog and secured around the neck. The length and tightness of the collar should be considered during restraint, as to work effectively the collar must project beyond the end of the dog's nose and should be snug enough to not easily allow the dog to back out of the collar. The collar can help impair the dog's ability to see because of their peripheral field of vision and may help calm down a difficult dog or allow a better restraint method, such as a muzzle, to be applied.

Use of Sedation

Some dogs that are too aggressive to handle safely may be sedated for the safety of the veterinary staff. Sedation is used to calm down a dog and allow it to safely be restrained. The use of sedation must be prescribed by the veterinarian and should be used as a final resort when all other options have been used. Sedation is an example of chemical restraint.

 MAKING THE CONNECTION

For a more general discussion of restraint techniques used for animals, please see Chapter 36. ■

Grooming

Grooming services may be provided by the veterinary facility, or veterinary assistants may be asked to provide grooming care to hospitalized patients. **Grooming** is the process of trimming and bathing in an appropriate manner for the best care of the patient (see Figure 8–27). Other grooming services may include ear cleaning, brushing the hair coat, brushing teeth, trimming nails, trimming hair, shaving, and ear plucking. Brushing is done first to allow the hair to be lying in the proper position and to remove any mats or items that may have collected within the coat. This allows the veterinary assistant to see any issues with the hair coat and have a neatly organized coat to work with.

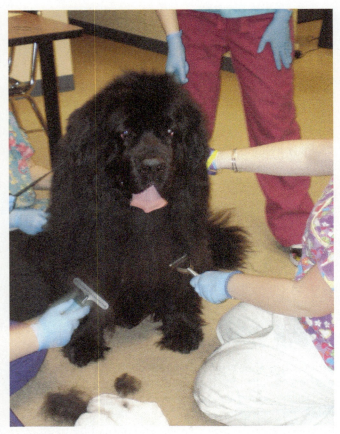

FIGURE 8–27 Some veterinary facilities provide grooming services.

TABLE 8–4
Clipper Blade Selection

BLADE SIZE	CUT LENGTH	USAGE
#50	1/125"	Surgical prep; poodle feet and face
#40	1/100"	Surgical prep; poodle feet; under snap-on combs
#30	1/50"	Poodle feet; between feet pads
#15	3/64"	Sensitive feet and faces; Poodles, Cockers, Terriers
#10	1/16"	Sanitary trim on genitals and rectum; cat clipping
#9	5/64"	Sporting breeds, such as Cockers and Schnauzers
#8.5	3/16"	Head, face, neck, and back
#7	1/8"	Matted dogs; body clipping
#5	1/4"	Body clipping
#4	3/8"	Body clipping and trimming wire-haired breeds; short trims
#3¾	1/2"	Medium to short cuts; puppy cuts
#5/8	5/8"	Longer cuts and puppy cuts
#3/4	3/4"	Body work on longer breeds

Tools

Clippers should be maintained with oil to keep blades lubricated, cool lube to prevent the clippers from overheating, and blade wash to keep the clippers and blades sanitary and properly disinfected. Clippers and blades are commonly cleaned with a toothbrush or wire brush to remove all hair and debris. Special chemicals are used to clean and disinfect the clippers and blades to prevent rusting and damage. Clipper blades can be sharpened if they become dull. Clipper blades come in many sizes (see Table 8–4). The blade size is determined by the type of hair coat and the length of hair being clipped or shaved. Clipper blades can be removed from the clippers for cleaning and changed to different sizes depending on the area to be trimmed or hair coat length.

It is important for veterinary assistants to know how to properly clean and care for grooming equipment and instruments. Daily maintenance is a must to keep grooming supplies sanitary and working properly. Tools and grooming supplies should be routinely sanitized with disinfectants and cleaners that are specifically for use with veterinary equipment. The spread of parasites and diseases is common in grooming items if they are not properly sanitized and sterilized between uses. Some veterinary assistants may be interested in pursuing a career in grooming and learning how to trim and clip animals according to breed standards.

Trimming or Shaving

Trimming and shaving areas of the hair may be done with scissors, thinning shears, or clippers. Scissors should be used with caution, as they can easily injure the animal. It is best to use a blunt (not rounded) tip, which is not sharp. Clippers are used to shave down areas of the hair coat. For example, some long-haired dog breeds may require shaving around the rectum or urogenital area to prevent feces and urine from collecting on the hair coat (see Figure 8–28). It is also common to shave between the digital pads of the feet, where long hair may collect and become irritating to dogs (see Figure 8–29). The area of the ear flaps may also be shaved or trimmed to allow better air circulation for those breeds prone to infections (see Figure 8–30).

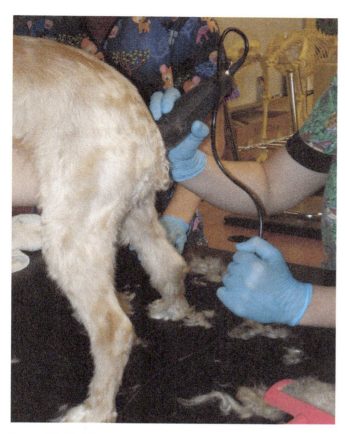

FIGURE 8–28 Some dog breeds may require clipping around the urogenital area to prevent infection and maintain cleanliness.

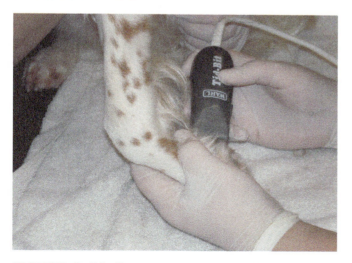

FIGURE 8–29 Dogs may grow hair between the digital pads of the feet that may require clipping for the comfort of the animal.

 SAFETY ALERT

Care must be taken when trimming or shaving animals. Be wary of cutting or injuring the animal, especially when using scissors. ■

FIGURE 8–30 The ear flaps are also a common area that may require clipping to prevent infection.

Bathing

A dog should be bathed in warm water with a gentle shampoo appropriate for the dog's skin. Most dog breeds need bathing every few months. Caution should be taken to avoid overbathing, which removes the natural oils in the body and causes the skin to dry out and develop dandruff, which can lead to other skin conditions. Before bathing, apply a protective eye ointment that lubricates the eyes to prevent injury from shampoo or water (see Figure 8–31). Shampoo and conditioner should be applied to the hair coat and worked into the hair and skin (see Figure 8–32). The shampoo and conditioner should be kept on the hair coat for 5 to 10 minutes before rinsing. Rinse with warm water and make sure all chemicals have been removed from the coat (see Figure 8–33). Dry the dog with a towel and place it under a dryer until the hair coat is dry. A blow dryer is often used to dry small dogs. Place a dryer in front of the cage to completely dry a dog (see Figure 8–34). Make sure the dryer is set at the appropriate temperature so as not to cause burns. When the dog is dry, use a **slicker brush** to brush and detangle the hair (see Figure 8–35). A spray-on **coat conditioner** should be applied to prevent the hair and skin from drying out (see Figure 8–36).

 SAFETY ALERT

Use dryers with caution. If the heat settings are too high, the patient can be burned. There have been instances in which the improper use of a drying device has resulted in the death of the patient. Never use a drying device without monitoring the patient while the device is in use. ■

136 SECTION II Veterinary Animal Production

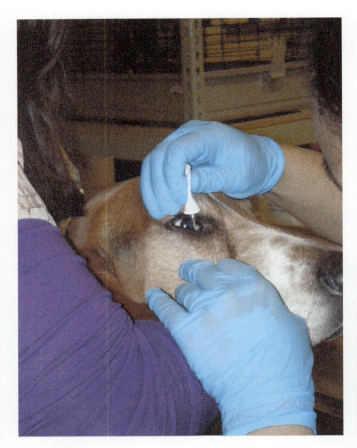

FIGURE 8–31 Application of eye ointment before bathing.

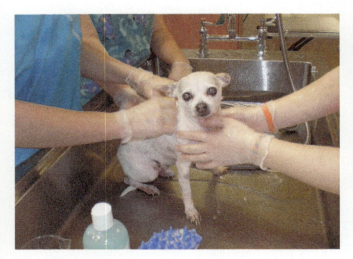

FIGURE 8–32 The dog should be lathered with shampoo and conditioner.

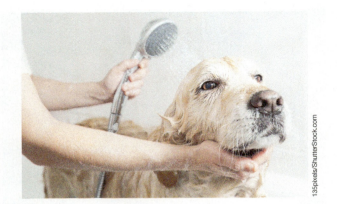

FIGURE 8–33 Rinse with warm water.

FIGURE 8–34 A dryer can be placed in front of a cage to completely dry the dog. It is critical to make sure the temperature is not too high.

Brushing

It is best to brush the hair coat of a dog on a daily to weekly basis to keep it healthy. Long-haired breeds may develop **mats**, which are tangled pieces of hair. Mats may need to be trimmed or shaved. Some long-haired breeds will require daily brushing. It is important to know how to properly care for the hair coat based on the breed of the dog. Some breeds will require regular bathing, while bathing others too often may cause damage to the hair coat.

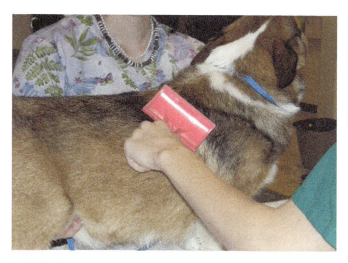

FIGURE 8–35 A slicker brush is used to brush the dog's coat once it is dry.

Ear Cleaning

Dogs should have their ears checked and cleaned as necessary. Some breeds will need more cleaning than others. Such cleaning help maintain breeds that are prone to ear infections. Ears can be cleaned with cotton balls, cotton-tipped applicators, or gauze pads (see Figure 8–37). Caution should be taken with applicators, as the wooden sticks can break off and become trapped in the ear canal. The veterinary assistant should clean only as far as can be seen in the ear canal (see Figure 8–38). Because of the shape of the ear, it can be damaged if not properly cleaned (see Figure 8–39). It is best to use an ear cleaner that is not water based. If only a small amount of debris or wax is noted, an ear cleaner may not be needed. Some breeds have large amounts of hair that grows within the ear canal that may need to be plucked (pulled) or

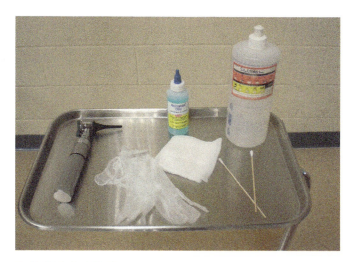

FIGURE 8–36 A conditioner is sprayed on the dog's coat to prevent the hair and skin from drying out.

FIGURE 8–37 Tools required to clean a dog's ears.

Basic Health Care and Maintenance

A dog should receive a yearly **physical examination (PE)** by a veterinarian. This exam will help determine the health status of the dog and whether it is developing any illnesses or conditions. Dogs require daily exercise, and some breeds will require more activity than others. Other regular maintenance needs include nail trimming (NT), ear cleaning, anal gland expression, and teeth brushing.

FIGURE 8–38 Cleaning the dog's ears.

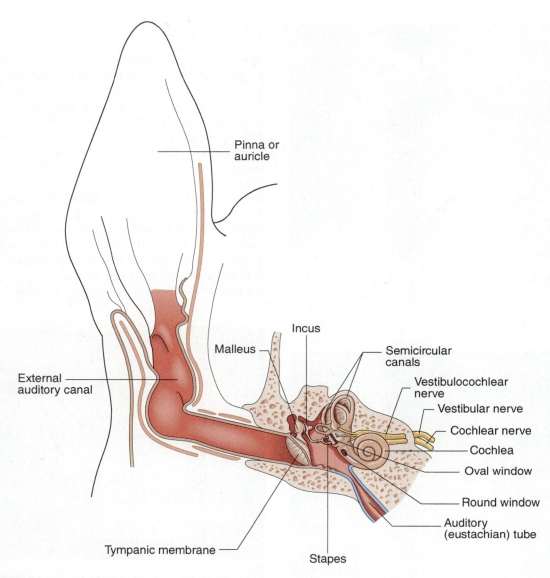

FIGURE 8-39 Anatomy of the dog's ear.

shaved. The ear canal should be examined for any dirt, wax, or debris that may build up within the inner ear. The opening of the ear is under the ear flap, with the ear canal lying internally within the ear in an L-shape. The end of the L-shape is located in the eardrum. This area can be damaged or punctured during ear cleaning. Debris can easily accumulate in the inner ear, causing severe irritation and possible infection. When cleaning ears, it is best to use gauze sponges or cotton balls rather than a cotton-tipped applicator, which can cause ear damage if used improperly. Begin cleaning the external ear flap by wiping the outer area with alcohol or an ear cleaner to remove any dirt or debris. Wipe away from the internal canal area. Next, clean the internal ear by placing an ear-cleaning solution on a gauze pad or cotton ball and wiping the ear backward and toward the outer ear to avoid pushing debris into the eardrum. Continue to clean until no debris is noted. After a physical examination, the veterinarian may recommend that severely dirty ears be flushed with ear cleaner. When the ears have been cleaned, use dry gauze sponges or cotton balls to dry out the ear canal. Protect the ears from water during a bath by placing a large cotton ball into each ear canal. Once the ear is clean, the veterinarian will be able to do a thorough ear exam and identify and abnormalities and, in some animals, may require medication administration by a veterinarian.

Nail Trimming

Nail trims are done based on the length of the nail and the type of surface the dog spends time on. Long nails can get caught and tear or may scratch people or damage furniture. To trim nails, you need some knowledge of the nail anatomy and practice clipping. Veterinary assistants should be able to use handheld nail trimmers as well as a **dremel** tool that grinds down nails (see Figure 8–40). The nail bed has a blood supply known as the **quick**. If a nail is trimmed too short, it will start to bleed. There are several ways to **clot** or stop the bleeding. A veterinary assistant may use **styptic powder** to clot the blood (see Figure 8–41). If the bleeding is difficult to control, a **silver nitrate stick** may be used. This is a chemical that burns the blood vessel and causes the blood to clot. Clients who trim their dog's nails at home may pack cornstarch, flour, or a bar of soap into the nail to stop the bleeding. If the nail is cut or torn very short, a bandage may be necessary.

Ear Plucking

Ear plucking removes hair visible to the eye from the inside of the ear canal. The hair may be removed with **forceps**, which grasp and pull the hair out. This is done for certain dog breeds, like the poodle, that have excessive amounts of hair within the ear canal. The growth of hair within the ear canal can cause irritation and hold in dirt and debris. Plucking the hair out can prevent dirt and wax from collecting in the ear canal.

Expression of Anal Glands

Anal glands are the scent glands located on either side of the rectum (see Figure 8–42). The **scent glands** allow dogs to determine "who is who" in the dog world. Dogs greet each other by smelling each other's scent glands. These sacs hold small amounts of fluid from the dog's bowel movements. Over time, the sacs fill with the fluid and cause pressure on the rectum. Some dogs are capable of releasing the pressure on their own during a bowel movement. Other dogs cannot and will show signs of needing the glands expressed, including excessive licking at the rectum, discomfort in the rectal area, and scooting the rear over the floor. Anal glands can be expressed both internally and externally. Veterinary assistants should be trained to express anal glands externally. This procedure is done with the dog standing. Exam gloves should be worn and paper towels and a bowl of warm soapy water should be available for cleanup. To express anal glands externally, locate the sacs on either side of the rectum. (If the glands were on the face of a clock, they would be at 3 o'clock and 9 o'clock.) Using the thumb and pointer finger, gently apply pressure to the sac area by massaging the site (see Figure 8–43). The sacs should press against each other and release the fluid. Be warned not to stand directly behind the dog when performing this procedure. The glands tend to release and spray a smelly fluid that can be projected several feet. If the glands become **impacted** or difficult to express due to thickening, a veterinarian or veterinary technician will need to express the glands internally. Please note that *external* means on the outside of the rectal area, while *internal* means on the inside of the rectum.

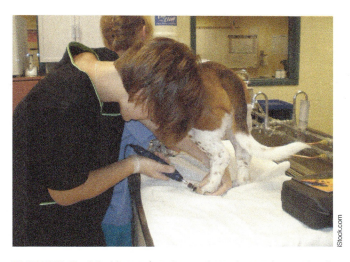

FIGURE 8–40 Use of a dremel tool to trim a dog's nails.

FIGURE 8–41 Styptic powder is used to stop bleeding that may occur when nails are cut.

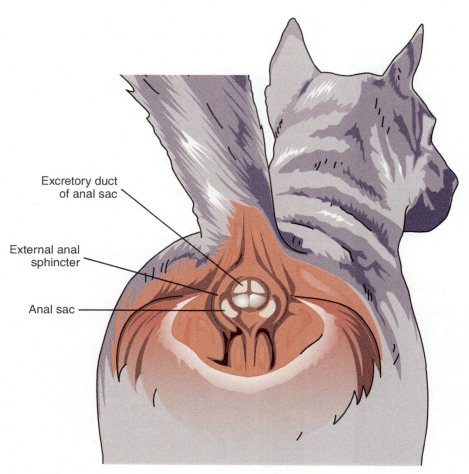

FIGURE 8–42 Location of the anal glands.

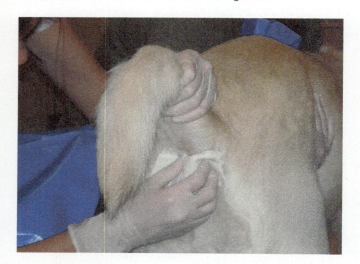

FIGURE 8–43 Expression of the anal glands.

Teeth Brushing

Brushing a dog's teeth helps prevent tartar buildup and reduce bad breath odor. Veterinary assistants should be able to brush a dog's teeth and note any dental problems (see Figure 8–44). Assistants should also be able to show clients how to brush their dog's teeth at home. Dogs require special toothpaste that has specific digestive enzymes that are absorbed in the body. Human toothpaste is not digestible and can cause toxic effects in dogs. There are several types of toothbrushes that may be used. One is a long-handled toothbrush, similar to a human toothbrush. Another is a **finger brush** that fits on the end of the pointer finger and allows for ease of brushing. There are also dental sprays that help reduce bacteria that cause bad breath. Typically, the outer surfaces of the teeth are brushed at a downward angle that also brushes the gums. The gums are sensitive and should be brushed gently. The lips should be lifted to make it easy to see the teeth. Note any broken, missing, or deciduous (baby) teeth while you are brushing.

Vaccinations

A **vaccine program** should begin for the puppy during 6 to 8 weeks of age (see Table 8–5). The vaccines will

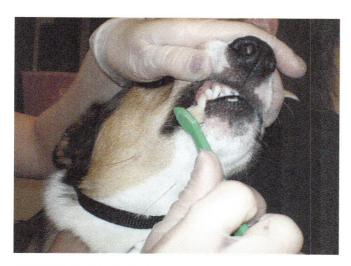

FIGURE 8–44 Tooth brushing is an important part of the dental care needs of a dog.

TABLE 8–5
Vaccination Schedule

AGE	VACCINE
6–8 weeks old	DHLPPC or DAPP, Bord
10–12 weeks old	DHLPPC or DAPP, Lepto
14–16 weeks old	DHLPPC or DAPP, Lepto, Lyme
12–16 weeks old	RV (first vaccine is good for one year), Lyme
Yearly to every three years	DHLPPC or DAPP, RV

protect the puppy from common canine diseases. Most puppies receive a vaccine series, or multiple vaccines, in one dose. The most common canine vaccine series is the DHLPPC or DAPP combination, also known as the distemper series. Each letter represents the disease from which the vaccination protects the puppy:

- D—Distemper
- H—Hepatitis or A—Adenovirus
- L—Leptospirosis
- P—Parainfluenza
- P—Parvovirus
- C—Corona virus

Vaccines may be considered to be core or noncore vaccines. The American Animal Hospital Association (AAHA) guidelines offer recommendations on which vaccines should be administered based on an animal's location and risk of disease. **Core vaccines** are considered necessary for all animals within that species due to the high-risk potential for disease. **Noncore vaccines** are considered only for those pets at high risk for certain diseases. A rabies vaccine (RV) is given between 12 and 16 weeks of age and is initially valid for one year; some rabies vaccines are good for one year, while others are good for three years. The rabies vaccine must legally be given to all dogs that are owned. A rabies tag and a proof of certificate are issued at the time the vaccine is administered. These vaccines are commonly given into a muscle [**intramuscular (IM)**] or under the skin [**subcutaneous (SQ)**]. The rabies vaccine is considered to be a core vaccine; the recommended site of administration is the right hind limb.

The DHLPPC or DAPP is given in **boosters**, or multiple times, to build up the protection of the immune system. Boosters are typically given a month apart (or every three to four weeks) three to four times, ending at about 16 to 18 weeks of age. The DHLPPC or DAPP vaccine is considered to be a core vaccine. The first series of completed vaccine boosters are good for one year; boosters are given every three years thereafter based on the veterinarian's recommendations. The leptospirosis vaccine component is a yearly vaccine, and a booster is recommended yearly. The recommended site for administration is subcutaneous (SQ) in the right front shoulder. The corona virus vaccine component is typically not recommended, as the vaccine is not shown to significantly reduce the disease.

The Lyme vaccine (*Borrelia*) may be a core or noncore vaccine based on the location of the pet. Many states in the eastern United States are at high risk for Lyme disease. The Lyme vaccine booster begins at 12 to 16 weeks of age, with a booster administered in three to four weeks. The vaccine is recommended on a yearly basis in high-risk areas and the recommended site for administration is subcutaneous (SQ) in the left front shoulder.

The *Bordetella* (kennel cough) vaccine is considered a noncore vaccine but should be recommended in dogs that are in high-risk situations, such as boarding kennels, dog shows, obedience training, or dog parks. The bordetella vaccine may be administered intranasally (IN) or by injection (SQ or IM). The injectable vaccine route requires a booster in two to four weeks; the intranasal route is good for one year, with a vaccine recommended yearly. The intranasal vaccine is administered by applying drops into the nasal cavity.

The canine influenza virus is a noncore vaccine for dogs that may be considered high risk. It is administered between 6 to 8 weeks of age, requires a booster in three to four weeks, and is recommended yearly.

Reproduction and Breeding

Canine reproduction and breeding techniques require a general knowledge of a dog's reproductive system and the **estrus cycle**, commonly known as being in **heat**. This is the time the female is receptive to the male and will allow breeding. **Receptive** means that the female dog will allow the male dog to breed with her. Estrus normally begins between 6 and 24 months of age, depending on the size and breed of the dog. This time frame is known as **puberty**: the dog is nearing adulthood and its reproductive organs are completely developed. Small-sized dogs typically begin puberty at about 6 to 12 months old; large-sized dogs begin at about 8 to 24 months old. Dogs cycle twice a year, typically every six months. Estrus typically lasts 14 to 21 days. Signs of a female dog in estrus include the following:

- Swollen vulva (see Figure 8–45)
- Vaginal discharge—usually yellowish to bloody in appearance
- Licks at vulva
- Swollen mammary glands
- Tail is held high and to the side
- Behavioral changes

Breeding usually begins 10 to 11 days after seeing signs of estrus. Vaginal cytology is helpful in determining where the female is within her cycle. The female will allow contact with the male, and breeding should occur every other day until the female will no longer accept the male. During the natural mating process in dogs, the male and female will form a "tie" for an additional 10 to 30 minutes after mating has taken place. This tie is formed due to the swelling of the bulbus glands within the penis during mating. The male and female will separate on their own; manually separating them can cause serious injury to both animals. It is common for the male and female to end up standing rear-to-rear during this time. Once a dog is bred, the gestation, or length of pregnancy, is an average of 63 days. The labor process of dogs is called "whelping." A whelping box may be provided to the female two to three weeks before the due date. The box should be placed in a quiet, warm, and darkened area to reduce stress on the dog. It is a good idea to monitor the dog for signs of labor and to allow for privacy when she begins the whelping process. Signs of whelping include the following:

- Restless and anxious behavior
- Stops eating 24 hours prior to labor
- Body temperature decreases to 98 degrees
- Nesting habits (collect items and make a nest)
- Licks vulva
- Circles and is uncomfortable
- Milk develops at mammary glands 12 to 24 hours prior to labor
- Pants

Normal labor occurs within 8 to 12 hours depending on the litter size. On occasion, some dogs may be in labor for up to 24 hours. Normal pup presentation is the head first. A difficult labor is called **dystocia**. Dystocia may occur for several reasons: the female tires during labor, a puppy becomes stuck within the birth canal, or the puppy is too large to enter the birth canal. When a female dog shows signs of having a difficult labor, it is an emergency and a veterinarian should be contacted immediately. After the birth, the mother will lick the puppies to stimulate them to breath and to help dry them. Human intervention may be necessary if the mother fails to break open the sac upon delivery. The puppies may need to be removed from the sac and the mucous removed from the nostrils using a bulb syringe. The umbilical cord may need to be tied off and cut, leaving at least an inch space between the belly of the puppy and the tie. The puppies may need to be vigorously rubbed to stimulate breathing and to dry the puppies. Once the puppies are born, it is important that they begin nursing as soon as possible to obtain the mother's **colostrum**: this is the milk produced during the first 24 hours and contains **antibodies** to protect the immune system (see Figure 8–46). The antibodies will protect the puppies from disease until a vaccine program is started. The temperature should be kept at 85 degrees Fahrenheit for the first seven days; it can then be gradually lowered to about 70 to 75 degrees Fahrenheit. The pups will nurse every 2 hours, and the mother will lick each puppy to stimulate urine and waste production. If the mother is unable to stimulate

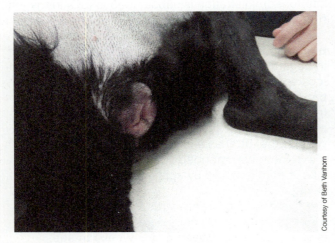

FIGURE 8–45 A swollen vulva, indicating a female dog in estrus.

FIGURE 8–46 Chihuahua giving birth, one baby is about to be born, while two are nursing.

waste or urine production, a warm moist cloth may be rubbed gently over the urogenital and rectal area of the puppies to stimulate production. The eyes and ears of the pups will open at 10 to 14 days old. They begin walking at about 2 weeks old; tail wagging and barking will begin at about 3 weeks old. At this time, they should be socialized in order to help form behaviors and personalities.

Common Diseases

Dogs may develop many types of infectious diseases. There are five classes of diseases in animals: bacterial, viral, fungal, protozoan, and rickettsial.

Bacterial Disease

Bacterial diseases are caused by bacteria that invade the body. Some examples of bacterial diseases in dogs include **tetanus**, commonly known as lockjaw, and **Lyme disease**, spread by the bite of a tick (commonly the deer tick) that transmits bacteria into the bloodstream.

Leptospirosis is a bacterial disease that affects dogs and is transmitted in the urine. Signs of leptospirosis include fever, **lethargy** (inactivity), blood in urine, skin bleeding, and liver and kidney damage. Vaccination and treatment are available. This disease is zoonotic and care should be taken to wear gloves when handling or cleaning urine.

Viral Disease

Viral diseases are caused by a virus that invades the body; the virus is not treatable, but rather must run its course. Examples of viruses in dogs are **rabies**, caused by saliva transmitted in bites and scratches, and **hepatitis**, a virus that causes an infection in the liver.

Canine distemper is a virus that can be fatal, especially in puppies. There are different levels of severity, and **incubation**, or the time that the dog is infected, takes 9 to 14 days. Sanitation is very important to control this virus. A vaccine is available, but treatment is limited to controlling symptoms. Signs of distemper include fever, lethargy, **anorexia** (not eating), eye discharge, nasal discharge, vomiting, diarrhea, and coughing. In some cases, a dog may have seizures. Dogs that recover from distemper can have **neurological** problems that affect the way they move.

Parvovirus is another viral infection that occurs in dogs. It is usually fatal in puppies. Dogs that transmit this disease are highly contagious and should be kept in isolation. Sanitation is the best way to control parvovirus. Signs of parvovirus include vomiting, bloody and smelly diarrhea, anorexia, lethargy, and fever. A vaccine is available.

Rabies is a fatal virus that affects all mammals and humans. It is transmitted through saliva into an open wound, most commonly from a bite. Each U.S. state has **rabies laws** that should be reviewed and understood by all veterinary staff members. All dogs are legally required to be vaccinated for rabies. Signs include a change in behavior, aggression, foaming at the mouth, drooling, fever, **paralysis** (loss of leg and body movement), and death. Any dog that is suspected of having rabies should be immediately quarantined in a veterinary facility. Veterinary professionals should consider a pre-exposure vaccination against rabies, which is given before an animal comes into contact with a diseased animal or contaminated area. This will hopefully prevent the animal from getting the disease.

Fungal Disease

Fungal diseases are caused by a fungus that lives on the outside of the body. One example is **ringworm**. All mammals are susceptible to the fungus, and it is contagious. It has the potential to be highly **zoonotic**, or transmitted to humans. Signs of ringworm include skin irritation, scratching, hair loss, and circular, crusty skin lesions. A site must be tested to determine the presence of the fungus. All bedding and grooming items should be disinfected to prevent spreading. Proper sanitation is necessary for control.

Protozoan Disease

Protozoan diseases are caused by single-celled organisms that invade the body. Examples include **giardiasis**, caused by water contamination, and **coccidiosis**, caused by contaminated bird droppings in water or soil.

Rickettsial Disease

Rickettsial diseases are caused by parasites, such as fleas and ticks. An example of a rickettsial disease is **Rocky Mountain spotted fever**, which is caused by a tick bite. (Note that Lyme disease is not a rickettsial disease but rather a bacterial disease, as it is caused by a tick bite that transmits bacteria into the bloodstream.)

Common Parasites and Prevention

Parasites found inside the body are commonly called "worms." Parasites may live inside the body, usually in the intestinal tract, or outside of the body, such as on the skin. Canines are prone to several types of internal and external parasites.

External Parasites

The veterinary assistant should be able to identify common external parasites, such as fleas and ticks. It is also important to know the signs that may suggest a dog has external parasites. Signs of external parasites include scratching, chewing at the skin, skin sores, visible parasites, and flea dirt on the hair coat.

Fleas

Fleas are jumping, wingless insects that seek heat to survive and feed on the blood of animals. They are brown to black in color. They are known to transmit certain diseases. The bite of a flea may cause an allergic reaction in dogs due to their sensitivity to the flea's saliva. Fleas will also bite humans. A female flea can lay up to 50 eggs a day; the eggs will hatch in 48 hours and the young become adults within 15 days of hatching. This life cycle continues and can become a large problem within a household. The term *life cycle* of a parasite explains how the parasite begins and the stages it goes through when an animal becomes infected. Dogs that have fleas will scratch, bite at the fur, and may develop sores in the area of the bites. There are several ways to determine if a dog has fleas. Fleas produce fecal droppings that appear to look like dirt on the hair coat. This is a warning sign of a flea problem.

To check for fleas, use a flea comb and brush the hair coat both with and against the growth of hair. If any dirt particles are removed from the animal's body, apply water to them and watch for a red or rust color. Red will indicate the blood meal within the flea's feces. Another way to check for fleas is to place a warm white towel on the floor of your house. Turn off any lights and

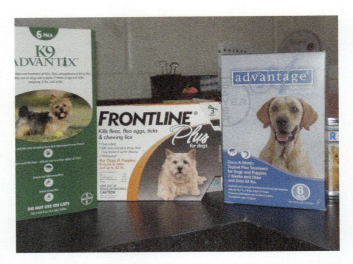

FIGURE 8–47 Various flea control products.

wait several minutes. Turn the lights back on and check the towel for fleas. Remember: fleas seek heat!

There are several ways to prevent fleas. There are **topical** medications that are applied to the skin and hair coat, **oral** medications that are given by mouth, flea collars, and flea shampoos and dips that kill or repel fleas (see Figure 8–47 and Table 8–6). There are also sprays and foggers that can be used inside a house to control fleas. Popular topical medications include First Shield, Frontline, and Advantage.

Ticks

Ticks are wingless insects that seek movement. They attach to animals and people by embedding their mouth into skin and then feed on blood (see Figure 8–48). During feeding, disease-causing bacteria may be transmitted. As the tick feeds on the blood meal, its body swells in size. It is important for veterinary assistants to know how to remove a tick: it must be done carefully and without detaching the head from the body. If the head is detached, the skin may become infected. Ticks lay their eggs on the ground and reach the adult stage in two years. You may detect a tick bite by finding a live tick; other symptoms include itching at the site, skin irritation, and redness. A skin infection causes redness, heat, pain, and pus at the site of the problem. Control is through routine brushing and checking skin for ticks and applying topical products that kill or repel ticks. Popular topical medications include First Shield Trio, Frontline Plus, and Advantix.

Mites

Mites are insects that can live on the skin, hair coat, or within the ears. Some mites are barely visible to the naked eye; others are **microscopic**, or only visible

TABLE 8-6
Common Flea Product Comparison Chart

PRODUCT NAME	APPLICATION TYPE	PARASITES CONTROLLED	AGE OF USE	WATERPROOF OR RESISTANT	LENGTH OF APPLICATION
First Shield Trio	Topical	Fleas, ticks, lice, biting flies, mites, mosquitoes	7 weeks	Water proof	Monthly
Advantage	Topical	Fleas	7 weeks	Water resistant	Monthly
Advantix	Topical	Fleas, ticks, flies, mosquitoes	7 weeks	Water resistant	Monthly
Bravecto	Oral	Fleas, ticks	6 months		Every 3 months
Capstar	Oral	Fleas	4 weeks	Waterproof	Daily or as needed
Comfortis	Oral	Fleas	14 weeks		Monthly
Frontline	Topical	Fleas	8 weeks	Waterproof	Monthly
Frontline Plus	Topical	Fleas, ticks, lice	8 weeks	Waterproof	Monthly
NexGard	Oral	Fleas, ticks	8 weeks		Monthly
Program	Oral/Injectable	Fleas	4 weeks	Waterproof	Monthly: oral 6 months: injection
Promeris	Topical	Fleas, ticks	8 weeks	Waterproof	Every 4–6 weeks
Revolution	Topical	Fleas, ticks, mites, heartworms	6 weeks	Waterproof	Monthly
Sentinel	Oral	Fleas, hookworms, roundworms, whipworms, heartworm	4 weeks	Waterproof	Monthly
Simparica	Oral	Fleas, ticks	35 days		Monthly

FIGURE 8–48 Tick.

with the use of a microscope. **Ear mites** live within the ear canal and cause a thick, dark, crusty waxlike material that builds up within the ear. Ear mites are not visible to the eye. They cause itchiness within the ear canal, and dogs will scratch their ears and sometimes cause open sores within the ear flap due to scratching. Ear mites are **contagious** and transmitted to all animals. The term *transmission* or *transmitted* means passing a disease by some route or method. A disease may be transmitted by direct contact, in the air, or by being bitten by an infected animal. Ears should be cleaned and evaluated under a microscope for mites. Ear medications are used to treat and kill the mites.

Skin mites occur within the hair and skin. There are two types of skin mites that cause a condition called mange. The sarcoptes mite causes **sarcoptic mange**. This mite is contagious and can spread to humans, in which the condition is known as **scabies**. The demodex mite causes **demodectic mange**. The proper medical term for hair loss is **alopecia**. Demodectic mange causes areas or patches of skin where the hair coat is thin or missing. Demodex is naturally occurring in small amounts on all dogs' skin, but the condition can become problematic and cause a skin irritation and infection. Both types of mange will cause severe itchiness that leads to scratching, hair loss, skin lesions and sores, skin infection, and discoloration of the skin. Both types of mites are not visible to the eye and the skin must be scraped by a veterinarian or technician and evaluated under a microscope. It is important for veterinary assistants to disinfect all grooming items and equipment since this is a common route of transmission for mange mites. All skin lesions should be evaluated for mange mites.

Lice

Lice are tiny insects that live on the hair. Lice are species specific and only contagious within a species. Signs of lice include scratching, hair coat damage, and visible spots of white on the hair. The small spots of white seen on the hair coat are the lice larvae, called **nits**. They appear to stick to the hair and can be removed with a piece of tape for microscopic evaluation. Lice must be treated with topical insecticides to kill the insects. All bedding and areas of the house where the dog has access should be treated.

Internal Parasites

The veterinary assistant should be aware of signs of internal parasites. Some common internal parasites occur within the body, usually within the intestinal tract. They feed on the blood of animals by surviving inside the body. Some are microscopic and some may be visible after passing through the dog's intestinal system. Regular preventative care is recommended for all internal parasites.

Heartworms

Heartworm disease is a serious issue among dog owners and veterinary professionals. It is caused by an internal parasite that lives in the heart, lungs, and bloodstream. It is transmitted by the mosquito, which is known as the **vector**, or route of transmission. The young heartworms, called **microfilaria**, are a serious stage in heartworm disease, as they can cause a blockage in the heart or bloodstream that may result in death. The heartworms reach the adult stage in six months. They can be as long as 12 inches and resemble a long piece of white spaghetti. Prevention is the best method. Several products are available to prevent heartworm and other parasites (see Table 8–7). Veterinarians require a negative blood test before they will prescribe a product. This should be done yearly for proper prevention and to verify the preventative is working. Dogs that test positive for heartworm disease should be treated with caution. Signs of heartworm include shortness of breath, coughing, exercise intolerance, easily tired, and possible death. Heartworm disease is treated with a drug called Immiticide, which contains arsenic and can potentially cause death during the treatment stages. During the lengthy treatment, heartworm-positive dogs must be caged at all times, with activity restrictions to reduce the risk of microfilaria lodging in the heart of blood vessels. Prevention is the key in this disease, and year-round treatment is recommended.

TABLE 8-7

Canine Heartworm Product Comparison Chart

PRODUCT NAME	APPLICATION TYPE	PARASITES CONTROLLED	AGE OF USE	LENGTH OF PROTECTION
Advantage Multi	Topical	Heartworms, roundworms, hookworms, whipworms, fleas	7 weeks	Monthly
Heartgard	Oral	Heartworms	6 weeks	Monthly
Heartgard Plus	Oral	Heartworms, roundworms, hookworms	6 weeks	Monthly
Interceptor	Oral	Heartworms, roundworms, hookworms, whipworms	4 weeks	Monthly
Iverhart Max	Oral	Heartworms, roundworms, hookworms, tapeworms	6 weeks	Monthly
Proheart 6	Injection	Heartworms, hookworms	6 months	Every 6 months
Revolution	Topical	Heartworms, fleas, ticks, ear mites, sarcoptic mange mites	6 weeks	Monthly
Sentinel	Oral	Heartworms, roundworms, hookworms, whipworms, fleas	4 weeks	Monthly
Tri-Heart Plus	Oral	Heartworms, roundworms, hookworms	6 weeks	Monthly

Roundworms

Roundworms, or **ascarids**, are the most common intestinal parasites found in dogs and puppies. They live in and feed off of the small intestine. Common causes are dogs ingesting rodents, contaminated soil, and eggs being passed from the placenta or to nursing puppies. When passed in waste material, roundworms appear as long, thin white pieces of spaghetti. Signs of roundworm infection include a swollen abdomen, diarrhea, vomiting, weight loss, and poor hair coat appearance. Roundworms are zoonotic to humans and can cause a condition known as **visceral larval migrans**, which is commonly caused through ingesting contaminated soil. The eggs can invade the lungs and intestines. Regular deworming with such products as Strongid or Pyrantel is recommended yearly. Puppies should be dewormed every three to four weeks between the ages of 6 and 18 weeks.

Hookworms

Hookworms are another internal parasite found in the intestinal tract that feed on blood. Transmission may occur through ingesting contaminated soil, eggs penetrating through the pads and skin, and puppies acquiring the infection when nursing. Signs of infection include **anemia** (low red blood cell count), diarrhea, vomiting, weight loss, and poor coat appearance. Hookworms are zoonotic to humans and cause a condition known as **cutaneous larval migrans**, in which hookworm eggs penetrate the skin and enter the intestinal tract. Regular deworming with such products as Strongid or Pyrantel is recommended yearly. Puppies should be dewormed every three to four weeks between the ages of 6 and 18 weeks.

Whipworms

Whipworms are internal parasites that attach to the colon and feed on intestinal blood and tissues. They are transmitted from contaminated soil that has been ingested. The eggs survive in soil for several years and are very hard to control. A **fecal egg analysis** is done to determine if parasites are infecting an animal (see Figure 8–49). This procedure may be completed in a variety of ways, as discussed in Chapter 38, Laboratory Procedures. Signs of whipworm infection include anemia, weight loss, diarrhea, and poor coat appearance.

Tapeworms

Tapeworms are intestinal parasites that live in the small intestine. They are commonly transmitted by ingesting rodents or a flea. The tapeworms form segments called **proglottids** that shed as the tapeworms grow in size (see Figure 8–50). These segments may

FIGURE 8–49 Conducting a fecal egg analysis test.

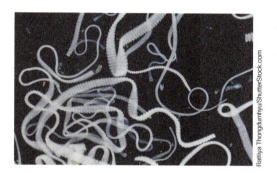

FIGURE 8–50 Tapeworm.

be seen in waste material or around the anus; they look like rice. Signs of tapeworm infection include flea problems, diarrhea, weight loss, and scooting on the floor. Diagnosis of internal parasites is through fecal egg analysis and treatment is with **anthelmintics**, commonly called dewormers. The best way to prevent parasites is through product control and removing all waste materials from the yard area.

Common Canine Surgical Procedures

Veterinary assistants will routinely discuss canine surgical procedures with dog owners. The veterinary assistant should understand common procedures in dogs: the **ovariohysterectomy (OHE or OVH)**, also called a spay; the castration; the dewclaw removal; and the tail docking.

In a spay surgery, a female dog's uterus and ovaries are removed to prevent reproduction. In castration surgery, a male dog's testicles are removed to prevent reproduction. "Neutered" refers to both male and female dogs that have been altered surgically to prevent breeding. Neutering is highly recommended for dogs that are not of breeding quality and are not intended for breeding or show situations: neutering is healthier for the pet and prevents unwanted animal populations.

Dewclaw surgery is the removal of the dewclaw or first digit on the foot pads, which doesn't have contact with the ground. This nail often becomes caught on objects and can be torn, causing a painful injury.

Tail docking, a cosmetic surgery to shorten the length of the tail, may be offered by the veterinarian. This is best done within three to five days of birth. It is important that the veterinary assistant has a basic understanding of the procedures in order to discuss the benefits of the surgery to the client.

Spay and Castration

Dogs are typically neutered at about 6 months old. Some veterinarians will neuter dogs at a much earlier age. The OHE or OVH is performed in the female dog with an incision made over the abdominal midline, or the center of the lower abdomen. Female dogs have two seasonal estrus cycles a year (with the exception in the Basenji, which has one seasonal estrus cycle). They are capable of having two litters of puppies a year. The OHE will decrease the dog population, reduce the attraction of male dogs, decrease mammary gland tumors, and decrease the incidence of spraying and marking territory. The spay is performed under general anesthesia. The uterus and ovaries are removed and the incision is sutured. Sutures are removed in 7 to 10 days, with a limit on exercise.

The **orchidectomy** (castration) in the male dog is much less complex than in the female dog. The testicles are exposed, **ligated** (tied off), and excised. There are typically a few sutures in a dog castration; sutures are removed in 7–10 days, with a limit on exercise. The castration procedure in dogs decreases the likelihood of marking, wandering for females, and chances of testicular tumors. The castration surgery is usually done under general anesthesia.

SUMMARY

Veterinary assistants work with dogs on a regular basis. It is important for them to have a basic knowledge of canine restraint, grooming, care and maintenance, and the information on health care required to educate clients on proper dog care. Veterinary assistants will also use this basic information to educate clients within the veterinary facility. They may need to show dog owners how to carry out a variety of procedures at home for the well-being of their pet. This is an important part of hands-on skills, as many other animals may have these same types of basic health care needs and maintenance.

Key Terms

aggressive   the behavior of an animal exhibiting an attempt to do harm, such as biting

alopecia   hair loss

anal glands   scent glands or sacs located on each side of the rectum

anatomy   the study of body structures

anemia   low red blood cell counts due to the inability of the body to make or replicate blood cells

anorexia   not eating

anthelmintic   a medication used to treat and prevent parasitic infection, also known as a dewormer

antibody   a substance that acts to protect the immune system

ascarid   also known as roundworm; an intestinal parasite that looks like a long white worm

bacterial   disease caused by bacterium

booster   a series of vaccines given multiple times to build up the immune system

brachycephalic   a short, nonexistent nasal area that gives a flat-face appearance

canine distemper   a viral infection that can be fatal to puppies; dogs who recover may have long-term neurological affects

Canis familiaris   genus and species of domesticated dog

carnivore   meat eating

cephalic venipuncture   a puncture to draw blood from the vein inside the front of the legs

chemical restraint   the use of sedatives or tranquilizers to calm and control an animal

clot   to stop or control bleeding

coat conditioner   a spray placed on the hair coat to eliminate drying of skin and hair

coccidiosis   a protozoan disease cause by contaminated bird droppings in water and soil sources

colostrum   milk produced by a female within 24 hours of giving birth; provides antibodies that protect the immune system

contagious   capable of spreading disease

core vaccine   a vaccine considered necessary for all animals within that species due to the high-risk potential for disease

cutaneous larval migrans   a hookworm infection in humans caused by penetrating the skin

demodectic mange   a disease caused by the demodex mange mite on the skin and hair of dogs

designer breed   the cross breeding of two breeds to form a new breed not yet recognized by the AKC

dewclaw   the first digit on the paw, which doesn't make contact with the ground

domestication   taming of an animal to coexist peaceably with humans

dominance aggression   aggression presented against a dog's pack family or human family

dorsal recumbency   a restraint position with the animal placed on its back

dremel   a tool used to grind nails

dystocia   a difficult labor process

ear mite   an insect that lives in the ear canal

ear plucking   removal of hair from the ear canal

endothermic   the ability to regulate body temperature within the body

estrus cycle   the period of time when a female dog is receptive to a male dog for mating; also known as the heat cycle

fecal egg analysis   a test to determine the type of parasite found in the fecal matter of an animal

finger brush   a tool used to clean a dog's teeth

flea   a wingless external parasite that feeds off the blood of an animal

forcep   a tool used to remove hair or parasites from the body

fungal   a disease caused by a fungus that lives or grows on the body

gauze   a thin cotton-like material used for bandages and to make restraint muzzles

giardiasis   a protozoan disease caused by water contamination

grooming   the process of trimming and bathing in an appropriate manner

gruel   a mix of equal parts of food to equal amounts of water, forming a soup-like consistency that can gradually be made into an oatmeal-like consistency

heartworm disease   a disease that is caused by an internal parasite, transmitted by mosquito, that lives in the heart and bloodstream

heat   the period of time when a female dog is receptive to a male dog for mating; also known as the estrus cycle

hepatitis   inflammation of the liver

hookworm   an internal parasite that may pass through the skin and pads from contaminated soil

housebreaking   training a dog to eliminate outside

hybrid   the cross breeding of two breeds to form a new breed not yet recognized by the AKC

impacted   difficult or unable to express

incubation   the time in which an animal becomes infected with disease

intramuscular (IM)   into a muscle, as in an injection

jugular venipuncture   a puncture to draw blood from the vein in the neck

lateral recumbency   a restraint in which the animal is placed onto its side

leptospirosis   a bacterial disease transmitted through contact with contaminated urine

lethargy   inactive; tired

lice   tiny insects that live on hair, are species specific, and are contagious

life span   the length of time an animal will live based on body size and health

life stage   particular periods in one's life characterized by specific needs during that time

ligated   to tie off

Lyme disease   a bacterial disease transmitted through the bite of a tick

maintenance diet   nutrition that allows animal to stay at the same weight

mat   tangled pieces of hair

microfilaria   young heartworm stage

microscopic   very small; unable to view with the naked eye

mite   an insect that lives on the skin, in the hair, or in the ears

mixed breed   an animal of unknown parentage that is a combination of two or more breeds

monogastric   a single stomach to digest food

muzzle   a covering for the mouth of an animal to prevent biting

nail trim   cutting the length of the nail to prevent injury and promote comfort

neurological   pertaining to the brain and spinal cord

neuter   surgical removal of the reproductive organs or either gender

nit   larvae of lice

noncore vaccine   a vaccine considered only for those pets at high risk for certain diseases

obedience   training a dog to obey commands

oral   pertaining to the mouth

orchidectomy   castration; surgical removal of the testes

orphaned   a young animal that has been rejected by or has lost its mother

ovariohysterectomy (OHE/OVH)   surgical removal of the uterus and ovaries

paralysis   unable to move one or more body parts

parasite   internal or external insects that live off an animal

parvovirus   a potentially fatal viral infection causing severe and bloody diarrhea

physical examination (PE)   the assessment of an animal to determine its overall health status

physical restraint   the use of a person's body to control the position of an animal

physiology   the study of body functions

proglottid   a tapeworm segment that is shed as the tapeworm grows

protozoan   a single-celled microorganism

puberty   sexual maturity

purebred   an animal with known parentage of one breed

quick   blood supply within the nail bed

rabies   a fatal viral infection spread through saliva

rabies laws   state legal standards to prevent the spread of rabies infections

rabies pole   a tool used to restrain and control an animal

ration   a specified amount of food

receptive   accepting of

recumbency   a lying position

redirected aggression   behavior that stimulates a response from another source, such as a dog outside of the yard or a car driving down the road, in which the dog turns on the owner

reduced-calorie diet   meals low in calories to promote weight loss

rickettsial   more complex multi-cellular organism that are tick- or flea-borne

ringworm   a fungus that grows on the skin and is contagious

Rocky Mountain spotted fever   a rickettsial disease caused by a tick bite

roundworm   an intestinal parasite that is long and white

saphenous venipuncture   a puncture to draw blood from the outer thigh area

sarcoptic mange   a contagious disease caused by a mite that lives on the hair and skin; known as scabies in humans

scabies   a contagious disease caused by a mite that lives on the hair and skin; known as sarcoptic mange in animals

scent glands   sacs located on either side of the rectum; also called anal glands

sedative   medication used to calm an animal

senior diet   nutritional requirements for an older, less-active animal

silver nitrate stick   an applicator used to apply a chemical that will stop bleeding

skin mite   an insect that lives within the hair and skin and is not visible to the naked eye

slicker brush a tool used to brush the hair coat

snare pole a tool used to restrain and control an animal

socialization an interaction with other animals and people to become used to them

sternal recumbency a restraint position in which the animal is lying on its chest and abdomen

styptic powder a chemical used to stop bleeding

subcutaneous (SQ) under the skin, as in an injection

tail docking surgical removal of all or part of the tail

tapeworm an internal parasite with a long segmented body caused by ingesting fleas

tetanus bacterial disease caused by a wound; commonly called lockjaw

tick wingless insect that feeds off blood of animal

topical the application of medication to the skin or the outside of the body

training teaching an animal to understand specific desired habits

tranquilizer medication used to calm an animal

vaccine program injections given to protect animals from common diseases and build up the immune system

vector an insect or organism that transmits a disease

verbal restraint voice commands used to control an animal

viral a disease caused by a virus

visceral larval migrans a roundworm infection in humans

weaning the process in which a young animal stops nursing and begins to eat solid foods

whipworm an internal parasite ingested from soil

zoonotic a disease that is transmitted from animals to humans

Review Questions

1. What is the difference between a purebred dog and mixed-breed dog?
2. List the seven groups of dogs and give three examples of breeds in each group.
3. What is the difference between lateral recumbency and sternal recumbency?
4. What are factors to consider when selecting a dog?
5. What are signs of a female dog in estrus?
6. When should a dog be fed a maintenance diet?
7. What is the ideal vaccine program for a puppy?
8. What are some general health practices that should be maintained for a dog?
9. What is the difference between an internal and an external parasite?
10. What are signs that a dog may have parasites?

11. What is the term for a disease passed from an animal to a human?
12. What are the five classes of diseases? Give an example of each type.
13. What is the importance of training a puppy or dog?
14. What grooming services should veterinary assistants be familiar with?
15. What is the importance of educating clients about zoonotic diseases and parasites?

Clinical Situation

A 5-year-old male basset hound is presented to the veterinary facility for a castration surgery. The dog has a history of bad breath and difficulty eating hard food. Last week, when the dog had a physical exam with yearly vaccines, the veterinarian discussed a dental procedure with the clients. The owners are dropping the dog off in the morning and ask if they can pick the dog up that afternoon. The veterinarian had noted in the medical chart that the dog will require general anesthesia and needs to stay overnight. Food and water are to be withheld the night before the surgery and the day of the surgery. The owners appear very distracted and in a hurry.

- What are some questions the veterinary assistant should ask in this situation?
- How would you answer the client's question about whether they can pick the dog up in the afternoon?
- What will you discuss regarding the castration surgery and the dental procedure?

Chapter 9: Cat Breed Identification and Production Management

Objectives

Upon completion of this chapter, the reader should be able to:

9.1 Identify common breeds of cats recognized by the Cat Fanciers' Association (CFA)
9.2 Describe the biology and development of the cat
9.3 Define common veterinary terms relating to the cat
9.4 Describe normal and abnormal behaviors in the cat
9.5 Describe how to properly and safely restrain cats for various procedures
9.6 Discuss proper cat selection methods
9.7 Describe health management practices for cats
9.8 Describe nutritional factors of cats
9.9 Explain cat breeding and reproduction methods
9.10 Discuss basic vaccine programs in cats
9.11 Discuss common feline diseases and prevention
9.12 Describe common internal and external parasites of cats
9.13 Discuss common feline surgical procedures
9.14 Discuss proper grooming needs of cats

Introduction

Cats are the most popular companion animal in today's society (see Figure 9–1). They are relatively easy to care for and tend to be self-sufficient. Cats have evolved from their wild counterpart, the lion. There are over 500 million domestic cats throughout the world and more than 30 different recognized breeds. The United States has over 70 million cats as pets—there are 10 million more cats than dogs. Cats are known to be playful, active, and alert. They have not been domesticated as long as the dog but have become an important part of the pet industry.

Veterinary Terminology

In veterinary medicine, it is essential to know species-specific terms and when to properly use them in a facility. Table 9–1 contains common feline terms and abbreviations

FIGURE 9–1 There are more cats as pets than dogs.

CHAPTER 9 Cat Breed Identification and Production Management

that are used frequently in veterinary medicine. Figure 9–2 shows the location of each body part on the feline and the corresponding term.

Biology

The cats descended from a wild cat, called **Felis catus**, that is commonly known as the lion. There are over 75 breeds of cats; over 40 of those are recognized as pure breeds. The average lifespan of a cat ranges from 12 to 18 years old. The oldest known living cats have been in their late 20s to early 30s. Cats are mammals that are carnivores, or meat eaters. Cats are similar to dogs in that they are endothermic, or capable of regulating their own body temperature internally, and they have a monogastric digestive system. *Monogastric* means

TABLE 9–1

Feline-Specific Terminology

FELINE	CAT
Queen	Female cat of breeding age
Tom	Male cat of breeding age
Kitten	Young cat that is less than 1 year of age
Gib	A male cat that is no longer able to reproduce
Spay	A female cat that is no longer able to reproduce
Bevy	Group of cats in a household
Queening or kittening	The labor process of the female cat
DLH	Domestic longhair
DMH	Domestic medium hair
DSH	Domestic shorthair

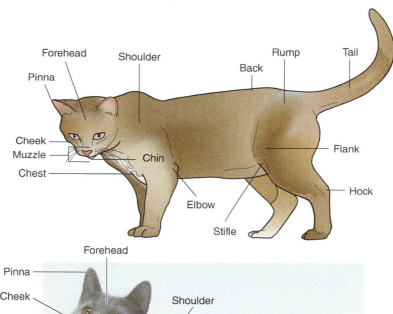

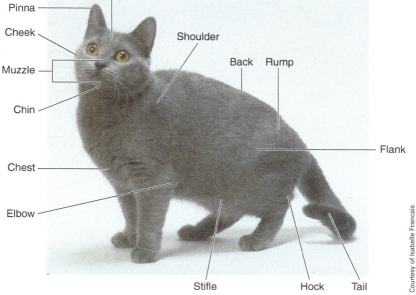

FIGURE 9–2 The anatomical parts of the cat.

a simple and single stomach within the digestive system. Much of the similarity ends there. Cats are known for their well-developed sensory systems. They have a sensitive nose, sensitive paws, extra smell and taste organs, sensitive ears that detect the slightest of sounds, 3-D and night vision, and specialized whiskers that serve as feelers and as a balance mechanism. Cats' feet differ from any other species of domestic animal. They have five toes on each paw, and each toe has a **retractable** claw. This allows the cat to place the nail outside or inside the toe. The cat has four long toes and one dewclaw. Many cats have extra toes, which means they are **polydactyl** (see Figure 9–3). The pads of the feet are soft layers of flesh that serve as cushions on the soles of the feet, which reduces the noise when a cat stalks its prey.

The cat's body and structure are designed for speed and quickness. Cats have an average of 250 bones in the body (see Figure 9–4) along with 500 muscles, the largest of which are in the rear legs to aid in climbing and jumping.

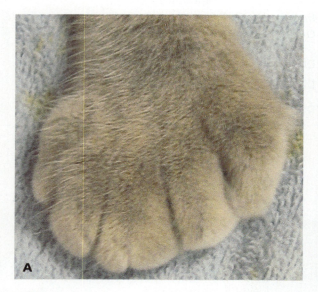
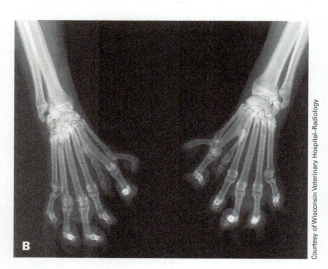

FIGURE 9–3 (A) A polydactyl cat has extra toes. (B) A radiograph of a polydactyl cat.

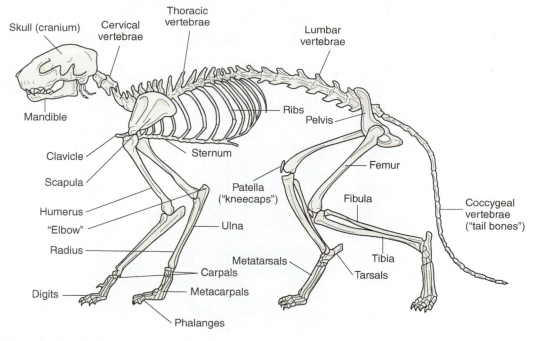

FIGURE 9–4 The skeletal structure of the cat.

Breeds

The **Cat Fanciers' Association (CFA)** recognizes more than 40 breeds of cats (see Table 9–2). The CFA is a breed registry that promotes the health and responsible breeding of cats and sets forth rules and regulations based on an individual breed's health, body structure, and physical beauty. Cats that are not registered and have unknown parentage are usually called **domestics** (see Figure 9–5).

There are a variety of different hair coat types, but most cats are distinguished by their **long hair**, **medium hair**, or **short hair**. Shorthair breeds tend to be low maintenance. Longhair breeds are high maintenance: they require daily brushing and routine grooming and are more susceptible to **hairballs**,

FIGURE 9–5 A domestic cat.

which are clumps of hair ingested from grooming that build up and collect in the digestive tract. Hairballs can cause obstructions and may lead to a severe health problem.

Cats come in a variety of sizes and range from 5 pounds to 25 pounds as an adult weight.

Breed Selection

Clients may ask a veterinary assistant for help choosing a cat. There are many factors to consider: Does the client want a kitten or an adult cat? Kittens will require more training (see Figure 9–6). Is the client interested in a purebred cat or in adopting a cat from a local shelter? Other considerations include the following:

- Indoor or outdoor cat
- Longhair or shorthair cat
- Grooming needs
- Health care needs
- Breed health problems
- Nutritional needs
- Breeding or show purposes

The best advice to offer clients is that they research several breeds and visit local animal shelters to determine what type of cat will best suit their lifestyle. Keep a list of reputable cat breeders in the area to share with clients. Offer reference materials to help future cat owners become educated and responsible pet owners as well. The CFA provides breed information on their website, cfa.org.

TABLE 9–2 Cat Breeds

SHORTHAIR BREEDS	
Abyssinian	Havana Brown
American Bobtail	Japanese Bobtail
American Shorthair	Korat
American Wirehair	Manx
Bombay	Ocicat
Burmese	Oriental Shorthair
British Shorthair	Ragamuffin
Chartreux	Russian Blue
Colorpoint	Scottish Fold
Cornish Rex	Snowshoe Breed
Devon Rex	Siamese
Egyptian Mau	Singapura
European Burmese	Sphynx
Exotic Shorthair	Tonkinese

LONGHAIR BREEDS	
Balinese and Javanese	Norwegian Forest
Birman	Persian
Cymric	Ragdoll
Himalayan and Kashmir	Somali
Maine Coon	Turkish Angora

MISCELLANEOUS BREEDS	
American Curl	Selkirk Rex
Burmilla	Siberian
Chinese Li Hua	Turkish Van
LaPerm	

FIGURE 9–6 Kittens require more training than an adult cat does.

Nutrition

Cats require specialized nutrients in their diet. They should be fed a high-quality dry food that helps prevent tartar buildup. Cats should never be fed dog food: the nutrient needs of both species are very different. Cats require a high-protein, high-fat diet that includes the nutrients taurine and thiamine. Cats need a high-protein diet because they break down protein very slowly. And as they do not conserve amino acids, they cannot use them for additional energy. Cats need **taurine** for proper development and body system functions. **Thiamine** is a compound of vitamin B that promotes a healthy coat. Cat foods are available in dry, semi-moist, and moist forms. Cats should not be given milk: it upsets the digestive system and causes diarrhea.

Feeding Kittens

Kittens require colostrum from the mother in the first 24 hours after birth. This will protect them from diseases until they begin to receive vaccines. Kittens will nurse from the queen for three to four weeks. At about 4 to 5 weeks of age, kittens should be offered canned or moist food. Over the next two weeks, they should begin dry food that is moistened with warm water. By 6 to 8 weeks of age, the kittens should be weaned and on solid kitten food. They should have several meals each day. It is important to make sure that you give a kitten a size-appropriate diet. A kitten's diet must include the following:

- Protein: 35 to 50 percent
- Fat: 17 to 35 percent

A kitten's energy needs are identified as follows:
- 2 months of age—175 kcal/day
- 3 months of age—260 kcal/day
- 6 months of age—280 kcal/day

Kittens that may be orphaned should be fed a commercial formula, such as KMR or pasteurized goat's milk. Kittens require 30 mL per pound body weight daily and can be given three to four feedings a day until they reach 4 weeks of age. Kittens may be fed with a nurser bottle or a syringe (see Figure 9–7). Record a kitten's daily weight to determine whether the kitten is growing. To increase its body temperature, a kitten should be placed on a heating pad for the first two weeks of life (see Figure 9–8). At about 4 to 5 weeks of age, a kitten can begin eating a soft, solid kitten food diet. The diet can be further softened with warm water, if needed.

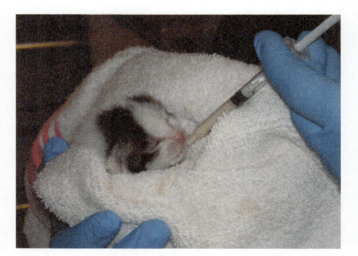

FIGURE 9–7 Orphaned kittens can be fed formula using a syringe.

FIGURE 9–8 Kittens should be kept warm by placing them on a heating pad.

Feeding Adult Cats

Adult cats should be fed a dry diet in a ration according to their ideal weight range. Healthy cats between 1 and 7 years old should be fed a maintenance diet. Senior cats should be fed a senior diet that is reduced in calories. Overweight cats should be fed a weight management diet. Adult cats should be fed twice daily. Some specialized cat foods to consider include tartar control to prevent buildup on the teeth and hairball formula diets to aid in digesting hairballs in the digestive tract. Adult cats on a maintenance diet require about 280 kcal/day. Pregnant queens should be fed **ad lib**, or as much as they want, with food available at all times during the last 20 days of gestation. This can continue post-queening until the kittens begin weaning. During weaning, the queen should have her food reduced by one-quarter rations until the ideal amount is reached. This will allow lactation to cease.

Adult male cats that are neutered and overweight may be predisposed to urinary tract issues. Careful attention to diet can improve an overall cat's urinary health by eliminating certain minerals and adding fiber and water as appropriate.

FIGURE 9–9 The happy cat.

Behavior

Cats have inherited many natural instincts from their wild cat ancestors, including hunting skills, marking behaviors, purring traits, and social skills. As the cat has been domesticated, it has also developed traits of companionship. Similar to the dog, a cat has three distinct behaviors. Each is displayed in the cat's body language. Most cats will display their behaviors relatively quickly. There is the happy cat, the angry cat, and the fearful cat. When handling cats, often "less is more"—meaning that it is may be helpful to begin with a minimum amount of restraint.

FIGURE 9–10 The angry cat.

The Happy Cat

Signs of a happy cat include rubbing against objects or people, holding the ears up and forward, holding the tail held high, purring and meowing, and appearing relaxed (see Figure 9–9).

The Angry Cat

The angry cat is likely to become aggressive. Signs of an angry cat include having a hair that looks puffed up, flicking or moving the tail rapidly, hissing, growling, flattening the back on the head, and having the eyes wide and pupils **dilated** (widened) (see Figure 9–10). Cats may become aggressive for several reasons. They may show territorial aggression (this is similar to what dogs do, but much more pronounced in cats). Cats are very protective of their territory, especially in multiple-cat households. **Intermale aggression** is common in adult male cats. These behaviors may be due to sexual dominance or social status. The occurrence is higher in reproductively intact males than those that are neutered. **Defensive aggression** commonly occurs in the veterinary facility. This is a how a cat protects itself. Cats will display this behavior toward people or other animals. **Redirected aggression** is a certain behavior

FIGURE 9–11 The fearful cat.

toward another animal or person that didn't provoke the cat in any way. This is most likely due to the social status of the cat.

The Fearful Cat

The fearful cat is often seen in the veterinary facility. Cats that are scared have a tendency to turn aggressive. Signs include having a lowered body position, tucking the tail under the body, flattening the ears on the side of the head, moving the hair up over the body, hiding, and widening the eyes (see Figure 9–11). When cats become upset, they will try to scratch and bite to protect themselves.

Basic Training

Cats should be **litter trained**: this means they will use a **litter box**. Place the cat or kitten inside the litter box frequently and help them make a scratching motion. Place several litter boxes throughout the house and in quiet and private spots. Cats need an isolated area when eliminating. The litter boxes should be cleaned daily and fresh litter provided on a routine basis.

Inappropriate Elimination (House Soiling)

Cats may eliminate or spray areas to mark territory. Many cats begin eliminating in the house due to issues with the litter box. It is important to determine if a cat is having difficulties accepting the litter box and possibly its location in the house. This is a behavior problem in which the cat chooses not to use the litter box for any one of a variety of reasons, choosing an alternative area for elimination of urine, feces, or both instead. Affected cats simply avoid the litter box and select a quiet spot behind a chair or in the corner of a room. A cat may dislike the litter box for the following reasons:

- Too few litter boxes
- Inappropriately positioned boxes (damp cellar, high-traffic area)
- Inconvenient location (basement, second level)
- Hooded box (most cats dislike hoods)
- Dirty box (not scooped/cleaned often enough)
- Clean box (cleaned with harsh-smelling chemicals, such as bleach)
- Liners (some cats are intimidated by plastic liners)
- Plastic underlay (convenient for the owner but not always accepted by the cat)
- Wrong type of litter (crystals or pellets)
- Litter not deep enough
- Animosity between cats in the house (competition/guarding)
- Difficulty getting in/out of the box, especially for elderly, arthritic cats

Cats with soiling problems can sometimes be easily accommodated; at other times behavior modifications, such as behavior or anxiety medications, must be prescribed by a veterinarian. It can be difficult to determine the causes of anxiety-related problems and they may require additional therapy and treatment. Stress can be a large part of anxiety. Stressors such as moving, changes in routine or in the makeup of the family, or holidays can result in inappropriate elimination. Reducing these stressors or decreasing their impact on the household will benefit the cat.

Equipment and Housing Needs

Cats should have a food and water bowl in the household. Other equipment that is recommended for cat owners includes the following:

- Carrier
- Litter pans
- Litter
- Scratching post
- Toys

Restraint and Handling

The most important consideration when handling a cat is to avoid being bitten or scratched. Cats can cause severely infected bites and scratches; therefore,

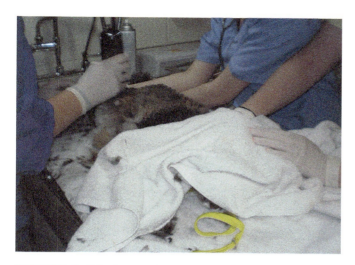

FIGURE 9–12 A towel can be used to help calm and control a cat.

FIGURE 9–13 A towel can be used to create a "kitty burrito" to restrain the cat.

care must be taken when restraining them. Veterinary assistants will need to be properly trained in cat restraint. Cats are quick and agile and can bite or scratch in an instant. You may use physical, verbal, or chemical restraints or a combination of methods. A towel is helpful in the handling of a cat. Cats feel less threatened when they are hiding under a towel; the towel can serve as a barrier if the cat becomes difficult to handle (see Figure 9–12). A towel comes in handy when you need to remove a cat from a cage: gently place the towel over the cat or, in certain circumstances, throw the towel over the cat. The towel can also be used to wrap the cat in a **kitty burrito**. The kitty burrito is made by placing a towel on a table and placing the cat on the towel. The sides of the towel are wrapped around the cat to contain the legs and body (see Figure 9-13). The head is left outside the towel and gives the appearance of a shell surrounding the cat. This allows better control of the cat to access and control the head.

Another way to control a cat is with the use of a **cat bag.** The cat bag is similar to the kitty burrito but has zippers to provide access to the cat's limbs or body parts in order to perform certain procedures (see Figure 9–14). This allows better control of the head and legs, preventing injury. To insert the cat into the restraint bag, open the bag wide and lay it out on a table. Place the cat in the middle of the bag and quickly wrap the neck strap securely around its neck, just snug enough to stay in place but not too tight as to block the airway. Then, holding a hand over the back and shoulders, zip the bag. Be careful not to catch any skin or hair in the zipper. The cat's limbs can be placed in the zippered openings for exam and venipuncture procedures.

Cat muzzles are similar to dog muzzles. They are placed over the entire face of the cat and typically cover the eyes, nose, and mouth. The cat can breathe through a small hole at the end of the nose area. Cats are more relaxed when they cannot see anything: this gives them a sense of security. Like the dog muzzle, the top of the cat muzzle is shorter than the bottom. Cat muzzles are closed with a Velcro strip. To best apply, stand to the back or the side of the cat (see Figure 9–15). Cats will try to scratch off a muzzle, so take care to control the legs. Care should also be taken

162 SECTION II Veterinary Animal Production

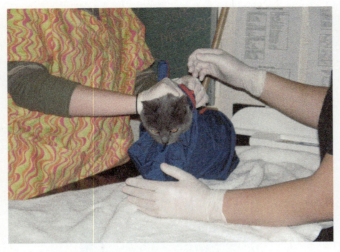

FIGURE 9–14 A cat bag can be used to restrain the cat while still allowing access to limbs (when necessary).

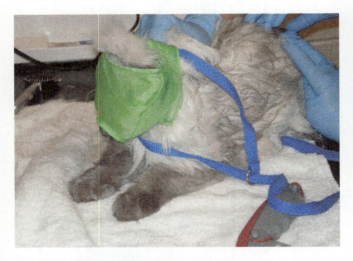

FIGURE 9–15 The use of a cat muzzle helps calm the cat and keep it from biting.

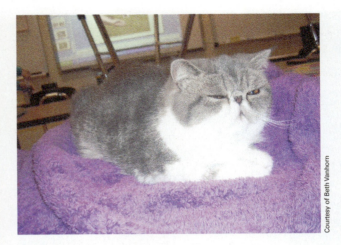

FIGURE 9–16 A brachycephalic cat breed.

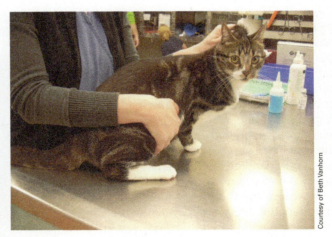

FIGURE 9–17 The scruff technique.

when muzzling brachycephalic cat breeds to ensure the cat is capable of breathing. *Brachycephalic* means a flat face that may prevent the animal from breathing properly (see Figure 9–16).

Cats can be restrained using a technique called the "**scruff**." The scruff is the loose skin over the back of the neck. Grasp the scruff firmly with the fist of one hand (see Figure 9–17). This will often cause the cat to become trancelike and will allow for better control of the cat's head and body. When "scruffing" the cat, hold the cat against a table to prevent injury. Use a free hand to hold the front legs of the cat to prevent scratches, if necessary.

Cats may also be placed in a **stretch technique**, in which the body of the cat is stretched. First, scruff the neck of the cat. Place the cat on its side in lateral recumbency, keeping a firm grasp on the scruff. Hold the rear limbs to avoid being scratched and to control the body (see Figure 9–18). This position often immobilizes the cat and keeps it quiet and still.

Distraction techniques may also be helpful when restraining cats. Some useful distraction techniques include blowing air onto a cat's face, scruffing while lightly bouncing the cat, and using gentle pats over the cat's head. Blowing small puffs of air directly into a cat's face can act as a distraction method using various amounts of air. The scruff technique along with light rapid bouncing can also easily distract a nervous or anxious cat. This can help the cat concentrate on something other than what procedure is being done to it. Small gentle pats over the head or shoulder

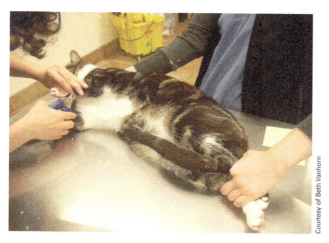

FIGURE 9–18 The stretch technique.

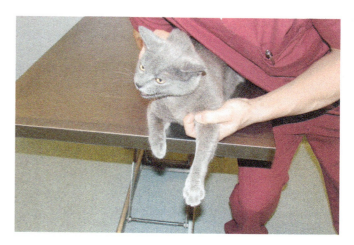

FIGURE 9–19 Restraint for cephalic venipuncture.

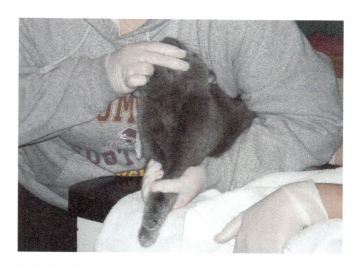

FIGURE 9–20 Restraint for jugular venipuncture.

area can also serve as a distraction. The pats can be rapid or slow with varying pressure based on how the cat responds to restraint. Pheromone products, such as Feliway spray or wipes, may also help in aiding to calm down a stressed-out cat. Fear-free and low-stress handling are recommended by the American Association of Feline Practitioners. Some examples of fear free methods include the process of traveling to the veterinarian often sets the stage for intensifying anxiety. One of the most powerful ways cat owners can reduce this stress buildup is to teach their cats to love their carriers. Leave the carrier out and place bedding and safe cat toys inside well in advance of a scheduled veterinary visit, so that when the time comes to head to veterinarian, the cat will already have a positive association with the carrier. The goal for the visit is prevent and alleviate fear and stress with respect of the cat's emotional well-being. This can be utilized by learning pressure points in the body, and alleviating odors, such as those from dogs, on the clothing, hands and other objects within the facility.

For a cephalic venipuncture restraint, scruff the cat with one hand to control the head and use the free hand to reach around the opposite side to extend the front leg (see Figure 9–19). Pressure can be applied around the limb to aid in blood collection. Sometimes the cat will sit quietly with less restraint if you restrain the cat under the jaw.

The jugular venipuncture restraint for a cat resembles the way you handle that for a small dog. Grasp the cat under the jaw and extend the head upward (see Figure 9–20). Use the free hand and arm to hold the front limbs to avoid getting scratched. Extend the front limbs over the edge of a table.

The saphenous venipuncture restraint is done using the scruff technique with one hand in combination with the stretch technique. A karate chop pressure hold is placed over the limb to allow the blood collector to visualize the vein.

Cats may also be restrained using a squeeze chute or an anesthesia chamber. The **squeeze chute** is a wire cage that restrains **feral** (wild) cats, rabies suspect cases, and very aggressive cats that are too difficult to handle. The boxlike cage holds a cat in place for examinations and procedures. The wire sides of the cage slide open, allowing the cat to be pulled inward against the side of the cage. Vaccinations or sedative injections to further restrain the cat may be while they are in the squeeze chute. The **anesthesia chamber** is used in a similar way. A cat may be placed in this glass enclosure for examination or further sedation. The chamber is an aquarium-like box with openings at the top that can be hooked up to an anesthesia machine. Anesthesia can be filtered in for further restraint. **Welding gloves** are also often worn in the act of restraining animals in a veterinary facility. The gloves are made of layers of leather that protect the hands and arms when handling aggressive cats. Because the gloves are bulky, be careful not to handle the cat too harshly, possibly causing injury. Bear in mind that the cat may be able to escape if the restrainer loses his or her grip.

Removal from a Carrier

Most cats will not willingly walk out of a carrier. Removing them may be done in one of two ways. A cat that is overall easy to deal with once it is removed from the carrier may be physically removed from the carrier by the "dump" method. Often, reaching in and grasping the cat gently by the body or the scruff causes the cat to hold onto the carrier. In the dump method, you simply elevate the rear of the carrier and let the cat slide out. If this doesn't work, or if the cat is aggressive or fearful, the best method to remove the cat is by taking the carrier apart. Most carriers are dismantled easily and quickly. Once the carrier is undone, place a towel over the cat and use the necessary restraint to remove it.

Grooming

Cats have the same grooming needs as dogs. Longhair and shorthair cats should be brushed on a daily basis using a medium- to soft-bristle brush, especially if they have medium to long hair (see Figure 9–21). Use a flea comb on a monthly basis to inspect for fleas and flea dirt. Cats should be bathed on a routine basis (see Figure 9–22). Cats may also be groomed in the same way dogs are. Longhaired cats may need trimming or shaving, especially if the hair coat develops mats (see Figure 9–23). Cats may be given a traditional bath with shampoo and water, or if the cat tends to be difficult, a waterless shampoo product may be used (see Figure 9–24). **Waterless shampoo** is applied to the hair coat and worked into the skin and allowed to dry. It is not rinsed out. The shampoo naturally disinfects the hair coat. Many cats are not cooperative at bath time and caution should be taken in order to avoid getting bitten or scratched. It is best to trim a cat's nails before bathing to reduce scratches.

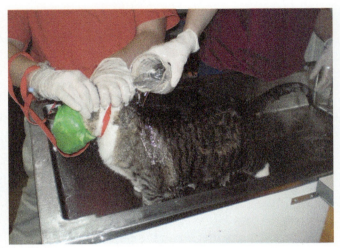

FIGURE 9–22 Cats should be bathed routinely.

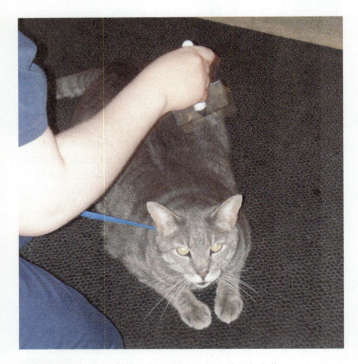

FIGURE 9–21 Cats should be brushed daily.

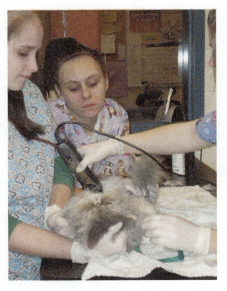

FIGURE 9–23 Longhair cats may need to be shaved if they develop mats.

CHAPTER 9 Cat Breed Identification and Production Management 165

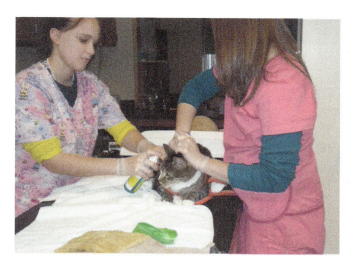

FIGURE 9–24 A waterless shampoo can be used on a cat that will not tolerate a bath with traditional shampoo and water.

tions that may be developing (see Figure 9–26). Every cat be tested for feline leukemia, feline AIDS, and feline infectious peritonitis. A simple blood test, the **SNAP test**, may be done in the veterinary facility (see Figure 9–27) to look for the antigens and antibodies of the diseases within the patient's bloodstream. Results are available in 10 minutes or less. It is important to know the health status of every cat that enters the veterinary facility.

A cat's ears, eyes, and teeth should be monitored on a regular basis (see Figure 9–28), as should their weight (see Figure 9–29). Veterinary assistants should be properly trained to perform basic grooming skills on cats, such as shaving hair, nail trims, cleaning ears, brushing teeth, and anal gland expression.

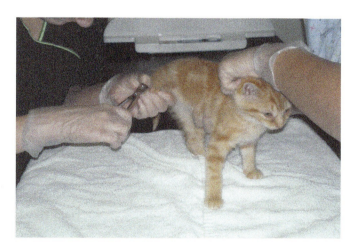

FIGURE 9–25 Trimming a cat's nails.

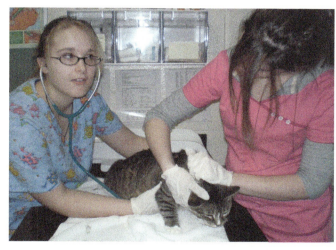

FIGURE 9–26 Cats should be seen by the veterinarian for a physical examination every year.

Brushing, nail trimming, ear cleaning, teeth brushing, and anal gland expression should be done on a regular basis as outlined in "Canine Production Management." Nails may be trimmed with cat-sized nail clippers or human fingernail trimmers (see Figure 9–25). Trim a cat's nails first before beginning any other procedure to reduce the likelihood of scratching injuries.

Basic Health Preventative Care and Maintenance

A cat should be seen by the veterinarian for a physical examination every year. This examination will evaluate the health status of the cat and any illnesses or condi-

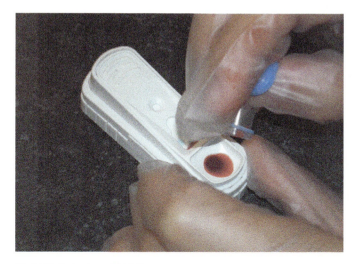

FIGURE 9–27 SNAP test.

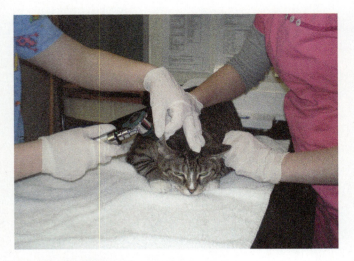

FIGURE 9–28 Checking the cat's ears.

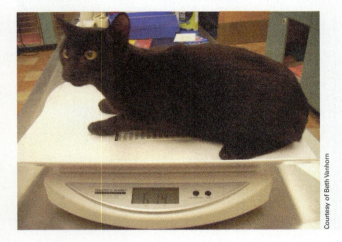

FIGURE 9–29 Monitoring the cat's weight.

Vaccinations

Similar to puppies and dogs, cats received booster vaccines that may be core or noncore based on the risk of disease for each cat. The following recommendations are based on AAFP and AAHA's vaccination standards guidelines for cats and kittens. Some vaccines are considered core vaccines and recommended for every cat regardless of environment and endemic areas. Noncore vaccines are limited for discussion with the veterinarian as to how high risk the cat may be and thus may not be necessary based on environment, health status and endemic regions.

Kittens should begin a vaccine program between 6 and 8 weeks of age (see Table 9–3). The most common feline vaccine series is the **FVRCP** combination, also known as the feline distemper series. Each letter stands for the disease it protects the kitten against.

TABLE 9–3
Vaccination Program for Cats

AGE	VACCINE
6–8 weeks old	FVRCP
10–12 weeks old	FVRCP
12 weeks old	FeLV
14–16 weeks old	FVRCP
16 weeks old	FeLV
16 weeks old	RV
Yearly to every three years	FVRCP, FeLV, RV

- FVR—Feline Viral Rhinotracheitis
- C—Calicivirus
- P—Panleukopenia

The FVRCP is given in boosters once a month three to four times until the kitten reaches 16 to 18 weeks of age. This vaccine is initially good for one year and thereafter may be given every years based on the veterinarian's protocol and risk of disease. The FVRCP should be administered subcutaneously (SQ) in the area of the right front shoulder. The FVRCP is a core vaccine.

A cat should be given a rabies vaccine between 12 to 16 weeks of age. The first initial rabies vaccine is legally good for one year; thereafter, it is good for one or three years based on the type of vaccine administered and state regulations. The rabies vaccine should be given intramuscularly (IM) or subcutaneously (SQ) in the right hind limb. The rabies vaccine is a core vaccine.

Some veterinarians may suggest other vaccines against such feline diseases as feline leukemia or feline infectious peritonitis. The feline leukemia virus vaccine, or **FeLV** vaccine, is given between 8 to 12 weeks of age; it requires one booster in three to four weeks and is then given annually. A negative FeLV/FIV test is required prior to vaccination and only negative cats should be vaccinated. The FeLV vaccine should be administered subcutaneously (SQ) in the left hind limb, preferably below the hock. The FeLV vaccine is noncore and recommended for at-risk cats.

Reproduction and Breeding

Female cats are known as **polyestrus**, which means that they have multiple heat cycles during a season. Typically, their heat cycle occurs every 15 to 21 days and will last for 5 to 7 days. Felines reach puberty between 5 and 8 months of age. Signs of a female cat in estrus include the following:

- Growling
- Behavior changes
- Vocal
- Very friendly and rubbing on objects
- Rolling and laying upside down (see Figure 9–30)
- Chewing on objects
- Assuming the **lordosis** position—prayer position of lowering the front of the body and raising the rear

Cats are **induced ovulators** and will become pregnant only after breeding has occurred. Gestation in cats is between 60–65 days with an average of 63 days. The labor process is known as "queening" or "kittening." Cats generally do not encounter problems during labor. Some breeders will provide a maternity box, but more female cats will find a dark, quiet location in the house to give birth, typically a bedroom or closet. Cats are private creatures, and noise should be kept at a minimum.

Signs of queening include the following:

- Restless and anxious
- Nesting behaviors
- Panting
- Stop eating 12 to 24 hours prior to labor
- Licking at vulva
- Body temperature decreases below 99 degrees

Cats should be monitored for dystocia, which is an abnormal or difficult labor. A normal labor process is front feet first followed by the head and body. The average labor process of queening typically lasts about 14 hours, from start to finish. The female will lick the kitten to stimulate breathing and to help dry the

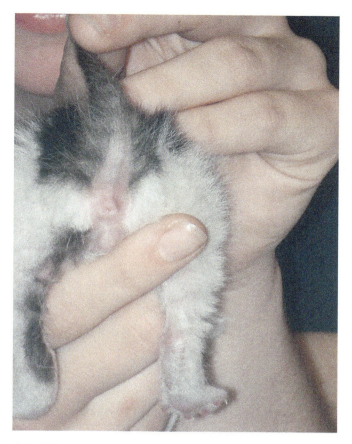

FIGURE 9–31 It is possible to determine the sex of a kitten within weeks of birth.

newborn. The kittens should not be handled for three to four days. A typical litter size is between three and four kittens. Kittens should be kept at a room temperature of 85 degrees for several weeks. Kittens' eyes and ears open in 10 to 14 days. Kittens can be sexed within weeks of birth (see Figure 9–31). Kitten gender can be determined by the distance of the anogenital area from the anus to the genital opening. Females have a shorter distance than do males. The female genital opening forms a slit, while the male genital opening is round. Males that have not yet been castrated may have already developed testicles.

Common Diseases

Cats may develop many types of infectious diseases. Some of the more common feline diseases seen in the veterinary facility include panleukopenia, commonly called feline distemper; rhinotracheitis, calicivirus, and the rabies virus. Several fatal diseases are of a concern for cats as well, including feline leukemia, feline immunodeficiency virus (FIV), and feline infectious

FIGURE 9–30 Rolling around or lying upside down is a sign of a cat in heat.

peritonitis (FIP). Clients of any cats diagnosed with one of these fatal diseases that are extremely ill should be counseled about treatment versus euthanasia.

Panleukopenia

Panleukopenia is also known as "**feline distemper.**" This systemic virus causes a decreased white blood cell count. It is transmitted through urine, feces, and direct contact. In kittens, it is usually fatal. The cause of this disease is a parvovirus infection. There is a vaccine available to prevent the disease. Signs of feline panleukopenia include the following:

- Vomiting
- Diarrhea
- Fever
- Abdominal pain
- Depression
- Dehydration
- Anorexia
- Seizures
- Death

Rhinotracheitis

Rhinotracheitis is a virus that affects the upper respiratory tract. It is spread through direct contact of saliva and nasal discharges. The disease onset is acute and is caused by a feline herpes virus. The disease typically runs a two- to four-week course until the cat shows signs of improvement. A vaccine is available for prevention. Signs of feline viral rhinotracheitis include the following:

- Nasal discharge (purulent or thick pus)
- Ocular discharge
- Sneezing
- Conjunctivitis
- Fever
- Drooling/salivation
- Anorexia

Feline Calicivirus

Feline **calicivirus** is a highly contagious virus that affects the upper respiratory tract. Cats that survive the disease usually have a permanent head tilt. The disease is spread through direct contact and body secretions. A vaccine is available for prevention. Signs of feline calicivirus include the following:

- Nasal discharge
- Ocular discharge
- Oral ulcers (stomatitis)
- Fever
- Lethargy
- Anorexia

- Pneumonia
- Head tilt

Feline Leukemia

Several fatal diseases occur in cats. These conditions are highly contagious; once a cat is diagnosed with the disease, there is no treatment. The **feline leukemia virus** is the most common of these diseases. It is a virus that is passed through direct contact and body secretions. A vaccine and test are available. Cats that test positive for the disease should be isolated from all other cats to prevent transmission. Cats may show signs of illness or they may be a **carrier** of the disease, which means they show no signs of disease but are still capable of passing the disease to another cat.

Feline Immunodeficiency Virus

Feline immunodeficiency virus (FIV) is a lentivirus that compromises the immune system. It is species-specific to cats but acts in the same manner as the human AIDS virus. It is highly contagious in cats and is typically spread through direct contacts and bite wounds. The immune system is greatly affected, and many cats die because of other illnesses. There is a test available as well as a vaccine, but the vaccine has shown not to protect cats against all strains of the virus. Cats that test positive should be placed in strict isolation. As with feline leukemia, cats may show any sign of sickness or be a carrier. Many cats affected with this disease develop chronic infections of the oral cavity, skin, or respiratory tract and have a chronic fever. A simple SNAP test may also be done in a clinic to determine a cat's FIV status (see Figure 9–32).

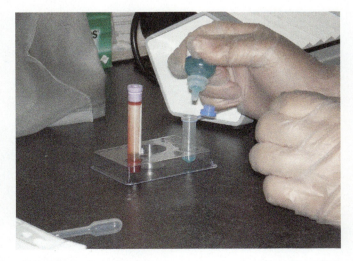

FIGURE 9–32 Conducting a SNAP test to determine FIV status.

Feline Infectious Peritonitis

Feline infectious peritonitis (FIP) is a highly contagious disease that may occur in a wet or dry form. This disease affects the lungs and chest cavity, causing severe respiratory problems that tend to lead to fatal pleural effusion. Pleural effusion is a buildup of fluid in the chest and abdominal cavities and is similar to pneumonia. **Pneumonia** is an abnormal condition that affects the lungs. There is a test and vaccine available.

Rabies

Rabies is also a fatal disease that may occur in cats. As in dogs, it is spread through saliva from bite wounds or scratches. It has a zoonotic potential, and state laws should be reviewed for cat owners. Feral cats are at a greater risk for rabies. Veterinary professionals should consider a pre-exposure vaccination against rabies for their patients.

Common Parasites and Prevention

Cats may become infected with internal and external parasites similar to dogs. Cats may become infested with fleas, ticks, lice, mange mites, and ear mites (see Chapter 8 for more information). The veterinary assistant should be able to identify these external parasites. They should also be familiar with products that are available to prevent and treat external parasites (see Table 9-4).

Cats may also become infected with internal parasites. Cats are susceptible to roundworms, hookworms, and tapeworms. Cats rarely develop whipworm infections. Cats also become infected with heartworms. Internal parasites of felines occur in the same manner as those in dogs. Refer to the "Canine Production Management" chapter to review the information on internal parasites. Veterinary assistants should be knowledgeable in products that are used to prevent and control internal parasites in cats (see Table 9-5).

TABLE 9–4

Feline Flea Product Comparison Chart

PRODUCT NAME	APPLICATION TYPE	PARASITES CONTROLLED	AGE OF USE	WATERPROOF OR RESISTANT	LENGTH OF APPLICATION
Advantage	Topical	Fleas	8 weeks	Water resistant	Monthly
Advantage Multi	Topical	Fleas, hookworms, roundworms, ear mites, heartworms	8 weeks	Water resistant	Monthly
Bravecto	Topical	Fleas, ticks	6 months	Water resistant	12 weeks
Capstar	Oral	Fleas	4 weeks	Waterproof	Daily or as needed
First Shield	Topical	Fleas, ticks, lice, biting flies, mosquitoes	7 weeks	Waterproof	Monthly
Frontline	Topical	Fleas	8 weeks	Waterproof	Monthly
Frontline Plus	Topical	Fleas, ticks, lice	8 weeks	Waterproof	Monthly
Program	Oral/Injectable	Fleas	4 weeks	Waterproof	Monthly-oral 6 months-Inj
Promeris	Topical	Fleas	8 weeks	Waterproof	Monthly to every 6 weeks
Revolution	Topical	Fleas, hookworms, roundworms, ear mites, heartworms	8 weeks	Waterproof	Monthly

TABLE 9–5
Feline Heartworm Product Comparison Chart

PRODUCT NAME	APPLICATION TYPE	PARASITES CONTROLLED	AGE OF USE	LENGTH OF APPLICATION
Advantage Multi	Topical	Heartworms, fleas, ear mites, hookworms, roundworms	9 weeks	Monthly
Heartgard	Oral	Heartworm, hookworms	6 weeks	Monthly
Interceptor	Oral	Heartworm, hookworms, roundworms	6 weeks	Monthly
Revolution	Topical	Heartworms, fleas, ear mites, hookworms, roundworms	8 weeks	Monthly

Cats and kittens should be placed on a regular de-worming prevention program. This includes regular fecal samples. Kittens should begin the de-worming process at the same time as their vaccination program begins, between 6–8 weeks of age. A fecal exam and de-worming product, such as pyrantel, should be done at each booster vaccine visit. Then a yearly fecal exam and de-worming should be initiated within the preventative care program.

Another parasite that affects cats and is zoonotic to humans and a large concern especially for pregnant women is called **toxoplasmosis**. Toxoplasmosis is a protozoan parasite that is passed in infected cat feces. An infected cat will shed the parasite in a 12–24 hour time frame. It is important for female clients who are cat owners to be educated about this parasite especially if they are pregnant. Pregnant women should not handle cat feces or clean litter boxes due to the possibility that the infection may cause spontaneous abortion or miscarriage in women or cause birth defects in the baby. Cats typically become infected with the organism through soil contamination.

Common Surgical Procedures

Veterinary assistants will routinely discuss feline surgical procedures with cat owners. (see Chapter 8, "Canine Production Management," to review the ovariohysterectomy (spay), castration, and declaw surgery). These surgical procedures are common in cats as well. The procedures are similar in both dogs and cats. In a **declaw** surgery in cats, the retractable claw or digit on the feet is removed. It is important that the veterinary assistant has a basic understanding of the procedures in order to discuss the benefits of the surgery with the client.

Spay

Cats are typically neutered at about 6 months of age. Some veterinarians will neuter cats at a much earlier age. The ovariohysterectomy (OHE/OVH) procedure is the same for dogs and cats. Female cats go in and out of estrus seasonally until they are bred. They are capable of having two to three litters each year. The OHE will decrease the cat population, reduce the attraction of male cats, and decrease the incidence of spraying and marking territory. The spay is performed under general anesthesia with a midline incision on the abdomen. The uterus and ovaries are removed and the incision is sutured. Sutures are removed in 7–10 days, with a limit on exercise.

The **orchidectomy** (castration) in the male cat is much less complex than in the male dog. The testicles are exposed, ligated, and excised. There are typically no sutures in a cat castration, so suture removal is not necessary. The castration procedure in cats decreases the likelihood of spraying, wandering for females, contracting diseases or wounds, and strong urine odor.

Onychectomy

Cats are often seen in the veterinary facility for a procedure known as **onychectomy**, commonly called a declaw surgery, which the surgical removal of the digits or claws. The procedure involves the amputation of the distal digit where the toenail begins. The procedure may be done in the front paws, rear paws, or both. After the digits are removed, the toes are either sutured closed or a glue-like **tissue adhesive** is applied to the area. The paws are then bandaged to control bleeding. The bandages are usually removed in 48 to 72 hours, depending on the age of the cat. Owners are instructed to keep their cat away from the litter in the litter box, as it will enter into the declaw sites and cause infection. Many owners purchase litter pellets made specifically for declaw surgeries or shred newspaper to place in the litter box. Cats are usually healed in two weeks. Declawed cats must be kept indoors, as they have a less likely chance of defending themselves outside. Some veterinarians will not perform this procedure, as it may be a painful and requires increased amounts of pain management. Cat owners should be counseled on the advantages and disadvantages of the declaw procedure. An alternative to declawing is to place plastic tips over the nail to reduce scratching and damage in the household.

SUMMARY

Cats have become popular companion animals. Veterinary assistants must be knowledgeable in feline health care and maintenance needs. Cats require an experienced restrainer, and the veterinary assistant should have a good working knowledge of normal and abnormal cat behavior. Veterinary medicine is becoming more specialized each day with feline-only practices.

Key Terms

ad lib   giving something, such as food or water, as often as necessary or so that it is available at all times

anesthesia chamber   a glass or plastic aquarium-like equipment with a lid and an opening that attaches to an anesthesia machine to sedate cats and small animals

calicivirus   a highly contagious virus that affects the upper respiratory system of cats and may cause a head tilt

carrier   a boxlike object that safely holds a cat for transport; also a cat that spreads a disease but shows no signs of having the disease

cat bag   restraint equipment used to hold and control a cat's body and limbs during certain procedures, such as a nail trim

Cat Fanciers' Association (CFA)   a cat breed registry association that promotes the health and responsible breeding of purebred cats

cat muzzle   restraint equipment placed over the face and mouth of a cat to prevent biting

declaw   the surgical removal of the digits or claws on the feet of a cat

defensive aggression   natural behavior that occurs when a cat is attempting to protect itself from people or other animals

dilated   the widening or opening of the pupils

domestic   a feline that is of two or more breeds and not registered

feline distemper   see *panleukopenia*

feline immunodeficiency virus (FIV)   a fatal virus in cats that attacks the immune system

feline infectious peritonitis (FIP)   a fatal virus in cats that has a dry and wet form and affects the lungs and chest cavity

feline leukemia virus (FeLV)   a fatal virus in cats that affects the white blood cells and immune system

Felis catus   genus and species of cats

feral   a cat that is wild but has a domesticated appearance

FVRCP a feline vaccine combination series also known as the feline distemper series; the acronym stands for feline viral rhinotracheitis, calicivirus, and panleukopenia

hairball a large amount of hair that collects in the digestive system of cats because of excessive grooming

induced ovulator when a female ovulates following mating with a male, as the release of an egg occurs at the time of mating

intermale aggression a behavior tendency that is common in adult male cats due to sexual dominance or social ranking

kitty burrito a restraint procedure in which a cat is wrapped in a towel

litter box a tray or pan filled with a pebble-like material used for elimination

litter trained the process of teaching a cat to eliminate in a litter box

long hair a cat's coat that is long and requires regular brushing and grooming

lordosis the prayer position of a female cat in heat; the front of the body is lowered, and the rear is raised

medium hair a cat's coat that is of medium length and requires some brushing and grooming

onychectomy the veterinary term for the surgical procedure that declaws a cat

orchidectomy a veterinary medical term for the removal of the testicles, also known as castration

panleukopenia commonly known as feline distemper; a systemic viral disease that causes a decrease in white blood cells in cats

pneumonia a lung infection in cats

polydactyl having many or multiple toes or digits

polyestrus having multiple heat or estrus cycles during a season

redirected aggression aggressive behavior toward another animal that did not provoke it but caused it to become upset

retractable capable of projecting inward and outward, as in the nail of a cat

rhinotracheitis a virus that affects the upper respiratory system of cats

scruff the loose skin over the base of a cat's neck that can be held to restrain the cat

short hair a cat's coat that is short and smooth

SNAP test a simple blood test completed in the clinic that tests for certain disease antigens in the bloodstream

squeeze chute a wire cage with sliding doors that is used to control a cat for procedures such as exams and injections

stretch technique a restraint procedure in which the loose skin of the cat's neck is held with one hand and the rear limbs are pulled backward with the other hand

taurine a nutrient required by cats that allows for proper development

thiamine a nutrient compound of vitamin B that promotes a healthy coat

tissue adhesive a glue-like substance used to close incisions

toxoplasmosis a protozoan parasite passed in cat feces that may cause abortion in pregnant women

waterless shampoo a cleansing product applied to the hair coat that does not require rinsing with water

welding gloves gloves of heavy layers of leather worn to restrain cats and small animals

Review Questions

1. Explain the differences between the cat and dog.
2. List the signs of the happy cat, angry cat, and scared cat.
3. Why is it important to be experienced and knowledgeable in cat restraint methods?
4. What factors should be considered when choosing a cat?
5. What are the signs of a female cat in estrus?
6. What are the nutritional requirements of the cat?
7. What should an ideal feline vaccine program include?
8. What fatal diseases of cats should clients be educated about?
9. What are some common internal parasites that occur in the cat?
10. What common surgical procedures are performed in cats?

Clinical Situation

A 3-week-old kitten is presented to the veterinary facility. The people who brought the kitten to the clinic found it this morning and were concerned because it seems so young. They gave it milk from their refrigerator, but it didn't want to eat. It appears bright, alert, and vocal. The veterinarian examines the kitten and feels it should be placed on a commercial formula and fed three to four times a day. When the veterinarian leaves the room, the owners begin questioning the veterinary assistant, Anne.

"We weren't expecting this today. We don't have any pets and don't want a kitten!" says the man.

Anne replies, "Dr. Smith went to get you some formula. We'll show you how to feed the kitten. It's easy and shouldn't be much work."

The woman says, "I don't think we can afford a kitten. Plus, we are planning on going out of town next weekend. I just don't think we can care for the little guy."

- What would you do in this situation?
- What are some options to discuss in this clinical situation?

CHAPTER 10: Avian Breed Identification and Production Management

Objectives

Upon completion of this chapter, the reader should be able to:

10.1 Define common veterinary terms relating to avian species
10.2 Describe the biology and development of avian species
10.3 Describe the classes of birds
10.4 Identify common breeds of exotic avian species
10.5 Discuss the nutritional requirements of birds
10.6 Describe normal and abnormal behaviors of birds
10.7 Describe how to properly and safely restrain small and large birds for various procedures
10.8 Discuss breeding and reproduction in birds
10.9 Discuss the health care and maintenance of avian species

Introduction

Birds are one of the animal species that have only been domesticated as companion animals in the last 50 years. They are also the third-most popular pet in the United States (see Figure 10–1). Birds are considered exotic and make excellent pets in homes or apartments. Each breed of bird has different needs and requirements. Proper care and health of a bird requires special equipment and facilities. For this reason, the veterinary assistant should have a basic understanding in and knowledge of the avian species.

Veterinary Terminology

There are a few terms that relate specifically to the avian species. It is essential that the veterinary assistant be familiar with a few terms and know when to properly use them in the facility (see Table 10–1).

Avian Biology

A bird is an animal that has feathers, two legs, and two wings. Many birds can fly, but not all breeds do. Birds have a **beak** that serves as a mouth. Birds stand upright on two limbs, and each limb has four toes for grasping and climbing. Some birds are very small (the hummingbird), while others are very large (the ostrich). This chapter mainly discusses veterinary medicine related to pet caged birds.

A bird's external body parts are very different from other companion animals (see Figure 10–2). The structure of the bird and its terms vary greatly. The veterinary assistant should be familiar with the terms and their locations shown in the diagram of the bird's external body parts (see Table 10–2). Birds have different types of feathers that

CHAPTER 10 Avian Breed Identification and Production Management   175

FIGURE 10–1 Birds are a popular pet in the United States.

TABLE 10–1

Avian Veterinary Terms

AVIAN SPECIES	BIRD
Hen	Female bird of breeding age
Cock	Male bird of breeding age
Brood	Young birds in a nest at one time
Clutch	Total number of eggs in the nest
Chick	Newborn bird
Juvenile	Young bird that has not yet developed feathers
Fledgling	Young bird that has left the nest but is not eating on its own
Weanling	Young bird that has left the nest and is feeding on its own
Flock or company	A group of birds

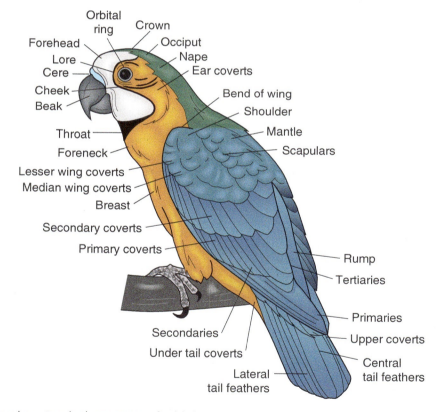

FIGURE 10–2 External anatomical structure of a bird.

play a role in flight, body covering, camouflage, and regulating body heat. Types of bird feathers include:

- Contour feathers: wing and body feathers used for flight and body cover
- Primary flight feathers: wing feathers located on the outer end of the wings
- Secondary flight feathers: wing feathers located between the body and the primary feathers
- Body feathers (coverts): feathers covering body for surface cover
- Down feathers: soft, fluffy feathers that provide insulation and may create a dusty white powder
- Blood feathers: new feathers that are developing and have a blood supply at the shaft

The feathers of a bird have anatomical parts, such as the barb, barbule, and calamus (see Figure 10–3).

176 SECTION II Veterinary Animal Production

TABLE 10–2
Common Avian External Anatomical Terms

Barb	Individual section that projects from the wing and forms the feather
Barbule	Small projection on the edge of the barb of a feather
Beak	Hard structure that forms the mouth of a bird
Breast	Front area of the chest
Calamus	Quill of the feather that attaches to the skin
Cere	Thick skin at the base of the beak that holds the nostrils; may be in different colors according to the gender
Crown	The top of the head
Feather	Similar to hair parts that form the wings and allow flight in certain species
Keel	Breast bone
Nape	Area at the back of the head
Orbital ring	Area that encircles the eye
Primary feathers	Lower feathers of the wing that form the first row
Secondary feathers	Upper feathers of the wing that form the second row
Shaft	Part of the feather where the barbs attach
Tail feathers	Feathers that form at the lower back of the bird
Talons	Claws or nails of the feet
Throat	Area below the beak
Wings	Formed by feathers on both sides of a bird and serve as arms

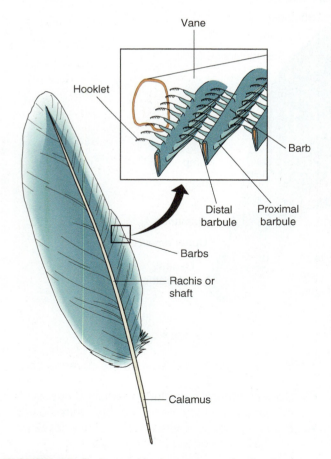

FIGURE 10–3 Anatomical structure of a feather.

The anatomy of the feather is important when carrying out restraint procedures and clipping the wings. A bird's health can be determined by examining the external parts of the bird for signs of disease.

Birds are **omnivores**: they eat both meat and plant sources. Some breeds in the wild are considered **herbivores**: they only eat plant sources. Birds in **captivity** (living in a cage and cared for by humans) learn to eat many types of foods. Birds have a specialized digestive system known as an avian digestive tract (see Figure 10–4). They have specialized organs that hold food and then break it down to make it easier to digest. Their digestive system includes the **crop**, which is located below the esophagus and acts as a holding tank for food (see Figure 10–5). The **gizzard**, located below the crop, acts as a filter system that breaks down hard foods, such as seed shells or bones. The **proventriculus** is a tube that connects the crop to the gizzard. Food moves through the intestines and passes waste materials via the **cloaca**, which serves as the rectum. The cloaca, also called the **vent**, is the external area of the digestive system.

A bird may have a long life span. Some larger species have been known to outlive humans. Birds reproduce by laying eggs.

The skeletal system of birds is somewhat similar to other companion animals (see Figure 10–6). The two legs of a bird act are equivalent to the rear limbs in such animals as cats and dogs. A bird's two wings are

CHAPTER 10 Avian Breed Identification and Production Management 177

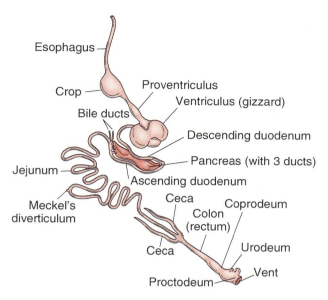

FIGURE 10–4 Avian digestive tract.

FIGURE 10–5 A young chick with a full crop.

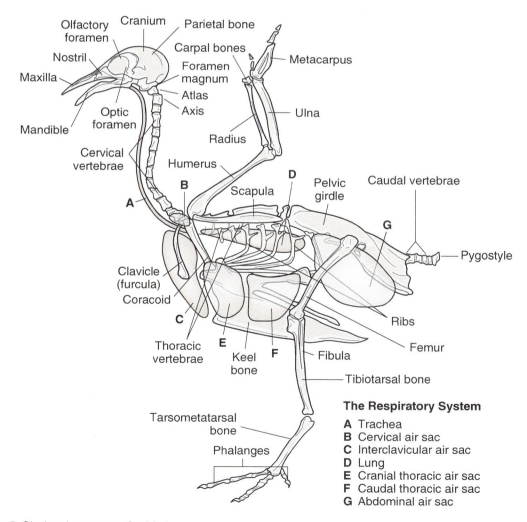

The Respiratory System
A Trachea
B Cervical air sac
C Interclavicular air sac
D Lung
E Cranial thoracic air sac
F Caudal thoracic air sac
G Abdominal air sac

FIGURE 10–6 Skeletal system of a bird.

equivalent to the front limbs of other animals. A bird's bones are hollow in the center and made up mostly of air. Birds have vertebrae bones in the neck and the tailbone only. A keel bone is located over the lower chest and abdomen area and serves as a breastbone to protect internal organs (see Figure 10–7).

The respiratory system of birds is very different from other animals. Air enters through the nares, located on each side of the beak over the **cere**, which is the fleshy area at the top of the beak. Air passes through the sinuses and into the **choana**, which is the opening to the oral cavity (see Figure 10–8). The choana is a V-shaped opening at the roof of the bird's beak that allows air to pass from the mouth and nasal area to the glottis. The glottis fits directly into the choana, creating a closed connection from the nostrils to the windpipe. Air moves down the trachea and into the lungs. Birds' lungs do not inflate, as they have no alveoli or lung lobes. In addition, birds lack a diaphragm: this means they must be able to move their chest muscles in order to breathe in and out. In order to breathe properly, birds must complete two breath cycles. The beak remains closed during breathing.

The heart rate of birds is 250 to 350 beats per minute in large species and up to 1400 beats per minute in small birds. The red blood cells in birds are different from those in mammals. In birds, the red blood cells are oval in shape and contain a nucleus.

Breeds

Birds are classified into 28 different orders. There are thousands of breeds of birds (see Figure 10–9). This chapter will discuss only the most popular breeds among breeders and bird owners. Each group has different characteristics, behaviors, attitudes, and a uniqueness that make them popular with people. The **psittacine** and **passerine** bird family breeds are the most common breeds kept as pets. Psittacine members include the larger parrot species that have a curved upper beak and four toes on each foot, two pointing forward and two backward. Passerine members include the smaller bird species with a beak that is pointed or only slightly curved and four toes on each foot, three pointing forward and one backward.

FIGURE 10–7 Keel bone of a bird.

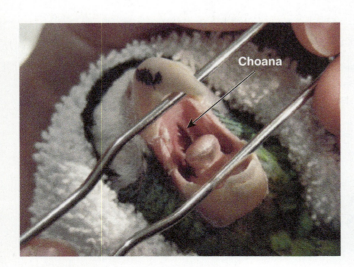

FIGURE 10–8 Choana of a parrot.

FIGURE 10–9 Gouldian finch.

Cockatiel

The **cockatiel** is a common breed, small in size and relatively easy to care for and train. This is a great beginner bird for someone just learning how to care for a member of the avian species. Adult cockatiels have a variety of feather colors, including yellow, gray, white, and orange (see Figure 10–10). Cockatiels originated in Australia. They are friendly, lovable, and appear to enjoy singing and whistling. Cockatiel eggs hatch in 25 to 26 days after being laid. The babies begin to mature at about 55 days old. Adult cockatiels eat cereal grains, fruits, seeds, and green foods. They do well on a high-quality commercial bird food. They range in size from 10 to 18 inches.

Parakeet

A **parakeet** is also known as a **budgie** or **budgerigar**. This small breed is easy to care for and excellent for beginners. Parakeets come in a variety of body colors, such as yellow, white, and blue (see Figure 10–11) and have black bars (lines) over the wing and tail feathers. They originated in Australia. Parakeets are very active. They develop extensive talking habits but tend to talk very rapidly. The eggs hatch in 25 to 26 days. They eat cereal grains, seeds, and fruits and do well on a high-quality commercial diet. Sizes range from 8 to 10 inches. They do tend to have some health problems.

FIGURE 10–11 Parakeet.

Finch

The **finch** is a very small breed whose appearance and colors vary (see Figure 10–12) and is a great beginner bird for breeders. Some species originated in Australia, others in Africa. Finches are easy to care for and raise. Eggs hatch in 12 days, and the birds begin to mature in 18 to 21 days. The males tend to be more vocal and have a superior singing ability. They eat seeds and small cereal grains. Adults range in size from 4 to 5 inches.

Lovebird

The **lovebird** is small in size and comes in a variety of colors (see Figure 10–13). The bird is easy to tame when young; adults are much more difficult to handle. They are

FIGURE 10–12 Society finches.

FIGURE 10–10 Cockatiel.

FIGURE 10–13 Lovebird.

excellent cage birds. Lovebirds originated in Ethiopia. They may sing but do not talk. They eat cereal grains, seeds, and fruits. Adults grow to be about 5 to 6 inches.

Canary

The **canary** originated in the Canary Islands, Spain, and is a small breed known for its solid bright colors (see Figure 10–14). The popular bird frequently sings. (The males tend to be better singers!) Canaries require larger cages because of their high-energy activity. Their eggs hatch in 14 days, and the birds begin to mature in 28 days. They prefer a seed diet with small amounts of green foods. Adults range in size from 5 to 7 inches.

Conure

The **conure** is a medium-sized active and playful bird that comes in a variety of colors (see Figure 10–15). They originated in South America. Eggs hatch in 23 days, and the birds mature in about a year. They eat cereal grains, fruits, vegetables, and green foods. Adults range in size from 12 to 14 inches.

Parrot

Parrots come in a variety of colors and sizes (see Figure 10–16). They are considered to be a medium-sized breed. Many originated in Central and South America. They are very active, playful, and

FIGURE 10–15 Conure.

FIGURE 10–14 Canary.

FIGURE 10–16 Parrots.

tend to bond to one specific family member. They are considered highly intelligent, are easy to train, perform tricks, and can develop an extensive vocabulary. Eggs hatch in 28 to 30 days, and the birds reach maturity in about a year. They eat cereal grains, nuts, fruit, vegetables, and pellets. They typically weigh several pounds and reach a variety of heights.

Macaw

The **macaws** is a large breed of bird and comes in a variety of bright colors (see Figure 10–17). Macaws are smart and very trainable. They are playful and very curious. They originated in Central and South America. Because of their size, they need a lot of room. They require more health care than smaller breeds. They are hardy but require protection from cold weather. Eggs hatch in 25 days, and adults mature in one to two years. They eat seeds, fruits, and green foods. Occasionally some breeds will eat small amounts of meat. They range in size from 15 to 36 inches.

Amazon

The **Amazon** is a large breed of bird that comes in a variety of bright colors (see Figure 10–18). Amazons originated in Central and South America. They are quite talkative and entertaining. They have social personalities and thrive on human company. They eat commercial pellets, fruits, and green foods. They need large cages and range in size from 15 to 20 inches.

FIGURE 10–18 Amazon.

Cockatoos

The **cockatoo** is a large breed of bird that typically has a white body color (see Figure 10–19). Cockatoos originated in Australia. They can develop a large vocabulary and tend to be loud—often screaming for attention. They are easy to train and thrive on human companionship. Eggs hatch in 25 to 30 days, and adults reach maturity in one to two years. They eat a variety of seeds, nuts, fruits, vegetables, and green foods. They need a large cage and range in size from 15 to 24 inches.

Table 10–3 presents an overview of bird breeds and their specific characteristics.

FIGURE 10–17 Macaw.

FIGURE 10–19 Cockatoo.

TABLE 10-3
Characteristics of Birds by Breed

BREED OF BIRD	WEIGHT (GRAMS)	AVERAGE LIFE SPAN (YEARS)	CHARACTERISTICS
Finch	10–15	5–10	Easy to care for and breed
Canary	15–30	10–20	Easy to care for
Parakeet	30–35	10–20	Easy to care for; great beginner bird
Cockatiel	80–90	10–20	Easy to care for and train
Lovebird	40–50	40–50	Taming and training more difficult
Parrot	400–500	60–70	Easy to train to talk and perform tricks
Macaw	1000–1400	60–70	Entertaining, can be trained to talk
Amazon	900–1200	60–70	Entertaining, enjoys talking
Conure	100–200	40–50	Easy to train and care for
Cockatoo	300–800	50–70	Very vocal; easy to train

Breed Selection

Before settling on a breed of bird for a pet, it is important to research the following points:

- Breed behaviors and characteristics
- Lifespan
- Cage needs
- Supplies and environmental needs
- Nutritional needs and diets
- Health care and veterinary needs
- Cost of bird
- Cost of care of the bird
- Common health problems or conditions

Cage birds have specific needs: a lot of attention, specific health care, and a variety of supplies. These items can be costly depending on the breed and size of the bird. Many new bird owners want a young bird, as older birds may have developed behavioral problems (such as biting or screaming). Young birds are easier to train and bond with new people. Adult birds have much more difficulty adapting to change. Choosing and purchasing a pet bird is a lifetime commitment and should be taken seriously. Some bird species can outlive humans; this must also be taken into consideration. The following websites may be helpful in researching various breeds of birds:

- Bird Breeds—www.bird-breeds.com
- BirdBreeders.com—www.birdbreeders.com

Nutrition

Birds eat a variety of different foods based on their breed and size and there are many commercial diets to choose from for a pet bird. The best type is a **pelleted diet**, which is very similar to what a rabbit eats. The pellets are small, pressed cubes that vary with the size of the bird. Each pellet contains the necessary amounts of protein, vitamins, and minerals to meet the bird's daily nutritional requirements. Another type of commercial diet is a **seed diet**. Seed diets come in a variety of types, colors, and sizes but are not the best nutritional choice, as they tend to be high in fat, low in calcium, and low in vitamin A. Many birds will select the best-tasting seeds, which tend to be high in fat but do not deliver the necessary nutrition. Many bird species require supplements of fruit and **green foods**. Green foods are non-dried vegetables with some juice substances still present. Examples include cabbage, carrots, fresh corn, peas, and bean sprouts. **Grit** is sometimes provided to a bird to help it break down foods—such as seeds and nuts—in the digestive system. The grit is a finely ground hard material similar to sand that rubs against food particles to make it easier for the bird to digest them. Pet birds should be provided with fresh food and water every day (see Figure 10–20). Do not feed a bird stale or spoiled foods. Take care not to feed a bird fresh vegetables or fruits that may have been sprayed for insects or weeds. When introducing a new food or diet to the bird, it is best to eat the food in front of the

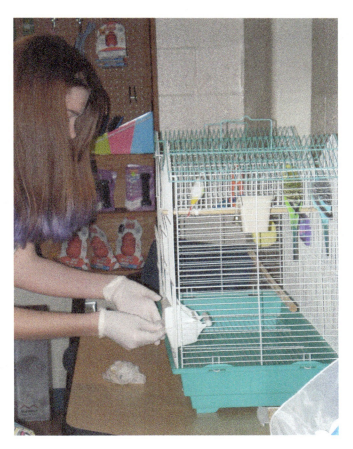

FIGURE 10–20 It is important to provide fresh food and water daily to birds.

bird and then offer the item. Birds learn by example! Some birds will prefer foods that are a certain color, substance, or temperature. Like dogs and cats, birds can easily become overweight. Smaller-size birds have a higher metabolic rate and thus need a higher-energy diet.

Suggested commercial diets include the following brands:

- Lafeber Diets
- Harrison's Diets
- Kaytee Diets
- Roudybush Diets
- ZuPreem Diets

Table 10–4 lists fruits and vegetables that are safe to feed pet birds.

These foods are unsafe and should not be fed to pet birds:

- High-fat junk food (potato chips, doughnuts, etc.)
- Avocado (guacamole)

TABLE 10–4
Safe for Birds

FRUITS	VEGETABLES
Apples	Broccoli
Apricots	Carrots
Banana	Cauliflower
Berries	Collard greens
Cantaloupe	Cooked sweet potatoes
Cherries	Cooled red potatoes
Cranberries	Corn
Grapefruit	Cucumbers
Grapes	Eggplant
Kiwi	Green beans
Mango	Kale
Oranges	Lettuce
Papaya	Peas
Plums	Radishes
Peaches	Tomatoes
Pears	
Pineapple	

- Chocolate
- Alcohol or caffeine
- Fruit pits
- Persimmons
- Table salt
- Onions
- Apple seeds
- Mushrooms

Behavior

Although birds have been domesticated, they still demonstrate many wild behaviors that influence how they interact with humans. In fact, many bird breeds are only one to three generations removed from the wild. To educate clients, the veterinary assistant must understand avian behavior basics—both normal and abnormal behaviors—and be knowledgeable about how training of the avian species. Table 10–5 lists factors that affect a bird's behavior.

Avian behaviors are designed for survival. In the wild, birds are natural prey and are at the bottom of the food chain and vulnerable to predators. They adapt behaviors for their own protection, including

TABLE 10–5
Factors Affecting Behavior in Birds

FACTOR	ITS EFFECT ON BEHAVIOR
Breed and genetics	Some birds are very vocal; some have gender appearances in color or characteristics
Socialization as a juvenile	Some birds will be easier to train; some adult birds will be less adaptable to change and to accepting a new owner
Rearing and environmental conditions	Daily handling will prevent birds from being territorial; some breeds will become territorial and difficult to handle
Post-weaning experiences	Learning to eat the proper diet, such as seeds, pellets, or green foods
Good training methods	Can be removed from the cage; can be carried; do not bite
State of health	Learning abnormal signs of bird health; birds mask signs of illness
Limited environment	Cage is too small; there are not enough toys; a bird becomes bored easily

biting, flapping their wings, and screaming (negative behaviors in a pet). In the wild, a flock of birds has a **pecking order**; each member has a social ranking. In a household, the bird will try to be the head. It is essential to use proper training methods to teach birds appropriate behaviors. Factors that lead to behavioral problems include the following:

- Over-bonding with one person
- Prolonged periods alone, absolute confinement
- No toys (which provide mental stimulation)
- Solitude
- Frequent environmental changes

It is crucial for bird owners to have a basic understanding of the behaviors of their particular species. The birds must be handled properly so they do not develop bad habits.

Need for a Routine and Stable Environment

Birds do not adapt well to change. In an unstable environment, the slightest change will cause the bird stress and behavioral problems. You may move the cage to a new location and the bird stops eating; you change its diet and it begins to bite; you have houseguests that the bird is not accustomed to and it begins screaming for attention.

Socialization and Stimulation

A bird has the intellect of a 5- or 6-year-old child. Emotionally, a bird is at the level of a 2-year-old. As do children, birds need boundaries, discipline, and guidance to learn appropriate behaviors.

One training method includes teaching a bird to step up and step down on command when inside or outside of its cage (see Figure 10–21). Besides following a command, the bird learns respect and that the owner is the leader. To teach the step-up process, hold one hand out vertically, like a perch. Then sweep that hand under the bird's feet while you say "Step up" and force the bird to perch (see Figure 10–22). Place the bird back in the cage with the command "Step down." Birds should not be allowed to climb over the body of the handler, as this can become a problem if the bird begins to bite. Place a thumb over the bird's digits to hold the talons to prevent it from climbing up your arm (see Figure 10–23). Birds that are being held should

FIGURE 10–21 Teaching a bird to step up and down off a finger upon command.

FIGURE 10–22 To teach a bird to step up, sweep the hand back toward the bird and force it to perch.

FIGURE 10–23 To prevent the bird from moving up the arm, put your thumb over the bird's digits to hold the talons.

be kept at chest level and never be allowed to sit on a shoulder or head. Dominant birds see that as the top of the tree, which gives them the greatest power in the flock.

Birds learn through repetition. Birds also need privacy. Perches should be placed low in the cage and kept below eye level to decrease the bird's feelings of dominance. It is recommended that the bird be placed on the bottom of the cage, which is their territory. During the night, the cage may be covered with a sheet or cloth so the birds feel secure, can sleep, and have some privacy. Some birds that are confined to the cage for long periods may become territorial and protective of their cage.

Vocalization

In the wild, screaming and other types of vocalization are normal behaviors for birds; in fact, screaming alerts the flock to danger. It is also a social ranking. In captivity, screaming is developed as an attention-seeking behavior—birds need frequent attention. Birds must adapt to a set schedule that rarely varies from day to day.

Loud screaming is best handled by placing a sheet over the cage. Do not attempt to scream or yell back at the bird. Birds enjoy a good argument, and your screaming only serves to reinforce the behavior. Instead, walk away and ignore the bad behaviors. Approach the cage when the bird is quiet and reinforce this good behavior with attention or a toy.

Biting

Biting is a defensive behavior. In the wild, birds bite to test a tree for perching suitability and to capture food. In captivity, biting is a way for a bird to get attention (see Figure 10–24). This can lead to further aggression if not dealt with properly.

Birds that bite should be handled cautiously. One technique to teach a bird perched on a hand that biting is not acceptable is called the "**earthquake**." The handler holds a thumb over the bird's toes to prevent the bird from moving up the arm or having too much freedom during handling. When the bird tries to bite, the handler quickly and carefully shakes the hand while holding on to the feet. The loss of balance distracts the bird. At the same time, the handler should say a firm "No!" each time the bird attempts to bite.

FIGURE 10–24 Biting is an instinctual behavior in wild birds but can become problematic in birds in captivity.

186 SECTION II Veterinary Animal Production

FIGURE 10–25 Using a perch can be helpful when working with birds that have a tendency to bite.

When handling a bird that bites, it is best to keep the cage and perch low to the floor, below the level of the human's head and shoulders. Birds that are kept in a high cage or that are capable of climbing up high have a feeling of dominance over humans, and this will increase behavioral issues. A perch can also be used to handle birds that bite to avoid injuries (see Figure 10–25).

Young juvenile birds are known to be "mouthy": they use their tongue to explore surfaces. Birds do not grow out of this phase. Rather, they tend to chew and grasp anything. Safety measures must be put in place.

Feather-Picking

Birds commonly pick their own feathers (see Figure 10–26). A bird that picks and bites its own feathers may also pull feathers out of its body. Some birds pick feathers out at a particular area of the body; others pick feathers over the entire body. Feather-picking should be evaluated by a veterinarian to rule out a physical problem, such as mites or other parasites.

Birds that have a behavior problem may engage in feather-picking out of boredom. Others may do it for attention. Sometimes stress leads to feather picking, and the bad habit becomes difficult to control or stop. It may not be possible to determine the cause of the feather-picking. An e-collar, or Elizabethan collar (similar to that used for dogs and cats), may be placed on the neck area of a bird to control feather-picking. Some birds become so destructive that they can traumatize their own skin, causing open sores that could become infected.

FIGURE 10–26 Feather-picking is a common behavior problem in birds.

Equipment and Housing Needs

Every bird needs a cage. The cage should be as large as possible (see Table 10–6). The cages should be wider than it is tall to accommodate the bird's flying and natural movements. Perches should be provided in the cage and should be of various heights and widths.

TABLE 10–6

Minimum Cage Size for Birds

SPECIES	CAGED PAIR LENGTH × WIDTH × HEIGHT	SINGLE BIRD LENGTH × WIDTH × HEIGHT
African gray	4' × 3' × 4'	3' × 2' × 2'
Amazon parrots	4' × 3' × 4'	3' × 2' × 2'
Budgerigars	24" × 14" × 8"	*
Canaries	18" × 10" × 10"	*
Cockatiels	4' × 2' × 3'	26" × 20" × 20"
Cockatoos	4' × 4' × 3.5'	4' × 3' × 4'
Conures	4' × 4' × 4'	4' × 3' × 4'
Finches	2' × 2' × 2'	12" × 12" × 12"
Lovebirds	4' × 4' × 4'	*
Macaws	6' × 6' × 6'	3' × 2' × 3.5'
Mynah birds	†6' × 3' × 3'	6' × 3' × 3'

*These birds prefer the company of other birds and should not be caged alone.

†The cage should be at least this size for a pair.

FIGURE 10–27 Offer a bird toys to keep it occupied and to alleviate boredom.

FIGURE 10–28 Do not place too much pressure over the bird's sternum during restraint, as the bird must be able to expand and contract its chest to breathe.

Newspaper is an appropriate floor material for the cage. Nothing toxic or harmful to the bird should be used as flooring. The cage should be placed in a draft-free area of the house where the bird will be near its owner. Toys can be provided to help keep the bird from becoming bored (see Figure 10–27).

Restraint and Handling

Veterinary assistants must learn the essentials of proper avian restraint. Birds are fragile animals and therefore must be handled with extreme care and with as little force as possible. Birds can become stressed over the shock of handling and die of heart failure if they are not accustomed to being handled. Small birds are more at risk of this. Before handling a bird, it is best to talk to it in a soft voice and place its cage in a quiet and low-lighted area for comfort. All doors and windows and escape routes should be closed. An appropriately sized towel may be helpful to aide in the capture and restraint. Using a towel to remove a bird from its cage will help reduce the bird's fear of human hands. When handling any bird, it is important not to apply large amounts of pressure over its chest or **diaphragm** (see Figure 10–28). The lungs and heart lie within the diaphragm, which encases the chest cavity. For the bird to breathe, its chest area must be able to expand (open) during **inspiration** and restrict (close) during **expiration**. This can be difficult to remember, as all other animals are capable of breathing while their chest area is being restrained.

It is important to note that medium- to large-breed birds that are being handled may become aggressive and bite if they feel threatened and are not used to people. Care must be taken to avoid severe injuries when handling certain birds. Large breeds, such as macaws and amazons, are capable of breaking or even amputating a finger. Their strong beaks can cause severe injuries. It is always a good idea to have a heavy towel at the ready when working with larger-breed birds (see Figure 10–29). Just as with cat restraint, it may be wise to use a pair of welding or heavy leather gloves when working with aggressive birds. Do not over-restrain the bird when wearing bulky gloves.

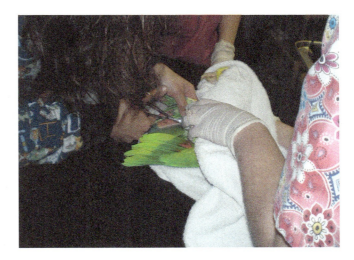

FIGURE 10–29 A towel can be used to control or restrain a large-breed bird.

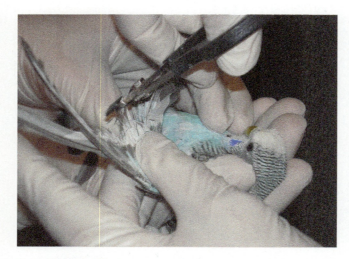

FIGURE 10–30 Restraint of a small-breed bird.

Small pet birds, such as the parakeet and cockatiel, can be picked up using one or two hands. This is done by grasping the bird from behind and securing both wings against the sides of the body. Use the thumb and forefinger to restrain the head and the other hand to cradle the bird's body (see Figure 10–30). Avoid the chest area. It is helpful to show clients how to train their bird to hop onto a finger (avoiding the need to pick up the bird). Small birds can also be restrained using a small hand towel or paper towel to prevent injury to the bird or restrainer.

Large birds can be held by perching them on your hand (if they are trained to do so). Always hold a thumb over the bird's toes to keep it from moving up the arm or flying. If a bird is unfamiliar with perching on a hand, cover your hand with a heavy towel. This will help protect your hands and arms as well as protecting the bird from injury. Never hold a large bird close to your face. Large birds can be restrained by holding the thumb and fingers on either side of the head to prevent biting. Place your pointer finger on the top of the head for added control and use your free hand to hold the wings at the sides of the body. The use of a towel makes this restraint much safer and easier. It is always wise to have a second person helping restrain a bird, especially in hospital procedures. In some cases, the veterinarian may opt to place the bird in an anesthesia chamber and sedate it for safer handling. Always remember to monitor a bird's respiratory rate and stress level the entire time it is being restrained.

You may need to remove toys and other items attached to the cage before attempting to remove the bird. This will make it easier to capture the bird during the restraint procedure. When you return a bird to its cage, it is important to place it on the bottom of the cage. This will keep it from falling from a perch or injuring itself by flying. After its return to the cage, a stressed-out bird may exhibit open-beaked breathing, increased respirations, wings held away from the body, and fluffing of the feathers. These signs should resolve shortly after restraint.

Grooming

Birds should have a daily bath, using a spray of water or a shallow bowl. This allows birds to keep their feathers clean, so they can then **preen**, or groom themselves.

Birds should be maintained with regular grooming practices, including nail, beak, and wing trims. Nail trims should be done with caution, as birds have the same type of "quick" or blood supply in their claws as dogs and cats. As the nail grows, the blood supply within the nail bed continues to grow, and the blood vessel is not visible in the nail bed, as it is in dogs. It is best to use a handheld **cautery unit** that burns through the nail bed and clots the blood. If this is not available, a cat nail trimmer or human nail clipper may be used; however, styptic powder or silver nitrate sticks should always be available to control bleeding.

A bird's beak continues to grow in the same fashion as its nails. Birds will require a **beak trim** based on their size and the type of foods in their diet. It is important to make sure the beak tips do not overlap or grow into each other, as this will prevent birds from eating properly. Beaks should be maintained using a dremel tool, which grinds down the edges of the beak (see Figure 10–31) and smooths the tip so it does not become sharp and overgrown. The smooth edges can be shaped along the natural lines of the beak. The sharp point of the beak's tip should be rounded to prevent injuries to

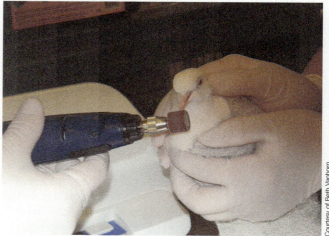

FIGURE 10–31 Beak trim.

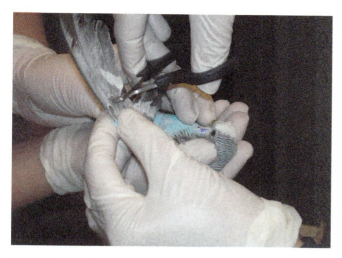

FIGURE 10-32 Wing trim.

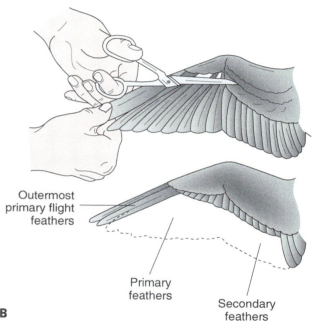

FIGURE 10-33 (A) Trimming the flight feathers. (B) Wing clipping in a pet bird.

the bird, reduce damage to the house, and lessen the risk of injuries to people handling the bird.

Wing trims should be done to prevent birds from taking flight (see Figure 10-32). This is important in households where pet birds are given freedom to fly about. Birds that take flight may fly into windows, fans, or walls, causing serious injury. They may also escape from the house and die from a lack of proper care in the wild. **Flight feathers** should be trimmed according to the size of the bird and how well it has adapted to flying (see Figure 10-33). The "flight feathers" are the adult feathers located over the edges of the wings. As these feathers begin to grow, they are called **blood feathers**, as there is a visible blood vessel at the end of the feather that allows it to grow. Caution should be taken to not trim a blood feather. Small breeds of birds can suffer severe blood loss from the trimming of a blood feather. Seasonally, adult birds will shed their flight feathers in a process called **molting**. Molting is similar to a dog shedding its winter coat. During this time, blood feathers will be present and wing trims should be avoided.

Basic Health Care and Maintenance

A bird needs several things to be healthy and stress-free. Birds do not adapt well to stress or change and should live in an environment that does not change. The cage should be at least 1½ times the bird's wingspan. The cage should include several perches so that droppings will not contaminate the food and water sources. Birds should be given toys with bright colors to keep them stimulated and prevent boredom. Birds that are bored tend to develop behavioral problems. Cages should be maintained and monitored, especially when they have several perches or many toys. It is not uncommon for birds to injure themselves on items inside the cage. Wings can be damaged, and toenails can be torn off (see Figure 10-34).

Birds should receive a yearly physical examination from a veterinarian, just as other companion animals. The physical exam helps evaluate each body system and any underlying changes or diseases that may be affecting the bird (see Figure 10-35). Birds have adapted in the wild by masking signs of illness or disease, so a proper physical examination will reveal any hidden diseases or health problems. Birds often appear healthy even when they are not feeling well. It is important to teach bird owners that when a bird shows

FIGURE 10–34 Birds can injure themselves if there are a lot of materials in the cage.

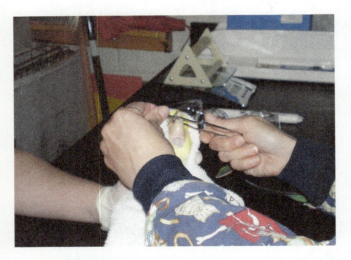

FIGURE 10–35 Birds should receive a yearly physical examination from a veterinarian.

The following items are dangerous to pet birds:

- Ceiling fans
- Cigarette smoke
- Teflon cooking pans
- Cleaning products
- Perfume
- Flowers and houseplants (poinsettias, holly, mistletoe, azalea, calla lily, philodendron)
- Lead and zinc
- Pressure-treated wood materials
- Cedar, wild cherry, and some oak woods

The following wood is safe for pet birds:

- Apple
- Maple
- Elm
- Ash
- Peach

Cages should be cleaned on a regular basis. This means that all areas of the cage must be cleaned, as bird droppings can accumulate anywhere in the cage (see Figure 10–36). The cage should be disinfected using a pet-safe cleaner.

Vaccinations

Vaccines for bird species are very rare. There is no specific vaccine program for birds (as there is for other animals, such as dogs or cats). Veterinarians will provide information to bird owners if they have a bird that is at high risk for specific diseases and suggest if the bird should be vaccinated.

FIGURE 10–36 Thoroughly clean all areas of the cage with a pet-safe cleaning product.

signs of not feeling well, this means it has been sick for a long time. As a result, many birds that show signs of illness do not recover well. Many veterinarians who specialize in aviary medicine suggest yearly blood work to determine a bird's internal health and changes in the bird's physical health due to this occurrence.

Veterinary assistants play a vital role in avian restraint and handling for hospital procedures and in educating bird owners on the needs and required health care of their pets. It is important to educate pet bird owners on the dangers that exist for pet birds. Many household items can cause injury or death to birds. The same goes for items placed in the bird's cage.

Fecal Exams

Fecal exams should be performed yearly during the annual physical exam. A fresh sample should be used and must include a fecal floatation as well as a direct smear. A gram stain will determine any bacterial overgrowth or yeast. A bird with a diet including fruit and grains should have gram-positive bacterial flora with occasional yeast. A bird with a diet including meat sources should have gram-negative bacterial flora. Budding yeast in any bird is abnormal. A cloacal swab can be collected for testing (see Figure 10–37).

Radiology

Most birds who are having a radiograph will require anesthesia to obtain proper restraint and positioning for diagnostic purposes. Standard whole-body ventral/dorsal (VD) views and right lateral views are necessary for proper evaluation. Plexiglas flat boards help to position the patient; when a bird is placed on its back with its wings stretched out to the sides, a piece of masking tape can be applied over the wings and limbs to secure the bird.

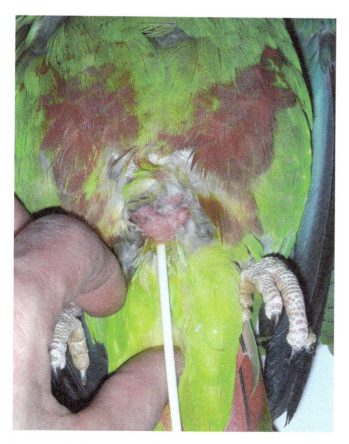

FIGURE 10–37 Cloacal swab.

Reproduction and Breeding

In general, birds reproduce by the females laying eggs. All female birds are capable of laying eggs, but a male must be present with the female to **fertilize** the eggs. If a male is not housed with the female and she lays an egg, it is called **infertile**. In some species, the male and female have the same characteristics and coloring. It is advisable to **DNA test** any birds that will be used for breeding purposes to determine their gender. The DNA test will determine if the bird is male or female. This is done with a simple blood sample or feather sample submitted to an outside lab.

When a male and female are breeding, the male will mate with the female several days before the female lays a fertilized egg. Some breeders will allow the eggs to remain in the nest and the birds to **incubate**, or keep the eggs warm until hatching, naturally. Other times, the eggs may be removed for **artificial incubation**. In this case, the eggs are placed in an **incubator**, a machine that keeps the eggs at a constant temperature until hatching. The best method is **natural incubation**, although some birds do not take care of the eggs and they may not hatch. Nests should be placed within the cage.

The incubation times of each bird breed vary from two to four weeks. It is recommended that bird owners who wish to breed research the type of bird they own and learn as much as possible about the reproduction needs and cycle of the bird. Some female birds will require diet supplements that increase the calcium intake to allow for egg production.

Basic and essential facts about breeding avian species include the following:

- The breeding pair must be mature, typically between 1 and 3 years old, depending on the species.
- Use caution when housing a male and female together; not all breeding pairs will bond.
- House the breeding pair in a large enough cage, with a nest or nesting box.
- Provide adequate food and water and necessary supplements for the female, such as a mineral block or calcium supplement.
- Provide a quiet, warm area with slight humidity.
- Provide a light source during the day and cover the cage at night.
- Clean the cage daily.
- Do not handle the eggs.
- Monitor the chicks as they hatch, making sure none are being pushed away.
- Monitor the chicks' weights daily until weaned.

One possible problem that breeding females may experience is known as "egg binding," or **egg bound**. This is when an egg does not pass through the reproductive system at a normal rate. It can occur in any female bird, regardless if a male is present. It occurs most commonly in small bird breeds and can be treated in the early stages. Some signs of the egg-bound female include the following:

- Abdominal straining
- Wide stance with legs
- Depression
- Loss of appetite
- Swollen abdomen
- Feathers fluffed
- Droppings stuck on vent
- Egg noted in vent

Respiratory Emergencies

A bird in respiratory distress may present for numerous reasons, including foreign body ingestion or aspiration, inhaled toxins, masses or tumors, or a respiratory infection. Normal respiration patterns in birds should be almost unnoticeable, with slow and regular breaths varying between 10 to 40 breaths/minute. Increased respiration can lead to trouble breathing, gasping for air, and tracheal collapse. Upper airway emergencies tend to be obstructions from ingesting foreign matter, such as seeds, toys, or pieces of material. Tumors and masses can also affect the upper airway. Lower airway problems include inhalation of toxic fumes, such as cigarette smoke, Teflon fumes, paint fumes, wood-burning stoves, scented candles, or perfumes.

Injuries and Fractures

Injuries to birds can occur in many different scenarios, such as bite injuries from other birds or animals, mishaps in the cage involving a perch or toy, flying into a window or fan, or being stepped on. Injured legs or wings may be splinted; in severe injuries, amputation may be the only option. Many birds do well after losing a leg or digit, as they often use their beak to help them move about. Wing amputations often take some adjustment because of a resulting loss of balance, but most birds seem to manage rather well. Broken wings typically take three to five weeks to heal and often are accompanied by muscle atrophy due to loss of the use of the wing. Limb fractures usually take four to six weeks to heal; the splint or bandage needs to be changed every one to two weeks. Circulation of the limb is important, and many birds will pick at the splint or bandage material and may require an avian e-collar.

Common Diseases

Avian species contract a variety of contagious and fatal diseases as well as internal and external parasites. The most common and important diseases include psittacosis, psittacine beak and feather disease (PBFD), fatty liver disease, and knemidokoptes mite infestation.

Psittacosis

Psittacosis is commonly called "parrot fever," or chlamydia. It is a disease that has zoonotic potential especially in young children and anyone with a poor immune system. In humans, it causes severe flulike symptoms. The disease is transmitted in the air or by direct contact. Birds shed the disease during times of stress and may be carriers of the disease. There are no specific signs of psittacosis: any abnormal appearance could be a sign of the disease. Diagnosis is made through blood testing. Prevention occurs through quarantine of all new birds and avoiding keeping birds in close quarters. This disease can be fatal to birds.

Psittacine Beak and Feather Disease

Psittacine beak and feather disease (PBFD) is a contagious and fatal viral disease among birds that affects the beak, feathers, and immune system. It is transmitted in the air and through direct contact or contamination. There are two forms, a short-term **acute** form and a long-term **chronic** form. Signs of PBFD include the following:

- Depression
- Diarrhea
- Weight loss
- Anorexia
- Abnormal feather development
- Beak growths and deformities
- Oral lesions
- Death

Fatty Liver Disease

Fatty liver disease, also called **hepatic lipidosis**, is a condition in which large amounts of fat are deposited in the liver. It is a serious condition that can result in death. Causes of the condition include a high-fat diet, overfeeding, nutritional deficiencies, hereditary factors, and toxic substances. A low-fat diet and medications are helpful if the condition is diagnosed early. Signs of fatty liver disease include the following:

- Sudden loss of appetite
- Lethargy
- Swollen abdomen
- Green droppings
- Obesity

Common Parasites

Bird species are also prone to both internal and external parasites, like cats and dogs. Certain parasites are more common in pet cage birds. Knowing the signs of parasites is helpful to the veterinary assistant.

Knemidokoptes

Knemidokoptes mites are commonly referred to as scaly leg and face mites. These external parasites occur on a bird's skin, beak, and feathers. The mites spend the entire life cycle on the bird, burrowing into its top layer of skin and are transmitted through direct contact and contamination of cages and equipment. White to gray lesions occur over areas of the skin, legs, face, and beak. Birds may show signs of itching and feather loss. Treatment is with ivermectin wormers. Indications of a sick bird include the following:

- Fluffed feathers
- Sitting on bottom of cage
- Depression
- Anorexia
- Lethargic
- Head tucked under wings
- Eyes closed

Common Surgical Procedures

Many bird owners do not typically have cage birds neutered. However, it is possible to spay a bird that may be a chronic egg layer or that becomes easily egg bound. Many avian veterinarians will have specialized skills in avian surgical procedures. General anesthesia is typically used with bird species. Sometimes sedation with general anesthesia is required for blood collection. Blood collection is commonly done at the jugular vein on the neck; in birds, it is located easily on the right side of the neck on a featherless tract, which is an area where feathers do not grow naturally. Once the sample is collected, it is important to apply pressure at the site of collection to prevent large hematomas from forming and possible blood loss.

SUMMARY

Birds have become popular companion animals. There are over 40 million pet birds in the United States. Veterinary assistants play an essential role in restraining and handling birds, educating bird owners on the care and needs of birds, and in their basic hospital care. Birds can offer the veterinary facility a challenging experience. It is important to have a basic knowledge and understanding of the behaviors and requirements of avian breeds.

Key Terms

acute a condition that occurs short term

amazon a large breed of cage bird from South or Central America that is talkative, entertaining, and very trainable

artificial incubation providing a controlled temperature in which eggs hatch, such as an incubator

beak the mouth of an avian or bird species used for chewing and grasping food

beak trim a procedure to shorten or round the sharp edges of the beak

blood feathers growing feathers that have a blood vessel in the center

budgerigar   commonly called a parakeet; a small bird breed that comes in a variety of bright colors with bars located over the wings and back

budgie   an abbreviated name for a budgerigar or parakeet

canary   a small breed of bird in solid bright colors that originated in the Canary Islands and is known for its singing ability

cautery unit   handheld equipment that heats to high temperatures to simultaneously burn through nails and clot them

cere   the fleshy area at the top of the beak that contains the nares or nostrils

choana   the opening to the oral cavity

chronic   a long-term condition

cloaca   the rectum area of a bird, where waste materials pass

cockatiel   a small bird breed that is a combination of yellow, gray, white, and orange colors and is very easy to train

cockatoo   a large bird breed that is commonly white and has a developed speaking ability

conure   a medium bird breed that is brightly colored from Central America that is active and playful

crop   the sac in the digestive system that acts as a holding tank for food

diaphragm   the chest area of a bird that must expand during the breathing process

DNA test   a test using blood or a feather that helps determine a bird's gender

"earthquake"   a procedure used in training a caged bird that teaches the bird to not bite; the bird sits on a hand and when it tries to bite, the hand is shaken slightly to cause a balance check

egg bound   a female bird that is attempting to lay an egg that has become stuck in the cloaca

expiration   to breathe out and expand the chest

fatty liver disease   a condition of birds in which large amounts of fat are deposited in the liver

fertilize   the ability of a male to mate with a female to create an egg with an embryo

finch   a small songbird that comes in a variety of colors and patterns and is easy to raise and breed

flight feathers   adult feathers located over the edges of the wings

gizzard   the digestive organ located below the crop of a bird that serves as a filter system to break down hard foods, such as seed shells or bones

green foods   foods such as fruits and vegetables that are green in color and have additional calcium

grit   a substance provided to birds to help break down hard substances, such as seed shells

hepatic lipidosis   a condition in which large amounts of fat are deposited in the liver

herbivores   animals that eat a plant-based diet

in captivity   housed in an enclosure or with people

incubate   to keep at a constant warm temperature

incubator   equipment used to keep eggs at a constant warm temperature for hatching

infertile   not able to reproduce; also known as sterile

inspiration   breathing in air and allowing the chest to depress

knemidokoptes mites external parasites that occur on the skin, beak, and feathers of birds; commonly called scaly leg mites

lovebird a small songbird of bright colors that originated in Ethiopia and is a popular cage bird

macaw a large bird breed that is brightly colored, originated in South and Central America, and is playful and easily trained

molting the natural seasonal dropping of feathers

natural incubation the process of a female or male bird nesting on eggs to hatch them

omnivores animals that eat both plant- and meat-based diets

parakeet a small songbird from Australia also known as a budgie that is brightly colored with bars located over the wings

parrot a medium-sized bird breed that comes from Central America and is bright in color, intelligent, playful, and easy to train

passerine a family of bird breeds that are small, have a pointed or slightly curved beak, and have four toes on each foot, three facing forward and one facing backward

pecking order a social organization of a group of birds where some are dominant over others

pelleted diet food fed to a bird in a size-appropriate pellet that is composed of the necessary ingredients

preen to clean the feathers

proventriculus the tube that connects the crop to the gizzard

psittacine a family of bird breeds that are large parrots with a curved upper beak that have four toes on each foot, two facing forward and two backward

psittacine beak and feather disease (PBFD) a contagious and fatal viral disease that affects the beak, feathers, and immune system of birds

psittacosis commonly called "parrot fever" or chlamydia; a disease that has zoonotic potential especially in young children and anyone with a poor immune system and causes severe flulike symptoms

seed diet a food composed of natural seeds eaten by birds

vent the external area of the digestive system; also called the cloaca

wing trims the procedure of clipping the wings to prevent a bird from flying

Review Questions

1. List the major classes of birds that are the most common pets.
2. List the factors that affect a bird's behavior.
3. What are the most common bird behavior problems?
4. What are the signs of egg-bound birds?
5. What are some foods that should not be fed to birds?
6. What grooming practices are necessary with pet birds?
7. What is psittacosis?
8. What are the causes of fatty liver disease?

Clinical Situation

Mrs. James has called to make an appointment for her Amazon, Goldie. Goldie has been known to be difficult to handle, biting and screaming when handled by veterinary staff members. The veterinarian, Dr. Black, has recommended that Goldie be brought in for a drop-off appointment and sedated for a physical exam.

Mrs. James feels she can safely handle Goldie and refuses to have her sedated. The veterinary assistant, John, is trying to discuss the situation with her.

"Mrs. James, we really feel that Goldie will be less stressed by a drop-off appointment. Dr. Black has suggested sedation for the exam. This will be helpful as we may also need to collect blood."

"I don't want Goldie to be sedated! She will be fine if I hold her. Please, let's schedule her for tomorrow morning and I will hold her for the exam."

- What would you suggest John do in this situation?
- What potential safety issues could be harmful in this situation?
- How would you handle this situation?

Chapter 11: Pocket Pet Health and Production Management

Objectives

Upon completion of this chapter, the reader should be able to:

11.1 Define common veterinary terms relating to rodent species
11.2 Describe the biology and development of rodent species
11.3 Identify common breeds of rodent or pocket pet species
11.4 Explain the nutritional requirements of rodents
11.5 Describe normal and abnormal behaviors of rodents
11.6 Explain how to properly and safely restrain rodents for various procedures
11.7 Discuss breeding and reproduction in rodent species
11.8 Develop a health care and maintenance plan for rodents
11.9 Describe common diseases affecting rodents

Introduction

A **rodent** is a mammal that has large front teeth designed for chewing or gnawing. Common rodents are mice, rats, hamsters, and guinea pigs. Rodents—also called pocket pets, as they can fit into one's pocket—have increased in popularity as companion animals (see Figure 11–1). Several species are more commonly kept. However, some rodent species are illegal to house as pets in California and other locations. Several states are currently considering changing their laws regarding ownership of species of hamsters, guinea pigs, and gerbils. States such as Hawaii and California have laws against owning certain species of animals, such as hedgehogs and ferrets. The veterinary assistant should have a basic knowledge in proper restraint and handling of rodent species, as many rodents will bite to protect themselves.

Mice and Rats

Mice and rats are curious and friendly pocket pets. They are usually tame animals, especially when acquired at a young age. The younger the rodent, the easier it is to tame. The best way to tame a rodent is by hand-feeding it. Both rats and mice can be trained to do tricks. Many times, these animals are used in research and nutritional studies. Unfortunately, both have received a negative reputation from their wild ancestors, which cause damage and spread disease.

FIGURE 11-1 Rodents, like this hamster, have become popular companion animals.

Veterinary Terminology

murine = mouse or rat
sire = intact male mouse or rat
dam = intact female mouse or rat
pup = young mouse or rat

Biology

Rats and mice typically have an average lifespan of 1 to 3 years. Rats and mice are in the family *Rodentia*: the veterinary term that refers to mice and rats is **murine** (see Figure 11–2). Rats and mice have poor vision but a well-developed sense of smell. Species that have red- or pink-colored eyes have reduced vision. Many species of rodents are **nocturnal**, which means they sleep during the day and are active at night. Many rodent species share common anatomical features. Common rat and mouse external body terms include the following:

- Claws—nails on feet
- Snout—nose and mouth area
- Tail—long part that extends off the back and is hairless and scaly
- Whiskers—located on nose and used as sensory devices

Some types of mice and rats are bred specifically for research, while others are bred as pets. Mice weigh between 20 and 40 grams and have a heart rate of 500 to 600 beats per minute. Mice have a high metabolic rate and are constantly active. Rats weigh between 250 and 500 grams; females are smaller than males.

FIGURE 11-2 (A) Rat with hooded markings. (B) Albino mouse.

Breeds

Mice and rats come in a variety of colors and types, more so than breeds. Rats are usually white, brown, black, or a variety called **hooded** (see Figure 11–2A). The body of a hooded rat is white in color, with black or brown coloring on the head and shoulders. This gives the rat a "hooded" look. White rats are commonly bred as research and laboratory species. Mice also come in a variety of colors, including black, white, tan, and spotted. There are species of fancy mice, as well as white mice, much like rat species.

Breed Selection

Pet rats and mice should be purchased at a young age and handled often to allow them to be socialized. Most rats and mice that are handled frequently and with care seldom bite. These timid, social pets can be entertaining. They make excellent pets for children if cared for and handled properly.

Nutrition

Mice and rats have the same basic nutritional requirements. They will eat almost any type of food and do well on commercial rodent pellets. They enjoy human foods, such as seeds, nuts, bread, cereals and grains, and raw vegetables. They can be fed *ad lib* or as much as they want, within reason. It is best to provide food in small dishes, which should be kept clean and free of bedding. Typical maintenance diets contain about 14 percent protein and 4 percent to 5 percent fat; diets for growth and reproduction contain 17 percent to 19 percent protein and 7 percent to 11 percent fat. Seed diets are also formulated for mice and rats, but these diets should only supplement the basic rodent pellet as a treat item. Rodents prefer sunflower-based diets to pellets, but these seeds are low in calcium and high in fat and cholesterol. When fed exclusively, seed diets can lead to obesity and nutritional deficiencies.

Water should be provided in bottles that are secured to the side of the cage. Fresh water should be provided daily.

Behavior

Rats and mice are timid animals that must learn to trust people (see Figure 11–3). When they become accustomed to people, they become very curious pets. They are always looking to escape and should be housed in well-confined cages. They are chewers and will chew through wood or plastic. Both species can be easily frightened and should be handled quietly and calmly. They should be kept in well-lit areas, as they can become stressed in dark or unlit rooms. A mouse may chew at the fur of another mouse, often in the facial area. This behavior is called **barbering** (see Figure 11–4). Male mice housed together often fight, but females housed together are often compatible. A stressed or upset rat will show anger by whipping its tail. A happy rat will chatter or click its teeth and occasionally make a vocal squeaking noise. Some rats and mice may become territorial of their cage and should be handled frequently to prevent this occurrence.

FIGURE 11–4 Barbering in a mouse.

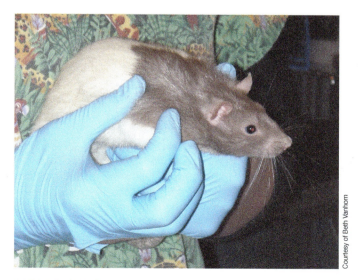

FIGURE 11–3 It is important to handle rats and mice at an early age so they become accustomed to people.

Basic Training Methods

Mice and rats are intelligent animals that can be easily trained. They will do almost anything for food and respond willingly to noises. The more they are handled and trained with consistency, the easier they will be to handle.

Equipment and Housing Needs

Cages should be made of heavy wire or metal because rodents will chew through wooden or plastic cages. Items placed in the cage should be small enough that mice and rats cannot use them to escape or climb out of the cage. Curious, mice and rats need toys that help keep boredom at bay. Perches, tunnels, and exercise wheels are common. Bedding should be free of dust and could be sawdust, cedar shavings, shredded paper, pellets, or cat litter (see Figure 11–5). At least one inch of bedding should be provided.

FIGURE 11–5 Rats and mice require clean bedding materials.

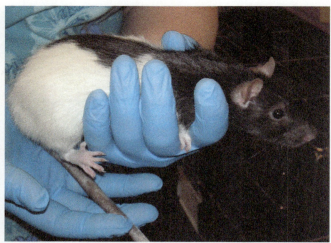

FIGURE 11–6 A tame rat or mouse can be picked up and will sit in the palm of the hand.

FIGURE 11–7 Scruffing a mouse.

The cages should be kept clean and dry. Adult mice require a minimum floor area of 15 square inches and a cage height of 5 inches. Rats need at least 40 square inches of floor space and a minimum of 7 inches in height. Breeder mice and rats require much larger areas.

Restraint and Handling

When handled frequently, a mouse or rat will typically sit in the hand and be easily held. If they are not handled a lot or used to people, they do have a tendency to bite. Rats and mice should never be touched or picked up when they are sleeping. Rats should be handled by the body and never picked up by the tail. The tail can be "stripped" of its skin or broken if used to pick up the rat. When the body is secure, grasp the rat by the tail base and place it on a firm surface. Then, with the opposite hand, encircle the body over the shoulder and neck region. It is necessary to perform this maneuver expeditiously, or the rat may turn and bite. Do not grip the chest so tightly that respiration is hindered. Once the head and body are secured, you may lift the rat (see Figure 11–6).

Mice can be picked up and handled by their tails, making sure to secure the base of the tail and not the end. Rats and mice should always be handled in a quiet, calm, and confident way. Most rats and mice will bite when a person shows fear or handles the animal incorrectly. To correctly handle a mouse, grasp it by the tail base and place it on a surface it can grip with its feet. Then, pull gently on the tail as the mouse holds onto the surface with its forelimbs. While maintaining traction on the tail, grasp the loose skin over the neck, immediately behind the ears, with the thumb and forefinger of the opposite hand (see Figure 11–7). Do this quickly to avoid being bitten. If the skin over the shoulders or an insufficient amount of loose skin is grasped, the mouse will be able to turn and bite. If you grasp too tightly, the animal's breathing can be compromised. Once the loose skin over the neck is appropriately secured, the mouse may be lifted, and the tail secured by the fourth and fifth fingers of the same hand (see Figure 11–8).

FIGURE 11-8 Mice can be picked up carefully by the base of the tail.

FIGURE 11-9 Mice pup litter.

Grooming

Mice and rats, like other members of the rodent family, do not require bathing or any additional grooming. They are fastidious groomers and will keep their coats shiny and clean if they are kept healthy. If the coat appears untidy, this may be a sign of disease.

Basic Health Care Maintenance

Mice and rats should have items in their cages to chew on to keep their teeth healthy and in good condition. Some teeth will get long and overgrown, which could cause eating problems or sore mouths. Mice and rats have four continuously erupting sharp incisors. On occasion, they may need their teeth trimmed.

Their coats should be monitored for external parasites, such as fleas and mites. These may be treated with topical parasite medications labeled for rodents. Do not use dog or cat products as they may be toxic.

Rats have a Harderian gland located behind the eye that secretes red tears used to lubricate the eyes. When stressed, tear production may increase, causing red staining over the face and nose.

Vaccinations

Rats and mice do not require vaccinations; moreover, there are no currently labeled vaccines for rats and mice.

Reproduction and Breeding

Mice reach puberty at an early age, usually by 8 to 10 weeks old. Rats usually reach puberty at 3 months of age and can be bred at this time. Both species have a gestation length of 21 days. They will have an average of six to eight babies in a litter (see Figure 11-9). Males and females may be housed together until labor occurs; the male should be removed until the babies are weaned, at about 3 weeks of age. Baby mice are active and learning to climb and jump. Housing should be considered carefully so that mice do not escape from their cage. It is important to separate the babies at weaning and determine their gender, as they will continue to breed. Sexing mice and rats is done through viewing their **anogenital** area (see Figures 11-10 and 11-11). This is the area located around the far stomach between the rear legs and the base of the tail. The distance is measured from the anus to the genital area. The anogenital distance in a male is much greater than that of a female. In a female, the distance is measured from the **urethra** (area where females urinate) to the anus; in the male, the distance is measured from the **penile area** (area where males urinate) to the anus. A male will also have a **scrotum**: this holds the **testicles**, where sperm is produced. Determining gender can be done when the mouse is 7 to 10 days old.

Common Diseases

Rats and mice are relatively healthy animals. They rarely get sick or acquire diseases. They will occasionally develop respiratory infections, with such signs as nasal discharge, sneezing, coughing, loss of appetite, and weight loss. They are also prone to tumors that may be malignant or cancer causing.

Common Parasites

Mice and rats may commonly develop external parasites, such as fleas, fur mites, and lice. The signs of external parasites include hair loss, itching,

202 SECTION II Veterinary Animal Production

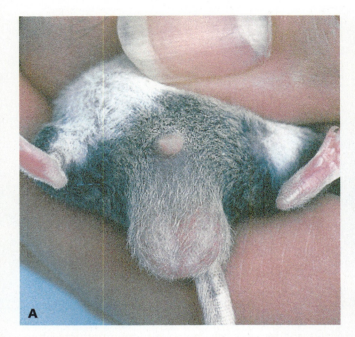

FIGURE 11–10 (A) Anogenital area of a male mouse. (B) Anogenital area of a female mouse.

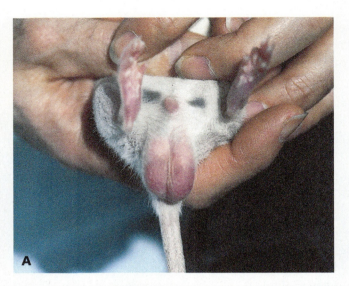

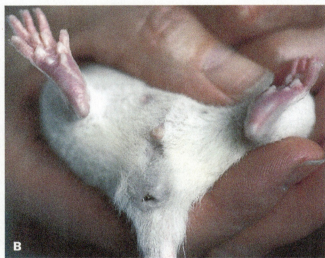

FIGURE 11–11 (A) Anogenital area of a male rat. (B) Anogenital area of a female rat.

and occasional scratches on the skin from the claws. Rats and mice may also be affected by internal parasites, such as tapeworms and protozoans, similar to those parasites that affect dogs and cats. Fecal analysis and treatment with deworming medications are necessary. It is important that products used to treat internal and external parasites are labeled for use in mice and rats.

Common Surgical Procedures

Surgical procedures of rats and mice are uncommon, mostly due to the short lifespan of the rodents. Rats and mice tend to develop tumors that may be surgically removed. Many tumors are **benign**, or noncancerous, but occasionally the tumors may be **malignant**, which means they are cancerous. Many tumors grow so large that the animal will have difficulty moving. Most rats and mice are euthanized when tumors develop (see Figure 11–12).

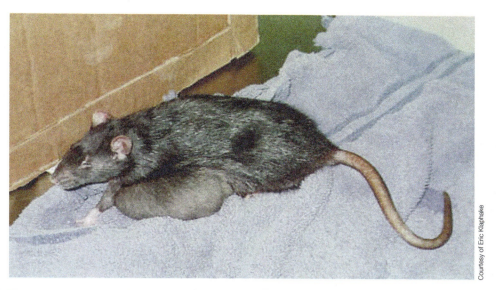

FIGURE 11–12 Tumors can be a problem in mice and rats. They can become large enough to restrict movement. This is an example of a vascular tumor in a rat.

Hamsters

The hamster is a popular pet. It is a very active and playful companion (see Figure 11–13). Hamsters are also a commonly used species in laboratory research. They are native to warm, humid desert climates and temperatures. The species originated in areas of Asia and Europe.

Biology

Hamsters have small round bodies with lots of skin, short legs, and a short tail. Like mice and rats, they are nocturnal and active at night. Common hamster external body terms include the following:

FIGURE 11–13 Hamsters are very active and playful.

- Cheek pouch—open area inside the cheek used to store food
- Ears—small hairless flaps on either side of the head
- Incisors—long teeth located on the upper and lower jaw
- Whiskers—long hair located on the face next to the nose

The average lifespan of a hamster is 18 to 24 months. Hamsters are best kept alone and are naturally solitary animals. They weigh between 35 and 120 grams; the females are usually larger than the males. Females tend to fight more and be more aggressive than males.

Breeds

Hamsters come in a variety of coat types and colors. The Syrian hamster—also called the golden hamster or teddy bear hamster—is one of the most popular breeds of pet hamsters (see Figure 11–14). They may grow to as large as 5 inches in length. The Chinese hamster is smaller in size and most often used as a research species. Several variations and mixtures of the breeds have been developed for different color characteristics and size variations. Dwarf hamsters are small in size and may only be an inch or smaller (see Figure 11–15).

Breed Selection

When choosing a pet hamster, it is important to observe the animal to make sure that it is alert,

Nutrition

Hamsters eat about ½ ounce of food a day. They should be fed in the evenings, when they are most active. Most commercial feeds include a variety of seeds, grains, and cereals; common ingredients are corn, oats, and wheat. They enjoy treats and will often store food and treats in their **cheek pouches**, using them as a storage area. They then **burrow** or tunnel the food items—often in an area of their cage—for later use (see Figure 11–16).

Behavior

A hamster's habitat should be kept at a temperature between 60 and 80 degrees Fahrenheit: a hamster's instinct is to hibernate if the temperature goes below 45 degrees Fahrenheit. **Hibernation** is a natural condition in which an animal goes into a long sleep and its

FIGURE 11–14 A teddy bear hamster.

FIGURE 11–15 Note the small size of the Chinese hamster.

FIGURE 11–16 (A) Hamsters store food in their cheek pouches to burrow it later. (B) Hamsters cheek pouches are large and may expand to several inches as items are stored.

active, and healthy. Some hamsters are susceptible to disease if they spend their first few weeks of life at a pet store. A hamster that is not active may be ill. Finding a healthy hamster with a winning personality will lead to years of enjoyment and companionship.

It is also important to select a breed or variation that will fit your cage space. Dwarf hamsters and some mixtures are more susceptible to disease. Keep in mind that a younger hamster will better bond with a new owner. It is best to select a hamster that enjoys being picked up and handled. Note that golden hamsters generally should be housed alone.

body systems slow down until the animal's body temperature rises. If this occurs, it is best to slowly warm the hamster and offer it warm milk when it is awake until it revives.

Basic Training

Training a hamster is similar to training rats and mice. It can be trained with consistency, proper handling, and treats (food). It is easier to train a young hamster (between 4 and 7 weeks old) than an older animal. Hamsters enjoy exercising on wheels and in balls for additional activities.

Equipment and Housing Needs

A hamster can live in an aquarium or rodent cage. Keep your eye on your hamster's dwelling, as it can squeeze through very small openings. Hamsters should not be kept in a wooden or plastic cage, as they love to chew and can easily escape. They are great climbers as well. They enjoy toys such as exercise wheels and tunnels, just like rats and mice. Clean a hamster's cage routinely and provide soft, dust-free bedding (see Figure 11–17).

Restraint and Handling

Hamsters are easy to handle if they are accustomed to being around people. A hamster is a member of an animal species that must be handled frequently to stay tame. It is best to begin by petting a hamster's back to allow it to adjust to a human's touch. Hamsters are sound sleepers. Do not attempt to pet or pick up a sleeping hamster: it *will* bite in a defensive reaction.

FIGURE 11–17 Hamsters can be housed in a clean aquarium with clean bedding materials.

FIGURE 11–18 Scruffing and supporting the body of a hamster.

A **defensive reaction** is when an animal protects itself from danger: examples are biting or scratching. Pick up a hamster by placing your hand underneath its body for support (see Figure 11–18). Many hamsters enjoy climbing into shirt or coat pockets. Older hamsters are sometimes considered to be difficult to handle. However, if simple precautions are taken, hamsters can routinely be handled with minimal stress to the animal and handler. Nonetheless, care must be taken to avoid being bitten. Factors that affect hamster restraint include a large amount of loose skin, which must be gathered to control the animal, and the tendency to bite when startled, such as being touched while sleeping. To remove a hamster from its cage, place a can or cup into the cage and gently encourage the hamster to enter. Then cover the opening and move the hamster from its cage. Alternatively, you may pick up a hamster by the loose skin over its back: when the hamster is on a flat surface, place the palm of your hand over the animal. Press down to control the hamster while you gather up the large amount of loose skin over its neck and back (see Figure 11–19). When you have secured enough of the skin, the hamster will be restrained and unable to bite.

Grooming

Hamsters are easy pets to maintain. They require little in the way of grooming aside from the occasional fur brushing and checking of teeth and nails. Like rats and mice, hamsters groom themselves to keep clean. Animals that are not grooming themselves, appear unkempt, or appear to have areas of wetness over the coat may be ill.

FIGURE 11-19 A hamster should be picked up firmly between the thumb and forefinger.

Basic Health Care and Maintenance

Hamsters can be given regular exams, like dogs and cats (see Figure 11–20). They should be monitored for any sign of illness, such as breathing problems, skin infections, or decreased activity. They may require regular trimming of nails and teeth: both continuously grow and can become overgrown if they are not worn down. The teeth should be monitored to make sure the hamster can eat appropriately.

Vaccinations

Hamsters do not require vaccinations, and no vaccines are currently labeled for hamsters.

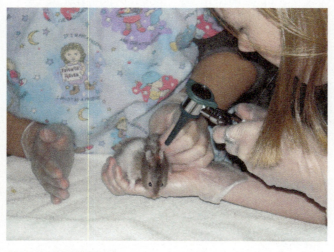

FIGURE 11-20 Physical examination of a hamster.

Reproduction and Breeding

Hamsters reach maturity at about 2 months of age. Males and females should only be housed together for breeding purposes. Place the female in the male's cage at night, when breeding takes place. If the male is placed in the female's cage, she is likely to kill him, feeling threatened and territorial. And if the female is not in estrus, they may fight. Place the female with the male at night, when breeding typically takes place. Wear gloves to protect your hands from being bitten. More information as follows:

- The estrus cycle lasts four days.
- Female hamsters should not be bred before 3 months of age.
- Many females do not produce **offspring**, or young, after a year of age.
- The gestation length is 16 days.
- A typical litter size is six to eight babies.
- The babies should not be touched for about 7 to 10 days after birth.

After the hamster has given birth, the cage should not be cleaned, and all contact with the female and young should be avoided. Female hamsters are very protective and feel threatened; if disturbed, they may kill their young. This is known as **cannibalism**. Hamsters can be sexed in the same manner as rats and mice (see Figure 11–21). The young should be separated and weaned within 2 to 3 weeks.

Common Diseases

Hamsters are prone to several diseases and conditions. One common condition is called **wet tail** and is caused by a bacterial disease that spreads rapidly through direct contact or bacterial spores. Causes include overcrowded cages, poor nutrition, poor sanitation methods, and stress. Hamsters with wet tail can be treated with **antibiotics** or medications and fluid therapy. They should be isolated from other hamsters and their cages thoroughly sanitized. They are most susceptible to this condition between the ages of 3 and 8 weeks old. Signs of wet tail include the following:

- Watery diarrhea
- Dehydration
- Weight loss
- Eye discharge
- Nasal discharge
- Lethargy
- Anorexia
- Wet tail appearance
- Irritable attitude

Another common condition is respiratory infections and diseases caused by bacteria, which may result

CHAPTER 11 Pocket Pet Health and Production Management

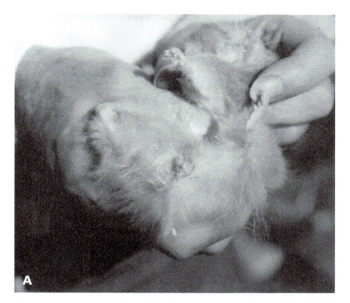

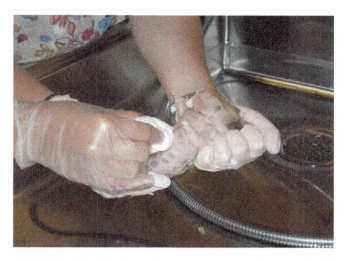

FIGURE 11–22 Hamsters can become infected with parasites.

FIGURE 11–21 (A) The genital area of a female hamster. (B) The genital area of a male hamster.

from poor sanitation or excessive dust from bedding. If diagnosed in time, the animal may respond to antibiotics. The cage conditions must be improved. Signs of respiratory disease in hamsters include the following:

- Nasal discharge
- Eye discharge
- Sneezing
- Labored or difficult breathing
- Anorexia
- Depression
- Weight loss
- Dehydration

Common Parasites

Hamsters may also become infested with mange mites, similar to those found in dogs (see Figure 11–22), usually from bedding or from other animals within the household. Signs include hair loss, especially large clumps, and scratching. Hamsters may also be afflicted with other external parasites, such as fleas, lice, and ticks. They should be treated with topicals labeled for rodents—never with dog or cat products. A **topical** is a type of medicine or chemical applied externally to the skin or hair coat.

Guinea Pigs

A guinea pig is docile, has lots of personality (is even known for greeting people with a whistle) and is an excellent starter pet for older children. Larger than hamsters, but smaller than rabbits, guinea pigs can weigh a couple of pounds and generally live for 5 to 7 years.

Veterinary Terminology

cavy = guinea pig
boar = intact male guinea pig
sow = intact female guinea pig
kit = young guinea pig

Biology

The guinea pig is a rodent with a short, heavy body; short legs; and no tail. Common external body terms include the following:

- Cheek pouch—area inside the mouth for storing food
- Fore feet—have four claws
- Hind feet—have three claws

Guinea pigs have an average lifespan of 4 to 5 years but have been known to live longer than 10 years.

Guinea pigs are social animals that prefer to live in small groups. Two or more females will usually get along well in the same living quarters. If housing two males together, it is smart to choose two babies from the same litter. As guinea pigs—like all rodents—multiply rapidly, keeping males and females together is not recommended.

Guinea pigs weigh between 700 and 1200 grams and have a normal body temperature of 99 to 103 degrees Fahrenheit. Their teeth are open rooted and erupt continuously.

Breeds

There are several breeds of guinea pig. They come in a variety of colors and hair coats: some are short and smooth, some are long and silky, and there is also a hairless breed. The three most common breeds are the smooth or American, with short, glossy fur; the Abyssinian, whose hair grows in fluffy tufts all over the body; and the Peruvian, with long, silky hair that flows to the ground (see Figure 11–23).

Breed Selection

Guinea pig selection should be based on coat type and grooming needs. Some longer-haired breeds require more maintenance and brushing (see Figure 11–24).

Nutrition

Guinea pigs should be given a high-quality commercial food that is typically a mix of pellets formulated for the needs of the species. Guinea pigs have a special need for **vitamin C** in their diets, as they cannot make it naturally (as other animals do). Their diet must include foods rich in vitamin C: fruits, vegetables, and greens, such as apples, carrots, lettuce, celery, spinach, and alfalfa hay. Food should be available *ad lib*. Water should also be made available; however, bear in mind that but many guinea pigs drink less water than other rodents, as they derive large amounts of water from their food sources.

Behavior

Guinea pigs are very social and can be housed together. They are not known to fight, jump, or climb like other rodent species. They are usually mild and timid animals but will bite if excited, scared, or in pain. Guinea pigs love to hide when they play, so be sure to place cardboard tubes and/or empty coffee cans with smoothed edges in the enclosure. Plastic pipes and rocks allow them to climb. All guinea pigs need an area for sleeping and resting. When they are happy and content, guinea pigs will make a whistling noise.

FIGURE 11–23 Examples of guinea pigs: (A) Abyssinian; (B) American shorthair; and (C) Peruvian.

Basic Training

Like mice, rats, and hamsters, guinea pigs can be trained. Time and correct handling are important.

Equipment and Housing Needs

Guinea pigs can be housed in a wire cage or an aquarium. They need at least 1 square foot of floor space

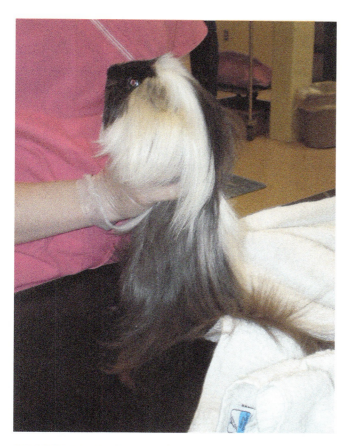

FIGURE 11-24 Long-haired guinea pigs will need to be brushed to maintain their hair coat.

FIGURE 11-25 A guinea pig enclosure must be cleaned on a regular basis.

FIGURE 11-26 Use both hands to secure the guinea pig.

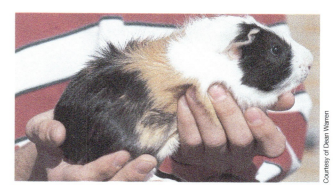

FIGURE 11-27 When picking up a guinea pig, one hand should support the torso and one hand should support the hind end.

per animal. They should have dust-free bedding, and cages should be cleaned on a regular basis (see Figure 11-25). Room temperatures should be kept between 70 and 80 degrees Fahrenheit. The young will stop growing if the temperature goes below 65 degrees or becomes excessively hot. Guinea pigs tend to be messy compared to other rodents and will scatter their food and bedding. Cages must be cleaned frequently.

Restraint and Handling

Most guinea pigs are easy to handle and rarely bite, but they can scratch with their nails. Guinea pigs should always be restrained using both hands (see Figure 11-26). One hand should gently secure the animal around the torso, while the second hand supports the hindquarters (see Figure 11-27). This is particularly important with large or pregnant animals in order to avoid internal injuries. If animals become scared or fractious, turning down the lights and covering the eyes may have a calming effect.

Grooming

Grooming practices in rodents should include trimming the teeth if they become overgrown, nail trims, and bathing as necessary. A guinea pig's teeth, like other rodents, grow continuously. Providing hard items for them to chew on will keep their teeth healthy. Routine evaluation is important to make sure the teeth are not becoming overgrown (see Figure 11-28). Rodents can be bathed using a safe waterless shampoo. Some guinea pigs require regular brushing, especially long-haired breeds.

FIGURE 11-28 A guinea pig's teeth should be checked regularly to make sure they are aligned and wearing down properly.

Basic Health Care and Maintenance

Guinea pigs should be examined on a regular basis (see Figure 11-29). Common signs of illness include sneezing, coughing, diarrhea, and lethargy. Guinea pigs are also susceptible to external parasites, such as mites and lice.

Vaccinations

Guinea pigs do not require vaccinations, and no vaccines are currently labeled for guinea pigs.

Reproduction and Breeding

Guinea pigs reach maturity between 3 and 4 months of age. They should not be bred before this time.

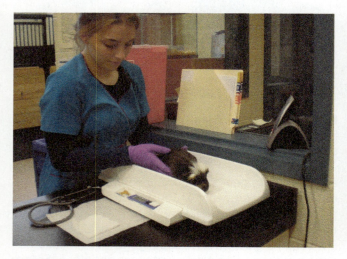

FIGURE 11-29 Routine exams are encouraged for guinea pigs.

A female guinea pig should be bred at about 6 months old; if not, the pubic bone begins to fuse and will cause dystocia, which is difficult labor. The male, or boar, should be placed in the female's cage (refer to the information on breeding hamsters). They may be left together for 3 weeks, during which the estrus cycle will occur. Sows are polyestrous and cycle every 15 to 17 days. Gestation occurs between 63 and 72 days. Sows will cycle and breed again within a few hours of the labor process. If they are not bred during this time, the female should not be bred until the litter is weaned. Babies are born with hair, eyes open, and the ability to eat solid food within the first day of birth. The sow will nurse the young. Weaning should occur at about three weeks of age and males and females separated to prevent breeding. Sexing guinea pigs is done by examining the genital area (see Figure 11-30). The female guinea pig has a Y-shaped anogenital region. The **anogenital** region is the area of the anus where bowel movements are produced; the genital area is where urine is produced. The male

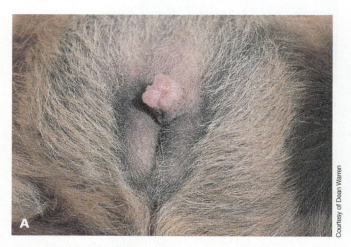
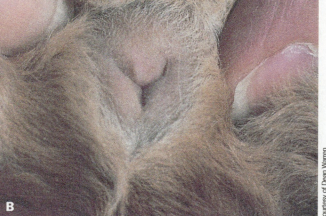

FIGURE 11-30 (A) Genital area of the male guinea pig. (B) Genital area of the female guinea pig.

has a straight slit and gentle pressure will expose the penis. Male guinea pigs will develop larger testicles than other rodents.

Common Diseases

Guinea pigs tend to be healthy pets. Some signs that an animal is not feeling well include sitting very still, sitting hunched up, a ruffled or messy coat, anorexia, weight loss, and watery droppings. If any of these symptoms appear, the guinea pig should be seen by a veterinarian. Guinea pigs are commonly known for developing respiratory problems, usually from dusty bedding; signs include wheezing, sneezing, eye and nasal discharge, and a decreased appetite.

Common Parasites

Occasionally, guinea pigs become infected with internal and external parasites, such as fleas, mites, or lice. They should be treated for external parasites with medications labeled for use in guinea pigs. Internal parasites are not a common problem.

Gerbils

A gerbil is a fascinating pet because it is a quick and curious animal (see Figure 11–31). Gerbils are clean and easy to maintain and handle. They produce little waste, making them one of the easiest pocket pets to care for. Most gerbils are **agouti** in color, which is a mixture of two or more colors. Other popular colors include white and black.

FIGURE 11–31 Gerbil.

Veterinary Terminology

Gerbils are known as **jirds**, or burrowing rodents. They have short bodies and a hunched appearance.

- sire = male gerbil
- dam = female gerbil
- kit = young gerbil

Biology

Gerbils are different than other rodents because they are **diurnal**: they sleep during the night and are active during the day. This makes the gerbil a great pet. About 4 inches long, the gerbil looks like a hamster but has a long tail that is typically as long as its body. Adults weigh about 3 ounces, or 100 grams. The average lifespan of a gerbil is 3 years. Gerbils are usually a dark or light-brown color and may have white markings. They can be shy and should be handled gently and regularly. Scratching a gerbil behind the ears may help it relax. Common external terms include the following:

- Eyes—large, black, and round and located on either side of the face
- Snout—the nose and mouth area on the face
- Tail—a very long extension off of the back, can be long as the body
- Tuft—dark hair at the end of the tail

Breeds

Mongolian gerbils are native to areas of Mongolia and China. They live in a variety of arid terrains, including deserts, low plains, grasslands, and mountain valleys.

Breed Selection

Gerbils should be housed in pairs. Selection is similar to that of other rodents. Gerbils are social animals and should usually be purchased in bonded, same-sex pairs. While personalities vary, and not every gerbil will run happily into your hand right away, an emotionally healthy gerbil will be curious and display a friendly demeanor. Healthy gerbils should be easy and willing to be handled and free of any signs of illness.

Nutrition

Gerbils may be given a commercial food that contains a variety of seeds, grains, corn, sunflower seeds, and oats. They also eat fruit seeds, apples, lettuce, and fresh grass. Gerbils will usually eat about a tablespoon of food once a day. Water should always be available.

Behavior

Gerbils will use one area of their cage to eliminate. Cages must be cleaned on a regular basis. Gerbils should be housed in temperatures between 65 and 80 degrees Fahrenheit. They do not do well in cool temperatures. Gerbils are very social animals; it is not a good idea to keep them singly. Pair-bonded or family units of gerbils are usually quite affectionate with each other. They will play—chase each other, wrestle, and box. They will also groom one another, sleep in piles, and cuddle together. When experiencing stress, a gerbil will drum its hind legs on the cage floor.

Basic Training

Gerbils are intelligent and may be easily trained. The more they are handled and trained with consistency, the easier they are to handle.

Equipment and Housing Needs

Gerbils are best housed in a 10-gallon aquarium. A glass fish tank is better than a wire hamster cage because gerbils like to dig in deep substrate and will kick the litter out of a wire enclosure. With a pair of gerbils, it can be helpful to put a wire cage topper above the aquarium to give them more room to run around. They are curious and will attempt to escape if given the opportunity. They like to climb, jump, and chew. A block of wood can help keep their teeth healthy. Gerbils like to build tunnels, so put at least 3 inches of substrate at the bottom of their habitat. A popular bedding is cedar chips or dust-free shavings.

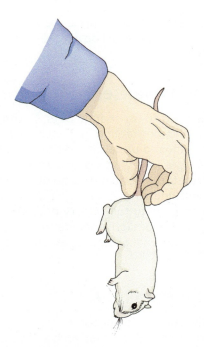

FIGURE 11–32 Never restrain or pick up a gerbil by the tail.

Restraint and Handling

To restrain a gerbil, grasp the base of the tail with one hand and the skin on the back of the neck with the opposite hand. Be very careful not to pull on the tail because gerbils are especially prone to tail injuries from improper handling. Never restrain a gerbil by the tail (see Figure 11–32). If additional restraint is desired, grasp the gerbil from above, securing its head between your index and middle fingers and the body with your thumb and remaining fingers. Alternatively, scruff the neck securely (see Figure 11–33).

Grooming

Gerbils do not need a lot of grooming care. They will groom each other and keep each other clean. Monitor the nails and teeth for overgrowth. Signs of a sick gerbil include unkempt hair coats and decreased amounts of self-grooming.

FIGURE 11–33 Scruffing a gerbil.

Basic Health Care and Maintenance

Gerbils should be examined regularly, as with any other pets. They come from a dry natural habitat have therefore adapted to conserve water. As a result, they produce scant amounts of urine and dry droppings, making it fairly easy to keep their cage fresh and clean. Provide toys and items to chew to make sure their teeth do not become overgrown.

Vaccinations

Gerbils do not require vaccinations, and no vaccines are currently labeled for gerbils.

Reproduction and Breeding

Gerbils mature at about 3 months of age. They will select one mate for life. They can be housed together but should be given time to get acquainted. The gestation length is 24 to 25 days. Most females have a litter of five babies. Both the male and female can be housed with the babies, which are typically weaned at about 6 weeks of age. Young gerbils should not be handled until their eyes are open. Sexing is done as in all other rodents, with the male anogenital distance being greater than that of the female gerbil (see Figure 11–34).

Common Diseases

Gerbils are relatively free of disease. They should be monitored for any signs of respiratory diseases: signs include sneezing, nasal discharge, increased breathing, and wheezing.

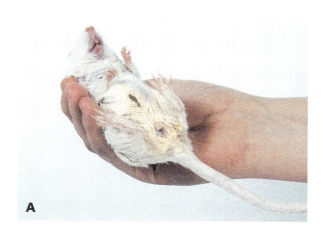

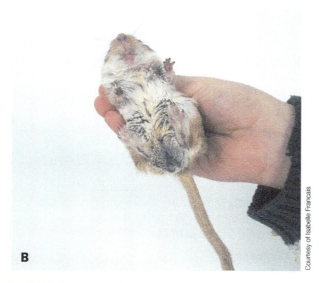

FIGURE 11–34 (A) The genital area of the male gerbil. (B) The genital area of the female gerbil.

Common Parasites

Common external parasites of gerbils include fleas and mites. Internal parasites are rarely seen, but tapeworms may occasionally develop. Gerbils should be medicated with label-approved medications.

Ferrets

The ferret is a member of the Mustelidae family, which classifies them similarly to weasels, mink, and martens. Domestic ferrets are small, furry mammals with an average size ranging from 1 to 5 pounds at maturity (see Figure 11–35). Ferrets are not wild animals. They do not belong to the rodent family but are still considered "pocket pets." They are very curious and entertaining pets. The ferret originated from a larger species of weasel known as the polecat, which is wild and undomesticated and is closely related to the black-footed ferret. Ferrets originated in Europe and were originally used for hunting rabbits.

Veterinary Terminology

hob = intact male ferret
jill = intact female ferret
gib = neutered male ferret
sprite = spayed female ferret
kit = young ferret
kindling = giving birth to ferrets

Biology

Ferrets have a long, slim body and a long tail. They come in several colors, with sable the most common. Common body terms include the following:

- Body—a long and thin central area
- Mask—black coloring around the eyes on the face
- Tail—a bushy area extending off the back

FIGURE 11–35 Sable ferret.

Ferrets have scent glands that give them a strong musky odor, which is much more notable in a male. They are primarily nocturnal. They have an average lifespan of 5 to 8 years. Ferrets combine the best features of dogs and cats with some unique features of their own. Like cats, ferrets are small and quiet. Like dogs, they are affectionate, playful, and enjoy human interaction. They are independent yet enjoy being with people. Their mischievous and playful nature, retained well into old age, makes them entertaining companions.

Ferrets weigh between 0.8 and 1.2 kg, with males slightly larger than females. A normal body temperature is 100 to 104 degrees Fahrenheit, with a heart rate of 180 to 250 beats per minute and a respiratory rate of 33 to 36 breaths per minute. Ferrets lose weight in the summer and gain it back in the winter and shed their hair in the spring and fall.

Breeds

There are no distinguishable breeds of ferrets. Ferrets are classified by their colors, such as sable, white, silver, and white-footed (see Figure 11–36).

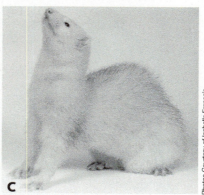

FIGURE 11–36 (A) White ferret. (B) Silver mitt ferret. (C) Sterling silver ferret.

Breed Selection

Ferret selection is similar to that of other pocket pets. A ferret should be playful, social, and enjoy being handled. Some ferrets may be de-scented and neutered, which makes their health care and maintenance more ideal.

Nutrition

The nutritional needs of a ferret include a commercial ferret food or high-quality cat food. Ferrets are true carnivores. They will eat both dry and canned food. They need a high-protein diet and fresh water daily. Table scraps should be limited. An appropriate commercial diet for ferrets might include the following:

- Totally Ferret (Performance Foods, Inc.)
- Ferret Forti Diet (Kaytee Products)
- Ferret Chow (Purina)
- Marshall Ferret Diet (Marshall Products)

Behavior

Ferrets are intelligent animals: what they can do and learn is amazing. They recognize their name and respond to verbal and visual commands. A healthy, well-trained ferret should not bite. Like all pets, ferrets need to be taught what acceptable behavior is. Ferrets have a lower bite rate than other household pets, such as hamsters. Ferrets are very curious and love to explore. They will easily get into trouble and may chew on items that can harm them, so care must be taken to ensure their safety. Ferrets are "pack rats": they will take anything that interests them and drag it off to their hiding place, usually under a couch or chair.

Basic Training

Similar to cats and kittens, ferrets can be litter trained. They can learn to perform tricks with consistent training and rewards. Like dogs, ferrets can be trained to come when their name is called. They can be trained to sit, beg for treats, and ride on an owner's shoulder. They may also be walked on a leash and harness (see Figure 11–37).

Equipment and Housing Needs

Ferrets should be kept in secure cages. Wire or metal cages are best. Make sure they cannot squeeze through any openings. When they are outside of a cage, they should be watched very closely.

Restraint and Handling

Ferrets vary greatly in temperament. Although some may be nonaggressive, others may be aggressive,

FIGURE 11–37 Ferrets can be trained to walk on a leash.

FIGURE 11–39 To hold a ferret, place one hand under the front of the body, supporting the head, and one hand under the hind end of the body.

FIGURE 11–38 Ferrets can be picked up by the nape of the neck.

FIGURE 11–40 Ferrets can benefit from brushing.

especially when restrained. Ferrets should be initially grasped around the neck and shoulders, similar to the scruff method in cats (see Figure 11–38). The handler should hold the ferret with one hand under the shoulders with a thumb under the jaw and the other hand supporting the animal's hindquarters (see Figure 11–39). Ferrets can move quickly: with their long and thin body, they can be a challenge to handle and control. Ferrets that are handled often become easier to handle. Ferrets can be taken outdoors on a leash and under close supervision. A leash can be attached to an H-type harness.

Grooming

Ferrets have a natural musky odor, even when descented. Frequent baths are not recommended, as the naturally oily skin will produce an even muskier smell. A health care program should include regular nail trims. Regular brushing may be necessary, but most ferrets will groom themselves and keep their hair coat clean (see Figure 11–40).

Basic Health Care and Maintenance

Ferrets should have an annual visit to a ferret-knowledgeable veterinarian, which helps identify potential problems early. This yearly visit should include a careful physical exam, inspection of the ears for mites, and inspection of the teeth (see Figure 11–41).

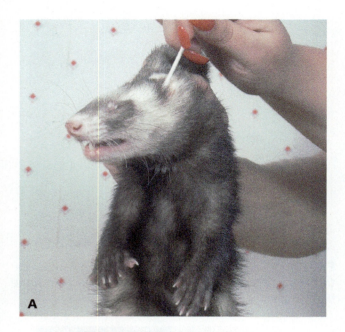

FIGURE 11–41 (A) Cleaning the ears of a ferret. (B) Trimming a ferret's nails.

Dental cleanings should be performed as necessary. A test for the Aleutian disease virus should be done at least once a year to ensure that the ferret is not carrying this highly contagious and potentially fatal disease.

Vaccinations

Ferrets are susceptible to canine distemper and are typically vaccinated to prevent the disease. The distemper vaccine (Fervac-D) is generally given between 6 to 8 weeks of age, and a booster is repeated monthly until 14 weeks of age. The vaccine is then given yearly. Some states that allow for legal housing of a ferret require a veterinarian to vaccinate a ferret for rabies.

Reproduction and Breeding

Ferrets reach maturity at about 10 months of age. The females are usually bred in the spring: gestation lasts an average of 42 days. Ferrets usually have between six and eight young in a litter. The open their eyes in 3 to 4 weeks. The young should start weaning at this time and should be weaned by 8 weeks old. They reach their adult weight by 14 weeks. Females should be spayed if not being bred. If a female goes into the estrus cycle and is not bred, she will continue to cycle, and bleeding may cause death. Spaying or castrating a ferret will decrease its musky body odor to some extent. Ferrets have scent glands located at either side of the rectum, similar to dogs and cats. These glands create a musky odor and are usually removed during neutering. The ferret will still have some odor. Ferrets should be neutered at about 5 to 6 weeks of age. Sex determination of the ferret can be done as in other rodent species by examination of the anogenital area (see Figure 11–42).

Common Diseases

Ferrets are relatively healthy animals but are susceptible to canine distemper and Aleutian disease. They can also develop flulike symptoms or respiratory illnesses similar to the "common cold," which can be transmitted by human companions. It is important to handle a ferret with extreme caution should you become ill. Older ferrets can develop diseases. The most common are diseases of the adrenal glands and pancreas. Signs of an adrenal gland disorder include hair loss, muscle atrophy, urinary blockage in males, and enlarged vulva in females (see Figure 11–43). Signs of pancreatic disease include lethargy, nausea, and seizures.

Any digestive problem (changes in bowel routine, extreme weight gain or loss, vomiting) that a ferret experiences is potentially serious. The best way to prevent these problems is to keep the ferret in an environment that is clean and free of dangerous objects. Keep foam packaging peanuts, rubber chew toys, erasers, rubber bands, latex, or plastic items away from ferrets.

Common Parasites

Ferrets are susceptible to external parasites common to those in cats, such as fleas, ticks, and skin and ear mites. They can be treated with medications safely labeled for use in ferrets. Ferrets are also capable of getting internal parasites, such as tapeworms. A fecal sample can be evaluated for internal parasites. Ferrets are also susceptible to heartworm disease.

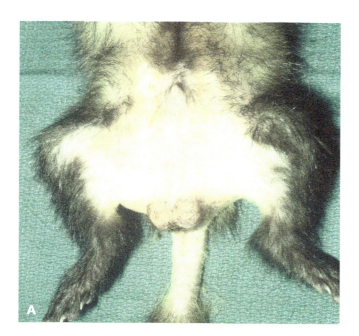
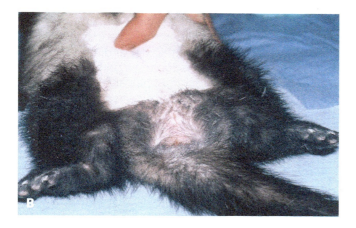
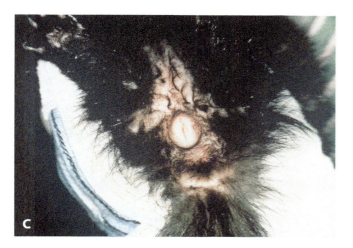

FIGURE 11–42 (A) Genital area of the male ferret. (B) Genital area of the nonestrus female ferret. (C) Genital area of an estrous female ferret.

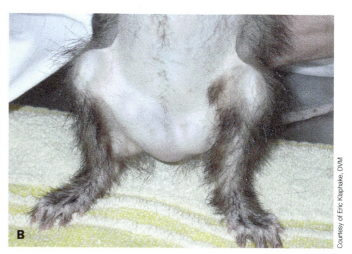

FIGURE 11–43 (A) Adrenal disease in a ferret. (B) Alopecia and edema in a ferret with adrenal gland disease.

Hedgehogs

A hedgehog is an interesting animal from the Erinaceinae family. This behavior of this animal can be unpredictable: a hedgehog tends to curl into a ball when handled. It has short spines that protect its body and can cause injury to a handler. Hedgehogs are as common in Europe and Britain as skunks are in the United States.

Veterinary Terminology

- hoglet or pup—young hedgehogs
- herd—group of hedgehogs
- self-anointing—the act of frothing at the mouth, arching the head back over the shoulders, and depositing the frothy saliva onto the quills; this is an action triggered by pungent smells and new tastes

Biology

Hedgehogs have **quills** or spines over their back and body that are between 1 and 1½ inches in length (see Figure 11–44) and number between 5000 and 7000. The muscles over a hedgehog's back cause the quills to stick out and protect it from predators. The hedgehog does not release or throw out its quills. Hedgehogs are a "salt and pepper" color. They are nocturnal and live an average of 4 to 7 years. When they are frightened or stressed, they will curl up into a ball, with the quills serving as their protection. They will also roll into a ball to sleep. Occasionally, hedgehogs make quick movements or dart to get away from danger. Common external terms include the following:

- Muzzle—mouth area located on the face
- Neck—located behind the ears
- Nose—a long area located at the point of the face
- Quills—spines located over the back and body
- Underside—the belly area that has hair and no spines

A hedgehog is about the size of a guinea pig and weighs between 1 and 3 pounds.

Breeds

There are no specific breeds of hedgehogs, but there are 17 different species found in New Zealand, Asia Europe, and Africa. Common colors are a mixture of brown, black, and white.

Breed Selection

Hedgehogs are naturally shy and tend to curl into a ball when handled. They require early socialization and should be handled often when they are young.

Nutrition

Hedgehogs are meat eaters: they enjoy a diet of mealworms and insects with a small amount of vegetables and fruits in their diet. They will eat a commercial hedgehog diet that is high in protein. They will also eat high-quality cat foods. As they are prone to obesity, take care not to overfeed them.

Behavior

The hedgehog is a relatively shy, **solitary** animal and does best when housed alone. A nocturnal animal, it starts its nightly activities, such as foraging, as the sun goes down. In the wild, hedgehogs may hibernate depending on the weather. During hibernation, the heart rate of the hedgehog will drop almost 90 percent to save energy. Hibernation periods may last from a few weeks to 6 months, depending on the severity of the winter. A hedgehog's primary defense mechanism is to roll into a tight ball and protrude its quills (see Figure 11–45). The hedgehog is able to form a ball by means of muscles located around the body. These muscles pull the skin around the sides

FIGURE 11–44 The hedgehog is covered in spines.

FIGURE 11–45 When frightened, the hedgehog will roll up in a ball as a defense mechanism.

and down over the feet and head. The hedgehog's spines can be moved individually by a complex layer of muscles beneath the skin. It erects its spines for protection from potential predators. Also, when the hedgehog feels threatened, it may hiss or make clicking sounds. Many hedgehogs are independent.

Basic Training

Hedgehogs are not typically trained due to their naturally shy behaviors. They can be trained to use a litter box.

Equipment and Housing Needs

An aquarium is ideal housing for hedgehogs (see Figure 11–46). They require dust-free bedding on the bottom of the cage. This may be shredded newspaper, pellets, or dust-free shavings. Ideal room temperatures are between 75 and 85 degrees Fahrenheit. Lower temperatures may make hedgehogs sluggish, as they are naturally hibernating animals. Hedgehogs need places to hide, especially during the day, when they sleep. Provide a plastic container with a hole in its side, a wide-diameter section of PVC pipe, or a cardboard tube. They do like to exercise and will run in wheels similar to hamsters.

Restraint and Handling

The most important thing to remember when beginning an examination of a hedgehog is that if it rolls up, it will be nearly impossible to examine. The more you fight with the pet, the fewer your chances of getting it unrolled. However, most pets are fairly docile and can be handled with latex exam gloves with relative ease and a minimum of discomfort. An open hedgehog should be scooped up from underneath and lifted off the table. When supported by two hands off the table, they will seldom attempt to roll up. A less cooperative hedgehog may be scruffed by grasping the skin between the ears and lifting the animal off the table. A latex glove will add a greater measure of control than a bare hand. A third technique is a leg hold: grasp the rear legs firmly and gently lift the rear end of the hedgehog slightly off the table. Some animals will squirm and try to pull away from the hold, but many pets will simply remain still when held in this manner.

Another method of unrolling a hedgehog is to wait a few minutes for it to uncurl on the examination table and then place the animal abdomen-side down. Heavy stroking of the spines toward the rump may help. Sometimes, none of these methods will work; as a last resort, the best examination technique becomes the use of isoflurane. A large cone may be placed over the entire animal until it starts to relax, then a small cone placed over the face.

Grooming

Hedgehogs may become dirty and need an occasional bath. This can be done with a toothbrush and warm water. Some hedgehogs will allow a light spraying of water to remove dirt and debris. They should be towel-dried well in a warm location. They may also need regular nail trims.

Basic Health Care and Maintenance

A yearly physical exam is ideal, but consideration must be taken that the hedgehog may need to be sedated for an adequate exam.

Vaccinations

Hedgehogs do not require vaccinations, and no vaccines are currently labeled for hedgehogs. Many states do not allow people to legally house hedgehogs.

Reproduction and Breeding

Breeding hedgehogs should be done after maturity, at about 7 weeks of age. They have a short gestation period of 35 days and typically a small litter. The male and female should be housed together only for breeding. The female is an induced ovulatory, and estrus lasts about 2 to 5 days. Males are known to kill and eat the babies if left with the female. The female will nurse the litter for about 5 weeks, after which they can be weaned (see Figure 11–47).

FIGURE 11–46 An aquarium makes a good enclosure for a hedgehog.

FIGURE 11–47 Young hedgehogs can be weaned by 5 weeks of age.

Common Diseases

Hedgehogs are prone to obesity as discussed in the nutritional section. They also are prone to bacterial skin infections and injuries from their spines. Dental disease may also be a problem: hedgehogs that are having difficulty eating should be examined for dental issues. They may also develop respiratory conditions due to dusty bedding materials, so care should be taken to monitor their breathing patterns.

Common Parasites

Hedgehogs are prone to mange mites, fleas, and ticks. Signs of mange mites include heavy dandruff, quill and/or hair loss, crusty thickened skin, thickening of ear margins, easy plugs in ear canals, and overall itchiness. Hedgehogs are not prone to internal parasites.

Chinchillas

A chinchilla is a small silky-haired rodent (see Figure 11–48). Chinchillas are bred as pets and for their fine hair. Chinchillas make excellent pets but require more care and maintenance than other small rodents.

FIGURE 11–48 Chinchilla.

Veterinary Terminology

- kit—baby chinchilla
- litter—group of newborn chinchillas

Biology

Chinchillas have a round body with a squirrel-like tail. They come in a variety of colors, with slate gray being the most common. Common external body terms include the following:

- Ears—large flaps on either side of the head
- Forelimbs—short front legs used for grasping
- Hind limbs—long back legs
- Pads—bottom surface of the feet
- Tail—a large bushy area extending off of the back

When stressed, chinchillas will throw out fine clumps of their hair. They are nocturnal. They enjoy being active while quickly jumping and bouncing. They are unpredictable in their movements and can easily escape. They live between 9 and 10 years of age. They weigh about a pound at adult size. A normal body temperature is between 99 and 100 degrees Fahrenheit.

Breeds

There are no specific breeds in chinchillas. There are several color variations, such as gray, white, beige, and black (see Figure 11–49). The chinchilla originated in South America.

Breed Selection

Chinchillas are social animals that may be selected in a similar way to other rodent species. A chinchilla is usually housed alone, as it may not tolerate other chinchillas. Chinchillas should tolerate being handled and not bite. The more they are handled, the tamer they become. They do demand a lot of attention.

CHAPTER 11 Pocket Pet Health and Production Management

FIGURE 11–49 Common chinchilla colors. (A) White. (B) Beige. (C) Black.

Nutrition

Chinchillas are usually fed a commercial diet that consists of pellets (similar to rabbits). They will also eat fruits, leafy vegetables, and raisins. They should be fed high-quality grass or alfalfa hay.

Behavior

Chinchillas are excellent jumpers and very quick. They are social but do not tend to like to be held and cuddled. Sudden moves and loud noises will startle and frighten them. Chinchillas will not bite unless they smell food on the fingers. They will sometimes nibble on clothing and belts. They are quite curious and should have as many places to climb into and on top of as possible. When they are stressed, they will lose a large amount of their fine hair, much like a dog or cat that is shedding.

Basic Training

A chinchilla may be trained, but it may take several weeks to months for it to trust a human. Consistency and frequent handling are the best training methods for chinchillas.

Equipment and Housing Needs

Chinchillas are acrobatic animals and therefore require a lot of space. Since they climb and jump in both horizontal and vertical directions, a large multilevel cage is recommended. Ideally, the larger the cage, the better off the chinchilla will be. The cage should be constructed of a small welded wire mesh to prevent leg or foot injury and include an area of solid flooring. Drop pans below the cage are ideal for easy cleaning. Since chinchillas are shy animals, they need a place to hide. Nondestructive boxes work well, such as wooden or metal boxes.

A chinchilla will suffer heatstroke in an environmental temperature at or above 80 degrees Fahrenheit coupled with humidity. A recommended temperature range is 50 to 68 Fahrenheit. Chinchillas need to chew hard items to keep their teeth in ideal shape, so it is best to provide wooden chew items.

Restraint and Handling

Chinchillas should never be captured or handled by their tails. This may cause severe injury or a broken tail. It is best to carry and lift a chinchilla by having a firm grip on the body with one hand under the abdomen or around the scruff and one hand holding the base of the tail (see Figure 11–50). Holding the base of the tail will help prevent the chinchilla from jumping. Using a towel can help the chinchilla feel more secure, as though it is in hiding.

Grooming

A chinchilla's hair is fine and delicate. When it is stressed or scared, a chinchilla will lose hair. This condition is called fur slip: it takes 6 to 8 weeks for the hair to grow back. Chinchillas require a **dust bath** every day to every other day. The chinchilla will roll in a particular type of dust to cleanse its delicate fur (see Figure 11–51). The dust bath hydrates and cleans their hair coat and protects it from damage. Place about one inch of the dust on the bottom of the cage. Commercial dust baths are available.

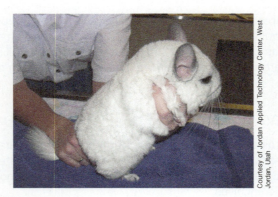

FIGURE 11–50 To restrain a chinchilla, place one hand under the torso while the other hand grasps the base of the tail.

FIGURE 11–51 A chinchilla taking a dust bath.

Basic Health Care and Maintenance

Yearly physical exams are recommended for chinchillas. They should have frequent dental exams, as their teeth continuously grow and they can develop difficulty with eating. As they groom themselves like cats, they should be monitored for hairballs, called **trichobezoars**. Chinchillas can develop respiratory problems, so signs of difficulty—breathing, wheezing, and sneezing—should be monitored.

Vaccinations

Chinchillas do not require vaccinations, and no vaccines are currently labeled for chinchillas.

Reproduction and Breeding

Chinchillas have a spontaneous estrus cycle that ranges from 30 to 50 days. The gestation length is 111 days.

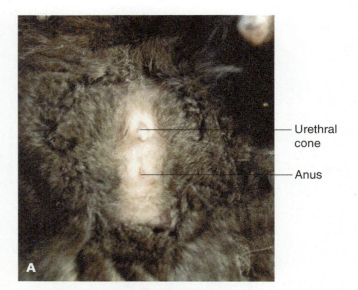

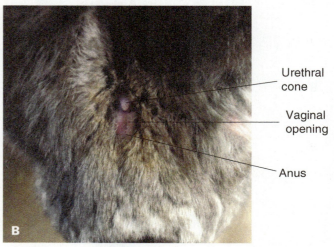

FIGURE 11–52 (A) Genital area of the female chinchilla. (B) Genital area of the male chinchilla.

The babies should be weaned between 3 and 8 weeks of age. Chinchillas are sexed in the same manner as in other rodents (see Figure 11–52). The female has a large urinary papilla that can be confused with a penis. In males, the penis can be protruded with manual pressure. Males do not have a true scrotum, and the testicles are contained within the abdomen or the inguinal canal.

Common Diseases

Chinchillas are healthy pocket pets but should be monitored for dental disease, signs of heatstroke, and respiratory problems.

Common Parasites

Chinchillas may have flea issues, but both external and internal parasites are rare.

SUMMARY

Rodents include some of the most interesting and popular companion animals. Gerbils, rats, guinea pigs, mice, and hamsters are the most popular of the pocket pets. It is important for the veterinary assistant to understand proper handling, health care, housing needs, nutrition, and reproductive care of common rodent species kept as pets. This area of veterinary medicine is becoming more popular in the animal industry as well as in specialty medicine.

Key Terms

agouti a mixture of two or more colors

anogenital the area located around the far stomach between the rear legs and the base of the tail

antibiotics medications used to treat infection and disease

barbering the tendency of mice to chew at the fur of another mouse

benign noncancerous

burrow to dig a tunnel into an area

cannibalism the act of rodents eating their young

cheek pouch an area within the mouth where food is stored

defensive reaction a behavior shown by an animal to protects itself from danger

diurnal being active during the day and sleeping at night

dust bath a method of grooming in which a chinchilla rolls in a dusty material to condition its skin and hair coat

hibernation a natural condition in which an animal goes into a long sleep and its body systems slow down until the animal's body temperature rises

hooded a variety of rats that are white in body color with black or brown coloring on the head and shoulders, giving it a "hooded" look

jird a burrowing rodent

malignant cancerous

murine a veterinary term for mice or rats

nocturnal sleeping during the day and being active at night

offspring the product of the reproductive processes of an animal

penile area in males, where urine is passed to the outside of the body

quills spines that project from a hedgehog's body to protect it

rodents mammals that have large front teeth designed for chewing or gnawing, such as rats, mice, and hamsters

scrotum in males, the sac of skin that holds the testicles

solitary alone and independent behavior

testicles in males, two glands where sperm is produced

topical applied to the skin

trichobezoar a hairball

urethra in females, where urine is passed to the outside of the body

vitamin C an essential nutrient and supplement

wet tail in hamsters, a condition caused by a bacterial disease that spreads rapidly through direct contact or bacterial spores

Review Questions

1. Explain the differences in the estrus cycles of the hamster and the guinea pig.
2. Why should ferrets be spayed if they are not going to be used for breeding?
3. What is the difference between nocturnal and diurnal?
4. Why do rodents develop respiratory problems?
5. What nutritional needs do guinea pigs have, and why?
6. Why are rodents commonly called "pocket pets"?
7. What common bacterial disease in hamsters causes diarrhea?
8. Which pocket pet species may be given vaccines as part of a health program?

Clinical Situation

Mr. and Mrs. Kline bring in their daughter's ferret, Mikey, a 6-year-old NM that has been showing signs of lethargy and has not been not eating well. The Klines are concerned that the ferret may not survive. Nicole, the veterinary assistant, is getting a weight on the animal and noting its listless appearance.

"How long has Mikey not been feeling well?" she asks the Klines.

"He has been sick for 2 or 3 days," they answer.

"Well, Dr. Davies will be in shortly to examine him. I am certain he will be fine.

Ferrets rarely get sick; he probably just has a cold." As Nicole leaves the exam room, the Klines feel confused and start talking about how they thought that ferrets *did* have several possible health problems. Wasn't that the information they had been given?

- What was done incorrectly in this situation?
- How would you have handled the situation?

CHAPTER 12: Rabbit Identification and Production Management

Objectives

Upon completion of this chapter, the reader should be able to:

12.1 Define common veterinary terms relating to rabbits
12.2 Explain the biology and development of the rabbit
12.3 Identify common breeds of rabbits and breed selection
12.4 Discuss the nutritional requirements of rabbits
12.5 Define normal and abnormal behaviors of rabbits
12.6 Explain how to properly and safely restrain rabbits for various procedures
12.7 Discuss the health care and maintenance of rabbits
12.8 Detail breeding and reproduction in rabbits
12.9 Describe common diseases affecting rabbits

Introduction

Rabbits have become very popular indoor pets and companion animals (see Figure 12–1). They are also commonly bred for meat and pelt production. Rabbits require special handling and have distinct health care needs compared to rodent species.

Veterinary Terminology

lagomorph = rabbit
buck = intact male rabbit
doe = intact female rabbit
lapin = neutered male rabbit
kit = young (blind, deaf) rabbit
kindling = giving birth to rabbits
herd = group of rabbits
junior = rabbit under 6 months of age
senior = rabbit over 6 months of age

226 SECTION II Veterinary Animal Production

FIGURE 12–1 A rabbit is a popular pet and companion animal.

is hardy and must continue to digest constantly to break down heavy foods, such as hay and grass. Common external body part terms are identified below (see Figure 12–2):

- Ears—long, floppy flaps located on top of the head
- Flank—lower thigh area of the rear legs
- Hock—joint or point of the rear leg
- Loin—the middle back area
- Neck—area behind the head
- Rump—lower back area just above the tail
- Tail—small cotton ball–like structure extending off of the back
- Toe—individual digits located on the feet

Figure 12–3 illustrates the skeletal structure of the rabbit.

Rabbits were developed for many purposes, including meat production, fur production, and research. They are used to show and to keep as pets. Rabbits have an average lifespan of 5 to 6 years. A normal body temperature is 101 to 104 degrees Fahrenheit; a healthy heart rate is 130 to 325 beats per minute and respiratory rate 30 to 60 breaths per minute.

Biology

Rabbits are mammals that are herbivores. They have long ears, a round body, a very short tail, and four large teeth. They are related to hares, which have a larger body and longer ears with black tips on the end. Rabbits are endothermic (they control their body temperature internally) but have a different digestive system compared to dogs and cats. Like horses, rabbits are **nonruminants**, which means they have one simple stomach that is made to filter and break down grasses and roughages. They do not vomit and lack a gall bladder. The digestive tract

Breeds

Rabbits come in a variety of breeds, colors, and sizes. There are about 45 breeds of rabbits classified by body shape and size. Sizes range from miniature to giant.

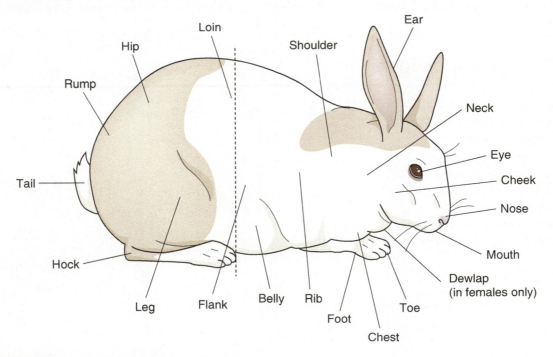

FIGURE 12–2 External anatomy of the rabbit.

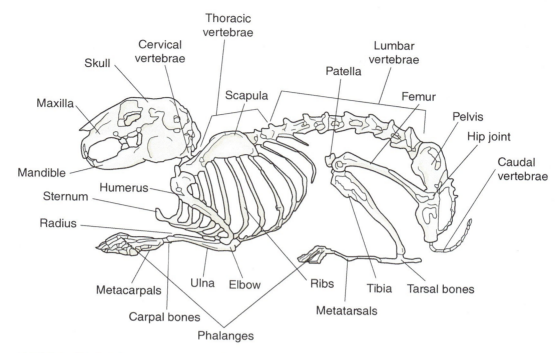

FIGURE 12–3 Skeletal anatomy of the rabbit.

Miniature rabbits weigh less than 2 pounds as adults and are sometimes also called **dwarf** rabbits. The Mini Lop and Netherland Dwarf are two examples (see Figure 12–4). Small-breed rabbits weigh between 2 and 7 pounds as adults and include Dutch and Polish breeds (see Figure 12–5). Medium rabbits weigh between 8 and 12 pounds and are sometimes referred to as **commercial** breeds. Examples include the New Zealand and Rex rabbits (see Figure 12–6). Large-breed rabbits weigh between 12 and 14 pounds. They are also known as **semi arch** breeds and include the Californian and Chinchilla breeds (see Figure 12–7). The largest group is the **giant** or **full arch** breeds, which weigh over 14 pounds. An example is the Flemish Giant (see Figure 12–8).

Breed Selection

Rabbit selection is dependent on the size, hair coat, and health and maintenance needs of the breed. Rabbits come in a variety of sizes and hair coat types. Prior to purchasing them as pets, potential owners should research breeds to determine the amount of care and maintenance they will require. Each animal should be evaluated for temperament; they should be docile and easy to handle. Rabbits are active and should be monitored for good health.

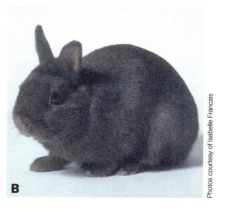

FIGURE 12–4 Miniature rabbits: (A) Mini Lop. (B) Netherland dwarf.

FIGURE 12-5 Small-breed rabbits: (A) Dutch. (B) Polish.

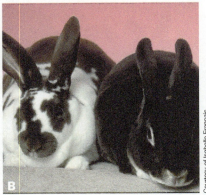

FIGURE 12-6 Medium or commercial breed rabbits: (A) New Zealand. (B) Shorthaired Rex.

FIGURE 12-7 Large or semi arch breeds: (A) Californian. (B) Chinchilla.

FIGURE 12-8 Giant or full arch breed: Flemish Giant.

Nutrition

Rabbits are herbivores and eat a plant-based diet. They are usually fed a commercial rabbit food in pellet form (see Figure 12-9). They should be fed high-quality hay, such as timothy or orchard grass that is available **free choice** (or at all times). Changes in food will cause an upset in the digestive system. Rabbits enjoy fruits,

FIGURE 12-9 Commercial rabbit feed.

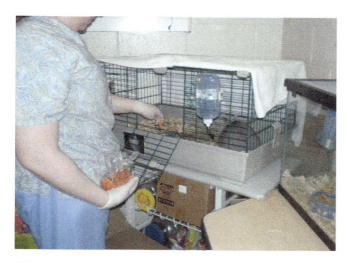

FIGURE 12-10 Rabbits can be fed fresh vegetables in small amounts.

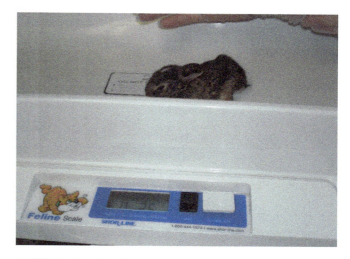

FIGURE 12-11 Orphaned bunnies should be weighed daily to monitor weight gain.

vegetables, and green foods that should be given in small amounts (see Figure 12–10). Too much of these items will cause digestive problems. Rabbits should not be overfed, as they will become obese. Rabbits should eat about 4 percent of their body weight daily as a maintenance diet.

Rabbits are coprophagic (dung eating): they will continue to feed during the night by eating their own feces, which become slightly softer at night. The feces contain broken-down plant materials and thus provide nutrition for the digestive system. These **night feces** are rich in vitamins and protein.

Sometimes young rabbits will become orphaned and require formula feeding. If this occurs, bunnies should be weighed daily to monitor weight gain (see Figure 12–11). Bunnies may be given a similar formula to what kittens and puppies eat, or a commercial product, such as KMR

TABLE 12–1
Formula Requirements for Young Rabbits

Newborn	5 ml	Divide into two daily feedings
One week old	12–17 ml	Divide into two daily feedings
Two weeks old	25–27 ml	Divide into two daily feedings
Three weeks old	30 ml	Divide into two daily feedings
Three weeks to weaning	30+ ml	Divide into two daily feedings

formula. If a commercial formula is used, an egg yolk should be added for protein. Newborns should receive 5 ml of formula daily, divided into two feedings. Table 12–1 outlines the formula requirements for young rabbits.

Behavior

Rabbits are nice, quiet, and gentle pets; most have docile attitudes. On occasion, a rabbit may produce a warning scream or growl to let you know that it is unhappy. Rabbits do have an instinct to protect themselves by biting, kicking, and striking with the front limbs. They will thump a foot as a warning sign that they are upset. Rabbits can be hypnotized for short periods of time by holding them upside down and rubbing their abdomen (see Figure 12–12). The term **hypnotized** means put into a trancelike (still) state.

Basic Training

Like cats, rabbits are trainable and can be litter-box trained (see Figure 12–13). Many pet rabbit owners allow them to have some free range in the household. This will require training the rabbit not to chew and demonstrate destructive habits. Rabbits can be trained fairly easily with consistency and time. It is important to note that rabbits will chew on anything, so if they are to be given free range, check that area for any safety issues, such as telephone cords or electrical wires.

Equipment and Housing Needs

Rabbits are often housed in a cage known as a **hutch:** this cage is designed for a rabbit and has a covered area that allows the rabbit to seek shelter

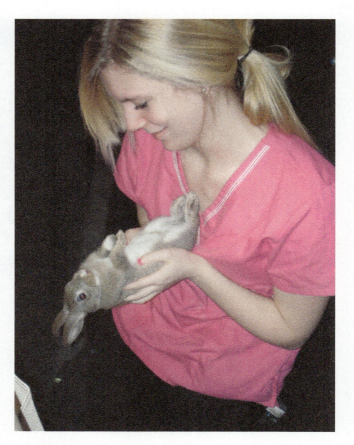

FIGURE 12–12 Holding a rabbit upside down and rubbing its abdomen will cause hypnosis for a short period of time.

FIGURE 12–13 Rabbits can be litter-box trained.

from the weather and an open area that provides proper ventilation. This pen is typically made of wire and has a solid wood bottom. A hutch should be free of drafts and provide 1 square foot of space per rabbit. Today, many people keep rabbits as pets

CHAPTER 12 Rabbit Identification and Production Management

FIGURE 12–14 Rabbit housing.

FIGURE 12–15 Carriers should be used when transporting a rabbit.

FIGURE 12–16 When handling a rabbit, always support the rear limbs.

inside the household. Rabbits must be monitored closely, as they are curious, love to chew, and can get injured if they bite into electrical wires or ingest certain items. Rabbit breeders will typically provide nest boxes, feeders, and water bottles inside a cage or the hutch.

A rabbit may also be housed in a commercial pen made of metal, wire, plastic, or wood. Cages should be made of solid materials and provide enough space for the rabbit (see Figure 12–14). When choosing a cage, remember that rabbits chew! The bottom of the cage may be metal or wire with a tray that collects waste materials. Cages should be cleaned on a regular basis, using shredded paper, pelleted litter, or dust-free shavings.

Restraint and Handling

Improper handling and restraint of rabbits can result in injury to the rabbit or to the restrainer. Rabbits should never be lifted or restrained by the ears. Cat carriers should be used when traveling with a rabbit (see Figure 12–15). Rabbits can be "hypnotized" by gently rubbing the belly for short periods of time. When carrying or lifting a rabbit, support the rear legs and back (see Figure 12–16). If a rabbit kicks its rear limbs, it may break its back or injure its spine. Wrapping a rabbit in a towel can help keep it calm and properly restrained (see Figure 12–17).

Grooming

Rabbits require regular brushing, which may include bathing and grooming depending on the hair type of the rabbit (see Figure 12–18). They should also be evaluated

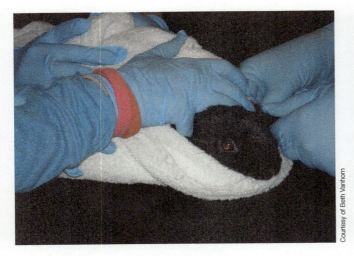

FIGURE 12–17 A towel can help restrain and keep a rabbit calm.

FIGURE 12–19 A rabbit's teeth should be checked regularly and trimmed if necessary.

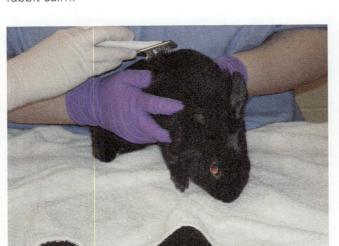

FIGURE 12–18 Rabbits require regular brushing.

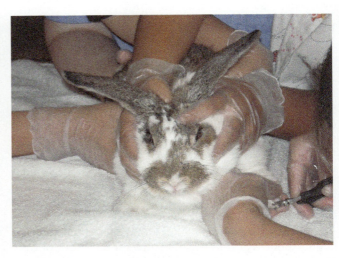

FIGURE 12–20 It is necessary to periodically trim a rabbit's nails.

for dental problems (see Figure 12–19). A rabbit's nails and teeth continue to grow during its lifetime and may require regular trimming and care (see Figure 12–20), requiring small handheld nail trimmers, styptic powder, and an oral speculum. Rabbits are self-groomers and clean their coats on a daily basis.

Basic Health Care and Maintenance

Rabbits are relatively healthy animals but can easily develop respiratory problems if they are not properly housed. Rabbit cages must be cleaned on a regular basis. Dust-free bedding should be used to keep cages clean and well bedded to prevent sores on feet.

Avoid cool and drafty cage locations, as rabbits do get colds.

Rabbits are prone to diarrhea from food changes and stress. They can also become **constipated** and unable to produce a bowel movement. Therefore, changes in diet should be made slowly over a few weeks to allow the digestive system time to adjust. Feline laxative medicines are commonly used. A **laxative** (also called a stool softener) can be given to allow the animal to have a bowel movement. Like cats, longhaired breeds are prone to hairballs (trichobezoars). Administering hairball medications or pineapple or papaya juice can help break down and prevent hairballs.

Rabbits should be monitored for sores on the legs and feet, especially in the area of the **hocks** (the back of the feet). Rabbits have long rear feet that they use

CHAPTER 12 Rabbit Identification and Production Management 233

FIGURE 12-21 Rabbits should be weighed as part of their annual examination.

FIGURE 12-22 Rabbits build nests and can be provided a nest box in the hutch.

to hop around, and pushing on the wires or hard surfaces of a cage can cause sores.

A change in a rabbit's behavior may indicate illness. Rabbits should receive a physical exam on a yearly basis and be taken to a veterinarian that specializes in rabbits if any signs of illness occur. It is important to note that if a rabbit seen in a veterinary facility requires blood work or surgery, the hair should be **plucked**, not shaved, from that area of the body. Plucking is pulling hair to remove it from the skin and is preferable to shaving, which will tear a rabbit's very delicate skin.

A regular weight check should also be taken (see Figure 12-21).

Vaccinations

Rabbits are not commonly given vaccines. In cases where certain diseases are more common or if show regulations require certain vaccinations, it should be discussed with the veterinarian.

Reproduction and Breeding

Rabbits reach maturity at about 6 months of age. The female rabbit should be taken to the male's cage for breeding, as she is more territorial of her environment and will fight and possibly injure the male. Mating should be monitored and the male and female separated after breeding has occurred. The gestation length of rabbits is 30 to 32 days. Females will begin building nests just prior to the labor process. Nest building typically includes materials such as straw, paper, or shavings, or a nest box can be provided (see Figure 12-22). The female will pluck hair from her abdomen to line the nest. Most rabbit breeds have an average litter size of six kits. Like puppies and kittens, baby rabbits are born hairless, with the eyes and ears closed. The young are known as **precocial**, which means they grow very quickly (see Figure 12-23). Baby rabbits are able to move away from danger in a short time, which is

FIGURE 12-23 (A) Baby rabbits grow quickly. This one is being examined in the veterinary facility. (B) Rabbits are born hairless, and their eyes and ears are not open.

an adapted instinct. The kits will nurse only one to two times daily. The eyes open in 10 days. When a rabbit's eyes open, it can begin eating pellets, hay, and leafy vegetables. Rabbits are usually weaned between 6 and 8 weeks of age.

Sexing rabbits is accomplished by placing gentle pressure over the skin at the genital opening. One may need to press the skin back to see the opening. Females have an elongated vulva, with a slit opening; males have a rounded, protruding penile sheath. The dewlap, or excess loose skin under the neck, is more prominent in females.

Common Diseases

Rabbits are prone to several types of diseases and health conditions. A healthy rabbit will be active, eating well, and showing signs of a bright and alert appearance. It is important for the veterinary assistant to be aware of common diseases that affect rabbits.

Respiratory Disease

Rabbits are prone to respiratory disease. They tend to develop excessive nasal discharge, often called "runny nose," from poor cage ventilation, dusty bedding, high temperature and humidity, dental disease, and foreign objects in the nasal passage. The respiratory cause must be determined to treat the condition. Signs of respiratory disease in rabbits include the following:

- Nasal discharge
- Sneezing
- Clumping of hair on the front paws

Pasteurella

Respiratory disease can sometimes be wrongly diagnosed when the actual cause is the bacteria **Pasteurella**, a naturally occurring bacteria in the respiratory tract of rabbits. Pasteurella is commonly called "snuffles": it is similar to a cold and may be an acute or chronic inflammation of the mucous membranes in the air passages and lungs. Mucus is discharged from the nose and eyes. Afflicted rabbits will rub their eyes and noses. The fur on the face and paws may become matted and caked with dried mucus (see Figure 12–24). The infected rabbit typically has bouts of sneezing and coughing. Rabbits with this disease may suffer from a depressed immune system and increased amounts of stress. This condition is treatable with antibiotics. Rabbits

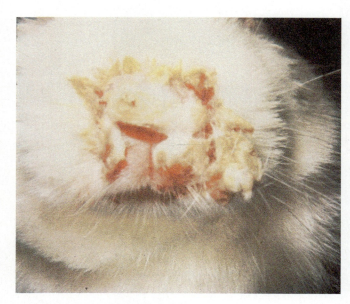

FIGURE 12–24 Snuffles in a rabbit.

that survive often remain carriers of this disease. It is recommended that any rabbits showing signs of this condition be isolated and a strict sanitation program be followed.

Tyzzer's Disease

Tyzzer's disease is another common bacterial disease. This infection affects the intestinal tract of rabbits and is spread through spores in the air. With no treatment or vaccine, prevention is the key and includes frequent sanitizing of the cage, eliminating dust in the bedding, and avoiding overcrowding of the rabbits. Signs of Tyzzer's disease in rabbits include the following:

- Watery diarrhea
- Feces or staining on rectum
- Anorexia
- Lethargy
- Dehydration
- Weight loss
- Death

Pododermatitis

Pododermatitis is a common condition in rabbits that affects the bottom surface of the rear feet and is commonly referred to as "sore hocks" or "bumblefoot." It is seen more often in obese and giant-size rabbits. Pressure sores over the bottom surface of the feet develop and become necrotic. Proper husbandry and

FIGURE 12–25 The rabbit's fur should be checked for signs of parasites.

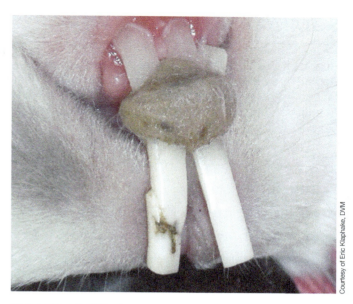

FIGURE 12–26 Overgrown teeth in a rabbit.

sanitation are key to the prevention and treatment of this condition.

Common Parasites

Rabbits commonly develop external parasites, such as fleas, ticks, and ear mites. Regular grooming practices will decrease the occurrence of these parasites. Fleas and ticks may appear on the hair or skin coat and cause itching and skin irritation. Ear mites produce a thick, dark, and crusty waxlike substance in the ear. Rabbits will scratch excessively at the ears, which may cause blood vessels to break, resulting in a **hematoma**, or blood-filled mass, within an area. Ear mites are treatable with topical ear medications. The skin and hair should be evaluated for signs of parasites (see Figure 12–25).

Internal parasites may also occur in rabbits and can be detected by a yearly evaluation of a fecal sample. A common intestinal protozoan in rabbits is coccidia. It is difficult to rid rabbits of these parasites. The organism may infect a rabbit's stomach or liver. Signs of the parasite in the rabbit include a potbelly appearance, diarrhea, weight loss, and a loss of appetite. A rabbit with a mild case may show no signs of the parasite.

Common Surgical Procedures

Common surgical procedures in the rabbit include castration, spay, and dental extraction procedures. Rabbits are frequently neutered to prevent reproduction and to decrease aggression and behavioral problems. Most rabbits can be castrated at 3 to 4 months old, although some veterinary professionals have recommended even earlier castration because rabbits may become sexually mature before 4 months of age. Castration decreases urine spraying as well. Most rabbit castrations are done in a similar way to cat castration surgeries.

Rabbit ovariohysterectomy or spay surgery is helpful in decreasing aggression and breeding as well as false pregnancies and reproductive tumors. The surgical procedure is similar to spaying a dog or cat. The surgery is typically recommended between 5 and 6 months of age but may be done earlier due to sexual maturity.

Dental problems and extractions are another common occurrences in rabbits. A rabbit's teeth continuously grow. Rabbits have six incisor teeth in the front and several cheek teeth. The cheek teeth sometimes overlap, which can cause trauma to the mouth and make it difficult for the rabbit to eat properly. Signs of dental disease in rabbits include the following:

- Dropping food
- Grinding teeth
- Salivating excessively
- Anorexia
- Nasal discharge

The teeth may need to be trimmed or extracted depending on the extent of the disease. **Malocclusion** is a condition in which the teeth become overgrown and dental problems develop (see Figure 12–26). When this occurs, the teeth need to be trimmed every 2 to 3 weeks.

SUMMARY

Rabbits make excellent pets: they can be housebroken, litter trained, and allowed to roam inside the house. Rabbits are also used in production programs for meat and fur. Rabbits are often referred to as rodents but are in fact an entirely different species. Rabbits require special handling and have particular health care and nutritional needs. Veterinary assistants should understand the basic care that is required and be able to properly retrain and handle rabbits in a clinical setting.

Key Terms

commercial refers to medium-sized rabbits that weigh between 8 and 12 pounds

constipated unable to pass waste materials and feces

dwarf this small rabbit breed weighs less than 2 pounds as an adult; see *miniature*

free choice food made available at all times or as necessary

full arch a giant rabbit breed that weighs over 14 pounds

giant a rabbit breed that weighs over 14 pounds

hematoma a collection of blood under the skin

hocks the tarsal joints at the point of the rear leg

hutch a special cage designed for a rabbit that has a covered area that shelters the rabbit from the weather and an open area that provides proper ventilation

hypnotized put into a trancelike state

laxative a stool softener

malocclusion the overgrowth of teeth, causing dental problems

miniature also called dwarf; a small rabbit breed that weighs less than 2 pounds as an adult

night feces waste that a rabbit passes at night that is softer, rich in vitamins and proteins, and consumed by the rabbit

nonruminant an animal that has a specialized digestive system with a simple stomach built for slow digestion

Pasteurella a naturally occurring bacteria in the respiratory tract of rabbits

precocial growth that occurs very quickly in bunnies that allows them to move away from danger in a short time

plucked hair pulled out by hand

semi arch a large rabbit breed that weighs between 12 and 14 pounds

Tyzzer's disease a bacterial infection in rabbits spread through spores in the air that affects the digestive system

Review Questions

1. Explain the nonruminant digestive system.
2. What are night feces, and what is their importance in rabbit nutrition?
3. What is the length of gestation in rabbits?

4. What is a trichobezoar, and how is it prevented in rabbits?
5. What are the nutritional needs of an adult rabbit?
6. How are rabbits housed?
7. What are the grooming and health needs of rabbits?
8. Why is it important to "pluck" the fur on a rabbit?
9. What injuries can occur to a rabbit's hocks?
10. What are signs of respiratory disease in rabbits?

Clinical Situation

Miss Crawford, a new rabbit owner, calls the veterinary clinic to ask several questions about her new bunny. She has recently purchased an Angora rabbit. She asks what health program the rabbit needs, what kind of care she will need to provide for the rabbit, and if the rabbit should be examined like her kitten "Louie" was after she bought him.

Jin, the veterinary assistant, takes the phone call. He says, "Hello, Miss Crawford. It would be ideal to have your bunny examined. Are you noticing any problems or concerns?"

"No, he seems very healthy. Actually, I don't even know if it is a boy. I have never owned a rabbit before and have very little information. Can we set up an exam for sometime next week?"

"Yes, that sounds good. Can you come Friday at 1 pm?" asks Jin.

"Yes. In the meantime, can you give me some information of what I should be doing or what supplies I need to buy?"

- How would you answer Miss Crawford's questions?
- What information should Miss Crawford be given immediately?
- Was this situation handled effectively?

Chapter 13: Reptile and Amphibian Breed Identification and Production Management

Objectives

Upon completion of this chapter, the reader should be able to:

- 13.1 Define common veterinary terms relating to reptiles and amphibians
- 13.2 Describe the biology and development of a reptile and amphibian
- 13.3 Identify common breeds of reptiles and species selection
- 13.4 Identify common breeds of amphibians
- 13.5 Identify the nutritional requirements of reptiles and amphibians
- 13.6 Describe normal and abnormal behaviors of reptiles and amphibians
- 13.7 Explain how to properly and safely restrain reptiles for various procedures
- 13.8 Explain how to properly and safely restrain amphibians for various procedures
- 13.9 Describe the health care maintenance needs of reptiles and amphibians
- 13.10 Explain the process of breeding and reproduction in reptiles and amphibians
- 13.11 Explain common diseases affecting reptiles and amphibians

Introduction

Reptiles and amphibians are becoming more popular as pets, which make knowledge and understanding of the health care of these companion animals crucial. These delicate pets have particular needs and require a certain environment.

Reptiles and amphibians are alike in several ways. Reptiles include species of snakes, lizards, turtles, crocodiles, and alligators. The most commonly kept reptile species are lizards and snakes (see Figure 13–1). There are over 6500 species of reptiles. Common examples of amphibians are frogs, toads, and salamanders.

Veterinary Terminology

Herpetology is the study of reptiles and amphibians. Some species are **aquatic** or live in the water (see Figure 13–2). Others are **terrestrial** or live on the land. Others are **semi-aquatic** or live both in the water and on land (see Figure 13–3). They need proper environmental conditions and temperatures to survive.

Snakes are commonly called **serpents**. Turtles are commonly called **chelonians** or **terrapins**. A group of reptiles or amphibians is called a **brood**. When an animal is carrying live young or eggs the term is **gravid**, or pregnant. A group of eggs that have been laid is called a **clutch**. Reptiles and amphibians may also lay eggs or bear live young, depending on the species.

CHAPTER 13 Reptile and Amphibian Breed Identification and Production Management

FIGURE 13–1 A lizard is a popular pet reptile.

FIGURE 13–2 Some reptile and amphibian species live in aquatic conditions.

FIGURE 13–3 Some reptile and amphibian species live in water and on land.

Reptiles and amphibians are known to hibernate when their body temperatures decrease. In these species, this process is called **brumation**. In this semi-hibernation state, the animal sleeps a majority of the time and eats very little.

Many of these species have specific terms for their anatomy. For example, the **cloaca**, or rectum, is also commonly known as the **vent**. This is the area that contains the genital organs and excretes wastes. The **crest** is the top of the animal's head.

Reptiles and amphibians are housed in a **terrarium**, a sealed glass container with plants and soil inside. The **substrate** is the type of floor material that is provided in the cage.

Both reptiles and amphibians may be **venomous** (poisonous) or **nonvenomous** (nonpoisonous). They may be herbivores, carnivores, or **omnivores** (eat both meat and plant sources).

Biology of Reptiles

A **reptile** is an animal with dry, scaly skin that can adjust its body temperature according to the outside temperature and environment. This type of process is called **ectothermic**. Reptiles are commonly described as *cold-blooded*, because they feel cool to the touch. Reptiles have scales that serve as a protective layer over their skin. They shed the outer layer on a regular basis. Snakes shed their skin all at once; before this process begins, the skin becomes sensitive and opaque in color. Lizards shed their skin in pieces or sections.

Snakes

Snakes have long, thin bodies and tails, but no legs, ears, or eyelids (see Figure 13–4). They have a smooth skin texture but are not slimy as many people think. Common external body part terms include the following:

- Body—long, thin central area of the snake
- Fang—long teeth in the front of the mouth
- Forked tongue—thin area in the mouth that has a slit in the center and is used to smell food
- Head—front part of the snake, shaped like a triangle
- Neck—area behind the head
- Rattles—located on the tail of certain species and used to sound a warning of danger
- Scales—outer skin covering the body
- Tail—end of the body

Figure 13–5 illustrates the anatomical structure of the snake.

FIGURE 13–4 A snake has a long thin body and tail, but no legs, ears, or eyelids.

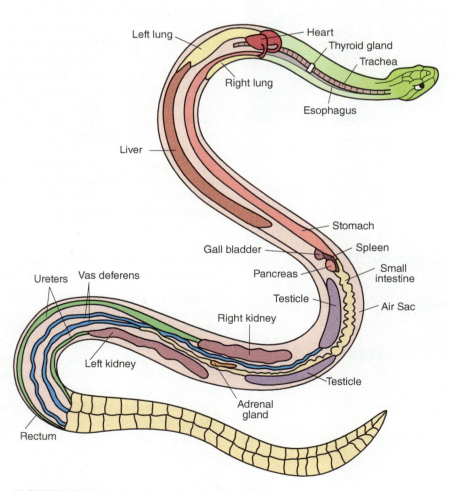

FIGURE 13–5 Internal anatomy of the snake.

Lizards

Most lizards have four legs, long tails, movable eyelids, and ear openings. Iguanas tend to be the most popular lizard species (see Figure 13–6). Common external body part terms include the following:

- Caudal crest—spines over the lower back
- Crest—neck area behind the head
- Dewlap—loose skin fold under the chin
- Dorsal crest—spines over the middle of the back
- Ear—hole in either side of the head behind the eyes
- Jowel—area of the head below the ear holes
- Spines—long projections over the back
- Tail—long extension off of the back
- Vent—opening on the underside of the body behind the rear legs

Figure 13–7 illustrates the anatomical structure of the lizard.

Turtles

The turtle has a shell and four legs and can pull its head, limbs, and tail into the shell for protection (see Figure 13–8).

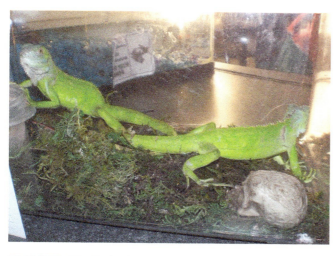

FIGURE 13–6 An iguana is a popular lizard species often kept as a pet.

Courtesy of Isabelle Francais

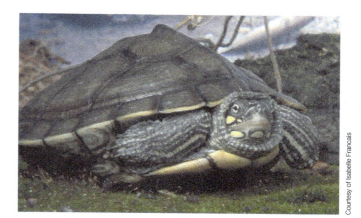

FIGURE 13–8 Some species of turtles can be kept as pets.

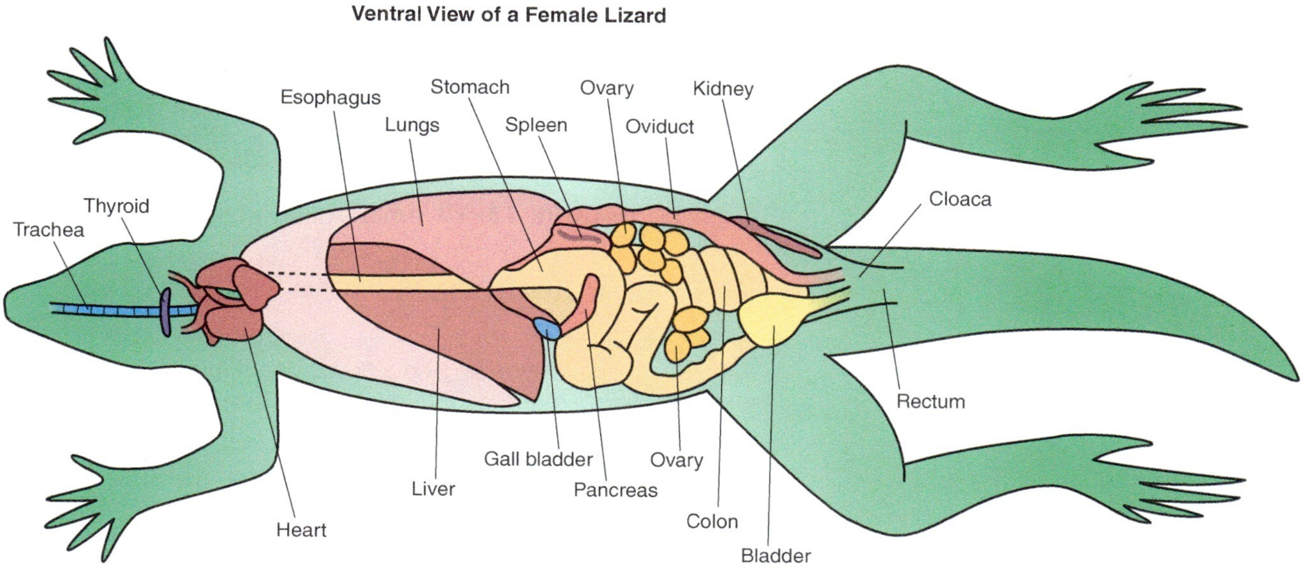

FIGURE 13–7 Internal anatomy of the lizard.

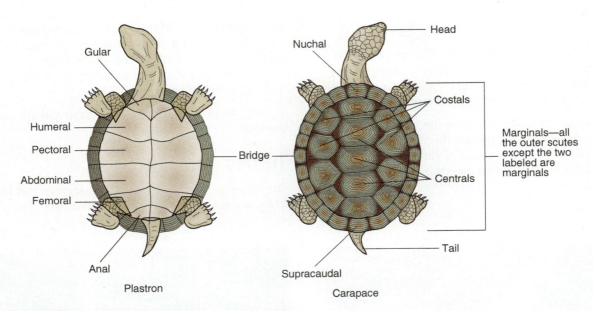

FIGURE 13–9 External anatomy of the turtle.

Turtles can live on land or in the water. Common external anatomy terms include the following (see Figure 13–9):

- Carapace—hard covering over the upper body; upper shell
- Head—located at the front of the body and shell
- Hinge—area where the upper and lower shells meet on either side of the body
- Plastron—hard covering over the bottom of the body; lower shell
- Scutes—individual plates over the shell
- Shell—hard covering over the body
- Tail—located at the back of the body and shell

Biology of Amphibians

An **amphibian** is an animal with smooth skin that spends part of its life on land and part in the water.

FIGURE 13–10 American bullfrog.

Frogs and Toads

Frogs and toads are very similar. They spend their **juvenile** (young) stages in the water. As adults, they have four legs and no tail (see Figure 13–10). Common external body terms include the following (see Figure 13–11):

- Dorsal surface—back area of the body
- Nares—nostril openings on the face
- Tympanum—ear opening behind the eyes on either side of the head
- Ventral surface—belly area of the body

Salamanders

Salamanders have a long body and tail and two or four legs, depending on the species (Figure 13–12). Common external body part terms include the following:

- Body—midsection of the salamander
- Crest—neck area behind the head
- Tail—long extension off of the back
- Vent—opening on the underside of the body behind the rear legs

Figure 13–13 illustrates the anatomical features of the salamander.

CHAPTER 13 Reptile and Amphibian Breed Identification and Production Management 243

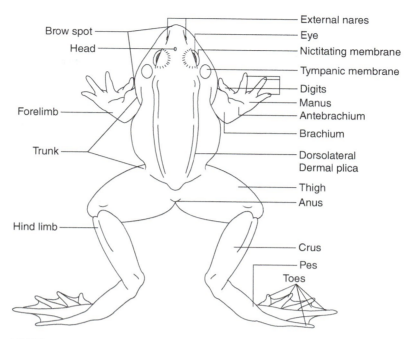

FIGURE 13–11 External anatomy of a frog.

FIGURE 13–12 Salamander.

Reptile Breeds

Reptile species come in a variety of sizes and colors. They can live for 10 to 20 years, depending on the species and the kind of health care they receive. Amphibian species also come in a variety of sizes and colors and have an average lifespan of 5 to 20 years or longer, depending on the species.

Snakes

Snakes come in a variety of breeds. Those that are typically kept as pets include the boa constrictor, python, and green snake. These snakes grow to varying sizes: some can be over 12 feet long. Many larger species of snakes constrict to eat live prey. Snakes enjoy heat sources, and cage temperatures should range from 85 to 90 degrees Fahrenheit. Snakes will climb. Most breeds of snakes make good pets, but the larger species are not recommended for children. Figure 13–14 shows a picture of a juvenile ball python.

Lizards

Lizards come in a variety of sizes and colors. Popular pet lizard breeds include iguanas, geckos, chameleons, and anoles: the green iguana is the most popular pet iguana breed (see Figure 13–15). Lizards tend to be high maintenance because they

244 SECTION II Veterinary Animal Production

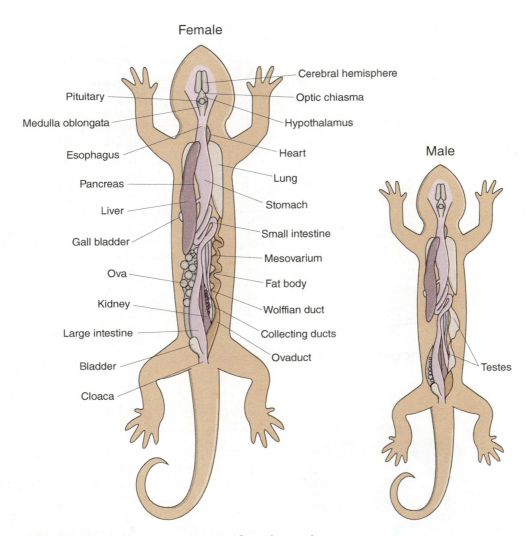

FIGURE 13–13 Internal anatomy of a salamander.

FIGURE 13–14 Juvenile ball python.

require strict diets and particular environmental care: if their needs are not met, they may suffer nutritional and health problems. The best starter lizards are the geckos and anoles. The most commonly kept gecko breeds are the leopard gecko (see Figure 13–16), fat-tailed gecko, and crested gecko. The green anole is

FIGURE 13–15 Green iguana.

CHAPTER 13 Reptile and Amphibian Breed Identification and Production Management

FIGURE 13–16 Leopard gecko.

a common small lizard breed that is readily available, requires little tank space, and does not need special lighting. Bearded dragons and some varieties of skinks are also popular lizard breeds. Chameleons tend to be slightly higher maintenance, with the veiled chameleon, Jackson's chameleon, and the Panther chameleon the most common.

Turtles

Turtles are ideal pets that require little maintenance and are relatively easy to care for. Some commonly kept turtle breeds include the Red-Eared Slider, map turtles, and box turtles (see Figure 13–17). Any turtle or hatchling under 4 inches in size is illegal to purchase and house in the United States.

Amphibian Breeds

Amphibian species come in a variety of sizes and colors and have an average lifespan of 5 to 20 years or longer, depending on the species.

Frogs

Frogs are easily acquired as pets, very entertaining, and relatively easy to care for. Common pet breeds include the bullfrog, green frog, and tree frog (see Figure 13–18).

Salamanders

Salamanders are less frequently kept as pets but relatively easy to care for. The most common breeds are the tiger salamander (see Figure 13–19) and the spotted salamander.

Salamanders have a head, body, limbs, and tail but spend part of their time on land and in water. They resemble lizards but have smooth skin versus scales.

FIGURE 13–17 (A) Red-eared turtle. (B) Map turtle. (C) Box turtle. (D) Tortoise.

FIGURE 13–18 Green tree frog.

FIGURE 13–19 Tiger salamander.

Reptile and Amphibian Breed Selection

People who have reptiles as pets must research the species and understand its environmental and health needs. Many of these species require specialized diets, proper housing, warm temperatures, and adequate cage space. They must be handled carefully and properly housed to prevent escape. Reptiles have important nutritional needs. Some reptiles reach very large sizes and may outgrow their cages. They can become aggressive. Some species will require more care and time than others. It is important to note that reptiles and amphibians are natural hosts for salmonella bacteria, which may be a potential health problem for some people.

Nutrition

The nutritional requirements of reptiles and amphibians must be researched and met: it is vital to their well-being and survival.

Snakes

Snakes are relatively carnivorous. Many people try to train snakes to eat dead prey, especially when they feed them rodents. A rodent can bite a snake, causing trauma or wounds. However, some snakes will not eat dead prey. It is best to observe the snake until it has killed its prey so that injuries do not occur. Food should be placed on a smooth surface so that particles from the substrate (sand, pebbles, pellets, water, cedar chips) are not ingested. The amount and frequency of feedings is based on the size and species of the animal. Snakes have developed a digestive system that is capable of digesting whole prey and defecating body parts that are not digested, such as hair or fur. The entire carcass provides added nutrients, such as calcium from bones.

Lizards

Lizards that are herbivores include the iguana species. They will eat dark green leafy vegetables that are high in calcium. All plant materials should be washed, chopped, and served at room temperature or slightly warmer. Newborn lizards should be fed twice a day. Juveniles up to 2.5 years of age are typically fed once a day. Adult species will vary from once-a-day feeding to every 48 hours. Most lizards eat in the early morning or late afternoon. Some common iguana foods include the following:

- Dark green leafy vegetables
- Grapes
- Kale
- Collard greens
- Dandelions
- Parsley
- Spinach
- Alfalfa pellets
- Turnip greens

Lizards that are carnivores should be fed dead prey. Live rodents will bite and cause trauma to lizards. Most carnivorous lizards are fed once a day. Many lizards that are fed a meat-based diet will require calcium supplements (ground beef sources do not provide adequate amount). Common meat-based foods include the following:

- Mealworms
- Crickets
- Grasshoppers
- Katydids
- Ants
- Pinkie mice
- Fish
- Snails

Turtles

Many turtles are mostly herbivores as adults. In the juvenile stages, most species are more carnivorous than they will be later in life. As a turtle ages, its need for a meat-based diet decreases and its need for a calcium-based diet increases. Many species remain omnivores and eat a variety of foods in their diets. Most tortoise species are herbivores, eating grasses, clover, leaves, and flowers. Snapping turtles are carnivores, eating insects, fish, worms, and snails. Box turtles are omnivores, eating a variety of fruits, vegetables, flowers, insects, worms, and small fish. Red-eared sliders, map turtles, and painted turtles eat water plants, worms, insects, and small fish.

Frogs and Toads

Amphibians in general need food from a variety of sources. They may require more food each day in the winter months. Frogs and toads that are terrestrial require insects, flies, crickets, and mealworms. Large toads may eat pinkie mice and adult rodents. Aquatic species eat a variety of insects, worms, fish, and commercial pellets. They should be fed in the water. Tadpoles should have access to **algae**, is a plant-based source of food that grows spontaneously in water when left in the sunlight. They can also be fed boiled dark green leafy vegetables. Tadpoles also need vitamin C to grow. They should be fed several times a day or they will cannibalize (eat) each other. They should stop feeding as their legs begin to form. When they reach adulthood, they should gradually be introduced to their new diet.

Salamanders

Salamanders eat insects, worms, flies, shrimp, and insect larvae. Many commercial diets have been formulated into pellets for reptiles and amphibians.

Behavior

Most reptiles and amphibians are not relatively social animals. They may become accustomed to humans, but excessive handling may cause stress and result in their refusing to eat.

Snakes

Snakes are relatively docile when handled in their early years and can grow accustomed to people. Many species will bite when they are upset or scared. Constrictors will strike and apply pressure by curling around parts of another body. Each species should be researched for other normal behaviors. Snakes will go through a skin shed cycle known as **ecdysis** (see Figure 13–20). When a snake begins the shedding cycle, the eyes will appear cloudy, and the skin may develop a milky coloration. At mid-cycle, the eyes and

FIGURE 13–20 Snake shedding its skin.

skin turn clear. The skin will shed about 1 to 4 days after this occurs. Snakes should not be handled during the shed cycle, as their skin is very fragile and the new skin may be damaged. For some snakes and some lizards, the tongue has a sensory function: they can use it to smell and hear vibrations through the air and ground. The tongue will flick as they sense their surroundings.

Lizards

Lizards are highly territorial, or protective of their cages. Males are more so than females. Males will be more aggressive with other males than with females. Women who are menstruating should use caution when handling lizards. Male lizards are capable of detecting hormonal changes and odors and their pheromones can lead to behavioral changes, such as aggression.

Turtles

Most species of turtles and tortoises are docile and easy to handle. When scared, they will pull their head, legs, and tail into the shell for protection. Contrary to popular belief, several species of turtles are fast and can move readily when necessary.

Frogs and Toads

Frogs and toads are docile and easy to handle. Some species have the ability to emit a bad odor when frightened or scared as a protective behavior. They tend to be more vocal during the evening and during breeding season.

Salamanders

Salamanders are relatively docile and easy to handle. Salamanders are lively, excellent climbers and very curious. Some salamanders are nocturnal; others are active during the day.

Basic Training

In general, reptiles' and amphibians' basic training includes being handled and accepting of humans. This is achieved by frequent and proper handling techniques. Socializing young reptiles and amphibians will allow them to become more accepting of handling.

Equipment and Housing Needs

All reptiles and amphibians should be kept in secure enclosures and provided with appropriate environments. Substrates should be selected carefully. Bark and wood-chip products are not safe for most species. They may ingest these items and the digestive tract may become obstructed. Substrates for cages include the following:

- Newspaper
- Wood or pine shavings
- Synthetic bedding
- Branches
- Sand
- Lab animal bedding pellets

Glass aquariums and homemade Plexiglas cages are ideal for most species. Snakes are escape artists and should have a well-sealed cage. Branches of various heights and sizes should be provided. Many species are **arboreal** and require items to climb on. Some containment areas for hiding, such as logs, rocks, or boxes, may be included for privacy. Food and water bowls should be able to be easily accessed. Cages should be sanitized on a regular basis and food and water bowls cleaned daily. It is important to regulate temperatures for reptiles and amphibians. Most species of reptiles require a cage temperature of 75 to 85 degrees Fahrenheit, with a relative humidity of 60 to 80 percent. This can be provided through the use of heat rocks, heating pads, and incandescent light bulbs (see Figure 13–21). A thermometer should be placed in the cage to monitor daily temperatures. Some heat items can cause burns.

Daytime temperatures may need to be increased and nighttime temperatures decreased to provide a natural environment. Sunlight or a light source should

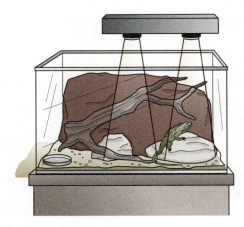

FIGURE 13–21 A glass aquarium with a heat source is an ideal home for reptiles or amphibians.

be provided for 12 to 14 hours on a daily basis. UV light sources may be required for species that cannot metabolize nutrients or synthesize vitamin D. The light should be full spectrum and kept at about 18 to 24 inches away from the pet. UV bulbs should be changed every 6 months. Systems that require water for semi-aquatic animals will require a **filtration system** to clean the water and provide oxygen. The optimum pH level of the water should be 6.5 to 8.5 and clear. The pH level of water is the acidic balance that is required for an animal to survive.

Amphibians are commonly kept in an aquarium that has an area covered in water and an area above the waterline that is dry and often covered in pebbles, stone, soil, or wood. Many pet amphibians have been captured from the wild. The quality and care of the water are important for the health of amphibians. Frequent water changes and a good filter system eliminate nitrogen waste buildup in the water.

Restraint and Handling

It is important to be familiar with the manner in which to safely handle and restrain reptiles and amphibian species.

Snakes

The most important factor to consider with snakes in a veterinary setting is whether they are poisonous. Many veterinary facilities are not equipped to deal with venomous species. Snakes may be handled by hand, with restraint poles, or within anesthesia chambers (see Figure 13–22).

Snakes that are accustomed to human handling should be restrained by holding the head with one hand and the body with the other hand (see Figure 13–23). Larger species of snakes may require more restrainers. It is best to avoid allowing larger species of snakes to wrap around any body parts of the restrainer. Some snakes may need to be placed in a bag or pillowcase for capture and carrying. Small species should be handled with care to avoiding holding them too tightly and causing injury (see Figure 13–24).

Lizards

Lizards should be well restrained at both the front end of the body, including the head, and the tail end.

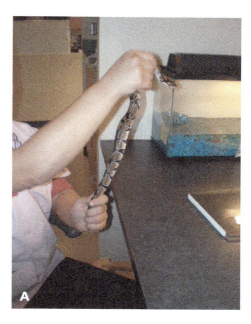

FIGURE 13–22 The use of an anesthesia chamber can help visualize and examine a pet snake.

FIGURE 13–23 (A) Proper method for handling a snake. (B) Proper handling includes secure control of the head and body.

FIGURE 13–24 Care should be taken not to hold the snake too tightly.

FIGURE 13–25 Proper handling technique for a lizard.

Many species of lizards can whip their tail and cause injury to the handler. It is best to control the front limbs and head with one hand and the rear limbs and tail with the other hand (see Figure 13–25). Larger species of lizards may require two or more restrainers. Lizards are capable of biting, and some iguanas and larger species can cause a serious bite wound. Lizards that are not accustomed to restraint may be difficult to handle. Lizards should never be picked up or handled by the tail. Many species can strip their tail or break off the tail stump as a defense mechanism for protection. Small lizards should be held carefully, as holding them too tightly can cause injury.

Turtles

Relatively docile, turtles are easy to handle; the most irritating instance in restraint is their ability to retract into their shell (see Figure 13–26). When lifting and restraining a turtle, hold the body of the shell firmly between one or both hands. To prevent a turtle from retracting into the shell, place the index finger under the shell and in front of the hind limbs. The shell will not close but may cause a pinch, depending on the size of the animal. Some turtle species may bite, causing injury.

Frogs and Toads

Frogs and toads may be picked up and restrained using one or both hands. One hand holds and restrains the head and the other hand restrains and holds the body. Larger species may be suspended in the air with the rear limbs hanging. Small species should be handled with care to avoid causing injury.

Salamanders

Salamanders can be quick. It may be necessary to use a net for capture if they are in water or a tank. When picking up and restraining a salamander, one hand should control the head and front limbs and the other hand should control the body and rear limbs. Aquatic species may need a water source to soak in. Larger species may cause injury with their tail. Salamanders should never be lifted or handled by the tail: it will be stripped and break off.

Grooming

Some reptile species will need regular bathing; the animal can be cleansed using a water source in the cage, a water bath, or a spray bottle (see Figure 13–27). Some species with nails may require routine nail trims.

FIGURE 13–26 A turtle's ability to retract into its shell can make handling and restraint a challenge.

FIGURE 13–27 Some reptile or amphibian species will require a water source for bathing.

Basic Health Care and Maintenance

Reptiles and amphibians should be observed daily for changes in behavior or appearance.

- Check the teeth for dental problems.
- Examine the vent for waste material buildup.
- Note **urates** (urine) and waste output.
- Take a fecal sample at the yearly physical examination.
- Monitor weight for any changes.

All new animals should be **quarantined** (isolated from others) for 2 to 3 months to make sure they are disease free. All reptiles and amphibians should be handled with caution: they have natural amounts of **salmonella** bacteria on their skin, which can create a potential zoonotic concern. Wear exam gloves and wash hands after handling and cleaning the cage (see Figure 13–28).

FIGURE 13–28 Care should be taken to wear gloves and wash hands after handling a reptile or cleaning its enclosure.

Vaccinations

Reptiles and amphibians do not require vaccinations.

Reproduction and Breeding

Reproduction and breeding in reptiles and amphibians is a popular hobby. Many species are easy to breed. Some animals may become aggressive and should be monitored for aggression and injuries. It is easy to determine the gender of some reptile and amphibian species; in others it is not. It is important to know that a male and female are paired together when breeding.

Snakes

Before breeding snakes, it is important to determine the gender, as many species of snakes look alike. Snakes are sexed by the **probe method**. A small metal probe is inserted into the area of the **hemipenis** (male reproductive organ) on the edge of the cloaca, or rectum (see Figure 13–29). In males, the hemipenis opening is much larger than in females, which only have scent glands in this area. The depth of the probe method is based on the length of the snake's scales. Males will probe between 7 and 12 scales, depending on size and species. Females will probe only 2 to 4 scales. Males and females can be kept together for breeding purposes. Many species are aggressive and should be separated after breeding. Some species of snakes bear live young; others lay eggs that must be **incubated**, or warmed until hatched, similar to birds. Incubation of snake eggs may be done in a nest, or the eggs may be laid in the ground and naturally incubated. Garter snakes and boa constrictors bear live

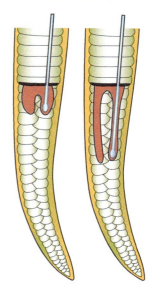

FIGURE 13–29 Determining the sex of a snake.

young. King snakes and pythons lay eggs. Gestation or incubation lengths vary from species to species. Garter snakes range from 90 to 100 days' gestation; boa constrictors range from 120 to 240 days. Burmese pythons have an incubation period of 56 to 65 days. A ball python's incubation is 90 days.

Lizards

Lizards can be sexed by visual exam. Males have taller dorsal spines over the back and tail, a larger **dewlap** (loose flap of skin under the chin), and bulges at the base of the tail. In males, the hemipenis can be **everted** (expressed or exposed to the outside of the body) with light pressure over the base of the tail. Some lizard species have horns or plates that distinguish male and female. Many lizard species undergo color changes during the breeding season (see Figure 13–30).

Males and females should be kept together for breeding purposes only, as they will become aggressive with each other. Iguanas, geckos, and anoles are egg layers; chameleons bear live young.

Table 13–1 outlines the incubation and gestation of common lizard species.

TABLE 13–1

Species and Incubation and Gestation Period

SPECIES	INCUBATION/GESTATION
Green Anole	60–90 days; eggs
Green Iguana	73–93 days; eggs
Jackson's Chameleon	90–180 days; live young
Leopard Gecko	55–60 days; eggs

Turtles

Turtles and tortoises can be sexed using a visual exam. Males have a longer tail and a wider tail base than females. The cloaca is more caudal and the shell area around the cloaca is **concave** (rounded inward) (see Figure 13–31). In some species, the males have longer claws on the front limbs. Females are usually larger in size. The male hemipenis can be expressed using gentle pressure and warm water over the vent area. Most male and female turtles can be kept together without fighting. Most turtle

FIGURE 13–30 Color change occurs in males when it is ready for breeding.

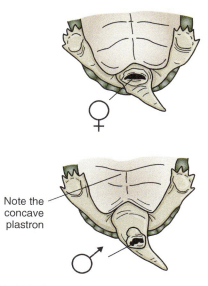

FIGURE 13–31 Determining the sex of turtle.

species lay eggs and have a variety of incubation periods. The box turtle has an incubation length of 50 to 90 days and the red-eared slider turtle 53 to 93 days. The incubation period for most tortoise species is about 60 days. Amphibians typically have some **dimorphism** characteristics, which means that the male and female are different. In most species, the female is larger in size. During the breeding season, many male species have an evident skin-color change.

Frogs and Toads

Frogs and toads begin as **tadpoles**. This is the larval stage of the newborn. Each species has a different time frame for when it reaches the adult stage.

Common Diseases

The most common concern in reptiles and amphibians is the possibility of salmonella infection. An estimated 3 percent of the population owns a reptile or amphibian. An estimated 70,000 people contract salmonella from contact with a reptile. Salmonella is passed to humans most frequently from such animals as snakes, lizards, and turtles. **Salmonellosis** is the bacterial infection that affects people. All reptile and amphibian owners should be educated on the possible risks and prevention of salmonella. Children and people with a compromised immune system are at increased risk and should avoid contact with these animals. Anyone who handles an animal or touches a cage should wash their hands after thoroughly, using soap and water. Reptiles and amphibians should be kept in a cage and not be allowed to freely roam the house. Any area that the animal comes in contact with should be properly sanitized.

Infections from wounds and burns are common among reptiles and amphibians because of poor breeding and inadequate housing. Keeping multiple reptiles and amphibians together can lead to fighting and aggression, and bite wounds become easily infected. Bacterial infections may also occur from injuries within the cage. Prevent animals from rubbing against sharp objects by setting up the cage with safety in mind. Thermal burns can occur from direct contact with hot surfaces, usually hot rocks, light bulbs, or direct sunlight. Provide a physical barrier to such objects within a cage. When such injuries occur, an animal will need treatment with antibiotics and topical medications to prevent further health issues.

Common Parasites

Reptiles and amphibians can become infected with internal and external parasites, including hookworms, roundworms, tapeworms, and pinworms. Also of concern is **coccidia**, a one-celled parasite that affects all animals and is passed in contaminated water sources, commonly from bird feces. Care should be taken when handling pets and cleaning their cages. Remember that mites and ticks are visible to the eye. Mites may come from substrate or items placed in the cage.

Common Surgical Procedures

Most reptiles and amphibians do not require routine elective surgical procedures. However, many species may be curious and ingest foreign objects that do not pass through the digestive system. These situations may require surgical repair. It is recommended that only veterinarians specializing in reptiles and amphibians attempt these procedures.

SUMMARY

Reptiles and amphibians are becoming more popular as companion animals. Their health and environmental needs are the responsibility of the pet owner; owners may well rely on veterinary staff members to educate them on the proper care of these animals. Many of these species are delicate and can easily become injured or sick if the owners are negligent. Therefore, it is important that the veterinary assistant be knowledgeable about the behavior, nutritional needs, health care, breeding, and disease control of these creatures.

Key Terms

algae a plant-based source of food that grows spontaneously in water when left in the sunlight

amphibian an animal with smooth skin that spends part of the time on land and part of the time in water

aquatic an animal that lives in the water

arboreal an animal that lives in trees

brood a group of reptiles or amphibians

brumation a process in reptiles in which their body temperature decreases; similar to hibernation

chelonian a veterinary term for turtle

cloaca the reptile's rectum; also called the vent

clutch a group of eggs

coccidia a single-celled parasite and protozoan passed in contaminated water sources

concave rounded inward

crest the top of an animal's head

dewlap a flap of skin under the chin

dimorphism the male and female having different body parts/appearances

ecdysis the process in which a snake sheds its skin

ectothermic the body temperature of an animal, controlled externally by the environment

evert to turn inside out

filtration system a mechanism to clean water and air

gravid pregnant, in reptiles

hemipenis the male reproductive organ, in reptiles and amphibians

herpetology the study of reptiles and amphibians

incubated eggs kept warm for hatching

juvenile young stage of growth

nonvenomous not poisonous

omnivores animals that eat a meat- and plant-based diet source

probe method the process of passing an instrument into the cloaca to determine the sex of a reptile or amphibian

quarantined isolated

reptile an animal with dry, scaly skin that adjusts its body temperature according to the outside temperature and environment

salmonella a bacterial infection that causes severe diarrhea

salmonellosis a bacterial infection that occurs from touching reptiles or amphibians

semi-aquatic living part of the time in the water and part of the time on land

serpent a veterinary term for snake

substrate the material on the bottom of a cage

tadpole a young newborn frog or toad

terrapin a veterinary term for turtle

terrarium a plant-sourced housing for reptiles and amphibians

terrestrial living on land

urate the urine waste material of reptiles and amphibians

venomous poisonous

vent the rectum in reptiles; also called the cloaca

Review Questions

1. Describe the characteristics of the snake, lizard, and turtle.
2. What are the differences between a reptile and an amphibian?
3. Explain the probe method of sexing a reptile.
4. What is the difference between gestation and incubation?
5. What are the signs of a reptile going through a shed cycle?
6. Why should reptiles and amphibians not be given food that is alive?
7. What are some good examples of iguana foods?
8. List some ideal substrates for reptiles or amphibians.
9. What is the ideal temperature for reptiles?
10. How long should quarantines occur for reptiles or amphibians?

Clinical Situation

Mr. Charles, an elderly client, calls the veterinary clinic about a turtle that he has just purchased as a gift for his grandson. He has some questions and concerns about his grandson handling and playing with the turtle.

"Hello, I'm calling about a red-eared slider turtle I just got for my grandson Billy. I was wondering if there are any health concerns with the turtle. And how should Billy hold it?"

The veterinary assistant, Katie, has had several turtles. She replies, "Mr. Charles, I have owned many turtles. They make excellent pets. I never had any issues with mine. Billy will love the turtle! As far as picking the turtle up, he should hold it by its shell. He can also let it walk around the house if he likes."

"Oh, I am so glad to hear that. Someone at the pet shop mentioned some concerns about handling it, and after I left the store I started to worry. Thank you so much for answering my questions."

- Did Katie provide the proper information?
- How would have you answered Mr. Charles's questions?
- What could potentially happen in this situation?

CHAPTER 14
Ornamental Fish and Aquaculture Identification and Production Management

Objectives

Upon completion of this chapter, the reader should be able to:

14.1 Explain the common veterinary terms relating to ornamental fish

14.2 Describe the biology and development of ornamental fish

14.3 Identify common species of ornamental fish and species selection

14.4 Explain guidelines for feeding ornamental fish

14.5 Discuss safe methods for capturing and handling ornamental fish

14.6 Explain the water sources and housing requirements of ornamental fish

14.7 Explain how to establish an appropriate water environment for ornamental fish

14.8 Discuss a health management plan for ornamental fish

14.9 Describe breeding and reproduction in ornamental fish

14.10 Discuss common diseases and health conditions of ornamental fish

14.11 Define veterinary terms relating to aquatics

14.12 Discuss aquatic biology

14.13 Identify common species of aquatic animals

14.14 Discuss the nutritional needs of aquatics

14.15 Explain the importance of aquaculture

14.16 Discuss the reproductive methods of aquatics

14.17 Describe common diseases and conditions of aquatics

14.18 Discuss aquaculture production systems

14.19 Discuss and detail aquaculture management

14.20 Describe water quality

Introduction

Ornamental fish are a large and popular industry in the United States. Many homes have a variety of sizes of aquariums that house all types of fish. These fish are more valuable as pets than food fish due to their small size and low weight. Ornamental fish require specialized housing and supplies. The popularity of ornamental fish production has evolved from people discovering that watching fish swim and interact in their aquatic environments provides stress relief. **Aquatic** species are decreasing in number in their natural environments, as water sources are drying up or becoming polluted. *Aquaculture* has become an industry that provides the resources for such animals and also allows a market for the species of aquatics used for human consumption. Aquaculture is a large industry for major food sources in the United States. Aquatic animals include such species as **fish**, shrimp, alligators, and lobsters.

Veterinary Terminology

Ornamental fish are primarily **omnivores**. Omnivores are animals that eat both plant and meat sources of food. There are no specific terms relating to the male and female fish. However, newborn fish are called **fry**. As fish mature, they are referred to as **yearlings** at one year of age and **fingerlings** in their second year of life. A group of fish is called a **run**, or **school**.

Ornamental Fish Biology

Ornamental fish are fish species that are kept for their color and appearance and because of a special appeal to people. Many species and breeds of ornamental fish have been developed based on color, fancy fins, interesting characteristics, and overall enjoyment. These fish are not raised as a food source. **Companion fish** are species kept in people's homes as pets or companions. They are kept for entertainment, amusement, and leisure. **Tropical fish** form another group of ornamental fish. They are small, brightly colored fish that require warm water and are naturally found in warm tropical locations. Most ornamental fish are not native to the United States or North America. Many species are found in South America, Asia, and the Pacific Islands. Hawaii is the largest U.S. state that has the largest amount of native fish species found in a tropical environment. Several states have the temperature and environment for producing quality fish species that have been imported from other locations. Florida is the leading U.S. state in ornamental fish production. Releasing nonnative fish species into the wild is illegal. Many fish species are captured in their natural locations and raised in confined breeding programs. There is a large **diversity** (having distinct qualities) of ornamental fish. The overall body shape has adapted fish to their specific habitat. **Surface-dwelling fish** live and feed near the surface or top of a body of water. **Bottom-dwelling fish** live and feed on the bottom or floor of a body of water. Surface-dwelling species have an upturned mouth and a flat back. Bottom-dwelling fish have flat bodies and small mouths. Tall, laterally compressed species, such as the discus or angelfish, are made for slow-moving waters. Slender streamlined fish are made for fast-moving waters.

Fins have been adapted for movement, floating, and breeding purposes (see Table 14–1). They may be single or paired. Many aquarium fish have long fins that are ideal for tanks. They have been developed through selective breeding. The **caudal fin** is also the tail fin. It is used to move the fish through the water. The **dorsal fin** is located on the back of the fish and is used to steer it through the water.

Most fish species are covered by **scales**. These body coverings serve as protection. The color of fish is determined by pigment and light reflection. **Pigment** represents the color of the skin. Some species are dark in color to blend into their environment. Others are brightly colored to attract mates. **Gills** are the organs that allow fish to breathe. Through the gills, fish absorb oxygen to exchange gases within the water (see Figure 14–1). Fish have a **lateral line** that is an organ located just under the scales that picks up vibrations in the water. This is what allows fish to detect danger, food, and navigate the water. Fish also

TABLE 14–1

Tail Fin Shapes

Forked tail fin	Fast swimmers
Round tail fin	Quick movements
Long tail fin	Attract mates

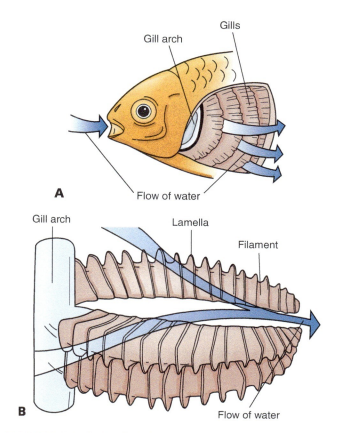

FIGURE 14–1 (A) Fish breathe through organs called gills. (B) Each gill consists of many filaments.

258 SECTION II Veterinary Animal Production

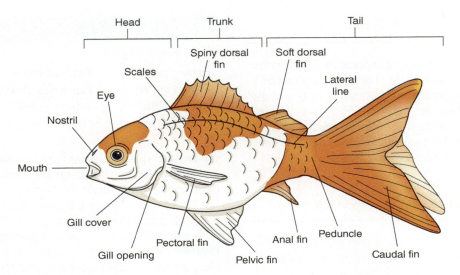

FIGURE 14–2 External anatomy of the fish.

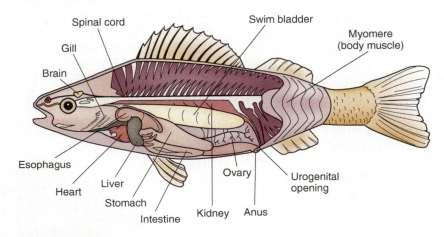

FIGURE 14–3 Internal anatomy of the fish.

have an organ called a **swim bladder** that allows them to float. The swim bladder is an air-filled sac that prevents fish from sinking. Common external body terms include the following:

- Anal fin—cartilage area located on the underside of the back of the body
- Caudal fin—cartilage area located at the base of the body, acting as a tail
- Dorsal fin—spiny cartilage area located in the center of the upper back
- Ear flap—opening located on either side of the head behind the eyes
- Gills—openings holding the lungs located on either side of the body in front of the pectoral fins and behind the head
- Lateral line—located on either side of the body, running along the entire body and lying just under the scales
- Pectoral fin—cartilage area located on both sides of the body, acting as arms
- Pelvic fin—cartilage area located under the pectoral fins on the underside of the front part of the body

Figures 14–2 and 14–3 illustrate the external and internal anatomy of the fish.

Ornamental Fish Breeds

Fish may live in freshwater or saltwater. **Freshwater fish** are species that live in freshwater that has no or very little salt. They are the most popular ornamental fish. **Saltwater fish** require salt in their water source. This water is a complex mixture of salt and various minerals. They require a synthetic seawater environment commonly called a marine aquarium.

Goldfish

Goldfish are one of the oldest breeds of ornamental fish (see Figure 14–4). They have been raised and bred for over 2000 years. Goldfish are small and commonly a bright orange color, but many new, fancy colors and interesting characteristics have been developed. Fancy fins and large eyes, as well as a variety of colors, have been bred. They are hardy and adapt well to a wide range of environments. Their size depends on the size of the tank they are housed in and their nutrition. Some are 1 to 2 inches in size and others can grow as large as 2 feet. Goldfish require about 2 gallons of fresh water to grow 2 inches in size. They are great beginner fish. They require regular cleaning, as they are a messy species compared to other fish. They have been known to live as long as 15 years. They can lay up to 10,000 eggs, which hatch in 3 to 6 days.

Koi

Koi are a variety of carp that look similar to large goldfish (see Figure 14–5). They have been developed by selective breeding for certain colors. Colors range from bright red, black, white, and gold to a combination of colors. They are freshwater fish that may reach a length of 3 to 4 feet and weigh up to 10 pounds or more. Most koi are kept in ponds or large tanks that may be housed indoors or outdoors.

Labyrinth Fish

Labyrinth fish are nest builders; the males maintain the nest and incubate the eggs. The nests are built from bubbles that the male builds within plants. Eggs hatch in 24 to 30 hours. Labyrinth fish come in a variety of sizes.

Gourami

Gouramis come in a variety of breeds (see Figure 14–6). They are large freshwater fish with interesting characteristics. They grow to 12 inches or more. Examples include the kissing gourami, blue gourami, and three-spot gourami. Gouramis lay eggs.

Barb

Barb fish are a popular species (see Figure 14–7). They are easy to raise and very hardy. They are related to goldfish and require fresh water. They typically range in sizes of 2 to 4 inches. They need plenty of light. Caution should be used with breeding, as some barbs eat their eggs after spawning.

FIGURE 14–4 Bubble-eye variety of goldfish.

FIGURE 14–5 Koi.

FIGURE 14–6 Kissing gourami.

FIGURE 14–7 Tiger barb.

FIGURE 14–8 Neon tetra.

Tetra

Tetras tend to require more care than other species of fish. The most common breed is the neon tetra, known for an electric-blue stripe running the entire length of the body (see Figure 14–8). Many grow to lengths of 1 to 3 inches. They lay eggs that hatch within 30 hours.

Catfish

Catfish have been developed into ornamental fish in small sizes and varieties. Most species prefer dark areas and like to hide in objects within the freshwater tank. They feed off the bottom of the tank and can be used to keep the tank clean. Favorite breeds include the upside-down catfish, which swims upside down, and the glass catfish, which is clear and may have a colored appearance (see Figure 14–9). Catfish lay about 100 eggs, which hatch in 4 to 5 days.

Guppy

Guppies are the most popular live-bearing fish. They come in a variety of colors and fancy-shaped tails and fins. They need larger amounts of food than other species. They usually reach lengths of 2 to 3 inches. Females tend to give birth to 200 babies at a time; only about 50 babies will survive. They give birth every 4 to 6 weeks. Many adults will eat the young if they are not separated. They are easy to raise and tend to be a hardy type of fish.

Swordtail

Swordtails are named for their long caudal fin. They come in a variety of colors and range in lengths of 3 to 5 inches. They bear live young and live in fresh water. The adults tend to eat the babies and should be separated from their young. They eat a variety of food.

FIGURE 14–9 (A) Glass catfish. (B) Upside-down catfish.

Molly

There are two types of Mollies: small-fin and large-fin breeds. Some large-fin fish can grow so large that they have difficulty swimming. Most mollies are black, but breeders breed for selective colors (see Figure 14–10). They live in large freshwater groups and bear live young.

Platy

Platys are a popular aquarium fish that come in a variety of bright colors. They grow to about 3 inches in length and live in fresh water. They are easy to keep and very hardy. They also bear live young.

FIGURE 14–10 Black molly.

FIGURE 14–11 Angelfish.

Angelfish

Angelfish come in a variety of colors, shapes, and sizes (see Figure 14–11). Most live in saltwater, but a few species can live in freshwater. They appear delicate but are hardy and can live a long time in a well-managed tank. They grow to an average of 6 to 10 inches. They lay eggs that hatch in about 30 hours. They are mouth brooders and will place the eggs on plants or sand sources prior to hatching.

Basslets

Basslets are small colorful fish that are popular in saltwater tanks. They are hardy fish and excellent for beginners just starting with saltwater fish. They tend to do best with other species: they can be aggressive with their own kind.

Butterfly Fish

Butterfly fish are beautiful, colorful saltwater fish. They require a large aquarium space. They range in size from 6 to 10 inches. They are territorial, best isolated from other species, and do best when housed alone.

Breed Selection

When putting together an aquarium, it is important to think about whether the fish will be compatible. Many pet shops and ornamental fish breeders offer guidance about passive and aggressive fish breeds and which should not be placed in the same tank. Size is another consideration: how big will the fish become, and how much space will they need? Water type and plant sources are also important and will be covered in this chapter.

Nutrition

It is easy to overfeed fish. The amount of food a fish should be given each day is called the *ration*. Fish should be fed rations in a diet that meets the needs of their species. The amount of food a fish eats daily is directly related to the temperature of the water in its tank, the type of fish, and its life stage. As water temperatures rise, fish need more food. Fish should finish eating their food within a few minutes: careful observation will allow an owner to know how much food to give their fish. Overfeeding creates more waste products in the water. Most people feed their fish a commercial fish food made from flakes or pellets. Most of these foods include protein and grain sources. It is important to read the list of ingredients on the label to determine the recommended amount for certain species. Table 14–2 summarizes the typical nutritional needs for common breeds of fish.

Equipment Needs

Ornamental fish require a specialized environment to survive and stay healthy: it must be artificially created to act as their natural habitat. The needs will vary for each species. For example, a goldfish can survive in a basic system, whereas an angelfish or gourami requires a more complex system.

TABLE 14–2

The Nutritional Needs of Fish

SPECIES	DIET NEEDS	SUPPLEMENTS
Cichlids	Commercial flakes	Seaweed
Angelfish	Commercial flakes; frozen bloodworms	Brine shrimp
Barbs	Commercial flakes	Brine shrimp
Bettas	Betta flakes; freeze-dried bloodworms	Brine shrimp
Catfish	Commercial flakes	Brine shrimp; sinking wafers
Discus	Discus flakes; frozen discus formula	Freeze-dried bloodworms
Goldfish	Goldfish flakes	Commercial flakes
Gouramis	Commercial flakes	Brine shrimp
Loaches	Commercial flakes	Freeze-dried brine shrimp
Tetras	Commercial flakes	Brine shrimp

Basic equipment begins with a water container. These items can be decorated with gravel or pebbles, plants, and other creative structures to add eye appeal. Many types of fish water containers are available commercially. The most common type for ornamental fish is the **tank**, or **aquarium**. These containers hold water and come in a variety of shapes and sizes. The most basic tank or aquarium is a small fish bowl, which is a rounded glass bowl that holds about a gallon of water. It may be used for a single fish that is hardy and can survive in water that does not need added oxygen. The next size of a basic tank is the rectangular or square glass aquarium. This ranges in size from 10 gallons to 100 gallons or more. All aquariums should be watertight and sealed to prevent leaking. Glass is preferred, especially with saltwater tanks, and is easier to clean with a lower risk of leaks. The glass should be between ¼ and ⅜ inches thick. Figure 14–12 shows a typical aquarium setup. The most important terms in the operation of an aquarium are the following:

- Air hose—allows flow of oxygen into the tank
- Air pump—controls the flow of oxygen into the tank
- Cover—lid that lies on top of the tank
- Filtration system—removes gases and solid materials from the water
- Gravel—pebbles located on the bottom of the tank that hold plants and other objects
- Heater—electric source that warms the water
- Light—electric source that lies within the cover to provide light into the tank
- Thermostat—area that provides the water temperature in a number scale

Basic and complex tank systems require proper equipment and supplies to maintain the tanks and the quality of the water. It is also necessary to maintain the proper health of the fish.

An indoor or outdoor **pool** or **fountain** may be used to house ornamental fish. They are usually decorative and may contain several species of fish, such as koi and goldfish. The species that live in these types of water containers must be hardy. The structures are usually made of plastic, fiberglass, or concrete materials. They must be watertight and easy to clean.

A **vat** is a large tank of water commonly used for reproduction and breeding systems. Vats are made of concrete or fiberglass heavy enough to hold hundreds to thousands of gallons of water. They must have proper water circulation (movement), aeration (oxygen or air), and drainage to keep the fish healthy.

Ponds are mostly found outside for larger species of hardy fish. They must be constructed for the climate and usually require proper heating sources and a way to protect fish from predators.

Oxygenation is the process of adding air to water through dissolved oxygen (DO). With their gills, fish remove and use the oxygen that has been dissolved in the water to breathe. Water that has little to no movement has very little oxygen. Fish that do not

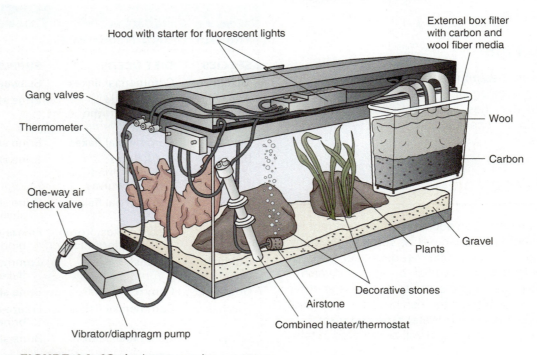

FIGURE 14–12 A glass aquarium system.

have adequate amounts of oxygen will die. There are several ways to add oxygen to a tank. The first is **aeration**, a process that creates bubbles within the water to create movement and allows gases to transfer from the water to the air. Oxygen is thus created in the water. Another way to get oxygen into the water system is with an air pump. The pump forces air into the tank through a plastic tube located in the bottom of the tank. In some ponds or vats, an air pump may lie on the surface of the water. Finally, some tanks can be oxygenated through the presence of live plants. This is known as **biological oxygenation**. Tiny plants called **plankton** and other types of live plants grown within the tank provide a natural source of oxygen to the water. This growth is called **photosynthesis** and allows oxygen to be released into the water naturally.

Filtration is a way to rid the tank water of solid waste materials and gases to provide a healthy environment for fish. There are several filtration methods that may be used. First, **biological filtration** uses bacteria and other living organisms in the water to change harmful materials into safer forms. Bacteria feed on fish wastes, uneaten food, and gases created in the water. Snails, crawfish, loaches, and catfish may also be used to scavenge or remove foreign matter from the water. Through this method, the bacteria change nitrogen and ammonia gas forms into dissolved oxygen. **Mechanical filtration** consists of equipment that removes harmful particles and keeps the water clear. Water flows over a filter placed in the water. Filters may be made of gravel, charcoal, and fiber materials called **floss**. Filters must be cleaned often or they will become clogged with waste matters and may stop functioning. Most of these systems have combination oxygen systems. Figure 14–13 shows various

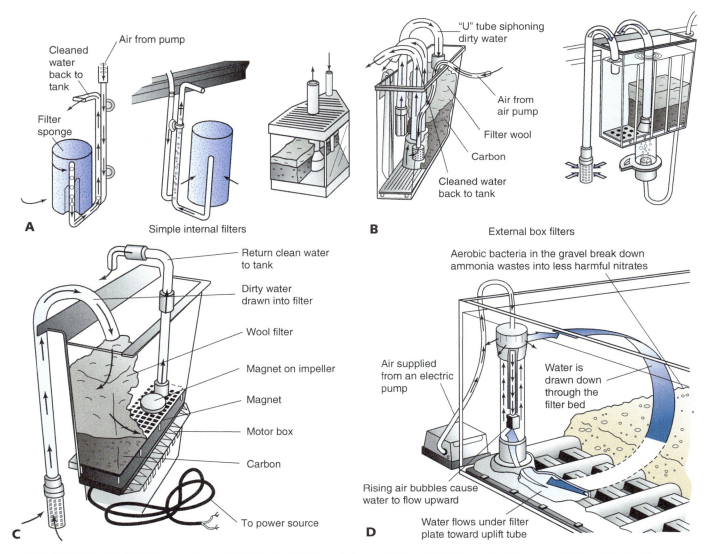

FIGURE 14–13 (A) Types of internal filters. (B) External box filters. (C) Motorized or power filter. (D) Undergravel filter system.

TABLE 14–3
Types of Filters

Undergravel filter	Placed on the bottom of the tank; water pulled through the bed of gravel to screen out materials
Canister filter	Located outside of the tank; includes pumps and tubes that move water through a cuplike tube to remove materials; must be cleaned or replaced every other week
Outside filter	Hang on the back of the tank; water is moved through a tube holding filter materials; common for small tanks
Nitrifier/dentrifier filter	Supplements the action of bacteria; expensive and only used in large specialized tanks

filtration systems. Table 14–3 summarizes various types of filters.

Chemical filtration uses special chemicals, such as ozone and activated charcoal, to keep the water clear and from becoming a yellowish color.

Other pieces of equipment needed in more complex systems include a **thermometer**, which measures the temperature of the water. Thermometers are often attached to or float within the tank. A **thermostat** can be set to maintain the water to above or below a certain temperature. Heaters keep tropical waters warm enough for certain fish species. Many fish require water temperatures ranging from 70 degrees to 85 degrees Fahrenheit. Lights within the tank allow visibility and provide a light source for fish and plant growth. A light source should be placed on the tank for 10 to 12 hours a day. Use caution with electricity around any water sources.

Water Sources and Quality

Water sources require a **habitat** that mimics the natural environment in which an animal lives. It is important to research the proper climate, plants, rocks, fish breeds, and other necessary items for a particular species. With this knowledge, a person can create an artificial environment that provides a healthy setting for fish. Water quality depends on its source and the substances in it. Freshwater sources include tap water, rainwater, and well water. **Tap water** comes from faucets in homes and businesses; it is readily available but may not be adequate for most tanks. Many public water systems have excessive amounts of chlorine, sulfur, and other chemicals. These substances are safe for humans but **toxic** (poisonous) to fish. Tap water can be prepared for tanks by allowing it to age. **Aging** is done by collecting water and allowing it to stand in an open container for at least 5 days. Some larger tanks will require large amounts of water. The containers that hold this water must be clean. Special tablets that clean the water can also be purchased to put in tap water sources. **Rainwater** can be used as long as it is not polluted or from a contaminated source. **Acid rain** and other substances will be toxic for fish. **Well water** is used by breeders and people who have large vats or ponds. The water source should be tested to determine its contents, as some well water may be unsafe for fish. Well water can be aged in the same manner as tap water. Saltwater can be provided from actual saltwater sources or **synthetic seawater** sources, which are man-made. Natural seawater can be collected from oceans and lakes. Synthetic seawater is made by mixing freshwater with a salt-based mix. Water quality is determined by the temperature, the pH levels, the nitrogen cycle, the amount of dissolved oxygen, and the amount of salt in the water. Water temperature should be within the appropriate range for the species kept in the tank. In some circumstances, the water will need to be warmed. Fish should be placed in water that varies only by 1 to 2 degrees from their previous water source. Fish do not do well in sudden temperature changes and will stress and likely die. Most ornamental fish species require water temperatures between 70 and 85 degrees Fahrenheit. Saltwater varieties are usually slightly warmer. Heaters may be used to warm water temperatures. The **pH level** of water determines its **alkalinity** or **acidity**. Water pH is measured on a scale of 0 to 14, with 7.0 being neutral. Alkaline water is measured at levels above 7.0; acid water is measured at levels below 7.0. The lower the number of the pH rating, the more acidic the water. Most ornamental fish require pH levels of 7.8 to 8.3. Materials in the water can change its pH.

The **nitrogen cycle** is a process that converts wastes into ammonia, and then into nitrites, and then into nitrates. This is a natural occurrence in water. Ammonia is toxic to fish and produces an odor that comes from the water. When fish are overfed or waste materials are not filtered from the water, there may be a buildup of excess ammonia. Commercial test kits may be used to determine these levels. They must be quickly, as ammonia evaporates rapidly in the air. **Dissolved oxygen (DO)** forms a gas state of oxygen in the water. Fish use the DO to breathe. Most fish require 5.0 ppm of DO. Below 3.0 ppm DO, the fish will likely die. Oxygen meters can be used to measure the DO levels in the water. **Salinity** is the amount of salt in the water. Freshwater species of fish cannot survive in water with salt contents. Salt content is measured as **specific gravity (SG)** using a hydrometer. Water has a specific gravity or weight of 1.0. Saltwater has a higher SG, and saltwater species require water values between 1.020 and 1.024. Salinity can be increased by adding prepared salts to the water. Water sources should be changed periodically. This can be a huge job in large tanks, vats, or ponds. Water management should be monitored to keep water clean and healthy. Fish will need to be temporarily relocated any time the water is changed.

Catching and Handling Ornamental Fish

When a fish must be caught and taken out of an aquarium, it is best to first remove some water from the tank. As there will be less water, the fish will have less room to move around, which means less time chasing it in the tank. The first and most common way to capture a fish is with a net. Nets come in a variety of sizes and colors. The best color for catching fish is black or dark green. A white net may be used to direct fish away toward another net for capture. Nets should be soft and move freely in the water. A net that has become coarse and stiff should be replaced. Place a net in the water for a few seconds before use to allow it to soften. This will also give fish a chance to grow accustomed it. Move the net slowly in the water. When the fish swims into the net, immediately remove the net and the fish from the tank. Sometimes, fish may become tangled in the net. If this happens, simply put the net and fish in a large water source and gently turn the net inside out to free the fish. If the fins are caught and cannot be freed, you might have to cut the net with scissors. Catch a fish with a net using these tips:

- Guppies can be best caught by bringing a net up from underneath them. If they cannot see the net coming, they will not try to escape.
- Use two nets to catch fast-moving fish or bottom dwellers. One net will herd them and one net will catch them.
- Stand a large net in one corner of the tank. Use a stick or straw to herd the fish into the net.
- Use a plastic baggie to catch angelfish and other delicate tropical fish to avoid damaging their fins.

Another method of capturing a fish is to use a Styrofoam cup, plastic baggie, or bottle. These objects should be dark in color, clean, and free of chemicals. Heavy items will sink more easily and should not float to the top of the tank. Herd fish into the object and carefully remove it from the tank. It may be necessary to use your hands to catch larger species of fish. This must be done carefully, as some fish can bite or have spines or whiskers used for defense. Use one hand to grasp the fish gently in front of the tail and the other hand to support the fish from underneath the body. This requires patience and care. The fish can then be removed and placed in the necessary area. Wash your hands thoroughly after handling fish and equipment.

Basic Health Care and Maintenance

Following a regular **aquarium maintenance schedule (AMS)** will ensure that the tank system is being kept in the appropriate manner (see Table 14–4). The AMS is a list of important duties that should be completed at a particular frequency, whether that is daily, weekly, monthly, or annually.

TABLE 14–4

AMS Fish Checklist

Daily Items	Provide food
	Check water temperature
	Monitor aeration
	Monitor filtration
	Remove any dead fish
	Observe fish behavior
	Note any unusual behavior
Weekly Items	Check water level
	Check pH levels
	Add water as needed
	Add chemicals as needed
Monthly Items	Change and add water
	Clean material on tank bottom
	Remove algae
	Observe plants
Quarterly Items	Clean filter
	Check electrical connections
	Check pump
	Check hoses
Annual Items	Clean tank thoroughly
	Rinse bottom gravel or pebbles
	Change bottom every other year
	Replace light bulbs

Breeding and Reproduction in Fish

The reproduction and breeding of fish has become a popular hobby as well as a profitable industry. Male and female fish are housed together for spawning or mating. Many fish will breed spontaneously, and others will pair up. Sometimes fish will need to be separated prior to mating to develop an interest in each other. Some fish will bear live young, while others are egg layers. **Egg-laying fish** reproduce when the female fish releases eggs that are fertilized by the male's sperm. Incubation of the eggs lasts only a few days, depending on the species. Some examples of egg-laying fish include goldfish, koi, gouramis, tetras, catfish, and barbs. The majority of ornamental fish species are egg laying. However, there are still many variations of how the eggs are incubated. Some breeds are **egg scatterers**. This means the fish spontaneously scatter their eggs around the tank as they swim. When housed in a tank with numerous fish, these eggs are often eaten by the other adult fish (in goldfish, for example). This can be avoided by placing pebbles, marbles, or plants within the tank. The eggs will rest among them for safety.

Some species of fish deposit their eggs on plants, rocks, or other objects. Examples of such **egg depositors** include cichlids and tetras. This protects the eggs from predators. Other species will build a nest for egg protection.

Examples of nest builders include bettas and gouramis. A few species of fish will carry their eggs in their mouths: they are known as mouth brooders. A few fish species are **live-bearing fish** and give birth to live young. These fish usually live in a group of five or more breeders, called a **shoal**. Live young bearers mate, and the male deposits the sperm inside the female. The fertilized eggs will develop into fully formed baby fish. Some females will store the sperm of the male until it is needed for reproduction. Examples of live-bearing fish include mollies, guppies, swordtails, and platys. These fish are easy to breed and make great beginner fish for new breeders.

Fish Disease and Health

Fish need positive bacteria within their environment to remain healthy. The bacteria are needed within the filtration system to keep the water quality proper and clean. Many tanks do not have enough good bacteria within the water source. In those cases, products that contain live bacterial sources can be added to the water. Natural bacteria growth occurs over several days to several weeks. Maintaining adequate amounts of bacteria within a fish's environment also helps keep it healthy by reducing the chances of disease outbreaks. Fish that are handled frequently or experience a sudden change in their environment experience stress. Fish that show a change in behavior may be ill or have a disease. Signs of diseases in fish include the following:

- Not eating
- Increased gill movement or respiration
- Gasping at the water surface
- Scratching on objects
- Folded fins
- Frayed fins
- Cloudy eyes
- Blood spots on body
- Fuzzy patches over body
- White spots on body
- Color changes
- Protruding eyes
- Large, swollen abdomen
- Material hanging from anus

Sick fish should be removed from the tank and held in isolation and a veterinarian who has fish knowledge should be contacted. It is recommended to treat the sick fish as well as the healthy tank where the fish was housed. When purchasing or receiving new fish, a quarantine tank should be used. All new fish should be observed for signs of disease before being placed with healthy fish. Some common diseases occur in ornamental fish. Veterinary assistants should be familiar with some basic information relating to these diseases. The following are some common fish diseases and conditions.

Fin Rot

Fin rot is a bacterial disease that begins when fin tissues start to erode and the tissue becomes inflamed. This may be caused by poor water quality, stress, or aggressive fish that have bitten or that attacked the fish. Treatment includes a bath immersion in saltwater if the fish species is salt tolerant, topical antibiotics, and antibiotics applied to the water.

TERMINOLOGY TIP

A *topical* product, such as a medication, is applied onto the skin or hair coat on the outside of the body. ■

Lymphocytosis

Lymphocytosis is commonly called **cauliflower disease**. This viral disease affects certain species of fish. Signs include white to gray growths over the fins and body. Causes may be stress or that a carrier fish is spreading the disease within the tank. There is no treatment, and death is rare. The only means of control and prevention of the disease is through water quality.

Ich

Ich is a parasitic protozoan disease caused by a one-celled organism. It is commonly known as **white spot** because of white spots that develop over the fins and body. Ich may be caused from stress, poor nutrition, and poor water quality from contamination. All new fish and plants should be quarantined. It is addressed by treating the water for 10 to 14 days.

Fish Fungus

Fish fungus is also known as **cotton-wool disease** due to the appearance of tufts on the skin of the face, gills, or eyes that look like cotton or wool. It is caused by fungus spores in the water. It is treated by isolating infected fish and administering antifungal medications. Salt can be added to the water of salt-tolerant species (2 grams of salt per liter of water).

Cloudy Eye

Cloudy eye is a bacterial infection that may affect one eye or both eyes of tropical fish. It can cause damage to the nervous system. The cause is poor water quality. Prevention is through monitoring and changing the water frequently while using proper filtration methods. It is treated with antibacterial medications and by adding salt to the water.

Ascites

Ascites, also called **dropsy**, may be caused by a bacterial or viral infection. The body of the fish swells because of fluid buildup: this causes the fish to lie on its side or have difficulty swimming. The scales of the fish develop a pinecone-like appearance. The causes of ascites are poor water quality and a buildup of ammonia in the water. It is treated with antibacterial medications and by adding salt to the water.

Swim Bladder Disease

Swim bladder disease is caused by a bacterial infection, poor diet, and genetic deformities. The disease causes a fish to have difficulty swimming to the water's surface and in depths of water. It is common in goldfish. Some fish will not recover from the disease, but treatments may include a diet change, improved water quality, and a diet of flakes instead of pellets.

Aquaculture Industry

Aquaculture is the cultivation of aquatic plants, animals, and other species of living organisms that grow and live in or around a water source. Most aquaculture in the United States involves species of fish and **crustaceans** (animals with a bony outer shell). Each species of aquatic organism is different and has specific requirements, but all need water for proper living conditions. Aquaculture creates **aquacrops**, commercially produced plant or animal species marketed for income and profit. They are produced as food, sport, recreation, and ornamental pets. Climate, water, and environmental conditions are important factors in aquaculture, and several species have adapted to grow and live in certain areas. The market is also an important consideration when deciding to raise an aquatic species, as the demand will control whether the species is profitable. Species in high demand in the United States include shrimp, oysters, lobster, salmon, trout, and catfish.

TERMINOLOGY TIP

The term *aquatic* means pertaining to water.

There are some general requirements that must be met in order to establish an aquaculture program. There must be an available market as well as a way to transport products to consumers. A suitable species must be selected that can adapt to the environment. The water quality and type must favor the species. The nutritional needs of the species must be met. Sanitation and disease control must be accounted for. The temperature of the water and environment must be monitored for proper growth and health.

Aquaculture requires an appropriate **production intensity**, which is the assurance that the amount of water will contain the corresponding amount of food sources and be large enough for a high-production level of products. The **intensity biomass**, or the number of aquatic species in a specified volume of water, is also important. Too many or too few fish in a body of water will affect production. Some aquatic species may overpopulate a body of water or land and not be able to survive. For example, in fish production, every 6000 to 8000 fish should have 1 acre of water for proper survival. Reproduction also factors into the size of the water source. In this example, the fish would be capable of producing 8000 pounds of fish per acre of water per year.

Veterinary Terminology

- Broodstock—adult fish retained for spawning
- Clutch—number of eggs incubated at one time
- Cultured fish—farm-raised fish or shellfish
- Fingerling—stage of life characterized by 1 inch in length to length at 1 year of age
- Fry—stage of life that begins at hatching and ends when the fish reaches 1 inch in length
- Roe—eggs of fish
- Spawn—mass of eggs deposited in water

Aquatic Biology

Fish are **vertebrates**, which means they have a backbone. They must live in water to survive and have such characteristics as a fragile bony skeleton, a body covered in scales, gills, and **fins**. The scales are individual waterproof segments that act as the skin. Some fish have a thin skin rather than scales. The gills act as lungs: as water passes over them, oxygen is filtered from the water to allow breathing and then dissolves into the water. The fins of a fish act as arms and legs and allow movement and balance in the water. Some fins have sharp ends that serve as protection from predators.

Fish have a streamlined body structure that helps propel them through water. They have a variety of colors and markings, mostly to help them blend into their surroundings. Fish have three segments that make up their body: the **head**, **trunk**, and **tail**. The head contains the eyes, mouth, and gills. An **operculum**, or gill cover, divides the head from the trunk. The trunk consists of the body, fins, and anus. The tail contains the caudal fin, which acts as a propeller to move the fish through the water.

The internal anatomy of fish is somewhat like other animals; however, some organ systems have more developed features for aquatic life. A fish's body is made of soft bones and cartilage that maintain the body structure and serve as protection in the water. The muscular system is the largest body system in fish, with the largest muscles located in the tail. Fish are adapted with barbels, or **barbs**, which act as sensory devices to detect food, danger, and movement. Fish have a monogastric digestive system. Female fish may lay eggs or give birth to live young, depending on the species. The most unusual and important feature of fish is the swim bladder. This structure acts like a small balloon that inflates and gives the fish buoyancy to float and balance in all levels of water. Without a swim bladder, a fish would not be capable of moving through deep water.

Crustaceans are an aquatic species that have a shelled body with legs. Crustaceans are considered **arthropods**, as they have an **exoskeleton** made of cartilage. Arthropods also have a segmented body and jointed legs. (Insects also belong to this class of organisms.) **Decapods** are crustaceans with five pairs of legs. The body structure of both arthropods and decapods consists of a head, the **carapace** (thorax), an abdomen, and limbs. They also have a pair of **pincers** located on the head for protection, smell, and feeling. Figures 14–14 and 14–15 illustrate the internal and external anatomy of crustaceans.

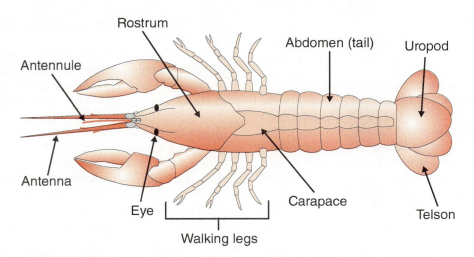

FIGURE 14–14 External anatomy of a crustacean.

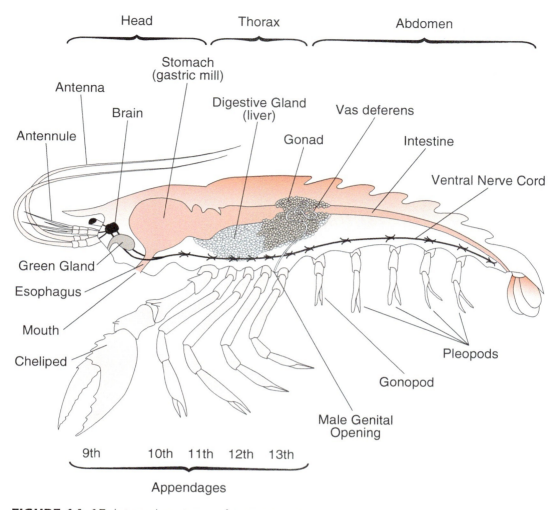

FIGURE 14-15 Internal anatomy of a crustacean.

Mollusks are aquatic species that have only a thick, hard shell. They have no limbs but instead have **valves** like gills for breathing. Their shells are hinged to open and feed and breathe. Some types of mollusks are **bivalves**, meaning they have two shells. Bivalves have a strong **abductor muscle** that allows the shell to open and close. They are filter feeders and remove particles from the water to feed. Other mollusks are **univalves**, meaning they only have a partial shell. One example of a mollusk is an oyster; oysters are farmed for their popularity in pearl production, which develop naturally from a single grain of sand. The internal anatomy of the mollusk is illustrated in Figure 14-16. Table 14-5 lists examples of aquatic species.

Aquaculture Production Systems

Aquaculture requires a **water facility**, including ponds, rivers, lakes, streams, tanks, vats, and aquariums. These facilities are developed to resemble the natural environment and accommodate each specific type of organism.

Pond

A **pond** is a contained area of water surrounded by human-built dams or levees; it is not a naturally occurring body of water. Ponds range in size of less than 1 acre to up to 50 acres. Ponds do not drain or have an active moving water source, so management of the water is important. Ponds are typically rectangular in shape and range from 3 to 5 feet in depth (see Figure 14-17).

Raceway

A **raceway** is a long, narrow water structure with flowing water. Raceways are common in fish hatcheries for breeding purposes and are used to increase the number of fish that may be stocked in other

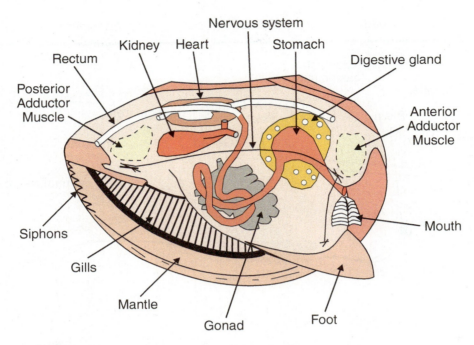

FIGURE 14–16 Internal anatomy of a mollusk.

TABLE 14–5	
Examples of Aquatics	
Crab	Crustacean–arthropod
Shrimp	Crustacean–decapod
Lobster	Crustacean–decapod
Oysters	Mollusk–bivalve
Clam	Mollusk–bivalve
Mussels	Mollusk–bivalve
Abalone	Mollusk–univalve
Snail	Mollusk–univalve
Bass	Fish
Tilapia	Fish
Salmon	Fish

bodies of water during fishing seasons. They may be constructed out of different types of materials and are usually built in naturally occurring waterways. They have the advantage of holding a larger number of fish, as flowing water creates more oxygen. Pumps may be needed to correct the water flow, and electricity costs can be expensive (see Figure 14–18).

Tank, Vat, or Aquarium

A tank, vat, or aquarium is a smaller body of water that may be used to house aquatic species. A tank is a large, round area (see Figure 14–19). A vat is similar in shape but smaller in size. An aquarium is much smaller in size and typically holds ornamental fish.

FIGURE 14–17 Crawfish ponds.

FIGURE 14–18 Earthen raceway for trout.

FIGURE 14–19 Tanks for raising catfish.

All three may be made of glass, wood, fiberglass, metal, or concrete. They require proper water flow and oxygenation.

Pen or Cage

A **pen** or **cage** is a small structure placed in a body of water to confine aquacrops. A pen is a structure that connects land with water (see Figure 14–20). It may have posts set in the ground for support and is usually placed in shallow water less than 3 feet deep. It rests in the muddy bottom of the water source to provide nesting and breeding areas. A cage is a structure that floats on the surface of the water. Typically made of netting, it does not touch the bottom of the water bed. It is frequently used in large bodies of water to confine aquatic species.

Water Sources

There are three types of water sources used in aquaculture production. The first is **freshwater**. Freshwater has little to no salt content and is the most common water source in North America. Some examples of freshwater sources include wells, streams, and surface runoff areas. A second water source is **saltwater**. Water is measured by its salinity, or salt content. Salt content is measured in **ppt** or parts per thousands; saltwater is 16.5 ppt or greater. Examples of saltwater sources are oceans and seas. The third type of water source is **brackish water**. This is a mixture of freshwater and saltwater. Brackish water is found where a stream enters a larger body of water, such as an ocean or sea. Its salt content is higher than freshwater but lower than saltwater. Only a few species of aquatics are capable of living in brackish water. Crabs are the most common.

Water Quality

When raising aquatic species, it is important to consider the temperature of the water, which is important for the organism's metabolism and regulation of body temperature. **Cool water**

FIGURE 14–20 Ocean net pens.

species grow best in water that is 50 to 60 degrees Fahrenheit. Water that is too cool or warm will impact the overall health and growth of the animal. Examples of cool water species are salmon and trout. **Cold water** is best for species that require temperatures below 50 degrees Fahrenheit. **Warm water** is defined as between 65 and 85 degrees Fahrenheit. While there are species that survive at both the high and low ranges of water temperatures, warm water species are the most abundant. Table 14–6 lists temperatures required for common aquatic species.

Other water-related factors that must be taken into account include the water quality, possible pollutants, pH levels, mineral contents, gases, and any plants and animals that live in the water. Water management is necessary in some aquatic production sources in which surface water and wells may cause contamination. **Surface water runoff** is excess water drainage from precipitation. The water must be pollution free, and there are legal regulations on the usage of this type of water. Well water is pumped from aquifers deep within the Earth where water is found naturally. This is a source of high-quality water that is free of pollutants; however, the need for drilling permits and the amount of labor involved make it is costly. It is also usually cold to cool water. A more common natural source of water is a **spring**, which is a natural opening in the ground that provides a natural, clean water source. Most springs develop at high elevations. These water sources may easily dry up and are thus not reliable.

Water quality involves how suitable water is for a particular source or use. Living organisms require fresh clean water that is pollutant free and has a sufficient amount of dissolved oxygen (DO). *DO* is oxygen present in water that living organisms use for breathing. It is found in a free gaseous form rather than in a molecular form such as humans breathe. Plants help transfer the oxygen into the water. (For information on water quality and management with regard to pH levels, alkalinity and acidity, and filters, see the sections on ornamental fish.) This type of production is very similar and should be reviewed for more information on water management and control.

Aquatic Reproduction and Breeding

Fish and aquatic animals reproduce and grow through production cycles. A **production cycle** is complete when the young are full grown and have reached market size. The cycles follow the seasons of the year. Reproduction takes place in the spring, and the young are born and grow over the summer. It takes most species two years to reach adulthood. Aquatic animals have four production cycles: broodfish and spawning, hatching and raising fry, producing fingerlings, and growing to adult size or market weight.

Broodfish and Spawning

The first production cycle includes **broodfish**, which are mature breeding fish used for reproduction. Broodfish may be male or female. Broodfish produce young so that fish may be stocked and used as food fish for other aquatics. Breeding, also known as **spawning**, usually begins at about 3 years of age. Adult fish are separated from the young, as many species of fish and aquatic animals eat their young. Spawning occurs when pairs or groups of fish are isolated in a water source. Broodfish actively lay a group of eggs called **spawn**. The spawn are laid in a nest with the eggs; the eggs vary in size, the smallest akin to a plant seed. Female broodfish produce thousands of eggs per spawning. Many fish can produce 2000 eggs per pound of body weight (see Figure 14–21). Spawning seasons vary with species. Catfish spawn when the water becomes warmer, in the late spring to early summer. Coho salmon will spawn once and then die. After spawning, some fish—males or females—will care for the nest. Other species are **mouth brooders**, which means they incubate the eggs in their mouth.

Artificial Fertilization

Artificial fertilization and incubation is used with some species, especially those that die after spawning. Fertilization occurs by **expressing** the female, or placing pressure on the abdomen and squeezing to remove eggs. The male is expressed in the same manner so that sperm, known as **milt**, is removed. Both eggs and sperm are then mixed in a container so that fertilization and hatching can occur (see Figure 14–22). This process is carried out at a **hatchery**. Fertilized eggs

TABLE 14–6
Common Species and Water Temperatures

Catfish	75 to 85 degrees Fahrenheit
Tilapia	75 to 85 degrees Fahrenheit
Bass	77 to 88 degrees Fahrenheit
Crawfish	65 to 85 degrees Fahrenheit
Atlantic char	40 to 50 degrees Fahrenheit
Salmon	50 to 60 degrees Fahrenheit
Trout	50 to 60 degrees Fahrenheit

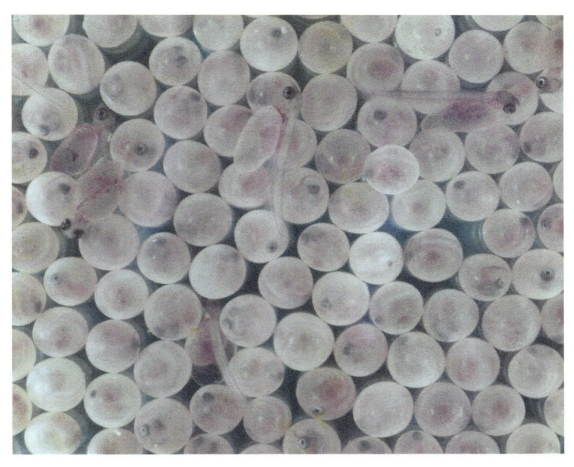

FIGURE 14–21 Trout eggs.

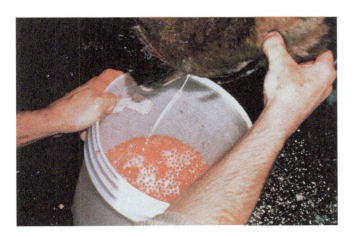

FIGURE 14–22 Fertilization of salmon eggs.

hatch after incubation, which is the period between spawning and hatching. The most important factors for incubation are water temperature and circulation. Specialized equipment is required, as eggs are incubated in trays, jars, or troughs (see Figure 14–23).

FIGURE 14–23 Incubation of trout eggs.

Hatching and Raising Fry

The newborn fish larvae that hatch are known as fry. Fry are very small and delicate and must be kept in ideal conditions for growth and to prevent disease. After the fry hatch, a **sac fry**, or yolk sac, is attached for a day (see Figure 14–24).

 TERMINOLOGY TIP
Fry is the term for young fish larvae; however, salmon young are known as **alevin**. ■

Fry are fed a **slurry**, or soft soup-like substance, that consists of a finely ground fish meal that is organ based, such as heart or kidney, and mixed with plankton, yeast, and chicken yolks. The fry grow to an inch long and are then moved to a new location. At this stage, they are known as fingerlings, or immature fish that are greater than 1 inch in size (but not yet at stocking or market size). As they grow, they are moved to large tanks or small ponds.

 TERMINOLOGY TIP
Fingerlings are young immature fish that are over 1 inch in size; however, salmon fingerlings are called **smolt**. ■

Producing Fingerlings

Fingerlings should be fed a high-protein diet (about 36 percent of the food should be protein) that is compounded as a commercial granule feed. It takes fingerlings one growing season before they meet stocking size. Fish species must be at least 5 inches long to be stocked in a natural water source.

Readying Fish for Market

Food fish production includes raising fingerlings to market size: so that they gain weight as quickly as possible, they should be fed 3 percent of their body weight. This type of production cycle requires a diet that is 28 to 32 percent protein. It typically takes 6 months from harvest time for fingerlings to reach market weight.

Breeds or Species

Many different species and breeds of aquatic animals are found natively in the United States, and many species are raised for production purposes. Species are divided into the following groups: finfish, ornamental fish, mollusks, crustaceans, and aquatic mammals. Popularly produced finfish in the United States include salmon and catfish; crustacean and mollusk production includes shrimp, clams, and oysters.

Breed Selection

Species and breed selection are based on many of the same factors as livestock. Successful production with aquatics species must take the following into consideration:

- Environmental needs
- Nutritional needs

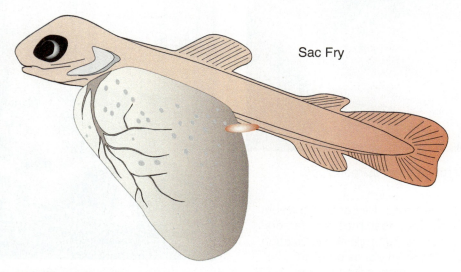

FIGURE 14–24 Fry with yolk sac.

- Reproductive habits
- Adaptability to housing
- Disease and health resistance
- Market demand and profitability

The ability of species to reproduce and survive is an important factor in aquatic production. Aquatic species require specific handling, care, and knowledge to be a successful operation.

Nutrition

Aquatic species require nutrition to meet their energy needs of swimming and regulation of their body temperature. Most need a balanced daily ration of protein, carbohydrates, fats, vitamins, and minerals. Protein is necessary for growth, and their diet should consist of at least 28 percent protein that comes from an animal protein source, such as shrimp, fish, or chicken. Carbohydrates—including soybeans, corn, and grains—should make up at least 10 percent of the diet; aquatic species will convert these carbs into energy. Fats, making up from 4 to 15 percent of the diet, come from by-products and are necessary for waste production.

Vitamins and mineral supplements such as calcium, iron, and sodium should also be included in the diet. Aquatic animals may eat a type of water plant nutrient called plankton, which grows naturally in the water. Vegetation sources that may be added to a commercial diet include rice, plant stems, leaves, and grains. Many commercial diets are formulated for specific species. Diets come in several formulas that are made to float in the water and easily break apart. **Meal** is finely ground food for small fish and aquatic animals. **Crumbles** are flakes or blocks of food. **Pellets** are meal bound into larger particles. Pellets vary in size and are made to float. Feeding rates vary with species and life stage and may occur every hour to one to two times a day. All aquatics and fish should be fed 3 percent of their body weight. Fry are fed every hour to several times a day as they grow. Fingerlings are fed two to three times a day. Food fish and broodfish are fed once or twice daily. Many aquaculture programs use an automatic feeder system. Such a feeder provides a specific amount of food on demand as the fish touches a *trigger* that releases the food (see Figure 14–25). Feeders can be set on a timer to release a specific amount of food at a certain time of the day.

FIGURE 14–25 Automated self-feeder used in trout production.

SUMMARY

Ornamental fish are kept for many reasons, from breeding programs to entertainment to personal appeal. Most of these species are bred in fish tanks and are not found in their natural habitats. Some of these fish bear live young and others lay eggs. Their nutritional and health care needs should be researched for optimal care. Fish are high maintenance when it comes to water sources, water quality, and housing conditions. Fish should be monitored closely and observed daily for any changes in behavior that may be a sign of disease. A veterinary assistant may work in a facility that tends to fish care or may become an ornamental fish breeder. Aquatic species are typically raised for human consumption. The most widely produced animals in aquaculture are fish, crustaceans, and mollusks. These animals may be raised in freshwater, saltwater, or brackish water. The water source and quality are important to the production of aquatic animals. Aquaculture is becoming a popular area within the veterinary industry.

Key Terms

abductor muscle a muscle that allows a shell to open and close

acid rain polluted rainwater

acidity water pH levels below 7.0

aeration oxygen or air created in the water by making bubbles

aging allowing a water source to stand in a container for at least 5 days

alevin young salmon

alkalinity water pH levels above 7.0

aquacrop a commercially produced plant or animal living in or around a water source

aquaculture the production of living plants, animals, and other organisms that grow and live in or around water

aquarium a small rectangular area that holds water to contain small fish

aquarium maintenance schedule (AMS) a list of important fish tank duties that should be completed at a particular frequency

aquatic pertaining to water

arthropod a crustacean with an external skeleton

ascites a bacterial or viral infection, also called "dropsy," that causes a fish to swell with fluid

barb a device on the fish that detects food, danger, and movement; also known as a barbel

biological filtration the use of bacteria and other living organisms in water to change harmful materials into safer forms

biological oxygenation the use of plants to create oxygen in the water

bivalve a mollusk with two shells

bottom-dwelling fish fish that live and feed on the bottom or floor of a body of water

brackish water a mixture of freshwater and saltwater

broodfish mature male and female breeding fish used for reproductive purposes

cage a structure for holding aquatics that floats on the water

carapace a dorsal section of the exoskeleton or shell similar to the thorax section of a mammal.

caudal fin the tail fin that acts as a propeller

cauliflower disease another name for *lymphocytosis*

chemical filtration the use of special chemicals such as ozone and activated charcoal to keep water clear

cloudy eye a bacterial infection that may affect one eye or both eyes of tropical fish

cold water water temperature below 50 degrees

companion fish a fish people keep as a stress reliever or for entertainment

cool water water temperature between 50 and 60 degrees

cotton-wool disease another name for *fish fungus*

crumbles flakes or blocks of fish food

crustacean an aquatic animal with an outer body shell and legs

decapod a crustacean with five pairs of legs

dissolved oxygen (DO) oxygen in a gas state in water

diversity having distinct qualities

dorsal fin a fin on the back of a fish used to steer it through the water

dropsy another name for *ascites*

egg depositers fish that lay eggs on areas such as plants, rocks, or the bottom of a water source for protection

egg scatterers fish that spontaneously scatter their eggs around the tank as they swim

egg-laying fish fish that reproduce when the females release eggs that are fertilized by the males' sperm

exoskeleton external cartilage on the outside of the body

expressing placing pressure on the abdomen to remove eggs or sperm

filtration a way of removing gases and solid materials from water

fin rot a bacterial disease that begins when fin tissue starts to erode and the tissue becomes inflamed

fingerling immature fish that is 2 years of age

fin a structure that acts as arms and legs and allows movement in water

fish aquatic species that have scales, gills, and fins

fish fungus a fungal disease known as "cotton-wool disease" because of tufts of growth resembling cotton or wool that occur on the skin of the face, gills, or eyes

floss a material within a filter made of gravel, charcoal, and fiber materials

food fish production raising fingerlings to market size to be sold as food product

fountain a watertight structure made of plastic, fiberglass, or concrete that hold fish species

freshwater water with little to no salt content

freshwater fish species of fish that live in freshwater with very little salt content

fry young newborn fish

gills organs that act as lungs and allow fish to breath by filtering oxygen in the water

habitat an area where an animal lives

hatchery place where fish are fertilized, bred, hatched, and raised

head the part of the body containing the eyes, mouth, and gills

ich a parasitic protozoan disease caused by a one-celled organism

intensity biomass the number of species in a volume of water

lateral line an organ located just under the scales that picks up vibrations in the water

live-bearing fish fish that bear or give birth to live young

lymphocytosis a viral disease that causes white to gray growths to develop over the skin; also known as *cauliflower disease*

meal fine-ground food for young fish and aquatics

mechanical filtration equipment that filters the water to remove harmful particles and keep it clear

milt fish sperm

mollusk an aquatic species that has only a thick, hard shell

mouth brooder fish that incubates eggs inside the mouth

nitrogen cycle a process that converts wastes into ammonia and then into nitrites and then into nitrates

omnivore eats both plant and meat sources of food

operculum a gill covering that divides the head from the trunk

ornamental fish popular pet kept in aquariums, small in size and varying in color and appearance

oxygenation the process of keeping air in the water by using dissolved oxygen

pellets fish meal bound into larger particles

pen a small structure placed in a body of water to confine aquacrops that connects land to water

pH level the quality of water on a scale of 0 to 14, with 7.0 being neutral

photosynthesis the process of sunlight allowing plants to grow in a water source

pigment a coloring matter in animals and plants

pincers a pair of structures on the head used for smelling, feeling, and protection

plankton tiny plants that grow in water sources

pond a contained area of water that is not naturally occurring

pool a structure made of plastic, fiberglass, or concrete that holds a large amount of water and fish

ppt parts per thousand; a measurement that determines the amount of salt in water

production cycle the time it takes for young aquatic species to reach full market size

production intensity an assurance that the size of the body of water will hold the appropriate amount of food sources and be large enough for a high-production level of products

raceway a long and narrow water structure with flowing water

rainwater a source of water from the sky that can be used if not contaminated or polluted

run a group of fish

sac fry a yolk sac attached to a fry for one day after the hatch

salinity the amount of salt in a water source

saltwater water with a salt content of at least 16.5 ppt

saltwater fish species of fish that require salt in their water source

scales a body covering that serves as skin and protects the outside of the fish

school a group of fish

shoal fish that usually live in groups of five or more breeders

slurry a soft, soup-like substance fed to fry

smolt immature salmon greater than an inch in size

spawn a group of fish eggs in a nest

spawning breeding or reproduction in fish

specific gravity (SG) the weight of a liquid measurement

spring a natural opening in the ground that provides a clean natural water source

surface-dwelling fish fish that live and feed near the surface or top of a body of water

surface water runoff excess water drainage from rainfall or precipitation

swim bladder an air-filled sac that prevents fish from sinking and allows them to float and move through depths of water

swim bladder disease a bacterial disease in fish that makes it difficult to swim to the water's surface and in depths of water

synthetic seawater water that is man-made

tail the part of the body containing the caudal fin that allows movement and guidance in the water

tank a basic container that holds water and fish

tap water water that comes from a faucet

thermometer a device used to measure the temperature of the water

thermostat a device used to control the temperature of water

topical a product applied to the outside of the body

toxic poisonous

tropical fish small, brightly colored fish that require warm water and are naturally found in warm, tropical locations

trunk a part of the body that contains the body, fins, and anus

univalve a mollusk with one shell or part of a shell

valve a structure in mollusks that acts like gills

vat a large metal or concrete structure that holds large amounts of water and is commonly used for breeding and reproduction of fish

vertebrate an animal with a backbone

warm water water between 65 and 85 degrees

water facility an area built to house aquatic animals in natural environments

well water water that is sourced from a well

white spot another name for *ich*

yearling a fish that is 1 year old

Review Questions

1. Where do ornamental fish occur naturally?
2. Give six examples of ornamental fish and note if they are freshwater or saltwater fish and if they bear live young or lay eggs.
3. List equipment that is needed for an ornamental fish tank.
4. What are oxygenation and filtration? Why are they important to fish?
5. What are the types of filters used in aquariums?
6. What are some sources of water for tanks?
7. What are the areas of importance in water quality?
8. What are some signs of disease in fish?
9. Describe a regular maintenance schedule for aquariums.
10. What are some guidelines for feeding fish?
11. What common species of aquatic animals are produced in the United States?
12. What are the general requirements for aquaculture production?
13. Explain production intensity.
14. What are the three types of water?
15. What types of production facilities are available for aquaculture or aquacrops?
16. What is dissolved oxygen?
17. What are production cycles in aquaculture?
18. What is ich?

Clinical Situation

Allie, a veterinary assistant at Feline Hospital, greets a client who walks in the front door. "Hi," she says. "How may I help you?"

"I'm not a client here but I have some tropical fish that have been sick. Some are now dying. They were very expensive, and I'd like to talk to someone about what can be done," he states.

The man seems very upset about the situation; however, the veterinarian at the hospital doesn't work with tropical fish.

Allie explains this. She feels bad that she can't help the man.

"Dr. Wills doesn't treat fish, he sees cats and some small animal species. I'm not even sure if he has any experience or knows anything about fish. Plus, he's currently in surgery, so he's not available. I am sorry we can't help you."

"Well, thanks anyway," says the man as he leaves the hospital.

- What could have been done in this situation to better serve the client?
- How would you have handled this situation?
- What was done correctly in this situation?

Clinical Situation

Jimmy, a veterinary assistant working in an aquaculture facility, is monitoring the tanks and vats, as he does every day. He checks the temperature and quality of the water in each tank. Jimmy notices that fish in several tanks are dead and floating at the top of the water source.

On closer examination, Jimmy sees that the fish in several tanks look diseased. He begins writing down the tank numbers, the signs of disease, and the number of fish that have died. He also decides to take some water samples of the tanks containing the diseased and dead fish.

- Did Jimmy handle this situation correctly? Why or why not?
- What would you do as a veterinary assistant in this situation?
- What is the importance of properly handling a food fish situation?

Chapter 15: Zoo, Exotic, and Wildlife Identification and Production Management

Objectives

Upon completion of this chapter, the reader should be able to:

15.1 Identify common wildlife species
15.2 Explain the importance of wildlife in veterinary medicine
15.3 Explain the classes of wildlife
15.4 Discuss important health management practices in wildlife
15.5 Explain how to implement wildlife management practices
15.6 Describe the career of wildlife rehabilitation
15.7 Identify common exotic and zoo animal species
15.8 Provide the definition of an exotic animal
15.9 Explain the purposes and uses of exotic animals
15.10 Describe the major classes of exotic animals
15.11 Discuss the health management practices in zoo and exotic animals

Introduction

Wildlife describes wild animals that can be found in their natural environments. People enjoy interacting with wildlife in many ways, such as observation, photography, and hunting. Wild animals may become sick or injured and need veterinary attention. Some careers focus on the care of wild animals. Some facilities receive wild animals that have been found sick or injured. **Wildlife management** has become an area of increased interest for veterinary medicine as the natural environments of many species are being destroyed.

Exotic animals are popular in zoos and, depending on the species, may be kept as pets. Many people enjoy watching and learning about these many different types of animals, which they do not normally see in a natural setting. These animals require specialized health care, diets, handling and restraint, and environmental resources that allow them to remain healthy and stress free.

New species of animals have been developed from several of these exotic animals. Many times, veterinary professionals do not have the information required to properly care for them. These new species depend on humans to survive; learning about their needs may be by trial and error. They also require more care and attention than the average domestic animal.

The Exotic Animal

An exotic animal is not native to the area where it is raised and may be rarely found in its natural habitat. Exotic animals are now being bred for new characteristics, such as color or hair coats, and as exotic pets. Many of these species require specialized care and additional resources to survive and be healthy. Most exotic animals are kept as companion animals or breeding and production animals. Most are newly domesticated animals that still have some wild instincts. Some may be used as food sources or for products they provide to humans. They may also be used for entertainment purposes, such as at the zoo and in circus performances. Others may be used for investment purposes, such as sales or breeding.

Purposes and Uses of Exotics

A variety of exotic animals are raised as food sources or other products used by people. These animals may be considered "agriculture" in respect to food sources. In the United States, llamas, alpacas, bison, deer, elk, alligator, and ostrich are raised for food and other **by-products**. These animals are raised, fed, and cared for by humans in an environment that is similar to a beef or dairy farm. They are produced for the benefit of humans and must be raised in a controlled environment. Some species are valuable and costly. They can also be challenging to care for, as there are limited veterinary resources for certain species of exotic animals.

Llama

The **llama** is a mammal that originated in South America and is used for its meat and hair. Llamas are pack animals. Many are now becoming "alternative" pets, meaning they are being used in a way that they were not bred for. Llama meat is low in fat and has a desirable taste. Few llamas are eaten in the United States. Llamas that are bred and can sell from several hundred to several thousand dollars. Another alternative use for llamas is as a guard animal. They can be aggressive and will protect property and other animals. They are used for guarding flocks of sheep and goats. Llamas range in size from 250 to 450 pounds and between 5 and 6 feet tall (see Figure 15–1). Their average lifespan is 15 to 25 years. Llamas are ruminants and can spit regurgitated food when upset or scared. A *ruminant* is an animal that has a stomach with four sections for breaking down food as it passes in the digestive tract. *Regurgitation* is the process in which

FIGURE 15–1 (A) Llamas are bred and raised for meat and hair and as pack animals. (B) Young llama.

food that has been chewed and swallowed is brought back into the mouth to be re-chewed for further breakdown. This is a natural process in ruminants.

Alpaca

An **alpaca** is a mammal similar to a llama that also originated in South America. Alpacas are smaller in size by about half and have much finer and silkier hair coats than llamas (see Figure 15–2). Alpacas are raised for their hair, which is used to make clothing and other linen products. The hair coat of the alpaca is very soft and provides insulation and warmth for winter clothing and other products. Alpacas are profitable animals. They weigh between 100 and 175 pounds. The alpaca and llama have a similar lifespan. **Breeding stock** alpacas—meaning that males and females are raised specifically for reproductive purposes—are very expensive because it costs a lot to achieve high hair quality. A female breeding alpaca may sell for $20,000 to $30,000; a male can range from $50,000 to $60,000, depending on the bloodlines and quality of the animal.

284 SECTION II Veterinary Animal Production

FIGURE 15–2 (A) Alpacas are bred mainly for their hair. (B) Alpacas are bred as pets and show animals and for products.

FIGURE 15–3 Bison are raised for beef production.

FIGURE 15–4 Beefalo are a cross between a cow and a bison.

Bison

The **bison**, also known as the American buffalo, is a ruminant mammal that is closely related to cattle. Bison are raised for meat and by-products, such as skin pelts and horns (see Figure 15–3). Bison are large animals that may become aggressive and dangerous. They require large amounts of land, secure types of fencing, and specialized handling and restraint equipment. Bison average around 1000 to 1500 pounds, stand between 6 and 7 feet tall as adults, and live to an average age of 30 years. They are normally quiet animals but do have the ability to jump over 6 feet high and run at speeds faster than many horses. The average price for breeding-quality bison is several thousand dollars. Many states have regulations on bison farming and reproduction. Bison ranches have become increasingly popular over the past several years due to bison meat being low in cholesterol and lower in fat content than beef.

Thus, the bison industry has become an alternative profitable business method to the beef industry. Several producers have begun crossbreeding bison and domesticated beef cattle breeds to produce beef animals—the **beefalo**—that offer healthier and tastier meat products (see Figure 15–4). **Crossbreeding** is producing an animal with a combination of characteristics from two different breeds of related animals.

Elk

Elk breeding and marketing has become a rewarding and profitable business. Elk are relatively easy to breed and maintain. They need adequate amounts of land but are adaptable to a variety of habitats. Elk are bred for their meat and by-products, including the **velvet** that is shed from the antlers and used in medications and as nutritional supplements (see Figure 15–5). The by-products of an animal are those parts that are not the main profit or production source, such as the antlers, hooves, hair coat, or internal organs. Elk antlers can

FIGURE 15–5 Elk are bred for meat and by-products.

produce a half pound of new growth in a single day. Elk are also raised for hunting-and-release programs, which breed and release elk back into the wild where populations are decreasing. They are ruminants that are related to deer and cattle. In fact, elk are raised in a way that is very similar to cows. Elk are considered the livestock of the future.

Deer

Deer are raised for their meat, which is called **venison** (see Figure 15–6). They are also raised for breeding stock, scents for hunting, and their antlers. Deer are ruminants that graze on pasture. They require specialized fencing due to their ability to jump great heights. Many different species of deer are raised, such as red deer, mule deer, and white-tailed deer. Deer are relatively easy to raise; however, a concern over the spread of diseases that the deer contract or carry has been an issue within the industry. Production management must be thorough to keep all deer healthy within the farm site.

Ostrich

Ostrich are large, flightless birds raised for their meat, feathers, oils, and other by-products. Ostrich stand about 7 feet tall and may live up to 70 years (see Figure 15–7). They can run at speeds of 40 mph and their kicks can cause severe injury. Ostrich do not require large amounts of land but do need proper fencing and handling training. They can produce between 30 and 50 eggs per year. They are adaptable to a variety of habitats. There has been a considerable decrease in ostrich production, as the popularity of ostrich farming in the past several years has caused an overabundance of ostrich, and the needs in the industry have decreased. However, rearing ostrich stills remain a marketable business.

Alligators

Alligators are large reptiles raised for their meat, skin, and by-products and for their release into the wild (see Figure 15–8). There are over 1.5 million alligators in the wild in Florida alone. In some locations, alligators are still an endangered species. Alligator production became popular in the 1980s in the southern United States: ever since, their numbers in the wild have been increasing. In fact, several southern states have organized alligator hunts. Alligator meat is a delicacy in parts of the United States and other countries. It has a firm texture and distinct flavor. Alligator skin is used to make leather products, such as boots, belts, and purses. The needs of alligators vary in regard to nutrition, housing, and climate. They grow and reproduce well in captivity. Alligators are dangerous wild animals and handling and restraint are critical for safety purposes, both to the handler and the animal.

FIGURE 15–6 Deer are raised for their meat, known as venison.

FIGURE 15–7 Ostrich are raised for meat, feathers, and other by-products.

FIGURE 15–8 (A) Alligators are raised for meat, skin, and by-products. (B) Alligators often bask in the water.

Rare Breeds

Other **rare breeds** of animals are raised for a variety of purposes. Some are considered alternative livestock; others are exotic species. These include mammals, reptiles, birds, and aquatics, such as caiman, flamingos, zebu cattle, and zebra (see Figure 15–9). Rare breeds are animals that are newly developed and not common; they are not breeds that were once popular but have now decreased in numbers. Producers and breeders take pride in having a rare breed thrive and exist. Many rare and exotic animals are domesticated and live well in the care of humans. Efforts are made to continue quality breeding and ensure that these animals stay rare. The popularity of breeds usually leads to poor breeding practices and genetic health problems that reduce the quality of the species. Genetic health problems occur in animals that are not carefully bred; these animals inherit unwanted characteristics or diseases from their parents.

Other species that are slowly becoming popular as exotic alternative species for reproduction, food, and by-products include eels, mink, turtles, antelope, snakes, sharks, and several species of game birds.

FIGURE 15–9 Flamingos are considered exotic species.

Exotic Animal Businesses

Exotic animals are obtained from a variety of sources. Some are captured in the wild; others are purchased from existing breeding programs. Yet still others are animals that perform in zoos or circuses (see Figure 15–10).

Each type of business requires that the animals be cared for by an experienced veterinary staff. Many small and large game animals that are marketed from wild animals, such as deer, elk, and alligators, require specialized health care. Exotic animals that are purchased from breeders as future investments include llamas and alpacas. These animals are raised for profit. The general purchase, care, housing, nutrition, and health needs are usually much costlier than those of a domestic animal. Pot-bellied pigs were once considered exotics; because of a lack of interest in the pigs, profits were not rewarding for breeders. Some exotic animal species become pets, as well. Monkeys, tigers, deer, bear, and other species have been raised in captivity as pets (see Figure 15–11). These animals are not always safe around humans. Keeping any exotic species of animal requires a **permit**: this is usually regulated by a state's department of agriculture or the U.S. Department of Agriculture.

Exhibition (Zoo) and Performance Animals

A **zoo**, also called a *zoological garden*, houses wild and exotic animals and plants in a park-like setting. The public sees the animals in an area that closely resembles the animal's native home (see Figure 15–12). Many of the species are from other countries or locations and kept in enclosures. Many of these

CHAPTER 15 Zoo, Exotic, and Wildlife Identification and Production Management

FIGURE 15–12 (A) Wild animals are housed and raised in zoos for study and entertainment. (B) Many wild animals are viewed by people who would otherwise not have the opportunity to learn about or see such wildlife.

FIGURE 15–10 (A) Bear playing with a ball in the zoo. (B) Bear enjoying a treat.

FIGURE 15–11 Many large cat species, such as this lion, have been raised in captivity.

animals are housed to breed in captivity to allow the species to continue to thrive. Zoos serve several purposes: entertainment, educational resources, research, and conservation. Many zoo employees have a background in **zoology**, which is the study of animal life. A **zookeeper** manages the zoo and the animals that are kept there. Zookeepers train the staff, oversee the health care needs of the animals, decide on proper nutritional needs of the animals, clean the animal enclosures, and give educational tours to people.

Zoos acquire animals in several different ways. Some animals are donated by people who cannot properly care for them or who are not licensed to house them. Animals that have retired from circuses or other performance industries will be donated to zoos. Many zoos exchange animals with other zoos to improve the breeding programs throughout the country. Zoos also purchase animals from **exotic animal dealers**. These are professional people who buy and sell exotic species of animals and help zoos meet the needs of certain species. Exotic animal dealers must have a license to purchase, sell, house, and handle a majority of these species. Table 15–1 outlines large zoos found in the United States.

TABLE 15–1
Famous Zoos of the United States

Baltimore Aquarium	Baltimore, Maryland
Bronx Zoo	Bronx, New York
Busch Gardens	Williamsburg, Virginia
	Tampa Bay, Florida
Cincinnati Zoo	Cincinnati, Ohio
Henry Doorly Zoo	Omaha, Nebraska
Philadelphia Zoo	Philadelphia, Pennsylvania
Pittsburgh Zoo	Pittsburgh, Pennsylvania
San Diego Zoo	San Diego, California
Sea World	San Diego, California
	Orlando, Florida
	San Antonio, Texas
Smithsonian Institute's National Zoo	Washington, D.C.
St. Louis Zoo	St. Louis, Missouri

FIGURE 15–13 The elephant is a popular zoo and circus animal exhibit.

Zoo animals require specialized health care through zoo veterinarians. Zoo vets are trained in the same way as other veterinarians but specialize in exotic and zoo animal species. They typically graduate and receive training and additional education in zoo species, learning how to treat them, provide proper health management programs, provide anesthesia and surgical services, and engage in proper restraint and handling. Veterinary technicians and assistants are also trained in zoo and exotic animal care.

An **animal refuge** is a large protected area where exotic animals are housed in a constructed setting that resembles their natural habitat. These areas house all types of exotic species, from mammals, reptiles, birds, and amphibians to species that are naturally found in the surrounding location. Animal refuges may include a variety of habitats, from swamplands to plains to deserts. The United States has over 475 national wildlife and exotic animal refuges. Exotic animals are **feral**, meaning they are wild, not domesticated, and live on land that is maintained by humans. Typically, the staff and people maintaining the refuge do not feed or provide routine health care for these animals. They only provide the environment. If an animal becomes sick or injured, the staff will contact a wildlife or zoo veterinarian or wildlife rehabilitator to care for it.

Some animal refuges have biologists on staff who have backgrounds in and knowledge of animal behavior as well as a degree in biology. Animal refuges are protected by law: hunting and fishing are prohibited. Many refuges are open for public visitation, and people may hike, go boating, and take photographs there.

Performing animals are exotic animals that are raised and trained to perform an unusual activity that is not natural for the animal. Many people feel that performing animals are mistreated or forced to do a job that they do not want to do. This has become a controversial topic over the years. The circus is a variety show that houses many species of animals that perform tricks and acts that entertain crowds of people. Circus animals are raised and trained and carefully cared for by humans. Some have been domesticated; others remain wild. The people who work with circus animals are usually trained in animal behavior, are animal trainers, and have degrees in an area of animal science or biology. The elephant is one animal that has been trained as circus performers a few generations from the wild (see Figure 15–13). The elephant is a species of animal that can be trained to work alongside humans. Elephants have been used in many countries in which they are native. They have been trained to remove trees and pull large loads, very much like a draft animal. Elephants weigh over 12,000 pounds and may reach 12 feet in height. They are strong enough to push over 30-foot trees. They may live up to 60 years in captivity. Because of these qualities, the elephant must always be treated as a wild animal with the potential to be dangerous.

Exotic Animal Care and Management

The use of exotic animals in breeding production programs, zoos, circuses, or as pets and alternative livestock means that they must be profitable. A business is **profitable** when it has gains (money) left over after paying its expenses, in this case, of purchasing and caring for a product (the exotic animal). The startup costs can be immense. It makes sense for anyone involved

FIGURE 15–14 Many wildlife enclosures mimic the habitat of animals in the wild, and in turn, can be quite costly to construct.

in this business venture to have a background in animal science or livestock production to understand the industry. Exotic animals require specialized equipment and housing as well as fencing needs, which can cost several thousand dollars (see Figure 15–14). Some items to consider are listed in Table 15–2.

It is important that people owning or operating an exotic animal–type of business research the animal species they are considering housing. Business owners may face certain obstacles, such as the environment, climate, state and local regulations, and **marketability**—how well something sells—of the species. Many exotic species of animals have environmental and climate requirements. They may need a tropical, wooded, or arctic habitat. It is important to determine if the animal will be able to live and thrive in the surroundings and climate that occur there seasonally. If the species will not do well in the natural area, it must be determined whether the right environment can be created. Fortunately, there are several exotic species that are not sensitive to unnatural environments. Some animals will need to be kept indoors year-round or at certain times of the year. These limitations must be known before purchasing the exotic animal. It is also important to research the federal, state, and local regulations for the state in which the animal will live. There are many laws pertaining to exotic animals. Some places require permits or licenses. Regulations can be very strict regarding owning and housing exotic animals that may be dangerous to other exotic animals or domestic pets. States such as Florida, Texas, and Louisiana have had problems with wild and exotic animals and have placed regulations and restrictions on owning and housing them (see Table 15–3). Exotic animals—such as tigers, lions, bears, and wolves—that are placed in the wrong hands do not mix well in the general public. There are regulations to prevent the outbreak and spread of disease among exotic animals. Some exotic species are closely related to canine, felines, equine, and bovines; they can easily spread a disease to domesticated animals, some of which are intended for human consumption. Research should be done on the species' marketability in the local area. It is important to know that there is a market for this type of exotic investment.

There are many similarities between exotics and livestock or other agricultural animals; however, there are even more differences. Raising an animal that requires specialized care and housing is the ultimate responsibility of the owner. Breeders, dealers, and private owners must be responsible for the animal's care and needs. The animal should be provided with the proper housing, fencing, diet, sanitation, health care, and environment. Exotic animals' nutritional needs will vary among species. The amount of food that domestic livestock and companion animals need has been thoroughly researched. There may not be food on

TABLE 15–2

Operating Expenses of Exotic Animals

Cost of the exotic animals and their availability	Varies between locations	$1,000–$100,000
Equipment	Sometimes difficult to locate	$5,000–$50,000
Housing and fencing	Extensive supplies, upkeep, may require permits	$10,000–$50,000
Nutrition and food supplies	May be difficult to locate	$1,000–$20,000
Veterinary medical care	May be difficult to locate veterinary specialists	$10,000–$80,000
Staff and labor	Must locate trained and knowledgeable employees	$15,000–$50,000
Marketing and advertising	May need to educate the public	$10,000–$25,000
Utilities: lighting, electricity, water, telephone, etc.	Varies with species needs	$5,000–$20,000
Permits and licenses	Depend on state and federal laws	$1,000–$10,000
Insurance and liability	Depend on state and federal laws and locating insurance companies with experience	$20,000–$100,000+

TABLE 15–3
States with Regulations on Exotics

Alaska	Arizona
Arkansas	California
Colorado	Connecticut
Delaware	Florida
Georgia	Hawaii
Illinois	Indiana
Iowa	Kansas
Kentucky	Louisiana
Maine	Maryland
Massachusetts	Michigan
Minnesota	Mississippi
Nebraska	New Hampshire
New Jersey	New Mexico
New York	North Dakota
Oklahoma	Oregon
Pennsylvania	Rhode Island
South Dakota	Tennessee
Texas	Utah
Vermont	Virginia
Washington	Wyoming

the market for some exotic animals. In those cases, the owner must provide the nutrients that the animal would eat in the wild. Many animal diets have been studied by biologists and zoos; the results of their research should be available. Housing and space requirements will also vary among exotics. To understand an animal's needs, owners must research the following:

- Indoor, outdoor, or a combination?
- Temperature range?
- Housed alone or in groups?
- Amount of space per animal?
- Housing materials?
- Fencing materials?
- Restraint equipment?

Finally, the most important aspect of owning an exotic animal is seeing to its health and veterinary care. A limited number of veterinarians and other health care providers are familiar with exotics. Even then, they may not understand the particular needs of the exotic species. A majority of exotic veterinary professionals are located at universities, zoos, aquariums, and animal rehabilitation centers. These staff members must know how to sedate and anesthetize exotics, perform various surgical procedures, and provide diagnosis and treatments. When handling and restraining exotic animals, veterinary staff should treat these animals as wildlife and act accordingly. The stress of restraint may cause aggression, and injury may occur even in the gentlest species of exotic animals. Special restraint equipment and sedation may be necessary.

The Importance of Wildlife

Animal wildlife consists of any living animal that has not been domesticated and includes mammals, birds, insects, fish, rodents, reptiles, and many other types of animals (see Figure 15–15). Humans may benefit from these animals in several ways. The animals may provide food, products (such as fur), and/or entertainment.

Wildlife as a Food Source

A **game animal** is an animal that is hunted for food. Common game animals in the United States include deer, elk, bear, ducks, and fish (see Figure 15–16).

FIGURE 15–15 The American crocodile in its natural habitat.

FIGURE 15–16 The white-tailed deer is often hunted in the United States.

Game animals may be raised in captivity for release into the wild where populations are decreased or sparse. Some areas have increased and overpopulated areas of game animals, which causes environmental damage and spread of disease. Wildlife is hunted as food sources and for their by-products, such as skin pelts, antlers, or hooves. Many parts of the United States have specific hunting seasons, depending on the animal species.

Wildlife as a Resource

Animals may also be hunted or trapped for fur or hides. **Trapping** is typically done with smaller mammals that have furs and pelts that are of economic value. Trapping involves the capture of small animals within wire cages or traps. Some of these animals may be used as food sources as well. Many areas of the United States have trapping seasons based on the animal species. Examples of animals that may be trapped include beaver, muskrats, and weasels.

Wildlife as Entertainment

Many people enjoy animal observing, photographing, and drawing wildlife. People enjoy the beauty of wild animals. People visit specific locations the world over to view animals living in their natural environment; this is a popular tourist, vacation, and recreational activity. Many parks and conservation areas dedicate their land and services to protecting the animals that live there. Zoos and wildlife parks have increased in popularity. These places give wild animals that may not survive in the wild a safe place to live and be observed (see Figure 15–17). Many of these facilities have veterinary staff who see to the well-being of the animals. They may also have breeding and reproduction programs

FIGURE 15–17 Zoos provide people with the opportunity to see wild animals in a manner that is safe for both the animal and the spectator.

that serve to increase populations in species that are endangered or extinct in their natural habitats. An **endangered animal** is a member of a species that is decreasing in numbers in the wild and is close to becoming extinct (see Figure 15–18). **Extinct** means that a species of animal no longer exists in the wild and may only be found in zoos or wildlife parks.

Most Endangered Wildlife in the United States

- Polar bear
- Humpback whale
- Grizzly bear
- Gray wolf
- Florida panther

FIGURE 15–18 The American alligator is endangered.

- Jaguar
- Otter
- Chinook salmon
- Bighorn sheep
- Manatee
- Puma (cougar)
- Green sea turtle
- American alligator

Control of Wildlife

Hunting is another way to control wildlife population and disease. Some species become overpopulated in the wild. This causes breakouts of disease to pass quickly throughout the species. Fishing also serves the same means for certain fish species. Each U.S. state establishes hunting and fishing seasons to control when people may fish for or hunt a specific species. These methods are a humane way of controlling large populations of animals.

The Role of Veterinary Medicine in Wildlife Management

The habitats and food sources of wildlife species are disappearing, as land is developed in these habitats due to **urbanization**, that is, creation of cities. Building cities and homes destroys the environments of animals that cannot survive in a human world and is a threat to many wild animals. Wildlife veterinarians may work in research, zoos, and wildlife parks or in government areas to care for wild animals that may become sick or injured. Their role is to serve the state or country by discovering disease outbreaks, handling and caring for wildlife, investigating the decrease in numbers of species, and working with public concerns regarding wildlife.

Classes of Wildlife

There are several classes of wildlife. Game animals are used for food and other products and may be large or small. Many states have designated hunting seasons for a variety of animals. The U.S. Fish and Wildlife Department and **Game Commission** are agencies that monitor these seasons by providing hunting rules, regulations, and licenses. Every state within the United States has its own Game Commission and U.S. Fish and Wildlife Department. The U.S. Fish and Wildlife Department governs the laws of each state. The Game Commission also serves educational and conservation purposes for both wildlife and people. Popular species of game animals include deer, bear, rabbit, and squirrel (see Figure 15–19). **Nongame animals** are wild species that are not hunted for food. They offer beauty and entertainment only. Some of these animals are food sources for predators or act as scavengers. Many of these animal species are found in urban and residential areas. Skunks, opossum, mice, raccoons, and porcupine are examples of nongame animals (see Figure 15–20).

Game fish are fish species that live in the wild—in both fresh water and salt water—and are used for food and sport. State agencies issue fishing licenses and monitor the fishing seasons. Freshwater game fish include bass, trout, and catfish (see Figure 15–21).

FIGURE 15–19 The bear is a game animal often hunted for food or fur.

FIGURE 15–20 The raccoon is a nongame animal often found in residential areas.

FIGURE 15–21 The brown trout is a freshwater game fish.

FIGURE 15–22 Tuna is a saltwater game fish.

Saltwater game fish include tuna, flounder, and snapper (see Figure 15–22). **Nongame fish** are species of fish that are too small or too large to be caught on a line. They may be found in both fresh water and salt water. Nongame fish include sunfish, bluegills, and sturgeon.

Game birds are wild bird species hunted for food and sport. They are divided into three categories. **Migratory** birds move from location to location, sometimes out of the local area or state. Examples are the crow and woodcock. **Waterfowl** swim and spend a large amount of time on the water. They include a variety of species of ducks and geese (see Figure 15–23). **Upland game birds** are wild species of birds that spend most of their time in the woods and do not move out of the area. They include the pheasant, quail, and turkey (see Figure 15–24). State and federal government agencies regulate hunting game birds.

Birds of prey or **raptors** are large species of birds that are protected by law; it is illegal to hunt or shoot them. They are hunters that prey on other wildlife, such as rabbits and squirrels. They have large **talons**, or claws, used for capturing prey and grasping or perching onto branches. Examples of birds of prey include eagles, owls, and hawks (see Figure 15–25). **Nongame birds** include all other species of birds and songbirds. Many are found in residential and

294   SECTION II   Veterinary Animal Production

FIGURE 15–23 Geese are migratory birds.

FIGURE 15–24 Turkeys are upland game birds.

FIGURE 15–25 The eagle is a bird of prey.

FIGURE 15–26 The cardinal is an example of a nongame bird.

urban areas. They live in a variety of environments, from woodlands to waterways. Nongame birds include robins, pelicans, blue jays, doves, cardinals, and wrens (see Figure 15–26).

Wildlife Health and Management Practices

A wildlife habitat is where animals live and can find food, water, space, and cover. The habitat must provide all four of these components. Any component that is missing is known as a **limiting factor**. Many animals eat *browse*, or the shoots, twigs, and leaves of trees and shrubs used as forage. When homes are built in wooded areas, browse is removed and becomes a limiting factor. Even when an area meets the needs of the wildlife, there are limits to how much of a population can survive and be sustained. The number of animals that a habitat can support is known as its **carrying capacity**. This number increases and decreases over the years.

When an animal claims an area as its own, it becomes its **territory**. Wildlife will protect this area and keep other wildlife from entering. Space can easily become a limiting factor.

Wildlife management is the practice of researching the needs of wildlife, providing them with the essentials of life, and monitoring their survival. Many people begin their career in wildlife management by taking courses in **mammalogy**, or the study of mammals. Many college programs offer **wildlife biologist** degrees. Wildlife biologists research and study data on all types of wild animals to determine how wildlife can be saved and protected (see Figure 15–27). Many biologists study a particular species of wildlife for a number of years to determine the animal's needs. Wildlife health practices include monitoring wild species for diseases and focusing on the control of those diseases. That may mean determining a treatment for the disease and treatment of the population of the species as a whole. It may also mean how to humanely control the population by increasing hunting or capturing and destroying sick animals affected by the problem. Other practices include manufacturing vaccines to control the spread of disease and capturing and tagging wildlife to monitor their health, provide treatment, and monitor the range the animal travels in the area.

Capture and Restraint of Wildlife

Capturing and handling wildlife can be a risk for both the animal and the wildlife team. Most wildlife is captured by trapping in humane cages that contain the animal safely or through the use of **tranquilizers** given to sedate the animal for safe handling. Tranquilizers are administered with a dart gun so that people do not have to get close to wildlife and risk injury. Both of these restraint methods are safe for the animal and for the people working with them. However, any wild animal can suffer from stress and be a safety risk during its capture and restraint. Thus, it is important that anyone working with wildlife be properly trained. An animal can be *sedated*, or put to sleep, by a drug known as a *sedative*. Then the animal can be handled without fear of it moving. Animals under sedation must be monitored for complications; they may also wake up before the procedure is over. A veterinarian must determine what is a proper sedative and how much an animal should be given. It is also important to keep accurate medical records of these situations.

Wildlife Rehabilitation

Wildlife and people can live together in certain areas. However, animals and people living in the same habitats means more injuries, more disease, and more death to the wildlife. Areas that have populations of wildlife usually have wildlife rehabilitation centers or wildlife rehabilitators. A **wildlife rehabilitation center** houses and cares for sick or injured wildlife (see Figure 15–28). The veterinary staff of such centers are knowledgeable and experienced in handling and caring for wild animals and usually consists of veterinarians, veterinary technicians, and veterinary assistants who have been trained to properly handle

FIGURE 15–27 Wildlife biologists study animal species to find ways to control and protect them.

FIGURE 15–28 Wildlife rehabilitation centers will take in and care for sick, injured, or young abandoned animals.

wildlife. Most centers also rely on volunteers to help with basic duties and raise money for these programs. These centers are not businesses run like a veterinary practice where owners bring a pet for services and pay for the care of them. Rather, a rehab center cares for wildlife and then returns it to the wild if possible; if the animal can no longer survive in a natural setting, it may send it to a zoo or wildlife park. A **wildlife rehabilitator** is someone who cares for wild animals that are sick or injured. Rehabilitators may offer this care in their homes, in private veterinary facilities, or at local nonprofit centers that they have developed specifically for wildlife. Many states require that wildlife rehabilitators take a course and pass an exam to become licensed. Many wild animals are protected within the states; housing them is illegal, and government agencies have been created to oversee wildlife rehabilitators and require their licensure. Handling wildlife can be dangerous to the handlers and the people in the local area. For this reason, most states also require a permit to house any wild animal on private property. Most rehabilitators work with a veterinarian who has experience in wildlife and can offer a diagnosis of illness and injury and recommend the proper treatment for the animal. Wildlife rehabilitators should be able to administer basic first aid care and basic physical therapy treatments to wildlife species.

Agencies for wildlife rehabilitation include the following:

- Game commissions or Fish and Wildlife departments
- U.S. Fish and Wildlife Service
- U.S. Department of Agriculture

Wildlife rehabilitators' education should begin with studies in biology and ecology. **Biology** is the study of life. **Ecology** is the study of the environments in which animals live. It is important that wildlife rehabilitators take courses in animal behavior in order to learn about normal and abnormal animal behavior. Other important areas of study and knowledge include wildlife nutrition, natural history, environmental needs, and proper caging requirements. A wildlife rehabilitator does not need a college degree but must be educated about the proper handling and health care techniques that wild animals require. Many veterinary technology schools and veterinary schools offer courses in wildlife management and rehabilitation.

The goal of wildlife rehabilitation is to provide professional care for diseased, injured, and orphaned wildlife so they can ultimately be returned to their natural habitats. Wildlife rehabbing is not an attempt to turn wild animals into pets. Fear of humans is a necessary trait that these animals must have in order to survive.

SUMMARY

Exotic animals are costly investments that may be used for a variety of purposes. The major types of exotics include food and fiber animals, production and breeding animals, zoo and performing animals, and investment animals. No matter the type of business or private use, exotic animals have specialized needs. There are many differences between exotics and domesticated species of animals. It is necessary to research the nutritional, environmental, health, and housing needs of the animal. It is also important to learn about regulations that may be in place about owning exotic animals.

Wildlife is an important part of the environment and means a lot to many people. Wildlife offers our society with beauty, entertainment, food, and other products. Wildlife management has become an important part of today's world, as necessary resources for wildlife, including food, water, space, and cover, are being lost. Wildlife is being threatened as more people move into their environments. For this reason, wild animals face more diseases and injuries than ever before. Wildlife rehabilitation and veterinary care has become an important part of the veterinary field. For additional information on the field of wildlife rehabilitation or wildlife services, see Appendix C.

CHAPTER 15 Zoo, Exotic, and Wildlife Identification and Production Management

Key Terms

alligator a large reptile raised for its meat, skin, and by-products and for release into the wild

alpaca a ruminant mammal similar to but smaller than a llama and that originated in South America; it has a silky, fine coat and is raised for its hair

animal refuge a large protected area of land that houses exotic animals and is set up to resemble their own habitat

beefalo a crossbreed of bison and domesticated beef cattle that produces a beef animal with healthier and tastier meat products

biology the study of life

bird of prey a large bird that is a hunter and eats small mammals and rodents; is protected by law and not legal to hunt

bison a ruminant mammal closely related to cattle

breeding stock the male and female animals of a species raised specifically for reproductive purposes

by-product a part of an animal that is not the main profit or production source, such as the hooves, antlers, hair coat, or internal organs

carrying capacity the number of animals living in an area that can be supported and sustained

crossbreeding crossing two varieties or breeds within the same species

deer a ruminant animal that grazes on pasture and has antlers

ecology the study of the environment that animals live in

elk a ruminant animal, much like a deer, that grazes on pasture and has antlers

endangered animal species decreasing in numbers in the wild and close to becoming extinct

exotic animal species not native to the area where it is raised and rarely found in its natural habitat

exotic animal dealer someone who buys and sells exotic animals to private collectors and zoos

extinct animal species that is no longer found in their natural habitats

feral wild and not domesticated to live in an area maintained by humans

game animal wildlife species hunted for food

game bird wild bird species hunted for food and sport

game fish aquatic species caught for food or sport

hunting the tracking and killing of animals for sport, food, or other by-products

limiting factor something that constrains a population's size and slows or stops it from growing

llama a ruminant animal native to South America raised for its hair and meat and used as a pack animal

mammalogy the study of animals

marketability a measure of how well a product will sell in an area

migratory moves from location to location throughout the year

nongame animal species that is not hunted for food or other products

nongame bird species that is not hunted for food

nongame fish species that is too small or too large to catch and eat as a food source

ostrich large, flightless bird raised for meat, feathers, oils, and other by-products

permit a written license granted by an authority

profitable having gains (money) left over after paying expenses

raptor a large bird of prey

rare breed an animal bred and raised for a variety of purposes but not found locally

talon claw on a bird of prey

territory an area of land claimed by an animal

tranquilizer a drug used to sedate animals for safe handling

trapping capturing and restraining of wild animals

upland game bird wild species that spends most of its time in the woods and does not move out of an area

urbanization the development of cities over an area

velvet material shed from the antlers of a deer used in medications and nutritional supplements

venison deer meat

waterfowl bird that swims in water and spends a large amount of time on the water

wildlife any living animal that has not been domesticated

wildlife biologist a person who researches and studies data on all types of wild animals to determine how wildlife can be saved and protected

wildlife management the practice of researching the needs of wildlife, providing them the essentials of life, and monitoring their survival

wildlife rehabilitation center an area that houses and cares for sick or injured wild animals, often run by a veterinary staff

wildlife rehabilitator an individual who cares for sick, injured, or orphaned wild animals

zoo a zoological "garden" that houses wild and exotic animals and plants for people to visit and observe their behaviors and beauty

zookeeper a person who manages and cares for animals in a zoo

zoology the study of animal life

Review Questions

1. What animal species are considered wildlife?
2. What is the difference between endangered and extinct animals?
3. How do animals become endangered or extinct?
4. What are the classes of wildlife? Give some examples of each class.
5. What are the four requirements of a wild animal's habitat?
6. What is a wildlife biologist?
7. What is a wildlife rehabilitator?
8. What are the requirements of a wildlife rehabilitator? What is an exotic animal?
9. List the similarities and differences between exotics and domestic animals.
10. What are some purposes and uses of exotic animals?

11. Discuss several species of exotic animals and how they are used.
12. What are the purposes of zoos?
13. How do zoos acquire their animals?
14. What must be considered before selecting an exotic animal?
15. What are the housing needs of exotic animals?
16. What is the difference between profitability and marketability?
17. What are feral animals?

Clinical Situation

Dr. Jefferson is a vet who works with a local zoo. He routinely has his veterinary assistants, Marcus and Jillian, and his veterinary technician, Bethany, accompany him on any medical and surgical cases. Today, they are going to the zoo to look at a mountain lion that may have a tooth abscess. The entire staff is excited to work at the zoo, as it makes for an interesting and challenging time.

When they arrive, the zookeeper, Anthony, shows them to the veterinary medical center so they can prepare for the day's work. The vet and Anthony prepare to tranquilize the mountain lion with a dart gun, which Anthony has been trained to use. The assistants and technician set up the surgical area for an exam and a possible dental extraction. The anesthesia machine is ready for use. The team goes together to sedate the lion.

"Okay, meet Romeo," says Anthony. "He is a 6-year-old male mountain lion and has been here for about 3 years. He can be aggressive and difficult, which is probably due to his feeling territorial and protective of his mate, Juliet. I have put her outside so we don't have to worry about her during the capture."

"I will dart him, and when he is down, everyone move quickly to lift him on the stretcher and move him to the surgery area," says Antony.

The lion is darted and down, and the team moves him quickly to the surgery area. When they arrive at the veterinary surgical center, Romeo is beginning to wake up.

- What is the problem with this situation?
- What could have caused this issue?
- How should the veterinary staff handle this situation?
- How could this problem have been avoided?

Clinical Situation

Alison, a veterinary assistant and wildlife rehabilitator, works at Davis Animal Hospital. The veterinarians at the hospital are also wildlife specialists. Alison has just received a call about a young fawn that was hit by a car along a local road. The people who saw the fawn are bringing it to the facility immediately. They stated it had some bleeding injuries and seemed dazed and scared. They would be arriving within 10 minutes. Alison alerts the staff of the emergency and they prepare the treatment area.

When the clients arrive with the young injured deer, they say the fawn has been trying to stand.

"The deer is very scared. We've been trying to make sure it stays lying down, but it's been fighting us," says one of the clients as they carry the deer into the treatment area.

The staff begins treatment. Alison shows the clients to the front door and thanks them for bringing in the animal. They ask, "What will happen to the animal now?"

- How would you answer the client's question?
- Who assumes the responsibility for the animal?
- What would you have done if the facility were not equipped to care for wildlife?

CHAPTER 16: Laboratory and Animal Research Production Management

Objectives

Upon completion of this chapter, the reader should be able to:

16.1 Identify common breeds and species of animals used in laboratory medicine
16.2 Describe methods and types of laboratory animal research
16.3 Discuss the use of animal models and computers in laboratory medicine
16.4 Describe the career opportunities in veterinary laboratory animal medicine
16.5 Discuss the history of animal research in relationship to disease discoveries
16.6 Explain the health management practices of laboratory animals
16.7 Explain how to implement a management plan for laboratory animals

Introduction

Laboratory animal science is a field of veterinary medicine devoted to the production, care, and study of laboratory animals used in biomedical research and education. It is a large field comprised of professionals from many scientific, educational, and veterinary disciplines. Professionals within this field have a deep sense of purpose and are dedicated to helping society and improving the lives of people and animals alike. Much controversy has evolved over the years with a difference of opinion regarding the use of animals in research. Animal well-being and high standards of care have been a goal in animal research over the past 10 years.

Animal Research

Scientists sometimes use animals in **research** to answer questions about medical conditions that affect humans and animals. From rats to dogs, pigs to fruit flies, a variety of animal species contribute to medical breakthroughs that save millions of human and animal lives each year. Through research on these animals, scientists have discovered cures and preventions for a number of human and animal ailments, as well as medications to treat these conditions. The **American Medical Association (AMA)** represents the human side of medicine and conducts many studies that benefit the advancement of human medicine and surgical procedures. In 1950, the **American Association for Laboratory Animal Science (AALAS)** was created for the exchange of information and expertise in the care and use of laboratory animals, to promote the humane care and treatment of laboratory animals, and to support the quality research that leads to scientific gains that benefit people and animals. Benefits of AALAS include the following:

- Promotion of the humane care and use of laboratory animals (animal welfare)
- Encouragement of responsible research
- Education of research animal personnel
- Establishment of standards and licensure of laboratory animal personnel

TABLE 16-1

Animals Used in Research

ANIMAL	USED IN
Armadillo	Development of a leprosy vaccine; current research into a cure for leprosy
Cat	Treatments for ocular conditions such as lazy eye, cross-eye, glaucoma, and cataracts
Dog	Discovery of diabetes; development of insulin; development of open-heart surgeries; development of organ transplants; development of anesthesia
Horse	Development of an antitoxin and vaccine for diphtheria; Treatment for tetanus
Ferret	Menstrual cycle and reproductive research; development of a bird flu vaccine; development of methods to control ovulation
Guinea Pig	Discovery of vitamin C; development of a vaccine for tuberculosis (TB); development of anticoagulants
Hamster	Study of brain disorders
Drosophila (fruit fly)	Development of drugs to treat skin infections, pneumonia, and meningitis; current research into breast cancer treatment
Jellyfish	Study of chemicals used in treating cancer and Huntington's disease
Kangaroo	Researching obesity and high blood pressure
Rabbit	Development of treatments for cystic fibrosis and asthma; development of polio vaccine; treatment of arthrosclerosis
Mouse	Development of the human genome
Pig	Transplant of organs; development of treatments for strokes; growing heart valves for humans
Rat	Development of medicines used in cancer, stroke, and bone repair; discovery of DNA
Monkey Tamarin	Development of measles vaccine develop treatments for colon cancer

FIGURE 16-1 Rats are commonly used for laboratory and research purposes.

A **lab animal** is raised for use in research laboratories. Table 16-1 outlines some uses and discoveries that have come from animal research. Lab animals are bred, raised, and cared for in controlled environments to ensure their genetic suitability for the work as well as the accuracy of results. Mice, rats, and other rodents make up 90 percent of the lab animals that are used in today's research studies (see Figure 16-1). Dog or cat species make up less than 1 percent. Primate use in labs is also on the decrease. One of the main goals of laboratory animal scientists is to make sure that research animals are not exposed to any unnecessary pain or stress. While the **Animal Welfare Act (AWA)** strictly regulates experiments on animals, statistics show that most experiments are not painful to animals. The act lists regulations on the unnecessary use of animals and recognizes that animals have rights and feelings and should not suffer in any way. According to a 2001 report by the U.S. Department of Agriculture, over 90 percent of research was not painful to animals. In a majority of cases, animals are not exposed to or involved in any painful procedures. Most animals receive anesthesia or pain-relieving medications during procedures that could possibly involve some pain or distress. The protocols that are used in animal labs many times exceed the standards of veterinary medicine.

Animals are also used for their **by-products tissue**—this is tissue removed from an animal to be used for other purposes in research or for human consumption. Many tissues are collected from animals produced for human meat sources. Other animal products that are commonly used include chicken eggs, which are used in manufacturing vaccines and medications.

Types of Research

Animals are used in three types of research (see Table 16-2). The first is **basic research**, which is carried out in a lab setting to determine knowledge and understanding in life processes and diseases. This type of research uses observation, measurements,

TABLE 16–2
Types of Research

Basic Research	• Laboratory setting • Conducting wide variety of experiments • Increase knowledge in life processes and diseases • No predetermined facts • Use of observation, measurements, descriptions • Provides data as answers • Studies a species of animal
Applied Research	• Any type of controlled setting • Specific topics • Builds on existing knowledge • Uses all types of animal models
Clinical Research	• Medical or veterinary research lab setting • Focus on human or veterinary medicine • Uses specific species based on the topic of research • Supports facts that have been proven

and experiments to determine data. An example of basic research is the discovery of anthrax: it was identified as bacteria when spores were isolated in a research experiment. **Applied research** is undertaken for a specific purpose, such as developing a new vaccine or medication. This research is carried out with existing information. Finally, **clinical research** is conducted in a lab setting and focuses on human- or veterinary-related issues. An example is using a dog or guinea pig to study specific diseases on canines or rodents.

Animal Models and Laboratory Medicine

The use of animals in a research setting must be completely controlled and monitored for accuracy. Laboratory medicine has developed several acceptable **animal models** that help researchers acquire knowledge and information on human and veterinary medicine. There are four basic models:

- Living animals
- Animal tissues
- Nonliving systems
- Computer or simulated programs

These models are used to study and better understand human and animal bodies, behavior, diseases, surgical procedures, treatments, and the effects of medications on the body. Research on **living animals** includes any species that is alive and will respond to a stimulus in some way (see Figure 16–2). Another research model uses **animal tissue**, which is a living part of an animal that is cultured in a lab and grown for specific research (see Figure 16–3), most often relating to cancer and chemotherapy. A **nonliving system** is a model that is mechanical in nature and mimics the species of animal that is being studied. This research typically monitors movement or injuries of living animals. An example is the artificial hip in humans or the artificial ball-and-socket joint in canine hip dysphasia. These models help develop a pattern or design without the use of a living animal.

FIGURE 16–2 Guinea pigs can be used as living animal models.

FIGURE 16–3 Animal tissue can be cultured and used in experiments for cancer and chemotherapy treatments.

The last research model is the **computer system** or the **simulated system**. This model mimics an animal's behavior as well as internal and external structures. It requires a significant amount of research on animal and human subjects to develop programs that work in the same manner as a live animal. Conducting experiments on a computer or simulator may not be perfect, but it is a step in the right direction toward not needing a live animal in research one day. One example is using a simulation of a frog dissection for biology classes.

Species of Animals Commonly Used in Research

Many types of animals are used in laboratory research. The species vary greatly and are commonly companion animals and livestock so that research conducted for human and veterinary medicine is true to the need of the experiment. Animals that are used in experiments should be as closely related to the desired outcome as possible.

Mouse

Mus musculus is the species name for the "house mouse"; mice (along with rats) make up the majority of animals used in medical research. Its small size, ease of handling, and low cost make the mouse an ideal participant in laboratory experiments (see Figure 16–4). Furthermore, with their efficiency in breeding, scientists can breed different strains of mice with natural genetic problems to achieve specific models of human and animal diseases. Researchers have developed mice with leukemia, breast cancer, and many other types of cancer through breeding and genetics. This allows new treatments to be tested on animal models, instead of humans. Scientists are always searching for the best animal model for the study of human AIDS. Much that we know about this disease has come from research using mice. The use of mice has also helped develop vaccines for influenza, polio, yellow fever, and rabies. Mice are popular for studies that involve large numbers of animals. Researchers have developed more than 100 bloodlines.

Rat

The larger cousin of the mouse is the rat, known as **Rattus norvegicus** (see Figure 16–5). Rats are commonly selected for use in studies of behavioral and nutritional problems. Rats are known for naturally developing tumors; thus, cells from human cancer tumors can be placed in rats and studied to further cures (see Figure 16–6). Five bloodlines of rats have been developed in research from the wild brown Norwegian rat.

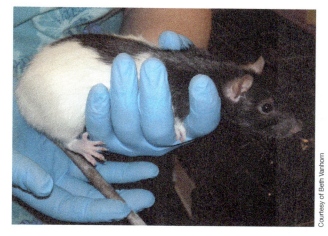

FIGURE 16–5 Rats are often studied for behavioral and nutritional problems.

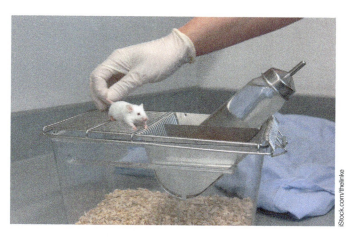

FIGURE 16–4 Mice are commonly used in labs for research purposes.

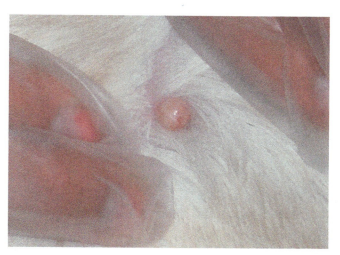

FIGURE 16–6 Tumor on a rat.

Rabbit

The rabbit is known in a lab setting as **Oryctolagus cuniculus**. There are several similarities in the physiology of rabbits and humans, which makes the rabbit a good model for research into human disease (see Figure 16–7). The term **physiology** refers to the functions of the body and how the systems within an animal work. Like the rat, the rabbit can produce tumors; this makes it a useful model for the study of chemotherapy treatments and the prevention of certain cancers. Rabbits are also used to study ear infections, which affect millions of infants and children each year. In cases of veterinary medicine, the rabbit has proven useful in finding treatments for diseases and problems involving the eyes, such as cataracts and glaucoma (see Figure 16–8).

FIGURE 16–7 Rabbits have similarities to humans that make them good candidates for living research models.

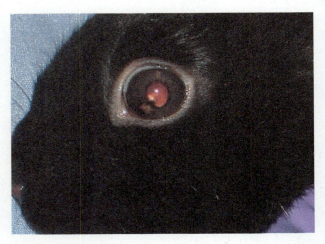

FIGURE 16–8 Rabbits are used to study disorders of the eye.

Primates

Primates are mammals, such as monkeys and apes, that have movable thumbs; in many ways, they resemble humans. Because of this resemblance, they are often referred to as **nonhuman primates** in the lab setting. Several species of monkeys are used in research. Primates are studied based on their wild behaviors and ability to communicate through verbal and nonverbal communication (Figure 16–9). Monkeys can be divided into two classes: **New World monkeys** and **Old World monkeys** (see Table 16–3). New World monkeys originated in South America, while Old World monkeys originated in Africa and Asia. Certain features distinguish them from each other.

Research Regulations and Guidelines

Animal research studies are closely monitored and regulated by various government agencies and humane associations. Despite the many benefits and breakthroughs born of animal research, a great debate has ensued regarding the use of animals in research and testing. Consequently, guidelines have been put in place to protect the rights and use of animals. One agency that monitors and approves studies for research is the **Institutional Animal Care and Use Committee (IACUC)**. All research studies involving animals must submit a proposal to the IACUC that states the type of study, how animals will be used and cared for, the personnel and veterinary staff and their credentials, and any drugs or anesthetics that may be used. The IACUC reviews the proposal and accepts or denies the study (see Figure 16–10). In 1985, the Animal Welfare Act (AWA) was developed to further regulate animal use in research on a federal level. This legislation was brought about due to the debate over

TABLE 16–3

Characteristics of Monkey Types

NEW WORLD MONKEYS	OLD WORLD MONKEYS
• Nostrils open to the side	• Nostrils are close together and open downward
• No cheek pouches	• Check pouches to carry food
• No pads on buttocks	• Pads on buttocks
• Long tails for grasping	

CHAPTER 16 Laboratory and Animal Research Production Management

FIGURE 16–9 Primates are used in research studies because of their similarities with humans.

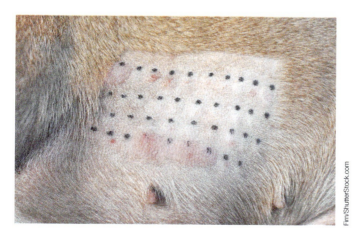

FIGURE 16–10 All studies—such as this animal undergoing skin irritancy testing—must be approved by the IACUC.

the need for animal testing and the need for proper use of animals in research. The U.S. Department of Agriculture enforces the AWA, which regulates and inspects all facilities that conduct animal-related studies.

Lab Animal Management and Health Practices

Lab animals require controlled conditions for their housing as well as a standard of care that includes proper hygiene and an **aseptic**, or sterile, environment. These facilities must follow proper sanitation practices. Studies may increase the risk of disease outbreaks or the contamination of a study source. Cages and equipment must be properly sterilized. Blood and tissue samples must be handled with extreme care. All bedding, food, and waste materials must be disposed of properly. Lab animals must be fed a strict, high-quality, balanced diet that follows what was set out in the proposal. Research can in fact be compromised if a proper diet is not followed. Diets will vary according to animal species, age, size, maturity level, and environment. Lab animal facilities require a specialized environment that meet specific conditions, the most important being space. All rodent species must be housed in at least 6 square feet of space. Livestock require at least 144 square feet of space. Every animal species has a specific space requirement that is available from the **National Institutes of Health (NIH)**. The ideal temperature is between 65 and 85 degrees Fahrenheit, with humidity between 60 and 70 percent. Light (and darkness) must be provided on a regular, set schedule. Proper ventilation allows fresh air to circulate, which eliminates odors and vapors and reduces the risk of airborne diseases. Noise barriers must be in place for both the animals and the outside area.

Lab Animal Careers

The animal research and lab areas offer several career opportunities. One option is a **laboratory animal technician**. Lab animal techs are trained to care for animals used in laboratory research (see Figure 16–11). They are responsible for feeding the animals; cleaning their living areas; monitoring their overall health; and providing basic care, such as administering medicine, feeding them, and assisting other employees. Lab animal techs have a basic knowledge of the needs and behaviors of animals. They play an important role before and after a scientific experiment, from being involved in the planning stages of an experiment to helping determine the care an animal will need before, during, and after a procedure. A technician is also involved in the analysis or study of the research data following an experiment.

Another career is **laboratory veterinarian**. Some veterinarians provide medical care to animals involved in research; others help improve surgical techniques for humans and animals. A number of veterinarians work for government agencies, universities, or corporations. Laboratory vets perform a variety of jobs, ranging from management to product development and research (see Figure 16–12).

FIGURE 16–11 Laboratory technicians play an important role in research, helping with the investigations and taking care of the animals.

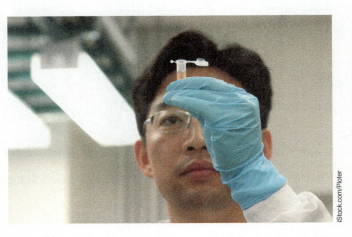

FIGURE 16–12 Laboratory veterinarians provide medical care to animals involved in research.

Other extremely important work for veterinarians is the humane care and treatment of laboratory animals. Through the U.S. Department of Agriculture, these doctors visit research laboratories and ensure the animals are being treated according to federal law.

Another area of the profession is animal behavior. An **animal behaviorist** studies animals to collect data on their behavior in captivity. Some may help rehabilitate animals from zoos or research facilities. Behaviorists also play an important role in research by discovering health programs and housing methods to promote a high quality of laboratory animal medicine.

Finally, there is the career of **research scientist**. Researchers are committed to human and animal health. They are responsible for planning experiments carried out on animals and ensuring that the animals are not in pain. Researchers are responsible for obeying the protocol set up by the federal government.

Veterinary medicine and animal research have led to cures, treatments, vaccines, or advancements in diseases and conditions such as the following:

- Allergies
- Arthritis
- Asthma
- Birth defects
- Bronchitis
- Cancer
- Deafness
- Diabetes
- Epilepsy and seizures
- Glaucoma
- Heart disease
- Hypertension
- Influenza
- Kidney disease
- Leukemia
- Lyme disease
- Measles
- Tetanus vaccine
- Tuberculosis
- Ulcers

SUMMARY

Animals and people live longer and better lives thanks to animal research. This research is a systematic and controlled attempt to answer questions on a specific topic that is based on a carefully planned procedure. Laboratory animals are bred and raised in controlled environments and used in a variety of research studies. Many types of animal species are used in these studies. Ninety percent of the animals used today are from the rodent family. Advancements in animal research have contributed to discoveries in the causes of disease, treatment and prevention of disease, and cures for former fatal diseases—both in human and veterinary medicine. Because of society's attitude toward animal research, the standards in health care and management practices of lab animals many times exceed those of doctors and veterinarians.

Key Terms

American Association for Laboratory Animal Science (AALAS) an association of professionals that advances the responsible use and care of laboratory animals to benefit people and animals

American Medical Association (AMA) the largest association of physicians and medical students in the United States; it conducts studies that benefit the advancement of human medicine and surgical procedures

animal behaviorist someone who studies animals to collect data on their behavior in captivity

animal model a live animal used for the study of behavior, diseases, procedures, treatments, and the effects of medications on the body

animal tissue a living part of an animal that is cultured in a lab and grown for specific needs

Animal Welfare Act (AWA) legislation that regulates experiments carried out on animals

applied research a study done for a specific purpose

aseptic free from contamination

basic research research carried out in a lab setting with the purpose of understanding of life processes and diseases

by-products tissue tissue that is removed from an animal when it is used for other purposes, such as research or human consumption

clinical research research conducted in a lab setting that focuses on human- and veterinary-related issues

computer system a computerized model that mimic animal behaviors and/or internal or external structures; also known as a *simulated system*

Institutional Animal Care and Use Committee (IACUC) an organization that monitors and approves proposed research studies

lab animal an animal raised for use in research laboratories

laboratory animal technician a professional trained to care for animals used in research

laboratory veterinarian a doctor who provides medical care to animals involved in research and performs surgical procedures used in improving techniques for humans and animals

living animal any species of animal that is alive and, in some way, responds to a stimulus

Mus musculus the genus and species of the mouse

National Institutes of Health (NIH) a federal agency responsible for biomedical and public health research

New World monkey a monkey species that originated in South America

nonhuman primate refers to primates in a lab setting that are being used for research

nonliving system a model that is mechanical in nature and that mimics the species of an animal under study

Old World monkey a monkey that originated in Africa and Asia

Oryctolagus cuniculus the genus and species of the rabbit

physiology the study of living organisms

primate an animal, such as a monkey or ape, that has movable thumbs and, in many ways, resembles a human

Rattus norvegicus the genus and species of the rat

research studious inquiry or examination aimed at the discovery and interpretation of information

research scientist a scientist involved in research, whether on animals or humans

simulated system a model that mimics an animal's behavior and/or internal or external structures for the purposes of research; see *computer system*

Review Questions

1. What organizations are involved in setting standards and regulations on animal research?
2. What species of animals are commonly used in animal research studies?
3. What are the three kinds of research types? Explain each type.
4. What are the four animal models used in animal research? Briefly explain each type.
5. What are the two types of monkeys? What are the differences between each type?
6. What are the career options in laboratory animal medicine?
7. What are some conditions or disease advancements that have been made in veterinary medicine because of animal research?
8. How is animal research regulated?

Clinical Situation

Mary and James, both laboratory animal technicians, are working in a research facility that is currently housing rats, mice, and rabbits for both veterinary and human medicine research. When the techs enter the lab housing area, they note the health status of each animal in the medical record. Today, the techs notice that several animals are showing signs of respiratory disease. Each animal is being used for several different studies. Some are housed separately, and others share a cage with two to three other animals.

- What are possible signs of respiratory disease in these animal species?
- What should the techs do as far as noting respiratory conditions?
- What protocols should the techs take in this situation?

CHAPTER 17
Beef and Dairy Cattle Breed Identification and Production Management

Objectives

Upon completion of this chapter, the reader should be able to:

17.1 Define common veterinary terms relating to cattle
17.2 Identify common breeds of beef cattle
17.3 Identify common breeds of dairy cattle
17.4 Discuss the nutritional requirements of cattle
17.5 Discuss normal and abnormal behaviors of cattle
17.6 Explain how to properly and safely restrain cattle for various procedures
17.7 Discuss the health care and maintenance of cattle
17.8 Discuss breeding and reproduction in cattle
17.9 Discuss common diseases affecting cattle
17.10 Explain the importance of beef cattle production
17.11 Explain the importance of dairy cattle production
17.12 Discuss the types of beef production systems
17.13 Discuss dairy production and milking
17.14 Discuss management practices with cattle

Introduction

Cattle are among the most widely raised livestock in the United States (see Figure 17–1). Americans enjoy both meat and milk, which makes raising both beef and dairy cattle a large, profitable business. The cattle industry provides many opportunities in the veterinary industry, from herd health management to reproduction and products for human consumption.

Veterinary Terminology

In veterinary medicine, cattle are known by the term **bovine**. The adult female is a **cow**, and the adult male is a **bull**. A castrated male is a **steer**. A young female cow that has not yet been bred is a **heifer**. A young female cow that is pregnant with her first calf is called a **springing heifer**. A young cow is known as a **calf**. The term for

CHAPTER 17 Beef and Dairy Cattle Breed Identification and Production Management

FIGURE 17–1 The cattle industry provides many opportunities for the practice of veterinary care.

a cow giving birth is **calving**; **freshening** also means the labor process in dairy animals. A group of cattle is known as a **herd**. Occasionally, an adult cow is sterile and cannot produce: it is known as a **freemartin**. Additional important cattle-related terms include the following:

- **Polled**—no horn growth
- **Forager**—animal that eats grass and pasture
- **Marbling**—the appearance of intramuscular fat within the meat as striations or lines through the meat layers
- **Cutability**—quality and quantity of meat from a beef animal
- **Dual-purpose breed**—a breed that serves more than one purpose, such as producing meat and dairy

Biology

Beef and dairy cattle may look alike, but they are raised for very different purposes. **Beef cattle** are produced for meat and the expectations are to have increased body size to produce more quality meat per cow (see Figure 17–2). **Dairy cattle** are raised for quality milk, with the expectation of producing large amounts of milk per cow (see Figure 17–3). Cows were one of the first domesticated livestock species; today there are over 1.5 billion cattle in the world. In many parts of the world, cattle are considered sacred animals: They are worshipped and valued greatly both for their milk and meat and as living animals.

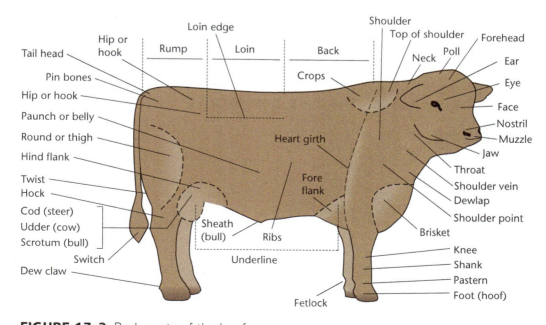

FIGURE 17–2 Body parts of the beef cow.

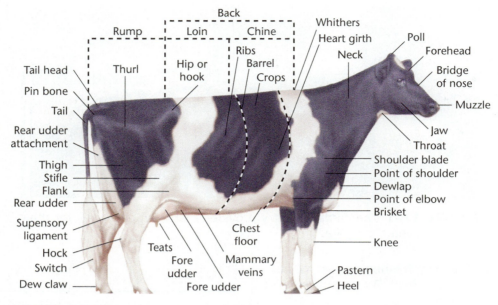

FIGURE 17–3 Body parts of the dairy cow.

Breeds

Cattle breeds vary greatly in size, structure, and color. Some cattle are miniature in size, such as zebu cattle; others are very large, such as Limousine and Brahman cattle. Beef breeds are bred to be heavier in appearance and body weight (see Figure 17–4). Beef cattle should have large amounts of muscle over the hindquarters, body, and pelvis. Dairy cattle are more angular in shape and, because of less body weight and muscle, have a bony pelvis and hind end (see Figure 17–5). Some cattle breeds are **dual purpose**; for instance, some cattle breeds—such as South Devon and Salers—are produced for both meat and milk.

FIGURE 17–5 Dairy cattle are slender in body composition.

FIGURE 17–4 Beef cattle are raised for a large body size to maximize meat production.

Beef Cattle

Beef cattle are raised specifically for meat that is called *beef*. The goal of beef cattle production is to select a beef animal that produces a large amount of high-quality meat that meets a particular standard of health care and a nutritional program. There are over 95 million beef cattle raised on farms in the United States. These farms bring in a profit of $74 billion a year, making the industry an important part of today's agriculture. Beef is consumed over 75 billion times a day in America alone, and the market for beef is increasing daily worldwide. The demand on the beef industry has increased over time; however, the number of cattle producers continues to decrease. Beef is one of the most frequently eaten foods due to its nutritional value of protein, vitamins, iron, and other essential nutrients. There has been an improvement in the genetics and overall health care of beef cattle as well as current changes in research to make beef quality even better.

Dairy Cattle

Dairy cattle are raised specifically for milk. The goal of dairy production is to produce the highest quality and quantity of milk per cow. Dairy cattle production is one of the oldest agriculture programs, dating back almost 7000 years. There are over 90,000 dairy farms located throughout the United States. Cows produce over 175 billion pounds of milk annually (see Table 17–1). A single cow can produce about 18,000 pounds of milk a year. With more nutrients than calories, milk continues to be one of the most wholesome foods there is. Similar to the beef industry, the demand for dairy cattle is on the rise but the number of dairy cattle is decreasing.

Breed Selection

The selection of a high-quality cow is based on the purpose or use of the cow, the breed, and the industry standards that are placed on cattle. The cow should bring the best market price based on the product.

Beef Cattle

Beef cattle are rated on cattle **selection guidelines** (see Figure 17–6). These guidelines set the standards for the cow's beef type and how the cow is chosen for a production program. It is important for breeders to research the standards of the beef breed and select a breeding animal that demonstrates quality traits of the breed. The **pedigree**, or parents' breeding lines, of the beef cattle should be evaluated for any known **genetic flaws**. A genetic flaw is an undesired trait or characteristic

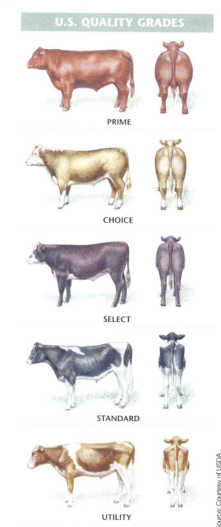

FIGURE 17–6 Quality grades of beef.

TABLE 17–1
Milk Production by State

STATE	AMOUNT IN MILLION POUNDS
1. California	40,000
2. Wisconsin	29,000
3. Idaho	14,100
4. New York	14,000
5. Pennsylvania	10,800
6. Texas	10,200
7. Michigan	10,000
8. Minnesota	9400
9. New Mexico	7800
10. Washington	6600
11. Ohio	5400
12. Iowa	4800
13. Arizona	4700
14. Indiana	4000
15. Colorado	3700

Source: United States Department of Agriculture, 2016

that is passed from one or both parents to the **offspring**. Quality beef cattle should have no genetic flaws. Superior animals that show ideal weights and health status should be selected. Cows should be **sound**, which means they show no signs of lameness or injury to the legs and hooves. Beef cattle should have excellent **conformation**, which is their body shape and form. The breed should be able to survive well in the type of environment in which they will be raised. The cow should be maintained easily with the proper diet. The marketability of the cow and the product should be at the high end of the production scale. The cow should be healthy and disease- and parasite-free.

Dairy Cattle

Dairy cattle standards are based on specific conformation requirements. Dairy cows should have an angular body shape that is rectangular in appearance (see Figure 17–7). They lack muscle mass in comparison to beef cattle. Dairy cows should have well-attached **udders**, or mammary glands. Depending on the breed, the udders should hold a total capacity of 50 to 70 pounds of milk. Each udder should contain four **teats**, which are sections of the mammary gland used to produce milk. The production of milk is known as **lactation**. All dairy cattle are bred to produce high-quality milk (see Figure 17–8).

The dairy industry has developed the **Dairy Herd Improvement (DHI) program**, a production testing and record-keeping system for dairy herds. The U.S. Department of Agriculture (USDA) works with individual dairy farms to compare information on local, state, and national herds. All cattle within the program have permanent identification numbers, and the milk is tested every 15 to 45 days to determine the quality and quantity of the product. The DHI program also has

FIGURE 17–7 Dairy cows should have an angular body shape that is rectangular in appearance.

FIGURE 17–8 A cow should produce up to 70 pounds of milk.

a registry that records information about dairy herds' milk production rates as well as breeding records of bulls. The DHI program and registry have rules that dairy producers must follow to be part of the program. These rules include the following:

- All cattle must belong to a registered breed association.
- All cattle must have permanent IDs.
- All cattle must have a pedigree on record.
- All cattle within the herd must be tested.
- Testing is required every month.
- Testing must be completed within 24 hours.
- Milk or fat samples above normal levels must be retested.
- The USDA must supervise testing.
- Surprise testing may be done at any time.
- Inaccurate files are liable for fraud charges.

DHI records also monitor the **progeny** or offspring of the cattle. The progeny records help identify any genetic flaws in breeding lines of cattle. The genetic flaws can then be traced to the **sire** (father) or the **dam** (mother). Each cow has a permanent ID number; when a calf is entered into the system, the progeny data is recorded under this identification number. This can help provide information to dairy breeders who wish to research which bulls to breed to their dairy cows to increase milk production rates. This information can also be used to estimate the traits that a bull or cow will pass along to offspring in order to determine genetic superiors and inferiors. The estimation record is called **predicted transmission ability**.

Culling

When beef or dairy cattle no longer produce a high-quality product, they may be **culled**, or removed, from the herd. Culling depends on the production rates of the cow, genetic flaws that are being produced, and the quality and quantity of beef or milk products being produced.

Nutrition

Cattle require nutrients that work to convert their food into the product that they are being bred for, such as meat or milk. Cattle are fed between three and four times a day. Cattle producers feed their herds a **total mix ration (TMR)** that consists of all the nutrients that they need in a day in a mixture of high-quality products (see Figure 17–9).

Lactating Dairy Cows

Lactating cattle require the most nutrients. Lactating dairy cattle require increased amounts of food on a daily basis. They also require large amounts of water. Both food and water nutrients are converted to milk. All lactating cattle are bred so their bodies produce milk. The first four months of the gestation period are the most critical for nutrition. Over this time frame, cows need increased calcium and fat for energy to produce increased amounts of milk. Lactation feeding can be the most challenging nutritional stage. Several factors must be considered when feeding the lactating cow, including the size of the cow, the stage of lactation, the quality of milk, the labor capability, the food quality, the costs of food, and the food **palatability** (how well it tastes and is eaten). Cows are usually fed 1 pound of high-quality grain for every 2.5 to 3 pounds of milk produced a day. Cows are fed high-quality hay at 1.5 to 2 percent of their body weight. High-quality grain products consist of:

- 20 percent protein
- 2 percent calcium
- 2 percent phosphorus

Cows produce milk for a 10-month period; after calving, they are given time for their reproductive systems to relax. This period of two months of not producing milk is called the *dry period*.

Feeding Dairy Calves

After the cows freshen, or give birth, the calves are allowed to nurse for the first 24 hours to receive **colostrum**; this is milk secreted for a few days after childbirth that is characterized by high protein and antibody content that keeps the calves free of disease until vaccines are given. A calf should gain 12 percent of its body weight in its first 24 hours. A calf is then removed from the mother and fed by hand. For the first three days, milk is collected from the cow and hand-fed to the calf in a nursing bottle (see Figure 17–10). The calf learns to nurse from

FIGURE 17–9 The daily feed and rations of cattle are important to the outcomes of milk and meat production.

FIGURE 17–10 Calves are hand-fed by bottle following the first 24 hours of nursing from the cow.

a bottle. Gradually, the calf will be placed on milk replacer for a 2- to 3-week period. Milk replacer has increased amounts of fat and protein for growth. The calf will be weaned to begin a solid food source once its digestive system has developed enough to support solid materials. This food source is usually a calf **starter food** that is easily digestible. The calf is fed a starter food diet beginning at four to six weeks of age; it is typically given a pound and a half per day. At five weeks of age, calves will begin eating **roughage**, or a hay or grass source. Most producers place calves on a high-protein diet—such as **alfalfa**, a hay high in protein—to increase the size of the calf. Calves are placed on free choice hay, which means it is available at all times (see Figure 17–11). During this time, the calf starter food is increased to about six pounds a day per calf. By 8 to 12 weeks of age, the calf should begin eating a grain source with concentrated supplements that will promote continued growth.

Body condition scoring (BCS) is used from the time a calf is born through its entire purpose as an adult, whether that is as a lactating dairy cow or a beef cow used for meat production. The BCS is rated on a scale from 1 to 9 (see Table 17–2). Producers use this information to estimate how much food an animal should be fed for optimal growth and maintenance.

Feeding the Dry Cow

The nutrition for the **dry cow period** varies from the nutrition given during lactation. A cow is placed on good-quality grass hay only for 2 to 3 weeks. At about days 25 to 30 of the dry period, the cow is given 4 to 7 pounds of grain per day. During the last week of the dry period, this is increased to 7 to 14 pounds per day. The cow resumes the lactation feeding once the cow has been bred.

TABLE 17–2
Cattle Body Condition Scoring

Score 1	Emaciated appearance with little muscle left
Score 2	Very thin appearance; no fat noted; bones visible
Score 3	Thin appearance; ribs visible
Score 4	Borderline—most ribs not visible
Score 5	Moderate—neither fat nor thin
Score 6	Good—smooth body appearance
Score 7	Very good—smooth appearance with fat noted over back and tail head
Score 8	Fat—blocky appearance; bone over back not visible
Score 9	Very fat—tail base buried in fat
Score 3.5–4	Dairy cattle ideal at end of lactation (dry period)
Score 2–2.5	Dairy cattle peak lactation @ 60 to 90 days

Beef Cattle

Beef cattle convert their nutrients to meat and muscle mass. The adults are typically fed high-quality grains or pellet-based diets that are high in protein and fat for energy and growth. Beef cattle should be fed 2 to 3 pounds of grain per day, which is 2 to 3 percent of their body weight daily. They also need to have free access to hay or a pasture source and water. The calves are typically kept with their mothers to nurse until weaning. Calves typically begin a starter food and then a grain supplement in much the same manner as dairy calves. When the weaning process begins, the adult cattle will eat the grain if they are given the opportunity. Calves are typically fed in **creep feeders**, small food troughs with bars that allow only calves to eat. Calves are fed 2 to 3 percent of their body weight until they reach market weight.

Behavior

Cows scare easily and are difficult to work with once frightened. It may be difficult to settle a cow down once it is spooked. Cattle can kick, stomp, and trample people. When cattle are around people or things that they are not used to, they often become nervous.

FIGURE 17–11 By five weeks of age, calves are fed free choice hay diets.

They may run, jump, or kick; their size and actions can cause severe injury to people. Cattle are very different from other types of livestock, such as horses, and should not be handled in the same way. Cattle are herd animals and considered prey. When they are placed in a threatening or frightening situation, they will go into a defensive reaction and use their speed, body size, head, and hooves to defend themselves. Cattle often head-butt and kick as a first means of protection. Cattle often kick out to the side rather than straight back, so standing near their rear legs is discouraged. Most dairy cattle are used to being handled by humans, as they are often milked several times a day. Beef cattle may not be as accustomed to people. Regardless, care must be taken when working around cattle. Signs of aggression in cattle include the following:

- Direct staring with head lowered
- Pawing at the ground with the front feet
- Lowering and shaking the head
- Snorting
- Short charging motions with the body
- Tail swishing quickly

Basic Training

Cattle that are shown or handled regularly for milking procedures are trained to follow routines and may be led with a halter and lead rope (see Figure 17–12). Cattle are often trained to herd to move them between locations. Cattle must be trained in a calm manner.

Cattle have a pressure point at their shoulder, so moving past the shoulder and going toward the rear of the cow will prompt it to move forward. You do not need to be close to the cow to make this happen. Moving from the head to the shoulder pressure point will cause a cow to stop, turn, and move away from you. This herding must be done slowly and calmly.

Equipment and Housing Needs

Chutes or stanchions are used to restrain cattle in order to administer medications or vaccines or to collect blood. These large pieces of equipment safely control the cow and prevent it from kicking and injuring the handler (see Figure 17–13). There are many different types of metal chutes and stanchions; most have a headgate that locks in and controls the head of the cow. They also have horizontal bars located on the sides to prevent kicking. Cows may need to be haltered and led or tied in certain situations, requiring a cow halter and lead rope. Sometimes anti-kick bars are placed over the hips or pelvis area of a cow to prevent it from kicking. These are metal devices in the shape of a V that clamp around the animal's flank.

Housing Needs

Cattle may be housed in barns or pastures depending on the type of weather and environment (see Figure 17–14). There are two types of housing for cattle: **warm housing** and **cold housing**. Warm housing, used in the winter, is a heated barn or building

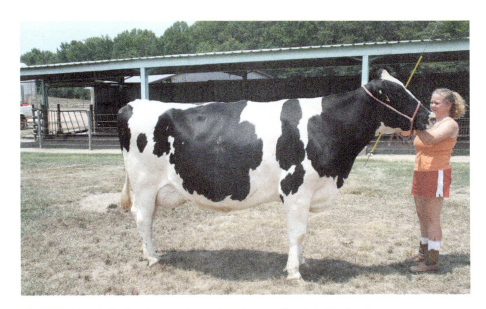

FIGURE 17–12 Cattle can be trained to walk on a lead rope.

FIGURE 17–13 A chute helps safely restrain the cow for performing medical procedures or exams.

FIGURE 17–14 It is common for cows to be housed in barns.

FIGURE 17–15 Bulk tank where milk is collected.

Pasture

Feeding management requires rotating cattle from pastures to allow the grass to grow and to provide sanitation methods to the fields to decrease diseases and parasites. Pastures are usually rotated every 48 hours. This practice improves nutrition and forces good grazing methods. It will help maximize the number of cows in the herd and maintain good health practices. Many pastures should have access to shelter so that cattle can get out of inclement weather (see Figure 17–16).

Feed Storage Needs

Proper storage of cattle feed products is important for good health practices. Grains are typically stored in a

with an insulated area with dairy stalls for each cow. Cold housing is a barn or building with no heat where the natural air circulates to keep out moisture. Cold housing is usually a large open area where the herd is kept together. This is a more common housing method for calves and dry period cows.

Milking Needs

Dairy cattle are milked in stalls or milking parlors. Stall milking entails milking each cow in the area in which it is housed. This requires a pipeline to transport the milk to an outside bulk tank for collection (see Figure 17–15). Milking parlors enable cows to be milked using improved efficiency and sanitation methods. The parlors typically include an elevated platform where the cow is restrained for milking. The floor area, known as the pit, is where the staff prepares the cows for milking.

FIGURE 17–16 Pastures should have structures or areas where the cows can be protected from the weather.

silo, which is a tall storage building that keeps the food free of moisture. Silos can also be controlled to feed specific amounts of food to the cattle during the day.

Waste Material Systems

Manure management is a concern for the health of the cattle herd and the health and sanitation of humans. Each cow produces about 8 percent of its body weight in waste materials on a daily basis. In one year, a cow can produce 15 to 20 tons of waste. The waste is called **manure**; it is high in nitrogen, phosphorus, and potassium. Waste material systems must be in place on all cattle production farms to ensure quality management and sanitation. The systems should provide efficient collection and removal methods. Some farms use a low-cost **solid manure system**, in which the manure is collected and hauled away on a daily basis. The wastes are stored on piles away from the cattle and removed periodically. A more costly **liquid manure system** is used in larger production systems and must meet environmental standards. This system requires large storage tanks that may be below or above the ground. Water is typically added to the system, as it requires pumping to remove the wastes on a regular basis of every 4 to 5 months.

Restraint and Handling

Beef cattle are more difficult to restrain than dairy cattle, mainly because they tend to be handled by people less frequently. Dairy cattle grow accustomed to humans during the milking process. Cattle are sometimes shown and must learn how to be handled with a **halter** and **lead**. The halter made of nylon or leather fits over the head and face of the cow, while the lead acts as a leash that is held to walk the animal. Cattle should be taught to tie and trained to stand while tied for both showing and milking purposes (see Figure 17–17). Cattle are typically tied using a **square knot**. The square knot is easy to make and will contain the cattle; it is also easy to untie when necessary.

It is easy to move cattle from place to place without using your hands: this is called **pushing** cattle. When you walk toward them, they move away from your movement. So, when you wave your arms and move behind cattle, they will move away from you in the direction in which you want them to move. When people are working around cattle, they should always stand at the cattle's shoulder area and never to the side of their rear legs. Cattle can kick to the side: this

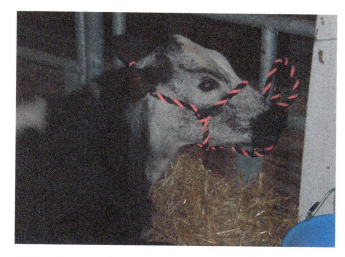

FIGURE 17–17 Tying of cattle.

is called **cow kicking**. Cattle can also be restrained using a **squeeze chute** or **stanchion**. The squeeze chute is a cage-like structure made of metal pipes that holds the cow and prevents it from kicking when being restrained. The cow can be tied in the chute or left free. Squeeze chutes are useful when veterinary staff are giving physical exams, administering medication, collecting blood, or conducting reproductive procedures. The stanchion is commonly called a *headgate*. This equipment may restrain the head or the entire body of the cow. The cow is walked into the stanchion, and a headgate is closed to contain its head and restrain it so it does not move (see Figure 17–18). Some stanchions squeeze the cow's body to restrain the entire cow.

Cattle may also be restrained using hands-on restraint methods. The **tail switch restraint** prevents cattle from moving and kicking while being handled. The tail is twisted at the base, which prevents the cow from moving and kicking (see Figure 17–19).

Cattle can also be restrained by placing a rope around a front leg and then lifting the leg off the ground. This prevents the cow from moving and provides minimal restraint. **Reefing** is another restraint procedure: a heavy rope is tied around the cow in such a way that when it is pulled, the cow is gently forced to the ground, and a handler can kneel on its neck to prevent it from standing (see Figure 17–20).

Cattle are sometimes restrained and led by **nose tongs**, which place pressure on the **nasal septum** (the cartilage located in the nostrils). Some cattle may be led by a **nose ring** placed through the septum. The pressure on the septum allows the animal to be handled in a similar way to a halter and lead (see Figure 17–21).

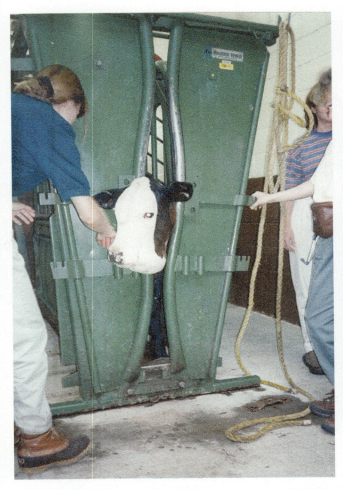

FIGURE 17–18 A headgate is used to hold cattle for treatment.

FIGURE 17–19 Tail switch restraint.

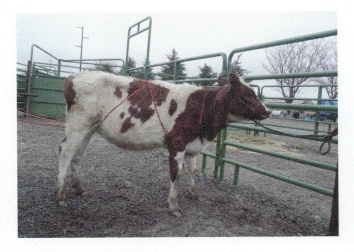

FIGURE 17–20 Reefing restraint of cow.

Cattle Behavior and Restraint Considerations

Aspects of cattle behavior that should be considered in the handling, restraint, or general work within their pastures or stall areas include the following:

- Cattle can see 360 degrees around them but can easily be startled by quick movements (see Figure 17–22).
- Cows may seem content in a pasture but are one of the most nervous domesticated herd animals.
- Cattle are easily startled by strange noises, people, or changes in the environment.
- Always announce your presence near a cow in a quiet and calm manner; gently touch a cow when next to it rather than allowing it to touch or move into you.
- When working with a cow, allow it to settle down and become used to the environment.
- Cows tend to kick forward and to the side rather than back.
- Always make sure you have an exit when you are working with a cow in an enclosed area.

CHAPTER 17 Beef and Dairy Cattle Breed Identification and Production Management

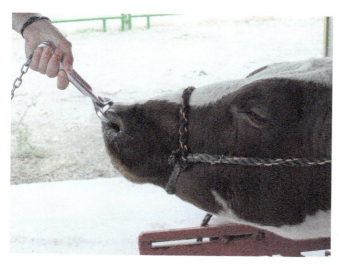

FIGURE 17–21 Nasal tongs are used to restrain a cow.

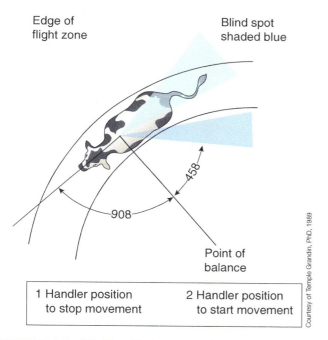

FIGURE 17–22 The flight zone of a cow. Note the presence of a blind spot and the positions where the handler can influence the animal's movement.

Grooming

Cattle that are shown need to be groomed on a regular basis. Show cattle are regularly given baths, dried using high-powered vacuums, and brushed. This requires close contact with humans. Show cattle are usually handled at a young age and learn to be groomed and handled by people. The grooming practices in cattle are similar to those of a dog or horse.

Basic Health Care and Maintenance

Because both dairy and beef cattle products are primarily used for human consumption, cattle production involves specific management and health practices of the herds. There are strict guidelines that must be followed when caring for cattle. First, it is important that the producers and staff have knowledge and experience in cattle production. Veterinary staff with a background in bovine medicine must be available. As cattle require human interaction, they should be accustomed to people so that health care and maintenance are relatively stress-free for the herds. Staff should include a **herd health manager** who records all health program and production information. The herd health manager needs a background in bovine veterinary care and reproduction and maintains herd health records, vaccine programs, deworming schedules, and reproductive care and records and works closely with the veterinarian to establish a high standard of care for the entire herd. The goal of the herd health manager is to provide a parasite- and disease-free herd.

Cattle should be placed on a regular daily schedule to provide a stress-free environment. They should be fed at the same time each day. Dairy cattle should be milked at the same times each day. All cattle should be given adequate space and pasture. A regular health care program should be established. Cattle should have a controlled climate, whether they are housed indoors or outdoors, that includes appropriate temperatures, proper shelter, appropriate humidity, and ventilation. Cattle should be provided shelter that is free from wind, rain, and cold weather. Water and feed troughs should be kept clean. Fencing should be monitored for repair and maintenance.

A herd health manager monitors the following:

- Vaccine program
- Deworming program
- Heat-cycle maintenance
- Body condition scoring
- Breeding log
- Veterinary maintenance
- Nutritional program
- Sanitation program
- Fly-control system
- Pasture control
- Housing
- Fencing

Each cattle should have a type of permanent identification. Cattle ID methods include **ear tag**

identification, **neck tag identification**, **microchips**, **tattoos**, and **branding**. Ear tags are applied in a way that is similar to ear piercing and have a number system that identifies each cow within the herd (see Figure 17–23). Cattle wear a neck tag (like a necklace); each cow is given a number that is recorded on the neck tag. A microchip is an electronic identification computer chip that is placed under the skin. Each chip has a number that identifies a cow within the herd. A scanner reads the microchip number to record data information. Cattle can be tattooed by writing a number on the skin with an electronic needle and ink. The tattoo is placed on an area of the body that is easy to locate, such as an ear or the gums. Branding is a method that uses extreme heat or cold to burn or freeze a number on the hair coat of each cow (see Figure 17–24). Branding methods cause the hair to change color and record the information on the hair. Occasionally, cattle are identified by **earmarks**, which are notches made on the edges of the ear. Today, some cattle producers use new technology identification methods, such as **DNA testing** and **nose printing**. DNA testing uses a blood or hair sample to identify each cow and its breeding lines. Nose printing is similar to fingerprinting: As each cow nose has its own skin lines, which vary from cow to cow, the nose print can be recorded.

FIGURE 17–24 Freeze branding in a cow.

Dehorning

Dehorning is the removal of horns to prevent injury to humans and other cattle. Methods include using **caustic** chemicals that burn the horn buds before the horns begin to grow. This is the most common procedure and causes less pain, less stress, and easier aftercare. Cattle that have developed horns must have them surgically removed with the use of sedation and dehorning equipment. Dehorners are specialized tools that have blades that remove the **horn bud**, or area of horn growth. The most important part of the aftercare is treating for flies and monitoring the site for infection.

Castration

Castration of cattle is the surgical removal of the testicles and is carried out to prevent reproduction. Cattle can be castrated in one of two ways. In the **banding** method, a tight rubber band is placed over the testicle area that cuts off the circulation to the area. In time, the tissue will **slough**, or fall off. The band is put in place using an **elastrator**, which is a tool that stretches the band over the testicle area. Banding is usually done in young calves. Older, developed bulls require a surgical procedure, in which veterinarians cut the spermatic cord with an **emasculator** and then remove the testicles

Vaccinations

Cattle vaccination programs should be discussed with the herd veterinarian: in some locations, cattle may be at higher risk of certain diseases. Young beef calves are typically vaccinated 14 to 21 days prior to weaning. Many calves are castrated and dehorned at this time.

FIGURE 17–23 Calf identified with an ear tag.

TABLE 17–3
Common Vaccines in Cattle

IBR (INFECTIOUS BOVINE RHINOTRACHEITIS)	BVD (BOVINE VIRAL DIARRHEA)	PARA INFLUENZA (FLU)
Bovine Respiratory Virus	Clostridium	Leptospirosis
Pasteurella	Bordetella	E-coli
Rotavirus	Coronavirus	Rabies

Dairy calves are typically vaccinated with boosters at 3 to 4 months of age and at 5 to 6 months of age and then yearly, as determined by the veterinarian. Adult breeding cattle are usually vaccinated prior to breeding and prior to calving. Table 17–3 lists the common vaccines used in cattle.

Common Diseases

In the face of cattle diseases, veterinary staff and cattle producers must work together to provide quality health care to establish a cattle herd that is disease- and parasite-free. It is important to monitor diseases within the herd that are common or of concern to the public; a robust vaccine program must be developed. Veterinarians and herd health managers should work together to determine which diseases and parasites are prevalent in their area and what health program should be in place. Some diseases affect both dairy and beef cattle equally; others are more prevalent in one cattle type.

Brucellosis

Brucellosis, informally known as **Bang's disease**, is a reproductive disease in cattle. The disease is passed through breeding and causes **abortion** (termination of the pregnancy) in cows and **infertility** (the inability to reproduce) in both bulls and cows. A vaccine is available to prevent this disease, but there is no treatment. All breeding animals should be tested prior to reproduction. This disease is zoonotic to humans, and all bodily fluids should be handled with caution. Bodily fluids include saliva, urine, feces, milk, semen, or vaginal discharge.

Bovine Viral Diarrhea

The many symptoms of the cattle virus **bovine viral diarrhea (BVD)** include severe diarrhea, fever, coughing and other respiratory problems, nasal discharge, possible lameness, and a poor hair coat. There is a vaccine for BVD but no treatment. The virus must run its course, with supportive care as therapy. Pregnant cows that develop this viral infection may abort.

Infectious Bovine Rhinotracheitis

Infectious bovine rhinotracheitis (IBR), commonly known as "red nose," is a virus that affects cattle. IBR is prevented through vaccination; however, like BVD, there is no treatment and the virus must run its course along with supportive therapy. Clinical signs include respiratory problems, coughing, nasal and ocular discharge, fever, and weight loss. The term *nasal* means pertaining to the nose; *ocular* means pertaining to the eye. In severe cases, it may cause staggering, seizures, and eventually death.

Bovine Respiratory Disease Complex (BRDC)

Bovine respiratory disease complex (BRDC), also known as "shipping fever," is a general term for pneumonia commonly seen in stressed calves, often while they are being shipped. Stress—caused by such events as weaning, dehorning, castration, and shipping—can make cattle susceptible to viruses and bacteria. The best way to reduce the risk of BRDC is through a routine vaccination.

Leptospirosis

Leptospirosis is a bacterial disease in cattle transmitted in the urine. There is a vaccine to prevent the disease, and treatments are available. There is a zoonotic potential with urine contamination. Signs of this disease include bloody urine, fever, rapid respiratory rate, and occasional lameness or stiffness.

Campylobacter

Campylobacter is a bacterial infection that affects either the intestinal tract or the reproductive system of cattle. The intestinal form causes diarrhea and digestive problems in cattle; this form is zoonotic in people. The reproductive form is more serious, as infertility may occur; its symptoms include possible abortion, decreased conception rates, and erratic heat cycles. Breeding animals should be vaccinated yearly.

Mastitis

Mastitis is an inflammation of the mammary gland that can occur in any female cow but most commonly affects dairy cattle (see Figure 17–25). Typically, bacteria enter the teat and cause an infection within the mammary glands. This can be prevented through good sanitation practices of the barn, pasture, and milking equipment. There is no vaccine for mastitis, but it can be easily treated. It is estimated that the average cost to treat mastitis in a herd is $200 per cow per year. The **California mastitis test (CMT)** is used to diagnose and measure the level of cell infection within the glands (see Figure 17–26). Most cows will need to be placed in a dry period during treatment.

Bloat

Another condition that can affect all cattle is **bloat**. Air is ingested into the stomach and causes the stomach to swell and possibly rotate, which is known as a **displaced abomasum**. Bloat may be caused by overeating or by eating a very rich pasture. Signs include restlessness, not eating, unable to have a bowel movement, and unable to regurgitate. Most stomachs rotate and are displaced to the left.

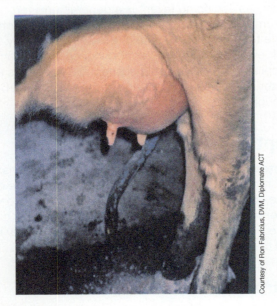

FIGURE 17–25 Mastitis in the rear quarter of a cow udder.

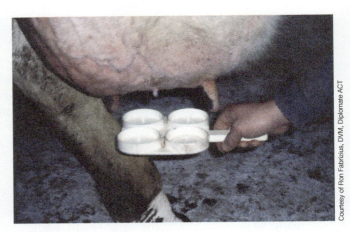

FIGURE 17–26 California mastitis test (CMT) paddle.

Retained Placenta

A **retained placenta** is a reproductive condition that can affect any female cow within 8 hours after labor. *Placenta* is the afterbirth; if it is not passed it can cause a severe infection within the female's body. This condition can lead to **sepsis**, a toxic condition caused by the spread of bacteria. In the case of a retained placenta, the toxins are the waste materials of the newborn. Signs include fever, anorexia, low milk production, foul odor, discharge from the vulva, and membranes hanging from the vulva. This condition should be treated as an emergency; veterinary attention is necessary.

Grass Tetany

When cattle are placed on rich green pasture that is high in nitrogen that they are unaccustomed to eating, the result can be low magnesium and calcium levels in the bloodstream. This condition, known as **grass tetany**, is more common in older cows during the lactation phase and occurs in the springtime. It can be prevented by feeding the cattle forage or placing them on new pastures for limited amounts of time. Signs include abdominal pain and seizures. The condition will occasionally be fatal.

Metabolic Disorders in Dairy Cattle

Dairy cattle are affected by other **metabolic diseases** that cause a chemical change in the body, usually from stress. Stress may occur with calving, milking, and changes in the nutrition or environment.

Hypocalcemia

Hypocalcemia, or **milk fever**, occurs in lactating cows and is due to low blood calcium. This condition may occur from improper calcium supplements in the diet or too much calcium in the lactating animal. It commonly occurs near calving time but may happen at any time. The condition can be avoided by providing phosphorus in the diet and a vitamin D supplement.

Ketosis

Ketosis is a condition in dairy cattle that causes low blood sugar, similar to diabetes. It may occur early in the lactation phase or be related to a nutritional imbalance. Signs include a poor appetite, a dull coat, depression, incoordination, increased licking motion, and a fruity odor to the breath.

Blackleg

Blackleg is a fatal disease in cattle caused by bacteria that can live in soil for many years. The bacteria are ingested through grazing and enter the digestive system or invade the body through wounds that become infected with the bacteria. The bacteria that cause this disease are a type known as *Clostridium*. Cattle between 6 months and 2 years old that are in pastures and eating high-protein diets are most susceptible to the disease. Signs of include lameness, loss of appetite, difficult or increased breathing, depression, fever, swelling, and pain. Cows suffering from blackleg usually die within 12 to 48 hours. As the disease progresses, the swelling causes the limbs to become infected and the tissue to turn black. This disease is not contagious from cow to cow and is preventable through vaccination.

Anaplasmosis

Anaplasmosis is a parasitic protozoan disease that affects dogs, cats, and ruminants—most often the cow. The disease is not contagious; it is most often transmitted by ticks, but also by flies and mosquitoes. Anaplasmosis invades the red blood cells. Signs include anemia, fever, weight loss, difficulty breathing, uncoordinated movements, and abortion. It can be fatal. This disease can cause cattle to become carriers, which can be a huge concern to dairy and beef producers. Anaplasmosis is a major problem in many southern states but is frequently diagnosed in the Midwest. A strict interpretation of disease control regulations for both interstate and intrastate movement of cattle precludes the sale of cattle from herds that are infected with anaplasmosis. A vaccine is available, and appropriate insect control methods are recommended.

Common Parasites

Because all parasites affect all cattle, a dewormer and sanitation program are vital. The most common external parasites in cattle include flies, ticks, mites, and lice. Fly control is an important part of cattle production and includes **insecticides**—sprays or pour-on chemicals that repel flies and other insects and prevent them from biting the cattle; sanitation control—the removal of body wastes to decrease the number of flying insects; and **back rubbers**—rolling bars covered with fly-control chemicals that are placed in a barn or pasture that cows rub their backs on. Internal parasites in cattle typically cause diarrhea, weight loss, poor hair quality, and a dull hair coat. Cattle may suffer from many types of external parasites—the most common being strongyles, flatworms, and roundworms. Cattle are treated with **anthelmintics** or dewormers once or twice a year in oral, injectable, pour-on, or topical varieties. A different product should be used each time to prevent a buildup of chemical resistance in the body system.

Common Surgical Procedures

Male cattle are often castrated if they are not going to be used as breeding bulls. Castration allows beef cattle to put on weight much faster for beef production systems. Older cattle that are castrated must be sedated; standing castration is done as a surgical procedure, similar to that in dogs or horses.

Reproduction and Breeding Beef Cattle

There are three primary types of beef cattle production systems that determine how beef cattle are raised and used for profit: the cow–calf system, the backgrounding system, and the finishing system.

The Cow–Calf System

The goal of the **cow–calf system** is to raise cattle and breed them as adults. Therefore, this type of production is based on beef cattle reproduction. One or more cattle breeds may be raised in this system, but only high-quality cows of a specific breed are used. The cows begin breeding at about 2 years of age; the calves are born in late winter or early spring and then sold as a profit. The cows are bred each season as long as they are producing a quality calf. The cows are maintained on pasture. The cow–calf system requires a large amount of land to hold the number of cattle in the production system.

The Back-Grounding System

The goal of the **back-grounding system** is to raise a calf to market size for a profit. Cows are bred to produce calves that are weaned and then fed to an ideal weight that increases their size and profit. This ideal calf size is known as **feedlot size**. The calves are fed quality pasture, hay, and grain sources that increase their body mass. They may also be fed supplements that allow their body to convert more food to muscle. When they reach feedlot size, they are sold for meat.

The Finishing System

The **finishing system** produces a calf that grows through the entire adult stage. The adult cattle are sold for profit either as breeding animals or for meat. Most adult cattle weigh between 1100 and 1300 pounds as adults at **market weight**. These cattle are fed **concentrates**, or a mixture of foods high in energy and fat, as well as minerals that help promote growth. This system is costlier and takes a longer time to reap a profit; however, the profit is much larger due to the size of the animal.

Reproduction and Breeding of Dairy Cattle

Dairy production is based on the cow producing a high-quality milk product. In order for lactation to occur, the cow must be bred. Most milk production systems have a large number of cows that are producing milk for about 305 days of the year. A cow's **gestation cycle**, or length of pregnancy, is 283 days. The cow produces milk through this entire phase. Milk production is typically halted for 50–60 days. This is known as the **dry cow period**: it gives the cow's reproductive system a rest before the cow is bred and the milking process begins again. The milk content consists of about 87 percent water and a mixture of fat, protein, and sugar nutrients. Milk is produced in the udder, or the bag-like structure that contains four mammary glands, or **quarters**, that are the sections of the glands that store the milk (see Figure 17–27). Each quarter has a teat that is used to remove the milk from the cow. Milk nutrients are converted to milk by tiny

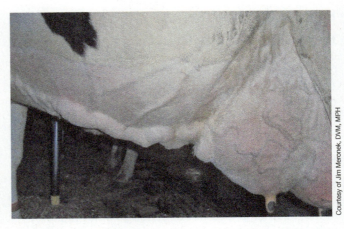

FIGURE 17–27 The veins found on the mammary gland of a cow udder.

structures within the udder called **alveoli**. Dairy production is a 24-hour-a-day job with a long-term financial growth possibility. However, dairy cattle are a costly investment with many risks. The size and type of the dairy herd may vary between farms. A **purebred business** consists of a breed of dairy cattle that is registered with a known pedigree. A **commercial business** consists of a grade type of dairy cow that is a mixed breed. Sizes range from a **family-size herd** that is fewer than 100 head of cattle, or a **large-size herd** that is more than 100 head of cattle. Many dairy producers have both a **milking herd** and a **total herd**. The milking herd is the group of cattle used to produce a high quality and large quantity of milk. A total herd uses replacement heifers when the adult cattle become too old to reproduce, thus, the heifers replace the adult aged cattle in the milk production system. Each farm herd also needs bulls to reproduce with the cows and continue the profitability of the herd. Small herds need between five and eight bulls for breeding purposes. Large herds need between 10 and 15 bulls.

A **dairy** is the building where cattle are milked two to three times a day. The milking process must be sanitary, as the product is for human consumption (see Figure 17–28). Before the milking process, the udders of each cow in the herd must be disinfected. The udder is cleaned and dried and then **primed**, which means that the milking machine devices are applied to the udders and begin collecting milk. The milking machine is attached to tubes that connect to pipes that transport the milk through a system and into a cooling tank. When all of the milk has been removed from the udder, iodine solution is applied to the udder. This prevents disease and bacteria from entering the mammary glands and helps close the muscle of the udders, preventing milk from continuing to drain.

CHAPTER 17 Beef and Dairy Cattle Breed Identification and Production Management

FIGURE 17–28 (A) Milking parlor. (B) Milking system.

SUMMARY

Beef and dairy cattle production systems are different in their types and practices; however, both require strict sanitary practices and have regulations that must be followed, as both products are for human consumption. There are several types of cattle production systems based on the market profit and resource used by the breeder. All cattle must have proper nutrition, health care, and management practices to ensure a healthy outcome. It is important for veterinary assistants to have experience in cattle behavior, restraint, and handling when working with these large animals.

Key Terms

abortion a termination of pregnancy

alfalfa hay high in protein

alveoli tiny structures within the udder where nutrients are converted to milk

anaplasmosis a parasitic protozoan disease that affects dogs, cats, and ruminants

anthelmintic a dewormer medication used to control parasites

back rubber equipment that dispenses fly-control chemicals to the backs of a cow

back-grounding system a production system that raises calves to market size for profit

banding method of castration in which a tight rubber band is placed over the testicle area of livestock; circulation is cut off and tissue falls away

Bang's disease see *brucellosis*

beef cattle cows specifically raised for meat

blackleg a fatal disease in cattle caused by bacteria; as the disease progresses, swelling causes limbs to become infected and tissue to turn black

bloat a swollen stomach from ingested air

Body condition scoring (BCS) body weight rated on scale from 1-9

bovine a veterinary term for a cow

bovine respiratory disease complex (BRDC) a stress-related disease in young cattle, also known as *shipping fever*

bovine viral diarrhea (BVD) a virus in cattle with symptoms that include diarrhea and respiratory distress

branding using extreme heat or cold to mark the skin of livestock with a number or symbol

brucellosis a reproductive disease in mammals spread through breeding practices

bull an adult male cow of breeding age

calf a newborn cow of either gender

California mastitis test (CMT) a test used to diagnose and measure the level of cell infection within mammary glands

calving the labor process of cows

campylobacter a bacterial infection in cattle that affects the intestinal tract or reproductive tract

castration the surgical removal of the testicles to prevent reproduction

caustic chemicals the chemicals used to burn horn buds to prevent horn growth

cold housing a building with no heat in which an entire herd is usually kept and where air circulates out moisture

colostrum mother's milk produced in the first 24 hours after giving birth that contains protein and antibodies to protect the immune system of the calf

commercial business a type of business activity that may distribute goods or provide services

concentrate a mixture of foods high in energy and fat

conformation the body's shape and form

cow an adult female cow of breeding age

cow kicking the way in which cows kick with their rear legs, which is off to the side

cow–calf system a production system that raises cattle to breed them as adults

creep feeder a small trough that allows calves to eat while preventing the adult cattle from doing so

culled removed from the herd

cutability the quality and quantity of meat from a beef animal

dairy a building where cows are milked

dairy cattle cows raised specifically for milk

dam a mother; female parent

dehorning the process of removing horns to prevent injury to people and other animals

Dairy Herd Improvement (DHI) program a way to maintain records of dairy herd information

displaced abomasum a condition in cattle that causes the stomach to rotate out of place

DNA testing using blood or hair samples to identify an animal's parents

dry-cow period a time in which milk production is stopped to give the reproductive system a rest

dual-purpose breed a breed that serves more than one purpose; such as milk and meat producing

earmarks notches made on the edges of the ear flaps to identify cattle

ear tag identification tags with numbers on them fastened by piecing the ear flaps

elastrator an instrument that stretches a band over the testicles during castration

emasculator a surgical tool used to cut the spermatic cord during castration

family-size herd a group of fewer than 100 head of cattle

feedlot size the ideal weight of calves that increases size and profit

finishing system a production system that produces a calf through its entire adult stage and then sells it as meat or for breeding

forager an animal that eats grass and pasture

free choice food that is offered in large quantities and at all times

freemartin an adult cow that is sterile and unable to reproduce

freshening the labor process of dairy animals

genetic flaw an undesired trait or characteristic passed from one or both parents to the offspring

gestation cycle the length of pregnancy

grass tetany a condition in cattle caused by eating rich pasture high in nitrogen gases; causes abdominal pain

halter a rope or strap that fits over the head and is used to control the animal

heifer a young female cow that has not yet been bred

herd a group of cattle

herd health manager someone who maintains the records and health care of a herd of cattle

horn bud the area of horn growth on top of the head

hypocalcemia a deficiency of calcium in the blood; also known as "milk fever"

infectious bovine rhinotracheitis (IBR) a respiratory virus affecting cattle; also known as "red nose"

infertility the inability to reproduce

insecticide spray or pour-on chemical used to control flies and other insects

ketosis a condition in dairy cattle that causes low blood sugar

lactation the process of milk production

large-size herd a group of more than 100 head of cattle

lead a rope that attaches to a halter and is used to walk an animal

leptospirosis a bacterial disease transmitted in the urine of infected animals

liquid manure system a costly waste-removal system used on large farms that requires above- or below-ground storage tanks with water added for ease of pumping

manure waste material high in nitrogen, phosphorus, and potassium

marbling appearance of intramuscular fat within the meat

market weight adult weight of cattle when they are sold for beef

mastitis an inflammation of the mammary gland

metabolic disease a condition that causes a chemical change within the body, usually from stress

microchip electronic identification contained in a computer chip placed under the skin

milk fever see *hypocalcemia*

milking herd a group of cattle used to produce high-quality amounts of milk

nasal septum cartilage between the nostrils

neck tag identification a tag with numbers on it worn around the neck of cows to identify them

nose printing a method similar to fingerprinting in which the lines of the nose are imprinted for identification

nose ring a metal ring placed through the septum of the nose that allows a cow to be led

nose tongs equipment placed in the nasal septum to apply pressure and allow a cow to be lead

offspring the product of the reproductive processes of an animal

palatability how well food tastes and is eaten

pedigree an ancestral line

polled no horn growth

predicted transmission ability estimation record that determines the possible traits that a bull or cow will pass along to offspring

primed milking machine devices are applied to the udders and begin collecting milk

progeny offspring

purebred business registered breed of cattle that has pedigrees

pushing walking toward cattle quietly and calmly to get them to move to another area

quarters sections of the mammary glands that store milk

reefing using rope to place a cow on the ground for restraint

retained placenta reproductive condition in female animals where the afterbirth materials have not passed within eight hours after labor

roughage hay or grass source fed to livestock and ruminants

selection guidelines set of rules that state the standards of the type of cow and how it is chosen for a production program

sepsis a toxic condition caused by the spread of bacteria

silo a tall storage building that keeps food free of moisture

sire male parent

slough tissue that has died and falls off

solid manure system a low-cost waste material system in which manure is collected and removed on a daily basis and stored away from cattle

sound showing no signs of lameness or injury

springing heifer a young female cow that is pregnant with the first calf

square knot a certain kind of knot that is easily tied and untied

squeeze chute a cage-like structure made of metal pipes that holds a cow and prevents it from kicking during restraint

stanchion a headgate that holds the head of a cow in place during restraint

starter food food for calves as they begin eating that is easily digested

steer a castrated male cow

tail switch restraint a method of twisting the tail at the base to prevent a cow from kicking and moving

tattoo a mark or number placed on the skin with an electronic needle and ink

teat a nipple of the mammary glands of a female animal, which milk is sucked by the young

total herd large group of cows kept together and replaced when milk production decreases

total mix ration (TMR) a mixture of high-quality nutrients

udder a large bag-shaped organ enclosing two or more milk-producing glands, with each draining into a separate nipple on the lower surface

warm housing a heated building with individual stalls that holds cattle in the winter

Review Questions

1. What are the differences between beef and dairy cattle?
2. What are the types of beef cattle production systems? Explain each one.
3. What is the basic concept of dairy cattle production?
4. What are the types of dairy production systems? Explain each one.
5. What is the DHI program?
6. What is the importance of the DHI program?
7. What are some methods of restraining cattle?
8. What is the role of the herd health manager?
9. What are some management practices in cattle?

10. What are some common diseases that may affect cattle?
11. What are some methods of cattle identification?
12. How much should adult dairy cattle be fed on a daily basis?
13. How much should adult beef cattle be fed on a daily basis?
14. What is the difference between warm and cold housing for cattle?
15. What are some management practices for waste materials of cattle?

Clinical Situation

Brittany, a veterinary assistant at the Delmont Bovine Practice, answered a phone call from Mrs. Shore, the wife a local dairy producer. Brittany remembered that Mr. Shore's Holstein dairy farm was one of the largest in the area.

"Hello, Mrs. Shore, how can I help you?"

"Hi, my husband asked me to call and set up a farm call appointment for his cattle herd to be vaccinated and examined. This is the yearly visit. He also said he needs about 100 head of cattle checked for pregnancy."

"Okay, Mrs. Shore, Dr. Evans is not in the office at the moment, so I will need to check with him to see what day works best for his schedule. How many head of cattle are in the herd?"

"I really don't know; it's my husband's business and I was just asked to call and set up an appointment."

"Well, let's tentatively set it up for next Wednesday. It looks like Dr. Evans has a light schedule. I will call you back if that day doesn't work."

- How did Brittany handle the situation?
- What would you have done in this situation?
- How could Brittany have obtained a history on this client and the dairy herd?

CHAPTER 18
Equine and Draft Animal Breed Identification and Production Management

Objectives

Upon completion of this chapter, the reader should be able to:

- **18.1** Define common veterinary terms relating to the horse
- **18.2** Describe the biology of the horse
- **18.3** Identify common breeds of horses and draft animals
- **18.4** Distinguish the different classes of horses
- **18.5** Discuss the purpose and uses of draft animals in society
- **18.6** Discuss the nutritional needs of horses
- **18.7** Describe normal and abnormal behaviors of horses
- **18.8** Describe housing, facility, and equipment requirements for horses
- **18.9** Explain how to properly and safely restrain horses for veterinary procedures
- **18.10** Discuss the necessary health care and veterinary practices of horses
- **18.11** Describe breeding practices with horses
- **18.12** Discuss common diseases of horses
- **18.13** Discuss common parasites of horses
- **18.14** Define common veterinary terms relating to draft animals

Introduction

Horses have become one of the most popular types of livestock for both breeders and animal lovers alike. Horses have a grace and beauty that are unlike any other animal (see Figure 18–1). They also have a great ability to bond with humans and serve many purposes to humans. Most horses are owned for recreation and entertainment but some are also used for working and in some countries as a food source. Today, there are over 9.5 million horses in the United States. Draft animals include several species of large animals developed to work and pull heavy loads. These animals were domesticated to help serve humans in today's society. Draft animals include mules and oxen.

Veterinary Terminology

Horses are classified based on age, gender, size, and type of use. Several different terms are used to describe horses. An adult female horse is called a **mare**; an adult male horse that is capable of reproduction is called

334 SECTION II Veterinary Animal Production

FIGURE 18–1 Horses are owned and raised for recreation, entertainment, and work.

FIGURE 18–3 A light horse.

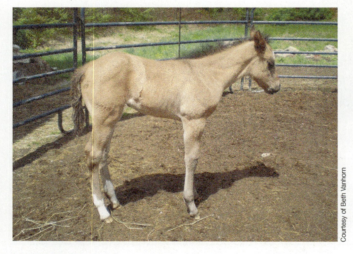

FIGURE 18–2 A young horse is known as a foal.

FIGURE 18–4 A draft horse.

a **stallion**. A female horse that has not been bred is called a **maiden** horse. The male parent is known as a **sire**; the female parent is called a **dam**. The male horse used in a breeding program is called a **stud** horse. When a male horse is castrated, it is called a **gelding**. A young newborn horse is called a **foal**; the female is specifically called a **filly**, and the male is known as a **colt** (see Figure 18–2). The labor or birth process of horses is known as **foaling**. In veterinary medicine, horses are called **equines**. They are classified by their sizes and purposes or uses. There are **light horses**, **draft horses**, and **ponies**. Equine species are measured in hands. A **hand** is equal to 4 inches and measures the height of a horse at the highest point of the shoulder. Light horses are typically riding horses that stand between 14.3 and 17 hands in height and weigh between 900 and 1400 pounds (see Figure 18–3). Draft horses are known as large, muscular animals that weigh over 1400 pounds and stand over 15 hands in height (see Figure 18–4). Ponies are small in size, standing under 14.2 hands in height and weighing less than 900 pounds. They are usually ridden by small children. Other types of equines are donkeys and mules. A **donkey** is a member of the equine family and sometimes called a *burro* (see Figure 18–5). The female donkey is called a **jenny,** and the male is called a **jack**. A **mule** is the cross of a female horse and a male donkey. A male horse crossed with a female donkey will produce a **hinny**.

FIGURE 18–5 A donkey.

Equine species also extend to draft horses and other similar types of livestock that are muscular, strong, and powerful. They are animals that were originally developed to help people with their work. People still use some draft animals to help them pull wagons, plow fields, and carry out farming activities. They are also kept as pets and recreational entertainment. Animals that were originally used in farming, travel, and the military still have a major role today.

Biology

Horses are large animals that are classified as livestock, similar to cattle. Horses are from the genus and species **Equus caballus** and developed from prehistoric types of horses. The purposes and use of horses have changed over the last several hundred years. Many breeds and varieties of horses have been developed.

Horses are herbivores and feed on a plant-based diet (see Figure 18–6). They are mammals that have a specialized digestive system called **nonruminant**. This system is similar in appearance to humans but requires small amounts of food to be eaten and digested throughout the day, similar to cattle and other grazing animals. Horses have a relatively small stomach with a long intestinal tract that allows food to flow all day. The horse's digestive system also has an enlarged **cecum**, which is an area within the intestinal tract that breaks down grasses and roughage. Horses do not have a gall bladder and are not capable of vomiting. A horse's digestive system is broken down into the fore-

FIGURE 18–6 Horses primarily eat plants and are known as herbivores.

gut and the hindgut. The foregut includes the mouth, esophagus, stomach, and small intestine. The hindgut includes the large intestine, cecum, and rectum. Food moves quickly through the foregut but greatly slows down when it reaches the hindgut. A majority of the nutrients are absorbed in the foregut. Because most people feed their horses once or twice daily, a large amount of food passes through the digestive system at one time. A horse should eat small amounts of roughage all day long for proper digestion and typically spends 75 percent of its day grazing.

Horses are hoofed animals; the hoof wall is made of a hard material similar to that of a fingernail (see Figure 18–7). Legs and hooves are critical parts of the horse and may easily become injured. Hooves continually grow and must be trimmed and cared for on a regular basis. A horse's legs are very fragile and can suffer severe damage.

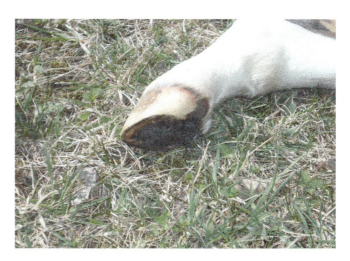

FIGURE 18–7 Horses are hoofed animals.

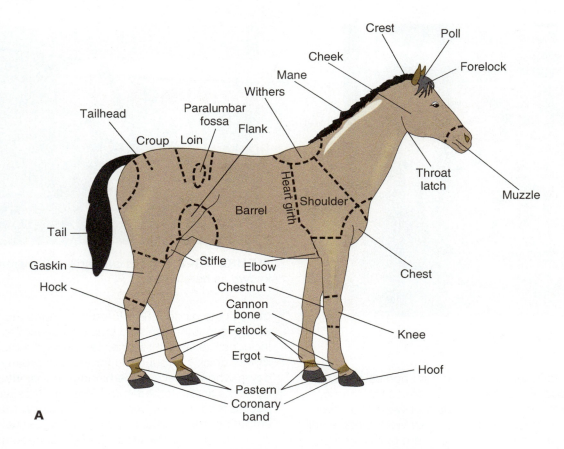
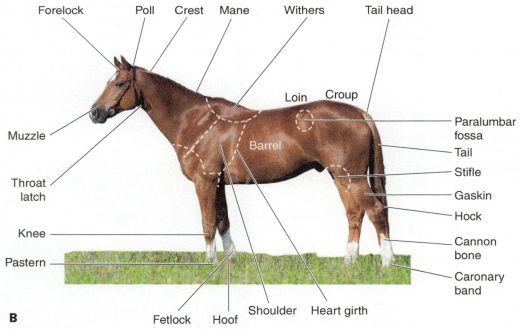

FIGURE 18-8 External anatomy of the horse.

Horses are endothermic, or **warm-blooded**, meaning they control their body temperature internally. The body and structure of a horse are based on the type of work they are used for (see Figure 18–8). Horses have about 205 bones in their skeletal system (see Figure 18–9). They can typically live into their 20s with proper health care.

Breeds of Horses

Horses are classed according to their body type, size, and purpose or use. Horse breeds come in a variety of sizes and colors; riding horses include the Quarter Horse, Paint, Thoroughbred, and Appaloosa (see Figure 18–10); draft horses include the Clydesdale and Percheron (see Figure 18–11); and ponies include the Shetland and Welsh (see Figure 18–12).

Riding Horse

Riding horses are ridden for pleasure and showing purposes. They are further classed by their type of gait and what people use them for. Some of the most popular riding horses are the Quarter Horse, Paint, Appaloosa, and Tennessee Walking Horse (see Figure 18–13).

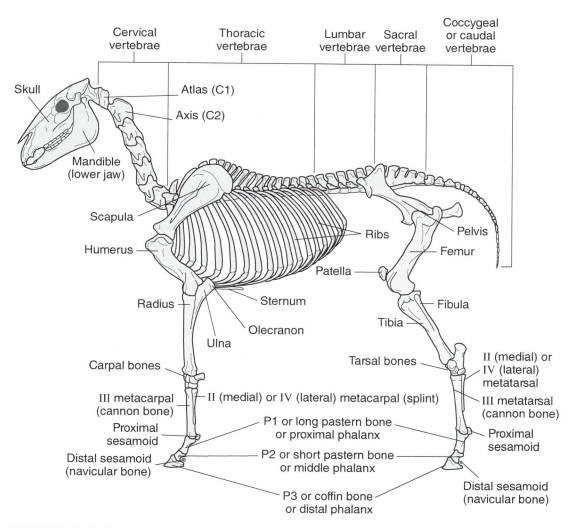

FIGURE 18–9 Skeletal anatomy of the horse.

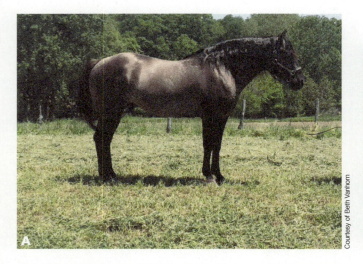

FIGURE 18–10 (A) American quarter horse. (B) American paint horse. (C) Thoroughbred. (D) Appaloosa.

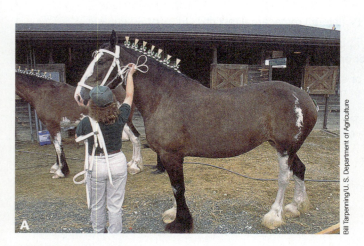
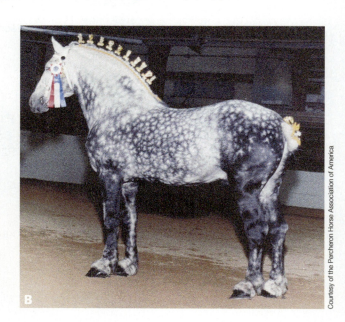

FIGURE 18–11 (A) Clydesdale. (B) Percheron.

Gaited Horse

Gaited horses are classified by the way they move in their **gait**, which is the way they walk and run. Gaited horses have a rhythm to their movement that is easy to sit and provides the rider with little movement or fewer rough rides. These horses are popular with older riders and people who may have back problems. Different speeds at each gait have been developed by people who spent a lot of time in the saddle and chose these horses due to the easy gait and disposition of the animal. Popular gaited horse breeds include the Tennessee walking horse, American saddlebred, and the Rocky Mountain horse (see Figure 18–14).

Stock Horse

A **stock horse** was once used on ranches for herding and working cattle. They are hardy, athletic, able to serve many purposes, agile and surefooted, fast, and durable. They are also known for their muscular and athletic body structure. Popular stock horse breeds include the quarter horse, Appaloosa, and paint horse (see Figure 18–15).

Hunter and Jumper

Hunters and **jumpers** are bred for fox hunting and cross-country riding. They tend to be tall, athletic, and muscular and have speed and endurance for jumping large fences. Hunter horses are typically used for cross-country fox hunting and jumping smaller fences; jumpers are used for jumping large fences on a set course (see Figure 18–16). Popular horse breeds used for hunters and jumpers include thoroughbreds and a variety of warmbloods.

Racehorse

Racehorses are bred for speed and used for racing (see Figure 18–17); this category includes horses raced under saddle and in harnesses. Racehorses may be used for other sports-related activities. Horse racing has become a popular sport: famous races include the Kentucky Derby and Belmont Stakes. Popular racehorses are the quarter horse and thoroughbred. Popular harness racing breeds include the standardbred and Morgan horse.

Pony

A pony is a small horse typically ridden by children. Ponies stand under 14.2 hands in height and tend to be gentle and intelligent. The most popular breeds include the Shetland, Welsh, and pony of the Americas (see Figure 18–18).

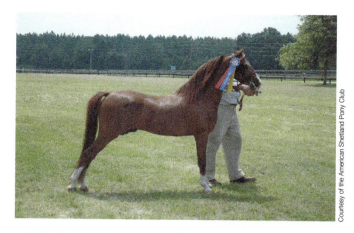

FIGURE 18–12 American Shetland pony.

FIGURE 18–13 Riding horses are used in shows and competitions.

Draft Horse

Draft horses are large and muscular and bred to pull heavy loads. They are also known as workhorses, as many are used today in helping work fields and pull large wagons. Draft horses were developed to weigh over 1500 pounds; several breeds weigh well over 2500 pounds. Their body structure is wide, deep, and large-boned with a low center of gravity, which holds a heavier muscle build. There are five recognized draft horse breeds still used as work animals in the United States: the Belgian, the Clydesdale, the Percheron, the Shire, and the Suffolk.

The Belgian is a Flemish breed that developed in the Middle Ages as a farm horse. They are light brown to sorrel in color with a lighter **flaxen** (blond) mane and tail. They are popular draft animals due to their large size, tremendous power, and excellent muscle build. They are the most popular and numerous draft horse breed today.

FIGURE 18–14 The Tennessee walking horse is a breed of gaited horse.

FIGURE 18–15 Stock horses are sturdy horses often used for working and showing.

FIGURE 18–16 Jumpers are used on courses in competitive riding events.

FIGURE 18–17 Racehorses are bred for speed.

FIGURE 18–18 Pony of the Americas.

The Percheron is the second-most-numerous draft horse. They originated in France and have a more refined build than the Belgian. They were used in the Middle Ages as battle mounts and then became popular carriage horses. The Percheron's colors range from black to dapple gray to white (see Figure 18–19).

The Clydesdale is the third-most-numerous draft horse breed. They have become notable for pulling heavy wagons and because of their role as mascot for a national beverage corporation. The Clydesdale is a Scottish breed that is usually brown or black in color with white faces and legs. They have long hair on their legs known as **feathers** (see Figure 18–20).

The Shire is a draft horse breed that developed in England and is the tallest of the draft horse breeds. They are typically black or bay in color and have feathers on their legs.

The Suffolk was developed in England solely for agricultural purposes. They are chestnut in color and tend to have very little white. They are easier keepers than other draft breeds and require less feed than other heavy breeds. They are smaller is size than the other draft breeds.

FIGURE 18–19 Percheron draft horses pulling a heavy wagon.

FIGURE 18–20 The famous Budweiser Clydesdales pulling a heavy wagon.

Donkey and Mule

Donkeys and mules are used as draft animals and are members of the equine family. A donkey is also called a *burro* or ass and was originally domesticated from a wild version of the African wild ass. Donkeys are not hybrids of horses. Donkeys that are bred to each other tend to be sterile; some are capable of being bred. A donkey is gray, pony-sized, has long ears, and has a bristle-like coarse coat. Donkeys have a vocalization sound known as a **bray** that can be heard from a long distance. Donkeys tend to be tough, versatile animals, which makes them hardy and handy as draft animals. They also tend to have a stubborn attitude and move at a slower pace than a horse. The female donkey is called a jenny, and the male is called a jack (see Figure 18–21).

The mule is a hybrid cross of a female horse and a male donkey. Mules are almost always sterile. A mule is larger than a donkey—similar in size to a horse—and looks like a donkey. Donkeys have long ears and move at a moderate pace (somewhere between a donkey and horse). Mules come in a variety of colors: bay, brown, or black. They are tough draft animals, powerful, and tend to be docile with people. They can be trained to ride like a horse. A female donkey crossed with a male horse is called a *hinny*. Compared to the mule, the hinny is rare. Mules and donkeys both endure heat well and are much less sensitive to digestive problems and leg injuries.

FIGURE 18–21 Two wild jacks registered as standard donkeys/wild burros, owned by Elmer Zeiss, Valley, Nebraska.

Breed Selection

Horse selection is based on several factors. Prospective horse owners need to take the costs involved in owning a horse into consideration: the purchase price along with its care and maintenance. Horse breed selection is based on size, disposition, breed, gait, purpose, gender, and color.

The price of a horse ranges greatly, depending on the breed and purpose. Show horses or specially trained horses may cost more than another type of horse. It is important to know the costs of both purchasing and owning a horse. Other factors to consider related to cost are boarding, hay and feed, veterinary care, shoeing, fencing and building, and supplies. The disposition of the horse must be evaluated: It is important that anyone working with the horse can get along with and handle the animal. Beginners require a horse that is calm, quiet, and gentle—and well broke for its purpose or use.

The size of the horse will depend on the size (height and weight) of the rider as well as the ability to maintain control over the horse. Breed, color, and gender are personal preferences that should be considered in the purchase.

Draft animals are mostly used for work purposes but are also bred as show animals. Horses, donkeys, mules, and oxen can all be shown within their respective breeds or species. They may be judged based on conformation (body structure), power, or athletic ability. They may compete in such events as horse pulls, draft-hitch classes, farming demonstrations, or log pulling. A **horse pull** is a contest of power to see which animal can pull the heaviest load with the least damage to the land. **Draft-hitch classes** are based on the number of animals used to pull a wagon and how the team works together to complete the job. **Farming demonstrations** involve draft animals pulling plows to work the land in farming activities (see Figure 18–22). **Log pulling** is a contest to see which animal or team can pull the greatest number of logs the greatest distance. The term **horsepower** was developed in England to describe the work done by large horses and other draft animals. It now refers to the unit of power of an engine or motor. It is defined as the amount of work done by lifting or moving 550 pounds at a distance of one mile per second. One horsepower equals 746 watts of electricity. Many draft animal contests measure this horsepower.

FIGURE 18–22 Horse pulling load of hay on a sled.

Draft capacity refers to the force an animal must exert to move or pull at a constant rate of 1/10 of its body weight. Power is the combination of draft capacity and speed. Draft horses are the most popular animal for use in these contests of power, as they have better skill, behavior, and coordination for team working events. Draft animals are selected according to the type of job or work they are meant to carry out. Other factors include the needs of the animal, gender, age, cost and upkeep, and health care. A social rank is established when several draft animals are kept together. This rank is their place or order within the group and depends on the animal's size, strength, and age. Table 18–1 lists factors to consider when selecting a draft animal.

TABLE 18–1

Selection Factors of Draft Horses and Animals

DRAFT HORSES	OXEN	DONKEYS/MULES
High purchase price	Low purchase cost	Moderate purchase cost
Require large amounts of food	Less food intake	Moderate food intake
Work fast	Work slowly	Work slowly
Require large space	Require moderate space	Require moderate space

Nutrition

Horses have a nonruminant digestive system that requires them to continually digest food to keep their system from binding up. The digestive system includes a cecum (a sac with an opening) located between the small and large intestine that helps break down roughage and protein. A horse's stomach is always digesting food and should have a notable gurgling sound at all times. Horses should be fed grains that include oats, corn, wheat, or barley mixtures. Grains supply the horse with adequate amounts of energy and protein. Most commercial quality feeds include 10 to 12 percent protein, 5 percent fat, and about 80 percent carbohydrates. Some horses require additional supplements or concentrates, depending on their health and activity levels. Common supplements include soybean meal or linseed oil for skin and coat conditioning. Horses can be fed in grain or pellet formulas. Pellet feed is easier to digest and break down in the intestinal system. Grains can be ground down, cracked, or shelled for better digestion. All foods should be free of dust and mold. The usual maintenance diet for a horse requires 1 pound of protein per 1000 pounds of body weight. Horses are typically fed twice daily with adequate amounts of pasture or hay (see Figure 18–23). **Forages**, or hay and pasture, may include timothy, orchard grass, or alfalfa sources. Hay should be kept dry and monitored for signs of mold. Hay should be fed at 1 to 2 pounds per 100 pounds of body weight and may be provided several times daily. Horses on pasture should have at least 3 acres of space for grazing. Some horses and foals may require minerals and vitamins for growth and development of teeth, bones, and tissues. Horses ingest about 2 percent of their body weight per day while grazing. The horse averages about 60,000 chews per day.

The most common nutritional deficiencies include calcium and phosphorus. Feed supplements should be given based on an individual basis. Horses require additional sources of salt that can be supplied in a **mineral block** or **salt block**. Horses with poor or inadequate pasture or hay sources may need to be given Vitamin A. Horses stalled indoors on a regular basis may need Vitamin B. Horses that are underweight may be fed higher amounts of protein or flaxseed oil for additional fat sources. All horses require daily freshwater sources (see Figure 18–24). The average horse requires 10 to 12 gallons of water a day. Increased amounts of water are necessary during high activity, during lactation, and in the winter. Water intake should be monitored, as horses can easily dehydrate. **Electrolytes** or powdered supplements can be placed in the water or food source to encourage horses to drink more and prevent dehydration. Horses should not be allowed to drink immediately after exercise or when sweated. This can lead to severe stomach pain. Any changes in diet should be done gradually over a 7- to 10-day time frame to avoid digestive problems. Table 18–2 outlines the recommended rations for horses.

Draft animals require at least 8 to 10 gallons of water every day. Draft horses have a nonruminant digestive system like other equine species. They require increased amounts of protein, salt, calcium, and phosphorus in their diets. They may also need additional vitamins and minerals based on the quality

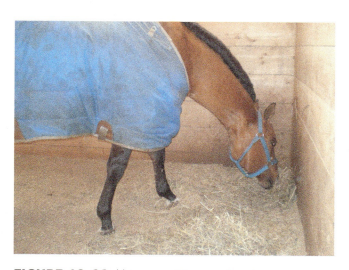

FIGURE 18–23 Horses eat hay and grains.

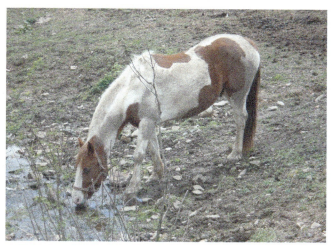

FIGURE 18–24 Horses should be supplied with water regularly.

344 SECTION II Veterinary Animal Production

TABLE 18–2

Recommended Maintenance Diet Rations

Protein	10–12%
Fat	6–8%
Fiber	12–15%
Calcium	1–3%
Phosphorus	0.5–1%
Sodium	0.2–0.7%

of food and their overall health. The feeding schedule should allow for 1 to 2 hours of rest before food or water after periods of work. Draft horses should not be worked for at least 45 minutes to an hour after being fed.

Behavior

Horses are herd animals and prefer to live in groups. Within a group or herd, each horse has its place; this structure is known as a **hierarchy**. It is also known as a "pecking order," in which certain animals' behaviors direct others and allow for more control over the group. Horses are also comforted by being near or with other horses. Depending on the situation, this can determine their behavior. Horses are suspicious by nature and can quickly detect a nervous person working with or handling them. This decreases the control a person has over the horse. Because horses were domesticated thousands of years ago, they have learned to trust people. The horse is not always accustomed to loud noises and will react in various ways, such as **spooking** or startling. Horses naturally have a "fight or flight" instinct. A horse's field of vision is almost 360 degrees, with an eye placed on each side of the head. A horse can see about 60 to 70 degrees in front and must move its head to see any other area outside of that range. Horses have also adapted to standing up while sleeping, ensuring a quick escape from any predators. Horses are known to display many types of behaviors that are normal for them but can be dangerous for humans handling or working near them. Such normal behaviors for a startled horse may include kicking, **bucking** (lifting and kicking with the rear limbs), **rearing** (lifting the front limbs off the ground and raising the body into the air), striking with the front legs, or biting. A horse will show if it is happy, angry, or scared through body language. A happy horse looks relaxed, with ears forward and alert, and with a relaxed tail and a relaxed expression in the eye. An angry horse has its ears pinned back on its head, its teeth bared, a flicking tail, stomps its feet, and kicks or bucks (see Figure 18–25). A scared horse is wide-eyed, with the whites around the eyes visible, its head is held high in the air, its ears flick back and forth, the tail may be held high in the air or clamped down, it may shake or move around nervously, and it may snort.

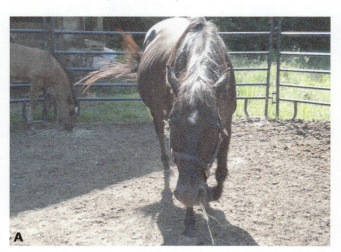

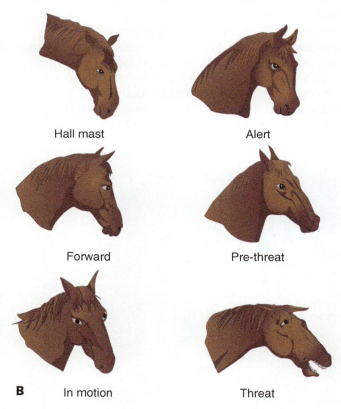

FIGURE 18–25 (A) This horse is displaying signs of anger. (B) Ear positions of horses.

Normal signs of horse behavior include the following:

- Behavior is consistent with other herd animals
- Displays fight or flight instinct
- As prey animals, feel threatened by animals that look at them as a food source
- Show vocal communication (neighing, snorting, squealing)
- Groom on other horses
- Nuzzle
- Ears are forward and alert
- Tail is down and quiet

Some abnormal signs of horse behavior include the following:

- Biting
- Kicking
- Pawing
- Chewing
- Striking
- Head tossing
- Ears moving quickly
- Ears flattened on head
- Tail swishing

Draft animals are typically **gregarious** in behavior, which means they have a natural instinct to join with the herd or group. Some draft animals that work alone may feel stress while they are away from the herd, which makes them difficult to handle and decreases work and production. Draft animals that work in teams must be able to get along to be safe to handle and prevent injuring each other. When working with a team or group of draft animals, make sure that they have established a pecking order and place them in the team according to their ranking in the group. Some animals will kick or bite at each other, possibly causing injury to the team or the handler.

Basic Training

Horses are trainable in many disciplines. Horses may be ridden, driven, used as pack animals, and kept as companions. There are many areas of training with horses, depending on their purpose and use. Professional trainers train young or inexperienced horses, as many problems—such as kicking, rearing, or bucking—may occur with someone who lacks experience. All horses should be taught to **lead** on a rope, walking calmly and quietly with a person (see Figure 18-26). They should be taught to be tied on a post or chain and also on cross ties, which are two ropes placed on either side of the halter and then

FIGURE 18-26 Horses can be trained to walk on a lead with a person.

secured to a wall or post (see Figure 18-27). Horses should also be trained to work on a **longe line**, which is a long rope that allows a horse to work at various gaits while circling the handler. Riding and driving horses should be taught basic commands, such as walk, jog or trot, and lope. They should also be trained to stop and back up. A large amount of basic training is gaining the horse's respect and trust.

The health of draft animals is based on **conditioning**. This is the practice of working an animal to build up its body and strength for its work needs. Conditioning begins slowly over time, with the amount of work increasing at regular intervals. This is similar to the way an athlete trains for a race or a competition. If an animal is not conditioned properly, it may pull a muscle or otherwise physically injure itself.

Equipment and Housing Needs

Horse facilities should consist of a barn or shelter area (see Figure 18-28). Shelter may be a run-in shed with at least three sides that will protect the horse from

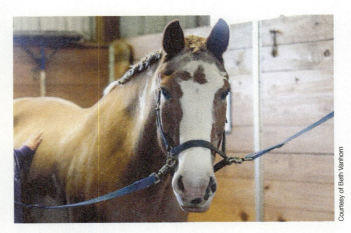

FIGURE 18–27 Horses can be trained to be tied or secured to a wall or post by cross ties.

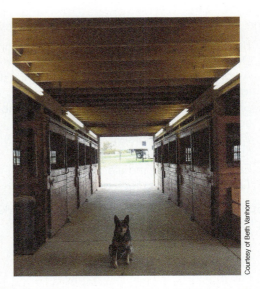

FIGURE 18–28 A horse barn lined with individual box stalls and aisle way.

the weather. Barns may consist of a large open area or single stalls for individual horses. Stalls should be at least 10 feet by 10 feet, and at least 8 feet high (see Figure 18–29). Cross tie areas or walkways should be about 5 feet wide (see Figure 18–30). The lower areas of the barn should be reinforced to prevent injury and damage should a horse kick. Paddocks and pastures should be built of high-quality fencing safe for horses and provide at least 1 acre of space per horse (see Figure 18–31). Barns should have to be equipped with water and electricity.

Working draft animals require specialized equipment; this includes an equine harness, which consists of leather straps that fit the horse and attach to a wagon or plow used for pulling. The harness attaches to a collar that lies around the neck and shoulders and holds the harness in place. **Hame straps** attach to the harness and collar through pieces of leather. A bridle is placed on the head and attaches to reins that are used to guide and control the animal (see Figure 18–32).

Restraint and Handling

Restraining horses requires experience, knowledge, and a good eye for reading a horse's body language. Horses can be calmed using a low, even tone of voice. It is best to allow a horse to evaluate the restrainer by smelling their hands; the restrainer should scratch or pat the horse's neck to show it is meant no harm. A **halter** should be placed on the horse to control its head; this will allow a handler to properly lead the animal (see Figure 18–33). Always approach a horse from the front and slightly near the left or **near side**. Never make your approach from directly behind, as

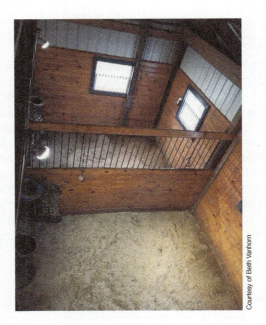

FIGURE 18–29 A box stall is often 10 × 10 feet.

this is a horse's blind spot. Horses are accustomed to being handled and trained from the left side. It is important to stay to the front of the horse and within its range of vision. A horse doesn't see well directly in front and directly behind it. Approach the horse quietly and calmly while watching its body language. If the horse tries to move away, stop and talk calmly to it and resume moving with slow movements toward its head. Once close enough, pat the horse on the neck before applying the halter. A lead rope

CHAPTER 18 Equine and Draft Animal Breed Identification and Production Management 347

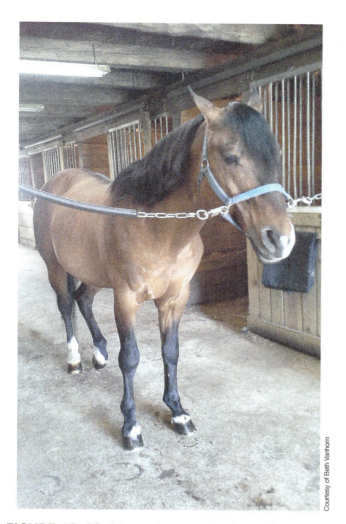

FIGURE 18–30 A horse in cross ties in an aisle way.

FIGURE 18–31 A pasture provides space and area for grazing for horses.

FIGURE 18–32 A Clydesdale draft horse in full tack and harness.

can be slipped over the horse's neck with a knot tied to prevent it from slipping off. Stand on the left side, to the side of the head and neck, gently place the halter over the muzzle, and slip the strap behind the ears. Then properly secure the halter and apply a lead rope to the center ring on the bottom of the halter. When handling and restraining a horse, never wrap the lead rope around the hand or wrist. Hold the lead in neat loops with the left hand and hold the lead rope close to the halter with the right. This allows the restrainer or handler to move freely from one side to another to control the horse. **Leading** a horse is done by walking with the horse near its left shoulder about 1 foot from the horse. Caution must be taken to avoid being stepped on by the horse. Tying a horse should be done with a **slip knot** that is used in case of a quick release. Tie the horse to allow 2 to 3 feet of lead rope to allow some movement and allow it to hold its head even with its withers. Horses can also be placed on **cross ties**, which are short ropes placed on either side of the horse's halter to allow safe movement around all areas of the animal. The ties are placed on the lateral cheek ring of the halter. When handling a horse, distractions can be helpful while the horse is examined, given an injection, or having blood collected. Distraction techniques include grasping the ear at the base and gently bending it back and forth; slowly sliding a hand up from the cheek to cover the eye; rubbing or grasping the skin at the shoulder (see Figure 18–34); or lifting the leg to prevent a horse from moving. Another technique to control a nervous or difficult horse is to use the **twitch**. A twitch is a

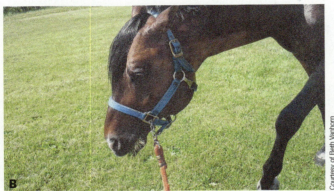

FIGURE 18–33 (A) Haltering allows the handler to properly lead the horse and control its movement. (B) Halters should be safe and secure.

the horse lifts the foot, grasp it at the hoof and move slightly forward to stretch the leg out and allow the hoof to rest on your knee.

Some other methods of equine restraint include **tail tying** to adjust a horse's weight and movement or the use of **hobbles** to prevent a horse from kicking. Tail tying is generally used with a sedated horse to move it or to keep it out of the way for certain procedures. The tail should be tied to the horse's body by securing a rope below the base of the tailbone (see Figure 18–37). The rope can then be secured to the horse's front leg or neck. Hobbles are leg restraints placed around the limbs to prevent horses from kicking. The safest method of restraint on a scared, aggressive, or difficult to

chain or rope loop attached to a long handle that is placed over the lip of the horse (see Figure 18–35). This will control a horse during certain procedures that may cause discomfort to the horse. It allows the horse to stand under better control. A **chain shank** can be used for better control as well. The chain is attached to a lead rope and placed on the halter to apply pressure over the muzzle, under the chin, under the lip, or in the mouth (see Figure 18–36). This is used as a control method as well as an additional distraction when working with a horse. To lift a horse's feet, the restrainer stands with their body facing the rear of the horse. When working with the front feet and legs, stand at the shoulder of the horse and run a hand down the rear aspect of the limb and gently squeeze the fetlock joint, just above the hoof. When the horse lifts its foot, hold it by the hoof and balance it on the knee or firmly in the hand. When lifting the rear feet, stand facing the rear of the horse and fairly close to the side of the hindquarters. Run the hand down the leg in the same manner as with the front limb. When

FIGURE 18–34 Pinching the skin roll provides a distraction when the horse is undergoing a procedure.

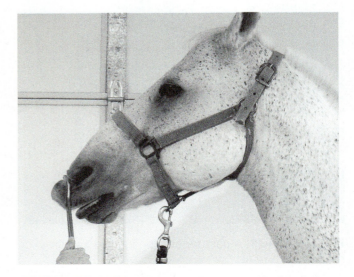

FIGURE 18–35 Application of a mechanical twitch.

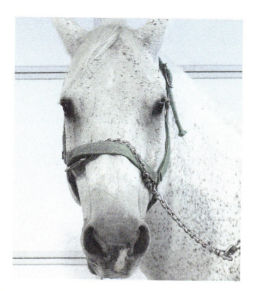

FIGURE 18–36 Restraint by using a chain over the nose.

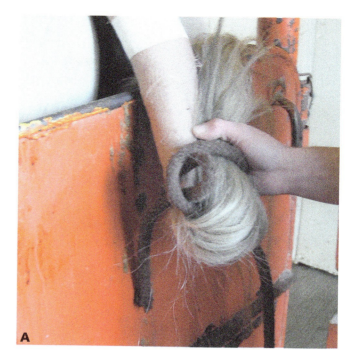

handle horse is sedation. Many sedatives allow the horse to remain in a standing position. Care must be taken to monitor the horse and body language, as the anesthesia can easily wear off. Foals can be restrained by grasping one arm around the front of the chest and the other arm around the rump and keeping it against a stall wall (see Figure 18–38). The tail can be grasped and pulled upward and forward over its back to prevent movement. When working with any horse, remember to talk and comfort the horse throughout the restraint and handling. Use caution when walking or standing anywhere near the back end of the horse, as kicking could easily injure the restrainer. A handler should never turn his or her back on a restrained horse.

Horse Behavior and Restraint

Equine behavior that should be considered during handling, restraint, or general work within their pastures or stall areas include the following:

- Horses have blind spots in which they can't see well, so work quietly and calmly around a horse.
- Allow a horse to know where you are at all times by moving slowly and talking gently.
- Horses point their ears in the direction they are focused on. Ears that are flattened back or pinned on the head indicate anger and may indicate that a horse is going to kick or bite. Horses' ears present different types of messages.
- Being nervous around a horse will make the horse nervous and difficult.

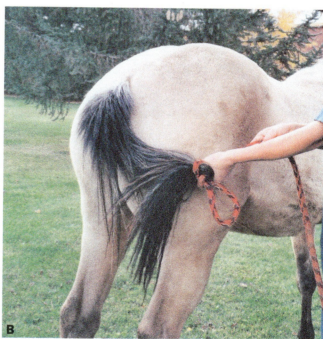

FIGURE 18–37 (A) Pass the bight through the tail loop. (B) Finished tail tie.

- Approach a horse from its left shoulder, moving confidently and calmly and watching the horse's body language.
- Horses that run toward you may be chased away by raising and waving your hands and arms above your body.

Chewing

It is common for horses to chew excessively. The consumption and chewing of wooden fences, walls, and boards may be due to a nervous habit, boredom, or a nutritional deficiency. Chewing can damage buildings and fences. It is not good for the health of the animal, as it may result in broken teeth, cuts in the mouth, or colic. Some horses exhibit a condition known as **cribbing**, in which they chew on an object and then suck in air (see Figure 18–39). This is a vice or bad habit that can easily be picked up because of stress or modeling.

Grooming

Horses need regular and consistent grooming, such as brushing, bathing, and hoof maintenance. Grooming equipment tools and supplies are essential for all horse owners. Basic tools and equipment needed are the following:

- Body brush
- Mane/tail brush
- Curry comb
- Hoof pick
- Body sponge
- Shedding blade
- Sweat scraper
- Conditioning spray
- Fly spray
- Hoof oil
- Clippers and blades
- Scissors

Horses require regular brushing over the body, mane, and tail. Each horse should have its own halter and lead rope. The halter should be well fitting over the face and throatlatch and not be too tight. Horses should have their hooves cleaned and picked out daily (see Figure 18–40).

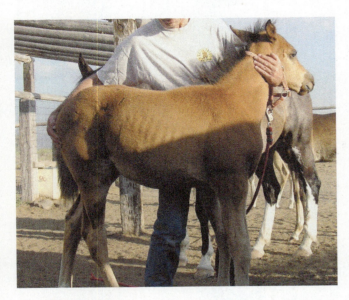

FIGURE 18–38 Cradling a foal in order to restrain it.

FIGURE 18–39 Cribbing.

CHAPTER 18 Equine and Draft Animal Breed Identification and Production Management

FIGURE 18–40 Horses require daily care and attention to their hooves.

Basic Health Care and Maintenance

The key to quality health and management of horses is prevention. Horses should be on a regular health program that includes a knowledgeable and experienced equine veterinarian. Horses should be monitored on a daily basis for any health problems or lameness. They should be on a quality exam, vaccine, and dewormer schedule. A horse's stall must be cleaned on a daily basis. Their pastures should be rotated on a regular basis and have a regular sanitation and disinfectant program to control parasites. Every horse should be on a high-quality nutritional program that meets its energy, health, and age needs. A horse should have a complete physical exam at least once a year. New horses should have a prepurchase or post-purchase exam before entering the barn area. It is important to determine the soundness of a horse, as the legs and hooves are its most important parts.

Health Testing

Horses are highly susceptible to contagious diseases and organisms, so new horses should receive a health exam before they are allowed on a farm. Exams may include a complete physical, blood work, and radiographs. Horses should be given a **Coggins test** every six months to a year, depending on state requirements. This blood test determines a horse's status for **equine infectious anemia (EIA)**. This deadly virus has no vaccine and no cure and spreads rapidly in horses via biting insects. All horses that are transported to shows, events, and travel throughout the country are required to have a negative test.

Dental Care

Horses require regular dental care. Horse teeth continue to grow as the horse ages. The **points** of the teeth, the corners and edges, become sharp and can make a horse's mouth sore and prevent it from eating properly. Horses can live well into their 30s with proper health care. A horse is in its prime years from 3 to 12 years of age. A horse's teeth can help determine its age through their appearance and wear (see Figure 18–41). Young horses have temporary teeth that are smaller in size and length and whiter than adult teeth. Young horses begin to lose their temporary teeth between 2 and 3 years of age and should have all of their adult teeth in place by 5 years of age. The shape and wear of the adult teeth change as the horse ages. A **bit** is a metal bar placed over the horse's tongue that provides pressure and control while riding. The **cusp**, or center of the tooth, wears. Until the age of 12, horses have oval-shaped teeth. After 12, the teeth become triangular in shape and begin to slant forward. Adult horses have between 36 and 40 teeth. Females have 36 and do not usually have canine teeth. Males usually have 40 teeth, which includes canines. Some horses will develop **wolf teeth**: small teeth that are about the size of an end of a pencil and located in front of the first molar. They most commonly occur on the upper jaw in about 15 percent of horses. They have only one root, so they are easy to extract, but they can cause many problems, including pain. The American Association of Equine Practitioners (AAEP) recommends that a horse have an oral exam every year. Routine maintenance of a horse's teeth

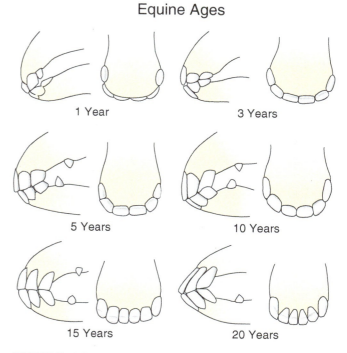

FIGURE 18–41 A horse's age can be determined through an examination of its teeth.

352 SECTION II Veterinary Animal Production

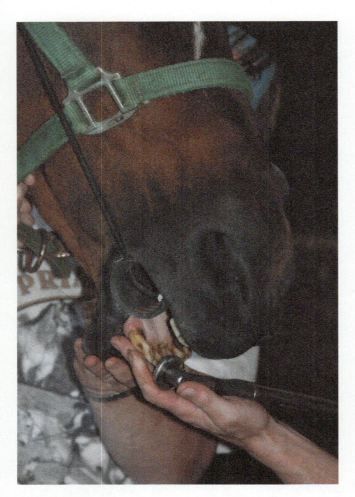

FIGURE 18–42 A float or rasp is used to file a horse's teeth.

has been historically referred to as **floating**. Floating removes the sharp enamel points. **Floating** is the filing of the teeth as they tend to wear down unevenly due to the upper jaw being wider than the lower jaw. A **float** or **rasp** is used to file the teeth and its sharp edges (see Figure 18–42). Sharp edges cause discomfort when a horse is eating or has a bit placed in its mouth. *Occlusal equilibration* is the term now used to describe smoothing enamel points, correcting malocclusion, balancing the dental arcades, and correcting other common dental problems. Some horses require more dental care than others. The following diagram details dental characteristics for horses of different ages.

Hoof Care

A horse will also require routine hoof care (see Figure 18–43). This is typically provided by a **farrier**. A farrier, sometimes called a blacksmith, is a professional who places horseshoes (aluminum or another metal) on the hooves. Hooves need to be

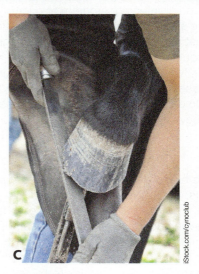

FIGURE 18–43 (A) Trimming a horse's hoof. (B) Using nippers to trim the hoof wall. (C) Using a rasp to shape the hoof.

CHAPTER 18 Equine and Draft Animal Breed Identification and Production Management 353

FIGURE 18–44 A horseshoe.

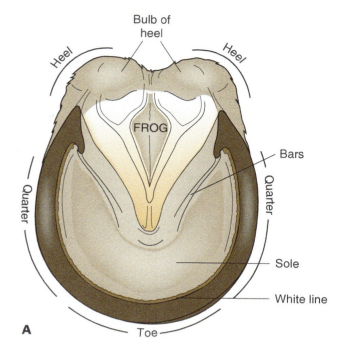

trimmed every 4 to 6 weeks; some horses will require shoes depending on their purpose and use (see Figure 18–44). Hooves grow continuously, like fingernails. They need to be a proper length and shape for the balance and comfort of the horse. The hooves should be cleaned out or picked daily. A **hoof pick** removes dirt and debris from the hoof. Rocks and other objects can become wedged into the hoof and sole and can cause lameness and infection. The **frog** is the soft padded V-shaped area at the causal aspect of the hoof (see Figure 18–45). It should not get too wet or too dry or it can become sore. The **hoof wall**, or outer covering of the hoof, may become dry and begin to crack. These injuries will cause a horse to become unsound.

Castration

It is common for male horses to be castrated to prevent reproduction and allow for ease of handling. Castration is usually done at about 6 months of age. Most castration procedures are done with the horse standing. The incision usually remains open because of the location and size of the testicles. This decreases swelling and allows the surgical site to drain. However, care must be taken to prevent bacteria from entering the incision. Castrations performed in late fall or early spring will help prevent flies from irritating the incision.

Assessment of Lameness

Horses should be evaluated for lameness by a veterinarian. The veterinary assistant may be asked to walk or jog the horse in hand with a lead rope to enable

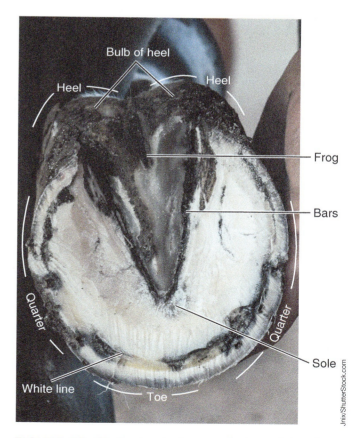

FIGURE 18–45 Anatomical parts of a horse's hoof.

the vet to evaluate the gaits. It is important to know the horse's gaits and natural movements. Many vets will also ask to see the horse on a longe line and under saddle. A horse's movements include the walk, jog, trot, lope, and canter (see Figure 18–46). The **walk** is a relaxed slow four-beat gait. The **jog** is a slow two-beat gait of Western horses. (A Western horse is ridden in ranch or rodeo events using a Western saddle or heavy saddle.) The **trot** is also a two-beat gait with a longer stride or distance between legs and is seen in English horses. (An English horse is ridden in light horse events using an English or light saddle.) The **lope** is a three-beat gait of Western horses, and the **canter** is a three-beat gait in English horses. The **gallop** is a four-beat fast gait or run where each foot hits the ground at a different time. There is a brief time between each stride where all four feet are off the ground.

Vaccinations

The equine vaccine program should begin in foals at about 4 to 5 months of age, with boosters given once a month for two to three series and then yearly as recommended by the veterinarian. The location and weather of the environment may dictate the type of vaccines and schedule. A regular dewormer program should begin in foals shortly after birth and continue

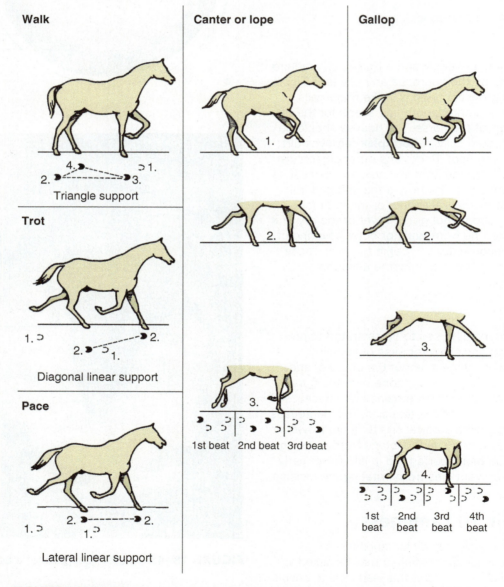

FIGURE 18–46 The basic gaits of a horse: walk, trot, pace, canter, and gallop.

TABLE 18–3
Common Equine Vaccines

VACCINE NAME	FREQUENCY	DISEASE/CONDITION
EEE/WEE/VEE	yearly	Encephalomyelitis
Tetanus	Yearly	Tetanus toxoid
Flu	Every six months	Influenza
Rhino or EHV	Every six months	Rhinopneumonitis Herpesvirus
Rabies	Yearly	Rabies virus
Strangles	Every six months	Strangles intranasal virus
West Nile	Every six months	West Nile virus
Potomac	Yearly to every six months	Potomac horse fever

TABLE 18–4
Common Equine Dewormers

CLASS	TRADE NAME	PARASITE CONTROL
Ivermectin	Zimecterin, Eqvalan, Rotation	Stongyles, stomach worms, pinworms, ascarids, bots, threadworms
Moxidectin	Quest	Stongyles, hairworms, pinworms, ascarids, bots
Fenbendazole	Safe-Guard	Strongyles, pinworms, ascarids
Pyrantel Pamoate	Strongid	Strongyles, pinworms, ascarids
Praziquantel	EquiMax, Zimectrin Gold, Quest Plus	Stongyles, hairworms, pinworms, ascarids, bots, stomach worms, threadworms, tapeworms

monthly until 6 months of age. Adult horses should be on a regular dewormer every 6 to 8 weeks. It is important to rotate wormers so that a horse's immune system does not become immune to the product. This is done by changing the class of wormer with each worming. Tables 18–3 and 18–4 highlight the necessary vaccines and dewormers for horses.

Reproduction and Breeding

Reproduction in horses can be difficult if a person is not experienced and knowledgeable in the reproductive system. Horses have the lowest conception rate of all livestock species. Typically, 50 percent of all horses bred will become pregnant during a breeding season. The breeding season in horses is usually mid-January through early summer. A mare's estrus cycle begins on average between 12 and 15 months of age but can begin as early as 6 months old. Breeding should not begin before 3 to 4 years of age. Horses in quality health can be bred into their late teens or early 20s. The estrus cycle of a mare is every 21 days and occurs from 4 to 10 days in length. Signs of estrus include the following:

- Relaxed vulva
- Increased urination
- "Winking" of the vulva
- Increased vocalization
- Slight clear mucous vaginal discharge
- Tail held high and to the side

A mare that is used for breeding purposes is called a **broodmare**. Broodmares should be kept in shape with adequate amounts of pasture and exercise (see Figure 18–47). Many broodmares are housed together. Broodmares should be examined prior to each breeding for health issues and cultured for possible bacterial infections or diseases that could be transmitted to the stallion. Exams include a **palpation** or feeling of the ovaries for **follicle** development, which determines when eggs will be released for fertilization to occur. This timing is critical for the mare to be bred. Some veterinarians will use **ultrasound**, or the use of radio and sounds waves to

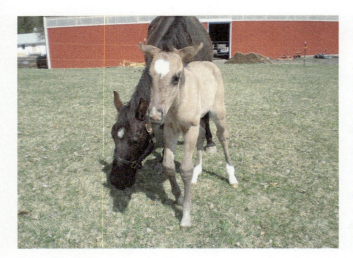

FIGURE 18-47 A broodmare.

view the ovaries for follicle development. Stallions should be evaluated for breeding purposes based on breed quality, ease of handling, and semen quality. Semen should be evaluated for **fertility** to ensure breeding will be successful. The semen should have a high sperm count, good **motility** (movement), and proper **morphology** (correct appearance). Stallions should be kept from other horses due to aggressive tendencies and the potential for accidental breeding.

Breeding can be achieved by several methods. **Live cover** breeding is when the stallion actually breeds with the mare. This may be done through **in-hand breeding**, where the stallion and mare are both restrained to allow natural breeding. This is typically done when the mare is showing signs of estrus and occurs every other day until the mare ovulates and signs of estrus are no longer present. **Pasture breeding** occurs when the mare and stallion are turned out together to allow natural breeding. This is typically over a specific period of time to ensure breeding. **Ovulation** is the release of the egg to unite with the sperm. Breeding can also occur by using **artificial insemination**. This is the process of collecting semen from the stallion and inseminating or placing the semen into the mare at the appropriate time of estrus during the ovulation phase. Artificial insemination must be timed correctly through an experienced reproductive equine veterinarian.

Semen can be shipped through several methods, usually cooled or frozen. **Cooled semen** is shipped in special containers within a 24-hour period and kept cool with ice packs. **Frozen semen** is shipped in special tanks with liquid nitrogen to keep it frozen. The semen is evaluated and prepared for shipping with specific medications that help preserve it during transport. Typically, mares are inseminated once or twice prior to ovulation, so timing must be accurate. The gestation length of mares averages 340 days or 11 months. The process of **parturition** (labor) should occur with a 15- to 30-minute time frame. Broodmares may be kept in a large stall or in a large pasture to foal. Signs of a mare going into labor will begin 12 to 24 hours prior to the parturition process. It is important that mares be kept in a dry and clean area during labor. Some signs will be noted 7 to 10 days prior to the labor process, especially in experienced broodmares. Signs of labor include the following:

- Large distended mammary glands
- Milk present at udder (lactation)
- Restless and anxious
- Stops eating 12 hours prior to labor
- Lays down
- Sweating
- Vulva distention and loosening
- Muscles over back and hind end drop

The labor process begins when the mare's water breaks, releasing the fetal membranes, and a large amount of fluid will be expelled. **Expelled** means released or pushed out, usually within the body. Foaling should occur shortly after the water breaks. The foal is born with normal presentation: the front legs with the feet pointed down are followed by the head and rest of the body (see Figure 18-48). The

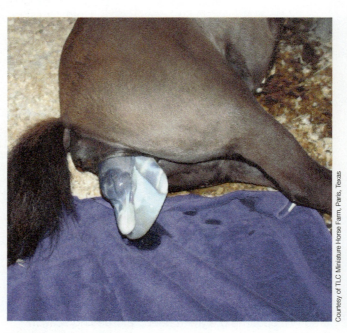

FIGURE 18-48 Normal presentation of the foal at birth.

foal should be monitored to see that breathing begins shortly after birth. The mare will begin to lick and clean the foal by nudging and working at the sac or membrane that encases the foal during development. The membranes are removed, which should stimulate the foal to begin breathing (see Figure 18–49). The mare will continue to lick and clean the foal to dry it (see Figure 18–50). The **placenta** may be attached to the foal or expelled shortly after birth. The placenta is the waste materials of the foal from the gestation period (see Figure 18–51). Some mares will attempt to eat the placenta and afterbirth membranes that contain the foal's waste products. This is a natural instinct to remove the membranes, which may attract predators to the foal. The placenta and membranes should be removed and examined to make sure all tissues have been expelled from the mare. A retained placenta will cause severe infection and should be expelled within hours of the labor process. The foal should rise within 30 minutes to 2 hours after the birth. Their movements are wobbly and take several days to become coordinated. Foals should begin nursing shortly after they stand (see Figure 18–52). The foal's umbilical cord should be treated with **iodine** to prevent bacteria from entering the body and causing an infection; this should be done shortly

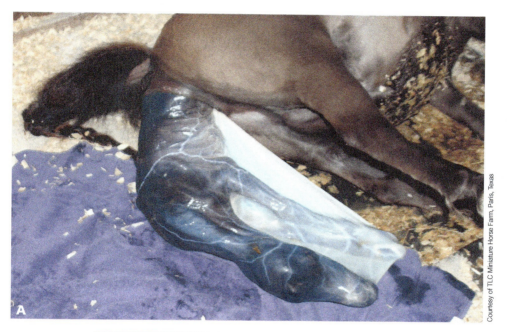

FIGURE 18–49 (A) The amniotic membranes may be pulled away from the foal. The mare will sometimes do this on her own. (B) Part of the amniotic sac on a newborn foal.

358 SECTION II Veterinary Animal Production

FIGURE 18–50 The mare will lick the foal after birth to clean it.

FIGURE 18–52 The foal should begin nursing shortly after birth.

Common Diseases

Horses are known to be fragile animals when it comes to diseases and injuries. They have delicate digestive systems and fragile bones that can be easily damaged. It is important for anyone working with horses to know the signs of pain, lameness, and illness. Horses that become sick can become gravely ill very quickly.

Colic

A common equine medical condition is **colic**: a severe stomach pain that can lead to the death of the horse if not cared for properly. A horse with colic tends to lie down and roll due to the abdominal pain and discomfort (see Figure 18–53). This can cause the stomach and intestines to twist and cut off circulation, leading to the death of the horse. Colic can also cause the horse to become obstructed and **constipated**—not able to have a bowel movement. This will slow down or stop digestion. A horse may develop colic due to many reasons, such as internal parasites, overfeeding or overeating, mold or poor food sources, ingesting too much water, eating or drinking shortly after exercise or being overheated, high fever, hay or protein sources that are too rich, and a sudden change in food. Horses that show signs of colic should be walked to prevent them from lying down and rolling. This is an immediate emergency and a veterinarian should be contacted. Signs of colic include the following:

- Flanking or looking at the abdomen
- Bloated or distended abdomen
- Frequent lying down and standing
- Rolling

FIGURE 18–51 The placenta should be examined to be sure it has been completely expelled.

after birth. When the foal begins to stand, an **enema** consisting of warm soapy water should be given rectally to begin the foal's bowel movements. The mare can be given small amounts of hay and water.

- Kicking at abdomen
- Biting at abdomen
- Sweating
- Restlessness
- Constipation or no bowel movements
- Anorexia (refusing food and water)

Laminitis

Another common condition in horses is **laminitis**, commonly called "**founder**." Laminitis is an inflammation of the lamina within the hoof bone and tissue connecting to the hoof wall. It can be caused by many factors, including overfeeding, rich grain or hay, a rapid change in diet, excessive amounts of cold water, fever or high body temperatures, inflammation of the uterus following labor, and overworking or stress. Laminitis can cause severe lameness and a rotation in the **coffin bone** located within the hoof joint. This can cripple a horse and causes severe pain. A horse that is foundering will

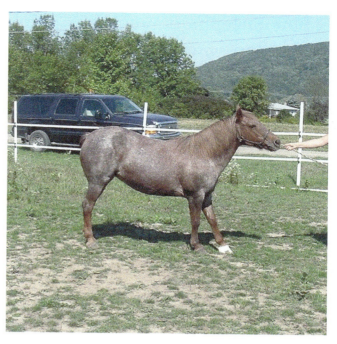

FIGURE 18–54 Laminitis.

show signs of not wanting to move, standing with the front and rear feet close together, pain noted in the hoof area, lameness, fever, and a parked-out appearance (see Figure 18–54). A **parked-out** horse stands with the front legs and back legs spread wide apart, which causes the back to be lowered or sunken in.

Equine Infectious Anemia

Another condition in horses that will lead to death is equine infectious anemia (EIA) or **swamp fever**. This virus is caused by biting insects, such as flies or mosquitoes, and can also be spread by contaminated needles. It is highly contagious, and there is no vaccine and no cure. All horses should be tested yearly by a licensed veterinarian: the test, known as a Coggins test (see page X), requires submitting a blood sample to the state veterinary lab and the U.S. Department of Agriculture. Many states require a negative Coggins test for show and travel purposes. Some horses show no clinical signs and are only carriers of the disease. Horses that test positive for EIA are usually euthanized and the property placed under quarantine. Signs of EIA include the following:

- High fever
- Stiffness
- Weakness
- Weight loss
- Icterus (yellow color of tissues)
- Anemia
- Swelling of limbs

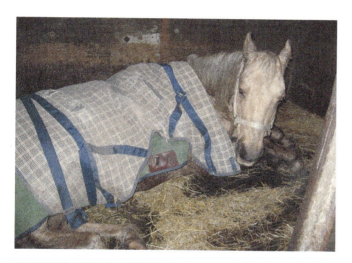

FIGURE 18–53 (A) Colic in a horse lying down. (B) Colic in a horse rolling because of abdominal pain.

Equine Encephalomyelitis

Equine encephalomyelitis, also called equine sleeping sickness, is a virus caused and transmitted by mosquitoes. It can be zoonotic. There are several strains of the virus: Eastern (EEE), Western (WEE), and Venezuelan (VEE). There is a vaccine available but no treatment. Insecticides and pesticides are recommended for insect control. Signs of EEE/WEE/VEE include the following:

- Lethargy
- Sleepy appearance
- Anorexia
- Fever
- Depression
- Local paralysis of lips and bladder
- Unable to swallow

Equine Influenza

Equine **influenza** (flu) is a highly contagious virus. Outbreaks can cause statewide emergencies. A vaccine is available as part of the yearly horse health program. Signs include typical flu symptoms, such as fever, anorexia, depression, rapid breathing, coughing, weakness, and nasal and ocular discharge.

Tetanus

Horses tend to injure themselves easily through wounds, punctures, and cuts. The possibility of infection from bacteria is likely. Bacteria, known as **Clostridium**, are naturally found in horse manure and can survive long periods in the ground. Bacterial infection causes the disease known as **tetanus** or lockjaw. Prevention and sanitation are key, and a vaccination is recommended as part of the yearly horse health program. Booster vaccines are usually given after a severe wound or injury to a horse, especially those involving metal or nails. A horse that develops tetanus will eventually die. Signs of tetanus include stiffness, difficulty chewing, inability to swallow, and reluctance to eat. These signs worsen over time.

Common Parasites

Common internal parasites include roundworms, bots, strongyles, and pinworms. Many of these parasites feed off of the blood in the intestinal tract (see Figure 18–55). The best control method of internal parasites is a regular dewormer, daily stall sanitation and cleaning, rotating pastures, cleaning and disinfecting buckets, and routine yearly fecal exams. External parasites that should be controlled include flies, mosquitoes, mites, ticks, lice, and ringworm fungus. External parasites carry many diseases, some of which are zoonotic. The best control methods for external parasites include insecticide sprays to control biting insects, regular stall cleaning, and regular manure removal.

FIGURE 18–55 Larvae of the horse bot fly that attached to the stomach lining of a horse.

SUMMARY

Horses are mainly used today as recreational and income sources. Many types of horses have been developed; they have become a popular livestock to own, show, and breed. They require exceptional health care, an experienced handler, and a knowledgeable veterinary staff. Routine maintenance and preventative care are the keys to overall good equine health. Veterinary assistants can play a vital role in the care and handling of horses. Draft animals have been developed to pull heavy loads and assist in work tasks such as farming, pulling wagons, and competitive competitions. Draft horses, mules, and donkeys are the more common domesticated species.

Key Terms

artificial insemination   the process of collecting semen from a stallion to insert into a mare for the purposes of breeding

bit   equipment placed in a horse's mouth when it is being ridden

bray   a loud vocal sound that a donkey or mule makes

broodmare   a female horse used for reproduction and breeding purposes

bucking   the lifting and kicking action of the hind limbs

canter   the three-beat running gait of English horses

cecum   a sac with an opening located between the small and large intestine that helps break down roughage and protein

chain shank   a chain placed around the head and halter of a horse for added control

Clostridium   bacteria that causes tetanus and is found naturally in manure and soil

coffin bone   a bone located within the hoof joint that rotates when laminitis occurs

Coggins test   a blood test for equine infectious anemia

colic   severe stomach pain

colt   a young male horse

conditioning   the process of working an animal to build up its body and strengthen the animal for work

constipated   unable to have a bowel movement

cooled semen   semen cooled on ice and shipped in a special container

cribbing   a condition in which horses chew on an object and then suck in air

cross ties   short ropes placed on either side of the horse to make it safe for those working near the animal

cusp   the center of the tooth that wears with age

dam   the female parent of the horse

donkey   a member of the equine family with large ears; also known as a *burro*

draft capacity   the amount of force an animal must exert to move or pull at a constant rate of 1/10 its body weight

draft hitch class   a contest based on the number of animals used to pull a wagon and how well the team works together

draft horse   a large and muscular horse standing over 17 hands in height

electrolytes   powdered supplements placed in food or water to encourage horses to drink more water

enema   warm soapy water used to produce a bowel movement

equine   of or related to a horse

equine encephalomyelitis   a viral disease of horses found in different strains transmitted by mosquitoes; also called *sleeping sickness*

equine infectious anemia (EIA)   a deadly virus in horses caused and transmitted by mosquitoes

Equus caballus   genus and species of the horse

expelled   removed

farming demonstration   draft animals pulling plows to work the land in farming activities

farrier   a professional who trims and shoes horses' hooves; also known as a *blacksmith*

feathers long hair over the lower legs of horses located just above the hoof

fertility the ability to reproduce

filly a young female horse

flaxen blond in hair color

float an instrument used to file down teeth; also called a *rasp*

floating the filing of horse teeth

foal a young newborn horse of either gender

foaling the labor process of horses

follicle development in the ovary that releases the egg

forage hay or pasture source

founder the common term for *laminitis*

frog a V-shaped soft pad in the bottom of the hoof

frozen semen semen kept frozen with liquid nitrogen and shipped in a special container

gallop a fast four-beat running gait in which each foot hits the ground at a different time

gait movement or way a horse moves

gaited horse a horse that moves with a specific rhythm and motion

gelding a castrated male horse

gregarious instinct or behavior to join a herd or group of like animals

halter the equipment applied to the head of the horse for control and handling

hame strap a leather strap that attaches the harness and collar

hand a measurement unit for the height of a horse; equal to 4 inches

hierarchy the place a horse has in the herd

hinny the cross of a male horse and female donkey

hobbles leg restraints that prevent a horse from kicking

hoof pick equipment used to remove rocks and debris from horse hooves

hoof wall the outer covering of the hoof

horsepower a unit of power in an engine or motor that equals 746 watts of electricity or the amount of work done by lifting or moving 550 pounds at a distance of one mile per second

horse pull a contest of power to see which horse can pull the heaviest load

hunter a horse typically used for cross-country fox hunting and jumping smaller fences

influenza the equine flu virus

in-hand breeding a method of holding the mare and stallion for breeding purposes

iodine a chemical disinfectant used to destroy bacteria

jack a male donkey

jenny a female donkey

jog the slow two-beat gait of Western horses

jumper a horse ridden over large fences on a set course

laminitis inflammation of the lamina within the hoof bone and surrounding tissue

lead a rope applied to a halter to handle, control, and move a horse

leading the act or motion of moving or walking a horse

light horse a riding horse that stands between 14.3 and 17 hands

live cover a natural breeding method of the mare and stallion

log pulling a contest to see which team of draft animals can pull the heaviest load of logs

longe line a long lead line used to exercise a horse in circles

lope the three-beat running gait of Western horses

maiden a female horse that has not been bred

mare the adult intact female horse capable of reproduction

mineral block minerals supplied in a block for horses to lick

morphology the form and structure of something

motility movement

mule a cross with a female horse and male donkey

near side the left side of the horse

nonruminant the specialized digestive system of the horse that requires small amounts of food to be digested throughout the day

ovulation the time at which the egg is released from the mare's ovary

palpation feeling

parked out standing with the front and rear legs wide apart

parturition a veterinary term that means the labor process

pasture breeding uncontrolled mating within a flock or herd

placenta membranes that contain waste materials of a foal

points sharp edges or corners of teeth

pony a small equine under 14.2 hands in height typically ridden by children

racehorse a horse bred for speed and racing purposes

rasp an instrument used to file down horse teeth; also called a *float*

rearing rising up on the hind legs

riding horse a horse ridden for show or recreational purposes

salt block salt provided in a block for horses to lick

sire the male parent of a horse

slip knot a way of tying that has a quick release

spooking a startling action

stallion an adult intact male horse capable of reproduction

stock horse a strong, athletic horse well suited for work

stud an adult male intact horse of breeding age

swamp fever a common name for equine infectious anemia

tail tying a method of tying the tail for the purposes of keeping it out of the way or to move a sedated horse

trot the slow two-beat gait of English horses

tetanus a bacterial condition known as *lockjaw*

twitch a chain or rope loop placed over the lip of a horse as a restraint

ultrasound the use of using the reflections of high-frequency sound waves to construct an image of a body organ

walk a relaxed, slow four-beat gait

warm-blooded able to control body temperature internally

wolf teeth small teeth that may develop in front of the first molars

Review Questions

1. What does the term "hand" mean in relationship to a horse?
2. List and describe the various classes of horses.
3. What are some normal behaviors of horses?
4. Describe the horse's digestive system.
5. What are pieces of equipment that can be used to restrain a horse?
6. What methods can be used to breed a horse?
7. How can a horse's teeth determine its age?
8. What are the signs of colic in a horse?
9. List and describe the movements of a horse.
10. What is EIA? What is the testing procedure for EIA?

Clinical Situation

Emily, a veterinary assistant at Lineville Equine Center, is accompanying Dr. Reese on a farm call visit. The particular horse is not eating; has been showing signs of not having bowel movements; and has been lying down and rolling in its stall, in noticeable pain. This has been going on for several hours today. The owner is not sure if he can pay for the treatment, so he has been resistant to having the horse examined. Finally, the owner agrees to an exam.

Dr. Reese attempts to explain how important it is that the horse receives care. "Sir, the horse is in a lot of pain and really should be seen at the clinic for possible surgery. At least allow me to give him some pain medication to make him more comfortable," he says.

"Well, I don't really have any money to spend on him. If he's really that bad, then I guess he should be put down."

Dr. Reese and Emily discuss the situation and possible causes for the horse's pain. The owner allows only a minimal exam of the horse and is not willing to give much background history.

- How would you handle the situation as a veterinary assistant?
- What are some likely issues with the horse's health?
- What are some options to give the horse owner?

CHAPTER 19: Swine Breed Identification and Production Management

Objectives

Upon completion of this chapter, the reader should be able to:

- **19.1** Define common veterinary terms pertaining to swine
- **19.2** Explain swine biology
- **19.3** Identify common breeds of swine
- **19.4** Discuss the nutritional needs of swine
- **19.5** Discuss normal swine behavior
- **19.6** Explain how to practice proper swine restraint methods and handling safety
- **19.7** Discuss the vaccine program of swine
- **19.8** Discuss swine reproduction factors and methods
- **19.9** Describe common health problems and diseases of swine
- **19.10** Discuss the swine production industry
- **19.11** Discuss swine selection factors
- **19.12** Discuss swine body types
- **19.13** Describe swine production methods
- **19.14** Discuss common health practices used in the swine industry

Introduction

Pigs, *hogs*, and *swine* are all terms for the popular animal that provides humans with the meat known as *pork*. Swine have become a popular production animal due to their ability to reproduce at a high rate and increase their body weight rapidly. Swine production requires low amounts of labor and has a high investment and profit rate. The swine industry is governed by strict regulations in how pigs are housed and cared for due to the use of the meat for human consumption. This means the care of swine plays an important role in the veterinary industry.

Veterinary Terminology

Swine, or **porcine**, is the veterinary term for pigs and hogs. The female pig is a **sow**; the male a **boar**. The labor process in swine is known as **farrowing**. Young pigs are **piglets**. A **barrow** is a young castrated male pig. A male pig castrated after maturity is a **stag**. A female pig that has not yet been bred is a **gilt**. The amount of meat produced by one pig is known as the **dressing**.

Biology

Swine have a system of organs that is similar to other animals and a monogastric digestive system similar to dogs, cats, and humans. Their food must be more concentrated. They have **cloven-shaped** hooves, which means there is a split in the toe of the hoof that separates the hoof into two parts. Most swine breeds weigh between 220 and 240 pounds, but some bacon breeds can top 400 pounds. Swine used for food products are usually young pigs, as the meat of older hogs, especially the boars, has a strong flavor. Most hogs in meat production systems yield a dressing of 65 to 80 percent of their body weight. Pigs should farrow about 7 to 12 piglets a year and have a gestation length of 114 days. The anatomical features of the swine are illustrated in Figure 19–1.

Breeds

Several swine breeds are found throughout the United States (see Figure 19–2). Swine provide pork and other meat products for human consumption. There are many different types of swine: they are classified as meat-type or bacon-type producers. Many breeds are now being developed through hybrid breeding or crossbreeding two or more purebreds based on

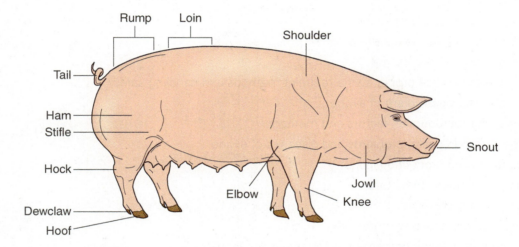

FIGURE 19–1 External anatomy of the swine.

specific traits. The conformation of the breed is the most critical factor. Swine breeds vary in size from 50 pounds, such as the Vietnamese potbellied pig, to several hundred pounds, such as the Yorkshire or Landrace.

Breed Selection

Swine have several requirements for ideal production systems that are based on the overall appearance and profitability. There are two types of swine: the **meat-type hog** and the **bacon-type hog**. Meat-type hogs have a large amount of meat, known as ham, on the body. Bacon-type hogs have a large amount of fat, known as bacon, on the body. Both types grow fast. The meat-type hog is muscular; the producers use the cuts of meat as ham, loins, and roasts that are high in cost. These meat cuts have low amounts of fat. Bacon-type hogs are fatter than muscle-based meat, such as bacon and sausage. Producers want to select a hog that meets the ideal type of swine for the product produced. Most producers will select a hog that has a verified pedigree and will measure the amount of meatiness on the pig's body. There are several ways to measure body fat versus body muscle. The probe is a tool used to measure the thickness of back fat, which is found over the mid-back and loin area. A lean meter measures the difference in fat versus muscle thickness over areas of the body. Both fat and muscle can be measured with an ultrasound in which the sound waves travel through the tissues and reflect the pulses of the hog. Registered breeds must meet the meat certification program guidelines of a production program. Sow selection factors include selecting gilts from large litters that have no genetic flaws and have back-fat measurements of less than 1.2 inches of fat. Figure 19–3 illustrates the grading of hogs.

Sows should be reevaluated when they reach between 150 and 200 pounds to ensure that they are growing adequately and will be excellent producers for breeding or meat production programs. How well the sow gains weight will determine if it may be replaced or put into a reproduction program.

Boar selection is based on temperament and body structure. Boars should be selected from large litters, have a medium to large body structure, should reach adult size by 155 days of age or less, and have less than one inch of back fat. Ideal weight gain for breeding boars is 2 pounds of body weight gain per day. It is necessary for reproduction programs to have one boar for every 15 to 20 sows. The boar should enter the breeding program between 6 and 7 months of age and begin breeding sows around 8 months of age.

FIGURE 19–2 (A) Saddleback. (B) Large black. (C) American Landrace.

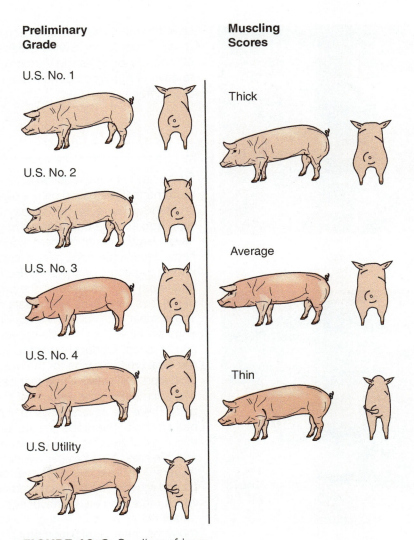

FIGURE 19–3 Grading of hogs.

Nutrition

Swine nutrition is based on life stage and production type. Mature boars are usually fed similarly to pregnant sows; that is, about 4 to 5 pounds of feed per day. Young boars under a year of age may need more nutrients than adults. Piglets that have been weaned can be placed on a starter food. Growing pigs will generally need more nutrients based on their intended type of production. Breeding sows should be fed increased amounts in the last 2 weeks of the gestation phase, typically 4 to 5 pounds of feed per day. Many producers feed *ad lib*, as pigs will not overeat. The increased nutrients help stimulate milk production. Food should slowly be decreased for the sow about three days after farrowing for a short period to prevent digestive problems. Sows can be managed on a lactation diet begun 2 weeks after farrowing that consists of 10 to 12 pounds of feed per day to maximize milk production. Many commercial feeds are balanced for the life stage of the pig.

Nutrition for general maintenance is classified by the production method as well as the size and life stage of the pig. Pigs that have been weaned and are between 40 and 125 pounds require a 16 to 18 percent protein diet. Pigs over 125 pounds require a 14 to 16 percent protein food source. Most producers feed a manufactured food source that meets the needs of their production system. Many are corn and soybean meal based. Supplements may be added as necessary. Swine that are used as meat sources for human consumption as well as breeding systems are typically fed a high-energy diet of fats and carbohydrates. The biggest source is cereal grains, such as corn, barley, and wheat (see Figure 19–4). Pigs do not eat roughage; it does not provide adequate sources of energy. Protein is an important part of a swine's diet: it is necessary for muscle development, fat development, and tissue quality and builds and repairs the body's cells. Protein is broken down in the stomach into amino acids that further help in muscle development. The most common protein is soybean meal. A corn-based diet requires protein supplements of 35 to 40 percent, as corn is primarily fat (see Table 19–1).

Swine are the most common livestock species to suffer a mineral deficiency. They need large amounts of salt, calcium, and phosphorous minerals. Copper, iodine, iron, zinc, and selenium are needed in small amounts. Vitamins A, D, E, and B_{12} are usually supplemented in the food source as well. Water is essential for the health of the pig, but the requirements

TABLE 19–1

Protein and Diets

PROTEIN DIET SOURCES	PERCENTAGE OF PROTEIN
Growth diet	12–16 percent protein
Finishing diet	12–16 percent protein
Boar breeding diet	13–16 percent protein
Lactation diet	13–16 percent protein

TABLE 19–2

Water Requirements

CLASS OF PIG	WATER INTAKE (GALLONS/PIG/DAY)
Sow and litter	8
Nursery pig	1
Growing pig	3
Finishing pig	4
Gestating sow	5
Lactating sow	7
Boar	5

are less than for most other types of livestock. Pigs need ¼ to ⅓ of a gallon of water per pound of dry feed (see Table 19–2). A 60-pound pig needs ¾ gallon of water daily. A 60- to 100-pound pig should drink about 2.5 gallons per day. A 100- to 250-pound pig needs 4 gallons per day. A mature boar and a pregnant sow typically drink about 5 gallons per day. A lactating sow needs 7 gallons a day, as milk consists mostly of water.

Behavior

Pigs are generally stubborn, contrary, and vocal; however, they are also very intelligent. They are also known for being aggressive, especially if a sow has piglets or a boar is present. These various characteristics make pigs one of the most difficult animals to handle and restrain. Pigs have poor eyesight and are easily scared, and their legs are susceptible to injury if handled improperly. It is important to move slowly and calmly when working around swine.

A pig's defense is its teeth. Swine—even piglets—have fine, needle-sharp teeth (see Figure 19–5). A pig's neck and shoulder muscles are very strong, and it can use them to push and pin people against objects. Larger pigs can be dangerous because of their body weight and size. Pigs will chase people if they feel threatened or cornered.

FIGURE 19–4 Hogs are fed grains such as (A) corn, (B) wheat, and (C) oats.

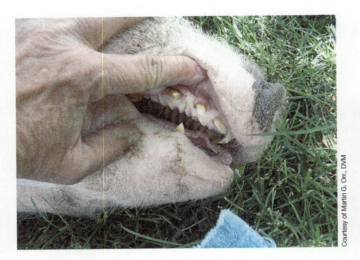

FIGURE 19–5 Swine have very sharp teeth.

Basic Training

Specialized equipment that protects a handler's lower body can be used to train pigs to move from location to location. Handlers can herd or push pigs to certain areas. The more often swine are handled by humans, especially at a young age, the more accepting they will be of handling.

Equipment and Housing Needs

Swine are usually kept in large groups. Housing needs are based on the production system. Gestation facilities house sows for breeding. Many producers are using free-access gestation housing, which allows the sows to move about and socialize and access an individual stall when they want to eat or rest (see Figure 19–6).

Sows are moved to farrowing facilities shortly before the birth of the piglets. Farrowing usually takes place in pens or crates (see Figure 19–7). During the pigs' growing period (weaning to 100 pounds), they may be

FIGURE 19–6 Free-access gestation stalls allow sows to gather in a socialization area and enter and exit individual stalls as desired.

FIGURE 19–7 (A) Farrowing crates are used when sows give birth. (B) Piglets nursing in a farrowing crate.

FIGURE 19–8 A modern swine finishing facility.

housed on pastures or in permanent facilities. Pigs over 100 pounds will be housed in finishing facilities, often in confinement housing or on pasture. Confinement housing is temperature and moisture controlled (see Figure 19–8).

Restraint and Handling

The principles of swine restraint are not different from those of other large animal species. The method used to handle pigs is determined by the size and age of the animal and the nature of the procedure being performed. As stated previously, swine are intelligent animals that may be stubborn to handle. They do not have a strong herd instinct but prefer to be with other pigs rather than being alone. They become very vocal when angry, stressed, or scared. Some pigs have large amounts of fat that—when they are stressed or handled in high temperatures—can cause overheating. People must learn to move slowly and deliberately when handling pigs—gentle handling and talking in a soft voice are necessary. Pigs, even piglets, can bite and cause serious injury with their teeth. They can squeeze easily through openings and escape. It is best to move pigs using a hurdle, a flat panel board that serves as a wall and is kept between the handler and the pig (see Figure 19–9). A **hurdle** is usually a solid piece of wood, plastic, or metal that is large enough to cover the legs of the handler. A **paddle** with a long handle (that acts as an extension of the handler's arm) can also be used.

Pigs that weigh less than 50 pounds can be restrained for exams, vaccines, or medication administration by lifting them by the rear feet and placing the head between the legs for control. The front legs can touch the ground to support the body (see Figure 19–10). To catch a pig, it is best to grasp one rear leg. Piglets that weigh less than 5 pounds can be handled in the same way as a dog. Quickly place one hand and arm under the piglet's chest and the other over its shoulders. Lift while supporting and controlling the head and back end. Pigs must be held firmly but gently. Pigs will likely squeal and become vocal. This can be dangerous when a sow with piglets is nearby,

FIGURE 19–9 Use of a hog hurdle.

FIGURE 19–10 Manual pig restraint technique.

FIGURE 19–11 Use of a hog snare.

- Young piglets should be isolated from the sow when veterinary attention is necessary.
- Use a panel to protect your lower body when moving swine.
- Pigs will bite when threatened or scared and typically go for the legs and feet.

Grooming

Pigs do not routinely need grooming. They tend to bathe themselves in mud and water sources to keep their skin moist. Show pigs may require bathing. The hooves of swine may need routine trimming and care. Teeth may need occasional trimming if sharp edges appear.

Basic Health Care and Maintenance

Swine producers typically raise pigs in aseptic and controlled environments to be disease free. Herds that are disease free and in excellent health are known as **specific-pathogen free (SPF)**. A *pathogen* is a specific causative agent (such as a bacterium or virus) of disease. Each pig should be disease free at birth; be raised in a sterile environment; and have no physical exam findings or internal or external parasites, swine diseases, or snout distortion (poorly shaped).

as the sow may become aggressive. In these cases, it is best to separate piglets and sows when handling. Restraint equipment includes a **hog snare**, **snubbing rope**, or **trough**. The hog snare, the restraint tool of choice, is a long pipe with a cable loop attached to the end. The loop is thrown over the upper jaw of the pig and tightened around the snout (see Figure 19–11). The snare is kept elevated above the pig to control the animal for 20- to 30-minute periods for blood collection, exams, and injections. A snubbing rope can be used to tie a pig. The rope is used in the same way as the hog snare, over the snout. The rope can be tied to a post for similar restraint. Smaller pigs can be restrained using a trough, a V-shaped table that holds the animal on its back for surgery and other procedures. A rope may be used to keep the pig still.

Swine Behavior and Handling

Tips on swine behavior to consider while handling, restraint, or general work within pastures or stall areas include the following:

- Sows are very protective of their young.
- Moving pigs is best done with panels or chutes.

Tail Docking

Tail docking is a common practice carried out shortly after piglets are born. Its purpose is to control feces from collecting on the tail and maintain proper sanitation as well as prevent piglets from biting each other's tails and causing serious injury. (Pigs that are confined may engage in tail biting.) The tail is cut about one inch or less from the tail bone (see Figure 19–12). Tails are docked when the newborn's teeth are trimmed and the umbilical cord is treated.

Ear Notching

A simple way to identify breeding stock and replacement pigs is **ear notching**: the practice of making a small cut into the ear in different locations to mark the pig and litter number (see Figure 19–13). A **V-notcher** is the tool used to make the notches. Identification is especially necessary in large herds. The right ear identifies the litter number; the left identifies the pig number. Each ear notch position has a specific value. Other methods of identification include ear tags, tattoos, and branding. However, ear tags tear easily and tattoos and brands are difficult to read on many breeds of pigs.

Castration

Male swine that are not used for breeding purposes are typically castrated. Those raised for meat are usually castrated before weaning. After the castration procedure, the pigs should be kept in a clean and sanitary environment. This practice is commonly carried out in the late fall or winter to prevent fly contamination.

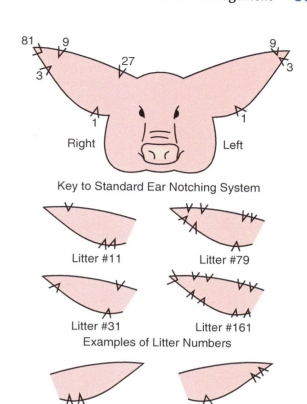

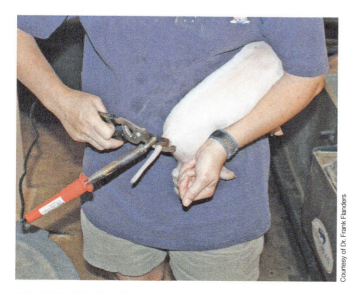

FIGURE 19–12 Tail docking.

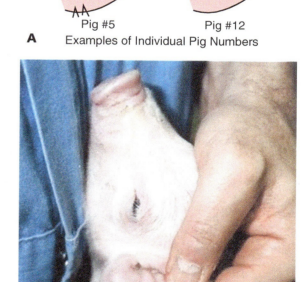

FIGURE 19–13 (A) Ear notching identification in swine. (B) Pig with notched ears.

Additives

Swine management includes feed and medication additives to meet the health needs of a pig. This is a standard practice for ideal health that is not related to nutrition and may include antibiotics, anthelmintics, or hormones. Antibiotics may be used to increase weight by 10 percent and help decrease the feed consumption by 15 percent. A growth hormone frequently used in swine production is **porcine somatotropin (pST)**. This hormone is used to increase protein synthesis that helps the body increase food use by 20 to 30 percent. The average daily weight gain is 15 to 20 percent of body weight. This increases the muscle mass by 10 to 15 percent. Additives are not fed for a specific time prior to slaughter. The term *slaughter* means to kill humanely for the meat source. The removal of additives is called **withdrawal time**.

There is a specific time that each medication or hormone is not to be used in the body to ensure that the meat will be drug free. The label of each additive states the necessary withdrawal time, which is regulated by law. Product withdrawal times range from 2 to 70 days, depending on the product. Injections are commonly given by intramuscular (IM) and subcutaneous (SQ) routes. The IM injection should be given into the front shoulder just behind and below the ear (see Figure 19–14). The ham and loin areas should *never* be used for injections. The SQ injection can be given into loose flaps of skin over the flank or behind the elbow. Injections should be given only into clean and dry locations.

Trimming Needle Teeth

Pigs are born with eight **needle teeth**, which are very long and sharp (like needles). Two teeth are located on each side of the upper and lower jaw. The teeth serve no purpose and can cause severe injury to the sow during nursing. The teeth are trimmed with forceps or nippers shortly after birth at the same time the umbilical cord is treated with iodine. Forceps or nippers are instruments that are used to grasp teeth and trim them short or completely remove them from the mouth (see Figure 19–15).

Vaccinations

Swine vaccine programs vary with location and high incidence of diseases. Swine are typically vaccinated for necessary intestinal, respiratory, and reproductive diseases. Swine are commonly vaccinated for protection from the following conditions:

- Pneumonia
- Rhinitis (inflammation of the nose)
- Flu
- E. coli
- Salmonella
- Leptospirosis
- Pseudo rabies
- Brucellosis
- Parvovirus

Swine vaccine programs begin when piglets receive an E. coli vaccine within 12 hours after birth. They should also have an iron injection for nutrition at this time. At 2 to 3 weeks of age, they should receive a vaccine for

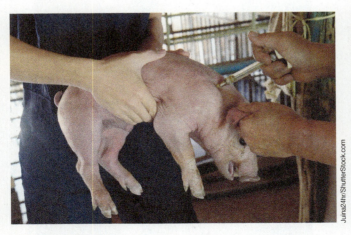

FIGURE 19–14 Location of an (intramuscular) IM injection.

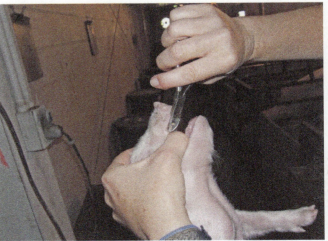

FIGURE 19–15 Trimming of the needle teeth.

pneumonia, salmonella, and flu. Piglets begin to produce natural antibodies between 5 and 6 weeks of age. At 7 to 8 weeks of age, all necessary vaccines should be given depending on the veterinarian's recommendation. Booster vaccines should be given within 30 days and then revaccinated yearly. All breeding pigs should be vaccinated 4 to 6 weeks prior to breeding and again in 3 to 4 weeks. Sows should be vaccinated 2 weeks prior to farrowing and again 2 weeks after farrowing.

Reproduction and Breeding

Swine reach puberty between 4 and 8 months of age. Gilts usually begin breeding at about 11 or 12 months of age, and boars at about 8 months old. Gilts and sows are in estrus from 1 to 5 days, with 2 to 3 days the average. Signs of estrus in swine include the following:

- Restlessness
- Mounting other pigs
- Vaginal swelling
- Vaginal discharge
- Increased urination
- Increased vocalization

Gestation occurs for approximately 114 days, or 3 months, 3 weeks, and 3 days. Swine are either purebreds or crossbreeds. Crossbreeding is a favorable method, as producers can develop an ideal type of hog. This allows for heterosis or breeding lines that outperform the parents. Crossbreeding also allows for increased production and increased profits. Over 90 percent of breeding operations are crossbreeds. Pigs may be bred naturally or through the use of artificial insemination; breeding naturally is the most common method. Breeding crates may be used to reduce fighting and injuries. Sows are checked for pregnancy between 30 and 45 days after mating. Ultrasounds may be done to determine pregnancy and are about 85 percent accurate. The sows are kept in confinement in individual pens, usually with hands-on care. This makes them easier to manage, especially after labor. Sows should be monitored closely during the labor process, as they tend to have difficulties during birth. Sows that tend to have difficulty with farrowing may be induced into labor at days 111 to 113. Farrowing should then occur within 18 to 36 hours. Only about 70 percent of piglets reach weaning, due to disease, difficulty during farrowing, poor nutrition, or injuries by the sow. Sows should be placed in an area to farrow at about day 110. Farrowing or gestation crates may be used to decrease injuries to the piglets. The sow and piglets may be kept in crates for several weeks until the piglets are large enough to avoid being injured. Signs of labor in swine include the following:

- Nervousness
- Anxiousness
- Restlessness
- Swollen vulva
- Vulva discharge
- Nesting
- Lactation

Sows that are kept in a nonsterile environment may be bathed prior to farrowing. Nesting material should include wheat, barley, rye, or oat straw. Chopped hay or corncobs are also ideal for gestation crates. The area should be disinfected daily. After the sow has farrowed and the piglets have been removed, the area should be sanitized and left empty for 5 to 7 days. Sows and piglets are sensitive to weather and should be housed in areas with adequate protection and ventilation: ideal temperatures are 60 to 70 degrees Fahrenheit. In winter or cold weather, heat lamps or hot pads may be necessary to keep piglets warm (see Figure 19–16). Boars should be kept in individual areas and have plenty of exercise. They should be de-tusked to prevent injury to each other and humans and should be handled regularly so they become easy to manage. It is important to keep reproductive records to determine if breeding problems may exist.

Common Diseases

Swine are highly susceptible to diseases and parasites and thus typically raised in a controlled environment when produced for human consumption. Pigs

FIGURE 19–16 Heat lamps may be used to keep piglets warm.

are also easily stressed. Stress can be brought on by transport, overcrowding, disease, or as a genetic factor.

Porcine Stress Syndrome

Pigs are prone to a condition known as **porcine stress syndrome (PSS)**. This is a nonpathogenic disorder, meaning it is not a disease that is spread but rather a condition that occurs due to the physical conditions and environment of an animal. PSS occurs most often in animals that are heavily muscled; it is usually genetic or spread through breeding lines. The stress condition can cause sudden death.

Brucellosis

Brucellosis, or **Bang's disease**, is a concern among swine breeders. This reproductive disease causes a virus known as "**hog cholera**," which is highly contagious and has no treatment. Pigs affected by this disease must be destroyed, as they are no longer capable of breeding and their meat will cause contamination that is zoonotic to humans. Signs of this virus include fever, weakness, loss of appetite (anorexia), and increased thirst.

Leptospirosis

Another disease of concern to the swine industry is **leptospirosis**, a bacterial condition spread by urine contamination. Signs of the disease include fever, poor appetite, and blood in the urine, or **hematuria**. *Hemato–* is the root word for blood, while *–uria* means pertaining to the urine. A vaccine is available for this disease, and antibiotics may be helpful. Breeders and producers will be affected if this disease is found in their program. Female sows with leptospirosis end to abort if pregnant.

Pneumonia

Swine are prone to a bacterial condition known as **pneumonia**. This condition affects the lungs and causes a secondary condition to the respiratory tract and lungs through lesions, or areas of the tissue that are damaged. The lesions cause a chronic cough. The term *chronic* means occurring long term; *acute* means occurring suddenly or for a short time. There are no control methods for the spread of this bacterium in a swine program, but antibiotics may be helpful to control or prevent a secondary infection. Lungworms and lungworm larvae may also cause lesions to form in the lungs.

Pseudo Rabies

Swine may show clinical signs or be carriers of the disease. A **carrier** may not show symptoms of the disease but is still contagious and capable of passing it along. Young nursing pigs with **pseudo rabies** may show signs of fever, nasal discharge, paralysis, or become comatose. Adults may show similar signs and pregnant sows may also abort or have stillbirths. Nasal and oral secretions spread the disease. A vaccine is available.

Swine Dysentery

Another common condition in pigs is **swine dysentery**, or *bloody scours*. This infectious bacterial disease is spread in fecal matter that is typically ingested by young pigs. Pigs tend to get infected in the late summer and early fall. This condition is hard to control and requires quality contamination methods as well as proper antibiotic control. Signs of the disease include soft and watery feces, bloody diarrhea, loss of appetite, and a slight fever.

Common Parasites

Parasites—both internal and external—are of concern in swine programs. They can be caused by soil contamination, poor sanitation practices, food contamination, and poor health practices. Parasites are more likely found in a pasture situation versus an aseptic controlled environment. Roundworms and tapeworms are the most common internal swine parasites. Regular deworming and sanitation methods are required. Signs of internal parasites include the following:

- Weight loss
- Poor skin and body condition
- Anemia
- Diarrhea

Common external parasites include lice and mites. Lice are common and spread rapidly in closely confined pig herds. Lice are to be suspected in pigs that commonly rub on trees, posts, or fences. Most cases occur in winter and can cause severe itching, skin lesions, and anemia. **Mange dermatitis** is a condition caused by mites. Mange mites cause itching and hair loss over the head and neck and spread rapidly over the body, from pig to pig. If not treated, severe skin infections can cause death. Back rubbers can be provided, along with the use of insecticides. Dipping and high-pressure sprays can also be used to treat and prevent external parasites.

Small herds that are housed outside are best kept at a ratio of 10 pigs per acre. Sows with piglets should be kept at seven pigs per acre. Boars should be isolated on one-quarter acre per pig. Fencing—made of strong wire and at least 3 feet high—should be used to keep pigs contained safely. Shelter must be provided so pigs can escape the weather. Ventilation is important to control temperatures, reduce disease, control odor, and remove moisture. Manure disposal is important in order to uphold government regulations on avoiding pollution and maintaining good health practices. Pigs housed in controlled environments need 2.25 square feet of space per 30 pounds of body weight.

Swine Production Industry

Swine domestication began thousands of years ago and was introduced to the United States by Christopher Columbus in 1493. Wild hogs, known as *wild boars*, were already found in North America and were hunted as a meat source. Domestic swine were bred with each other as well as the wild boars to create new breeds. Many of today's swine breeds are the result of crossbreeding two or more pure breeds of swine. Swine are used as meat producers for either pork or bacon sources. Pork is sometimes referred to as the "other white meat" and is a tasty and tender type of meat. Today there are about 100 million pigs in the United States. Swine production is the second-largest livestock production industry in North America. Pigs are known for rapidly converting the food they eat to meat and muscle. Compared to other livestock species, pigs can eat smaller amounts of food and convert and produce larger amounts of meat. Fewer than 5 pounds of food can produce 1 pound of pork. A lot of technology and research has been put into the swine industry, thus highlighting the importance of swine production in veterinary medicine.

Corn is an industry that parallels that of swine production. The "Corn Belt" is the area in the Midwest where corn yield is high, the price is low, and the swine industry is growing. The top-producing swine states in the country are in the Corn Belt and include Illinois, Indiana, Iowa, Minnesota, Missouri, Nebraska, and North Carolina. More than 75 percent of the swine produced in the United States are raised in confinement. Many producers are contracted by large companies or corporations that process their own meat. There are many factors in the swine industry that make it a favorable business. Swine are efficient in converting their food to meat. Swine are prolific, which means they are capable of producing a large amount of young in a short time frame. Labor requirements are low because swine are self-sufficient. The startup costs for swine programs are low because they require less land and space depending on the type of production method. A profit is made in a relatively short amount of time, usually less than 10 months. Although many factors make swine production a favorable business, other unfavorable factors should also be considered:

- Susceptibility to disease and parasites
- The need for large amounts of concentrates to be added to food
- The need for close monitoring during labor
- Economic issues and conditions
- Waste management laws and regulations
- Production permit requirements
- High cost of swine feed; swine do not eat pasture or forage sources

Swine Production Methods and Systems

Most swine producers have become corporations, from small farms to large factory farms. They are contracted by companies, usually markets or grocery stores, for their meat sources. A *contract* in this case is an agreement in writing between the producer and the buyer that lists the breed of swine, type of production system, and price of each hog. This is typically arranged with feeder pigs or market hogs. Because of the susceptibility of disease and parasites, most swine are raised in isolation, with limited human contact. All employees must decontaminate their clothing and equipment and be sterile before entering the swine's environment. This includes changing clothes and sanitizing footwear.

Feeder Pig Production Systems

Feeder pig production systems are typically used for breeding purposes as well as meat market sources. A herd of brood sows is bred and farrowed. The litter is raised to 40 pounds at weaning and sold to continue feeding until market size. There are boars used for breeding usually through artificial insemination to continue an aseptic environment.

Finishing System

The **finishing system** includes feeder pigs that are fed until market weight. Market size is typically 200 to 240 pounds based on breed and hog type. The ideal outcome of this system is the best weight gain at a minimal cost. This is the most common type of swine production method.

Farrow to Finish Systems

Farrow to finish systems are less specialized production systems. The sows are bred and farrowed out and the piglets are raised to market weight on the same farm (see Figure 19–17). This system requires multiple housing areas for breeding and feeding.

FIGURE 19–17 Farrow to finish systems raise piglets to adults.

SUMMARY

Swine production is important to the veterinary industry as well as to consumers and producers. Meat production is a large part of the swine industry and is regulated by the government; therefore, health care practices and a controlled environment are essential for proper swine management. More than 60 percent of all swine production systems are in controlled and aseptic environments. Veterinary assistants play a large role in the care of and sanitation for large herd operations and programs. Knowledge and experience in restraint and handling, management practices, and sanitation are essential for proper swine production.

Key Terms

bacon-type hog   a hog raised for the large amount of meat, known as bacon, on their body

Bang's disease   see *brucellosis*

barrow   a young castrated male pig

boar   an adult male pig of reproductive age

brucellosis   a highly contagious reproductive disease for which there is no treatment; also known as *Bang's disease*

carrier   an animal that is infected by a disease but shows no signs

cloven shaped   a split in the toe of the hoof that separates the hoof into two parts

dressing   the amount of meat produced on one pig

ear notching   the practice of making a small cut on the ears to identify the pig number and litter number

farrowing   the labor process of pigs

farrow to finish systems   a production system in which sows farrow out and the piglets are raised to market weight

finishing system   a production system where pigs are fed to market size

gilt   a young female pig that has not yet been bred

hematuria blood in the urine

hog cholera the virus that causes brucellosis

hog snare a metal pole with a loop on one end used to restrain a pig by putting pressure over the snout

hurdle wood, plastic, or metal board used to direct and move pigs

leptospirosis a bacterial condition spread by urine contamination

mange dermatitis a condition caused by mange mites that cause itching and hair loss over the head and neck and that can lead to severe skin infection

meat-type hog a hog raised for the large amount of meat, known as ham, on their body

needle teeth eight long, sharp, needle-like teeth located on each side of the upper and lower jaw

paddle a long-handled board similar to a hurdle that acts as an extension of the hand to move pigs

piglet a newborn pig

pneumonia an acute disease that is marked by inflammation of lung tissue

porcine a veterinary term for pigs, hogs, and swine

porcine somatotropin (pST) a hormone used to increase protein synthesis and food use that produces weight gain

porcine stress syndrome (PSS) a condition affecting heavily muscled pigs that can result from stress and may cause death

pseudo rabies a condition caused by a virus that has similar symptoms to rabies and is spread through oral and nasal secretions

snubbing rope a rope placed over the snout and tied to a post for the purposes of restraint

sow an adult female pig of reproductive age

specific-pathogen free (SPF) disease free at birth

stag a male pig castrated after maturity

swine pigs or hogs

swine dysentery a condition in pigs that causes diarrhea and spreads rapidly through contamination

tail docking cutting a pig's tail short to prevent chewing and injury to the tail and to keep it free of feces

trough a V-shaped table that holds the animal on its back for surgery and other procedures

V-notcher a tool used to make cuts in a pig's ear

withdrawal time a period of time before slaughter when pigs are given no additives

Review Questions

1. What are the different types of swine?
2. What are common restraint methods used with swine?
3. What are factors used in selecting swine?
4. How should a sow be managed during farrowing?
5. Describe the nutritional requirements of swine.

6. What are the production system types in swine programs?
7. What is the importance of swine production?
8. What management practices are necessary with young pigs?
9. What is the difference between a barrow and a stag?
10. What are the requirements of an SPF herd?

Clinical Situation

Mr. Townson, a local dairy farmer, has decided to purchase several Yorkshire piglets. He has phoned the Davidson Large Animal Clinic and is talking to Devin, the veterinary assistant.

"I've just acquired 10 Yorkshire piglets and the sow. I've never raised pigs before and don't know a lot about them. Normally, Dr. Richards comes out to check my cattle herd every 2 to 3 months. I guess I need to know what care is needed for the pigs."

"First, Mr. Townson, we should schedule an exam for the sow and the piglets. They may need vaccines and some routine health care. Dr. Richards is available this Friday afternoon. How is that day for you?"

"That is fine. Thank you."

- What other information could Devin have given Mr. Townson?
- What production factors should be considered in this situation?
- What are some basic areas that Dr. Richards will probably have to address with the sow and piglets?

Chapter 20: Sheep Breed Identification and Production Management

Objectives

Upon completion of this chapter, the reader should be able to:

20.1 Define veterinary terms relating to sheep
20.2 Discuss sheep biology
20.3 Identify common breeds of sheep
20.4 Discuss the nutritional needs of sheep
20.5 Describe normal sheep behavior
20.6 Discuss the proper restraint and handling of sheep
20.7 Discuss the health care and maintenance of sheep
20.8 Describe the reproduction of sheep
20.9 Discuss diseases and conditions that affect sheep
20.10 Discuss the purposes and uses of sheep
20.11 Discuss the production types of sheep

Introduction

Sheep have been raised for the purposes of food and clothing for thousands of years. Sheep were domesticated more than 8000 years ago in Europe and Asia. They were brought to America by settlers over 400 years ago. Sheep can be used as meat animals, for wool and fiber, and as pets. There are six major breeds that have been instrumental in developing the different breeds of today. The United States is seeing a decline in the number of sheep producers. Australia and New Zealand are the leading sheep producers in the world.

Veterinary Terminology

Sheep are known as **ovine** in veterinary medicine. The **ewe** is the adult female and the **ram** is the adult male. A **lamb** is a young sheep under a year of age. The labor process of sheep is called **lambing**. A **wether** is a castrated male sheep. Breeding types include the ram, the ewe, and the dual-purpose sheep. A dual-purpose sheep is bred for both meat and wool production, and occasionally milk. Sheep meat from an animal over a year of age is known as **mutton**. The meat from a sheep under a year of age is referred to as lamb.

Biology

Sheep are ruminants, have cloven hooves, and are members of the *Bovine* family. Sheep breeds range in size from 100 to 225 pounds. Sheep have an average life span of 7 to 13 years. The wool of some breeds may weigh 15 pounds or more. Sheep have a gestation length of 148 to 150 days. Sheep are a commonly herded livestock; the herd is best moved from behind. Figure 20–1 illustrates the anatomical features of sheep.

Breeds

Six major breeds of sheep have been the basis for the development of over 200 total breeds—and over three-quarters of all sheep breeds—seen today. These breeds are the Dorset, Suffolk, Hampshire, Rambouillet, Border Leicester, and the Columbia (see Figure 20–2). These breeds have been crossbred over time. Sheep are classified by purpose and wool quality (see Table 20–1). Many of these breeds of sheep have become popular in North America.

Breed Selection

Sheep are used for several purposes, mostly as meat and wool products (see Figure 20–3). However, they can also be used for milk products. Lamb is popular, as it is very delicate meat, whether from a male or female. Mutton tends to be tougher, with a stronger taste, and is less desirable. Sheep's milk is not popular, as it tends to be less digestible than other milk

TABLE 20–1
Sheep Classified by Wool Types

Fine wool	Medium wool
• Jacob	• Suffolk, Dorset, Hampshire, Oxford, Columbia
Long wool	**Crossbred wool**
• Rambouillet, Border Leicester, Cheviot, Merino	• Southdown
Carpet wool	**Fur bred**
• Corriedale	• Katahdin

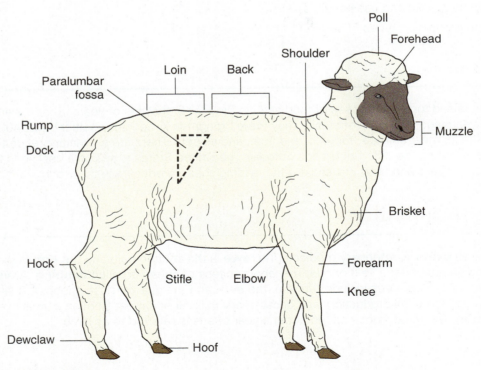

FIGURE 20–1 External anatomy of the sheep. (*Continues*)

CHAPTER 20 Sheep Breed Identification and Production Management

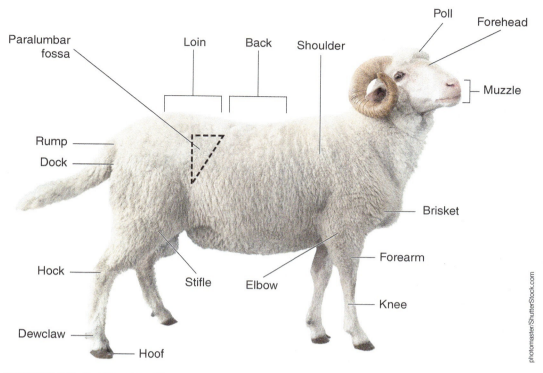

FIGURE 20–1 (*Continued*).

sources. Another popular product of sheep is their hair or **wool**, a soft fiber that acts as a coat. A by-product is the skin of sheep, known as **chamois** (pronounced "chammy"), a leatherlike material frequently used to polish cars.

Sheep selection by the breeder requires research into the characteristics of the breed, the size of the flock, the time of the year for breeding purposes, the overall health status, age, cost, whether it is a crossbred or a purebred, its purpose, and the space and facilities.

Nutrition

Sheep are ruminants; they eat a roughage-based diet and require quality hay and pasture sources (see Figure 20–4). Typically, most sheep will only require

FIGURE 20–2 (A) Dorset ram. (B) Hampshire ram. (*Continues*)

384 SECTION II Veterinary Animal Production

FIGURE 20–2 (C) Suffolk ram. (D) Leicester sheep. (E) Columbia ram. (*Continued*)

FIGURE 20–3 Wool samples. Longhair wool on left; shorthair wool on right.

provided through pasture, hay, silage, or grains. Their protein needs are met through alfalfa hay, linseed oil, or peanut oil supplements. Sheep need vitamins A, D, E, and K and get them through quality hay and pasture sources. Vitamin B is naturally produced in the *rumen*, or the filter system, of the animal ruminant digestive system. This filter helps break down fiber sources, such as hay and grass. Sheep require some minerals in small amounts, such as salt, phosphorus, calcium, and potassium. Sheep need about a gallon of water per day.

Behavior

Sheep have an instinct to remain with a flock; when separated, they can be easily stressed. Veterinary assistants must know how to handle sheep as well as understand their behavior. Many veterinary

a grain source during gestation. Other health issues may require additional supplements. Sheep use carbohydrates and fats for energy that are usually

FIGURE 20–4 Pasture or other roughage is the basic feed for sheep.

professionals work with sheep that are still in the flock (see Figure 20–5). Sheep behavior and body language are often the result of a *defense mechanism*, which is an animals' automatic reaction to protect itself, usually from prey or a stressful situation. Sheep will show upset or anger by stomping their front feet or **butting** with their heads. Rams will head butt when angry or during the breeding season.

Equipment and Housing Needs

Sheep are usually maintained in pastures or in barns with adequate ventilation. Flocks require shelter as protection from the weather, especially sheep raised for wool. Stalls should be kept well bedded and dry. Muddy and moist surfaces are not ideal. They need electricity and water sources, especially in wintertime. Water troughs should be kept from freezing and fresh water provided daily. Indoor housing for ewes should include free stalls and large open areas for young lambs to exercise. The horns of sheep or rams should be trimmed to prevent injury to other sheep or people. Fencing requirements include wire that is 60 inches or higher, with 4 to 5 inches between strands.

Restraint and Handling

When working with sheep, it is important to stay calm, quiet, and gentle and act with assurance. Sheep can be easily injured by a careless handler: their bones are fragile and may break. It is important to keep this in mind when attempting to grasp and catch a sheep. Sheep that have a full wool coat can easily become overheated due to the stress of handling. Handling can damage the wool, reducing the quality and profit of the product. Sheep are otherwise easy to work with; they rarely bite or kick during restraint. Capturing a sheep is best done within the herd by approaching the animal and swinging one arm around the neck and front shoulders and another around the **dock** (tail area). Grasping the dock and then steering the animal to the destination is relatively easy. Halters can be placed on sheep, but because of the animal's short nose and odd-shaped head, they can slip off and occlude the nostrils, causing breathing problems. There are several methods of restraint depending on the procedures being completed. One of the most common and easiest is the **rump method**: which is to sit the sheep up on its hind end. Sheep rarely struggle, and this allows for examination, hoof trimming, shearing, or injections. When holding the sheep, stand on the left side, Place one hand on the chin and pull the sheep close to your legs. Using head control and a quick twist, one person can usually sit a sheep on its rump. Larger sheep can be held with the restrainer's back against a wall for support. A restraint used for medicating is typically to hold the sheep against a wall or use the limbs to restrain and hold the head in place. Newborn lambs can be held like small dogs.

Grooming

Many breeds are raised for wool. Few breeds shed their own wool and require regular **shearing** or shaving. Shearing is done carefully with clippers or shears that easily remove and prevent damage to the wool quality (see Figure 20–6). Most producers will shear several times a year, but care must be taken so sheep do not overheat in the summer or become sick in the winter because of the length of the wool coat. Shearing equipment includes electric shearers, shearing blades, trimmer disinfectant, shearing tables or chutes, and blankets and hoods.

Basic Health Care and Maintenance

Sheep are highly susceptible to diseases and parasites and require a quality health care program. There are many signs of illness that should be of concern to veterinary assistants and sheep producers, including

FIGURE 20–5 Sheep tend to remain in flocks and are best handled if allowed to remain with the flock.

FIGURE 20–6 Shearing sheep.

poor wool or coat quality, loss of wool, discharge from eyes and nose, pale mucous membranes, anorexia, no signs of chewing cud, diarrhea, and signs of standing alone from the flock.

When new sheep are added to a program, it is important to quarantine and isolate all sheep for 30 days to determine their health status. All sheep flocks should receive yearly physical exams, updated yearly vaccines, and dewormers every three months. All sick animals must be isolated and stalls cleaned and sanitized daily.

Lamb tails are usually docked by the banding method, which involves placing an elastic band tightly over the tail area, causing it to slough off. This is done 3 to 5 days after birth and is not painful. Tail docking prevents a buildup of feces on the wool. Adult sheep may be docked using a tail docking instrument that cuts the tail (see Figure 20–7). Young male sheep can be castrated through the banding method while they are still in the juvenile stage. Adult sheep should be castrated through the use of an *elastrator*, similar to cattle and goats.

Vaccinations

Vaccine programs will vary from location and veterinarian recommendations. Numerous sheep vaccines are available and should be discussed with the veterinarian. Many lambs receive a combination vaccine of the most common diseases for that particular area. Lambs should begin a vaccine program between 6 and 8 weeks of age, with booster vaccines in 2 to 4 weeks. Vaccines for breeding programs may require boosters every 6 months to a year; other sheep production programs will require yearly vaccine boosters.

Breeding and Reproduction

Sheep producers strive to have a high number of lambs born each season (see Figure 20–8). The average lamb crop is two lambs per ewe per breeding. Ewes reach puberty at about 8 to 10 months old, while rams reach puberty at about 5 to 7 months old. Ewes are typically bred for the first time at about 2 years old. Ewes can sometimes be difficult to breed, as they do not normally show any signs of heat. The estrus cycle lasts an average of 30 hours, sometimes making it difficult to determine a breeding time, and occurs every 16 to 17 days. The gestation length is between 148 and 150 days. The labor process in sheep is known as *lambing*. It is important to ensure that each ewe has been bred, because a sheep that does not lamb is not profitable. Rams are normally kept apart from ewes. One may be placed in a flock of ewes for short periods of time to allow for breeding, but the timing must be right. When not breeding, rams should be kept isolated from ewes to prevent fighting and aggression. They require large amounts of exercise and are best kept in a pasture. They do not require large amounts of grain but

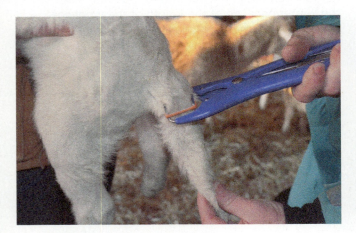

FIGURE 20–7 Tail docking.

FIGURE 20–8 Ewe with lamb.

should have quality grass or hay sources. The pregnant ewe should have quality pasture and hay sources and adequate exercise and be given grains until close to labor. When the lambing process nears, the ewe should be placed in a dry and sanitary area. The udders should be sheared for proper sanitation and to allow lambs to more easily locate the milk source. The lambing process is divided into four phases: dilation, accelerated contractions, expulsion of the lamb, and the expulsion of the placenta. *Expulsion* refers to removal from the body by internal muscular contractions.

Sheep typically have a quick, easy labor but should be monitored for difficulty, especially when twins are expected. Many ewes will have twins; this is the ideal goal of a reproductive program (see Figure 20–9). When giving birth to twins, the ewe may need some help in reviving and drying the lambs. During the **dilation phase**, the **cervix** (birth canal) will **dilate** or expand. This prepares the birth canal for the lamb, which should be born within the next 12 to 24 hours. The dilation phase is completed when a clear to whitish color discharge appears at the vulva. This leads to the **contraction phase**. The contractions or wavelike movements of the uterus allow the lamb to move into the birth canal. This phase occurs over a 6- to 12-hour period and is completed when the water breaks and a clear, water-like fluid is passed. This indicates the lamb is beginning to be born. The labor phase should be monitored for **dystocia**, or difficult birth signs. In a normal birth, the tip of the nose and front feet deliver first. When the lamb is born, the ewe will clean the head and nasal passage first to stimulate breathing. If a twin is then born, the ewe will lose interest in the first lamb as contractions begin. The placenta or afterbirth is expelled within 30 minutes to an hour after the labor process is completed. It is important to note the passing of the placenta, as sheep will consume the membranes as a natural instinct in order to not attract predators. A **retained placenta** that is not expelled within 24 hours after birth will lead to a serious and possibly fatal infection. A veterinarian should be contacted immediately. Occasionally, a ewe may die or reject a lamb, and the lamb becomes known as an **orphaned lamb**. Some sheep will become a foster mother if they have a lamb at their side. Otherwise, a milk replacer or cow's milk may need to be supplemented through bottle feeding.

Common Diseases

Sheep are easily susceptible to diseases. This is important to note in production programs where sheep are raised for meat products and human consumption.

Foot Rot

Sheep tend to be affected by a common condition that affects their hooves and is caused by a bacterial infection. Bacteria can enter a wound or the hoof wall and cause a condition known as **foot rot** or an **abscess**. Foot rot is the decay and damage of the soft tissues within the cloven hoof. An abscess is a buildup of pus caused by an infection and is typically localized to one area. The infection causes the sheep to become lame and limping will occur. The joints and tendons may also be affected. Quality treatment and prevention includes trimming away any areas of damage, applying a topical disinfectant and antibiotic, and keeping the sheep in a clean and dry location until healed. Sheep kept in moist, muddy, or wet areas are more prone to bacterial infections. Abscesses may need to be opened and drained to relieve pressure buildup from the pus and infection. A vaccine is available for bacteria prevention from foot rot. Signs of foot rot or abscesses include the following:

- Limping and lameness
- Standing without putting weight on hoof
- Weight loss
- Anorexia
- Fever
- Foul odor from the hoof area
- Pus drainage from the hoof
- Swollen joints

Mastitis

Another common condition that affects ewes is **mastitis**, or inflammation of the mammary glands. This udder infection occurs during lactation when bacteria enter the teat glands and cause an infection.

FIGURE 20–9 Twins are common in sheep.

The udder may feel hard, hot, and tender to the touch. A veterinarian should prescribe antibiotics; the sheep may need to stop nursing to allow the udder to dry up.

Actinobacillosis

Another infection that must be considered is called **actinobacillosis**, commonly called *lumpy jaw*. This disease is of importance to producers, as there is an economic loss when this occurs: the meat is condemned and cannot be used for human consumption. This disease affects the head and jaw and causes tumorlike lumps of yellow pus to build up within the jaw. The infection spreads to the muscles, organs, and tissues of the body. The affected areas are not legal to slaughter for human use. This condition does not cause death but can spread to other herd animals.

Actinomycosis

Another disease that affects sheep is called **actinomycosis**, or "wooden tongue." This bacterial infection causes lesions on the head that spread to the lymph nodes located in the neck, causing swelling. Hard lesions will also form on the tongue, causing difficulty eating and drinking. Weight loss and death will occur as the disease progresses. This is caused from contamination of food and water sources. Infected animals should be isolated from the herd to decrease spread of the disease.

Blue Tongue

Blue tongue is a virus that is spread by gnats, which are tiny flying insects that bite and feed off of the blood meal of animals and humans. Blue tongue has no treatment; the virus rapidly weakens the sheep's immune system. A secondary infection typically causes the death of the animal. A vaccine is available and should be administered at the time of shearing to boost the immune system. Signs of blue tongue include the following:

- Fever
- Weight loss
- Anorexia
- Difficulty eating
- Lethargy/sluggish behavior
- Swelling on the mouth and ears
- Ulcers on the mouth and tongue

Enterotoxemia, the "overeating disease," is a bacterial disease that affects lambs and show animals as a result of overfeeding of grain. Bacteria within the rumen develop and produce toxins that cannot be absorbed by the body. This leads to seizure activity and most often death. There is no treatment of this condition. Vaccination is recommended, and good health practices lessen the chance of infection. Vaccines should be administered 2 and 6 weeks prior to lambing.

Johne's disease is a digestive condition that causes the intestinal wall to thicken. Signs include diarrhea, weight loss, and possible death. No treatment has been developed, but research continues to look into the causes of this disease. It is important to conduct an exam and review a medical history before purchasing new sheep.

Soremouth is a highly contagious zoonotic viral disease that causes sores around the mouth and lips of lambs (see Figure 20–10). Scrubbing the sores with iodine will help them heal. Scabs should be removed; scrubbing should continue until sores have cleared. Isolation and sanitation are keys to treatment and prevention. Exam gloves should be worn when treating young lambs. A vaccination is available and should be given shortly after birth.

Lamb dysentery is caused by bacteria that affect newborn lambs 1 to 5 days after birth. This disease spreads rapidly through newborn lambs and requires proper sanitation and control management. Signs include loss of appetite, diarrhea, and depression; sudden death may occur.

Polyarthritis

Young lambs may be affected by **polyarthritis**. This condition can occur at about 3 to 5 weeks of age and causes pain and swelling in joints. Affected lambs are reluctant to move and inactive and experience weight loss or no weight gain. They can be treated with tetracycline antibiotics.

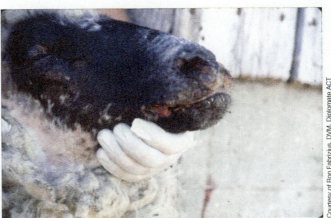

FIGURE 20–10 Soremouth in a sheep.

Common Parasites

Parasite prevention programs are a vital part of a sheep production program, as many sheep are housed outdoors and feed off of pasture. This makes them more susceptible for internal and external parasites. Common external parasites include lice, mange mites, ear mites, and ringworm. External parasites severely affect the wool quality of the coat. Ringworm is highly contagious to other sheep and humans. Mites can be treated with common parasitic medicine and disinfectants, such as Fulvicin, nolvasan, or a bleach solution. Common internal parasites include lungworm, whipworm, tapeworm, coccidian, liver fluke, and stomach worms. Parasite prevention and treatments include topical dips, insecticide sprays, anthelmintics (dewormers), waste control and removal, and pasture rotations. Complete eradication of parasites is impossible. Dewormers should be given every 3 months through **drenching**: a method of giving medicine by mouth with a dose syringe that is helpful in administering large amounts of medicine at one time for large numbers of animals. Signs of parasites in sheep include the following:

- Diarrhea
- Weight loss
- Rough coat
- Wool loss
- Anemia
- Respiratory problems
- Anorexia
- Bloated appearance
- Depression

Sheep Production

Sheep production has its advantages and disadvantages in relation to other types of livestock programs. Sheep are the best-suited livestock for grazing, are excellent scavengers, are very efficient at converting food to meat, serve a dual purpose, produce a low amount of waste material, are easy to raise and handle, and require little space. Disadvantages of sheep include industry competition, low wool prices, low meat production, susceptibility to disease, and an existence as easy prey.

There are several types of systems used in sheep production. The most popular is the **farm flock method**. Its goal is to produce market lambs whose primary product is a lamb meat source; wool is the secondary product (see Figure 20–11). Typically, producers may raise the sheep with other types of livestock in a pasture setting.

The **purebred flock method** is usually a small production system of purebred sheep. The goal is to produce quality rams and ewes for sheep breeding programs. This requires a high level of management and a high standard for the breed. Producers are raising ideal types of sheep that meet the standards of that particular breed. This is a full-time business that requires more labor compared to other production systems. Pasture space is also necessary.

The **range band method** is seen more in the Midwest, where larger amounts of land are used to

FIGURE 20–11 Example of a farm flock sheep production system.

and rainfall. The focus of the range band method is wool or lamb production.

The **confinement method** is becoming more popular as an aseptic system for raising sheep for human consumption. Sheep are raised in a sterile and controlled indoor environment to prevent little to no health or parasite problems (see Figure 20–12). There is no need for land. This system requires more skill and care in the production management of the sheep.

Lamb Feeding

Lamb feeding is a specialized production system in which lambs are raised to weaning and then sold to feedlots. Proper breeding management is crucial. This system is typically seen on a large farm with quality pasture space (see Figure 20–13).

FIGURE 20–12 Lambing pen.

raise large sheep herds. The herd is known as a **band**, a large group that ranges free over a vast amount of land and is moved and cared for by **herders**. It is important that the land features prominent vegetation

FIGURE 20–13 Lamb feeding systems raise lambs to the age of weaning before selling them.

SUMMARY

Sheep production is an important part of the United States and essential for economic success due to its economic impact on agriculture. However, the past several years have seen a decrease in the number of producers and sheep flocks. Sheep are used for meat, wool, and milk sources. They tend to be an easier, more efficient livestock source to raise and manage. Because of their high susceptibility to disease and parasites, a quality health care program must be followed to ensure quality products.

Key Terms

abscess   a bacterial infection that causes pus to build up in a localized area

actinobacillosis   a disease of the head and jaw that causes tumorlike lumps of yellow pus to build up within the jaw; also called "lumpy jaw"

actinomycosis   a bacterial infection of the head and neck lymph nodes that causes hard lesions to form over the tongue; also called "wooden tongue"

band a large group

blue tongue a virus spread by gnats that weakens the immune system and causes secondary infections

butting using the head to hit an object or a person

cervix the birth canal

chamois a by-product of sheep skin used as a leather product for polishing

confinement method a system of raising sheep in an aseptic environment for the purposes of human consumption

contraction phase the second phase in the labor process in which wavelike motions in the uterus move the fetus into the birth canal

dilate to open, expand, or widen

dilation phase the first phase of the labor process in which the birth canal opens and expands

dock the tail area

drenching giving medicine by mouth with a dose syringe to large numbers of animals

dystocia a difficult or abnormal birth

enterotoxemia a bacterial disease from overeating that causes toxins to build up within the rumen that are not absorbed by the body

ewe the adult female sheep capable of reproduction

farm flock method the most popular sheep production system: meat sources are the primary system and wool production the secondary system

foot rot a bacterial condition of the hoof that causes decay or damages the hoof tissue

herder the person in charge of a group of sheep

Johne's disease a contagious, chronic, and usually fatal infection that primarily affects the small intestine

lamb a sheep less than 1 year old; also the meat from a lamb less than 1 year old

lamb dysentery a bacterial diarrhea condition in newborn lambs

lamb feeding a specialized production system of raising lambs to weaning and then selling to feed lots

lambing the labor process of sheep

mastitis an inflammation of the mammary glands

mutton meat from a sheep that is over 1 year old

orphaned lamb a young sheep that has lost or been rejected by its mother or and must be bottle fed

ovine the veterinary term for sheep

polyarthritis a condition in young lambs that causes multiple joints to become inflamed and painful

purebred flock method a small production system that raises purebred sheep for reproductive purposes

ram the adult male sheep capable of reproduction

range band method raising large sheep herds on large amounts of land

retained placenta fetal membranes that have not been passed from the uterus after the labor process

rump method a restraint method in which a sheep is placed on its hind end for handling

shearing a method of shaving wool from sheep

soremouth a highly contagious zoonotic viral disease that causes lesions over the mouth and lips of sheep

wether a male castrated sheep

wool the soft fiber of sheep that acts as a coat

Review Questions

1. How are sheep classified?
2. What are the advantages and disadvantages of sheep production?
3. What are the types of sheep production systems? Describe each one.
4. What is necessary for ewe management in regard to reproductive needs?
5. What are the nutritional requirements for sheep?
6. What are the health needs of sheep?
7. What are signs of parasites in sheep?
8. What treatments and preventions are necessary for foot rot and abscesses?

Clinical Situation

Jim and Chris Smith, clients at Millers Animal Clinic, have an appointment for their dog to have its yearly exam and vaccines. During the visit, the couple asks Pedro, the veterinary assistant, who is taking a patient history, about their newest pets, twin lambs.

"We just got twin lambs that are bottle babies. We got them for our kids. Would Dr. Anderson see them in the same way that he sees our dog?"

"Dr. Anderson does see both small and large animals. He would schedule a farm call visit to examine the lambs. We usually only see large animals in the clinic if they need to be hospitalized."

"What would the lambs need as far as veterinary care?" Mrs. Smith asks.

- What would be the best answer that Pedro could give Mrs. Smith?
- How would you go about scheduling the lambs for a farm call appointment?
- What educational information should you give to the Smiths about the lambs?
- What information will Pedro need to schedule the farm call visit?

Chapter 21: Goat Breed Identification and Production Management

Objectives

Upon completion of this chapter, the reader should be able to:

21.1 Define veterinary terms relating to goats
21.2 Discuss goat biology
21.3 Identify common breeds of goats
21.4 Discuss the nutritional needs of goats
21.5 Explain normal goat behavior
21.6 Demonstrate the restraint and handling of goats
21.7 Discuss the health care and maintenance of goats
21.8 Describe the reproduction of goats
21.9 Discuss diseases and conditions that affect goats
21.10 Discuss the purposes and uses of goats

Introduction

Although **goats** are similar to sheep in several ways, different characteristics do set them apart. A goat's coat is hair rather than wool. Goats and sheep have similarly shaped heads and ears; however, goats tend to be slightly larger. Both are cloven-hoofed. Some goat breeds have horns. Goats have beards. Goats tend to be more intelligent, independent, and are more capable of protecting themselves.

Developed more than 8000 years ago, goats were one of the first types of livestock to be domesticated. They can be used as grazers to keep grass trimmed, to clean up plants and weeds, and to provide company to other types of livestock. They are used as a source of fiber and kept as pets. Goats are classified by their purpose and use. Today, China and India lead the world in goat production. As a food source, goats provide milk, meat, and by-products.

Veterinary Terminology

In veterinary medicine, **caprine** means resembling or relating to goats. A female goat is a **doe** or **nanny**; the male goat is a **buck** or **billy**. A young goat under a year of age is a **kid**. The labor process of goats is referred to as **kidding**. A group of goats is a **herd**.

Biology

Goats, like sheep, are ruminants. Goats weigh anywhere from 20 pounds to well over 150 pounds and can be 1.5 to 4 feet tall. The average life span of a goat is 8 to 10 years. Male goats have a strong odor from hormones that are present on the top of the head, usually behind the horns. These **scent glands** create an unpleasant odor; goats typically rub them against objects or areas to mark their territory. The scent glands of young male goats are usually removed during dehorning. Figure 21–1 illustrates the anatomical features of the goat.

394 SECTION II Veterinary Animal Production

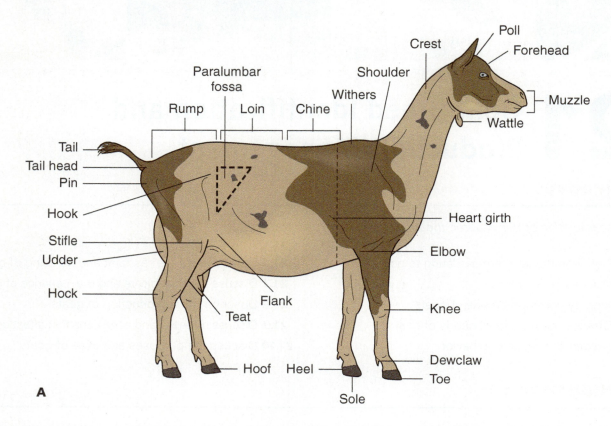

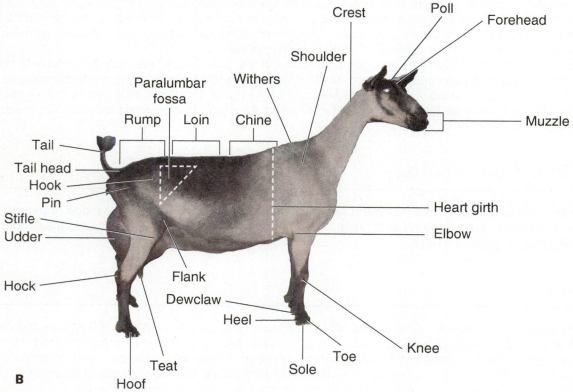

FIGURE 21–1 External anatomy of the goat.

Breeds

There are over 300 breeds of domesticated goats. Wild goats are still found in the United States and other countries. There are five groups of goat breeds: Angora, dairy, meat, **cashmere**, and pygmy.

Angora

Angora goats are produced for their hair coat, which is used to make clothing and fiber products (see Figure 21–2). The fiber produced from Angora goats is called **mohair**. Angora goats are also used for meat.

Dairy

People who suffer from digestion problems and dairy sensitivities may drink the milk of **dairy goats**. It is easy to digest, has more vitamin A than the milk from other animals, and has less fat because of **globules**, which are small round particles of a substance. There are six common breeds of dairy goat in the United States: French alpine, Oberhasli, Lamancha, Nubian (see Figure 21–3), Saanen, and Toggenburg.

Meat

Meat goats are raised as a meat source and bred to develop quality muscles. Most meat goats are from Spanish breeds. A popular meat breed is the Boer goat (see Figure 21–4).

FIGURE 21–3 Nubian goat.

FIGURE 21–4 Boer goat.

Cashmere

Cashmere is a soft down or undercoat produced by most breeds of goat. (Angora goats do not produce cashmere.) There is no registered breed that produces cashmere; however, goats have been bred to improve on this characteristic.

Pygmy

Pygmy goats are miniature-sized goats often kept as pets and for research (see Figure 21–5). They are very

FIGURE 21–2 Angora goat.

FIGURE 21-5 Pygmy goat.

FIGURE 21-6 Goats eat browse or woody plants.

entertaining and playful. They can be produced for milk and meat. Because of their small size, these goats are relatively easy to handle.

Breed Selection

Goat selection should be based on production or purpose (such as dairy or meat). Hair and coat characteristics are important considerations for fiber goat production. The main purpose of dairy goats is milk production.

Goat Nutrition

Goats eat anything, especially weeds and mostly **browse**: shrubs, weeds, woody plants, and briars (see Figure 21-6). A goat's diet consists mostly of carbohydrates, fats, protein, vitamins, minerals, and water. Like sheep, goats have a ruminant digestive system. Their diet is mainly roughage and plants. They should be fed according to their use and purpose. Goats chew cud to break down the plants and roughage they eat so it can filter through their digestive system. Many goats do not get adequate nutrients, especially does during pregnancy and lactation. It is important to note that some weeds and plants are toxic to goats, such as azalea, cherry, hemlock, and laurel.

The best maintenance diet for goats includes a diet that is 14 to 16 percent protein, with added salt, phosphorous, and calcium. Goats may be fed grains or a pellet-based diet. Growing kids and lactating females require calcium supplements. Goats require minerals in small amounts, including cobalt, copper, iodine, iron, manganese, selenium, and zinc. Goats need copper more than other ruminant species. Most commercial grain or pellet diets are given at 1 to 1.5 pounds per day with high-quality roughage. Hay, grass, and browse requirements range between 3.5 and 4.5 pounds per day. Kids nurse for about 60 days and should then be weaned onto a starter diet that consists of soft pellet-formulated food. Young goats should begin a regular diet as soon as possible.

Behavior

Goats do not have the same instincts as sheep to join a group; they tend to scatter when herded. In a herd of goats, there is a lead goat, usually a doe. Most other goats follow this lead goat. Goats are typically docile and friendly, which makes them easy to handle. They can become agitated and butt in a similar way to sheep, usually rearing up on their hind legs. Signs of anger and aggression include their hair rising over the back (similar to the hackles on a dog), holding the tail close to the back, sneezing, stomping, and snorting. Goats are naturally playful and are often kept as pets. They become less playful with age. Nanny goats remain calmer as adults than do billy goats.

Basic Training

Goats may be trained to halter and lead like sheep and cattle. When shown, goats must be able to lead in hand and walk with a handler. They should be taught

CHAPTER 21 Goat Breed Identification and Production Management

to stand quietly during judging. Goats can be groomed like cattle, with regular brushing, bathing, and hoof trimming.

Equipment and Housing Needs

Goats may be housed indoors or outdoors. Indoor barns or sheds should serve as shelter from the weather; any open sides should open to the south. Stalls should be kept well bedded and dry. Proper ventilation is necessary to allow air to circulate through the open area. The barn should include running water, electricity, and food and water troughs. If several goats and kids are housed together, an open area should be large enough to contain the herd comfortably. Goats should be dehorned to prevent injury to people and other animals. Outdoor fencing should be at least 60 inches high, as goats are excellent jumpers. Fence panels should be at least 4 to 5 inches between strands or boards. Goats that are shown or regularly transported should have loading chutes and weight crates to determine their weight. Other equipment that may be necessary includes blankets, hoof trimmers, clippers, shearing table, blankets, halters, leads, and fans.

Restraint and Handling

Goats struggle if handled roughly or improperly. A minimum amount of restraint is better for both the goat and the handler. Goats should not be handled nor treated like sheep.

The best method of catching a goat is to grasp one leg and lift it. Most goats will stand still and allow many types of procedures to be completed in this manner. The best way to carry out this restraint is to reach over the back of the goat and grasp one front leg. This is a common restraint for hoof trimming. Goats that have not been handled or are scared may need to be caught by grasping a hind leg. It is important to be cautious, as goats may kick and struggle. Goats with horns pose a safety issue: horns can cause severe damage to humans during restraint (see Figure 21–7). Like sheep, goats can be placed against walls and held against a solid structure. A goat should never be restrained by its rump. As with a dog, lateral recumbency is another way to restrain a goat: grasp the head and muzzle with one hand, reach over the back and grasp one front leg, and then move the other hand to a front leg. Gently raise the rear legs and lay the goat on its side, holding down the limbs to prevent it from standing. Goats can be trained to

FIGURE 21–7 Goats with horns can pose a safety issue to humans during restraint.

be haltered and will lead. Halters should be fitted so they do not fall off the nose or occlude the nostrils. Head restraint for a goat is similar to the way a dog is held. Venipuncture restraint may also be done, as with a dog or cat: extend the front limb for cephalic blood collection and the head upward for jugular blood collection (see Figure 21–8).

Grooming

Goats can be washed and kept clean using water and a mild soap. Brushing them on a regular basis with a stiff brush helps remove dirt and shedding hair. Goats

FIGURE 21–8 Proper restraint technique for jugular access.

can be sheared like sheep. Hooves require routine trimming and maintenance. Goats that are shown commonly are bathed, brushed, trimmed, or sheared before a show. A blanket placed over the body keeps the animal clean and dry. Some goats have hair. Others have mohair, which may need to be sheared rather than clipped.

Basic Health Care and Maintenance

Goats are relatively healthy and can live a long life with good care and management. Healthy goats graze on pasture and plants throughout the day, chew on cud, have a shiny and healthy hair coat, have strong legs and hooves, appear social, and have bright and clear eyes. Table 21–1 outlines normal vital signs for a goat.

Goats require the same health practices and programs as sheep. Their scent glands are usually removed with a caustic ointment or electric dehorner when they are kids. Males are also banded as kids at 3 to 5 days of age.

Vaccination

Goats should begin to receive vaccines at 6 to 8 weeks of age, following the recommendations of a veterinarian. The vaccine program is similar to that of sheep. Most goats are routinely vaccinated for overeating disease, tetanus, soremouth, and goat pox, but programs vary with location and occurrence of disease. Kids should begin a deworming program 4 to 8 weeks after they begin to graze.

Reproduction and Breeding

As with sheep, the ideal number of young kids produced each year is two to three per breeding season. This means that the ideal breeding doe will have twins or triplets each season. The breeding seasons are late summer and late fall. Does have a gestation length of 148 to 151 days; the breeding is timed for kids to be born in the early spring or late fall. Does are typically bred at about 2 years of age. The estrus cycles are regulated by the climate and weather and occur every 16 to 17 days. A doe is usually receptive to the ram for 24 to 48 hours. Kids average about 5 pounds in weight. Rarely do the does have reproductive or dystocia issues. A normal labor presentation is similar to sheep, with the head and front limbs presented first.

Common Diseases

Goats are relatively healthy but can acquire many of the diseases to which sheep are susceptible. That said, a review of Chapter 20 is valuable for learning common conditions and possible diseases in goats. Signs of an unhealthy or diseased goat include a poor or rough hair coat, hair loss, abnormal temperature, swelling of joints and body, pale mucous membranes of the mouth and eye mucosa, and thickened nasal and ocular discharge.

TERMINOLOGY TIP

The term *mucosa* refers to the mucous membranes located throughout the body, in the gums, eye tissues, nasal tissues, and rectal tissues. The mucous membranes allow for gum color determination that identifies blood circulation and oxygen flow through the body. ■

Joint Ill

Young kids may be affected by **joint ill**. This is a condition in which the umbilical cord site becomes contaminated and bacteria enters the blood stream, causing severe inflammation and pain in the growing joints.

Bloat

Goats that ingest foreign objects may suffer stomach pains, digestive upset, and **bloat**: an accumulation of air or gas in their stomach that causes it to swell. Bloat in goats is handled in a similar way as in cattle. Goats with poor nutrition or that ingest objects, such as garbage, may suffer from indigestion.

Goat Pox

Goat pox is a viral disease that affects mostly young goats or those with compromised immune systems and spreads rapidly through herds. Infected goats

TABLE 21–1	
Normal Vital Signs in Goats	
Temperature (T)	103–103.6 degrees Fahrenheit
Heart rate or pulse (HR)	80–85 bpm
Respiratory rate (RR)	25–30 br/min

should be isolated. The virus causes flulike symptoms as well as lesions over the head and face. These lesions eventually open up and drain; they must be cleaned with hydrogen peroxide daily and topical antibiotics should be applied.

Hypocalcemia

Hypocalcemia, or low blood calcium, is also known as "*milk fever*." It often occurs in does that have been lactating. Does with hypocalcemia require calcium supplements in order to produce milk. Milk fever in goats may lead to death; signs include the following:

- Inability to stand
- Decreased respiratory rate
- Gasping for breath
- Coma

 MAKING THE CONNECTION

Other common conditions are mastitis, foot rot, and hoof abscesses. These conditions are further discussed in Chapter 20.

Common Parasites

Parasites that affect sheep may also affect goats. The most common external condition in goats is the **ringworm** fungus. Goats that are housed outside and have access to wet and muddy areas are more susceptible to this fungus. Ringworm causes hair loss in patches on areas of the coat. This condition is highly contagious to all animals and humans.

SUMMARY

Sheep and goats are closely related and have similar needs. Goats have relatively minor differences in physical characteristics. Goats can be used for milk, meat, fiber, and by-products. They have become popular as pets. Their importance as livestock and products for human consumption has made their health care a vital part of veterinary medicine.

Key Terms

Angora goat a group of goats that produce hair

billy a common name for the adult male goat

bloat air in a goat's stomach that causes it to swell

browse a woody plant

buck an adult male goat

caprine a veterinary term meaning of, relating to, or being a goat

cashmere a fine down undercoat of goats that is warm and used in clothing

dairy goat a goat that produces milk

doe an adult female goat

globule a small round fat particle

goat any of various hollow-horned ruminant mammals related to the sheep but of lighter build and with backwardly arching horns, a short tail, and usually straight hair; long domesticated for its milk, wool, and flesh

goat pox a viral disease that affects a goat's immune system and causes flulike symptoms

herd a group of goats

hypocalcemia low blood calcium

joint ill a bacterial infection caused by contamination of the umbilical cord that results in pain and inflammation within a joint

kid a young newborn goat

kidding the labor process of goats

meat goat a goat that produces meat

mohair long silky locks of hair from the Angora goat used in clothing and bedding

nanny a common name for the adult female goat

Pygmy goat a small goat kept as a pet

ringworm a contagious external fungus found on the skin and hair coat

scent gland an organ located on the head behind the horns that produces a strong hormonal odor in male goats

Review Questions

1. How are goats and sheep similar in appearance? How are they different?
2. Describe how goats are grouped.
3. What is the normal behavior of a goat?
4. What is the best restraint method to use when catching a goat?
5. What is the estrus cycle of a female goat?
6. How long is the period of gestation in goats?
7. What are the normal vital signs of goats?
8. What are some restraint methods used with goats?

Clinical Situation

Mr. Reyes, a local goat farmer, calls to schedule a farm call appointment. The veterinary assistant, Greg, tells Mr. Reyes that the veterinarian, Dr. Johnson, is on vacation. Greg says that he can schedule exams for Mr. Reyes's goat herd in two weeks.

"I really need my meat goat herd seen as soon as possible. Some of the herd is showing signs of nose and eye discharge. Can you set up an appointment for tomorrow?"

"I am sorry, but Dr. Johnson is off until next week. I would recommend calling another veterinarian to see your goats." Greg then hangs up the phone.

- What was handled incorrectly in this situation?
- How should this situation been handled?
- Why is it important for Mr. Reyes's meat goat herd to be seen as soon as possible?

CHAPTER 22: Poultry Breed Identification and Production Management

Objectives

Upon completion of this chapter, the reader should be able to:

- **22.1** Define common veterinary terms used in the poultry industry
- **22.2** Describe the biology of poultry
- **22.3** Explain egg anatomy, color, and quality
- **22.4** Identify common breeds of poultry species
- **22.5** Describe the different species and classes of poultry
- **22.6** Discuss the nutritional needs of poultry
- **22.7** Discuss common health practices and management of poultry
- **22.8** Describe common poultry diseases and parasites
- **22.9** State the importance of poultry production
- **22.10** Explain poultry production systems
- **22.11** Discuss poultry reproduction and chick development

Introduction

Poultry species include chickens, ducks, geese, turkeys, and other bird species that may be seen in the wild or have been domesticated for a particular use or product to serve humans (see Figure 22–1). Poultry production needs have increased greatly over the past 10 years, with chicken being the most popular poultry. The average person eats over 75 pounds of chicken per year. Poultry production is both economical and a large part of the veterinary industry.

Veterinary Terminology

Poultry are also called birds or **fowl**. Many terms are used in veterinary medicine that relate specifically to chickens. These terms are often used interchangeably with other poultry species. The adult female is a **hen**, and the adult male is a **rooster** or **cock**. A young chicken is a **chick**. A male chicken under a year of age is a **cockerel**. A young chicken raised to lay eggs is a **pullet**. The adult female hen that lays eggs is a **layer**. As the layer hens age and no longer produce eggs, they are **spent hens**. Spent hens are typically used as a meat source in processed foods. Other terms are used to refer to chickens used for food purposes. A young chick between 6 and 8 weeks of age used for meat production is a **broiler**. It weighs 4 pounds or less and may be male or female. The meat is tender and easy to cook and may also be known as a **fryer**. A larger chick over 4 pounds and usually over 8 weeks of age is a **roaster**. A male chicken that has been castrated (and thus gains more weight) usually over 6 pounds and 6 months of age is a **capon**. A group of similar poultry is a **flock**. Additional poultry terms are listed in Table 22–1.

402 SECTION II Veterinary Animal Production

FIGURE 22-1 Chicken is the most popular poultry production species.

TABLE 22-1	
Additional Terms Related to Poultry	
Poult	Young turkey
Drake	Adult male duck
Duckling	Young duck
Goose	Adult female goose
Gander	Adult male goose
Gosling	Young goose
Peacock	Adult male peafowl
Peahen	Adult female peafowl

Biology

Poultry have an avian digestive tract. Thus, they are much like caged pet birds. Modern poultry do not have a need to fly, and many breeds grow too large to fly. They use their beaks to break food particles into small pieces for ease of digestion. Like other avians, poultry have a **gizzard**, a muscular organ used to break down food particles. They also ingest **grit** to help break down hard substances, such as seed shells. The beak can grow sharp and may become overgrown, making it difficult to eat. Most producers will **debeak** young chicks to prevent them from pecking and injuring each other. Debeaking involves trimming the sharp end of the beak with small clippers. The external features of most poultry are similar to other bird species (see Figure 22–2). Poultry have certain head and neck features that help identify the species and breed. Chickens have **combs** on the tops of their heads, which are flesh-like projections (see Figure 22–3). Some combs are complete, while others have **serrations**: spaces between fingerlike sections called *points*. Some poultry species also have a **wattle**, a flesh-like projection under the chin. Turkeys have a small amount of hair beneath the wattle, known as a **beard**. Some species of geese and ducks have a projection located on the top of the beak called a **knob**.

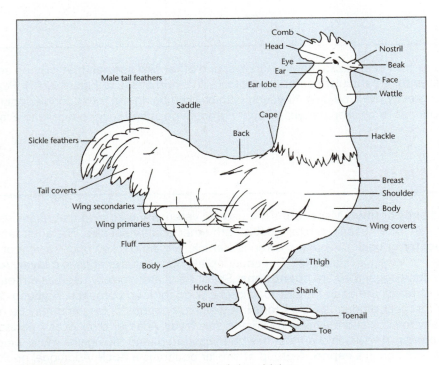

FIGURE 22-2 External anatomy of the chicken.

CHAPTER 22 Poultry Breed Identification and Production Management

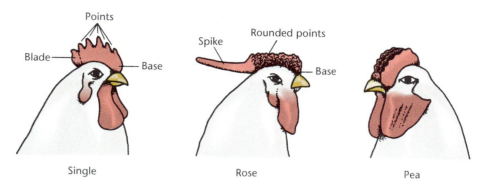

FIGURE 22–3 Comb types found on poultry.

Males tend to have larger bodies and head features (such as the comb and wattle). The eyes, skin, wattle, and comb are useful in determining the overall health and condition of the bird. A sign of good health for most breeds is a large red comb (see Figure 22–4).

Poultry Species and Classes

Poultry are considered domesticated birds with feathers, two legs, two wings, and a beak. Poultry species are raised primarily for eggs, meat, feathers, and by-products.

Chickens

Chickens are the most important and most popular type of poultry. They are raised for eggs and meat; some breeds are better producers in their respective products. There are four classes of chickens: American, English, Asiatic, and Mediterranean.

The **American class** developed in the United States and was created based on the need for top egg and meat production requirements. Examples of breeds in this class include Rhode Island Red and Plymouth Rock (see Figure 22–5).

FIGURE 22–4 A large red comb is a sign of a healthy bird.

FIGURE 22–5 (A) Rhode Island Red. (B) White Plymouth Rock.

The **English class** developed in England as a meat group that is commonly crossbred for improved meat quality. Cornish is one example of this class.

The **Asiatic class** developed in Asia primarily as a show quality group bred for size and appearance. Cochin is one example of this class (see Figure 22–6).

The **Mediterranean class** developed in the Mediterranean as an egg production group bred for large size and quality eggs.

Chicken production is the most economical program in the poultry industry. Chicken provides humans with a healthy food source in eggs and meat and as well as feathers and by-products for numerous other markets. Chicken eggs are useful in research for the development of medications and vaccines. Chickens are typically raised in confinement and are under strict regulations and guidelines from the U.S. Department of Agriculture. Over 8 billion chickens are raised each year in the United States for production purposes. Raising chickens has also become a popular hobby.

FIGURE 22–7 Beltsville small white turkey.

Turkeys

Turkeys are a large poultry breed raised specifically for meat (see Figure 22–7). The goal of producers and consumers is to raise turkeys with a large amount of white meat. Today, about 200 million turkeys are raised each year in the United States, which is a decrease in production over the past several years. Most turkey producers are on commercial farms raising over 100,000 birds a year on a single farm. Turkeys are usually marketed at about 20 weeks of age, with males being more popular due to their greater size.

Ducks

Ducks are water birds raised for meat, eggs, feathers, and soft down (see Figure 22–8). **Down** is a soft

FIGURE 22–6 Cochin hen.

FIGURE 22–8 Ducks are raised for eggs, meat, down, and feathers.

feather covering that grows under the primary feathers of young ducks. Ducks are not raised as commonly as chickens and turkeys. A little over 15 million ducks are raised in the United States each year. Ducks are economical in that they grow faster and become heavier than chickens.

Geese

Geese are water birds that are larger in size than ducks and are usually raised for meat, eggs, feathers, and down (see Figure 22–9). A little over 1 million geese are raised in the United States each year. Geese enjoy eating grass and weeds. They may be kept as ornamental pets to keep weeds under control and are

CHAPTER 22 Poultry Breed Identification and Production Management 405

FIGURE 22–9 Geese.

also known to be good watch animals. They are vocal, can be aggressive, and are resistant to many common poultry diseases. They can also become destructive to crops in large numbers.

Peafowl

Peafowl are poultry raised for their beautiful colors and large feathers. The male peafowl is known for the spread of its tail feathers, called a *train* (see Figure 22–10). It is not uncommon for the train of the male to grow to five times the length of the body. Most peafowl are raised for ornamental pets and eggs.

Swans

Swans are water birds similar to ducks and geese. They are known for having a long, thin, and graceful neck and come in a variety of colors (see Figure 22–11). They are raised for ornamental purposes.

FIGURE 22–11 Swan.

Other Poultry Species

Several species of wild birds have been domesticated for poultry production purposes. The small and hardy **Guinea fowl** are raised for meat, eggs, and hunting purposes (see Figure 22–12). They produce eggs that have a thicker shell than chickens and are often used for dying and decorations, as they are not easily broken. **Pigeons** are small birds that are common in the wild and raised in captivity for meat and competition purposes. At one time, many people raised carrier pigeons, which are bred and trained to carry small messages or identification bands from one location to another, common in competitive events. **Quail** are another common wild species raised in captivity. They are small round-bodied birds known for their "*bobwhite*" whistle sound. They are raised for both meat and eggs. **Pheasant** is also a wild species raised in captivity. They are larger in size and raised primarily for meat and for hunting purposes (see Figure 22–13).

FIGURE 22–10 Peacock.

FIGURE 22–12 Guinea fowl.

FIGURE 22–13 Pheasant.

FIGURE 22–15 Emu.

Ratites

Ratites are a group of large flightless birds that have become popular in commercial poultry production programs. The largest and best-recognized ratite is the ostrich. The adult **ostrich** can weigh over 350 pounds and stand over 10 feet (see Figure 22–14). They are known to have a long life span, capable of living over 70 years. Ostrich are raised for meat, eggs, feathers, and by-products, such as oil and skin. Their feathers, *plumes,* are used in decorations. **Emus**, smaller than ostrich but with a similar body structure, are native to Australia (see Figure 22–15). Another common ratite is the **rhea**, which is the smallest of these three and native to South America (see Figure 22–16).

FIGURE 22–14 Ostrich.

FIGURE 22–16 Rhea.

Selection of Species or Breed

A bird that produces the largest number of eggs or meat as quickly as possible is desired. For layers, this means producing one egg every day. Chicks should be in excellent condition and grow to a large size, based on breed standards. Meat breeds should reach market weight in 6 weeks, an efficient time that will reduce the cost of production. Other factors that should be considered when selecting poultry include how well the bird eats and whether it uses the food efficiently, a fast return on the investment, a high profit compared to the cost of feed, the amount of space needed, and the amount of labor involved.

Nutrition

Adult poultry should receive a commercial feed formulated for its species of fowl: a maintenance diet should be grain consisting of 12 to 15 percent protein. Young chicks begin on a starter food that is between 18 and 20 percent protein; they should be placed on a commercial diet as quickly as possible. Poultry will eat off of the ground or in pans and dishes. Most producers use an automated system to prevent contamination or disease in aseptic environments. Automated feeder systems consist of a trough or tray and a conveyor belt: the feed is weighed and a specific amount is distributed at certain times (see Figure 22–17). Water may be provided in a similar manner; however, poultry must be monitored to ensure they are consuming an appropriate daily amount. Water troughs must be cleaned on a daily basis to prevent contamination.

Behavior

Poultry are known for their adaptable behavior to any type of environment and changing conditions. The domestic fowl by nature is a wary, shy animal. Poultry have excellent vision and hearing, but other senses tend to be poorly developed. Many poultry species are gregarious in nature and have a social status within the flock. Fowl have specific body postures, associated with the position of the head, that allow communication with other birds. They maintain contact with flock mates by sight and vocal communication. They often become nervous and more vocal when flock mates are out of sight. A male rooster will establish hens as his territory and become aggressive and fight with other roosters that may attempt to enter the flock. Poultry can become aggressive with humans: they may become territorial and give chase, flapping their wings and using them as a weapon.

Basic Training

Poultry species that are shown can be trained to be handled and restrained rather easily. Handling should begin when the fowl are young and continue routinely. The wings should be extended for observation and examination (see Figure 22–18). The feet and legs should be handled and feathers checked for proper appearance. With regular handling, poultry will easily become accustomed to humans.

Equipment and Housing Needs

Housing depends on the type of poultry production and the number of birds within the flock. Housing may be as basic as a cage or shed or an elaborate building that houses 40,000 birds (see Figure 22–19). Housing must meet the needs of the birds as far as temperature and humidity go. Day-old chicks require 1/4 to 1/3 square feet of space per bird. Space requirements increase as the chick grows. Growers will commonly use portable wire fencing to contain chicks and remove it as the birds grow. Floor and cage coverings must be gentle and dust-free. **Litter** made of fine wood shavings that absorb droppings is often used as bedding. Poultry need proper lighting for ideal growth and health; this may include artificial

FIGURE 22–17 Automated feeder system.

lights set on timers to boost growth and egg production. Temperatures should be set at 85 to 95 degrees Fahrenheit for young growers; after 6 weeks of age, the temperature can be reduced to 70 to 75 degrees Fahrenheit. Poultry require adequate ventilation. A relative humidity of 50 to 75 percent is ideal. Emergency generators should be available in large commercial programs in case of power failures.

Restraint and Handling

To restrain and handle poultry, hold the wings against the bird's body to keep them from flapping, which can injure the handler or the animal. Restrain the body securely with one arm and hold the head around the neck with your other hand (see Figure 22–20). Poultry breathe using their diaphragm and chest and should never be held tightly in this area. Young chicks often lie quietly in the hand and are thus easily examined (see Figure 22–21).

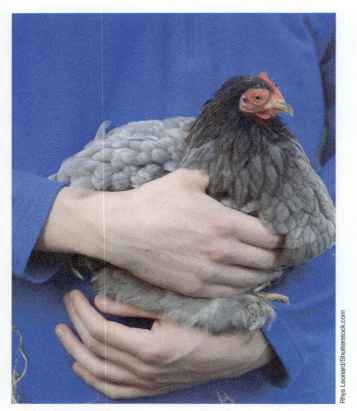

FIGURE 22–18 Poultry can become accustomed to being handled with care.

FIGURE 22–19 Enclosure for a small flock of chickens.

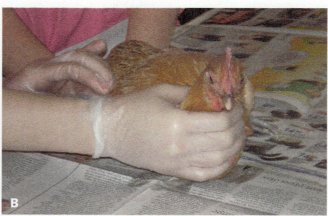

FIGURE 22–20 Proper methods for restraining poultry.

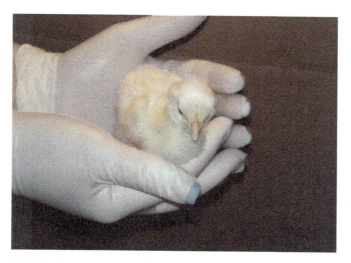

FIGURE 22–21 A young bird will sit quietly in the hand.

Grooming

Poultry do not typically require grooming; however, species that are shown should be clean and free of external parasites, dirt, and debris. Feathers can be gently sprayed with water to remove dust and debris. Fowl will preen and clean their wings and feathers many times throughout the day. Some owners may prefer to have the nails trimmed, wings clipped to prevent flying, and beaks trimmed and shaped to avoid overgrowth (see Figure 22–22).

Basic Health Care and Maintenance

All birds shed their feathers in a process known as **molting**. This allows for new feather growth. During molting, hens decrease their egg production or may stop laying eggs all together. The body is depleted of calcium as new feather development and growth occurs. Most producers will rest hens during molting to allow time for the body to replenish its level of calcium. Birds typically begin to molt at about 1 year of age; the process begins in the late fall. Molting takes place over a 4-month period until new feathers develop. Producers may do a forced molting of some hens so that egg production profits do not stop for all layers. A **forced molting** triggers the molting phase to begin at a time of the year when it does not normally occur. This is done by decreasing the amount of light that the birds receive to 8 hours per day. After the molting, the amount of light is restored to 14 to 16 hours per day.

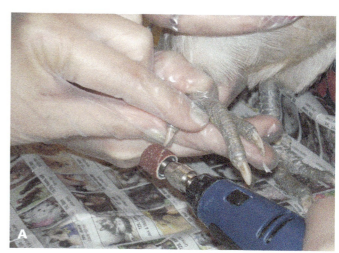

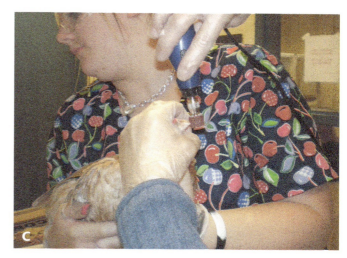

FIGURE 22–22 (A) Trimming the nails. (B) Clipping the wings. (C) Filing the beak.

Sanitation is essential for all poultry systems, as one sick bird can quickly kill an entire flock. Sound sanitation control methods are required to ensure disease-free flocks. Insect control and removal of waste materials are important. Sanitation control methods include the following:

- Disinfection of cages, equipment, and clothing
- Use of insecticides
- Waste control
- Disposal of dead birds
- Removal of litter
- Limited access for people and vehicles

The disposal of dead birds must adhere to guidelines and regulations because of the potential for human contact and the spread of disease in wild bird species. Dead birds should be removed immediately. The best control method of disease is through incineration, or burning the **carcass** (the dead animal's body) to ashes. Other methods of disposing of a dead bird include **composting**, or layering the carcass with waste materials and bedding to allow it to decompose or wear away. **Burial** means placing the carcass in the ground and covering it with soil.

Vaccinations

Poultry should be placed on a regular vaccination program to help develop an immunity to specific diseases (see Figure 22–23). **Fertile** eggs can be vaccinated using the *in vivo* method: the embryo is injected with a vaccine during the incubation phase. This is usually done on the 18th day of incubation. Chicks will hatch with immunity; this decreases the stress of regular vaccination and allows for a healthier chick that is ready for market 2 days earlier than with conventional vaccination. A specialized injection system is used to vaccinate eggs at a high rate of speed through the use of a needle that punctures the egg shell and delivers a set dose of vaccine. This machine is capable of injecting 20,000 to 30,000 eggs per hour. If the *in vivo* method is not used, vaccines should be administered on day 1 of the hatch at the time of the debeak procedure. The vaccine can be administered into a wing, within the eye membrane, or via a nostril. The program should vaccinate for common diseases. All vaccines and boosters should be given by 12 weeks of age. Breeders are typically vaccinated every 3 months, or as necessary.

Reproduction and Breeding

Female hens lay eggs that are sterile if no male is present and fertile if a male rooster is available within the flock. The roosters and hens must mate for the **fertilization** of eggs to occur. The eggs may be incubated or left on the nests to be hatched out. The **incubation period** varies from species to species and occasionally by breed and egg size (see Table 22–2). Hatching also depends on the temperature and humidity of the environment. Most producers will hatch out eggs in an **incubator**, which is a large enclosed secure piece of equipment with controlled temperature and humidity settings (see Figure 22–24).

Hens do not need to mate for eggs to be produced. There is no difference in taste or quality to the consumer between a fertile and sterile egg. Fertile eggs must be kept in an environment with controlled temperatures for eggs to hatch. Fertile eggs placed in a refrigerator will not produce an embryo. Artificial insemination is not used in chicken production but is commonly used in turkey production systems. Turkeys often have difficulty breeding due to their size.

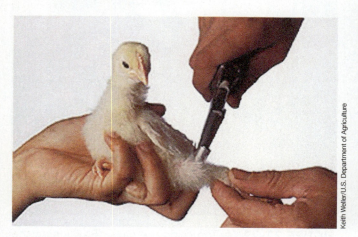

FIGURE 22–23 Chick receiving a vaccine.

FIGURE 22–24 Eggs in an incubator.

TABLE 22-2

Incubation Lengths of Poultry

Chickens	Average of 21 days
Ducks	28–35 days
Geese	29–31 days
Turkeys	27–28 days

The incubation of chicks varies between species. Most incubation temperatures are between 99 and 100 degrees Fahrenheit (see Figure 22–25). The humidity should remain at between 82 and 84 percent. Ideally, eggs should be rotated daily for uniform temperature and development.

Common Diseases

Poultry are susceptible to 33 different disease pathogens. They are highly susceptible to viruses and bacterial infections; the most common problems occur from nutritional problems. Poultry are costly to treat, especially when large flocks are affected. When fowl become sick, many will die. When they are sick, they become less efficient and decrease or stop egg production or cannot be used for human consumption. This causes a loss in profits.

Marek's Disease

One disease that affects poultry is **Marek's disease**, also known as **range paralysis**. This condition is caused by a **herpes virus** that results in diarrhea, weight loss, leg or wing paralysis, or perhaps death.

FIGURE 22-25 Temperature must be monitored in order for the eggs to hatch.

There is currently no treatment, but a vaccine is available. Some poultry species are genetically predisposed and have a poor resistance to the virus.

Newcastle Disease

Newcastle disease is caused by a virus that affects the respiratory system and causes wheezing, gasping for breath, open-mouthed breathing, possible paralysis, soft-shelled eggs, and possibly no egg production. One telltale sign is that birds will attempt to twist their necks around from the respiratory distress. There is a vaccine, but no treatment to date.

Infectious Bronchitis

Infectious bronchitis is a respiratory virus that affects only chickens and mostly young birds. Older birds can also become infected and may suffer a decrease or halt in egg production. A vaccine is available. The disease can be prevented through proper sanitation and isolation of sick birds. Signs of infectious bronchitis include the following:

- Bronchitis
- Wheezing
- Nasal discharge
- Gasping for air
- Poor appetite
- Ruffled feathers
- Depression

Fowl Cholera

Fowl cholera is caused by a bacterial infection that infects all species of poultry. Signs include fever, purple coloration to heads and combs, yellow droppings, and sudden death. A vaccine is available, and it is possible to treat the disease with a class of antibiotics known as *sulfonamides*.

Avian Pox

Avian pox is a virus spread by mosquitoes. It is a slow-spreading disease that has several strains; it is species specific. This virus is difficult to control, and there is no treatment. Vaccination is the key to prevention. Signs of the virus include respiratory distress and wart-like growths on the skin and beak.

Avian Influenza

The most recent and worrisome disease affecting the poultry industry is **avian influenza**, also commonly known as **bird flu**. This virus affects all breeds of poultry and occurs from a naturally occurring virus within the intestinal tract. It is transmitted directly through body fluids, such as nasal discharge or droppings. It is zoonotic to humans, but no human-to-human contact transmission has been recorded. There are several strains, including a highly fatal one. So far, the United States has not had any fatal outbreaks. This virus is of alarming concern, as studies show that it is capable of *mutating*, or changing forms to survive. A human vaccine has been developed, but currently there are no avian vaccines or treatments. Isolation of all sick birds is key.

Common Parasites

Poultry are susceptible to 10 parasites. Internal and external parasites are common. Isolated cages, confinement methods, and specialized production systems help reduce the possibility of infection. Proper sanitation controls are the key to prevention. Control of insects and wild birds is helpful.

Coccidiosis

Coccidiosis is a parasitic disease of the intestinal tract of animals caused by coccidian **protozoa**. *Coccidia* are the most common internal parasite affecting all birds. They are commonly seen in the droppings of wild birds and are spread by feces in the soil, food, or water sources. Companion animals, livestock, and people may easily be infected from contaminated sources. Treatment is in the form of antibiotics and dewormers given by mouth or in a water source. Signs of coccidiosis include the following:

- Lethargy or depressed appearance
- Bloody droppings
- Pale skin
- Weight loss
- Loss of appetite
- Ruffled feathers

Large Roundworms

Large roundworms may occur in the intestinal tract and can reach up to 3 inches in length. Signs of roundworm infestation include diarrhea, weight loss, an **emaciated** (severely dehydrated and appearing very thin) appearance, and droopy wings. Prevention and treatment are through sanitation and dewormers.

Tapeworms

Tapeworms are more common in poultry with access to the outdoors: most birds are infected through ingesting hosts that include snails, earthworms, beetles, and flies. Prevention is through diet control, dewormers, and proper sanitation. Signs of tapeworms include pale skin over the body, head, and legs.

Mites

Mites are the most common external parasite affecting poultry. Birds may also have lice, ticks, and chiggers. The key to prevention is through the use of approved insecticides and limited exposure to the outdoors and wild birds. External parasites are visible; other signs include droopy wings, a listless appearance, itching, feather loss or damage, and a pale skin color.

Poultry Production

Poultry production began over 5000 years ago in Asia. Poultry was then introduced to America in the 1600s. Families began raising small flocks of chickens for both eggs and meat sources. Turkeys were later introduced by the Aztec Indians, and settlers began raising and domesticating wild turkeys. In the 1950s, poultry production became a large and important industry in the United States. The confinement method of production began with standards set by the U.S. government. Currently, California is the largest state poultry egg producer. Because of the popularity of poultry and egg production, many agricultural and animal science college programs offer a program in **poultry science**. This is the study of poultry and includes breeding, incubation, raising, housing, and marketing (see Figure 22–26). The goal of poultry science is to learn to produce a quality product at a reasonable price. Most poultry houses hold 40,000 birds. One hen can produce over 300 eggs a year. Over the last 50 years, poultry science has developed methods for increasing production rates of commercial fowl.

Poultry uses have gone beyond the conventional food and fiber sources. Eggs are currently being used in the research and development of vaccines for veterinary and human medicine. Many types of human and animal medications are now egg based.

Poultry Production Systems

There are four main poultry production systems, mostly relating to chickens: broiler production, egg production, pullet production, and broiler egg production. These systems are used in raising poultry, eggs, meat, feathers, or other products for human use.

FIGURE 22–26 Poultry science courses focus on the study of breeding, incubation, raising, housing, and marketing.

Pullet Production

In **pullet production**, pullet hens are raised for the purposes of fertile egg production. They are selected as day-old chicks and raised in a similar way to broilers. Only the females are selected, and they are selected for maximum egg production. The pullets are raised for 20 weeks and then placed in breeding systems. The hens usually begin laying eggs at about 24 weeks of age and reach maximum production rates between 30 and 35 weeks of age. Pullets are also called **replacement hens**. This is usually a contracted commercial industry that raises breeding poultry. One cockerel is placed per 8 to 10 pullets and raised together for future breeding. As they mature, they may be placed in a larger flock.

Broiler Production

The goal of a **broiler production** system is to produce the largest amount of meat in the shortest amount of time. Chicks are raised to 6 weeks of age and should weigh about 4 to 4.5 pounds in size. They are raised in large poultry houses in confinement to prevent disease and contamination (see Figure 22–27). They must

FIGURE 22–27 Broiler chicks in a commercial broiler operation.

stay healthy to keep growing at an adequate rate. The chicks should grow efficiently in order to reduce costs. The mortality (death rate) should be kept at less than 5 percent. The more livestock and production animals that are produced, the higher the death rate.

Broilers should grow at a rate of 1-pound weight gain for every 2 pounds of food. Young newborn chicks are given starter food, which is finely ground grain meal. Starter food should be between 18 and 20 percent protein for ideal growth. As the chicks age, the protein amount should decrease to 12 to 15 percent. Most starter food is soybean meal with fish or a meat-meal base. As they age, they can begin eating a corn- and grain-based diet with added vitamins and minerals for continued growth.

Egg Production System

The goal of an **egg production** system is to produce a high-quality egg for human consumption. Each hen should produce one egg per day. The eggs are not fertile and are not hatched or incubated. They are graded by their size and quality: grades range from small to jumbo. The eggs are placed in cartons with the small end down and the large end up to protect the air cell of the egg, which helps protect the quality of the egg. Ninety percent of all egg production systems use layers kept in specialized cages with nests that have slatted trays that collect the eggs as they are laid. The eggs may be collected at regular times or through the use of an automatic conveyer belt (see Figure 22–28). The facilities are kept sanitary to protect the eggs from contamination and damage.

Egg Anatomy

The parts of an egg are important to egg production and in determining the quality of the egg (see Figure 22–29). The **eggshell** is the outer layer of

414 SECTION II Veterinary Animal Production

FIGURE 22–28 Eggs roll to the front of the layer cages and are moved on an automated system to the packing house.

the egg and is made of calcium. The inner yellow membrane is the **yolk**, which is a cell created by the hen's ovary. The **albumin** is the egg white layer that surrounds the yolk. The **air cell** is the empty area located at the large end of an egg, where oxygen is stored. Within the egg yolk is a **germinal disc**, which is a white spot within the egg where the sperm enters the cell to allow fertilization to occur. The egg also has a **chalaza**, which is the part of the egg that anchors the yolk into the egg white. The egg parts are held together by two membranes: the **inner shell membrane**, which is a thin area located outside of the albumin, and the **outer shell membrane**, which is a thin area just inside of the shell.

Egg Quality

Egg quality is determined by the internal and external appearance of the egg (see Figure 22–30). The size, shape, color, and condition are important factors to egg production. Cracks, blemishes, and dirt are signs of a poor-quality egg. The shell should be of good quality and consistency, as it is made of calcium and is what protects the egg. Ninety-five percent of eggs sold in markets are white; however, the demand for brown eggs is increasing as more chicken breeds produce brown eggs than white. There is no difference in taste or anatomy.

The internal structures of the egg can be viewed through a process called **candling**. Candling is done by shining a light source through the air cell to visualize the inside egg parts (see Figure 22–31). Fresh eggs will have small air cells. Older eggs will have a larger air cell. The yolk should look yellow because of high amounts of **carotene**, a protein that promotes the health of organs, such as the human eye. The albumin should cling near the egg yolk when rotated. If the albumin moves or spreads out with movement, that indicates age and deterioration. Occasionally, a blood cell may show up in the yolk of a fertilized egg that is a sign of embryo development. Eggs that are transparent or show signs of dark spots are damaged.

Broiler Egg Production System

In a **broiler egg production system**, pullets and cockerels are raised together specifically to produce broilers. The females should have quality eggs and large production numbers and the males should be large in size. The chicks are hatched out and selected for broiler production systems.

LAYERS OF ALBUMEN
- Outer liquid
- Dense (albuminous sac)
- Inner liquid
- Chalaziferous
- Ligamentum albuminis
- Ligamentum albuminis
- Egg shell
- Chalazae
- Ligamentum albuminis
- Air cell
- Egg membrane
- Ligamentum albuminis
- Shell membrane
- Yolk

Structure of the egg

FIGURE 22–29 Parts of an egg.

CHAPTER 22 Poultry Breed Identification and Production Management **415**

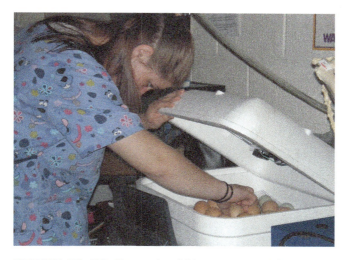

FIGURE 22–30 Eggs should be examined for cracks and blemishes.

FIGURE 22–31 (A) Device used for candling eggs to inspect structure. (B) Candling eggs to reveal cracks in the shell or to see embryo development.

SUMMARY

The poultry industry has become one of the most marketable and economical animal production systems in the United States. Poultry species include chickens, turkeys, ducks, geese, and ratites. Chicken and turkey are the most popular species for poultry production. Poultry science programs have been developed because of the popularity and need for further development of the industry. Poultry meat, eggs, feathers, and numerous by-products have become popular because of the health benefits that these products offer humans.

Key Terms

air cell   an empty area located at the large end of an egg that provides oxygen

albumin   the egg white surrounding the yolk

American class   a chicken class developed in the United States for egg and meat production

Asiatic class   a chicken class developed in Asia for show and ornamental breeds

avian influenza   a virus that affects the intestinal tract of both poultry and humans; also known as *bird flu*

avian pox   a virus affecting poultry that is spread by mosquitoes and slowly lowers the immune system

beard   hairs located under the wattle in turkeys

bird flu   an influenza virus found in avian species that may be zoonotic to humans

broiler   a young chick that is between 6 and 8 weeks of age and under 4 pounds used for meat

broiler egg production system   a system in which chicks are raised for large production numbers and selected for broiler programs

broiler production   a system used to produce the most meat possible

burial   the process of placing a carcass in the ground and covering it with soil to allow decomposition

candling   a technique of using a light to view the internal structures of an egg

capon   a male castrated chicken over 6 pounds and 6 months of age

carcass   the dead body of an animal

carotene   protein that promotes the health of organs

chalaza   the part of the egg that anchors the yolk to the egg white

chick   a young chicken

chicken   the most important type of poultry; has four different classes

coccidiosis   a disease caused by an internal parasite found on poultry and wild birds that is spread through water, food, and soil contamination

cock   an adult male chicken

cockerel   a male chicken under a year of age

comb   a flesh-like projection on the top of a chicken's head

composting   layering the carcass of a dead bird with waste materials and bedding to allow it to decompose or wear away

debeak   to remove the tip of a chick's beak to prevent pecking damage to other chicks

down   a soft feather covering that grows under primary feathers

drake   an adult male duck

duck   a small water bird used for meat, feathers, and down

duckling   a young duck

egg production   a system used to produce a quality egg for human consumption

egg quality   a standard determined by the egg's internal and external appearance

eggshell   the outer calcium layer that protects an egg

emaciated   severely dehydrated and appearing very thin

emu a large flightless bird native to Australia that is smaller than the ostrich

English class a chicken class developed in England for better meat production

fertile an egg that has an embryo because it has been fertilized by a male

fertilization the process of the egg and sperm uniting to create an embryo

flock a group of poultry

forced molting triggering feathers to be shed and regrown at specific times of the year

fowl poultry

fowl cholera a bacterial infection of poultry that causes a purple coloration to the head

fryer broiler

gander an adult male goose

geese large water birds used for meat, eggs, feathers, and down

germinal disc a white spot in the egg yolk where the sperm enters the egg cell to allow fertilization

gizzard a muscular organ used to break down food to allow digestion

goose an adult female goose

gosling a young goose

grit a substance used to break down hard materials, such as seed shells, to allow food to digest

guinea fowl a small wild bird domesticated for meat, eggs, and *hunting*

hen an adult female chicken

herpes virus a virus in bird species that causes diarrhea, weight loss, wing or leg paralysis, and death

incubation period the length of time until eggs hatch

incubator equipment used to house eggs at a controlled temperature and humidity

infectious bronchitis a respiratory virus that affects chickens only as well as egg production

inner shell membrane a thin area located outside of the albumin

knob a projection on the top of the beak

layer an adult female hen that lays eggs

litter wood shavings used as bedding that absorb droppings

Marek's disease a poultry disease caused by a herpes virus; also known as *range paralysis*, as it can cause wing or leg paralysis

Mediterranean class chicken class developed in the Mediterranean for large quantities of egg producers

molting the process of shedding feathers to allow new feather growth

Newcastle disease a respiratory virus that causes poultry to twist their necks

ostrich the largest and most common ratite, which grows to over 10 feet tall

outer shell membrane the thin area just inside of the shell

peacock an adult male peafowl

peafowl poultry raised for their colorful, large feathers

peahen an adult female peafowl

pheasant a wild species of game bird raised for meat and hunting purposes

pigeon a small wild bird raised in captivity for meat and competition purposes

poult a young turkey

poultry domesticated birds with feather, two legs, and two wings that is used as a source of eggs, meat, feathers, and by-products

poultry science the study of poultry breeding, incubation, production, and management

protozoa a single-celled organism or parasite

pullet a young chicken raised to lay eggs

pullet production a system in which hens are raised for fertile egg production

quail a small wild bird raised for eggs that makes the whistle sound "bobwhite"

range paralysis a common name for Marek's disease

ratite a large flightless bird raised for poultry production purposes

replacement hens a pullet that takes the place of an older hen that is no longer capable of egg production

rhea a moderate-sized flightless bird native to South America

roaster a chick over 8 weeks of age and 4 pounds that is used as meat

rooster an adult male chicken

serration a space on the top of the comb

spent hens an adult female hen that has aged and whose egg production has decreased or stopped that is used as a meat source in processed foods

swan a large water bird with a long, thin neck that is raised as ornamental poultry

turkey a large size poultry breed with a large body and a long, thin neck that is specifically raised for meat

wattle a flesh-like area located under the chin

yolk the inner yellow cell membrane created by the hen's ovary

Review Questions

1. What species of birds are considered poultry?
2. What are the classes of chickens, and where did they develop? What were they developed for?
3. What are some purpose and uses of poultry?
4. What are the four types of poultry production systems? Explain each type.
5. What is poultry science? What is its importance to the poultry industry?
6. What are the parts of an egg?
7. Explain the difference between normal molting and forced molting.
8. What methods can be used to dispose of dead birds?
9. What is candling? In what ways can it be useful in determining egg quality?
10. What are the needs of eggs during incubation?

Clinical Situation

Dr. Ross, a poultry veterinarian, and Harry, a veterinary assistant, have just arrived at a poultry egg production facility. Today is a monthly scheduled health inspection visit for the several thousand chickens housed on the property.

Dr. Ross and Harry put on uniforms—which have been sterilized—and face masks before they enter the facility. They are just about to enter the first poultry barn when Harry realizes he has not put on sterile booties over his shoes. He tells Dr. Ross about this mistake and goes back to the veterinary truck to get the booties.

- What was done correctly in this situation?
- What was done incorrectly in this situation?
- What should have been done when the doctor and assistant arrived at the facility?

Section III
General Anatomy/Physiology and Disease Processes

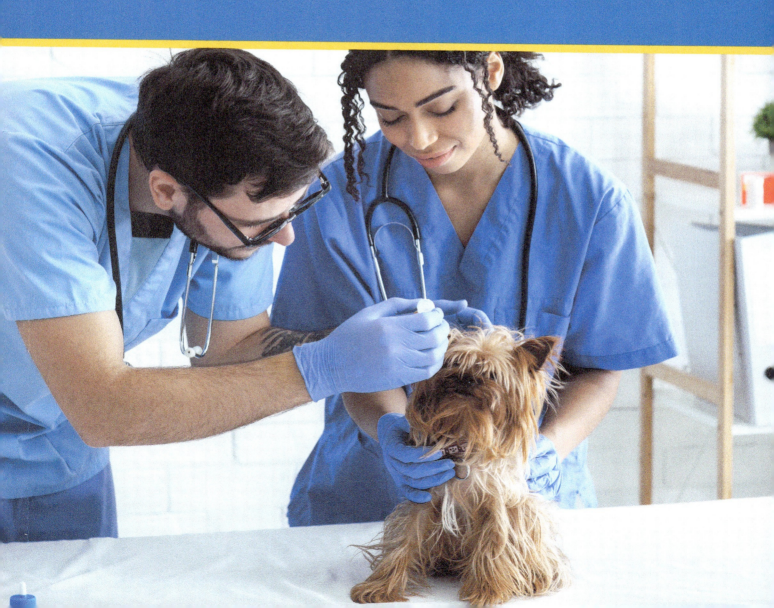

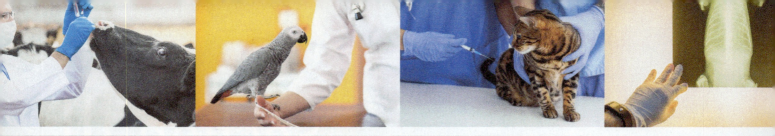

CHAPTER 23: The Structure of Living Things

Objectives

Upon completion of this chapter, the reader should be able to:

23.1 Define anatomy, physiology, and pathology
23.2 Identify the structural units of living things
23.3 Describe the functions of cells and cell division
23.4 Identify the function of various types of tissues
23.5 Explain the purpose of organ systems

Introduction

Knowledge and understanding of normal anatomy is critical to the veterinary professional who assists in restraint, handling, and care of a variety of animals. **Anatomy** is the study of the internal and external body structure. **Physiology** is the study of the functions of those structures and how they work. A sound understanding of what is normal will aid in the identification of disease processes when something abnormal occurs. **Pathology** the study of diseases, especially of the structural and functional changes produced by them.

While animal species vary in appearance, there are similarities in external body structures and the functions of those structures. Sometimes the structures share a common name; other times the name for the structure may be specific to a particular species of animal. Each species of animal is discussed in the production chapters of this text. It is recommended that you refer to each chapter for an external anatomy diagram and proper terminology of the external body parts related to a specific species. All animals have a head, body, and tail area. Some animals have feet, paws, or hooves. It is important that veterinary assistants become familiar with the terminology of external anatomical structures of the species they commonly work with.

Cells

The most basic structural unit of an animal is the **cell**. Cells are microscopic in size, and each living thing is made up of trillions of cells. Cells come in a variety of shapes and sizes, and each cell has a specific function. The study of cells is called **cytology**. Animals are made up of many different cell types that allow the animal to function.

Cell Makeup

Cells are made up of molecules. **Lipids**, also known as fats, combine hydrogen, oxygen, and carbon into a form that does not dissolve well in water. Therefore, fats float easily to the top of water. Fats are made up of a molecule of fatty acids and glycerol and are stored in the body as a source of energy.

Phospholipids are similar but have fewer fatty acids in their makeup. One end of their molecule is **hydrophilic**, or dissolved in water, and one end is **hydrophobic**, or repelled by water. These characteristics play an important role in the structure of the cell wall.

Carbohydrates supply energy and aid in the structure inside a cell. **Monosaccharides** are the simplest form of these molecules and include the sugar known as glucose that are stored as energy within the bloodstream of animals. The amount of glucose in the bloodstream may be monitored: too much or too little may compromise an animal's health. A condition known as diabetes may occur when blood sugar levels increase, and the animal's body does not use it correctly. Diabetes treatment requires a lowering of glucose in the bloodstream. **Polysaccharides** are starches and made up of many monosaccharides. One starch, known as *glycogen*, stores energy within cells. These molecules work together to build a cell's structure.

Proteins are another important part of cell structure and function. Proteins make up 50 percent of an animal's dry weight. Proteins are large molecules that work with 22 different amino acids that may join with sugars, starches, and lipids to form the strong structure of a cell. Muscle is an example of protein that enables cells to move. **Enzymes** are protein molecules that enable chemical reactions and changes to occur in the body and are part of the metabolism process. Enzymes break down cell components to enable the cells to be absorbed and used in the body—similar to how medications work within the body.

Proteins also carry oxygen through the blood and aid in blood clotting. Certain proteins, **antibodies**, help fight off bacteria, infections, and diseases.

A complex chain or proteins make up **nucleic acids** that are made from nucleotides, each containing a five-carbon sugar backbone, a phosphate group, and a nitrogen base. **Deoxyribonucleic acid (DNA)** provides the code for the cell's activities, while **ribonucleic acid (RNA)** converts that code into proteins to carry out cellular functions. The amino acid and protein combination will then be stored in a cell's genetic material. DNA is a double-strand molecule with loose binding bases. Its structure is twisted in the shape of a double helix (see Figure 23–1). DNA bases can be converted to the molecule RNA and play a role in protein synthesis. The order of DNA bases plays a part in the genetic code of an animal.

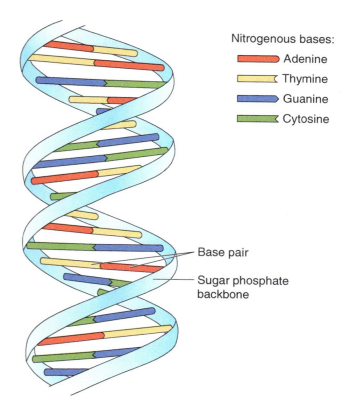

FIGURE 23–1 Double helix of DNA.

Clinical Concepts

Cells contain enzymes that aid in breaking down medications so they can be absorbed by the body and then metabolized. This explains why many medicines must be given repeatedly over a specific amount of time. Animal species may react differently to the same type of medicine. Many medications that are safe for humans are not digested well by animals. This can lead to reactions and toxic effects. Cats have fewer of certain enzymes that aid in breaking down medicines. Acetaminophen, which is the active ingredient in Tylenol, is found in many pain relievers. It is very dangerous to cats: as little as a quarter of a tablet can make a cat extremely sick. Since cats do not have the necessary enzymes to break down acetaminophen, it can easily build up in a cat's body and bloodstream and cause severe toxicity and even death. It is very important to determine if medications are safe for use in animals before administering them.

Cell Structure

A cell consists of many **organelles** (small units within the cell responsible for a specific function; see Figure 23–2). The structure of a cell includes a **cell membrane**, also

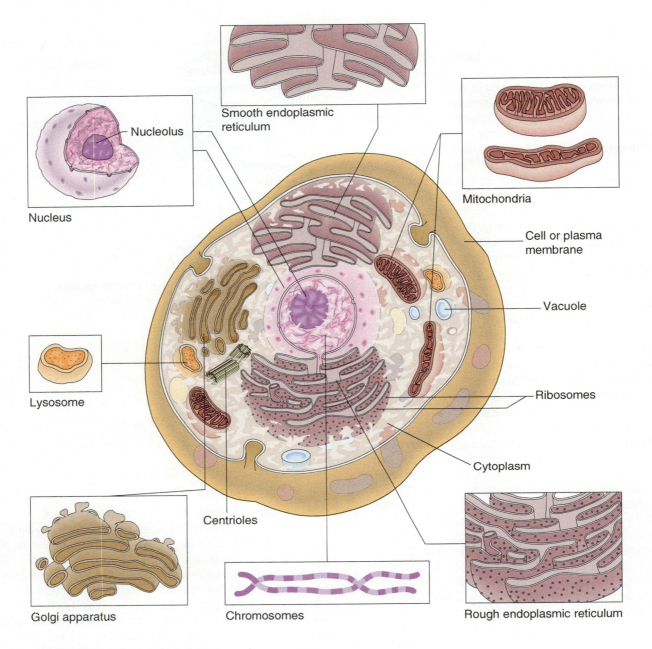

FIGURE 23–2 Parts of the cell.

called a cell wall or plasma membrane, which holds the cell together. This membrane is half lipid and half protein. One end is attracted to water and the opposite end is repelled by water. Cell membranes are semipermeable, meaning they allow some substances to pass through but not others. Some molecules, like water, pass through easily, whereas other molecules, like proteins and starches, are not able to pass. The **cytoplasm** is the fluid part of the cell that allows the cell's internal structure to move. The **nucleus** is the brain of the cell and is located in the center of the cell. (Note that red blood cells do not have a nucleus.) DNA is found in the nucleus of a cell.

As the cell divides, chromosomes begin to form, and the genetic material is passed on to all daughter cells. Within the nucleus is the **nucleolus**, which forms genetic material. This is where RNA is produced that then forms ribosomes. The **ribosomes** are structures within the cell that make protein. Ribosomes are macromolecular machines, found within all living cells, that perform biological protein synthesis. Ribosomes link amino acids together in the order specified by the codons of messenger RNA molecules to form polypeptide chain. The **endoplasmic reticulum** is a collection of folded membranes that attach to the nucleus. Endoplasmic reticulum that is lined with ribosomes that give the appearance of bumps along the edge is called *rough endoplasmic reticulum*. *Smooth endoplasmic reticulum* has no ribosomes attached to it and occurs less often throughout the body. The **Golgi apparatus**, where lysosomes are produced, is formed with large amounts of folded membrane similar to smooth endoplasmic reticulum. **Lysosomes** are enzymes that digest food and proteins. They help break down other molecules. Cells that die within the body are removed when enzymes within the lysosomes are released into the cytoplasm. This is called **autolysis** and allows for room for new cells to develop. This is often how the body fights infection. The **mitochondria** are the part of the cell that makes energy and are often referred to as the "*powerhouse of the cell.*" These rod-shaped cells have a double membrane. The inner membrane is folded and forms ridges, while the outer membrane is smooth. Mitochondria convert food particles into energy.

Cell Functions

Cells allow for **metabolism**, a reaction that occurs within the body in which chemicals are broken down and used by the body. Cells serve many functions within the body (see Table 23–1).

There are two main types of metabolism: anabolism and catabolism. **Anabolism** describes reactions in which smaller molecules are combined into larger molecules. **Catabolism** describes reactions in which larger molecules are broken down to form smaller molecules (the opposite of anabolism). *Extracellular fluid* is a liquid that surrounds all living cells, supplying them with all they need to function.

The skin is the only part of the body that does not contain this fluid. Extracellular fluid begins within the blood. It helps lubricate the body (such as in the eyes, where tears are produced, it helps moisturize the eyelids). Water is the major component of extracellular fluid. Oxygen and carbon dioxide are transported via this fluid. Many particles within this fluid must be maintained at a constant normal concentration, a state of equilibrium known as **homeostasis**.

Materials must be able to move in and out of cells. A process called **diffusion** allows molecules to move from a higher to a lower concentration. Molecules are always moving and are more likely to move to an area of lower concentration until they are equalized. The smaller the molecule, the more easily it is diffused. **Osmosis** is a special type of diffusion and allows particles to move across a membrane. Some larger molecules are unable to diffuse through a membrane and must move into higher concentrations. This process is called **active transport**. It requires a lot of energy and uses enzymes to aid in changing the concentration amounts. The process of moving these large molecules through a membrane is called **endocytosis**. The cell membrane wraps itself around a particle and pinches it off, allowing movement into the cytoplasm. Lysosomes and enzymes aid in this process of breaking down the cell into fragments.

Clinical Concepts

Edema is the buildup of excessive fluid within tissue. *Pitting edema* is demonstrated when finger pressure placed on the area leaves a "pit" impression that

TABLE 23–1
Cell Functions

Active Transport	A process in which a substance found in a lower concentration will move to an area where it is higher in concentration
Anabolism	A process in which smaller particles combine to form larger particles
Catabolism	A process in which larger particles break down into smaller particles
Diffusion	A process in which a substance found in a higher concentration will move to an area where it is lower in concentration
Endocytosis	Process in which a cell can take in a particle
Extra-Cellular Fluid (ECF)	Fluid found outside the cell, such as blood
Homeostasis	The tendency toward a relatively stable and balanced equilibrium between interdependent elements, especially as maintained by physiological processes
Osmosis	The movement of water or other solvent through a plasma membrane from a region of low solute concentration to a region of high solute concentration
Phagocytosis	Process in which dead cells and waste materials are eaten or removed from the body

426 SECTION III General Anatomy/Physiology and Disease Processes

slowly resolves. Protein levels within the bloodstream are often affected with edema, causing small blood vessels called *capillaries* to wall off the protein and not allow movement into the extracellular fluid. Water, however, is able to move freely. Osmosis occurs across the walls of the capillaries. As the protein levels decrease, the osmotic pressure allowing water to remain within the blood vessels also decreases. Water thus begins to accumulate within the tissues, creating edema.

Cell Division

Cell division occurs when one cell splits into two. Two types of cell division occur within the body: mitosis and meiosis. **Mitosis** is the cell division of **somatic cells** (nonreproductive), which aid in the body's growth and repair. There are five stages of mitosis (see Figure 23–3). One cell divides into two cells that are identical to the original cell. The rate of cell division depends on the needs of the animal. Some cells divide more rapidly than others. When uncontrolled mitosis occurs, the result is a condition known as *cancer*. New cells are produced more rapidly than necessary; the accumulation of these new cells in one area is known as a **tumor**.

In the first stage of mitosis, **interphase**, a single cell forms genetic material called *chromatin* loosely in the nucleus. The cell doubles its DNA. **Prophase** begins when the chromatin thickens and forms a chromosome.

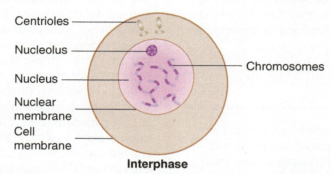

Interphase
The nucleus and nuclear membrane are distinct. The chromosomes are contained in the nucleus in a thread-like mass.

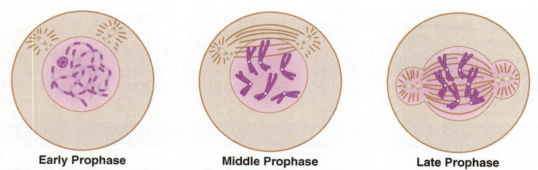

Early Prophase   **Middle Prophase**   **Late Prophase**
The centrioles begin to move toward opposite poles creating fibers between them. The chromosomes begin to condense. The nuclear membrane becomes less distinct.

Metaphase
The chromosomes begin to attach to the fibers between the centrioles.

FIGURE 23–3 The phases of mitosis. (*Continues*)

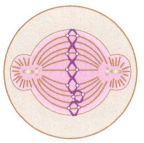 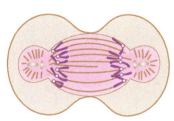

Early Anaphase   **Late Anaphase**

The chromosomes separate and move to opposite poles.

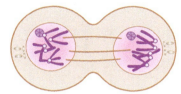

Telophase

The chromosomes are at opposite poles and a nuclear membrane begins to form.

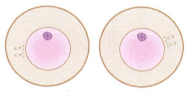

Interphase

Cell division is complete. Two new cells are identical to the original.

FIGURE 23–3 (Continued)

An "X" shape forms in the chromosome as the cell begins to separate. **Metaphase** begins when a spindle forms between the separation site. In the **anaphase** stage, the chromosomes split at the center of the spindle and begin to move outward in a "V" shape. **Telophase** is the reverse of prophase. Chromatin is a complex of DNA and protein found in eukaryotic cells. Its primary function is packaging long DNA molecules into more compact, denser structures. Chromatin loosely forms as a groove down the center of a cell, thus creating the development of a second identical cell structure. This continuous process repeats.

Meiosis is cell division in breeding and reproduction. This process involves a cell that divides into four unique **daughter cells** (see Figure 23–4). Each daughter cell has half the number of chromosomes of the parent cell. The process of fertilization provides for the remaining formation of chromosomes after the egg and sperm cells have united. All new cells contain identical genetic materials from the parents. The stages of cell division are outlined in Table 23–2. All chromosomes come in pairs. Each paired cell divides, resulting in four cells and then undergoing stages similar to those of mitosis. This allows the offspring to acquire traits from both parents.

Meiosis I in Males

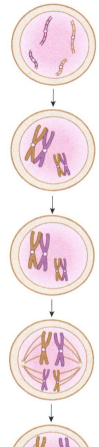

Prophase I
Chromosomes begin to condense and chromosomes similar in structure pair and cross over.

Metaphase I
Spindle fibers attach to chromosomes and chromosomes line up in center of cell.

Anaphase I
Chromosomes start to move to opposite ends of cell as spindle fibers shorten.

Telophase I
Chromosomes reach opposite ends of cell and nuclear membrane forms.

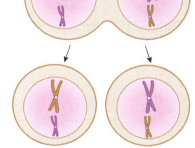

Cytokinesis
Cell division occurs producing two identical cells.

FIGURE 23–4 The phases of meiosis. (Continues)

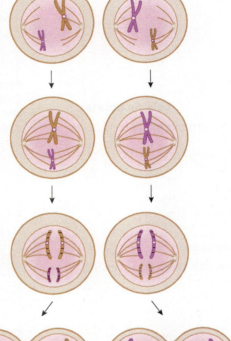

Meiosis II in Males

Prophase II
Chromosomes begin to condense and the nuclear membrane dissolves. Spindle fibers form.

Metaphase II
Spindle fibers attach to chromosomes and the chromosomes line up in center of cell.

Anaphase II
Centromeres divide and sister chromatids move to opposite ends of cell as spindle fibers shorten.

Telophase II
Chromosomes reach opposite ends of the cell and the nuclear membrane forms.

Cytokinesis
Cell division occurs producing four cells each with half of the original parent cell.

FIGURE 23-4 (*Continued*)

TABLE 23-2	
Stages of Mitosis and Meiosis	
Interphase	Cells in their normal state
Prophase	Chromatin forms and begins to form an "X" shape
Metaphase	A spindle forms at the center
Anaphase	Chromosomes split
Telophase	Final phase of cell division to create two or four new cells

Genetics

All animals pass on characteristics to their offspring through genes. These genes contain the information necessary to form a new animal. Animals with the most desirable traits should be evaluated for breeding through a process called **selection**. Selection is the process in which environmental or genetic influences determine which types of organism thrive better than others, regarded as a factor in evolution. This helps determine which animals should be allowed to reproduce and pass their genes on to future generations. **Geneticists** are scientists who study genes and genetics and use mathematical equations to predict the outcome of breeding.

 TERMINOLOGY TIP

Fertilization is the union of the sperm to the egg to create an embryo. A *chromosome* is the genetic material inherited from each parent and passed on to the offspring. ■

Clinical Concepts

Animals often develop lumps and bumps. It may be necessary to take a biopsy sample or completely remove a mass. This procedure allows a veterinarian to diagnose cell division of tumors and determine if a mass is benign or malignant (see Figure 23–5). A **benign** mass is localized to one area, does not spread to other areas, and is not cancerous. A **malignant** mass is more likely to invade other tissues, more likely to and spread to other parts of the body, and is considered cancerous. A mass may also be referred to as a *tumor* or *neoplasm*. A **pathologist**, one who interprets and diagnoses the changes caused by disease in tissues and body fluids, will evaluate the sample and provide a treatment plan based on the cellular findings. The sample is sent out for cytology analysis, which is the study of cells. This analysis will determine cell origin, structure, function and cause (see Table 23–3). Tumors develop because of rapidly growing and dividing cells that are abnormal. When a cell is determined to be cancerous, it will likely be abnormal, be of varying shapes and sizes, have enlarged nuclei, develop multiple nuclei, and have clumped chromatin. It is important that samples be sent to a laboratory to determine the mass's cellular makeup. Pathologists can now not only determine between cancerous and noncancerous cells, but they can also identify the type of tumor and the likelihood of reoccurrence. This is important in the veterinarian's treatment of the animal.

TERMINOLOGY TIP

Tissue can form normally or abnormally. The suffix *–plasia* is used to describe the formation, development, and growth of tissues and cells in excessive amounts. The suffix *–trophy* describes the formation, development, and increase in size of tissues and cells. The suffix *–oma* refers to a tumor or neoplasm. ■

Enzymes are chemicals that change other chemicals and are part of the metabolic process. Enzymes break down the components of a cell to allow them to be absorbed and used in the body. (Medications too are absorbed within the body; that is how they work.)

Tissues

Tissue is composed of a group of cells that are alike in structure. Tissues perform specific functions, such as forming skin and bone. The study of tissues is called **histology**. There are four types of animal tissues: epithelial, connective, muscular, and nervous. **Epithelial tissue** covers the body's surface, lines internal organ structures, and serves

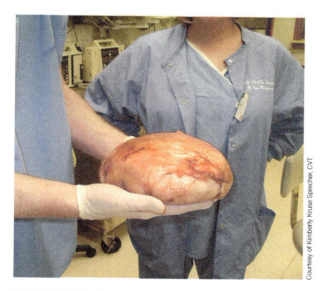

FIGURE 23–5 A lipoma tumor removed from a dog. This is a benign mass and is made up of fat cells.

TABLE 23–3

Terminology Relating to Abnormal Tissues/Cells

Anaplasia	A change in the structure of cells and their orientation to each other
Aplasia	A lack of development of a tissue or cell
Dysplasia	The abnormal growth or development of a tissue or cell
Hyperplasia	An abnormal increase in the number of tissues or cells
Hypoplasia	The incomplete development of a tissue or cell
Neoplasia	The abnormal new growth of tissue, resulting in the rapid multiplication of cells
Atrophy	A decrease in the size of a tissue or cell
Dystrophy	The defective growth in the size of a tissue or cell
Hypertrophy	An increase in the size of a tissue or cell

as a protective layer. **Connective tissue** holds and supports body structures by connecting cells, such as ligaments and tendons (see Figure 23–6). There are six main types of connective tissues: adipose (fat), loose, dense, elastic, blood, and bone/cartilage. **Muscular tissue** allows movements of body parts (see Figure 23–7). **Nervous tissue** contains particles that respond to a stimulus and cause a reaction in the body.

Epithelial Tissue

Epithelial tissue consists of tightly packed cells layered in sheets that line the body's surface and cover the openings of the urinary tract, intestinal system, and reproductive system. Epithelial tissue performs many functions, including protection. The largest organ system is the skin, which is made up entirely of epithelial tissue. The skin protects the body and internal organs from trauma, disease, extreme temperatures, and sunlight. Epithelial tissue also provides secretions—such as tears, saliva, and mucous—which protect and moisten surfaces that can become damaged if they dry out. Epithelial cells allow absorption of materials through the lining of cells internally and exchange materials from the blood and extracellular fluid. These cells can also prevent excessive fluid loss.

Epithelial tissue is classified by the shape of the cells and the number of layers in which it is housed (Figure 23–8). The layers of the epithelium are classified as simple, stratified, and transitional. **Simple epithelial tissue** consists of a single layer. **Stratified epithelial tissue** consists of multiple layers of cells, while the **transitional epithelium** is multiple layers of cells that have different shapes. The shape of epithelial tissues includes **squamous** (very flat); **cuboidal** (cube shaped); and **columnar** (tall and thin). The epithelium can be identified by its shape and type of layer. The

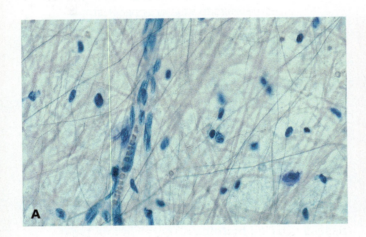
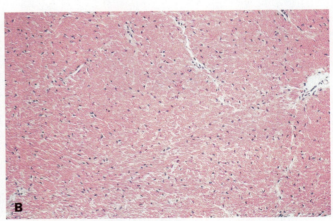
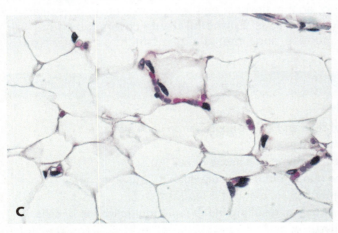
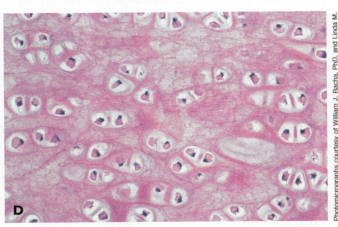

FIGURE 23–6 Types of connective tissue: (A) Mesentery (loosely arranged connective tissues attached to abdominal organs) of a cat, which shows (pink) collagen fibers and (dark, thin) elastic fibers; (B) densely packed elastic fibers found in the ligament of a sheep; (C) adipose (fat) tissue; (D) cartilage found in a pig.

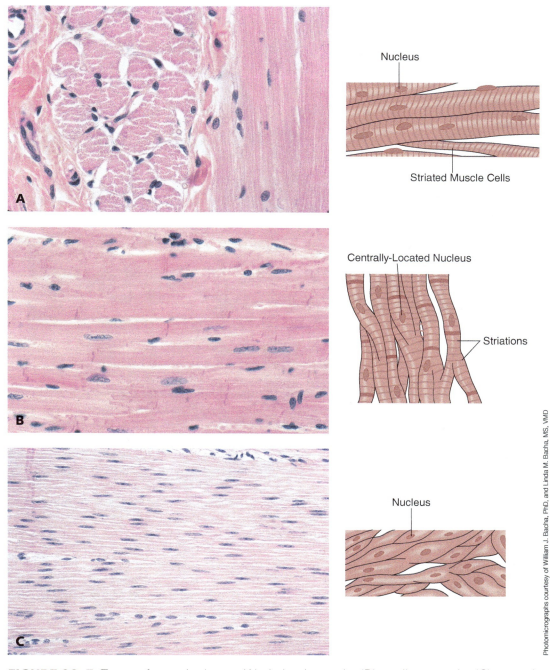

FIGURE 23-7 Types of muscle tissue: (A) skeletal muscle; (B) cardiac muscle; (C) smooth muscle.

simple squamous epithelium contains a single layer of flat cells and is found in the lining of both blood vessels and the respiratory tract. This tissue is very fragile and exists deep within the body in protected areas. *Simple cuboidal epithelium* contains a single layer of cubed-shaped cells. This type of cell is found in areas where absorption and secretion are necessary, such as in the lining of the endocrine glands and the kidneys. It is most commonly found in areas responsible for transport of bodily substances. *Simple columnar epithelium* contains a single layer of cells that are tall, thin, and fingerlike in appearance. This type of tissue lines the stomach and intestines. *Stratified squamous epithelium* consists of many layers of flat cells and is found in the external skin. Over time, these cells become flatter and are

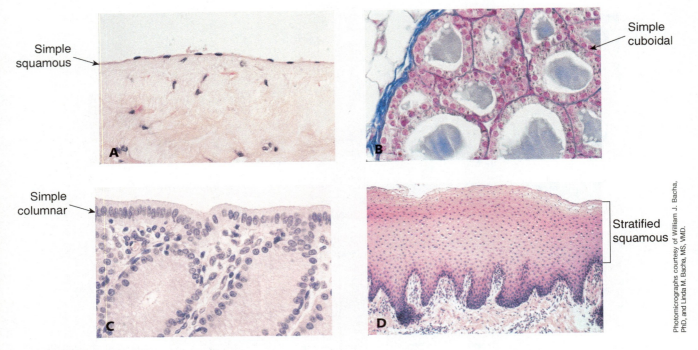

FIGURE 23–8 Types of epithelial tissue.

constantly being shed. This tissue is also found in the mouth and the lining of the rectum and vagina. *Transitional epithelium* is found specifically in the lining of the urinary tract. This tissue has multiple layers of cells and can stretch.

Connective Tissue

Connective tissue consists of specialized cells within large amounts of extracellular material that connect one organ or tissue to another, serve as protection, and provide structural support. Examples include tendons, ligaments, bone, and cartilage. Connective tissue looks smooth to fibrous. **Tendons** are bundles of protein—known as *collagen*—that provides strength and connects muscle to bone. **Ligaments** are layers of protein fibers that connect bone to bone. Ligaments have both collagen and elastin that provide strength and the ability to stretch.

Connective tissue also includes cartilage and bone. Cartilage has no blood supply and is maintained by the surrounding extracellular fluid. It is a thin lining that serves as padding located between each bone and moveable joint area throughout the body. Cartilage is also found in the rings of the trachea and holds its shape, within ear flaps, and within the growth plates of immature, growing animals. Cartilage is flexible and somewhat elastic. Bone consists of minerals that provide hardness for the protection of internal organs and allow the body movement, structure, and shape. Bone has a blood supply that provides nourishment for bone cell growth and repair.

Connective tissue is also found around the lining of internal organs for support. Much of this tissue consists of **adipose** tissue, or fat that is filled with lipid cells. This type of tissue is found within the abdomen, behind the eye, inside the bone marrow, and between muscles. The amount of adipose tissue present depends on the amount of stored fat that is reserved for later energy use. This tissue provides insulation during extreme temperatures. Blood is also considered connective tissue in the liquid substance called *plasma*.

Muscle Tissue

Muscle tissue primarily allows for movement. There are three types of muscle tissue found in an animal's body: smooth, skeletal, and cardiac. *Smooth muscle* is involuntary muscle, meaning it is controlled by the brain stem. This muscle type is located in hollow body organs, such as the urinary bladder, blood vessels, and intestinal tract. Smooth muscle is arranged in sheets and can contract and make the openings of organs and vessels smaller. In blood vessels, this contraction is known as *constriction*. *Skeletal muscle* is a voluntary muscle attached to the bones that allows movement sent from nerve signals delivered by the nervous system. This muscle type enables an animal to control which muscles it will move. *Cardiac muscle* is found within the heart. This muscle type is involuntary and continuous and functions without thought, such as during heartbeats, where the heart muscle initiates its own contraction.

Muscles consist of multiple strands of fibers. A group of muscle cells is called a **myofiber** and allows muscles to contract and relax. Two proteins—actin and myosin— work together to allow muscles to contract. **Actin** is a thin filament fiber; **myosin** is a thick filament fiber. Both of these proteins have bridges between them that allow them to slide and move. Muscles contract when a nerve cell sends a signal to release calcium and causes the proteins to slide across each other. When muscles relax, the cell actively transports calcium back to its origin. The number of fibers in a muscle depends on the size of the muscle. A muscle grows larger depending on how much it is used. When muscles are used more frequently, and strengthened, the myofibers become larger. When a muscle is used less (often from inactivity), it may decrease in size. This is often seen in animals that have injuries and cannot use the leg for a certain period of time.

Clinical Concepts

Dairy cattle are known for developing a condition called *hypocalcemia*, or "milk fever," in which blood calcium levels become too low. This may happen near calving time, when a cow excretes large amounts of calcium in its milk, causing calcium levels in its blood to quickly drop. These cattle become weak; they are often unable to stand up, as their muscles cannot contract to allow movement. Veterinarians treat this condition by giving the cow a calcium supplement directly into the bloodstream. As the calcium supplement enters into the muscles, the cow responds, and its body allows the muscles to contract. Within minutes, the cow is back on its feet and moving.

Nerve Tissue

Nerves allow the body to communicate by receiving and transmitting of signals. This requires nervous tissue, which is found within the brain, spinal cord, and peripheral nerves located throughout the body (see Figure 23–9). Nerve tissue contains large cells called **neurons** that react to stimuli and conduct

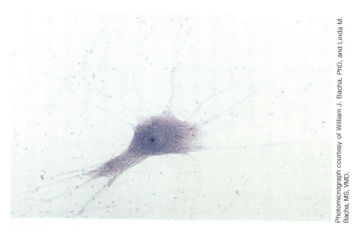

FIGURE 23–9 Nervous tissue.

electrical pulses. A hairlike extension from the neuron called an **axon** carries a **nerve impulse**, or message. Axons from multiple neurons form a **nerve**. A nerve impulse occurs when sodium and potassium particles stimulate and move the neurons rapidly down the axon to the nerve.

Clinical Concepts

When nerves become damaged, they show abnormal signs based on where the injury is and the group of nerves affected. Dysfunction to eye and surrounding facial nerves in the autonomic nervous system may cause a condition called *Horner's syndrome*. The sympathetic nerves are located at the base of the skull and controls several functions of the eye. Symptoms may include a constricted pupil, a droopy upper eyelid, an eye sunken back into the socket, and the upward protrusion of the third eyelid. These signs indicate which nerve is damaged and help a veterinarian better determine the diagnosis and treatment of the underlying cause.

Organs

A group of similar tissues form **organs**, including the liver, heart, skin, kidneys, and pancreas. Each organ performs a specific function. Organs do not work separately; rather, they act together to perform specific related functions. A group of organs working together is known as an *organ system*. Each organ system serves a particular function in maintaining life. Every living thing is made up of many organ systems. Each system coordinates to make up a complete living, functioning organism.

Disease and Injury

Body systems and their functions can be disrupted by disease processes or injury. **Infection** is the invasion of a foreign substance, causing disease. Infection usually results in **inflammation**, a protective response to an injury that results in pain, swelling, redness, and warmth.

External forces—such as a fall from a certain height—can cause trauma, which can result in tissue damage or injury. Emergency management of trauma is necessary to prevent complications such as excessive blood loss or infection.

SUMMARY

Cells are the most basic structure of living organisms. Cells divide and multiply, forming tissues. Tissues form organs, which carry out specific functions. Groups of organs work together to provide all vital life functions for living things.

Key Terms

actin a thin protein filament fiber found in muscle tissue

active transport the process in which molecules in a lower concentration move to a higher concentration

adipose fat tissue

anabolism a process in which smaller molecules combine into larger molecules

anaphase the stage of mitosis and meiosis in which the chromosomes move toward the poles of the spindle

anatomy the study of the internal and external body structures and parts of an animal

antibody a protein that fights off disease

autolysis the breakdown and removal of dead cells

axon a hairlike projection from the neuron that carries a nerve impulse

benign a noncancerous tumor

carbohydrate provides energy to cells and aids in cell structure

catabolism a process in which large molecules are broken down into smaller molecules

cell the basic unit of life and the structural basis of an animal

cell division a process in which one cell splits into two cells

cell membrane holds a cell together; also called the cell wall or plasma membrane

columnar a tall and thin shape

connective tissue holds and supports body structures by connecting cells

cuboidal cube shaped

cytology the study of cells

cytoplasm the liquid or fluid portion of a cell that allows the internal cell structures to move

daughter cells one of four cells created from one cell during meiosis; each cell has half the number of chromosomes of the parent cell

deoxyribonucleic acid (DNA) a self-replicating material that is present in nearly all living organisms as the main constituent of chromosomes. It is the carrier of genetic information

diffusion a process in which molecules move from higher to lower concentrations

endocytosis a process in which large molecules move from a higher concentration to a lower concentration

endoplasmic reticulum a collection of folded membranes that attaches to the nucleus

enzyme a chemical that changes other chemicals during metabolism

epithelial tissue a protective layer that covers the body's organs and acts as a lining of internal and external structures

geneticist a scientist who studies genetics and genes and uses mathematical equations to predict the outcome of breeding

Golgi apparatus a chemical processor of the cell

histology the study of tissues

homeostasis the tendency toward a relatively stable and balanced equilibrium between interdependent elements, especially as maintained by physiological processes

hydrophilic having a tendency to mix with, dissolve in, or be wetted by water

hydrophobic tending to repel or fail to mix with water

infection an invasion of the body by foreign matter, causing illness and disease

inflammation a protective mechanism causing redness, pain, swelling, and warmth

interphase the interval between the end of one mitotic or meiotic division and the beginning of another

ligament a fibrous strand that attaches from one bone to another

lipid any of a class of organic compounds that are fatty acids or their derivatives and are insoluble in water but soluble in organic solvents

lysosome a structure within the cell that digests food and proteins

malignant cancerous

meiosis a type of cell division that results in four daughter cells each with half the number of chromosomes of the parent cell, as in the production of gametes and plant spores

metabolism a reaction within the body in which chemicals break down and are used by the body

metaphase the stage of mitosis and meiosis in which the chromosomes become arranged in the equatorial plane of the spindle

mitochondria the "powerhouse of the cell"; the part of the cell that makes energy

mitosis a type of cell division that results in two daughter cells each having the same number and kind of chromosomes as the parent nucleus, typical of ordinary tissue growth

monosaccharide simple sugar

muscular tissue allows movement of body parts

myofiber a muscle cell

myosin a thick protein filament fiber found in muscle tissue

nerve where axons from multiple neurons meet

nerve impulse a message sent to a nerve

nervous tissue contains particles that respond to a stimulus and cause a reaction in the body

neuron a nerve cell

nucleic acid a complex organic substance present in living cells, especially DNA or RNA, whose molecules consist of many nucleotides linked in a long chain

nucleolus a small dense spherical structure in the nucleus of a cell during interphase. It is the largest structure in the nucleus of eukaryotic cells and is best known as the site of ribosome biogenesis

nucleus the brain of the cell, located in the center of a cell

organ a group of tissues similar in structure and function

organelle a structure found within a cell that has a specific function

osmosis the movement of water or other solvent through a plasma membrane from a region of low solute concentration to a region of high solute concentration.

pathologist one who interprets and diagnoses the changes caused by disease in tissues and body fluids

pathology the study of diseases, especially of the structural and functional changes produced by them

phospholipid similar to fat, with less fatty acids in its makeup

physiology branch of biology that deals with the normal functions of living organisms and their parts

polysaccharide a starch made up of simple sugars

prophase a stage in where chromatin thickens and begins to form chromosomes

protein a large molecule that works with 22 different amino acids that may join with sugars, starches, and lipids to form the strong structure of a cell

ribonucleic acid (RNA) a nucleic acid present in all living cells. Its principal role is to act as a messenger carrying instructions from DNA for controlling the synthesis of proteins, although in some viruses RNA rather than DNA carries the genetic information

ribosome a structure within the cell that makes protein

selection a natural or artificial process that results or tends to result in the survival and propagation of some individuals or organisms but not of others with the result that the inherited traits of the survivors are perpetuated

simple epithelial tissue single layer of cells with every cell in direct contact with the basement membrane that separates it from the underlying connective tissue

somatic cell a cell that divides in mitosis and help in the growth and repair of the body

squamous flat

stratified epithelial tissue several cell layers composed of multiple cells and many layers of thickness

telophase the final stage of mitosis and of the second division of meiosis in which the spindle disappears and the nucleus reforms around each set of chromosomes

tendon a fibrous strand of connective tissue that attaches muscle to bone

tissue an aggregate of cells usually of a particular kind together with their intercellular substance that form one of the structural materials of a plant or an animal

transitional epithelium consists of multiple layers of cells with varying shapes

tumor a mass of tissue made up of abnormal cells

Review Questions

1. What is the term for cells that are alike in size, shape, and function?
2. What type of tissue lines internal structures?
3. What is the study of tissues called?

4. What structure is known as the "powerhouse of the cell"?
5. Explain the term "homeostasis."
6. What are the parts of a cell?
7. What are the stages of mitosis?
8. What is the difference between anatomy and physiology?
9. What is the difference between a benign and a malignant tumor?
10. What is another name for the cell membrane?
11. Where does extracellular fluid derive from?
12. What is the difference between smooth and rough endoplasmic reticulum?
13. What disease can result from too much blood sugar in the body?
14. What type of epithelial tissues line the surface of the urinary bladder?
15. Name the hairlike extension from the nerve cell that carries the nerve impulse.
16. Describe the shape of squamous cells.
17. Discuss the difference between a tendon and a ligament.
18. List the three types of muscle tissue.
19. What type of tissue is affected by Horner's syndrome?
20. List two types of involuntary muscle types.

Clinical Situation

Karena, a veterinary assistant at Green Lane Animal Clinic, is preparing for surgery on Callie Ames, a 12-year-old SF cocker spaniel. Karena has read the medical history and notes that Callie has a 4 cm mass on her L lateral abdomen that Dr. Colón is scheduled to surgically remove. The tumor is to be biopsied at an outside lab to determine the cause and if it is malignant.

Karena sees that Callie is to have presurgical blood work and that the owner has left several questions with the receptionist. Mrs. Ames, the owner, would like to know what will happen if the mass is cancerous and what care Callie will need at home.

- What items and supplies will Karena need to prepare for Callie's surgical procedure?
- What are some treatments to consider if the tumor is cancerous?
- What are some factors that should be addressed in Callie's home care instructions?
- Who should discuss the lab results with Mrs. Ames?

Chapter 24: The Musculoskeletal System

Objectives

Upon completion of this chapter, the reader should be able to:

24.1 Name the basic structures that make up the musculoskeletal system
24.2 State the functions of the musculoskeletal system
24.3 Discuss the structure of bone
24.4 Discuss bone development and growth
24.5 Detail muscle and bone groups and their relationship to movement
24.6 Identify types of joints and where they are found in the body
24.7 Describe common diseases and conditions of the musculoskeletal system

Introduction

The **musculoskeletal system** consists of bones, muscles, tendons, and ligaments. The skeleton makes up the bony structure of the animal, providing shape, support, and protection for vital organs. Bone and muscle are tissues that adapt to the needs of the animal. A majority of this system is composed of connective and muscle tissue. The muscular system allows the animal to move. The skeleton is adaptable to allow for bending and growth.

Bones

Bone is a hard, active tissue that consists mostly of calcium and forms the skeleton of the animal. Bones give the animal support and structure while protecting the internal organs. Bones also serve as a mineral reserve to the body, storing calcium and phosphorous for use when necessary and providing nutrients to aid in red blood cell formation. Young and growing animals need calcium and other essential minerals in their diets to allow the bones to form and grow. Bone is also the location of blood cell production, specifically within the medullary cavity of bones within the marrow.

There are different types of bones that are classified based on shape. Long bones are straight and long and have a shaft at either end. The shaft ends are called **diaphysis**. The long center portion between each shaft is called the **epiphysis**. Examples of long bones are the femur and humerus. Flat bones expand in two different directions and serve as an area for muscle attachment. Examples of flat bones are the skull and scapula. Small bones are cube-shaped and equal in all directions. They are often located at joints, for example, the carpal and tarsal bones. Irregular bones are odd or irregular in shape, such as the vertebrae. Sesamoid bones are tiny in size and found along tendons to help reduce friction and stress on tendons. An example of a sesamoid bone is the patella.

The number of bones varies with the age and species of animal. Bone structure is classified as either **compact** or **cancellous**. Compact bone is a thick tissue that forms the outer layer of bone. This part of the bone can be repaired, and it gives elasticity and rigidity to the bone. Cancellous bone is spongelike and softer and is located inside the end of bones. Bones are formed through the process of **ossification.** A hollow center within the bone is called the **medullary cavity** and produces blood cells within a fluid known as **bone marrow**. Additional structures of bone are defined in Table 24–1. Figure 24–1 illustrates the anatomical features of a long bone.

CHAPTER 24 The Musculoskeletal System

TABLE 24–1
Other Structures of Bone

BONE STRUCTURE	DEFINITION
Osteoblast	Particles that begin the ossification process in young, developing bones
Osteocyte	Bone cell that begins to develop mature bone
Osteoclast	Mature bone particle that forms minerals and compact bone
Periosteum	Thin connective tissue covering outer bone
Endosteum	Thin connective tissue covering inner bone

Bones are covered externally with a thin connective tissue called **periosteum**. The periosteum binds ligaments and tendons to the bone and supplies nutrients to the bone tissue. The open spaces within bone are covered with a similar connective tissue called **endosteum**. Both types of connective tissue provide cells for bone repair.

Bone Development

There are three types of bone cells: osteoblasts, osteoclasts, and osteocytes. **Osteoblasts** are responsible for the formation of bone developed from collagen fibers. They are immature bone cells that produce bony tissue. Osteoblasts then develop into **osteocytes**, which are mature bone cells. **Osteoclasts**

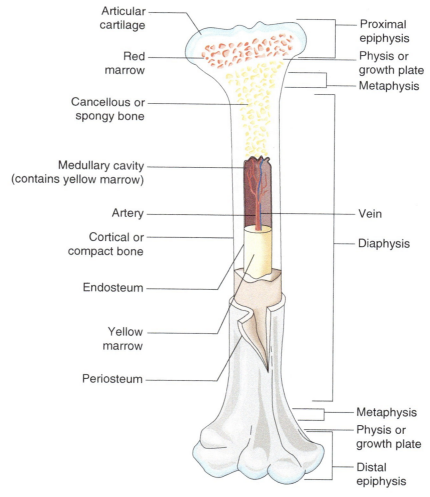

FIGURE 24–1 Anatomy of a long bone.

are large cells that release minerals into the bone and eat away at the bony tissue within the medullary cavity. Bone is living tissue that is constantly repairing and remodeling in response to the body's need for calcium.

Joints

The area where two bones meet forms a **joint**. Joints allow the animal to bend in areas with a level-like movement. There are three main types of joints located in the body. A **fibrous joint** is a fixed joint with little to no movement. Thick connective tissue links the structures of a fibrous joint. The area where two bones of a fibrous joint meet is called a **suture**. This is a fine line that connects the bones, but little to no movement occurs there. Examples of this type of joint can be found throughout the skull. A **cartilage joint** connects where two bones meet and where **cartilage** forms to protect and cushion the area where two bones meet. When an animal is young and growing rapidly, this area is called the **growth plate**; over time and as the animal ages, the cartilage joint becomes bone (see Figure 24–2). The third type of joint is a **synovial joint**, in which a thick layer of bone at the joint is covered by a layer of cartilage. The joint is enclosed by the *joint capsule*, which contains an outer layer of strong connective tissue. This area is lined with a *synovial membrane* that produces lubrication and nutrients from *synovial fluid*. This type of joint is movable and may be a **hinge** joint that opens and closes in a simple one-way movement, a **pivot** joint that rotates around a fixed point, or a **ball and socket** joint that rotates in numerous directions and provides the largest amount of motion. Various joints are illustrated in Figure 24–3 and listed in Table 24–2.

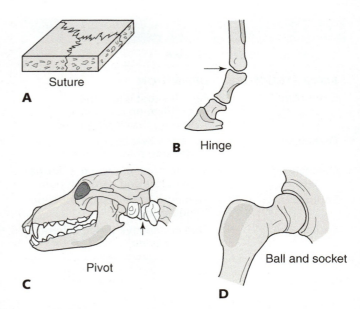

FIGURE 24–3 Examples of joints: (A) suture; (B) hinge; (C) pivot; (D) ball and socket.

TABLE 24–2
Types of Joints and Locations

LOCATION	TYPE OF JOINT
Skull	Fibrous joint
Growth plate at end of bone	Cartilage joint
Elbow	Synovial joint (hinge)
Vertebrae	Synovial joint (pivot)
Hip	Synovial joint (ball and socket)

Joints move in different directions depending on the type of joint. Joints that bend move with **flexion** or closing causing the joint to shorten; **extension** or opening causing the joint to lengthen. Animals also are capable of joint movements known as **abduction** and **adduction**. *Abduction* is movement away from the body and *adduction* is movement toward the body (see Figure 24–4). *Rotation* occurs when a joint spins on its axis.

Muscles and Connective Tissues

Muscles are located throughout an animal's body and allow movement. They attach to different locations of the body, serve as protection, and provide the animal's strength. **Ligaments** are fibrous strands of tissue that connect one bone to another bone and serve as

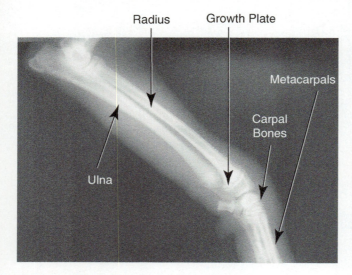

FIGURE 24–2 Example of a growth plate.

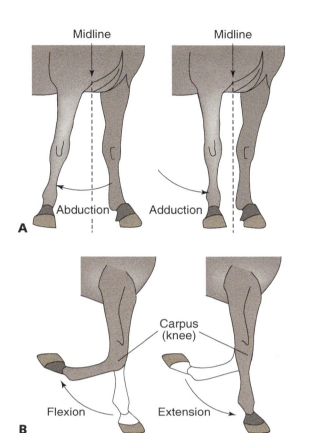

FIGURE 24–4 (A) adduction versus abduction; (B) flexion versus extension.

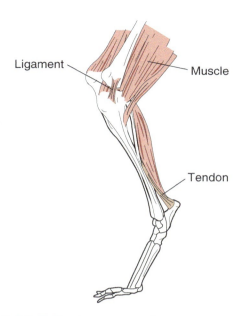

FIGURE 24–5 Tendons versus ligaments.

support and strength to all areas of the body. They also allow bending and movement of the limbs. **Tendons** are fibrous strands of tissue that connect a bone to a muscle. They also aid in movement of the limbs. Some joints have a **meniscus**, which is a hard cartilage pad and acts as a cushion between the ends of bones. An example of this is the knee joint that has ligaments both inside and outside the joint capsule. Menisci also serve as a cushion for the wide range of motion within this joint. Figure 24–5 shows the difference between a tendon and ligament.

 TERMINOLOGY TIP

The root word for cartilage is *chondr/o*. The root word for joint is *arthr/o*. The root word for ligament is *ligament/o* and for tendon is *ten/o* or *tend/o*. ■

Axial Skeleton

The skeleton is divided into two parts, the **axial skeleton** and the **appendicular skeleton**. The axial skeleton contains the bones of the body that lie perpendicular or lengthwise. This includes the skull, spinal column, ribs, and sternum. The **skull** holds and protects the brain and is made up of many bones. The **maxilla**, or upper jaw, and **mandible**, or lower jaw, make up part of the skull (see Figure 24–6). The jaw is a synovial joint that allows animals to secure and chew their food. Small open spaces located within the skull and nasal canal are called **sinuses**. The eye sockets are known as the **orbits**. The **spinal column** contains the **vertebrae**, which are individual bones that surround and protect the **spinal cord**. The spinal column extends from the base of the skull to the end of the tail and allows the animal to move. It may also be referred to as the vertebral column or spine. The spinal column is divided into sections according to location.

 TERMINOLOGY TIP

The term *vertebrae* is plural and refers to the entire spinal column. The singular form is *vertebra*, which refers to one bone. ■

 TERMINOLOGY TIP

The *cranium* is the area inside the skull where the brain is housed. The root word *crani/o* means "skull." ■

The body of the vertebra is surrounded by a bony arch, and the entire vertebral canal consists of a series of arches that forms a long tube that holds the spinal cord. The *spinous process* is the projection at the top of a vertebra along the dorsal aspect, and the *transverse process* is the wings of the vertebra

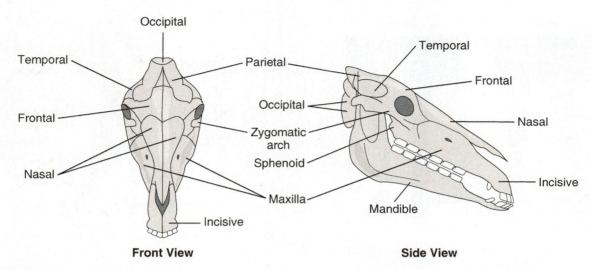

FIGURE 24-6 Selected bones of the skull and face.

that are located at the base and project laterally left and right. Both the spinous and transverse processes have tendons and ligaments that form attachments. The center of the vertebra is an opening called the *vertebral foramen*, which allows the passage of the spinal cord (see Figure 24–7). Between each vertebra (except the first two cervical vertebrae) lie **intervertebral discs**. These discs offer a protective cushion with a soft spongy center.

Clinical Concepts and Critical Thinking

A common occurrence in long-backed dog breeds is a condition known as *intervertebral disc disease (IVDD)*. Certain breeds of long-backed dogs, such as the Basset Hound, Dachshund, and Corgi, are more predisposed to this condition. IVDD is caused when pressure occurs between the discs due to inflammation from an injury. The center of the disc becomes less spongy, which causes the center to rupture. The inflammation and swelling exert pressure between the affected vertebrae, causing pain via the spinal nerves and possible trauma to the spinal cord. Most common signs of the condition include severe pain and possible paralysis. The paralysis can affect the back, head, neck, and limbs. The paralysis may affect the animal's ability to control urination and defecation. Steroids may be used to help decrease inflammation, and muscle relaxers may be necessary to reduce pain. Depending on the extent of the condition, dogs may be treated medically or surgically.

The first section of the spinal column consists of the **cervical** vertebrae and extends through the head and neck area. The first vertebra, located at the base of the skull, is the **atlas** and is the first cervical vertebra (C1) and allows the up-and-down nodding motion of the head. The second cervical vertebra (C2) is the **axis** and allows for the rotation and shaking motion of the head. The second section of vertebrae consists of the **thoracic** vertebrae, which lie over the shoulders and chest area. Each **rib** attaches to an individual thoracic vertebra. The ribs protect the heart and lungs, and they allow for expansion of the lungs during breathing. The **sternum**, or breastbone, helps to protect the organs of the chest. Some ribs attach to the sternum at the ends with cartilage called **costal cartilage**. Ribs that are not attached to cartilage are called **floating ribs**.

The next section of the spinal column is the **lumbar** vertebrae, which lie over the lower back or **loin** area. This area allows an animal to extend and flex. It also helps support the organs in the abdomen. The **sacral** vertebrae form over the pelvis or rump area.

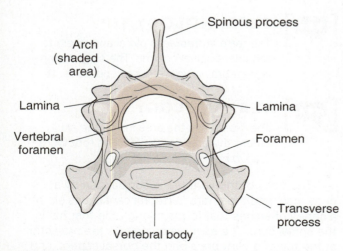

FIGURE 24-7 Vertebra, a bone of the spinal column.

Several vertebrae are fused together directly over the pelvis at the highest point of the spine in an area called the **sacrum** and support the weight over the back end of the animal. A firm joint on each side of the sacrum connects to the **pelvis**, or hip bones. This area is easily damaged, causing the pelvis to separate from the sacrum, which is common in animals hit by a car. Pelvic fractures are very painful, but many heal on their own with restricted activity and cage rest. Severe trauma may require surgical repair. The last section of the spinal cord is the **coccygeal** vertebrae and forms the tail area. Much of this section doesn't contain the spinal cord. Figure 24–8 illustrates the various parts of the axial skeleton. Each species has a different number of vertebrae in each section of the spinal column (see Table 24–3).

TERMINOLOGY TIP

Spondylitis is inflammation of the vertebrae. *Spondylosis* is the vertebral condition that causes degeneration and bridging between vertebrae. ■

Clinical Concepts and Critical Thinking

Different animal species vary on how many bones they have regardless of their size. A dog has about 320 bones, 134 of those in the axial skeleton and 186 in the appendicular skeleton. A dog's skull consists of 50 flat bones. A cat has a total of 244 bones, with 29 flat bones in the skull. A horse has fewer total bones at about 250, with fewer bones in their limbs compared to smaller animals. A cow has 207 total skeletal bones.

Appendicular Skeleton

The appendicular skeleton is the part of the body that contains the appendages or limbs. The forelegs or **forelimbs** are the front legs of an animal. The front limbs include the **scapula**, commonly called the shoulder blade, which lies flat against the rib cage and connects to the axial skeleton by a group of muscles. This allows for a range of motion in the shoulder of as much as 25 degrees with such activities as running and jumping. Some species have a **clavicle**, commonly known as the collarbone, that connects the sternum to the scapula. The upper bone of the forelimb, called the **humerus**, joins the scapula via a ball and socket joint that allows for a wide range of motion. The lower end of the humerus forms the elbow, or **olecranon**, and connects to the lower long bones of the **radius** and **ulna**. The *radius* is the larger bone located in the lower front limb. The *ulna* is a smaller bone and lies directly behind the radius. Both bones are fused at the top and bottom and connect to the humerus to form the elbow joint. The lower leg forms the **carpal** bones, similar to the wrist bones in humans.

There are several small carpal bones that are arranged in two rows collectively called the **carpus** joint. The **metacarpal** bones form the long bones

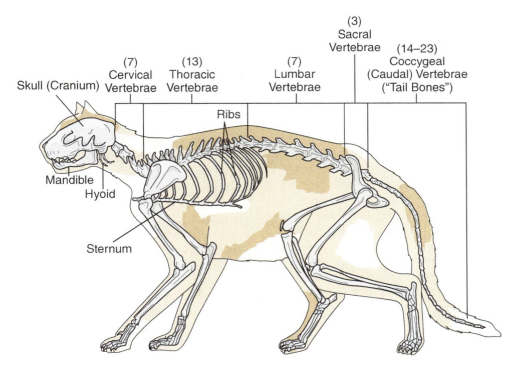

FIGURE 24–8 Parts of the axial skeleton.

TABLE 24–3
Vertebrae Formulas for Common Species

SPECIES	CERVICAL	THORACIC	LUMBAR	SACRAL	COCCYGEAL
Dog	7	13	7	3	20–23
Cat	7	13	7	3	5–23
Horse	7	18	6	5	15–21
Cow	7	13	6	5	18–20
Pig	7	14–15	6–7	4	20–23
Avian	14	7	Fused	14	6

of the front feet. In hoofed animals, these bones form the **cannon** bone that is one large single bone. Horses have two small bones that form at the back of the cannon bone called **splint bones** that can become inflamed with large amounts of work due to the body absorbing movement over the front legs. The **phalanges** form the toes or digits. The short inside bone that forms the first digit and acts as an opposable thumb is called the **dewclaw**.

The last section of the *phalanx* is covered by the nail or hoof. Horses have *sesamoid* bones within their tendons that help reduce wear on the tendons. These bones are located between the metacarpal bone and the first phalanx in a joint called the *fetlock*. Another sesamoid, the *navicular* bone, lies in the joint between the last two phalanges.

TERMINOLOGY TIP
The term *phalanges* refers to more than one digit. In hooved animals that have only a single digit, the singular form is *phalanx*. ∎

The hind limb or the **pelvic limb** forms the rear leg of an animal. The pelvis is formed by three sections, the **ilium**, **ischium**, and **pubis**. The ilium forms most of the pelvis. The pubis is located in the center of the pelvis where the halves come together, meet, and are fused by cartilage. During labor this cartilage softens due to hormonal responses, allowing the newborn to pass through the pelvic canal. The pelvis holds the rear limb in place via the **acetabulum**, or ball and socket joint that is located on each side of the hip bone, or pelvis. The **femur**, or thigh bone, is the large bone of the upper rear leg. A strong ligament within the acetabulum holds the hip joint in place and seats the femur firmly in position. The **patella** forms the joint between the upper and lower bones of the rear leg. This is commonly called the kneecap. This joint is supported by two ligaments and a tendon externally called the lateral and medial *collateral* ligaments. Internally lie two ligaments that form an X over the knee joint and are called the *cruciate* ligaments. They provide stability to the joint, as this is an area that is easily damaged and often torn. The entire joint around the knee is commonly called the *stifle*. The bones in the lower leg are called the **tibia** and **fibula**. The *tibia* is the larger bone in the lower rear leg, lies in the front, and bears most of the weight over the hind limb; the *fibula* is smaller, lies to the rear of the leg, and provides for muscle attachment. The point of the rear leg where the tibia and fibula meet are commonly called the **hock**. The hock forms several bones similar to the ankle bone called **tarsal** bones; the collective term for the entire joint is **tarsus**. This joint is very similar to the carpal joint in the front limb. The **metatarsal** bones form the long bones of the rear feet, and the digits of both the front and rear limbs are all called the *phalanges*. Figure 24–9 illustrates the bones that make up the appendicular skeleton.

Common Musculoskeletal Diseases and Conditions

Hip dysplasia is a common genetic condition of large-breed dogs. The ball and socket joint of the pelvis and femur becomes diseased and does not sit properly (see Figure 24–10). This condition varies in degree of severity. The socket of the pelvis is shallow, and the ball of the femur **subluxates** or comes partially out of the joint. Over time the cartilage that lines the acetabulum and the head of the femur wears down. This is called *degenerative joint disease*. Common signs of the condition include reluctance to lie down or rise, inability to move on stairs, and lameness and pain in mild to severe stages. Some breeds are more predisposed to the condition, and large and giant breed puppies that grow too quickly are more prone to this condition. Anti-inflammatory

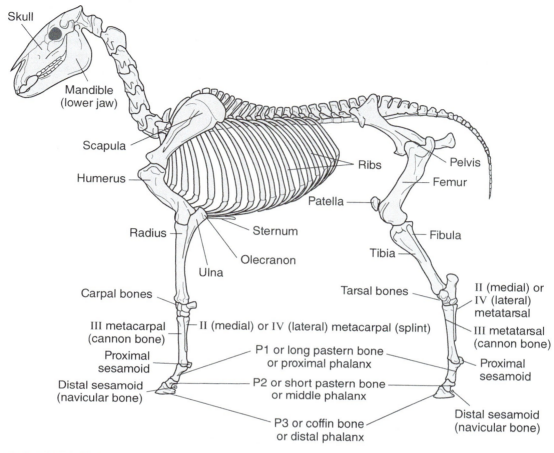

FIGURE 24–9 Bones of the appendicular skeleton.

medications may be given to relieve the pain and inflammation related to the condition. Severe stages of hip dysplasia may require surgical hip replacement. Careful breeding of some dogs can decrease the occurrence of hip dysplasia. Breeders should evaluate their breeding dogs for hip dysplasia by having a hip X-ray certification through the Orthopedic Foundation of Animals (OFA) or through the University of Pennsylvania's PennHIP program. If two dogs with hip dysplasia are bred, the puppies from their litters are likely to have it as well.

Dog Breeds with Genetic Disposition to Hip Dysplasia

- German Shepherd
- Rottweiler
- Labrador Retriever
- Great Dane
- Golden Retriever
- Saint Bernard

Arthritis is a common condition of older animals in which the joints become inflamed, and other changes in the joints may be noted. This condition causes lameness and pain, and it worsens over time. Signs of arthritis in animals include the following:

- Favoring a limb or limping
- Difficulty sitting or standing
- Sleeping more
- Seeming to have stiff or sore joints
- Hesitancy to jump, run, or climb stairs
- Weight gain
- Decreased activity or less interest in play/activity
- Attitude or behavior changes
- Being less alert

Pain relievers and anti-inflammatory medications as well as dietary supplements may be helpful in reducing pain. Many arthritis medications are now available to aid the joints of affected animals.

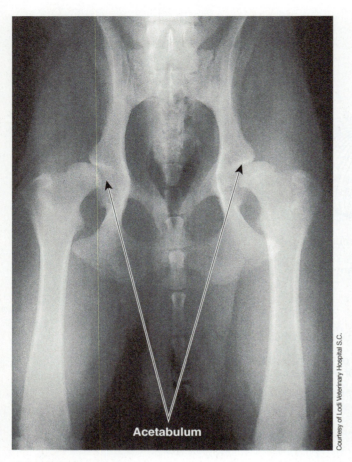

FIGURE 24–10 Radiograph of a dog with hip dysplasia. Note the shallow acetabulum and the subluxation (partial dislocation) of the femur.

 TERMINOLOGY TIP

Osteitis is inflammation of bone. *Osteoarthritis* is a degenerative joint disease affecting both the bones and joints. *Osteochondrosis* is a degenerative condition affecting bones and cartilage. *Osteomalacia* is the softening of bone, and *osteosclerosis* is the hardening of bone. ■

Cruciate ligament tears are a common injury in active dogs and tend to be more complicated in large-breed dogs. It is possible for any type of animal to have a ligament injury. Ligaments are located throughout the limbs, but the most common ligament injury is to the cranial cruciate ligament, which is an intraarticular ligament located within the stile or knee joint. The cruciate ligament provides stability to the kneecap and stifle joint and is commonly torn during vigorous exercise or trauma to the knee. Ligaments may also deteriorate with age and break down and finally tear. Lameness and sometimes complete loss of use of the limb will result. A torn ligament is diagnosed via the **cranial drawer sign**, which involves moving the knee forward and back like a drawer opening and closing (see Figure 24–11). This is done by pressing backward on the end of the femur and forward on the tibia. If the ligament is torn, the tibia will slide forward, much more than in a normal knee. In small-breed dogs, this condition can often be managed with pain medication and rest, but in larger dogs, surgical correction is often recommended. Often while one cruciate tear is healing, the opposite limb is damaged and requires treatment. The tear may also involve the **meniscus**, a cartilage cushion within the patella that aids in lubrication of the joint. There are two pieces of connective tissue forming a "C" shape that are located in front of the patella, including the meniscus. This cartilage layer acts as protection to prevent rubbing on the patella. When a cruciate ligament and meniscus injury occur, owners often describe hearing a clicking sound within the knee when a dog is walking. This is due to the movement of a free section of the torn cartilage. Restricted movement, pain medications, and surgical repair are necessary to reconstruct and repair the ligament tear. Cranial cruciate tears are a common cause of degenerative joint disease within the stifle.

Bone **fractures** are a break in the bone (see Figure 24–12). A bone may fracture in many different ways. A bone may come partially out of place, causing a subluxation, or it may completely move out of place, causing a **dislocation** or **luxation**. A **simple fracture** is a single break in a bone that stays in place;

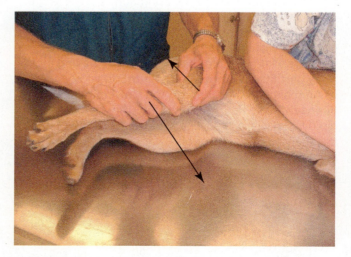

FIGURE 24–11 A veterinarian tests for cranial drawer sign in a dog's knee joint. The arrows indicate the direction of force being applied.

immobilize the injury. A **cast** is a material placed on the area of the body where the fracture occurred and is made of a hard substance to keep the bone in place. The support must be able to maintain the weight of the animal, as once a bandage or splint is in place, the animal must still have some use of the limb. It may be difficult for some active pets to keep a splint, bandage, or cast in place, and it must remain clean and dry at all times. Surgical repair may require an **intramedullary** or **IM pin** (see Figure 24–13). This is a stainless steel pin or rod placed into the center of a broken bone to hold the fracture site together while the bone heals. A **bone plate** may be required and is a surgical steel plate that is placed around the fracture site that holds the bone together with surgical screws (see Figure 24–14). If bone fragments occur, they may be held together with **cerclage wire** to support the fragments as they heal (see Figure 24–15). Bone healing is similar to bone development. Cartilage rebuilds and forms a **callus** over the fracture site. This area becomes thickened, within two weeks osteoblasts begin to develop, and gradually the callus is replaced by bone. This bony callus fills in between the fracture sites and extends into the bone marrow and the outside edges of the bone. The duration required for complete healing depends on the fracture site, type of trauma, and healing care. Any bandage changes and postoperative care are essential for proper healing.

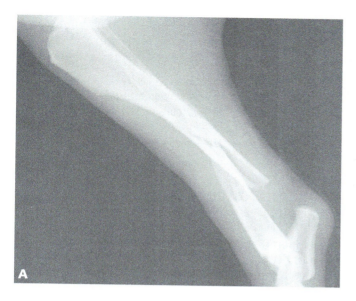

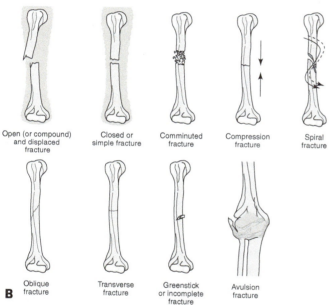

FIGURE 24–12 (A) Radiograph of a fractured leg. (B) Fracture types.

this is often called a clean break. A **comminuted fracture** occurs when a bone breaks in several different locations, and there are bone fragments. A **compression fracture** occurs when the broken bones are pressed together. A **compound fracture** is a break in the bone in which the bone penetrates the skin. The rate of infection in compound fractures is high and may cause treatment complications. Fractures can be repaired in several ways, but for the bone to heal, it must be realigned and not allowed to move. Some fractures can be healed with the use of a **splint**, which is a support applied with padding and bandages to

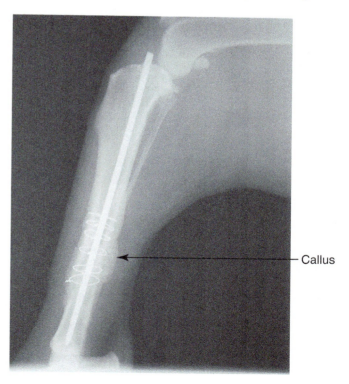

FIGURE 24–13 Radiograph showing placement of intramedullary pin.

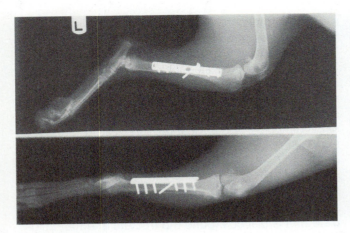

FIGURE 24–14 Radiograph showing use of bone plate to treat a fracture.

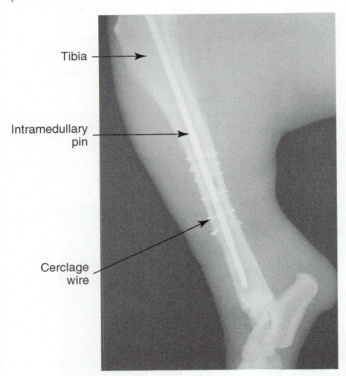

FIGURE 24–15 Radiograph showing use of cerclage wire in treatment of a fracture.

 TERMINOLOGY TIP

The terms *crepitation* and *crepitis* describe the cracking sound that is felt and heard when broken bones rub together. *Immobilization* is the term used to describe holding or fastening a bone in a fixed position, such as with a splint or bandage. *Amputation* means the removal of a body part. *Ostectomy* is the surgical removal of bone. *Osteopexy* is the surgical fixation of bone to the body wall. *Osteoplasty* is the surgical repair of bone. ■

 MAKING THE CONNECTION

For additional reading relating to disorders of the musculoskeletal system, see the following:

Chapter 18—laminitis in horses
Chapter 19—porcine stress syndrome
Chapter 20—foot rot and polyarthritis
Chapter 21—joint ill ■

Clinical Concepts and Critical Thinking

A condition that occurs in horses that is one of the most common causes of front limb lameness is known as *navicular syndrome*. Damage or degeneration of the navicular bone in the hoof is the cause for this lameness. This condition can also cause damage to surrounding structures of the hoof, including tendons, ligaments, and joints. When clinical signs first develop, a horse often becomes mildly lame. The lameness may initially improve with exercise but over time the lameness increases, and the horse is often unable to move. Navicular disease often occurs in both front feet, with the problem being more severe in one bone than the other. The horse often shortens its stride, causing a limp in an attempt to reduce the pain. A hoof tester is often used over the region of the navicular bone to determine the degree of pain (see Figure 24–16). Mild cases can be controlled with regular hoof trimming, rest, corrective horseshoes, and anti-inflammatory medicines. Severe cases require surgical correction to aide in the comfort of the horse's feet.

FIGURE 24–16 Example of a hoof tester for pain assessment.

Muscular System

Muscles are organs that contract to produce a movement. These contractions aide in walking, running, and jumping; movement of internal organs and tissues; movement of bodily fluids; and heat generation to keep the body warm. Muscles are composed of strands of fibers. Muscle is covered by connective tissue called **fascia** that covers, supports, and separates muscles. The *linea alba* is a fibrous band of connective tissue located on the ventral abdomen where the abdominal muscles attach (see Figure 24–17). This means "white line" in Latin and describes the appearance of where the muscles come together, causing a thin white line down the center of the abdomen below the umbilicus. This is the line that veterinary surgeons follow when they make an incision into the abdomen.

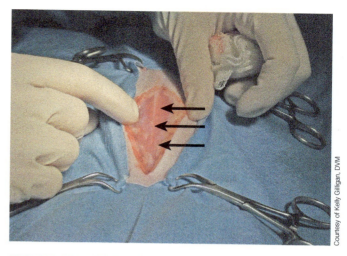

FIGURE 24–17 The linea alba of the abdomen.

 TERMINOLOGY TIP

The root word for muscle is *my/o*. *Myopathy* means a diseased muscle. *Myositis* is the inflammation of muscle. *Myoplasty* is the surgical repair of muscle. *Myotomy* is a surgical incision into a muscle. ■

Some muscles work in pairs and others work against each other. Muscles that work together are called **agonistic** muscles and those that work against each other are called **antagonistic** muscles. Agonistic muscles contact at the same time as another muscle and help to move and support the body. Antagonistic muscles work with one muscle producing a *contraction* (tightening) with another muscle in *relaxation* (loosening). During contraction the muscle becomes shorter and thicker, and during relaxation the muscle becomes longer and returns to its original shape. Muscles form at the **origin** and end at the **insertion**. The muscle origin is the place where a muscle begins and is well attached. The muscle insertion is where a muscle ends and is more movable. Muscles move by a variety of ranges of motion. An *abductor* muscle moves away from the midline, where an *adductor* muscle moves toward the midline. A *flexor* muscle bends at a joint and decreases its length, whereas an *extensor* muscle straightens at a joint and increases in length. A *rotator* muscle turns a body part on its axis.

Specific muscles may be named for their location or according to the direction in which their muscle fibers are arranged. *Pectoral* muscles are located in the chest; *epaxial* muscles are located above the pelvis; *intercostal* muscles are located between the ribs. Muscles that are straight are called *rectus* muscles; slanted muscles are called *oblique*; *transverse* muscles lie crosswise to the midline; and *sphincter* muscles form a tight band for constricting around an opening. Some muscles are named for the number of divisions or sections they have, such as the biceps (two sections), triceps (three sections), and quadriceps (four sections).

SUMMARY

The musculoskeletal system consists of bones, muscles, joints, and tendons. The axial skeleton consists of the skull, spine, and rib cage. The appendicular skeleton consists of the upper and lower extremities. This system provides support and protection for vital organs. It also allows the animal to move. Common disorders of the musculoskeletal system include disorders affecting joints. Trauma can also result in bone fractures. Animals with musculoskeletal injuries may present with muscle pain or weakness, paralysis, and deformity. The animal may be less active and gaining weight.

Key Terms

abduction   movement away from midline or the axis of the body

acetabulum   area of the pelvis where the femur attaches to form the hip joint

adduction   movement toward midline or the axis of the body

agonistic   muscles that work in pairs with each other

antagonistic   muscles that work against each other

appendicular skeleton   contains bones of the body that hang; the limbs or appendages

arthritis   inflammation of the joint

atlas   first cervical vertebrae located at the base of the brain; allows for up-and-down motions of the head

axial skeleton   contains bones of the body that lie perpendicular or lengthwise

axis   second cervical vertebrae; allows for rotation and shaking motions of the head

ball and socket   joint that rotates in numerous directions

bone   hard, active tissue that consists mostly of calcium and forms the skeleton of the animal; gives the body support, structure, and protection

bone marrow   soft substance within the medullary cavity where blood cells are produced

bone plate   surgical steel plate placed around a fractured bone that holds the bone together with surgical screws

callus   cartilage that rebuilds and thickens over a fractured bone that has healed

cancellous bone   softer spongelike layer of bone located inside the end of bones

cannon   long bones in hoofed animals located above the ankle joint

carpus   joint formed by several carpal bones and arranged in two rows in the area of the wrist

cartilage   forms at the end of bones to protect and cushion the bone

cartilage joint   connects at the end of a joint where two bones meet; acts as a cushion

cast   material made of a hard substance; placed over a broken bone to keep the bone in place as it heals

cerclage wire   surgical wire used to hold fragments of broken bone together while healing

cervical   first section of the spinal column located over the neck area

clavicle   bone that connects sternum to scapula; commonly called the collarbone

coccygeal   last section of the vertebrae that lies over the tail area

comminuted fracture   bone breaks in several different locations; causes fragments

compact bone   thick tissue that forms the outer layer of bone

compound fracture   a break in a bone that causes the bone to penetrate the skin

compression fracture   when a broken bone presses into another bone

costal cartilage   tissue located at the end of the ribs where they attach to the sternum

cranial drawer sign   in and out movement or motion of the knee when the cruciate ligament has been torn

cruciate ligament tears   common knee injury to large-breed dogs that tears the cruciate ligament in the stifle

dewclaw   the short inside bone that forms the first digit that acts as an opposable thumb

dislocation   displacement of one or more bones at a joint

endosteum   thin connective tissue covering inner bone

extension   bending that causes a joint to open and to lengthen

fascia   connective tissue that covers, supports, and separates muscles

femur   large upper bone of the rear leg; thigh bone

fibrous joint   fixed joint with little to no movement

fibula   smaller bone that lies in the lower rear leg in the back of the limb

flexion   bending that causes the joint to close and shorten

floating ribs   ribs that are not connected to the sternum by cartilage

forelimb   front leg of an animal

fracture   break in a bone

growth plate   area of bone in young animals that allows the bone to grow and mature as the animal ages and the cartilage joint turns to bone

hinge   joint that opens and closes

hip dysplasia   common genetic condition of large-breed dogs where the ball and socket joint of the femur and pelvis becomes diseased

hock   common term for the point of the rear leg where the tibia and fibula meet

humerus   large upper bone of the forelimb

ilium   first section of the pelvis

IM pin   intramedullary pin

insertion   where a muscle comes to an end

intervertebral disc   soft, spongy cushioning that protects the vertebrae

intervertebral disc disease (IVDD)   common injury of the back in long-backed breeds; inflammation and swelling occur when one vertebra puts pressure on another

intramedullary pin   stainless steel pin or rod placed into the center of a broken bone to keep the bone in place as it heals

ischium   second section of the pelvis

joints   where two bones meet and allow a bending motion

ligament   fibrous strands of tissue that attach bone to bone

loin   the lower back or lumbar area

lumbar   third section of the vertebrae; lies over the lower back

luxation   occurs when a bone or body part comes out of place

mandible   lower jaw

maxilla   upper jaw

medullary cavity   hollow center within bone where blood cells are produced

meniscus   cartilage within the patella; forms an X shape over the cruciate ligament

metacarpal one of the bones that forms the long bones of the feet in the forelimbs

metatarsal one of the bones that forms the long bones of the feet in the rear limbs

muscle one of the structures located throughout the animal's body that attaches to different locations and serves to protect, allow bending and movement, and aids in the strength of the animal

musculoskeletal system body system consisting of bone, muscles, tendons, and ligaments

olecranon bony process behind the elbow joint

orbits eye sockets within the skull

origin where a muscle begins or originates

ossification the formation of bone

patella kneecap; protects the front of the stifle joint

pelvic limb rear leg of an animal

pelvis hip bone

periosteum thin connective tissue covering outer bone

phalanges toes or digits

pivot joint that rotates around a fixed point

pubis third section of the pelvis

radius larger bone on the front of the lower forelimb

rib bone that attaches to individual thoracic vertebrae and protects the heart and lungs

sacral the fourth section of the vertebrae; lies over the pelvic area

sacrum area over the pelvis that is fused together and forms the highest point of the hip joint

scapula shoulder blade

simple fracture a break in the bone that may be complete or incomplete, but does not break through the skin; also known as a closed fracture

sinuses small, open spaces of air located within the skull and nasal bone

skull the bone that holds and protects the brain and head area

spinal column extends from the base of the skull to the end of the tail and allows movement; also called the spine

spinal cord canal of nerves that run through the center of the vertebrae (the spinal column)

splint support applied to a broken bone with bandage and padding to allow a bone to heal

splint bone one of two small bones on the back of a horse's cannon bone that may become inflamed from large amounts of work and stress on the legs

sternum breastbone that protects the organs of the chest

subluxate causes bone to be partially out of the joint

suture where two fibrous joints meet; fine line where little or no movement occurs

synovial joint moveable joint at the area where two bones meet

tarsus joint of the ankle of the rear limb

tendon fibrous strands of tissue that attach bone to muscle

> **thoracic** second section of vertebrae located over the chest area
>
> **tibia** larger bone in the lower rear leg; lies in front
>
> **ulna** smaller bone on the back of the lower forelimb
>
> **vertebrae** individual bones of the spine that surround and protect the spinal cord

Review Questions

1. What minerals are commonly stored in bones?
2. What does the term *flexion* mean?
3. What are the fibrous bands of tissue that connect one bone to another bone called?
4. In which section of the spinal column are the atlas and axis bones located?
5. What is the common name for the bone at the point of the rear leg where the tibia and fibula meet?
6. What type of disease is hip dysplasia?
7. Name the parts of the axial skeleton.
8. Name the parts of the appendicular skeleton.
9. Name and describe the musculoskeletal condition that large-breed dogs are prone to.
10. Name and describe a common disorder that often affects older animals.
11. What type of joint is the pelvis?
12. What type of joint is the elbow?
13. What is the scapula?
14. What is the difference between the tibia and fibula?
15. Describe a compound fracture.
16. Name the sections of vertebrae from head to tail.
17. What is the term for the toes or digits?
18. What is the patella?
19. What is the name of the open hole in a vertebra called?

(Continued)

20. Label the following diagram:

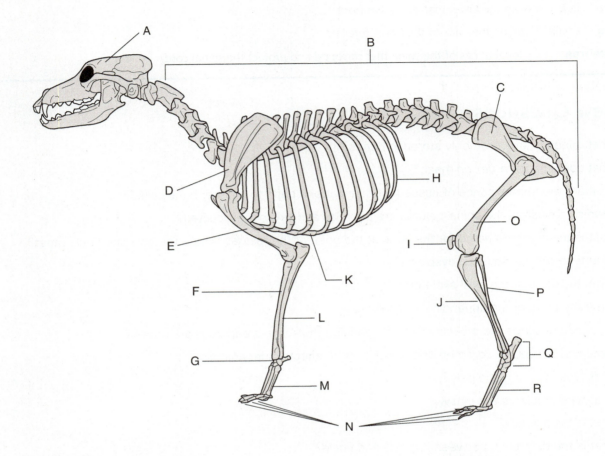

Clinical Situation

Jake Anderson, a 6 yr old NM English Springer Spaniel is rushed into Bell Animal Hospital. The dog has been HBC and is not bearing any weight on his R front limb. It is very painful. The veterinary assistant, Mark, helps transport Jake into an exam room and notes severe swelling over the leg and thinks the leg appears to be broken. Dr. Johns does a PE and some blood work, which appear to be normal, except for the leg injury. Mark and the veterinary technician, Julie, are asked to take radiographs of the leg. They perform two X-ray views and take the X-rays to Dr. Johns for evaluation.

Dr. Johns determines the R front radius and ulna are fractured. It appears to be a simple fracture over both bones. Dr. Johns takes the X-rays into the exam room to discuss his findings with the owner.
- What treatment options might Dr. Johns discuss with the owners?
- What supplies or equipment may be necessary to treat Jake when the vet is done talking to the owners?
- What home care will Jake require when he is discharged?

CHAPTER 25 The Digestive System

Objectives

Upon completion of this chapter, the reader should be able to:

25.1 State the structures that make up the digestive system
25.2 Describe the function of the digestive system
25.3 Discuss dentition in various animal species
25.4 Differentiate between the monogastric, ruminant, nonruminant, and avian digestive systems
25.5 Describe common disorders associated with the digestive system

Introduction

The **digestive system** or **gastrointestinal (GI) system** is responsible for **digestion** of food. *Digestion* is the breaking down of food particles into nutrients that are used by the body to allow the animal to live. Nutrients are converted to a form that cells can use and are transported through the digestive system. The process in which nutrients are absorbed and used by the body varies from species to species and is complex. The major structures of the digestive system are the mouth (with or without teeth), the stomach, and the intestines.

 TERMINOLOGY TIP
The digestive system is the common name for the **alimentary canal**, which is the veterinary medical term for the GI system. ■

Teeth

Food enters the mouth and is chewed through the use of the teeth. Animal teeth are adapted and structured according to the type of food they eat. Carnivores are animals that eat a meat-based diet. Their teeth are made for grasping food, tearing it up, and chewing it into smaller particles (see Figure 25–1). Dogs and cats are carnivores and have long **canine** teeth, also known as fangs, that are used to tear apart food. **Herbivores** are plant-based food eaters, and their teeth are structured for grinding down plant particles. Cows, sheep, goats, and horses are examples of herbivores. **Omnivores** are animals that eat both a plant- and meat-based diet. Their teeth are adapted for a variety of foods. Pigs and humans are omnivores.

Teeth are made of **enamel**, the hardest substance found in the body. Enamel is the covering of the tooth known as the **crown** and protects it from damage. The crown is the top part of the tooth that lies above the gum line and is visible in the mouth. The second layer of tooth is called the **dentin** and is similar to bone. The **root** is the portion of the tooth that is located below the gum and holds the tooth in place with one or more roots. Within the center of a tooth is the **pulp cavity**, which holds the nerves, arteries, and veins. Figure 25–2 illustrates the anatomical structure of the tooth.

Teeth are either **deciduous** or **permanent**. Deciduous teeth are known as baby teeth, are developed in the newborn animal, and appear curved in shape. They are shed around the time the animal reaches maturity and replaced by the permanent teeth or the adult set, which are straighter in shape. The teeth are arranged in the mouth in a specific

456 SECTION III General Anatomy and Disease Processes

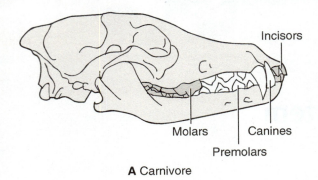

A Carnivore

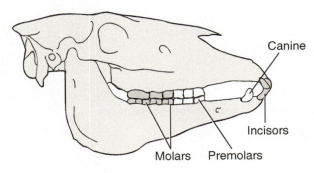

B Herbivore

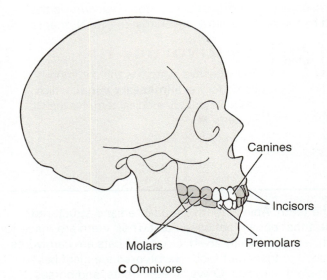

C Omnivore

FIGURE 25–1 (A) carnivore teeth, (B) herbivore teeth, (C) omnivore teeth.

way known as **dentition**, and the specific location in the mouth with teeth in the upper arcade in the numerator position and teeth in the lower arcade in the denominator position is called the **dental formula** (see Figure 25–3 and Table 25–1). The teeth are arranged the same in every domestic animal, but the number of teeth differs according to the species. The front teeth located on the upper and lower jawbones are the **incisors**. They are used to bite and grasp food. The canines or fangs are the next set of teeth and are the longest and sharpest teeth in the mouth. They are used to tear food apart. The root of the canine tooth is twice as long as the exposed crown. The next set of teeth are the **premolars** and are wider teeth used to grind down and crush food. Dogs and cats have an upper fourth premolar and a lower first molar that have a tendency to become infected and abscessed. This tooth is called a **carnassial tooth** and when abscessed causes a swelling in the cheek under the eye (see Figure 25–4). The last sets of teeth are the **molars**. The molars are large teeth located in the back of the mouth. Horses use their teeth in a grinding motion that breaks down food. Over time, the teeth wear with age, helping to determine a horse's age due to the shape of the tooth and the amount of surface wear.

The Mouth

The mouth or oral cavity consists of the lips, tongue, hard and soft palate, salivary glands, and teeth. Animals have four **salivary glands** within the mouth that produce **saliva**, a fluid used to break down food and line the digestive tract for ease of swallowing and moving food along the intestinal tract (see Figure 25–5). Saliva contains the enzyme *amylase* that begins the process of digestion. The **tongue** is a muscle that lies within the mouth and is used to hold food within the mouth. The tongue is where saliva begins to mix with food. Some animals have **papillae** or hair on the tongue that act as taste buds (see Figure 25–6). The next part of the digestive tract is the throat or the area at the back of the mouth called the **pharynx**. Food passes into the **esophagus**, the tube from the throat that passes food into the stomach. The size of the esophagus is regulated by a sphincter muscle consisting of muscular rings that serve as a valve to seal the esophagus off from the stomach. This sphincter muscle allows for opening when food is swallowed to pass into the stomach and closing and relaxes in animals that can vomit or ruminate to allow passage back into the esophagus. The entire digestive system is lined with a thin connective tissue called **mucosa**, which helps ease the passing of food. As food enters the stomach and intestinal tract, it moves through the system in wavelike motions called **peristalsis** (see Figure 25–7). These contractions move food through the entire GI system.

Stomach Systems

There are four types of animal digestive systems: monogastric, ruminant, nonruminant, and avian. Each system serves the animal by digesting particular types of food.

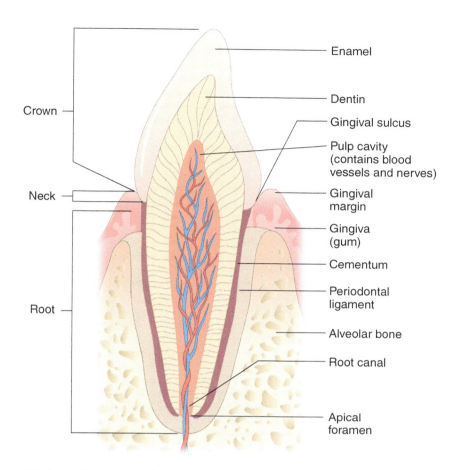

FIGURE 25–2 Anatomical structure of teeth.

TABLE 25–1	
Dental Formulas of Adult Domestic Animals	
Dog	2 (I 3/3 C 1/1 P 4/4 M 2/3) = 42
Cat	2 (I 3/3 C 1/1 P 3/2 M 1/1) = 30
Horse	2 (I 3/3 C 1/1 P 3–4/3 M 3/3) = 24–42
Goat	2 (I 0/3 C 0/1 P 3/3 M 3/3) = 32
Sheep	2 (I 0/3 C 0/1 P 3/3 M 3/3) = 32
Pig	2 (I 3/3 C 1/1 P 4/4 M 3/3) = 44
Cow	2 (I 0/3 C 0/1 P 3/3 M 3/3) = 32
Rabbit	2 (I 2/1 C 0/0 P 3/2 M 3/3) = 28

Note: I—Incisor; C—Canine; P—Premolar; M—Molar

Monogastric Systems

The **monogastric** stomach is a single, simple sac that is divided into three parts: the **cardia**, the **body**, and the **pylorus** (see Figure 25–8). The cardia is the entrance to the stomach and filters food into the entrance through an opening called the **fundus**. The fundus is a pouchlike area at the end of the stomach that allows for expansion. The body of the stomach enlarges as food enters. The pylorus is the exit passage of the stomach. When the stomach is empty, it lies in folds known as **rugae**. Sphincter muscles allow food to enter and exit the stomach, and peristalsis motions continue to move the food through the stomach as digestion occurs. Stomach acids are produced by the **liver**, the organ that lies behind the stomach and creates **bile**, secretions that help break down food to aide in digestion and absorption of food. Bile can be emptied into the small intestine or is stored in the gall bladder, a pear-shaped organ lying between the lobes of the liver. Only monogastric animals have a gall bladder. Please note that bile created by the liver is not the same type of yellow bile that occurs in vomiting animals.

Food then moves from the stomach into the small intestine, where most of the digestion and absorption occur. The small intestine is actually longer than the large intestine, but the width is smaller. The small intestine is divided into three sections: duodenum, jejunum, and ileum. The **duodenum** is the short first section of the small intestine. The duodenum is associated with the pancreas, an exocrine gland. The **jejunum** is the second or middle section of the small intestine. The last and final section is the **ileum**.

SECTION III General Anatomy and Disease Processes

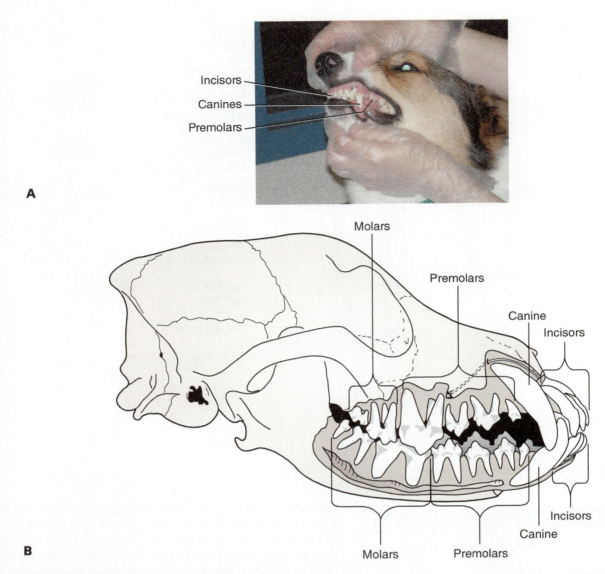

FIGURE 25-3 (A) Dentition in dog. (B) Types of teeth.

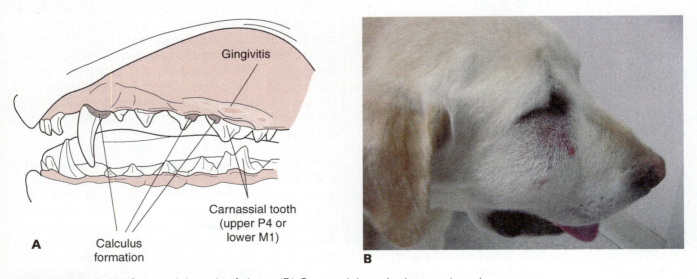

FIGURE 25-4 (A) Carnassial teeth of dogs. (B) Carnassial tooth abscess in a dog.

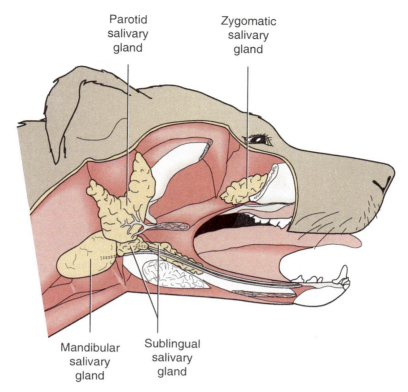

FIGURE 25–5 Salivary glands.

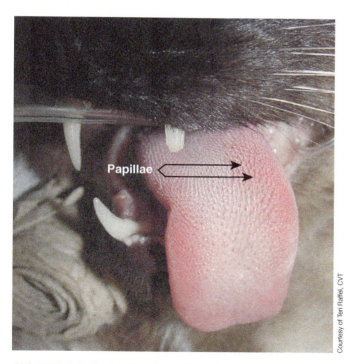

FIGURE 25–6 Papillae found on the tongue.

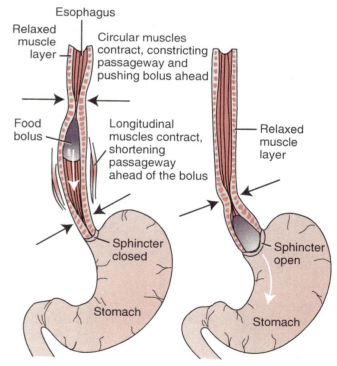

FIGURE 25–7 Peristalsis activity.

The small intestine then enters the large intestine by way of a small sac called the **cecum**. This pouch aides in further digestion of food and begins the widening of the large intestine, commonly called the **colon**.

The cecum does not play a vital role in monogastric animals. The large intestine is also divided into three sections: the **ascending**, **transverse**, and **descending** colon. The ascending is the first section, the transverse

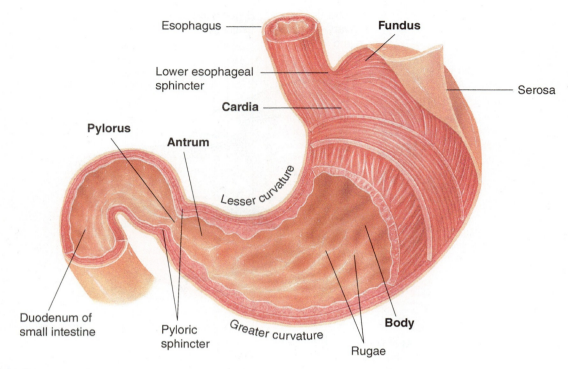

FIGURE 25-8 Example of the monogastric stomach.

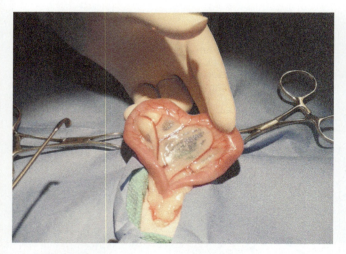

FIGURE 25-9 Mesentery fold that attaches to the small intestine.

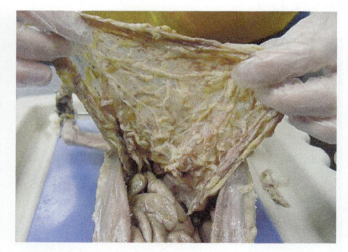

FIGURE 25-10 Omentum.

is the second or middle section, and the descending is the last section, which leads into the rectum. The rectum is the end of the intestinal tract and exits the body at the anus. The anus has a muscular sphincter that controls defecation.

The stomach and intestinal system are lined with different types of connective tissues. The **peritoneum** is a smooth, thin, clear lining of the entire abdomen that allows the organs to move freely within the abdominal cavity. The **mesentery** is a connective tissue that extends from the peritoneum and allows blood vessels and nerves to supply the small intestine (see Figure 25-9). The **omentum** is a thin lining that surrounds all abdominal organs and helps decrease the spread of infection and inflammation within the abdominal area (see Figure 25-10).

Several organs are responsible for digestion and metabolism occurring in the digestive system. The **pancreas** lies next to the intestinal tract and secretes enzymes that help aid in digestion of food. The three types of digestive enzymes produced by the pancreas are **trypsin**, which digests proteins; **amylase**, which breaks down starches; and **lipase**, which breaks down fats. The pancreas is also the organ that helps maintain blood sugar levels and produces **insulin** that is released into the bloodstream and helps to regulate the body's use of blood sugar, also called **glucose**. The liver lies under the pancreas, specifically within the loop of the duodenum, and is constantly producing bile, which is a digestive enzyme that is transported to the gall bladder, where it is stored for later use. Bile helps break down and digest fat. Bile contains bile acids that aide in digesting and absorbing fat. Many drugs and other toxins are metabolized and broken down by the liver. When too little insulin is produced, the body finds it difficult to regulate, which results in a condition known as **diabetes**. The liver converts excessive glucose into glycogen and stores it as energy. When sugars other than glucose are available, the liver converts those to glucose and then into glycogen. When blood glucose levels decrease, the glycogen is converted back into glucose and released in the bloodstream.

Figure 25–11 illustrates the organs associated with the monogastric digestive system.

Clinical Concepts and Critical Thinking

Animals are known for ingesting items that they should not eat, such as toys, clothing, metal, and other solid objects. These items that are ingested may be unable to pass, forming a foreign body obstruction. Not all foreign bodies are seen on X-ray but may show a pattern in the stomach and intestinal tract that is abnormal, causing suspicion of an *obstruction*. When an obstruction is suspected, an *exploratory laparotomy* is performed to determine the location of the blockage. An excision is made into the abdomen, and the veterinarian explores the abdominal cavity for a blockage with a focus on the stomach and intestines. When a foreign body is removed from the stomach, this procedure is called a *gastrotomy*. When an object is removed from the intestines, the procedure is called an *enterotomy*. Sometimes a blockage causes the intestine to telescope back on itself, cutting off the circulation and blood supply to the intestine and becoming an *intussusception*.

This tissue is no longer healthy and must be addressed surgically by removing a section of the intestine. This procedure is known as an *anastomosis*. Clamps are gently placed on either side of the intussusception to prevent gastric contents from leaking. When the area is removed, the veterinarian will resect the healthy ends that remain back together carefully, creating a leak-proof connection. During this procedure it is vitally important that all abdominal contents be kept hydrated and moist to prevent further damage. Recovery time from such a surgery is several weeks, with the animal requiring a soft and easily digestible diet to allow proper healing of the intestinal lining.

The Ruminant System

Another type of stomach in the digestive system is the **ruminant** stomach (see Figure 25–12). A ruminant is an animal such as a cow, sheep, or goat that has one large stomach with multiple compartments or sections. The ruminant stomach occupies about three-quarters of the abdomen. Ruminant animals have no upper front teeth, only lower incisors, and are known as "cud chewers." **Cud** is the mixture of grass or hay sources and saliva. Food is broken down by the ruminant system and regurgitated for further breakdown to allow digestion to occur. **Regurgitation** is the process of bringing food into the mouth after it has been swallowed. The first section of the ruminant stomach is called the **rumen**. The rumen acts as a storage vat and soaks food to soften it during a process called **fermentation**, which allows food particles to break down by the use of bacteria for easier digestion. The rumen consists of about 80 percent of the stomach and is the largest compartment, lying mostly on the left side of the abdomen. The surface of the rumen is covered with tiny hairlike projections called *papillae* that increase the surface area and allow for the absorption of nutrients. The rumen regularly

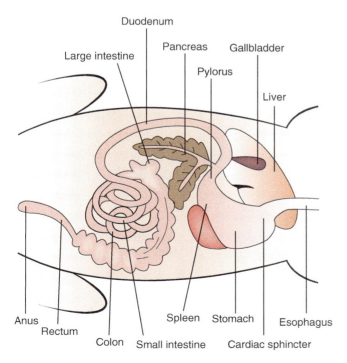

FIGURE 25–11 Structures of the monogastric digestive system.

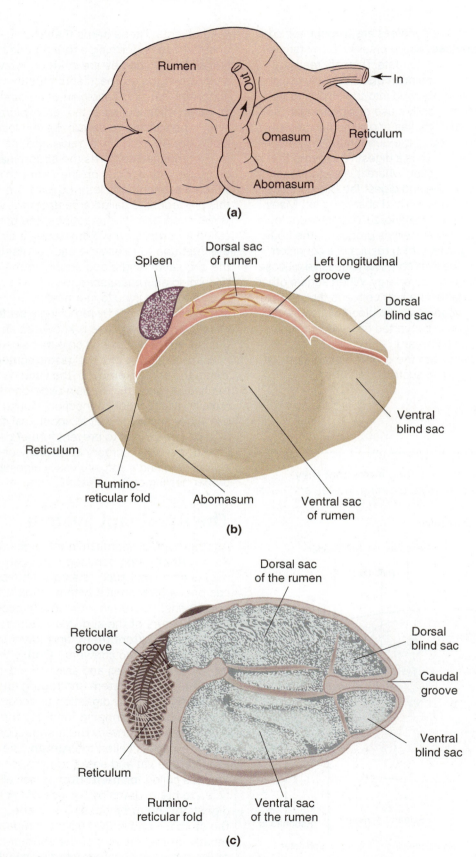

FIGURE 25-12 (A) Example of the ruminant stomach. (B) Linings of the four sections of the ruminant stomach. (*Continues*)

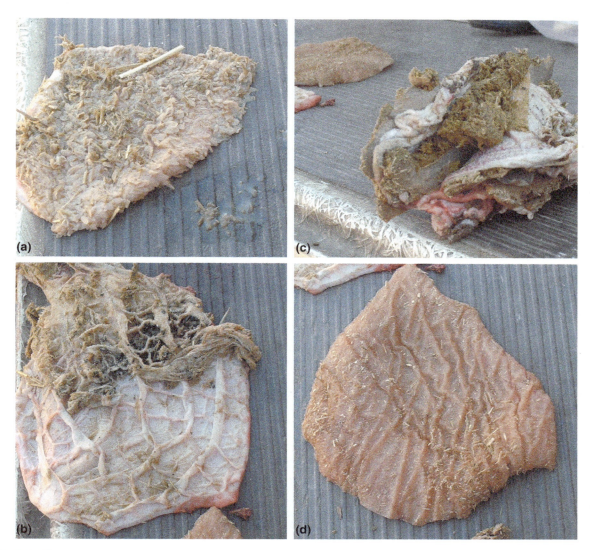

FIGURE 25-12 (Continued)

contracts to help mix the contents within it. These contractions result in a rumbling noise that can be heard on auscultation of the abdomen. On evaluation, a normal ruminant should have two to three contractions per minute. The contents of the rumen separate, with liquid portions in the lower area and gas in the uppermost section. The second section is called the **reticulum**. This compartment acts as a filter and composes about 5 percent of the stomach. The reticulum looks similar to a honeycomb pattern. This helps to trap dangerous materials that are not digestible and prevents them from passing through the digestive tract. Items that are not capable of passing this compartment are regurgitated for further breakdown. Often metal components that are swallowed may penetrate the thin lining of the reticulum. It is common practice to place magnets in the stomach to collect metal objects, preventing sharp objects from penetrating the wall. The third compartment is the **omasum**. The omasum composes about 8 percent of the stomach and absorbs nutrients and water. It is lined with long, thin folds that allow an increase in surface area that absorbs water and nutrients. It also helps to grind down roughage and grains as they pass through the system. The last section is the "true" stomach called the **abomasum**. The true stomach is about 10 percent of the GI system and acts similar to the monogastric stomach. This section contains acids and enzymes that begin digestion of food particles. Food passes through each compartment, then is regurgitated to allow further chewing to break down food into smaller particles, thus causing an increase in saliva production and food needing to be reswallowed. This occurrence continues until food is able to pass into the abomasum.

Ruminants eat food quickly, chew briefly, and then swallow. A process called *rumination* creates contractions that force the swallowed material back into the mouth through *regurgitation*. The ruminant chews the food again, combining more saliva, and then swallows the material. This process repeats itself until the material is finely ground. This area holds a large number of bacteria that helps break down fiber and plant material. These microorganisms release nutrients into the rumen, are absorbed, and produce a gas. Ruminants are known to chew with open mouths, and air easily enters the rumen, mixing with the gas. As the gas builds up, belching occurs to rid the rumen of the gas in a process called **eructation**. Young calves have a poorly developed rumen and reticulum and initially function as monogastrics. A specialized area within the reticulum contracts as a calf suckles. A groove in the esophagus directs the material to the abomasum, bypassing the rumen. It takes several months for the rumen and reticulum to fully develop, and gradually the calf transitions from a monogastric to a ruminant.

The Nonruminant System

Another type of digestive system is the **nonruminant system**. This system is noted in horses, rodents, and rabbits. This system is quite similar to the monogastric digestive tract in appearance except it has a well-developed intestinal tract with a large, long cecum located between the small and large intestines (see Figure 25–13). Plant fibers are broken down and digested in the cecum. Small amounts of healthy bacteria also break down plant and hay fibers through fermentation. Nonruminant animals tend to eat slowly and require digestion to occur through most of the day. A horse's intestinal system is about 70 feet in length, with the cecum measuring about 4 feet. Nonruminant animals are not capable of vomiting, as they do not have a **gall bladder**, which is an organ that stores bile. Nonruminant animals make bile in their intestinal systems and have no need for a gall bladder.

The Avian System

The last type of digestive system is the **avian system** (see Figure 25–14). This is a specialized GI system in birds and poultry species characterized by the inclusion of several organs not found in other species that are used to break down and grind hard food particles. Avian mouthparts have no teeth and form what is called a **beak**. Salivation occurs to help soften food for swallowing. The esophagus passes food into a small holding tank called the **crop**. The crop softens and stores food for later use and slows digestion. The crop can be palpated to determine if digestion is occurring. Similarly to ruminants, birds can regurgitate food from the crop to be filtered and broken down for later digestion. When the food particles are small enough to pass into the stomach, the food passes into two specialized organs located in the stomach: the **proventriculus** and the **gizzard**. The proventriculus acts like the monogastric stomach and begins the digestion

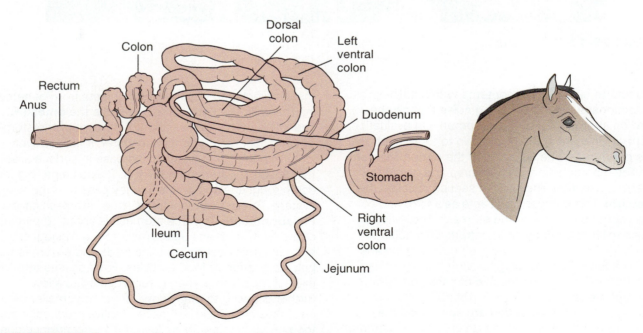

FIGURE 25–13 Example of a nonruminant digestive system.

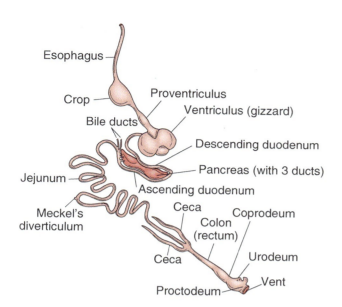

FIGURE 25–14 Avian digestive system.

process by releasing excretions to soften food. The gizzard is a muscular organ located after the proventriculus and grinds down hard food particles, such as bones or seed shells. The food filters through the small and large intestines similar to other animal species. The end of the digestive tract is the **cloaca**, which is the area where waste materials pass. Avian species pass both urates and feces at the same time from the cloaca. The cloaca is commonly called the **vent** and is the external area similar to the rectum.

Common Digestive Conditions

Digestive upset can be caused by many factors. The digestive system's reaction to this upset is typically **vomiting** or **diarrhea**. *Vomiting* is the act of bringing up partially digested or undigested food that has been in the stomach. Vomiting occurs in monogastric animals only. Ruminants and avian species have a similar normal action in regurgitation, but this is due to how the animal needs to have food sources filtered and broken down for digestion. Vomiting may begin with nausea, which can slow down the *motility* (movement) of the stomach and esophagus. Nausea may cause retching or strong rapid abdominal contractions that cause the animal to gag. Vomiting then occurs as the stomach's contents are forced out of the mouth. Animals that are vomiting often need to be taken off food and water for up to 8–12 hours to allow the GI system to rest. When vomiting has resolved, it is often recommended to introduce small amounts of water and then begin a bland diet, such as boiled turkey and boiled rice, which are low in fat and high in fiber. Fiber in the diet helps the body absorb water in the large intestine and slows down the output of feces. *Diarrhea* is the occurrence of waste materials becoming soft and watery, which can lead to an uncontrollable bowel movement. Diarrhea originating in the large intestine often causes a high frequency of bowel movements, often in small amounts with blood, mucous, or both. Diarrhea originating in the small intestine often has an increased volume of liquid material without an increase in frequency. Animals that have occurrences of vomiting or diarrhea should be examined to determine the cause.

Dehydration

Both vomiting and diarrhea can lead to **dehydration**, the loss of fluids and electrolytes in the body. Dehydration can be established by the evaluation of **skin turgor**, which can be evaluated in the area of skin at the base of the neck and over the shoulders and back that is elastic-like. Hydration status can be determined by lifting the skin up and monitoring how quickly it "snaps" back in place. Skin with normal turgor should snap back in place immediately like an elastic band. When dehydration occurs, the skin will return more slowly to its normal position over the neck and shoulders. Hydration status can also be determined by evaluating the gums or **mucous membranes (mm)** (see Figure 25–15). The color and amount of moisture can help identify if dehydration is occurring. Dehydration can also be seen in the eye sockets; the eyes will appear sunken into the head.

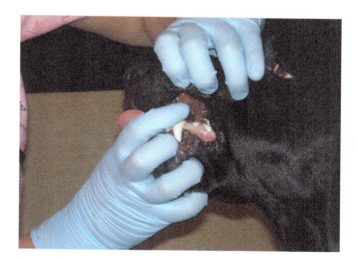

FIGURE 25–15 Examination of mucous membranes to determine hydration status.

 TERMINOLOGY TIP

The gums or other mm can be labeled normal or moist, **tacky** or slightly dry, and dry or no moisture.

Dehydration may cause the need to replace bodily fluids with **intravenous (IV)** or **subcutaneous (SQ)** fluids. Intravenous fluids are injected into the vein by placement of an IV catheter, which is a small, sterile plastic tube that is inserted into the vein. Some supplies are different for IV and SQ fluid therapy. This procedure is typically done by the veterinarian or veterinary technician and is necessary for severely dehydrated animals that may need long-term fluid therapy. Subcutaneous fluids are given under the skin, usually at the base of the neck or an area where the skin is plentiful and elastic-like (see Figure 25–16). This is a procedure that may be done by the veterinary assistant and requires a bag of fluids, a fluid line, and an appropriately sized needle. This type of fluid replacement is for animals that are not too dehydrated and is a short-term therapy. The need for and type of fluid replacement must be decided by the veterinarian. The type of fluids will depend on the amount of dehydration and the cause of the dehydration. It will also depend on the animal's overall physical health. Many fluids are sodium based, as thirst control is provided to the animal. When vomiting occurs, the veterinarian may require an animal to have food and water withheld to allow the digestive system a period of rest.

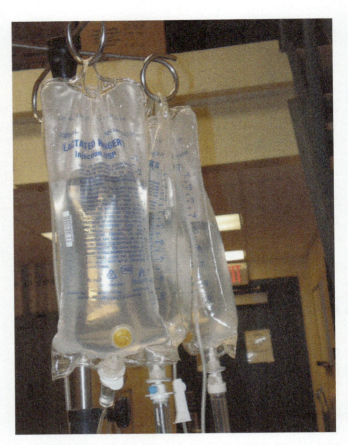

FIGURE 25–17 Bag of LRS used in SQ and IV fluid administration.

 TERMINOLOGY TIP

The abbreviation **NPO** means "nothing per os" or nothing by mouth, meaning no food or water. The opposite of this abbreviation is **PO** or "per os" or by mouth.

Fluids can be given to replace water in the body until the animal is ready to eat and drink. When sodium-based fluids are given, it encourages the animal to drink. **Sodium chloride** or NaCl is a saltwater fluid. **Normal saline** is an isotonic solution, meaning it has the same concentration level as table salt, and is often used to trigger an animal to drink water. **Lactated Ringer's solution (LRS)** is a fluid that is commonly used to replace lost fluid (see Figure 25–17). The animal's dehydration status helps to determine how much replacement fluid is necessary. Table 25-2 shows different amounts of dehydration and the corresponding clinical signs. The calculation used to determine the amount of replacement fluids is the percentage of dehydration multiplied by the animal's body weight in kilograms (kg) multiplied by 1000. This determines the replacement volume of fluids in milliliters (ml).

% dehydration × body weight in kg × 1000
= ml replacement volume

FIGURE 25–16 Administering SQ fluids under the skin to treat dehydration.

TABLE 25–2

Dehydration Percentages

6–7 percent dehydration	Moist mm; slight skin turgor; slightly sunken eyes
8–9 percent dehydration	Tacky mm; obvious sunken eyes; slow skin turgor with noted tenting
10–12 percent dehydration	Dry mm; very sunken eyes; very slow and very tented skin turgor

TABLE 25–3

Causes of Colic

Ingesting large amounts of grain or pasture	Ingesting fresh grass or hay sources
Stress	Excessive gas
Internal parasites	Excessive amounts of water after exercise or high temperatures
Dehydration	Ingesting sand
Sudden change in diet	Constipation or impaction
Abdominal tumors	Hernia
Medications	Unknown origins

Constipation

Constipation is another occurrence in the digestive system that can cause little to no bowel movement production. A constipated animal may show signs of stomach pain, straining to produce a bowel movement, and decreased appetite. Many factors can cause constipation, such as poor diet, eating items that are not digestible, not drinking enough, or a low-fiber diet. Constipated animals should be evaluated by a veterinarian to determine the cause. Patients should be treated by the veterinarian according to how constipated they are. Some animals may require an **enema** or a procedure that passes fluids into the rectum and colon to soften feces to allow the animal to have a bowel movement. Many times, this is done using warm, soapy water. Animals may need **stool softeners** or **laxatives** given by mouth to help keep the waste materials soft for ease of passing. A high-fiber diet may also be recommended to produce normal bowel movements. Extreme cases of constipation may require surgery to remove large, firm stool from the intestinal tract.

Colic

Colic is a common condition in horses. Colic is severe stomach pain that may be caused by many factors, though sometimes the factor remains unknown (see Table 25–3). Horses are not able to vomit, and when they begin showing signs of colic, it is important to prevent them from rolling. The stomach pain caused by colic can cause a horse to lie down and roll to try to relieve the pain and pressure. When a horse rolls, it is likely that the stomach and intestinal tract can rotate upon itself. This is a condition known as **intussusception**. This causes the intestine to telescope upon itself, thus cutting off circulation through the digestive tract (see Figure 25–18). When circulation is cut off to an area of the body, the tissue begins to die, and this can be fatal to the animal. Horses, being nonruminant animals, must slowly digest food particles throughout the entire day. When colic or intussusception occurs, this prevents digestion from occurring. Colic is an emergency and a veterinarian should be notified immediately. A physical exam and rectal palpation are necessary to determine how severe the condition may be. Feces should be monitored for amount and consistency to determine if the animal is capable of passing stool. Laxatives are often used to keep feces soft and passable. The pain is typically treated, and the cause may or may not be determined. A common pain reliever used for colic is Banamine. Intussusception is corrected through surgical repair, but the longer the condition has been occurring the less likely the surgery will be successful. Following surgery, horses require intensive care, antibiotics, and IV fluid therapy. Signs of colic include frequent lying down and rising, rolling, **flanking** (looking at or biting at the sides of the abdomen), kicking at the sides of the abdomen, little to no bowel movement production, anorexia, depression, and little to no drinking.

Bloat

Bloat is a condition that affects dogs and some ruminants, such as cattle, sheep, and goats. Bloat causes the abdomen to become swollen and painful due to air and gas within the stomach and intestinal tract. Bloat in dogs is more common in large and giant breeds, but it can occur in any type of dog. Dogs that become bloated should be handled as an emergency and seen immediately by a veterinarian. Dogs may become bloated for several reasons, usually by ingesting food very quickly, exercising shortly after eating, getting into garbage or food they are not accustomed to eating, or drinking excessive amounts of fluids. This condition is called **gastric dilatation**. Dogs that become bloated could progress to a condition known as **gastric dilatation volvulus** or GDV, in which the stomach or intestinal tract rotates and causes a similar condition to that of horses with colic. Bloat or gastric dilatation may be treated by placing a stomach tube into the GI system to remove excessive air or gas. Dogs that have GDV require immediate surgery to correct the rotation. In cattle and other ruminants, the rumen can become gas filled, and eructation can't occur to remove the gas. Bloat develops, causing a **distended** or swollen abdomen. This typically occurs on the left side of the rumen (see Figure 25–19). Two types of bloat may occur in ruminant animals: free gas or frothy bloat. **Free gas** accumulates in the dorsal rumen, and the animal usually shows signs of choking when the esophagus becomes obstructed with ingesta, causing the gas to not be able to escape. The rumen stops contracting and regurgitation ceases. **Frothy bloat** is caused by gas being trapped in small bubbles within the rumen. This is typical of animals that eat lush green pastures and build up gas that becomes trapped. Large ruminants are typically treated by the veterinarian by placing a **trocar** within the rumen to allow the gas to escape and relieve the pressure on the animal's stomach. A trocar is a plastic or metal instrument with a pointed, sharp end that is placed within the rumen through the abdominal wall and that allows the air to drain (see Figure 25–20).

COMPETENCY SKILL 28

Converting pounds (# or lb) to kilograms (kg)

Equipment and Supplies:
- Scale
- Pen or pencil
- Paper
- Calculator

1. Determine the weight of the animal in pounds using a scale. Record the weight of the animal in pounds.
2. Set up the formula of determining kg by using the following:

   weight in pounds divided by 2.2 = _____ kg

   # or lb/2.2 = _____ kg

3. Record the body weight in pounds and kg in the medical record.

Converting kilograms (kg) to pounds is done in reverse.

1. Record the weight of the animal in kilograms (kg).
2. Set up the formula of determining # or lb by using the following formula:

   weight in kg multiplied by 2.2 = _____ # or lb

   kg × 2.2 = _____ # or lb

EXAMPLE:

20# animal = 9.0 kg
20#/2.2 = 9.0 kg

EXAMPLE:

10 kg animal = 22 pounds
10 kg × 2.2 = 22 #

Ruminants can also have dilation volvulus occur in which the stomach rotates and becomes displaced, called a **displaced abomasum**. This requires a surgical incision into the side of the abdominal wall in which the veterinarian can place an arm and rotate the stomach back to its place (see Figure 25–21).

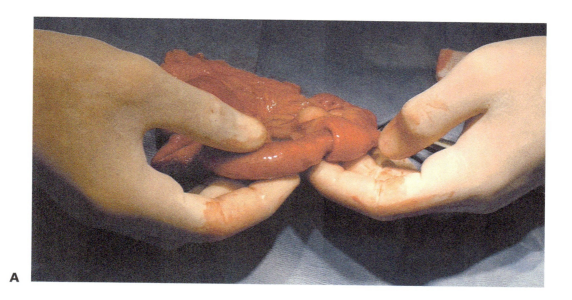

A

Foreign Body Obstructions

Foreign body obstructions may occur in any type of animal that ingests an item that is not digestible or may become impacted within the digestive system. Such items may include toys, household items, or items found outside (see Figure 25–22). Diagnosis of a foreign body within the digestive tract may be done through radiographs or through the use of an **oral barium study**. Oral barium is given by mouth and radiographs are taken over time as the barium passes through the digestive system. The barium is **radiopaque** and fluoresces as it passes through the GI system (see Figure 25–23). The radiographs will

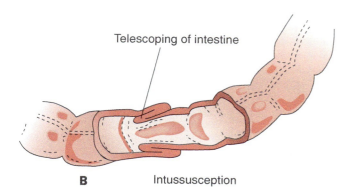

FIGURE 25–18 (A) Intraoperative photograph of intussusception, where a region of the intestine has telescoped into another section. (B) Telescoping of the bowel upon itself.

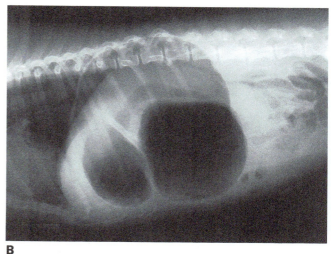

FIGURE 25–19 (A) Bloat in a cow. Note the distended appearance on the left side. (B) Gastric dilation volvulus formation.

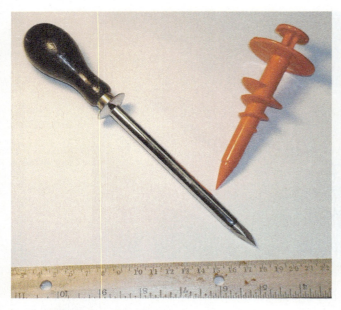

FIGURE 25-20 Examples of trocars.

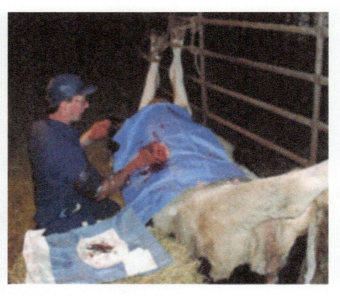

FIGURE 25-21 Displaced abomasum surgical repair in a cow.

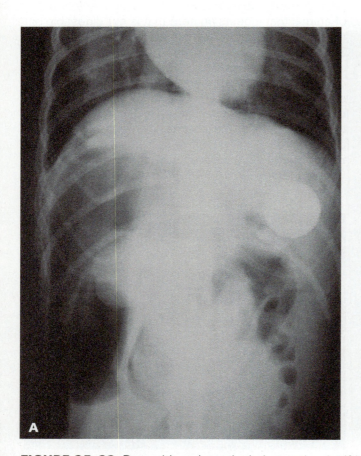
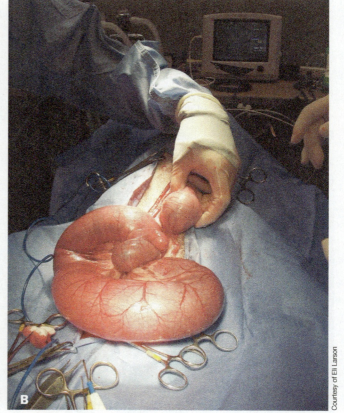

FIGURE 25-22 Dog with an intestinal obstruction (golf ball).

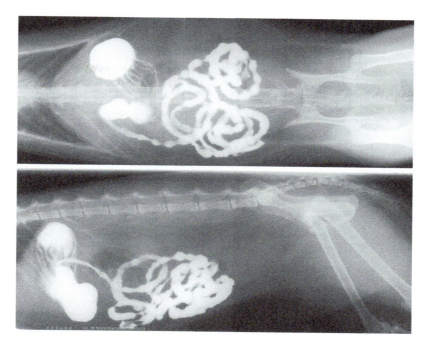

FIGURE 25–23 Radiograph showing a barium series in a cat. The bright white area is the barium within the gastrointestinal tract.

help determine if an object is moving through the system or if it has become obstructed and requires surgical removal. A foreign body obstruction may cause the intestinal tract to telescope upon itself, causing an intussusception. This causes the blood supply to be cut off. When a foreign body is surgically removed from the stomach or intestinal tract and the blood supply has been cut off, causing an area of dead or **necrotic** tissue to occur, there may be a need for an **anastomosis** surgery. This is the surgical removal of an area of dead tissue along the intestinal tract, which requires resection of the intestines.

Signs of an animal with a foreign body obstruction include the following:

- Vomiting
- Retching or gagging
- Diarrhea
- Anorexia
- Abdominal pain
- Dehydration
- Depression

 MAKING THE CONNECTION

Additional information on disorders affecting the digestive system can be found in the following:
Chapter 8—bloat
Chapter 10—fatty liver disease
Chapter 12—Tyzzer's disease
Chapter 17—bovine viral diarrhea, campylobacter, bloat, grass tetany, and ketosis
Chapter 18—colic
Chapter 19—swine dysentery
Chapter 20—enterotoxemia, Johne's disease, and lamb dysentery.

SUMMARY

The digestive system provides nutrients to the body. Four different types of digestive systems can be found in animal species: monogastric, ruminant, nonruminant, and avian. The variation in the types of digestive systems found in animals is related to the types of foods the animals eat. Some of the common conditions veterinarians will treat related to the digestive system are vomiting, diarrhea, constipation, dehydration, bloat, colic, and foreign body obstructions. Some of these conditions present as medical emergencies, and immediate care is required to save the animal's life.

Key Terms

abomasum last section of the ruminant stomach that acts as the true stomach and allows food to be digested

alimentary canal veterinary medical terminology for the GI system

amylase enzyme produced by the pancreas that breaks down starches

anastomosis surgical removal of a dead area of tissue along the digestive tract and re-sectioning the area back together

ascending colon first section of the large intestine

avian system specialized digestive system of birds

beak avian mouth with no teeth that forms an upper and lower bill

bile yellow fluid secretion that helps break down food for digestion and absorption

bloat condition that causes the abdomen to become swollen and painful due to air and gas within the intestinal tract

body central part of the stomach that expands as food enters

canine teeth also known as fangs that are used to tear apart food

cardia entrance of the stomach that filters food

carnassial tooth upper fourth premolar and lower first molar in dogs and cats that tends to become infected and abscessed

cecum the small sac that lies between the small and large intestines

cloaca end of the digestive tract where waste materials pass

colic condition in horses that causes severe stomach pain

colon common term for the largest section of the large intestine

constipation occurrence in the digestive tract that can cause little to no bowel movement

crop small sac that acts as a holding tank for food as it is passed from the esophagus in birds

crown the upper part of the tooth that lies above the gum line

cud mixture of grass sources and saliva that is chewed and regurgitated to break down food for digestion

deciduous baby teeth that are developed in newborn animals and eventually shed when adulthood is reached

dehydration loss of fluids in the body

dental formula arrangement of teeth in the upper arcade in the numerator position and teeth in the lower arcade in the denominator position

dentin second layer of teeth; similar to bone

dentition the way teeth are arranged in the mouth

descending colon third or last section of the large intestine

diabetes condition that is produced when too much or too little blood sugar is produced, and the body finds it difficult to regulate

diarrhea process of waste materials and feces become soft and watery

digestion breaking down food particles into nutrients to be used by the body to allow an animal to live

digestive system the body system that contains the stomach and intestines

displaced abomasum condition in ruminants where the stomach twists and rotates out of place

distended swollen, as in the abdomen

duodenum short, first section of the small intestine

enamel hardest substance in the body; covers and protects the teeth

enema passes fluids into the rectum and colon to soften feces to produce a bowel movement

eructation gas buildup where belching occurs to rid the rumen of air

esophagus tube that passes food from the mouth to the stomach

fermentation process of soaking food that allows bacteria to break down food for easier digestion

flanking looking at or biting at the sides of the abdomen due to stomach pain, often in a colic situation in horses

foreign body obstruction occurs when an animal ingests a foreign object that is not digestible and it becomes impacted within the intestinal tract

free gas air accumulates in the dorsal rumen of a ruminant's stomach, causing the animal to choke when the esophagus becomes obstructed with food and saliva and the gas is not able to escape

frothy bloat caused by gas being trapped within small bubbles within the rumen, causing the abdomen to become swollen and painful

fundus opening of the stomach

gall bladder organ that stores bile

gastric dilation the condition known as bloat in which air or gas fills the stomach, causing the abdomen to become swollen and painful

gastric dilatation volvulus (GDV) condition where the stomach and intestinal tract rotate after becoming swollen due to air or gas in the GI tract, causing the intestinal tract's circulation to be cut off

gastrointestinal (GI) system the digestive system; contains the stomach and intestines

gizzard muscular organ located after the proventriculus in birds that grinds down hard food substances

glucose more technical term for blood sugar

herbivores animals that eat plant-based food sources

ileum third or last section of the small intestine

incisors the front teeth located in the upper and lower jaws

insulin chemical produced by the pancreas that is released into the bloodstream and regulates the body's blood sugar

intravenous (IV) within the vein

intussusception condition when the stomach or intestine telescopes upon itself, cutting off circulation of the digestive tract

jejunum second or middle section of the small intestine

lactated Ringer's solution (LRS) fluid of lactic acid that is commonly used to replace fluids, as in dehydration

laxative stool softeners or medicine given to soften feces to produce a bowel movement

lipase enzyme produced by the pancreas that breaks down fats

liver organ behind the stomach that makes bile and produces glucose

mesentery connective tissue from the peritoneum; allows blood vessels and nerves to supply the small intestine

molars last set of teeth; are large and located in the back of the mouth

monogastric refers to the digestive system of an animal with one simple stomach

mucosa thin connective tissue that lines the intestinal tract

mucous membranes (mm) membranes that line body cavities; the gums are an example

necrotic dead area of tissue

nonruminant system digestive system similar to monogastric animals but with a larger, well-developed cecum for breaking down plant fibers

normal saline solution with the same concentration level as salt

NPO nothing by mouth, as in food or water

omasum third section of the ruminant stomach; absorbs nutrients and water

omentum thin lining that surrounds the organs within the abdomen

omnivores animals that eat both plant- and meat-based food sources

oral barium study barium solution given by mouth to pass through the digestive system to allow X-rays to be taken over time to view internal structures of the GI tract

pancreas organ that lies next to the stomach and secretes enzymes that aid in digestion

papillae hair on the tongue that acts as taste buds

peristalsis wavelike motion of the stomach that moves food through the intestines in contractions

peritoneum clear thin lining of the abdomen

permanent adult teeth that are formed after the deciduous teeth are shed

pharynx throat or area at the back of the mouth

PO by mouth

premolars wider teeth at the back of the mouth used to grind and tear food

proventriculus acts as a monogastric stomach and begins the digestion process in birds by releasing excretions to soften food

pulp cavity center of the tooth that holds the nerves, veins, and arteries

pylorus exit passageway of the stomach

radiopaque solution that fluoresces and allows radiation to pass through to view internal body structures during X-rays

regurgitation process of bringing food into the mouth from the stomach to break it down

reticulum second section of the ruminant stomach; acts as a filter for food

root part of the tooth located below the gum line; holds the tooth in place

rugae folds within the stomach when it is empty

rumen first section of the ruminant stomach; acts as a storage vat and softens food for fermentation

ruminant animal with a digestive system that has a stomach with four sections or compartments

saliva fluid that helps soften and break down food for ease of swallowing and digestion

salivary glands area within the mouth that produces saliva

skin turgor assessed during the process of evaluating an animal for dehydration by lifting the skin over the base of the neck or shoulder blades

sodium chloride NaCl; saltwater fluid

stool softeners medication given to produce a bowel movement by softening feces

subcutaneous (SQ) under the skin

tacky slightly dry, as the gums may be

tongue muscle within the mouth used to hold food within the mouth

transverse colon second or middle section of the large intestine

trocar plastic or metal pointed instrument placed into the rumen of a ruminant animal that has bloated to relive the air pressure on the animal's stomach

trypsin enzyme produced by the pancreas that digests proteins

vent external area of an avian that passes waste materials; also called cloaca; similar to the rectum

vomiting process of bringing up partially digested or undigested food that has been in the stomach of monogastric animals

Review Questions

1. The involuntary muscle action of the digestive system is called:
2. What is an animal that eats a meat-based diet called?
3. What are the teeth used to grind down food called?
4. What is the enzyme that breaks down starches called?
5. What is a common gastrointestinal condition in horses that causes severe pain and may be fatal called?
6. What are the four types of digestive systems?
7. List two examples of species with each type of digestive system.
8. Compare and contrast bloat and colic.
9. List the organs associated with the digestive system.
10. Describe the function of the ruminant digestive system.
11. Which type of diarrhea typically occurs very frequently?
12. What does the abbreviation NPO mean?
13. Which portion of the tooth lies above the gum line?
14. Which section of the ruminant stomach has a honeycomb-like appearance?
15. Which tooth tends to become infected and causes an abscess and swelling under the eye?
16. What are the small hairlike projections along the tongue called?
17. Which ruminant compartment is known as the "true stomach"?
18. Discuss possible treatments used in the condition known as constipation.
19. What is the surgical procedure to explore the abdomen called?
20. What is the surgical term for removing a section of the intestine called?

Clinical Situation

Harley is a 4 yr NM lab known for ingesting clothing, toys, and anything he can get in his mouth. He has been vomiting for 2 days and is lethargic. He appears depressed and in some discomfort when Allie, the vet assistant, takes him into an exam room. Harley has a normal RR and HR but has an elevated temperature of 103.0 degrees Fahrenheit.

Dr. Osborne completes a PE and blood work, and orders radiographs. The X-rays show a buildup of gas in the stomach and small intestine, but no foreign body is noted. Dr. Osborne suggests that Harley be admitted for a barium study.

Allie and Kathleen, another vet assistant, admit Harley into the ICU. The owner has several questions for the assistants as they place Harley into a kennel. Her main concern is asking about the barium study procedure.

- What should the assistants explain to the owner about the barium study?
- What should be discussed with the owner if the vet finds surgery is necessary?
- What supplies and equipment will Allie and Kathleen need to complete the barium study radiographs?
- What steps should the assistants take when admitting Harley into the ICU?

CHAPTER 26 Circulatory System

Objectives

Upon completion of this chapter, the reader should be able to:

26.1 State the structures that make up the circulatory system
26.2 Describe the functions of the circulatory system
26.3 Describe the functions of the various types of blood cells
26.4 Describe blood flow through the heart
26.5 Describe common disorders of the circulatory system

Introduction

The **circulatory system** is essential for life and involves the heart, blood, arteries, veins, and capillaries. The functions of the circulatory system include oxygen flow, blood circulation, transport of nutrients, waste removal, and movement of hormones.

Blood

The circulatory system includes blood cells and blood flow. **Blood** is composed of 40 percent cells and 60 percent **plasma**. *Plasma* is formed of various proteins: albumin, globulins, and fibrinogen. **Albumin** draws water into the bloodstream and helps in providing hydration to the body. **Globulins** provide antibodies to help prevent disease. **Fibrinogen** aids in clotting blood. Plasma consists mainly of water.

 TERMINOLOGY TIP

The term to **clot** or clotting refers to the process of stopping blood flow, as in after a cut or puncture that causes bleeding. ■

Hematology is the study of blood and an essential practice in the veterinary facility. The veterinary assistant serves an important role in setting up for blood collection, running blood analysis samples, restraining patients for blood collection, and completing blood testing kit samples. Blood samples are collected by the veterinarian or technician. Blood tube samples should be labeled with the patient's name, owner's name, date, and test procedure requested. Some testing procedures require the blood tube to be spun at high speeds using a piece of equipment called a **centrifuge** (see Figure 26–1). The centrifuge uses speed to separate blood elements, causing the cells to separate from the **serum**, the liquid portion of a blood sample that is used to analyze chemistry values to determine the functions of the body's organs (see Figure 26–2). The serum portion of blood makes up 30–45 percent of the total contents, which may vary between species. The blood is composed of red cells and white cells. **Red blood cells (RBCs)** are known as **erythrocytes** and are the most abundant blood cell in the body (see Figure 26–3A). Their main function is to transport oxygen throughout the body. The red blood cells are produced in the bone marrow found within the medullary cavity of bones through the process of **erythropoiesis**. Red blood cells are constantly being produced and replaced. The main component of red

477

FIGURE 26–1 Centrifuge.

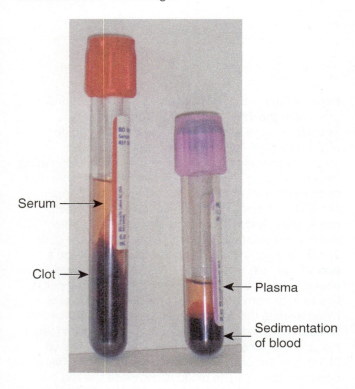

FIGURE 26–2 Serum versus plasma.

aim to destroy any microorganisms in the tissues. A mature neutrophil has a nucleus with segments or divisions. An immature neutrophil is called a **band cell** and indicates that there is an infection in the body. Band cells have a nucleus that is shaped like the letter *U* (see Figure 26–3B). If the body is invaded by infection, neutrophils will move to the site of infection within the tissues to act as an army against infection. An infection typically has a higher number of abnormal cells that may include band cells. **Lymphocytes** are the largest cell part within the bloodstream. They are the largest of the five types of WBCs and aid in immune functions to help fight diseases by producing antibodies in the blood. Lymphocytes have one large single nucleus that makes up most of the blood cell (see Figure 26–3C). **Eosinophils** fight against allergic reactions, help control inflammation, and help prevent parasite infections within the body. They look similar to neutrophils in that they have a segmented nucleus, but they have a large number of granules in the cytoplasm that give them a dotted appearance (see Figure 26–3D). **Monocytes** are the largest white blood cells and help the neutrophils by removing organisms, dead cells, and foreign particles. As monocytes age they become **macrophages**, which are cells that eat and destroy organisms at different locations within the body. **Basophils** are cells that are involved with allergic reactions, and they have a segmented nucleus with granules that stain very dark (see Figure 26–3E). The granules contain **histamine**, which is released during an allergic reaction. Another type of cell found within the blood is a **platelet**. Platelets aid in the clotting time of blood after an injury occurs to a blood vessel, causing the vessel to constrict. Platelets attach to the vessel site and plug the hole, thus decreasing bleeding. Platelets are also known as **thrombocytes** and appear in a blood smear as small dots. When not enough platelets are formed in the blood, a condition called **anemia** may occur. This may cause a low red blood cell count that doesn't allow the red blood cells to replenish. An **autoimmune disease** involves the animal's red blood cells being destroyed by its immune system. Figure 26–4 displays the components of blood.

blood cells that allows for oxygen transport and has an iron component that allows red blood cells to be replicated or continue being produced is a protein called **hemoglobin**. **White blood cells (WBCs)** are called **leukocytes** and are the body's main defense against infection. The blood cells are commonly evaluated under a microscope to determine the type of cell and how many are noted in the blood. (Figure 26–3B, C, D) There are five types of white blood cells found in an animal's bloodstream: neutrophils, lymphocytes, eosinophils, monocytes, and basophils. Each type of white blood cell serves a specific function. **Neutrophils** are the most commonly seen white blood cells and

 TERMINOLOGY TIP

Terms used in bloodwork evaluation to determine the diagnosis of an animal's condition include the following:

- hyper—above normal or excessive
- hypo—below normal or reduced
- emia—refers to the blood
- penia—refers to a blood count that is below normal
- cytosis—refers to a blood count that is above normal

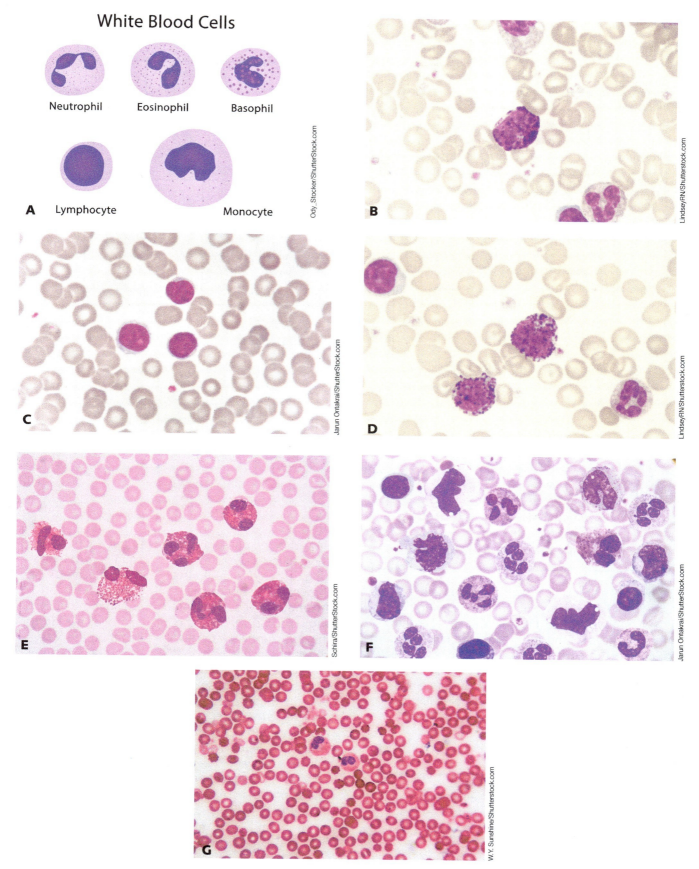

FIGURE 26–3 (A) Types of White Blood Cells, (B) Neutrophil, (C) Eosinophil, (D) Basophil, (E) Lymphocyte, (F) Monocyte, (G) Example of a differential blood smear.

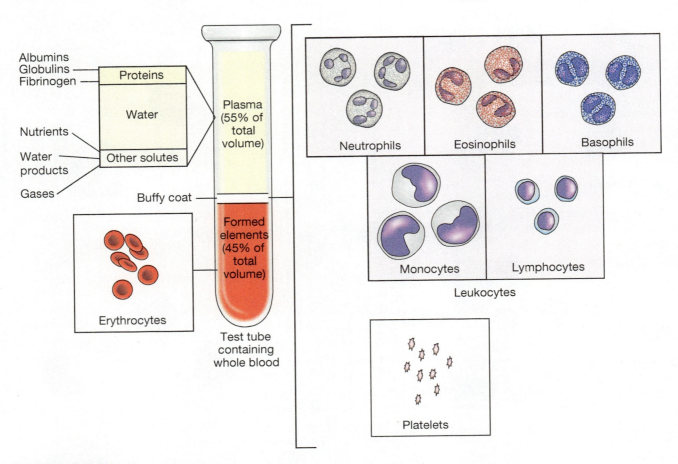

FIGURE 26–4 Blood components.

The Heart

The **heart** is an organ with four chambers and is located in the chest between two lungs. The wall of the heart is made of thick cardiac muscle called **myocardium**. A thin inner layer of muscle is called the **endocardium**. The outside thin covering of the myocardium is called the **epicardium**. The outside of the heart is lined by a sac called the **pericardium**, which is a thin membrane that covers, protects, and maintains the beating action of the heart (see Figure 26–5).

Blood Flow

Blood flows throughout the body, allowing for **systemic circulation** and oxygenation. The systemic flow of blood delivers nutrients to the entire body. Oxygen is delivered to the lungs and is exchanged with carbon dioxide. **Veins** are vessels located throughout the body that carry blood to the heart. **Arteries** are vessels located throughout the body that carry blood away from the heart. Arteries are larger the closer they are to the heart and then become smaller branches of blood vessels, called **arterioles**, the further away from the heart they lie. They then branch into tiny vessels called **capillaries**, as they eventually course throughout the body. The capillaries serve as a transport system to exchange nutrients and hormones, and to help rid the body of waste products and gases. Veins are much thinner and less muscular than arteries, and they contain valves that keep blood flowing in one direction. Arteries have thick, muscular walls and can be used to determine the **pulse** of an animal. The pulse is the **heart rate**, which is the

CHAPTER 26 Circulatory System   481

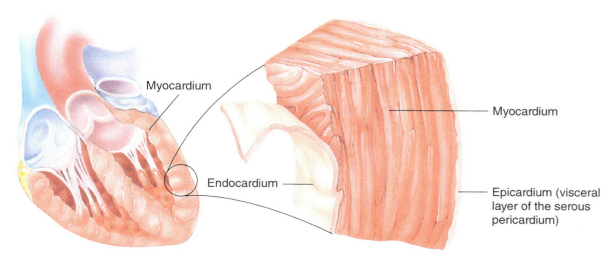

FIGURE 26–5 Layers of the heart.

pumping action of blood through the heart that can be counted to determine a number of beats per minute. Blood flows into the heart through a large vessel called the **vena cava**.

As blood flows into the heart, it enters the thin-walled chamber of the **right atrium** (see Figure 26–6).

It then passes into the **right ventricle**, a thicker-walled chamber that pumps blood to the lungs. The right atrium and right ventricle are separated by the **AV valve**, which opens and closes to allow blood to flow through in one direction. The **pulmonary arteries** carry oxygen-poor blood and prevent blood flow

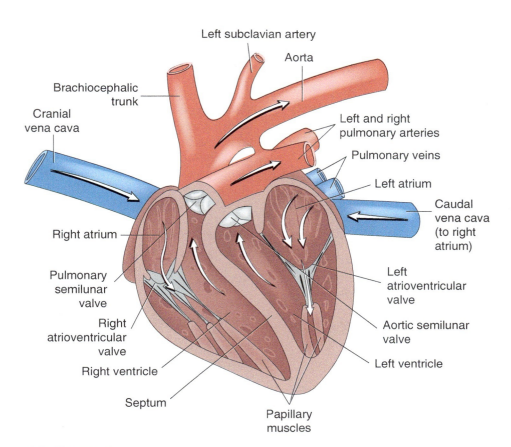

FIGURE 26–6 Internal structure and blood flow through the heart.

back to the heart. The pulmonary veins allow blood to flow back to the heart and into the **left atrium**, a thin-walled chamber, then into the **left ventricle**, the thickest-walled chamber of the heart. They left atrium and left ventricle are also separated by an AV valve. The **aorta** is the large vessel that allows blood flow out of the heart and back into systemic circulation. The aorta is an artery that is located in the mid-chest and abdomen. The AV valve opens when the ventricle contracts and is closed by the backup of blood when the ventricle relaxes, preventing blood from flowing back into the heart. Table 26–1 lists the normal heart rates for various species.

As blood flows through the heart, the body produces **blood pressure (BP)** that can be measured for contraction and relaxation. The contraction is caused by **vasoconstriction**, in which the diameter of the vessel decreases and causes the blood pressure to increase. Relaxation is caused by **vasodilatation**, in which the vessel diameter increases and causes the blood pressure to decrease. Blood pressure measures one complete contraction of the heart in the cardiac cycle. The contraction phase is called **systole**, which is the top number that is normally higher. The relaxation phase is called **diastole**, which is the bottom number that is normally lower. Blood pressure is measured in millimeters (mm) of mercury (Hg). Dogs typically have a BP of 115/65 to 150/90, and cats' normal range is around 125/88.

Example
Blood pressure = 120/80
120 is systole
80 is diastole

TABLE 26–1
Normal Heart or Pulse Rates

SPECIES	NORMAL HEART RATE (BEATS PER MINUTE)
Dog	60–120 bpm
Cat	160–240 bpm
Rabbit	130–325 bpm
Guinea Pig	240–250 bpm
Horse	30–40 bpm
Cow	60–80 bpm
Pig	60–80 bpm
Chicken	200–300 bpm
Sheep or Goat	70–80 bpm

Veins

Veterinary assistants in some states may not be allowed to perform venipuncture techniques; however, they do provide a significant role in assisting with successful blood collection. The main role of the veterinary assistant is to provide proper animal restraint, and it is important to know which veins are commonly used and be able to locate them using proper terminology. Dogs and cats are frequently restrained for blood collection from the following locations:

Cephalic veins—moderate-sized vessels located on anterior surface of the front limb running medially along the inner leg surface

Jugular veins—large vessels located in the neck on either side of the trachea

Saphenous veins—superficial vessels running diagonally over the lateral and medial surfaces of the distal tibia

Femoral veins—small vessels located on the medial aspect of the thigh near the groin

Heart Sounds

Heart sounds are created as the valves close. Heart sounds can be heard through the use of a **stethoscope**, an instrument used to listen to the heart, lungs, and other areas of the chest (Figure 26–7). These sounds are heard when the valves within the heart wall vibrate during closing. These vibrations spread through the chest wall and are detected with the use of the stethoscope. Heart sounds are controlled by the **pacemaker** of the heart, known as the **SA node**, or sinoatrial node. The pacemaker system controls the heart's rhythm. The first heart sound, which sounds like "lub," is created by the closure of the AV valves. The second heart sound, or the "dub," is created by the closure of the aorta and pulmonary valve. The **cardiac cycle** includes one complete contraction and the relaxation that follows. During the relaxation period, the atria fill with blood to prepare for the next contraction. Normal heart rates vary between species and health and fitness levels.

Heart Rates

When listening to the heart, it is important to note the patient's heart rate or pulse. First, the rate in beats per minute (bpm) is counted. The heart rate then is compared with what is normal for its species. Many factors may influence heart rate, especially in the veterinary hospital. Stress, fear, anxiety, and

CHAPTER 26 Circulatory System

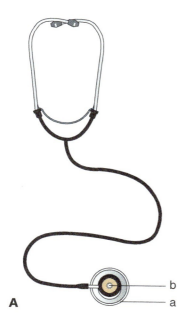

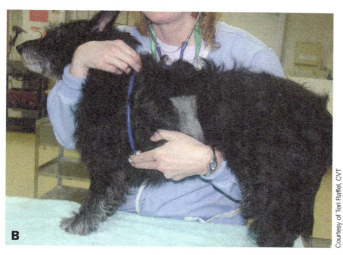

FIGURE 26–7 (A) Stethoscope. (B) Listening to heart sounds.

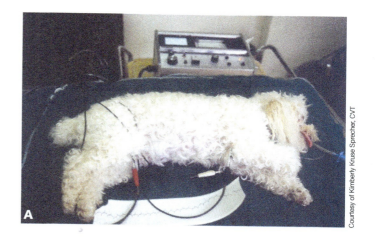

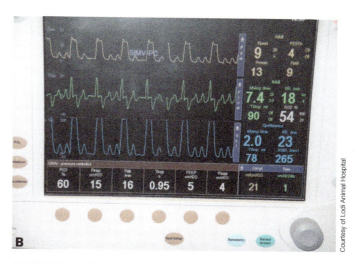

FIGURE 26–8 (A) ECG monitor. (B) A dog's electrocardiograph.

nervousness, as well as disease, may increase the heart rate, and this must be taken into consideration during the evaluation.

Electrocardiography

The veterinary facility needs to be able to monitor heart sounds and rhythms of animals that are under anesthesia or may have preexisting heart conditions. This may be done with the use of **electrocardiography** or the evaluation of the electrical currents of the heart through the use of machines. An **electrocardiogram**, also known as an ECG, traces the electrical activity of the heart (see Figure 26–8).

An electrocardiograph is a machine that records the electrical currents of the heart. The rhythm of the heart is shown on a screen and is depicted in waves or lines that form peaks according to the heartbeats. Each peak and drop is identified by a letter that relates to the heart's activity and functions. The ECG helps veterinarians identify problems with the heart. The normal heart rate and rhythm is called **sinus rhythm**.

The first peak is called the **P wave**, which forms at the SA node (pacemaker) and reflects the flow of electricity and blood through the atrium and that a contraction has occurred. The QRS wave or **QRS complex** is a series of peaks and drops that show that the electrical current and blood have flowed through the AV node and have caused a contraction in the ventricles. The **T wave** is the last peak and the most important part of the ECG. It shows that an electrical impulse has traveled through the entire heart and has completed the contraction and repolarization of the heart (see Figure 26–9).

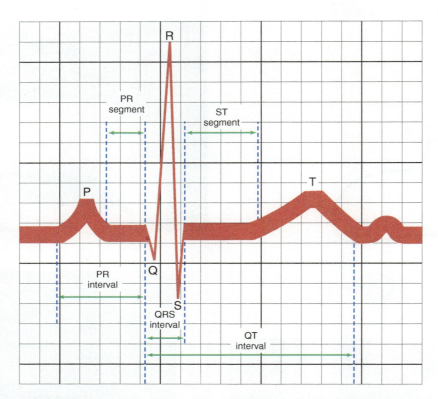

FIGURE 26-9 (Anatomy of an electrocardiogram. The first deflection, the P wave, represents excitation (depolarization) of the atria. The PR interval represents conduction through the atrioventricular valve. The QRS complex results from excitation of the ventricles. The QT interval represents ventricular depolarization and repolarization. The ST segment represents the end of ventricular depolarization to the onset of ventricular repolarization. The T wave results from recovery (repolarization) of the ventricles.

TERMINOLOGY TIP
The term **repolarize** means to reset or restart, and when describing the heart, it means to get ready for the next electrical impulse to start. ■

TERMINOLOGY TIP
Auscultation or to auscultate means to listen to, as in using a stethoscope to auscultate the heart. ■

Common Diseases and Conditions of the Circulatory System

Many abnormalities can occur with the heart. These conditions may be diagnosed through the use of physical exam, radiology, ECG, or ultrasound evaluation. Some common types of abnormal heart conditions include a **heart murmur**, which is caused by an abnormal valve that produces an abnormal flow of blood that creates a "swishing" noise upon auscultation.

Arrhythmias

An **arrhythmia** is a change in the heart's rhythm, rate, or conduction (see Figure 26–10). **Tachycardia** refers to the heart beating faster than normal. **Bradycardia** is the term that refers to the heart beating slower than normal. **Sinus arrhythmia** is a condition when the heart rate increases with inspiration and decreases with expiration, the complexes appear normal, but a change in breathing rhythm is noted. When **cardiac arrest** occurs, the heart is not contracting appropriately and is similar to a heart attack in humans. Cardiac arrest may be from **atrial fibrillation**, commonly called A-fib. This is a condition that occurs

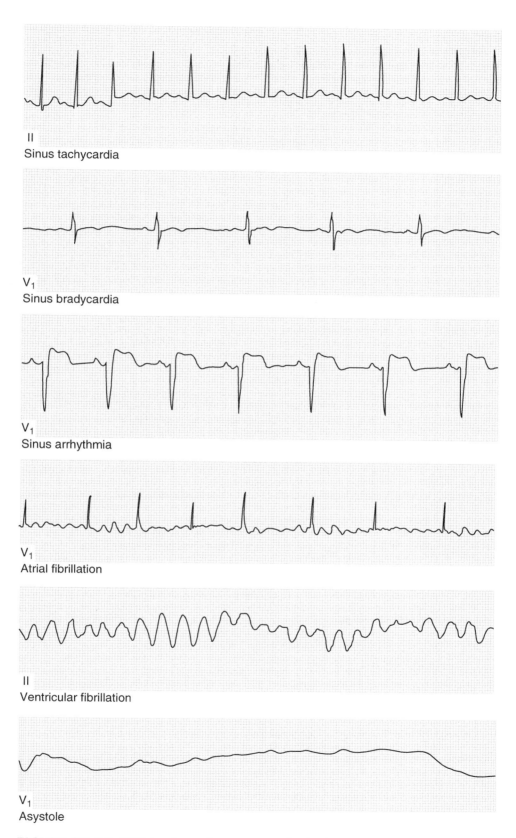

FIGURE 26–10 ECG tracings of common arrhythmias.

when the pacemaker of the heart or SA node is not working correctly. **Ventricular fibrillation**, commonly called V-fib, is a condition that causes the ventricles to fire electrical currents rapidly and is the most serious cause of cardiac arrest. When this occurs, the heart goes into a condition called **asystole**, meaning the heart stops contracting and heart failure occurs. **Cardiopulmonary resuscitation** (CPR) can then be used to stimulate the heart to deliver oxygen to the lungs.

Shock

Shock is a condition that occurs when an animal does not have enough blood and thus oxygen reaching the tissues. This may occur with any type of trauma or heart condition. Shock is a medical emergency and care should be provided immediately. Animals in shock are weak, are depressed, and have a rapid heart rate but weak pulse. Respiration rates are often increased, and body temperature drops below normal. Mucous membrane (gum) color often is white to gray in color, and the capillary refill time (CRT) is decreased. The CRT can be established by pressing on the gums with a finger, causing the gums to turn white. When the finger is released, the gum color should return within 1–2 seconds (this is the normal condition). Normal gum color is a shade of healthy pink. Treatment for shock involves addressing the underlying cause. IV fluids are often given to replace any fluid loss, raise blood volume, and improve the flow of blood and thus oxygen to the tissues. In addition, medications may be administered.

Hypovolemic shock occurs when there is not enough blood volume due to blood loss internally or externally. This may be from hemorrhaging, vomiting, or diarrhea. Septic shock occurs when a severe infection invades the body, such as a bacterial infection, causing too many blood vessels to dilate (widen) and open, resulting in decreased blood output, as an insufficient amount of blood is not available to fill every vessel in the body.

SUMMARY

The cardiovascular system consists of the heart, blood, and blood vessels. The function of the system is to transport oxygen and nutrients to all body structures through the blood. Arteries carry blood away from the heart, and veins carry blood back to the heart. Diseases and disorders of the cardiac system include alterations in the heart's rate and rhythm. Severe blood loss resulting from trauma is a medical emergency.

Key Terms

albumin   part of the blood that draws water into the bloodstream and provides hydration

anemia   low red blood cell count where the red blood cells are not replenished

aorta   large vessel that allows blood to flow out of the heart and back into systemic circulation

arrhythmia   change in the heart's rate or rhythm

arterioles   smaller vessels that branch off of an artery

arteries   vessels that carry blood away from the heart

asystole   heart stops contracting and heart failure occurs

atrial fibrillation   A-fib; condition that occurs when the SA node or the pacemaker of the heart is not working

auscultation   the procedure of listening to a body part

autoimmune disease   animal's red blood cells are destroyed by the immune system

AV valve   opens and closes between the atrium and ventricle to allow blood to flow through each chamber of the heart

band cell   immature neutrophil that indicates an infection within the body; shaped like a *U*

basophil   white blood cell with a segmented nucleus and granules that stain very dark; aids in allergic reaction control

blood   red liquid within the circulatory system that contains 40 percent of cells in the body and transports oxygen throughout the body

blood pressure (BP) the heart's contractions and relaxations as blood flows through the chambers

bradycardia heart beating slower than normal

capillaries smallest blood vessels in the body that branch off of arterioles

cardiac arrest occurs when the heart is not contracting appropriately

cardiac cycle one complete contraction and relaxation of the heart

cardiopulmonary resuscitation process known as CPR that stimulates heart to provide oxygen to the lungs

centrifuge piece of equipment that uses high speeds to separate liquid portions from solids

circulatory system body system essential for life that includes the heart, blood, veins, arteries, and capillaries with functions that include oxygen flow, blood circulation, transport of nutrients, waste removal, and movement of hormones

clot formation of a plug in the blood to stop bleeding

diastole relaxation phase of blood pressure or the bottom number, which is normally lower than the top number

electrocardiogram (ECG) a machine that records the electrical currents of the heart

electrocardiography the evaluation of the electrical currents of the heart through the use of machines

endocardium thin inner muscle layer of the heart

eosinophil white blood cell that fights against allergic reactions, controls inflammation, and protects the body from parasite infection; has a large nucleus with segmented granules within the cytoplasm

epicardium outer thin covering of the myocardium

erythrocytes red blood cells

erythropoiesis the production process of red blood cells within the bone marrow

fibrinogen protein that aids in blood clotting

globulins provide antibodies to help prevent disease

heart organ with four chambers located in the chest between the lungs

heart murmur occurs when an abnormal valve produces an abnormal flow of blood that creates a "swishing" noise upon auscultation

heart rate pumping action of the blood through the heart creating a beating action

hematology the study of blood

hemoglobin main component of red blood cells that allows oxygen transport; includes an iron component that allows cells to multiply

histamine chemical released during an allergic reaction

left atrium thin-walled chamber of the heart that is situated below the pulmonary artery on the left side of the heart

left ventricle thickest-walled chamber of the heart

leukocytes white blood cells

lymphocytes largest of the white blood cells; aid in immune functions that protect the body from disease; have one large single nucleus that makes up most of the cell

macrophage cell that eats and destroys organisms throughout the body

monocyte largest white blood cell; helps neutrophils rid the body of wastes and cell debris

myocardium thick muscle that forms the wall of the heart

neutrophils most common white blood cell; destroy microorganisms within tissues; has a nucleus with segments

pacemaker system of the heart that controls heart sounds and rhythm

pericardium thin membrane that covers, protects, and maintains the beating action of the heart

plasma formed of various proteins in the body; comprises 60 percent of the blood system; remaining portion of blood after red and white cells have been removed

platelet blood cell that serves to clot blood by constricting (closing) the vessels and allowing bleeding to stop

pulmonary arteries carry oxygen-poor blood and prevent blood flow back to the heart

pulse the heart rate of an animal, or each beat created from the heart

P wave first peak on an ECG that forms at the SA node (pacemaker) and reflects the flow of electricity and blood through the atrium and that a contraction has occurred

QRS complex series of peaks showing that the electrical current and blood have flowed through the AV node and have caused a contraction in the ventricles

red blood cells (RBC) most abundant cells within the body; are produced in the bone marrow and transport oxygen throughout the body

repolarize reset or restart the heart to get ready for the next electrical impulse to start

right atrium thin-walled chamber of the heart where blood enters

right ventricle thicker-walled chamber of the heart that pumps blood to the lungs

SA node the pacemaker of the heart

serum liquid portion of blood

shock condition that occurs when an animal does not have enough blood and thus oxygen reaching the tissues

sinus arrhythmia condition when the heart rate increases with inspiration and decreases with expiration

sinus rhythm normal heart rate and rhythm

stethoscope instrument used to listen to the heart, lungs, and other areas of the chest

systemic circulation blood system that delivers nutrients and oxygen to all areas of the body

systole contraction phase of blood pressure or the top number, which is normally higher

tachycardia heart beating faster than normal

thrombocyte platelet

T wave last peak and the most important part of the ECG; shows that an electrical impulse has traveled through the entire heart, has completed the contraction, and has repolarized the heart

vasoconstriction diameter of a vessel decreases as blood pressure increases

vasodilatation diameter of a vessel increases as blood pressure decreases

veins vessels that carry blood to the heart

vena cava large vessel that transports blood into the heart from systemic circulation

ventricular fibrillation V-fib; condition causing the ventricles to fire electrical currents rapidly; is the most serious cause of cardiac arrest

white blood cells (WBCs) five types of blood cells that protect the body from infection and aid in immune system protection

Review Questions

1. What is the outer layer of the heart called?
2. What is the blood vessel that transports blood to the right atrium called?
3. What is the device used to measure the electrical activity of the heart called?
4. What is the blood vessel that carries blood away from the heart toward the lungs called?
5. What is the name of the protein that aids in blood clotting?
6. Name the organs of the circulatory system.
7. Name the three major types of blood cells.
8. Describe the path of blood through the heart.
9. Explain the difference between tachycardia and bradycardia.
10. Differentiate between an artery and a vein.
11. What is the main function of leukocytes?
12. What directions in relationship to the heart do veins carry blood? Arteries?
13. What purpose do heart valves serve?
14. Which of the four heart chambers has the thickest wall?
15. What sound does a heart murmur make?
16. Discuss two different types of shock.
17. What number is the top number or numerator in a blood pressure reading?
18. Which are smaller, capillaries or arterioles?
19. What is a cardiac cycle?
20. What is the membrane that maintains the beating action of the heart?

Clinical Situation

Dr. Jansen and Dale, a veterinary assistant from Ashton Oak Equine Center, are on a farm call to examine a Thoroughbred named Roscoe. The horse has been lethargic, has been weak after exercise, and has had a decreased appetite for several days.

The vet completes a PE and determines the horse has pale mucous membranes, has a decreased HR and RR, is 5–10 percent dehydrated, and appears weak and to have lost some weight since the last exam. The vet is suspicious of parasites or an autoimmune disease, since some of the signs appear to point toward anemia.

Dr. Jansen explains to the owner that anemia causes a low red blood cell count, causing less oxygen to be transported to the tissues. Thus, the body is not producing enough red blood cells. Upon completing the exam, Dale assists the vet in obtaining some blood samples and a fecal sample for further diagnosis.

- What are some possible causes for the anemia?
- What are some factors that could prevent the horse from having anemia?
- What treatment options might Dr. Jansen discuss with the owner?

CHAPTER 27
The Respiratory System

Objectives

Upon completion of this chapter, the reader should be able to:

- **27.1** State the structures that make up the respiratory system
- **27.2** Describe the function of the respiratory system
- **27.3** Describe common disorders associated with the respiratory system

Introduction

The **respiratory system** works closely with the circulatory system to supply the body with oxygen and rid the body of carbon dioxide (see Figure 27–1). This provides an exchange of gases between the animal and the environment.

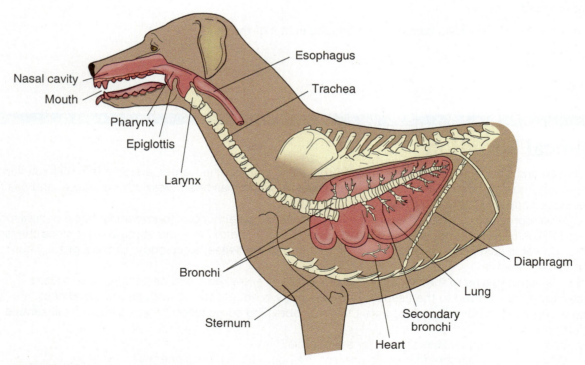

FIGURE 27–1 Structures of the respiratory system.

Structures of the Respiratory System

The respiratory system begins at the nostrils or **nares**. The nostrils are covered by a lining of mucous membranes and have numerous blood vessels. They also contain thin bonelike scrolls or plates of cartilage. The nostrils create an airway and help trap dust and foreign particles within the upper airway to prevent their entry into the lungs. The nostrils can raise the temperature of cold air before it enters the lungs. The mouth also serves as an airway that leads to the respiratory system. The **pharynx** or throat area is the common passageway for air and food. The upper area of the pharynx consists of the soft palate. Air passes through the pharynx, whether it is brought in through the nose or mouth. The pharynx functions during both swallowing and breathing; thus, the structures in the airway consist of special features to prevent food from being inhaled. The **larynx** lies within the throat and is firm cartilage that opens into the airway. This area can be palpated at the top of the neck. The larynx contains folds that allow vocalization in animals, such as barking or meowing. The **hyoid apparatus** found at the base of the tongue is a series of small bones that allow muscle attachment to aid in the process of swallowing. The **glottis** opens into the larynx, lies between the vocal folds, and has a covering called the **epiglottis** that hinges at the base of the larynx and prevents food from entering the airway and lungs during swallowing (see Figure 27–2). When breathing occurs, the epiglottis does not cover the opening and allows for air exchange. The **trachea** is the windpipe and is the airway that needs to be established during anesthesia. The trachea is made up of a series of cartilage rings of connective tissue. These rings are C-shaped and can be palpated in the neck. An **endotracheal tube** is placed in the trachea to help maintain control during anesthesia and to allow the animal to be sedated during surgery (see Figure 27–3). Endotracheal tubes or E-tubes are also used to establish an airway when an animal is not breathing or in cardiac arrest. It is important to inflate the cuff on the end of the tube to prevent anesthesia gas from escaping or any objects, such as vomit or saliva, from entering the lungs. It is also essential to deflate the cuff prior to removing it from the trachea, as failure to do so can cause serious damage to and lacerations of the trachea.

TERMINOLOGY TIP

Anesthesia is the process of placing an animal under a loss of consciousness state to perform surgery. Anesthesia prevents pain sensation.

The trachea enters the chest near the region of the heart. The base of the trachea forms two branches, one leading to each lung. The branches are called **bronchi**. The bronchi are large passages connected to the lungs through tiny tubes called **bronchioles**. The bronchioles are made up of smooth muscle that can allow airways to open further or close more tightly. This helps in keeping irritants out of the lungs. And when breathing is difficult, the bronchioles can cause the airway to open more fully (bronchodilation). They resemble tree branches that connect through tiny sacs filled with air called **alveoli**. The alveoli exchange oxygen and carbon dioxide from the blood. These branches connect to the lungs (see Figure 27–4).

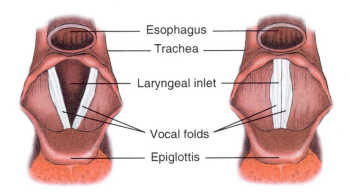

FIGURE 27–2 Epiglottis and larynx.

FIGURE 27–3 Endotracheal tubes and a laryngoscope.

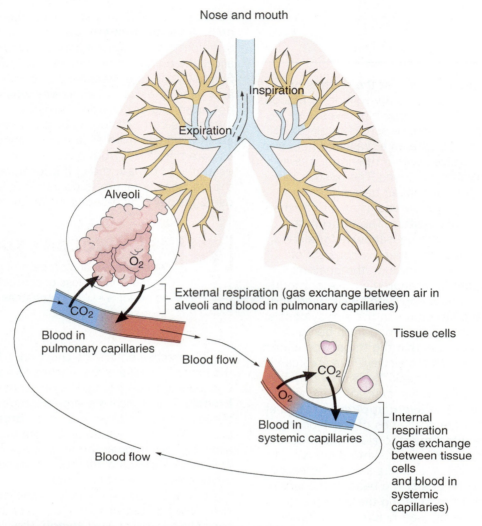

FIGURE 27-4 External and internal respiration.

The two **lungs** surround the heart and inflate with air during breathing. The lungs are the main organs of respiration (see Figure 27-5). They are elasticlike sacs that act like balloons filled with air, and the lungs act like a sponge and absorb water vapor. The lining of the lungs is called the **pleura**, a double-membrane covering of the **pleural cavity**, which holds the lungs. The pleura contains a fluid that allows the lungs to function and exchange oxygen. This allows the process of respiration to occur.

Respiration

Respiration is the act of inhalation and exhalation that creates a breath. **Inhalation** occurs when the nostrils

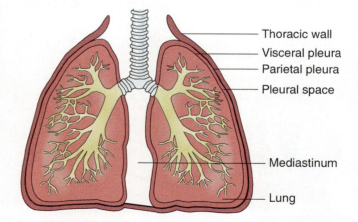

FIGURE 27-5 Respiratory structures of the thoracic cavity.

and trachea open, and air is taken into the lungs. As the rib muscles expand or contract and the diaphragm contracts, the **thoracic cavity** or chest cavity enlarges and air rushes in (see Figure 27–6).

 TERMINOLOGY TIP

The **diaphragm** is the area within the chest cavity that creates a thin, domelike sheet of muscle that separates the thoracic cavity from the abdominal cavity. It allows expansion and relaxation during the breathing process.

Exhalation is the act of air flowing out of the lungs. Relaxation of the trachea and nostrils occurs, causing the rib muscles to relax and the size of the thoracic cavity to decrease as air flows out. Respiration or breathing is an involuntary action controlled by the respiratory center of the brain. The respiratory rate (RR) is controlled and affected by several factors, including excitement, body temperature, exercise and activity, fever or pain, and oxygen levels in the blood. Table 27–1 lists normal respiratory rates of some species. Respiration rates change in response to demands on the body. When an animal is more active, receptors in the brain stimulate

TABLE 27–1

Normal Respiratory Rates of Animals (breaths per minute)

Dog	16–20 br/min
Cat	22–26 br/min
Rabbit	30–60 br/min
Guinea pig	42–105 br/min
Horse	12–15 br/min
Cow	30–35 br/min
Pig	12–15 br/min
Chicken	36–40 br/min
Sheep/Goat	20–22 br/min

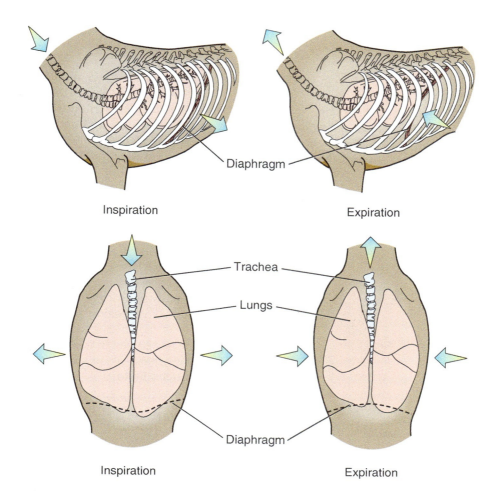

FIGURE 27–6 The mechanics of breathing.

a faster respiration rate. This allows the bronchioles to dilate (open up) and improves the delivery of oxygen to the alveoli. Oxygen levels must fall to very low levels before they stimulate a respiratory response.

Diseases and Conditions

Respiratory disorders are common and are often the result of bacterial or viral infections. Some respiratory conditions may be the result of trauma. Some of the most common disorders seen in veterinary practice are discussed here.

Upper Respiratory Infection

A common disorder of the respiratory system is an **upper respiratory infection** or URI. Signs of a URI include sneezing, coughing, nasal discharge, ocular discharge, wheezing, and difficulty breathing (see Figure 27–7). **Sneezing** is a common action that removes dust and foreign particles from the respiratory tract through the use of physical force. Clinical signs of an URI often begin in the nose. The nose filters incoming air, and organisms become trapped in the mucus and lining of the nasal cavity.

Pneumonia

Animals can also develop a condition of the lungs called **pneumonia** (see Figure 27–8). This is a disease

FIGURE 27–7 Young kitten suffering from an upper respiratory infection. Note the sore eyes and thick nasal discharge.

FIGURE 27–8 Heifer suffering from pneumonia as exhibited by thick nasal discharge and labored breathing.

that causes the lungs to become inflamed. White blood cells move into the lungs to attack invading organisms, and these cells and tissue fluid accumulate in the alveoli, causing the animal to breathe harder and faster. This can cause damage within the lungs and chest cavity. There are two types of pneumonia: bacterial and viral. The causes can be related to a virus, bacteria, fungus, parasitic, trauma related or aspiration from vomitus, food, medication or other type of liquid being directed into the lungs and chest cavity. Advanced stages of pneumonia can lead to the development of a **pleural friction rub**, an abnormal lung sound. It may be a crackling noise called **crackles** or **rales**, which sound like cellophane paper being folded. With each breath the pleural folds rub together, causing pain and inflammation.

Asthma

Asthma is a condition in which the lungs have difficulty taking in enough oxygen. Asthma can lead to **cyanosis**, a condition where a lack of oxygen causes the animal's mucous membranes to turn blue. During an asthma attack the sound of **wheezing** can be heard. A wheeze sound has a long, high-pitched musical note quality. Medications may be used to open up the bronchioles and lungs during these situations. These medications are known as **bronchodilators**.

Kennel Cough (Bordetella)

Bordetella or tracheobronchitis, commonly called kennel cough, can live in the respiratory tract of animals. This bacterium causes a severe and chronic cough, mostly in dogs. Some dogs also develop nasal discharge. Contact and transmission often occur with coughing and the transfer of nasal discharge. Vaccination is typically done using a nasal spray in the nostrils. However, several vaccines are available and may be administered in various ways. It is important to vaccinate dogs that are frequently kenneled, are show dogs, or are taken to locations where many dogs gather, such as dog parks.

Shipping Fever

Shipping fever is a condition that occurs in livestock, such as horses and cattle, and that can be highly contagious and spread rapidly during the stress of transport. Shipping fever causes a respiratory infection that affects the lungs and chest cavity, causing a severe problem that can be fatal. Shipping and transporting livestock can be stressful and can cause overheated conditions in which bacteria commonly grow. In addition, poor sanitation within a trailer can cause the disease to spread rapidly through livestock. Antibiotics and proper sanitation are key to treating the disease. Signs of the disease in livestock include fever, nasal discharge, and coughing. A vaccine is available for cattle and can be given three to four weeks prior to transport. Proper trailer ventilation and sanitation are the best methods to prevent the disease.

Diaphragmatic Hernia

A **diaphragmatic hernia** is a condition of the respiratory system caused by severe trauma, such as being hit by a car (HBC). This affects the diaphragm by causing a tear in the chest muscle that internal organs may protrude through (see Figure 27–9). The lungs cannot properly expand, and breathing is labored, which results in difficulty breathing. Surgical correction is necessary to repair the tear and return any misplaced organs to their original position. It can be an emergency if the tear in the muscle is extensive and affects numerous chest and abdominal organs, such as the stomach, intestines, or spleen. During surgical repair, the animal can't breathe on its own. Once an incision is made into the chest, the technician must do the breathing for the animal, providing pressure on the air entering the lungs. This is done with air from the anesthesia machine being manually forced into the lungs.

Heaves

Horses are affected by several respiratory conditions. One condition is called **heaves** and is caused by an allergy usually related to dust. Heaves is a common term for COPD, or **chronic obstructive pulmonary disease**.

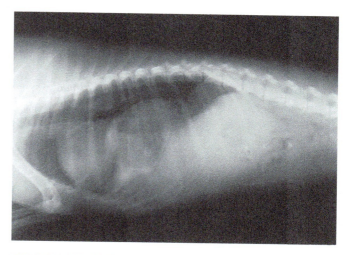

FIGURE 27–9 Radiograph of a diaphragmatic hernia. Intestinal organs are visible in the chest cavity.

This is a lifelong condition that may be controlled by antihistamines and steroids. Horses that are affected may need to be housed outside or in an area with good ventilation, as dust and moisture can cause serious breathing difficulty. They may also need to have their hay source watered down to reduce the amount of dust spores in the air (see Figure 27–10). Horses with COPD tire easily, have labored and difficult breathing and exhibit nasal discharge, coughing, and wheezing.

FIGURE 27–10 Horses with heaves should be fed in a well ventilated area and hay may need to be watered down to reduce dust spores in the air.

MAKING THE CONNECTION

For additional information on the respiratory system, please see the following chapters:

Chapter 9—rhinotracheitis, feline calicivirus, feline infectious peritonitis

Chapter 12—respiratory disease in rabbits, pasteurella

Chapter 17—infectious bovine rhinotracheitis

Chapter 22—Newcastle disease, infectious bronchitis ■

Roaring

Another common condition in horses is called **roaring**. Roaring is a condition that causes a sound (like a roar) due to the larynx only opening a small amount due to a trauma to the nerves of the throat area. As a horse exerts itself, the air forced though a half-open larynx creates an increased muffled roaring noise. It may be controlled with activity limitations and medications. Severe trauma and breathing problems may be corrected surgically to hold the vocal folds open and allow air to pass through without obstruction.

Pneumothorax

The pleural lining of the lungs is not attached to the lining of the chest cavity but has a very small area that may become larger. In a healthy animal, this small space allows the lungs to stay inflated within the thorax. In an animal with lung issues, the lungs may become partially or fully deflated, and the space may fill with air or fluid that may leak into this pleural space. This is a condition called **pneumothorax** (see Figure 27–11). This condition is often associated with severe trauma or punctures, such as an HBC or gunshot wound. Trauma can also cause bleeding within the pleural space, and air is often present on both sides of the chest cavity. This accumulation of blood within the chest cavity is called **hemothorax**. This is often associated with minor trauma or animals ingesting rodenticides, causing a clotting disorder. Infection may also enter the pleural space, resulting in a **pyothorax**. This is an accumulation of white blood cells (WBCs), pus, and cellular debris in the pleural space. This is often seen secondary to pneumonia, when bacteria escapes from the lung tissue and invades the pleural space. All of these conditions involving the pleural space result in decreased space available for the lungs, affecting breathing and the oxygenation of blood and tissues. Often these animals are uncomfortable, have an increased rate of respiration, and are depressed and reluctant to move. As the condition progresses, the animal may become unable to breathe and begin gasping or open-mouthed breathing in an attempt to expand the lungs. Untreated, these cases are fatal. Initial treatment involves removing the fluid or gas from the chest cavity by inserting a chest tube or aspirating with a needle. This provides immediate relief and allows the animal to expand its lungs more fully. The underlying condition then must be treated, and repeat aspirations are often necessary.

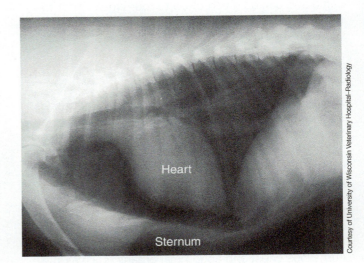

FIGURE 27–11 Pneumothorax in a dog.

SUMMARY

The respiratory system regulates breathing and the exchange of oxygen and carbon dioxide in the body. This oxygen exchange is critical to maintaining life processes. Most of the diseases of the respiratory system are related to viral or bacterial invasion. Trauma can also be a source of respiratory injury.

Key Terms

alveoli tiny sacs filled with air at the end of the bronchioles

asthma condition of lungs in which there is difficulty taking in enough oxygen

bordetella tracheobronchitis; commonly called kennel cough

bronchi branches at the base of each lung connected by large passages

bronchioles tiny tubes that connect to the lungs and bronchi; resemble tree branches

bronchodilator medication used to open up the bronchioles and lungs

chronic obstructive pulmonary disease (COPD) respiratory condition similar to heaves and asthma that causes difficulty breathing

crackle noise that sounds like cellophane

cyanosis blue color of mouth and gums due to a lack of oxygen

diaphragm area of the chest that separates the thoracic cavity from the abdominal cavity

diaphragmatic hernia condition that affects the diaphragm by causing a tear in the chest muscle; internal organs may protrude through

endotracheal tube E-tube; tube placed into the windpipe to create and establish an airway during anesthesia or CPR

epiglottis covering or flap that prevents food from entering the airway and lungs during swallowing

exhalation act of exhaling or air leaving the lungs

glottis opening into the larynx

heaves respiratory condition in horses due to an allergic reaction, usually to dust

hemothorax blood within the chest cavity

hyoid apparatus series of small bones at the base of the tongue that attach to muscles and aide in swallowing

inhalation act of inhaling or taking air into the lungs

larynx cartilage that opens into the airway and lies within the throat area

lungs organs that surround the heart and inflate with air for breathing

nares the nostrils

pharynx throat area for the passage of food or air

pleura double-membrane covering that lines the lungs

pleural cavity double-membrane covering that holds the lungs

pleural friction rub abnormal lung sound that develops with pneumonia and causes a crackling noise

pneumonia condition of the lungs that causes inflammation; may have viral or bacterial source

pneumothorax air within the chest cavity

pyothorax pus and cellular debris within the chest cavity

rale noise or crackle caused by the lungs during pneumonia

respiration the process of breathing through inhalation and exhalation

respiratory system system of the body that controls oxygen circulation through the lungs and the rest of the body

roaring respiratory condition in horses that sounds like a roar due to the larynx only opening a small amount due to a trauma to the nerves of the throat area

shipping fever a condition that occurs in livestock, such as horses and cattle, that are transported

sneezing act of removing dust or other particles from the respiratory tract through physical force

thoracic cavity the chest

trachea the windpipe or tubelike airway within the throat

upper respiratory infection (URI) condition causing sneezing, coughing, nasal discharge, ocular discharge, wheezing, and difficulty breathing

wheezing long, high-pitched sound produced during asthma, which causes difficulty breathing

Review Questions

1. What system works closely with the respiratory system in the transport of oxygen to organs?
2. Which of the following contains organs of the respiratory system?
    a. periosteum, medullary cavity, cartilage
    b. fundus, mesentery, peritoneum
    c. pericardium, plasma, erythrocytes
    d. pharynx, pleura, trachea
3. What abnormal lung sound develops from pneumonia?
4. What type of medication is used to open the bronchioles and lungs?
5. What respiratory condition may be related to a dust allergy?
6. Describe the process of respiration.
7. What is the shape of the tracheal rings?
8. What is the reflex action that occurs when irritation occurs in the nasal cavity?
9. What species can develop a respiratory condition known as roaring?
10. What is the covering at the base of the larynx that prevents food from entering the lungs called?
11. What are three factors that affect respiratory rate?
12. What is a hemothorax?

Clinical Situation

Dr. Joseph and José, a vet assistant at Redmont Animal Hospital, are on a farm call at a local dairy farm. Mr. Calvins is a local farmer who has a large herd of Holstein dairy cattle and has a cow that he has been treating for about a week. The cow has appeared sick to him, and he has been giving injectable penicillin that he purchased at a local farm supply store.

José takes a history of the cow while Dr. Joseph gets his equipment. Mr. Calvins says the cow has been sick about a week with signs of labored breathing, poor appetite, and decreased milk production. José also notes in the history the penicillin information on the bottle.

Dr. Joseph completes a PE and upon auscultation of the chest, hears a crackling sound in the lungs. The cow also has an increased HR and RR, as well as a slight fever.

- What does Dr. Joseph likely determine is the cow's medical problem?
- Was self-treatment by the farmer a good idea? Why or why not? Explain.
- How might have this situation been prevented?

CHAPTER 28: The Endocrine System

Objectives

Upon completion of this chapter, the reader should be able to:

28.1 State the structures that make up the endocrine system
28.2 Describe the functions of the endocrine system
28.3 Explain the glands of the endocrine system and their functions
28.4 Explain the hormones secreted by the endocrine system
28.5 Describe common disorders of the endocrine system

Introduction

The body needs to coordinate and integrate all of its functions into one harmonious whole. This is called **homeostasis**. The maintenance of homeostasis involves growth, maturation, reproduction, and metabolism. The body system responsible for this coordination is the **endocrine system**. This system provides chemicals that control hormone functions from numerous glands throughout the body.

Structures of the Endocrine System

The endocrine system is composed of numerous glands throughout the body that secrete hormones into the bloodstream and **excrete** chemicals to rid the body of wastes (see Figure 28–1). These chemicals and hormones are passed throughout the body through blood and tissues. The **chemicals** and **hormones** provide changes in the body that regulate growth, regulate sexual production and development, metabolize nutrients in cells, and maintain homeostasis in the body, allowing for a balance in the body systems. Hormones may affect the entire body or may target a specific organ or location. Once delivered, the hormones influence the specific area and activities of cells. Hormones are divided into four chemical classes: fatty acids, steroids, amino acids, and peptides.

Fatty acids include the *prostaglandins*, which are involved in the estrous cycle in females. Steroids are small molecules derived from cholesterol and are an essential compound in animal physiology. Amino acids are the simplest of chemicals that include the compounds carbon, nitrogen, hydrogen, and oxygen. Peptides are the largest of the hormones and may be a short chain of molecules or a large protein.

Hormone receptors are present on the cell membrane or within the cytoplasm or nucleus. An analogy of how a key fits into a lock may be used to describe the receptor and the hormone. The hormone molecule has a distinct shape, as does the receptor to accept this specific shape. Once a hormone is delivered, it signals cells how it is to be used and the body responds. In the endocrine system, hormones are secreted in response to a change within the internal environment and when the response is completed, the secretion stops.

Clinical Concepts and Critical Thinking

The release of insulin occurs within the body when blood sugar levels increase. As the blood sugar is used by the body and begins to decrease, insulin secretion stops. In the endocrine system, the effect of insulin is rather slow. To prevent blood sugar levels from falling too low, another hormone, *glucagon*, is released to counteract the

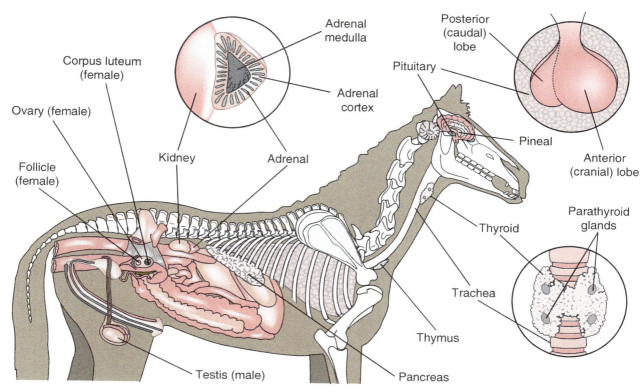

FIGURE 28–1 Locations of the endocrine glands. The relative locations of the endocrine glands are shown in this horse.

effects of insulin and elevate the blood sugar. Thus, the sugar level is—ideally—kept within a specific range. For example, following a meal rich in carbohydrates, blood sugar levels increase rapidly. Insulin is then released and transported through the bloodstream. The liver is stimulated to convert glucose into glucagon and fat. This allows the sugar to be metabolized and the effect is a decrease in blood sugar levels. This cycle begins again following the next meal.

Endocrine Glands

The endocrine glands secrete hormones directly into the bloodstream. They are then transported to all areas of the body. The endocrine glands are the following:

- Pituitary gland
- Thyroid gland
- **Parathyroid glands**
- Adrenal glands
- Thymus
- Pancreas
- Pineal
- Gonads (ovaries in females and testes in males)

Pituitary gland

The main endocrine gland is the **pituitary gland** and is often referred to as the master gland. The pituitary gland works with the **hypothalamus** to control the endocrine system and create a link with the nervous system. The pituitary gland lies at the base of the brain and is separated into an anterior and a posterior lobe. The **posterior lobe** connects to the hypothalamus. The hypothalamus develops from brain tissue while an animal is in the embryo stage. The hypothalamus serves as a reservoir for hormones made in the posterior pituitary and allows for the release and regulation of hormones (see Figure 28–2). The posterior or back lobe releases two peptide hormones: oxytocin and antidiuretic hormone (ADH). **Oxytocin** plays a vital role in **parturition**, or the labor process. The hormone is released and causes the muscles of the uterine wall to contract and milk production to begin in the mammary glands. An injectable form can be given to increase these functions. **Antidiuretic hormone** or ADH is a hormone that promotes urine formation and water absorption, controls blood pressure, and conserves water volume in the body. The release of ADH is controlled by receptors in the hypothalamus

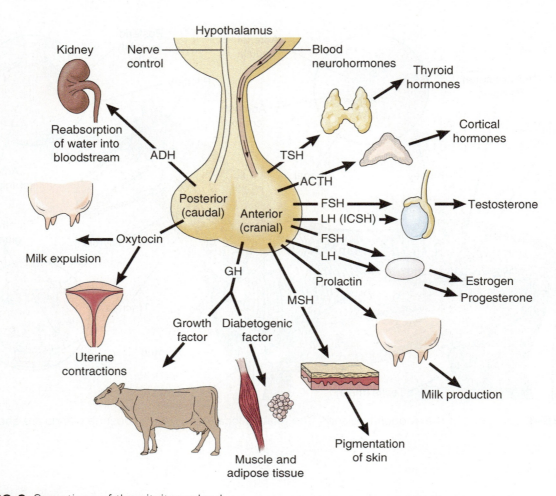

FIGURE 28–2 Secretions of the pituitary gland.

when changes in osmotic pressure within the bloodstream are detected. A condition known as **diabetes insipidus** affects the water content in the bloodstream, causing the urine to become very dilute. This decreases ADH levels, causing the animal to have polydipsia and polyuria.

TERMINOLOGY TIP

Polyuria means excessive and frequent urination; **polydipsia** means uncontrolled thirst. The terms are abbreviated PU and PD and often occur simultaneously. ■

The anterior lobe of the pituitary gland is controlled by the hypothalamus. The anterior lobe's main function is to release hormones into the bloodstream and produce and regulate growth hormones within the body. The major growth hormone **somatotropin** increases protein synthesis within the body, thus increasing the size of the animal. The anterior lobe also produces the hormone **prolactin**, which controls milk production. The pituitary gland controls the **thyroid gland**, which produces another hormone made by the anterior pituitary gland known as **thyroid-stimulating hormone or TSH** (see Figure 28–3). Therefore, the anterior pituitary stimulates or controls the thyroid gland. The hypothalamus acts as a messenger that links the brain and anterior pituitary to send messages to release hormones when necessary.

Thyroid Gland

The thyroid gland, located in the neck with a lobe located on each side of the trachea, is the only endocrine gland that can be palpated. In many animals, though, this is possible only if the gland becomes enlarged. The thyroid gland produces two hormones, **thyroxine** and **calcitonin**. Thyroxine acts similar to growth hormones and helps regulate metabolism in an animal's cells and also helps regulate body temperature. Calcitonin regulates the body's calcium level and is secreted

FIGURE 28–3 Thyroid and parathyroid glands from a cat. The white areas are very enlarged parathyroid glands. Normal parathyroid glands would not be visible.

when blood calcium levels are elevated. The thyroid gland is stimulated by TSH or thyroid-stimulating hormone made in the anterior pituitary gland. This signals the thyroid gland to produce thyroxine. There are two forms of thyroxine produced, known as **T3** (triiodothyronine) and **T4** (tetraiodothyronine). T3 is a more potent and active form of the hormone. T4 breaks down fats and helps control cholesterol. Assessing both T3 and T4 levels is important in diagnosing diseases of the thyroid gland; blood levels are often evaluated.

Parathyroid glands

The parathyroid glands are several small nodules located on the surface of the thyroid gland. They secrete **parathormone**, which helps regulate blood calcium and is secreted when blood calcium levels become decreased.

Adrenal Glands

The **adrenal glands** are also a part of the endocrine system and lie cranial to the kidneys. The adrenal glands produce and release adrenaline and other hormones. **Adrenaline** is a chemical that is released by the nervous system in times of stress to instigate response of an animal's fight-or-flight reaction. There is a release at this time of a hormone called **epinephrine**, which is short acting. A longer-acting hormone is **norepinephrine**, which increases heart rate, blood pressure, blood flow, blood glucose, and metabolism. The adrenal glands also regulate the function of **ACTH** or adrenocorticotropic hormone. ACTH controls blood pressure, releases cholesterol, and controls the body's production of steroids. Stress is a factor in ACTH production and increases the amount of ACTH in the body, which in turn decreases its production.

Pancreas

The **pancreas** is another organ within the endocrine system. It serves a dual role in endocrine and exocrine production and functions. The pancreas produces and regulates **insulin** and **glucagon** production. Insulin allows blood sugar from the bloodstream to enter the body's cells for energy. Insulin is regulated by the **blood glucose** level, also known as blood sugar. The release of insulin following a meal is created by a response from the pancreas. This response causes sugar metabolism, which is a conversion of insulin to glucose to glycogen to fat. Sugar metabolizes as the blood level decreases as food digests, and this is repeated after each meal. Without insulin, the blood sugar levels stay elevated and the body's cells starve. The pancreas also produces the hormone glucagon, a peptide hormone produced by alpha cells that raises the concentration of glucose in the bloodstream, thus increasing the blood sugar level. Its effect is opposite that of insulin, which lowers the glucose concentration. Figure 28–4 shows how insulin works.

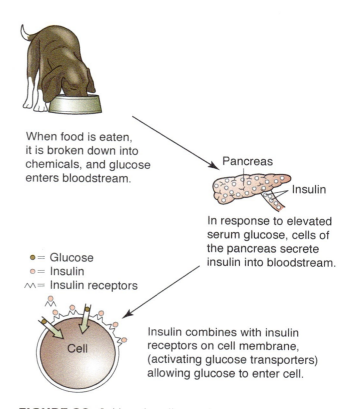

FIGURE 28–4 How insulin works.

Thymus

The **thymus** is a gland in young animals. It has immunologic functions through maturation of T lymphocytes.

Pineal Gland

The pineal gland is located in a central portion of the brain. The functions of the pineal gland are not fully understood. It plays a role in the regulation of body rhythms.

Gonads

The gonads are the glands associated with reproduction and are cell-producing organs. The male gonads are the testes, and the female gonads are the ovaries. The main hormone produced by the testes is **testosterone**, which is produced at constant levels. A small amount of the female sex hormone **estrogen** is also produced by males. In females, the two main hormones produced by the ovaries are estrogen and **progesterone**. Female animals' cycle is based on species, and levels of hormones produced are based on this cycle. As follicles form within the ovaries, estrogen is produced and is responsible for signs of estrus, also called heat. Once a follicle ruptures and an egg is released, progesterone is released. Progesterone is responsible for maintaining pregnancy. If the animal is pregnant, the **corpus luteum (CL)** is maintained. If an animal is not pregnant, the corpus luteum lasts only a short time and then becomes smaller in size and the estrus cycle begins at the next time frame.

Hormones

Hormones are made up of four major chemical groups that regulate parts of the body. The **fatty acids** control hormones involved in estrus in the female animal's heat cycle. **Steroids** occur naturally in the body and regulate chemicals, such as cholesterol, in the body that control essential life functions. **Amino acids** are the simplest compounds and the building blocks of proteins, which form hormones that are stored within the endocrine system until needed by the body. **Peptides** are the largest hormones and control the proteins in the body. Hormones are delivered to target cells located throughout the body and fit into each cell like a key fits into a lock. The body will send a signal for the necessary hormone function, which responds by creating an **enzyme**. Enzymes are chemicals that create a reaction that changes within the body. The enzymes create and release the hormones that are needed until the function is complete. An example of hormone and enzyme function is insulin production as the blood sugar level rises. The blood sugar or glucose level begins to drop, causing insulin secretion to slowly stop. The hormone glucagon is released to increase the blood glucose level. Diabetes results when this function is not properly controlled by the hormones in the body.

Hormones and Estrus

The endocrine system is active in producing and controlling hormones involved in male and female reproduction. Estrus hormones are produced to control the heat cycle of female animals. One estrus hormone is called **luteinizing hormone (LH)**, which allows the production of testosterone, allows for ovulation to occur, and begins the formation of the corpus luteum (CL) during the reproductive cycle. **Ovulation** is the release of the egg in the female to allow for reproduction to occur when sperm is present. The space vacated by the released egg, the corpus luteum (CL), forms in the female during reproduction and is present through the entire gestation period. This is what maintains the pregnancy.

TERMINOLOGY TIP

The term *gestation* means the length of pregnancy of an animal and development from embryo to fetus. Each species of animal has a different gestation period. ■

Another estrus hormone, **follicle-stimulating hormone (FSH)**, allows for sperm production, regulates the female estrus cycle, forms the follicle during the breeding process, and produces estrogen. Estrogen is the hormone that begins the estrus cycle. Both LH and FSH are controlled by the hypothalamus. Another hormone, **gonadotropin-releasing hormone (GnRH)**, is produced to regulate and maintain a normal estrus cycle. LH, FSH, and gonadotropins are produced in the anterior pituitary gland.

TERMINOLOGY TIP

Gonadotropin is often called chorionic gonadotropin or gonadotropin-releasing hormone, abbreviated GnRH. ■

Diseases and Conditions

Many diseases and conditions are caused by the glands that secrete chemicals in the body and are part of the endocrine system. These conditions often cause notable signs and symptoms in animals, such as weight problems, skin problems, and changes in thirst.

Hypothyroidism

Hypothyroidism is a condition that decreases thyroxine hormone production by the thyroid gland. This condition causes the body to slow down, causing weight gain, lethargy, hair and skin problems, weakness, and intolerance to cold. Skin conditions that are common in this disease include **alopecia** (hair loss) and **dermatitis** (inflammation of the skin). Treatment is a lifetime therapy of thyroxine supplementation. This condition is common in dogs and is inherited in many breeds (see Figure 28–5).

Dog Breeds Prone to Hypothyroidism

- Boxer
- Great Dane
- Golden retriever
- Dachshund
- Doberman
- Schnauzer

Hyperthyroidism

Hyperthyroidism causes increased thyroxine production in the thyroid gland. This condition most commonly occurs in elderly cats. Signs of this disease include weight loss, a ravenous appetite, polyuria (PU), polydipsia (PD), increased activity, increased heart rate, and an enlarged, palpable thyroid gland. As the disease progresses, vomiting and diarrhea may occur. Treatment options include a **thyroidectomy**, or the surgical removal of the thyroid gland, or thyroxine therapy and supplementation. Cats that have had a thyroidectomy usually require medication to prevent the condition from becoming hypothyroidism. The owner should be cautioned before the surgery: Any damage or accidental removal of the parathyroid gland will cause severe **hypocalcemia**, or a decrease in blood calcium. Another treatment option is a **radioactive iodine** treatment known as Radiocat. This involves iodine being injected into the bloodstream to treat any areas that are overactive in thyroxine production by damaging the thyroid tissue. Less active areas of tissue are left undamaged. Another medication therapy option is **Methimazole**, an oral medication that blocks the synthesis of thyroxine; its dosage must be adjusted over the lifetime of the cat.

Diabetes

Diabetes mellitus is a condition that causes a high blood glucose level (see Figure 28–6). This condition of high blood glucose is referred to as **hyperglycemia**. Animals with this condition tend to have increased appetite but lose weight, thirst, water consumption, and urination (PU/PD). Animals that develop diabetes have cells that deteriorate and break down the pancreas, which in turn causes the body's immune system to attack the cells. It is estimated that about 75 percent of the cells are destroyed by the time signs of the disease begin to appear. The treatment is supplementing the body's natural insulin via injection with an insulin needle. The insulin is measured in IUs or insulin units. Diabetes can be challenging to regulate and control. It requires a regular and consistent diet that may be a therapy diet necessary to control weight. The insulin dose is determined based on the blood glucose levels. The goal of diabetes control is to keep insulin levels consistent and as close to normal on a daily basis, and it requires regular attention and blood level evaluations. Insulin resistance can occur in some animals due to immune system responses, other underlying disease conditions, or poorly controlled insulin amounts. It may lead to nerve damage to the kidneys and thus can lead to kidney failure. Side effects of the disease include vision problems that can lead to **cataracts**, or lens opacity, or retinal problems. Insulin overdose or a defect in insulin production can occur, resulting in **hypoglycemia** or a low blood glucose level. Signs of low blood glucose include

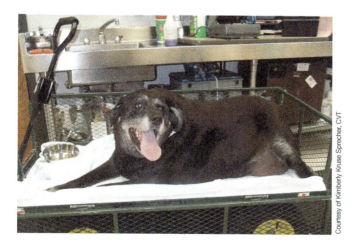

FIGURE 28–5 Dog with hypothyroidism.

Hyperadrenocorticism and Hyposadrenocorticism

Hyperadrenocorticism, often called **Cushing's disease**, involves a production problem within the adrenal gland (see Figure 28–7). This results in a pituitary gland tumor when ACTH is increased or an adrenal gland tumor when cortisol production is increased. This disease can be **iatrogenic**, meaning it may be caused by humans, such as in the situation of using steroids. Diagnosis is done by measuring cortisol levels. This is completed through a dexamethasone suppression test. A baseline blood sample is acquired and then the corticosteroid **dexamethasone** is administered by injection. A blood sample is then collected 4–8 hours postinjection. Normal results will show decreased ACTH and cortisol levels. If a pituitary tumor is present, there ACTH levels will be increased, with no change in cortisol levels. If an adrenal gland tumor is present, there will be no change in ACTH levels and an increase in cortisol levels. Signs of Cushing's disease include the following:

- PU/PD
- Increased appetite
- Thin skin and hair coat
- Panting
- Swollen abdomen
- Weakness
- Lethargy
- Hair loss

Treatment of Cushing's disease includes surgical removal of the tumor and medication to decrease cortisol levels in the bloodstream. This is a lifelong therapy that may lead to decreased production of corticosteroids by the adrenal glands if they fail to function. This condition is known as **hypoadrenocorticism** or **Addison's disease**. Signs of this condition include lethargy, weakness, weight loss, vomiting, diarrhea, poor appetite, and GI problems. **Hyponatremia** may

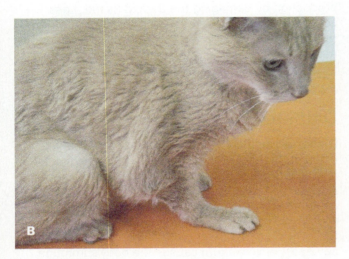

FIGURE 28–6 (A) Cat with diabetes mellitus. (B) Stance of a cat with diabetes.

weakness, lethargy, ataxia, a trancelike appearance, seizures, and possibly coma. Young newborn animals can be treated with sugar water or Karo syrup placed on the gums. Severe cases will need treatment with IV fluids and glucose supplements.

Diabetes insipidus is caused by the release of antidiuretic hormone (ADH) from the posterior lobe of the pituitary gland. This peptide hormone causes constriction of arterioles, which increases blood pressure and causes the distal tubules within the kidneys to become more water permeable. Thus, more water is retained, and the urine becomes more concentrated. The release of ADH is controlled by receptors in the hypothalamus and in turn detects the blood pressure changes. Most animals with this form of diabetes have uncontrolled thirst and increased urination.

FIGURE 28–7 Dog with hyperadrenocorticism.

occur, in which sodium (Na) concentration in the blood becomes too low and the body holds onto too much water. **Hyperkalemia** may also occur, which is an increase of potassium (K) in the blood. Diagnosis occurs with the use of an ACTH stimulation test. A blood sample is obtained to determine baseline cortisol levels, and then an injection of ACTH gel is administered. The cortisol blood level is sampled 1–2 hours postinjection. Normal results will show an increased cortisol level. The cortisol level will not change with potential Addison's disease. Treatment of Addison's disease includes long-lasting corticosteroid therapy and IV fluids.

MAKING THE CONNECTION

Chapter 17—hypocalcemia and ketosis

SUMMARY

The endocrine glands work with the nervous system to regulate homeostasis in the body. This is accomplished through the secretion of hormones, which act on specific tissues and organs to produce a desired effect.

Key Terms

ACTH adrenocorticotropic hormone released by the adrenal glands; controls blood pressure, releases cholesterol, and regulates the body's production of steroids

Addison's disease occurs due to a decreased production of steroids in the adrenal glands, causing the glands to fail to function; common term for hypoadrenocorticism

adrenal glands located cranial to the kidneys; produce and release adrenaline and other hormones

adrenaline chemical that is released by the nervous system in times of stress to instigate response of an animal's fight-or-flight reaction

alopecia hair loss

amino acid simplest hormone that controls thyroid gland functions

antidiuretic hormone (ADH) hormone that promotes urine formation, water absorption, blood pressure control, and conserves water volume in the body

blood glucose blood sugar or insulin produced by the pancreas

calcitonin hormone that regulates the body's calcium level and is secreted when blood calcium levels are elevated

cataracts opacity of the lens in the eye

chemical change in the body that affects growth, sexual reproduction, and development

corpus luteum (CL) known as the yellow body; forms during the gestation period of a female to maintain pregnancy

Cushing's disease condition where an increase in ACTH or cortisol causes a production problem within the adrenal glands; common term for hyperadrenocorticism

dermatitis inflammation of the skin

Dexamethasone corticosteroid administered by injection to determine pituitary problems

diabetes insipidus condition that affects the water content in the bloodstream, causing the urine to become dilute

diabetes mellitus condition that causes high blood glucose levels

endocrine system   excretory system that rids the body of waste materials

enzyme   chemical reactions that change within the body and create and release hormones

epinephrine   short-acting chemical released during the fight-or-flight response

estrogen   female reproductive hormone that begins the estrus cycle

estrus   heat cycle that releases hormones for reproduction

excrete   to remove and rid the body of waste

fatty acids   control hormones involved in estrus in the female animal's heat cycle

follicle-stimulating hormone (FSH)   estrus hormone that allows for sperm production, regulates the female estrus cycle, forms the follicle during the breeding process, and produces estrogen

gestation   length of pregnancy

glucagon   hormone produced in the pancreas that increases blood sugar levels

gonadotropin-releasing hormone (GnRH)   hormone produced and released to maintain a normal estrus cycle

homeostasis   balance within the body

hormones   chemicals in the body that regulate growth, regulate sexual production and development, metabolize nutrients in cells

hyperadrenocorticism   increased production problem within the adrenal gland that results in a pituitary gland tumor when the ACTH hormone is increased or an adrenal gland tumor when cortisol hormone production is increased; also called Cushing's disease

hyperglycemia   high blood glucose level

hyperkalemia   increased potassium in the blood

hyperthyroidism   condition that increases thyroxine hormone production from the thyroid gland; common in cats

hypoadrenocorticism   decreased production of steroids within the adrenal glands if the glands fail to function; also called Addison's disease

hypocalcemia   decreased blood calcium

hypoglycemia   low blood glucose level

hyponatremia   decreased sodium in the blood; sodium concentration in the blood becomes too low and the body holds onto too much water

hypothalamus   develops from brain tissue while an animal is in the embryo stage; the gland serves as a reservoir for hormones and allows for the release and regulation of hormones

hypothyroidism   condition that decreases thyroxine hormone production from the thyroid gland; common in dogs

iatrogenic   condition caused by humans

insulin   hormone produced in the pancreas that decreases blood sugar levels

luteinizing hormone (LH)   estrus hormone that allows the production of testosterone, allows for ovulation to occur, and forms the corpus luteum (CL) during the reproductive cycle

Methimazole   medicine for cats to treat hyperthyroidism; blocks the synthesis of thyroxine

norepinephrine   long-acting chemical hormone that increases heart rate, blood pressure, blood flow, blood glucose, and metabolism

ovulation   release of the egg during estrus in the female to allow reproduction to occur when united with sperm

oxytocin hormone that causes the muscles of the uterine wall to contract and milk production to begin in the mammary glands

pancreas organ with a dual role in endocrine and exocrine production and functions; produces insulin and regulates insulin production

parathormone hormone that regulates blood calcium and is secreted when blood calcium levels become decreased

parathyroid glands glands located below the thyroid gland

parturition the process of giving birth

peptides largest hormones; control proteins in the body

pituitary gland "master gland" that lies at the base of the brain and controls hormone release from the endocrine glands

polydipsia increased thirst; PD

polyuria increased urine production; PU

posterior lobe back lobe of the pituitary gland; controls peptide hormones

progesterone hormone in females that maintains pregnancy

prolactin hormone that controls milk production

radioactive iodine involved in treatment in which iodine is injected into the bloodstream to treat areas that are overactive in thyroxine production; damages thyroid tissue

somatotropin growth hormone that increases protein synthesis in the body, causing an increase in the animal's size

steroid occurs naturally in the body; regulates chemicals, such as cholesterol, in the body that control essential life functions

testosterone hormone in males produced in the testes

thymus a gland in young animals; has an immunologic function

thyroidectomy surgical removal of the thyroid gland

thyroid gland endocrine gland located in the neck area; produces thyroxine and calcitonin

thyroid-stimulating hormone (TSH) hormone produced in the anterior pituitary gland; controls the chemical thyroxine

thyroxine hormone that produces growth hormones and helps regulate metabolism in an animal's cells; also helps regulate body temperature.

T3 most potent and active thyroid gland hormone; measured in the bloodstream to diagnosis thyroid problems

T4 thyroid hormone known as thyroxine; breaks down fats and helps control cholesterol

Review Questions

1. What organ is known as the "master gland"?
2. What is the hormone responsible for stimulating ovulation called?
3. Hypoadrenocorticism, a disorder caused by deficient adrenal cortex production of glucocorticoid, is known by what other name?
4. What is an abnormally low blood glucose level called?

5. What is the difference between hypothyroidism and hyperthyroidism?
6. What hormone causes milk production?
7. Which gland is the only endocrine gland that can be palpated?
8. Which gender produces testosterone?
9. Which two hormones work together to regulate blood sugar?
10. What condition can cause poor vision and cataracts?

Clinical Situation

Amelia, a veterinary assistant at Small Creatures Vet Clinic, is answering the phones. She is talking to Mrs. Carl about her cat, Ruby, a 16 yr SF DSH. Ruby has been having some problems such as weight loss, excessive drinking, and increased vocalization. Mrs. Carl also notes that she sees Ruby in the litterbox more frequently than normal.

Amelia schedules an appointment with Dr. Hall and asks Mrs. Carl to bring Ruby right in. Mrs. Carl asks if she should be concerned, and Amelia replies, "With Ruby's age and clinical signs, this may be an emergency."

- What may be wrong with Ruby?
- What other signs may be present with Ruby on a PE?
- What will Dr. Hall most likely suggest to the owner?

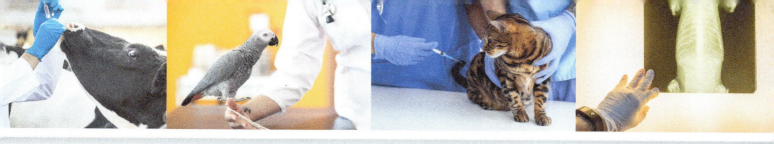

CHAPTER 29: The Renal System

Objectives

Upon completion of this chapter, the reader should be able to:

29.1 State the structures that make up the renal system
29.2 Describe the functions of the renal system
29.3 Explain the formation and regulation of urine
29.4 Describe common disorders associated with the renal system

Introduction

Food is transformed through the process of digestion, absorption, and metabolism. The blood and lymph fluids transport products of digestion to the tissues. The cells of the tissues use the oxygen and food they need for growth and repair. The waste products are then taken away to be removed from the body. The primary body system involved in this process is the **renal system**. The functions of the renal system include excreting waste products, such as nitrogen; regulating the body's water balance; regulating chemicals and electrolytes, such as potassium, sodium, and chloride; producing hormones that control blood pressure; and producing a chemical called **creatinine**. *Creatinine* is a chemical waste product filtered by the kidneys and indicates the kidney function levels within the blood. The major waste product of protein metabolism is called **urea**, which is filtered by the kidneys and used in some diagnostic tests to evaluate the health status of the kidneys.

Structures of the Renal System

The structures of the renal system include the kidneys, ureters, bladder, and urethra. The renal system controls urinary production to rid the body of waste products within the bloodstream (see Figure 29–1). Urine is formed in the kidneys, flows through the ureters to the bladder, is stored in the bladder, and flows through the urethra and outside the body.

Kidneys

The **kidneys** are two organs located in the dorsal abdomen on either side of the spine. The right kidney lies more cranial than the left. The kidneys produce urine. They are reddish brown in color and are bean shaped and smooth (see Figure 29–2). The right kidney of horses is heart-shaped. Cows are the only animal species that have lobed kidneys, rather than organs with a smooth surface (see Figure 29–3). Each kidney has a **renal artery** and **renal vein** that are located in the center of the kidney and allow blood to flow through the renal and urinary structures. The area where these blood vessels are located is an indented location called the **hilus**. About 20 percent–25 percent of the body's blood flows through the kidneys.

The kidney's anatomy includes the **cortex**, which is the rough outer section of the structure. The **medulla** is the smooth center section of the kidney. The **renal pelvis** is the innermost section and filters blood and urine through the veins and arteries. The **nephron** is the structural and functional unit of the kidney that is composed

512 SECTION III General Anatomy and Disease Processes

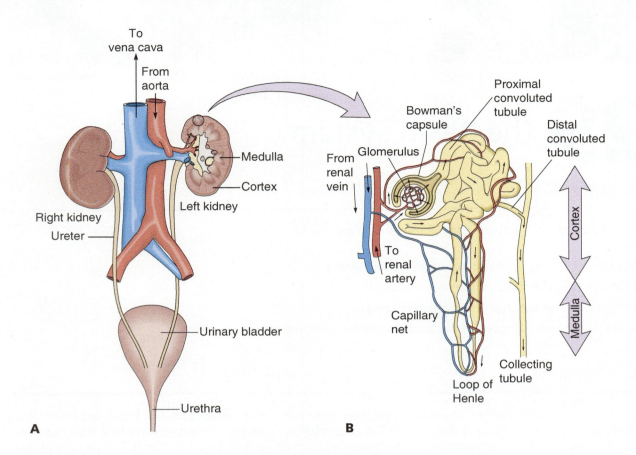

FIGURE 29–1 (A) Structures of the renal system. (B) Structures of a nephron.

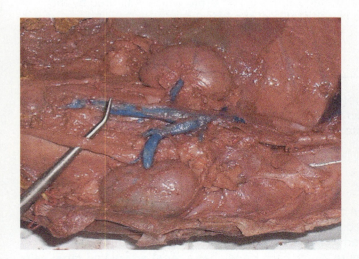

FIGURE 29–2 External appearance of the feline kidney. This photograph is of a prepared specimen. The blue vessels are the renal veins and caudal vena cava injected with latex. The ureter passes over the probe.

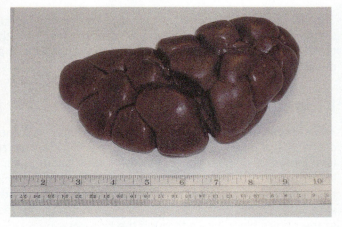

FIGURE 29–3 External appearance of a bovine kidney.

of a renal corpuscle and a renal tubule. The renal corpuscle consists of a group of capillaries called a glomerulus and Bowman's capsule. **Bowman's capsule** is a cuplike sack at the beginning of the tubular component of a nephron in the kidney that performs the first step in the filtration of blood to form urine. A **glomerulus** is enclosed in the sac and filters substances like water, salt, sugar, urea, and other nitrogenous wastes from the blood. This is where urine is first formed. Urine production begins through **osmosis** or water being absorbed by the body. Water mixes with the blood and absorbs sodium, causing the water to become diluted. The glomerulus then filters waste materials from the blood and begins urine formation, which involves 80 percent of the body's water. The filtered substances from blood collect in Bowman's capsule and then pass into the renal tubules. Substances in the renal tubules that can be used again by the body are reabsorbed back into the bloodstream. Any substances not able to be used by the body are excreted in the form of urine. Formed urine travels to the renal pelvis prior to entering the ureters and eventually traveling to the urinary bladder.

Ureters

In the center of each kidney is a tube that carries urine to the bladder. This tube is called the **ureter**. Each ureter is made of smooth muscle and pushes urine out of the kidneys through contractions. The ureters enter the bladder, forming valve-like openings that prevent backflow of urine into the ureters as the bladder fills.

Urinary Bladder

The ureters connect to the **urinary bladder**, which is a hollow muscular sac that changes size based on the amount of urine being held. The bladder holds a large amount of urine based on the size of an animal. For example, a 25-pound dog has a bladder that can hold around 100 ml of urine. The kidneys constantly produce urine. As the urine accumulates, the bladder wall receptors are activated to release the urine at a certain point in a process called urination. A spinal reflex initiates a contraction of the smooth muscle in the bladder wall. A voluntarily controlled sphincter muscle around the neck of the bladder allows for control of urination.

Urethra

The bladder connects to the urethra, which is a muscular tube that allows urine to flow to the outside of the body for elimination. Males typically have a longer, more curved, and narrower urethra than females. Male animals are more prone to urinary obstructions due to this anatomy. Male dogs have a bone within the urethra known as the **os penis**. The end of the urethra is typically S-shaped and is the site of the narrowest portion of the tube. Urine is passed to the outside of the body through the **penis**. The female urethra is shorter and wider. Female animals are more susceptible to urinary **incontinence** and uncontrolled bladder leakage.

Urinary Evaluation

The evaluation of urine, called a urinalysis, is vital in assessing the health status of an animal. Many products are metabolized and excreted by the urine, and a urinalysis can provide information about many body systems. The urinalysis is detailed extensively in Chapter 39. In addition to urinalysis, blood testing is available to evaluate kidney functions. With an understanding of how the renal system functions, the interpretation of a urinalysis and blood values is vital in assessing an animal's renal and urinary health.

Common Diseases and Conditions

The process of eliminating waste from the body is critical to maintaining health and homeostasis.

Urinary Incontinence

Urinary incontinence is the uncontrolled leaking of urine from the bladder. This condition is more frequent in females that are seniors and have been spayed. It commonly occurs during sleep and causes accidents to occur in the household due to the inability to control the sphincter muscles within the bladder. Many animals respond to treatment with medication, a natural hormone supplement of estrogen, to help control the bladder. Females are also more prone to urinary tract infections (UTI) called **cystitis** or inflammation of the urinary bladder. Bacteria migrate into the urethra through contaminants from the environment or increased licking of the **vulva** or **prepuce** areas. Signs of cystitis include **hematuria** (blood in the urine), increased or frequent urination, accidents within the house, and **dysuria** (difficult or painful urination).

 TERMINOLOGY TIP

The *vulva* is the external female anatomy where urine is excreted; the *prepuce* is the skin covering the penis where urine is excreted. ■

Urinary Blockage

Urinary blockage is caused by several conditions. One condition is **uroliths** or bladder stones that form within the urinary bladder, ureter, or urethra and cause an obstruction (see Figure 29–4). Diet is the common reason for bladder stone formation. Radiographs can be helpful in diagnosis of the condition, and many animals may need a **cystotomy**, surgical removal of the bladder stones (see Figure 29–5). The components of the uroliths should be analyzed to allow for proper treatment with medications or dietary control to prevent further occurrences (see Figure 29–6). Dalmatians are a breed of dog known for a genetic urolith condition from uric acid bladder stones. Dalmatians, as a breed, can't metabolize uric acid, as their bodies do not create enough enzymes to allow for its metabolism. This creates a buildup of uric acid that

FIGURE 29–6 Bladder stones.

then leads to bladder stone formation. This condition can be controlled by a specialized diet. Many animals are also predisposed to struvite and oxalate crystal formation, among others, that can lead to bladder stone formation.

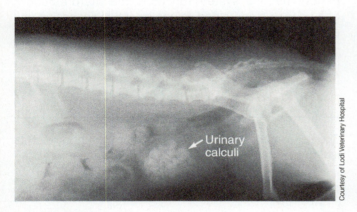

FIGURE 29–4 Radiograph of the urinary bladder showing urinary calculi in the bladder.

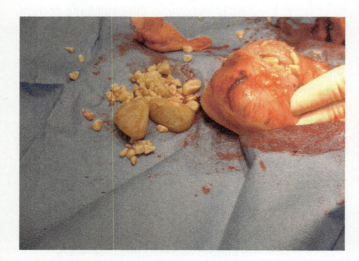

FIGURE 29–5 Cystotomy procedure.

Feline Urologic Syndrome

Cats are, as a species, are known for urinary tract disease. Male cats that are castrated have an increased incidence for a condition called **feline urologic syndrome** or FLUTD (feline lower urinary tract disease). Male cats that are overweight or obese are more at risk for this condition. The condition causes a cat's urethra to become blocked, and it is not able to urinate. This is an emergency situation that must be treated quickly. Cats have urine that tends to build up **sediment**, which is sandlike **crystals** that form in the urine. The sediment easily causes a blockage within the urethra. Cats with FLUTD must be treated immediately, as the inability to urinate causes kidney failure to develop within hours and can easily be fatal. Cats with FLUTD are sedated, and a urinary **catheter** is used to flush out the bladder and urethra. A urinalysis should be evaluated to determine treatment options. Typically, veterinarians will keep FLUTD cats catheterized for 48 hours to allow urination and flushing of sediment along with fluid therapy and medication treatment for underlying infection. The catheter is then removed, and the cat is monitored for an additional 24 hours for proper urination. Many FLUTD cats can be placed on a specialized diet to reduce blockages. Those cats that have increased occurrences of FLUTD blockages may be candidates for a **perineal urethrostomy** or

PU surgery. This surgery essentially removes the urethra and makes the male cat more anatomically like a female, eliminating the long narrow tube that allows sediment buildup and blockage. Signs of FLUTD include the following:

- Vocalization due to pain
- Painful abdomen
- Hard, palpable bladder
- Dehydration
- Weakness and lethargy
- Panting, along with increased HR and RR

When urine can't be properly eliminated from the body, a condition known as **azotemia** often occurs. This causes elevated phosphorus, blood urea nitrogen, and creatinine in the bloodstream. It may be the result of inadequate blood flow to the kidneys. Prerenal azotemia occurs when a pet is severely dehydrated, such as with vomiting or diarrhea. Animals with urinary obstruction develop postrenal azotemia. As the pressure increases within the bladder, the kidneys are unable to continue producing urine. Eventually, this causes damage to the kidneys and when the obstruction is relieved, the animal may continue to be azotemic. Kidneys that are not functioning properly may develop primary renal azotemia. For this to occur, more than 75 percent of the nephrons must be damaged. Once a kidney is damaged, it can't be repaired, but the disease process can be slowed. A kidney may be removed, and an animal may survive normally with one kidney, as long as the existing kidney remains healthy. A kidney transplant is also an option and has increased in companion animal medicine in recent years.

Toxicity

Antifreeze toxicity or **ethylene glycol** poisoning is a condition that affects the renal system and causes a severe condition in an animal. Due to its fruity taste and smell, making it inviting to animals, dogs and cats may ingest antifreeze that is found outside. Ethylene glycol is the active ingredient in antifreeze and is not capable of being metabolized by the liver, and toxins are then circulated through the body to the kidney, eventually causing renal damage and failure. This buildup becomes toxic to the renal system, causing it to fail and shut down. About 80 percent of animals that ingest antifreeze will die. Time is of the essence with this condition. Signs of antifreeze toxicity include central nervous system (CNS) conditions such as ataxia or loss of body control, disorientation, inability to stand, weakness, difficulty breathing, oliguria or anuria, PU/PD, painful abdomen, vomiting, seizures, coma, and death. Bloodwork and urinalysis evaluation help in determining the diagnosis and treatment of antifreeze toxicity, with the main therapy being fluids to flush out the renal system.

A condition known as **uremia** may occur, with toxicity from the urine invading the bloodstream. The animal then presents with clinical signs of azotemia due to an increased concentration of waste products, with the elevated amounts of phosphorus, urea, and creatinine getting into the lining of the GI system, which often causes ulcers within the intestinal tract and in the mouth. This condition is common in renal failure. Signs of uremia include vomiting, diarrhea, anorexia, bad breath with a fruity odor, and lethargy.

TERMINOLOGY TIP

The **central nervous system** or CNS is the part of the body under involuntary brain control, such as the brainstem and spinal cord. **Ataxia** is a term used to refer to an animal that has lost control over its body, causing wobbliness and inability to control movement. **Oliguria** is decreased amounts of urine production, whereas **anuria** is no urine production. ∎

Renal Failure

Renal failure may occur as an **acute** condition, which is one that occurs suddenly and is short term. It may also be a **chronic** condition in which the signs appear over a long period of time. Chronic renal failure is typical in senior pets due to age and other disease conditions that cause problems with the renal system. Signs of chronic renal failure include increased thirst (PD), poor appetite, weight loss, loss of urinary control (incontinence), vomiting, and diarrhea. Blood work typically reveals increased blood urea nitrogen (BUN), creatinine, and phosphorus. Anemia is also seen in a complete blood count or CBC. Fluid therapy along with dietary control with reduced protein and phosphorus are the usual treatment. Routine monitoring of blood levels helps with managing the condition.

MAKING THE CONNECTION

Chapter 8—leptospirosis
Chapter 17—leptospirosis ∎

SUMMARY

The renal system's main responsibility is the removal of wastes from the body. The removal of wastes is achieved by filtering the blood. This mainly occurs in the kidneys. The renal system also helps to maintain the proper balance of body fluids, which aids in the maintenance of homeostasis. Many nutritional manufacturers —such as Hills, Purina, and Royal Canin—now offer prescription diets that include urinary diets to aid in treating and decreasing the potential of many urinary conditions.

Key Terms

acute   short-term condition

antifreeze toxicity   poisoning that affects the renal system and shuts it down, causing renal failure; also known as ethylene glycol toxicity

anuria   no urine production

ataxia   term used to refer to an animal that has lost control over its body, causing wobbliness and inability to control movement

azotemia   elevation of blood urea nitrogen and creatinine in the bloodstream

Bowman's capsule   cuplike sack at the beginning of the tubular component of a nephron in the kidney that performs the first step in the filtration of blood to form urine

catheter   thin plastic or rubber tube that is passed into the urinary bladder

central nervous system (CNS)   part of the body under involuntary brain control, such as the brainstem and spinal cord

chronic   long-term condition

cortex   outer section of the kidney structure

creatinine   chemical waste product filtered by the kidneys; serum creatinine levels can be analyzed to assess renal function

crystals   solid formations within a urine sample that form from increased components within the urinary tract or renal system

cystitis   term for a bladder infection; also called urinary tract infection

cystotomy   surgical incision into the urinary bladder to remove urinary bladder stones

dysuria   difficult or painful urination

ethylene glycol toxicity   poisoning by antifreeze that causes renal failure

feline urologic syndrome (FLUTD)   condition that commonly affects obese neutered male cats in which the urethra becomes blocked and are unable to urinate; also known as feline lower urinary tract disease

glomerulus   capillaries located in a bundle at the renal pelvis

hematuria   blood in the urine

hilus   the indented center of the kidney where the renal vein and renal artery enter and exit

incontinence   uncontrolled bladder leakage

kidneys   two organs located in the dorsal abdomen on either side of the spine that produce urine, are reddish brown in color, and are bean shaped and smooth

medulla   center section of a kidney

nephron structural and functional unit of the kidney; produces urine

oliguria decreased amounts of urine production

osmosis water absorption by the body through membranes

os penis bone within the urethra of male dogs

penis male sex organ; allows urine to be excreted externally

perineal urethrostomy (PU) surgical creation of a new opening at the urethra and penis to eliminate the occurrence of urinary blockages

prepuce skin covering the penis where urine is excreted

renal artery located in the center of the kidney; allows blood flow out of the renal system

renal pelvis innermost section of the kidney; filters blood and urine through the veins and arteries

renal system body system that involves the kidneys and urinary production to rid the body of waste products within the bloodstream

renal vein located in the center of the kidney; allows blood flow into the renal system

sediment sandlike buildup in the urine composed of crystals and casts

urea waste product of protein metabolism by the kidney

uremia toxic condition caused by the kidneys when an increased concentration of waste products gets into the bloodstream

ureter tube made of smooth muscle that pushes and filters urine out of the kidneys through contractions

urinalysis breakdown and evaluation of the components of urine

urolith bladder stone that forms within the urinary bladder

vulva external female anatomy where urine is excreted

Review Questions

1. What is a UTI? What are the signs of a UTI?
2. The pathway of urine formation is
    a. kidney, ureter, urethra, bladder.
    b. ureter, pelvis, urethra, bladder.
    c. kidney, urethra, bladder, ureter.
    d. kidney, ureter, bladder, urethra.
3. What does the term *oliguria* mean?
4. What are the parts of the urinary system in both male and female animals?
5. List and describe the functions of the renal system.
6. Explain the difference between urinary incontinence and urinary blockage.
7. Describe how urine is formed.
8. What tubelike structure connects the kidney to the bladder?
9. Do males or females have a higher incidence of urinary incontinence?

10. What breed of dog has a predisposition for difficulty in metabolizing uric acid?
11. Can azotemia result from dehydration?
12. What is urinary incontinence? Explain.
13. What is the medical term for difficult urination?
14. What is the difference between chronic and acute?
15. What is the name of the test that evaluates urine?

Clinical Situation

Kelly, a vet assistant at High Pines Animal Hospital, is performing a UA for Max, a 4 yr NM Dalmatian. Kelly has just recently been trained to run lab work samples. Max's owner, Mr. Kline, brought the urine sample in this morning and Kelly placed it in the fridge until she had a chance to analyze it. Max has a history of bladder stones and frequent UTI. Kelly notes the color of the urine is brown and it appears cloudy and somewhat thick. She performs a clinical stick analysis with the following results that she provides to the vet.

Color: brown
Appearance: cloudy, thick, viscous
SG: 1.245
Glucose: NEG
Ketones: NEG
Bilirubin: TRACE
Protein: TRACE to +
Blood: POS +++
pH: 8.5

- What is most likely wrong with Max?
- What other parts of a UA might Kelly complete?
- What factors may be involved with the sample and how the owner collected the specimen?

CHAPTER 30: The Reproductive System

Objectives

Upon completion of this chapter, the reader should be able to:

30.1 State the structures that make up the male reproductive system
30.2 State the structures that make up the female reproductive system
30.3 Describe the hormonal functions of the male reproductive system
30.4 Describe the hormonal functions of the female reproductive system
30.5 Understand the female estrus cycle
30.6 Explain the steps in establishing pregnancy
30.7 Discuss the stages of parturition
30.8 Describe common disorders of the male reproductive system
30.9 Describe common disorders of the female reproductive system

Introduction

The **reproductive system** is responsible for producing offspring. Both male- and female-specific organs are required to complete offspring production. The animal reproductive system, in general, is basically similar in many species. The reproductive organs, whether male or female, are called genitals or **genitalia**. The male produces sperm that is delivered into the female, who is responsible for maintaining pregnancy.

The Female Reproductive System

In the female, the **vulva** is the external beginning of the reproductive system. The vulva is the external opening of the **vagina**, or the internal female reproductive organ, that allows for breeding, labor, and urination processes to occur. At the base of the vagina is the **urethra**, which is a narrow tube that connects to the urinary bladder. The end of the vagina holds the **cervix**, which is the opening to the uterus. The **uterus** is the main area for allowing an embryo to begin the gestation phase. The uterus is Y-shaped and has two tubes that are called **uterine horns**. It is a large, hollow organ that expands through the gestation period to allow the young to grow. Some animal species carry single or multiple young in the uterine horns, and other species carry the embryo or embryos within the uterus. Each end of the horn tapers into a small tube called an **oviduct**, which holds the ovary and is the transport system for sperm and eggs. The oviduct is where the sperm and egg meet during fertilization. Then the fertilized egg continues to move down the oviduct to the uterus. At the end of the oviduct is a thin membrane called the **infundibulum**, which wraps around the ovary to catch the egg after it is released from the ovary. In non-egg-laying animals, the infundibulum guides the released unfertilized egg into the oviduct to continue its journey to the uterus. The **ovary** is the primary female reproductive organ and produces eggs in the female through the process of meiosis. There are two ovaries, one located at the end of each uterine horn. Some variations exist between species; for example, chickens have only one fully developed left ovary. The ovary contains **follicles** that

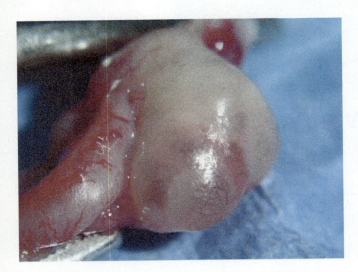

FIGURE 30–1 Follicle on the surface on an ovary.

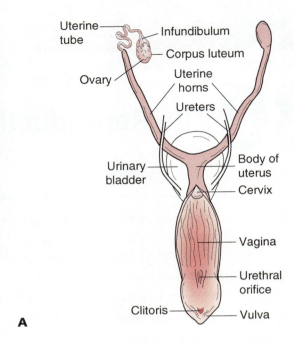

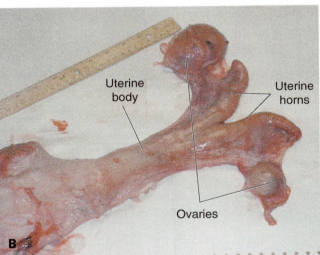

FIGURE 30–2 (A) Reproductive tract of a female dog. (B) Reproductive tract of a cow.

are tiny structures where the eggs are produced (see Figure 30–1). Each animal species has a set number of follicles and eggs that are present when the animal is born. The eggs are used slowly throughout the life of the animal. As an egg matures, it will move to the edge of the ovary for release. The time frame when an egg is released is called **ovulation**. Figure 30–2 shows female reproductive organs in a dog and cow.

Some species of animals, such as avian species, some fish, and some reptiles, have a reproductive system for laying eggs rather than bearing live young. The egg develops in the **infundibulum**, which is a funnel within the uterus. Species that lay eggs have an oviduct that has four sections or segments: the **ovum** (egg yolk), **albumen** (egg white), **isthmus** (shell membrane), and uterus. The external reproductive area where the egg is laid is called the **cloaca** or **vent**. The time frame of ovulation until an egg is laid is about 24 hours. Figure 30–3 shows the reproductive organs in a bird. In non-egg-laying species, the infundibulum guides the released egg into the oviduct, where it continues its journey to the uterus.

Estrogen is the main female reproductive hormone that causes behavior changes in females during the estrus cycle and also allows follicle production and development to occur. **Progesterone** is a hormone that is produced to maintain pregnancy.

The Male Reproductive System

The male reproductive system begins with the external structure known as the **penis**. The penis is the male reproductive organ used in breeding and is also part of the urinary and renal systems. The penis is covered and protected by skin called the **prepuce** (see Figure 30–4). The primary reproductive organs in males are called **testicles**. Before birth, the testicles develop in the abdomen and then descend through slits in the abdominal muscles called **inguinal rings**. The testicles produce sperm and are held in place by a sac known as the **scrotum**. The scrotum consists of skin with two lobes that hold and protect each testicle. The scrotal sac holds the scrotum, which is temperature controlled for ideal sperm production, and thus must be cooler than the overall body temperature. **Sperm** production begins at puberty and continues throughout the life process. The sperm is the male reproductive cell.

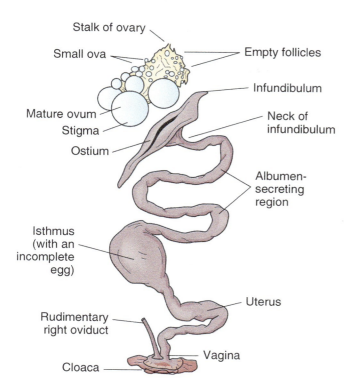

FIGURE 30-3 Reproductive tract of a bird.

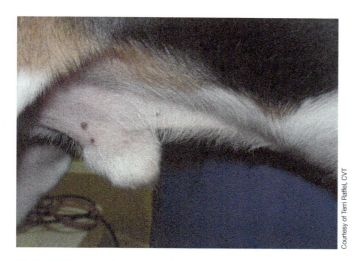

FIGURE 30-4 Prepuce of a dog.

Sperm are formed in the **seminiferous vesicles**, which are U-shaped tubes located within the testicles. Sperm pass through the **epididymis**, a long tube lying on the surface of the testicles, and into the **vas deferens**, the transport systems for sperm from the testicles to the outside of the body. Sperm are stored in the epididymis until ejaculation and if not expelled, will die and be absorbed. Both organs also carry **semen**, which is fluid that transports the sperm during breeding. In mammals, the sperm travel through the urethra. In avian species and reptiles, the sperm travel through the cloaca. Avian species do not have a true penis due to an undeveloped organ. In fish species, the sperm travel through the urogenital pores, which also pass feces and urates. Figure 30–5 shows the reproductive system of a stallion. **Testosterone** is the main male reproductive hormone that produces male characteristics and is essential for sperm production. Male characteristics developed by testosterone include the following:

- Large body size
- Muscles and powerful structure
- Aggressive tendencies
- Deep vocal sounds

There are several accessory sex glands. One is the **prostate gland**, which is found in all common mammal species. The prostate gland is located at the base of the urinary bladder and produces an accessory secretion that releases prostatic fluid that produces some of the ingredients of semen and is necessary for survival and motility of sperm. The **seminal vesicles** and **bulbourethral glands** are present in only some species, such as cattle and horses. These are paired glands that contribute to the fluid portion of semen.

Semen

Semen morphology is the process of determining if sperm are normal and adequate for breeding. Sperm cells multiply by meiosis. The sperm have a head and a tail, known as **flagella** (see Figure 30–6). The flagella allow for movement and motility. Chromosomes are found in the sperm's head and are what identify the DNA of the offspring. A normal sperm cell has one head and one straight tail (see Figure 30–7). Sperm quality is noted using a microscope, and all sperm should move rapidly. Sexual reproduction occurs when an egg and a sperm unite. This requires a male and female of the same species. When egg and sperm unite, the process is called **fertilization** (see Figure 30–8).

Fertilization

In mammals, fertilization occurs directly in the reproductive tract. In avian species, the female develops an egg and passes it externally, and incubation must then occur to allow the egg to hatch. **Incubation** is the process of increasing the temperature to allow the embryo to grow. Fish fertilization occurs outside of the reproductive tract through either egg, as live young, or as mouth brooders. **Mouth brooders** carry

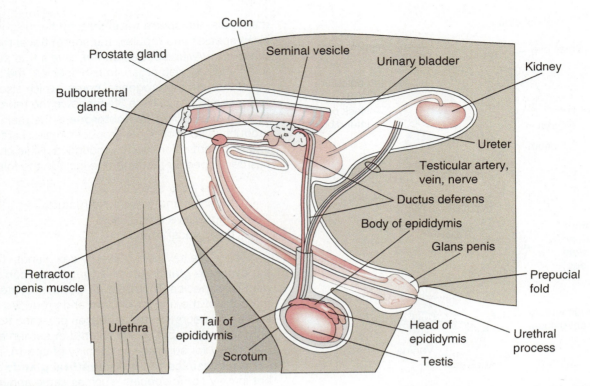

FIGURE 30-5 Reproductive tract of a stallion.

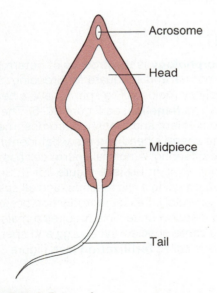

FIGURE 30-6 Parts of sperm.

their eggs in their mouth until hatching. Fertilization leads to **conception**, which is the process of creating new life in the form of an embryo. An **embryo** is a cell that develops into a newborn animal. Pregnancy occurs; **gestation** is the length of time until the animal is born. **Hormones** are chemicals that are produced by the body that allow for sexual reproduction to occur in both males and females.

Estrus

The **estrus cycle** is also called the heat cycle in animals. This is the period of time that animals are receptive to breeding. The estrus cycle is the heat cycle and when discussing the active heat cycle, the term is estrous. During the heat cycle a female animal is bred. There are several stages within the heat cycle, and each animal species has a different length of cycle and duration. Many animals have a seasonal estrus, such as many large animal species, which have an estrus cycle that depends on the length of the day. **Anestrus** is the period of time when no ovarian activity is occurring, and the animal is not actively in heat. **Proestrus** is the stage that occurs just prior to the heat cycle. Hormones are released in this stage and cause initial follicle development. Estrogen is produced and causes genital and behavioral changes that attract the male and prepare the female reproductive tract for mating. **Estrus** is the time when the female animal is receptive to breeding or mating with a male. Some estrus cycles are as short as 12 hours and others as long as 14 days. The estrus cycle is triggered by an increased release of estrogen. Signs of estrus vary from animal to animal, but many female animals in estrus will have a swollen vulva (see Figure 30–9), a change in behavior, vaginal discharge, increased urination, and increased vocal behaviors. **Metestrus**

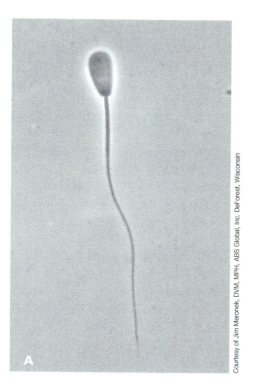

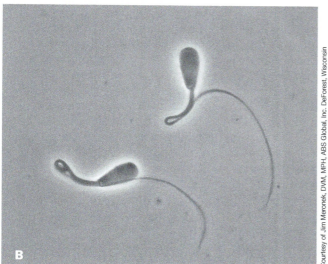

FIGURE 30–7 (A) Normal sperm. (B) Defective sperm with bent tails.

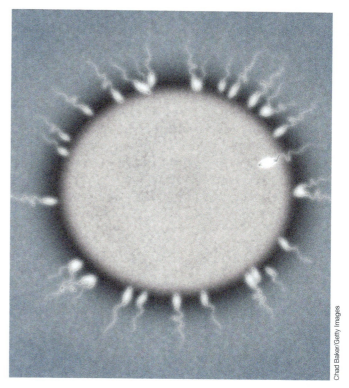

FIGURE 30–8 The outer membrane of an egg must be dissolved before the sperm can enter the egg.

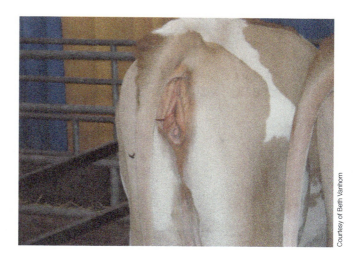

FIGURE 30–9 Swollen vulva and vaginal discharge in a cow.

is the short stage just following the heat cycle when ovulation occurs, and progesterone begins to be secreted. The female is no longer accepting of the male. Some species are **induced ovulators** and must be mating for the release of the egg to occur. Examples of induced ovulators are cats and rabbits. Some species release one egg during ovulation, whereas other species release multiple eggs. Table 30–1 summarizes the types of ovulation that occur in various species.

 TERMINOLOGY TIP

Induced ovulation is the occurrence of the release of the egg at the time sperm is introduced. This causes immediate fertilization to occur at the time of mating.

TABLE 30–1
Animal Species Ovulation Types

Dog	Multiple eggs released during ovulation
Cat	Induced ovulation with multiple eggs released
Horse	One egg released during ovulation
Cow	One egg released during ovulation
Sheep	One egg released during ovulation
Pig	Multiple eggs released during ovulation
Goat	One egg released during ovulation
Rabbit	Induced ovulation with multiple eggs released

Proestrus produces an ovary that contains a **corpus luteum (CL)**, which produces the hormone progesterone. Prostaglandin hormone is released into the bloodstream from the uterus and causes the CL to degenerate and in turn decrease the amount of progesterone produced. **Follicle-stimulating hormone (FSH)** is then released from the pituitary gland and results in the development of a follicle in the ovary. The follicle is a fluid-filled structure and surrounds the egg. The follicle begins to secrete estrogen in response to the FSH. Estrogen produces the behavioral changes in females and causes **luteinizing hormone (LH)** to increase production and further mature the follicles. Estrogen also causes the release of an egg during ovulation. After ovulation, the follicles collapse and metestrus begins. Estrogen levels begin to decrease, and LH stimulates the follicles to develop a CL, also called the yellow body. The cells of the CL begin to produce progesterone and stimulate the uterus to prepare for an embryo, and the egg remains in the oviduct. **Diestrus** is the stage when pregnancy occurs and is characterized by a functional CL that releases increased progesterone to maintain pregnancy. This occurs over the course of 9 to 12 days and prepares the uterus for gestation. If pregnancy doesn't occur, diestrus allows the CL to break down, and the animal goes back into either proestrus or anestrus, depending upon the species. When **gestation** is occurring, the CL becomes smaller in size and hormone production decreases. If breeding does not occur or pregnancy is not maintained, the CL breaks down to prepare the animal to both go back into proestrus and begin a new cycle. Depending on the species of animal, the female will go back into a new cycle or remain in anestrus. Figure 30–10 illustrates the estrus cycle in a cow. Figure 30–11 illustrates the stages of the estrus cycle. Table 30–2 summarizes estrus cycles in various species.

TERMINOLOGY TIP

Many terms relate to breeding animals. The following are several common terms:

Lordosis—exhibiting a prayer position where the front end of the body is lowered to the ground with the back end held high in the air

Polyestrous—many heat cycles that occur at regular intervals

Monestrous—one single heat cycle

Standing heat—occurs when the female animal stands for breeding when a male is present

Tail flagging—holding the tail high in the air during the estrus cycle

Winking—movement of the vulva with frequent vaginal discharge and urination.

Establishing Pregnancy

During mating, semen is deposited into the vagina. Sperm move rapidly through the cervix, into the uterus, and up the oviducts via a combination of swimming actions and contractions of the female reproductive system. When the sperm unites with the egg and fertilization occurs, the resulting cell implants in the wall of the uterus. Following implantation, the placenta of the developing fetus develops and is the life support system consisting of multiple layers of a fluid-filled sac that attaches to the uterine wall and transfers nutrients and gases from the mother to the growing offspring. The first stage of pregnancy, the first trimester, is when the implanted cell is developing and is called the embryo stage. During the second stage or second trimester, the developing animal is called a **fetus**. This is the stage when the anatomy is developing and taking shape. The third stage or trimester is when the most fetal growth occurs dramatically, preparing the fetus for birth.

Gestation and Incubation

Pregnancy or gestation is the period of development of the fetus in the uterus from conception to birth. Gestation periods vary in length in different species (see Table 30–3). Signs of pregnancy include increased appetite, weight gain, and increased size of mammary glands.

In egg-bearing animals, once an egg is laid the embryo continues to grow inside the egg. This period of growth is known as the incubation period. This period varies by species (see Table 30–4). Proper temperature and humidity are important to incubation.

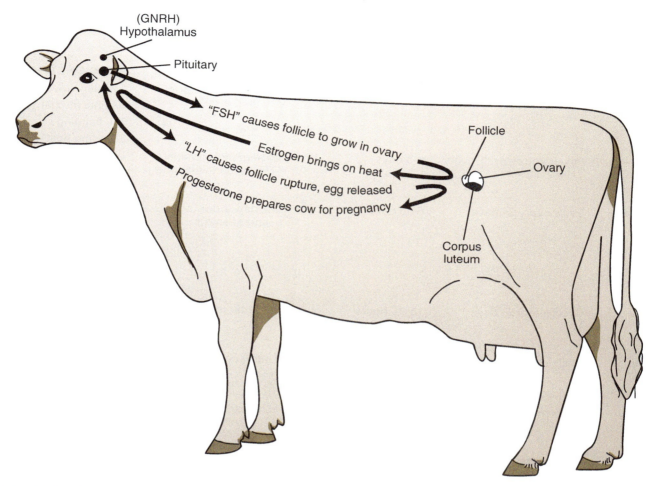

FIGURE 30-10 Hormones of the bovine estrus cycle. GnRH = gonadotropin-releasing hormone.

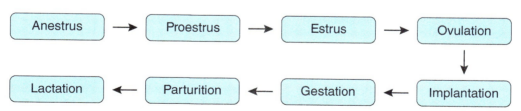

FIGURE 30-11 Anestrus. (1) Proestrus. (2) Estrus. (3) Ovulation. (4) Implantation. (5) Gestation. (6) Parturition. (7) Lactation.

TABLE 30-2

Estrus Cycles in Various Species

SPECIES	CYCLE OCCURRENCE	LENGTH OF ESTRUS	OVULATION TIME	SIGNS OF ESTRUS
Dog	Cycle twice a year*	9–10 days	1–2 days after estrus	Swollen vulva Vaginal discharge Tail flagging Standing heat
Cat	Polyestrous—cycle every 14–21 days	4–5 days	24 hours after mating (induced ovulation)	Lordosis Friendly Vocal
Horse	Cycle every 21 days	6–10 days	1–2 days prior to the end of estrus	Tail flagging Winking Increased urination Vaginal discharge Swollen vulva Vocal
Cow	Cycle every 21 days	12–18 hours	10–14 hours after estrus	Standing heat Restless Mounting other cows
Goat	Cycle every 21 days	30–40 hours	End of cycle	Rapid tail wagging Tail held in air Mounting other goats
Sheep	Cycle every 17 days	24–36 hours	Late estrus	Minimal signs Difficult to detect Mounting other sheep
Pig	Cycle every 21 days	40–72 hours	Mid-estrus	Swollen vulva Standing heat Lordosis
Rabbit	Constant cycle—polyestrous	Polyestrous	8–10 hours after mating (induced ovulation)	Aggression Lordosis

*The exception in the dog is the Basenji, which has only one estrus cycle per year.

TABLE 30-3

Gestation Length of Various Species

SPECIES	GESTATION LENGTH
Dog	Average 63 days
Cat	Average 63 days
Horse	Average 330 days
Cow	Average 285 days
Goat	Average 50 days
Sheep	Average 150 days
Pig	Average 114 days
Rabbit	Average 30 days

TABLE 30-4

Incubation Table

SPECIES	INCUBATION PERIOD
Chicken	21 days
Duck	28 days
Turkey	28 days
Goose	28–30 days
Pheasant	24 days

The Labor Process

The **parturition** process, also known as birth or labor, has three stages. During the labor process levels of progesterone decrease and estrogen increases. Estrogen prepares the newborn for the delivery process. Labor begins when cortisone is released by the fetus. The first stage of labor begins with **contractions** of the uterus as the fetus begins to move around. Signs of first-stage labor include restlessness, anxiety, nervousness, and abdominal discomfort. The cervix begins to dilate or open thus causing the cervix tissues to soften. The rupture of the fetal membranes ends this stage. The second stage of labor occurs when the fetus moves into the pelvic opening of the birth canal. Strong abdominal contractions increase as each fetus enters the vaginal area for birth. The fetus exits the birth canal during this stage (see Figure 30–12). Most female animals will lick the newborn immediately after birth to help clean and dry the baby and also to stimulate breathing. This process removes the fetal membranes over the mouth and nose. After the active labor stage is complete, the third stage occurs, and the placental tissues are expelled from the uterus. The **placenta** is the tissue and membrane collection that connects the fetus to the mother to allow nutrients to be absorbed during gestation (see Figure 30–13). The placenta is also the collection area for fetal wastes. After the placenta is removed, the uterus begins to shrink to its normal size through a process called **involution**. This usually occurs 1–3 weeks after birth depending on animal species. After delivery of the offspring, the umbilical cord is separated from the placenta and the newborn (see Figure 30–14). Milk production, called **lactation**, begins at the end stage of pregnancy and allows for nursing of the newborn. The hormones prolactin and oxytocin allow for milk letdown and production. The initial milk secretion following parturition produces **colostrum** and is vital in transferring antibodies from the mother to the newborn. It is important for the newborn to nurse as soon after birth as possible, as the colostrums is produced for only a short period.

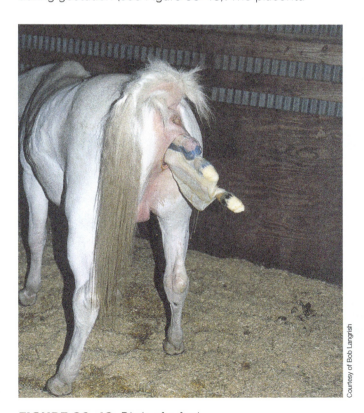

FIGURE 30–12 Birth of a foal.

FIGURE 30–13 Placenta.

FIGURE 30–14 Equine fetus and placenta after the labor process of a newborn foal.

Semen Collection

Semen collection is accomplished using specialized veterinary equipment to collect sperm from male animals to be used in breeding practices. Semen collection is done in numerous species and breeds of animals, especially horses, dogs, and cattle. Collection methods vary between each species of animal, but proper sterile collection methods are important to the success of the practice. The most common collection method uses an **artificial vagina (AV)**, which is a sleeve or tube used to collect the semen during the ejaculation process. It is important that veterinary assistants and other professionals have advanced handling and restraint skills when working with breeding animals. This includes knowledge of and experience with body language and behavior of the species. Once semen is collected, it should be evaluated for any abnormal appearance. It must then be cooled slowly, preserved, and properly stored. Semen that is to be stored and shipped may be cooled or frozen. **Cooled semen** must be shipped and used within a 24- to 48-hour time frame for sperm survival. **Frozen semen** may be stored in liquid nitrogen and may be preserved for up to 40 years.

Artificial Insemination

Artificial insemination (AI) is a procedure that is popular in many animal species and breeds and is used in many different situations. This procedure is the practice of breeding female animals through the use of veterinary equipment rather than the traditional mating method (see Table 30–5). Timing is of vital importance when using AI methods. The female must be in heat and near ovulation. Determining this requires proper estrus cycle evaluation and observation. The most common failure of the AI procedure is failure to properly detect and determine ovulation. There are many reasons to use AI, such as the male animal is located in another part of the country, the male and female are not compatible, the animal is too aggressive to safely breed, or the female's estrus cycles are difficult to regulate. The first step in the artificial insemination process is having quality sperm. The sperm should be evaluated on arrival for motility and quality. Cooled and frozen semen must be warmed to 95–98 degrees Fahrenheit. Fresh semen should be used within 24 hours. The insemination process should be done at intervals at least two times prior to ovulation. This requires knowledge of the species and proper timing of the cycle.

Some AI procedures place the semen in the uterus, while others place the semen within the cervix (see Figure 30–15). This depends on the animal species and where the embryo grows within the reproductive system. Many female animals will stand for the AI procedure, while other species may need to be sedated for surgical AI. It is important to prevent the female from urinating or leaking any semen from the vagina for 30 minutes after the AI process is completed. An X-ray or ultrasound procedure is done after gestation has started to ensure the AI breeding process has occurred and to determine the number of offspring.

TABLE 30–5
Common AI Procedures

SPECIES	ESTRUS DETERMINATION	AI TIMEFRAME	AI METHOD
Dogs	Estrus smear to evaluate cells and stage in cycle	AI during last stages when cornified epithelial cells are present	AI tube placed in cervix
Horses	Palpate or ultrasound to determine stage in cycle	AI on days 3, 5, and 7 in every-other-day method	AI speculum used with caution due to delicate tissue; AI tube placed into uterus
Cows	Most estrus activity occurs at night	AI during the last two-thirds or 1–2 hours after estrus ends	Palpate to feel ovaries; place AI tube in cervix
Sheep	Rare; difficult to detect estrus	AI twice during estrus cycle	AI tube placed in uterus
Pigs	Monitor for standing heat	AI 24 hours after estrus cycle begins; AI again 40–48 hours after estrus begins	AI tube placed in cervix
Turkeys	Used due to difficulty breeding	Once a week for 2 inseminations	Storage glands for semen in hens

CHAPTER 30 The Reproductive System

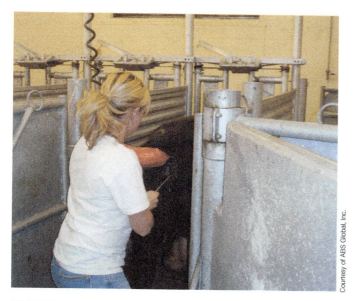

FIGURE 30–15 Artificial insemination of a mare.

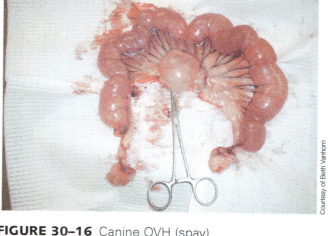

FIGURE 30–16 Canine OVH (spay).

FIGURE 30–17 Canine castration.

Neutering

Neutering is the surgical removal of the reproductive organs. The **ovariohysterectomy (OHE/OVH)** or spay is the surgical procedure in females that removes the uterus and ovaries. A surgical incision is placed on the midline of the abdomen or the center of the belly near the umbilicus or navel (see Figure 30–16). This procedure is done to prevent overpopulation of animals, eliminate aggressive or difficult behaviors, and prevent cancer development in animals. It is not recommended to perform surgery in females that are in estrus, as increased bleeding may occur, causing complications.

The **castration** surgery is the removal of the testicles in male animals (see Figure 30–17). It is done by making a small incision near the testicles over the scrotum. This is a surgical procedure under anesthesia in small animals. Sutures or staples are usually used to close the incision. In large animals, such as cows and horses, castration is usually done standing with sedation. The area is left open to heal. Both surgeries require postoperative care such as reduced activity, monitoring for swelling and infection, and suture removal if necessary.

Some male animals may not have both or one testicle descend into the scrotum. This is known as cryptorchid. A **unilateral cryptorchid** is one retained testicle, and a **bilateral cryptorchid** occurs when both testicles are retained. Retained testicles may be in the abdomen, in the inguinal canal, or under the skin near the scrotum. Most testicles that don't descend properly into the scrotum are smaller in size. Animals with this condition are at an increased risk for tumor development. The condition can also cause pain. Cryptorchid animals have a genetic tendency to pass this condition on to their offspring.

Reproductive Problems

Female animals may develop a uterine infection known as a **pyometra**, where the uterus fills with pus (see Figure 30–18). This condition is an emergency and must be treated immediately. The buildup of bacteria in the system is toxic when it gets into the bloodstream, causing a **septic** condition. An immediate ovariohysterectomy should be done to remove the infected uterus. Another reproductive condition in females is a **prolapsed uterus**. This is more common in cattle but can occur in any species of female following labor (see Figure 30–19). It is more common in females that have

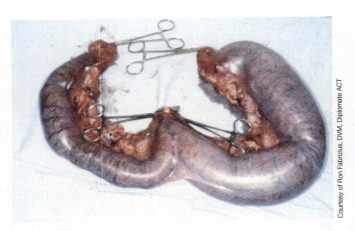

FIGURE 30–18 Pyometra in a female dog.

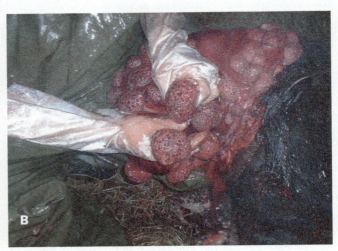

FIGURE 30–20 Ovarian cysts on a canine uterus.

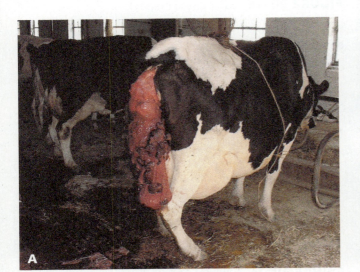

FIGURE 30–19 (A) Cow with a prolapsed uterus. (B) Close-up of prolapsed uterus showing placenta.

large newborns or numerous young for which labor requires excessive straining. Difficult deliveries involving **dystocia** also may cause a uterus to prolapse. The uterus and vaginal tissues separate from the body and allow the uterus body to be passed outside the body due to the stretching of the vagina and vulva. This is a life-threatening emergency situation, as exposing the uterus to the environment outside the body can cause severe septic infection; possible tearing can cause a uterine artery to hemorrhage.

Ovarian cysts also occur in intact females and may occur on the follicles or ovaries of the reproductive system (see Figure 30–20). They usually are larger than 1 cm in size and begin when the follicle or ovary doesn't release an egg. This condition typically causes the female to be **sterile**. Signs of ovarian cysts include irregular and decreased estrus cycles. Many females with cysts also produce an excessive amount of estrogen.

Male animals that are not castrated have an increased risk of prostatic medical conditions that can lead to neoplasia. The most common condition in intact males is **benign prostatic hyperplasia (BPN)** or an enlarged prostate gland. This typically affects intact male animals between 2 and 6 years of age and clinically may not have any symptoms or may have a bloody discharge from the prepuce and pain or straining with urination. Castration is the best control method.

Mammary Gland Problems

Mammary gland tumors are a common occurrence in females with intact reproductive systems. These tumors may be cancerous and should be evaluated by a veterinarian. Swelling may occur on one or more mammary glands. The surgical removal of the mammary glands is known as a **mastectomy**. There is a decreased occurrence in females that have had an OHE prior to having the estrus cycle begin.

SUMMARY

The reproductive system allows animals to breed and create new life. Male and female reproductive organs are required for fertilization and incubation to take place. Reproduction is an important part of the animal industry. Disorders of the reproductive system can cause sterility, rendering an animal incapable of giving birth. Some reproductive disorders can be life-threatening.

 MAKING THE CONNECTION

All chapters in Section II contain information on reproduction and breeding of various animal species.

Key Terms

albumen   egg white

anestrus   time when a female animal is not actively in the heat cycle

artificial insemination (AI)   the practice of breeding female animals through the use of veterinary equipment rather than the traditional mating method

artificial vagina (AV)   sterile sleeve or tube used to collect the semen of an animal during the ejaculation process

benign prostatic hyperplasia (BPN)   enlarged prostate gland in males

bilateral cryptorchid   both testicles are retained and do not descend into the scrotum

bulbourethral glands   paired glands that contribute to the fluid portion of semen

castration   surgical removal of the testicles in male animals

cervix   end of the vagina that opens to the uterus

cloaca   external reproductive area where egg-laying animals pass urine, feces, and eggs; also called the vent

colostrum   antibodies produced in the mother's initial milk supply

conception   process of creating a new life that forms an embryo

contraction   first stage of labor when the fetus begins to move in the uterus to prepare for birth by moving into the birth canal

cooled semen   process of storing and shipping sperm at a cool temperature for use within 24–48 hours

corpus luteum (CL)   known as the yellow body of the ovary; forms to allow the ovary to increase in size as an egg develops

diestrus   stage of the heat cycle when pregnancy occurs; is characterized by a functional CL that releases progesterone to maintain pregnancy or if pregnancy doesn't occur, the corpus luteum breaks down, preparing the animal to cycle again

dystocia   difficult labor

embryo   early stage of development of a multicellular organism

epididymis   tube that transports sperm from the seminiferous vesicles to the vas deferens

estrogen main female reproductive hormone that causes behavior changes in females during the estrus cycle and also allows follicle production and development to occur

estrus the time when the female animal is receptive to breeding or mating with a male

estrus cycle heat cycle in female animals when the female is receptive to mating with a male

fertilization uniting of egg and sperm to allow an embryo to form

fetus unborn offspring of an animal that develops from an embryo

flagella tail of sperm that facilitates movement

follicle tiny structures within the ovary that enlarge with egg development and allow ovulation to occur

follicle-stimulating hormone (FSH) hormone released to allow a follicle to grow and ovulation to occur

frozen semen process of storing and shipping sperm in liquid nitrogen which may be preserved up to 40 years

genitalia also called *genitals*; reproductive organs

gestation length of pregnancy

hormones chemicals that are produced by the body that allow for sexual reproduction to occur in both males and females

incubation the process of increased temperature to allow an embryo to grow

induced ovulation occurrence of the release of the egg at the time sperm is introduced when fertilization occurs at mating

infundibulum funnel within the uterus that catches eggs as they are released from the ovary

inguinal rings slits in the abdominal muscles where the testicles descend into the scrotum

involution the process of the uterus shrinking after labor

isthmus shell membrane of an animal that lays eggs

lactation milk production from the mammary glands

lordosis exhibiting a prayer position where the front end of the body is lowered to the ground with the back end held high in the air

luteinizing hormone (LH) allows an egg to develop and be released in the ovary during metestrus

mastectomy surgical removal of the mammary glands

metestrus stage of the heat cycle following estrus; when ovulation occurs

monestrous one single heat cycle

mouth brooder fish that carry their eggs in their mouth for incubation to protect and allow hatching

ovarian cyst small masses that develop on the ovary of an intact female animal, causing sterility and eggs not to develop

ovariohysterectomy (OHE/OVH) term for a spay; surgical removal of the ovaries and uterus

ovary the part of the uterine horn that allows egg development and is the primary female reproductive organ in egg growth

oviduct small tube at the end of the uterine horn that is tapered and allows the egg to pass from the ovary

ovulation the release of the egg from the ovary

ovum egg yolk

parturition the labor process

penis the male external reproductive organ; used in breeding and mating

placenta tissue and membrane collection that connects the fetus and the mother to allow the fetus to absorb nutrients during gestation

polyestrous having more than one heat cycle in a year

prepuce the skin that covers and protects the penis in male animals

proestrus beginning stage of the heat cycle before estrus; causes follicle development and genital and behavioral changes that attract the male to the female and prepare the female for mating

progesterone hormone produced to maintain pregnancy

prolapsed uterus condition when the uterus and vaginal tissues separate from the body and allow the uterus body to be passed outside the body

prostate gland produces accessory secretion to add to the sperm to produce semen

pyometra severe infection of the uterus in female intact animal that causes the uterus to fill with pus

reproductive system the organs in the body responsible for both male and female breeding and reproduction

scrotum skin over the two testicles that protects and controls the temperature for sperm production

semen fluid that transports sperm during reproduction

semen collection procedure using specialized veterinary equipment to collect semen from male animals to be used in breeding practices

semen morphology process of determining if semen and sperm is adequate for transport and breeding by analyzing the quality and quantity

seminal vesicles paired glands that contribute to the fluid portion of semen

seminiferous vesicle tubes located within the testicle that hold the sperm

septic refers to buildup of bacteria in the system that causes a toxic effect when it gets into the bloodstream

sperm male reproductive cell developed in the testicle

standing heat situation in which a female animal stands for breeding when a male is present

sterile not able to reproduce

tail flagging holding the tail high in the air during the estrus cycle

testicle the primary reproductive organ in the male that holds sperm

testosterone main male reproductive hormone that produces male characteristics and is essential for sperm production

unilateral cryptorchid condition in which one retained testicle has not descended into the scrotum

urethra the tube by which urine is passed out of the body from the bladder, and which in male vertebrates also passes semen

uterine horn two parts of the uterus that hold single or multiple embryos that expand as the embryos grow

uterus large, hollow reproductive organ that holds an embryo during the gestation period and is Y-shaped

vagina internal female reproductive organ that allows mating and labor to occur

vas deferens small tube that transports sperm from the testicles to the outside of the body

vent the external area where eggs are laid in egg-bearing animals; also called the cloaca

vulva external beginning of the reproductive system in the female; also where urine is excreted

winking movement of the vulva with frequent vaginal discharge and urination

Review Questions

1. What is the heat cycle in females known as?
2. What is the term used for both male and female external reproductive organs?
3. What is the veterinary medical term for a spay?
4. What are the male hormones? Describe each.
5. What is the time frame when an egg is released in a female called?
6. Explain the estrus cycle.
7. Describe the stages of labor.
8. How does fertilization and incubation of an egg occur?
9. Explain artificial insemination and why it may be practiced.
10. Explain neutering and why it may be performed.
11. Where is sperm produced?
12. What organ becomes infected during a pyometra?
13. In which trimester does the fetus experience the most growth?
14. Can females continue to produce eggs throughout their life?
15. What does cryptorchid refer to?

Clinical Situation

Leslie, a veterinary assistant at Whiskers and Tails Veterinary Clinic, is setting up for a C-section surgery. A 3 yr F English Bulldog is scheduled for surgery. Leslie reviews the patient's medical record history and notes that the dog has had a C-section a year ago and had a normal routine surgery with 6 puppies. The medical record notes show that the dog was seen about 3 weeks ago and radiographs were performed, showing 8 puppies in the uterus.

Dr. Dawson and the vet tech, Sara, arrive in the prep area and prepare to anesthetize the patient. Leslie continues to set up for surgery.

- What equipment and supplies will Leslie need to provide for the surgical procedure?
- What concerns and factors should the veterinary medical team consider with this patient?
- What client education may be provided to the owner of the animal?
- What factors in this situation may or may not show the owners of this dog are responsible breeders?

CHAPTER 31: The Immune System

Objectives

Upon completion of this chapter, the reader should be able to:

31.1 State the structures that make up the immune system
31.2 Describe the functions of the immune system
31.3 Describe common disorders of the immune system

Introduction

The **immune system** is responsible for keeping the body healthy and protecting animals from disease. **Immunity** is the term used to mean *protection*. Immunity begins at birth when an animal begins nursing from its mother and acquiring **antibodies** from the first 24 hours of milk known as **colostrum**. Antibodies are specific proteins produced to protect a newborn against disease and attack cells relating to any diseases, known as **pathogens**. Antibodies protect young animals from possible disease until vaccines start to be administered to help build up the immune system. Antibodies form as chains within the bloodstream.

Antibodies

Antibodies are also referred to as **immunoglobins**. There are several forms of antibodies within the immune system and each type is responsible for a specific activity. Protein molecules are produced from **B lymphocytes** when a foreign substance is recognized. The B lymphocytes are white blood cells that develop within the bone marrow. Antibodies are referred to as Ig, for immunoglobin, and include IgG, IgA, IgD, IgE, and IgM. Each antibody is produced at a specific time and in relationship to an immune and cellular response. Some are produced when certain antigens are involved, such as when an animal has parasites. Antibody production occurs primarily in the **lymph nodes**, as well as in the spleen and bone marrow. These lymph nodes are collections of tissue located throughout the body and allow for a rapid immune response when diseases enter the body (see Figure 31–1). Lymph nodes contain large amounts of lymphocyte cells that filter the lymph fluid that removes foreign materials from the body. Typically, lymph nodes should be small in size and not palpable, but when a pathogen invades and an immune response is initiated, lymph nodes may become enlarged and palpable in areas affected by the disease.

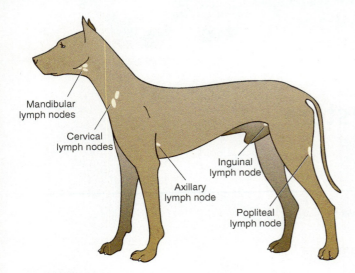

FIGURE 31–1 Location of lymph nodes throughout the body of a dog.

Innate Versus Adaptive Immunity

The innate immune system is able to work quickly to begin an immune response and is the body's first line of defense. The immunity is not specialized for specific pathogens, and it does not need a long start-up phase. Because of this broad effect, there is only a certain degree to which it can stop germs from entering and spreading in the body.

The innate defense consists of several elements:

- The skin and all mucous membranes in the body openings, which form external barriers
- Different defense cells from the white blood cell group (leukocytes)
- Various substances in the blood and in body fluids

The adaptive immune system takes longer to come into play, typically 4–7 days after the innate immune system is introduced to a pathogen. The adaptive immune system is activated if the innate immune system is not successful in protecting the body. It can remember the antigens by producing memory cells, which are capable of responding to a particular antigen or disease pathogen should the body be exposed in the animal's lifetime, thus allowing the body to become immune and protected from developing the disease.

The adaptive defense consists of the following:

- T lymphocytes
- B lymphocytes
- Antibodies as soluble proteins in the blood
- Cytokines in the blood and tissue as hormonelike messenger substances

Vaccines

Vaccinations are placed within the body as **antigens**. Vaccinations are foreign materials used to create and stimulate an immune response. Vaccinations are started in young animals as a series of **booster shots** that slowly build up the immune system and help prevent infection of the body by helping it become resistant to diseases. Vaccine booster shots provide a primary response and a secondary response. The **primary response** to a vaccine is the production of antigens. Between 3 and 14 days are needed to build and stimulate an immune response that begins by protecting against antibodies and antigens. The **secondary response** to a vaccine is due to quick, repeated exposure to the antigen that creates immunity to prevent disease development. Some booster shots are required two to three times to build up an immune response. This depends on the type of vaccine, age, species of animal, and disease type.

 TERMINOLOGY TIP

A *vaccine* is used to build up immunity and resistance to disease within the body. Vaccines are also called *shots*, *vaccinations*, *boosters*, and *immunizations*. ∎

Vaccine types include modified live and killed vaccines. **Modified live vaccines** are made from altered or weakened antigens created from pathogens and place small amounts of a disease into the animal's body (see Figure 31–2). Modified live vaccines stimulate an immune response and may cause the disease, but typically only mild clinical signs are observed. **Killed vaccines** are manufactured from dead pathogens and placed into the animal's body in an inactive form. Dead organisms cannot divide within the body, so repeated doses are needed to produce a protective level of antibodies. A **recombinant vaccine** is made up of a live nonpathogenic virus into which the gene for a pathogen-related antigen has been inserted. Thus, the virus is injected into the animal and the animal will produce antibodies without ever having been exposed to the disease. Immunity thus provides different types of protection. **Active immunity** is developed from exposure to a pathogen through the process of vaccination. **Passive immunity** is developed through antibodies acquired from an animal or other

CHAPTER 31 The Immune System

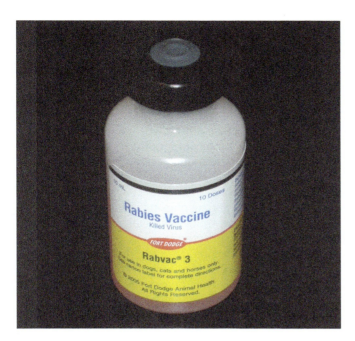

FIGURE 31–2 Rabies is a disease that can be controlled with a vaccine.

source. Passive immunity can be provided through colostrum, commercial diets or formulas, and plasma donated from related animals.

 TERMINOLOGY TIP

Plasma is a part of the blood that is spun down and separated from the serum. The plasma contains natural immunity antibodies that may be given to a newborn in cases in which the young have not received adequate amounts of colostrum.

Vaccine Storage and Use

Proper storage of vaccines is imperative to help reduce vaccine failure. Vaccines must be kept refrigerated and shipped on ice at 35.6–39.2 degrees Fahrenheit. Microorganisms within a live vaccine will be killed by warm temperatures. Potential causes for vaccine failure include the following:

- Expired vaccine
- Improper storage temperature
- Improper administration
- Inaccurate timing of vaccine schedule
- Excessive use of alcohol or disinfectant on the skin prior to vaccination

Vaccination Routes

Vaccines are given by various routes. The most common routes are **subcutaneous (SQ)** or under the skin and **intramuscular (IM)** or into the muscle. Vaccination programs are the most important part of the clinical health of an animal and its immune system. Some vaccines may be labeled for use as **intranasal (IN)** or by applying in the nostril. This is a common route of administration of the bordetella or "kennel cough" vaccine. Another route is **intraocular** or into the eye as in an eye drop. *Lingual* or *oral* routes mean by mouth. **Intradermal (ID)** means into the dermis layer of skin. Figure 31–3 illustrates the various routes of injections.

Infection

The body's immune system controls body temperature (see Table 31–1) and helps prevent **infection**. Inflammation is the immune system's response to an abnormal condition. **Inflammation** is a sign of infection within the body. Inflammation causes white blood

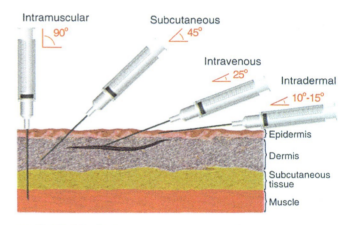

FIGURE 31–3 Examples of routes of injection.

TABLE 31–1
Normal Body Temperatures

Dog	101–102 degrees Fahrenheit
Cat	101–102.5 degrees Fahrenheit
Cow	101–101.5 degrees Fahrenheit
Horse	99–100 degrees Fahrenheit
Pig	102–102.5 degrees Fahrenheit
Goat	102 degrees Fahrenheit
Sheep	103 degrees Fahrenheit
Rabbit	102–104 degrees Fahrenheit

cells to build up at the affected site and may cause pus to form, redness around the area, warm to hot skin temperatures, increased body core temperatures, edema, and pain.

> **TERMINOLOGY TIP**
> **Edema** is a buildup of fluid under the skin (see Figure 31–4).

Spleen

The spleen is an organ located in the central abdomen, filters foreign material from the blood, and stores lymphocytes that are released when an immune response is necessary (see Figure 31–5).

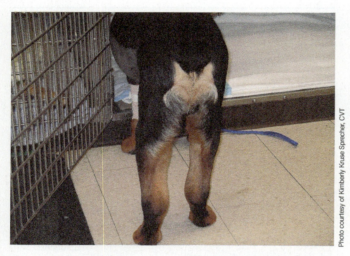

FIGURE 31–4 Edema (swelling) in the hind legs of a dog.

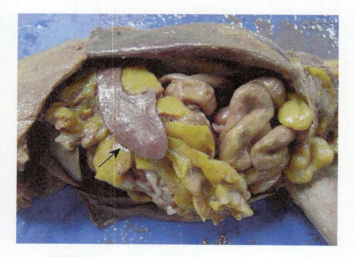

FIGURE 31–5 The spleen is a mass of lymphatic tissue located in the cranial abdomen.

Allergies

Allergies are a common clinical condition seen in animals. Allergies develop from **allergens**, which are substances in the environment that cause an allergic reaction. Some examples of common allergens are ragweed, trees, grass, and flowers. Animals that have allergies have a **hypersensitivity** within their immune system to certain allergens. This sensitivity causes a release of chemicals known as **histamines** when an allergic reaction occurs. This response may cause signs of allergy such as itchiness, eye or nasal discharge, sneezing, rubbing at eyes, or redness of the skin. Many allergic reactions are treated with **antihistamines** or drugs used to prevent and control the allergic reaction. Skin allergies are known as **atopy**. Skin allergies may be reactions to items in the air, the household, the outside environment, or the animal's diet. Atopy can lead to pyoderma, which is a secondary skin infection from scratching. Skin testing may be done to determine the cause of the atopy. Skin testing injections consist of **intradermal** injections into the layers of the skin. The skin becoming swollen and red at the site of the injection indicates a positive reaction to an allergen. Another method of allergy testing is through **blood titer** levels. The titer is a measured amount of antibody within the bloodstream. Blood titer levels can be measured with a simple blood draw to detect diseases and allergies within an animal's body. High blood titer levels mean that an allergy or a pathogen causes the allergic reaction.

ELISA Testing

Some clinical tests that are used in-house to detect pathogens and diseases are enzyme-linked immunosorbent assay **(ELISA)** tests, which are enzyme-linked immunosorbent assays. These simple tests are used to measure antigen or antibody levels in a blood sample. Many ELISA tests come in a test kit that contains all the necessary items to run a blood sample in a facility's lab. Most tests can be completed within 5 to 10 minutes. They are relatively inexpensive and easy to run. Some examples of ELISA test kits are heartworm tests, feline leukemia tests, and feline immunodeficiency virus (FIV) or Feline infectious peritonitis (FIP) tests. Most of these test kits are called **SNAP tests** because they are packaged as a piece of equipment that when the blood sample and conjugate are mixed together, the test is snapped to activate the test (see Figure 31–6). Antigens are often present in the blood, so the sample used for these types of tests is often a blood sample.

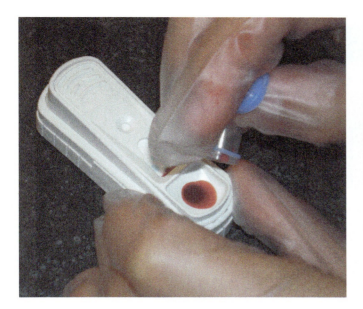

FIGURE 31–6 SNAP test.

Neoplasia (Cancer)

Neoplasia, commonly called cancer, can occur in all animals, and there are many types of cancers that affect every area of an animal's body. The rate of cancer increases with age. Certain animal species and breeds are more prone to certain types of cancer, such as osteosarcoma or bone cancer (see Figure 31–7). In general, animals get cancer at roughly the same rate as humans. Cancer accounts for almost half of all deaths in animal over the age of 10. Some cancer is easily preventable, such as with early neutering. When an animal is suspected of having cancer, such diagnostics as blood work, X-rays, ultrasound, and biopsy (tissue sample) can be helpful in diagnosis and possible early treatment. Common signs of cancer in animals include the following:

- Abnormal swelling that persists or continues to grow
- Sores that do not heal
- Weight loss
- Loss of appetite
- Bleeding or discharge from any body opening
- Offensive odor
- Difficulty eating or swallowing
- Hesitating to exercise or loss of stamina
- Persistent lameness or stiffness

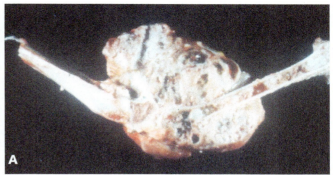

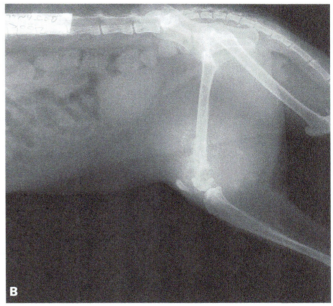

FIGURE 31–7 (A) Osteosarcoma, a malignant bone tumor. (B) Radiograph of an osteosarcoma.

- Difficulty breathing, urinating, or defecating
- Overall poor general appearance
- Depression

Each type of cancer requires specific care, and some cancers may be treatable at the early stages, while others may be difficult to treat at any point. Cancer treatment may include **chemotherapy** (treatment with chemicals), surgical removal, radiation, **cryosurgery** (freezing tissues), medications, or a combination of these (see Figure 31–8). Table 31–2 lists commonly occurring cancers in animals.

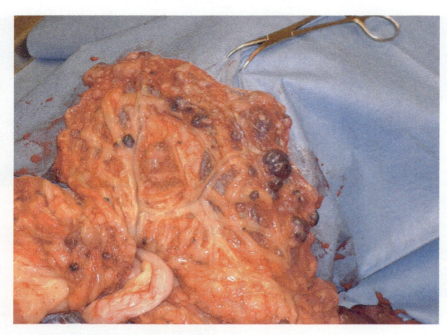

FIGURE 31–8 Intraoperative photograph showing numerous dark red tumors in the omentum of a dog with a splenic tumor. The tumors had also spread to the liver.

TABLE 31–2				
Common Cancers in Animals				
CANCER TYPE	COMMON SPECIES	OCCURRENCE	SIGNS
Bone cancer	Large/Giant breed dogs	Leg bones near the joints	Lameness, swelling, pain
Skin cancer	Dogs, horses	Body and limbs	Change in color, swelling, drainage, pain
Lymphoma	Dogs, cats, horses, cattle	Enlarged lymph nodes throughout the body, spleen, bone marrow	Palpable lymph nodes, anorexia, pain, weight loss, lethargy
Abdominal cancer	Dogs, cats, horses, rodents	Stomach, intestines	Weight loss, abdominal swelling, pain
Mammary cancer	Dogs, cats	Female: uterus, mammary glands	Swelling, palpable hard mammary glands, pain
Head and neck cancer	Dogs, horses	Mouth, nose, throat	Bleeding, odor, oral masses, difficulty eating and drinking, difficulty breathing, facial swelling, pain

SUMMARY

The immune system functions to protect the body from harmful substances. The immune system produces antibodies that respond to a foreign substance invading the body. Immunity can be provided through the use of vaccinations.

Key Terms

active immunity   developed from exposure of a pathogen through the process of vaccination or the disease itself

allergen   substance that causes an allergic reaction

antibody   specific proteins produced to protect against disease and attack cells relating to any diseases

antigen   foreign materials used to create an immune response

antihistamine   drugs used to prevent and control an allergic reaction

atopy   allergy causing a skin infection

blood titer   measured amount of antigen within the bloodstream

B lymphocytes   white blood cells that develop and mature in the bone marrow

booster series   vaccines introduced to the immune system to build up protection and immunity over a period of time

chemotherapy   therapy treatment with chemicals or drugs

colostrum   the mother's first 24 hours of milk after labor; passes antibodies to the newborn

cryosurgery   procedure involving freezing parts of tissue

edema   buildup of fluid under the skin; swelling

ELISA   enzyme-linked immunosorbent assays; simple test used to measure an antigen or antibody level within the blood

histamine   chemical released during an allergic reaction

hypersensitivity   increased reaction to allergens; causes an allergic reaction

immune system   responsible for keeping the body healthy and protecting an animal from disease

immunity   protection that begins at birth during the nursing process

immunoglobulins   another term for antibodies

infection   the immune system's response to an invasion of an organism

inflammation   process that causes white blood cells to build up in an area within the body; and may cause pus to form, redness around the area, warm to hot skin temperatures, increased body core temperatures, edema, and pain

intradermal (ID)   into the layers of skin

intramuscular (IM)   into the muscle

intranasal (IN)   into the nasal cavity or nostrils

intraocular   into the eyes

killed vaccine   vaccine manufactured from dead pathogens and placed into the animal's body in an inactive form

lymph node   small collection of tissue that produces lymph fluid

modified live vaccine   vaccine made from altered antigens created from pathogens that places small amounts of the pathogen into the animal's body

neoplasia   veterinary term for cancer

passive immunity   developed through antibodies acquired from another animal or source, as in colostrum, formula, or plasma

pathogen a disease-causing agent

primary response process that occurs when an antigen comes in contact with the immune system for the first time

recombinant vaccine made up of a live nonpathogenic virus into which the gene for a pathogen-related antigen has been inserted

secondary response process that occurs when an antigen comes in contact with the immune system for the second time and provides a repeated exposure to an antigen that creates immunity to prevent disease development

SNAP test test kit equipment that when mixed with the blood is snapped to activate the test

subcutaneous (SQ) under the skin

vaccination antigen placed into the body to build up resistance in the immune system to disease

Review Questions

1. Explain the difference between antibodies and pathogens.
2. Explain the difference between a live vaccine and a killed vaccine.
3. Describe the primary and secondary responses to a vaccine.
4. Explain the difference between inflammation and infection.
5. Describe allergies and how they can be treated.
6. Is it normal for lymph nodes to be easily palpated?
7. What is another term for antibodies?
8. Why is blood often used in ELISA or SNAP tests?
9. What are two ways to acquire passive immunity?
10. Where are B lymphocytes produced?

Clinical Situation

Dr. Huang, a large animal veterinarian, is on a farm call to examine a young sheep that has recently been banded for castration. The banding was done by the owner about a month ago. The sheep has not been doing well for several days. When Dr. Huang examines the sheep, he notes the sheep's stance is very wide, and it is stiff and reluctant to move. Dr. Huang performs a PE and notes when he attempts to perform an oral exam that the sheep is unable to open its mouth. The area of the testes has developed an open wound due to poor circulation from the band. Bacteria had most likely entered the wound and caused the clinical signs that Dr. Huang was seeing. Treatment of the sheep at this stage would not be successful.

- What is most likely wrong with the sheep?
- How could this condition have been prevented?
- What other signs may Dr. Huang see in this sheep as the condition progresses?
- What options might Dr. Huang give the owner?

CHAPTER 32: The Nervous System

Objectives

Upon completion of this chapter, the reader should be able to:

32.1 State the structures that make up the nervous system
32.2 Describe the major structures of the brain
32.3 Compare and contrast the central nervous system (CNS) and the peripheral nervous system (PNS)
32.4 Describe the functions of the nervous system
32.5 Describe common disorders of the nervous system

Introduction

The **nervous system** is the body's communication control network. It coordinates and controls body activity. It does this by detecting and processing internal and external information and responding in an appropriate manner.

The nervous system is made up of **neurons**, or specialized cells within the nervous system, that control impulses. The body has three types of neurons: sensory, motor, and interneurons. The **sensory neurons** deliver signals to the central nervous system (CNS). **Motor neurons** deliver signals from the central nervous system to the muscles to give a response (see Figure 32–1). The **interneurons** deliver signals from one neuron to another from the brain to the spinal cord.

Neurons

The basic structural unit of the nervous system is the neuron. Neurons respond to stimuli and send impulses from one part of a cell to another. Two fiberlike processes extend from the neuron, the **axon** and **dendrite**. There is usually just one axon, and it conducts impulses away from the cell body. There is often more than one dendrite, and they conduct impulses received from neurons located toward the cell body. The junction of an axon with another nerve cell is called the **synapse**. The synapse is a junction between two nerve cells, consisting of a small gap across which impulses pass, sending a message that creates a response from the stimuli (see Figure 32–2). The messages that are sent across the synapse are called **neurotransmitters**. These molecules move across the synapse to relay information to other nerve cells.

Neurons do not reproduce, they require a lot of oxygen, and they have limited healing capabilities when damaged. Neurons suffer permanent damage when deprived of blood for more than a few minutes, which is why CPR must begin within minutes of cardiac arrest.

Structures of the Nervous System

The major structures of the nervous system are the brain, spinal cord, peripheral nerves, and sensory organs. The nervous system has two major divisions: the **central nervous system** and the **peripheral nervous system**.

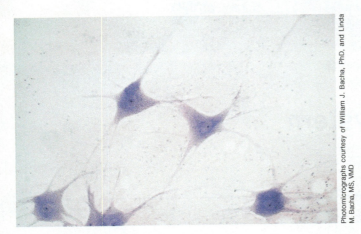

FIGURE 32-1 Multiple motor neurons from the spinal cord of a cow.

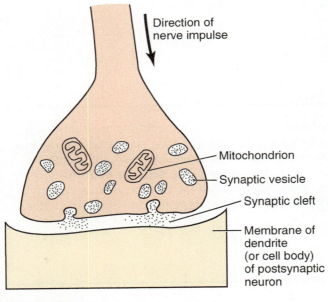

FIGURE 32-2 Synapse.

Central Nervous System

The central nervous system or CNS consists of the brain and spinal cord. The major organ in the nervous system is the **brain**. The brain is an organ of soft nervous tissue contained in the skull of vertebrates, functioning as the coordinating center of sensation and intellectual and nervous activity. The brain is divided into three regions or sections: the cerebrum, cerebellum, and brainstem. The **cerebrum** controls the voluntary movements of the body and thought processes and is the largest region of the brain. The **cerebellum** provides control over coordination and movement. The **brainstem** controls functions that maintain life. Within the brainstem is a small area called the **medulla oblongata**, which controls the body's functions, such as the heart rate, respiratory rate, and blood pressure. Damage to this region causes instant death. The **midbrain** area controls the senses, which include sight, smell, hearing, taste, and sensation. Damage to the midbrain causes a **coma** condition, where there is a loss of consciousness and lack of awareness. Other organs that are housed within the brain include the **thalamus**, which is located at the top of the brainstem; the **hypothalamus**, which is located in front of the thalamus and controls hormone production; and the **pituitary gland**, which is located between the thalamus and hypothalamus and controls hormone release and body functions. The brainstem attaches to the **spinal cord**. Figure 32–3 shows the structures of the brain.

The spinal cord begins at the base of the brain and continues to the sixth or seventh lumbar vertebrae. The spinal cord is held in place by the vertebrae and is a hollow tube made of strands and fibers that run through the middle of the vertebrae. Nerves, called **tracts**, are a major passage in the body that contain large bundle of nerve fibers and run through the spinal column, allowing for body sensations. These are sensory nerves that allow for pain detection. The spinal cord nerves also allow reflexes and coordination functions. The spinal cord serves as a pathway for impulses to go to and from the brain.

 TERMINOLOGY TIP

The **lumbar vertebrae** lie over the lower back and are designated L when noting spinal cord and vertebral conditions. The other vertebrae are cervical (C), thoracic (T), sacral (S), and coccygeal (Co) (see Figure 32–4).

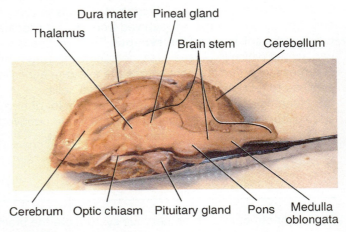

FIGURE 32-3 Structures of the brain.

CHAPTER 32 The Nervous System 545

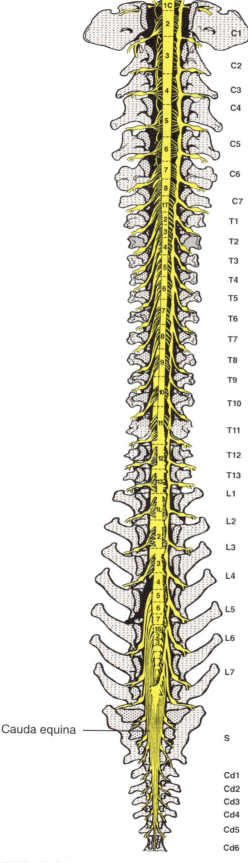

FIGURE 32–4 Spinal cord.

The Peripheral Nervous System

The peripheral nervous system or PNS consists of the cranial and spinal nerves, as well as the autonomic nervous system.

There are 12 pairs of cranial nerves. They originate from under the surface of the brain. The cranial nerves are named for the function they serve and are represented by Roman numerals (see Figure 32–5 and Table 32–1).

The spinal nerves arise from the spinal cord and are paired. Each segment of vertebrae has two nerve branches: the dorsal root and the ventral root. The **dorsal root** contains sensory nerves, and the **ventral root** contains motor nerves.

The autonomic nervous system is part of the peripheral nervous system and functions to innervate smooth muscle, cardiac muscle, and glands. It is further divided into the parasympathetic system and the sympathetic system. The **sympathetic system** is responsible for the "fight-or-flight" reaction, increased vital signs, and certain drug reactions. The **parasympathetic system** is responsible for opposite reactions, such as decreased vital signs, regulating the body system back to normal. The main nerve in these systems is the **vagus nerve**.

 TERMINOLOGY TIP

Fight or flight is a natural instinct of animals in which they take care of themselves by either running away or protecting themselves by fighting. **Vital signs** are the body signs such as heart rate (HR), respiratory rate (RR), and temperature (T). **Peristalsis** is the wavelike action that allows food to digest in the stomach. ■

Coordinated Function of the CNS and PNS

The CNS controls the brain and spinal cord by receiving signals from the PNS. The PNS controls the nerves, detects a stimulus, sends signals to the CNS, and causes a response or action to occur. The body is made up of electrical units of measurements called **millivolts**. These electrical millivolts create a force within a cell in specific amounts. Millivolts help in creating **reflexes**, which are units of function that produce a movement. A signal is sent to the brain to create a reflex, and the movement occurs. Reflexes may be voluntary or involuntary responses. **Voluntary reflexes** occur when an animal asks its body to perform a function, such as a dog jumping up to greet its owner or a

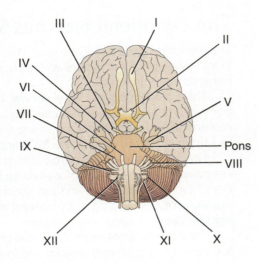

FIGURE 32-5 Cranial nerves.

I. Olfactory
II. Optic nerve
III. Oculomotor nerve
IV. Trochlear nerve
V. Trigeminal nerve
VI. Abducent nerve
VII. Facial nerve
VIII. Vestibulocochlear nerve
IX. Glossopharyngeal nerve
X. Vagus nerve
XI. Accessory nerve
XII. Hypoglossal nerve

TABLE 32-1
Cranial Nerves

CRANIAL NERVE	NAME	FUNCTION
I	Olfactory	Conducts sensory impulses from the nose to the brain (smell)
II	Optic	Conducts sensory impulses from the eyes to the brain (vision)
III	Oculomotor	Sends motor impulses to the external eye muscles (dorsal, medial, and ventral rectus; ventral oblique; and levator superioris) and to some internal eye muscles
IV	Trochlear	Sends motor impulses to one external eye muscle (dorsal oblique)
V	Trigeminal	Three branches: ophthalmic = sensory to cornea; maxillary = motor to upper jaw; mandibular = motor to lower jaw
VI	Abducent	Motor innervation to two muscles of the eye (retractor bulbi and lateral rectus)
VII	Facial	Motor to facial muscles, salivary glands, and lacrimal glands and taste sensation to anterior two-thirds of tongue
VIII	Acoustic or vestibulocochlear	Two branches: cochlear = sense of hearing; vestibular = sense of balance
IX	Glossopharyngeal	Motor to the parotid glands and pharyngeal muscles, taste sensation to caudal third of tongue, and sensory to the pharyngeal mucosa
X	Vagus	Sensory to part of the pharynx and larynx and parts of thoracic and abdominal viscera; motor for swallowing and voice production
XI	Accessory	Accessory motor to shoulder muscles
XII	Hypoglossal	Motor to the muscles that control tongue movement

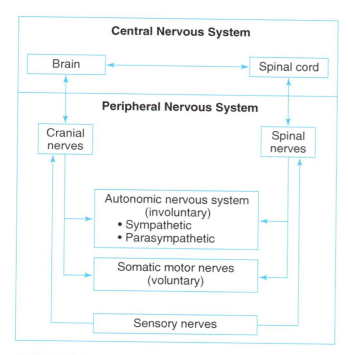

FIGURE 32–6 Divisions of the nervous system.

horse kicking at another horse. **Involuntary reflexes** occur without thinking and serve as many of the body system functions, such as the heart beating and breathing. Figure 32–6 summarizes the divisions of the nervous system and their coordinated function.

Receptors

Receptors are found within the nerves located throughout the body and allow for detecting changes within the body and the environment. The receptors cause blood vessels to dilate or constrict depending on the body's temperature. Skin receptors help control the body temperature of an animal; pain receptors detect pain; pressure receptors determine feeling and sensations; and mechanical receptors determine **proprioception** or movement. **Chemoreceptors** are chemical receptors that allow animals to smell, taste, and detect sounds. The taste buds in the mouth have four types of receptors: bitter, sour, salty, and sweet.

 TERMINOLOGY TIP
The term *proprioception* relates to the body's ability to recognize body position for posture or movement. ■

Nerve Testing

The **cranial nerve test** is done to determine how well the animal's reflexes and nervous system are working. This exam helps to determine if a brain-related injury has occurred as a result of some trauma.

The **spinal nerve reflexes** should also be evaluated for proprioceptive reflexes of the muscles and tendons. The animal should stand to be evaluated. One test to determine proper body balance is turning one foot to the opposite surface to see if the animal can place it in the correct position. Normal animals will right the foot immediately. Nerve damage causes a prolonged response to appropriately position the foot. The **knee jerk reflex** is also used to assess spinal nerves. This involves the use of the **reflex hammer**, a tool used to tap the area of the knees and joints to elicit a response. The patella ligament is tapped lightly to determine any reflex movements (see Figure 32–7). Normal reflexes should have the motor nerves extend the knee joint. A lack of response is a sign of nerve damage to the lower lumbar region of the vertebrae.

Common Diseases and Conditions

Diseases of the nervous system can originate from bacterial or viral invasion, other disease processes, or traumatic injury.

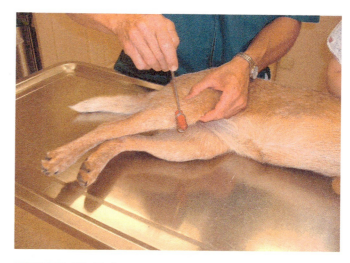

FIGURE 32–7 A veterinarian tests the patellar reflex on a dog. By striking the patellar tendon, receptors sense the sudden lengthening and quickly contract the muscle in response.

Intervertebral Disc Disease

Many many specific animal breeds are predisposed or inclined to a condition that causes pressure to be placed on an area of the vertebral column, causing compression on the discs. This is called **intervertebral disc disease** or IVDD. Enough pressure can cause compression of the spinal cord (see Figure 32–8). This condition causes severe back pain and in extreme cases partial or complete **paralysis**. In dogs, the long-backed breeds are more susceptible to injury. Some of these breeds include the Basset Hound, Dachshund, and Corgi.

TERMINOLOGY TIP

The term *paralysis* means a loss of motor skills. This may be partial loss in an area of the body or complete loss of motor skills throughout the body. ■

Epilepsy

Epilepsy is the medical term for neurological disorders often characterized by seizures. A **seizure** is a loss of voluntary control of the body that causes a sudden attack or fit (see Figure 32–9). Seizures may occur due to disease or heredity, or they may be a toxin-related response. In many animals, abnormal behavior is noted prior to seizure activity and may also be noted after the seizure. Signs of seizures include the following:

- Paddling motion of legs
- Opening and closing of the jaw (chewing gum seizures)
- Muscle twitches
- Excessive drooling/salivation
- Uncontrolled urination
- Uncontrolled defecation
- Vocalization—crying, barking, howling
- Loss of consciousness
- Stiff appearance to body
- Ocular dilation

The severity of signs will depend on the length and severity of the seizure activity. Care should be taken for any animal suffering a seizure. Animals may tend to bite, as they have no control over their nervous system. They should be monitored around stairs or on furniture, as seizure activity can cause trauma and further injury. Each episode of seizure activity should be timed; 30 minutes of seizure activity becomes an emergency situation. It is important for a veterinarian to examine any animal that has had a seizure to determine the cause of the episode. Some common causes of seizures are:

- Epilepsy
- Hereditary factors
- Trauma or nervous system injury
- Tumor—brain or nervous-system related
- Toxins or chemicals
- Infectious diseases affecting the nervous system

Hereditary epilepsy is the most common cause of seizure activity. It is often difficult to determine a specific cause, as the seizures are caused by genetic factors passed from one or both parents. Genetic seizure conditions usually begin to occur between 6 months and 5 years of age. Several breeds of dogs are known for genetic epilepsy: Cocker Spaniel, Beagle, Dalma-

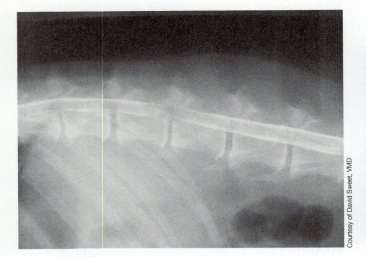

FIGURE 32–8 Myelogram showing intervertebral disc disease. Dye injected into the spinal column shows the compression of the spinal cord.

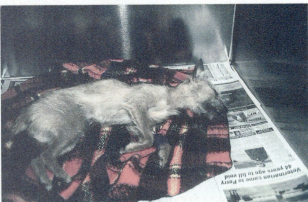

FIGURE 32–9 Seizing dog. During a seizure, dogs typically are unresponsive, lying on their sides, paddling their legs, and chomping their jaws.

tian, Labrador Retriever, and Springer Spaniel. Epilepsy begins with infrequent occurrence of seizure episodes that increase with age and time. Medication treatments may be used to control frequent and severe episodes. Some examples of drugs used to control and prevent seizures include phenobarbital and potassium bromide (KBr). These medicines must be monitored over time; doses may need to be adjusted, and liver function must be assessed because this is the main organ in which these drugs are metabolized. Toxicity and seizures can occur when animals ingest chemicals such as organophosphates, which are ingredients in insecticides and weed killers. These chemicals are toxic to the nervous system and often produce a seizure activity response, as they block normal activity between the CNS and PNS.

Rabies

Rabies is a deadly virus that can affect any mammal, including humans. Rabies is usually transmitted by the saliva of an infected animal and then moves into the bloodstream. Most cases of rabies occur in wild animals, with bats, skunks, foxes, and raccoons being the most common. Rabies vaccinations should be given to all dogs, cats, ferrets, and livestock species to help control and prevent the rabies virus. Vaccines should be kept up to date, and state and local rabies laws should be evaluated, as many rabies laws exist for certain animal species and differ from location to location. Limiting animals—especially companion animals—to the indoors will help decrease the potential for exposure. Also, attracting wild animals to residential or business properties is discouraged; not feeding wild animals and not leaving garbage where it can be accessed will decrease the risk of wild animals coming in contact with pets and livestock. Signs of rabies in animals vary. Animals may become aggressive, but many wildlife species become calm and friendly, venturing out during the daytime and approaching people. Other signs include excessive salivation, foaming at the mouth, lack of coordination, stumbling, circling, and other changes in behavior. Any pet or livestock suspected of being bitten by another animal should be seen immediately by a veterinarian. The rabies virus is fatal, and testing is done through use of brain tissues, so any animal suspected of having rabies must be euthanized and the brain submitted to a state laboratory testing facility.

West Nile Virus

West Nile virus is of concern for birds and horses. West Nile is transmitted through mosquito bites and causes inflammation or swelling of the brain and spinal cord. The life cycle starts when an infected bird is bitten by a mosquito, and the mosquito carries the virus to a human, bird, or horse and transmits it through another blood meal. Signs of infection may only take 5–15 days. A West Nile vaccine is available for prevention. Signs of West Nile virus include the following:

- Stumbling or tripping
- Muscle weakness or twitching
- Partial paralysis
- Loss of appetite
- Depression or lethargy
- Head pressing or tilt
- Impaired vision
- Wandering or circling
- Inability to swallow
- Inability to stand up
- Fever
- Convulsions
- Coma
- Death

SUMMARY

The nervous system is the body's control center. It regulates and coordinates body functions as well as movement. There are two divisions of the nervous system: the central nervous system and the peripheral nervous system. Disorders of the nervous system can arise from disease processes, viral or bacterial invasion, or traumatic injury.

 MAKING THE CONNECTION

Chapter 8: rabies
Chapter 9: rabies
Chapter 19: pseudorabies
Chapter 22: Marek's disease

Key Terms

axon single process of a neuron that conducts impulses away from the cell body

brain the major organ in the nervous system that controls the body's actions

brainstem part of the brain that controls functions that maintain life

central nervous system (CNS) controls the brain and spinal cord by receiving signals from the PNS

cerebellum part of the brain that controls coordination and movement

cerebrum part of the brain that controls the voluntary movements of the body and thought processes; is the largest region of the brain

chemoreceptor chemical receptor that allows an animal to taste, smell, and detect sounds

coma damage to the midbrain that causes a loss of consciousness and awareness

cranial nerve test reflex test of the eye where a hand is moved quickly toward the eye without touching the eye or any hair around the eye

dendrites multiple processes that extend from a neuron and conduct impulses received from other neurons located toward the nerve cell body

dorsal root nerve branch of the vertebrae that contain sensory nerves

epilepsy veterinary term for neurological disorders characterized by seizures

fight or flight natural instinct of animals faced with a threat in which they take care of themselves by either running away or protecting themselves by fighting

hypothalamus organ within the brain located in front of the thalamus; controls hormone production

interneurons cells that deliver signals from one neuron to another from the brain to the spinal cord

intervertebral disc disease (IVDD) injury to the spine that causes pressure on the discs and/or spinal cord and may cause partial or complete paralysis

involuntary reflex occurs without the need for thinking and serves as a necessary body function, such as the heart beating

knee jerk reflex reflex used to assess the spinal nerves

lumbar vertebrae lie over the lower back and are designated L when noting spinal cord and vertebral conditions

medulla oblongata part of the brain that controls the body's involuntary functions

midbrain part of the brain that controls the senses

millivolt electrical units of measurement within the body that create a force within cells

motor neuron cells that deliver signals from the central nervous system (CNS) to the muscles to give a response

nervous system the body's control network; the network of nerve cells and fibers that transmits nerve impulses between parts of the body

neurons specialized cells within the nervous system that transmit nerve impulses

neurotransmitters molecules that travel along the synapse relaying information like a message transport system

paralysis partial or complete loss of motor skills

parasympathetic system spinal cord system responsible for reactions of the sympathetic nervous system, such as decreased vital signs and regulating the body system back to normal

peripheral nervous system (PNS) controls the nerves, detects a stimulus, sends signals and informs the CNS, and causes a response or action to occur

peristalsis wavelike motions of the stomach that allow digestion of food

pituitary gland organ located between the thalamus and hypothalamus; controls hormone releases in the body

proprioception relates to the body's ability to recognize body position for posture or movement

rabies a deadly virus that can affect any mammal, including humans; usually transmitted by the saliva of an infected animal and then is transferred into the bloodstream

receptor found within the nerves located throughout the body; allows for detecting changes within the body and the environment

reflex units of function that produce a reaction to a stimulus

reflex hammer tool used to tap the area of the knees and other joints to elicit a response

seizure loss of voluntary control of the body with an amount of unconsciousness causing uncontrolled violent body activity

sensory neuron cell that delivers signals from the central nervous system (CNS)

spinal cord strands of fibers that connect the brainstem to the PNS to send and receive signals through the body; begins at the base of the brain and continues to the lumbar vertebrae

spinal nerve reflexes evaluated for proprioceptive reflexes of the muscles and tendons

sympathetic system division of the autonomic nervous system that is responsible for the fight-or-flight reaction, increased vital signs, and certain drug reactions

synapse junction of an axon and another nerve cell

thalamus organ within the brain; located at the top of the brainstem

tract a major passage in the body that contains large bundles of nerve fibers and runs through the body, providing sensations and feelings

vagus nerve main nerve within the nervous system

ventral root nerve branch of the vertebrae that contains motor nerves

vital signs body signs such as heart rate (HR), respiratory rate (RR), and temperature (T)

voluntary reflex occurs when an animal asks its body to perform a desired function, such as walking or running

West Nile virus a mosquito-borne disease that causes inflammation or swelling of the brain and spinal cord

Review Questions

1. Which division of the autonomic nervous system is concerned with body functions under emergency or stress?
2. What type of nerve carries impulses away from the CNS and toward muscles?
3. What is the largest portion of the brain involved with thought and memory?
4. What part of the brain is associated with coordinated muscle movement?
5. What type of reflexes are involved in the heart and breathing?
6. Describe the structures of the central nervous system and explain their functions.
7. Describe the structures of the peripheral nervous system and explain their functions.

8. Differentiate between the sympathetic and parasympathetic divisions of the autonomic nervous system.
9. Describe how an impulse travels to cause a response.
10. Differentiate between the cranial nerves and the spinal nerves.

Clinical Situation

"Sally," a 3 yr SF Cocker Spaniel, was brought into the emergency clinic due to multiple seizure episodes. Alice, the vet assistant, obtained information from the owner and noted in the medical record that "Sally's" seizures began the night before. As Alice was taking the vital signs of the dog, "Sally" began to have another seizure. She immediately called for the veterinarian, Dr. Howe. "Sally" was lying on her side with her legs thrashing wildly. Her jaws were opening and closing quickly. The seizure episode lasted around 60 seconds.

The owner was very upset and concerned. He said that "Sally" had had 4 episodes overnight, each lasting from 1 to 2 minutes. Dr. Howe performed a PE and drew some blood work, which showed normal results. Dr. Howe concluded that "Sally" had epilepsy.

- What causes epilepsy?
- What signs may occur to alert the owner that seizure activity may be occurring?
- What signs or symptoms may occur after a seizure?
- What treatments may be offered for "Sally"?
- What precautions might Dr. Howe discuss with "Sally's" owner about epilepsy in dogs?

CHAPTER 33: The Sensory System

Objectives

Upon completion of this chapter, the reader should be able to:

33.1 State the structures that make up the sensory system
33.2 Describe the structures of the eye
33.3 Describe the functions of the eye
33.4 Explain the purpose of reflex testing the eye
33.5 Describe the structures of the ear
33.6 Explain the function of the sense of hearing
33.7 Explain equilibrium and how it is controlled by the ear
33.8 Explain the function of the sense of smell
33.9 Explain the function of the sense of taste
33.10 Describe common disorders of the sensory system

Introduction

The sensory system consists of those structures and organs associated with touch, vision, hearing, smell, and taste. The function of the senses is to receive stimuli from sensory receptors and transmit those impulses to the brain for interpretation. This chapter will cover the senses of sight, sound, taste, and smell.

The Eye

Vision receptors within the eye allow the lens to focus on light-sensitive cells and allow for day and night vision in animals. The entire eye—including surfaces of the sclera and cornea—is the **globe**, often referred to as the eyeball, and is formed by a connective tissue layer that gives shape to the eye. The eye is made of several parts that work together to allow vision to occur. There are three layers of the eye: retina, lens, and sclera (see Figure 33–1). The **retina** is the inner layer of the eye and contains two types of cells that aid in seeing color and determining depth. The **rods** are the long, narrow sensitive cells that allow the eye to detect light and depth and make up 95 percent of the cells within the eye. The **cones** are cells that allow an animal to see certain colors. Both rods and cones work together to allow each animal species to react to light, color, and detail. The **posterior chamber** of the eye, which holds the nerves, blood vessels, and optic vessels, is located between the iris and the lens. The posterior chamber contains the **vitreous humor**, a transparent gel-like substance that helps regulate pressure within the eye.

The middle layer of the eye is called the **iris**. The iris gives the eye its color and holds the pupil in the center, which dilates and constricts in darkness and light to regulate the amount of light entering the lens. Light causes the pupil to **constrict** (close), and darkness causes the pupil to **dilate** (open). In front of the iris is the **anterior chamber**, which holds the **aqueous humor**, a watery fluid. Behind the iris lies the **lens**. The lens is responsible for focus as well as near and far vision accommodation. The lens is connected to the **ciliary body** by tiny suspensory

554 SECTION III General Anatomy and Disease Processes

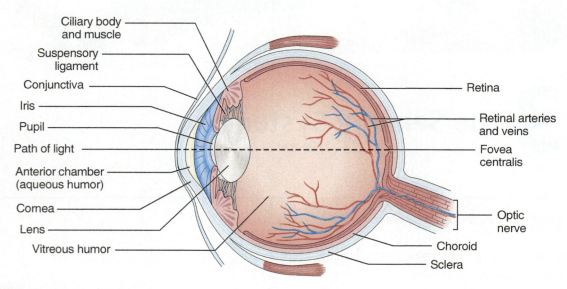

FIGURE 33–1 Structures of the eyeball.

ligaments. The ciliary body contains the muscles responsible for changing the shape of the lens to allow light to focus and differentiate near and far vision. The muscles react when the animal is looking at something nearby, by contracting and allowing the lens to assume a natural round shape and focus on the close-up image. When focusing on distant objects, the muscles relax and pull the lens flat to allow distant focus.

The outer white layer of the eye is the **sclera**. The sclera is covered by a clear layer called the **cornea**. This is a common site for scratches and trauma in the eye (see Figure 33–2). The cornea is a transparent layer through which light rays enter the eye. A dye can be placed into the eye to determine whether a corneal ulcer is present (see Figure 33–3).

The eye is a sensitive organ that is protected by accessory structures called the **conjunctiva**, **lacrimal glands**, and **eyelids**. The conjunctiva is a thin membrane that lines the underside of the eyelids and covers the outer area of the eyeball. It is transparent and allows the sclera and blood vessels to show through. This is an area that can become inflamed, a condition called **conjunctivitis**. The color of this membrane can help in the assessment of jaundice and anemia. The lacrimal glands, or tear ducts, are located in the **lateral apparatus** of the eye. The angle where the upper and lower eyelids meet is called the **canthus**. The **lateral canthus** is the outer corner of the eye farther away from the nose and helps produce tears to lubricate the eye surface and aid in the blink reflex. Tears flow down the surface of the eye and drain in the inner corner of the eye known as the **medial canthus**. The eyelids are folds of skins located on the upper (dorsal) lid and lower (ventral) lid of the eye surface that cover and protect the eye when an animal blinks and sleeps. The third

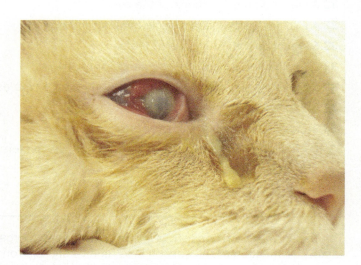

FIGURE 33–2 Corneal ulceration of an eye.

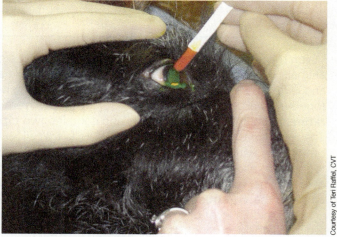

FIGURE 33–3 Dye stain used to detect ulcers or scratches in the eye.

CHAPTER 33 The Sensory System

FIGURE 33–4 Third eyelid, or nictitans, in a cat.

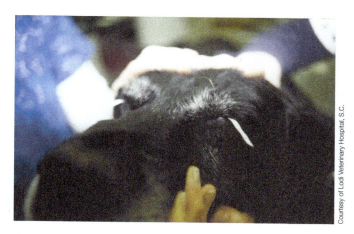

FIGURE 33–6 The Schirmer tear test is used to measure tear production in a dog.

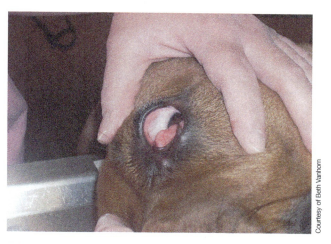

FIGURE 33–5 Prolapsed nictitating membrane, often called cherry eye.

FIGURE 33–7 Eye exam in a horse.

eyelid lies below the lower lid surface and is a cartilage membrane called the **nictitating membrane** (see Figure 33–4). It becomes abnormally visible when a pet is not feeling well or when the membrane prolapses in a condition often called cherry eye (see Figure 33–5).

Tear production may be abnormal in animals, causing a condition known as KCS, or **keratoconjunctivitis sicca**. This condition causes decreased to no tear production in one or both eyes. Animals with this condition develop inflammation of the conjunctiva that leads to thickened eye discharge, painful corneal ulceration, and corneal scarring. This condition is also called dry eye and occurs in dogs, cats, and horses. A **Schirmer tear test** is used to measure tear production (see Figure 33–6). The test includes a paper test strip for each eye.

Reflex Testing

The **menace response test** is a reflex test of the eye where a hand is moved quickly toward the eye without touching the eye or any hair around the eye. A normal response to this reflex is to blink or close the eye. An abnormal response would show no change in the eye and not elicit a blink. The **pupillary light response test** uses a light source shown into the animal's eyes to note constriction or shrinking of the **pupil** and dilation or expanding when the light source is removed from the eye (see Figure 33–7). Both pupils should react in the same manner. Normal reflexes show that the optic nerves are working correctly. The eyes should also be examined for any signs of **nystagmus**, which is a condition where the eyes jump back and forth in rhythmic jerks due to damage in the inner ear, brainstem, or cranial nerves. This movement

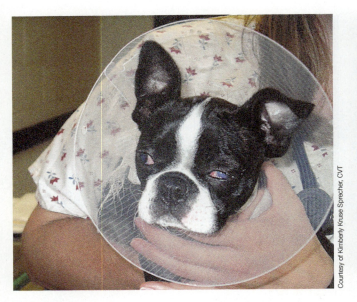

FIGURE 33-8 Glaucoma in a dog.

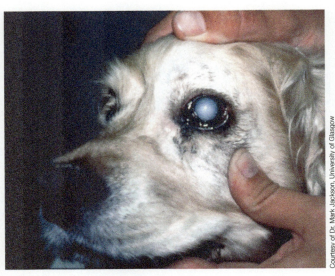

FIGURE 33-10 Cataract in a dog.

FIGURE 33-9 Tonometer to measure eye pressure.

causes the animal to have symptoms similar to motion sickness, such as vomiting and lack of coordination. Normal cranial nerves will show no signs of uncontrolled ocular movement.

Animals can develop increased pressure behind the eye called **glaucoma** (see Figure 33-8). Eye pressure measurements are taken using a **tonometer**, which is a piece of equipment used to measure the pressure within the eye (see Figure 33-9). The tonometer is applied to the eye with a light tap, and this measures corneal resistance to the applied force.

A **cataract**, or cloudiness or opacity of the lens, is common in older animals but can occur at any point in an animal's life (see Figure 33-10). A slight change in the lens with age is called **nuclear sclerosis**, which causes the lens to dry out.

Ear

The **ear** is an organ that enables hearing and helps to maintain balance. The ear has outer, inner, and middle portions. The outer ear consists of the pinna and external auditory canal. The **pinna** (ear flap) is the external portion of the ear that catches sound waves and transmits them to the external auditory canal. The **external auditory canal** is the tube that transmits sound to the tympanic membrane. The middle ear consists of the tympanic membrane, the auditory ossicles, Eustachian tube, oval window, round window, and tympanic bulla. The **tympanic membrane** (eardrum) is thin connective tissue that separates the outer and inner ear structures by stretching across the opening to the middle ear. Sound waves cause this membrane to vibrate. The **auditory ossicles** are three bones of the inner ear (malleus, stapes, and incus) that transmit sound vibrations. The **malleus** attaches to the medial surface of the tympanic membrane. It forms a tiny joint with the second bone, the **incus**, which in turn forms a third joint with the **stapes**, which forms a contact point with the cochlea in the inner ear. The **Eustachian tube** is a narrow duct that leads from the ossicles to the nasopharynx and serves to maintain air pressure in the middle ear. The **oval window** is a membrane that separates the middle and inner portions of the ear. The **round window** is a membrane that receives sound waves, and the **tympanic bulla** is an osseous chamber at the base of the skull. The inner ear consists of the bony labyrinth, which is divided into three parts: the vestibule, semicircular canals, and cochlea. The **vestibule** contains receptors for balance and position. The **semicircular canals** contain sensory cells that detect changes in position. The **cochlea** is a spiral-shaped passage that vibrates and relays vibrations that allow sound to be heard. The cochlea is a fluid-filled space that is similar in appearance to a snail shell. Most of the auditory structures are located within the temporal bone of the skull. The structures of the ear are illustrated in Figure 33–11.

An **otoscope** is used to examine the internal ear canal and can help determine if an inner ear infection is present (see Figure 33–12). Ear infections commonly cause an animal to shake its head and scratch at the affected ear. They often also cause an odor, inflammation, redness, and discharge. Many animals have pain along with these symptoms. A swab is often taken of the ear discharge, and a cytology exam is done under the microscope to determine the cause of the ear infection, known as **otitis**. Otitis is a clinical sign of otic inflammation, which often causes a head tilt. Bacteria, yeast, and mites are common sources of ear problems (see Figure 33–13).

Hearing

Sound waves enter the ear through the pinna, travel through the auditory canal, and strike the tympanic membrane. The tympanic membrane vibrates and moves the ossicles. The ossicles conduct the sound

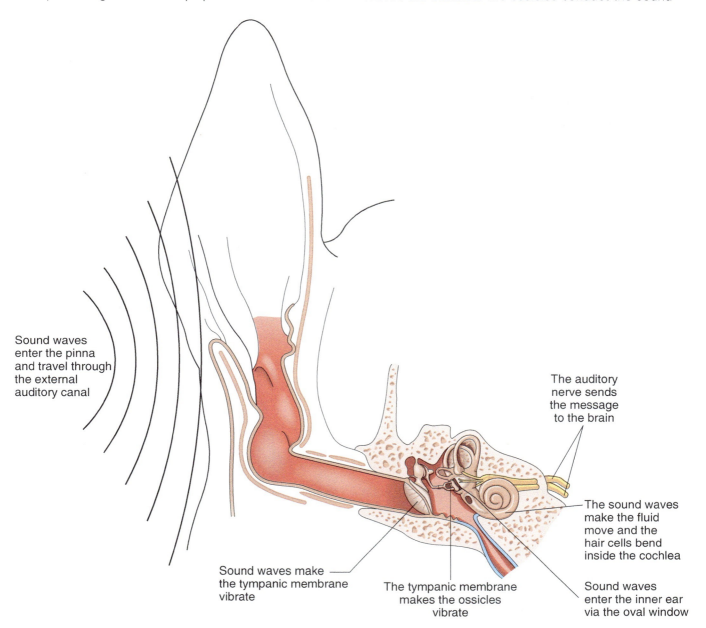

FIGURE 33–11 Structures of the ear.

FIGURE 33-12 Otoscope used to visualize the inner ear canal.

the back of the nasal cavity. Sound vibrations reach the inner ear through the round window and displace fluid within the structures of the inner ear, specifically the cochlea, which is responsible for hearing. Receptor cells of the cochlea receive transmissions of sound wave vibrations and initiate a nerve impulse that is relayed to the brain.

Equilibrium

The ear canal aids in **equilibrium**, which is a state of balance. **Static equilibrium** is controlled by the organs of the inner ear, specifically the vestibule. The structures of the inner ear bend with the movement of the head in response to gravity. **Dynamic equilibrium** is controlled by the semicircular canals. These structures maintain balance in response to rotational or angular movement of the head.

An **aural hematoma** is caused by a blood vessel bursting within the ear flap or pinna (see Figure 33-14). This creates a large buildup of blood in the ear flap, and the ear flap itself appears swollen. In some mild cases, the ear flap may be drained, but in most circumstances the ear flap must be opened with an incision and then flushed, drained, and surgically repaired.

Smell

The **olfactory sense** is a chemical sense that forms the perception of smell. Olfaction has many purposes, such as the detection of hazards, pheromones, and food. Olfaction occurs when odorants bind to specific sites on

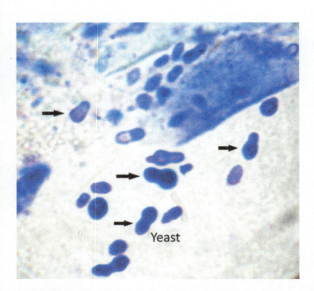

FIGURE 33-13 Ear cytology of a dog with otitis. Yeast is found in the ear of this dog.

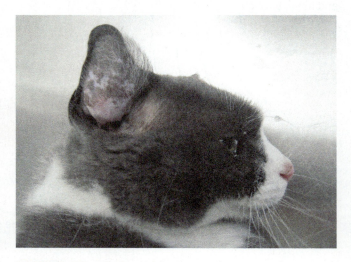

FIGURE 33-14 Aural hematoma in a cat.

waves through the middle ear. Air pressure within the middle ear must create equilibrium with the atmospheric pressure of the external air to prevent bulging of the membrane. This is accomplished by the Eustachian tube, a canal that connects the middle ear to the nasopharynx, which consists of the upper throat and olfactory receptors located in the nasal cavity. The receptor cells are located within the nasal passages, which are covered in hair overlying the mucous membrane within the nasal epithelium. The mixture of mucus and a chemical substance provides information about odors that is transmitted to the brain via the olfactory nerve.

Taste

The sense of taste is also a chemical sense. Chemical substances mix within the mouth and dissolve within the saliva. The receptor cells are located in tiny taste buds located on the tongue. Each taste bud has an opening much like a pore. Hairlike projections over the tongue hold chemical substances that then dissolve and enter the pores, creating an interaction that generates an impulse to the brain that leads to the detection of different tastes.

SUMMARY

The eyes and the ears are organs of the sensory system. The eyes allow sight and the ability to see in dark and light situations. The ears allow hearing and help to maintain equilibrium. The nose is part of the olfactory system. The sense of taste is also a sense in animals.

Key Terms

anterior chamber area of the eye in front of the iris that holds the aqueous humor

aqueous humor watery fluid located in the anterior chamber

auditory ossicle bones of the middle ear that transmit sound vibrations

aural hematoma swelling of the ear flap caused by a broken blood vessel in the pinna

canthus angle where the upper and lower eyelids meet

cataract cloudiness or opacity of the lens

ciliary body area consisting of muscles that change the shape of the lens to allow near- and far-sighted vision

cochlea spiral-shaped passage in inner ear that receives sound vibrations and initiates an impulse to the brain for translation

cones cells in the eye that allow animals to see colors

conjunctiva thin membrane that lines the underside of the eyelids and the outer surface of the eyeball

conjunctivitis inflammation of the conjunctiva eye membrane

constrict to narrow (as in the pupil in the eye)

cornea clear outer layer of the eye; is commonly easily damaged or scratched

dilate to open (as in the pupil in the eye)

dynamic equilibrium a state of balance in response to rotational or angular movement

ear sensory organ that enables hearing

equilibrium state of balance

Eustachian tube narrow duct that leads from middle ear to nasopharynx and maintains air pressure

external auditory canal tube that transmits sound from the pinna to the tympanic membrane

glaucoma   increase in pressure behind the eye

globe   the eyeball; formed by connective tissues that give shape to the eye

incus   tiny ossicle bone that forms second joint within inner ear

iris   middle layer of the eye; gives the eye its color and holds the pupil

keratoconjunctivitis sicca   condition that causes decreased to no tear production in one or both eyes

lacrimal glands   tear ducts that produce tears on the surface of the eye

lateral apparatus   structure that produces tears and holds tear glands

lateral canthus   outer corner of the eye; farther from the nose

lens   transparent elastic structure located behind the iris; allows for focus

malleus   small ossicle bone; attaches to the medial surface of the tympanic membrane

medial canthus   inner corner of the eye; closer to the nose

menace response test   a reflex test of the eye where a hand is moved quickly toward the eye without touching the eye or any hair around the eye

nictitating membrane   third eyelid cartilage membrane located under the lower surface of the eye

nuclear sclerosis   drying of the lens with age

nystagmus   condition where the eyes jump back and forth in rhythmic jerks due to damage in the inner ear, brainstem, or cranial nerves

olfactory sense   the chemical sense that controls smell

otitis   inflammation of the ear canal

otoscope   equipment used to visualize the inner ear

oval window   membrane that separates the middle and inner portions of the ear

pinna   external portion of the ear

posterior chamber   area behind the eye that holds the nerves and is filled with vitreous humor; helps regulate pressure within the eye

pupil   area in the center of the eye that dilates and constricts with light responses

pupillary light response test   test in which light is shined into the animal's eyes and examiner notes constriction (shrinking) of the pupil and dilation (expanding) when the light source is removed

retina   inner layer of the eye; contains two types of cells that aid in seeing color and determining depth

rods   cells that allow the eye to detect light and depth

round window   membrane that receives sound waves through fluid passed through the cochlea

Schirmer tear test   diagnostic test that uses a paper strip to measure tear production in the eye

sclera   outer white layer of the eye

semicircular canals   sensory cells that generate nerves impulses to regulate position

stapes   tiny third ossicle bone that forms contact with the cochlea in the inner ear

static equilibrium   a state of rest in the body

tonometer   equipment used to measure intraocular pressure

tympanic bulla   osseous chamber at the base of the skull

tympanic membrane   eardrum; tissue that separates the outer and middle portions of the ear

vestibule   portion of the inner ear that contains specialized receptors for balance and position

vitreous humor   transparent gel-like fluid located behind the lens of the eye

Review Questions

1. What is the term for "a state of balance"?
2. What part of the ear separates the outer or external ear from the middle ear?
3. What is the colored muscular layer of the eye that surrounds the pupil?
4. What is the region of the eye where nerve endings of the retina gather to form the optic nerve called?
5. What is the name of the spiral-shaped passage that leads from the oval window to the inner ear?
6. Explain the pupillary light response.
7. Explain the process of hearing.
8. Explain how the inner ear aids in maintenance of equilibrium.
9. What is the common term for a prolapsed nictitating membrane?
10. What inner ear organ is responsible for hearing?

Clinical Situation

"Sophie," a 10 yr SF Persian, is being examined by Dr. Lister. Mrs. George, the cat's owner, has noticed changes in "Sophie's" eyes recently. Dr. Lister notes that both eyes have developed a thin whitish appearance of both lenses. Mrs. George tells the vet that the cat has been having difficulty jumping up on furniture and often walks into items.

Dr. Lister performs a PE on "Sophie" and notes in the progress notes the bilateral opaque appearance to the lenses. The pupil responses are decreased.

Dr. Lister discusses some treatment options with Mrs. George and states that he recommends a CBC and blood chemistry panel due to the patient's age and that often cataracts in cats are caused by another underlying condition. Mrs. George agrees to the blood work and allows the vet tech, Asha, to take "Sophie" to the prep area for blood collection.

- What causes or factors may begin cataracts in cats?
- What treatment options may be discussed with Mrs. George?
- What is most likely the cause of "Sophie's" cataracts?

CHAPTER 34: Animal Nutrition

Objectives

Upon completion of this chapter, the reader should be able to:

34.1 Discuss the major nutrient needs of animals
34.2 Describe the purpose of quality animal nutrition
34.3 Contrast feed requirements with digestive systems types and structures
34.4 Describe the ways that animals use nutrition and nutrients
34.5 Describe the types of animal feed
34.6 Explain how animals are fed
34.7 Discuss the importance of rations
34.8 Explain the resting energy requirements (RER) and daily energy requirements (DER)

Introduction

Animal nutrition is a part of veterinary medicine that provides knowledge and scientific evidence about food. It addresses both the necessary quality and specific quantity for different animals. Many scientists, veterinary professionals, and other researchers devote large amounts of time to improving animal nutrition and health care. Every species of wild or domesticated animal has its own specific nutritional requirements. Producers want animals to grow big and strong as quickly as possible, pet owners want animals to be healthy and happy, and veterinary professionals want to provide proper information regarding every aspect of animal health, especially one as important as nutrition. Veterinary assistants can provide detailed information to clients regarding appropriate nutrients and diets. They may also help owners select favorable diets.

Nutritional Needs of Animals

All animals require food to grow, live, reproduce, and work. **Animal nutrition** is the science of determining how animals use food in the body and all body processes that transform food into body tissues and energy for activity. Each animal species has requirements based on the animal's environment, work and activity level, age and stage of development, genetic makeup, and health. A **nutrient** is any single class of food or group of like foods that aids in the support of life and makes it possible for animals to grow or provides energy for physiological processes in life.

Animals require a specific amount of food daily. This amount is known as a **ration** and is the total amount of food an animal needs within a 24-hour time frame. A ration may be provided once daily or in divided amounts throughout the day. A **balanced ration** is a diet that contains all the nutrients required by an animal in correct specific amounts. Nutrients may be broken down, digested, absorbed, and used by the body. Some nutrients may be given in decreased or increased amounts that may be harmful to the animal's health or growth. Therefore, it is important to have an understanding of the correct rations for various species of animals.

Nutrients

Nutrients are the components of food that are meant to sustain life. There are six essential nutrients that animals require: water, carbohydrates, fats, proteins, vitamins, and minerals. Nutrients are divided into categories based on their chemical makeup and the purpose they serve in the body.

Water

Water makes up more than 75 percent of an animal's body. The body weight of newborn animals may be up to 90 percent water. Water serves several functions within the body, including controlling body temperature, maintaining body shape, transporting nutrients within the body's cells, aiding in digestion of food, breaking down food particles, and serving as a carrier for waste products. Water makes up the major part of all body fluids, such as urine, blood, feces, sweat, and lung vapors. Water is the most critical nutrient in an animal's diet. An animal can live longer without food than without water. Animals require 3 pounds of water for every 1 pound of food they consume. The amount of water an animal needs beyond what is contained in its food is determined by the amount of activity, gestation or lactation occurrences, and the environment. Water loss through the process of **dehydration** can cause serious problems, especially when 10 percent or more fluid is lost by the body. At 15 percent–20 percent of water loss, an animal will die. The following factors contribute to water loss in an animal:

- Diarrhea
- Vomiting
- Urination
- Panting
- Sweating
- Defecation
- Lactation
- Gestation

Water allows nutrients to be transported through the body and also helps to break down cells into smaller parts through a chemical process called **hydrolysis**. This means that water is added to a molecule in a process that breaks nutrients down into smaller particles, allowing for them to be transported throughout the body. The water requirement for animals is expressed in ml/day and is equivalent to the animal's energy requirements expressed in kcal/day.

Carbohydrates

Carbohydrates are nutrients that provide energy for body functions and allow for body structure formation. Carbohydrates are the largest part of an animal's food supply, at about 75 percent of the diet. Carbohydrates include starches, sugars, and fiber materials. They are made of combinations of carbon, hydrogen, and oxygen. These nutrients are not stored in the body and are required from a food source on a daily basis. Carbohydrates can be converted to fats. Some of the functions of carbohydrates include maintaining the body's blood sugar levels, creating lactose in milk, storing fat, and aiding metabolism. **Starches** are plant or grain materials that provide fiber and bulk in an animal's diet. Starches convert to glucose (sugar) during digestion. Examples of starches are cereal grains, silage, oats, and corn (see Figure 34–1).

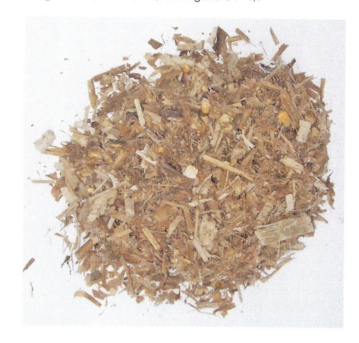

FIGURE 34–1 Silage is an example of carbohydrates that can be provided in the diet of livestock.

Sugars are another type of carbohydrate that forms the simplest example of a nutrient in an animal's diet. Sugars include such food items as fruits and milk. Sugars are further classified as **simple sugars** (or **monosaccharides**) and **double sugars** (or **disaccharides**). A simple sugar is glucose and a double sugar is sucrose or table sugar. A long chain of simple sugars forms a **polysaccharide**. These carbohydrates form the building blocks of other major nutrients and are easily digested within the stomach and intestines. **Fiber** is material from plant cells that is left after other nutrients are digested. Fiber helps produce beneficial bacteria and aids in digestive health. Examples of fiber include hay and grass (see Figure 34–2). Fiber helps slow down the process of digestion and helps protect the lining of the stomach and intestinal tract.

Fats

Fats or **lipids** furnish animals with a concentrated source of energy. Fats are found in every cell within the body and contain the highest amount of energy of all nutrients. Fats function within the body by providing insulation from the environment, protection for vital organs, energy reserves, and flavor in food. Fats form cholesterol, steroids, and other hormones found naturally within the body. When fat is absent from the diet, it can lead to hair and skin problems. In increased amounts, it can lead to obesity and other health problems. Many diets contain **fatty acids**, which are oils that are products of fat sources that may be used as nutrients or supplements within a diet. Fatty acids are helpful in skin and hair coat hydration. Fats are measured by **calories**, the unit of measurement that defines the energy in food. Animal feed is measured in **kilocalories**, the amount of energy needed to raise 1 gram of water by 1 degree. The abbreviation is written as a kcal. Calories and kilocalories are measured with a calorimeter, which is an instrument that measures the amount of heat released by a food product. This measurement determines how long it takes the product to be metabolized, which in turn predicts the amount of energy in the food particle. This is the information that is placed on food labels for as part of nutrition information. Examples of fatty acids include fish oil, linseed oil, fish meal, and vegetable oil. Fats form solid particles, and oils or acids form liquid particles. One benefit for animal feeds is that fat increases the **palatability** of food, or how good it tastes, which directly affects whether it will be eaten by an animal. Fatty acid sources may be **essential** or **nonessential**. The main difference between the two is that essential fatty acids cannot be produced by the body and have to be consumed through food or dietary supplements. Nonessential fatty acids can be produced by the body, although they can still also be ingested from some of the food that we eat. Fats are much more energy dense than carbohydrates, with 2.25 times more calories per gram.

Proteins

Proteins are nutrients that are essential for growth and tissue repair. Protein is used for forming and developing muscles, internal organs, skin, hair, wool, feathers, hoofs, and horns. Protein is the basis of the structure and function of cells. It is also vital to growth and development of young animals and to reproduction and breeding. Protein-based diets aid animals that require weight gain for health and market situations. **Amino acids** form in chainlike structures and are the building blocks of proteins. Twenty-three amino acids are available from food sources, and 10 are essential to the health of animals. Essential amino acids are used to make other amino acids required by the body for life. Essential amino acids must be provided in the diet because they can't be metabolized by the body. Examples of essential amino acids include arginine, leucine, lysine, methionine, tryptophan, taurine, and glycine. Nonessential amino acids do not need to be provided in the diet, as they can be manufactured by the body by other amino acids. Proteins are made of amino acids that work together to help food digest in the body via the animal's digestive system. Protein sources may come from soybean meal, skim milk, fish meal, or alfalfa hay (see Figure 34–3). Proteins are listed as a percentage on food labels based on the **biological value**, which describes the quality of the source. The higher the biological value of the protein, the more easily digested the food will be. Many animal food sources provide protein through such ingredients as eggs, milk, fish, meat, and corn.

FIGURE 34–2 Pasture provides fiber for some species of animals such as horses.

FIGURE 34–3 Proteins are added to foods of animals.

Protein deficiency tends to be one of the most common nutritional problems in animals. Animals that do not receive adequate amounts of protein will have both developmental and health problems that may affect their lives. Young animals require more protein than do adult animals. Signs of protein deficiency include the following:

- Poor or slow growth
- Anorexia or decreased appetite
- Anemia
- Poor appearance and coat
- Low birth weight
- Low milk production
- Dental problems
- Edema
- Brittle bones

Minerals

Minerals are needed in every area of the body but are found mostly in the bones and teeth. Minerals are used by the body based on the animal's needs and mineral availability. Minerals do not contain carbon and are therefore inorganic. Calcium is a mineral that makes up about 50 percent of the body's mineral sources and is found in many organic body fluids, such as blood and tissues. Phosphorus also plays an important role in metabolism and the composition of teeth and bones. The body's functions rely on the appropriate amounts of mineral to maintain regular rates and duties, such as the heart rate and respiratory rate. There are two types of minerals: macrominerals and microminerals. **Macrominerals** are needed in large amounts. Examples are calcium, phosphorus, and magnesium. **Microminerals** are needed in small amounts. Examples are iron, potassium, and zinc. These minerals are often called **trace minerals**.

Vitamins

Vitamins are organic nutrients needed in small amounts to maintain health and life. Vitamins help provide a defense against disease (vitamin E), promote growth (vitamin D) and reproduction (vitamin B12), and contribute to the general health of the animal via regulation of body functions. Vitamins react with enzymes to allow body functions to occur. Some vitamins act as **antioxidants** within the body's system to boost the immune system. There are two types of vitamins: **fat soluble** and **water soluble**. Fat-soluble vitamins are stored in fat and released when needed within the body. Vitamins A, D, E, and K are fat soluble. Vitamin A is the most common vitamin *not* available in commercial animal foods. Excessive fat-soluble vitamins may be provided, which can result in a toxic reaction that poisons the system and that can cause damage to the kidneys and heart. Water-soluble vitamins are not stored within the body, are dissolved by water, and therefore need to be provided in daily doses. Vitamins B and C are water soluble. Vitamin C is available in many types of forage. Vitamins are usually supplied to the body through the use of **supplements**, which are additives provided via the diet in solid or liquid form when needed by the animal.

Animal Nutrition and Concentrations

The **concentration** of food is dependent on how it is delivered to the animal. Food concentration is based on **dry matter**, or the amount of nutrients without the water. The concentration is determined by the amount of food fed divided by the percentage of dry matter. Dry matter is critical in how much the animal needs to be fed. The concentration helps determine the amount of food to feed and helps ease digestion. **Digestion** is the breakdown of food from large molecules to smaller molecules for use by the body.

Palatability

Palatability refers to the degree to which an animal likes the taste of food. A balanced diet must be palatable to ensure an animal will eat it. Several factors determine how a food is eaten, including temperature, odor, and texture in the animal's mouth. Fat, salt, and water content also affect taste. The palatability of food can be increased through pouring water over it or increasing the temperature.

Types of Diets

Animal foods come in a variety of types (see Figure 34–4). Different species as well as the animal's age, activity level, and health status all must be considered when selecting a diet. Factors to consider when planning an animal's diet include the following:

- Age
- Environment
- Species
- Size
- Health condition
- Breed
- Medical history

Growth Diets

Young animals may be placed on a **growth diet** (see Figure 34–5). Growth diets are specifically formulated to increase body weight—including the size of

FIGURE 34–4 There are many pet foods on the market to choose from.

FIGURE 34–5 Puppies should be fed a diet that will encourage growth.

muscles, bones, and organs—of young offspring. Each animal species has specific requirements. Livestock and meat production animals are raised in programs that provide quick body mass increases in a short amount of time to achieve market weight and profitability. Growth diets include large components of such nutrients as protein, vitamins, and minerals. These types of diets are often called life stages diets and are not best for all pets.

Maintenance Diets

Maintenance diets are given to adult animals that are in their prime age and health condition. These animals may be working or competing, and the aim of a maintenance goal is to maintain that animal at an ideal weight and health condition. Animals on these types of diets require energy to work and be active, to maintain the body's temperature, and for good health. Maintenance diets have high percentages of fats and carbohydrates and small amounts of proteins, vitamins, and minerals.

Reproduction Diets

Reproduction diets are given to breeding animals, which have additional nutrient needs. Specialized reproduction diets are used to address the energy needs of female animals with developing embryos and those that are beginning the lactation phase. The first **trimester** of a pregnancy in animals is the most critical time for nutrition. This is the first three months of pregnancy when the embryos require more protein, vitamins, and minerals for proper development. Male breeding animals require a reproduction diet for quality sperm production. Breeding animals that don't get adequate nutrition show signs of poor overall health, underweight offspring, and possible abortion.

 TERMINOLOGY TIP
The term **abortion** refers to when a pregnant female animal loses a fetus.

Lactation Diets

Lactation diets are provided to females that have completed the gestation phase and are currently producing milk for young to nurse. These diets require large amounts of water, protein, vitamins, and minerals. The most important minerals are calcium and phosphorus. These nutrients improve the milk's quality and quantity.

Work Diets

A **work diet** is typically provided to livestock that use a large amount of energy for some type of work or strenuous activity. All activity requires nutrients to provide the energy to fuel work. Examples of work animals include plow animals, draft or wagon animals, racehorses, hunting animals, and show or competition animals. Work diets include increased carbohydrates, fats, vitamins, and minerals.

Reduced Calorie Diets

Reduced calorie diets are often given to animals that are overweight or less active due to a health issue. These diets are formulated specifically for an animal's low energy needs, with decreased amounts of carbohydrates, fats, and proteins. They have moderate vitamins and minerals to provide for normal body functions.

Senior Diets

Another type of diet that is specific to age and health is the **senior diet**. This type of food is formulated for geriatric or senior animals over a specific age for their species that may require certain increased nutrients and some decreased nutrients. Senior diets are usually low in carbohydrates and fats, moderate in protein for healthy bones and skeletal mass, and higher in vitamins and minerals to protect the body and immune system as it ages. Most reduced calorie and senior diets are species specific and formulated for each animal type based on their digestive systems.

Feeding Animals

Veterinary medicine requires all veterinary professionals to have some basic knowledge of animal nutrition. This is helpful in determining quality diets and proper amounts to feed the animals under their care. **Feed** is what animals eat to obtain nutrients. **Foodstuff** refers to the ingredients in animal food that determine the nutrient content, such as cornmeal, barley, or oats. Animal feed is divided into classes based on the most prevalent type of nutrient: forage, pasture, concentrates, and supplements.

Forage

Forage is also called roughage, which are high-fiber plant-based sources, such as grass, stems, and leaves. Most forage is low in protein, with the exception of alfalfa hay. Hay and grass forage sources tend to be relatively inexpensive and are usually plentiful, though they are dependent on the weather. Silage is a type of forage commonly used in feed and is a fermented form of a plant. Legumes are a plant that can convert nitrogen into protein by use of nodules within their roots. They are high in protein, up to 18 percent in some plants. Common legumes used in animal feed include alfalfa, soybeans, and clover.

Pasture

Pasture is grown as either **temporary** or **permanent** food sources. Temporary pastures are planted each season by reseeding the grass source, such as corn, hay, or millet seed. Permanent pastures are grass sources that regrow each year; they do not need to be replanted or seeded and include clover and bluegrass (see Figure 34-6). Pastures include land covered with grass. Grasses are relatively low in protein; timothy, orchard grass, and Bermuda grass are common animal feed sources.

FIGURE 34-6 Permanent pastures don't require replanting year after year.

Concentrates

Concentrates are provided to an animal as an additional nutrient source when the primary food source is not adequate or abundant. Concentrates are high in protein and energy but low in fiber and tend to be provided to animals when other food sources, such as hay, are not of the best quality. These foods include many of the bagged commercial animal feeds from Purina, Hills, and Agway. Energy feeds are products with less than 20 percent protein (dry basis) and less than 18 percent crude fiber (dry basis), for example, fish, grain, and mill by-products. Most grains are energy diets and include corn, oats, and barley. Protein feeds are high in protein (typically more than 20 percent) and may be from animal or plant sources. Soybean and cottonseed oil meal are common.

Supplements

Supplements are the last class of nutrients that will be discussed in this chapter. They are provided when necessary as a diet additive to address specific health or conditional requirements. Supplements may be a vitamin, mineral, or mineral block such as salt. Supplements are based on individual needs and add to a balanced diet based on species. It is important for the veterinary assistant and veterinarian to provide client education and recommendations on types of acceptable diet and feeding amounts. Supplements should be added to an animal's diet only based on veterinary recommendation and health requirements.

Ideal Weight

When feeding an animal its required food source to provide adequate nutrients, it is important that the foodstuff be researched and of high quality. When feeding for ideal nutritional value, it is best to feed the animal according to the manufacturer's recommendations and the animal's ideal weight. The **ideal weight** is the breed standard based on the animal's age, species, breed, purpose or use, and health condition.

Example

An animal that is 80 lb and considered overweight for its age and breed should be assessed by the veterinarian. The veterinarian recommends the animal lose 10 lb, so the animal's ideal weight is 70 lb. The owner would feed the animal based on the 70-lb ideal weight range according to the food label and manufacturer's recommendations.

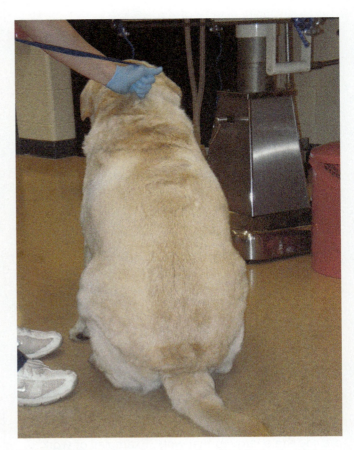

FIGURE 34–7 Example of a dog that is overweight.

Body Condition Scoring

Weight is assessed by an ideal body appearance called the **body condition scoring**. This rates how an animal appears based on an animal's ideal weight. An animal above that ideal weight may appear to be overweight or obese. This animal will have excess amounts of fat and will appear wide over the body, including at the hips (see Figure 34–7). On a scale from 1 to 9, 5 is average and considered to represent ideal weight. The numbers 6 to 9 are overweight and the numbers below 5 are underweight and thin (see Figure 34–8). Animals that are thin lack sufficient body fat and substance (see Figure 34–9).

Weight Management

Obesity is the most common disease seen in pets and the most common form of malnutrition. Between 25 percent and 30 percent of pets are overweight to obese. In pets older than age 5, this increases to 40 percent. Pets that are 10 percent–20 percent above their optimal weight are considered overweight, and

CHAPTER 34 Animal Nutrition 569

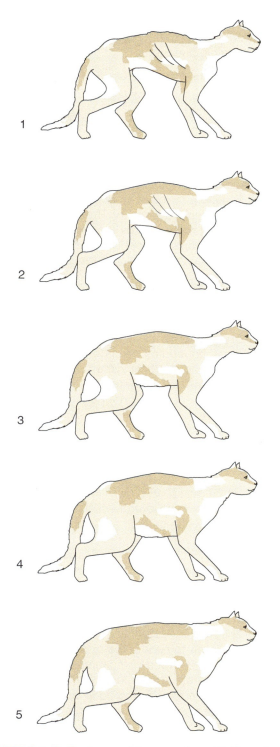

FIGURE 34–8 Body condition scoring. (1) Emaciated. (2) Thin. (3) Ideal (4) Overweight. (5) Obese.

those that are 20 percent above their ideal weight are considered obese. Obesity increases the risk for other diseases.

Pets that are overweight are at increased risk for other diseases, such as arthritis, heart disease, diabetes, and respiratory issues. In turn, life span decreases.

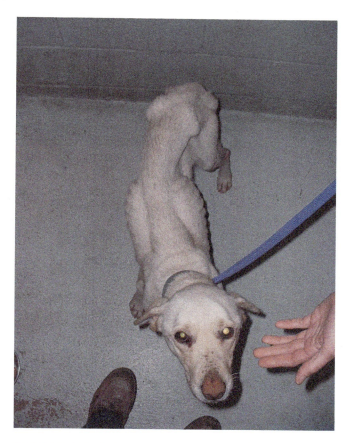

FIGURE 34–9 Example of a dog who is emaciated due to neglect.

Pets become overweight when their energy balance is too high—they are taking in too many calories and not expending enough in exercise. Spaying and neutering, genetics, age, physical activity, and type of food all factor into imbalanced energy intake. Pets that are overweight should begin a gradual and realistic weight loss program. A journal describing how the pet is fed is recommended and should include the following:

- Type of food
- Amount of food
- Calorie content of food
- When pet is fed
- How the pet is fed (free choice versus individual feedings)
- Treats (type and amount)
- Calorie content of treats
- Exercise schedule and frequency

It is important to emphasize to clients the importance of setting realistic goals before implementing their pet's weight loss program. Also, feeding small amounts more frequently will help keep a pet that is on a weight loss program from feeling hungry. Owners should be encouraged to exercise their pets gradually, increasing intensity and duration only as tolerated by the pet.

Food Analysis

Animal feed companies complete a process called **food analysis**. This process requires specific information acquired through testing that allows the company to document the food is safe for the species of animal the food is intended for. It determines the nutrients in each food item and prepared mixes to ensure it serves to provide a balanced ration. The food analysis is legally required to be placed on the animal food label. Items that are displayed include dry matter, crude protein, ash, fat, crude fiber, nitrogen-free extracts, and various other nutrients.

Information about the nutritional content of animal food is prepared by the National Research Council (NRC) and addresses all foodstuff ingredients, including an international feed number or IFN. This information is provided to allow for the formulation of proper rations (i.e., how much to feed according to the size of the animal). The NRC of the National Academies of Sciences, Engineering and Medicine and the Association of American Feed Control Officials are the two most influential bodies when it comes to feeding domestic cats and dogs. Since the 1940s, the NRC has released reports on the nutrient requirements of cats and dogs, based on available literature and research. The following information is required on all feed labels (see Figure 34–10):

- Product name
- Nutrient list
- Bar code
- Manufacturer name and address

Many pet food labels now provide caloric information to help owners better determine their pets' ration needs. Resting energy rates (RER) and daily energy requirements (DER) are important to understand to achieve proper feeding amounts. The caloric content of pet foods is typically expressed in kilocalories or kcal, which is 1000 calories. A simple method for healthy dogs and cats starts with calculating the RER. The RER is the energy required for a healthy animal at rest in its environment. It includes energy expended during eating and recovery from physical activity. You can locate how many calories your dog or cat's food contains (usually per cup)—along with the other required nutrition information—on the container of food.

Example

If your dog food contains 500 calories per cup and your dog requires 1000 calories per day, he will need 2 cups per day. If you feed your dog twice per day, he would get 1 cup per feeding to sustain his weight at rest. To determine how many calories per day (RER) a dog requires to maintain body weight, follow these simple directions:

1. Weigh the dog on a scale.
2. Convert to kilograms (kg) by dividing the weight in lb by 2.2.
3. Multiply the weight in kg by 30, then add 70.

DER refers to the energy requirements of a moderately active animal in its environment. It includes energy needed to obtain, digest, and absorb food in amounts sufficient to maintain body weight, as well as energy for spontaneous activity. The formulas to calculate MER (metabolic energy requirements) consider age and neuter status. An estimated MER calculation is $1.6 \times$ RER. The veterinarian determines MER based on several factors.

An example of a basic calorie calculator table is shown in Table 34–1.

It is recommended that veterinary assistants and all veterinary support staff utilize veterinary nutritional manufacturers' continuing education options and prescription diet educational systems, which typically include certificates of completion. One such example is the Hills Veterinary Nutritional Advisor program (http://vna.hillsvet.com).

AAFCO

Pet foods should be tested by the AAFCO (Association of American Feed Control Officials). This is especially important when selecting a quality prescription diet food versus over-the-counter pet food. AFFCO's responsibility is to provide uniform regulations and policies for labeling pet food and listing ingredients. AFFCO provides standards on diets' feeding trials to determine if they can be approved for their intended use, that is, for the pet's species and life stage. This resource is useful for veterinary professionals and pet owners who want to review the diet ingredients and nutritional value. Pet food companies are regulated, but many diets are not ideal for every animal. Companies can make the food sound good when it is not, in fact, healthy for a particular pet. Clients may not fully understand ingredients such as by-products, bone meal, and mixed tocopherols. By-products are defined by AFFCO as the fresh, clean parts—other than meat—derived from slaughtered animals. Examples of by-products include spleen, kidneys, liver, lungs, blood, bone, undeveloped eggs (in the instance of poultry by-product meal), and intestines. They do not include indigestible materials such as hair, hooves, teeth, horns, and feathers.

100% Complete and Balanced

Holistic Select® Lamb Meal & Rice Formula provides complete and balanced nutrition for maintenance and is comparable in nutritional adequacy to a product which has been substantiated using AAFCO feeding tests.

ingredients: lamb meal, ground brown rice, oatmeal, chicken fat (preserved with natural mixed tocopherols and citric acid), dried beet pulp, flaxseed, brewers dried yeast, rice bran, dried egg product, salt, potassium chloride, air dried peas, ground carrot cubes, inulin, glucosamine hydrochloride, dl-methionine, vitamin a acetate, vitamin d3 supplement, vitamin e supplement, riboflavin supplement, vitamin b12 supplement, d-pantothenic acid, niacin supplement, choline chloride, pyridoxine hydrochloride, thiamine mononitrate, folic acid, ascorbic acid, biolin, rosemary extract, inositol, dehydrated kelp, polysaccharide complexes of zinc, iron, manganese, copper and cobalt, potassium iodate, sodium selenite, *Yucca Schidigera Extract, Lactobacillus acidophilus, Lactobacillus casei, Enterococcus faecium, B. Subtillus, Bacillus lichenformis, Bacillus coagulins, Aspergillus oryzae,* and *Aspergillus niger.*

ingredients: Farine d' agneau, riz brun moulu, flacon d' avoine, gras de poulet (conservateurs: mélange de tocophérols et acide citrique), pulpe de betterave déshydraté, lin, levure de biére déshydraté son de riz, oeuf entier déshydraté, sel, chlorure de potassium, pois sèche a l'air, cube de carottes broyées, inuline, hydrochlorure de glucosamine, DL méthionine, acélate de vitamine A, supplement de vitamine D3, supplément de vitamine E, supplément de riboflavine, supplément de vitamin B12, d-pantothenic acide, supplément de niacine, chlorure de choline, chlohydrate de pyridoxine, mononitrate de thiamine, acide folique, acide ascorbique, biotine, extrait de romarin, inositol, varech déshydraté, polysaccharide complexe de zinc, fer, manganèse, cuivre et cobalt, potassium iodate, sodium sélénite, extrait de yucca schidigera, lactobacilles acidophiles, lactobacilles casai, faecium d'entérocoque, subtillus-b, bacilles de lichenformis, bacilles de coagulins, d'oryzae et Niger d'aspergille.

Guaranteed Analysis / Analyse Garantie:

Crude protein/protéines brutes	Min.	22%
Crude fat/matières grasses brutes	Min.	15%
Crude fiber/fibres brutes	Mix.	4.0%
Moisture/humidité	Mix.	10%
Calcium/calcium	Min.	1.3%
Phosphorus/phosphore	Min.	0.85%
Vitamin A/vitamine A	Min.	22,000 IU/kg.
Vitamin E/vitamine E	Min.	125 IU/kg.
Omega 6 fatty acids*/Oméga 6*	Min.	1.6%
Omega 3 fatty acids*/Oméga 3*	Min.	0.17%
Glucosamine hydrochloride*	Min.	400 ppm
*Total Lactic Acid Producing Live Microorganisms (Lactobacillus acidophilus, Lactobacillus casei, Enterococcus faecium)		240 million CFU/LB.
*Total Bacillus Organisms (Bacillus Subtillus, Bacillus lichenformis, Bacillus coagulins)		7 million CFU/LB.
*Protease (from Aspergillus oryzae and Aspergillus niger)[1]		280 HUT/LB.
*Cellulase (from Aspergillus oryzae and Aspergillus niger)[2]		100 Cellulase Units/LB.
*Amylase (from Aspergillus oryzae and Aspergillus niger)[3]		5 Dextrin Units/LB.

*Not recognized as an essential nutrient by the AAFCO Dog Food Nutrient Profiles.
*Ne figure pas à la liste des éléments nutritifs essentiels reconnus par l'AAFCO en matière d'alimentation pour chiens. Enzyme Functionality Statement: Product contains enzymes. Protease for protein hydrolysis; Cellulase for cellulose hydrolysis; Amylase for starch hydrolysis.

(1) One HUT unit of proteolytic (protease) activity is defined as that amount of enzyme that produces, in one minute under the specified conditions (40 deg Celsius, pH 4.7), a hydrolysate whose absorbance at 275 nm is the same as a solution containing 1.10 ug per ml of tyrosine in 0.006 N hydrochloric acid.

(2) One Cellulase Unit (CU) is that activity that will produce a relative fluidity change of one in 5 minutes in a defined carboxymethylcellulose substrate under the conditions of an assay (40 deg Celsius, pH 4.5.).

(3) One Dextrinizing Unit (DU), or (SKB), unit of alpha-amylase activity as defined as that amount of enzyme that will dextrinize soluble starch ... at the rate of 1 g per hour at 30 deg Celsius.

FIGURE 34–10 Example of a commercial pet food label.

Feeding Schedules

Feeding animals may be done through **free choice** or **scheduled feedings**. Free choice is also called free access. This is a popular feeding method for large herds of livestock. Free choice allows animals to eat whenever they want food. The food is provided in a bulk amount and is always available for the animal. This is common with hay and grass sources for livestock. Water is also commonly allowed in free choice feedings. Scheduled feedings are more common for companion animals and livestock that are housed separately or in reduced numbers. Scheduled feedings provide a set amount of food given at specific times during the day. This is the best method when the goal is to provide

TABLE 34–1

Life stages and corresponding factors used to estimate daily energy needs for dogs

Neutered adult	= 1.6 × RER
Intact adult	= 1.8 × RER
Inactive/prone to obesity	= (1.2 to 1.4) × RER
Weight loss	= 1.0 × RER for ideal weight
Weight gain	= (1.2 to 1.8) × RER for ideal weight
Active, working dogs	= (2.0 to 5.0) × RER
Puppy, 0–4 months	= 3.0 × RER
Puppy, 4 months to adult	= 2.0 × RER

ideal nutrition, ensuring each animal receives and takes in an adequate amount of food. Most animals on a feeding schedule are fed at least two to three times a day. The foodstuff and schedule are based on the management and practice of the animal program. This includes individual rations given to each animal. Some larger producers and animal owners may purchase specialized equipment that is computerized and set to release certain amounts of food at specified times throughout the day. This equipment is called an **automatic feeder** and may also be available in water form, as an automatic waterer. Factors to consider when working to balance rations include the following:

- Knowledge of the needs of the animal species
- Knowledge of the food, including nutrient content
- Research and documentation

SUMMARY

Nutrition is the process by which animals ingest and utilize their food for survival. The food is used for various purposes according to several factors involving the individual animal. Nutrients are substances found within the food that are necessary for animals to function properly. Each animal species has different nutritional requirements. It is essential for veterinary assistants to have a basic understanding of animal nutrition and the needs of animals regarding a quality diet and proper ration.

Key Terms

abortion   occurs when a pregnant female animal loses the fetus

amino acids   building blocks of proteins that form chainlike structures

animal nutrition   science of determining how animals use food in the body and all body processes that transform food into body tissues and energy for activity, and the process which by animals grow, live, reproduce, and work

antioxidants   vitamins that boost the body's immune system

automatic feeder   specialized equipment that is computerized and set to release certain amounts of food at specified times throughout the day

balanced ration   diet that contains all the nutrients required by an animal in correct and specific amounts

biological value   percentage on a food label that describes the quality of the food source

body condition scoring   rating of how an animal appears, including weight based on its ideal body weight

calories   unit of measurement that specifies the amount of energy in food

carbohydrates   nutrients that provide energy for body functions and allow for body structure formation; make up the largest part of an animal's dietary needs

concentrates   food sources provided to an animal as an additional nutrient source when the primary food source is not adequate or abundant

concentration   determined by the amount of food fed divided by the percentage of dry matter and how it is delivered to the animal

dehydration   water loss by the body

digestion   the process of breaking down food from larger particles into smaller particles for use by the body in order to function

disaccharides   double sugars

double sugars   two particles of carbohydrates that form the building blocks of nutrients and are called disaccharides

dry matter   the amount of nutrients in a food source without the water content

essential fatty acids   nutrients that are not produced by the body and must be consumed

fats   concentrated source of energy; also called lipids

fat soluble   refers to vitamins that are stored in fat and released when needed within the body; include vitamins A, D, E, and K

fatty acids   oils that are products from fat sources; may be used as nutrients or supplements within a diet

feed   materials that animals eat to obtain nutrients

foodstuff   ingredients in animal food that determine nutrient content

fiber   material from plant cells that is left after other nutrients are digested

food analysis   process of determining the nutrients in food and prepared mixes to ensure they serve as a balanced ration

forage   plant-based sources of nutrition that are high in fiber; also called roughage

free choice   feeding method that allows animals to eat when they want; also called free access

growth diet   specialized food formulated to increase the size of muscles, bones organs, and body weight of young offspring

hydrolysis   chemical process of breaking cells down into smaller particles

ideal weight   breed standard based on an animal's age, species, purpose, or use and health status

kilocalories   amount of energy to raise 1 gram of water by 1 degree; written as a kcal

lactation diet   specialized food provided to breeding females that have completed the gestation phase and are currently producing milk for their offspring

lipids   concentrated source of energy; contains the highest amount of energy of all nutrients

macrominerals   minerals needed in large amounts, such as calcium

maintenance diet   specialized diet fed to animals that may be working or competing; the goal is to keep the animal at a specific ideal weight

microminerals   minerals needed in small amounts, such as iron

minerals   nutrients needed in every area of the body; are found mostly in the bones and teeth; used by the body based on the animal's needs and mineral availability

monosaccharides   simple sugars

nonessential fatty acids   nutrients that are both produced within the body and added to the diet when necessary

nutrient   any single class of food or group of like foods that aids in the support of life and makes it possible for animals to grow or provides energy for physiological processes

palatability how food tastes and is eaten by an animal

permanent pasture grass source that regrows each year without replanting or reseeding

polysaccharide long chain of simple sugars

proteins nutrients that are essential in growth and tissue repair

ration total amount of food an animal needs within a 24-hour time frame

reduced calorie diet specialized food given to animals that are overweight or less active due to health status

reproduction diet specialized food given to breeding animals to meet additional nutrient needs

scheduled feedings occur when a set amount of food is given at specific times during the day

senior diet specialized food given to geriatric or older animals older than a specific age (this is species specific) that require certain nutrient levels due to age

simple sugar single molecule of carbohydrates that form the building blocks of nutrients and are called monosaccharides

starches plant or grain materials that provide fiber and bulk in an animal's diet; convert to glucose or sugar during digestion

sugars type of carbohydrate that forms the simplest type of that nutrient

supplements additives incorporated into the diet in solid or liquid form when needed by the animal

temporary pasture grass source that does not regrow each year; needs to be replanted or reseeded

trace minerals needed in small amounts; also called microminerals

trimester period of 3 months of pregnancy in the gestation cycle

vitamins nutrients needed in small amounts for maintenance of life and health

water nutrient that makes up 75 percent of the body and provides several functions within the body, including controlling body temperature, maintaining body shape, transporting nutrients within the body's cells, aiding in digestion of food, breaking down food particles, and serving as a carrier for waste products

water soluble refers to vitamins that are not stored within the body, are dissolved by water, and are therefore needed in daily doses for the body to work; these include vitamins C and D

work diet specialized food given to animals that use a large amount of energy for work or strenuous activity

 MAKING THE CONNECTION

For nutrition related to specific species, review the content in each chapter in Section II of this book. Also review Chapter 25: The Digestive System.

REVIEW QUESTIONS

1. What is animal nutrition?
2. What is the importance of proper animal nutrition?
3. What is a ration?
4. What is a balanced ration?
5. What are nutrients?
6. What are the six nutrients necessary for animals?
7. What is the function of lipids?

8. What is the function of proteins?
9. What is the difference between vitamins and minerals?
10. What ways are animals fed?
11. What are the types of diets available for animals?
12. What are the RER and DER for a 15-lb dog?

Clinical Situation

Keiko, a veterinary assistant at Happy Dog Vet Hospital, is discussing some nutritional information with Mr. Good, who owns Snoopy, a 10-yr-old NM Beagle that is used for hunting and field trials. Mr. Good has noticed that Snoopy has been losing weight over the past few months. His medical records show that he has lost 4 pounds since his visit last year. Mr. Good is concerned about the food he is feeding Snoopy. "Do you think he should be on another type of food?" he asks.

Keiko responds, "Well, that depends. There are several options that you may want to consider. You have to think about Snoopy's age, his activity level, and his health. It appears from his medical record that he has been in good health. Dr. Wilson will be examining him today, so you may want to talk to him about a diet change."

"What do you think he should eat?" asks Mr. Good.

"Well, if he were my dog, I'd feed him a high-quality diet that has a lot of fat to put some weight back on him quickly," replies Keiko.

- Did Keiko handle educating the client correctly?
- What other options might Keiko have considered as you responded to Mr. Good?
- What are some factors to consider when selecting an ideal food for Snoopy?

COMPETENCY SKILL 29

Food and Water for the Presurgical or Preanesthesia Patient

Objective:
To properly prepare a diet and provide water for an animal undergoing surgical procedures

Preparation:
- Appropriate charts related to amount of water to provide
- Appropriate foodstuffs
- Water
- Containers to hold water
- Cleaning materials for the environment

Procedure:
1. Withhold food from certain species of animals (dogs and cats) for 12 hours before anesthesia is administered.
2. Withhold water 8 hours prior to administration of anesthesia.
3. Water can be provided postoperatively when the patient is fully awake and standing.
4. Food should be offered within hours post-surgery in ruminant and nonruminant animals.
5. Food should be offered within 12–24 hours post-surgery in monogastric animals.

COMPETENCY SKILL 30

Watering a Hospitalized Animal

Objective:

To properly prepare water for an animal under the care of the veterinarian

Preparation:

- Water
- Container for holding water
- Cleaning materials for the environment

Procedure:

1. Provide water as appropriate.
2. Refill the water bowl or bucket throughout the day.
3. Monitor the amount of water that is consumed by the patient.
4. If the water container is spilled, clean and dry the area as soon as the spill is noticed.
5. Replace the water bowl or bucket.
6. Clean the bowl or bucket as needed.

COMPETENCY SKILL 31

Feeding a Hospitalized Animal

Objective:

To properly prepare a diet for an animal under the care of the veterinarian

Preparation:

- Appropriate foodstuff
- Container for food
- Measuring devices
- Cleaning products

Procedure:

1. Prepare the appropriate diet. Consult with the veterinarian for specialized dietary needs.
2. Determine the amount of food to be consumed in 24 hours.
3. Place the food in the cage or animal holding area. Use a clean bowl or bucket.
4. Remove the bowl or bucket when the animal is done eating.
5. Clean and disinfect the bowl or bucket.

Chapter 35: Microbiology and Parasitology as Disease Processes

Objectives

Upon completion of this chapter, the reader should be able to:

35.1 Describe the various types of microorganisms that contribute to disease processes
35.2 Describe common animal external parasites
35.3 Describe common animal internal parasites
35.4 Explain the importance of zoonosis in humans

Introduction

The gastrointestinal (GI) tract of animals is a complex system of normal flora and microbiota that aids in the health of the animal. The health of an animal can be altered by invasion of unhealthy microorganisms from its environment. There are also smaller insects and animals that may play a role in passing disease to companion or production animals. These parasites can only survive by feeding off of the host animal, and as a result they pass disease to the animal. These types of diseases in production animals can greatly impact the success of the business and health of humans who ingest products from these animals. A sound understanding of how microorganisms and parasites contribute to disease will enable the veterinary assistant to identify these situations and help to provide care and control of the disease processes.

Microbiology

The study of microorganisms (*microbes*) is called **microbiology**. Many microorganisms are beneficial and can be used to make antibiotics and food products. However, some microorganisms are harmful; these are called **pathogens**. Pathogens can cause disease that may be contagious or noncontagious. Contagious diseases are easily spread between animals and humans. Noncontagious diseases are not passed to others.

Contagious diseases may be spread through direct or indirect contact. **Direct contact** means through contact with an animal or a body fluid. **Indirect contact** means spread through ways other than touching an infected animal, such as airborne or through bedding sources. Tissues and body fluids can be evaluated to determine the presence of specific disease-causing organisms and to aid in treatment. Samples may be collected by various methods, including swabs, scraping, and aspiration. The collection methods vary depending on the infection and location on the patient's body. Careful attention to aseptic techniques with samples is vital to obtaining accurate results.

Types of Microorganisms

Disease may be defined as a change in structure or function within the body. It is usually identified when abnormal symptoms occur in the health or behavior of an animal. There are five primary disease-causing microorganisms: viruses, bacteria, fungi, protozoa, and rickettsia.

Viruses

Viral diseases are caused by viruses, which are particles that are contagious and spread through the environment. Viruses can only live inside a cell. They cannot get nourishment or reproduce outside of a cell. Viral diseases are often not treatable with antibiotics, must run the course of the illness, and the animal may be maintained through supportive therapy or care. **Supportive care** or therapy includes such treatment as fluid replacement, medications to decrease vomiting and diarrhea, and pain medication. Common viruses include influenza strains, rabies, parvovirus, and West Nile virus.

 MAKING THE CONNECTION

See the following chapters:

Chapter 8—discussion of viral disease
Chapter 9—discussion of panleukopenia, rhinotracheitis, feline calicivirus, feline leukemia, feline immunodeficiency virus, and rabies
Chapter 10—discussion of psittacosis
Chapter 14—discussion of lymphocytosis
Chapter 17—discussion of bovine viral diarrhea
Chapter 18—discussion of equine infectious anemia, equine encephalomyelitis, and equine influenza
Chapter 19—discussion of brucellosis and pseudorabies
Chapter 20—discussion of blue tongue and sore mouth
Chapter 21—discussion of goat pox
Chapter 22—discussion of Marek's disease, Newcastle disease, infectious bronchitis, avian pox, and avian influenza
Chapter 31—discussion of lymphocytosis and ascites ■

Bacteria

Bacterial diseases are spread through single-celled organisms called bacteria (singular: *bacterium*). Bacteria are located in all areas of an animal's environment and are small cells with few cellular organs and no nucleus. Bacteria need an environment that will provide food for them to survive. Bacterial diseases can cause diarrhea, pneumonia, sinusitis, and various infections. Bacterial diseases are often treated with antibiotics that are specific to the bacteria being treated and are also called *antibacterial agents*. **Rickettsial** diseases are spread by biting insects, such as fleas and ticks. *Rickettsia* need to be in a living cell to reproduce and are considered to be bacteria spread by biting insects.

Bacteria occur in different shapes. Rods, cocci, and spirals are the most common and can occur individually, in pairs, and in clusters. Rods are long and straight, cocci are circular, and spirals are bent and squiggly.

 MAKING THE CONNECTION

Chapter 8—discussion of bacterial and ricksettial diseases
Chapter 12—discussion of Pasturella and Tyzzer's disease
Chapter 13—discussion of salmonella
Chapter 14 and Chapter 24—discussion of fin rot, cloudy eye, ascites, and swim bladder disease
Chapter 17—discussion of leptospirosis and Black Leg
Chapter 19—discussion of leptospirosis and pneumonia
Chapter 20—discussion of foot rot, mastitis, actinomycosis, enterotozemia, and lamb dysentery
Chapter 21—discussion of joint ill
Chapter 22—discussion of fowl cholera ■

Fungi

Fungal diseases are spread by simple-celled organisms or spores called *fungi* (singular: *fungus*) that grow on the external body and other areas of the environment, mostly in humid conditions. A fungal disease generally occurs in an animal that is immunologically impaired. Fungal diseases often cause hair loss, skin irritation, and itchiness. Fungal diseases are often treated with *antifungal agents*. The study of fungi is called **mycology**. Fungal cultures are often obtained to determine the type of **dermatophyte**, or the type of fungal growth, that is present. This is often done with a dermatophyte test medium (DTM). The medium is typically solid and placed into a slant-typed container or tube. Superficial fungal infections, such as dermatophytosis, occur on the outer skin, hair, and nail bed surfaces. Systemic fungal infections occur internally, often as a result of inhaling spores from the soil or other areas of the environment.

MAKING THE CONNECTION

Chapter 8—discussion of fungal disease
Chapter 14—discussion of fish fungus

Protozoa

Protozoa are the simplest forms of life—single-celled organisms that are known as parasites and that may live inside or outside the body. Most protozoa feed off of dead or decaying matter and cause infection in animals that ingest them. **Protozoan** diseases often cause vomiting, diarrhea, weight loss, and overall poor body and coat condition. These diseases are often treated with *dewormers* or *anthelmintics*.

MAKING THE CONNECTION

Chapter 8—discussion of protozoan disease
Chapter 14—discussion of ich
Chapter 17—discussion of anaplasmosis

Parasitology

Parasitology is the study of organisms that live on or in other organisms to survive. Parasites may invade the internal or external parts of an animal. Most parasites have one location that they live in and feed off the **host**, or the animal that is infected. The host may be a **definitive host** for the adult stages of the parasites or an **intermediate host** for the immature or larval stages of a parasite. Parasites go through a life cycle in which they are born as **larvae** and continue through growth stages until they become adults and are capable of reproduction. Each parasite has a different life cycle and growth stages. Each life cycle may be simple or complex and involve one or more **vectors**, which are the modes of action via which a parasite is transmitted. The term *life cycle* refers to the maturation and developmental stages in one or more hosts. Many parasites invade the digestive tract, skin, or muscles of an animal. They may enter the animal's body system through many different methods, for example, through ingestion, as in food or water sources. Other routes of contamination include contact with soil, through penetration of the skin, through nursing young, and contact with infected feces. Parasites can be controlled and prevented through proper sanitation and disinfection methods.

Endoparasites

Common internal parasites called **endoparasites** occur in small and large animals and include roundworms, whipworms, hookworms, heartworms, coccidian, tapeworms, and strongyles. Many other types of internal parasites occur in various species of animals and should be evaluated in a fecal analysis (see Figure 35–1). Many internal parasites invade the intestinal system, causing vomiting and diarrhea. Although the infective stage is internal to the host, many endoparasites spend at least some part of their life cycle outside the body of the host.

Roundworms

Roundworms are also called **ascarids** and are the most common intestinal parasite in small and large animals. They occur most often in young animals. All roundworms live in the small intestine of animals. There are several different species of roundworms. Female roundworms can lay up to 200,000 eggs a day.

Toxascaris leonina has the simplest life cycle. After an animal ingests infective eggs, the eggs hatch and the **larvae** mature within the small intestine. The adult female worm lays eggs that are passed in the feces. The eggs become infective after remaining in the environment for at least 3–6 days. Animals become infected if they eat something contaminated with infected feces.

Toxocara canis has a more complicated life cycle and an effective way of making sure the species will survive from generation to generation. They are very hardy and resistant to eradication efforts. There are several ways by which an animal can acquire a *T. canis* infection: ingestion of eggs, ingestion of a transport host, or larvae entering the animal while in the uterus or through the milk. The larvae migrate through the circulatory system and go to the **respiratory** system or other organs or tissues in the body. If they enter body tissues, they can **encyst** (become walled off and inactive). They can remain encysted in tissues for months or years.

Toxocara cati is similar to *T. canis*. The infective eggs are swallowed, and the larvae hatch and penetrate the stomach wall. From there, the larvae migrate through the liver, other tissues, and lungs. Some larvae may encyst in the tissues. Larvae that enter the lungs are coughed up and swallowed. The larvae mature in the stomach and small intestine, and the adult female worms start laying eggs.

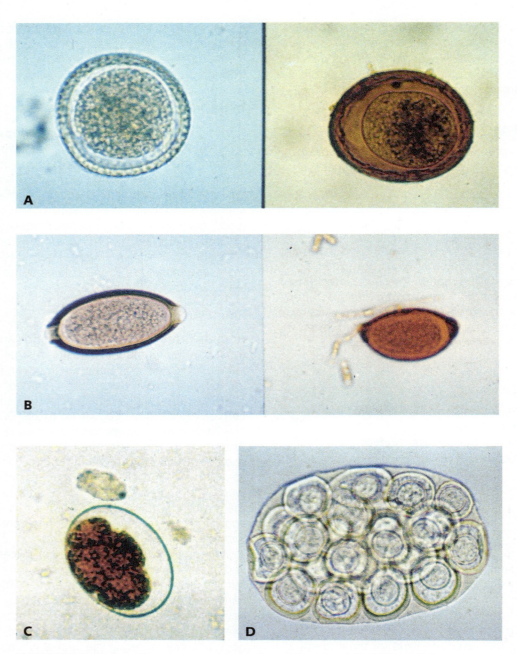

FIGURE 35–1 Parasite eggs commonly found in a fecal flotation. (A) Roundworm eggs (*Toxocara canis*, left; *Toxascaris leonina*, right). (B) Whipworm (*Trichuris vulpis*, left) and lungworm (*Capillaria aerophila*). (C) Hookworm (*Ancylostoma*). (D) Tapeworm (*Dipylidium caninum*).

Parascaris equorum in large animals are swallowed after they have contaminated hay or water, and they hatch in the intestinal tract. The young worms burrow through the intestinal wall, taking about a week to make their way to the lungs. From there the young worms travel up the trachea to the mouth, to be swallowed a second time. They mature in the intestine in 2 to 3 months, then lay eggs that are passed in the feces to start the cycle anew.

Small animals infected with roundworms will show signs of vomiting and diarrhea; a bloated stomach appearance; and visible roundworms in the feces, which look like long, thin pieces of pasta (see Figure 35–2). Large animals will show signs of abdominal pain, coughing, and diarrhea, as well as visible roundworms in the feces that may be 6–12 inches in length. The eggs of roundworms are circular in shape with dark circular centers that can be

CHAPTER 35 Microbiology and Parasitology as Disease Processes

FIGURE 35–2 Common roundworm found in dogs (*Toxicara canis*).

seen when viewed under a microscope. Table 35–1 summarizes types of roundworms and their common hosts.

Whipworms

Whipworms are another common intestinal parasite and occur mostly in dogs. Whipworms get their name from the whiplike shape of the adult parasite. The front end of the worm is thin and the back end of the worm is thick, similar to a whip handle. Whipworms live in the large intestine of infected animals. **Trichuris vulpis** species infect animals after they ingest food or water contaminated with whipworm eggs. Once the eggs are swallowed, they hatch and—in a little less than 3 months—the **larvae** mature into adults in the cecum and large intestine, where they burrow their mouths into the intestinal wall and cecum, and feed on the host's blood. Adult worms lay eggs that are passed in the feces. The eggs must remain in the soil for about a month to mature and be capable of causing infection. The eggs are resistant to eradication, hardy, and difficult to control. Animals infected with whipworms may exhibit diarrhea, weight loss, possible blood in the feces, and anemia. Whipworms are not visible in the feces and are diagnosed through a fecal analysis under a microscope. The eggs of whipworms are football shaped (see Figure 35–1B).

Hookworms

Hookworms are another common intestinal parasite of dogs and occasionally cats. Hookworms have either teethlike structures or cutting plates with which they attach themselves to the wall of the intestine and feed on the animal's blood. Hookworms can cause a skin disease in humans called **cutaneous larval migrans**. Infections of the intestines in people can also cause severe abdominal pain. **Ancylostoma caninum** lives in the small intestine of its host, where it attaches itself and feeds on the host's blood. The adults lay eggs that pass in the feces. In 2–10 days, the eggs hatch and the larvae are released. These larvae are excellent swimmers that travel via raindrops or dew on leaves and vegetation as they wait for a suitable host animal to come along. The larvae enter a host either by being ingested or by burrowing through the host's skin. Animals infected with hookworms show signs of vomiting, diarrhea, anemia, weakness, black darkened feces, a dull coat appearance, and occasional coughing. Hookworms are not visible in the feces and are diagnosed with a fecal analysis using a microscope. Hookworm eggs are oval and clear (see Figure 35–1C).

Heartworms

Heartworm disease was first identified in the United States in 1847 and occurred most frequently in the southern United States. In recent years, heartworm disease has been found throughout the United States. The movement of infected animals that could serve as sources of infection for others is probably a significant contributing factor to heartworms spreading throughout North America. The actual number of infected dogs and cats in the United States is unknown. Heartworm disease is spread by mosquitoes and is of most concern in dogs and cats, and occasionally ferrets. Heartworms (**Dirofilaria immitis**) belong to the same class as roundworms and look similar in structure—like long,

TABLE 35–1
Roundworm Hosts

ROUNDWORM	PRIMARY HOST	TRANSPORT HOST
Toxascaris leonina	Dog, cat, fox, other wild carnivores	Small rodents
Toxocara canis	Dog, fox	Small rodents
Toxocara cati	Cat	Small rodents, beetles, earthworms
Parascaris equorum	Horse	Small rodents, flies

thin pieces of pasta. *D. immitis* affect the heart and circulatory system of infected animals. Adult heartworms in the heart lay tiny **larva** called **microfilaria**, which then settle in to live in the bloodstream. The microfilaria enters a mosquito when it sucks blood from an infected animal. In 2–3 weeks, the microfilaria develops into larger larvae in the mosquito and migrate to the mosquito's mouth. When the mosquito bites another animal, the larvae enter the animal's skin. The larvae grow and after about 3 months finish their migration to the heart, where they grow into adults, sometimes reaching a length of 12–14 inches (see Figures 35–3 and 35–4). The time from when an animal was bitten by an infected mosquito until adult heartworms develop, mate, and lay microfilaria is about 6 months in dogs and 8 months in cats. Dogs that have heartworm disease will show signs of exercise intolerance, difficulty breathing, coughing, decreased appetite, weight loss, and lethargy. Cats typically show no signs and may die suddenly. Some cats show signs similar to those of dogs. Simple heartworm test kits are commonly used in diagnosing the disease (see Figure 35–5). Treatment of heartworm disease is a concern; when the heartworms die within the heart and/or bloodstream, this can cause blood flow to be blocked, which may cause the pet to die. Prevention with monthly medication is the key to this avoiding disease.

Strongyles

Large and small **strongyles** are a common parasite of large animals, especially horses. Large strongyles are a group of internal parasites, also known as

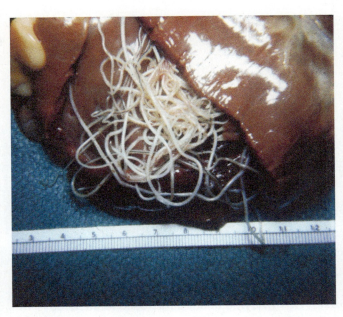

FIGURE 35–4 Heartworms can lead to blockage of the heart of an animal.

bloodworms or redworms. Eggs in manure hatch into larvae that are consumed by the grazing animal. The larvae mature in the intestinal tract and burrow out into blood vessels, where they migrate throughout various organs and eventually end up back in the intestine. The larvae can cause extensive damage to the lining of blood vessels. Signs of large strongyles are weight loss, anemia, abdominal pain, and—in extreme cases—sudden death. Small strongyles

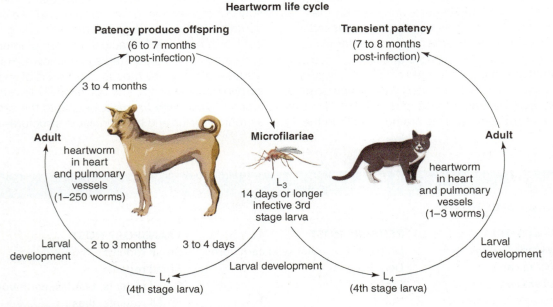

FIGURE 35–3 The life cycle of heartworm.

CHAPTER 35 Microbiology and Parasitology as Disease Processes

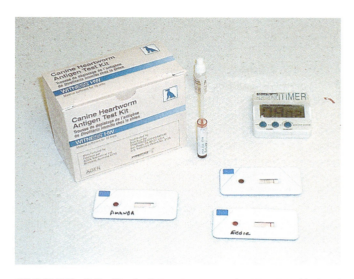

FIGURE 35–5 A heartworm (enzyme-linked immunosorbent assay (ELISA) test.

differ from large strongyles in several ways. First, small strongyles do not migrate through tissues as do large strongyles. Second, small strongyle larvae may become encysted. This means that they burrow into the intestinal wall and lay dormant as they wait for the proper conditions to emerge. During this encysted period, unlike adult parasites, small strongyle larvae are not susceptible to most dewormers. Signs of small strongyles include diarrhea, weight loss, poor growth, poor coat quality, and abdominal pain. Frequent deworming is recommended with all livestock, as the environment increases the possibility of parasite infection. Yearly fecal analysis should be done to identify any internal parasites (see Figure 35–6). Strongyle eggs appear as large, long, oval shapes under a microscope.

Coccidia

Coccidia is a common protozoan internal parasite that is a simple one-celled organism that is present in all mammals. Most animals that have **coccidiosis** (i.e., they are infected with coccidia) are infected by *Iospora canis*, although several species of coccidian do occur. Animals with coccidiosis will shed **oocysts**, or single-celled eggs, in the feces. In the infected host, the eggs live within the intestinal tract and the infected animal may show no signs of illness. Other animals may develop chronic diarrhea, and blood and mucous may be present. Sometimes coccidiosis is difficult to diagnose, even with fecal analysis. The microscopic cells appear small and circular and may be missed on a fecal smear (see Figure 35–7). Most animals become infected with coccidian from contaminated food, water, or soil sources. Bird droppings are the most common source of the parasite.

Giardia

Giardia is a protozoan or single-celled organism that lives in the intestinal tract of animals. Animals acquire **giardiasis** (i.e., become infected by giardia) via ingestion of contaminated water or soil sources. This is a common occurrence in water sources that are outside or are contaminated when pipes break. Many people receive boil-water advisory information when giardia is present in a water source. Animals frequently get diarrhea when giardia is present in the feces. Other signs of giardiasis include vomiting, weight loss, and poor overall appearance.

Tapeworms

Tapeworms are flatworms that are segmented or have individual parts that grow and shed as the parasite ages. Tapeworms live within the small and large

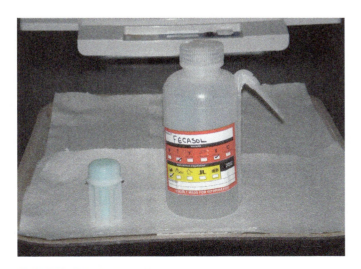

FIGURE 35–6 Fecal floatation materials.

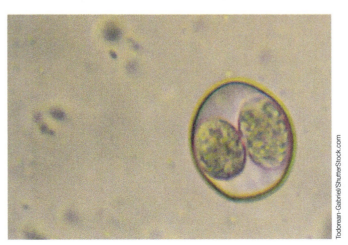

FIGURE 35–7 Coccidia egg.

intestines. The segments, known as **proglotids**, are often be seen near the anus of the infected dog or cat. These segments may move if recently passed, or if dried, they look like grains of uncooked rice or cucumber seeds (see Figure 35–8). Tapeworm infections are usually diagnosed by finding these segments on the animal. Most animals develop tapeworms from ingesting a flea, rodent, or mite. Flea tapeworms (***Dipylidium caninum***), which can be up to 20 inches long, live in the small intestine. The segments, full of eggs, are passed in the feces. While warm, the segments are active, but as they dry, they break open and liberate the eggs inside. Either an adult louse or a flea **larva** ingests the eggs. Each egg develops into an immature form in the insect. When a dog or cat eats the insect, the immature form develops into an adult worm, and the life cycle is completed (see Figure 35–9). Large and small animals may become infected with **Taenia** species, which have a similar life cycle (see Figure 35–10). Fleas may begin the life cycle in small animals. Mites living in a pasture may consume tapeworm eggs from the feces of infected animals. Grazing animals may then swallow the mites and become infested with tapeworms. Signs of tapeworms include poor hair coat, abdominal discomfort, or visible segments in the feces.

Flukes

Flukes are **trematodes**, or flat worms that lack a body cavity. They are unsegmented and appear leaflike. They have suckerlike attachments and loose tissue that holds the organs. One sucker is located anterior on the body and serves as the mouth, and another sucker is located ventrally near the middle to caudal aspect and serves as the rectum. Many trematodes have both male and female reproductive organs. The life cycle of trematodes is often complicated, with the organism passing through several larval stages and typically requiring one or more hosts (which are mainly mollusks, e.g., snails and slugs).

Bots and Pinworms

Botflies lay yellowish eggs on a horse that are then ingested by the horse. The eggs hatch in the stomach, mature during the larval stage, and later are passed out through the feces, where they complete maturation into an adult fly.

Pinworms are small parasites in horses that travel through the large intestine and settle in the anus where they cause irritation and itching, causing

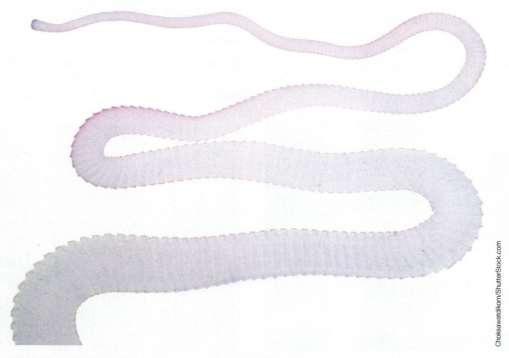

FIGURE 35–8 Tapeworm. Note the segmented body.

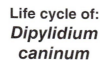
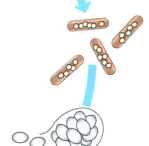

Life cycle of: *Dipylidium caninum* (dog/cat-flea tapeworm)
INTERMEDIATE HOST: The Common Flea

An important point to remember about this tapeworm species is that any pet that has fleas also has a very real possibility of having tapeworms

1. Pet passes egg-filled tapeworm segments in feces
2. The segments rupture, releasing individual eggs that are eaten by flea larvae
3. During grooming or biting, pet ingests tapeworm-carrying fleas
4. Tapeworm matures in pet
5. Adult tapeworm releases mature, egg-filled segments and, without treatment, the cycle begins again

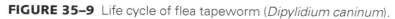

FIGURE 35–9 Life cycle of flea tapeworm (*Dipylidium caninum*).

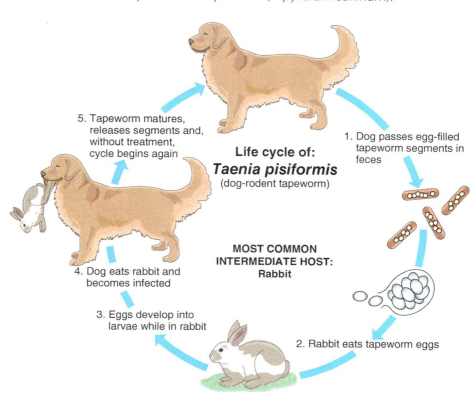

Life cycle of: *Taenia pisiformis* (dog-rodent tapeworm)
MOST COMMON INTERMEDIATE HOST: Rabbit

1. Dog passes egg-filled tapeworm segments in feces
2. Rabbit eats tapeworm eggs
3. Eggs develop into larvae while in rabbit
4. Dog eats rabbit and becomes infected
5. Tapeworm matures, releases segments and, without treatment, cycle begins again

FIGURE 35–10 Life cycle of rodent tapeworm (*Taenia pisiformis*).

animals to rub their tails. The rubbing of the tail hair can lead to hair loss, one of the most frequent signs that pinworms are present.

Ectoparasites

Several external parasites, or **ectoparasites**, invade all types of small and large animals. Mites, fleas, ticks, lice, and biting flies are common, especially in warmer weather. External parasites may live in the hair coat, on the skin, or within the ear canals.

Fleas

Adult female fleas can lay up to 50 eggs a day. The eggs are laid on the host, from which they fall to hatch in the environment. Eggs incubate best in high humidity and temperatures of 65–80 degrees Fahrenheit. The eggs will hatch after around 10 days, moving into the larval stage. Larvae are like little caterpillars crawling around and grazing on the flea "dirt"—black pepper-like appearance in haircoat—that is generally in their vicinity. This is the stage when large animals pick up tapeworms while they graze. Larvae go through several molt stages and then become **pupae**, at which point they spin a cocoonlike structure inside which they become an adult flea. At this stage, they can remain dormant for several months to a year before emerging as an adult flea. Under ideal temperatures and conditions, this stage may occur in as little as 21 days from the time eggs hatched. The pupae emerge when they feel the vibrations and heat of a host to feed off of. Adult fleas may survive up to several months without a food source, that is, a blood meal (see Figure 35–11). Once a female flea has a blood meal, it will produce eggs within 24 to 48 hours and then continually lay eggs until it dies. The average life cycle of an adult flea is 4–6 weeks.

Fleas are easily noted on an animal by applying a flea comb to the coat. The veterinary assistant may also look for flea dirt, which consists of small dark black to brown particles that make it look like the animal is dirty. This flea dirt is actually flea feces. If the veterinary assistant isn't sure if what he or she is seeing is flea dirt or real dirt from the outside environment, some of the particles can be placed on a paper towel or other light-colored surface, then a drop of water can be applied to the particles. If the particles turn red rust to brown in color, they are flea feces, which determine the blood meal of the host animal. Owners can also place a warm towel in an area the animal often accesses, such as a living room, with the lights turned off for several minutes. After turning the lights on, the owners can inspect the towel for fleas. Fleas will quickly seek heat (on the warm towel) if they are in the household.

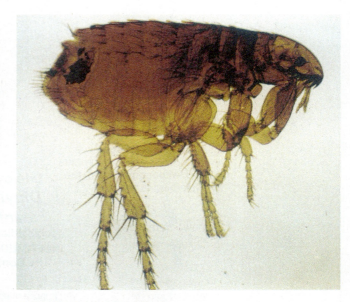

FIGURE 35–11 Flea.

Signs of an animal having fleas include itching and scratching of the skin, biting at the skin, hair loss, scabs, and visible fleas or flea dirt. Fleas can cause severe skin irritation and itching, and in some animals, the flea saliva causes an allergic reaction when an animal is bitten. This is called **flea allergy dermatitis** or **FAD**. Signs of FAD include itching and scratching of the skin, biting at the skin, hair loss, scabs or bumps, and red areas on the skin. Fleas can also cause an animal to have **anemia** or a loss of blood. An animal that has anemia from flea bites will have pale gums, exhibit signs of weakness, and be lethargic. Animals that ingest a flea or a stage of the life cycle of the flea are susceptible to developing tapeworms. The most common sign of tapeworms is itching around the anal area or visible segments shedding from the tapeworm noted around the rectum or on bedding. Another less common condition caused by flea bites is **plague**. This disease affects animals and humans, causing high fever, dehydration, and enlarged lymph nodes. It can eventually lead to death. **Bartonella** is a bacterium spread by fleas and ticks and is most common in cats. Cats transmit the organism when they are infested by fleas, scratch themselves, get infected flea dirt in their claws, then scratch a person or another cat with their dirty claws. Cats can also harbor *Bartonella* in their mouths and transmit the infection via bites or groom their feet and self-infect their claws. Cats also are prone to a disease called **feline infectious anemia** (which was once called **hemobartonella**). This disease is caused by the bite of a flea or tick with bacteria that infects the animal's red blood cells and immune system. Infected cats are often pale, jaundiced, anemic, lethargic, and febrile.

Ticks

Ticks are arthropods that seek heat and movement (see Figure 35–12). There are two types of ticks, hard and soft ticks. Hard ticks are the vectors of protozoal, rickettsial, bacterial, and viral diseases. Hard ticks are flat in appearance with a hard covering on the dorsal body surface and often have grooves or notches visible on the body surface, which helps identify them. They often feed on one to three hosts throughout their life cycle. Soft ticks lack a covering, they have a round body shape, and their mouthparts are not visible from the dorsal surface. Soft ticks are hardier and have a longer life span. Ticks undergo four life stages until they become adults. There are several species of ticks that transmit different types of serious diseases. Adult female ticks lay eggs depending on how much of a blood source they have from a host. The larger the blood meal it has ingested, the more eggs the female will produce. After an adult has completed a blood meal, it will drop off of the host and lay its eggs in the environment. Eggs are laid over several days to several weeks. The female dies shortly after laying the eggs. The eggs develop into the larval stage then molt into a nymph stage, where they become an adult tick. During each stage the tick needs a blood meal from a host to develop into the next life cycle stage. The entire life cycle process from egg to adult typically takes 2 years. Ticks feed off any animal species, including humans. In some species, the female tick produces toxic saliva that may cause various degrees of paralysis, referred to as **tick paralysis**. When they have a blood meal, they can transmit viral, bacterial, and rickettsial diseases. The most common tick species in animals are the American dog tick, the deer tick, the brown dog tick, the lone star tick, and the *Ixodes* species.

FIGURE 35–12 Ticks.

The American dog tick transmits a disease known as **Rocky mountain spotted fever**. Signs include fever, joint pain (arthrodynia), depression, and anorexia. This disease can be fatal if not treated. The lone star tick transmits a disease known as **ehrlichiosis**. This bacterial disease causes fever, joint pain, depression, anemia, anorexia, weight loss, and some edema. If not treated, this disease may be fatal. The deer tick, also known as the black-legged tick, transmits a common disease called **Lyme disease**, which is transmitted from a bacterium carried by the tick called *Borrelia burgdorferi*. This disease causes joint pain, lameness, fever, depression, anorexia, lethargy, and swelling of the joints. In more severe cases, it may lead to significant hepatic, cardiac, and neurological problems. Some animals show no signs of the disease in the early stages of infection. Another less common disease caused by the deer tick is **babesia**. This disease causes anemia, jaundice (yellow coloring of the skin and **mucous membranes [mm]**), fever, and vomiting. Severe stages can lead to kidney failure. This disease may be difficult to control, and relapses often occur. Table 35–2 summarizes the various types of diseases caused by ticks.

TABLE 35–2

Tick Species and Diseases

TICK SPECIES	APPEARANCE	DISEASE POTENTIAL
American dog tick	• Chestnut brown • Light to white-colored markings on back	Rocky mountain spotted fever
Brown dog tick	• Shades of light to dark brown • Solid color with no markings	None
Deer tick (black-legged tick)	• Dark brown body • Black legs	Lyme disease Babesia
Lone star tick	• Reddish brown • White spot or "star" on back	Ehrlichia (ehrlichiosis)
Ixodes cookie (groundhog tick)	• Light brown to tan body • Dark brown head and neck area	Encephalitis (rare)

Lice

All animals and humans can get lice. Each animal species, including humans, can get lice that are transmitted only to that species; thus, lice are species specific. Humans can't get lice from dogs, cats, horses, or other animals, and people can't transfer lice to animals. The two types of lice are biting lice and sucking lice. Animal lice are almost motionless, and some species attach to the hair by sticking to and feeding off of the hair and skin cells. These are the sucking lice, which feed off blood using their long, narrow heads and often cause anemia. Other species of lice bite and remove blood from the animal. Biting lice also feed on epithelial tissue and move more rapidly over the animal's body. Biting lice have a broad rounded head and are irritating to the host. Animal lice are gray to light beige in color and are wingless parasites that are about one-twelfth of an inch in length. They are small in size but can be seen with the naked eye (see Figure 35–13). The lice and eggs easily stick to a piece of tape, which can be used to remove them for better identification. Lice can spread by direct contact or contact with bedding and other areas upon which animals frequently lay. Lice eggs, called **nits**, may fall off and be passed to another animal of the same species. Grooming supplies and clippers are a common source of infection if sanitation is not done properly. Lice cause severe itching, areas of hair loss, and anemia. Sometimes adult lice and nits can be seen with the naked eye. Lice can lay up to 100 eggs, which will reach adulthood in about 21 days.

Mosquitoes

Mosquitoes are small flying insects that bite humans and animals, ingesting a blood meal to survive. When they bite, they are capable of spreading diseases from animal to animal. One disease transmitted by mosquitoes is heartworm disease. Mosquitoes transmit microfilaria, or immature heartworms, from animal to animal when they have a blood meal from an infected animal.

West Nile virus is also a mosquito-borne disease that causes inflammation and swelling of the brain and spinal cord. Birds and horses are at the greatest risk.

Biting Flies

Biting flies can cause animals, especially large animals, lots of irritation. One disease that is of concern from biting flies, especially the horsefly, is **equine infectious anemia**, or **EIA**. Equine infectious anemia, also known as swamp fever, is a viral disease for which there is no vaccine and no cure. This is of huge concern in some areas, and many horses that travel are required to have a **Coggins test**, which screens for the EIA virus. The Coggins test is performed by a state-approved laboratory on a blood sample. Many states require horses to have a negative Coggins test, and veterinary assistants should be familiar with state laws on equine travel. The best prevention is the use of fly sprays to help reduce the number of biting flies that have contact with a horse. Control and removal of manure and other waste products are helpful.

Flies cause painful bites, suck blood, produce sensitivity reactions, and may deposit eggs in sores and on body surfaces. Flies can lay eggs that burrow, with larval stages migrating through the tissues of the host and emerging through holes in the skin. Botflies are beelike flies that glue their eggs to hair on the legs or other torso of a host. The larvae hatch and penetrate the skin of the host—which may cause large pockets in the subcutaneous tissue locally—or they may migrate to other areas. The larvas survive due to air holes in the skin that enable them to breathe. These larval stages produce warbles, or *Cuterebra* species, that cause pain and irritation and can lead to an underlying skin infection.

Mites

Mites come in many forms and invade several areas of an animal's body. Mites may live on the skin or hair coat, or they may invade the ears. **Ear mites** are common in all animals (see Figure 35–14). They are tiny infectious organisms resembling microscopic ticks. Infection usually produces a characteristic black ear discharge that is dry and crusty and is commonly said to resemble coffee grounds. The mite lives on the surface of the ear canal, though it sometimes migrates out onto the face and head of its host. Eggs are laid and hatch after 4 days of incubation. The larvae hatch

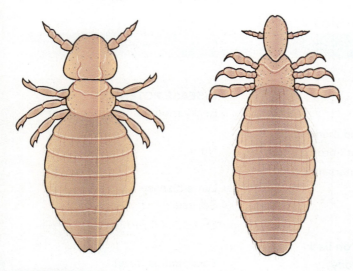

FIGURE 35–13 Lice: biting louse (left) and sucking louse (right).

FIGURE 35–14 Ear mite (*Otodectes* mite).

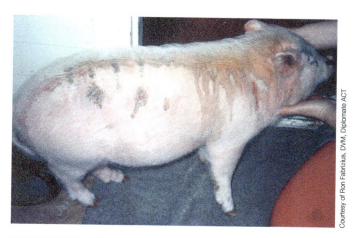

FIGURE 35–16 Alopecia in a pig with sarcoptic mange.

from the eggs, feed on ear wax and skin oils for about a week, then go through several molts to become adult mites. The entire life cycle takes about 3 weeks to complete. Adult ear mites live about 2 months. Ear mite infections can be easily treated by cleaning the ear canals and applying medications that kill the ear mites (see Figure 35–15). Other signs of ear mites in animals include scratching the ears, shaking the head, and open sores within the ear.

Mange mites are another common occurrence, especially in dogs and pigs. Mange mites live on the skin and hair coat. There are two types of mange mites: sarcoptic mange and demodectic mange. **Sarcoptic mange** is the name for the skin disease caused by infection with the *Sarcoptes scabei* mite (see Figure 35–16). Sarcoptic mites are burrowing mites that tunnel into superficial layers of skin and feed on the tissue. Infestations begin locally as areas of hair loss and inflammation then spread rapidly. Pruritis and hair loss are the most common signs. Mites are not insects; instead, they are more closely related to spiders. They are microscopic, so cannot be seen with the naked eye. Adult mites live 3–4 weeks in the host's skin. After mating, the female burrows into the skin and deposits three to four eggs in a tunnel behind her. The eggs hatch in 3–10 days, producing larvae that then molt into adult mites within 10 to 15 days. This type of mite is zoonotic and causes the condition known as **scabies** in humans. Sarcoptic mange is highly contagious via direct contact or contact with infected bedding or other items such as grooming tools that have not been properly disinfected. Signs of sarcoptic mange in animals include excessive itching, hair loss, damage to the hair coat, open lesions on the skin, and skin infection. Diagnosis is typically achieved through a skin scraping and identifying the mites under a microscope (see Figure 35–17). **Demodectic mange** is another mite that invades the skin by burrowing into the hair follicles and sebaceous glands, and all dogs raised typically by their mothers possess this mite, as mites are transferred from mother to pup via cuddling during the first few days of life (see Figure 35–18). Most dogs live with their mites without any problems, never suffering any consequences. However, the mites can overpopulate in some dogs and cause severe skin irritation, especially when an immunodeficiency is present. This mange is not contagious. Signs of the disease include localized hair loss, pustules, increased itching, and dry thickened areas of skin. A skin scraping is used to identify the microscopic mites, which are shaped like cigars. Both types of mange mites are treatable with medicated dips and other medications.

Maggots

Maggots are insect larvae, usually flies, that live on animal tissue. A maggot infestation results when eggs

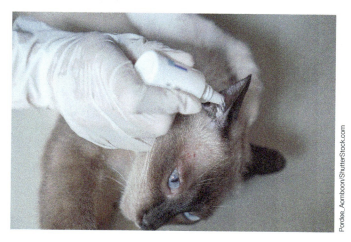

FIGURE 35–15 Eardrop products for control of mites.

SECTION III General Anatomy and Disease Processes

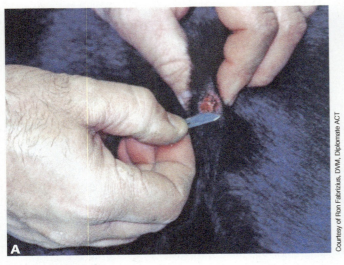
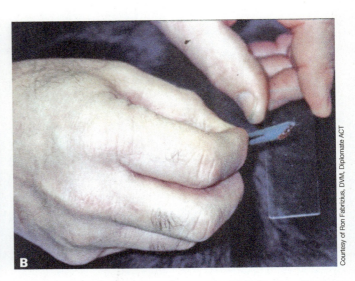

FIGURE 35–17 Skin scraping. Skin scrapings are used to detect mites. (A) The use of a surgical blade to collect a skin scraping sample. (B) Several areas of the skin lesion should be sampled with a surgical blade.

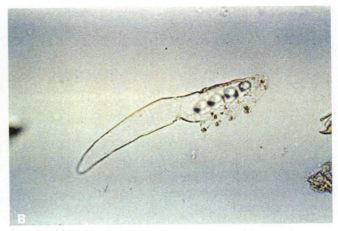

FIGURE 35–18 (A) Demodex mange in a dog. (B) Demodex mite.

are laid on an animal, typically on a wet area or where skin is damaged. The eggs will then hatch and feed off of the tissue (see Figure 35–19). This can cause severe infection and skin damage.

Ringworm (Dermatophytosis)

Dermatophytosis, commonly known as **ringworm**, is a fungal disease of animals and can be transmitted to humans (see Figure 35–20). This condition can be passed from any infected animal to another species, including humans. It is the most common contagious skin disease of animals. Dogs, cats, and all livestock species commonly get ringworm from the environment. Spores from infected animals can be shed into the environment and live for up to 24 months. Humid, warm environments encourage growth of the fungus. Spores can be on brushes, bedding, furniture, stalls, or anything that has been in contact with an infected

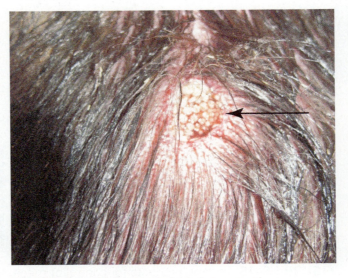

FIGURE 35–19 Maggots in a dog wound.

other locations. The most common sites in animals are on the head, face, ears, and tail. The most common method for diagnosis is by skin scraping and looking at the hair and skin under a microscope. Another method is by use of a Wood's lamp, which fluoresces the area with a black light. Several species of the ringworm fungus will glow a fluorescent color when exposed to a Wood's lamp. However, many species do not show up with this method. Oral and topical treatment is used to cure the animal, but the environment must also be disinfected to prevent reoccurrence.

Anthelmintics

Small and large animals alike should be on a regular deworming program. Large animals should be wormed every 6–8 weeks due to the increased exposure to parasites. Small animals should be wormed yearly or as recommended by the veterinarian. Dewormers, or anthelmintics, should be given according to the common parasites that affect the specific species of animal. Many dewormers are available for treatment of these parasites. Unfortunately, none of the over-the-counter or prescription dewormers will kill all of these parasites. Thus, dewormers must be chosen according to the type of parasite that is present. Table 35–3 summarizes the recommendations by the Companion Animal Parasite Council (CAPC) and details control methods for common internal (GI) parasites for dogs and cats in the United States. Table 35–4 lists common deworming products used for companion animals.

Large animal deworming recommendations include deworming livestock every 2 months for common internal GI parasites. Additional recommendations for parasite control include the following:

- Remove manure daily from stalls and weekly from pastures.
- Be sure pastures and paddocks are well drained and not overpopulated.
- Compost manure rather than spreading it on fields where animals graze.
- Use a feeder for hay and grain, and avoid ground feeding.

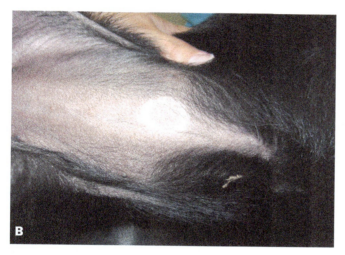

FIGURE 35–20 (A) Holstein heifer with ringworm. (B) Ringworm lesion on the abdomen of a dog.

animal or the animal's hair. Ringworm is transmitted by direct contact with spores, which may be on the animal or in the environment. Skin lesions may vary in appearance and size. The classic sign is a small round area with no hair and scaly, dry skin located in the center. The lesion may start as a small spot and continue to grow. The area may or may not become itchy. In some cases, the fungus can spread over the body to

TABLE 35–3

Companion Animal Deworming Recommendations

1. Administer year-round broad-spectrum heartworm medications.
2. Deworm puppies and kittens at ages 2, 4, 6, and 8 weeks of age and then monthly until 6 months of age.
3. Deworm nursing dogs and cats at the same time as their offspring.
4. Conduct fecal examinations two to four times per year for adult animals, depending on health status and lifestyle factors.

TABLE 35–4

Common Companion Animal Deworming Products

PRODUCT NAME	PARASITES TREATED
Drontal	Roundworms, hookworms, whipworms, and tapeworms
Albon	Coccidia
Droncit	Tapeworms
Heartgard	Heartworms, roundworms, and hookworms
Interceptor	Heartworms, roundworms, hookworms, and whipworms.
Nemex	Roundworms and hookworms
Panacur	Roundworms, hookworms, whipworms, and some species of tapeworms
ProHeart 6/12	Heartworms and hookworms
Revolution	Heartworms, fleas, ear mites, ticks, roundworms, mange, and hookworms
Sentinel	Heartworms, roundworms, hookworms, whipworms, and flea eggs
Strongid T	Roundworms, hookworms, whipworms, and tapeworms

- Initiate effective fly control programs.
- Routinely examine all animals for signs of infestation.
- Establish a parasite prevention and monitoring program with your veterinarian.

 MAKING THE CONNECTION

All chapters in Section II discuss parasites and parasitic disease specific to the species discussed in the chapter. Please review these content areas in those chapters for more information on parasites. ■

Zoonosis

Diseases that are transmitted from animals to humans are called **zoonoses**. Many animals infected with a zoonotic disease may not show any signs of illness. The large majority of zoonoses are not clinically significant. However, it is important for veterinary professionals to be aware of zoonotic disease and take precautions to prevent the spread of such diseases. The most common zoonotic diseases of significance are discussed here.

Toxoplasmosis

Toxoplasma gondii is a single-celled organism that can occur in any mammal but is shed through the feces of cats, which are the only animals that pass the toxoplasmosis through the feces. This disease is of concern to humans, especially pregnant women who may contract the disease because it may lead to congenital defects or even abortion. Cats that have this condition shed oocysts (immature egglike particles) in their feces, and the oocysts must be in the environment 1–5 days before they are infective (see Figure 35–21). Cats pass the eggs in their feces only 2–3 weeks after becoming infective. The oocysts can survive several years and are difficult to remove from the environment. Cats generally become infected through eating rodents or birds. Signs of animals having toxoplasmosis include depression, anorexia, fever, and lethargy. Prevention is key, so it is recommended that pregnant humans not clean or handle feline litter boxes during pregnancy. When cleaning cat litter, working outside in gardens or other sources of soil, or playing in sandboxes, adults and children should wash their hands immediately upon going inside. These are common sites for cats to eliminate and pass oocysts.

Brucellosis

Brucellosis is a reproductive disease that can affect all mammals. It is of most concern in dogs and cattle. This is a contagious disease spread through vaginal discharge, usually through breeding programs. In cattle this disease is also known as Bang's disease. This disease causes infertility problems and increased abortions, usually with the animal becoming sterile. When humans come in contact with milk from a cow that has brucellosis, they will become infected with a disease called undulant fever, which causes a high fever and thus makes brucellosis a public health issue. There is no treatment for brucellosis, so testing of breeding

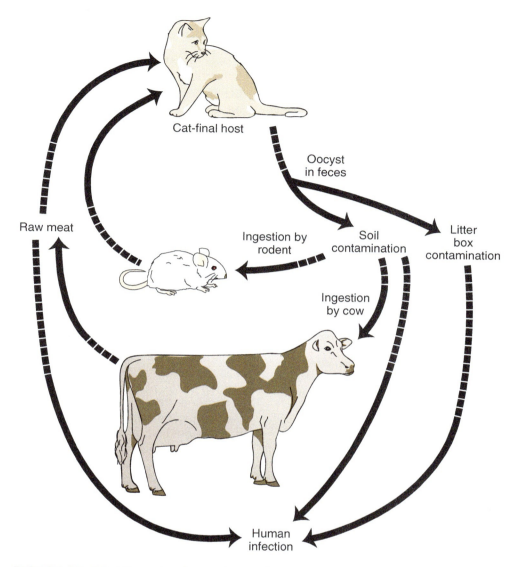

FIGURE 35–21 Life cycle of toxoplasmosis.

animals prior to reproduction is essential. A vaccine is available for cattle. Testing in breeding dogs is vital for control and prevention. The organism becomes infective in about 3 weeks.

Encephalitis

Encephalitis, commonly called sleeping sickness, is a viral disease that affects horses and people. Several strains of the disease are of concern to horses, such as eastern equine encephalitis (EEE), western equine encephalitis (WEE), and Venezuelan equine encephalitis (VEE). This disease is spread by mosquitoes, and the virus causes central nervous system infections. Birds are the reservoir host—a mosquito bites an infected bird that then spreads the disease on to horses. No treatment is available, but a vaccine is available to prevent these strains of disease. Signs of encephalitis in horses include possibly blindness, walking aimlessly, walking into things, fever, anorexia, depression, a sleepy appearance, difficulty swallowing, grinding teeth, paddling motion with front limbs, and eventually death.

Maintaining Good Animal Health

Many factors must be considered in maintaining ideal animal health. The environment is one of the essential factors. Vaccination programs are the best method of lowering disease risks (see Figure 35–22). Knowing

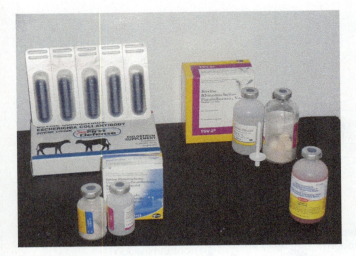

FIGURE 35-22 Examples of vaccines.

TABLE 35-5

Common Injection Routes

INJECTION ROUTE	LOCATION
SQ—subcutaneous	Under the skin
IM—intramuscular	Into a muscle
IV—intravenous	Into a vein
ID—intradermal	Into the skin (within the layers)
IN—intranasal	Into the nasal cavity
IP—intraperitoneal	Into the peritoneum or body cavity
IO—intraosseous	Into the bone

the common diseases that affect a species of animal or that are at high risk in a given area will help to decrease the chances that animals will become infected with disease. If animals do become infected, it is essential to isolate any sick animals, especially those housed in large groups, to decrease the spread of disease. Young animals can be preconditioned to reduce stress, which will lower the chances of disease. **Preconditioning** prepares young animals for possible stress factors, such as learning to load and travel, learning to be around large numbers of people or animals, and being in a show environment. Other factors related to keeping animals healthy include a clean and sanitary environment, a proper nutritional program, proper space, a yearly physical exam and health program, and a deworming program.

Treating Disease

When animals do become sick and need veterinary treatment, veterinary assistants should be familiar with medication treatments and routes of medication administration. Medications are called **biologicals**, that is, drugs used to treat disease. They are purchased from a pharmacy or a pharmaceutical company that manufactures and sells medications. Some biologicals are **antibiotics**, a group of medications used to kill organisms such as bacteria. **Pesticides** are a group of medications used to treat internal and external parasites. Some of these pesticides are known as **anthelmintics**, or dewormers. Medications may be given by many different routes, such as injectable, oral, or topical. Injections are administered in various ways (see Table 35-5). One route of injection is subcutaneous (SQ), or under the skin. Another method is intramuscular (IM), or into a muscle. Another common injection method is intravenous (IV), or into a vein.

Veterinary assistants are commonly supervised to perform SQ and IM injections. Oral medication is given by mouth. Topical medications are placed on the external body, such as the skin and hair coat. Some medications are additives that are mixed in with the food or water.

SUMMARY

Animal diseases are an important part of veterinary medicine, and veterinary assistants must have an understanding of normal and abnormal animal health. Some diseases are contagious, while others do not spread from animal to animal. Some diseases are zoonotic and create a public health hazard. Certain animal species are of concern, as they may be consumed by humans. Many factors must be considered when dealing with animal health and diseases. It is essential to know how to properly handle an animal that is ill and understand concerns related to animal diseases.

Key Terms

Ancylostoma caninum hookworm species that live in the small intestine of the host animal, where they attach themselves and feed on the host's blood

anemia loss of blood; causes blood cells not to replenish

anthelmintic a dewormer used to treat parasites in animals

antibiotic medication used to treat infection or disease

ascarids veterinary term for roundworms

babesia less common disease caused by the deer tick; causes signs of anemia, jaundice (yellow coloring of the skin and mucous membranes), fever, and vomiting; more severe stages can lead to kidney failure

bacterial class of disease spread through single-celled organisms called germs that are located in all areas of an animal's environment

bartonella bacterial infection caused by a flea bite that affects cats' bloodstream; infects the red blood cells and immune system

biological drugs used to treat disease

brucellosis reproductive disease of mammals; spread through breeding

coccidiosis common protozoan internal parasite that is a simple one-celled organism; present in all mammals

Coggins test blood test for horses that screens for equine infectious anemia

cutaneous larval migrans hookworm infection in humans caused by penetration of the skin

definitive host organism that provides life through the adult stages of the parasite

demodectic mange mite that invades the skin; all dogs raised typically by their mothers possess this mite, as mites are transferred from mother to pup via cuddling during the first few days of life

dermatophyte fungi

dermatophytosis fungal disease caused by an external parasite

Dipylidium caninum tapeworm species that can be up to 20 inches long and lives in the small intestine

direct contact contact with an animal or a bodily fluid from that animal

Dirofilaria immitis heartworm species that belongs to the same class as roundworms; they look similar in structure—like long, thin pieces of pasta

disease poor health that is a disturbance or change in an animal's function and structure

ear mites microscopic parasites that live within the ear canal, causing severe itching and a dark, crusty discharge

ectoparasites external parasites

ehrlichiosis bacterial disease caused by the lone star tick; causes fever, joint pain, depression, anemia, anorexia, weight loss, and some edema

encephalitis viral disease in horses and people that is also called sleeping sickness; several strains are spread by mosquitoes and affect the nervous system

endoparasites internal parasites

equine infectious anemia (EIA) highly infectious disease of horses; also known as swamp fever; a viral disease for which there is no vaccine and no cure

feline infectious anemia another name for hemobartonella, which causes infection within the red blood cells of cats

flea allergy dermatitis (FAD) allergic reaction to a flea bite caused by the flea's saliva that affects the animal's skin

fungal class of disease spread by single-celled organisms or spores that grow on the external body and other areas of the environment, mostly in humid conditions

heartworm disease disease spread by mosquitoes; of most concern in dogs, cats, and—occasionally—ferrets; causes long white worms to build up in the heart

hookworms common intestinal parasite of dogs and occasionally cats that have either teethlike structures or cutting plates with which they attach themselves to the wall of the intestine and feed on the animal's blood

host the animal that is infected by a parasite and that the parasite feeds off on

indirect contact method via which disease spreads through ways other than touching an infected animal, such as airborne or through bedding

intermediate host organism that provides immature and larval stages of a parasite

Iospora canis most common coccidia species affecting animals

larvae immature stage of a parasite

Lyme disease bacterial disease transmitted by the bite of a tick that causes joint inflammation in affected animals

microbiology the study of microscopic organisms

microfilaria tiny immature heartworm larvae that live in the bloodstream

mucous membrane (mm) the gums of an animal

mycology the study of fungus

nits lice eggs

oocysts single-celled eggs

Parascaris equorum roundworm species in large animals that are swallowed along with contaminated hay or water; hatch in the intestinal tract

parasitology the study of organisms that live off of other organisms to survive

pathogens a bacterium, virus, or other microorganism that can cause disease

pesticide chemical used to treat external parasites

plague disease that affects animals and humans; causes high fever, dehydration, and enlarged lymph nodes; can eventually lead to death

preconditioning building up an animal's body to prepare young animals for possible stress factors

proglotids segments that shed off a tapeworm and appear like small white pieces of flat rice

protozoan class of disease; protozoa are the simplest forms of life; they are single-celled organisms that are known as parasites that may live inside or outside the body

pupae immature larval stage that goes through several molting stages during growth by spinning a cocoon

rickettsial class of disease spread by biting insects, such as fleas and ticks

ringworm fungal condition that affects the skin of humans and mammals and is highly contagious

Rocky Mountain spotted fever rickettsial disease transmitted by the American dog tick; signs include fever, joint pain (arthrodynia), depression, and anorexia

roundworms visible long white worms; the most common intestinal parasite in both small and large animals

sarcoptic mange name for the skin disease caused by infection with the *Sarcoptes scabei* mite; causes a zoonotic mite skin condition in which mites burrow into the skin and cause infection

scabies sarcoptic mange condition in humans

strongyles common intestinal parasite in large animals; are long, thin, and white in appearance

supportive care therapy or treatment that includes such interventions as fluid replacement, medications to decrease vomiting and diarrhea, and pain medicine

taenia genus of tapeworm that affects companion animals and livestock

tapeworms intestinal flat worms that are segmented or have individual parts that grow and shed as the parasite ages

tick paralysis varying degrees of paralysis caused by saliva of a female tick

Toxascaris leonina roundworm species with the simplest life cycle

Toxocara canis roundworm species with a more complicated life cycle than some other roundworms; can ensure its species will be passed from generation to generation

Toxocara cati roundworm species similar to *T. canis*; may infect the lungs or abdominal tissue

Toxoplasma gondii single-celled organisms that can occur in any mammal but are shed through the feces of cats, which are the only animals that pass it through the feces

trematode flat worm lacking a body cavity; also called a fluke

Trichuris vulpis whipworm species; infects animals when they ingest contaminated food or water

vector mode of transmission of a parasite

viral class of disease caused by particles that are contagious and spread through the environment

whipworms common intestinal parasites; occur mostly in dogs; get their name from the whiplike shape of the adult parasite

zoonosis refers to disease passed from animals to humans

Review Questions

1. What are signs of normal animal health?
2. What are signs of concern in the abnormal health of an animal?
3. What are common internal parasites in animals?
4. What are common external parasites in animals?
5. What are some ways to treat or otherwise control parasites?
6. Name several diseases that are zoonotic.
7. What is preconditioning?
8. What are several routes of injection?

9. How does the tapeworm cycle begin in pets?
10. What disease is significant to prevent in pregnant women?
11. What mite has the potential for zoonosis in people?
12. Which test is required for interstate travel and showing of horses?
13. How long does it take heartworms to become infective in the adult stage in dogs?
14. What are tapeworm segments called?
15. What is another term for dewormers?

Clinical Situation

The Dhar family has just purchased a farm. They called the Alpine Veterinary Clinic to ask some questions about the sanitation of the farm and when they can start transporting their animals to the location. Amanda, a veterinary assistant at the facility, is talking with Mr. Dhar about his concerns. She asks, "I was wondering if you can tell me about the types of animals that were raised on this farm previously."

Mr. Dhar responded, "Well, the previous owners had some cattle, some horses, and several small animals. We saw most of their livestock and pets. I am just concerned about moving my animals here in case there are any health concerns I need to worry about. Is it possible to talk to the vet about this?"

Amanda said, "Sure, I will take your phone number and have Dr. Daniels call you."

- What are some concerns the new owner should have?
- What precautions might the veterinarian suggest he take with the farm?
- When might the veterinarian recommend he move his animals to the farm?

Section IV
Clinical Procedures

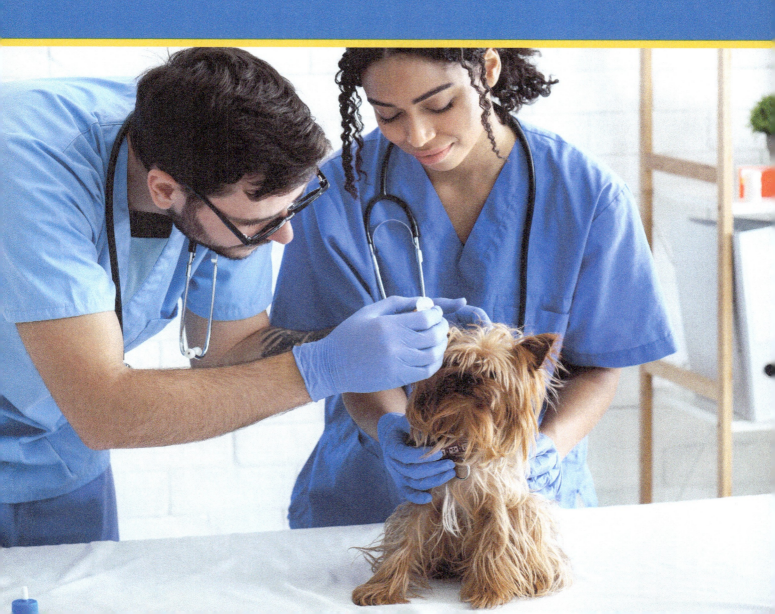

Chapter 36: Basic Veterinary Restraint and Handling Procedures

Objectives

Upon completion of this chapter, the reader should be able to:

36.1 Discuss safety concerns related to proper restraint and handling
36.2 List the equipment needs for animal restraint
36.3 Explain the circumstances necessary for animal restraint
36.4 Demonstrate how to properly tie knots used in animal restraint
36.5 Explain how to properly restrain small companion animals, such as dogs and cats
36.6 Explain how to properly restrain large animals, such as cattle and horses
36.7 Explain how to properly restrain exotic animals, such as birds and reptiles
36.8 Explain how to properly restrain rodents, such as mice and rats
36.9 Describe types of behavior
36.10 Describe how to recognize normal animal behavior
36.11 Describe how to recognize abnormal animal behavior

Introduction

Veterinary assistants, as well as the entire veterinary staff, should be trained in the proper safety and handling of animals for procedures involving animal restraint. This is important, as many veterinary patients may not be willing to have certain procedures done and their behaviors may change, making for an unsafe environment for both the animal and the staff. Therefore, it is extremely important that the veterinary staff learn how to properly restrain all types of patients. Practice is the most important part of training the staff in animal restraint and handling techniques.

Animal behavior relates to what an animal does and why it does it. The types of behaviors exhibited are varied. Some are genetically determined, or instinctive, while others are learned behaviors. Animals learn behaviors from their mothers, other animals, and people. Any behavior occurs due to a **stimulus**, an internal or external change that causes a reaction. Many behaviors are triggered due to hormones within the animal's body. The study of animal behavior is called **ethology**.

Instinctive Behaviors

Instinctive behaviors are an animal's natural reaction to a stimulus. Many behaviors in the early life of companion animals are instinctual. The kneading action of the paws of a puppy or kitten on the mother's mammary glands is an instinctual behavior that literally determines the young animal's survival. This behavior stimulates an equally instinctual behavioral response for the mother—lactation. Hormones are released from the brain, which leads to milk being released from the mammary ducts to the teats. This lactating response is seen in human beings as well.

Many instinctive behaviors can be observed between a mother and her offspring. Calling out to the mother as the young animal's eyes are still closed and nestling in with the rest of the litter to help maintain the body heat of such a small individual are instinctive behaviors (see Figure 36-1). The mother instinctively licking the young keeps their coat clean and aids in the process of feces and urine elimination. This licking action is comforting to the young and in the case of a cat elicits the purring response that is characteristic only of the cat. Further, many mothers with newborns will become protective and may become more defensive and difficult to handle. When possible, allow young animals to stay close to the mother during handling to reduce stress in both the mother and her young.

Marking is another instinctive behavior in animals, and it has both social and sexual purposes. **Pheromones** present in the urine give all organisms that encounter the marks left by others a variety of messages, including sexual receptiveness, eating habits, age, and overall health. Mating behaviors are also instinctual in nature. Female animals in estrus give off pheromones that indicate readiness for breeding. In return, male animals detect these odors and respond by locating the female and breeding with her.

Predatory behaviors are another example of instinct. Animals understand that to survive they must eat, and food must be hunted.

Submission and dominance postures also provide examples of behaviors that are instinctive. These postures allow animals to know who the leader is or that a particular animal doesn't pose a threat.

Instinctive behaviors include those that are acquired through an animal's genetic makeup, which is the set of instructions for building the parts of the body, including the brain. Each species' brain is built on a somewhat different pattern. When happy, a dog may wag its tail, a cat often purrs, and a human is likely to smile, while a turtle shows no outward signs at all. These variations are due to the different brain structures inherent to each species. The brain and other parts of the nervous system generate behaviors in response to environmental input and hormones. Some of these behaviors may be simple, immediate responses to simple stimuli, including the knee jerk reflex and the cry of pain when an animal is hurt.

Learned Behaviors

Learning is modifying behavior in response to specific experiences. Animals' **learned behaviors** can be classified in various ways. Understanding how an animal learns helps those in the veterinary field better serve clients with animals presenting with behavior problems. Some common behavior modification techniques include conditioning, modeling, and imprinting.

Conditioning

Animals can learn to associate one stimulus with another. This is a behavior that is learned through **conditioning**. Conditioning is the process of teaching an animal an action in relationship to another action. An example of conditioning is rewarding a dog for coming when a clicking sound is given. The dog hears the clicking sound and has learned that when it approaches its owner following this sound, it receives a reward such as food. Conditioning is used with a reward system and is the process of training or accustoming an animal to behave in a certain way or to accept certain circumstance. This clicking sound example presents an activity that has a punishment or reward. Classical conditioning involves linking a stimulus to an action, which results in a physical response. Operant conditioning involves linking a stimulus with a response, such as a reward.

Modeling

Another learned behavior is **modeling**. Modeling is when the animal learns a behavior through watching other animals perform the behavior. For example, in a pack animal such as the wolf, hunting behaviors fit this category.

Imprinting

Finally, **imprinting** is a learned behavior that comes to fruition due to a process that must occur within a distinct, usually short, time period within a few hours

FIGURE 36-1 Kittens will snuggle together to maintain body warmth.

of the birthing process. It involves an attachment to an object that emits adult behaviors and can be generalized to all examples of the object. For example, imprinting in a newborn horse can be achieved by rubbing its entire body and lifting and touching its feet to allow it to become accustomed to being touched and otherwise handled. These training techniques, if done properly, make an animal much easier to handle and more apt to adjust to changes in the environment. Much of this learned behavior allows animals to trust people.

Interpreting an Animal's Body Language and Behavior

It is important for veterinary staff to have knowledge of and experience in working around the types of animals that are seen in the facility. Each species has instincts and behavior characteristics that allow it to respond to the people around it, its environment, its mental and physical health status, and its natural instincts. When an animal's environment changes, its behavior patterns may change in response. A trip to the veterinary facility is an example of an environmental change that an animal may respond either positively or negatively to. Animals visiting the veterinary facility may exhibit characteristics indicating that they are happy, scared, or angry. This requires the veterinary assistant to gain knowledge and experience with the typical *body language* of the species they work with. Body language is communication by the animal about how it feels toward other animals, people, and its environment. Interpreting body language will help veterinary staff determine how easy or difficult an animal may be during times of handling and restraint.

The Happy Animal

The happy animal displays behavior that is typical for its species. A happy animal is relaxed; is alert; stands, sits, or lies with a comfortable appearance; and has ears up and forward (see Figure 36–2). Various species have other body language that displays happiness. A happy animal is easier to handle than one that is scared or angry. However, the veterinary assistant should be alert to changes in behavior depending on the procedures being performed.

The Scared Animal

The scared animal may be difficult to handle and become aggressive if not dealt with in an appropriate

FIGURE 36–2 A happy animal is relaxed, alert, and comfortable.

manner. Animals that are scared tend to have a stiff stance and tremor from nervousness, avoid direct eye contact, lay their ears flat or back on the head, and lower their body or tail to the ground (see Figure 36–3). They may become **submissive**, meaning they give in to humans due to an instinct that makes them feel threatened. This may lead to fear biting or some other protective instinct that makes them injure a human out of the instinctive desire to protect themselves.

The Angry Animal

An angry animal is aggressive. The term **aggression** refers to a behavior that makes an animal difficult and unsafe to handle. An animal may be aggressive for several reasons. Aggressive animals may display body language such as a stiff stance, teeth bared, head lowered to the ground, staring, and tail raised (see Figure 36–4). Aggressive animals may fall into the category of dominance aggression, fear aggression, territorial aggression, or redirected aggression. **Dominance aggression** refers to the "pack" animal instinct and social status within a group. **Fear aggression** refers to the defense reaction in response

CHAPTER 36 Basic Veterinary Restraint and Handling Procedures **603**

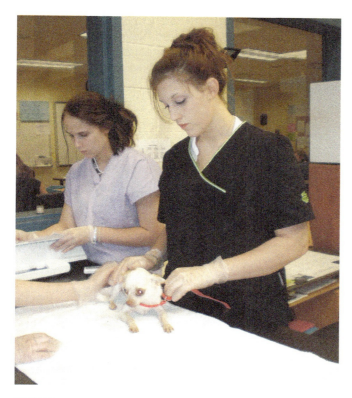

FIGURE 36–3 An animal that is scared will lower its body and put its ears back, close to its head.

FIGURE 36–4 An aggressive animal will bare its teeth and may vocalize or attack.

to being harmed and the animal's instinct to protect itself. Many animals are ruled by the "fight-or-flight" instinct, *fight* being protection and *flight* being to run away. **Territorial aggression** refers to an animal's protective nature over its environment, such as an owner, offspring, or food. **Redirected aggression** refers to an animal's predatory instinct whereby the animal turns its aggressive behaviors on its owner. This is the most serious situation involving animal aggression. It is important, once again, to note that the veterinary staff must gain experience of and knowledge about typical animal behavior and body language. It is very difficult to work with a species of animal and handle it safely without knowledge of what is typical or atypical.

MAKING THE CONNECTION

Refer to individual species chapters for more information regarding animal behavior and aggression.

Restraint Considerations

Restrain means to hold back, check, or suppress an action and to keep something under control using safety and some means of physical, chemical, or psychological action (see Figure 36–5). Various degrees of restraint are used by veterinary staff to allow an animal to be controlled for procedures in the interest of the safety of both the animal and people. Restraint may be accomplished in many forms: physical control using the hands, the body, or a piece of equipment; chemical control using a sedative or tranquilizer; psychological control using the voice, eyes, or body language of a person to help control of an animal.

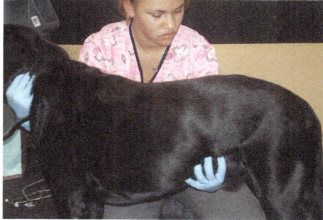

FIGURE 36–5 The veterinary assistant is maintaining physical restraint of this dog.

604 SECTION IV Clinical Procedures

 TERMINOLOGY TIP
A **sedative** or **tranquilizer** is a medication given to an animal to keep it calm during certain stressful procedures or circumstances.

Veterinary assistants should learn animal restraint techniques for the species they work with through experience and practice. The goal of any restraint situation is to minimize the effects of the handling by controlling the animal safely and with the least amount of stress possible—"less is more." It is important for all veterinary professionals to use good judgment, contain their temper, and be alert to the animal's signals at all times. Working as a team is an essential component of maintaining focus and safety (see Figure 36–6). All people involved with animal restraint should heed the guidance of those who have restraint experience and provide safe and expert coaching.

Animal Safety

Animal safety must be considered with every patient. Many animals are accustomed to humans, but those that are not or that are removed from familiar territory may become easily stressed. This consideration is especially important when working with very young or very old animals. Young animals have young bones that are cartilaginous, making them more pliable due to growth and formation requirements, and they should be handled with particular care. Older animals may be arthritic and in pain, and should also be handled with particular care. Rough handling may delay recovery or could even lead to death. Animals that are pregnant must also be treated with caution during times of stress. Every time restraint of animals is necessary, the safety of both the animal and the staff must be considered. Never allow nonveterinary staff or animal owners to restrain an animal; any mishap could potentially be a legal issue for the staff or veterinary facility.

Restraint Equipment

When animals become difficult and attempt to bite, scratch, or kick, it may be necessary to use other means of control to ensure safety, for example, muzzles, antikick bars, hobbles, or stanchions. Determining how to handle overly aggressive behaviors depends on the species of animal, situation involved, and behavior pattern of the animal. The hands can be used to soothe and calm an animal when it is placed in a safe location. The veterinary assistant should consider his or her hands to be fragile instruments that can easily be injured when placed in the wrong area. Most restraint equipment is designed for use on a specific species, with most being used to distract the animal from the pressure or pain of a procedure while being worked on. Equipment can cause injury if not properly used.

Muzzles

A muzzle is a device placed over the mouth and nose to prevent an animal from biting (see Figure 36–7). Muzzles are commonly made for dogs, cats, and horses.

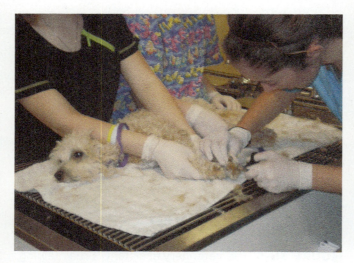

FIGURE 36–6 Teamwork is important when working with animals that require restraint.

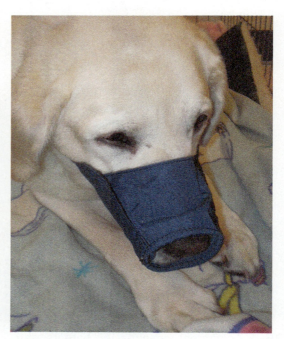

FIGURE 36–7 Muzzles can be used to prevent an animal from biting.

Muzzles may be made of nylon, leather, wire, or basket materials. Muzzles can also be made of gauze, tape, or leashes. Muzzles should be individually fitted to the animal and used for short-term procedures. Always make certain the animal is able to breathe easily while a muzzle is in place.

COMPETENCY SKILL 32

Applying a Commercial Muzzle on Dogs and Cats

Objective:

To properly and safely apply and remove a muzzle to prevent the animal from biting

Preparation:

- Appropriate-sized muzzle

Procedure for Dogs:

1. Hold the muzzle with the narrow side up and the wide side down. Grasp the muzzle on each side.
2. Stand to the side of the dog and place the muzzle carefully over the nose of the dog. The muzzle should fit snugly (see Figure 36–8A).
3. Place each end behind the ears and snap in place like a seat belt.
4. Pull the end snugly (see Figure 36–8B).
5. Release the clasp by pinching and pulling open.
6. Hold one end of the muzzle and allow it to slip off the nose of the dog.

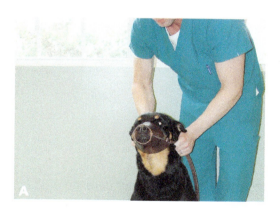
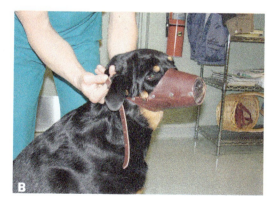

FIGURE 36–8 Placing a muzzle on a dog. (A) The muzzle is placed on the dog with the straps pulled to the back of the head. (B) The straps are clasped and pulled snugly.

Procedure for Cats:

1. Hold the muzzle with the wide side upward and the narrow side down.
2. Stand to the side or the rear of the cat and gently slip the muzzle over the face, including the eyes (see Figure 36–9A).

3. Secure the back of the muzzle behind the ears by closing the Velcro flap. Make sure the flap is pulled snugly and secure (see Figure 36–9B).
4. Restrain the limbs to prevent the cat from scratching off the muzzle.
5. Release the Velcro flap and pull the muzzle off the face.

 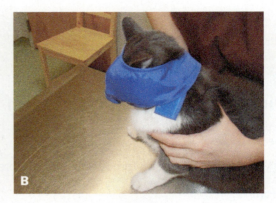

FIGURE 36–9 Applying a muzzle to a cat. (A) Cover the cat's face with the muzzle, pulling the straps to the back of the head. (B) Secure the straps.

COMPETENCY SKILL 33

Applying a Gauze Muzzle

Objective:

To properly and safely apply a muzzle to prevent the animal from biting

Preparation:

- Gauze

Procedure:

1. Make a loop using a piece of heavy gauze long enough to fit the muzzle circumference of the dog (see Figure 36–10A).
2. Slip the loop over the dog's nose while standing to the side of the dog. Pull the loop snugly (see Figure 36–10B).
3. Take each side under the jaw and tie in a square knot under the chin (see Figure 36–10C).
4. Take each side behind each ear and tie in a knot with a bow behind the ears (see Figure 36–10D).
5. Untie the bow and slip the muzzle off the nose of the dog.

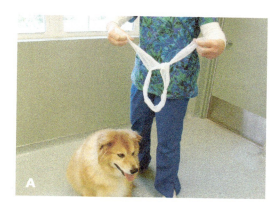

FIGURE 36–10 Applying a gauze muzzle. (A) Make a loop large enough to fit over the dog's muzzle. (B) Tighten the gauze on the muzzle. (C) Tie with a square knot under the muzzle. (D) Tie with a bow behind the dog's head.

COMPETENCY SKILL 34

Applying a Tape Muzzle

Objective:

To properly and safely apply a muzzle to prevent the animal from biting

Preparation:

- Tape

Procedure:

1. Make a muzzle using a piece of 1-inch tape that is long enough to fit the muzzle circumference of the dog.
2. Fold the entire piece of tape onto itself to make a long piece of nonstick tape.
3. Make a large loop with the tape.
4. Slip the loop over the dog's nose while standing to the side of the dog. Pull the loop tightly.
5. Take each side under the jaw and tie in a square knot under the chin.
6. Take each side behind each ear and tie in a knot with a bow behind the ears.
7. Untie the bow and slip the tape muzzle off the nose of the dog.

COMPETENCY SKILL 35

Applying a Leash Muzzle

Objective:

To properly and safely apply a muzzle to prevent the animal from biting

Preparation:

- Leash

Procedure:

1. Make a loop using a commercial nylon leash long enough to fit the dog (see Figure 36–11A).
2. Slip the loop over the dog's nose while standing to the side of the dog. Pull the loop tightly (see Figure 36–11B).
3. Take each side under the jaw and tie in a square knot under the chin.
4. Take each side behind each ear and tie in a knot with a bow behind the ears (see Figure 36–11C and D).
5. Untie the bow and slip the leash muzzle off the nose of the dog.

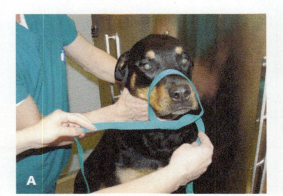
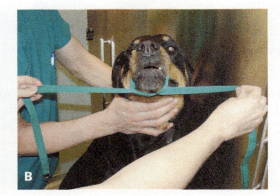
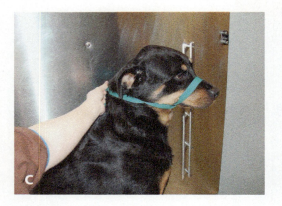
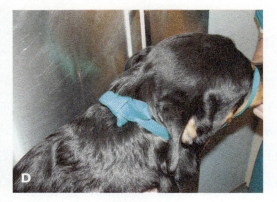

FIGURE 36–11 Applying a leash muzzle to a dog. (A) Make a loop large enough to go around the dog's face. (B) Tighten the loop around the dog's face. (C, D) Tie the leash behind the dog's head.

Towels

Towels can be used to restrain small animals by keeping them wrapped and contained within the material (see Figure 36–12). Many animals are more comfortable wrapped and secure, especially when they feel that they are hidden from view. Small pets can be wrapped in a towel with a front or back leg exposed for blood collection or injections. It's often easier to wrap a small animal in a towel while allowing the head to be exposed to give medications.

Squeeze Cages

Squeeze cages may also be used with small animals to help to contain the animal without placing a person's hands directly on the animal but on a cage with open areas that are used to sedate the animal or give vaccines. Squeeze cages are often used for wild or feral animals to prevent bites and allow ease of injections. These cages have a divider that moves to one side, allowing the animal to be gently squeezed to the side of the cage, thus decreasing the animal's movement.

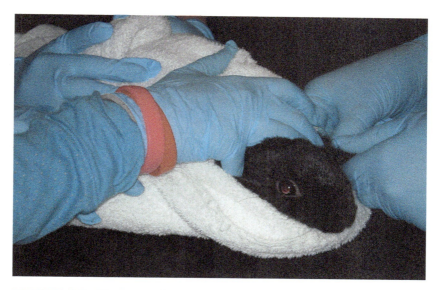

FIGURE 36–12 A towel can be used to restrain some animals and make them feel more secure.

COMPETENCY SKILL 36

Use of a Towel to Restrain a Cat

Objective:

To successfully use a towel to calm and restrain a cat

Preparation:

- Heavy towel

Procedure:

1. Gently apply a towel over the cat's body and head.
2. Reach under the towel and scruff the cat.
3. Pull the cat's body toward your body to keep secure.
4. Keep the towel over the cat's head to keep the cat comfortable.

COMPETENCY SKILL 37

Use of a Squeeze Cage

Objective:

To safely and properly restrain an animal using a squeeze cage

Preparation:

- Squeeze cage

Procedure:

1. Open the top of the squeeze cage.
2. Place the cat inside the cage and close the lid.
3. Use the side flaps to pull the cat forward against the side of the cage.
4. Release the sides of the cage.
5. Open the lid and remove the cat from the cage.

Leather Welding Gloves

Leather welding gloves may also serve as protection for the hands when handling certain species (see Figure 36–13). These restraint gloves only serve to slow down or decrease the severity of a bite, as many dogs, cats, and birds are still able to bite or pinch through leather. Care must be taken to not over-restrain with the use of the bulky gloves, as too much pressure can injure an animal, and the gloves tend to reduce one's sense of touch and strength.

Antikick Bars

Antikick bars are used to immobilize cattle when giving injections or medications.

Hobbles

A hobble can limit the movement of an animal by tethering its legs. Hobbles are made of various materials, such as leather, nylon, or rope. They look similar to a set of handcuffs that are placed on the animal's legs.

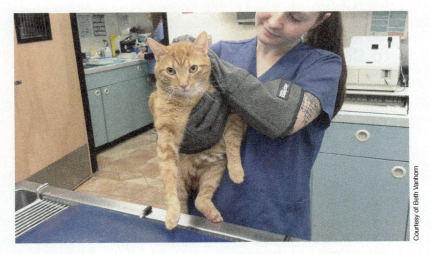

FIGURE 36–13 Leather welding gloves can be worn to protect the hands when working with some animals.

CHAPTER 36 Basic Veterinary Restraint and Handling Procedures   611

COMPETENCY SKILL 38

Use of Welding Gloves

Objective:

To properly use welding gloves to restrain animals for procedures

Preparation:

- Welding gloves

Procedure:

1. Apply a welding glove to each hand.
2. Grasp the animal firmly over the neck with one hand and around the body with the other hand. The scruff technique may be used in cats.
3. Use caution to avoid injuring the animal with too harsh of a grip. Use caution to not let the animal get away.
4. A towel can also be used to keep the animal covered and quiet.

COMPETENCY SKILL 39

Use of the Antikick Bar on a Cow

Objective:

To properly and safely restrain a cow

Preparation:

- Antikick bar

Procedure:

1. Locate the metal antikick bar within the veterinary facility or mobile vehicle.
2. Stand behind the cow or at the front or side of the cow closer to the cow's head.
3. Avoid standing to the side of the rear legs.
4. Position the bar around the cow's flanks.
5. Tighten the bar down over the flanks.

COMPETENCY SKILL 40

Placement of Hobbles on Horse

Objective:
To safely restrain a horse and keep it from wandering too far too quickly

Preparation:
- Hobbles

Procedure:
1. Locate the hobbles within the veterinary facility or mobile vehicle.
2. Stand to the left side of the horse about 1 foot from the middle of the horse's body, using caution so as not to be kicked.
3. Place the left hobble around the left rear leg and buckle in place.
4. Move to the right side of the horse and place the right hobble around the right rear leg and buckle in place.

Stanchions or Headgates

Headgates and stanchions are structures that are used to contain large animals so that procedures can be performed safely. A headgate is a large cratelike structure that an animal, such as a cow, can be loaded into. At the front of the structure, the device has bars that lock around the animal's head, securing it. On some models, the sides can then be adjusted to prevent the animal from moving around in the structure. On some, side panels can be removed to give the handler or veterinarian access to the animal so that he or she can administer procedures such as injections.

Planning the Restraint Procedure

The best place to administer restraints is an area that offers enough room for both the animal and restrainers and that is clean, dry, and well lit. This may be an exam room or a barn stall. Several aspects should be considered when planning restraint areas. It is best to discuss restraint plans with everyone involved in handling the animal. Such communication allows everyone to work together safely. Move any costly equipment out of the way so as not to damage it. Ensure a nonslip area for both staff and animals. Especially in warm areas and with certain procedures, temperature should be considered when working outside with animals. Some species may become hyperthermic (i.e., an excess increase in body temperature) easily when handled roughly or stressed in warm weather. When heat is a factor in handling animals, early morning may be the best time. When heat is a factor, allow for a shaded area, use fans, and provide water. In cold weather, hypothermia may be a factor. Anesthetized animals have difficulty maintaining their body temperature. Doors and windows should be monitored when working with unruly animals. Always plan what should be done if an animal happens to get away from a restrainer. Areas to hold and contain livestock should be inspected for safety before use. Regardless of how well a restraint is planned, unpredictable circumstances may cause problems. It is best to always anticipate what may happen and have a backup plan if something should occur that requires deviation from the plan. This encourages for safe planning and awareness. Do not become stressed when animals are difficult, as animals can sense such emotions and may become even more difficult to control.

Restraint Knots

Knots are made from one to two pieces of rope material, where one section of the rope prevents slipping of another. Knots allow animals to be tied and thus restrained for a temporary amount of time. Ropes

CHAPTER 36 Basic Veterinary Restraint and Handling Procedures

COMPETENCY SKILL 41

Use of a Stanchion or Headgate

Objective:

To safely restrain a cow for various medical treatments or procedures

Preparation:

- Headgate or stanchion chute

Procedure:

1. Open the headgate or stanchion chute.
2. Push the cow forward into the chute.
3. As the cow reaches the front of the chute, allow the head to enter the gate.
4. Close the gate at the base of the neck at the shoulders (see Figure 36–14).
5. If the chute has sidewalls, they should be squeezed in toward the cow to prevent movement.

FIGURE 36–14 Use of a headgate.

should be inspected for tears and the ability to hold the animal. Ropes may be used to tie animals in a standing position or in a lying position. It is important that the knots used to tie an animal are properly placed and tied.

Square Knot

A **square knot** is commonly used to secure an animal. It is a nonslip knot that doesn't come untied or easily loosen. The saying "right over left, left over right" can be remembered as the common method for making a square knot. Two ropes may be used to make a square knot. When complete, it looks like two intertwined loops and is easily untied when the opposite ends are pushed together. A single rope can be used to make the square knot by passing one shorter end behind the opposite longer end, folding a loop, and pushing it through the loop created by the ends of the rope (see Figure 36–15).

FIGURE 36–15 Square knot. (A) Loop in the short end held over the long end or the rope attached to the animal's halter. (B) Place the fold of rope into the loop.

 COMPETENCY SKILL 42

How to Tie a Square Knot

Objective:
To tie a knot that will not slip or easily loosen

Preparation:
- Rope

Procedure:
1. Take a single length of rope and apply the right over the left.
2. Then apply the left over the right.
3. The end product will appear intertwined.
4. Untie by pushing the opposite ends together.

COMPETENCY SKILL 43

How to Tie a Slip Knot or Quick-Release Knot

Objective:

To tie a knot that can be quickly and easily removed

Preparation:

- Rope (at least 2 feet)
- Tie area with a pole and space enough to safely contain a horse

Procedure:

1. Place the rope over a pole or tie area.
2. Allow about 2 feet of rope for the animal to move slightly.
3. Make a loop in the rope close to the pole or tie area.
4. Pass the loop under the lead rope near the pole and pass it through the loop. The resulting configuration should look like a pretzel.
5. Pull the loop to tighten the knot.
6. Untie by pulling the end of the lead rope for a quick release.

Reefer's Knot

A **reefer's knot** is a single bow knot that allows a nonslip, quick-release tie. It is the same as the square knot with the exception that the second throw is made upon itself, creating a hold that can easily be untied by pulling the end of the rope (see Figure 36–16). This is a common tie for large animals to prevent them from injuring their heads and necks during restraint.

FIGURE 36–16 Reefer's knot. (A) Form a loop with the short end crossing over the long end. (B) Wrap the short end of the rope around the long end. (C) Pass the short end through the loop and pull in the direction indicated by the arrow. (*Continues*)

FIGURE 36–16 (*Continued*)

COMPETENCY SKILL 44

How to Tie a Reefer's Knot

Objective:

To tie a knot that can be quickly and easily removed

Preparation:

- Rope
- Tie area

Procedure:

1. Place the rope around a pole or tie area.
2. Make a loop in the end of the long end so the short end overlaps the long end.
3. Pass the short end of the rope up through the loop.
4. Reach under the long end of rope and grasp the short end so that it wraps around the long end.
5. Pass the short end of rope back through the loop in the opposite direction of the first pass.
6. Pull securely.
7. To release, pull short end of rope back through original loop and pull to release.

Half Hitch

A **half hitch** is a tie that makes a loop around a stationary location, such as a post or fence. One pass is made around a location, with the next pass going through the resulting loop and pulling up on the pinch ends of the hitch between the standing part of the rope and the item it is tied to (see Figure 36–17). The half hitch is commonly used to secure an animal to a surgery table.

FIGURE 36–17 Half hitch. (A) Pull in direction shown by arrow. (B) Second half knot.

COMPETENCY SKILL 45

How to Tie a Half Hitch

Objective:
To tie a knot that will securely restrain an animal to a particular site

Preparation:
- Rope
- Pole or tie area

Procedure:
1. Place the rope over a pole or tie area.
2. Pass the short end of rope under the long end and then back over the top, thus creating a closed loop around a tie area.
3. Push the rope down between the tie area (e.g., a pole) and the loop that was formed.
4. Pull the loop tight.
5. Pass the short end of rope over and under the end forming a loop.
6. Pass the short end up through the loop and pull it tight.

Restraint Positions

Common restraint techniques in small animals include standing restraint, sitting restraint, sternal recumbency, lateral recumbency, dorsal recumbency, and restraint for blood collection. The term **recumbency** indicates a lying position. **Standing restraint** is used to keep an animal standing for a procedure, thus preventing it from sitting or lying down. **Sitting restraint** is used to keep an animal in a sitting position for ease of completing a procedure. **Sternal recumbency** involves placing the animal on its chest for restraint. **Lateral recumbency** involves placing the animal on its side for restraint and may be done in left or right lateral recumbency. In left lateral recumbency, the left side is down and in contact with the table or the floor. **Dorsal recumbency** is used to place the animal on its back for restraint and is common during surgical procedures and radiology techniques.

COMPETENCY SKILL 46

Standing Restraint of Dogs and Cats

Objective:

To safely restrain a dog or cat while it is in a standing position

Preparation:

No special equipment is required. Small dogs and cats may require only one restrainer; medium to large dogs will require two or more restrainers.

Procedure for a Dog:

1. Dog should be grasped in a bear hug around the head and neck with one arm.
2. The free arm should be placed under the dog's stomach, and the dog should be raised to a standing position (see Figure 36–18).
3. Do not allow the dog to sit down.
4. Larger dogs may require an additional restrainer to stand behind the dog to prevent it from backing up.

Procedure for a Cat:

1. Scruff the cat with the dominant hand and support the abdomen with the other hand (see Figure 36–19).
2. Keep the cat's body close to your body for better control.
3. Monitor the cat for changes in body language and divert the cat's attention as necessary.

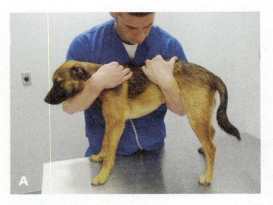 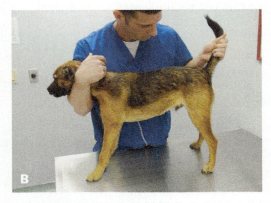

FIGURE 36–18 (A) Standing restraint. (B) Standing restraint holding the tail up.

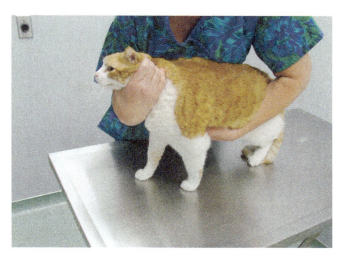

FIGURE 36–19 Standing restraint of the cat.

 COMPETENCY SKILL 47

Sitting Restraint

Objective:
To safely restrain a dog in the sitting position

Preparation:
No special equipment is required. More than one person may be necessary to properly restrain a larger animal.

Procedure:
1. Dog should be grasped in a bear hug around the head and neck with one arm.
2. The free arm should pull the dog's body close to the restrainer and gently place pressure over the back to place the dog in a sitting position (see Figure 36–20).
3. Small dogs may require only one restrainer; medium to large dogs will require two or more restrainers.

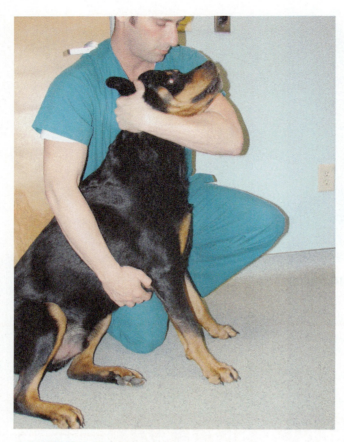

FIGURE 36–20 Sitting restraint.

 ## COMPETENCY SKILL 48

Sternal Recumbency of Dogs and Cats

Objective:

To properly and safely restrain a dog or cat for procedures

Preparation:

No special equipment is necessary. Small dogs may require only one restrainer; medium to large dogs will require two or more restrainers.

Procedure for Dogs:

1. The dog should be in the standing position.
2. Apply pressure over the hips, forcing the dog to sit.
3. Slide one hand under the front legs while applying pressure over the shoulders with the free hand.
4. Maintain pressure over the back and shoulders until the dog is lying on its chest.
5. Restrain the head in a bear hug with one arm.
6. Use the free arm to apply continued pressure over the back to prevent standing (see Figure 36–21).

CHAPTER 36 Basic Veterinary Restraint and Handling Procedures

FIGURE 36–21 Sternal recumbency.

Procedure for Cats:

1. The cat should be in standing restraint with the scruff method applied with the dominant hand.
2. Gently place the free hand over the back near the hips.
3. Reach around the cat's body with the free hand and gently use the elbow to place the cat on its chest and restrain both front and rear limbs to keep the cat from scratching (see Figure 36–22).
4. A towel can be helpful in cats that are aggressive or scared.

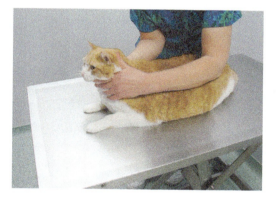

FIGURE 36–22 Sternal recumbency of the cat.

COMPETENCY SKILL 49

Dorsal Recumbency

Objective:

To properly and safely restrain a dog or cat for procedures

Preparation:

No special equipment is necessary. Small dogs may require only one restrainer; medium to large dogs will require two or more restrainers.

622 SECTION IV Clinical Procedures

Procedure for Dogs:

1. The dog or cat should be placed in lateral recumbency as seen in the following skill.
2. The dog or cat is gently rolled onto its back with the front legs extended forward and the back legs extended backward (see Figure 36–23).
3. The head should be kept between the front legs and held securely.
4. The rear legs should be kept still to avoid injury to the back.

Cats may be restrained with one or two people, depending on the attitude of the cat. The wrist and forearm can be used to hold down the head, but caution must be used to monitor the cat's mouth and avoid being bitten, such as use of a blanket or heavy restraint gloves!

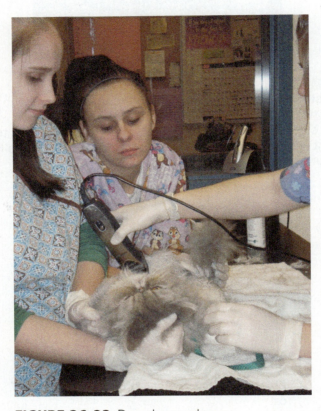

FIGURE 36–23 Dorsal recumbency.

 COMPETENCY SKILL 50

Lateral Recumbency

Objective:

To properly and safely restrain a dog or cat for procedures

Preparation:

No special equipment is necessary. Small dogs and cats may require only one restrainer; medium to large dogs will require two or more restrainers.

Procedure for Dogs:

1. Determine which side of the dog must be placed down on the floor or table.
2. Stand on the side of the animal that is to be the downside.
3. Reach both arms over the back and chest of the dog and firmly grasp the front and rear paws (see Figure 36–24A).
4. With a sudden jerk, flip the dog onto its left side. Caution should be used to avoid having the dog's head hit any objects or the floor.
5. The front and rear limbs that are on the down (left) side should be firmly grasped to prevent the dog from standing.
6. The upper body of the restrainer should be used to apply pressure over the back and chest of the dog to prevent standing.
7. The elbow of the restrainer should apply pressure over the neck of the dog to avoid biting (see Figure 36–24B).

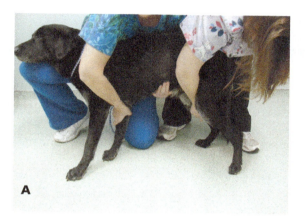 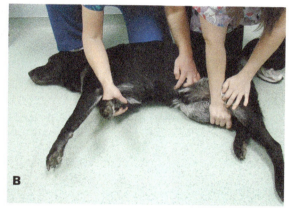

FIGURE 36–24 Placing a dog in lateral recumbency. (A) Place the arms over the dog's back and grasp the legs that will be on the floor side. (B) Gently lift and lower the dog to the floor and use pressure of your body to maintain control to prevent the dog from standing.

Restraint for Blood Collection

In small animals, blood collection is commonly performed in one of three locations: the cephalic vein, the jugular vein, or the saphenous vein. The **cephalic vein** is located in the medial aspect of the front limbs, and blood collection from this site is called cephalic **venipuncture**. The restraint for blood collection in the cephalic area is typically done in sitting or sternal recumbency. The **jugular vein** is located on either side of the neck in the lower throat area. The restraint for jugular venipuncture is commonly sitting or sternal recumbency. The **saphenous vein** is located on the lateral surface of the rear limbs just proximal to the hock. The restraint commonly used for saphenous venipuncture is lateral recumbency.

 MAKING THE CONNECTION

Many terms refer to directional locations on an animal's body and are useful in restraint procedures. Review Chapter 1 for discussion of anatomical and directional terminology.

COMPETENCY SKILL 51

Cephalic Venipuncture Restraint for Dogs and Cats

Objective:

To properly and safely restrain a dog or cat while obtaining a blood sample from the cephalic vein

Preparation:

No special equipment is necessary.

Procedure for Dogs:

1. Place the dog in sitting recumbency.
2. Stand on the opposite side of the limb to be restrained.
3. Restrain the head using a bear hug.
4. Use the free arm and hand to reach around the backside of the dog and hold the target leg at the elbow.
5. Extend the elbow forward.
6. Apply pressure above the elbow joint by making a fist and allowing the vein to pop up over the front limb.
7. Keep the leg extended throughout the procedure (see Figure 36–25).
8. When the venipuncture is complete, gently apply pressure gently apply pressure until a bandage is applied.
9. If blood is on the fur, clean with hydrogen peroxide; then fur with water and dry.

Procedure for Cats:

1. Place one hand around the head and under the jaw to prevent biting. Use the free hand to extend the elbow of the leg that is being used for blood collection.
2. Grasp the elbow and apply pressure to hold off the cephalic vein at the top of the target leg (see Figure 36–26).
3. Place a thumb over the puncture site after blood collection to stop bleeding or until a bandage is applied.
4. Clean the fur with hydrogen peroxide then wash the area with water.

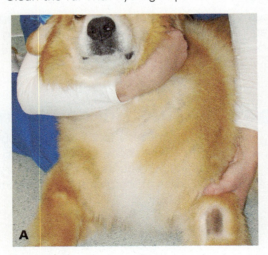
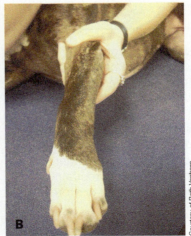

FIGURE 36–25 (A, B) Cephalic venipuncture in a dog.

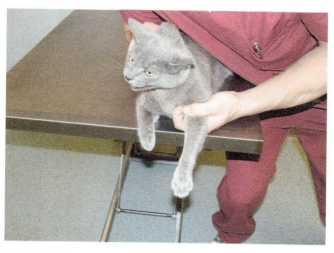

FIGURE 36-26 Cephalic venipuncture in a cat.

 COMPETENCY SKILL 52

Jugular Venipuncture Restraint for Dogs and Cats

Objective:

To properly and safely restrain a dog or cat while obtaining a blood sample from the jugular vein

Preparation:

No special equipment is necessary.

Procedure for Dogs:

1. A large dog should be placed on the floor in sitting or sternal recumbency. A small dog should be placed on a table with the front legs extended over the edge.
2. Collars should be removed to prevent injury.
3. The head is extended upward and away from the person collecting blood. The muzzle is held closed to prevent biting (see Figure 36–27).
4. The free arm is held over the body to prevent moving.
5. Once the needle is removed from the jugular vein, the restrainer should gently apply pressure until a bandage is applied.
6. Clean the fur with hydrogen peroxide; rinse with water and dry.

Procedure for Cats:

1. Place one hand around the head with the palm grasping the lower jaw to prevent biting.
2. With the free hand, hold the front limbs, placing the fingers between each paw and holding firmly to prevent scratching (see Figure 36–28).
3. Extend the front limbs over the edge of a table.
4. Extend the head upward securely.
5. Place a finger over the puncture site after blood collection to stop bleeding.
6. Clean the fur with hydrogen peroxide, then wash the area with water.

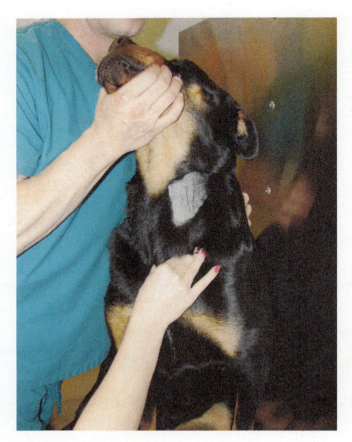

FIGURE 36–27 Jugular venipuncture in a dog.

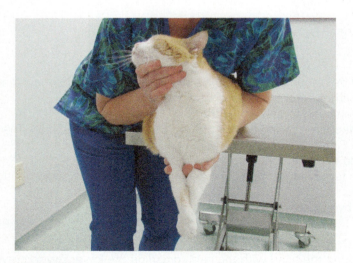

FIGURE 36–28 Jugular venipuncture in a cat.

COMPETENCY SKILL 53

Saphenous Venipuncture Restraint in Dogs and Cats

Objective:
To properly and safely restrain a dog or cat while obtaining a blood sample from the saphenous vein

Preparation:
No special equipment is necessary.

Procedure for Dogs:
1. Place the dog in lateral recumbency according to the limb being used. The hind limb being used is the one that should be on top.
2. Restrain the head in a bear hug.
3. Modify the restraint for lateral recumbency by placing the hand restraining the hind limbs around the top of the leg below the knee or thigh area.
4. Maintain pressure over the lateral surface of the limb so that the saphenous vein will pop up (see Figure 36–29).
5. Once the venipuncture is complete, place a thumb over the blood collection site and maintain pressure until the bleeding stops or a bandage is placed.
6. If blood is on the fur, clean with hydrogen peroxide; then rinse fur with water and dry.

Procedure for a Cat:
1. Place the cat in lateral recumbency using the scruff and stretch technique.
2. Modify the restraint for lateral recumbency by placing the hand restraining the hind limbs around the top of the leg below the knee or thigh area (see Figure 36–30).
3. Maintain pressure over the lateral surface of the limb so that the saphenous vein will pop up.
4. Once the venipuncture is complete, place a thumb over the blood collection site and maintain pressure until the bleeding stops or a bandage is placed.
5. If blood is on the fur, clean with hydrogen peroxide; then rinse fur with water and dry.

FIGURE 36–29 (A, B) Saphenous venipuncture in a dog.

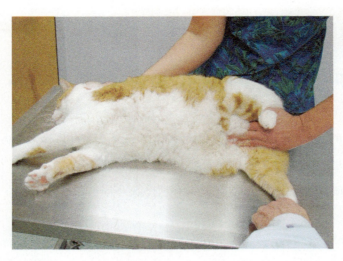

FIGURE 36–30 Saphenous venipuncture in a cat.

Restraint Procedures and Techniques

There are various methods of restraint for large and small animals. Many of these procedures include using equipment that is meant to protect both the animal and the restrainer and other veterinary staff. Basic restraint begins with removing an animal from its cage, stall, or pen. This is done with a type of head control device specific to the animal species. The control device should be properly fitted and placed over the animal's head or neck area according to the animal husbandry and production system methods discussed in previous chapters. For example, dogs are commonly controlled via a leash and collar, whereas horses are commonly controlled via a halter and lead rope. Many basic components are used for handling animals for physical exams.

Small Animal Restraint

Small animals typically are restrained for exams and veterinary procedures in which the animal is held in a safe manner through body control. Injuries that commonly occur to veterinary staff during small animal restraint include bites and scratches. Small animals include cats, dogs, rodents, rabbits, ferrets, reptiles, and birds. When working with small animals, it is helpful to talk with and reassure the animal and use several techniques that may divert its attention from the procedures being done. Such diversions that may be helpful include not only talking to the animal but also making soothing and calm noises, lightly blowing on the face, lightly rubbing the temple area, or a combination of these techniques. Sometimes it is necessary to be creative when distracting an animal's attention from a procedure that is being done so that the restraint can be safely undertaken. Learning about and monitoring animal body language helps ensure safe restraint outcomes.

 MAKING THE CONNECTION

Species-specific restraint techniques have been covered in previous chapters on individual species production management practices. Please see chapters in Section II. ■

Cats

Cats tend to be one of the most difficult domesticated animals to control during restraint when they become upset and aggressive from stress. Cats are likely to bite and scratch, which—due to increased bacteria (compared to other animals) on the teeth and nails—can cause severe skin infections quickly and easily. Caution should always be taken if a scratch or bite ensues; the wound should be cared for immediately. If any wounds caused by a cat appear to be infected, a health care provider should be consulted.

One of the most important parts of restraining cats is to properly and safely restrain and have control over the head. Towels placed over a stressed cat may allow it to feel less threatened, as cats feel more secure when they are hidden under an object (see Figure 36–31). Cats that become aggressive may be muzzled with special cat muzzles that cover the mouth

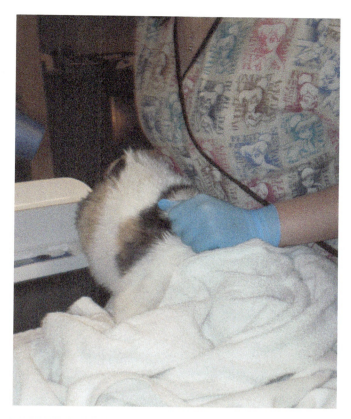

FIGURE 36–31 A towel can be used to help remove a cat from a cage.

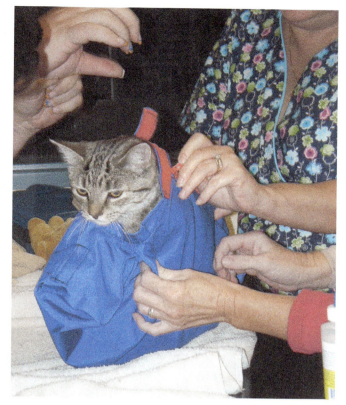

FIGURE 36–32 A cat bag can be used to help control and restrain the cat for procedures.

and eyes. **Cat bags** are also available to control the limbs and allow better control over the head, which is the only body part exposed when a cat is placed in the cat bag (see Figure 36–32). Cats may also be placed in squeeze cages or anesthesia chambers for better control and less need for hands-on restraint. **Squeeze cages** are wire boxes with small slats that allow injections such as vaccines or sedatives to be administered. They have handles that pull a cat to the side of the box for better access. An **anesthesia chamber** is used to sedate a cat or other small mammal under gas anesthesia that is pumped into the glass or plastic chamber. Holes in the chamber allow the anesthesia machine to attach to the chamber and filter gas and oxygen in to induce sedation. These chambers can also be used when examining and observing a cat, without the need for hands-on restraint. They can further be used to sedate a difficult or aggressive cat if need be.

Some hands-on restraint techniques that work specifically in cats include the scruff and the stretch. The **scruff technique** is used on the base of the neck over the area where the skin is elastic and able to be grasped in a clenched fist. This may divert the cat's attention from the procedure that is being done. It gives some control over the head, but safety must be monitored; a cat can easily turn and bite, even with the best grip. The **stretch technique** is done by scruffing the cat with one hand while in lateral recumbency and using the free hand to hold the rear limbs and pull them dorsally, thus putting the cat in a stretch position.

COMPETENCY SKILL 54

Head Restraint of a Cat

Objective:

To safely restrain a cat by manually holding its head

Preparation:

No specific equipment is required.

Procedure:

1. Place one hand around the side of the head while standing over and to the rear of the cat.
2. Place the thumb on the top of the head and the fingers under the chin to hold the mandible (lower jaw) shut.

See Figure 36–33. This technique can be done with the left or right hand.

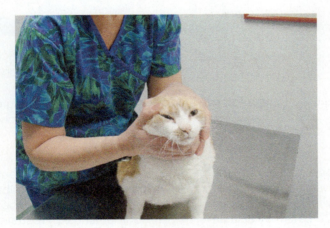

FIGURE 36–33 Restraint of the cat's head.

COMPETENCY SKILL 55

Use of a Cat Bag

Objective:

To properly use a cat bag to restrain and control a cat during veterinary procedures

Preparation:

- Cat bag

Procedure:

1. Open the cat bag and place it on a table.
2. Place the cat on top of the open cat bag.

3. While scruffing the cat, pull the bag over cat's body.
4. Zip the bag shut carefully to avoid catching any hair in the zipper.
5. Close the bag over the entire body with the head exposed.
6. Control the head to prevent biting.
7. Open the zipper areas of the cat bag to perform procedures.

See Figure 36–32.

COMPETENCY SKILL 56

Use of an Anesthesia Chamber

Objective:

To safely and properly use an anesthesia chamber for restraint or observation of an animal

Preparation:

- Anesthesia chamber

Procedure:

1. Remove the lid of the anesthesia chamber.
2. Place the cat inside the chamber and place lid securely on top and latch sides.
3. Make sure the openings over the top of the chamber are open and clear.
4. Open the lid of the chamber and remove the cat.

COMPETENCY SKILL 57

Scruff Technique

Objective:

To safely and successfully use the scruff technique to restrain a cat

Preparation:

No special equipment is required.

Procedure:

1. Use the dominant hand to grasp the loose skin over the base of the neck (see Figure 36-34).
2. Use the free hand to hold the rear limbs to prevent scratching.
3. Use a towel over the cat if the cat is difficult or trying to scratch with the front limbs.

632 SECTION IV Clinical Procedures

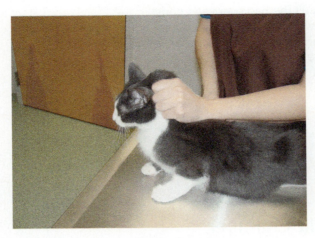

FIGURE 36–34 Cats can be restrained using a scruff technique.

 COMPETENCY SKILL 58

The Stretch Technique

Objective:
To safely and successfully use the stretch technique to restrain a cat

Preparation:
No special equipment is required.

Procedure:
1. Grasp the cat firmly at the scruff of the neck using one hand.
2. With the free hand, grasp the two rear legs by slipping a finger between the rear paws.
3. Roll the cat to the left side as in lateral recumbency.
4. Use the forearm of the hand holding the scruff to support the body and keep the cat from moving (see Figure 36–35).

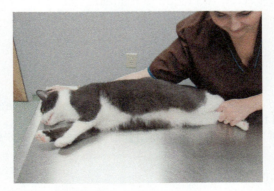

FIGURE 36–35 Another common restraint technique used for cats is the stretch technique.

COMPETENCY SKILL 59

Use of the "Kitty Taco"

Objective:

To successfully use a towel to calm and restrain a cat

Preparation:

- Heavy towel or blanket

Procedure:

1. Open a heavy towel or blanket on a table.
2. Place the cat on top of the open towel.
3. Scruff the cat and pull one side of the towel over the body of the cat and tuck under the cat.
4. Take the opposite side of the towel and wrap around the cat to make a tacolike appearance. The cat's four limbs will be enclosed in the towel and the head will be exposed (see Figure 36–36).
5. Properly control the head to prevent biting.

This restraint may be used to allow examination of the head and mouth and administer medications. One limb may be removed from the towel to examine or complete a nail trim.

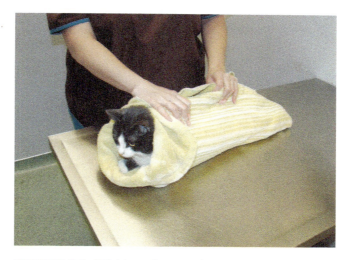

FIGURE 36–36 Use of a towel to create a kitty taco.

COMPETENCY SKILL 60

Removing a Cat from a Cage or Carrier

Objective:

To safely remove a cat from a cage or carrier for examination

Preparation:

- Towel

Procedure:

1. Close all windows and doors within the veterinary exam area prior to removing the cat.
2. Open the cage or pet carrier door and allow the cat to move to the front or step out.
3. If the cat is not willing to come to the front of the cage, reach into the area and scruff the cat with the dominant hand.
4. Place the free hand under the abdomen and lift the cat to your body.
5. A towel may be used for cats that are scared or aggressive so that they feel more secure. Use the towel by placing it over the cat and then follow the above steps (see Figure 36–37).

FIGURE 36–37 A towel can be placed over the cat to safely remove it from its cage.

CHAPTER 36 Basic Veterinary Restraint and Handling Procedures   635

COMPETENCY SKILL 61

Carrying a Cat

Objective:

To safely carry a cat to an examination room or cage

Preparation:

No special equipment is required.

Procedure:

1. Carry a calm and happy cat by placing one hand on the front of the body to control the head and front limbs.
2. Place the other hand under the abdomen and rump to control the rear limbs.
3. Pull the cat to your body for support.

If the cat is aggressive:

1. Place the dominant hand over the scruff of the cat's neck and apply the scruff technique.
2. Use the free hand to lift the cat under the rump and hold onto the rear limbs to prevent scratching.
3. Lift the cat in the air with the cat's limbs away from the body.

See Figure 36–38. A towel can be placed over the cat, and the same steps can be followed.

FIGURE 36–38 Technique for carrying an aggressive cat.

Dogs

Dogs that are difficult may be captured and restrained using a **rabies pole** or **snare pole** (see Figure 36–39). This pole is long and placed inside a cage or kennel to capture a dog around the neck. A noose on the end of the pole acts as a collar, and the pole acts as a leash. The end of the pole controls the noose by pulling in or out on the pole to loosen or tighten the noose to fit the dog's head. Once the noose is on the dog, the animal can be safely removed from the cage or kennel and walked to a safe area for further restraint or sedation. When walking a dog with a snare pole, the dog should be kept in front of the restrainer and pushed along, rather than pulled. Pulling a dog can cause severe head and neck injuries.

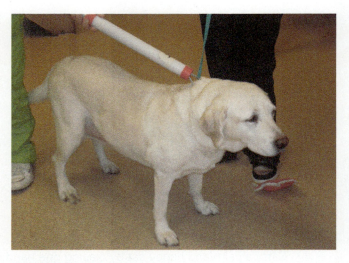

FIGURE 36–39 Use of a rabies control to restrain a dog.

COMPETENCY SKILL 62

Removing a Dog from a Cage or Kennel

Objective:
To safely and successfully remove a dog from a cage or kennel

Preparation:
- Leash

Procedure for Ground Level Cage:
1. Place a leash in one hand with a large loop open and ready to place over the dog's head (see Figure 36–40A).
2. Open the cage door enough to slip the hand holding the leash into the cage (see Figure 36–40B).
3. Slip the leash over the neck of the dog and gently tighten the leash around the patient's neck (see Figure 36–40C).
4. Open the door and allow the dog to exit the cage (see Figure 36–40D).
5. Keep the dog to your side while maintaining a slight tension on the leash.

Procedure for Elevated Cage:
1. A small dog in an elevated cage should be removed from the cage area. Carry a calm and happy dog by placing one hand on the front of the body to control the head and front limbs.
2. Place the other hand under the abdomen and rump to control the rear limbs.
3. Pull the dog to your body for support.

 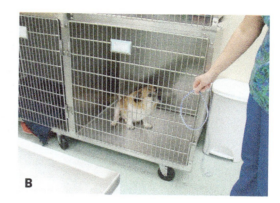

FIGURE 36–40 (A) Make a large loop at the end of the leash. (B) Open the door of the cage wide enough to place the hand with the leash in the cage. (C) Place the loop over the dog's head and gently tighten. (D) Allow the dog to exit the cage.

COMPETENCY SKILL 63

Walking a Dog in the Heel Position

Objective:
To safely walk a dog in the veterinary facility

Preparation:
- Leash

Procedure:
1. Place a leash in one hand with a large loop open and ready to place over the dog's head.
2. Slip the leash over the head while standing to the side of the dog.
3. Gently tighten the leash around the patient's head.
4. Stand with the dog on the left side of the body.
5. Walk forward with the dog at the left side of the body (see Figure 36–41).

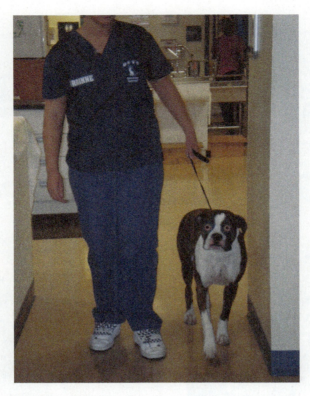

FIGURE 36–41 Walking the dog in the heel position.

 COMPETENCY SKILL 64

Applying a Commercial Leash to a Dog

Objective:
To properly and safely use a leash to control a dog

Preparation:
- Leash

Procedure:
1. Place a leash in one hand with a large loop open and ready to place over the dog's head.
2. Slip the leash over the head while standing to the side of the dog.
3. Gently tighten the leash around the patient's head (see Figure 36–42).
4. Stand with the dog to the left side of the body.

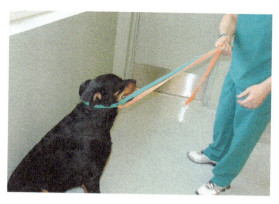

FIGURE 36–42 Application of a leash on a dog.

 COMPETENCY SKILL 65

Dorsal-Ventral (DV) Recumbency

Objective:

To properly and safely place an animal in a **dorsal-ventral** recumbency position

Preparation:

No special equipment is necessary.

Procedure:

1. The dog or cat should be in the standing position.
2. Cats should be scruffed.
3. Apply pressure over the hips, forcing the dog or cat to sit.
4. Slide one hand under the front legs while applying pressure over the shoulders with the free hand.
5. Maintain pressure over the back and shoulders until the animal is lying on its chest (see Figure 36–43).
6. Restrain the head in a bear hug with one arm with a dog. Cats should be scruffed.
7. Use the free arm to apply continued pressure over the back to prevent standing. Small dogs may require only one restrainer; medium to large dogs will require two or more restrainers. Cats may be restrained with one or two people.

This is a common position for X-rays when the light beam enters the dorsal area (back) first and the ventral area (stomach) second.

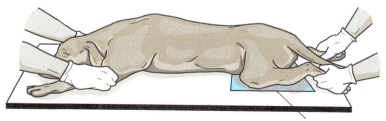

Ventral recumbency/sternal recumbency

FIGURE 36–43 Dorsal-ventral recumbency.

COMPETENCY SKILL 66

Ventral-Dorsal (VD) Recumbency

Objective:

To properly and safely place an animal in a **ventral-dorsal** recumbency position

Preparation:

No special equipment is necessary.

Procedure:

1. The dog or cat should be placed in lateral recumbency as outlined in the lateral recumbency competency skill.
2. The dog or cat is gently rolled onto its back with the front legs extended forward and the back legs extended backward.
3. The animal's head should be kept between the front legs and held securely.
4. The rear legs should be kept still to avoid injury to the back (see Figure 36–44).

Small dogs may require only one restrainer; medium to large dogs will require two or more restrainers. Cats may need two restrainers.

This is a common position for X-rays when the light beam enters the ventral area (stomach) first and the dorsal area (back) second.

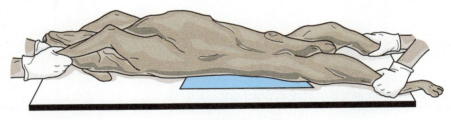

Dorsal recumbency

FIGURE 36–44 Ventral-dorsal recumbency.

COMPETENCY SKILL 67

Returning a Dog to a Cage or Kennel

Objective:

To safely move a dog from a procedure area to the cage or kennel

Preparation:

- Leash

Procedure:

1. Open the cage door completely.
2. Place the animal in the cage with the dog facing you.

3. Close the door with only one arm remaining in the cage.
4. Gently slip the loop of the leash over the dog's neck and pull the leash from the cage.
5. Completely close the cage door and latch securely.
6. Check the door to make sure it is properly closed.

COMPETENCY SKILL 68

Handling a Fractious or Scared Dog

Objective:

To properly and safely handle a dog that is behaving in an unruly manner

Preparation:

- Leash
- Muzzle

Procedure:

1. Approach and work with the scared dog calmly and slowly.
2. Talk with a soft and gentle voice and reassure the dog.
3. Move slowly toward the dog from the side. Do not approach from the front or rear of the dog.
4. Gently apply a leash using the loop method over the neck.
5. Continue to talk softly to the dog.
6. Slowly walk the dog from the kennel or cage area.
7. Use a muzzle as necessary, as fear biting may occur. Apply the muzzle prior to any procedures to allow for safe handling.

Handling of Other Small Animals

Many of the techniques discussed for dogs and cats can be used or slightly adapted for use with other small animals. For example, the scruff technique works well on ferrets. Birds and small mammals can be handled and secured by wrapping them in towels without restricting the chest cavity and the bird or other animal's ability to breathe. Note that the restraint of psittacine pet birds is different than that of birds of prey in both beaks versus talons and the potential for injury. Some small animals have some specific needs that the veterinary assistant should become familiar with. Rabbits, for instance, can become paralyzed if their hind limbs are not properly supported.

COMPETENCY SKILL 69

Lifting a Rabbit

Objective:

To properly and safely lift a rabbit

Preparation:

No special equipment is necessary.

Procedure:

1. When lifting a rabbit, keep its feet forward and its back toward your body.
2. Grasp the scruff of the neck with one hand and the hind legs with the other hand.
3. Extend the rear legs and lift the rabbit.
4. Use the palm of the hand that is holding the rear legs to support the hindquarters and body (see Figure 36–45).
5. Keep the back against your body.

FIGURE 36–45 Properly holding a rabbit.

Fear-Free Handling

The fear-free veterinary visit is best created with a multimodal approach that must involve the pet owner. The owner should be encouraged to take the pet to a fear-free veterinarian and thus may be instructed by the veterinarian on how to adapt to changes from previous veterinary visits. The owner should condition the pet to the carrier or restraint device and to car rides long before the day of the veterinary visit. Cats should be allowed to use their carriers daily as a resting place so that the carrier is familiar and comfortable, rather than something associated with a frightening car ride. Also, the veterinary team should learn to detect fear early on by asking owners about the anxiety their pets have experienced and can screen owners on pet history prior to the initial visit. If you don't ask, you may miss it.

To keep the pet calm on the day of the appointment, instruct pet owners to use pheromones, calming nutraceuticals, and carrier covers. Owners will need to remain calm themselves. They may find it helpful to play special calming music and use aromatherapy in the car. Some pets may require prescription anxiolytic medication before they leave home.

If the practice has enough exam rooms and the technology to do so, consider conducting check-ins for fearful pets in the exam room or a holding room, rather than the entrance and reception area. The pet owner could take the pet directly into an exam room, or the owner could check in alone and then return to the vehicle to wait until the pet can be ushered directly into an available exam room. It is best for the veterinarian and the team to have already entered the room before the pet arrives. If the waiting room is large, calming, and quiet, a fearful dog or cat can stay there for about 10 minutes. Even a small waiting room can be made more pleasant by including visual barriers, such as plants or shelves, so that clients and patients can wait and not have other animals right next to them. Cat carriers can be placed on shelves rather than on the floor, where the cat can be investigated by every dog that passes by. When possible, assign specific exam rooms to a single species to allow for species-specific places to examine pets that can also incorporate pheromones, calming music, and appropriate wall coverings. These exam rooms can also serve to limit stressful odors and sounds from other patients.

Large Animal Restraint

Large animals include livestock such as horses, cattle, goats, swine, and sheep. Large animal behavior is typically based on these species being in groups or herds. Livestock have the instincts of prey animals and many times when people approach them for restraint, if they are not used to human contact or being handled, their typical behavior is that of prey. The "fight-or-flight" instinct is likely to be part of their reaction to restraint. Horses tend to be more willing to be handled by people, as they have usually been trained and otherwise worked with by humans. Some may become frightened by certain movements or smells. Other livestock, such as cattle, sheep, and swine, may not be as accustomed to human handling and should be restrained with caution by an experienced hand. The large size of livestock can be a safety issue, and many species can kick, rear up into the air, bite, and use their large bodies to injure people. This is a safety situation that requires veterinary staff to be experienced with large animal behaviors and species-specific handling techniques. Many times, large animals that are aggressive or otherwise deemed dangerous to handle should be sedated for the safety of the staff and the animal.

COMPETENCY SKILL 70

Application of a Halter

Objective:

To properly and safely apply a halter to lead a large animal such as a cow or horse

Preparation:

- Halter

Procedure for Cows:

1. Stand on the left side of the cow.
2. Place the noseband of the halter over the face and around the nose.
3. The headstall should then be placed over the back of the head and behind the ears (see Figure 36–46).
4. Make sure the halter is a good fit and doesn't slip off.
5. Apply a lead to the bottom of the halter.

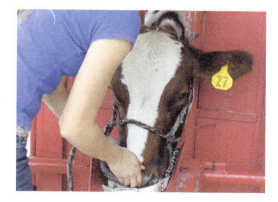

FIGURE 36–46 Haltering a cow.

Procedure for Horses:

1. Stand on the left side of the horse.
2. Place the noseband of the halter over the face and around the nose.
3. The headstall should then be placed over the back of the head and behind the ears.
4. Secure the halter in place (see Figure 36–47).
5. Make sure the halter is a good fit and doesn't slip off.
6. Apply a lead to the bottom of the halter ring.
7. Hold the lead loops in the left hand and hold the lead close to where it connects to the halter in the right hand.

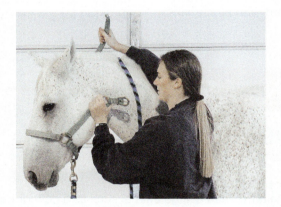

FIGURE 36–47 Haltering a horse.

 COMPETENCY SKILL 71

Leading a Cow or a Horse

Objective:

To safely lead a horse or a cow from location to location

Preparation:

- Halter
- Rope
- Pole or prod

Procedure:

1. Hold the lead loops in the left hand and hold the lead close to where it connects to the halter in the right hand.
2. Stand on the left side of the horse or cow near the shoulder (see Figure 36–48).
3. Stand about a foot from the horse or cow.
4. Face the same direction as the horse and click with the tongue to ask the horse to move forward. Cattle may need to be pulled or prodded with a pole.
5. Walk the horse or cow forward and watch the front feet to avoid being stepped on.

FIGURE 36–48 Leading a horse.

 COMPETENCY SKILL 72

Jogging a Horse on a Lead

Objective:
To safely exercise a horse while on a lead

Preparation:
- Rope
- Halter

Procedure:
1. Hold the lead loops in the left hand and hold the lead close to where it connects to the halter in the right hand.
2. Stand on the left side of the horse near the shoulder.
3. Stand about a foot from the horse.
4. Face the same direction as the horse and click with the tongue to ask the horse to move forward.
5. Jog with the horse in a forward motion and watch the front feet to avoid being stepped on.

COMPETENCY SKILL 73

Tying a Cow

Objective:

To safely and securely tie a cow for restraint

Preparation:

- Rope

Procedure:

1. Pull the cow's head to the side.
2. Place the lead over a bar or chute.
3. Apply a quick release knot by making half of a bow tie.
4. Pull the end of the rope to quickly untie.

COMPETENCY SKILL 74

Tying a Horse

Objective:

To safely and securely tie a horse for restraint

Preparation:

- Rope

Procedure:

1. Place the lead over a pole or tie area.
2. Allow about 2 feet of rope for the horse to move slightly.
3. Make a loop in the lead rope close to the pole or tie area.
4. Pass the loop under the lead rope near the pole and pass it through the loop. This should look like a pretzel.
5. Pull the loop to tighten the knot.
6. Untie by pulling the end of the lead rope for a quick release.

COMPETENCY SKILL 75

Cross Tying a Horse

Objective:

To safely and securely tie a horse for restraint

Preparation:

- Rope

Procedure:

1. Walk the horse to the cross-tie area.
2. Place the left cross tie on the lateral cheek ring of the halter.
3. Move in front of the horse to the right side.
4. Place the right cross tie on the lateral cheek ring of the halter (see Figure 36–49).

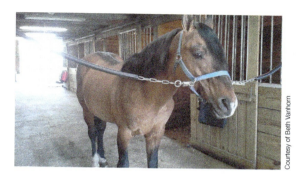

FIGURE 36–49 Horse in cross ties.

COMPETENCY SKILL 76

Longing a Horse

Objective:

To safely exercise a horse.

Preparation:

- Rope

Procedure:

1. Place a long longe line onto the horse's halter by snapping securely.
2. Allow the horse to move away from you in a large circle.
3. Work the horse at a walk, jog, and lope in both directions to allow the vet to assess any lameness problems.
4. The horse should continue to move in large circles around the handler (see Figure 36–50).
5. Hold the end of the longe line firmly in one hand.

FIGURE 36–50 Lunging a horse.

COMPETENCY SKILL 77

Tail Tie Technique

Objective:

To safely and properly place a tail tie to aid in maneuvering of a horse

Preparation:

- Rope

Procedure:

1. Stand to the rear of the horse and slightly to the side.
2. Grasp the tail about a third of the way from the tail base just below the tail bone.
3. Use both hands to gently lift the tail upward.
4. Place a rope around the tail making a loop.
5. Make a slip knot and pull the end of the rope toward the head (see Figure 36–51).
6. Tie the end of the rope to the front leg or neck using a slip knot tie.

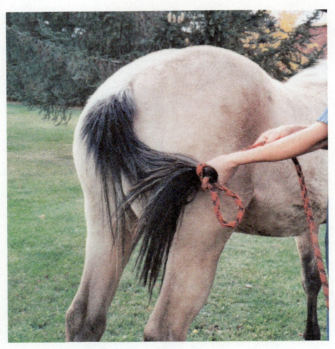

FIGURE 36–51 Tail tie.

COMPETENCY SKILL 78

The Twitch Restraint

Objective:

To use a device to elicit minor pain to distract an animal from procedures

Preparation:

- Twitch device

Procedure:

1. Stand to the left side of the horse's neck and head.
2. Place the end of the twitch loop over the left wrist.
3. Hold the twitch handle with the right hand.
4. Grasp the horse's upper lip with the left hand and press the edges together.
5. Quickly slip the twitch loop over the lip so that it lies around the upper lip (see Figure 36–52).
6. Tighten the twitch loop by twisting the handle clockwise before letting go of the lip.
7. Hold the twitch in the left hand and the halter in the right.
8. Release the twitch by untwisting and removing from the lip.

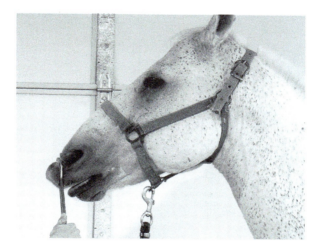

FIGURE 36–52 Application of a twitch.

COMPETENCY SKILL 79

Use of a Chain Shank

Objective:

To apply an increased amount of restraint

Preparation:

- Chain shank

Procedure:

1. Stand to the left side of the horse's neck and head.
2. Place the chain through the left lateral cheek ring.
3. Run the chain over the nose, under the chin, or under the lip (see Figure 36–53).
4. Move in front of the horse to the right side.
5. Place the chain through the right lateral check ring and snap in place.
6. Move in front of the horse to the left side to control the horse. Hold the chain and lead close to where it connects to the halter.

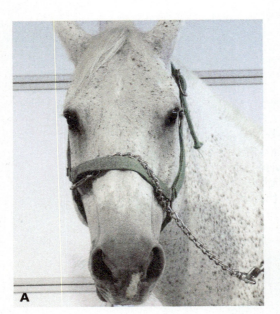
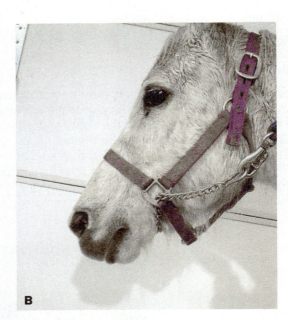
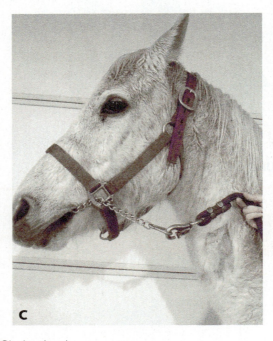

FIGURE 36-53 (A–C) Chain shank.

COMPETENCY SKILL 80

The Tail Jack Restraint

Objective:

To properly and safely restrain a cow

Preparation:

No special equipment is required.

Procedure:

1. Stand to the rear of the cow, just behind the legs.
2. Grasp the tail about 1/3 of the way from the tail base.
3. Use both hands to gently move the tail upward (see Figure 36–54).

FIGURE 36–54 Tail jack restraint.

652 SECTION IV Clinical Procedures

COMPETENCY SKILL 81

The Tail Switch Restraint

Objective:

To properly and safely restrain a cow

Preparation:

No special equipment is required.

Procedure:

1. Stand to the side and as close to the legs as possible.
2. Use both hands to grasp the tail switch.
3. Rotate the tail gently by the switch and turning it under the cow.
4. Make a small circle with the end of the tail.

COMPETENCY SKILL 82

Restraint of a Large Animal for Blood Collection or an Injection

Objective:

To properly and safely restrain a large animal for the purpose of collecting a blood sample or to give a medication by injection

Preparation:

- Halter

Procedure:

1. Apply a halter and lead to the animal.
2. Stand to the left side of the animal near the shoulder.
3. Hold the animal close to the halter with the left hand.
4. Place the right hand just behind the eye to distract the animal.
5. Reach to the neck area and pinch the skin with the right hand to further distract the animal.

COMPETENCY SKILL 83

Restraint of a Foal

Objective:

To properly and safely restrain a foal

Preparation:

No special equipment is required.

Procedure:

1. Hold one arm and hand around the neck to control the head and front of the body.
2. The arm may be moved lower to hold the chest as necessary.
3. Place the other arm around the rump, holding the foal in the arms (see Figure 36–55).
4. The tail may be lifted over the rump toward the head to control a foal from moving.

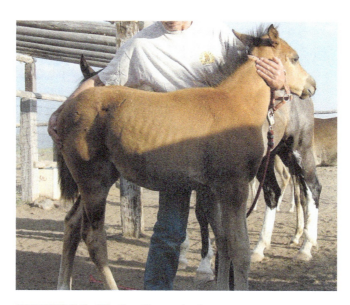

FIGURE 36–55 Cradling a foal.

 MAKING THE CONNECTION

Some species of livestock may require specialized restraint. Refer to the chapters on specific large animal production for techniques.

SUMMARY

Restraining small animals must be done with several safety issues being kept in mind at all times. The safety of the animal and the restrainers is most important. The restrainer must focus on the animal's body language during all procedures. The environment must be monitored for safety issues. Possible adverse outcomes must be anticipated to provide a safe restraint setting. Animals can be unpredictable, and behaviors may change with the slightest amount of stress, especially in a veterinary setting. It is important to provide for the most pleasant experience possible for both clients and patients. Large animal restraint requires a great deal of practice and skill. It also requires knowledge and experience in species behavior. Large animals can cause severe injury and possibly death in situations in which handling and restraint are difficult. Care and handling of animals may require additional restraint equipment that must be used properly to ensure the safety of the staff and the animal. Sedation may also be a possibility.

Key Terms

aggression behavior that makes an animal angry, difficult, and potentially unsafe to handle

anesthesia chamber restraint equipment used to sedate a cat or other small mammal under gas anesthesia that is pumped into the glass or plastic chamber

animal behavior relates to what an animal does and why it does it

body language communication by the animal about how it feels toward other animals, people, and its environment

cat bag material that covers the body and limbs of a cat to control it for various procedures; only exposed part is the head

cephalic vein blood vessel located on the medial front limb

conditioning process of teaching an animal an action in relationship to another action

dominance aggression behavior that refers to the pack instinct of an animal and its social status within a group

dorsal recumbency restraint position with an animal lying on its back

dorsal-ventral (DV) position where the animal is restrained on its chest, and the back of the animal is higher than the chest and abdomen.

ethology the study of animal behavior

fear aggression refers to the defensive reaction to being harmed and the animal's instinct to protect itself

half hitch tie that makes a loop around a stationary location, such as a post or fence

imprinting process of learning through an attachment to an object that emits adult behaviors and can be generalized to all examples of the object

instinctive behavior behavior that occurs naturally to an animal in reaction to a stimulus

jugular vein blood vessel located in the neck or throat area

knot two pieces of rope tied together so that they do not slip; tied to contain an animal

lateral recumbency restraint position with animal lying on its side

learned behavior response to a stimulus that develops as a result of experience

modeling learning a behavior through watching other animals conduct that behavior

rabies pole long pole used to capture a dog around the neck with a noose on the end of the pole that acts as a collar and a pole that acts as a leash; also called a snare pole

recumbency the state of leaning, lying, resting, or reclining

redirected aggression refers to an animal's predator instinct whereby the animal turns its aggressive behaviors on the owner

reefer's knot single bow knot that ties in a nonslip but quick release tie

restrain to hold back, check, or suppress an action and to keep something under control using safety and some means of physical, chemical, or psychological action saphenous vein blood vessel located on the lateral thigh of the rear limb just proximal to the hock

scruff technique used on the base of the neck over the area where the skin is elastic and able to be grasped with a fist technique

sedative medication used to calm an animal

sitting restraint restraint used to keep an animal sitting on its rump and prevent it from standing

snare pole long pole with a noose on the end that acts as a leash and collar to capture a difficult dog; also called rabies pole

square knot commonly used to secure an animal; a nonslip knot that doesn't come untied or loosen easily

squeeze cage wire box with small slats that allow injections to be given to a cat, such as vaccines or sedatives

standing restraint restraint used to keep an animal standing and prevent it from sitting or lying down

sternal recumbency restraint position with the animal lying on its chest

stimulus internal or external change that causes a reaction

stretch technique scruffing the cat with one hand while in lateral recumbency and using the free hand to hold the rear limbs and pull them dorsally in a stretch

territorial aggression refers to an animal's protective nature of its environment, such as an owner, offspring, or food

tranquilizer medication used to calm an animal during stress

venipuncture practice of placing a needle into a blood vessel

ventral-dorsal (VD) position where the animal is restrained on its back, and the chest and abdomen of animal is higher than the back

Review Questions

1. What is animal behavior?
2. What is the difference between instinctive and learned behavior?
3. What is body language?
4. What is a square knot?
5. What are some similarities in restraining small and large animals?
6. What are some differences in restraining small and large animals?
7. What is the proper way to restrain for cephalic venipuncture?
8. What is dorsal recumbency?
9. Why are towels often helpful in restraining small animals?
10. Explain the stretch technique.

Clinical Situation

Lila, the veterinary assistant at Companion Animal Vet Clinic, was working in the veterinary kennel ward. She began going down the cage rows, giving water to those animals that were allowed to have it. She noted that one kennel had some waste material that needed to be cleaned. When she opened the cage and began to remove the waste, Precious, a nervous Bichon Frise, began growling and raising his upper lip. Lila gently reached toward the dog and patted him on the head.

"It's okay, Precious, no one is going to hurt you," she said quietly.

The dog snapped at her hand, causing a scratch with his teeth. Lila quickly closed the cage door and went to clean the wound. The scratch was not deep, but to be safe, she checked with Dr. Todd to see if there was anything to be concerned about.

- How did Lila handle the situation correctly?
- What did Lila do incorrectly?
- What preventative measures should have Lila taken in this situation?

Chapter 37: Physical Examination Procedures and Patient History

Objectives

Upon completion of this chapter, the reader should be able to:

- 37.1 Demonstrate how to obtain a complete patient history
- 37.2 Explain the importance of the patient history
- 37.3 Discuss the tools and supplies necessary for a physical examination
- 37.4 Explain the importance of the physical examination
- 37.5 Describe normal and abnormal body system signs
- 37.6 Discuss each body system evaluated in the physical examination
- 37.7 Demonstrate common examination procedures
- 37.8 Describe common examination procedures with clients
- 37.9 Define common veterinary terminology and abbreviations relating to examination procedures

Introduction

The physical examination is an essential evaluation of an animal's overall health. The complete and accurate results of a physical exam begin with a full history of an animal's past medical problems and other health factors that will determine if an animal is healthy or if the animal is showing signs of disease. Veterinary assistants are a vital part of obtaining a patient history. Some veterinarians also expect the support staff to be able to record vital signs and complete a prephysical exam based on the owner's chief complaints and reason for visiting the veterinary facility. Examination procedures must be understood by the entire veterinary health care team. Even though liability and responsibility lie with the attending veterinarian, the veterinary team and support staff must be trained as reliable professionals that can perform auscultation, obtain vital signs, provide client education, and assist the veterinary technician and veterinarian in performing common examination procedures. It is important to note that cleanliness, good hand hygiene, and PPE, or personal protective equipment, should be used when necessary. The veterinary assistant plays a vital role in the examination room.

The Patient History

The **patient history** is background information of past medical, surgical, nutritional, and behavioral factors that have occurred over the life of the animal. A patient history should be obtained for every new animal and updated with each patient visit to the veterinary facility. A thorough knowledge of the history of an animal is necessary. Frequently, a careful description of the history or onset of a problem will simplify the diagnosis. Most owners will not provide additional information if they are not asked the appropriate questions, so the questions are important to the patient history. While it is usually best to deal with the person who spends a major portion of the day with the animal, information from other family members should be included if it contributes to a better understanding of the problem. Specific areas of history that should be included are as follows:

- Signalment (age, gender, breed)
- Past illnesses or diseases
- Past surgeries or medical procedures
- Previous injuries
- Genetic history of parents and other offspring
- Chief complaint
- Signs observed by the owner
- Changes that have occurred in the animal (behavior, attitude, exercise)
- Duration of signs
- Vaccine history
- Nutritional history

The medical history should be noted in brief comments within the chart, either on a patient history sheet or within the progress notes. It is best to not ask leading questions. Instead, it is best to ask "yes" or "no" questions to get an idea of any changes that have occurred with the animal. Asking open-ended questions will encourage owners to give a more descriptive answer about what they are seeing with their pet. When the opportunity arises, ask the owners to explain in detail what they have seen in the animal. Table 37–1 lists common questions used to obtain accurate patient histories in veterinary practice.

The veterinary assistant should record the patient history according to what he or she interprets from the owner. It is important to note the **chief complaint**, or the reason the animal is being seen, as this is the most important issue to address. It is also important to establish how long the chief complaint has been occurring. Other areas of importance include obtaining the patient's name, age or date of birth, gender, color, and breed or species. This patient information is known as the animal's **signalment**. This information provides statistics on the animal that may help indicate problems and identify the specific characteristics of the animal. A patient's health history is vital to the medical record and physical exam and includes past and current illnesses, surgeries, and medications. Vaccine history should include the last vaccination dates and the types of vaccines that were given. Nutrition is important to recognize, so the veterinary assistant should obtain diet information, how much is fed and how often, and if any changes in eating have occurred. Some problems are frequently associated with certain breeds, so it is important to establish an awareness of particular genetic problems associated with parents, grandparents, siblings, or progeny. *Progeny* refers to the offspring of the animal; genetic problems are easily passed on to offspring. While the degree of a genetic disease will vary from animal to animal, the problem may be present to some degree in the presenting animal. Such information from an owner can be helpful in diagnosing a disease or condition. In some cases, especially behavior-related situations, it is ideal to obtain a home environment history, such as if the animal stays indoors, outdoors, or both and what type of cage or confinement the animal has. Some other areas to consider with some patients are travel history, any medications that the animal is currently taking, and other animals in the household or those the pet otherwise comes in contact with.

The Physical Examination

A **physical examination (PE)** is an observation of each body system to determine abnormalities that can cause a health problem. This includes examining each body system of the animal from head to tail. The PE is one of the most important procedures completed on an animal. It is also one of the most challenging, as animals can't talk and tell us what is wrong or where it hurts. A proper PE can detect early signs of illness before they become serious health issues. A yearly physical exam should be completed on every patient to recognize any changes in health (see Figure 37–1). As animals age, the frequency of a PE may be

TABLE 37–1

Examples of Patient History Open-Ended Questions

PATIENT HISTORY COMPONENT	CORRECT QUESTION	INCORRECT QUESTION
Appetite	"Tell me how your pet's eating habits have recently changed."	"You haven't observed any change in appetite, have you?"
Chief complaint	"What is the reason for today's visit?" "How long has the pet been sick?"	"What do you need done today?" "Has this been occurring a long time?"
Behavior	"Tell me how your pet's behavior has been different."	"There hasn't been a change in behavior, has there?"

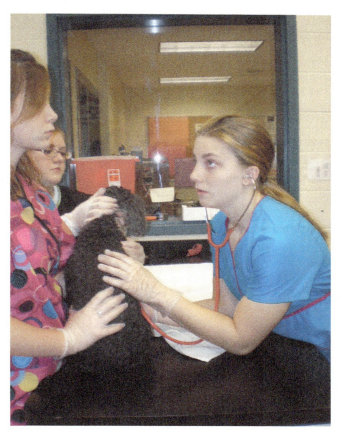

FIGURE 37–1 Physical examinations are an important part of maintaining a pet's health.

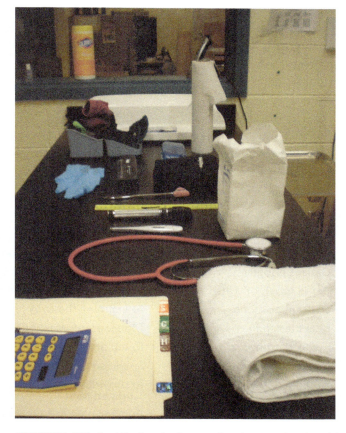

FIGURE 37–2 All physical examination tools should be set up before beginning any examination.

increased to every 6 months to detect changes. Animals that are having anesthesia or surgical procedures should have a PE to determine whether they are healthy enough for the procedure.

A consistent method should be developed and used with every patient exam. This helps to maintain an effective exam and not miss an area of importance. It also allows the staff member to assess areas quickly and easily. Although it is important to focus on the chief complaint, all other areas must be evaluated as well. A full PE should be given regardless of the immediate health problem. In some cases, multiple animals that need to be seen immediately for emergency situations may present to the facility. In these cases, the support staff may need to perform **triage**, which is a quick general assessment of an animal to determine how quickly the animal needs a veterinarian's attention.

Tools for the Physical Examination

It is important to have all the PE tools in the areas of the facility where physical exams are typically completed (see Figure 37–2). This would include the exam rooms and the treatment area. All tools should be maintained and kept in working order; this would include checking any batteries, electrical cords, and switches. In some cases, certain species may require additional restraint and safety devices that should be available in case of handling problems. This would include muzzles, towels, welding gloves, antikick devices, and hobbles, depending on the animal species. Certain procedures that may also need to be completed during a physical exam may require additional tools and supplies. Anticipation of these items is important in saving time and testing an animal's patience. It is also important to make certain there is enough space available for the PE in accordance to the size of the animal. Some animals may require additional restrainers, and enough people should be available to help as necessary. The tools and the exam location should be prepared and disinfected prior to an animal entering the exam area. It is important to perform the PE in a quiet area that is comforting to the animal.

Table 37–2 lists the tools commonly used in the physical examination and their functions.

TABLE 37-2	
Tools and Supplies for the Physical Examination	
TOOL	**FUNCTION**
Stethoscope	To listen to the heart, chest, and lungs
Exam gloves	To cover the hands to prevent contamination of the skin or transfer organisms to other animals
Paper towels	For quick cleanup of any messes or spills
Restraint equipment	To safely handle an animal
Vaseline or KY Jelly	Lubricant to ease discomfort when performing rectal exams or using a rectal thermometer
Otoscope	To examine the ears
Medical chart	To record patient history and document examination findings
Cotton-tipped applicators	Long thin wooden sticks with a cotton end; used to clean areas of the body or apply medications; commonly called Q-tips
Gauze sponges	Soft square mesh pads; used to clean areas of an animal
Clock, timer, or watch	To count respiration and pulse rates
Ophthalmoscope	To examine the eyes
Blue/black pen	To record information in a paper medical chart
Reflex hammer	To tap areas of the body to test the animal's reflex responses
Cotton balls	Small, soft, white circular balls; used to clean areas of an animal
Penlight	To provide a light source to observe the mouth or other body areas where there is limited light
Calculator	To determine drug dosages
Scale	To weigh an animal
Thermometer	To determine an animal's internal body temperature

Observation

The PE should begin with an observation of the animal. This is done before a hands-on exam is performed. An **observation** is an inspection of an animal at a distance to see what it looks like, how it acts, or how it moves naturally. The veterinary assistant should observe the animal as it walks to the exam area, observing the gait for any limping or signs of lameness. This is also an appropriate time to observe the animal's overall behavior for signs of aggression, shyness, fear, dominance, or calmness. This is important when the hands-on exam must begin, to predict how the animal will react. The observation helps to note how the animal acts prior to stressing it out and potentially becoming excited or agitated by the hands-on assessment. Some other areas to note on observation include the following:

- Mental status (e.g., alert, focused)
- Conformation of the animal (body shape)
- Body condition score (e.g., thin, obese)
- Neurological problems (e.g., wobbling, unable to control movements, head tilt)
- Overall general appearance (e.g., weak, labored breathing, depressed)

On completing an observation, the veterinary assistant should socialize and greet the animal by allowing the animal to approach and sniff a hand. This will allow the animal to familiarize itself with the person and allow the person to determine the animal's behavior and acceptance.

Weight Assessment

The weight of an animal should be established when the animal enters the exam area. A scale should be provided according to the size of the animal. Small animals may be weighed on a feline scale or a pediatric baby scale (see Figure 37–3). Large dogs may be weighed on a floor scale, and livestock should be weighed on a livestock scale or a weight tape measurement (see Figure 37–4). The weight should be recorded and assessed according to species and breed, as should age, gender, purpose or use, and health status to determine any abnormal conditions.

An animal should be weighed and evaluated for obesity or thin body conditions. An animal is weighed using a size-appropriate scale or weight tape. There are

FIGURE 37–3 Small animals can be weighed on a feline or pediatric scale.

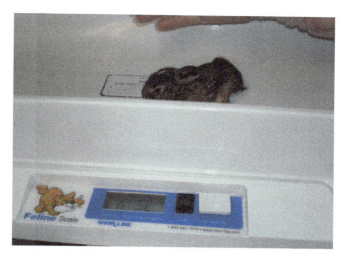

FIGURE 37–5 Weight is an important measurement in assessing the health status of animals.

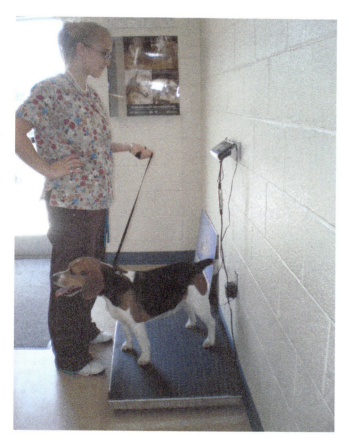

FIGURE 37–4 Larger animals can be weighed on a floor scale.

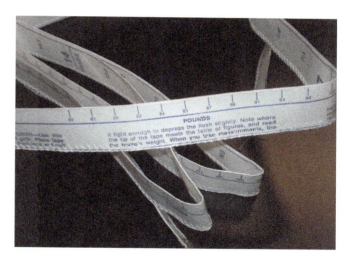

FIGURE 37–6 The weight of livestock can be measured using a weight tape.

scales for small animals, medium-sized animals, and livestock. These scales are typically meant for an animal to stand or sit on and weigh the body mass of the animal through digital measurement (see Figure 37–5). Livestock can also be weighed by a **weight tape** (see Figure 37–6). This is the use of a measurement tape around the girth of an animal that gives an estimate of its body mass. Weight is recorded in either pounds (lb) or kilograms (kg). Care must be taken when recording the weight, as some scales record the weight with a decimal and others record in pounds and ounces (oz). For example, 8.8 lb does not equal 8 lb, 8 oz—as there are 16 oz in a pound. If a weight in pounds needs to be converted into kilograms, it is done with the following equivalency: 1 kg = 2.2 lb. For example, a pet weighing 30 lb is 13.63 kg. This is found by the following equation: 30 lb × 1 kg/2.2 lb = 13.63 kg.

COMPETENCY SKILL 84

Weighing an Animal Using a Scale

Objective:

To properly use a scale to obtain an accurate measurement of weight

Preparation:

- Scale

Procedure:

1. Turn on the scale and balance it prior to placing the animal on the scale.
2. Allow the animal to walk on the scale or place the animal on the scale.
3. Maintain the animal by allowing it to stand, sit, or lie quietly and calmly on the scale with as little movement as possible (see Figure 37–7).
4. Read the digital weight measurement on the scale screen.
5. Remove the animal from the scale.
6. Turn off the scale.
7. Record the animal's weight in the medical record.
8. Disinfect the scale.

FIGURE 37–7 Try to keep the animal still while weighing it.

COMPETENCY SKILL 85

Weighing an Animal Using a Weight Tape (Livestock)

Objective:

To properly use a weight tape to accurately measure the weight of large animals and livestock

Preparation:

- Weight tape

Procedure:

1. Tie or otherwise restrain the large animal.
2. Unfold or unroll the entire weight tape. Note the side of the tape that measures weight.
3. Place the weight tape around the girth of the animal just behind the front limbs and encircling the back and chest area.
4. With the tape snugly around the girth, read the measurement at the end of the tape where the ends meet.
5. Remove the tape.

Vital Signs

Animals are evaluated for normal health status by taking their **vital signs**, which are the signs of life that are measurements to assess the basic functions of the body. Vital signs include temperature, heart rate, respiratory rate, blood pressure, mucous membrane color, capillary refill time, and weight of the animal. The first three vital signs are often referred to as **TPR**: temperature, pulse, respiration.

Temperature Assessment

The patient's temperature is an essential part of an animal's exam and health status. The patient's temperature should be evaluated immediately to achieve accurate results and should be recorded in degrees Fahrenheit (F). Many thermometers also record temperatures in degrees Celsius, so care must be taken to record the correct unit of measure and use that unit each time a temperature is taken on a patient (see Table 37–3). The body temperature can allude to signs of infection, disease, environmental conditions, and other factors. The **endothermic** temperature is evaluated in an animal that regulates its body temperature internally or within the body. An **ectothermic** temperature is evaluated in an animal that regulates its body temperature externally or through its environment (see Table 37–4).

TABLE 37–3

Converting Fahrenheit and Celsius Temperatures

Fahrenheit to Celsius (F – 32) / 1.8
Celsius to Fahrenheit (C × 1.8) + 32

TABLE 37–4

Examples of Endothermic and Ectothermic Species

ENDOTHERMIC SPECIES		ECTOTHERMIC SPECIES		
Dog	Cow	Mouse	Snake	Frog
Cat	Goat	Rabbit	Lizard	Turtle
Horse	Sheep	Rat	Fish	Toad

Hypothermia is a condition where the body temperature is below normal, causing the animal to become cold, and it may be a sign of environmental exposure to cold climates, cardiac conditions, metabolic disease, or shock. Hypothermia can be dealt with by placing circulating warm water blankets or pads under the animal; applying warm air devices; or using warmed blankets, a heat lamp, or warm IV fluids to increase body temperature. IV fluids should be warmed to 98.6 degrees F with a fluid warmer. Care should be taken not to overheat the patient, and when temperatures reach 99–100 degrees F, the heat source should be removed. A towel or pad between the animal and the heating object will help prevent burns and overheating.

Hyperthermia is when the body temperature is above normal and causes the animal to have a **fever**, which is an excessive amount of heat in the body. Hyperthermia can be a sign of infection, toxicity, or heatstroke. **Heatstroke** is a common occurrence in areas that have high environmental temperatures and can be a serious condition for animals. Heatstroke may be considered when body temperatures reach or exceed 105 degrees F and can produce severe damage to internal body organs, the nervous system, and the blood supply. Hyperthermia can be controlled with ice packs wrapped in towels, cold water baths, fans, alcohol applications to pads and limbs, and cool drinking water. Cooling should be discontinued when the patient's body temperature reaches 103 degrees F.

The core body temperature is best evaluated by taking a rectal temperature using a rectal thermometer (see Figure 37–8). The animal should remain standing or in a sternal position so as not to break the thermometer or cause injury to the animal's rectal area. The rectal thermometer is applied carefully within the rectum by lubricating the end of the thermometer that has a silver metal tip. The thermometer only needs to be inserted ¼ inch to ½ inch within the rectum. Some rectal thermometers are digital and will alert with an alarm sound when the temperature is recorded (see Figure 37–9). Some thermometers are mercury based and have to be read after a set amount of time to

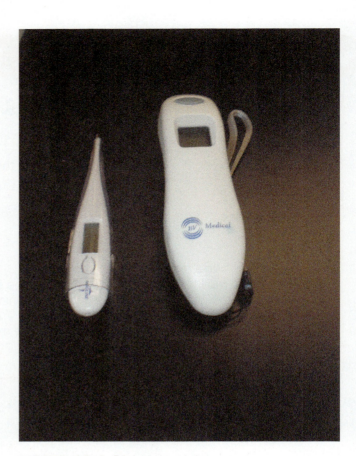

FIGURE 37–9 Digital rectal thermometer.

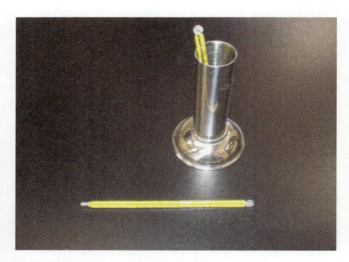

FIGURE 37–10 Mercury rectal thermometer.

determine the temperature (see Figure 37–10). Thermometers should have probe covers placed on them for each patient's individual use. The rectal thermometer should be left in the rectum for approximately 2 minutes and the temperature should be recorded in the medical notes throughout the patient's stay. Mercury thermometers are a safety hazard in that if

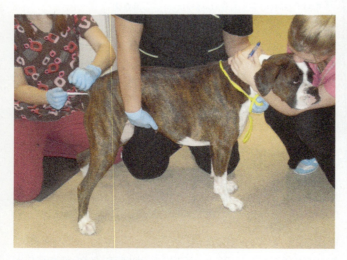

FIGURE 37–8 Rectal temperature is the most accurate measurement of body temperature.

broken, mercury is a toxic substance that will erode and burn. Care must be taken when handling mercury thermometers. Never touch mercury or glass with bare hands. Prior to use, a mercury thermometer must be carefully shaken down to decrease the mercury from the previous use. The mercury line in the glass thermometer is visible as a thin silver line located in the center of the glass when the number line is seen. This is also the area where the temperature reading is noted. The mercury will stop at the highest point of the internal body temperature and should be read on the number scale at that point. Ear thermometers fit into the external ear canal and record the body temperature through the ear. Use caution in that if the ear thermometer is not properly placed within the ear canal at the point of the tympanum, inaccurate readings will occur. Table 37–5 lists common normal body temperatures in various species.

TABLE 37–5
Normal Core Body Temperatures

SPECIES	NORMAL RECTAL TEMPERATURE
Dog	99.5–102 degrees F
Cat	100–102.5 degrees F
Rabbit	102–104 degrees F
Guinea pig	100–103 degrees F
Horse	99–100.5 degrees F
Cow	101–101.5 degrees F
Sheep	102–102.5 degrees F
Goat	103–103.5 degrees F
Pig	102–102.5 degrees F
Chicken	105–106 degrees F
Hamster	101–102 degrees F

COMPETENCY SKILL 86

Taking a Rectal Temperature

Objective:

To properly take a rectal temperature to accurately measure core body temperature

Preparation:

- Rectal thermometer (digital or mercury)
- Vaseline or KY Jelly lubricant
- Exam gloves
- Gauze sponges
- Paper towels

Procedure:

1. Apply exam gloves.
2. Shake down mercury thermometer with a flick of the wrist or turn on and calibrate digital thermometer.
3. Apply a small amount of Vaseline or KY Jelly to a gauze sponge.
4. Place the metal-tipped end of the thermometer into the lubricant.
5. Lift the animal's tail and secure one hand at the base of the tail. Keep animal in standing or sternal position.
6. With the other hand, gently insert the lubricated end of the thermometer into the animal's rectum.
7. Allow digital thermometers to beep when done or mercury thermometers to be in place for 1–2 minutes based on type of thermometer (mercury versus digital).
8. Remove thermometer gently from rectum and clean end of thermometer with a clean gauze sponge.
9. Read the thermometer for the core body temperature.
10. Record the temperature in degrees Fahrenheit in the medical record.
11. Disinfect the thermometer and place it in a safe area.
12. Clean up supplies and disinfect the work area.

Pulse

The **pulse** is the number of times the heartbeats per minute. The pulse is taken by palpating an artery. The pumping action that is felt is a shock wave that travels through the arteries and is generated by a contraction of the heart. When taking a pulse rate, one to two fingers may be placed over an artery to feel for the pulsing action. Each pulse that is felt is counted as one heartbeat. The most common site for establishing a pulse is on the femoral artery located on the inner thigh of a hind leg or the dorsal pedal artery of the paw. The **heart rate (HR)** is the number of times the heart relaxes and contracts per minute. The heart rate is taken by auscultating the heart with a stethoscope (see Figure 37–11). A heart rate should be recorded in beats per minute or bpm. Table 37–6 lists normal heart rates for various species. In a typical animal, the pulse and heart rate should be the same. When determining a heart rate or pulse on a patient, it is easier to count the number of beats over a shorter period of time, typically for 15 seconds, and then multiply the number of beats by the number of units of time per minute. For example, you would count the number of beats heard in 15 seconds and then multiply by 4. If counting an extremely fast heart rate, one may count the number of beats for 5 seconds and then multiply by 12.

Pulse strength should be recorded and any deficits noted. A pulse deficit is a difference in the number of heartbeats and the number of pulse beats calculated. Pulses may be described as the following: normal, regular, irregular, thready, weak, bounding, or absent. A normal sinus rhythm is determined when a heartbeat sound is regular. The pulse rate is the same as the heart rate. A sinus arrythmia occurs when the heart and pulse rates increase during inspiration and decrease during expiration.

TABLE 37–6

Normal Heart Rates

SPECIES	HEART RATE (BPM)
Dog	60–140
Small	90–140
Medium	80–120
Large/giant	60–100
Cat	140–250
Rabbit	130–325
Guinea pig	240–350
Horse	35–45
Cow	60–70
Sheep	60–80
Goat	70–80
Pig	60–80
Chicken	200–400
Hamster	250–500

Respiratory Rate

The respiration or **respiratory rate (RR)** is related to the breathing of an animal and evaluates how many breaths an animal takes in a minute. Respiration is the cycle of one inhalation and one exhalation (see Figure 37–12). The respiratory rate is recorded in breaths per minute or br/min. The respiratory rate can be obtained by watching an animal inhale and exhale or by using a stethoscope over the thoracic cavity while counting each breath heard for a specific amount of time. Again, one would establish the respiratory rate by counting the number of breaths heard in 15 seconds and multiplying by 4, as in establishing a heart rate. Table 37–7 lists the normal average respiratory rates in various species.

TERMINOLOGY TIP

The heart rate or pulse may be recorded **P** or **HR** and is labeled **bpm**, which stands for beats per minute. The respiration or respiratory rate is recorded **R** or **RR** and is labeled **br/min**, which means breaths per minute. The weight may be recorded as **Wt**. or the pound sign (#) to refer to pounds. Pounds are also abbreviated lb. ■

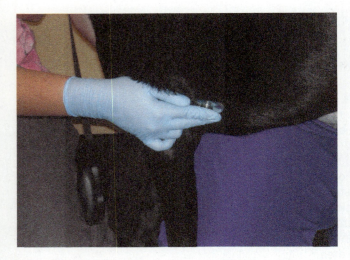

FIGURE 37–11 Listening to the heart to determine heart rate.

CHAPTER 37 Physical Examination Procedures and Patient History

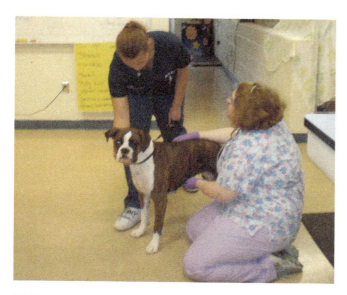

FIGURE 37-12 Counting the dog's respirations.

TABLE 37-7	
Normal Respiratory Rates	
SPECIES	RESPIRATORY RATE (BREATHS/MINUTE)
Dog	10–30
Cat	20–30
Rabbit	32–60
Guinea pig	40–150
Horse	8–16
Cow	10–30
Sheep	12–20
Goat	12–20
Pig	8–15
Chicken	15–30
Hamster	35–135

The depth of respiration should be noted to detail the amount of air volume needed with each breath. An increased depth of respiration requires a greater demand for oxygen. A shallow breathing pattern indicates disease or injury. Abnormal respiratory patterns should be noted for the veterinarian. These include *dyspnea*, or difficult or abnormal breathing noted on observation as distressed breathing. *Labored breathing* occurs when abdominal movements are noted at the same time as the breathing pattern. *Tachypnea* is very rapid breathing. *Hyperpnea* is breathing that is deeper and more rapid than normal. *Hyperventilation* is when increased shallow breaths are noted and often occur with pain or discomfort.

 COMPETENCY SKILL 87

Determining Pulse Rate

Objective:

To properly obtain a pulse rate when performing a physical examination

Preparation:

- Watch or clock with second hand
- Exam gloves
- Quiet location

Procedure:

1. Apply exam gloves.
2. Locate the femoral artery on the inner thigh of the hind leg. This is deep within the groin area. Apply the fingertips gently to the area.
3. Count the number of pulses felt in 15 seconds.
4. Multiply the number of pulses by 4 and note the pulse rate.
5. Remove gloves and discard. Record the pulse rate in the medical record as bpm.
6. Note the characteristic of the pulse as regular, irregular, thread, bounding, weak, or strong.

COMPETENCY SKILL 88

Determining Heart Rate

Objective:

To properly use a stethoscope to obtain a heart rate when performing a physical examination

Preparation:

- Stethoscope
- Watch or clock with second hand
- Exam gloves
- Quiet location

Procedure:

1. Apply exam gloves.
2. Locate a quiet place to listen to the chest cavity.
3. Locate the heart by taking the left front limb and placing the elbow at the lateral chest area. The area where the elbow meets the chest is the heart location.
4. Place each end of the stethoscope that moves outward into each ear canal.
5. Apply the bell end of the stethoscope to the lateral chest area and identify the heartbeat.
6. Count the number of heartbeats for 15 seconds.
7. Multiply the number of heartbeats by 4 and record your results in the medical record in beats per minute or bpm.
8. Record any abnormal sounds of the heart.
9. Disinfect the bell end of the stethoscope and the ends of the ear probes.

COMPETENCY SKILL 89

Determining Respiratory Rate

Objective:

To properly use the stethoscope to obtain a breathing rate when performing a physical examination

Preparation:

- Stethoscope
- Watch or clock with a second hand
- Exam gloves
- Quiet location

Procedure:

1. Apply exam gloves.
2. Locate a quiet place to listen to the chest cavity.
3. Place each end of the stethoscope that moves outward into each ear canal.
4. Apply the bell end of the stethoscope to lungs over the dorsolateral aspect of the thorax.
5. Count the number of breaths in 15 seconds.
6. Multiply the number of breaths by 4 and record the results in the medical record as breaths/minute or br/min.
7. Record any abnormal sounds of the heart and lungs.
8. Disinfect the end of the stethoscope and the ends of the ear probes.

Blood Pressure

Blood pressure is the tension exerted by blood on the arterial walls. Blood pressure is measured by a sphygmomanometer (blood pressure cuff). This device measures the amount of pressure exerted against the walls of the arterial vessels. Systolic pressure occurs when the ventricles contract. Diastolic pressure occurs when the ventricles relax. Hypertension is the term used to describe high blood pressure, while hypotension is the term used to describe blood pressure that is too low. Normal systolic blood pressure readings in dogs and cats are between 100 and 160 mm Hg; normal diastolic pressures are between 60 and 90 mm Hg. The **mean arterial pressure (MAP)** refers to the average of the systolic and diastolic pressures. Normal MAP in dogs and cats is 80–120 mm Hg.

Mucous Membranes

Mucous membrane color refers to the color of an animal's gums. The gums should be a nice light pink color. Gum colors such as grey, blue, brick red, or white are abnormal and should be reported to the veterinarian (see Figure 37–13).

Capillary Refill Time

The **capillary refill time (CRT)** evaluates how well an animal's blood is circulating in the body. This is assessed by placing a finger on the gums with a small amount of pressure, allowing the area to turn white. Releasing the finger and thus the pressure allows the color to return to the gum area (see Figure 37–14). The time it takes for this return of color to occur is the CRT. A normal CRT time is 1–2 seconds. Times that are longer than this should be reported to the veterinarian.

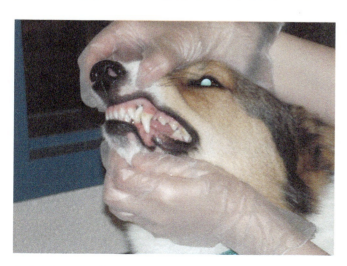

FIGURE 37–13 Checking mucous membranes.

The Oral Cavity

The **oral cavity** includes the gums, mouth, teeth, tongue, and throat. Examining the mouth area should include examining the teeth by lifting the lip, noting any signs of tartar on teeth, stains, odor, or broken or missing teeth. The mucous membrane (mm) or gum color can be noted at this time. Pink is the normal color of healthy gums. The capillary refill time (CRT) can be established by pressing on the gums and noting the time it takes for the color to return to the tissues. A normal CRT is 1–2 seconds. Then safely open the mouth to examine the tongue and throat area for any abnormal appearance or objects (see Figure 37–15). If an animal appears to be aggressive, do not attempt to pry open the mouth, as this may cause the animal to bite.

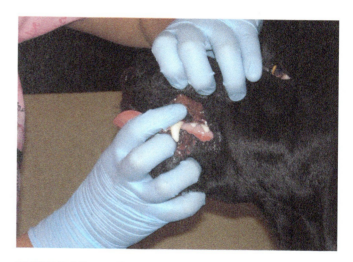

FIGURE 37–14 Performing a capillary refill test.

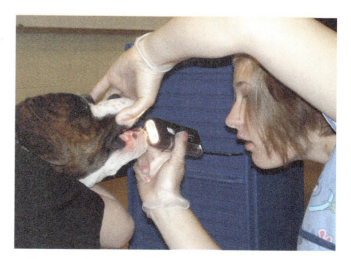

FIGURE 37–15 Look inside the mouth to note any abnormalities or objects.

670 SECTION IV Clinical Procedures

FIGURE 37–16 The eyes should be examined using an ophthalmoscope to note any signs of ocular discharge, redness, squinting, scratches, sores, and pupil reflexes in response to a light source.

The Head Exam

The **head exam** includes examining the eyes, ears, and nose. The eyes should be examined using an ophthalmoscope to note any signs of ocular discharge, redness, squinting, scratches, sores, cloudiness, and pupil reflexes in response to a light source (see Figure 37–16). When a light is shined directly into the pupil, the pupil should **constrict** or narrow and become smaller. When the light source is removed, the pupil should **dilate**, or open wide. It is important to note even pupil size and reflexes on each eye. The ears should be examined using an otoscope, and any signs of ear wax discharge, redness, sores, or odor should be noted (see Figure 37–17). The nasal area or nostrils should be examined with a penlight or the otoscope to evaluate the nasal passage for discharge, foreign objects, or swelling. The head, face, and neck should be examined for signs of sores, hair loss, or areas of swelling.

Eye Exam

Normal eyes appear clear and responsive. The white part of the eye, called the sclera, should have small blood vessels visible. The cornea, or the colored part of the center of the eye, should be clear. Any abnormal signs, such as a scratch, redness, swelling, discharge, or jaundice, should be noted. **Jaundice** is a yellow coloration to the skin or mucous membranes and may be a sign of liver disease or other illness. A light source should be placed on the eye with a

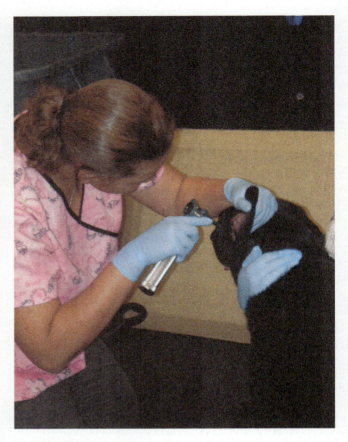

FIGURE 37–17 The ears should be examined using an otoscope, and any signs of ear wax discharge, redness, sores, and odor should be noted.

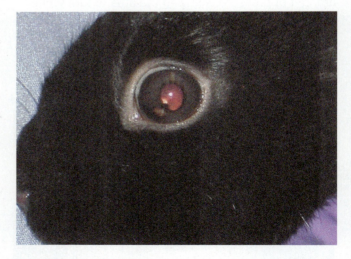

FIGURE 37–18 When examining the eye, note any abnormalities.

penlight or ophthalmoscope. Any abnormal signs such as poor pupil response, sensitivity to light, squinting, ocular discharge, or irregular eye movements should be noted (see Figure 37–18). Pupillary light response

(PLR) may be normal or absent. When the PLR is normal, it is called **direct PLR**. When the PLR is abnormal, it is called **indirect PLR**. Uneven-sized pupils seen in a condition known as **anisicoria**. Rapid and irregular eye movements that may cause the eyes to bounce from side to side or in circular motions indicate a condition known as **nystagmus**. This may suggest a balance problem, such as an ear infection or a central nervous system (CNS) disease. Blood noted inside the eye is a condition called **hyphema**.

COMPETENCY SKILL 90

Using an Ophthalmoscope

Objective:
To properly use an ophthalmoscope to assess the eyes and pupillary response to light

Preparation:
- Ophthalmoscope
- Exam gloves

Procedure:
1. Apply exam gloves.
2. Elevate the animal's head with one hand under the chin area to stabilize any movements. Each eyelid may need to be held open to observe the eye structures.
3. Turn on the light source of the ophthalmoscope.
4. Place the ophthalmoscope about 2–3 inches from one eye and look into the eye using the eyepiece located at the center of the ophthalmoscope.
5. Observe the opposite eye in the same way.
6. Note any abnormal signs of the eyes.
7. Turn off the light source of the ophthalmoscope.
8. The cone should be discarded and the instrument cleaned and recharged if needed. Place it in a safe location.
9. Record notes in the medical record.

Ear Exam

The ears should be examined using an otoscope (see Figure 37–19). Generally, the ear canal and the ear flap, also known as the pinna, should be a light pink color and free of wax and discharge (see Figure 37–20). Changes in the appearance of the ear should be noted, including any odor, redness, swelling, discharge, or debris. A common condition of the ear flap in animals that may scratch or shake the ears frequently is called an **aural hematoma**. A hematoma is a blood-filled swelling caused by the rupture of a blood vessel. The area fills up with blood, causing pressure on the area.

Dental Examination

The dental exam should be performed to note any signs of dental disease or other tooth problems (see Figure 37–21). Signs of dental disease include broken teeth, decayed teeth, inflamed gums, oral odor, and tartar or stains of the teeth. **Plaque** is a soft buildup of material found over the surfaces of the teeth that is composed of bacteria, food materials, and saliva. It may be easy to remove by brushing the teeth but often returns quickly (see Figure 37–22). Tartar or **calculus** is a mineralized plaque material that appears brown to dark yellow in color and is difficult to remove (see Figure 37–23). The **gingiva** or gums are the soft tissue that surrounds the teeth and are generally pink

672   SECTION IV Clinical Procedures

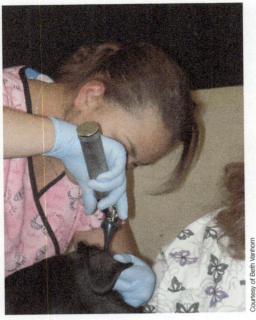

FIGURE 37–19 Examination of the ear is done using an otoscope.

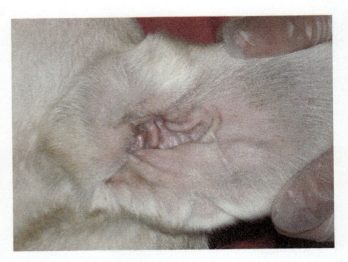

FIGURE 37–20 Note the normal color that should be observed on examination of the ear.

 COMPETENCY SKILL 91

Using the Otoscope

Objective:

To properly use an otoscope to assess the ears during a physical examination

Preparation:

- Otoscope
- Exam gloves
- Otoscope cone

Procedure:

1. Apply exam gloves.
2. Place a clean cone on the end of the otoscope and make sure it is secure.
3. Turn on the light source for the otoscope.
4. Pull back an ear flap to expose the internal ear canal. Hold the ear flap with one hand.
5. Using the free hand, place the otoscope cone near the internal ear canal.
6. Observe the ear canal by looking into the otoscope eyepiece at the center of the instrument.
7. Note any abnormal appearance.
8. Repeat the above steps with the opposite ear.
9. Turn off the otoscope light source.
10. Remove the cone and disinfect the cone.
11. Place the cone and otoscope in a safe location.
12. Record any abnormal findings in the medical record.

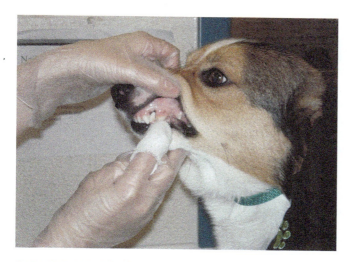

FIGURE 37–21 Examination of the oral cavity and dentition is an important part of the physical examination.

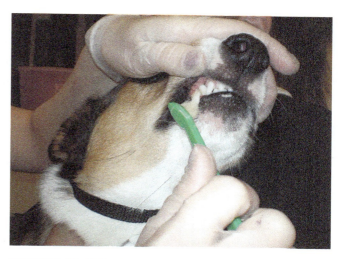

FIGURE 37–22 Brushing may help to prevent plaque buildup and promote good oral health.

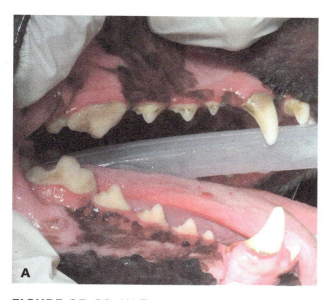
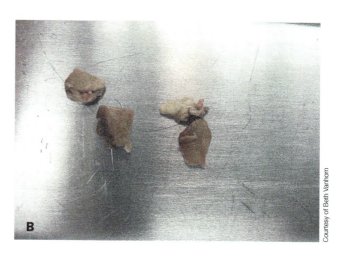

FIGURE 37–23 (A) Tartar buildup on teeth prior to a dental cleaning. (B) Tartar and calculus that have been removed from a tooth

in color, or they may be darkly pigmented in certain breeds. **Gingivitis** is inflammation of the gums; is the first stage of dental disease; and causes red, inflamed gums that are swollen and often bleed easily when touched. Animals that have dental disease may require anesthesia to clean the teeth through the process of dental **prophylaxis**, which is the term for the professional cleaning of a patient's teeth (see Figure 37–24).

A common condition in animals is a **tooth abscess**. The roots of teeth can abscess when a tooth is broken or damaged. This causes an infection within the root and gingival tissues. Injuries can occur when the animal chews on hard items. A root infection can also be caused by severe dental disease. The **carnassial teeth** provide a common site for root abscess. The carnassial teeth are the largest teeth in an animal's mouth and are the upper fourth premolars and the lower first molars. These teeth may have three to four roots that may be affected. The abscess of the root causes severe facial swelling and pain, typically located below an eye, as the teeth are located close to the sinuses and eye sockets.

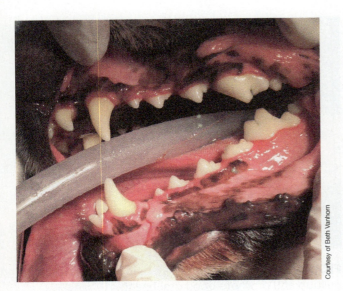

FIGURE 37–24 Teeth that have been cleaned; all tartar and calculus build-up has been removed.

Another common dental condition is excessive amounts of gingival tissue growth called **gingival hyperplasia**. These masses are common in some dog breeds, such as the Boxer, Great Dane, Doberman, and Dalmatian. Some animals, such as horses and rodent species, will require teeth to be floated or trimmed. **Floating** is the process of filing down long, sharp edges of teeth that can prevent horses from eating properly and can be painful during chewing or riding. Rodents' and rabbits' teeth continuously grow and may need to be trimmed for them to properly eat and not have the teeth grow into areas of the mouth, causing irritation and pain.

 COMPETENCY SKILL 92

Performing a Dental Exam

Objective

To properly perform an examination and assessment of the teeth and oral cavity

Preparation

- Exam gloves
- Gauze sponges
- Dental probe
- Dental mirror

Procedure

1. Apply exam gloves.
2. Carefully approach the animal and observe the facial area for any swelling.
3. Carefully restrain the head and gently lift the upper lip to observe the teeth. Note any abnormal appearance or odor.
4. Repeat the same process on the opposite side of the mouth.
5. The mouth can be opened by applying pressure at the area where the maxilla and mandible meet. Once open, carefully and gently pull the tongue out to one side to allow viewing of the mouth.
6. Animals may bite down on the tongue, causing wounds, so caution should be used when examining the mouth.
7. Note any abnormal appearances or odors in the medical chart.
8. The dental probe may be used to measure depths of tooth injuries, missing teeth, or areas of recession. It may also be used to check debris and tartar buildup.
9. The gauze sponges may be used to wipe away any bleeding, saliva, or debris from the teeth.
10. Clean up the area and disinfect any tools.
11. Add notes to medical record.

The Thoracic Cavity

The thorax or chest area should be evaluated with a stethoscope by listening to the heart, chest, and lungs for any abnormal sounds (see Figure 37–25). This process is known as **auscultation**. A breathing pattern should be determined and any abnormal signs noted, such as **dyspnea** (difficulty breathing); whether chest movement is normal, shallow, or deep; and abnormal sounds. Abnormal sounds may include wheezing (sounds like musical notes) or **rales** or **crackles** (sounds like cellophane paper). The heart and lungs sounds should be evaluated for abnormal sounds, such as a **murmur** (often heard as a swishing sound), an **arrythmia** (abnormal heartbeat), or any muffled sounds that may be caused by fluid buildup in the chest. The heart rate (HR) and respiratory rate (RR) should be determined and recorded (see Figure 37–26). The femoral pulse should be evaluated for strength and quality, with any signs of weakness or bounding noted.

FIGURE 37–26 Heart rate (HR) and respiratory rate (RR) should be evaluated and recorded.

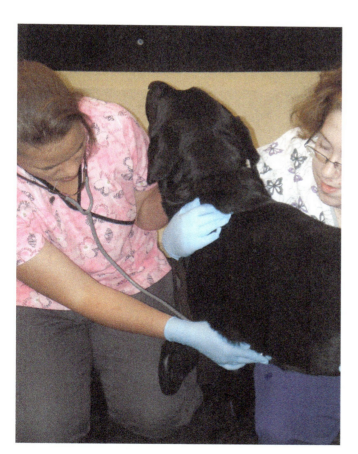

FIGURE 37–25 The thorax or chest area should be evaluated with a stethoscope by listening to the heart, chest, and lungs for any abnormal sounds.

Auscultation

Auscultation of the chest cavity includes listening to the heart and lungs. A stethoscope is used to listen to the chest area by placing the end of the stethoscope that is bell-like in appearance over the animal's heart area. The heart area can be located by taking the left front leg and pulling it toward the chest, and where the elbow meets the chest is the location of the heart. This area is also near the ribs. The stethoscope should be moved to at least three areas of the left side of the chest to listen to the heart and lungs to rule out abnormal sounds (see Figure 37–27). The same procedure should be completed on the right side of the chest. The heart can also be listened to through the sternum or breast bone on the lower ventral chest area. A strong heartbeat should be noted.

Irregular heart sounds such as decreased rhythm may be a sign of a mass or fluid located in the chest area. **Bradycardia** is the term for a decreased or slow heart rate, and **tachycardia** is the term for an increased or fast heart rate. Either abnormality may be a sign of heart disease.

Irregular lung sounds present as crackles or wheezing. Crackles, sometimes referred to as rales, sound like crumpling cellophane and many times mean fluid buildup in the chest. Wheezing is a raspy sound that produces a whistlelike noise and is common in narrow or constricted airways. Some sounds may be normal for breeds that have brachycephalic (flattened or shortened) nasal passages such as Pugs, Bulldogs, and Himalayans.

Cardiac murmurs may be detected during auscultation and should be evaluated when the pet is in a sternal and in a standing position. Murmurs are often heard in young animals and are often soft systolic murmurs originating in the mitral or aortic valves. They are innocent murmurs that the pet often outgrows and tend to disappear by the age of 3–4 months. Older pets typically have murmurs related to dysfunctional heart valves or septal defects.

Integumentary System

The **integumentary system** includes the skin, hair, pads, horns, beak, nails, hooves, and feathers. The skin and hair coat should be evaluated for signs of **alopecia** (hair loss), masses, dandruff or flaky skin, dryness, excessive oil, matting, redness, sores, or signs of parasites (see Figure 37–28). External parasites include fleas, ticks, mites (ear and skin), lice, and fly eggs called maggots. Signs of hair loss may indicate a fungus or another contagious condition. The hair coat should be evaluated for signs of cleanliness or unkempt appearances. The animal's **hydration status** should be evaluated over the neck and shoulder area of the skin, observation of the eye sockets, and gum moistness. Loss of elasticity of the skin (skin turgor) is the first sign of dehydration.

FIGURE 37–27 A stethoscope is used to assess heart and lung sounds.

COMPETENCY SKILL 93

Auscultating

Objective

To properly use the stethoscope to listen to heart sounds when performing a physical examination

Preparation

- Stethoscope
- Exam gloves
- Quiet location

Procedure:

1. Apply exam gloves.
2. Locate a quiet place to listen to the chest cavity.
3. Locate the heart by taking the front limb and placing the elbow at the lateral chest area. The area where the elbow meets the chest is the heart location.
4. Place each end of the stethoscope that moves outward into each ear canal.

5. Apply the bell end of the stethoscope to the lateral chest area and listen to the heart sounds.
6. Move the stethoscope to at least three areas of the chest and listen for abnormal sounds.
7. Move the stethoscope slightly caudal to the heart and listen to three areas of the lung field for abnormal sounds.
8. Repeat these steps on the opposite side of the chest.
9. Record any abnormal sounds of the heart and lungs.
10. Disinfect the bell end of the stethoscope and the ends of the ear probes.

This is indicated by tenting, which should be reported in subjective forms of body weight percentages from 0 percent to 20 percent. The amount of tenting should be reported in a percentage to identify dehydration. The nail beds, hooves, and pads, as well as beaks, horns, and antlers, should be evaluated for signs of injury, dryness, or wounds (see Figure 37–29). Feathers should be evaluated for similar signs relating to hair and skin quality.

The Nervous System

The **nervous system** is evaluated throughout the entire physical exam. Areas to note are gait, pupil reflexes, and other body reflexes such as those related to the limbs, paws, and digits. Reflexes that can easily be evaluated include the joint reflexes, which can be assessed using the reflex hammer to see the **knee jerk reaction**. This is the reaction to a slight tap over the knee joint that causes the knee to jerk in response. The limbs should be evaluated by bending at each joint to note any pain or swelling (see Figure 37–30). The paws can be assessed by placing the top surface of the paw in contact with the ground and noting whether the animal immediately rights the paw to the proper position. The digits can be pinched to elicit a response. The desired response is pulling away from the pressure (see Figure 37–31). This is called the **toe pinch response**. These reflexes allow the assistant to determine the animal's nervous system function. Small animals that have questionable nervous system signs may be evaluated using the **wheelbarrow technique**, which is slightly lifting the animal's hind end off the ground and pushing the body forward in a "wheelbarrow" motion. The animal should walk forward in a typical fashion.

The Musculoskeletal System

The **musculoskeletal system** includes the bones, joints, and muscles. Gait can be used to assess for signs of lameness or pain. Areas of the limbs and

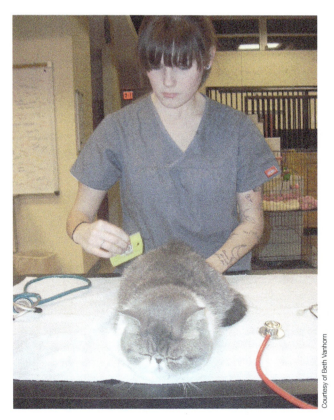

FIGURE 37–28 The skin and hair coat should be evaluated for signs of alopecia (hair loss), dandruff or flaky skin, dryness, excessive oil, matting, redness, sores, or signs of parasites.

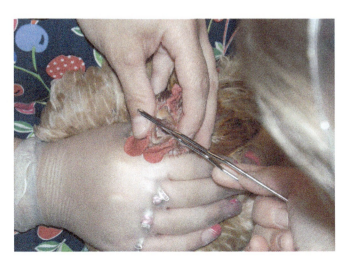

FIGURE 37–29 Checking the beak for injury or abnormalities.

shrinking of the muscle when it is not being used. The joints of the body should be evaluated for signs of swelling, pain, inflammation, or arthritis.

The Abdominal System

The **abdominal system** includes the stomach, intestines, and other internal organs, such as the liver, pancreas, and spleen. The abdomen should be palpated (i.e., felt with the hands) by the veterinarian for any signs of masses, distention, pain, or bruising. The stomach area should be palpated gently and slowly without grabbing any locations that may feel abnormal (see Figure 37–32). Using both hands and carefully allowing the fingers to move gently over the abdomen will allow the veterinarian to feel for any signs of masses, feces in the intestines, fluid, gas, or fetuses in a pregnant female. Large animals should be handled carefully to prevent injury from kicking or striking.

The Urogenital System

The **urogenital system** includes the kidneys, urinary bladder, external genitals, mammary glands, and rectal area. Palpation of the dorsal abdomen will allow evaluation of the kidneys and bladder. It is difficult to palpate the kidneys of some animal species. The mammary glands should be palpated for signs of masses, tumors, nodules, or cysts and any sign of milk discharge or pain and swelling. Female animals that have recently been bred or experienced labor may show signs of mastitis, including heat, redness, and discharge. Lactating animals should have their milk expressed to determine whether it has a normal appearance. Female animals should have the vulva examined

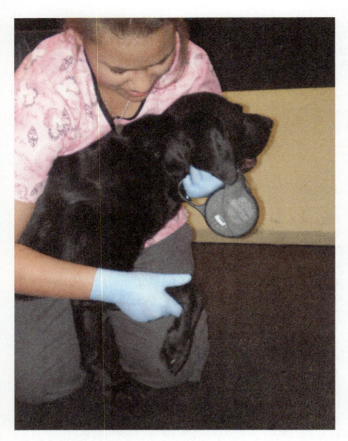

FIGURE 37–30 The limbs should be evaluated by bending at each joint to note any pain or swelling.

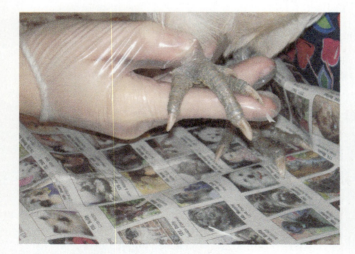

FIGURE 37–31 Toe pinch response.

other body parts should be evaluated for signs of swelling, heat, injury or wounds, atrophy, and pain. The muscles are areas in the body where movement occurs and should be assessed for quality. These areas are prone to pain, swelling, and **atrophy**, which is the

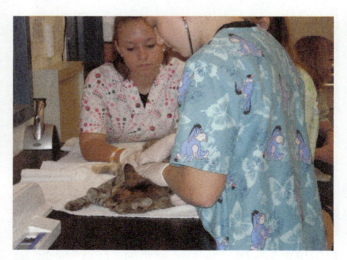

FIGURE 37–32 Palpating the abdomen for signs of swelling or discomfort.

for signs of discharge, swelling, inflammation, and masses. Male animals should have the prepuce, penis, and testicles evaluated for discharge, inflammation, swelling, and tumors. Urine samples should be collected and evaluated as necessary (see Figure 37–33). Normal daily urine output for dogs and cats is 20–40 ml/kg and should be determined as part of the PE.

The rectal area should be evaluated for signs of diarrhea, size of anal glands, odor from the anal glands or rectum, and any masses (see Figure 37–34). The rectal temperature of the animal can be taken at the beginning of the exam by lubricating a digital or mercury thermometer with Vaseline or KY jelly and applying a probe cover prior to use to ease insertion of the tool rectally and minimize any discomfort to the animal

FIGURE 37–33 Obtaining a urine sample from a dog.

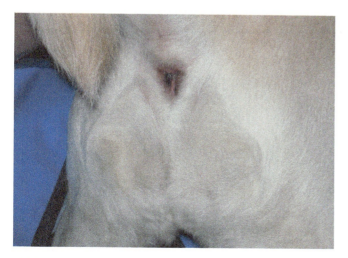

FIGURE 37–34 The rectal area should be evaluated for signs of diarrhea, size of anal glands, odor from the anal glands or rectum, and any masses.

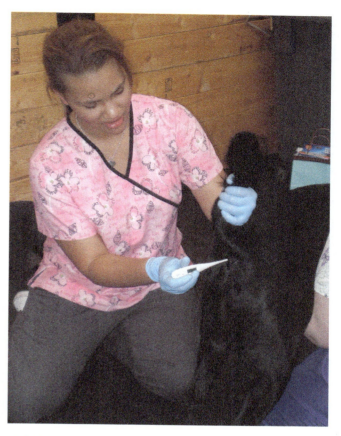

FIGURE 37–35 Taking a rectal temperature.

(see Figure 37–35). It normally takes about 1–2 minutes depending on thermometer type (mercury versus digital) for the thermometer to provide a reading. It is ideal to end with this so as not to stress or agitate the animal and then have a difficult animal during the rest of the PE.

Recording the Physical Examination

Many veterinary facilities have a physical exam (PE) report form that is used to note information about each body system of an animal. This form should be completed by the veterinarian, veterinary technician, and veterinary assistant as their duties allow (see Figure 37–36). Any abnormal findings may be noted in the progress notes so that follow-up reports can be monitored. It is important to note that when learning to complete PEs, one should concentrate on the hands-on skills and have another person record the findings until the veterinary assistant becomes experienced and familiar with each body system. This will

FIGURE 37–36 Completion of a physical examination report.

allow fewer mistakes and missing and abnormal notations in the medical record. When the assistant becomes familiar and experienced in evaluating patients, he or she should be able to complete a full PE and record the findings at the end of the exam, when the exam gloves can be removed and hands sanitized prior to recording in the medical record. This avoids potential contamination of the exam area and medical records. When the medical chart is completed, take a few moments to review the notes and make certain all items were covered, results were recorded accurately, and spelling and grammar are correct. It is important to use descriptive words and proper veterinary terminology when noting any normal or abnormal occurrences. Any problems noted that are a possible emergency should be immediately reported to the veterinarian for further evaluation. The evaluating veterinarian will then review the medical record and continue prompt assessment and examination of the patient.

SUMMARY

Completing a thorough physical examination even when there are no obvious problems allows veterinary staff to determine whether a problem is occurring or locate signs of diseases or changes in a patient. In addition, in animals with known abnormalities, such as fatty tumors or heart murmurs, regular exams allow the staff to chart the course of disease and determine whether it is progressing or remains the same. The veterinary assistant and support staff play a vital role in assessing the patient's overall appearance and condition. Exam room procedures entail a large amount of a veterinary facility's daily caseload. Various animals and appointments will occupy exam areas on a daily basis, with the potential for contagious organisms to come in contact with animals, staff members, and clients. Understanding common exam room procedures and proper exam room sanitation will ensure a clean and happy veterinary environment.

Key Terms

abdominal system   includes the stomach, intestines, and other internal organs such as the liver, pancreas, and spleen

alopecia   areas of hair loss

anisicoria   uneven pupil size

atrophy   refers to muscle that has shrunk due to disuse

aural hematoma   collection of blood within the cartilage of the ear

auscultation   process of listening to the heart, chest, abdomen, and lungs for any abnormal sounds using a stethoscope

bradycardia   decreased or slow heart rate

calculus   mineralized plaque material that appears brown to dark yellow in color and is difficult to remove from teeth

capillary refill time (CRT) indicates how well an animal's blood is circulating in the body; determined by pressing on the gums and monitoring how long it takes for the color to return to the area

carnassial teeth the largest teeth in an animal's mouth and are the upper fourth premolars and the lower first molars; common abscess sites

chief complaint the reason the animal is being seen by the veterinarian

constrict to close

crackles abnormal sounds within the chest that sounds like cellophane paper

dilate to open

direct PLR normal pupil light response assessment

dyspnea difficulty breathing

ectothermic temperature evaluated in an animal that regulates its body temperature externally or through its environment

endothermic temperature evaluated in an animal that regulates its body temperature internally or within the body

fever elevated body temperature due to excessive heat

floating filing down teeth that are overgrown and sharp

gingiva soft gum tissue

gingival hyperplasia excessive amounts of gingival tissue growth

gingivitis inflammation of the gums

head exam includes examining the eyes, ears, and nose

heart rate (HR) how many beats the heart produces in a minute

heatstroke serious condition where the body overheats at temperatures of 105 degrees F or more

hydration status evaluated over the neck and shoulders, eye sockets, and gums to determine how much water loss has occurred

hyperthermia condition where body temperature is above normal; causes the animal to have a fever

hyphema blood noted inside the eye

hypothermia condition where body temperature is below normal causing the animal to become cold

indirect PLR abnormal pupil light response assessment

integumentary system includes the skin, hair, pads, horns, beak, nails, hooves, and feathers

jaundice yellow coloration to the skin or mucous membranes

knee jerk reaction the reaction to a slight tap over the knee joint that causes the knee to jerk in response; used to assess the nervous system

mean arterial pressure (MAP) the average of the systolic and diastolic blood pressures

mucous membrane (mm) the gums of an animal

murmur abnormal heart sound that presents as a swishing sound

musculoskeletal system includes the bones, joints, and muscles

nervous system complex collection of nerves and cells; assessed through gait; the response of pupil reflexes; as well as other body reflexes such as those of the limbs, paws, and digits

nystagmus condition that causes the eyes to bounce from side to side or in a circular motion

observation inspection of an animal at a distance to see what it looks like, how it acts, and how it moves naturally

oral cavity includes the gums, mouth, teeth, tongue, and throat

patient history background information of past medical, surgical, nutritional, and behavioral factors that have occurred over the life of the animal

physical examination (PE) observation of each body system to determine issues that can cause a health problem

plaque soft buildup of material found over the surfaces of the teeth; composed of bacteria, food materials, and saliva

prophylaxis professional cleaning of a patient's teeth

pulse heart rate

rales abnormal sounds within the chest; sound like musical notes

respiratory rate (RR) how many breaths an animal takes in a minute

signalment patient's name, age or date of birth, gender, color, and breed or species

tachycardia increased or elevated heart rate

toe pinch response occurs when digits are pinched to elicit a response, which should be to pull away from the pressure; used to assess the nervous system

tooth abscess infection of the tooth root

TPR temperature, pulse, and respiration

triage quick general assessment of an animal to determine how urgently the animal needs a veterinarian's attention

urogenital system includes the kidneys, urinary bladder, external genitals, mammary glands, and rectal area

vital signs signs of life; measurements that assess the basic functions of an animal's body; include heart rate, respiratory rate, temperature, blood pressure, mucous membrane color, capillary refill time, and weight of the animal

weight tape a measurement tape used around the girth of an animal to estimate its body mass

wheel barrow technique slightly lifting the animal's hind end off the ground and pushing the body forward in a "wheelbarrow" motion to assess the nervous system

Review Questions

1. What is the importance of the patient history?
2. What are the parts of a patient history?
3. What is the importance of the physical exam?
4. What are the body system parts of a physical exam?
5. Why is observation performed first in a physical exam?
6. What is meant by the term *signalment*?
7. What is the chief complaint?

8. What is triage? Why is it performed?
9. What are vital signs?
10. What does the term *auscultation* mean?
11. What is the difference between a scale and a weight tape?
12. What is the difference between hypothermia and hyperthermia?
13. What is the importance of safe use of a mercury thermometer?
14. When obtaining the rectal temperature of a dog, what should the normal temperature be? What should the normal rectal temperature of a horse be?
15. Explain how to locate the heart in the chest area of an animal.
16. What is the term for increased or elevated heart rate? Decreased or slow heart rate?
17. What are some signs of dental disease?
18. What are some abnormalities that may be noted when using an ophthalmoscope?

Clinical Situation

Melissa, a veterinary assistant at the Ponderosa Veterinary Clinic, is setting up for a dog exam in one of the hospital's exam rooms. Melissa has worked in the facility only for 2 weeks and has been taking notes regarding the duties she must complete and certain expectations of the veterinarians and staff. Melissa has checked the appointment schedule, retrieved the medical chart for the patient, and located all of the necessary paperwork for the office visit and physical exam. The appointment notes that Charles Delmont is bringing in a dog named Taffy at 4:00 p.m. for a PE and anal gland expression. Taffy is a 4-year-old SF Pomeranian.

Melissa reviews the medical chart and notes that Taffy has a WILL BITE sticker on the front of the chart. Taffy also has a history of matted hair around the rectum and skin problems. Several times she has had fleas, ticks, and alopecia.

- What tools and supplies should Melissa prepare for the PE?
- What information should Melissa ask the owner for a patient history?
- Does Melissa need to do anything else to prepare the exam room for Taffy and Mr. Delmont?

CHAPTER 38 Veterinary Assistant Procedures

Objectives

Upon completion of this chapter, the reader should be able to:

38.1 Explain the need for dental care for patients.
38.2 Discuss basic teeth-brushing techniques
38.3 Perform the proper method of filling a syringe
38.4 Describe the proper labeling of a syringe.
38.5 Describe subcutaneous, intranasal, and intramuscular injections
38.6 Demonstrate how to administer SQ fluids
38.7 Explain the importance of monitoring IV catheters and fluids
38.8 Discuss the importance of socialization and exercise of a hospitalized patient
38.9 Explain the euthanasia procedure
38.10 Demonstrate veterinary laundering techniques
38.11 Discuss proper admission and discharge practices for hospitalized patients
38.12 Describe how to record observations of hospitalized patients
38.13 Determine the proper handling of emergency situations
38.14 Explain how to tell the gender of animal species
38.15 Explain how to tell the approximate age of an animal
38.16 Demonstrate how to properly mix and prepare a vaccine
38.17 Demonstrate wound cleaning techniques
38.18 Demonstrate basic bandaging techniques
38.19 Discuss proper procedures for feeding and watering hospitalized patients
38.20 Discuss how to perform proper sanitation of cages of hospitalized patients
38.21 Discuss proper isolation ward practices
38.22 Describe common grooming procedures performed in the veterinary facility
38.23 Maintain grooming equipment and tools through proper sanitation and care
38.24 Perform a brush-out of patients, including patients that may be matted
38.25 Demonstrate proper maintenance and care of clippers and blades
38.26 Demonstrate how to properly trim, clip, and shave patients using clippers and appropriate blades
38.27 Demonstrate how to properly clean normal ears of a patient
38.28 Demonstrate how to properly trim nails of a patient
38.29 Demonstrate how to properly express anal sacs using external expression techniques
38.30 Demonstrate how to properly bathe patients
38.31 Demonstrate how to properly dry patients

Introduction

Veterinary assisting for animal patients is similar to nursing care for human patients. This is a service provided by trained professionals and is essential to caring for patients and preventing illness and disease in hospitalized patients when the owners are not capable of caring for them at home. Veterinary assistant skills are essential to the health of the patient. Each state's practice act includes rules and regulations regarding what educational requirements and skills veterinary assistants versus credentialed veterinary technicians must have and what they are allowed to do. Every team member should be well versed in their state's regulations. Nursing care provided to patients is a team effort by the veterinarian and support staff. Veterinary nursing skills include dental care, providing injections for the health of the patient, monitoring IV fluids and catheters, and understanding the needs related to and process of euthanasia. Injuries and wounds are as much a disease condition warranting nursing care as are parvovirus and heartworm disease. Working within a veterinary facility requires being knowledgeable about and experienced in hospital skills, as well as being able to recognize an emergency situation when an animal is presented to the facility or is a hospitalized patient. The veterinary assistant must be aware of the needs of the patients and veterinary staff to provide successful work within the hospital treatment area. Several procedures may be necessary for hospitalized patients, and the veterinary assistant will be a contributor to the preparation phase, the procedure, and follow-up care. Grooming is a large part of the veterinary assistant's responsibilities. **Grooming** is the care of an animal's external body, including hair coat, ears, nails, and anal glands. This is different from grooming care provided by a professional groomer, whose training requires an entirely different skill set and perspective than that of the veterinary assistant. The focus of grooming in a veterinary facility includes improving the well-being—but not necessarily appearance—of the patient. The main goal in grooming tasks in the veterinary facility is improving the overall health and comfort of the animal.

The Hospitalized Patient

Animals that are admitted to the hospital may require constant care, thorough observation, critical care and support, or routine and regular treatments. It is important that the veterinary assistant know the proper intake procedures for hospitalized patients, how to note the treatments or procedures that are required, and where the patient should be placed for care. It is also necessary for the assistant to know the case history

FIGURE 38–1 Clients must sign a consent form before leaving the patient at the facility.

and when the patient is ready to be discharged so that the owner is given accurate and appropriate instructions for home care.

Admitting the Hospitalized Patient

The client should be aware of what treatments, procedures, and necessary care are required for the patient prior to admitting the animal to the hospital. Appropriate waiver forms and estimates should be provided to the client and then the client may leave (see Figure 38–1). This may be difficult for some people, and to ease the separation from the animal, they may need to be reassured that their pets are in excellent care and will be as comfortable as possible. The patient, medical record, cage card, and treatment instructions should be placed in the treatment area where all information is kept on hospitalized patients. Some facilities also immediately place a **neckband** on the patient for further identification. The neckband and cage card should contain patient information and reason for hospitalization. Table 38–1 summarizes the various forms of patient identification that may be used in the hospital.

The cage, kennel, or holding area should be prepared based on the size of the animal and the type of care that is needed. Towels, beds, blankets, or newspapers may

TABLE 38–1			
Identification Information for Hospitalized Patients			
Cage card	Neckband	Patient name	Patient age
Species	Breed	Color/Markings	Gender
Phone number	Reason for hospitalization	Client name	Personal items (toys, blanket, etc.)

FIGURE 38–2 Recording information on the hospital treatment board.

FIGURE 38–3 Note how the information is recorded on this hospital treatment board.

be used to line cages for small animals and companion animals. Clean bedding should be provided in a stall for large animals. When the patient is placed in the enclosure and properly secured, the assistant should locate the **hospital treatment board** and medical record bin. The hospital treatment board is an essential tool in the veterinary clinic (see Figure 38–2). This board contains patient information and displays treatments that need to be completed or have been completed, along with general health information that can be easily accessed by the entire veterinary and support staff. The information captured on the hospital treatment board includes the following:

- Client name
- Phone number
- Patient name
- Patient age
- Patient breed
- Patient gender
- Patient weight
- Reason for hospitalization
- Treatment needs or procedures
- Medications and dosages
- Warnings or other vital information

Many facilities record all hospitalized patients by the medical record filing method so that the medical chart is easily located and follows the same chart labeling and filing method for accuracy. Facilities that file records by client last name or client number place the information on the board in the same manner to identify the client, patient, and medical chart (see Figure 38–3).

Every hospitalized patient should have a treatment sheet with the veterinarian's treatment instructions written on it. It is important that every staff member documents every treatment, procedure, and observation on a daily basis as the work is completed. Every associate—especially those on the veterinary treatment team—should have a generalized understanding of the treatment plans and be able to discuss them with the client. This includes knowledge of procedures, medications, and laboratory diagnostics.

Discharging the Hospitalized Patient

When the veterinarian feels the patient no longer needs constant care and is able to safely go home, it is important that written home-care instructions be provided in the medical chart. This will be a guideline for the veterinary assistant and support staff as they provide information on patient care to the client. Home-care compliance is improved when the owner has the appropriate information and instructions for home care. The medical record should include home-care instructions, any medications that need to be dispensed and sent with the pet, any pending laboratory tests, and an invoice. Once any prescriptions are filled, the assistant should review with the owner

COMPETENCY SKILL 94

Admitting the Hospitalized Patient

Objective:

To efficiently and effectively process admission information for patients that are to be hospitalized

Preparation:

- Medical record
- Cost estimate
- Consent form or waiver
- Cage or holding area according to patient size and care needs (bedding, water, feed, litter box, etc.)

Procedure:

1. Review patient treatment plan and estimate with the client.
2. Obtain a signature from the client on a consent form or waiver.
3. Assure the client the patient will be well cared for.
4. Take the patient to the treatment area, along with the medical record.
5. Place a neckband and cage card on the cage for identification (see Figure 38–4).
6. Place patient in cage or holding area.
7. Close cage and check that cage is secure.
8. Place patient information on the hospital treatment board.
9. Place medical record in treatment bin.

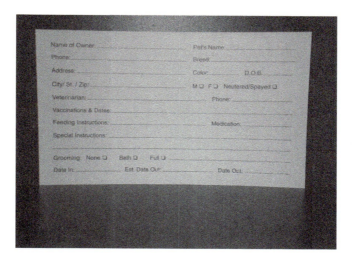

FIGURE 38–4 Cage card.

the dosage and directions for giving the medicine. The owner should be told when any pending lab test results will be available, if a follow-up appointment is necessary, and any health conditions to monitor at home. It is important to ask each client if he or she has questions regarding the patient or the home-care instructions. The veterinary assistant should stress that the owner should call if he or she has any problems or questions (see Figure 38–5). The client and patient should be escorted to the front desk or reception area for invoicing and checkout procedures. The assistant should note the conversation in the medical chart.

688 SECTION IV Clinical Procedures

FIGURE 38–5 The veterinary assistant should go over all discharge instructions with the client and allow the client time to ask questions.

Recording Observations

It is essential for all staff members to learn to observe patients. This is a necessary skill from the moment a patient enters the facility until the time of discharge. **Observation** is a key factor in successful patient care and treatment. Observation includes watching and noting an animal's behavior, appearance, mental status, and overall health (see Figure 38–6). These signs should be compared with what is normal for the species and breed of the animal. Any changes that occur should be monitored and noted in the medical chart. Some changes may be subtle and can be easily overlooked. Observations include visual monitoring, smelling odors, palpating changes, and auditory monitoring. Examples of observations that may be helpful in treating a patient include noting odorous gas or diarrhea, a tense abdomen, or ocular discharge. Daily observations should be noted in the medical record.

 COMPETENCY SKILL 95

Discharging a Hospitalized Patient

Objective:

To efficiently and effectively process the discharge information for patients that are able to return home after hospitalization

Preparation:

- Medical record
- Home-care instructions
- Medication
- Invoice
- Patient
- Any personal items

Procedure:

1. Prepare any prescriptions.
2. Review home-care instructions, medications, and other information with the client.
3. Demonstrate how to administer medications.
4. Provide information on lab tests or follow-up visits.
5. Review the patient discharge form with written instructions.
6. Obtain the patient for the client.
7. Escort the client to the front desk.
8. Give the medical record and invoice to the receptionist.
9. Inform the receptionist if a follow-up visit is needed.
10. Thank the client for coming and ask him or her to call if any problems or questions arise.

FIGURE 38–6 Observation skills are critical to the health and safety of the patient while under the care of the veterinarian.

Pain Assessment

The level of pain is vitally important to the comfort of the patient. Animals can't talk and describe their level of pain, so the recognition of pain and how it is managed will help improve a patient's well-being and recovery time. Pain is an unpleasant sensory or emotional experience that is associated with actual or potential injury. Nerve endings that conduct pain responses are found throughout the tissues and when stimulated, pain results. Pain can be classified as clinical, peripheral, neuropathic, or idiopathic. Pain can cause possible harmful effects on the body such as tissue breakdown or damage, immunosuppression, poor healing, and stress. All animals react differently to pain. Pain can be difficult to recognize in some patients. Signs of pain include the following:

- Changes in behavior
- Increased vocalization
- Licking/biting at painful areas
- Restlessness
- Increased heart rate (HR), respiratory rate (RR), temperature, blood pressure
- Shaking/tremors

Pain scales and sets of pain management guidelines have been developed for use in veterinary patients. They can be used to monitor pain levels in hospitalized patients, which facilitates optimal pain management for in-clinic patients, despite different personnel caring for them throughout their stay.

General Nursing Care

Patients should be kept clean and dry, as well as comfortable, while hospitalized. Pets that can't be taken outside or that have soiled their cage should be cleaned immediately. Patients that are recumbent and can't be moved may need management to help prevent from decubital sores, often called bedsores, as a result of urine scald. Recovering and ill patients that are recumbent for long periods of time and develop these types of sores can develop sepsis. Prevention is key. Decubital sores result from continual pressure on one site for extended amounts of time, which can lead to tissue damage. Foam pads, fleece blankets, and mattresses can be placed under these patients to allow better padding and reduce the tendency to develop sores. These items may need frequent changing, as moisture from feces or urine will only increase the chance of skin damage and infection. All recumbent patients should be rotated from side to side every 2–4 hours. Any sores that develop should be disinfected and cleaned with a sterile surgical scrub and completely dried after cleansing.

Exercise of hospitalized patients should be allowed as long as doing so is permitted by the veterinarian. Caution should be taken to allow leash walking only in fenced-in areas that are safe and secure. Walking will help improve muscle tone and reduce swelling and edema, as well as benefit the mind of the pet.

Monitoring urine output is another vital aspect of caring for the hospitalized patient. Note and evaluate the frequency of urine from cats and pocket pets by cleaning litter pans and bedding often. Dogs should be walked every 4 hours. Patients on IV fluids may require more attention, as are likely to have increased urine output. Normal urine output is 1–2 ml/kg/hr. Disposable pads are often placed under blankets to help reduce moisture in cages. The pads can be weighed. This weight can be compared to the initial pad weight to determine the weight of the urine, and this result can be converted into milliliters to approximate the total urine output. Patients with urinary catheters can have attached collection systems that aid in keeping the bladder empty and allow monitoring of how much urine is passed.

Evaluating Emergency Situations

Emergencies are a fact of life in all veterinary facilities. Patients may arrive in serious or critical condition from injuries, hemorrhaging, shock, respiratory distress, or poisoning. An **emergency** is a situation that requires immediate life-saving measures. When emergencies arrive, or when they occur within the hospital, it is important that all staff members work together as a team under the leadership of the attending veterinarian to provide adequate emergency care. The success of handling emergencies relies on the staff working as a team and staying calm under pressure. The veterinarian and veterinary technician should be providing hands-on care, triage, evaluation, telephone emergency, and communication situations, while the veterinary assistants are responsible for locating supplies, medications, and emergency equipment for use in the treatment area. A **crash cart**, which is a movable cart with enclosed drawers to hold emergency equipment and supplies, should be easily accessible when caring for a patient in serious condition. All items are prepared ahead of time for emergency situations. An effective and efficient veterinary assistant should be able to complete the following:

- Locate the crash cart
- Locate emergency equipment that may not be located on the crash cart or emergency kit due to size or limited supply
- Maintain emergency equipment, drugs, and supplies on the crash cart or kit
- Update the emergency supplies as newer equipment and fresh drugs are ordered
- Identify common emergency equipment
- Locate common general supplies
- Restrain and position patients properly
- Provide skin preparation as necessary
- Perform simple in-house laboratory procedures

Table 38–2 lists the common emergency equipment the veterinary assistant should be familiar with.

Once the patient has been stabilized, the assistant should provide veterinary care and patient monitoring as needed under the supervision of the veterinarian and veterinary technician. All staff members should be trained on what to do during an emergency so that each member of the veterinary team has a job to perform and is useful in these situations. Monthly staff meetings can be devoted to practicing such emergency procedures and protocols at the direction of the veterinarian. Knowledge of common emergency procedures is invaluable during an emergency, as this is not a time to learn and be trained.

Signs of an animal that is in a life-threatening emergency include the following:

- No signs of a heartbeat
- No signs of breathing
- Not alert
- Difficulty breathing
- Low body temperature and pale gum color
- Excessively high temperature
- Excessively hemorrhaging

It is important that assistants have knowledge of emergency equipment and how to clean and maintain it. A regular maintenance schedule should be followed to make sure all items are in working order. All drugs should be monitored to ensure they are up to date, and a checklist should be made for all items.

Common Hospital Procedures

The veterinary assistant should be able to perform common hospital procedures. These procedures include determining the gender of an animal, applying a basic bandage, mixing up a vaccine or medication, basic wound care, determining the approximate age of an animal, and feeding and watering.

Determining Animal Gender

Although it may seem humorous and odd, some owners may not know the gender or may incorrectly identify the gender of an animal. In some animals, such as kittens, rabbits, and pocket pets, it may be more

TABLE 38–2

Common Emergency Equipment

Laryngoscope and blade (see Figure 38–7)	Various endotracheal tube sizes (see Figure 38–8)	Heat pad or blanket	Clippers and blades
Instrument packs (see Figure 38–9)	IV stand (see Figure 38–10)	IV bags	IV lines
Pulse oximeter (see Figure 38–11)	Oxygen cage	Oxygen tank	Anesthesia machine (see Figure 38–12)
Stomach tube	IV catheter	Bandage materials	Urinary catheter
Stethoscope	Penlight	Syringes	Needles
Electrocardiogram (ECG)	Blood pressure monitor	Ambu bag	Emergency drugs

CHAPTER 38 Veterinary Assistant Procedures **691**

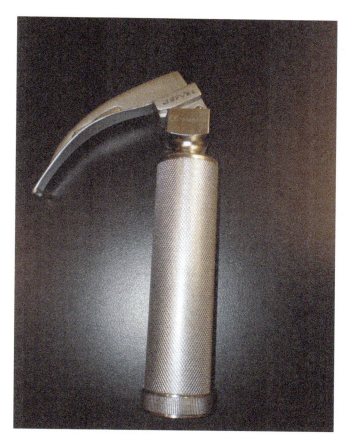

FIGURE 38–7 Laryngoscope.

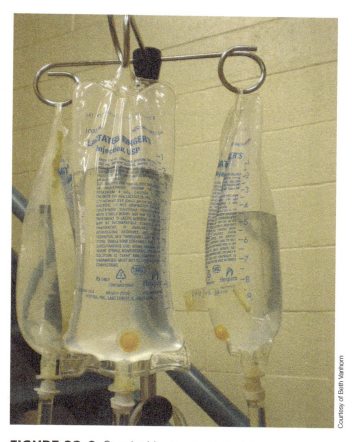

FIGURE 38–9 Surgical instrument packs.

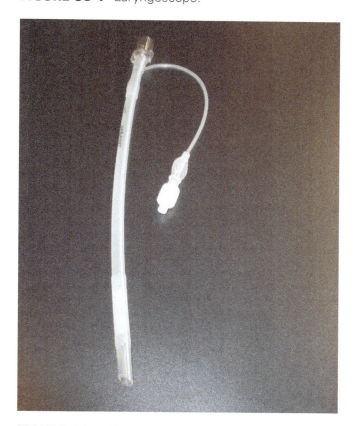

FIGURE 38–8 Endotracheal tube.

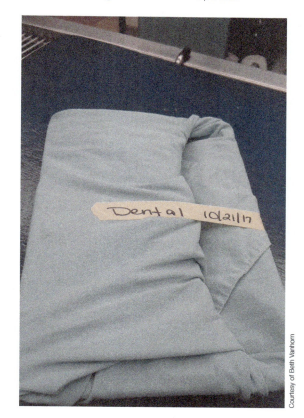

FIGURE 38–10 IV fluid stand.

692 SECTION IV Clinical Procedures

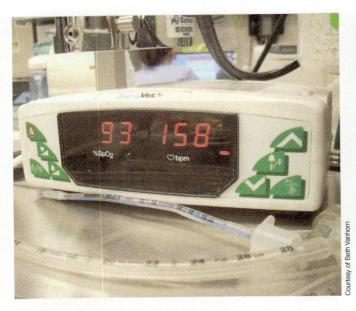

FIGURE 38–11 Pulse oximeter.

FIGURE 38–13 Determining sex in kittens. The female kitten is on the left and the male is on the right.

difficult to identify the sex. Some young animals may not have developed reproductive organs that can aid in easy identification of their gender. Some animals may have been neutered. The gender of most pocket pets, rabbits, and cats can be identified by the shape of the genital opening and the distance between the anus and genital area, commonly called the **anogenital distance**. Some animals can be sexed by the distance from the rectum or anus and the genital area where urination occurs. Many male rodent species, cats, and puppies have a wider space than in the female (see Figure 38–13). Gently lift the tail at its base and move it upward to get a visual appearance. Table 38–3 summarizes the shape of the genitals in various species.

Determining Approximate Animal Age

The age of an animal can be approximated in part by examining its **dentition**. The dentition is the arrangement of teeth in the mouth. The eruption of deciduous and permanent teeth occurs at a specific time in each animal species (see Table 38–4). The deciduous teeth, or the baby teeth, are sharper and smaller than the adult teeth. Also, as animals age, the teeth become more widely spaced as the jaw grows, and the baby teeth are pushed out and replaced by adult teeth. How the teeth look will also help determine an estimated age in an animal (see Figure 38–14). Signs of staining, tartar, and general wear will be noted in older animals. In the case of horses, the shape of the angle of the teeth changes, and the cups (center spot in the teeth) disappear with age (see Figure 38–15). Younger horses have more rounded teeth, middle-aged horses have

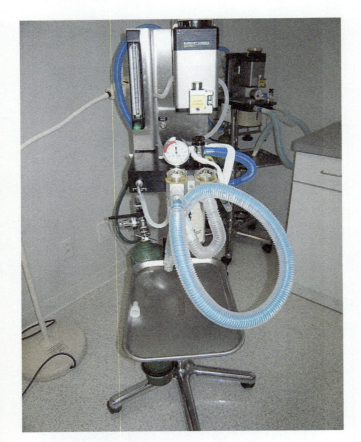

FIGURE 38–12 Anesthesia machine.

TABLE 38–3
Genital Shape of Various Species

SPECIES	MALE GENITAL SHAPE	FEMALE GENITAL SHAPE
Mouse	Penile shaft	Vaginal orifice or opening with flap of skin
Rat	Outward projection	Closed
Rabbit	Round opening	Slit opening
Hamster	Round opening	Long, slitlike opening
Guinea pig	*I*-shaped; able to expose penis	*Y*-shaped opening
Cat	Round; period shaped	Slit; comma shaped

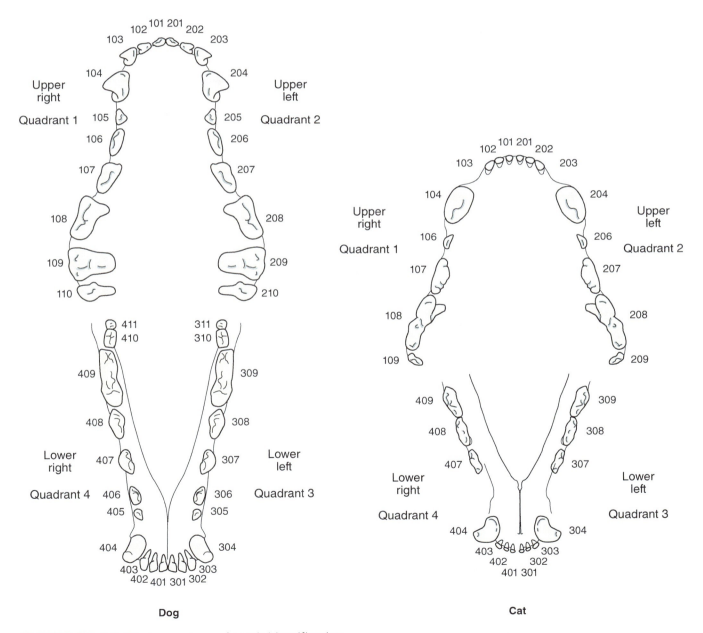

FIGURE 38–14 Triadan system of tooth identification.

TABLE 38–4
Tooth Eruption Timeframes in Dogs and Cats

TOOTH	DECIDUOUS, PUPPIES	DECIDUOUS, KITTENS	PERMANENT, DOGS	PERMANENT, CATS
Incisors	4–6 weeks	3–4 weeks	12–16 weeks	11–16 weeks
Canines	3–5 weeks	3–4 weeks	12–16 weeks	12–20 weeks
Premolars	5–6 weeks	5–6 weeks	16–20 weeks	16–20 weeks
Molars			16–24 weeks	20–24 weeks

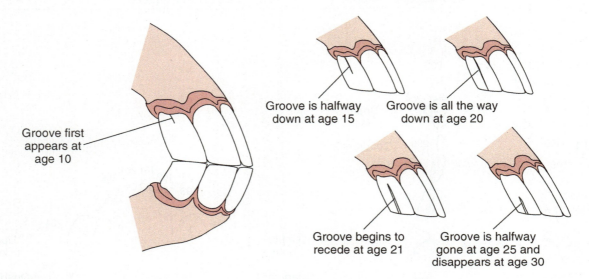

FIGURE 38–15 Grooves to determine age in horse teeth.

TABLE 38–5
Comparison of Dentition Between Species*

		DECIDUOUS				PERMANENT			
		INCISORS	CANINES	PREMOLARS	MOLARS	INCISORS	CANINES	PREMOLARS	MOLARS
Dogs	Upper	3	1	3		3	1	4	2
	Lower	3	1	3		3	1	4	3
Horse	Upper	3	0	3		3	1	3 or 4†	2
	Lower	3	0	3		3	1	4	3
Cattle	Upper	0	0	3		0	0	3	3
	Lower	3	1	3		3	1	3	3

*Some horses only develop 3 and some may have wolf teeth removed for pain relief during riding due to the bit location in the mouth

triangular-shaped teeth, and senior horses have rectangular-shaped teeth. Dogs and cats typically begin losing their deciduous teeth between 3 and 6 months of age and have all of their adult teeth between 6 and 8 months of age. Table 38–5 presents a comparison of the dentition in various species.

Dental Care

Dental care in animals is important and has an increased demand in all types of species, including companion animals, large animals, and pocket pets.

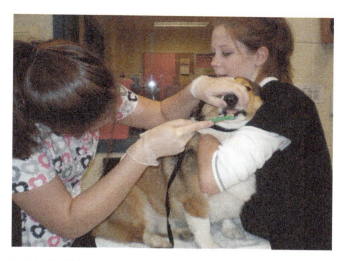

FIGURE 38-16 The veterinary assistant can perform teeth brushing.

Several aspects of dental care fall into the hands of the veterinary assistant. The veterinary technician performs dental care under anesthesia, and the veterinarian performs oral surgery and extractions. Preventive dental care and teeth brushing are an important part of the veterinary assistant's responsibility (see Figure 38–16). All of these factors make animal dental care an essential part of veterinary medicine and patient care. Promoting ideal dental health begins with educating clients on home dental care needs, including daily brushing of teeth, feeding a dry diet, and providing routine veterinary dental care such as dental prophylaxis or floating. It is also important to educate owners on how to properly keep teeth strong and healthy, and also prevent dental injury. Proper toys and dental chews should be provided, such as KONGs or veterinary dental bones (see Figure 38–17). Items such as raw bones, chains, or rocks should not be chewed on, as they may damage teeth and gums. Young animals should be monitored for proper tooth eruption as deciduous (baby) teeth are shed and permanent (adult) teeth begin moving into alignment. A yearly physical exam should include a dental exam by the veterinarian.

Daily Dental Care and Brushing

Daily dental care should include feeding a hard, dry diet that promotes dental health. Hard treats and dental toys or bones are highly recommended to further allow healthy teeth and provide animals that chew with appropriate items to prevent tooth injury. Daily dental care also means brushing the teeth in the proper manner. Many owners think of animal dental care as similar to that which they provide for their own teeth; however, there are several modifications in pet dental care that can cause difficulty or harm to the animal if

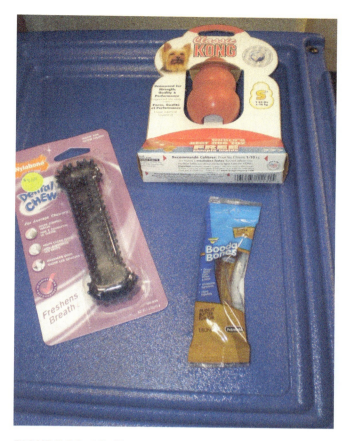

FIGURE 38-17 Chew toys promote good dental health.

not carried out properly. The veterinary assistant should teach owners how to properly brush their pet's teeth. Teaching the proper method and necessary tools and supplies will help owners accept the needs of the pet's dental care and make the process easier. When owners have difficulty brushing their pet's teeth, many soon give up and stop dental care altogether. Animals should never have human toothpaste placed in their mouth, as it has certain enzymes that are not digestible by animals; it may cause severe irritation and can even be toxic. Veterinary **dentifrices**, or toothpastes, come in a variety of flavors that increase the animal's acceptance of brushing the teeth (see Figure 38–18). These dentifrices have safe, digestible enzymes that easily break down in the animal's body. Flavors that are enjoyed by pets include fish, beef, malt, and poultry.

The toothbrush should also be a specialized instrument that fits the shape and size of the animal. Human toothbrushes are not ideal for all animals, such as cats and small dogs. Pet toothbrushes may have a small round head with soft bristles and a short narrow handle that fits comfortably into the mouth of an animal. Various sizes are available depending on the size of the animal. A **fingerbrush** is a small tool that fits on the end of a finger and is a plastic, thimblelike device with small, soft bristles that are rubbed over the teeth (see Figure 38–19).

The brushing procedure is similar to brushing one's own teeth with the exception that the lingual surface, or inner aspect of the teeth near the tongue, is not brushed. Only the buccal, labial, and occlusal surfaces of the teeth are brushed. The **buccal** surface of the teeth is located on the outer area near the cheek (see Figure 38–20). The **labial** surface of the teeth is the front area that is covered by the lips (see Figure 38–21). The **occlusal** surface is the top area of the teeth. It is difficult and not safe to open an animal's mouth to brush the inside surfaces of the teeth. Injury may be caused to the person or the animal. The areas of importance are easily accessed by lifting the lips to expose the outer tooth surface (see Figure 38–22). Animals will accept teeth brushing with time and consistency, as well as patience. Begin with a small amount of toothpaste placed on the tip of a finger or piece of gauze to allow the animal to smell and possibly taste it. A small amount of toothpaste can be rubbed over the outer buccal surface by using the fingertip or gauze and a rubbing motion on the tooth (see Figure 38–23).

As the pet accepts this, the owner can then proceed to using a toothbrush. By carefully lifting one area

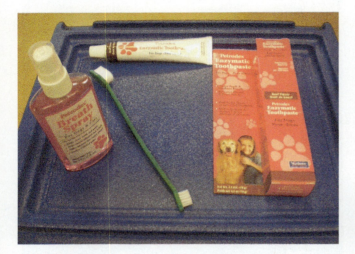

FIGURE 38–18 Proper tools for dental care in pets.

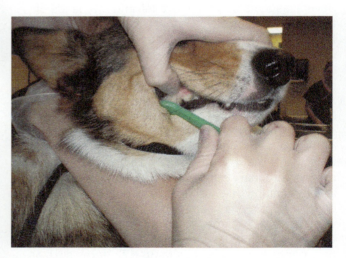

FIGURE 38–20 Buccal surface of the teeth.

FIGURE 38–19 Toothbrushes and fingerbrushes.

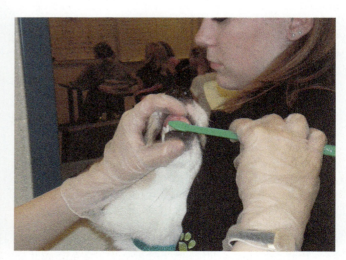

FIGURE 38–21 Labial surface of the teeth.

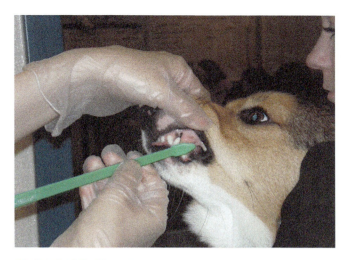

FIGURE 38–22 Lift the lips to expose the outer tooth surface.

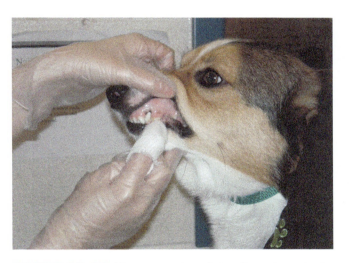

FIGURE 38–23 Use gauze to rub toothpaste onto the surface of the teeth.

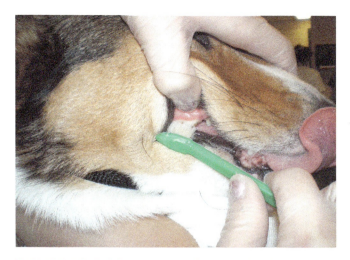

FIGURE 38–24 Proper angle for brushing the teeth.

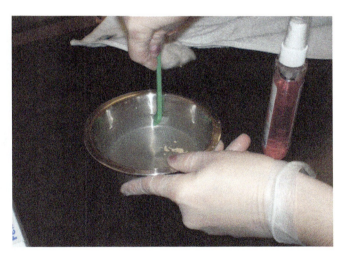

FIGURE 38–25 Use a small bowl of water to rinse the toothbrush.

of the lip and working on both the upper and lower aspects of the teeth, the teeth can be brushed. The toothbrush should be held at a 45-degree angle over the tooth and moved in a circular motion (see Figure 38–24). This process can be repeated until all outer aspects of the upper and lower teeth are brushed. Toothpaste should be reapplied as needed. A small amount of water in a bowl can be used to rinse the brush before applying more toothpaste (see Figure 38–25).

Pets should be introduced to teeth brushing at an early age, and the veterinary assistant should show owners how to properly brush the pet's teeth. This helps the process become more pleasurable to both the animal and the owner. Pets require daily teeth brushing and dental care. Sample kits that include a toothbrush, finger brush, and pet-safe toothpaste should be sent home to encourage compliance. A dental model may be used to show owners the proper brushing technique. Owners may need to be encouraged to provide home dental care. It is also important to discuss the possible need for **dental prophylaxis** or professional cleaning that includes scaling and polishing the teeth (see Figure 38–26). This may need to be done once a year or more depending on the health and age of the animal.

Large animals may require regular floating of the teeth to prevent sharp edges from forming and subsequently causing eating difficulties. Floating is an advanced technique that requires additional training and experience. Floating equipment often includes a tooth float, file blade, and mouth gag (see Figure 38–27). Horses commonly require floating when sharp edges form on the teeth that may cut into the gums, cheek, or tongue. When this condition progresses, they may have difficulty chewing hay and grass, and they often drop grain out of the mouth during chewing.

698 SECTION IV Clinical Procedures

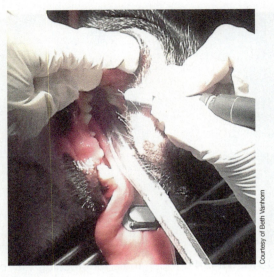

FIGURE 38–26 Dental prophylaxis (teeth cleaning).

FIGURE 38–27 A tooth float is used to file the teeth of horses. A gag is used to hold the mouth open for the procedure.

 COMPETENCY SKILL 96

Brushing Teeth

Objective:

To properly provide dental care and cleaning of pet's teeth

Preparation:

- Toothbrush or fingerbrush
- Pet-safe toothpaste
- Gauze sponges
- Small bowl with fresh water
- Dental spray
- Exam gloves

Procedure:

1. Apply exam gloves.
2. Place a small amount of toothpaste on the end of the index finger or piece of gauze.
3. Allow patient to smell or taste the toothpaste.
4. Lift the upper lip and rub the toothpaste over a tooth area. Repeat as necessary until patient is accepting of the process.
5. Apply a small amount of toothpaste to the toothbrush or fingerbrush.
6. Hold the mouth closed and lift the upper lip to expose the outer tooth surfaces.
7. Begin brushing the central incisors, holding the brush at a 45-degree angle.
8. Work posteriorly to the back of the mouth, ending at the molars.
9. Rinse the toothbrush with water and apply more toothpaste as needed.
10. Brush the left and right upper and lower dental arcade.
11. If the patient is willing, slip the toothbrush between the upper and lower arcades. Rotate the bristles so that they contact the inner surface of the upper arcade.
12. Repeat on the lower arcade.
13. Brush both surfaces of the upper and lower lingual arcade.
14. Clean the patient's face when done.
15. Clean up the work area and disinfect all items.

Dental Prophylaxis

Professional dental prophylaxis should be provided by the veterinary technician. To educate clients, the veterinary assistant must to understand the importance of the procedure. Some veterinary assistants may be asked to perform certain aspects of the dental prophylaxis under direct veterinary supervision. Several courses are available for advanced training in dental techniques and should be evaluated for requirements and state practice act guidelines. Dental prophylaxis includes placing the animal under general anesthesia. A dental exam should be completed first, noting any damaged or broken teeth, signs of gingivitis, excessive plaque, or any signs of missing or lose teeth (see Figure 38–28). A veterinarian may be asked to check any areas that are diseased or to extract teeth. The prophylaxis should then begin with scaling all surfaces of the upper and lower dental arcade. Care should be taken to not overheat any areas of the teeth or stay on one tooth surface too long without movement, as this can cause severe tooth injury and irreversible damage. Then all tooth surfaces should be polished. This process allows for the best dental care for the patient and should be discussed with all clients. Each animal should be evaluated for dental health care needs.

Preparing Vaccines

Vaccinations are a regular and routine occurrence in the veterinary hospital. All animals require booster vaccines while young or when beginning a new type of vaccine and are then placed on a regular vaccine program throughout the adult life. Animals that appear ill or diseased should not be vaccinated until they are healthy. Vaccines come in either single or multiple doses. Single doses may be premixed or in two vials, one with a liquid **diluent** and one with a powder (see Figure 38–29). The two vials must be combined to form a viable vaccine. The liquid diluent must be removed with a syringe and needle and may be an active vaccine component or sterile water. The diluent is then inserted into the powder vial and inverted to mix the components. Always mix the diluent provided for the vaccine, never one from another vaccine type. The vaccine is then ready to be administered in the animal species that it is labeled for. Once a vaccine is reconstituted, it must be used within an hour unless the label indicates otherwise. Vaccines are considered biologicals and must be kept refrigerated (see Figure 38–30). Multiple-dose vials will indicate the amount of vaccine required for each animal. This is usually 1 ml, but the label should be read for the directions. Multiple doses are premixed and ready for use. Care must be taken not to contaminate the vial, as doing so will make the entire contents unusable.

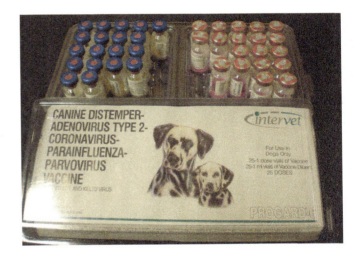

FIGURE 38–29 Vaccines.

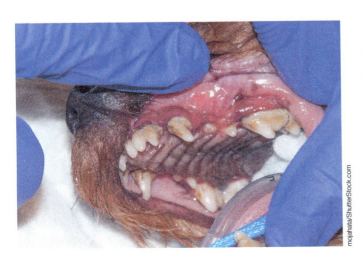

FIGURE 38–28 Dental calculus and gingivitis in a dog.

FIGURE 38–30 Vaccines must be kept refrigerated.

COMPETENCY SKILL 97

Mixing Vaccines

Objective:

To properly prepare vaccinations to be administered to patients

Preparation:

- Vaccine for the appropriate species
- Appropriate-size syringe and needle (usually 1–3 ml)
- Alcohol
- Cotton balls
- Exam gloves (see Figure 38–31)

Procedure:

1. Apply exam gloves.
2. Soak a cotton ball with alcohol.
3. Wipe the top of each vial with the alcohol-soaked cotton ball.
4. Obtain the liquid diluent vial and invert the bottle. (Multidose vial's volume should be withdrawn according to label; see Figure 38–32.)
5. Insert the needle into the center of the rubber vial (see Figure 38–33).
6. Remove entire contents of vial into the syringe. For a multidose vial, remove only the desired amount as specified on the label (see Figure 38–34).
7. Inject entire contents of liquid diluent into the powder vial.
8. Gently invert bottle 4–5 times until thoroughly mixed.
9. Withdraw the contents into the syringe and remove the needle from the vial.
10. Recap the needle.
11. Label each syringe with the peel-off label or a piece of tape indicating the contents.
12. Place the syringe in the treatment area.
13. Clean up work area.

FIGURE 38–31 Preparation for mixing vaccines.

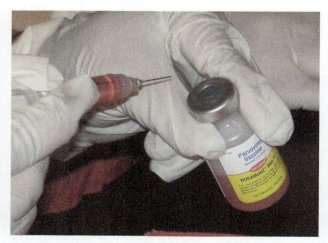

FIGURE 38–32 Look at the label for the correct amount to withdraw when working with a multiple-dose vial.

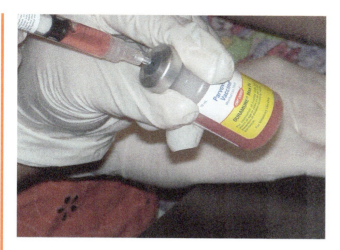

FIGURE 38-33 Insert the needle into the rubber top of the vial.

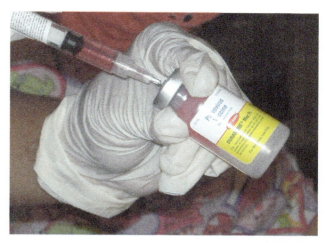

FIGURE 38-34 Only withdraw the amount indicated for a single dose when using a multiple-dose vial.

Administering Injections

Some veterinary practices will allow veterinary assistants to prepare and administer certain injections. Subcutaneous, intranasal, and intramuscular injections may be administered by the veterinary assistant in some states, depending on the veterinary practice act and the tasks delegated within the facility (see Figure 38–35). The **subcutaneous (SQ)** injection is administered under the skin. The **intranasal (IN)** injection is administered into the nose, usually in drops placed into the nostril rather than via a needle. The **intramuscular (IM)** injection is administered into a muscle. Commonly used muscles for injection sites are the *dorsal lumbar muscle,* located on either side of the midline, and the *semitendinosus muscles*, located in the rear leg and often called the hamstring muscles. Intravenous (IV) and intradermal (ID) injections are given by the veterinary technician and veterinarian, respectively.

Filling a Syringe

Several things must be considered before filling a syringe. The syringe and needle size are the first items to consider when preparing for an injection. Select a syringe that has a volume slightly larger than the dose being administered. Vaccines typically hold 1 ml, so an appropriate-sized syringe would be 3 ml (see Figure 38–36). When the syringe volume size is slightly larger than the amount to be administered, it allows

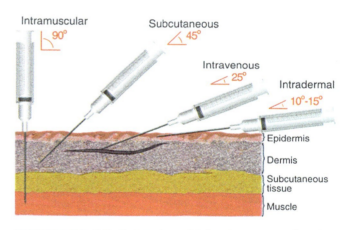

FIGURE 38-35 Examples of injection routes for drug administration.

for space to remove any air bubbles that may be drawn into the syringe. It also allows space for aspiration. The **aspiration** process occurs prior to administering the injection and requires the plunger to be drawn back slightly after the needle has broken the skin to make certain no blood vessel has been accidently penetrated. This is done immediately prior to injecting any medication into a patient, regardless of the route, to avoid accidental injection of a medication into the bloodstream. The needle size or **gauge** should be determined by patient size, thickness of the liquid being administered, and rate at which the injection is being administered. Gauge is related to the diameter of the needle. The greater the diameter of the needle, the

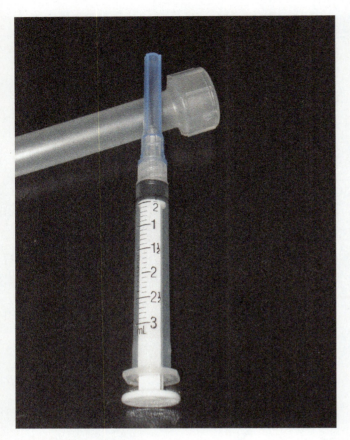

FIGURE 38–36 3-ml syringe.

TABLE 38–6

Labeling a Syringe

- Drug or vaccine name/type
- Amount or dose prepared
- Date
- Patient name
- Initial of person preparing syringe

lower the gauge size; for example, an 18-gauge needle has a greater diameter than a 25-gauge needle does. A lower-gauge needle is needed for more **viscous**, or thick, liquid drugs. A lower-gauge needle is needed for drugs that are to be administered more rapidly. The gauge selected should be appropriate for the injection type and the animal species; for example, cats have smaller veins and muscles than horses do. Needle length is also based on the type of injection and depth at which the medication will be administered. A short needle may be used in thin-skinned animals, such as cats and rodents, and a longer needle may be used when an injection is given into a muscle. The syringe needs to be handled with care to avoid any possible contamination. It is important to clean any injection vial with alcohol prior to inserting a needle into the vial. Medications and vaccines should never be mixed in the same syringe unless otherwise directed by the label, as combining different substances may result in an unwanted chemical reaction. It is important for the person preparing the syringe to always label it with the contents and initial it. The syringe should be labeled as in Table 38–6.

When ready to withdraw contents into a syringe, the vial may create a vacuum that makes it somewhat difficult to remove the substance from the container. When this occurs, detach the syringe from the needle so that it is still penetrating the top of the vial. This will allow some air to enter the container and eliminate the vacuum. Securely attach the syringe back onto the needle and continue withdrawal of the necessary amount of medication. The vial should be held upside down with one hand, with the other hand controlling the syringe, which is pointing upward toward the vial. The needle should penetrate the rubber stopper of the vial to the level of the medication. The plunger of the syringe should be pulled back to withdraw the proper amount of contents. Then the needle is withdrawn from the vial and the side of the syringe is tapped gently with a flick of the fingertip to remove any air bubbles. A gentle push on the end of the plunger may also help expel air bubbles (while not allowing any of the contents to escape the syringe). Recap the needle by gently sliding the needle into the needle cover and attach an appropriate label for the medication or vaccine. Always check the patient record or hospital treatment board to make certain the correct dose and medication are being administered. Read every drug label at least three times to make certain the correct drug is being used.

Subcutaneous Injections

The SQ injection is the easiest to administer and is the most frequently used injection for vaccines and antibiotics. Small animals have loose skin over the base of the neck and between the shoulder blade area, making an easy site for the SQ injection (see Figure 38–37). The site to be injected should be cleaned with alcohol. One hand should hold the syringe and the free hand should pinch the skin over the shoulder blades and lift gently to form a triangle, or tent. The needle is inserted at the base of the triangle or tent formation parallel to the body. If the needle is

COMPETENCY SKILL 98

Filling a Syringe

Objective:
To properly prepare a syringe with an accurate dose of medication or vaccine

Preparation:
- Proper-sized syringe
- Proper gauge needle
- Proper length needle
- Proper vial of medication or vaccine
- Alcohol
- Cotton balls
- Adhesive tape
- Exam gloves

Procedure:
1. Determine the drug or vaccine and amount to be placed in the syringe.
2. Apply exam gloves.
3. Select the proper-sized syringe, needle length, and gauge.
4. Prepare a label with the drug or vaccine name, amount to be withdrawn, date, patient name, and your initials.
5. Place the label on the distal barrel of the syringe.
6. Prepare a cotton ball saturated with alcohol.
7. Place the alcohol-saturated cotton ball on the top of the vial and wipe the rubber stopper area.
8. Place the vial upside down in one hand with the fingers curling around the vial securely.
9. Uncap the needle and insert it into the rubber top of the vial.
10. Withdraw the proper volume.
11. Remove the needle from the vial.
12. Gently tap or flick the edge of the syringe to remove any air bubbles, or slightly expel the air by pushing the end of the plunger.
13. Recap the needle.

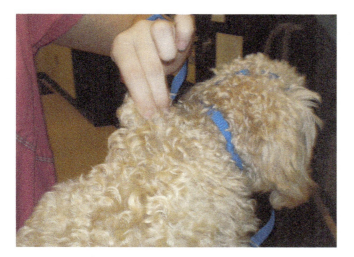

FIGURE 38–37 Site for subcutaneous injections.

short, it may need to be fully inserted; if the needle is long, it may only need to be partly inserted. Caution should be taken to avoid passing the needle outside of the skin. To determine proper needle placement, release the skin and use the index finger to feel for the end of the needle under the skin. If the needle is entirely under the skin, aspirate for proper placement. If no blood is noted, administer the injection. Withdraw the needle, briefly rub the area, and praise the patient. SQ injections are absorbed slowly over 30 minutes or longer and for larger volumes of fluids up to 6–8 hours. Most SQ injections are done using a 22-gauge needle.

COMPETENCY SKILL 99

Administering a Subcutaneous Injection

Objective:

To properly administer a subcutaneous injection

Preparation:

- Proper-sized syringe
- Needle of proper gauge and length
- Alcohol
- Cotton balls
- Exam gloves

Procedure:

1. Apply exam gloves.
2. Lift the skin between the shoulder blades using the thumb and forefinger of one hand to form a triangle or tent with the skin.
3. Wipe the area with an alcohol-soaked cotton ball.
4. Using the other hand, insert the syringe into the skin at the base of the tent or triangle parallel to the body.
5. Once the needle is placed, release the skin.
6. Use the free hand to palpate the needle below the skin to check for accurate placement, noting the needle is not through the skin.
7. Aspirate for placement, looking for any signs of blood entering the syringe. If no blood enters the syringe, administer the injection.
8. Withdraw the needle and place in sharps container.
9. Rub the injection site with one hand and praise the patient.

Intramuscular Injections

The IM injection is placed deeper into the body into a muscle. The muscles are more vascular than the SQ tissue and allow more absorption. Muscle tissue can only allow 2–5 ml of fluid to be injected into one location. There are many muscle sites on an animal's body to place IM injections. The common sites for IM injections in small animals are the **quadriceps** or **hamstring** group of muscles and the **epaxial** muscles located over the bands of muscles along either side of the spinal column near the back end of the animal. The quadriceps muscle is located on the hind limb at the cranial part of the thigh, and the hamstring muscle is located on the hind limb at the mid- to distal part of the thigh. In large animals, IM injections are administered in the brisket muscle over the front chest between the front limbs, the neck muscles, or the hamstring muscles.

IM injections that are given into the rear limb should be given while one hand stabilizes the limb so that the animal cannot readily move the leg. Caution must be taken when giving an IM injection in the hamstring or quadriceps area of the thigh, as it can be easy to inject the sciatic nerve, which can cause irreversible damage and potential paralysis. The fingers should be placed on the medial aspect of the leg, with the thumb on the lateral aspect at the mid-thigh. The free hand should hold the syringe, and the femur area should be palpated for correct muscle placement. The needle should be placed anterior to the femur bone approximately halfway between the hip and the knee and parallel to the femur. The site should be cleaned using an alcohol-soaked cotton ball. When injecting the needle, place the syringe slightly downward as it enters the muscle to control the depth of the needle placement (see Figure 38–38). Avoid any bone. Aspirate to ensure a blood vessel has not been entered. If

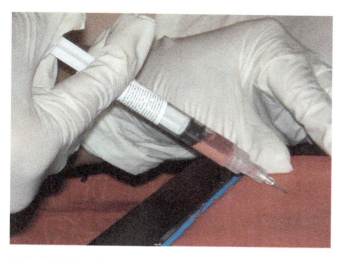

FIGURE 38–38 Correct angle for an IM injection.

there is no sign of blood, slowly inject the substance. Withdraw the needle and gently massage the injection area. If multiple or repeated IM injections are required, it is recommended to use alternating rear limbs to decrease pain and trauma to areas of the limbs. Needle sizes for IM injections typically range from 22 to 24 gauge. A 45- to 90-degree angle into the muscles is helpful and if multiple or repeated injections are necessary, injections should be alternated between muscle groups and different sides of the body.

Large animal injections are commonly placed in the muscles located at the base of the neck. This area forms a triangle shape from the muscles and ligaments in this area where the shoulder meets the crest of the neck. The center of the triangle is the best IM injection site. The pectoral muscles or brisket area over the lower front chest located at the top of the front limbs is another possible IM injection site in large animals. When placing the needle into the muscle of a large animal, use a quick and decisive thrust that moves the needle deep into the muscle. The site should be aspirated to make sure no blood vessel has been entered.

 COMPETENCY SKILL 100

Administering an Intramuscular Injection

Objective:

To properly administer an intramuscular injection

Preparation:

- Proper-sized syringe
- Needle of proper gauge and length
- Alcohol
- Cotton balls
- Exam gloves

Procedure:

1. Apply exam gloves.
2. Place one hand with the fingers located medially along the middle of the femur. The thumb is placed on the lateral aspect of the mid-thigh. The IM injection may also be given in the epaxial muscles located over the back end of the animal along either side of the spinal column.
3. Rub an alcohol-saturated cotton ball over the injection site.
4. Insert the needle cranially and parallel to the femur, directing the needle below the skin and into the muscle mass in front of the femur between the knee and hip joints. Do not go through the muscle or touch the femur bone.
5. Aspirate the plunger on the syringe. If no blood is noted, inject the substance slowly.
6. Withdraw the needle and place in a sharps container.
7. Massage the area where the injection was given and praise the patient.

Intranasal Injections

The IN injection is administered by placing drops of liquid into the nasal cavity or nares of an animal. This type of administration doesn't involve the use of a needle, but some vaccines may have a protective cover that can be placed over the end of the syringe. Many respiratory medications and vaccines are given by the IN route. Small animals may be standing, sitting, or in sternal recumbency, and large animals should be standing. The most important aspect of administering the IN injection is elevating the head and neck to prevent the nasal drops from being forced back out of the nostrils. A firm grip must be kept on the head for best control. The nose should be pointed upward long enough for the drops to coat the mucous membranes of the nares. Many animals do not like or tolerate nasal drops, so care must be taken when administering IN injections. It is often helpful to cover the eyes during this procedure, to help the pet remain calm and still.

COMPETENCY SKILL 101

Administering an Intranasal Injection

Objective:

To properly administer an intranasal injection

Preparation:

- Proper-sized syringe
- Exam gloves

Procedure:

1. Apply gloves.
2. Extend the head and neck of the patient with the nose pointed upward.
3. Apply entire amount of vaccine from vial into the nostril.
4. Keep the head and neck extended to allow drainage of the substance onto the mucous membranes.
5. Gently blow on the nasal area to eliminate sneezing.
6. Allow head to relax, and discard syringe in sharps container.

Fluid Administration

Fluids are administered to patients for fluid replacement therapy due to illness, surgical support, or dehydration. Some patients may need small amounts of fluid replacement given as an SQ injection. Other animals that are very sick and require large amounts of fluids may need an IV catheter and IV fluid therapy. SQ fluid therapy may be administered by the veterinary assistant. IV catheter placement is completed by the veterinary technician. However, the veterinary assistant should be able to monitor the catheter and IV fluid line while the patient is undergoing treatment. Fluids can be provided to address replacement or maintenance needs.

Subcutaneous Fluid Therapy

SQ fluids are placed into the loose skin at the base of the neck of small animals that require small amounts of fluid replacement. This is usually less than a 200-ml volume of fluid. These fluids are placed under the skin and slowly absorbed by the body. Common reasons for SQ fluids include treatment for kidney disease, the need for urine collection, and cases of minimal vomiting and diarrhea. The most common type of SQ fluid used for therapy is **lactated Ringer's solution (LRS)**.

SQ fluids are given using an IV bag with an IV line that is attached to the proper-sized needle according to the animal's size and species, just as in injections. Most SQ fluid administration needle sizes range from 18 to 16 gauge. The IV line or extension set is a flexible plastic tubing that connects to the fluid bag. The fluid bag is made of clear vinyl material. Most animals tolerate being given SQ fluids, which are similar to an SQ injection. One location can hold about 5–10 ml per pound body weight as a volume of fluid. Average absorption time of SQ fluids is between 6 and 8 hours. The fluid bag should be opened, and an IV drip set applied to the end of the bag, which is kept sealed by

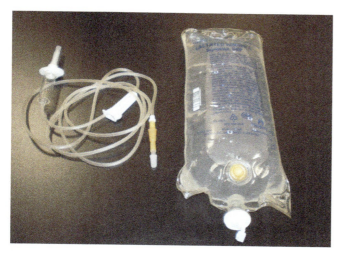

FIGURE 38–39 IV fluid and administration set.

at the unplugged site. The IV drip set control located on the plastic tubing should be maneuvered closed to prevent fluids from leaking from the tubing. The bag is then placed upright and hung on an IV pole. The tubing may be opened to allow fluids and air to flow out of the bag. A fluid bag has two ports located on the end: the injection port and the spike port. The **injection port** is used to remove and insert additional substances using a needle and syringe. The **spike port** is where the drip set tubing is placed. Most fluid bags hold 1000 ml of fluid.

The needle is inserted over the base of the neck and shoulder blade area, and fluids are administered by gravity flow. This allows for patient comfort during the process of administering large volumes of fluid. The patient should be sitting, standing, or in sternal recumbency. The needle should be held in place until the total fluid amount is administered. The needle is then removed, and the site can be gently pinched to prevent excessive leaking of fluid at the injection site. It is common for fluids to gradually migrate ventrally over the chest area before complete absorption occurs.

a plug that must be pulled and removed (see Figure 38–39). To prevent fluids from leaking, the bag should be inverted when the plug is removed. The sharp end of the IV set should be inserted into the open fluid bag

COMPETENCY SKILL 102

Preparing an IV Fluid Bag and IV Drip Set

Objective:

To properly prepare an IV bag and tubing

Preparation:

- IV fluid bag
- IV drip set
- IV stand or pole
- Appropriate-sized needle

Procedure:

1. Wash hands well, as gloves are not necessary with this task.
2. Remove the IV fluid bag from the plastic cover.
3. Remove the IV drip set from the plastic cover.
4. Close the IV drip set wheel located on the plastic tubing.
5. Remove the cover over the white spike of the IV drip set located at the end of the tubing. The spike is sharp and should be handled with care.
6. Remove the spike port cover on the IV fluid bag. Pull firmly to remove.
7. Hold the spike port in the left hand to guide the spike straight into the port. Push firmly. If the spike is pushed in at an angle, it may puncture the bag of fluids.
8. Hang the bag on an IV pump or IV pole.
9. Open the IV drip set clamp to allow fluids and air to be removed from the bag and tubing.
10. Close the IV drip set clamp when all air bubbles have been removed.
11. Apply the appropriate-sized needle to the end of the IV drip set tubing.

COMPETENCY SKILL 103

Administering SQ Fluids

Objective:
To properly administer SQ fluids

Preparation:
IV fluid bag, IV drip set, IV stand or pole, appropriate-sized needle, exam gloves

Procedure:
1. Place prepared IV fluid bag and line on an IV pole.
2. Apply gloves.
3. Tent the skin over the base of the neck between the shoulder blades as in an SQ injection (see Figure 38–40).
4. Insert the needle attached to the IV tubing into the base of the tented area slightly angled downward and just under the skin.
5. Insert along the long axis of the skin fold.
6. Let go of the tented area of skin.
7. Hold the needle in place with one hand.
8. Use the free hand to turn on the wheel of the IV drip set clamp. The higher the bag is placed on the pole, the faster the fluid rate. The bag can be squeezed gently to increase fluid flow.
9. Give the appropriate volume of fluid as directed by the veterinarian.
10. When complete, close the clamp on the IV drip set.
11. Remove the needle and apply pressure over the injection site.
12. Some fluid will leak from the injection site.
13. Place needle in sharps container.
14. Clean work area and disinfect supplies.

FIGURE 38–40 Preparation for the administration of SQ fluids.

IV Catheter Monitoring

An **IV catheter** is a small plastic piece of equipment that is placed within a vein to administer fluids and medications directly into a patient's bloodstream (see Figure 38–41). This task is done by the veterinary technician or veterinarian, as it requires advanced skills in vein location and properly inserting the catheter into the area, which is relatively small. The veterinary assistant is responsible for maintaining and monitoring IV catheter patency after it has been placed. **Patency** refers to maintaining the proper flow and purpose of the tubing, ensuring it remains intact and usable. Most IV catheters are placed in the cephalic vein; however, large animals and larger breed dogs may have a jugular catheter. The veterinary assistant should be able to recognize where the catheter is placed, whether it is patent and flowing properly, and whether the fluid rate is proper and flowing. A bandage is placed to cover and protect the catheter, and the bandage should be checked on a daily basis for signs of leaking, disconnection, bleeding, or swelling. This assessment may require a daily bandage change. Some catheter sites may become inflamed and require antibiotics or antiseptics around the placement site. The catheter site should then be re-bandaged. **Phlebitis** may develop around the site, so it should be monitored for signs of redness, swelling, pain, and inflammation. The IV catheter should be replaced every 3 to 4 days if the treatment is lengthy.

One of the most common problems with IV catheters is lack of patency, which may be caused from the IV tubing kinking or a clot occluding the vein and preventing the free flow of fluids. Catheters require flushing with several milliliters of sterile saline fluid after insertion and prior to any medication administration to assure catheter patency. This is done every time the catheter is detached from the IV set and prior to medication being administered directly into the catheter. Some animals with cephalic vein IV catheters may kink the tubing by pulling the leg in a direction that prevents fluid flow. A splint or bandage can be placed to help maintain the flow of fluids. Animals that chew or bite at the catheter site or bandage may need an E-collar to prevent them from damaging or removing the catheter. Catheter site bandages must be kept clean and dry. If the bandage becomes wet or soiled, it must be changed, and the catheter must be evaluated for any underlying problems. To prevent contamination, the catheter may need to be covered with plastic in incontinent patients.

If **pitting edema** is noted or swelling around the catheter site is causing the fluid to flow outside of the vein, the IV fluids should be stopped, the catheter removed, and a new catheter placed at another vein site.

IV Fluid Monitoring

IV fluids require constant and consistent monitoring by the veterinary assistant. Frequent observations will help detect improper fluid flow or occlusion. The veterinarian should determine the volume to be delivered to each patient on a daily basis. The amount of fluids to be delivered over the course of a day is called the **flow rate**. When a total volume is determined, the veterinary assistant may note how much fluid is in an IV bag at the start of treatment by placing a piece of adhesive tape along the edge of the bag or container lengthwise where no information is written. The initial volume and time are marked in pen along the tape at the level of the fluid line. The final volume desired and when the volume should be completed are marked on the opposite end of the tape at the proper fluid level. Ideally, an **infusion pump** should be available to adjust the rate and regulate the fluid flow (see Figure 38–42). This equipment is used to provide a constant flow of fluid at a specific rate throughout the day. The infusion pump has an alarm that sounds when fluids are no longer passing through the tubing due to an obstruction or an empty bag. Support staff should always consult with the veterinarian as to proper fluid rates, especially if any additives are to be placed into the IV fluid bag.

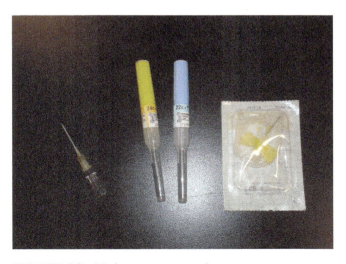

FIGURE 38–41 Intravenous catheter.

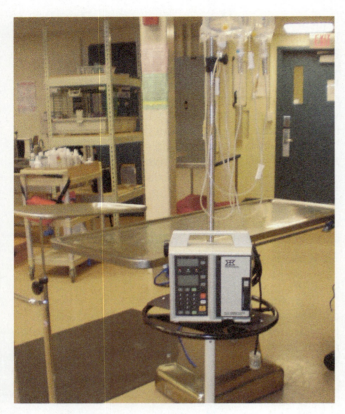

FIGURE 38–42 Infusion pump.

TABLE 38–7
Fluid Rate Calculation Examples
Example 1
Total volume: 1000 ml
1000 ml/24 hours per day = 41.6 ml/hr
Example 2
Total volume: 3500 ml
3500 ml/24 hours per day = 145.8 ml/hr
Example 3
Total volume: 9000 ml
9000 ml/24 hours per day = 375 ml/hr
Surgical Maintenance Rates
Dogs: 10 ml/kg/hr
Cats: 5 ml/kg/hr
5-lb dog = 5 lb/2.2 kg = 2.27 kg × 10 ml = 22.72 ml/hr maintenance
13-lb cat = 13 lb/2.2 kg = 5.90 kg × 5 ml = 29.54 ml/hr maintenance

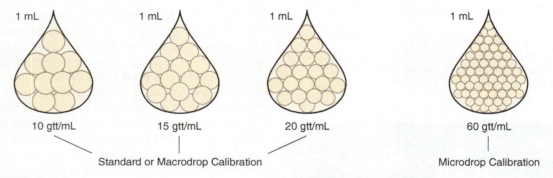

FIGURE 38–43 Comparison of drop size in macrodrip and microdrip infusions.

There are two types of IV drip lines that are used depending on patient fluid volume. A **macrodrip line** is a regular IV drip that delivers 15 drops of fluid per milliliter or 15 drops/ml. A **microdrip line** is a small line that delivers 60 drops/ml (see Figure 38–43). After the veterinarian has determined the total volume a patient requires in a day or 24-hour period, the veterinary assistant may determine how much fluid should be given each hour. This is done by taking the total volume and dividing by 24, the number of hours in a day. This number gives you the amount of fluid to be delivered to a patient each hour. Table 38–7 shows some examples of fluid rate calculations.

COMPETENCY SKILL 104

Ensuring Catheter Patency

Objective:

To properly assess the IV catheter and site of administration to ensure patency of the line

Preparation:

- 3-ml syringe
- Sterile saline
- Lactated Ringer's solution (LRS)
- Heparin

Procedure:

1. Observe fluid flow of catheter through IV tubing and IV bag.
2. If fluid flow has stopped, determine the cause of occlusion.
3. Extend the patient's leg by extending the foot and elbow forward.
4. If flow returns, place a splint or bandage to maintain extension of the limb.
5. If no flow returns, remove the bandage over the catheter. Note any swelling, redness, or irritation.
6. If no swelling is noted and the catheter appears to be properly placed in the vein, detach the IV tubing from the catheter.
7. For a total amount of 3 ml of fluid, flush a combination of saline, LRS, and a small amount of heparin into the catheter using a 3-ml syringe.
8. If the fluid flows through the catheter, reattach the tubing and monitor fluid flow.
9. When the fluid flow is appropriate, rebandage the catheter site.
10. If no fluid flow appears, contact the veterinary technician or veterinarian.

COMPETENCY SKILL 105

Monitoring the Infusion Pump

Objective:

To properly assess the IV infusion pump to maintain patency of the line

Preparation:

- Infusion pump, 3-ml syringe, sterile saline, lactated Ringer's solution (LRS), heparin

Procedure:

1. Check tubing to see that it is properly placed in the infusion pump mechanism. See that there are no kinks or obstructions.
2. If the clamp is controlling the fluid flow, make certain it is open.
3. Press the tubing under the clamp to make certain fluids can flow through the tubing.
4. If no fluid is flowing, for a total amount of 3 ml of fluid, flush a combination of saline, LRS, and a small amount of heparin into the catheter using a 3-ml syringe.
5. Monitor the catheter site for signs of swelling.
6. Note if the rate of fluid flow is correct and the amount of fluids that have been administered to the pet have been recorded in the medical record.

COMPETENCY SKILL 106

Monitoring Fluid Flow Rate

Objective:

To properly assess the infusion pump and administration set for proper flow rate

Preparation:

- Infusion pump
- IV bag and tubing

Procedure:

1. Note drops per milliliter multiplied by the total number of milliliters equals the total number of drops to be administered; for example, 15 drops/ml × 300 ml = 4500 total drops to be delivered per day.
2. Total volume of fluids delivered per day divided by 24 hours equals how much fluid should be administered per hour: 300 ml/24 = 12.5 ml/hr
3. Total number of drops to be administered divided by total number of minutes over which to be administered equals the drops per minute: 4500 drops/240 minutes (4 hours × 60 minutes/240 minutes) = 18.75 drops or approximately 19 drops/minute.

COMPETENCY SKILL 107

Replacing Catheter Bandage

Objective:

To properly assess the IV catheter and administration site, and replace the bandages as appropriate

Preparation:

- Cotton roll
- Gauze roll
- Vet wrap
- Adhesive tape
- Bandage scissors
- Exam gloves

Procedure:

1. Apply exam gloves.
2. Remove all layers of bandage.
3. Do not touch tape holding catheter in place.
4. Examine catheter for swelling or bleeding. Note tubing patency.
5. Flush catheter as necessary.
6. If catheter is patent, rebandage by applying a cotton roll as the primary layer.
7. Apply a gauze roll layer as the secondary bandage layer.
8. Place a loop of IV tubing without any kinks into the tertiary bandage layer.
9. Apply the vet wrap as the tertiary bandage layer.
10. Apply tape to the outer area as necessary.
11. Contact the veterinarian or veterinary technician if the catheter site is not normal or patent.

Urinary Catheter Maintenance

Urinary catheters provide access to the urinary bladder through the urethra. Urinary catheters may be placed for several reasons, including collection of urine for a urinalysis, to relieve urethral obstruction, to maintain urine flow, or administration of radiographic contrast material. Urinary catheters are available in a variety of sizes and materials. Care of urinary catheters includes attaching a closed urinary collection system to prevent infection and allow monitoring of urine volume. Urinary catheters must be placed aseptically and should be changed every 72 hours. Daily nursing care duties include inspecting the system for blood clots and kinked tubing, cleaning the vulva or prepuce with an antiseptic cleanser, flushing the urinary catheter for patency, and emptying the collection bag multiple times daily. All patients with a urinary catheter should be placed in an area where the collection system is lower than the patient to allow gravity aid in urine collection.

Wound Cleaning and Care

Wounds may be as superficial as a lesion or skin irritation or may be deep within the flesh such as a cut or puncture (see Figure 38–44). It is important that the wound be evaluated by the veterinarian to determine treatment options and the possibility of suturing. Many wounds may be on areas where excessive amounts of hair are located, which will need to be clipped away for better visibility and treatment. Depending on the treatment, the area will need to be cleaned and disinfected. The area may be contaminated with debris and bacteria, so it is best to flush the site after the hair has been clipped. This is called **lavaging** and is accomplished using a sterile flush to rinse the area and

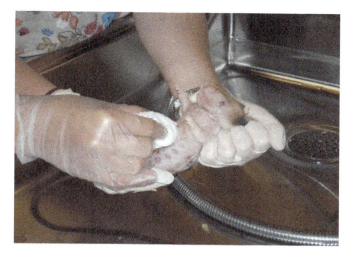

FIGURE 38–45 Clean the wound with a surgical scrub to disinfect the area.

an increased fluid jet pressure forcefully expelling the fluid from a 35- to 60-ml syringe through an 18-gauge needle. This will also flush out any hair that may have gotten into the site. Sterile water or normal saline is often used with a mixture of povidone–iodine solution at one part iodine to nine parts sterile fluid. A surgical scrub such as Nolvasan or chlorhexidine should be used to cleanse and disinfect the area (see Figure 38–45). Subsequent care will depend on the location of the wound, its size and depth, and how well the care is managed. Antibiotics are a common treatment to protect the animal against infection.

Second-intention healing of wounds may be necessary for areas that have significant tissue damage or loss, or for older wounds that have already begun to heal. This type of healing requires the area to be left open, without suturing. This is common in wounds occurring on the limbs where there may not be a lot of extra tissue to allow proper suture closure. When left open to heal, the wound must be cleaned daily to remove any exudates and debris. A layer of petroleum jelly may be placed around the outer area of the wound to prevent further irritation during healing.

Bandaging and Bandage Care

Bandages may be applied to an animal for several reasons, such as wound dressings, protection, immobilization (to prevent movement), or for covering splints. Bandages consist of several layers, each with a specific purpose. The veterinary assistant should have a general knowledge of the materials and equipment necessary for bandaging techniques and a basic knowledge of how to apply a simple protective

FIGURE 38–44 Wound in a cat.

714   SECTION IV Clinical Procedures

COMPETENCY SKILL 108

Cleaning a Wound

Objective:
To properly clean and prep the site of a wound for the veterinarian to assess and provide treatment for it

Preparation:
- Exam gloves
- Clippers
- KY Jelly
- Vacuum
- Gauze sponges
- Nonabsorbent pad
- Water bowl
- Sterile saline
- Betadine
- Nolvasan or chlorhexidine scrub
- 50-ml syringe and large-gauge needle
- Sponge
- Forceps
- Towels

Procedure:
1. Apply KY Jelly over the wound area.
2. Clip the hair from the wound area, going against the growth of the hair coat.
3. Clip wide margins with straight edges.
4. Vacuum loose hair from the area.
5. Place towels around the area of the animal where the wound is located.
6. Gently use a 50-ml syringe to flush the wound with sterile saline.
7. Repeat the flushing until the wound is free of dirt and debris.
8. Begin the scrubbing process by applying sterile water and Nolvasan or chlorhexidine scrub to gauze sponges. Use a water bowl to soak the sponges. Scrub over the wound and clipped margin to remove as much debris as possible. Repeat as necessary for three to four cycles.
9. Begin scrubbing at the center of the wound in circular movements, working outward to the hair coat. Repeat three to four times.
10. Use a new gauze sponge with each cleaning.
11. Apply a surgical disinfectant, such as betadine (iodine), over the wound.
12. Place a sterile nonabsorbent pad or gauze sponge over the site to keep it clean.

bandage (see Figure 38–46). The general materials needed for bandaging include the following:

- Rolled cotton
- Adhesive tape (1"–2")
- Rolled gauze or kling
- Co-flex or vet wrap bandage
- Nonadhesive pad
- Bandage scissors

The **primary bandage layer** is the layer closest to the skin. The primary layer should be soft and provide padding to prevent rubbing and skin irritation and act as a protective covering of the area. This may include cotton rolls, cotton bandages, or other soft fiber materials wrapped around the area. If a wound or suture line is present, the area should be covered with a nonabsorbent protective pad. The bandage layer application should be tight enough to remain in place but

FIGURE 38–46 A selection of bandage material and splints.

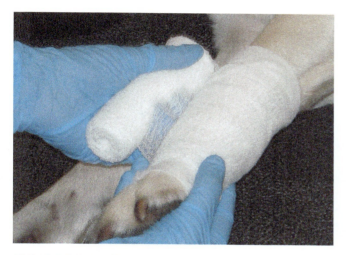

FIGURE 38–47 The secondary layer of gauze is placed over the primary layer of the bandage.

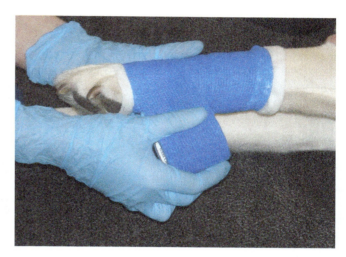

FIGURE 38–48 The tertiary bandage layer is usually waterproof and helps to protect and keep the underlying layers clean.

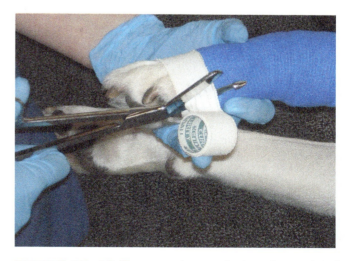

FIGURE 38–49 Tape can be applied to the ends of the bandage to help keep it in place.

the animal. It is usually waterproof to protect the underlying layers. It should extend proximally and distally beyond the edges of the secondary layer. The material for this layer is usually self-adherent, meaning it sticks to itself, such as co-flex or vet wrap (see Figure 38–48). The edges should be wide enough to prevent slipping but not so extensive that they cause discomfort to the patient. Adhesive tape can be applied to the outer edges of the last layer of bandage on the proximal and distal ends to further help prevent the bandage from slipping (see Figure 38–49). Half of the tape piece should attach to the bandage and half to the hair of the animal. It is important to note that bandage layers that are placed too tightly may cut off circulation, causing serious damage to tissue.

not so tight as to cut off circulation. Bandage wrapping should begin at the center of the area to be bandaged and move downward or forward of the area and then reversed to cover the upward or backward area. This allows the bandage to cover all areas and provides several primary layers of padding and protection. Many times, the bandage layers are applied at a 45-degree angle, which may help prevent bandage slipping. Each layer should lie smoothly against the previous layer with no wrinkling present and with overlap of the layer before it.

The **secondary bandage layer** is a thin fiber material that clings to the first layer and holds the primary layer in place. It typically consists of a nonadherent wrap, such as gauze or kling, extending proximally and then distally beyond the edges of the primary layer (see Figure 38–47).

The **tertiary bandage layer** is the final layer that acts as the outer covering and holds the bandage on

Bandage application and type depend on the purpose of the bandage and the location to which they are applied. If slipping is a concern, such as on the limbs, adhesive tape strips may be applied directly onto the skin on opposite sides of the limb (see Figure 38–50). The strips should extend beyond the end of the limb. The bandage is then applied, and the strips are incorporated into the secondary bandage layer. These tape strips are referred to as **stirrups**.

Bandage care and cleanliness are essential. This means that the bandage should remain dry and free of debris. This may mean frequent bandage changes for areas of wounds or if a bandage becomes soiled. Animals are frequent lickers and chewers of bandages, so care must be taken to prevent the animal from damaging or removing the bandage. Sprays that have nontoxic but bad-tasting chemicals may be applied to the outer layer to limit these behaviors. Examples of this type of spray include Bandguard and Bitter Apple. More drastic measures may be required for difficult patients, such as applying an E-collar around the head and neck to prevent chewing. Bandage care may also include restricted exercise, monitoring for signs of swelling or slipping, applying a plastic bag over the bandage to protect from wetness, and monitoring for odor.

Bandage removal is undertaken when the veterinarian determines the bandage may be removed. **Bandage scissors** are used to cut away the layers of the bandage (see Figure 38–51). Bandage scissors have an angled end called a dogleg (see Figure 38–52). This angled end of the scissor blades prevents cutting into the animal during bandage removal. The bandage should be removed from the proximal end, working distally and making certain that the end of the bandage scissors is always pointed upward and in contact with the bandage layers.

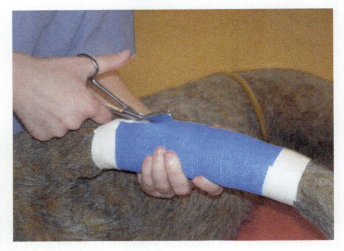

FIGURE 38–51 Use bandage scissors to cut away the bandage materials for removal.

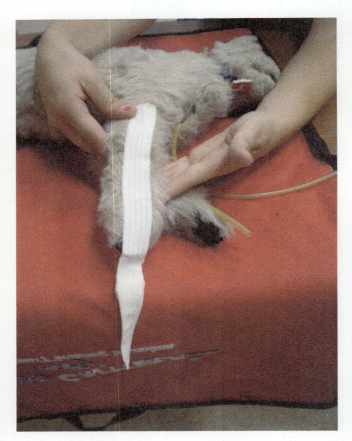

FIGURE 38–50 Application of adhesive tape to prevent slippage of the bandage.

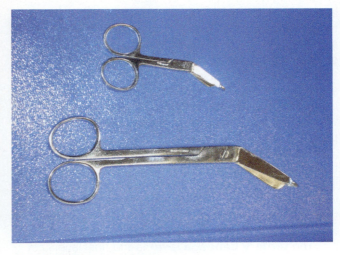

FIGURE 38–52 Bandage scissors.

COMPETENCY SKILL 109

Applying a Limb Bandage

Objective:

To properly apply a bandage to a wound

Preparation:

- Cotton roll,
- Gauze roll or kling
- 1"–2" adhesive tape
- Vet wrap or co-flex bandage
- Bandage scissors
- Exam gloves

Procedure:

1. Apply exam gloves.
2. Apply a strip of 1"–2" adhesive tape along the cranial aspect of the limb and one strip on the caudal aspect of the limb to prevent slipping. The tape strips should extend beyond the end of the limb.
3. Hold the roll of cotton in one hand while using the other hand to hold the vet wrap in place. *Do not* place too tight, cutting off circulation.
4. Apply the primary layer by applying the cotton roll distally to the end of the area and then proximally to cover the entire surface. Apply the cotton roll at a slight 45-degree angle and smoothly and evenly to prevent wrinkling.
5. Apply each layer of cotton so that it overlaps the previous layer. Continue wrapping to desired amount of padding and protection.
6. Apply the gauze roll as the secondary layer distally to the end of the area and then proximally to cover the entire surface. *Do not* place too tight, cutting off circulation.
7. Apply the gauze roll at a slight 45-degree angle. Apply each gauze layer smoothly and evenly, preventing wrinkling. Apply each layer of gauze so that it overlaps the previous layer. Pull each strip of tape upward into the gauze roll layers.
8. Apply gauze to desired amount with a slight extension beyond the end of the primary layers.
9. Apply the vet wrap tertiary layer by applying distally to the end of the area and then proximally to cover the entire area.
10. Apply the vet wrap at a slight 45-degree angle. Apply the vet wrap smoothly and evenly, preventing wrinkling. Apply each layer of vet wrap so that it overlaps the previous layer.
11. Apply to desired amount with the edges of the bandage extending slightly beyond the secondary layer. *Do not* place too tight, cutting off circulation.
12. Apply 1" or 2" adhesive tape to the proximal end and the distal end of the bandage edges. Attach half of the tape layer to the bandage and half of the tape layer to the hair.
13. A small amount of tape can be placed along the last edge of the vet wrap at the end of the bandage opening. You may make a triangle end of the vet wrap bandage to easily locate the opening.

COMPETENCY SKILL 110

Removing a Bandage

Objective:

To properly remove a bandage without causing any additional trauma

Preparation:

- Bandage scissors
- Exam gloves

Procedure:

1. Apply exam gloves.
2. Work from the proximal end to the distal end of the bandage.
3. Place the long, blunt blade of the bandage scissors against the skin and slightly under the bandage edge.
4. Keep the blade flat against the skin and the end raised slightly upward in contact with the bandage.
5. Place the bandage layers between the scissor blades.
6. Begin cutting proximally. Peel the bandage layers away from the patient using a firm motion.
7. Unpeel or cut layers moving toward the distal portion of the bandage.
8. Remove stirrups, if present, when the entire bandage has been removed.
9. Gently remove each layer of bandage.
10. Notify the veterinarian when the bandage has been removed.
11. Clean up work area.

Socialization and Exercise of Patients

Animals are accustomed to human interaction and when hospitalized may not receive enough human contact. It is important that the veterinary assistant provide positive social interaction to all hospitalized patients during treatment, cage cleaning, or exercise. This includes petting animals, talking to the patients, using a gentle and soothing voice, or holding them if possible (see Figure 38–53). These positive actions help put patients at ease.

Exercise may be provided to some hospitalized animals, depending on their condition and treatment needs. Exercise is important, as some animals, such as dogs, will not urinate in the confines of the cage. This requires a brief walk several times a day outside of the facility (see Figure 38–54). A designated walking area should be determined to control the spread of parasites and disease and allow proper sanitation measures. Always place the appropriate leash or other control needs on the patient. Areas inside the facility may be used for exercise and play areas. The time an animal is outside of its cage will vary from patient to patient. Be careful to indicate when an animal is roaming

FIGURE 38–53 It is important for hospitalized patients to receive human contact and attention.

CHAPTER 38 Veterinary Assistant Procedures

FIGURE 38–54 Some animals will need to be walked while in the care of the veterinary facility.

free within an area. Interaction between patients is an important part of their recovery and treatment.

Feeding and Watering Hospitalized Patients

There are several factors influencing patient feeding and watering in the veterinary hospital. The health status of the patient, procedures or surgeries to be completed, medications, and appetite must be factored into the patient's needs. Some animals may be placed on NPO, or no food or water by mouth, status. This may be due to the patient's condition, disease, or surgical needs. If an animal is allowed to eat and drink, it must be determined which diet is appropriate. Age and illness will influence the type of diet. Patients with certain conditions will need specialized veterinary diets or foods that may be prescribed by the attending veterinarian. Typically, patients with heart failure need a low-salt diet, and patients with renal failure need a low-protein diet. The diet should be noted in the chart so that consistency during the hospital stay is followed.

Some animals may feel comfortable eating in a veterinary setting. Other animals may feel insecure about eating in a caged setting or outside of their normal environment. This needs to be considered when trying to encourage patients to eat. Some animals may not be used to the type or consistency of food being offered. For example, some dogs eat only canned or dry food and may not eat another type. Some horses or cattle may eat a specific type of hay and not want another variety. For small animals, canned foods may be warmed up to increase palatability. Gentle heating of food also increases the aroma or smell, making it more appealing. Mixing in small amounts of water can sometimes increase acceptance of the food. Dry food can be watered down to make it softer. For large animals, water or applesauce can be added to grains to make them more appealing. Foods may also be force-fed to animals if necessary. **Force-feeding** is placing the food in the animal's mouth and making it swallow. Several methods may be used to accomplish this task. One can gently open the animal's mouth and use a finger against the roof of the mouth to insert a small amount of food. Another method is to liquefy the diet and feed it through a syringe or feeding tube. The syringe is filled with food, and the end of the syringe or the feeding tube is placed into the side of the mouth near the back teeth. The nose of the patient is tipped slightly upward. The end of the syringe is pushed to move the food contents into the mouth and the patient is allowed to swallow between administrations. The feedings are broken down into several meals throughout the day. Food should be provided slowly to prevent the patient from aspirating food into the lungs. With **aspiration**, food enters the lungs rather than the stomach. Some patients may begin to eat on their own after some amount of food is ingested. Some animals will fight the process of force-feeding. It may be necessary to use a **nasogastric tube** placed through the nose and into the stomach to feed the animal (see Figure 38–55). This is done by the veterinarian or veterinary technician. The nasogastric tube enters the nostril and passes through the pharynx and into the esophagus until the tip reaches the stomach. Liquefied food can then be administered through the tube as needed. The amount of food required should be evaluated by the veterinarian to make certain the patient's total calorie needs are met. Animals that are capable of eating should be monitored to determine the amount of food they eat. Animals should never go longer than 24 hours without eating. Feeding hospitalized patients often requires creativity, persistence, and patience.

720 SECTION IV Clinical Procedures

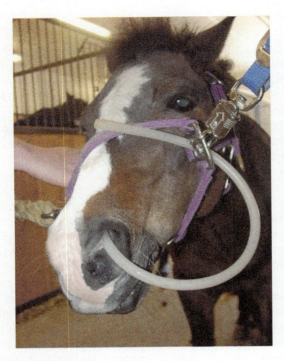

FIGURE 38–55 Nasogastric feeding tube.

Water is another important nutrient required by animals. Hospitalized patients may or may not be allowed to have water made available to them. Animals may become dehydrated if they are not consuming enough water. Water should be available in adequate amounts if allowed. Hospitalized patients readily spill their water, so cages should be monitored, or bowls should be placed on metal rings to prevent spilling. Some patients will require more water than others. As much water as the animal needs should be provided, and it should be kept fresh and clean.

Understanding Euthanasia

Euthanasia is the process of putting an animal to sleep using humane methods to allow for a painless death. Some hospitalized patients may not recover and continue to be ill, and the owners may decide to end the animal's suffering by means of euthanasia. This decision is based on personal values, religious beliefs, and previous experiences along with the veterinarian's guidance and recommendations. This process is often difficult for the owners as well as the

COMPETENCY SKILL 111

Feeding Hospitalized Patients

Objective:
To properly meet the dietary needs of hospitalized patients through feeding and watering

Preparation:
- Select the appropriate variety of food
- Note how much food is to be given to the patient
- Food bowl, bucket, or other utensils
- Exam gloves
- Microwave or heat source

Procedure:
1. Prepare the approved diet.
2. Offer the food to the animal. If the animal doesn't seem interested in eating, apply a small amount of water to the food. For soft foods, gently warm the contents to room temperature.
3. If the patient is not interested, place a small amount on the fingertip.
4. Gently open the patient's mouth and place the food against the roof of the mouth.
5. Allow the patient to swallow the food.
6. Repeat slowly, allowing the patient time to swallow.
7. Repeat until the food is refused or the animal eats the amount provided.

COMPETENCY SKILL 112

Force-feeding Hospitalized Patients

Objective:

To properly administer nutritional products to the animal by force

Preparation:

- Select the appropriate variety of food
- Note how much food is to be given to the patient
- Syringe
- Feeding tube
- Exam gloves
- Microwave or heat source

Procedure:

1. Apply exam gloves.
2. Prepare the food by liquefying the appropriate amount with warm water.
3. Draw the food into a syringe. A feeding tube can be applied to the end of the syringe if needed.
4. Place the syringe in the side of the patient's mouth toward the back teeth.
5. Tilt the head slightly upward.
6. Pull the corner of the lip outward to form a small pouch.
7. Gently administer the liquified food into the mouth between the cheek and back molars.
8. Allow the food to be swallowed.
9. Repeat the process throughout the day until total daily calorie needs are met.

---

veterinary staff. Many people see their pets as part of the family and have developed a strong human–animal bond that is an emotional connection between human and pet.

The process of euthanasia should be as free of pain and stress as possible. During this time, the veterinary staff should be supportive and sympathetic to the owner's needs and values. It is important to remember the need for patient and client confidentiality and limit discussions about the case to only those team members involved with caring for the patient. The owner should sign a waiver giving permission for the procedure to be done. The preparation for euthanasia should include discussing the procedure and sequence of events with the client. It is also important to determine what the owner would like to have done with the animal's remains. Many facilities offer burial or cremation services and some owners may opt not to take the animal's body or complete burial at home. Several items need to be considered in the preparation for euthanasia:

- Will the client be present with the animal during the procedure?
- Would the client like to pay before the procedure or be billed?
- Inform the client of legal restrictions in certain areas for private burial.
- Determine how the client would like to care for the animal's remains, such as a pet cemetery burial or cremation.
- Inform the client of the costs of the services.
- Complete all necessary paperwork and consent forms prior to the procedure.

An exam room should be prepared with a box of tissues, a blanket or towel, chairs for the owner and other family members, and any other items the family may need. A cadaver bag should be placed in an area of the room that is out of sight of the owner, such as a drawer or closet. The drug of choice should be available to the veterinarian, as should an accurate weight of the animal. Some facilities administer sedatives to calm an animal prior to the procedure. If the procedure requires IV catheter placement, the catheter and materials should be placed on the exam counter.

The euthanasia procedure itself should be performed in a quiet area, with the patient and family members placed in the area immediately upon arrival to the facility. The patient and family should be made as comfortable as possible. Any paperwork should be completed prior to the procedure and placed in the patient's medical record. Many family members may ask about the procedure so that they understand what to expect during the process. It is important to make certain the owners are prepared for the procedure and have some understanding of what the veterinarian will be doing and how the patient will react. The veterinarian may discuss the procedure with the owner and other family members or may ask the assistant to answer any questions. It is important that the family know the veterinarian will perform the procedure and the assistant will be available to restrain the patient and assist as needed. If a sedative or tranquilizer is administered, the patient and family members will remain in the exam area until the medication takes effect. The veterinarian will administer any sedatives and the assistant should check on the patient every few minutes during this period to determine when the drug has taken effect. The assistant should make certain the owners are comfortable and if anything can be done for them.

If an IV catheter is required, the veterinary technician will insert it with the assistant restraining the patient. If a simple venipuncture is used to administer the euthanasia solution, the assistant will restrain the patient. Once they are ready for the procedure to begin, the veterinarian will prepare the necessary amount of solution and administer the injection. Once the entire amount has been administered, the patient's body will relax completely. The needle is then withdrawn and placed in a sharps container. The patient should be placed in lateral recumbency, and the veterinarian will check for residual heartbeats or signs of a pulse and reflexes. During this time there may be some **agonal** respirations, which are gasps of breaths while the respiratory system shuts down. The patient's pupils will dilate, and the mucous membranes will become **cyanotic** or bluish gray in color. It is important to note that the patient may lose control of the bladder and bowel functions.

When no heartbeat or pulse is detected, the patient is pronounced deceased. Owners' reactions will vary, and the staff should be prepared at this point to handle a variety of grief reactions from relief to hysteria. Make the owners as comfortable as possible and ask if they would like to remain with the patient alone to say good-bye. The owners may wish to remove the patient's collar to keep as a memory. Some clients may benefit from information on a pet loss group or grief support number. When the owners have finished visiting with the pet, they are escorted outside—by a side door if possible. If the owners are not taking the patient's body with them, the assistant should then enter the exam area to place the body in a cadaver bag and label the bag with an ID tag containing the patient's name and owner's name, and what is to be done with the remains. The bag should be closed with adhesive tape and the remains placed in the appropriate area of the facility. The assistant should make certain that the sedatives and euthanasia solution are returned to the pharmacy area and any controlled substances are recorded in the controlled substance log. The exam area should be cleaned and disinfected.

COMPETENCY SKILL 113

Preparing for Euthanasia

Objective:

To properly prepare for and assist with euthanasia procedures

Preparation:

- Blankets or towels
- Tissue box
- Chairs or seats for the family
- Cadaver bag
- Tape
- ID tag
- Sedatives or tranquilizers
- Euthanasia solution
- Appropriate-sized syringe and needle

Procedure:

1. Prepare an exam room or area for the euthanasia procedure.
2. Place a blanket or towels on the exam table or floor, depending on patient size.
3. Place a box of tissues on the exam counter.
4. Place chairs near the area where the euthanasia will be performed.
5. Place a cadaver bag in a drawer or cabinet with a roll of adhesive tape and an ID tag to label the bag with patient name and client name.
6. Place any sedatives that may be used to sedate and calm the patient on the counter.
7. Place the euthanasia solution on the exam counter.
8. Place an appropriate-sized syringe and needle on the exam counter near the euthanasia solution.

Grooming Skills in the Veterinary Facility

Many patients require hair care due to poor overall hair coat quality, appearance, and lack of care. Long-haired animals may become matted; have waste materials collected on their hair coat; have hair become attached to the rectum or urogenital areas; or have overgrown nails, enlarged anal sacs, or dirty ears. Many of these factors can cause an unhappy, unhealthy, and uncomfortable animal. Routine care and maintenance may be provided by the veterinary facility, usually by the assistant.

A variety of grooming tools should be available at the veterinary facility. All tools that have been in contact with an animal should be properly disinfected to prevent transmission of external parasites or other contagious diseases. Grooming items tend to be the most common spread of parasites and disease in a veterinary facility.

Brushing and Combing

Brushing and combing of an animal's hair coat is required for all animals, especially those that may have long and thick hair. **Brushing** is the act of cleaning the hair coat using a soft brush (see Figure 38–56). This removes dead hair and dirt materials. **Combing** is the act of cleaning the hair with a thin comb that helps remove tangles and mats, as well as other debris and foreign substances located in the hair. All animals should have regular brushing and combing of the hair. Combing is important before a bath to untangle any hair or mats that may have developed (see Figure 38–57). Many shampoos and conditioners can't penetrate a matted hair coat, and this may cause further tangling and matting to occur. Mats in the hair coat may cause sores on the skin and are uncomfortable to animals. Brushes and combs come in various sizes, shapes, and material textures (see Figure 38–58). Any combs or brushes with wire teeth, such as a slicker brush, should be used with caution, as they can scratch or irritate the

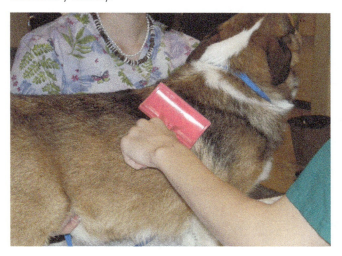

FIGURE 38–56 Brushing helps maintain a healthy hair coat.

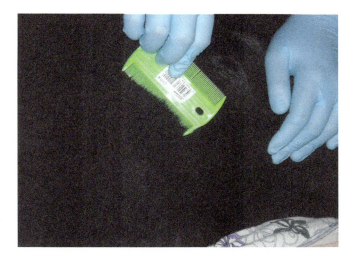

FIGURE 38–57 Combing helps remove matted hair.

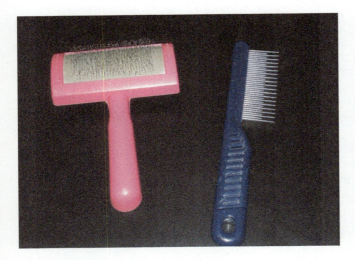

FIGURE 38–58 Types of brushes and combs.

FIGURE 38–59 Flea comb (left) and detangler comb (right).

FIGURE 38–60 Mat splitter.

arate mats into smaller ones, but care must be taken because these tools are sharp and can injure the person, animal, or hair coat (see Figure 38–60). It is best to use with a comb and gently work the areas of the mat using patience, as some animals will become irritated with excessive pulling on the hair coat. Severely matted areas may need to be clipped out with electric clippers (see Figure 38–61). Clipping out individual mats over the hair coat may result in a choppy appearance to the hair coat. When clipping mats, it is best to use the lowest blade number possible. For example, a number 5 blade leaves a longer coat length than does a number 10 blade. When using clippers, it is important that the clipper blade be kept flat or parallel to the skin and hair coat (see Figure 38–62). The blades should not be tipped or angled inward toward the hair coat. When mats have been removed, it is important to note the condition of the skin under the matted area. These areas may become inflamed or irritated, have open wounds, have signs of parasites, or have dry skin. Any observation of abnormal skin should be reported to the veterinarian.

skin or pull on mats and tangles, causing discomfort. Some brushes are so soft that they don't penetrate the hair and get through the hair coat. The teeth of combs and brushes may be close together or farther apart. Combs with teeth close together, such as a flea comb, are meant to pull debris from the coat (see Figure 38–59). Teeth that are farther apart are meant to break apart mats and tangles. Combs and brushes must meet the needs of each individual patient.

Basic brushing and combing should begin at the back of the body and work forward, moving with the direction of the hair coat. Also, it is best to begin at the feet and work upward. Any mats should be broken down by working across the mat with a wide-toothed comb until the mat is free from the underlying skin and hair. **Mats** are areas of hair that are interwoven and form a large clump that is irritating to an animal. A **mat splitter** may be used to break down and sep-

CHAPTER 38 Veterinary Assistant Procedures

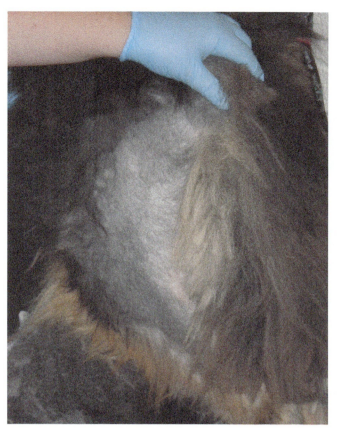

FIGURE 38–61 A severely matted hair coat may need to be shaved.

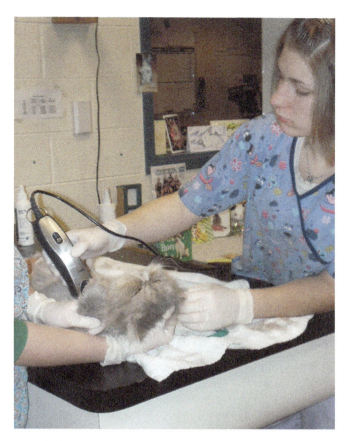

FIGURE 38–62 The clippers must be kept parallel to the skin or hair coat.

COMPETENCY SKILL 114

Brushing or Combing

Objective:

To properly brush or comb various hair coats as a health maintenance technique

Preparation:

- Appropriate brush (thin-spaced teeth)
- Appropriate comb (wide-spaced teeth)
- Scissors
- Clippers
- Clipper blades in assorted sizes
- Exam gloves
- Grooming table (if needed)

Procedure:

1. Begin combing medium to long hair areas over the back and sides of the animal, working from the tail toward the head (see Figure 38–63).
2. Part the hair and expose the skin to allow the comb to work through all hair.

FIGURE 38–63 When brushing, work from the tail forward.

3. Comb legs, working from the feet proximally up the leg.
4. Comb the medial aspect of each limb while standing on the opposite side of the body.
5. Comb lateral areas of the limb closest to you.
6. Comb the axillary (armpit) and groin areas with the patient standing. Do not pull excessively in these areas, as they are tender.
7. Comb the tail by working cranially from the tip.
8. Comb the ventral body last, working from the distal area to the proximal area.
9. Comb out or clip any mats as necessary. Be patient working on mats, being careful not to pull excessively on the hair.
10. Brush over the body to remove loose hair, brushing in the direction of the hair coat.
11. Clean and disinfect all tools and the work area.

Bathing

Bathing is done to clean the skin and hair coat of the animal or to apply medicated shampoos or dips to the skin and hair coat. Bathing removes dirt and debris from the skin and hair with the use of shampoo, conditioner, and water (see Figure 38–64). **Dipping** is the process of applying a chemical pesticide or medication to the skin and hair coat to treat a specific condition, such as mites or fleas. Dips usually remain on the skin and hair coat for a specific period of time to allow them to work as necessary.

Bathing should begin after the animal has been completely combed and brushed and is free of mats. Other grooming needs, such as anal gland expression, ear cleaning, or nail trims, should be performed prior to bathing. Large cotton balls should be placed into each ear canal to prevent water from entering the ear and potentially causing an inner ear infection. A protective ophthalmic ointment should be applied to the eyes to protect them from shampoo, conditioner, or any medications that may irritate the eyes (see Figure 38–65). A common ointment used is Puralube.

All supplies and tools should be located and prepared prior to placing the patient in the bathing area, either a washtub or wash stall. Tools used to bathe a patient are listed in Table 38–8. A shampoo and

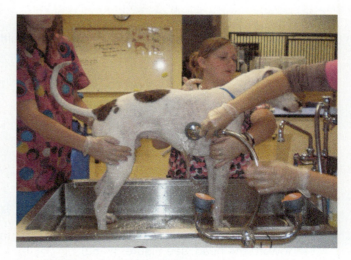

FIGURE 38–64 Bathing removes dirt and debris.

CHAPTER 38 Veterinary Assistant Procedures 727

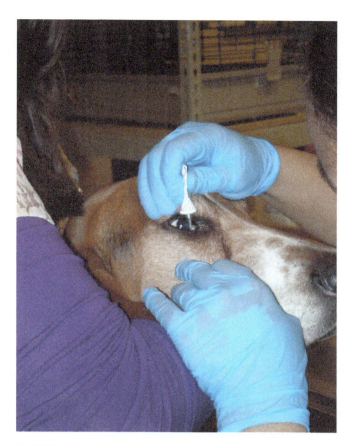

FIGURE 38–65 An eye ointment is used to protect the eye from cleaning products used when bathing.

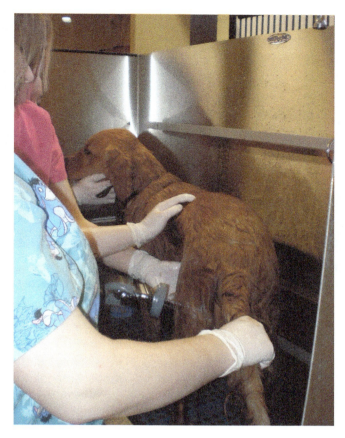

FIGURE 38–66 Wet the entire coat with warm water to begin the bathing process.

TABLE 38–8
Bathing Tools

Protective eye ointment	Cotton balls	Towels	Washcloths, sponges
Warm water	Dryer	Aprons/smocks	Goggles
Exam gloves	Shampoo	Conditioner, detangler	Medicated shampoo

conditioner or medicated shampoo should be selected based on the needs of the patient. Medications should be prescribed by the veterinarian. Towels should be available to dry the patient. A cage should be set up for drying or a clean area should be provided for tying large animals for drying. Washcloths, mitts, sponges, or other absorbable washing material should be used to bathe the animal, especially the head and face area. Aprons or water-resistant smocks should be worn to prevent excessive wetness of scrubs. Goggles or glasses may be worn to prevent water from getting into the eyes. All drains should have protection to prevent hair and other debris from entering. Make certain the water is warm and comfortable for the patient, usually around 70–72 degrees F.

The patient should be placed in the bathing area with a leash or halter and lead for added control. One end may be held or tied to the bath area to prevent the animal from escaping. The patient should never be left unattended when tied in a bath area. Begin the bath by thoroughly soaking the entire coat with warm water (see Figure 38–66). The shampoo should be mild, lathered into the hair coat, and rubbed deep into the hair and skin with a massaging motion. All areas of the body should be lathered well, including between the digits, around the rectum and the genital area, in the axillary areas, and behind the ears. The face area should be washed using a small amount of shampoo applied to a wet washcloth. Care must be used around the eyes to prevent any shampoo from entering and burning the eyes. Once the patient is fully lathered, allow the shampoo to stand on the body for about 5 minutes to thoroughly clean the skin and hair coat. Any medicated shampoos that are used should be handled according to the label; many require standing time in contact with the hair and skin. While standing,

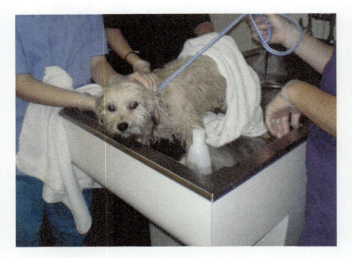

FIGURE 38–67 Use towels to begin the drying process and to remove excess water from the hair coat.

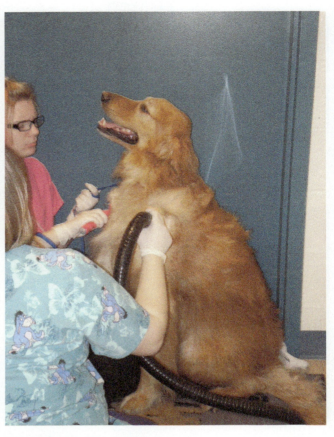

FIGURE 38–68 Handheld dryers can be used to more thoroughly dry the coat.

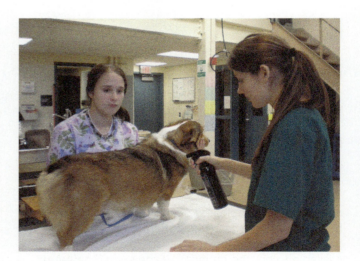

FIGURE 38–69 Applying a spray conditioner helps keep the hair coat neat and odor free.

It is important to continue massaging the shampoo into the skin for the time indicated. After shampooing is completed, begin rinsing the patient with warm water. All areas should be rinsed well, especially inside creases and folds of the skin. The conditioner is then applied much in the same manner as the shampoo. Conditioners do not generally lather up like shampoos, so it is important to use enough conditioner in all areas of the hair coat. A comb or massage brush can be used to work through the hair coat when the conditioner is on to help make certain the conditioner makes appropriate contact. The conditioner must then be rinsed well from the entire hair coat. Any dips should be applied according to the label directions. Rinse all materials from the hair coat, using the hands to remove excessive amounts of water by squeezing as much water off the hair as possible. Towels should be applied around the animal and rubbed over the body to begin the drying process (see Figure 38–67). Blow drying using handheld dryers, cage dryers, or high-powered vacuum-type dryers is a convenient way to dry the coat (see Figure 38–68). It is important to comb out the hair coat as the animal is drying, as this helps prevent the animal from overheating, reduces tangling of hair, softens the hair coat, and decreases the smell of wet animal. As in combing and brushing the animal's coat, work from the distal end to the proximal area and caudal to cranial. It is important to make certain all areas have been completely dried, such as between the digits, under the tail, at the abdomen, and in the ear flaps. Applying a spray-on conditioner also helps keep the hair coat neat and odorless (see Figure 38–69). When hand drying, do not leave the heat of the dryer on one spot very long. When cage drying, check on animals every 5 minutes to make certain they are not overheating. To prevent skin burns, use low temperatures to dry and make certain the dryer is not too close to the animal. Cotton balls can be removed from the ears.

After the patient is dry, a final comb-out and brush-out should be done and a spray-on conditioner applied to the hair coat. The patient should be placed in a clean area to avoid becoming soiled. The bath area should be cleaned and disinfected. It is important to remove all hair from the bath area and drain. A mop should be used to dry any floor areas that may be wet. Towels should be laundered and all items, including shampoo and conditioner bottles, cleaned and put away. Hair should be vacuumed, and the dryer disinfected and put away.

COMPETENCY SKILL 115

Bathing

Objective:
To properly bathe an animal in a safe manner for effective health maintenance

Preparation:
- Shampoo/conditioner/medicated shampoo
- Protective ophthalmic ointment
- Cotton balls
- Towels
- Washcloths
- Sponges
- Comb/brush
- Leash
- Apron/smock
- Exam gloves
- Goggles
- Dryer

Procedure:
1. Apply exam gloves.
2. Place leash on patient.
3. Place cotton balls in both ear canals.
4. Apply eye ointment to both eyes.
5. Place patient in wash area.
6. Secure patient in wash area.
7. Wet entire hair coat thoroughly with warm water.
8. Apply shampoo by lathering well through coat. For medicated shampoos, read and follow all labels. Allow to stand 5 minutes after being massaged into the haircoat.
9. Rinse entire coat well.
10. Apply conditioner by massaging into hair coat. Allow to stand 5 minutes after being massaged into the haircoat.
11. Rinse entire coat well.
12. Apply dip, if necessary, by following all directions.
13. Rinse well.
14. Squeeze excessive hair from coat.
15. Wrap patient in towels and begin towel drying.
16. Remove patient from wash area.
17. Place in cage with cage dryer set on low to moderate heat, checking every 5 minutes for overheating.
18. Hand dry at low temperature, moving over entire coat.
19. Comb or brush through hair coat as it dries.
20. Apply spray-on conditioner with combing or brushing.
21. When completely dry, apply final spray-on coat conditioner.
22. Place animal in clean cage.
23. Clean up work area and disinfect all tools.

Clipping, Trimming, and Shaving

Clipping of the hair is typically the process of removing small amounts of hair from one or several areas of the patient. **Trimming** refers to the process of removing a specific amount of hair from one or more locations. **Shaving** is the process of taking down an amount of hair close to the skin. Clipping and trimming may be done using scissors or electric clippers (see Figure 38–70). In veterinary medicine, this usually refers to areas of the body that need care and maintenance, such as the rectal area, genital area, or facial area. Clipping and trimming around the rectum is common in animals that may have long hair that may collect around the rectum, causing a buildup of feces (see Figure 38–71). This causes a sanitary problem that may cause bad odor or soiling of furniture. The genital area may need trimming for similar reasons. Many male and female animals may urinate with urine collecting around the hair of the genital area. Trimming and clipping hair from these sites is common when providing care for patients. Shaving is another skill in veterinary medicine that is necessary for several reasons. One is the removal of mats and another is shaving a surgical area to properly prepare it for a surgical procedure.

When clipping or trimming hair, blunt scissors should be used. Blunt scissors have a nonsharp end so that the possibility of injury is decreased. Smaller electronic clippers may be used, but care should be taken so that areas are not trimmed too short (see Figure 38–72). To prevent trimming areas too short while using electronic clippers, it is recommended that veterinary assistants use **clipper guards**, which fit over the end of the clippers. The clipper guards come in various sizes from ¼ inch to 2 inches to allow uniform

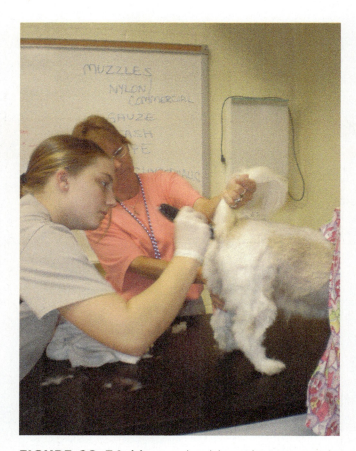

FIGURE 38–71 Many animal breeds may require trimming of the hair coat around the rectum and genital areas to prevent urine and feces from building up on the hair coat.

FIGURE 38–72 Electric clippers.

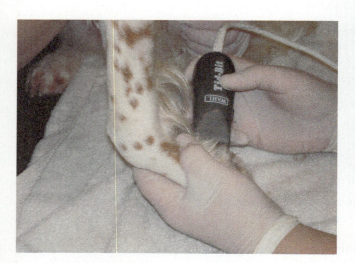

FIGURE 38–70 Electric clippers are used to clip and trim the hair coat.

trimming (see Figure 38–73). Clipper guards should be placed over a #40 clipper blade. When trimming a hair coat to a shorter length, work against the hair coat. This is best done by using a slicker brush and brushing

CHAPTER 38 Veterinary Assistant Procedures

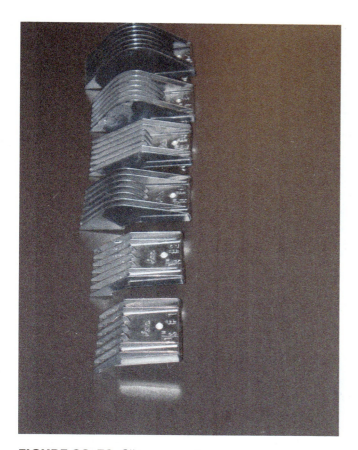

FIGURE 38–73 Clipper guards.

FIGURE 38–74 Cooling spray and blade cleanser.

against the hair coat, allowing the area to stand. Use long, even strokes when trimming long-haired coats. Shaving is typically done with surgery patients or animals that may have severely matted coats or problematic skin conditions. When shaving an animal, it is best to know how short the area must be shaved. Surgical patients will need the entire surgical preparation surface shaved to the skin, free of all hair and debris. This is typically done with a #40 or #50 surgical clipper blade. Longer shaves may be done with smaller-number clipper blades, which have wider spaces between the teeth, causing less hair to be removed. Clipper guards can also be used to keep the area uniform. When using clippers, one must keep the clipper blade surface parallel to the animal's body and flat, going in the direction of the hair coat. In surgical preparation, shave in the direction of the hair and then against the hair growth to remove all hair from the desired site.

Clipper and Clipper Blade Care and Maintenance

Clippers and blades receive a large amount of hard use. To prolong their performance, several things may be done. It is best to read the manufacturer's manual to know the proper use and care of the clippers. A backup clipper should always be available in case of damage or overheating. Electric clippers come in a variety of sizes, shapes, and speeds. Some are cordless, and others have electrical cords. Corded models should be checked on a regular basis for any damage to the cord or plug. Cordless models are kept in a charger and should be disinfected and returned to the charger immediately after each use. Some clippers have multiple speeds.

One area on the clipper that should be checked after each use is the air filter, which may easily become blocked by hair and debris. This will cause the motor to overheat and possibly become damaged. The instruction manual should note regular maintenance checks. Cleaning the clippers after each use and keeping clipper blades free of hair and debris are essential to prolonging the life of the clippers. When using the clippers, always use a cooling spray every 5 minutes and a lubricating oil every 15 to 20 minutes to prevent the clippers from overheating and excessive wear and tear (see Figure 38–74). Never use an oil lubricant other than the clipper oil. Turn off the clippers to apply a spray that cools down the clipper motor and lubricating oil to keep the clippers working properly.

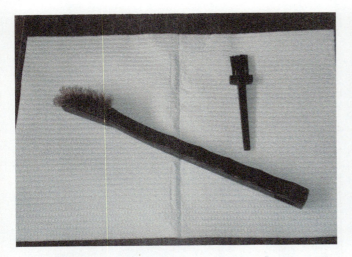

FIGURE 38–75 Wire brush used to clean clipper blades.

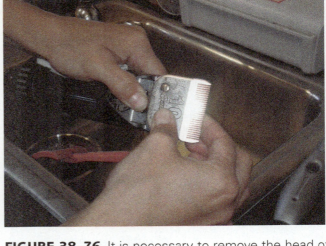

FIGURE 38–76 It is necessary to remove the head of the clipper to thoroughly clean the device to maintain it properly.

Items such as WD-40 will ruin the clippers. After using a blade, remove it from the clipper body and clean the teeth using a heavy wire brush (see Figure 38–75). Blade wash should be used to soak the blade. Blades should be stored face down to prevent tooth damage. Blades should be monitored for damaged or missing teeth. Broken teeth can cause skin tears and irritation. The blades should be dried well after each cleaning. Any wet areas on clippers or blades will cause corrosion. Blades will become dull with excessive use. Avoid using clippers on wet and dirty areas of the coat. Blades can be sharpened by professionals or using commercial blade sharpeners. Extra sharpened blades in various sizes should be available. The clipper blade attachment area of the clippers should be cleaned after each use by using a wire brush and clipper blade wash. Again, dry all areas well to prevent corrosion. Several times a year, the clipper head should be removed, and clipper grease should be applied to the drive gear (see Figure 38–76). This should be noted in the instruction manual. Some clippers may need to be serviced by the manufacturer if damaged or not working properly.

Ear Cleaning

The ear canals of animals may need to be cleaned on a regular basis. Floppy eared pets are prone to ear infections known as **otitis**. Some animals have hair growing within the ear canal as part of a breed or species characteristic. Some people will pluck out this hair, while others will shave it out to remove it (see Figure 38–77). Check with the veterinarian for his or her preferences. If the hair is to be plucked, try

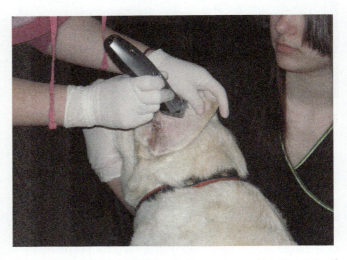

FIGURE 38–77 Some breeds may require shaving of the inside of the ear to keep the ear canal clean.

to remove as much of it from the inside ear flap with your fingers as you can. The hair growing within the ear canal can be removed using a pair of forceps or hemostats (see Figure 38–78). If clipping the ear flap and canal, use caution not to damage the folds of the ear. Blades should be kept flat on the ear surface. The ear anatomy includes three sections: the outer ear, middle ear, and inner ear. The **pinna**, or ear flap, can be cleaned using a cotton ball that has alcohol or a commercial ear-cleaning solution presoaked on the cotton ball. Wipe all areas of the ear flap and external ear canal. If signs of infection are present, such as redness, inflammation, strong odor, or discharge, notify

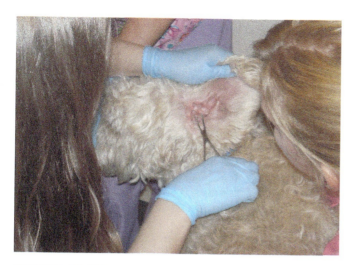

FIGURE 38–78 Hair inside the ear can also be removed using forceps to pluck it.

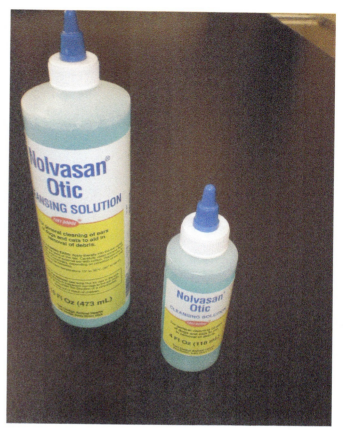

FIGURE 38–79 Solution used to clean internal structures of the ear.

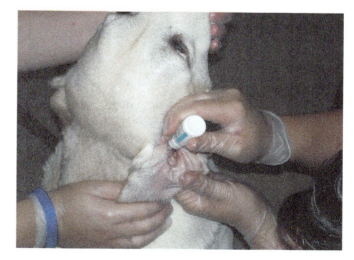

FIGURE 38–80 When instilling ear solutions, pull the ear flap caudally to open the ear canal.

the veterinarian. The ear opening is shaped like a *U*, whereas the ear canal is shaped like an *L* and bends at a 90-degree angle in the deeper canal of the ear. The eardrum, or **tympanic membrane**, is located at the end of the *L*. This area is easy to puncture during cleaning and is especially fragile if an infection is present. It is also easy to pack ear wax and discharge further into the canal around this area. It is not recommended to use cotton-tipped applicators in the internal ear canal. They may be used to clean the folds of the external ear flap.

Ears that are filled with wax or discharge should be cleaned by placing a small amount of ear-cleaning solution or ear antiseptic into the internal ear canal (see Figure 38–79). Pull the ear fold caudally to open the ear canal (see Figure 38–80). After applying the solution, fold the ear flap back into place and gently massage the base of the ear and cartilage to allow the solution to loosen and break up debris, which will allow the debris to easily come to the ear's surface. When done correctly, you should hear a sloshing sound as the fluids soften and loosen ear debris. Place a cotton ball or gauze sponge over the ear canal opening and tip the patient's head sideways ventrally. This will help any excess fluid drain out onto the cotton ball or sponge. Wipe the ear canal to remove excess discharge. Repeat this several times until the fluid is clean and the ear canal is free of debris. Dry the ear canal with a cotton ball or gauze sponge, reaching into the canal with the fingertip. Make certain all ear-cleaning solution has been removed from the ear, as moisture buildup can promote bacterial growth and an ear infection. An otoscope can be used to examine the internal ear canal (see Figure 38–81). Prior to administering any medication following an ear cleaning, the ear canal should be aired out for 10–15 minutes.

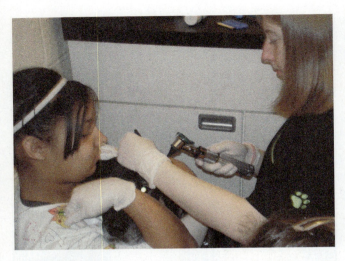

FIGURE 38–81 Use an otoscope to check the ear for buildup of wax or debris.

Nail Trimming

The nails on animals can easily become overgrown. They may require regular trimming so that they don't begin to catch on rugs, blankets, or other items or split from overgrowth. Sometimes long nails will begin to curl and grow into the footpad, causing severe pain and discomfort. A normal nail should be even with the bottom of the paw pads and during walking should not touch the ground. The purpose of nails is protection or defense and traction on smooth surfaces. In cats, they are also used for climbing. Some animals have hoofs, which need to be trimmed and protected from overgrowth as well. This section focuses on small animal nail trimming, as most hoofed animals are maintained by a farrier, who is a professional trained in trimming hoofed animals.

COMPETENCY SKILL 116

Cleaning Ears

Objective:

To properly trim or shave hair around the ears and remove excessive wax or debris from the ear canal

Preparation:

- Cotton-tipped applicator
- Cotton balls or gauze sponges
- Alcohol/Ear-cleaning solution
- Forceps
- Exam gloves

Procedure:

1. Apply exam gloves.
2. Examine ears using an otoscope.
3. Contact veterinarian if signs of redness, odor, inflammation, pain, or discharge are present.
4. Pluck or shave any hair from ear flap or ear canal if necessary.
5. Clean outer pinna and ear flap with alcohol or ear solution on cotton ball or gauze sponge. Wipe away from the internal ear canal.
6. Clean the outer pinna and ear flap creases and folds with cotton-tipped applicators. Work away from the internal ear canal.
7. Repeat until clean.
8. Pull flap caudally, and fill ear canal with ear-cleaning solution.
9. Massage the base of the external ear canal and cartilage.
10. Place cotton ball or gauze at external ear canal opening and tilt head. Drain fluid from ear, and wipe away any discharge.
11. Repeat until fluid is clean.
12. Wipe out external ear canal using a cotton ball or gauze sponge.
13. Check internal ear canal for moisture using an otoscope.

The anatomy of the nail includes the **nail bed**, the area where the nail growth occurs and that is located at the end of the digit near the hair growth of the foot. Within the nail bed lies the **quick**, or the blood and nerve supply to the nail. Similar to human nails, the outer nail covering is composed of **keratin**. Keratin is a protein that allows the nail to grow and strengthen. The goal of the nail trim is to cut the nail just distal to the quick (see Figure 38–82). Some nails are white or unpigmented, whereas others are black to brown in color and contain pigment. In white nails, it is much easier to see and determine where the quick lies. The quick is a faint pink triangular area within the end of the nail (see Figure 38–83). In dark-colored nails, it is not easy to see the quick. The goal is to cut about 1/8 inch in front of the quick. In dark nails, small cuts of nail must be done to look for the dark spot in the nail bed that surrounds the quick. At this point, you should stop trimming because it is the near the quick. Most animals use the hind digits more for traction and movement than they do the front digits, so typically the front nails need more taken off. Dewclaws may also be present and have no contact with the ground. They typically become long and tend to curl in toward the skin and hair coat. Every animal should be checked for signs of dewclaws on all four limbs. Some cats are known for being polydactyl and may have extra toes and digits that need to be trimmed.

If the quick is cut into or near the vessel, the nail will bleed, and the animal will react in pain. This is painful to the animal because the nerve supply also lies within the nail bed. When trimming the nails, always have **styptic powder** available to apply to any bleeding nail beds. The styptic powder is a loose yellow powder that packs into the nail bed and clots the blood. There are special well containers that can hold the powder, or the powder can be placed on a gauze sponge (see Figure 38–84). Caution must be used to not allow the powder to have contact with any skin or pads of people or animals, as the powder is caustic and burns materials. Sometimes the powder will not provide proper clotting, and a wooden **silver nitrate stick** is necessary. These are long slim wooden applicators with silver nitrate located on the end (see Figure 38–85). The silver nitrate burns the nail bed to induce a clot and stop the bleeding. Pets may find the silver nitrate uncomfortable when it is being applied and may briefly struggle. Caution must also be used when using these sticks due to the caustic risks and because the silver nitrate tends to stain skin, clothing, and tabletops. Animals that have severely

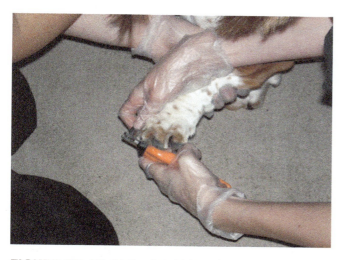

FIGURE 38–82 Nails should be trimmed just distal to the quick.

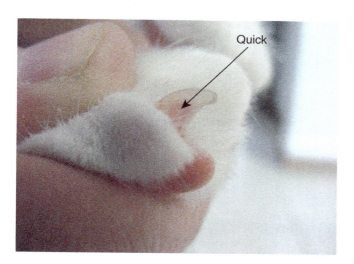

FIGURE 38–83 Quick of a cat nail.

FIGURE 38–84 Styptic powder can be used to stop bleeding that may occur when nails are trimmed.

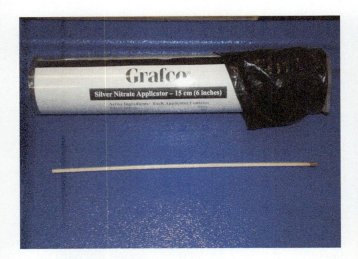

FIGURE 38–85 Silver nitrate sticks.

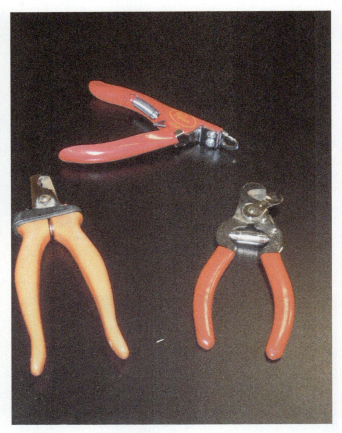

FIGURE 38–86 Nail trimmers.

FIGURE 38–87 Dremel tool used to trim nails.

overgrown nails may need to be sedated to trim the nails, as they may need to be cut back into the quick, which is painful for the animal and may require pain medication. The area should be examined for signs of infection, especially if the nail has grown into the surrounding pad. Antibiotics and pain medications are often prescribed, depending on the severity of the injury.

Nail trimmers come in a variety of shapes and sizes (see Figure 38–86). Most commonly used types include the White clippers, which are scissorlike and useful for cats, small dogs, birds, and pocket pets. They are useful in trimming overgrown nails that curl into pads. Another common type of clipper is handheld professional trimmers that have a blade and scissorlike handle and come in a variety of sizes depending on the animal. A less-used nail trimmer is the Rescoe, or guillotine nail clipper. A sharp blade in the center moves to cut the nail. Extreme care must be used with this type of clipper, as in small animals it could amputate a toe. Human nail trimmers can be used on small thin nails. The dremel tool is also useful with animals that may not sit well for nail trims. The dremel is a grinding tool with a sandpaper surface that rounds and smooths the end of the nail without as much possibility as hitting the quick (see Figure 38–87).

When trimming or dremeling nails, the foot should be held and pressure applied to the digit with the nail being trimmed (see Figure 38–88). This will extend the nail and hold it in place to prevent too much from being cut. The nail is then trimmed. The animal can be in a standing or sitting position, and smaller pets may need to be held for the nails to be trimmed. It is best to begin at the rear paws to allow the animal time to settle down before moving to the front paws. The front legs are then trimmed, pulling the limbs back away from the head. This allows animals that do not like their nails trimmed to feel less threatened.

Expressing Anal Glands

Anal glands or sacs are located on either side of the rectum and are the scent glands of dogs and cats (see Figure 38–91). The sacs themselves lie ventrally and slightly anterior to the skin around the rectum at the 4 o'clock and 8 o'clock positions. A duct travels to a small opening that allows the glands to be expressed. The material that builds up in the sacs is secreted into the sac every time the animal has a bowel movement. If the material becomes thickened and is not removed properly, the sacs may become abscessed and painful. Signs that the anal glands should be expressed include the dog or cat scooting across the floor on its hind end, excessive licking at the rectum, and odor noted from the rectum. Normal material within the glands is a yellow to tan color and may produce a fishlike or skunklike odor.

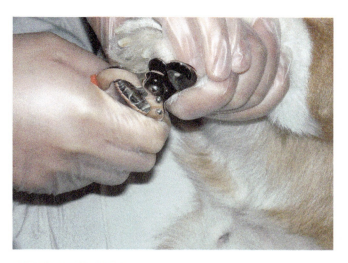

FIGURE 38–88 Proper method of holding the foot when trimming nails.

COMPETENCY SKILL 117

Trimming Nails

Objective:

To properly trim nails

Preparation:

- Nail trimmers
- Dremel
- Styptic powder
- Silver nitrate sticks
- Powder well
- Gauze sponge
- Exam gloves

Procedure:

1. Apply exam gloves.
2. Locate the proper trimmers or dremel tool.
3. Grasp a rear foot with one hand (see Figure 38–89).
4. Apply pressure to each digit to extend the nail (see Figure 38–90).
5. Use the other hand to clip or file the nail just distal to the quick. Dark nails in which it is difficult to see the quick should be trimmed in small amounts until the dark center circle of the quick is noted.
6. Repeat with each nail, remembering to check for dewclaws, if present.
7. If bleeding occurs, use the styptic powder by applying the yellow powder onto the nail and applying pressure using a gauze sponge. Severely bleeding nails should have a silver nitrate stick applied to the end of the nail to control bleeding.
8. Trim each footpad, beginning with the rear limbs and then moving to the front limbs.
9. Check all nails for signs of bleeding after completing the trim.

738 SECTION IV Clinical Procedures

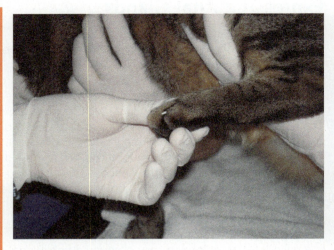

FIGURE 38–89 To begin trimming nails, start with a rear limb and grasp the foot firmly.

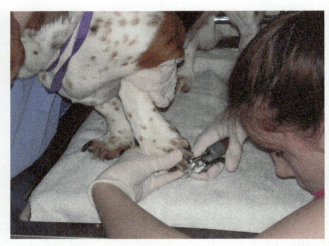

FIGURE 38–90 Apply pressure to the digit to extend the nail.

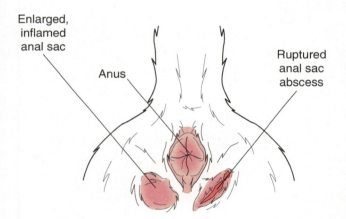

FIGURE 38–91 Location of the anal glands.

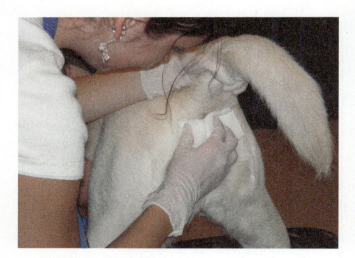

FIGURE 38–92 External expression of the anal glands.

There are two techniques for expressing anal glands. The first that should be used by the veterinary assistant is external expression of the anal glands (see Figure 38–92). The pet should be in a standing position. Gloves are worn while the thumb and index finger are used to gently palpate the ventral and lateral sides of the anus. The tail of the animal is held upward and out of the way. Stand to the side of the animal after locating the anal glands, place a paper towel over the area, and gently press the fingers medially while squeezing each sac against the other with slight pressure. The fingers can be moved in an upward and then inward semicircular motion while squeezing the glands against each other. This will allow the release of the anal secretions through each gland duct. The secretions will be caught in the paper towel. Note how much is expressed along with the color and consistency. This procedure may need to be repeated several times until all debris are expressed. Once it has been completely emptied, you should no longer be able to palpate the anal gland. After completing the procedure, place the paper towel and gloves in the trash. Apply a new pair of gloves and then clean the anus and hair around the anus with a damp paper towel or gauze sponges soaked in warm, soapy water. A pet-safe cologne or coat conditioner can be applied to the hair coat (see Figure 38–93).

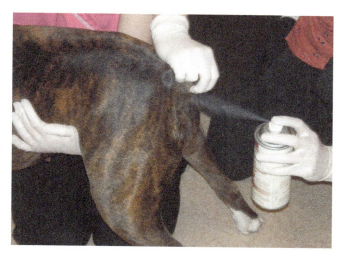

FIGURE 38–93 After expression of the anal glands, the animal should be cleaned, and a scented spray can be used.

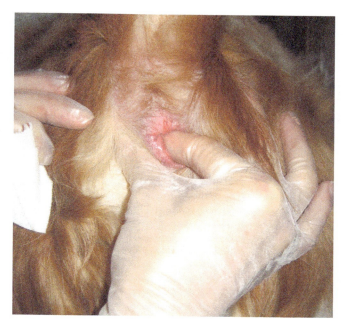

FIGURE 38–94 Internal expression of the anal glands.

If the anal glands are difficult to express using the external anal gland expression technique, the veterinary technician or veterinarian should be alerted, and an internal expression method should be used (see Figure 38–94). This technique exerts greater pressure on the anal sacs when the secretions may be thickened or a duct is blocked. If the secretions are expressed, it is important to note any abnormal findings, such as signs of blood, yellow or puslike secretions, red or swollen anal glands, or severe pain in the anal gland area. These may be signs of an infection or abscess. With a gloved hand, a lubricant is applied on the index finger, which is then inserted inside the rectum. The anal gland can be palpated with the index finger on the inside and the thumb on the outside. Gently squeezing each gland will allow the material to be milked out of the duct. The area should be cleaned as noted above and an odor-reducing spray applied. Anal glands are best expressed prior to bathing to help clean the pet following the expression procedure.

 COMPETENCY SKILL 118

Externally Expressing Anal Glands

Objective:

To safely and effectively express the anal glands using the manual method

Preparation:

- Exam gloves
- Vaseline or KY Jelly
- Paper towels
- Gauze sponges
- Small bowl
- Soap or scrub
- Cologne or conditioning spray

Procedure:

1. Apply exam gloves and stand to the side of the animal, lifting the tail upward. A small amount of lubricant can be applied to a paper towel for comfort.
2. Place the paper towel over the anus and, using the thumb and index finger, locate and palpate the anal sacs, located laterally and slightly ventrally at 4 o'clock and 8 o'clock.
3. Place the thumb over the lateral aspect of one gland and the index finger over the lateral aspect of the other gland.
4. Gently squeeze the fingers together in an upward and outward, semicircular motion, as if massaging the area.
5. Note the amount and appearance of the secretions.
6. Place paper towels in the trash.
7. Replace exam gloves if soiled.
8. Clean the rectum area and the hair around the anus with clean paper towels or gauze sponges soaked in warm, soapy water.
9. Dry the area.
10. Apply cologne or conditioning spray.

Laundering Materials

Laundry disinfection and cleaning is an important part of a sanitation and disease control plan. Many items must be laundered in a facility after being used with animals. These items include the following:

- Towels, blankets, and other bedding materials
- Muzzles and cat bags
- Collars and leashes
- Scrubs, surgical gowns, lab coats, and other clothing materials
- Surgical towels and pack covers

Sorting the laundry materials is an important part of laundering technique. In most cases, laundry should be pretreated by soaking in water with detergent or bleach materials as necessary. Laundry should be sorted according to surgical materials, regular hospital materials, and contagious items. The three types should not be mixed. This will help prevent and control the spread of disease. Any items with bloodstains, urine, feces, or other bodily fluids should be soaked and treated in water and detergent or bleach mixtures for at least 30 minutes before washing. Items that are bloodstained should also be pretreated with hydrogen peroxide to remove stains. All contagious items are presoaked in bleach and warm water.

Washing is more effective at high temperatures and with high-quality industrial-strength detergents. A fabric softener may be used during the rinse cycle or in the dryer via softener sheets. Each load of laundry should be set to the proper load size and monitored for overloading. The washer must be cleaned and disinfected on a regular basis to remove hair, debris, and detergent residues. Lint should be removed from filters after each load. Surgery laundry should be folded immediately and placed in an area to complete packs and resterilize materials. Other hospital laundry should be folded and stored in clean, dry locations.

COMPETENCY SKILL 119

Doing Laundry

Objective:

To properly clean and sanitize all soiled items that may be laundered

Preparation:

- Laundry baskets (3)
- Buckets (3)
- Industrial-strength laundry detergent

- Bleach
- Fabric softener
- Hydrogen peroxide
- Exam gloves
- Washer
- Dryer

Procedure:

1. Sort laundry into baskets labeled surgery, hospital, and contagious.
2. Place items that need soaking in buckets with detergent or bleach.
3. Pretreat any bloodstains with hydrogen peroxide.
4. Pretreat each laundry load in the washing machine. This and subsequent steps should be carried out three times, once for each category of wash.
5. Set wash cycle on hot and appropriate load size.
6. Add fabric softener in rinse cycle.
7. Dry according to load size.
8. Disinfect and clean washer.
9. Clean lint filters.
10. Disinfect and clean dryer.
11. Fold laundry.
12. Put laundry away in proper clean area.

SUMMARY

Hospitalized patients require specialized care, constant observation, and proper sanitation. Many patients are cared for by the veterinary assistant. Hospitalized patients may be contagious, be experiencing a serious health condition, or require many treatments throughout the day. The veterinary assistant must know what is being done with the patients, what is normal or abnormal for each patient, and what treatments are to be done by the assistant. This requires knowledge, experience, and a planned working schedule to meet the diverse needs of every patient. It is essential that the veterinary assistant properly handle and care for hospitalized patients. For the assistant to properly provide this care, he or she must be able to recognize the needs of the patient and the requirements of the veterinary care to be provided. Such needs include dental care, veterinary assistant nursing care, proper setup and maintenance of fluid therapy, understanding of euthanasia procedures, and the proper laundering of bedding and hospital materials.

Veterinary assistants play an important role in grooming and maintaining the care and cleanliness of hospitalized patients. Animals may require veterinary care that the owners have not been able to provide at home or have neglected for some other reason. Common basic grooming skills such as nail trims, ear cleaning, and bathing and anal gland expression are helpful to many clients that can't properly provide this care at home. The veterinary assistant should be able to complete these duties easily and efficiently.

Key Terms

agonal   refers to gasps of breaths while the respiratory system shuts down at the time of death

anal glands   scent glands located on either side of the rectum

anogenital distance   space between the area of the vulva or penis and the rectum by which gender can be determined

aspiration   process used with a syringe when the plunger is drawn back slightly to make certain no blood vessel has been accidently penetrated prior to administering an injection; also refers to swallowing food that enters the lung field

bandage covering applied to a wound, fracture, or area of injury to prevent movement and protect the wound

bandage scissors scissors with an angled blunt end used to carefully remove a bandage from an animal

bathing process of cleaning the skin and hair coat of the animal or applying medicated shampoos or dips to the skin and hair coat

brushing cleaning the hair coat using a soft brush

buccal surface of the teeth located on the outer area near the cheek

clipper guard tool that fits over the end of clippers to prevent trimming hair too short

clipping removing small amounts of hair from one or several areas

combing brushing the hair with a thin comb that helps remove tangles and mats

crash cart movable table that holds emergency equipment and supplies; is easily accessible to address a serious patient condition

cyanotic bluish gray color of the mucous membranes due to a lack of oxygen

dental prophylaxis professional cleaning that involves scaling and polishing the teeth under general anesthesia

dentifrices toothpaste used on animals

dentition arrangement of teeth within the mouth

diluent liquid mixture of a vaccine

dipping applying a chemical pesticide or medication to the skin and hair coat

epaxial located above the axis of a body part, such as above the spine

emergency situation that requires immediate life-saving measures

euthanasia process of putting an animal to sleep using humane methods

fingerbrush a small toothbrush that fits on the end of a finger; is a plastic, thimblelike device that has small, soft bristles that are rubbed over the teeth

flow rate rate at which ordered fluids are to be delivered over the course of a day

force-feeding placing food within a patient's mouth and forcing it to swallow

gauge refers to size of needle; related to its diameter

grooming caring for an animal's external body, including hair coat, ears, nails, and anal glands

hamstring muscle located on the hind limb at the mid- to distal part of the thigh

hospital treatment board area in the facility where hospitalized patients are listed along with the procedures that are to be completed on them

infusion pump equipment used to provide a constant flow of fluid at a specific rate throughout the day

injection port area on an IV bag used to remove and insert substances using a needle and syringe

intramuscular (IM) refers to injection given into the muscle

intranasal (IN) refers to medication administration into the nasal passage

IV catheter a small plastic piece of equipment that is placed within a vein to administer fluids and medications directly into a patient's bloodstream

keratin protein that allows the nail to grow and strengthen

labial surface of the teeth located on the front area that is covered by the lips

lactated Ringer's solution (LRS) used to hydrate animals by subcutaneous or intravenous fluid therapy

lavaging flushing a wound using jet pressure to forcefully clean an area

macrodrip line a regular IV drip that delivers 15 drops of fluid per milliliter or 15 drops/ml

mat area of hair interwoven together and form a large clump

mat splitter tool used to break down and separate mats into smaller ones

microdrip line a small line that delivers 60 drops/ml

nail bed area where the nail growth occurs; is located at the end of the digit near the hair growth of the foot

nasogastric tube tube passed through nostrils into the stomach to force-feed a patient

neckband paper or plastic slip placed around the neck that identifies the animal

observation act of watching and noting an animal's behavior while hospitalized

occlusal top surface area of the teeth

otitis ear infection

patency refers to maintaining the proper flow and purpose of IV tubing, ensuring it remains intact and usable

phlebitis causes the area around the IV catheter site to experience redness, swelling, pain, and inflammation

pinna external ear flap

pitting edema swelling around the catheter site that causes the fluid to flow outside the vein

primary bandage layer first bandage layer, against the surface of the animal, that acts as padding

quadriceps muscle located on the hind limb at the cranial part of the thigh

quick blood supply to the nail

secondary bandage layer holds the first layer in place

shaving taking away an amount of hair close to the skin

silver nitrate stick long slim wooden applicator with silver nitrate located on the end; is used to clot a bleeding nail by burning the quick

spike port site on an IV bag where the drip set tubing is placed

stirrups pieces of tape at the end of a bandage on a limb that hold the bandage in place and prevent slipping

styptic powder loose yellow powder that gets packed into the nail bed and clots the blood

subcutaneous (SQ) refers to injection given under the skin

subcutaneous fluids fluids placed under the skin and slowly absorbed by the body

tertiary bandage layer third bandage layer that acts as a protective covering

trimming removing a specific amount of hair from one or more locations

tympanic membrane eardrum; located in the inner ear

viscous thick liquid substance

Review Questions

1. What are dental care responsibilities of the veterinary assistant?
2. What tools and supplies are necessary for brushing a patient's teeth?
3. What type of injections may be completed by the veterinary assistant?
4. What does the term *aspiration* mean?
5. What is the difference between an 18-gauge needle and a 24-gauge needle?
6. What items should be placed on a syringe label?
7. What is the difference between using SQ fluid therapy versus IV fluid therapy?
8. What items should be monitored with patients that have IV catheters?
9. What are ways to socialize hospitalized patients?
10. What items are necessary when preparing for euthanasia?
11. What is the importance of the hospital treatment board?
12. What information should be placed on a hospital treatment board?
13. What are some veterinary assistant duties and responsibilities in emergency situations?
14. What are some common emergency tools and supplies?
15. What are the three layers of a basic bandage?
16. What is the purpose of each layer of a basic bandage?
17. What items are needed for applying a basic bandage?
18. What information should be addressed for basic bandage care?
19. What tasks are involved in the grooming care of an animal?
20. What is the difference between combing and brushing?
21. What is meant by trimming, clipping, and shaving patients?
22. What items and supplies are necessary for bathing?
23. What are signs of an ear infection?
24. What is the importance of styptic powder and silver nitrate sticks?
25. Why do dewclaws often become overgrown?
26. What are signs that a dog or cat needs the anal glands to be expressed?

Clinical Situation

Alicia, a veterinary assistant at Kind Heart Pet Hospital, is working as the exam room vet assistant. Dr. Moore has asked her to set up the treatment area for a bandage application for a cat that has an injured leg. The cat had X-rays that show no fracture, but the leg is swollen and sore, and the vet feels the cat should not use the limb for a week until it has some time to heal and the swelling is reduced. Alicia goes into the treatment area to set up the supplies and equipment and to assist Dr. Moore in applying the bandage.

- What supplies will be needed for this treatment situation?
- What may be needed if the cat is in pain and difficult to handle?
- When complete, what home-care instructions might Dr. Moore have Alicia prepare for the owners?

Clinical Situation

Veterinary assistant Kathryn is monitoring a cat's IV fluids during the patient's hospital stay. Kathryn notes the fluids are not flowing correctly and, at times, no fluid is dripping from the IV bag. Kathryn checks the hospital treatment board and the medical record and notes the cat should be receiving 1400 ml of fluid a day. She is concerned since the fluids are obviously not being received correctly.

- What should Kathryn do?
- What volume of fluid should the cat receive on an hourly basis?
- What may be causing the fluid flow problem?

Clinical Situation

Rhonda, the veterinary assistant, and Kelly, the veterinary technician at Harland Veterinary Hospital, are working with Mrs. Cramer's dog, Tilly. Tilly is a 5-year-old SF Boxer that is hyper and tends to get excited at the veterinary clinic. Tilly needs her ears cleaned frequently, as they are not cropped and are floppy, and she tends to have a lot of wax and discharge. When there is a lapse in the cleanings, she is prone to ear infections. Mrs. Cramer frequently visits the clinic for routine ear cleanings.

"I would really like to clean Tilly's ears at home. It would save me driving to the clinic so much. Is it possible for me to clean her ears at home?" she asks.

"Sure, you would just need to have the necessary supplies and know the correct procedure," says Rhonda.

Mrs. Cramer replies, "Wonderful, maybe that will make Tilly's ears less of a problem."

- What should Rhonda discuss with Mrs. Cramer about home ear cleaning?
- What items will Rhonda suggest Mrs. Cramer have at home for cleaning Tilly's ears?
- What signs of problems should Mrs. Cramer know about that should alert her to the need to schedule a veterinary appointment?

CHAPTER 39: Laboratory Procedures

Objectives

Upon completion of this chapter, the reader should be able to:

39.1 Identify common laboratory equipment used in a veterinary facility

39.2 Explain proper collection of a fecal sample

39.3 Demonstrate how to properly conduct a gross fecal examination

39.4 Demonstrate how to properly prepare a fecal smear

39.5 Demonstrate how to properly set up a fecal floatation

39.6 Explain how to properly prepare a blood sample

39.7 Demonstrate how to properly prepare a usable blood smear

39.8 Demonstrate how to properly stain a blood smear using Wright's stain

39.9 Explain how to properly complete a CBC

39.10 Demonstrate how to properly complete a blood chemistry sample using in-house analyzers

39.11 Demonstrate how to properly determine a plasma protein or total protein sample using a refractometer

39.12 Explain how to properly use assorted serologic testing kits

39.13 Demonstrate how to properly collect a voided urine sample

39.14 Explain how to properly conduct a gross urine examination

39.15 Demonstrate how to properly determine urine specific gravity

39.16 Demonstrate how to properly understand and conduct urine chemistries using test strips

39.17 Demonstrate how to properly prepare urine sediment for microscopic examination

39.18 Demonstrate how to properly collect a sample for culture and sensitivity

39.19 Demonstrate how to properly prepare a Gram stain smear

39.20 Demonstrate how to properly set up equipment and materials for a veterinary necropsy

Introduction

Laboratory skills are meant to aid the veterinarian in determining a patient diagnosis and thus determining the necessary treatment. Many testing procedures are done both inside and outside the veterinary facility. Most laboratory testing begins after the initial patient history and physical exam. The veterinarian will determine the testing needs of each patient, and the veterinary technician and assistant will work together to prepare samples, conduct in-house laboratory tests, and record laboratory data in medical records. The patient's well-being depends on the entire veterinary health care team providing accurate and timely results. Laboratory testing technology related to veterinary medicine changes regularly. Many protocols, equipment, and tests may have unique factors in each veterinary facility. These factors require flexibility in all staff members and understanding of the importance of continuing education and learning to keep up with the changes in veterinary medicine. It is important to remember safety when handling lab samples to ensure both protection of the sample and protection of the person handling the samples. Many samples may contaminate an area, a person, or a patient, so wearing appropriate personal protective equipment, such as exam gloves and goggles, is essential. Please note that any laboratory test performed improperly, imprecisely, or inconsistently will lead to errors and inaccurate results that may harm the patient.

Veterinary Laboratory Equipment

Many laboratory tests and services are performed within the veterinary facility and are referred to as **in-house testing**. Some tests that may not be completed at the facility are often sent to a commercial laboratory referred to as a **reference lab**. The equipment that is available for the test determines which services are offered in-house or through reference labs.

Microscope

Microscopes are a basic and key component to all in-house veterinary labs (see Figure 39–1). They are extremely useful for basic testing procedures and can help identify and diagnose problems quickly and accurately. Many veterinary facilities have a binocular microscope and a research microscope that allow a variety of magnification powers to view different test samples. They also allow slides to be observed under **oil immersion**, which is the use of a specialized oil substance that is placed over a sample to view the contents. It is important that the veterinary assistant has a general and clear understanding of how to appropriately use a microscope. All microscopes should be covered when not in use. Each microscope has several **lens objectives** that offer a variety of magnifications to view slide samples (see Figure 39–2). These objectives offer a variety of powers (see Table 39–1). The most common objective lenses are 4× for scanning, 10× for low power, 40× for high power, and 100× for oil immersion. A slide should be placed on the **stage**, which is the flat section under the lens. The stage is movable from left to right and forward and backward to allow movement of the slide for viewing. A slide should always be viewed using the lowest power magnification first and then moving to a higher power. A slide is viewed by looking through the **eyepiece**. To focus each lens to view a sample, the **focus knobs** should be turned gently and slowly while looking through the eyepiece. When the microscope is on, a light source will be present. The slide sample should be moved up and down and side to side to view the entire sample. This motion is done by moving the **diaphragm mechanism**. All lenses should be cleaned after each use, and this is accomplished by using lens paper and a small amount of lens cleaner or alcohol to wipe each lens. It is important to follow the manufacturer's instructions when using and

FIGURE 39–1 It is important to learn the proper use of the microscope.

FIGURE 39–2 Note the different lens objectives that can be used on this microscope.

cleaning any microscope. All slides and **coverslips**, which fit over the sample on the slide, should be clean and have no cracks or chipped edges. The microscope should be wiped clean after each use and kept covered when not in use. The eyepieces used for viewing the slide should also be cleaned with lens paper. Cleaning and adjustment should be scheduled annually with a microscope professional. Light bulbs need regular changing according to manufacturer recommendations. When changing a bulb, allow the microscope to cool before removal. Changing a light bulb requires the device to be powered off and unplugged. Avoid touching the replacement bulb, as oils from the skin can shorten the life span of the bulb.

TABLE 39–1
Magnification Powers

	EYEPIECE	OBJECTIVE	MAGNIFICATION
Low Power	Green	10×	100×
Medium Power	Blue	30×	300×
High Power	Yellow	40×	400×
Oil Immersion	Red	100×	1000×

COMPETENCY SKILL 120

Use of a Microscope

Objective:

To properly understand the mechanics of the microscope and use it in practice

Preparation:

- Microscope
- Microscope slide
- Coverslip
- Lens paper
- Immersion oil

Procedure:

1. Remove the plastic cover from the microscope. The arm of the microscope should face the user.
2. Plug in the microscope and turn on the power source including the light.
3. Place a slide on the stage of the microscope, securing it in place (see Figure 39–3).
4. Place the projection lens on a low power.
5. Use the adjustment knob to lower the objective power while looking through the eyepiece (see Figure 39–4).
6. Look at the stage through the eyepiece while focusing the slide. Adjust the focus accordingly. Raise the objective arm away from the slide when complete.
7. Apply a drop of immersion oil into the center of the slide specimen if using the oil immersion objective lens.
8. Lower the immersion objective slowly into the immersion oil.
9. Slowly adjust the focus knob while looking in the eyepiece.
10. When complete, raise the objective arm away from the slide.

11. Remove the slide from the microscope stage.
12. Use a kim wipe or lens paper piece to gently cleanse the objective.

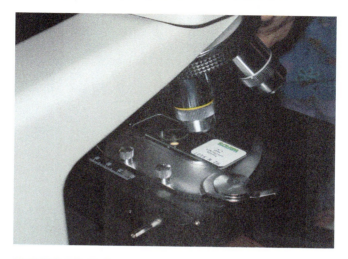

FIGURE 39–3 Be sure to secure the slide when placing it on the stage of the microscope.

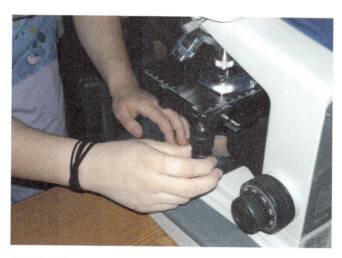

FIGURE 39–4 Lower the objective power while looking in the eyepiece.

Centrifuge

A **centrifuge** is another common veterinary tool and is used to spin lab samples at a high rate of speed and force (see Figure 39–5). This device is used to separate or concentrate materials suspended in a liquid form. Gravity helps to separate the different forms. When solid and liquid components are present in a sample, the liquid portion is called the **supernatant** and the solid portion is called the **sediment**. Centrifuges come in a variety of shapes and sizes and many facilities have two, one used to spin microhematocrit tubes and one standard machine used to spin larger volumes of fluid samples. All centrifuges have a variety of **rotors**, or wheels, which spin at a variety of speeds depending on the sample type (see Figure 39–6). Each time the centrifuge is used, it must be balanced before turning it on. This means the tubes must be counterbalanced with tubes of equal size and weight. This allows the machine to operate correctly and not

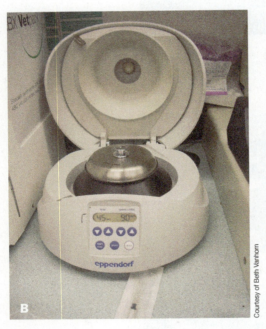

FIGURE 39–5 (A) Standard centrifuge. (B) Small commercial centrifuge.

FIGURE 39–6 Rotors in the centrifuge.

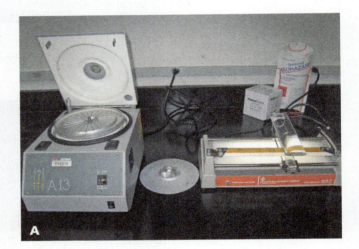

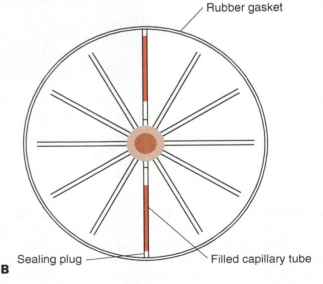

FIGURE 39–7 (A) Microhematocrit centrifuge. (B) Proper placement of sealed capillary tubes in a microhematocrit centrifuge.

shake or wobble, potentially spilling liquid from a tube. Water-filled tubes are often used to balance the centrifuge. Incorrect handling and balancing can cause damage to the machine. The centrifuge must be cleaned immediately if anything is spilled or broken within the machine. Samples that may be used in a centrifuge include blood tubes of various sizes, **centrifuge tubes** that have **conical** or pointed ends and may hold urine or fecal samples, or **microhematocrit tubes**, which are small thin glass tubes that hold blood (see Figure 39–7). When a sample is placed within a rotor, the lid should be closed and secured (see Figure 39–8). The machine should be set on the proper test sample setting and speed, and the timer should be set for the required spinning time. The centrifuge spins at a high rate of speed; no one should stand directly over the centrifuge while it is in use, and one should never try to stop or slow down the rotors. Most centrifuges have speed settings calibrated in revolutions per minute (rpm) times 1000. A setting of 5 thus represents 5000 rpm. Safety brakes are included on the centrifuge for when the device must be stopped, and this rapidly shuts down the rotors so that a sample can be safely removed or added.

Refractometer

A **refractometer** is a tool used to measure the weight of a liquid and determine a liquid's specific gravity (see Figure 39–9). The veterinary refractometer has two scales that are viewed in a fashion similar to how a prism is viewed. This is a lightweight handheld tool. A liquid sample is placed on the prism plate and the cover is closed (see Figure 39–10). The refractometer is then held up to a light source, and the sample is viewed through the eyepiece to determine the specific gravity of the liquid (see Figure 39–11). Each scale is labeled, one on the right side of the prism and one on the left side. The specific gravity and protein concentrations are measured in g/dL. The tool should be cleaned after each use by using lens paper

FIGURE 39–8 After placing a sample in the hematocrit, the rotor should be closed and secured.

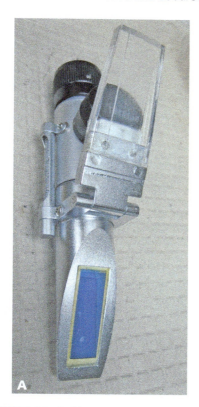

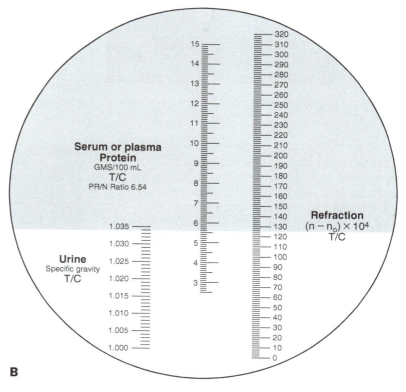

FIGURE 39–9 (A) Refractometer. (B) This refractometer scale shows a urine specific gravity of 1.034 (lower left scale) and serum or plasma protein of 5.6 (middle scale).

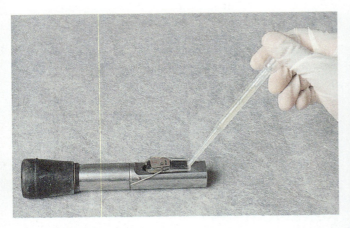

FIGURE 39-10 Adding liquid sample to the refractometer.

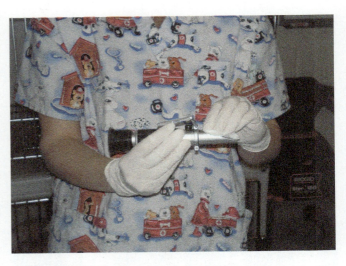

FIGURE 39-12 Cleaning the refractometer.

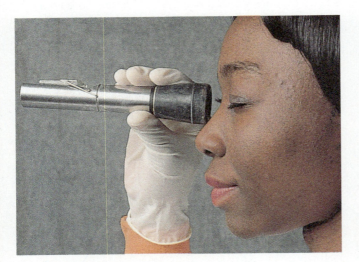

FIGURE 39-11 Viewing the sample in the refractometer.

Blood chemistries and CBCs may be run using samples of whole blood, serum, or plasma. The manufacturer's instructions should be followed vis-a-vis how to use and maintain the equipment, as there are a variety of types of analyzers. Commercial blood chemistry machines are more efficient, accurate and less time consuming, providing results for patient diagnosis in a timely manner. Many analyzer systems require slides or cartridges for use with each sample but may be time consuming to prepare and complete testing of a sample for results. Blood samples are commonly placed on a blood rocker, which allows the blood components to mix until it is analyzed (see Figure 39-13). These analyzer machines are sensitive pieces of equipment, so care and maintenance are important. Most machines have a warm-up period and require use of control samples to ensure the machine is properly calibrated.

and alcohol or another cleaner (see Figure 39-12). The manufacturer's instructions should be followed for cleaning and use. It is important to calibrate the refractometer on a regular basis using distilled water. To check whether the refractometer is correctly calibrated before use, place a drop of distilled water on the prism, close the lid, and hold the instrument up to the light while looking through the eyepiece. Check that the distilled water drop reads 1.000 on the specific gravity (U.G.) scale. If the tool is not properly calibrated, it will not provide accurate results.

Blood Chemistry Analyzers

Blood chemistry analyzers are machines that run blood samples and measure routine blood chemistries and electrolytes, including **complete blood counts** or **CBCs**.

FIGURE 39-13 Blood rocker.

Serological Test Kits

Commercial serologic test kits are commonly used in a veterinary facility to provide quick and accurate results related to common viruses and diseases (see Figure 39–14). There is an increasing number of test kits available by several manufacturers to detect a variety of diseases. These test kits usually have a set number of individual tests along with the items necessary to perform them. Step-by-step instructions accompany each test kit and should be followed when conducting the test and interpreting the results (see Figure 39–15). The instructions should be kept with the test kit, which is usually kept refrigerated and then warmed to room temperature prior to use. A variety of **reagents** or chemicals are used to run each individual test in the kit and should be kept with the test kit. They should not be thrown out until the entire kit has been used (see Figure 39–16).

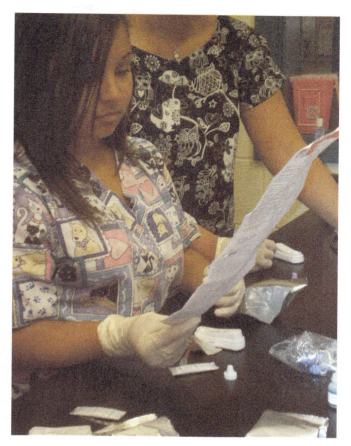

FIGURE 39–15 Instructions that come with each kit should be followed.

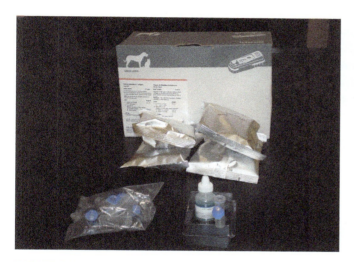

FIGURE 39–14 Serological test kit.

Several manufacturers produce testing kits that help diagnose conditions and diseases quickly and easily within the veterinary facility (see Table 39–2). These tests vary on the disease type being screened, species being screened, and type of sample needed. Some samples may require whole blood, serum, or a fecal sample. Many of these tests involve the presence of antibodies or antigens that react in the test kit. Some serology kits test for one specific disease, while others test for multiple diseases. Most of the test kits offer either positive or negative results. The majority of testing kits provide results within 10 to 15 minutes. The veterinary assistant is essential in performing these tests quickly and accurately.

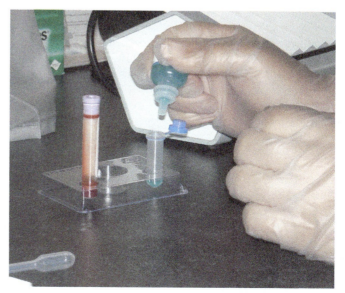

FIGURE 39–16 Reagent used in the serological test kit.

TABLE 39-2

Types of Serologic Test Kit

TEST TYPE	MANUFACTURER	DISEASES TESTED
SNAP FeLV	Idexx	Feline leukemia virus
SNAP FIV/FeLV Combo	Idexx	Feline immunodeficiency virus; feline leukemia virus
SNAP Giardia	Idexx	Feline giardia; canine giardia
SNAP Heartworm	Idexx	Canine heartworm disease
SNAP Parvo	Idexx	Canine parvovirus
PetCheck Heartworm	Idexx	Canine heartworm antigen
SNAP Foal IgG	Idexx	Equine IgG Antibodies
SNAP Antibiotic Residue	Idexx	Ruminant/Dairy penicillin; tetracycline, aflatoxin, gentamicin, sulfamethazine
SNAP cPL	Idexx	Canine pancreatitis
SNAP Feline Triple	Idexx	Feline immunodeficiency virus; feline leukemia virus; feline heartworm disease
SNAP 3Dx	Idexx	Heartworm disease; *Ehrlichia canis*; lyme disease
SNAP 4Dx	Idexx	Heartworm disease; *Ehrlichia canis*; lyme disease; *Anaplasma phagocytophilum*
D-Tec CB	Synbiotics	Canine brucellosis
Witness Relaxin	Synbiotics	Canine and feline pregnancy test
FeLV Assure	Synbiotics	Feline leukemia virus
OvuCheck	Synbiotics	Canine ovulation
Solo Step	Heska	Feline and canine heartworm disease

Test kits come packaged individually in boxes with all of the materials required to run each test. Step-by-step instructions are provided for setting up and reading the test. Read each testing kit's requirements for storage. The kits, including the chemicals provided, have expiration dates that should be monitored. Reagents within a test kit should always be kept with the kit and never removed, replaced, or thrown away. Most serologic test kits are based on a **SNAP test**, which is named for the snap action that begins the reaction to determine the test result (see Figure 39-17).

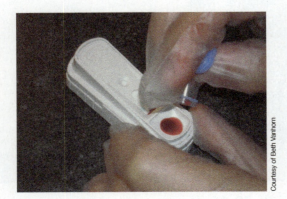

FIGURE 39-17 SNAP test.

Agglutination tests are used to detect antibodies. This type of test requires adding a specific antigen to the test sample. If the sample contains the antibody for that antigen, agglutination, or clumping, occurs. Some tests may be performed on a slide or within a kit. Sample liquids or latex beads cause agglutination reactions to occur. The most common type of tests using agglutination are blood typing and detection of brucellosis in dogs.

Stains, test strips, and other chemical reagents are also commonly used in a veterinary lab. Veterinary assistants should be familiar with the types of items used in the facility and how to properly interpret the results. Over time, some stains and reagents tend to crystallize or evaporate and should be monitored for age and expiration date. Some items are kept in **Coplin jars**, which hold the chemical for repeated use (see Figure 39-18). These items may need to be changed on a regular basis as the materials become aged and collect debris. Never add a new chemical to an old chemical. The items should be entirely discarded and fresh ones should be provided. Keep all items tightly capped and stored appropriately.

FIGURE 39–18 Coplin jars.

Hematology Sample Collection

Sample quality will affect testing results and accuracy. Careful attention must be paid to collection procedures and the type of tubes and equipment used. Many blood samples are taken after a dog or cat has fasted for several hours. These samples are called **preprandial** samples. Blood samples taken from animals that have recently eaten may include increased amounts of fats or lipids that affect the blood work results and can lead to hemolysis of the sample, that is, blood cells will be damaged. Some tests require a **postprandial** sample, or blood collected after an animal has had a meal. Blood is typically collected using a needle and syringe, with the needle size varying depending on species. It is important to use the correct needle and syringe size, as a syringe that is too large can collapse a patient's vein, and a needle that is too small can cause hemolysis of the blood sample. When placing a blood sample into a tube, it is important to remove the needle before placing the blood in the collection tube. Red blood cells may rupture if forced back through the needle.

It is important to fill collection tubes with the proper amount of blood. Most tubes must be filled two-thirds to three-quarters of the way with blood. This ensures a proper blood to anticoagulant ration. The blood tube should be mixed thoroughly by inverting the tube gently 10 times over 10–20 seconds after transferring the blood. Whole blood samples are often placed into purple- or lavender-top collection tubes that have the anticoagulant EDTA (ethylenediaminetetraacetic acid) coating the tube to prevent the sample from clotting. Blood samples should be analyzed immediately; if they can't be analyzed quickly, they should be refrigerated for later use, preferably within 2 hours following collection. Blood preserved in EDTA remains fresh for several hours and can be kept overnight if stored in a refrigerator at 39 degrees F.

Recording Laboratory Results

All laboratory results should be recorded or placed within the patient's medical record. Some laboratory tests are kept in logs in the lab area for easy and quick reference. As soon as test results are available, they should be recorded in the file. The veterinarian should be notified to interpret the results for patient diagnosis. All results are given in units of numeric value or may be a positive/negative result and should include normal reference ranges in parentheses to the right of the results. Many reference values vary between various laboratories and textbooks. Some test results are provided as either positive or negative. Laboratory logbooks should be maintained in a neat and clean condition, ideally in a three-ring binder with clear plastic sleeves. Medical charts should have an area to place all laboratory reports according to date, with the newest being on top.

Blood Chemistry Procedures

Blood collection procedures are often performed to obtain samples for various tests both in-house and at reference labs (see Figure 39–19). These samples must be placed in **vacutainer tubes**, which are a commercial product with a vacuum. The needle and syringe are inserted into a rubber plunger to deposit the blood sample (see Figure 39–20). Some samples may require the needle to be removed from the syringe and the blood tube plunger to be removed for the sample to be added directly to the tube to prevent hemolysis. The blood sample is then housed within the tube and prepared for analysis by being placed on a blood rocker or placed in a centrifuge for serum sampling. A variety of tubes are available for specific testing requirements. The most frequently used tubes are red-top tubes, tiger strip–top tubes, and lavender-top tubes. Blood tubes should be handled gently and carefully to prevent damage to the sample. Some blood chemistry analyzers have special small blood collection tubes that require the lids to be removed and the sample placed directly into the tube. This prevents **hemolysis**, which is the rupture of red blood cells that causes a pink coloration to develop in the plasma or serum (see Figure 39–21). This discoloration indicates the cells have been lysed, and this often interferes with test results.

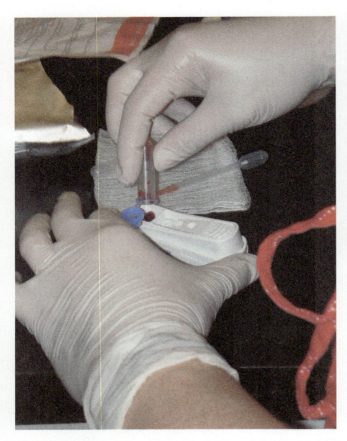

FIGURE 39–19 Blood samples are commonly used in diagnostic procedures.

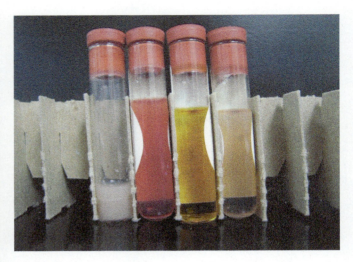

FIGURE 39–21 The second tube from the left shows the pinkish blood-tinged coloration that occurs when red blood cells are damaged upon improper handling of the specimen.

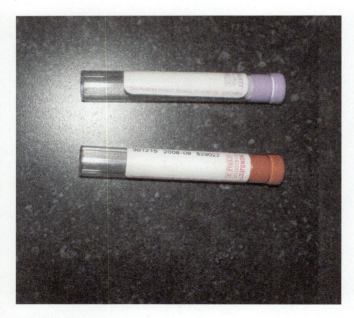

FIGURE 39–20 Vacutainers.

Each type of blood test requires a specific amount of blood, serum, or plasma to run the sample. The testing requirements should be noted before blood collection. A general rule is to collect at least 2.5 times the amount of whole blood needed for serum or plasma test samples. Blood samples that are not able to be evaluated immediately should be refrigerated to preserve freshness and sample accuracy. Blood samples that require centrifuging should be allowed to clot at least 30, but no more than 60, minutes prior to spinning. Each vacutainer tube should be labeled with the patient name, client name, testing requirements, and date. A laboratory test log should be kept to record any test performed. Any outside reference tests should be prepared, the accompanying paperwork completed, and the lab notified for pickup.

Most testing samples require **blood serum**, which is the liquid portion of the whole blood sample. Serum blood samples are usually placed in red-top or tiger stripe–top blood tubes. The samples are allowed to clot and are then spun within a centrifuge for 10 minutes to separate out the serum. **Blood plasma** samples may be needed and require the blood sample to be frozen and then centrifuged to separate out the plasma. Serum is most commonly used in blood chemistry profiles.

Whole blood samples are common for complete blood count (CBC) testing and are placed in lavender-top tubes (see Figure 39–22). The lavender-top tubes contain the **anticoagulant** chemical EDTA (ethylenediaminetetraacetic acid), which prevents the blood from clotting. When a whole blood sample is

- Examination of blood for blood-borne parasites
- Packed cell volume (PCV)
- Plasma or total protein (TP)
- Total WBC count
- Mean corpuscular volume (MCV)
- Hemoglobin concentration

Many of these tests are conducted in-house, simple to run, and performed by the veterinary assistant. Some facilities may need to send them to outside reference labs. In-house blood analyzer machines efficiently provide results and several can be run by hand (see Figure 39–23).

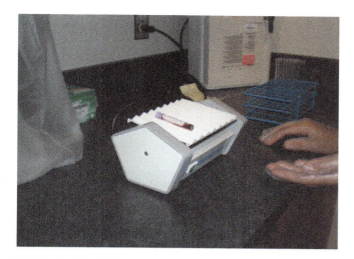

FIGURE 39–22 A whole blood sample placed in a blood rocker before testing.

placed in a tube, it should be inverted several times to allow the EDTA to react with the blood. Table 39–3 lists the various types of blood collection tubes and their uses.

The Complete Blood Count

The complete blood count or CBC is a group of tests that evaluate the different types of white blood cells (WBCs), red blood cells (RBCs), and platelets. The CBC is composed of the following:

- Examination of a blood smear
- Differential white blood cell count and morphology
- Platelet count
- Red blood cell morphology

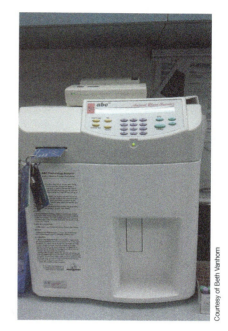

FIGURE 39–23 Blood chemistry analyzer.

TABLE 39–3
Blood Collection Tubes

Red-top tube	Sterile; no anticoagulant or additives. Contains gel separator. Collection of serum for chemical or serological and bacteriologic studies. May be used for any procedure requiring serum.
Tiger stripe–top tube	Contains no silicone, gel separators, anticoagulants, or additives of any kind. Can be used for collection of serum.
Lavender-top tube	Sterile. Contains EDTA as the anticoagulant. Primarily for collection of hematology studies, blood bank procedures, and certain chemistries with whole blood.
Green-top tube	Sterile. Contains lithium heparin as the anticoagulant. For collection of other miscellaneous studies. Electrolytes, glucose, blood urea nitrogen (BUN) can be performed more quickly than from a red top.
Light blue–top tube	Sterile. Contains sodium citrate solution as the anticoagulant. Tube calibrated to hold only 4.5 ml of blood. Primarily for collection of coagulation studies.
Gray-top tube	Sterile. Contains potassium oxalate and sodium fluoride as the anticoagulant. Used for the collection of glucose and lactate samples. Not suitable for enzymes or electrolytes.

The CBC requires unclotted whole blood collected in a lavender-top tube. The lavender-top tube should be filled as much as possible. An in-house sample is run on a CBC analysis machine, and the assistant should be proficient in understanding the results and sampling methods. The results should be printed out and placed in the patient's medical chart. The results should be provided to the veterinarian to aid in diagnosis and treatment.

Blood Smear

The **blood smear**, also known as a blood film, is used to look at the morphology of blood cells. The **morphology** includes the cell structure, shape, color, and appearance. The blood smear is made by the assistant using one slide, acting as a spreader, to pull a small drop of blood across a clean slide to make it into a thin film (see Figure 39–24). This is done by holding the spreader slide at a 45-degree angle. Two methods may be used to make a blood film: pushing or pulling the spreader slide. Each method requires consistent pressure applied during the entire spreading process. It takes patience and practice to develop a quality film. A suitable film is thin and tapers to a **feathered edge** before the far end of the slide (see Figure 39–25). The spreader slide ensures two straight edges parallel to the long edges of the slide. Immediately after the film is spread, it is dried quickly by fanning the slide in the air. The feathered edge is the area the veterinarian or veterinary technician reads to interpret the blood on the slide. It must be layered and forms a slight circular pattern at the end of the slide. Some recommendations to prevent errors in blood slides include the following:

- Use slides that are clean and free of breaks and chipped areas; no residue should be on the edges of any slide.
- Do not use large amounts of blood.
- Use a very small drop of blood at the center of one end of the slide.
- Place the spreader slide centrally into the drop of blood.
- Always keep the spreader slide at a 45-degree angle.
- Keep a constant and even pressure on the spreader slide.
- Do not push or pull the spreader slide slowly; this makes a poor feathered edge.
- Make a quick motion when using the spreader slide.

When the slide is completely dry, it should be stained with Wright's stain, which is a three-step staining method using commercial mixed solutions that allow blood cells to be easily viewed under a microscope. The solutions may also be in a commercial kit known as a Diff Quick set. These include three solutions labeled A, B, and C, which fix the stain with red and blue coloration (see Figure 39–26).

When preparing blood smear slides, it is best to make two slides: One is stained and viewed and the other remains unstained in case of further testing. Both slides should be labeled on the frosted end using a wax pencil or regular pencil with patient name, client name, and date (see Figure 39–27). Immersion oil should be placed next to the microscope and the slide placed on a paper towel near the microscope while awaiting evaluation. Prior to evaluation of the slide under the microscope, a small droplet of immersion oil should be placed on the feathered edge. The sample should then be viewed under the oil immersion lens.

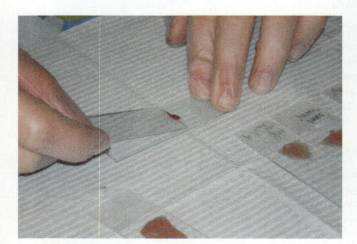

FIGURE 39–24 Creating a blood smear.

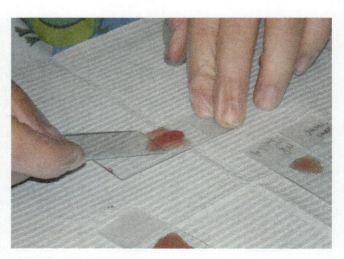

FIGURE 39–25 The edge of the blood smear should taper to a feathered edge.

CHAPTER 39 Laboratory Procedures

FIGURE 39-26 Stains used for blood smears.

The stain will allow the red blood cells to appear a pink, red, or salmon color. The nuclei of white blood cells will be stained with a dark blue to purple coloration. The platelets stain dark blue to violet in color. The cytoplasm and granules will vary in color from pink to purple. When evaluating the stained slide with the naked eye, it should have an overall even coloring of purple with pink hues. The stains should be maintained in separate Coplin jars marked with the appropriate order of use, such as A, B, and C. The jars should be filled at least ¾ full to allow slides to be covered by the stain. Distilled water should be used to rinse slides during staining.

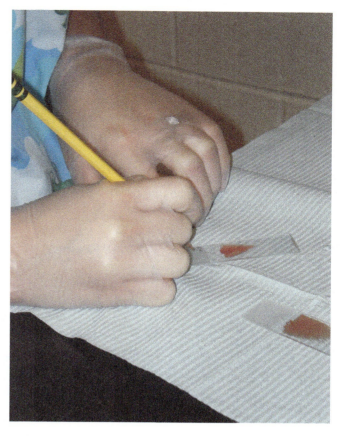

FIGURE 39-27 Properly label all blood smear slides.

COMPETENCY SKILL 121

Blood Smear Procedure

Objective:
To properly prepare a blood smear for microscopic evaluation

Preparation:
- Microscope
- Microscope slides
- Wax pencil or pencil
- Blood sample
- Syringe
- Exam gloves

Procedure:

1. Apply exam gloves.
2. Place a small drop of blood on the end of a slide using a clean syringe or from the lip of the blood top tube (see Figure 39–28A).
3. Apply a second slide as the spreader slide directly in front of the blood drop (see Figure 39–28B).
4. Place the spreader slide at a 45-degree angle.
5. Gently pull or push the spreader slide lengthwise along the slide (see Figure 39–28C).
6. Keep constant and even pressure on the spreader slide.
7. Use a quick motion, making an even feathered edge. Repeat this method until an even feathered edge is formed (see Figure 39–28D).
8. After the slide is completed, air dry by fanning or set aside to dry.
9. Make a second blood slide using the same steps.
10. Set slides aside to stain.

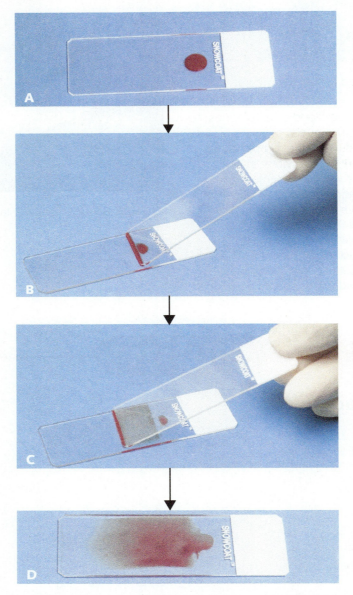

FIGURE 39–28 (A) Place a small drop of blood on the end of a slide. (B) Using a spreader slide, gently bring back the edge of the spreader slide into contact with the drop of blood. (C) Quickly push the spreader slide at a 45-degree angle forward to make the smear. (D) A properly prepared blood smear.

COMPETENCY SKILL 122

Blood Smear Stain Procedure

Objective:
To properly prepare a blood smear with the appropriate stain

Preparation:
- Blood smear slides (dried)
- Wright's stain or Diff Quick set
- Forceps
- Distilled water
- Immersion oil
- Tray
- Paper towels
- Exam glove
- Coplin jars (3)

Procedure:
1. Apply gloves.
2. Open each Coplin jar, marked A, B, and C.
3. Using forceps, the slide should be immersed completely inside jar A (the eosinate or red color stain) to apply the stain fixative. Time immersion according to kit directions.
4. Lift slide from jar and tilt and drain onto a paper towel, dabbing the end of the slide several times to remove excess stain.
5. Immerse slide in distilled water, lifting up and down to rinse.
6. Using forceps, the slide should be immersed inside jar B (the polychrome or blue color stain) (see Figure 39–29). Time immersion according to kit directions.
7. Lift slide from jar and tilt and drain onto a paper towel, dabbing the end of the slide several times to remove excess stain.
8. Immerse slide in distilled water, lifting up and down to rinse.
9. Using forceps, the slide should be immersed inside jar C (clear stain). Time immersion according to kit directions.
10. Lift slide from jar and tilt and drain onto a paper towel, dabbing the end of the slide several times to remove excess stain (see Figure 39–30).
11. Wipe off back of the slide using a paper towel.
12. Allow to air dry completely.
13. Place on a paper towel near microscope.
14. Place a bottle of immersion oil near the slide.

FIGURE 39–29 Staining technique for a blood smear.

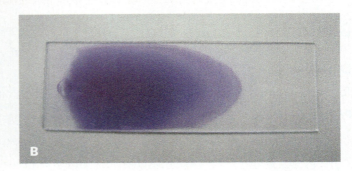

FIGURE 39–30 (A) Allow slide to drain onto a paper towel. (B) A properly stained blood smear.

Packed Cell Volume

The **packed cell volume**, or **PCV**, is also referred to as a **hematocrit**. This is a measurement of the percentage of red blood cells in whole or unclotted blood. The red blood cells, called **erythrocytes**, carry oxygen in the bloodstream. The PCV test is simple, is rapid, and requires a very small amount of blood. The equipment includes a centrifuge with a microhematocrit rotor, hematocrit tubes, and clay. The microhematocrit centrifuge has a lid that screws onto the rotor to prevent tubes from emerging. The rotor must be properly balanced. Plain tubes may be used with whole blood that contains an anticoagulant, such as EDTA, and untreated whole blood directly from a syringe should be in a heparinized hematocrit tube. Hematocrit tubes are sometimes called capillary tubes.

Blood is drawn into the hematocrit tube by **capillary action**, meaning the blood drains into the tube through gravity. One end of the tube is placed into the blood sample and is allowed to rise through the tube until it is three-quarters to completely full. Some tubes have a black fill line indicating how much blood should be placed into the tube, although achieving the mark is not necessary for the test to be accurate. Place a gloved finger over the end of the tube and wipe the tube clean using lens paper. Quickly and firmly place the end of the tube into a clay sealant. This creates a seal at the end of the tube to prevent blood from leaking out. The tube is then placed into the centrifuge rotor and balanced with another capillary tube filled with water. Multiple PCV tubes can be run in one rotor. The tubes should be placed opposite each other to balance the rotor, and the clay end of the tubes should be facing outward. This placement is important because if not properly placed, the sample will be lost as it spins. Then apply and tighten the metal lid and set the centrifuge to the proper setting and time.

When the centrifuge has completed spinning, the rotor is opened, and the tube is ready to read. The tube will have a dark red layer composed of red blood cells near the clay-sealed end. Just above this line is a yellow to clear area of plasma. Between both layers is a tiny area of visible white, known as the **buffy coat**, composed of white blood cells and platelets (see Figure 39–31).

To determine the PCV level or the percent of red blood cells in the patient's sample, a microhematocrit scale is used (see Figure 39–32). The scale can be a chartlike paper or it may be a plastic form that fits over the centrifuge. Using

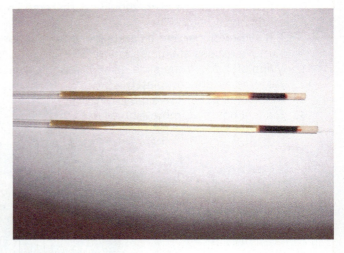

FIGURE 39–31 A blood sample spun to perform a packed cell volume. Note the yellow color of the serum showing icterus or jaundice.

CHAPTER 39 Laboratory Procedures

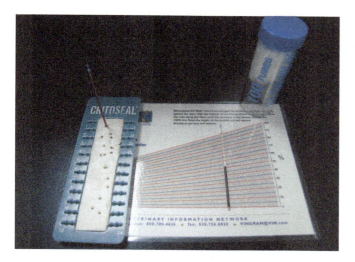

FIGURE 39–32 Hematocrit scale.

either type of scale, align the sealant at zero and the top of the plasma at 100. The line running through the interface of red blood cells and the buffy coat layer is the patient's percentage of red blood cells. This test helps the veterinarian determine if anemia is present in an animal. The percentage should be recorded in the medical chart.

COMPETENCY SKILL 123

PCV Procedure (Hematocrit)

Objective:

To properly prepare hematocrit samples for PCV testing

Preparation:

- Capillary tube
- Hematocrit tube or microhematocrit tube
- Clay sealant
- Lens paper
- Exam gloves
- Centrifuge
- Microhematocrit scale
- Blood sample

Procedure:

1. Apply gloves.
2. Fill one capillary tube with the blood sample.
3. Apply a finger to the end of the capillary tube that was in the blood tube.
4. Wipe the tube clean with lens paper.
5. Quickly apply the end of the tube firmly into the clay sealant.
6. Fill one capillary tube with water and apply end to clay sealant.
7. Apply the microhematocrit rotor to the centrifuge.
8. Place tubes on opposite sides within the centrifuge rotor. Both tubes should have the clay sealant facing outward.
9. Apply lid to the rotor and secure tightly.
10. Place setting on hematocrit and time according to manufacturer's instructions.
11. After the centrifuge has stopped spinning, remove the tube with the blood sample.
12. Place the tube on the microhematocrit scale. Align the top of the sealant at zero and the top of the plasma at 100.
13. Read the line crossing the red blood cells and the buffy coat.
14. Record the percentage of the PCV in the medical chart.

Plasma Protein

Plasma protein, also called **total protein (TP)**, is the ratio of protein within the blood and helps veterinarians determine a patient's hydration status and occurrence of inflammation. TP is determined using a refractometer or total solids meter. The TP sample is evaluated after the PCV test has been conducted. The microhematocrit tube holding the spun-down blood is used to evaluate TP by using the plasma located above the buffy coat layer in the tube. The clear to slightly yellow coloration at the top area of the tube holds the plasma. The tube is broken just above the buffy coat to allow a drop of plasma to be applied to the refractometer prism. The cover plate is lowered, and the refractometer is held toward a light source to read the scale for serum or plasma protein, often labeled TP. The scale is read where the dark and light colors intersect on the scale. Record the results in the medical chart in g/dL (grams per deciliter). This is a simple and quick test that the veterinary assistant should be able to proficiently perform.

COMPETENCY SKILL 124

Total Protein (TP) Procedure

Objective:

To properly prepare a sample for total protein evaluation

Preparation:

- Capillary or microhematocrit tube (spun)
- Refractometer
- Lens paper
- Exam gloves

Procedure:

1. Apply exam gloves.
2. Open the refractometer lid.
3. Break the capillary tube containing the patient's blood just above the buffy coat in the blood plasma (clear to light yellow fluid).
4. Apply one drop of fluid to the refractometer.
5. Close the lid to the refractometer.
6. Direct the refractometer toward a light source.
7. Focus the eyepiece and read where the light and dark areas interface on the TP scale.
8. Record the results in g/dL in the medical record.
9. Place tube in medical waste container.
10. Clean refractometer using alcohol and lens paper.

Blood Chemistry Samples and Electrolytes

Blood chemistry samples and electrolytes are often used to diagnose diseases and conditions of ill patients. Some of these tests may be run in-house and others may be sent to reference labs. In-house blood analyzers allow access to rapid results (see Figure 39–33). Many preanesthetic blood panels are performed on patients prior to surgery, as well as yearly health profiles on senior or geriatric pets. Many of these blood chemistries are useful in determining developing conditions before a patient becomes clinically ill. Approximately 10 percent of veterinary patients have a diagnosed condition determined during routine wellness or preanesthetic blood work.

FIGURE 39–33 Blood chemistry analyzer.

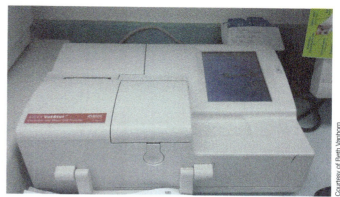

FIGURE 39–34 CBC analyzer.

Most in-house veterinary labs consist of analyzers used for serum blood chemistries, electrolytes, and hematology evaluation. Blood serum is obtained from a blood sample spun down in the centrifuge to allow serum to be analyzed. **Electrolytes** are minerals that maintain the balance of elements in the body, specifically potassium, sodium, and chloride. Electrolytes can be measured using serum or heparinized plasma. **Hematology** refers to the study of the blood, specifically with a CBC. Many analyzer machines may run a series of tests called **panels** or individual tests. These tests provide chemical information on specific organ functions that help evaluate a patient's health. It is important to note that blood machines are constantly upgrading in systems and software, so it is expected that current equipment may require frequent upgrades to further allow testing capabilities in-house. Testing becomes faster and more dependable, testing errors are reduced, and often the machines become easier to use. Veterinary assistants should be trained in proficiently operating, cleaning, and maintaining blood machine equipment. There are many different veterinary blood machine brands and there is much information to be learned about them (see Figure 39–34). The assistant should be able to understand and determine the following general information on blood machines:

- How to obtain and print out patient results
- How to determine reference value ranges for the machine's results
- How to properly calibrate machines and keep records of the calibration process
- How to change empty and expired reagent chemicals used in the machine
- How to properly run the testing panels
- How to troubleshoot problems that may occur with the machine and who to contact with manufacturer and technical problems

It is important that assistants know how to operate and maintain the blood machines. The machine must be kept clean, and recommended directions for maintenance must be followed. Valid results depend on the skill of the person performing the testing. Equipment must be turned on properly, allowed to warm up, calibrated, cleaned on a regular basis, and serviced by the manufacturer as necessary. The directions and testing instructions should be kept in a binder near the blood machines to refer to as necessary.

The Fecal Sample

Fecal samples are used to diagnose internal parasites and see the presence of blood within the stool sample. A fecal sample is assessed using a small amount of an animal's stool or bowel movement materials. Changes in fecal samples aid in determining animal health issues and signs of disease.

The fecal sample begins with the collection, which may be done at the facility or brought in by the animal's owner. The fecal sample should be placed in an airtight container or bag labeled with client name, patient name, sample test, and date. Samples that are collected fresh and not immediately tested should be kept refrigerated. At least 1–2 grams of feces are needed to complete a fecal analysis. Samples may be collected directly from the rectum using a fecal loop. This method only uses a small amount of feces and may not be desired based on the small quantity evaluated and may not be sufficient to evaluate parasites.

The first part of the fecal analysis begins with the **gross examination** or visible observation of the feces. The observations are recorded in the patient's medical record and possibly in a laboratory fecal log. The gross fecal observation should include the following:

- Color
- Consistency
- Odor
- Presence of blood, ranging from dark black to bright red
- Presence of observable parasites
- Presence of mucous
- Presence of foreign materials or debris

The next step is performing a **fecal direct smear**, which is prepared by placing a small amount of the fecal sample onto a microscope slide. This is usually done using an applicator stick to transfer a small amount of the sample directly onto a glass slide by rolling the material lengthwise along the slide (see Figure 39–35). Samples that are dry may require 1–2 drops of saline solution to mix with the feces before applying to the slide. A thin layer of feces should be placed across the slide surface. The veterinarian or veterinary technician should then read the fecal smear using a microscope. This procedure is used to detect the presence of eggs and larvae and is useful in visualizing the trophozoite stages of giardia and other protozoans.

The **fecal floatation** technique is done to sample the fecal material to determine if any parasite eggs are within the sample. This is done based on the concept that fecal ova (eggs) are lighter in weight than the solution they are submerged in. So any ova float to the top of the solution and become attached to a coverslip or slide applied over the sample. The sample should be undisturbed for 10–20 minutes to allow eggs to surface. The slide or coverslip is then viewed under the microscope for evaluation by a veterinarian or veterinary technician.

Commonly used fecal solutions include sugar, sodium nitrate, sodium chloride, and zinc sulfate. Selection is often based on cost, efficiency, availability,

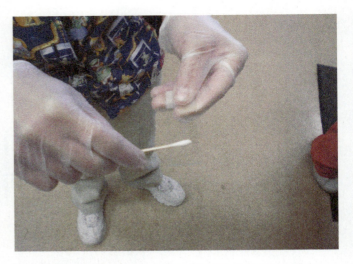

FIGURE 39–35 Preparation of a fecal sample slide.

ease of use, and common parasites encountered in the area. Some manufacturers have packaged kits with prepared solution that are convenient but more costly. The specific gravity of the egg, life cycle stage of the parasite, and type of solution will affect which eggs will able to float and be visible. Most eggs have a specific gravity (SG) of 1.10–1.20 g/ml. The SG refers to the weight of an object compared to the weight of an equal volume of distilled water. Fecal solutions are formulated with a SG higher than that of common parasite eggs. Most solutions vary, with SG ranging from 1.18 to 1.40 g/ml.

Fecal samples require fecal floatation solution, a bottle of Lugol's solution, and new methylene blue stain. Solutions commonly used to prepare a fecal sample are listed in Table 39–4. Stains can be applied to the slide to help view the eggs. Slides and coverslips should be available, along with applicator sticks, floatation system kits, centrifuge tubes, and a laboratory space designated for fecal samples (see Figure 39–36). A small tray may be

TABLE 39–4	
Fecal Sample Solutions	
Lugol's solution	Iodine tincture solution applied to fecal smear to determine parasites, commonly oocysts and single-celled organisms
New methylene blue	Blue stain solution applied to fecal smear to note bacteria and single-celled organisms
Saline	Saltwater solution used in fecal smears to determine presence of ova
Sodium nitrate	Most common solution; used in commercial fecal floatation kits; relatively expensive
Zinc sulfate	Used to diagnose giardia cysts; used with centrifuging fecal samples
Sucrose	Granulated sugar solution used when centrifuging fecal samples; more expensive and sticky
Magnesium sulfate	Epsom salts; inexpensive but crystallizes quickly and easily, causing ova distortion
Sodium chloride	Table salt solution; causes cloudy appearance of sample; inexpensive

used to decrease contamination and spread of disease within the lab. Many parasites and other pathogens within a fecal sample are zoonotic, and exam gloves should be worn at all times. The area should be cleaned and disinfected after every fecal sample analysis is conducted.

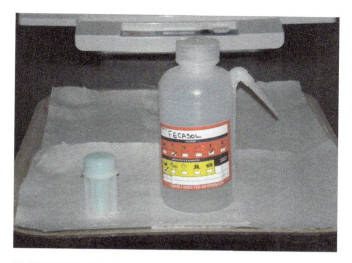

FIGURE 39–36 Preparation needed for fecal floatation.

 COMPETENCY SKILL 125

Fecal Analysis Procedure

Objective:

To properly view and prepare a fecal sample

Preparation:

- Exam gloves
- Fecal sample
- Tray
- Laboratory work area
- Fecal kit/container/centrifuge tube
- Fecal floatation solution
- Microscope
- Microscope slide
- Coverslip
- Applicator sticks
- Wax pencil
- Lugol's solution
- New methylene blue
- Saline
- Centrifuge
- Timer

Procedure:

1. Apply exam gloves.
2. Conduct gross fecal examination. Record results.
3. Prepare fresh fecal smear.
4. Label slide with wax pencil or pencil (patient name, client name, date).
5. Place small amount of feces in center of slide using applicator stick.

6. Add one drop of saline solution.
7. Mix well and spread thin layer lengthwise over slide.
8. Place coverslip over smear.
9. Place at microscope for evaluation.
10. Prepare another fecal smear as above.
11. Place 1 drop Lugol's solution or New Methylene blue on slide.
12. Place coverslip over slide.
13. Place at microscope for evaluation.
14. Set up fecal floatation (using commercial fecal solution kit such as Ovassay or Fecalyzer).
15. Label the container with patient name, client name, and date.
16. Place small amount of fecal sample in container.
17. Fill the container halfway with the fecal floatation solution.
18. Mix the sample and solution well.
19. Insert any filter and lock container in place.
20. Fill the vial to the top and allow solution to dome upward without overfilling (see Figure 39–37).
21. Apply a coverslip to the top of the solution to form a vacuum (see Figure 39–38).
22. Place timer on 15 minutes to allow ova to float (see Figure 39–39).
23. When timer sounds, remove coverslip and apply to microscope slide.
24. Place slide at microscope for evaluation.
25. Prepare a fecal floatation with a centrifuge if not using a commercial kit.
26. Mix fecal sample in centrifuge tube using 1 teaspoon of feces and small amount of water to soften feces. Mix using applicator stick.
27. Add fecal floatation solution to the top of tube, allowing an upward dome to form.
28. Place a coverslip on the tube to form a vacuum.
29. Place in centrifuge and spin 3–5 minutes on slow speed.
30. Remove coverslip and apply to a microscope slide.
31. Place slide at microscope for evaluation.
32. Clean work area well.
33. Dispose of feces in biohazard waste container.

FIGURE 39–37 Preparation of the fecal floatation.

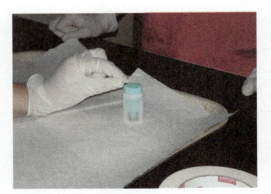

FIGURE 39–38 Placing a slide coverslip on the top of the fecal floatation preparation.

FIGURE 39-39 Time the sample.

Centrifugal Fecal Floatation

This fecal procedure is similar to fecal floatation, but once the sample and solution are mixed, they are then strained to remove excess debris. The straining can be done using a tea strainer or cheesecloth over a centrifuge tube. Once strained, the sample is placed in a centrifuge with a coverslip over the tube and centrifuged at 1500 rpm for 5 minutes. The centrifugal force holds the coverslip in position during spinning as long as the tube is counterbalanced. A drop of liquid is then placed on a slide to be examined under the microscope. This procedure is more sensitive than fecal floatation and recovers more eggs from a sample in less time.

Fecal Sediment

A sedimentation test is done when suspected eggs are too large to be detected via simple floatation. The fecal sample is mixed in a small amount of water and strained into a centrifuge tube. The sample is centrifuged at 1500 rpm for 5 minutes or may sit undisturbed for 20–30 minutes. The supernatant liquid is poured off and a sample of the sediment is placed on a slide and examined under the microscope. If a lot of fecal debris is seen, a drop of liquid detergent can be applied to remove excess debris from the sample. This procedure is most often used when fluke eggs are suspected. Most fluke eggs are too heavy to float.

The Urine Sample

Assessing the urine is essential to an animal's health, especially when urinary problems exist. The veterinary assistant's role in the urine sample involves collecting voided samples, restraining during other urine collection methods, performing gross examination of urine samples, using clinical strip tests, and preparing the microscopic urine examination. The complete urine exam is called **urinalysis.** This is a relatively simple, cost-effective, and rapid procedure to perform in the veterinary facility. The urinalysis should be completed immediately after sample collection. Samples left at room temperature for an hour will begin to change and disintegrate. If the urinalysis cannot be completed immediately, the sample should be refrigerated and then brought to room temperature before testing.

Urine Collection

There are several methods for collecting urine, depending on the urine test and condition of the animal. Most samples that are collected are **voided** samples, meaning they are collecting as the animal is urinating. This is the only type of urine collection method the veterinary assistant may perform. The voided sample should be collected **midstream** shortly after urination begins and just prior to the end of the process (see Figure 39–40). The container must be sterile when collecting a urine sample. Dog urine can be collected using a sterile plastic collection cup, and in difficult animals or those low to the ground, a wire attachment with a loop can be made to hold the container and give the assistant an extension without interfering with the patient's normal urination process. A similar method can be used when collecting urine samples from large animals. Cats should be provided a pelleted nonabsorbable litter that doesn't soak up the urine when the cat urinates in the litter pan. Some cats may need an empty litter pan. The urine sample is then transferred into a sterile container. It is best

SECTION IV Clinical Procedures

FIGURE 39-40 Midstream urine collection.

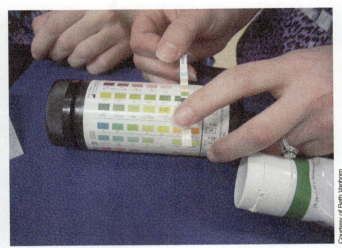

FIGURE 39-41 Reading the urine chemical strip.

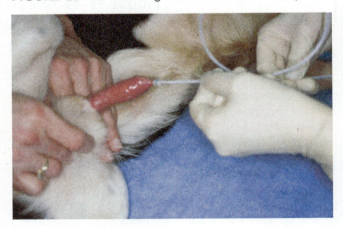

FIGURE 39-42 Catheterization of the male.

to not use urine samples collected from the floor or a cage area if at all possible, as these samples will be contaminated and give inaccurate results.

Other urine collection methods include palpating the bladder and expressing the urine by applying external pressure, which is a common method used with cats and small dogs. The veterinarian or veterinary technician expresses the bladder, and the assistant should hold the container and catch the midstream sample. Some animals may need to be **catheterized** for urine collection. This is inserting a long, thin sterile rubber or plastic tube through the external urinary opening and into the bladder. The assistant restrains the patient while the catheter is placed by the veterinarian or technician. A syringe is then attached to the end of the catheter and urine is aspirated. In a female animal, the catheter is placed directly into the external vulva and through the urethra to the bladder, and in a male animal the catheter is passed into the penis and then through the urethra to the bladder (see Figure 39-41). A male animal is generally much easier to catheterize than a female animal. Aseptic technique is required for catheter placement and urine collection. The animal may be restrained in standing recumbency, lateral recumbency, or dorsal recumbency depending on the veterinarian's preference.

A urine sample may also be collected via **cystocentesis**, which is surgical puncture of the bladder and using a needle to collect the sample (see Figure 39-42). This is typically used to collect a sterile sample that will be checked for bacteria or other microorganisms, such as with a culture and sensitivity **(C/S)** test. A **culture and sensitivity** will determine if a bacterial issue is occurring, what type of bacteria is causing the health condition, and what antibiotic should be used to treat the problem. A cystocentesis avoids bacterial contamination by avoiding having the urine come in contact with the urethra. This technique requires advanced skills on the part of the veterinarian or technician, as a needle attached to a syringe is passed through the abdominal wall directly into the bladder. Urine is aspirated into the syringe. The assistant restrains the patient, usually in lateral or dorsal recumbency.

Gross Examination of Urine

The urine exam is divided into several parts: physical appearance, chemical properties, and microscopic examination. The veterinary assistant is responsible for completing the physical examination and clinical test strip evaluation and preparing the sample for microscopic examination. The sample should be completely evaluated within 30 minutes of collection.

Assessing the physical appearance of the urine sample includes noting the color, clarity and consistency, odor, and presence or absence of foam. Most veterinary facilities have a urinalysis laboratory sheet that can easily be completed by noting the findings. Normal urine color in most animal species is various shades of yellow. The color of urine can range from colorless to red or brown. Certain medications and supplements can alter the color of urine, as can certain types of diseases. Most animal species' urine is clear, although some may normally be cloudy. Clarity can range from clear to cloudy to **flocculent**, which contains large amounts of particles called sediment suspended in the urine. The odor may not be diagnostic and may range in varying degrees by animal species. Certain large animals that may have more bacteria present in the urine tend to have the odor of ammonia noted in their urine. Another odor that may be detected in animal urine is due to the presence of ketones, which produce a sweet, fruitlike smell. When urine is shaken, a small amount of white foam should be produced. High levels of protein may create a large volume of foam, and bile pigments noted in the urine will produce a foam that is greenish-yellow in color.

COMPETENCY SKILL 126

Collecting a Voided Urine Sample

Objective:
To properly obtain a midstream urine sample from a patient

Preparation:
- Exam gloves
- Sterile container
- Label
- Pen
- Wire cup holder and extension

Procedure:
1. Apply gloves.
2. Leash or control the patient and take the animal outside.
3. Walk around the area, monitoring the patient for signs of urine elimination.
4. When the patient assumes the urination posture, prepare to slip the sterile container under the patient into midstream urine.
5. Withdraw the container prior to the patient ending urination.
6. Allow the patient to finish urinating.
7. Return the patient to the cage or enclosure.
8. Clean the outside of the sterile container and label the container with patient name, client name, and date.
9. Place the container in the refrigerator if it is not to be tested immediately.

Urine Specific Gravity

Urine **specific gravity (SG)**, or the weight of the liquid, is determined using a refractometer. The SG indicates the concentration of the urine and helps with evaluation of kidney function. Changes in the kidneys' ability to concentrate urine indicate renal tube damage. A drop of the urine sample is placed on the prism of the refractometer. The specific gravity should be noted on the urine SG scale on the refractometer. The SG number is recorded as a decimal, such as 1.025. The SG should be recorded on the urinalysis report.

Chemical Test Strips

The urine chemical test strips are used to evaluate the chemical properties of a urine sample using a **reagent strip**, also called a **chem strip** or **dip stick** (see Figure 39–43). Each test is on a long, thin plastic strip separated by individual square pads containing a chemically treated paper. There may be from one to 10 tests on the strip, depending on the manufacturer and test type. The bottle containing the testing strips provides a label for help with interpreting the color changes and results. Results of each test should

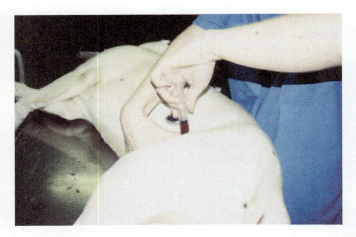

FIGURE 39-43 Cystocentesis of a dog.

be noted on the urinalysis sheet. The test strip is aligned with the label to determine what the color change means (see Figure 39-44). The manufacturer's instructions should be read and followed. The urine strip should be saturated with urine, which is usually accomplished by applying a drop of urine to each square pad with a plastic applicator or needle and syringe. Each test requires a specific amount of time until it should be read. All values should then be recorded on the urinalysis sheet. Some clinical urine test strip results are recorded in numbers and others in values such as +1, +2 pr +3, or negative (−).

Microscopic Exam

The purpose of the microscopic examination of urine is to determine abnormally formed elements in the sample, such as crystals, cells, or bacteria. Begin preparing urine for microscopic examination by placing it in a conical or tapered-end centrifuge tube, similar to a test tube. Typically, this requires 5 to 10 ml of fresh urine. The same amount of water should be placed in a second centrifuge tube, and both tubes should be placed on opposite sides in the centrifuge. The correct rotor size should be selected for the tubes. It is important to balance each tube placed in the rotor. The centrifuge should be placed on the urine spin cycle, which is typically between 1500 and 2000 rpm for 5 minutes. When the centrifuge has completed spinning, the test tube holding the urine should be removed and decanted over a sink area. The **decant** process involves pouring the urine out of the tube and allowing only the sediment on the bottom to remain. A drop of urine should cling to the bottom of the tube within the sediment. A drop of urine sediment stain, such

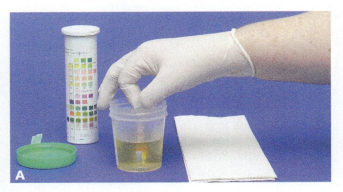
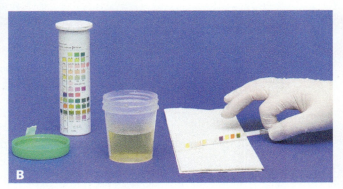
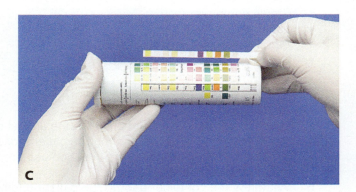

FIGURE 39-44 Urinalysis using chemical reagent strip.

as new methylene blue or a commercial urine stain, should be added to the sample, and a gentle tap on the test tube should help to mix the sample. The most common commercial stain is Sedi-stain. Some veterinarians may ask for two slides, one that is unstained and one stained. A clean slide should be labeled with patient name, client name, and date. A drop of urine should be placed on the center of the slide. A coverslip is placed over the drop. Place the slide next to the microscope for evaluation by a veterinarian or veterinary technician. Figure 39-45 shows crystals, cells, and casts as seen in microscopic evaluation of a urine sample.

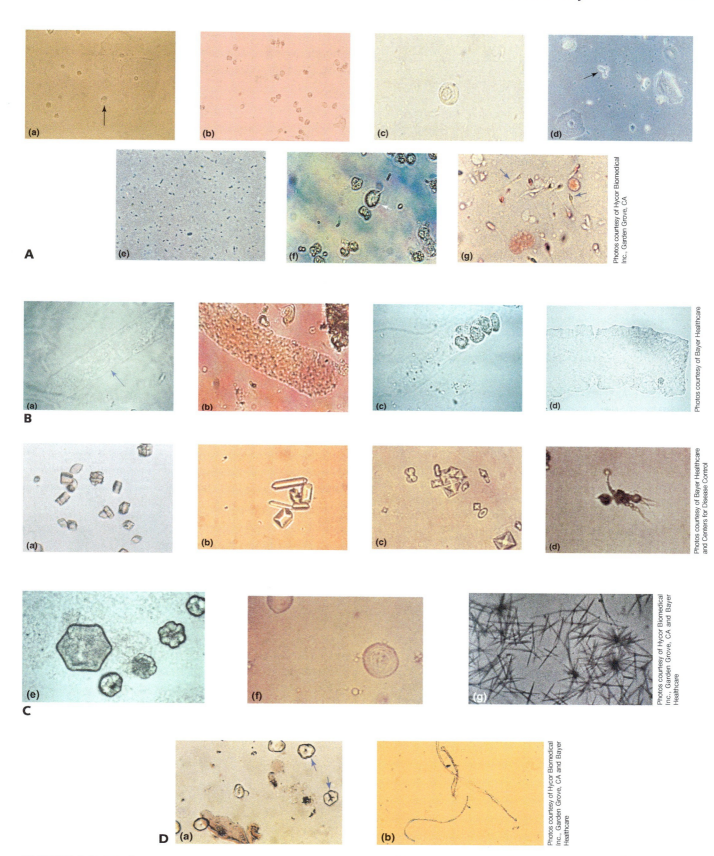

FIGURE 39–45 Crystals, cells, and casts found in urine. (A) Examples of cells found in urine, (B) Examples of casts found in urine, (C) Examples of crystals found in urine, (D) Examples of other substances found in urine.

Elements in Urinary Sediment

Many types of cells and elements can be seen in urinary sediment. The presence of red blood cells and white blood cells may indicate bleeding within the urogenital tract or a urinary tract infection. Squamous epithelial cells are often seen, as they often shed from the bladder wall, ureters, or urethra during normal urine expression. Transitional epithelial cells may be seen in clumps, and larger numbers of these cells may be an indication of carcinoma or inflammation and require close examination of the morphology to look for signs of malignancy. **Crystals** may be seen in sediment, and the type of crystal is dependent on the amount of substance in the urine and solubility of the crystal. The pH and SG of the urine affect the formation of crystals. Acidic urine forms calcium oxalate, cystine, uric acid, amorphous urate, calcium sulfate, and sodium urate crystals. Alkaline urine forms amorphous phosphate and struvite crystals. The shape of the crystals will help identify them. More common crystals, such as struvite, have a coffin lid–shaped appearance. Calcium oxalate crystals are square-shaped with an X in the center. The formation of crystals may indicate urinary tract inflammation or the formation of bladder stones. **Casts** are cylinder-shaped structures formed from protein secreted by the renal tubules. The cast takes on the shape of the tubule. Any component may form within the cast, such as bacteria or other cells. Several common types of casts include waxy, cellular, and fatty. The formation of casts may indicate renal tubule damage.

Gram Stain

Gram stain tests are evaluated to determine the presence of bacteria and—if present—the type of bacteria in a sample. **Gram-positive** and **Gram-negative** bacteria may be identified, and the type of bacteria is considered when determining the diagnosis of the animal. Gram-positive bacteria stain purple, and Gram-negative bacteria stain red. The bacteria may be in the shape of **rods** (oblong) or **cocci** (round). **Budding yeast** cells may also be noted, appearing as a budding flower. The sample tested may be feces, urine, or other body fluid such as pus or discharge. The sample must be collected aseptically so as to not introduce any organisms that are not present in the sample. If contaminants are introduced into the sample, the results will be invalid, and diagnosis will be difficult. The sample should be collected using a sterile cotton swab. The cotton swab tip is placed into a sample of feces collected from the patient and then transferred to a microscope slide, where a thin sample is spread over the center of the slide. The slide must be air dried or heat fixed prior to staining. When dry, a Gram stain kit is used to stain the slide. The kit consists of three stains—the crystal violet or purple stain, Lugol's iodine or orange stain, the safranin stain or red stain—and the decolorizer as a rinse. The stains should be placed in Coplin jars for easy staining. The veterinary assistant should be able to collect a sterile sample, perform a Gram stain, and alert the veterinarian or technician when the slide is ready to examine.

Culture and Sensitivity

A culture and sensitivity (C/S) is a test done through a sterile process to determine if any bacterial or fungal growth occurs in a sample. This requires a liquid or solid sample using a sterile cotton swab that is then placed within an applicator that may or may not contain a **medium** or gel where organism growth may occur. The cotton swab should not touch any site other than the sample area and the medium. A cap should be placed over the swab, which may entirely be submerged in the gel, or the stick can be broken off. The container may be a culturette system, agar plate, or dermatophyte test medium (DTM). The cotton swab can be swept or rolled over the Agar plates. This must be done gently so that the agar gel is not damaged. Agar plates and DTM tests should be placed in a safe place for 4–6 weeks to monitor growth and color change; the presence of bacteria or fungus should be noted. So that ventilation may occur, the top of the bottle should not be secured tightly.

COMPETENCY SKILL 127

Urine Sample Procedure

Objective:

To properly prepare a urine sample for analysis

Preparation:

- Sterile urine container
- Exam gloves
- Pen
- Conical test tubes
- Test tube rack
- Centrifuge
- Clinical urine test strips
- Pipette
- Microscope slides
- Coverslips
- Urine stain
- Refractometer
- Lens paper
- Alcohol
- Paper towels
- Tray and work area

Procedure:

1. Apply gloves.
2. Conduct physical appearance examination.
3. Note color, transparency, consistency, odor, and foam content.
4. Perform the specific gravity test by using the refractometer. Record value.
5. Clean refractometer with alcohol and lens paper.
6. Locate a clinical urine test strip kit.
7. Apply a drop of urine onto each pad of the test strip using a pipette.
8. Time each test according to the direction panel.
9. Read each test pad as indicated. Align each test strip with the correct label.
10. Record results based on the urine color chart on the bottle label.
11. Prepare urine sediment.
12. Fill one conical test tube with urine.
13. Fill another conical test tube with the same amount of water.
14. Place tubes opposite each other in a standard centrifuge.
15. Set centrifuge for urine spin and recommended speed and time.
16. After centrifuge stops, decant urine.
17. Place one drop of nonstained urine on microscope slide.
18. Place coverslip over drop.
19. Mix a drop of urine stain with sediment.
20. Tap bottom of tube to mix.
21. Place a drop of urine sediment onto a microscope slide (see Figure 39–46).
22. Place coverslip over drop.
23. Place slides near microscope for evaluation.
24. Clean work area.

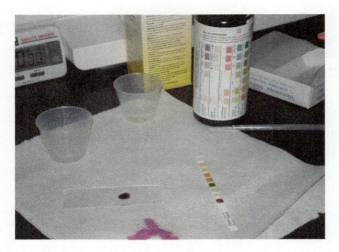

FIGURE 39–46 Preparation of a urine sample slide.

 COMPETENCY SKILL 128

Gram Stain Procedure

Objective:
To properly prepare a sample as a Gram stain

Preparation:
- Exam gloves
- Sterile cotton swab
- Microscope slide
- Open flame or match for heat source
- Gram stain kit
- Staining rack
- Paper towels
- Immersion oil

Procedure:
1. Apply gloves.
2. Collect sample using a sterile cotton swab.
3. Apply thin layer of sample to clean microscope slide by rolling sterile swab over the center of the slide.
4. Place over open flame to dry.
5. Apply crystal violet stain for 30 seconds.
6. Gently rinse slide with tap water.
7. Apply Lugol's iodine solution for 30 seconds.
8. Gently rinse slide with tap water.
9. Wash slide with decolorizer for 10 seconds or until purple color is no longer present.
10. Apply safranin stain for 30 seconds.
11. Gently rinse with tap water.
12. Air dry or blot dry using a paper towel.
13. Place slide near microscope.
14. Place immersion oil near slide.

COMPETENCY SKILL 129

Bacterial or Fungal Culture Procedure

Objective:

To properly prepare a sample for a culture and sensitivity test.

Preparation:

- Agar plate
- DTM bottle or culturette tube
- Exam gloves
- Sterile cotton swab

Procedure:

1. Apply gloves.
2. Apply sterile cotton swab to sample site.
3. Place swab in transport media (culturette tube, agar plate, or DTM bottle).
4. Place patient name, client name, and date on sample.
5. Send culturette tube to lab for analysis.
6. Close DTM bottle lid so that it fits loosely to allow ventilation.
7. Place DTM bottle in well-lit area and monitor for fungal growth or color change.
8. Place agar plate in incubator upside down.
9. Check daily for growth.

Cytology and Histology

Cytology is the study of cells or evaluation of inflammation and neoplasia. Knowing the type of cells and number of cells is helpful in determining diagnostic information. Samples for cytology are often collected quickly and are easily prepared within the facility. These samples are often sent to outside labs for pathologists to review and determine a diagnosis. Cytology samples may be from impression smears from an active lesion on an animal's body or from tissue removed from an animal. Scrapings may be prepared from external lesions or tissue. Swabs may be submitted with samples from different body locations. A fine-needle aspirate may be used to provide a sample directly from lesions or internal body structures.

Histology samples are often collected through a biopsy site, either excisional or incisional. An excisional tissue sample is taken when only a small amount of the tissue sample is removed from the patient, such as a **punch biopsy** site, where a small circular sample of tissue is removed for testing (see Figure 39–47). An incisional sample is when the

FIGURE 39–47 Punch biopsy of the skin.

entire area is removed and submitted for testing. Most tissue samples are placed in a formalin jar and sent to a pathologist for testing. The amount of solution within the formalin jar is typically 10–20 times the volume of the sample. Larger samples will require sectioning so that the solution can penetrate and preserve the entire tissue.

Necropsy Procedure

The **necropsy** procedure is done by examining the body of a deceased animal to determine the cause of death. It is similar to a human autopsy. A necropsy procedure is performed shortly after death, since prolonging the time will result in a breakdown of tissues and inaccurate results.

The body is examined, and all organs are viewed to note any abnormalities. Tissue samples are collected and sent to a veterinary pathologist for microscopic exam. Packaging and shipping these samples is often a duty of the veterinary assistant. This requires the assistant to know what reference lab is used and how the sample is to be submitted. Many labs provide containers and paperwork for submitting samples. Certain tests require special instructions and should be read to determine handling of the sample. The submission form should provide client name, patient name, and other identification information as well as a patient history and the type of tissue samples included. Most tissue samples are placed in **formalin** to preserve the tissue. The necropsy procedure is often performed in the treatment area away from general facility activities. It is best to place the animal on a rack over a sink for ease of area cleanup. The tools and supplies the veterinarian will need include the following:

- Scalpel blade and handle
- Forceps
- Sharp scissors
- Ronguers (bone cutters)
- Large, sharp knife
- Needle holders
- Suture needle
- Suture material
- Slides
- Sample containers

The veterinarian should wear an apron, mask, goggles, and heavy rubber gloves to prevent contamination. The necropsy procedure is thorough and detailed. Key observations are made much like in a physical exam. In the case of rabies suspects, the head must be sent for rabies testing immediately upon death of the animal, including euthanasia. The packaging and shipping should follow the state veterinary lab requirements.

On completing a necropsy, all tissues and organs should be replaced in their respective body cavities, and the animal is sutured closed. The cadaver is placed in a cadaver bag for disposal. The entire work area should be disinfected, and all tools rinsed and resterilized.

COMPETENCY SKILL 130

Necropsy Setup

Objective:

To properly prepare for a necropsy procedure

Preparation:

- Waterproof apron
- Goggles or face shield
- Exam gloves
- Heavy rubber gloves
- Mask
- Specimen containers
- Pathology forms
- Slides
- Large sharp knife
- Rongeurs
- Scalpel
- Scalpel blade
- Forceps
- Sharp scissors
- Needle holder

- Suture material
- Suture needle
- Cadaver bag
- Tape
- ID tag

Procedure:

1. Apply gloves.
2. Place cadaver over a sink on a drainage rack.
3. Move the animal in left lateral recumbency.
4. Place all tools and supplies next to the work area.
5. When necropsy is complete, place cadaver in cadaver bag.
6. Tape bag securely closed.
7. Label bag with client name, patient name, date, and how to dispose of remains.
8. Place in refrigerator or holding area for pickup.
9. Clean work area by disinfecting table, sink, and supplies.
10. Place all lab samples in lab area with forms prepared.
11. Place specimens in area for pickup.

Rabies Suspect Handling and Sample Submission

All persons involved in rabies testing should receive pre-exposure immunization with regular serologic tests and booster immunizations as necessary. Unimmunized individuals should not enter facilities where rabies sampling is conducted. All tissues processed in an infectious disease laboratory must be disposed of as medical waste, and all activities related to the handling of animals and samples for rabies diagnosis should be performed using appropriate biosafety practices to avoid direct contact with potentially infected tissues or fluids. Personnel working with rabies suspect cases are at risk of rabies infection through accidental injection or contamination of mucous membranes with rabies virus-contaminated material and by exposure to aerosols of rabies-infected material. All manipulations of tissues and slides should be conducted in a manner that does not aerosolize liquids or produce airborne particles. Barrier protection is required for safe removal of brain tissue from animals submitted for rabies testing. At a minimum, barrier protection during necropsy should include the following as personal protective equipment (PPE): heavy rubber gloves, laboratory gown and waterproof apron, boots, surgical masks, protective sleeves, and a face shield. Fume hoods or biosafety hoods are not required, but they provide additional protection from odor, ectoparasites, and bone fragments. Glass chips and shards from slide manipulations are also potential sources of exposure to rabies. Care should be taken to protect eyes and hands during manipulation and staining of slides and during cleanup of the microscope and surrounding area. Sample submission guidelines should be followed, as each state's veterinary laboratory requirements vary.

SUMMARY

The veterinary laboratory has a variety of testing procedures and clinical evaluations that are performed by the veterinary assistant. These practices, procedures, and protocols require knowledge of the procedure and the tools and equipment that are necessary to perform it. The equipment must be used properly, cleaned, and kept well maintained. Protective equipment must be used when obtaining and analyzing samples. The veterinary assistant plays an essential role in all aspects of lab medicine and obtaining results.

Key Terms

agglutination clumping

anticoagulant substance that prevents the blood from forming a clot

blood chemistry analyzer machine that runs blood samples that measure routine blood chemistries and electrolytes

blood plasma blood sample that is frozen and then centrifuged to obtain plasma

blood serum liquid portion of the whole blood sample

blood smear blood film placed on a slide that is used to look at the morphology of blood cells

budding yeast bacteria that are in the shape of a budding flower

buffy coat the layers within a hematocrit or PCV that form a tiny area of visible white composed of white blood cells and platelets

capillary action the action of blood draining into the tube through gravity

casts cylinder-shaped structures in urine formed from protein secreted by the renal tubules of the kidney

catheterized refers to animal that has a long, thin rubber or plastic sterile tube inserted into the bladder through the external urinary opening to collect a urine sample

centrifuge tube glass or plastic tube that holds samples within a centrifuge

centrifuge veterinary tool used to spin lab samples at a high rate of speed and force to separate or concentrate materials suspended in a liquid form

chem strip long, thin plastic strip separated by individual square pads containing a chemically treated paper

cocci round bacteria shape

commercial serologic test kit commonly used test kit; provides quick and accurate results related to common viruses and diseases

complete blood count (CBC) blood test

conical refers to the end of an object that is a pointed shaped

Coplin jar glass container that holds chemicals for use in staining slide samples

coverslip thin piece of glass that fits over the sample on the slide

crystals artifacts in urine based on the amount of substance in the urine and its solubility

culture and sensitivity (C/S) test that determines if a bacterial issue is occurring, what type of bacteria is causing the health condition, and what antibiotic should be used to treat the problem

cystocentesis surgical puncture into the bladder using a needle to collect a urine sample

cytology the study of cells

decant process of pouring urine out of a tube and allowing only the sediment on the bottom to remain in the tube

diaphragm mechanism part of the microscope that allows the slide sample to be moved both up and down and side to side to view the entire sample

dip stick long, thin plastic strip separated by individual square pads containing a chemically treated paper

electrolytes determine the balance of elements in the body, specifically potassium, sodium, and chloride

erythrocytes red blood cells

eyepiece the portion of a microscope that you look through

feathered edge the staggered area at the end of the slide where the veterinarian or veterinary technician reads and interprets a blood slide

fecal direct smear prepared by placing a small amount of the fecal sample onto a microscope slide

fecal floatation technique done by placing feces within a liquid to determine if any parasite eggs are within the sample

fecal sample used to diagnose internal parasites and the presence of blood within the stool sample

flocculent contains large amounts of particles called sediment suspended in the urine and appearing cloudy

focus knob the portion of the microscope that allows better visualization of the sample

formalin chemical used to preserve tissue samples

Gram-negative bacteria that stain red

Gram-positive bacteria that stain purple

Gram stain used to determine the presence of bacteria and type of bacteria in a sample

gross examination visible observation of feces

hematocrit measure of the percentage of red blood cells in whole or unclotted blood; also called a PCV

hematology the study of blood

hemolysis the rupture of red blood cells; causes a pink coloration to develop in the plasma or serum

histology the study of tissues

in-house testing occurs when laboratory samples are analyzed within the veterinary facility

lens objective microscope viewer that offers a variety of different powers

medium gel where organism growth may occur

microhematocrit measurement of the percentage of red blood cells in whole or unclotted blood; also called a PCV

microhematocrit tube small thin glass tube that holds blood within a centrifuge

midstream refers to urine collected shortly after urination begins and just prior to the process ending

morphology the structure, shape, color, and appearance in numbers of cells

necropsy procedure done by examining the body of a deceased animal to determine the cause of death

oil immersion process by which a specialized oil substance is placed over a sample to view the contents

packed cell volume (PCV) measurement of the percentage of red blood cells in whole or unclotted blood; also called a hematocrit

panel individual blood test

plasma protein test that measures the ratio of protein within the blood; helps veterinarians determine the hydration status and inflammation occurrence in patients; also called total protein (TP)

postprandial refers to blood sample taken after an animal has had a meal

preprandial refers to blood sample taken after a dog or cat has fasted to reduce the amount of fats or lipids present

punch biopsy a small amount of tissue is removed for testing from a circular site

reagent chemical used to run each individual test in the entire kit

reagent strip chemical test strip used to analyze urine or blood on a long, thin plastic strip separated by individual square pads containing a chemically treated paper

reference lab laboratory samples analyzed outside of the facility in a commercial lab

refractometer tool used to measure the weight of a liquid and determine its specific gravity

rods oblong bacteria shape

rotors wheels in a centrifuge that spin at a variety of speeds depending on the sample type

sediment solid portion of a sample

SNAP test a serologic test

specific gravity (SG) the weight of a liquid

stage part of the microscope that is the flat section under the lens

supernatant liquid portion of a sample

total protein (TP) test that measures the ratio of protein within the blood and helps veterinarians determine the hydration status and inflammation occurrence in patients; also called plasma protein

urinalysis breakdown of urine components to determine a diagnosis

vacutainer tube tube in which blood samples are placed for future sampling

voided refers to urine sample collected as the animal is urinating

whole blood blood sample placed in a lavender-top tube to prevent clotting of the sample

Review Questions

1. What is the difference between an in-house test and a reference lab test?
2. What are some important pieces of lab equipment used in the veterinary facility and what do they do?
3. What are the components of performing a fecal sample?
4. What is the difference between whole blood, serum, and plasma?
5. What is the importance of the feathered edge on a blood slide?
6. What are the components performed in a urinalysis?
7. What are the ways to collect a urine sample from an animal?
8. What is the importance of the Gram stain?
9. What lab tests are performed using a refractometer?
10. What lab tests are performed using a hematocrit tube?
11. How are crystals formed in urine?
12. What is preprandial? What is postprandial?
13. What does DTM stand for?
14. What is the difference between cytology and histology?
15. What is agglutination?

Clinical Situation

The veterinary assistant, Andrea, is attempting to collect a urine sample from a patient named Kimba. Kimba is a 3-year-old SM Doberman that has a history of eliminating in the house. Each time that Andrea walks the dog outside to collect a urine sample, the dog either will not eliminate or attempts to eliminate, and when Andrea tries to place a container for collection, the dog stops urinating.
- What may be clinical conditions to consider with Kimba?
- How can Andrea try to collect a voided urine sample?
- What other ways may a urine sample be obtained?

Chapter 40: Radiology and Imaging Procedures

Objectives

Upon completion of this chapter, the reader should be able to:

- **40.1** Explain radiation safety practices within the veterinary facility
- **40.2** Explain how to maintain a radiology log of radiographs taken within the facility
- **40.3** Describe directional terms and abbreviations pertaining to radiology
- **40.4** Discuss proper animal restraint and placement during radiographic procedures
- **40.5** Demonstrate how to use a caliper to measure the thickness of a body part in centimeters
- **40.6** Explain how to properly use a technique chart
- **40.7** Explain how to determine proper settings on an X-ray machine
- **40.8** Explain how to use identification markers for X-ray films
- **40.9** Explain how to properly develop X-ray films using a hand developing tank
- **40.10** Explain how to properly develop X-ray films using an automatic developer
- **40.11** Demonstrate how to properly load and unload film from a cassette
- **40.12** Demonstrate how to properly clean a film cassette
- **40.13** Explain how to properly file patient films
- **40.14** Explain how to properly clean and maintain the veterinary darkroom
- **40.15** Explain the importance of ultrasound and endoscopy equipment

Introduction

Radiology is the study of radiation and the proper way to take radiographs, commonly called **X-rays**. **Radiation** is a safety hazard in the veterinary industry and requires proper training and safety methods to allow a safe working environment for staff. The veterinary assistant plays an important role in properly performing radiographic procedures and other diagnostic techniques. **Diagnostic imaging** is the term used to refer to performing radiographs, ultrasounds, and endoscopy procedures. These tools are helpful in allowing the veterinarian to properly diagnose patients. It is important for the assistant to be familiar with these procedures and the proper handling and setup of the valuable equipment used to carry them out.

Radiology Generation

The production of radiation or X-rays occurs when fast-moving electrons passed from the **cathode** collide with the **anode**. This causes a negatively charged electrode to collide with a positively charged electrode. Thus, when charged particles such as electrons are slowed down during a collision, X-ray results. This process occurs within an X-ray tube to create a beam composed of a bundle of energy that travels in a wave called photons. This pure energy of photons is carried by the wave and exposes a film to create an image called a radiograph (see Figure 40–1).

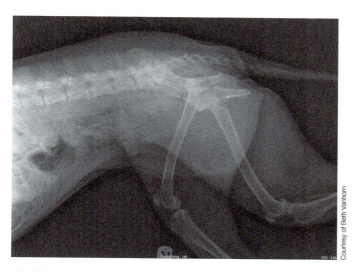

FIGURE 40–1 Radiograph image.

FIGURE 40–2 Radiology caution sign.

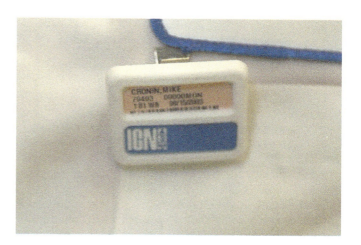

FIGURE 40–3 Dosimeter.

The cathode is within the X-ray tube and contains a tungsten filament which, when heated, creates electrons. The tungsten filament is housed within a cup that allows the beam to focus on the anode, which is a rotating target where X-rays are generated.

Radiation Safety

The veterinary assistant should have a basic understanding of radiation, the hazards of working around radiation, and how to properly reduce exposure and enact safety measures when working in radiology. Regulations regarding equipment and personal protective safety are governed by the Occupational Safety and Health Administration (OSHA) and your state's Department of Health. Each state has various laws governing radiation exposure and radiation safety principles.

Radiation causes generalized damage to cells throughout the body, with reproductive cells, thyroid cells, and eye cells being especially of concern because they can be easily damaged. Certain body cells are more sensitive to radiation than others. Damage from radiation occurs with small exposure over a lengthy time frame or from large exposure over a short period of time. It is important to note that radiation is invisible and cannot be felt, smelled, or tasted, so exposure is unknown, and safety precautions and use of radiation measurement equipment is vital.

Following safety measures greatly reduces the risk of radiation exposure (see Figure 40–2). This is achieved through the use of the as low as reasonably achievable (ALARA) method of taking radiographs. Achieving the lowest amount of radiation exposure will increase the safety of the patient and staff when obtaining X-rays.

A **dosimeter**, or X-ray badge, measures the amount of radiation a person is exposed to (see Figure 40–3). It should be worn every time a radiograph is performed, or a person is working around or near radiation. It measures the level of radiation exposure each staff member is exposed to. The standard exposure level for professionals in an occupation that uses radiation is 0.005 sievert per year. The general public should receive only a tenth of this amount and should never have access to the radiology area. All professionals should be educated about not placing any body part directly into the X-ray beam, where the radiation is the strongest. All individuals must be at least 18 years of age to work with radiation, and pregnant women should not be near any areas of radiation.

Whenever there is exposure to radiation, protection begins with the use of protective equipment (see Figure 40–4). This includes the use of a lead apron, lead gloves, thyroid shield, and lead glasses. Because

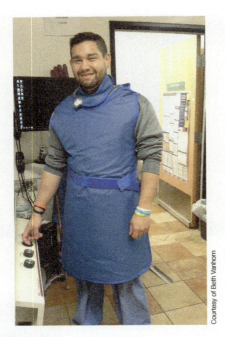

FIGURE 40–4 Radiology lead gown and gloves.

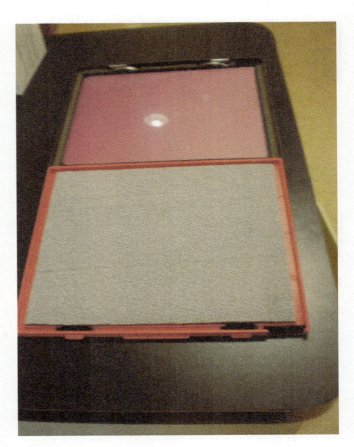

FIGURE 40–5 Screen on X-ray cassette.

lead items may not be folded, this equipment should be hung on a wall in the radiology area. Folding the lead items may cause cracks and tears in the material, which would reduce the level of protection they could provide. Ideally, the items should hang on specialized hangers on the wall, with the gloves on a glove rack for proper ventilation after each use. This equipment should be radiographed on a regular basis to inspect for leaks and damage that may occur in the lead shielding. The dosimeter should be stored outside of the radiation area. When dressing for radiographs, the thyroid shield should be placed over the neck first. The lead apron should fit properly over the shoulders and chest and tie or Velcro easily. It should cover the entire front of the person. The thyroid shield should completely cover the neck and be tucked under the gown. The goggles should then be applied over the eyes and fit well so that they do not easily slide off during movement. The dosimeter (X-ray badge) should fit onto the front of the gown; it is usually clipped onto the neck or pocket of the gown. It should face outward and lie flat.

All X-ray equipment should be maintained and cleaned after each use. The **cassette** that holds the film has a **screen** that lines the cassette and should be cleaned every month using a screen cleaner to prevent a buildup of dust and debris (see Figure 40–5). Any cracks or damage in the screen must be noted because they will necessitate replacement of the cassette. The exterior of the cassette is cleaned after each use with soap and water or with a mild disinfectant if it has been exposed to an animal. Screens may get dusty and collect hair, which can show up as an unwanted artifact on the image. Pressurized air can be used to remove debris, or a soft brush may be used, but care must be taken not to apply too much pressure during cleaning. Commercial screen cleaners are available, or mild soap and water can be used. Cassettes should be left open in a vertical position to dry before new film is placed inside. An **X-ray log** should be kept to include patient information, type of x-ray taken, measurement calculations, and machine settings (see Figure 40–6). The patient's name, client's name, date, X-ray number, X-ray position, thickness of body measurement, and area exposed should be noted in the log. The X-ray machine should be serviced by a field representative on a regular basis to make certain the equipment is in proper working order and properly calibrated. Dosimeters should be evaluated on a regular basis to determine employee exposure rates.

When working with patients, it is helpful to reduce radiation exposure as much as possible. If the patient is easily restrained and can be X-rayed safely, the smallest cassette size possible should be used to allow less radiation exposure. This will allow the **collimator** to reduce the field size and prevent

X-RAY LOG												
DATE	X-RAY #	CLIENT NAME	PATIENT NAME	SPECIES	SEX	AGE	WEIGHT	VIEW	CM	KVP	MAS	BODY PART	INITIALS
									----	----	----		--------
									----	----	----		--------
									----	----	----		--------
									----	----	----		--------
									----	----	----		--------
									----	----	----		--------
									----	----	----		--------
									----	----	----		--------
									----	----	----		--------
									----	----	----		--------

FIGURE 40–6 X-ray log.

scatter radiation. The collimator sets the size of the X-ray beam that produces radiation to take the picture, based on cassette size. The setting that allows the least amount of radiation should be used. Avoid having to retake films by using the correct exposure settings, proper patient positioning, and correct area to be radiographed. Items such as sandbags, foam wedges, and a commercial positioner may be used to help keep the patient still. Animals that are in pain or not cooperative may require anesthesia or other sedation to make the radiograph process easier. Some states have regulations regarding whether staff may be in the same room as the patient when the radiograph is being taken. While assisting in taking radiographs, it is important to turn your head away from the primary beam during exposure. Also close your eyes while taking the image to protect them and limit radiation exposure.

Strict safety measures must be implemented and followed by every veterinary team member to help reduce radiation exposure. Once radiation damage occurs, it is not reversible and can cause serious illness or injury. Simple safety precautions during every radiographic procedure will help prevent staff, clients, and patients from unnecessary exposure.

Radiology Terminology

There are many terms related to radiology equipment, procedures, and positioning of patients. When working in radiology, it is important for veterinary assistants to understand basic radiology terms and positioning. This is important for proper film exposure, film development, and patient diagnosis.

X-ray Terms

When a radiograph or X-ray is taken and developed, there are several terms that should be understood to appropriately provide notations in the X-ray log and medical chart. Contrast is the difference in density of adjacent areas of an image. Density is the degree of blackness on a radiograph. X-rays may be developed and must be properly exposed, avoiding such circumstances as causing an X-ray to become radiopaque or radiolucent. **Radiopaque** means appearing white to light gray in color and may indicate hard tissue such as bone or may be an exposure factor in that the machine settings were too low or the patient was not properly measured (see Figure 40–7). **Radiolucent** means appearing black or dark gray in color and may indicate soft tissue or air and may be an exposure factor relating to increased exposure settings or improper patient measurement (see Figure 40–8). The scale of contrast may be lengthened or shortened by increasing or decreasing the voltage.

X-ray Machine Terminology

When taking X-rays, the machine must properly be set and used. This requires knowledge of the procedure and understanding of terms. The patient should be measured using a caliper to determine the body part thickness in centimeters (cm) (see Figure 40–9). This helps determine **kilovoltage peak (kVp)** and **milliamperage (mA)** machine settings. The kVp is the strength or voltage of the X-ray beam. Thus, this is the strength of radiation. The mA represents the number of X-ray beams based on time. Thus, this is the amount of radiation

788 SECTION IV Clinical Procedures

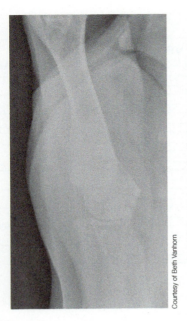

FIGURE 40–7 Radiopaque film.

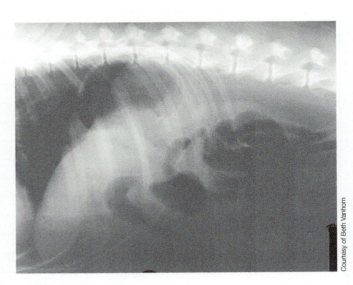

FIGURE 40–8 Radiolucent film.

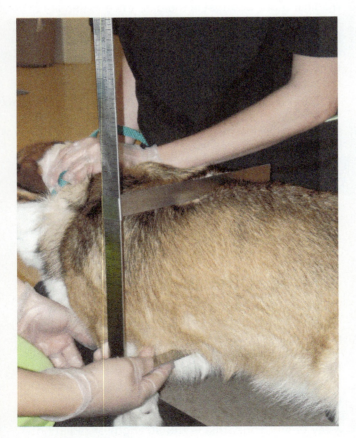

FIGURE 40–9 Measuring the patient to determine proper settings for taking an X-ray.

released during a set time. The settings of the machine will be based on the thickness of the animal's body part to be X-rayed; thus, the body part should be measured accurately, and the settings should be properly calculated.

X-ray Positioning Terms

Animals can be positioned in several different ways for X-rays, depending on the animal and body part being evaluated. The most common X-ray positions are lateral, ventrodorsal, dorsoventral, and oblique. The **lateral (LAT)** view requires an animal to be positioned on its side (see Figure 40–10). The X-ray beam passes the animal from side to side. The **ventrodorsal (V-D)** view requires the animal to be positioned on its back, with the X-ray beam going through the ventral area first (stomach) and then the dorsal area (back) (see Figure 40–11). The **dorsoventral (D-V)** view requires the animal to be positioned on its stomach, with the X-ray beam going through the back first and the stomach second. The **oblique** view is for areas that need to be positioned at an angle to prevent double exposure from other body organs. This is common with skull X-rays to prevent an area from causing shadows on the film. It is important to note the left and right sides of an animal on an X-ray view so that the proper side of the animal is identified on the film. An **anterior posterior (AP)** view is often used on limbs and is common for large dogs and other large animals. This means the X-ray beam penetrates the anterior portion of the body area first and then exits through the posterior portion of the body.

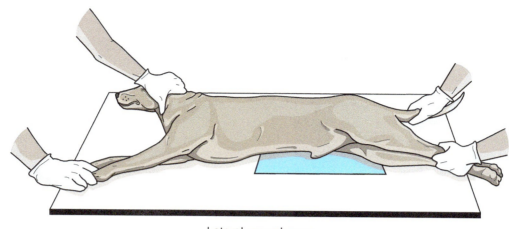

Lateral recumbency

FIGURE 40–10 Lateral X-ray positioning.

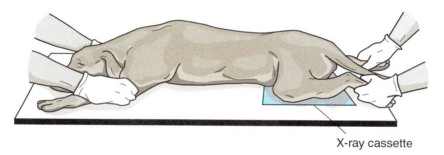

Ventral recumbency/sternal recumbency

FIGURE 40–11 D-V X-ray positioning.

Radiographic Detail

Radiographic detail is imperative to diagnosis and must include sharp detail of internal organs and tissues. Many factors—most commonly patient motion—can affect radiographic detail. Patient motion causes a loss of detail due to the blurring of the image. Decreasing the exposure time can improve image quality and if the patient continues to be too restless, the animal may require sedation for appropriate detail. It is important to use the smallest focal point when taking images. This reduces blurring and improves the sharpness of detail in the image. Distortion can also occur when the X-ray beam is not perpendicular to the surface. In some circumstances a small amount of padding or foam can be placed under the patient to better view an area. Care must be taken not to elevate the spinal column, as this will produce a narrowing of the intervertebral spaces.

When taking X-rays, it is important to reduce as much scatter radiation as possible. Scatter radiation projects in all directions and causes a decrease in contrast quality of the image, thus fogging the film. Of greater importance, scatter radiation is a safety hazard, and animals and staff members can be exposed. Exposure techniques that require increased kVp produce more scatter radiation. Any body part measuring larger than 10 cm also increases scatter radiation. A collimator can be used to decrease the X-ray beam distance and confine radiation waves to the area being filmed. Collimators are adjustable openings within the X-ray tube that allow a light to project on the area of the body being exposed. **Grids** also reduce scatter radiation and improve the contrast of the image. The larger the area being exposed, the greater the kVp setting needed. Grids are useful to help absorb scatter radiation, as they are made up of thin strips of radiolucent and radiodense materials. The radiolucent material is plastic or fiber, and the radiodense material is lead. The grid lies under the patient within the X-ray table and absorbs the radioactive waves when an X-ray is taken by passing through the patient and then into the grid strips, thus reducing scatter radiation.

Diagnostic Terms

There are several diagnostics performed in addition to general radiograph films. Radiographic **contrast mediums** are used to study parts of the body, and films are taken in sequence to monitor the contrast medium as it moves through the area. A contrast medium is a material or substance administered to the body to show structures on an X-ray film that would otherwise be difficult to view. Contrast studies help the veterinarian determine the size, shape, location, position, and function of an organ. Areas that are white are radiopaque, and positive contrast agents appear white. Radiolucent areas appear black, and negative contrast agents will be black. **Positive contrast agents** contain elements that absorb more X-rays, allowing for less radiation exposure to the patient, decreased exposure time, and creation of a white coloration on the film. **Barium sulfate** is an example of a positive contrast medium used in a **barium study**. A barium study is a series of films taken as the material passes through the digestive tract (see Figure 40–12). Barium sulfate is a solution administered by mouth or rectally and used to fluoresce parts of the digestive system. It is insoluble and not affected by gastric secretions, so it provides quality mucosal detail. Barium comes in a powder, suspension, or paste and is inexpensive. Barium may take up to 3 hours to adequately pass through the stomach to the colon. It should never be administered if perforation is suspected. An **upper GI** is a contrast study of the structures of the upper digestive tract in which the barium is given orally. A **lower GI** is a contrast film of the lower digestive tract organs and is often given via an enema, in which a large amount of a substance passed into the colon via the rectum. This is called a **barium enema.** Always record the type, amount, and administration route of the contrast agent. **Negative contrast agents** include air, oxygen, and carbon dioxide. They appear radiolucent on an image and are often used to inflate an area for better visibility and contrast. Care must be taken to not overinflate organs such as the bladder.

Veterinary Animal Restraint

When an animal is positioned and restrained during a radiographic procedure, correct positioning is important. A lateral view requires the proper side down depending on the reason for the X-ray and the location of the film and body part. Most patients are placed in right lateral recumbency unless otherwise requested by the veterinarian. The limbs are pulled forward or backward depending on the area to be viewed. A V-D film requires the patient to be placed on its back, with the front limbs extended parallel to each other and beyond the patient's head. The head should be stabilized between each leg (see Figure 40–13). The hind limbs are parallel and pulled firmly away from the patient's body until they are straight. A D-V film requires the animal to be placed on its sternum or chest, with the front limbs pulled forward and the head stabilized between each front leg. The hind limbs are placed comfortably under the patient and out of view of the body part being X-rayed. The X-ray beam is centered directly over the center of the body part being radiographed. After the proper position is determined, the veterinary assistant and technician must maneuver the patient into position. The use of sandbags, a trough positioner, or foam wedge may keep the patient in place. The goal is to establish and maintain as normal a position as possible while preventing movement or superimposed

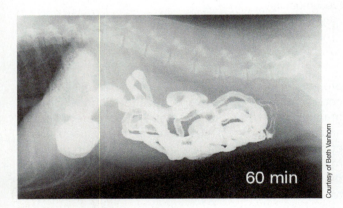

FIGURE 40–12 Barium study.

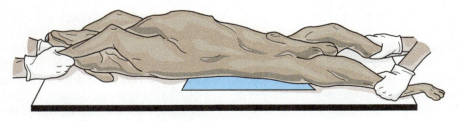

Dorsal recumbency

FIGURE 40–13 Proper positioning of the front limbs for a ventrodorsal X-ray.

body areas that may create a shadow. A minimum of two films should be taken to ensure proper film positioning, calculation, and development. All items should be set up and prepared prior to placing the animal on the X-ray table. This reduces patient and staff stress and allows the team to focus on the task at hand.

Radiology Log

The radiology or X-ray log is a legally required document that must be updated regularly and accurate and should be kept within the radiology area. The X-ray log serves several purposes, including meeting the standards of the veterinary profession, allowing comparison of techniques used on patient follow-up films, and recording the quality of each film taken as a reference to locate the film or improve film quality. The log is usually maintained in a binder with a blue or black pen secured to the binder. Individual entries are made during each radiographic procedure. The format is similar in all logs and is composed of columns and rows where information is entered for each patient. X-ray log papers may be created by the hospital or purchased commercially through an office supply store. Each state practice act outlines the required information kept within the veterinary X-ray log. These items include the following:

- Date
- X-ray or radiograph number
- Client name
- Patient name or identification
- Breed or species
- Gender
- Age
- Weight
- Body location of radiograph or study
- Thickness of body part in centimeters (cm)
- X-ray view or position
- Kilovoltage peak (kVp)
- Milliamperage (mA)
- Seconds of exposure time
- Quality of the film
- Veterinary diagnosis and comments

Patient Measurement

Patients are measured to ensure that the X-ray beam can sufficiently penetrate the thickness of the tissue requiring an X-ray. **Calipers** are used to measure the thickness of a body part (see Figure 40–14). This device or instrument is a simple rulerlike tool that slides on a fixed bar or arm that moves parallel to it. The upright portion along the moveable bar is marked in **centimeters** (cm) and inches. The centimeter scale is to be used when measuring for X-rays. This is due to the techniques chart that is based on centimeter readings and settings for the X-ray machine. The caliper is placed over the thickest part of the area to be X-rayed. The moveable bar is allowed to move and fit lightly around the area being measured (see Figure 40–15).

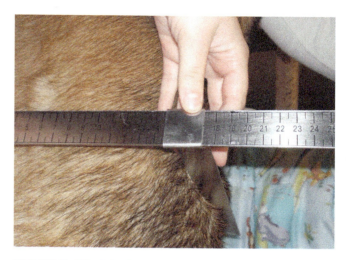

FIGURE 40–14 Calipers are used to measure the thickness of the body part to be X-rayed.

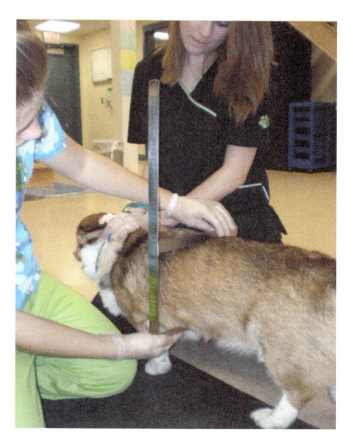

FIGURE 40–15 Use of a caliper to measure the thickness of a body part to be X-rayed.

TABLE 40-1
Example of Caliper Readings

BODY PART	SPECIES/BREED	MEASUREMENT	VIEW
Carpal joint	Feline/Domestic shorthair (DSH)	2 cm/2 cm	LAT/V-D
Chest	Canine/Dachshund	8 cm/10 cm	LAT/V-D
Skull	Canine/Mastiff	15 cm/11 cm	Oblique/D-V
Abdomen	Ferret	3 cm/5 cm	LAT/V-D
Tarsal joint	Equine	6 cm/9 cm	LAT/AP

The centimeter scale should be read at the point just *below* the moveable arm. Reading the scale above the arm will provide an inaccurate measurement. Table 40–1 shows some examples of caliper readings for various species and body parts.

The caliper should be kept close to the X-ray log so that it is easily accessible for noting the proper centimeter measurements and necessary views. Correct use of the caliper leads to quality films and less error.

COMPETENCY SKILL 131

Caliper Measurement Procedure

Objective:
To properly measure the thickness of a body part to ensure an accurate and useful X-ray is obtained

Preparation:
- Caliper
- X-ray views requested
- X-ray log book
- Pen

Procedure:
1. Place the fixed portion of the caliper at the site where the X-ray beam is to exit.
2. Slide the moveable arm over the site the beam is to enter.
3. Move the caliper along the body area to the thickest portion of tissue.
4. Read the centimeter (cm) scale at the portion of the moveable arm closest to the body. Read at the area just below the arm.
5. Record the measurement in cm along with the view in the X-ray log.
6. Repeat for any additional views.
7. Clean the caliper after each use by disinfecting.
8. Put caliper in storage area.

Technique Chart

An X-ray **technique chart** is a listing of settings on the X-ray machine based on the thickness of the area to be radiographed. It is set up in rows and columns and lists the thickness in centimeters (cm). Once the thickness of a body part is determined, the technique chart is used to determine the machine settings. This is done by noting the measurement in the column listed *cm* and moving across the row to determine the kVp, mA, and exposure time in seconds or mAs. Technique charts are different for a grid technique and a tabletop technique. In the **grid technique**, a plate holds the film below the X-ray table and causes the **X-ray tube** holding the radiation source to be lower. The film is not in contact with the patient. In the **tabletop technique**, the distance differs between the X-ray tube and the top of the X-ray table surface, and the film cassette is placed on top of the table in contact with the patient.

Technique charts are formulated for a specific machine. The constants of the machine are determined before the chart is created. These factors include the following:

- The distance from the source of the X-ray beam to the film. This is usually between 36 and 40 inches. This is a fixed distance based on the machine and changes when the grid is used versus the tabletop technique.
- A filter is located between the window of the X-ray tube and the collimator, which is usually between 2 and 2.5 mm in thickness. The filter absorbs the scatter radiation to prevent excessive exposure.
- The grid height and width within the table.
- Film speed, which depends on exposure time. Fast film requires less exposure time but lacks definition and detail, whereas slow film requires greater exposure time but allows for greater detail.
- **Intensifying screens** that are located within the film cassette to help produce better exposure to the film. They are rated according to speed and relate to exposure time, with high speeds requiring less exposure to radiation. Slow film speeds have greater detail but cause blurry images with any movement.

A technique chart should be developed for both the grid technique and the tabletop technique. Most common procedures include a tabletop technique used for views of the skull; extremities; and avian, and rodent, or exotic animals. The grid is used for larger animals and the thorax, abdomen, and chest.

Machine Setting Procedures

The control panel of the machine has several dials, switches, and knobs (see Figure 40–16). It is important to first locate the on/off power switch. This turns the machine on and off. It is important to make certain the X-ray machine is on during use. Some machines have a buzzing sound and others are silent. The three important selector knobs used for setting the machine are the kVp, mA, and timer. Some machines link the mA and timer, and may also provide the mA

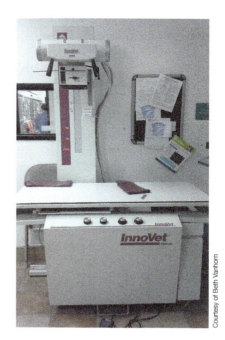

FIGURE 40–16 (A) Radiology machine. (B) Radiographic control panel.

setting. The goal is to use the highest mA setting and the shortest exposure time. The knobs should be set to the readings on the technique chart based on the patient's body measurement. It is important for the assistant and technician to locate the exposure button on the console panel, which may also be in the form of a foot pedal that enables the staff to take a picture while simultaneously restraining the patient in the proper position. When the exposure button is pressed, a red light flashes or a buzzer sounds to indicate the emission of radiation. The age of the machine will also determine how the machine is set up and used. All settings should be double-checked and confirmed for accuracy. All staff members working in radiology should be properly trained and experienced in taking radiographs.

Milliamperage

The mAs is a product of the milliamperage and exposure time. The milliamperage controls the number of electrons generated in the cathode by controlling the temperature. When the mAs is increased, the temperature increases, producing more electrons. Increasing the mAs increases the density of the tissues being evaluated in the patient. Time also affects proper radiographic technique. The amount of time electrons are allowed to flow from the cathode to the anode controls the number of X-rays produced. mAs and exposure time are related. As the mAs increases, the exposure time decreases. The best technique is to use the fastest exposure time to allow less movement on the image, which results in the largest mAs. Common mAs and exposure times include 300 mA at 1/60 sec and 100 mA at 1/20 sec. Consider the following when seeking to adjust density:

- Doubling the radiographic density doubles the mAs.
- Halving the mAs decreases the density.

kVp

The kVp is the voltage applied between the cathode and anode. It accelerates electrons flowing from the cathode toward the anode. Increasing the kVp increases the positive charge on the anode, causing the electrons to move faster with a shorter wavelength and more penetrating power. The correct kVp setting is determined by the thickness of the body part being radiographed. The thicker the body part, the higher the kVp setting because more penetration power is required. Consider the following when seeking to adjust density via kVp:

- Increasing the kVp by 20 percent increases density
- Decreasing the kVp by 16 percent decreases density

Radiographic Film

Film consists of three layers: a thin protective layer, an emulsion containing crystals, and a polyester base. The outer protective layer consists of a clear gelatin covering that protects the emulsion layer. The second layer is a sensitive emulsion containing fine silver halide crystals that cover both sides of the film. This allows for better film image and contrast. The film base is located in the center and gives the film its support. X-ray film is sensitive to heat, light, radiation, and pressure. Two types of film are used in radiology: screen type and direct exposure. Screen type film is available in green- or blue-sensitive film and is more sensitive to light produced by the intensifying screens. Direct exposure film is more sensitive to the direct X-rays than to light. The intensifying screens are not used with this type of film, and it requires higher mAs than screen film. Direct exposure film is most often used for increased detail in areas such as the mandible or maxilla, or dental X-rays. Film speeds range from high to average to slow. The faster the film, the more sensitive it is and the lower the mAs it requires. Thus, high-speed film requires less exposure time than low-speed film. Most veterinary applications use average film speed. The latitude of the film allows it to produce shades of gray. Increased latitude allows for more shades of gray in the contrast quality; longer film latitude is desirable.

Film Identification

Radiographs are part of a patient's medical record and are also considered legal documents. Most X-rays are filed separately from the medical record but must be properly labeled and identified so that they can be located and reviewed. Each film that is taken should be permanently marked with patient and client information, as well as the veterinary facility's information. Each state requires specific information to be placed permanently on the film, meaning it must be placed on the film prior to X-ray exposure or X-ray development. Information that must be on the label includes the following:

- Hospital name, address, and phone number
- Veterinarian's name
- X-ray number
- Client name
- Patient name
- Date

The film should also be marked with a directional label or marker to note which side of the animal is being viewed (see Figure 40–17). This is usually in the form of a lead L or R placed on the proper side of the cassette film prior to X-ray exposure. The markers identify the left and right side of the patient when using a V-D, D-V, or AP view and the downside when using a LAT or oblique view. Make certain the marker is placed properly on the film cassette and within the area of

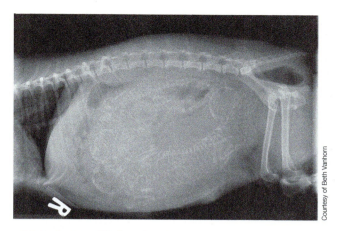

FIGURE 40–17 Positioning marker.

the X-ray beam. Some facilities use lead tape that contains the patient information and is placed on the X-ray film cassette to be exposed onto the film during the X-ray. Other facilities use an electronic film identification printer that stamps the information onto the film prior to film development. If an electronic flasher or printer system is used, an area must be blocked on the film cassette prior to exposure to prevent that area from being occluded.

Developing Film

There are two methods for developing film in the veterinary facility. The methods are **manual developing** and **automatic film processing**. The older method is manual developing through use of **hand tanks**. Manual development is cheaper to set up and maintain than an automatic processing system, but it takes longer to develop a film than it does using the automatic method. The manual method also requires temperature control of the hand tanks, which hold the chemicals used to develop the film. The automatic film processor is a costly piece of equipment that has advantages over hand developing, most notably the speed of development and quality of the film. The solutions' temperature is controlled by the machine, which reduces errors and increases the life span and quality of the film. The time for a person to develop the film is significantly reduced. Professional maintenance is also required with the automatic processor unit. Table 40–2 highlights the pros and cons of manual versus automatic developing.

Film processing involves several steps independent of the method of developing. These steps include film developing, film **rinsing**, film **fixing**, and film drying. **Processing** involves being in a darkroom that has no light source or light leaks from the outside area. Light will ruin the film and make it undevelopable. The darkroom is equipped with a **safe light**, which is a red light that is low in intensity and has a filter so that the light doesn't damage the film. This light allows enough visibility to work in the darkroom without damaging the film. The light should be at least 48 inches above the work area. The entrance door should be secured firmly to prevent anyone from entering during film development. A light or sign should be placed on the door when the room is in use. The undeveloped film is removed from the cassette before it is developed. A new film is replaced within the cassette.

Film should be stored in a cool, dry area away from chemicals and should be stored on end and not flat on its side. The storage boxes must be kept closed tight to prevent light exposure.

TABLE 40–2
Pros and Cons of Development Methods

	PROS	CONS
Manual developing	• Low cost • Easy to use • No backup system required • Low maintenance	• Lengthy developing time • Must maintain chemical temperatures • Must stir and prepare chemicals • Shorter life span of developed film
Automatic processing	• High speed of development • Ready-to-use temperature-controlled chemicals • Reduced developing errors • Increased life span of film • Higher quality film	• High cost • Machine warm-up and setup time • Requires maintenance and servicing • Requires backup method if equipment breaks or fails

Film Cassettes

Cassettes hold the film that is used to take an X-ray and prevent the film from being exposed to light (see Figure 40–18). They have latches that secure the cassette tightly to keep light from entering. It is important to work in the darkroom when loading and unloading film from a cassette. The cassette should be placed face down so that the back latches can be unlocked. The cassette is then turned over face up with the top open. Film is removed and should be handled with care. Only handle the film at the corners and as carefully as possible, avoiding dropping the film to prevent streaks, smudges, dirt marks, static electricity, and light exposure. When the cassette is refilled with film, it is removed from the film box, which is kept in a film bin that is light-tight. The appropriate film size should be selected, and the lid removed. Each sheet of film is covered by a paper or plastic cover that protects the film. A single sheet of film should be carefully removed and placed in the cassette. The cassette is closed and properly locked. The cassette is ready for use or storage in its customary place. A cassette should never be left without film. Cassettes should be handled carefully. They should not be dropped, and heavy objects should not be set on top of them.

Inside of the cassette are intensifying screens made up of fluorescent crystals that line both sides of the cassette. When exposed to X-rays, the screens emit light, provide protection to the film, reduce static electricity, and allow for decreased exposure time. The screens are available in various speeds, just as

FIGURE 40–18 Film cassette.

with the film: high, average, and slow. High-speed screens require less exposure time, but the detail is decreased. Shorter exposure time with high-speed screens is ideal for soft tissue imaging, such as with the abdomen. Better image detail is achieved with slow speed screens and longer exposure time. Routine cleaning of intensifying screens is necessary to ensure screens are free of dirt and debris. Dirt can block light emitted from the screens, leaving parts of the film unexposed. The screens should be cleaned with a soft cloth and screen cleaning solution or warm water. Screens must be dry before film is placed inside of the cassette.

 COMPETENCY SKILL 132

Unloading Film from a Cassette

Objective:

To properly remove film from a cassette while maintaining the integrity of the film

Preparation:

- Film cassette
- Film
- Darkroom
- Safe light

Procedure:

1. Place the cassette upside down and unlock.
2. Turn cassette face up and open cover.
3. Remove film from cassette by one corner, grasping firmly with the fingertips.
4. Leave the cassette open while developing film.

COMPETENCY SKILL 133

Loading Film to a Cassette

Objective:

To properly prepare a cassette to maintain the integrity of the film

Preparation:

- Film cassette
- Film
- Darkroom
- Safe light

Procedure:

1. Place cassette face down and unlock.
2. Open cassette.
3. Place the film box containing the same size film next to cassette.
4. Remove the box lid.
5. Open paper or plastic cover and remove one sheet of film.
6. Place film in cassette and close. Lock cassette.
7. Fold paper or plastic cover over film box and place in storage bin.
8. Place cassette in storage area.

 COMPETENCY SKILL 134

Cleaning a Cassette

Objective:

To properly maintain equipment for use in obtaining X-rays

Preparation:

- Film cassette
- Mild soap
- Warm water
- Paper towels

Procedure:

1. Open cassette.
2. Apply small amount of warm water and soap to paper towel and wipe cassette screens on both surfaces.
3. Dry each screen surface.
4. Allow cassette to remain open until dry.

Film Hangers

Film hangers are used to secure film for manual processing and development. These tools hold the film onto a metal frame with clips and hold the film as it is submerged in chemicals during developing (see Figure 40–19). The film is removed from the cassette, and the proper hanger size is selected to hold the film. The film hanger is held upside down as the film is attached to clips. Once the bottom clips are secured onto the corners of the film, rotate the film and secure the film to the top clips by pulling them downward and snapping them onto the corners. The film holder should secure all four corners of the film. The film is then handled by holding the bar at the top of the hanger. This bar allows the film to hang in the tank during the developing process.

Manual Developing

Developing tanks are required for hand-processing films. These tanks may be metal or heavy plastic, and they hold large volumes of chemicals. The solutions require periodic changing, at which time the chemicals are drained. The tanks are cleaned well to remove bacteria and algae, rinsed, and refilled. The frequency of this practice depends on how much use occurs and any debris and contaminants that may get into the chemicals. The tanks contain the developer solution, the fixer solution, and a wash tank that contains water. The **developer solution** is usually located on the left-hand side of the tank. The developer solution develops the X-ray film. The **fixer solution** is usually located on the right-hand side and is a smaller tank compared to that of the water. The fixer solution fixes the X-ray film and helps maintain the life span of the film, allowing it

FIGURE 40–19 Film hanger.

to stay readable. The water rinse tank is usually located at the far right side and is larger than the other two tank sections. When the processing method is complete, the film should hang to dry.

COMPETENCY SKILL 135

Manual Film Processing

Objective:

To properly develop X-ray film using the manual method

Preparation:

- Film cassette
- Film
- Darkroom
- Safe light
- Automatic film marker
- Patient identification card

- Pen
- Hand developing tanks
- Developer solution
- Fixer solution
- Water
- Timer
- Thermometer

Procedure:

1. Make patient identification card and place in automatic film marker.
2. Stir developer and fixer solution and make certain proper temperatures are noted.
3. Place film cassettes face down in the darkroom and unlock back.
4. Place film cassettes face up and open top.
5. Remove film by one corner.
6. Place unexposed corner of film into automatic film marker over card and lower top to expose film corner.
7. Begin attaching film to bottom corner of film hanger; attach to both clips.
8. Rotate hanger and attach top clips to both corners.
9. Place film hanger in the developer on the left side of the processing tank.
10. Lift film up and down several times.
11. Set proper time required for developer.
12. While film is in developer, refill cassette with film.
13. Put cassette away.
14. Place film box in storage bin.
15. When timer sounds, remove film from developer and rinse in water tank by moving film hanger up and down several times.
16. Place film hanger in fixer solution and lift up and down several times.
17. Set proper time required for fixer.
18. Cover tanks with the correct lids.
19. Exit darkroom.
20. When timer sounds, return to darkroom, remove film from fixer, and rinse in water tank.
21. Place in water for 30 minutes. After 30 minutes, hang film to dry.
22. When film is dry, unclip the film from the hanger.
23. Notify veterinarian when film is ready to read.

Automatic Film Processing

The automatic film processor unit should be started early to allow it to warm up and the chemicals to reach the desired temperature. All chemical tanks should be checked for proper levels. The roller racks should be removed and washed on a regular basis. The process of using the automatic process for film developing is simple. The film cassette is taken to the darkroom, and the exposed film removed from the cassette. The film should be grasped by the edges and placed on the feeder tray of the processor. The film is aligned with the roller bar and placed against it. The film will be pulled into the processor and developed automatically. The completed film will be removed on the opposite end of the feeder tray dry and ready to view. X-rays are viewed on a viewing box that illuminates the film and allows the veterinarian to make a diagnosis (see Figure 40–20).

Digital Radiology

Many facilities are beginning to switch from using chemical processing to produce radiographs to using digital radiology. Digital X-rays are transferred onto a computer disc rather than a film. This type of diagnostic imaging has multiple advantages, including ease of use, production of high-quality images, affordability, and improved neatness and cleanliness of images. This radiology practice is quite common in veterinary medicine and in the coming years will be considered the norm in radiology. Digital radiography uses an imaging plate composed of multiple detectors. These detectors translate and convert images into electrical signals that are then digitized by a computer into a radiographic image. The imaging plate is connected directly to the X-ray machine and a computer dedicated to producing a high-quality image, which eliminates the need for a

SECTION IV Clinical Procedures

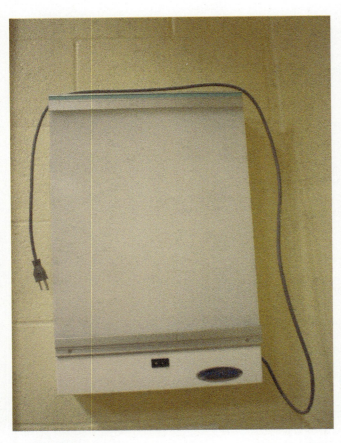

FIGURE 40–20 X-ray view box.

film cassette. The image can then be manipulated to enhance the contrast and brightness. In addition, the image can be enlarged and rotated, and selected areas can be measured. Digital radiology uses a DICOM or digital imaging and communications in medicine format for storing medical images. Images may be stored on a DVD or managed through a picture archiving and communications system (PACS). This allows storing images within a hospital or network of hospitals, making the images available at any time on any computer. It is important to note that the same concepts of proper digital image identification as discussed with film are applied to digital imaging. The use of markers to identify patient sides, proper ID of patient/client/veterinary facility/veterinarian, date of imaging, image ID number, and body location are required.

Filing Film

Just as medical records are filed for ease of location, so radiographs must also be filed. X-rays are too large to be filed with the medical record and are often stored in a nearby file area in large protective X-ray folders or paper envelopes (see Figure 40–21). Several systems have been used to file radiographs,

 COMPETENCY SKILL 136

Automatic Film Processing

Objective:

To properly develop X-ray film using the automatic method

Preparation:

- Film cassette
- Film
- Darkroom
- Safe light
- Automatic film marker
- Patient identification card
- Pen
- Automatic processor
- Developer solution
- Fixer solution

Procedure:

1. Make certain the processor is ready to use.
2. Make certain all chemicals are replenished.
3. Place cassettes face down in darkroom and unlock.
4. Turn cassettes over and open cover to remove film.
5. Grasp film by one corner and align film with feeder tray.
6. Gently place film against roller bar.
7. The film will enter the processor automatically.
8. While the film is being processed, refill the film cassette. Replace film box in storage bin.
9. Remove film from opposite end of feeder tray.
10. Give film to veterinarian to be read.

FIGURE 40–21 Filing X-ray films.

such as alphabetical or numeric filing. Most facilities file radiographs by the X-ray number and then store them in numeric order. All X-rays for a particular patient are stored in one file envelope. Once a film has been developed, an envelope and number should be recorded in the patient's medical record and a file folder labeled with the X-ray and patient information. The information label should include the following:

- X-ray number
- Patient name
- Client name
- Veterinarian name
- Dates of radiographs
- Type of study or view
- Diagnosis

The file folder is labeled properly, the film is placed within the file folder, and the folder is then properly filed in the storage location. Filing should be done accurately, as a misfiled folder creates disorder in a clinic.

Darkroom Care and Maintenance

The darkroom should be kept clutter free and clean at all times to help improve film quality and proper developing methods. The darkroom should be checked on a regular basis for any light leaks from the outside. The door should be observed for cracks around the edges. The safety light should be checked by closing the door and keeping the darkroom dark with only the safe light source on. An unexposed film should be placed on the counter with a metal object, such as a paper clip or nail, over the film. The film should be exposed for 2 minutes and then developed. If the object appears on the film, the safety light is not working properly, or some other light source is entering the room. This test evaluates the lighting environment of the darkroom.

The counter workspace should be cleaned on a daily basis, and all items should be placed in their storage locations for ease of working in the dark. There should be two areas to the darkroom: a wet side and a dry side. The wet side holds the processor and hand developing tanks. The dry side is the counter workspace where cassettes are loaded and unloaded with film. The film storage boxes should be stored in a bin under the work area to decrease light exposure.

Floors and work areas may become wet, and towels should be kept in the area to clean up leaks and spills. A mop should be used to clean the floor as spills occur. A cleaning regimen should be developed for darkroom maintenance, and materials and chemicals should be monitored to determine when it is time to reorder them. Figure 40–22 shows an example of a darkroom maintenance schedule.

TASK	CLEANING SCHEDULE	DATE	INITIALS
Clean processor	Daily		
Clean tanks	Weekly		
Replenish solutions	Daily, as needed		
Clean counters	Daily		
Clean/Mop floor	Daily		
Inventory supplies	Weekly		
Check light leaks	Monthly		
Check safe light	Monthly		

FIGURE 40–22 Darkroom maintenance schedule.

Radiograph Artifacts and Errors

An error or artifact is common on images and may be in the form of a blemish, incorrect positioning, incorrect techniques, improper machine settings, or poor processing techniques. Common error or artifact occurrences include black areas, white areas, fogging, lines, poor image detail, uneven development, streaks, clear film, and brown discoloration. Black areas on the film may occur due to light exposure in the film box or bin, or they may be due to light in the darkroom or damaged cassettes. White areas on the film often are due to dirty screens, debris or dirt on the film, chemical spills on the film, contrast medium getting on the patient, fingerprints on the film due to fixer on the hands, or grid damage. Fogging is due to excessive scatter radiation, exposure to radiation during film storage, film storage in excessive heat or humidity, a damaged safelight, low-grade light leakage in the darkroom, or film being past its expiration date. Decreased detail is typically due to patient motion, poor film or screen contact, or the film being too far away from the patient. Uneven development of film is often due to improper stirring of chemicals, uneven chemical levels, or excessive removal of the film from the tank during development. Spots or streaks are typically due to film sticking together during storage or during development, splashing of chemicals prior to development, or damage to rollers during automatic processing. When the entire film is clear, no exposure has occurred or the film was placed in the fixer before the developer. When you note film is brown in color, it is due to improper washing during the final rinse.

Ultrasound Diagnostics

Ultrasound is a diagnostic tool that uses ultrasonic sound waves to view images of internal organs and other structures. The sound waves bounce off the patient, create an echo within the tissues, respond back to the ultrasound machine, and are projected on a screen (see Figure 40–23). The stronger the ultrasound signal during the return phase, the brighter and whiter the image.

The veterinary assistant prepares and restrains the patient for the ultrasound procedure. The quality of the image is affected by how the patient is prepared and how calm the animal is kept while restrained. The image quality may be decreased if air, hair, dirt, or other debris is not properly removed, making for an unclean environment that may distort the image quality. The **transducer** is the wand used to scan the area being examined. The transducer sends low-intensity, high-frequency sound waves into the soft tissues. Some sound waves reflect back to the transducer, and some waves are transmitted deeper into the tissues. The sound waves that reflect back to the transducer are echoes that produce a gray image on the screen. Transducers are extremely fragile and expensive, and they are the most important part of the ultrasound equipment. The area where the transducer is placed should be clipped as for a surgery using a #40 or #50 clipper blade. The area is gently washed with sterile scrub materials and dried well to enhance the contact area of the transducer. A small amount of transducer gel is applied to the area to help visualize internal structures. The patient should be positioned depending on the area being examined. Abdominal examination in small animals requires dorsal or lateral recumbency, and dogs and cats should fast for better visualization. Cardiac examination requires lateral recumbency. Many large animal ultrasounds are done with standing recumbency.

The equipment should be safely placed in the examination or treatment area, plugged in, turned on, and prepared for use. The equipment should be set up prior to the patient being brought into the area. When the ultrasound is complete, the equipment should be disinfected properly according to the manufacturer's instructions. A printout of the ultrasound, or a **sonogram**, is made, the image is dated, and the report is placed in the patient's file. The sonogram is the image of the internal structure(s). Some images may be recorded on videotape or CD. The ultrasound diagnostic report should note this, and the tape or disc should be dated and stored within the medical record or other storage area.

Endoscopy

Endoscopy is the procedure of visually examining the interior of the body by using an endoscope, that is, a tool used to view the inside of the body (see Figure 40–24). The **endoscope** is made of a bundle of fine glass rods that are moved through the body to project an image using a light source. The use of the endoscope is limited by its diameter and length relative to the structure it is passing. A small opening within the endoscope allows for other diagnostic tools and instruments to pass for further sampling. Procedures such as biopsies are completed with an endoscope. The procedure requires sedation or anesthesia.

The glass rods within the endoscope are fragile and may break easily. The endoscope should be stored within a padded case provided by the manufacturer. The endoscope should be stored in a straight position, disinfected and dried after each use, and stored in a dry area. There are two types of endoscopes: rigid and flexible. Rigid endoscopes are less expensive and vary in size, but all have a hollow tube containing no fiber bundles. They should be handled by the eyepiece and not by the rod. Any slight bending of the rod causes damage that can affect angle and degree of visualization. There are two kinds of flexible endoscopes: fiberoptic and video. Flexible endoscopes use glass fiber bundles to transmit images. The bundles transmit light from the light source to the distal tip of the endoscope. Video endoscopes contain a microchip at the distal end of the scope that records and transmits images to a computer and then to a monitor.

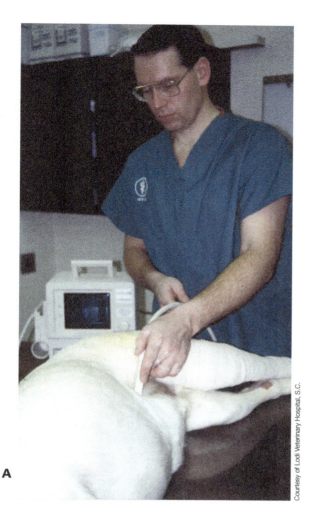

FIGURE 40–23 (A) Endoscopy procedure. (B) View of the intestine using an endoscope.

804 SECTION IV Clinical Procedures

endoscope lens cleaner. The scope should be stored in a hanging position in a clean, well-ventilated area.

The veterinary assistant should be able to set up the endoscope and monitor, understand proper cleaning and disinfection methods, and understand factors that affect storage. If tissue samples or biopsies are taken during the procedure, the assistant should prepare lab work forms and containers for sampling. The reference lab should be contacted for any special instructions related to handling guidelines.

CT and MRI Procedures

Computed tomography (CT) involves ionizing radiation, with computer assistance, passing through the patient to allow for the display of internal body structures in cross-sectional views (see Figure 40–25). It is commonly called a CT scan. The scans produce three-dimensional images that are downloaded to a computer, analyzed, and used as diagnostic tools. **Magnetic resonance imaging (MRI)** involves radio waves and a strong magnetic field passing through the patient to allow for the display of internal body structures in three-dimensional images. As cells in the body move, atoms respond to the magnetic field and emit radio waves that produce an image (see Figure 40–26). MRIs are often used to provide imaging of the brain, joints, and spinal column.

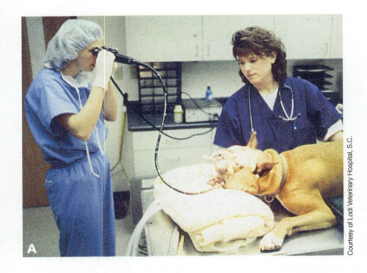
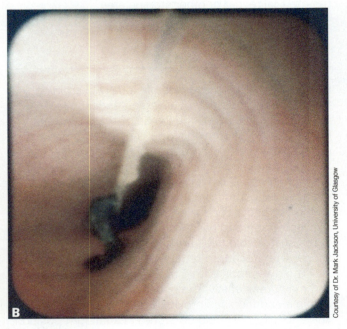

FIGURE 40–24 (A) Endoscopy procedure. (B) View of the intestine using an endoscope.

The endoscope equipment should be cleaned immediately after use, and the person cleaning the scope should wear latex gloves to help prevent damage to the equipment. The scope should be flushed with water and air, and the insertion tube should be gently washed with a soft cloth or gauze soaked in detergent. A detergent and water mixture should be used to flush the distal end of the scope. The outside of the scope can be cleaned with alcohol-soaked gauze pads. The lenses must be cleaned with an

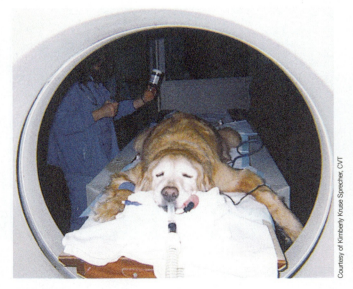

FIGURE 40–25 CT scan.

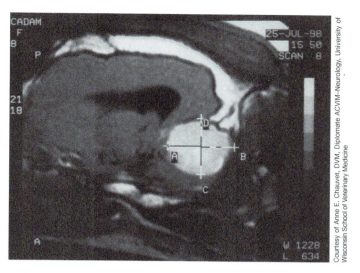

FIGURE 40–26 MRI image from a dog with a brain tumor.

SUMMARY

Radiology procedures are an important part of the veterinary assistant's duties. Proper training and experience are necessary in the areas of radiation safety, maintaining radiology logs and films, understanding directional terms, setting up and maintaining the X-ray machine, developing and handling film, and restraining patients during radiology procedures. Radiology is an area that is changing every day, and keeping up with current knowledge and education in this field is recommended.

Key Terms

anode rotating target that creates positively charged electrons

anterior posterior (AP) view that requires the X-ray beam to penetrate the anterior portion of the body area; the beam exits from the posterior portion of the body

automatic film processing equipment that develops and dries film automatically

barium enema involves a large amount of fluorescent substance passing into the colon via the rectum

barium study series of films taken as barium passes through the digestive tract

barium sulfate substance administered to the patient to show structures on an X-ray film that would otherwise be difficult to view

caliper device or instrument with a simple rulerlike tool that slides on a fixed bar or arm that moves parallel to it and is used to measure the thickness of a body part

cassette tool that holds X-ray film

cathode area within X-ray tube that creates negatively charged electrons when heated

centimeters (cm) unit of measurement used when measuring body parts before X-rays

collimator part of the X-ray machine that sets the size of the X-ray beam that produces radiation to take the picture

computed tomography (CT) procedure in which ionizing radiation, with computer assistance, passes through the patient to allow for the display of internal body structures in cross-sectional views

contrast medium   material or substance administered to the patient to allow imaging of structures that would otherwise be difficult to view on an X-ray film

developer solution   chemical that develops X-ray films

diagnostic imaging   procedures that involve radiographs, ultrasounds, and endoscopy

dorsoventral (D-V)   view requiring the animal to be positioned on its stomach; the X-ray beam goes through the back first and the stomach second

dosimeter   X-ray badge that measures the amount of radiation a person is exposed to

endoscope   fiberoptic instrument used to visualize structures inside the body

endoscopy   procedure used to visually examine the interior of the body via an endoscope

film hanger   used to secure film for manual processing and development

fixer solution   chemical that fixes X-ray film and helps maintain the life span of the film, allowing it to stay readable

fixing   using a solution to maintain the life span of X-ray film

grid   series of radiolucent and radiodense material that reduces scatter radiation

grid technique   plate that holds the film below the X-ray table

hand tanks   large containers that hold chemicals and water for developing film

intensifying screen   located within the film cassette to help produce a better exposure to the film

kilovoltage peak (kVp)   strength of the X-ray beam

lateral (LAT)   view that requires an animal to be positioned on its side

lower GI   contrast film of the lower digestive tract organs; in which contrast medium is often given via an enema

magnetic resonance imaging (MRI)   procedure in which radio waves and a strong magnetic field pass through the patient to produce three-dimensional images of internal body structures

manual developing   developing X-rays by hand using chemical tanks

milliamperage (mA)   represents the number of X-ray beams based on time

negative contrast agent   air, oxygen, or carbon dioxide that appear radiolucent on an image

oblique   view used on areas that need to be placed at an angle to prevent double exposure from other body organs

positive contrast agent   element that absorbs more X-rays; causes white appearance on an image

processing   developing film in a darkroom that has a safe light source and no light leaks from the outside area

radiation   safety hazard that allows X-rays to be produced to take a radiograph

radiology   study of radiation

radiolucent   related to black or dark gray appearance, which may indicate soft tissue or air; may be an exposure factor related to increased exposure settings or improper patient measurement

radiopaque   related to white to light gray appearance, which may indicate hard tissue such as bone; may be an exposure factor in that the machine settings were too low or the patient was not properly measured

rinsing   using water to remove any chemicals from an X-ray film

safe light   red light that is low in intensity and has a filter so that it doesn't damage the film

screen   area that lines the film cassette

sonogram printout of an ultrasound recording

tabletop technique the distance differs between the X-ray tube and the top of the X-ray table surface, and the film cassette is placed on top of the table in contact with the patient

technique chart list of settings on the X-ray machine based on the thickness of the area to be radiographed

transducer instrument used with an ultrasound; scans the body and transmits the waves back to a screen

ultrasound diagnostic tool that uses ultrasonic sound waves to produce images of internal organs and structures

upper GI contrast study of the structures of the upper digestive tract; barium is given orally

ventrodorsal (V-D) view that requires the animal to be positioned on its back with the X-ray beam going through the ventral area (stomach) first and the dorsal area (back) second

X-ray common term for radiograph

X-ray log records the patient's name, client's name, date, X-ray number, X-ray position, thickness of body measurement, and area exposed

X-ray tube part of the machine that holds the radiation source

Review Questions

1. What safety equipment items are used in the radiology area?
2. What is the kVp?
3. What is the mA?
4. What is the difference between a V-D view and a D-V view?
5. What information is kept in an X-ray log?
6. What tool is used to measure a patient before an X-ray? What is the correct way to use this tool to measure the animal?
7. What is the significance of the technique chart?
8. What is the difference between manual and automatic developing?
9. What information must legally be placed on a film?
10. What is an ultrasound?
11. What is an endoscope?
12. What is the difference between negative and positive contrast agents?
13. How does a transducer work?
14. What age must staff be before they are allowed to work near radiation?
15. What does ALARA stand for?

Clinical Situation

Molly, a veterinary technician, and Lee, a veterinary assistant, are working in radiology at the Best Friends Vet Hospital. They have several patients to X-ray, including two large dogs, a challenging cat, and a rabbit. The facility has both an automatic processor and manual hand tanks. The vet would like the animals restrained for the X-ray views and would like to complete each view without the use of anesthesia or other sedatives if possible.

- What should Molly and Lee do to prepare for the X-rays?
- In what order should Molly and Lee complete the X-rays?
- What restraint items may be useful with the patients?
- How should Molly and Lee utilize the developing equipment to process the films?

Chapter 41: Pharmacy Procedures

Objectives

Upon completion of this chapter, the reader should be able to:

- **41.1** Explain how to interpret a prescription
- **41.2** Explain how to read labels correctly
- **41.3** Explain how to properly label a dispensing container
- **41.4** Calculate the proper quantity of medication to dispense
- **41.5** Demonstrate how to properly use a pill counting tray to count medications
- **41.6** Explain how to process and log controlled substances according to Drug Enforcement Agency (DEA) regulations
- **41.7** Demonstrate how to administer oral medications to a patient
- **41.8** Demonstrate how to administer aural medications to a patient
- **41.9** Demonstrate how to administer topical medications to a patient
- **41.10** Demonstrate how to administer ophthalmic medications to a patient
- **41.11** Explain how to properly store medications in the pharmacy according to manufacturer labels
- **41.12** Discuss the importance of accuracy with pharmacy skills

Introduction

The **pharmacy** is the area where medications are stored and prepared for veterinary patients. Pharmacy skills are an essential part of the veterinary assistant's duties and must be accurate in several major areas, including reading the veterinarian's prescription, understanding common abbreviations used in the veterinary pharmacy, properly identifying the drug of choice, and properly calculating the amount of medication to dispense to a patient. Accuracy is essential and mandatory when working with medications and prescriptions. The study of drugs and their effect on living organisms is called **pharmacology**. Medications are composed of active ingredients that are mainly plant or mineral sources.

Medication Names

Medications are often referred to by three names: the **proprietary** name, the **nonproprietary** name, and the chemical name. The proprietary name is commonly called the trade name and is the name of the medication given by the manufacturer to its particular brand of a drug, such as Amoxi-Tabs, Clavamox, and Rimadyl. The trade name is always capitalized, as it is a proper noun and the registered trademark symbol is located after the name to signify that is cannot be used by other manufacturers and is typically only found on the medication bottle.

Many manufacturers produce similar products under different names. The nonproprietary name refers to the generic name, which is based on the active ingredient in the medication. Examples are amoxicillin, prednisone, and cefazolin. The chemical name describes the drug's chemical makeup. In veterinary clinics, most drugs are referred to by their generic name. Any drug used in veterinary medicine must be approved for use by the U.S. Food and Drug Administration (FDA).

Traditionally, drugs are based on mineral or plant sources. These include alkaloids, glycosides, oils, resins, and gums. Alkaloids typically end in *–ine*, and glycosides end in *–in*. Bacteria and molds make up many antibiotics and dewormer components. Electrolytes are made of mineral sources.

Medication Forms

Medications come in several forms, most commonly solid, liquid, or semisolid. Solid forms include tablets and capsules. Tablets are composed of powdered components pressed into pill form. Some tablets are coated to protect them from breaking down too rapidly in the stomach and dissolving until they enter the intestinal tract. Capsules are powdered drugs enclosed by a gelatin surrounding. Some medications are sustained release and slowly release small amounts of the medicine into the abdominal lumen over an extended period of time. Suppositories are solid forms inserted into the rectum to be dissolved and absorbed into the wall of the large intestine. Semisolid forms are ointments, creams, and pastes that are applied to the skin or given by mouth. Ointments and creams liquefy at body temperatures, but pastes keep their semisolid form.

Liquid medication forms include solutions, suspensions, syrups, elixirs, and tinctures. Solutions are dissolvable and do not change or settle when left standing. A suspension doesn't dissolve and typically has particles that settle at the bottom of the bottle; it must be shaken before use. Syrup is a solution composed of drugs with water and sugar components. Elixirs are solutions in sweetened alcohol but don't dissolve in water. Tinctures are solutions of alcohol base that are applied to the skin topically.

Injectable forms of medicine are administered by needle and syringe either into a patient or into a fluid bag. Syringes are available in different sizes and styles. Commonly used syringe sizes in veterinary medicine include 1, 3, 5, 12, 20, 35, and 60 ml (see Figure 41–1). Syringes may be supplied with a needle attached, or they may be supplied as the syringe only, requiring a needle to be placed on the end of the syringe.

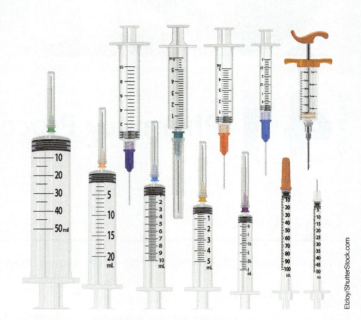

FIGURE 41–1 Various sizes of syringes.

Where the needle attaches to the syringe may be a luer-lock, slip tip, eccentric tip, or catheter tip. The syringe consists of a plunger, barrel, hub, and needle. The area where fluid remains within the plunger after the medicine has been deposited is called the dead space. A tuberculin (TB) syringe holds up to 1 ml of fluid. It is commonly used to administer amounts of medicine less than 1 ml. An insulin syringe is different from other syringes, as it is divided into units instead of milliliters and is used to administer insulin.

Needles are available in different sizes. The specified size is determined by the gauge and length. *Gauge* refers to the inside diameter of the needle shaft. The larger the number, the smaller the shaft diameter. Thus, a 22-g needle is larger than a 25-g needle. The needle length is measured from the hub to the shaft end. Lengths over 1 inch are commonly used in large animals and also for obtaining samples, as in a cystocentesis. The bevel is the opening in the needle tip. The bevel should be up when performing a venipuncture.

The VCPR and Prescriptions

The *veterinary client patient relationship (VCPR)* as discussed in Chapter 2 should be understood relating to each state's veterinary practice act requirements. Typically, veterinarians must establish a VCPR to prescribe medications and continue to refill medications. This requires a yearly physical exam by the veterinarian prescribing the medication.

It is unethical for a veterinarian to write a prescription or dispense a prescription drug outside a VCPR.

Reading a Prescription

Veterinary prescriptions, often abbreviated Rx, are written in a combination of words and abbreviations. The **prescription** is the type of medication, amount of medication, and directions for use of the medication. The prescription is determined and prepared by the veterinarian. There are several parts to the prescription, including the following (see Figure 41–2):

- Medication name
- Medication strength
- Method of administration
- Amount to be administered
- Frequency of administration
- Length of use
- Amount to be dispensed
- Special instructions
- Number of refills
- Veterinarian name, address, phone number

It is essential for the assistant to understand all parts of the prescription and have knowledge of common veterinary pharmacy abbreviations, as discussed in Chapter 1. The patient must legally have a VCPR established within 6 months to a year for the veterinarian to legally prescribe medications.

COMPETENCY SKILL 137

Reading a Prescription

Objective:

To properly read and interpret a prescription written by the veterinarian

Preparation:

- Patient file
- Prescription

Procedure:

1. Identify the drug name.
2. Identify the drug strength.
3. Determine how the medication is to be administered (oral, aural, ophthalmic, topical, etc.).
4. Determine how much is to be administered.
5. Determine the frequency of administration (SID, BID, TID, etc.).
6. Determine how long the medication should be used.
7. Determine how much medicine is to be dispensed.
8. Note any special instructions.
9. Determine the number of refills.
10. Recheck all items for accuracy.
11. Recheck any math used to determine items.

Labeling a Prescription

The purpose of a **prescription label** is to indicate to the owner how to properly give the medication and identify the medication within the container (see Figure 41–3). Facilities use printed or handwritten labels and place individual labels on each container. When a label is prepared it should be legible (which is especially a concern if it is handwritten), clean and understandable, and accurate; it must also contain all required label information. Legally, the following information is required on a pharmacy label:

812 SECTION IV Clinical Procedures

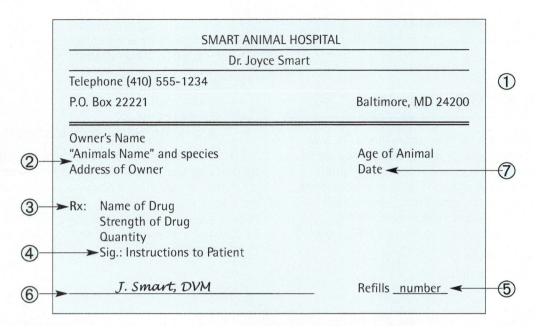

FIGURE 41–2 Parts of the prescription: (1) Name, address, and phone number of veterinarian. (2) Client's name and address; species and patient's name. (3) Name, strength, and quantity of drug. (4) Instructions for giving drug to patient. (5) Number of refills. (6) Veterinarian signature. (7) Date script was written.

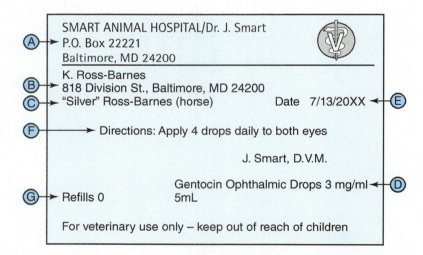

FIGURE 41–3 Parts of the dispensed drug label: (A) Veterinarian name and address. (B) Client name. (C) Animal name and species. (D) Drug name, strength, and quantity. (E) Date ordered. (F) Directions for use. (G) Refill information.

- Veterinary facility name, address, phone number
- Name of prescribing veterinarian
- Client name
- Client address if a controlled substance
- Patient name or ID number
- Medication name
- Medication strength
- Quantity dispensed
- Expiration date of medication
- Number of refills
- Amount per treatment or use
- Route of use
- Frequency of treatments
- Length or duration of use
- Special instructions

Additional labels may be required for the container and are often added as warning labels in sticker form, such as "Keep refrigerated," "Give with food," or "Shake well" (see Figure 41–4). Every container should be labeled with "For Veterinary Use Only" as a legal documentation of how the drug is meant to be used. All FDA-approved medications must contain the following statement: "Caution: Federal law restricts the use of this drug to use by or on the order of a licensed veterinarian." When labeling the directions for use on the medication, it is imperative that plain and understandable basic words are used. Never use pharmacy terms or abbreviations on a label. Recheck each label three times for accuracy, discuss the directions with the client, and make certain he or she has no questions about use of the drug. When applying the label to the container, make certain it is centered and straight, and no folds are located over any information.

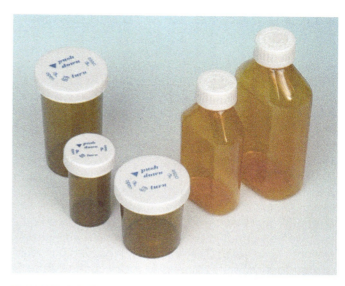

FIGURE 41–5 Various child-resistant caps.

Dispensing Medications

Every medication should be dispensed in a childproof container (see Figure 41–5). The only exception is if a client requests a nonchildproof container due to physical limitations that make it difficult to open such a lid. It is important to note that children or pets can easily get into containers, and safety is an issue. It is important to document in the medical record when a nonchildproof container is requested and dispensed.

Medication Containers

Plastic vials and bottles are commonly used with a twist-off or snap-off lid. The size of the container is measured in **drams**. Properly sized vial or bottle must be selected to hold the required amount of medicine to be dispensed. Vials and bottles are usually amber in color to protect the medication from degrading due to light exposure. Vials are used for tablets, capsules, and powders. Bottles are used for liquid medications.

Metric System

The metric system is the most common measurement system used in veterinary medicine. The metric system uses powers of 10 as a base for units and the decimal system with three basic units: weight, mass, volume, and length. Weight and mass use the unit gram (g). Volume uses the unit liter (L). Length uses the unit meter (m). Each unit uses multiples or powers of 10 to describe their numbers and is assigned prefixes based on the powers. Common prefixes the multiples of basic units include: *centi–*, *kilo–*, *milli–*,

FIGURE 41–4 Warning labels placed on prescription medication.

and *deci–*. A cubic centimeter (cc) is equal to a milliliter (ml) and designates the same volume. Any decimal number less than 1 (e.g., 0.60) should be designated as being part of a whole number, and a zero should be placed to the left of the decimal point as a placeholder. Fractions should not be used but instead be written in decimal form. For example, ½ g should be expressed as 0.50 g.

Drug Identification

Accuracy in selecting the proper medication to be dispensed is essential. The proper dosage and strength must also be accurate. A patient's condition may worsen or not improve if the wrong medicine or strength is selected. There are also legal ramifications related to incorrectly dispensing medications. Many drugs have similar names and spellings. Always compare the spelling of the prescription with that of the drug label on the bottle. Many drugs come in a variety of strengths, and the prescription should match the drug strength located on the drug label. Drug strengths may be written in milligrams (mg), kilograms (kg), milliliters (ml), cubic centimeters (cc), or grains (gr). Some may be a combination such as mg/kg or mg/tablet.

All drugs selected for dispensing should be evaluated for expiration dates. Never dispense a drug that has expired. The expiration date may be located on the drug label or the lid. Remember to check the following information for accuracy:

- Right patient
- Right drug
- Right strength
- Right quantity
- Right frequency

Amount to Dispense

The amount to be dispensed may be listed on the prescription or may need to be calculated by the veterinary assistant. This requires knowledge of pharmacy abbreviations and basic math. The amount to be dispensed is based on the prescription provided by the veterinarian. This requires knowledge of how much to give per treatment, how many treatments per day, and how long the duration of the treatment will be. Some medications may require determining a dose if the actual drug strength used is not equal to the required amount to be given. Recheck all dosages and dispensing amounts three times for accuracy.

Example 1:
Rx: Amoxicillin 250 mg PO BID × 14 d
1. Determine how often the drug is given (BID = twice a day).
2. Determine how long the drug is to be given (14 days).
3. Multiply the number of doses per day by the duration of treatment (2 × 14 = 28).
4. Locate the strength of drug that is as close as possible to the prescription (250 mg).
5. Determine how many units per dose are needed. (250 mg tablet per dose).
6. Multiply the number of tablets per dose by the number of doses (1 × 28 = 28). This is the number to be dispensed (28).
7. Determine the amount per dose needed. This is done by dividing the units per dose by the units per tablet available, which in this example is one tablet that is 250 mg.

Example 2:
Rx: Amoxicillin 250 mg PO SID × 10 d (Note: SID also may be written as q 24 hours.)
1. Determine how often the drug is given (SID = once a day).
2. Determine how long the drug is to be given (10 days).
3. Multiply the number of doses per day by the duration of treatment (1 × 10 = 10).
4. Locate the strength of drug as closely as possible to the prescription (500 mg).
5. Determine how many units per dose are needed. (500 mg tablet is broken in half so that 0.5 of the tablet is given, which equals an amount of 250 mg, or half of the 500 mg tablet.)
6. Multiply the number of tablets per dose by the number of doses (0.5 × 10 = 5). This is the number to be dispensed (5).

Dose Calculations

Doses should be calculated using the following steps:
1. Weigh the patient. If the weight is in pounds, convert to kilograms.
2. Based on how the drug is dosed, use the patient's weight to calculate the correct dose (mg/kg).

3. Based on the concentration of the medicine, determine the amount to dispense.

Example 3:

What is the amount to dispense if a 40-pound dog needs 5 mg/kg of a drug that has a concentration of 50 mg/ml?

1. Begin by converting pounds to kilograms: 40 lb/2.2 lb/kg = 18.18 kg
2. Calculate the dose: 5 mg/kg × 18.18 kg = 90.9 mg
3. Calculate the volume of the solution needed based on concentration: 90.9 mg/50 mg/ml = 1.8 ml

Always check the rounded dosage with the veterinarian, as some medications have a narrow margin of safety and may need to be rounded down rather than up.

Pill Counting Tray

When the amount to be dispensed has been determined, the assistant should locate the medication, and if tablets or capsules are used, the **pill counting tray** should be used for counting the required quantity. The pill counting tray is a device that has a flat area for medicine to be placed and a channel or funnel-like area to place medicine that has been counted and is to be dispensed to a patient (see Figure 41–6). This is helpful in limiting the amount of contact with the medicine and helps determine that the proper amount to be dispensed is accurate.

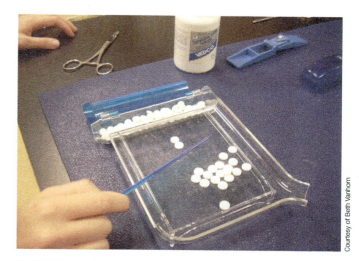

FIGURE 41–6 Pill counting tray.

COMPETENCY SKILL 138

Pill Counting Tray Procedure

Objective:

To properly prepare a prescription using a counting tray

Preparation:

- Medication bottle
- Pill counting tray
- Spatula or tongue depressor
- Medicine vial or container

Procedure:

1. Place the pill counting tray on the pharmacy counter.
2. Pour the medication tablets or capsules onto the tray plate.
3. Open the channel cover.
4. Using a spatula or tongue depressor, push groups of five tablets or capsules into the channel.
5. When you have counted the desired number of tablets or capsules, close the channel cover. Lift the tray and place the channel spout into the medicine vial or container.
6. Tilt the tray to pour the medicine into the vial or container.
7. Place the vial on the counter.
8. Place the vial on the pharmacy shelf.
9. Clean the pill tray using water and allow to dry.

Types of Medications

Medications are divided into three categories: **over-the-counter (OTC) drugs**; **prescription drugs**; and **controlled substances**, which require a special license for veterinaries to prescribe. OTC drugs are nonprescription drugs and can be purchased by anyone at any time, commonly at human pharmacies. Prescription drugs can only be ordered from a pharmaceutical distributor and prescribed by a licensed veterinarian.

Controlled Substances

Controlled substances are prescription drugs that have the potential for abuse or addiction and are regulated by the **Drug Enforcement Agency (DEA)**. The DEA has specific regulations and guidelines for ordering, storing, and dispensing controlled drugs. A veterinarian must have a **controlled substance license** to prescribe these drugs. This license provides a number separate from the veterinary license and must be kept posted within the pharmacy. A controlled substance is also referred to as a **scheduled drug**, as the class of a controlled drug is categorized by schedule type from Schedule I to Schedule V (see Table 41–1).

Handling controlled substances is common among all veterinary health care team members. Several guidelines and regulations must be known regarding controlled substances and how they are stored and dispensed. The only person who may order and prescribe controlled substances is a licensed veterinarian who has a DEA controlled substance license. If any veterinarian in the facility doesn't have a DEA license, he or she may not prescribe controlled drugs. These licenses are issued by the DEA and renewed every three years. The license should be posted within the pharmacy area in a reasonably visible area.

Controlled substances must legally be stored behind three locks (see Figure 41–7). This means they

FIGURE 41–7 DEA controlled substance locked cabinet.

must be stored within a locked box within a locked safe unit that is padlocked. The outer locked area must be securely and permanently fixed in place. Access to the controlled substance storage location should be limited to one or two people within the facility. This controls access to medications and potential theft. Manufacturers of controlled substances are required to identify the substance on its label with an uppercase *C*, followed by the class of drug noted via Roman numeral. For example, C-II.

Any controlled drug must legally be logged into the **controlled substance log** kept within the pharmacy. A written entry is required for every substance used or dispensed by the facility. The log must be recorded using blue or black ink and must be a permanently bound book maintained for at least two

TABLE 41–1

Scheduled Drug Classes

SCHEDULE TYPE	EXPLANATION OR EXAMPLE
Schedule I (C-I)	No medical use; illegal drugs, e.g., cocaine; limited use in research only; not found in veterinary facilities
Schedule II (C-II)	Potential for severe addiction, e.g., morphine; no refills allowed
Schedule III (C-III)	Moderate potential for addiction, e.g., Hycodan; limited to five refills every 6 months
Schedule IV (C-IV)	Low potential for addiction, e.g., Valium; limited to five refills every 6 months
Schedule V (C-V)	Low potential of addiction or abuse, e.g, Robitussin; no limits on refills

years, with requirements on record keeping varying from state to state. Information that must be entered into the log includes the following:

- Name of controlled substance
- Drug strength
- Drug form (tablet, capsule, liquid, injection, etc.)
- Quantity dispensed
- Quantity on hand
- Date dispensed
- Time dispensed
- Client name and address
- Patient name
- Animal breed/species
- Initials or signature of person dispensing medication

A running inventory account of all controlled substances is legally required for each facility. The amount dispensed and the amount left on hand must be accurate. A quarterly and end-of-year inventory should be maintained and be accurate according to the controlled substance log.

Educating Clients on Medication Use

The veterinarian may discuss medications with the client, but it is up to the staff members to make certain the client completely understands how to give the medicine, including when and how much. Certain terminology may confuse a client, or so much information may be provided at one time that people may easily become confused and may be afraid to ask about a topic. There is always the potential for a client to misunderstand or have incomplete information, which may serve as a failure for proper pet health. This may delay or impede a patient's recovery. The veterinary assistant should be able to review all medication instructions before the client and patient leave the facility. The client should also be urged to call with any questions or problems. Medications that may require the splitting of pills should be used only as recommended by the veterinarian and may require the use of a pill splitter that helps the client cut the pill equally. Many medications may not break down properly in the digestive tract after splitting. Medications that have an outer coating should never be split in half, as this can cause irritation to the patient and may decrease the absorption rate during metabolism. Care should be used to ensure proper cleaning of pill splitters and reuse of any medication vials to reduce any cross-contamination of different types of medications that may be a health risk to patients.

Some clients will receive a handout about how to give the medication, with examples or diagrams showing hand placement. It is important that the assistant demonstrates the procedure for the client and ask if he or she has any questions. A written copy of these instructions, or a review, should include the following information:

- Why medication is being administered
- How medication is to be administered
- How much medication is to be administered
- When the medication is to be administered

Administering Oral Medications

Oral medications are administered by mouth. Oral medications may be tablets, capsules, or liquids. Determination of which form of the medication to administer may be based on several factors. The species of animal determines how the mouth can be opened and what instrument may be used to administer the medication. Small animals, such as dogs and cats, will require the mouth to be opened and the medicine placed at the back of the mouth and swallowed. Large animals, such as horses and cattle, require more power to hold and administer medications. Some animals may require pet piller or dosing syringes to aid in giving medicine by mouth. A **pet piller** is a small device, usually made of plastic, that has a long thin handle with a plunger on the end. The plunger holds the medicine and the handle is used to throw the medicine to the back of the throat (see Figure 41–8). Large animals may require a **balling gun**, which is a metal device with a long handle that has a plunger at the base to hold medicines (see Figure 41–9). The disposition of an animal also plays a role in how medication is given. When administering medicine, it is important to consider the type of drug used. Tablets and capsules may be coated with butter, shortening, or vegetable oil to make them slippery and easier to pass down the throat. They may also be coated with foods that tempt the animal to eat, such as peanut butter or a piece of meat. Some animals may eat the food item and spit out the medicine, so caution must be used in this method. Tablets without an outer coating may be crushed into a fine powder and mixed with food or water and administered as a liquid medication. This is often done using a mortar and pestle, which grinds medications to a powderlike substance (see Figure 41–10). Liquids may be given using a syringe without a needle (see Figure 41–11). This allows proper measurement of the volume to be given and allows for slow administration to prevent the animal from spitting out the medicine. Pastes may be given from the original container; they are usually squeezed from the tube and placed within the inner cheek area of the animal.

SECTION IV Clinical Procedures

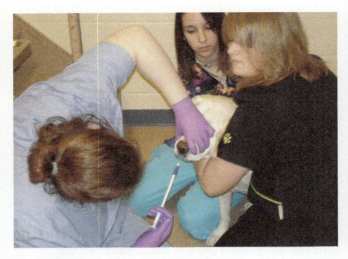

FIGURE 41–8 Use of a pet piller to administer medication to a dog.

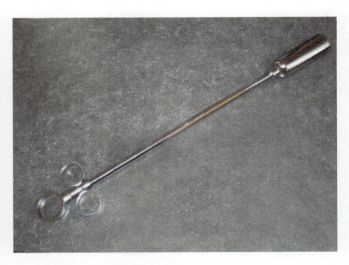

FIGURE 41–9 Balling gun.

FIGURE 41–10 A mortar and pestle can be used to grind medications.

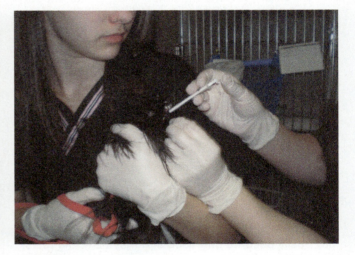

FIGURE 41–11 Use of a syringe to administer medications.

COMPETENCY SKILL 139

Administering Oral Medications

Objective:

To properly administer a medication vial the oral route

Preparation:

- Proper medication (right patient, right drug, right strength, right dose, right time)
- Pet piller or balling gun
- Syringe
- Exam gloves

Procedure:

Tablets or capsule (small animal)
1. Apply gloves.
2. Elevate head upward.
3. Open the mouth at the side by pressing between the upper and lower jaw joint by cheek (see Figure 41–12).
4. Place medicine between index finger and thumb of free hand.
5. Place middle finger of same hand in front of mouth and apply pressure to open mouth wide.
6. Drop the tablet or capsule at the back of the throat.
7. If safely able to use index finger of free hand, push the medicine deeply into the throat (see Figure 41–13).
8. Close the mouth and hold until the patient swallows.
9. Gently blow on face or rub throat to stimulate swallowing.
10. Monitor for any signs of the medication not being ingested.

FIGURE 41–12 Press on the joint between the upper and lower jaw to open the mouth.

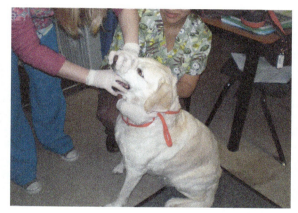

FIGURE 41–13 Gently press medication deep into throat, if possible to do so safely.

Pet Piller or Balling Gun

1. Apply gloves.
2. Place pill in end of plunger of pet piller or balling gun.
3. Apply the end of the piller or balling gun into the side of the mouth between the upper and lower teeth.
4. Once in mouth, push plunger to throw medicine to the back of the throat.
5. Remove balling gun or piller and immediately elevate head upward.
6. Blow on nose or rub neck to stimulate swallowing.
7. Check to make sure medicine was swallowed.
8. Disinfect items and put away.

Liquids and Pastes

1. Apply gloves.
2. Elevate head upward.
3. Insert syringe or paste tube into the side of the mouth between upper and lower teeth.
4. Press the plunger to the desired amount into the back of the throat.
5. Remove syringe or tube and continue elevating head upward.
6. Blow on nose or rub throat to stimulate swallowing.
7. Keep mouth held shut until swallowing occurs.
8. Disinfect any tools and put away.

Administering Aural Medications

Aural medications are placed within the ear canal and may be used for such treatments as ear infections, medicated ear cleaning maintenance, or ear mites. The opening of the ear canal structure includes the haired area of the **tragus**, located cranially on the ear flap (see Figure 41–14). The area located caudally and with no hair is called the **pinna**. The internal ear canal is shaped in an L fashion where at the beginning or exterior area the structure is wider and as the canal gets closer to the eardrum it gets shorter. The ear canal can be palpated through the skin, below the opening of the external pinna, and extending down toward the lower jaw. The ear canal is made of cartilage and when palpated feels similar to a rubber hose or tube.

Ears may be treated for many reasons, most commonly after developing an ear infection, which is more common in long-eared dogs and those with floppy ears. These ears do not allow for sufficient air ventilation, and moisture and bacteria may easily build up within the canal. Some dog breeds, such as the Poodle or Bichon Fries, have hair that grows within the ear canal and may prevent proper ventilation. All animals are susceptible to **ear mites**, a parasite that occurs deep within the ear canal. Flies and mosquitoes may bite the ear flaps, especially in livestock, causing crusting and bleeding around the ear edges. Wax and other debris may become lodged within the internal ear canal, causing trauma and irritation. Dogs and cats commonly inflict bite wounds over the ear flaps, causing damage to the pinna and surrounding tissue that may develop into a **hematoma**. This may also occur due to excessive head shaking and may be a secondary trauma to an ear infection. When an animal shakes its head severely it may cause a blood vessel to rupture and the pinna to fill with blood, known as an aural hematoma.

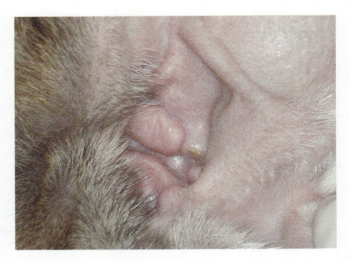

FIGURE 41–14 External anatomy of the ear.

Medications placed into the ear are usually in liquid form, either as a droplike solution or ointment (see Figure 41–15). Some medicine is packaged in tubes and others in bottles with droppers. Many tubes are meant for multiple uses. It is often necessary for the ear to be cleaned prior to aural medications being administered. When ears are a concern during a diagnosis, the veterinary assistant is asked to restrain the animal for an exam. Medications may be administered after the cleaning has occurred. The veterinary assistant may be asked to show clients how to properly administer aural medications at home. It is important that gloves be worn to prevent contamination. When a multiple-use ear dropper or container is used, alcohol should be used to clean the top after each use.

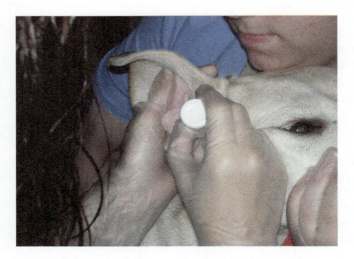

FIGURE 41–15 Administration of ear drops.

COMPETENCY SKILL 140

Administering Aural Medication

Objective:

To properly administer medications via the aural route

Preparation:

- Ear medication
- Exam gloves

Procedure:

1. Apply gloves.
2. Clean ears as recommended by the veterinarian. This procedure is discussed in Chapter 38.
3. Place medication dropper or tip deep inside the ear canal only to the depth of the start of the vertical portion of the L shape.
4. Apply the proper amount of medication into the ear as noted by the veterinarian. This is usually in drops that are counted according to the pharmacy label.
5. Remove the medication dispenser from the ear.
6. Massage the base of the outside of the ear canal. This will create a swishing sound caused by the medication moving around the ear canal.
7. Wipe away any solution that may have leaked onto the outside of the ear flap or hair.
8. Disinfect the medication dispenser and place in appropriate area.

Administering Topical Medications

A **topical** medication is a substance applied to the outside of the hair or skin coat on an external body surface. Topical medications may include antiseptics to clean the skin surface, flea and tick preventions, or wound treatments (see Figure 41–16). Some areas of the body must be cleaned prior to a topical being administered. Areas of wounds or abrasions may have scabs or crusts that need to be soaked prior to topical medication application. This is best done by soaking with warm water and a surgical scrub, sterile saline, or betadine solution. This may take several minutes, and soakings may need to be repeated. The area then must be dried. Application of topical medicines is then completed at the appropriate site following the veterinarian's recommendations.

Some topical medications come in a single-use container, and others are in a large volume meant for multiple uses. Topical medications in large volumes used in the treatment area should be kept sterile by removing the required amount from the container with an item that is sterile and has not touched the patient. This may be done using a spatula, tongue depressor, or other similar item. Wounded areas on the skin should be treated carefully to prevent further damage or trauma. Wounds and healing tissue are fragile.

Topical flea and tick medications should be applied according to the label instructions. There are a variety of products available that offer monthly treatments. Each product treats specific animals for specific purposes. Some topicals treat fleas only, whereas others treat fleas and ticks, and possibly mosquitoes or other biting insects. Gloves must be worn when applying these topicals, as they include a variety of chemicals that may be harmful to human skin. Most of these products require separating the hair and placing the entire contents onto one area of the skin or onto multiple areas of skin.

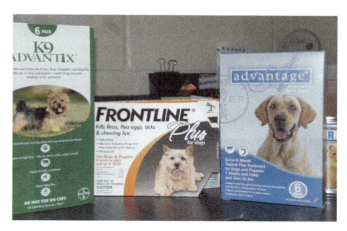

FIGURE 41–16 Flea and tick products are topical medications.

COMPETENCY SKILL 141

Administering Topical Medications

Objective:

To properly apply topical medications for safe and effective use

Preparation:

- Exam gloves
- Topical medication
- Tongue depressor

Procedure:

1. Apply gloves.
2. Clean area as necessary. This is discussed in Chapter 38.
3. Use a tongue depressor to transfer the appropriate amount of topical medication if removing from a multiple-use container.
4. Apply the topical medication onto the area in a circular motion, starting at the center of the area being treated and gently working outward.
5. Do not contaminate the medication by touching items that touched the animal. Flea and tick topicals should be applied according to the manufacturer's instructions.
6. Separate hair from skin and apply as directed, as in steps 3 and 4.
7. Disinfect work area and put items away.

Administering Ophthalmic Medications

Ophthalmic medications are ointments or solutions applied to the eyes (see Figure 41–17). This may be used to treat an eye condition or prior to bathing and grooming to protect the cornea from damage. Ophthalmic medicines are usually packaged for a single patient. In the veterinary treatment area, some items may be used on multiple animals and must be kept sterile. After removing the cap, wipe the end of the applicator with a cotton ball or gauze sponge with a small amount of alcohol. After using the medication, repeat this process. Avoid allowing the applicator to touch the animal's eye or other body parts when applying the medicine into the eyes. Touching the surface of the eye with the tip will not only contaminate it but may also cause trauma to the eye. Wear gloves when applying eye medicines. The eye will need to be held open to better view the eye and place the substance (see Figure 49–18). Eye drops can be dropped into the eye by holding the bottle above the open eye. Ointments should be applied as a thin layer over the lower eyelid; then allow the animal to blink several times to move the substance throughout the eye.

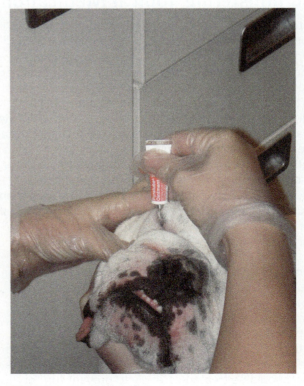

FIGURE 41–17 Application of ophthalmic medications.

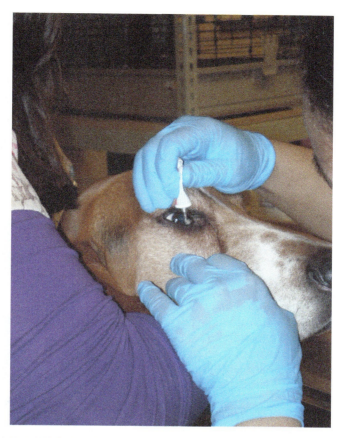

FIGURE 41–18 Hold the eye open to properly apply an ophthalmic medication.

COMPETENCY SKILL 142

Administering Ophthalmic Medication

Objective:

To properly administer medication to treat eye ailments

Preparation:

- Exam gloves
- Ophthalmic medicine
- Gauze sponges

Procedure:

1. Apply gloves.
2. Wipe any discharge from the patient's eye using a gauze sponge.
3. Open the end of tube or syringe holding the ophthalmic medicine and hold in one hand.
4. Using the free hand, use the index finger and thumb to pull the upper and lower lids apart to open the eye. The thumb pulls the lower lid down and the index finger pulls the upper lid upward.
5. The remaining unused fingers may rest on the head of the animal.

6. Gently tilt the head upward.
7. Apply the drops gently into the eye, counting each drop and applying the proper amount. Do not touch the surface of the eye with the dispenser.
8. Apply the ointment in a thin layer on the lower lid. Do not touch the surface of the eye with the dispenser.
9. Release the eyelids.
10. Allow the animal to blink to move the medication throughout the eye.
11. Clean the dispenser with a small amount of alcohol on a gauze sponge.

Medication Storage

Medications have an expiration date and should be monitored for proper dating prior to dispensing. All medicines in the pharmacy should be stored so that those with the oldest dates are closest to the front of the shelf in order to facilitate use of the products that will expire first (see Figure 41–19). Drugs expiring later are placed at the back of the shelf. The expiration date is located on both the product and the label of the bottle. This date must be included on the pharmacy label when dispensing a medication. Any outdated items are removed from the shelf.

Drugs typically arrive in an individual container. Each medicine has a drug insert within the container that details the actions and use of the drug. Some medications may have the insert attached to the outside of the bottle. Storage information is also located on the drug insert and details how the medication should be stored. Some items will need to be stored at room temperature and others may require refrigeration. Biologicals, especially vaccines, are stored under refrigeration. A separate refrigerator should be used for pharmacy items. Some drugs must be stored in a dark place and kept from direct light. These drugs will lose their **efficacy**, or the strength and life of the drug, when exposed to light sources. All drugs without information on storage will most likely be kept at room temperature and in a dry location.

Most veterinary pharmacies keep medications stored on shelves in alphabetical order for ease of locating. All controlled substances must be kept locked at all times. These items must be inventoried similar to other items in the hospital. Once the items have been opened, they should be sealed tightly and returned to their proper pharmacy location. Multiple bottles of the same medicine should be stored in a separate area to make certain they are not opened before the previous bottle is used. The pharmacy shelves must be cleaned on a regular basis to keep dust and debris from the bottles.

FIGURE 41–19 Proper storage of medications at the veterinary facility.

Pharmacokinetics

Pharmacokinetics is the term for how drugs move into, through, and out of the body. This allows knowledge of how the medicine is to be given and the options for using different routes of administration. This includes how the medicine is absorbed, distributed, metabolized, and excreted. The movement of drug particles from the site of administration into the body system is called **absorption**. After the drug is absorbed into the body it enters the bloodstream and travels to the intended area, called the **target tissue**. The proper route of administration is important, as the amount of drug reaching the target tissue may be greatly altered if not administered properly. **Distribution** describes the movement of the drug into the tissues. Many drugs become altered by the body before they are eliminated. This is known as the process of drug **metabolism**. The liver is the primary organ involved in drug metabolism; however, other areas such as the skin and intestinal tract may be part of the metabolic process. **Elimination** of the drug waste occurs when it is excreted and passed outside of the body, typically through the liver or kidneys. The liver removes drugs from the body by excreting waste material out of the bile and then into the feces. The kidneys eliminate waste materials through urine. Drug elimination can be affected by age, disease, dehydration, heart issues, and kidney or liver problems.

Drugs with systemic actions administered by mouth (PO or per os) must cross the gastrointestinal (GI) lining of the stomach or small intestine. Drugs given as an injection are administered parenterally and can be IV, IM, IP, or SQ. Intravenous or IV injections are given directly into a vein, which allows the blood containing the medicine to pass through the heart and mix with blood in general circulation before moving to the body's tissues. Intramuscular or IM injections are delivered directly into the muscle and begin working more quickly but don't have a long-acting effect. Subcutaneous or SQ injections are given under the skin and take longer to begin working but have a long-lasting effect. Intraperitoneal or IP injections are given directly into the peritoneal cavity of the abdomen and are used when IV or IM injections can't be used or when a large amount of solution is needed to be given to ensure rapid absorption.

Types of Medications

There are many types of drugs used in veterinary medicine. **Emetics** are drugs used to induce vomiting. A drug that causes emesis is controlled by the medulla in the brainstem and is used to induce vomiting in animals that have ingested toxic substances or foreign objects. These drugs should not be used in all situations of poisoning or ingesting objects. There is a risk of aspiration or inhaling stomach contents into the lungs during vomiting. Any corrosive substances or sharp objects should also never be induced. Some examples of emetics include hydrogen peroxide, apomorphine, and ipecac syrup. Emetics do not always work consistently. Apomorphine can be administered by injection or by placing a tablet in the conjunctiva of the eye. It tends to be less effective in cats but should produce a vomiting effect within 10–15 minutes. Ipecac syrup and hydrogen peroxide are both given by mouth and typically take 10–30 minutes to work, as they need to pass from the stomach into the intestines and then be absorbed.

Antiemetics are drugs that aid in preventing or decreasing vomiting. It is important to determine the cause of the vomiting, but several types of medication can help manage certain vomiting cases. Motion sickness is common in animals, and some medications can help prevent nausea and vomiting in these circumstances. The antihistamine meclizine (Dramamine) may be beneficial in decreasing motion sickness. Cerenia is an approved medication to help relieve acute vomiting and motion sickness. Metoclopramide (Reglan) can be helpful in cases of vomiting due to slowed intestinal motility. Cisapride may help reduce regurgitation in dogs with megaesophagus and in cats with hairballs.

Antidiarrheals are medications used to treat various types of diarrhea. Common medicines used to treat diarrhea include loperamide (Imodium), aminopentamide (Centrine), Bismuth subsalicylate (Pepto-Bismol), and kaolin and pectin (Kaopectate).

Adsorbents and protectants are often used when toxins enter the body and help substances to adhere to the outer surface of the drug, reducing contact with the intestinal wall. Common adsorbents include activated charcoal, which may include a substance called sorbitol to help eliminate the charcoal from the body. Sorbitol additives should not be used in patients with dehydration.

Laxatives are stool softeners that help to gently soften stool and evacuate the bowel. These drugs pull water into the bowel and retain water within the feces. Metamucil and milk of magnesia are common laxatives. Docusate sodium succinate (Colace) is a stool softener that acts as a water agent to help water penetrate the feces. Lactulose is a common stool softener that draws water into the colon, allowing fecal material to become more liquid.

Antacids reduce the acidity in the stomach and are typically composed of calcium, magnesium, or aluminum. Common antacids include cimetidine (Tagamet), ranitidine (Zantac), and famotidine (Pepcid).

Antiulcer medications are used to treat ulcers of the stomach and upper small intestine and to aid in prevention of ulcers in patients that are on multiple medications at one time. Sucralfate (Carafate) is a common antiulcer medication that creates a sticky paste that adheres to the ulcer site, protecting it from the acidic environment of the stomach. Omeprazole (Gastrogard, Prilosec OTC) binds to the surface of the stomach cells and decreases acid secretions to protect the stomach lining.

Anti-inflammatory drugs or nonsteroidal anti-inflammatory drugs (NSAIDs) are used to reduce pain and inflammation (see Figure 41–20). NSAIDs have few adverse effects but in large doses can cause GI irritation that can lead to gastric ulcers or decreased blood flow

FIGURE 41–20 Example of an NSAID.

to the kidneys. Most NSAIDs are metabolized by the liver and can cause liver damage with long-term use. It is recommended that animals on long-term NSAID use have regular blood work to monitor liver enzyme functions. Carprofen (Rimadyl), meloxicam (Metacam), deracoxib (Deramaxx), and etodolac (Etogesic) are commonly used NSAIDs in veterinary medicine.

Glucocorticoids are corticosteroids that produce an anti-inflammatory effect that is short acting, usually between 12 and 24 hours. These include prednisone, prednisolone, methylprednisolone, and triamcinolone (Vetalog). Longer-acting corticosteroids, such as dexamethasone, remain effective for 48 hours or longer. Weaning doses are used with these corticosteroids, as there can be many adverse effects if they are stopped abruptly. Overuse of corticosteroids can produce Cushing's disease, with signs including hair loss, swollen abdomen, muscle wasting and atrophy, polyuria (PU), polydipsia (PD), polyphagia (PP), and slow healing of wounds.

Antiarrhythmic drugs are used to treat abnormal patterns of electrical activity in the heart. There are two main types of arrhythmias: (1) those that result from increased heart rates (tachycardia) and (2) those that result from decreased heart rates (bradycardia). Lidocaine is an injectable drug used to decrease the rate of movement of sodium into the heart. Beta-blockers are drugs that cause the heart to contract with less force. Common beta-blockers are propranolol and atenolol.

Positive **inotropic** drugs increase the strength of contraction of a weakened heart. Digoxin is the common medication for long-term effects and makes more calcium available for the contractions within the heart muscle cells. Digoxin toxicity is possible, and owners should be instructed to watch for signs of vomiting, diarrhea, and anorexia. Pimobendan is an inotropic drug with vasodilator effects and helps treat congestive heart failure due to cardiomyopathy. Dobutamine is used in short-term cases of heart failure and shock.

Vasodilators are used to treat low blood pressure caused by heart failure. These drugs expand constricted vessels, making it easier for the heart to pump blood. Nitroglycerine relaxes venous blood vessels and helps dilate arterioles. The drug is absorbed through the skin and mucous membranes. Enalapril (Enacard) is an ACE inhibitor that relaxes the smooth muscles of veins and arterioles in patients with heart disease involving the ventricles.

Diuretics are drugs that increase urine formation and promote water loss (diuresis). Diuretics are often used in cases of congestive heart failure and help decrease or prevent occurrences such as water retention that leads to pulmonary edema, ascites, and increased cardiac workload. Removing water from the body helps to reduce these critical conditions. Diuretics must be used with caution in animals with **hypotension**, or low blood pressure, and **hypovolemia**, or low blood volume. A common diuretic is furosemide (Lasix), which acts by inhibiting sodium absorption and water retention from the loop of Henle in the neurons within the kidneys. When sodium absorption is delayed, osmosis causes additional water retention in the kidneys. Sodium is also retained in urine, which causes further water retention to occur, and the water is not resorbed. Mannitol is a sugar carbohydrate used in diuretics to promote osmosis. Mannitol is poorly resorbed and during osmosis, water retention occurs within the kidneys. This process allows medications used as a diuretic to help reduce cerebral edema that may be caused by head trauma.

Bronchodilators are drugs that decrease bronchiole constriction and help to open up the airway for improved breathing. These medications are available in injectable, oral tablet, and inhaler form. Common bronchodilators include aminophylline and theophylline. Aminophylline is often used, as it is available as an injectable. Albuterol is a common inhaler.

Expectorants are medications that increase the fluid content of mucous within the respiratory tract to allow the body to better rid itself of these secretions. Guaifenesin is a common oral expectorant. Many expectorants also have **decongestants** that reduce congestion within the respiratory tract and mucous membranes. A common decongestant to reduce nasal congestion is phenylpropanolamine.

Antitussive drugs work by blocking the cough reflex within the brainstem and help to suppress coughing that aides in removing mucous and other debris that accumulate within the bronchi. This type of medicine is typically used in patients with a nonproductive cough that is dry and low within the chest and produces little or no mucous. Antitussives are commonly used in cases of tracheobronchitis or kennel cough, which often create a gagging or retching type of reflex and are extremely irritating to the upper airway. Butorphanol (Torbutrol) is an opioid with antitussive cough suppressant properties and causes very little sedation. Hydrocodone (Hycodan) is a controlled substance with antitussive properties that causes mild sedation and with long-term use can cause constipation. Codeine is a weak opioid found in many cough suppressants and causes sedation similar to that of hydrocodone.

Reproductive drugs include hormones that are used to prevent pregnancy or alter the state of the uterus to improve breeding cycles and timing of breeding. Medications can be used to suppress the estrus cycle and prevent pregnancy from occurring. Megestrol acetate or Ovaban is an oral progestin that is used as a contraceptive in dogs and cats. Prolonged use of megestrol can lead to pyometra, endometritis, mammary gland tumors, and cysts within the reproductive tract. Estradiol cypionate (ECP) is commonly used as an injectable estrogen when mismating occurs in dogs. The estrogen levels prevent pregnancy by increasing the number and thickness of folds within the oviducts and prevent sperm and ova from entering the uterus. Hormones used to improve estrus cycles and timing for mating include gonadotropin-releasing hormone (GnRH), which stimulates the release of luteinizing hormone (LH) and follicle-stimulating hormone (FSH) from the pituitary gland, causing the ovary to develop follicles. These hormones are also called gonadotropins. Human chorionic gonadotropin (hCG) is produced by pregnant women, and equine chorionic gonadotropin (eCG) is produced by pregnant mares. hCG and eCG are agents occasionally used to induce estrus in dogs and cats. Oxytocin is a naturally occurring hormone used to increase uterine contractions in cases of dystocia due to a weak or fatigued uterus.

Hypothyroid drugs are used to treat hypothyroidism or the insufficient production of thyroid hormone. These medications supplement the hormone T4, which is delivered to various organs and tissues, where it is converted to T3. The most common hypothyroid medicines are levothyroxine (T4) and synthetic levothyroxine (T3) (see Figure 41–21). The synthetic medicine is beneficial in that T3 supplementation does not require conversion in the body; however, it is costly and requires administration

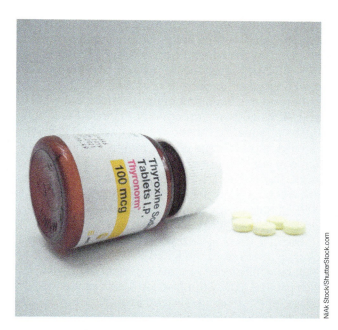

FIGURE 41–21 Example of a thyroid supplement.

three times a day. The most common drug of choice is Soloxine for T4 supplementation, as it is more cost effective and only requires once daily treatment. Hyperthyroid drugs are used to treat increased thyroid hormone production. Hyperthyroidism is most common in cats and is often caused by a hormone-secreting thyroid tumor. Methimazole (Tapezole) is often used to control hyperthyroidism in cats by blocking the tumor's ability to produce hormones T3 and T4. This medicine is available in oral tablet form, or it may be compounded into a topical medicine to be used on the ear pinna. Radioactive iodine may be used as an IV treatment by placing iodine as a normal component of the thyroid hormone into the active tumor cells, which are then destroyed by radiation. Care must be taken—including proper use of PPE—when handling radioactive or chemotherapeutic agents.

Hyperadrenocorticism or Cushing's disease affects the adrenal glands by producing excessive glucocorticoids within the body. The causes of this condition are pituitary tumors, adrenal tumors, or prolonged use or increased amounts of corticosteroids. The treatment is to suppress the adrenal gland either by surgical removal of the gland or discontinuing corticosteroid use. Mitotane (Lysodren) causes the adrenal glands to become necrotic, which decreases the release of steroids. It is used daily for initial treatment and then weaned to twice a week administration. Anipryl causes a decrease in dopamine, which increases the production of cortisol. All treatments can lead to the reverse effect of hypoadrenocorticism.

Hypoadrenocorticism or Addison's disease is caused by a decreased secretion of natural corticosteroids from the adrenal glands or the hormone ACTH. ACTH or adrenocorticotropic hormone causes the adrenal gland to secrete corticosteroid Thus, the treatment is supplementation of corticoid with medications such as Percorten or Florinef. Percorten requires adequately functioning kidneys and is given as an injectable in dogs once a month. Florinef is used as an oral supplement in dogs and cats. Glucocorticoid agents may be required in addition and include dexamethasone, prednisolone, and prednisone. Proper use of PPE should be implemented to ensure care in handling any type of hormone medication.

Insulin is used to treat patients with diabetes by moving glucose from the blood into the tissues. Lack of natural insulin release results in diabetes mellitus, which causes high blood glucose levels as well as glucose in the urine. Insulin administration can control blood glucose levels through daily injections (see Figure 41–22). Several insulin choices for animals include NPH insulin in dogs and longer-acting insulin in cats, such as Glargine or PZI. Humulin N and PZI insulin are often used in dogs and provide moderate duration.

Anticonvulsants are used to treat seizure activity. Seizures cause a loss of consciousness, altered muscle tone and movement, neurologic problems, and altered levels of sensation. Phenobarbital is the most commonly used anticonvulsant. It is used as a long-term treatment and is an inexpensive barbiturate given one to two times daily. Side effects of long-term use include increased liver enzymes and liver damage. Yearly phenobarbital blood levels and internal organ function blood work are necessary for adequate control of seizures and long-term side effects. Phenobarbital is measured in grains or gr. 1 grain = 60 mg. Potassium bromide or KBr is another long-term seizure control medicine and may be used alone or in conjunction with phenobarbital. It works by decreasing neuron activity but takes at least 1 month of use for adequate therapy. Potassium bromide blood levels are measured yearly for proper dosing. Valium is the drug of choice during active seizure episodes and is given IV.

Muscle relaxants are used to relax a patient's muscles during tremors or acute inflammation, as well as and trauma-related injuries of skeletal muscle and treatment of muscle spasms. Methocarbamol (Robaxin-V) is the most common muscle relaxant and is available in oral and injectable forms.

Antimicrobials are medications that kill or inhibit the growth of microorganisms, such as bacteria, protozoa, or fungi. Antimicrobials are often referred to as antibiotics. There are several classes of antimicrobials: **bactericidal** (kills bacteria), **bacteriostatic** (inhibits bacterial growth), **virucidal** (kills viruses), **fungicidal** (kills fungus), and **protozoastatic** (inhibits growth of protozoa). Antimicrobials work to kill or inhibit growth of microorganisms at the cell wall, cell membrane, enzymes, ribosomes, and nucleic acids. Some organisms develop a resistance to long-term use of medications. Bacteria are known to become resistant to certain drugs due to long-term use, genetic mutations of bacteria, and acquired resistance of chromosome makeup. The penicillins were among the earliest classes of antibacterial drugs. Penicillins are divided into subclasses based on chemical structure (e.g., penicillins, monobactams, and carbapenems), spectrum (narrow, broad, or extended), source (natural, semisynthetic, or synthetic), and susceptibility to β-lactamase destruction. Manipulation of some drugs has improved the spectrum, resistance to β-lactamase destruction, or clinical pharmacologic characteristics that enhance efficacy. They are typically absorbed well when they are injected or given orally. The only penicillin that should not be given by mouth is penicillin G, which is given as an injectable form due to inactivation by gastric acids. The penicillins are commonly used to treat or prevent local and systemic infections caused by susceptible bacteria. Several acute infectious disease syndromes are specifically responsive. Because of their synergistic interaction with other antimicrobials, they are often used as part of combination therapy. Penicillins also are used topically in the eye and ear as well as on the skin; intramammary administration is common for treatment or prevention of bovine mastitis. Amoxicillin with or without clavulanic acid is among the first-choice antimicrobials for treatment of canine or feline urinary tract infections. Tetracyclines

FIGURE 41–22 Example of types of insulin.

are bacteriostatic drugs used when treating Gram-positive organisms. Doxycycline and minocycline have a longer half-life and are less likely to cause resistance. Some organisms develop a resistance to long-term use of sulfonamide medications and other bacteriostatic and bactericidal drugs. Trimethoprim is a compound commonly combined with sulfa drugs and is effective against many Gram-positive organisms. Cephalosporins are bactericidal drugs that are used to treat both Gram-positive and Gram-negative organisms. There are multiple generations of cephalosporins categorized by when they were developed. They are absorbed well by the gastrointestinal tract. Common cephalosporins include cefazolin, Naxcel, Cefa-Tabs, Convenia, Simplicef, and Keflex. Bacitracins are polypeptide antibiotics with Bacitracin A as the main ingredient. Bacitracins are often composed in topical creams and ointments. More broad-spectrum activity is achieved when bacitracin is added to polymyxin B and neomycin. Common bacitracins include BNP ointment and Neosporin.

Aminoglycosides are bactericidal drugs used to treat aerobic bacteria, which require oxygen to survive. These medications may be toxic to the kidneys and inner ear at normal dosages. Common aminoglycosides include gentamicin, amikacin, tobramycin, and streptomycin. Fluoroquinolones are bactericidal drugs effective against specific diseases caused by both Gram-negative and Gram-positive bacteria. These medications should not be used in young and growing animals, as they may cause arthropathies in immature animals. Common fluoroquinolones include enrofloxacin (Baytril), ciprofloxacin, Orbax, and Zeniquin. Metronidazole (Flagyl) is a bactericidal that is effective against protozoa that cause intestinal diseases such as giardia and trichomoniasis. Macrolides are bacteriostatic drugs that help in cases of bacterial cross-resistance and include erythromycin and azithromycin. Lincosamides are bacteriostatic or bactericidal depending on the concentration and site of infection. They are typically used to treat Gram-positive aerobic cocci and include clindamycin or Antirobe.

Antifungals are drugs used to kill and treat fungus. Ketoconazole and fluconazole are antifungals with the least amount of side effects. Griseofulvin is a fungistatic drug used to treat dermatophytes—such as ringworm—that infect skin, nails, and hair follicles. Nystatin is an antifungal used to treat candidiasis on the skin, mucous membranes, and lining of the gastrointestinal tract. It is toxic to tissues and should be used with caution.

Preanesthetics and Anesthetics

Tranquilizers and sedatives are used as preanesthesia agents for pain relief and mild to moderate sedation. A **tranquilizer** acts as a drug to relax and calm an animal without significant drowsiness. A **sedative** provides a mild degree of drowsiness, but pain effects are still present. Acepromazine maleate is a common tranquilizer that reduces anxiety and causes a mentally relaxed state. It may be used to calm an animal for exam or transport but provides no pain relief. Other common tranquilizers include diazepam (Valium) and midazolam, which are used as preanesthesics due to their adequate calming and muscle-relaxing effects. Droperidol is a sedative that provides moderate relaxation and drowsiness. Xylazine (Rompun), dexmedetomidine (Dexdomitor), and detomidine (Dormosedan) produce calming effects and decrease an animal's excitability. They also provide a mild amount of pain relief. Xylazine often causes vomiting in dogs and cats. An **analgesic** is a drug that reduces pain without the loss of other sensations. Hydromorphone is commonly used as a preanesthesic as an analgesic agent. Butorphanol (Torbutrol, Torbugesic) and tramadol are used for generalized pain relief following surgery. Buprenorphine is often combined with sedatives and tranquilizers as a long-duration analgesic, which lasts between 6 and 8 hours.

Anesthetics are used to induce anesthesia and provide complete loss of pain, loss of consciousness, and all sensory sensations. A common injectable anesthetic is propofol, which is injected as an IV bolus for rapid induction of anesthesia with a short time frame of unconsciousness (see Figure 41–23). It may cause

FIGURE 41–23 Example of an injectable anesthetic.

mild pain during administration and should not be administered rapidly; otherwise, it may cause respiratory depression. Ketamine is a short-acting injectable anesthetic that causes the animal to feel dissociated from its body. This effect causes no corneal, laryngeal, or pharyngeal reflexes, a lack of muscle relaxation, and an increased heart rate. This makes the agent inappropriate as a sole agent for surgery. Ketamine is often coupled with another analgesic for added benefits, such as Telazol, and is often used as a restraint agent for procedures such as radiographs, nail trims, and exams of uncooperative patients. General anesthetic agents are for long-term use during surgery and are used as inhalants. Common inhalants include isoflurane and sevoflurane. Both provide smooth induction and rapid recovery.

Respiratory stimulants are used to stimulate respiration, especially in surgical patients. Doxapram (Dopram) is a central nervous system (CNS) stimulant that increases respiration in animals with apnea or bradypnea. **Apnea** is a condition when an animal is not breathing, and **bradypnea** is slow or decreased breathing. Dopram can be used to stimulate breathing in newborns following C-sections and may be administered via the umbilical cord or sublingually.

Anthelmintics and Antiparasitics

An **anthelmintic** is a dewormer used to kill or prevent internal parasites (see Figure 41–24). Antinematodal agents are used to treat nematodes infections such as roundworms, hookworms, whipworms, and strongyles. Anticestodal agents treat infections with cestodes such as tapeworms and segmented flatworms. Common anticestodals include Droncit and Cestex. Antitrematodal agents treat infections with trematodes such as flukes or unsegmented flatworms. Antiprotozoal agents treat infections with protozoa such as giardia, coccidia, and toxoplasmosis. The most common antiprotozoal agent is Albon. Common internal parasitic agents include pyrantel pamoate (Strongid, Nemex) as the active ingredient and fenbendazole (Panacur). It is used to treat and prevent a variety of nematodes in domestic animals.

Avermectins are broad-spectrum anthelmintics and are commonly used as internal parasite control. The most common is ivermectin. Ivermectin (Ivomec,

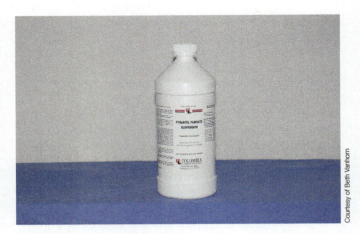

FIGURE 41–24 Example of an anthelmintic.

Heartgard) is used in many species as a dewormer and is an active ingredient in many heartworm preventatives, but must be used with caution in Collie and Collie-mixed dogs due to sensitivity and adverse reactions. Immiticide is the agent approved to treat heartworm disease as an adulticide. The active ingredient is melarsomine dihydrochloride, and it is given as an IM injection deep within the lumbar muscles.

External antiparasitics are commonly used to treat and prevent fleas, ticks, lice, and biting flies. Most topical applications are applied to the base of the neck around the shoulder blade area and used on a monthly basis. Imidacloprid is the active ingredient in Advantage, which is used as a topical to treat fleas, and Advantix, which is used to treat fleas and ticks. Fipronil is the active ingredient in Frontline and Revolution, both of which are used to treat fleas and ticks. Lufenuron inhibits the development of parasites, including egg formation and larval development. Program and Sentinel are oral medications that distribute throughout the animal's tissues; when the animal is bitten by a flea, it prevents the eggs from hatching. Amitraz is an insecticide used to treat demodectic mange and is also incorporated into a dip product called Mitaban and a tick collar called Preventic. Pyrethrins and permethrins are common ingredients in insecticide sprays. They have a fast-acting effect on insects and are often used as premise sprays. There are many new external parasite control medications available based on veterinary recommendation and patient health status. These products include Simparica, Nexgard, Seresto collar, and Capstar. Please refer to Chapters 8 and 9 for discussions and tables relating to internal and external parasite control options.

SUMMARY

Pharmacy techniques and skills require accuracy. When handling and dispensing medicines, it is important to make sure the right patient, right medicine, right strength, and right dose are determined. If you are unsure of any item on a prescription, check with the veterinarian for clarification. It is best to ask the veterinarian to clarify what is written rather than risk incorrectly filling a prescription. It is important when calculating and dispensing medications to always check your work at least three times for accuracy.

Key Terms

absorption the movement of drug particles from the site of administration into the body system

analgesic drug that reduces pain without loss of other sensations

anesthetic drug that causes loss of pain, consciousness, and all sensory sensations

antacid drug that reduces acidity of the stomach

anti-inflammatory drug used to treat inflammation and pain without corticosteroid components

antiarrhythmic drug used to treat abnormal electrical currents of the heart

anticonvulsant drug used to treat seizure activity

antiemetic drug used to stop vomiting

antifungal drug used to kill or treat fungus

antimicrobial drug that kills or inhibits growth of microorganisms; often called an antibiotic

antitussive drug that suppresses coughing and blocks the cough reflex within the brainstem

antiulcer drug used to treat or prevent gastric ulcers

apnea condition in which the patient is not breathing

aural pertaining to the ears

bactericidal kills bacteria

bacteriostatic inhibits growth of bacteria

balling gun metal device with a long handle that has a plunger at the base; used to hold medicines to administer to large animals

bradypnea decreased rate of breathing

bronchodilator drug that decreases bronchiole constriction and opens up the airway

controlled substance license paper given by the DEA to allow a veterinarian to prescribe scheduled drugs

controlled substance log requires written entry for every controlled substance dispensed by the facility

controlled substance prescription drug that has the potential for abuse or addiction; also called scheduled drug

decongestant medicine used to reduce nasal congestion

distribution movement of a drug into the body's tissues

diuretic drug used to increase urine production

dram measurement size of vials or bottles

Drug Enforcement Agency (DEA) government agency that sets regulations and guidelines for ordering, storing, and dispensing controlled drugs

ear mites   microscopic parasites that live within the ear canal

efficacy   strength of the drug

elimination   removal of a drug from the body

emetic   drug used to induce vomiting

expectorant   medicine that increases the fluid content of mucous

fungicidal   kills fungus

glucocorticoids   corticosteroids used to treat inflammation

hematoma   rupture of a blood vessel that causes a pocket of blood

hypotension   low blood pressure

hypovolemia   low blood volume

inotropic   drug that increases the contractile strength of the heart

laxative   stool softener

metabolism   alters drugs within the body prior to elimination

over-the-counter (OTC) drugs   medications that do not require a prescription

pet piller   small device, usually made of plastic, that has a long thin handle with a plunger on the end to administer medications to small animals

pharmacokinetics   how drugs move into, through, and out of the body

pharmacology   the study of drugs and their sources

pharmacy   the area where medications are stored and prepared for veterinary patients

pill counting tray   tool used to count out tablets or capsules to send home with a client

pinna   the ear flap

prescription (Rx)   type of medication, amount of medication, and directions for use of the medication prepared by the veterinarian

prescription drug   medication prescribed by the veterinarian

prescription label   indicates to the owner how to properly give the medication and identify the medication within the container

protozoastatic   inhibits growth of protozoa

scheduled drug   prescription drug that has the potential for abuse or addiction; also called controlled substance

sedative   drug that causes drowsiness but no loss of pain sensation

target tissue   location where a drug travels within the body; location of its intended use

topical   drug that is administered to the skin

tragus   opening of the ear canal structure; includes the haired area

tranquilizer   drug that causes relaxation and calmness without drowsiness

vasodilator   drug used to treat low blood pressure due to heart failure

virucidal   kills viruses

Review Questions

1. What items should a prescription include?
2. What items must be included on a label for a medication that is being dispensed?
3. What items should be checked at least three times for accuracy when dispensing medicines?
4. Complete the following pharmacy questions based on the information given:

    2 caps Cephalexin 500 mg BID × 14

    a. What does the prescription say in the above statement?

    b. How many items should be dispensed for this dosage?

    c. What total mg amount is given per day?

    ½ tab Amoxicillin 100 mg TID × 10

    d. What does the prescription say in the above statement?

    e. How many items should be dispensed for this dosage?

    f. What total mg amount is given per day?

5. Why is a pill counting tray used?
6. What items are recorded in a controlled substance log?
7. What is the difference between a pet piller and a balling gun?
8. What is an aural medicine?
9. What is a topical medicine?
10. What is the importance of storing items out of a direct light source?
11. A 14-pound dog is prescribed a medicine at 2 mg/kg. The drug comes in a 50 mg/ml concentration. How much will you prepare?

Clinical Situation

Amy, a veterinary assistant at Seaside Vet Clinic, is working in the pharmacy to dispense medications to hospitalized patient. She has several prescriptions to fill for Dr. Andrews. Amy has to fill the following prescriptions for the same patient:

Advantage Canine for a 50# dog, apply 1 dose topically once a month for 6 months
2 capsules Doxycycline SID for 3 weeks
1 tablet Cefa-Tabs BID for 1 week
1 ml Amoxi-Drops BID for 10 days

- What information does Amy need to fill each prescription?
- How should each prescription be written for the client?
- How many items of each prescription will Amy send with the patient?

CHAPTER 42 Surgical Assisting Procedures

Objectives

Upon completion of this chapter, the reader should be able to:

42.1 Describe how to maintain asepsis during surgical procedures
42.2 Explain how to maintain and record in a surgical logbook
42.3 State the importance of a sterile surgical suite
42.4 Explain how to assist with anesthesia preparations and inductions
42.5 Demonstrate how to properly restrain the patient for intubation
42.6 Demonstrate how to properly clip and prepare the patient for surgery
42.7 Demonstrate how to properly handle and open sterile surgical packs
42.8 Demonstrate how to properly prepare surgical drapes, gowns, instrument packs, and towel packs for surgical procedures
42.9 Demonstrate how to properly use an autoclave for sterilization procedures
42.10 Describe how to properly assist the surgeon in gowning
42.11 Explain how to properly position the patient for surgery
42.12 Describe how to place and maintain patient monitors
42.13 Explain the planes of anesthesia
42.14 Discuss how to properly monitor a patient during anesthesia and postoperative recovery
42.15 Explain how to set up, maintain, and disconnect the anesthesia machine
42.16 Describe the difference between rebreathing and nonrebreathing anesthesia systems
42.17 Discuss how to properly refill the vaporizer and soda lime components of the anesthesia machine
42.18 Identify components on the anesthesia machine
42.19 Explain how to select appropriate-sized rebreathing bags for the patient
42.20 Explain how to properly test the functions of the anesthesia machine
42.21 Explain how to make adjustments to the anesthetic flow of gases during induction, maintenance, and recovery phases
42.22 Describe the anesthesia and surgical report
42.23 Explain how to properly glove and gown for surgical assisting
42.24 Discuss how to properly extubate a patient during recovery
42.25 Explain how to review the postoperative care of a patient
42.26 Explain how to complete a suture removal appointment
42.27 Explain how to clean and maintain the surgical suite

Introduction

Surgical assisting requires the veterinary assistant and all other staff members to understand and maintain **aseptic techniques**. These are precautions taken to prevent contamination and infection of a surgical incision. This is essential in the surgical suite. This requires knowledge and attention to the patient and the staff members working in the surgical area. **Asepsis** governs the patient, the health care team, the surgical suite, the instruments, and all items housed within the surgical area. Cleaning and sanitation are a vital part of keeping the surgery suite sterile and aseptic. Any break in asepsis may lead to a potentially life-threatening patient infection, delayed healing, or patient death.

Surgical Asepsis

The goal of surgical aseptic technique is to prevent any organisms from entering a patient via the surgical incision, inhalation, or intravenous (IV) anesthesia. The first barrier to entry is the skin. Lengthy surgeries pose more of a potential risk for infection, as well do surgeries involving deep body cavities, immune-compromised patients, and orthopedic surgeries. Many preparation techniques have been developed to ensure patient safety.

Proper disinfecting and sterilization techniques help reduce the number of living organisms in the surgical environment. Any item that is not sterile is not usable. The goal is to maintain the utmost sterility of every object that comes in contact with the patient during surgery. This would include the patient, anesthetic equipment, the surgery table, surgical prep equipment, surgical instruments, and staff members.

Another area that must be considered in maintaining asepsis is the surgical suite's ventilation and airflow. The surgical suite should have a separate ventilation area from the rest of the hospital. Surgical doors should be kept closed during all surgery procedures. Staff member traffic should be kept at a minimum to reduce the risk of spreading organisms.

Only sterile surgical procedures should be completed in the surgery room. This means such procedures as flushing abscesses, unblocking cats with urinary obstructions, suturing wounds, and performing dentals should be done in the treatment area to reduce the spread of living organisms.

Surgical Logbook

All aspects of the surgery should be noted in a surgical log and report. This includes presurgical documentation, surgical notes, and postoperative recovery stages. The surgical consent form, surgical fee estimate, anesthesia report, surgical report, and recovery report are forms that should be included in each patient's medical record. The **surgical log** is a record of details recording all surgeries performed in the facility (see Figure 42–1). Similar to the radiology log, this book is used to meet legal and state regulations in noting all procedures completed within the facility. The information within the surgery log should include the following:

- Surgery date
- Patient name or number
- Client name
- Breed/Species
- Gender
- Weight
- Procedure(s) performed
- Preanesthetic medications administered, dosage, and route of use
- Anesthetic administered, dosage, and route of use
- Surgical assessment score
- Technician and assistant initials
- Veterinarian's initials
- Length of surgery time
- Laboratory specimens taken
- Surgical comments

The surgery logbook is kept in the surgical suite and entries are recorded prior to surgery and at the completion of the surgery.

Anesthesia Logbook

An **anesthesia logbook** may also be kept in addition to or included in the surgery logbook. The anesthesia log details the patient's status through induction, maintenance, and recovery stages. This should also include temperature, pulse, respiration, and blood pressure (TPR and BP) every 5 minutes, fluid type and amount received during the procedure, pain medicines used, and oxygen flow rates. These items should be noted every 5 minutes throughout the surgical procedure. Items that should be recorded in the anesthesia log should include the following:

- Preanesthetic drug and dosage
- Anesthetic drug and dosage
- Route of drug use (IV, SQ)

- Time medications given
- Pain relief drugs and dosage
- Vital signs
- Complications
- Length of anesthesia
- Technician initials

The anesthesia log is helpful in determining what medications were used during each patient's procedure and any adjustments that may be necessary for follow-up procedures. The log is also legal documentation of the anesthesia used during surgery.

ANESTHESIA/SURGERY LOGBOOK

Date	Patient	Client	Procedure(s) Performed	Drugs Administered	Dosage	Route(s) of Administration	Length of Procedure	Surgeon(s)	Anesthetist

Comments:
_____
_____
_____
_____
_____
_____

FIGURE 42-1 A sample anesthesia/surgery logbook.

Surgical Suite Maintenance

The surgical suite should be the cleanest area of the veterinary facility. The surgery area should only be used for surgical procedures. Patient preparation should be completed in the treatment area and the patient carefully moved to the surgical suite to reduce risk of contamination. The surgery room should be kept closed to prevent spread of organisms. Only surgical staff should be in the room during surgeries. The surgery room itself should have separate cleaning equipment from all other areas of the hospital. A scrub area should be provided outside of the surgery suite to allow the doctor and any assistants to complete a surgical scrub, room to gown and glove, and complete preparations for entering the sterile surgery area. Sanitation should occur after each surgery, and a final sanitation and disinfection procedure should be completed at the end of every surgery day.

Ceiling Sanitation

The ceiling of the surgical suite should be spot-cleaned daily. The entire ceiling should be mopped once a week using a sponge mop and bucket used only within the surgical suite. The ceiling should be cleaned prior to other areas, as dust particles and hair or other debris that have collected on the ceiling may fall onto other objects during cleaning. Dry vacuuming may be necessary depending on the ceiling surface material. When vacuuming areas within the surgery area, use a clean filter and bag each time to prevent spread of organisms. Ventilation fan filters should be changed on a weekly basis.

Wall Sanitation

The walls of the surgical suite should be spot-cleaned after each surgery. Each wall surface should be mopped down using a sponge mop and bucket on a daily basis. A vacuum may be used as necessary as outlined in the ceiling sanitation section. Spot-cleaning of walls should be done with a disinfecting cleaner and paper towels.

Counter and Shelf Sanitation

Counters, tabletops, shelves, sinks, and waste containers should be disinfected on a daily basis. Items that require spot-cleaning should be cleaned between each surgery using a disinfectant and paper towels. All flat surfaces and edges should be wiped down and kept clean. All items that have been used and are disposable should be placed within a medical waste container. The medical waste container should be emptied each day or when full between surgeries.

Floor Sanitation

Floors should be mopped on a daily basis and may require mopping between surgeries. The surgical mop and bucket should only be used in the surgery area. It may be necessary to label these items "for surgery room use only." Never use a regular hospital mop and bucket that have been used in other areas of the facility. This is an easy way to contaminate the surgery room from highly contagious sites.

Surgical floor mopping should include the dual mop method. This process involves one bucket containing fresh warm water for mop rinsing. A second bucket should contain a disinfectant solution for actual floor mopping. The mop is placed in the disinfectant solution and wrung out well. A section of floor is mopped in the furthest area of the surgical room. The mop is then placed in the rinse water and wrung out. It again goes into the disinfectant solution and is wrung out and mopping continues. This process continues until the entire floor area is mopped. At the end of the mopping, the mop is rinsed and disinfected and wrung out well and hung to dry. Empty both buckets immediately and refill them for each use. The mop head should be washed weekly. This should be done in a washing machine with hot water and bleach. Multiple mop heads should be available for regular changing and cleaning purposes.

Equipment

All surgical equipment should be cleaned and disinfected following the manufacturer's recommendation. Any surgical equipment that is located within the surgery suite may not be able to be cleaned depending on the part of the equipment and its power source. It is important to read the cleaning and care guidelines for all items. Permanent fixtures, such as surgical lights, should be wiped down and cleaned daily. The surgery table must be disinfected after each use and should include cleaning the top surface, edges, bottom surface, and base (see Figure 42–2). The same procedure applies to any **Mayo stand**, or the instrument tray that is elevated to hold the surgeon's instruments and supplies. Surgical ties should be washed weekly unless they are soiled or contaminated by bodily fluids, in which case they should be laundered immediately. Any surgical positioner and restraint devices should be disinfected after each use. It is important to note that any disinfectant used in the surgical suite be safe for patient contact as most items are used in close contact with the patient.

838 SECTION IV Clinical Procedures

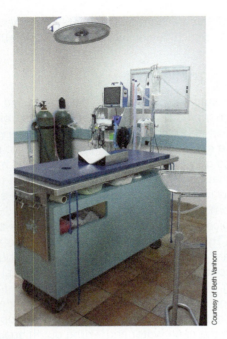

FIGURE 42–2 The surgical suite, table, and equipment must be cleaned and maintained on a daily basis.

Pressurgical blood work is frequently performed and may be required by the facility or the type of surgical procedure involved (see Figure 42–3). The information provided by the blood work helps to determine the anesthetics that will be administered to the patient and the anesthesia classification of the patient. The classification of each patient is a presurgical assessment score that rates the patient in a class between I and V. The age and physical status of the patient serve as a guideline for the classification. The older the patient, the higher the score and the more surgical protocols that will be in place during the procedure. Table 42–1 summarizes the classifications.

The choice of presurgical blood testing, preanesthetic drugs, anesthesia, and surgical monitoring will vary according to the classification assessment. Each facility will have its own protocol and doctor preference for presurgical and surgical procedures. The choice of preanesthetic medications will depend on several factors, including the following:

- Species/breed
- Age
- Anesthesia

Preanesthetic Patient Care

Any patient undergoing a surgical procedure should be admitted to the hospital early in the day to complete necessary presurgical blood work and physical exams prior to the surgery. All consent forms and fees, as well as an estimate and review of procedures, should be provided to the client to complete as soon as the patient is admitted. Emergency surgical procedures are worked into the schedule dependent on how soon the patient needs to be attended to. The early admission also allows patients to **fast**, or not have any food or water, prior to surgery depending on the animal species and required fasting time. For some species this means a 12-hour fast, and for others fasting is not required. This is dependent on the animal's digestive system. Dogs and cats typically are fasted 12 hours prior to surgery. Rodents, cattle and other ruminants, and horses do not require fasting. Monogastric animals are fasted so that no contents are in the stomach, reducing the chances of vomiting during anesthesia.

The physical exam for surgery patients should include an evaluation of each body system, evaluation of the vital signs, and the overall health status of the patient. The veterinary assistant should evaluate the vital signs, and the veterinarian should complete the physical exam (PE).

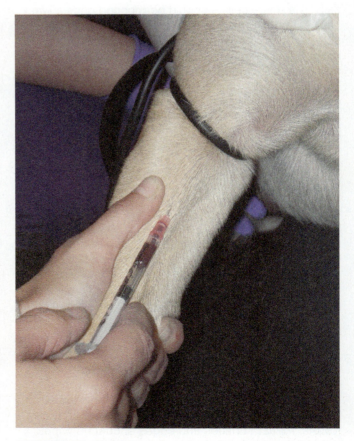

FIGURE 42–3 A blood draw is an important step in the presurgical assessment.

TABLE 42-1
Anesthesia Classification Assessment

Class	Description
Class I	Minimal risk: normal, healthy patient
Class II	Slight risk: slight systemic disease with no clinical signs; may include very young, geriatric, underweight, or obese patients
Class III	Moderate risk: moderate systemic disease; slight clinical signs noted
Class IV	High risk: severe systemic disease; severe clinical signs with a threat to life
Class V	Grave risk: moribund patient; may die with or without surgery within a 24-hour period

- Classification/Score
- Surgical procedure
- Length of surgery
- Veterinarian preference
- Patient weight

The purpose of preanesthetic drugs is to calm patients, reduce pain, reduce the amount of anesthesia necessary for the surgery, and reduce side effects of other drugs necessary during the procedure. All medications used for presurgical anesthetics should be recorded in the patient chart and any controlled substance noted in the log.

Fluid Therapy

A surgical protocol that may be used for every surgical patient or on a surgical case basis is fluid therapy. It is important that all surgical patients have an IV catheter placed during the preanesthesia preparation (see Figure 42–4). This provides access to the patient's bloodstream for easy administration of anesthesia medications and pain medicines. It is also access to the bloodstream should any complications occur during anesthesia or surgery for quick accessibility to provide emergency medications should the need arise. The IV catheter is commonly used as a fluid therapy site to provide fluids during surgery, which allows for correcting fluid deficits, electrolyte imbalances, and acid-base imbalances. Most IV catheters are placed in the cephalic vein by the veterinary technician and attached to an IV fluid bag and line. The fluids of choice are usually **isotonic crystalloid** fluids, such as lactated Ringer's solution (LRS) (see Figure 42–5). In small patients or young patients, a 5 percent dextrose solution may be used or added to the LRS. The veterinarian should select any fluids being used with the patient and determine the formulation of fluid therapy based on the level of hydration and patient assessment. The flow rate of the fluids is typically 10–20 ml/kg/hr. Dehydrated patients will require a higher flow rate. Patients with renal or cardiac damage should receive a reduced rate of flow, which should be established by the veterinarian. The veterinary assistant should be able to set up an IV bag and line along with the correct size of IV catheter necessary for the technician to place the catheter. The assistant should also be able to properly monitor the fluid rate and flow.

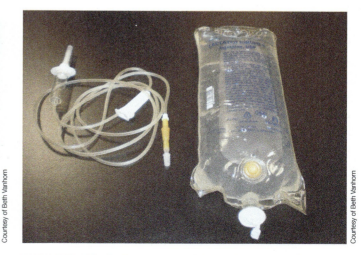

FIGURE 42–4 Examples of IV catheters; butterfly catheter.

FIGURE 42–5 Fluid bag and line (lactated Ringer's solution).

This includes monitoring the catheter site for signs of phlebitis or SQ fluid accumulation and monitoring the digits for signs of swelling. The catheter bandage must be kept clean and dry to prevent contamination and be changed immediately if wet or soiled with debris. Catheter bandages may require a plastic covering for protection in patients with incontinence. Many patients that move around with an IV line in place may cause the line to kink and occlude, thus preventing the flow of fluids. Catheters should be flushed several times daily to prevent occlusion and maintain patency, especially if any medication is administered IV through the catheter site.

Anesthesia Induction

The veterinary assistant will provide restraint for all patients being induced with anesthesia. The **induction phase** is the time when the patient is being given anesthesia to make it sleep and remain unconscious and free of sensation during the surgery. The induction phase is short term. The veterinarian or credentialed technician will administer the anesthesia. The assistant should properly restrain the patient depending on the type of anesthesia being administered. Anesthesia may be given as an injectable, such as IV or IM, or may be given as an inhalant, in which the patient inhales the anesthetic gases. The inhalant route may cause the animal to struggle, and enough staff members should be available for additional restraint. The inhalant route uses an anesthesia machine to continue to keep the patient anesthetized (see Figure 42–6).

Vital signs should be monitored throughout the procedure. A pulse oximeter is used to evaluate heart rate and oxygen levels (see Figure 42–7). Once an animal has been induced and has achieved an initial brief period of unconsciousness, the patient should be intubated with an endotracheal tube. **Intubation** is the process of placing a tube into the trachea to establish an airway that allows the patient to continue to inhale gases that keep it under general anesthesia as long as necessary. The **endotracheal tube** is a flexible tube that should fit snugly into the trachea.

The **plane** of anesthesia determines how awake or how asleep a patient is under anesthesia. A **blink reflex** can be evaluated to determine which plane of anesthesia a patient is in. The blink reflex should not be visible in an adequate plane of anesthesia for surgery. An eye ointment lubricant is placed into each eye to protect the cornea and prevent the eyes from drying out during anesthesia and surgical procedures.

Intubation Procedure

The veterinarian or veterinary technician performs the endotracheal tube placement as the assistant restrains for the procedure. Intubation consists of passing a tube through a patient's mouth and pharynx and into the trachea. This establishes an airway and allows the patient to be attached to the anesthesia machine during a procedure. It also prevents the patient from inhaling and aspirating saliva, vomit, or other fluids into the lungs. The endotracheal tube comes in a variety of lengths and diameters to fit securely into the trachea of various animal species (see Figure 42–8). The selected size should be the approximate diameter of

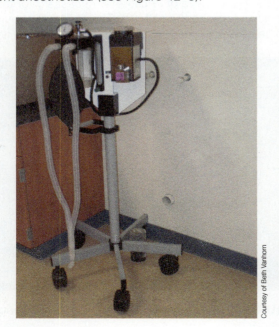

FIGURE 42–6 A machine used to provide inhalation anesthesia.

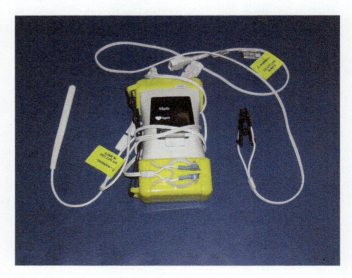

FIGURE 42–7 Pulse oximeter.

CHAPTER 42 Surgical Assisting Procedures 841

lubricant that is placed directly on the trachea prior to tube placement. An example of this type of solution is lidocaine gel. This is most commonly used in cats due to a very active reflex that opens and closes the trachea during intubation. Lidocaine gel helps relax the reflex.

The patient is placed in sternal or lateral recumbency with the head and neck extended upward to allow the throat to be viewed during placement. The mouth is held open with the tongue pulled forward and down over the patient's front teeth. This requires gloves, and the tongue can be held using a gauze sponge to prevent slipping. The veterinarian or technician will use a **laryngoscope** to place the tube into the trachea (see Figure 42–9). This is a tool made of heavy metal that has a light source on the end to help light the airway view. The laryngoscope is placed into the throat area, and a blade that extends off the tool is used to hold down the cartilage of the larynx. As the patient breathes, the larynx reflex opens and closes, and the laryngoscope helps in visualization of the laryngeal cartilage and trachea. Once the tube is placed within the trachea, it should be checked for proper placement. This may be done by feeling the end of the endotracheal tube during a breath to feel

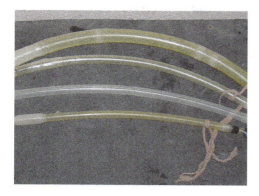
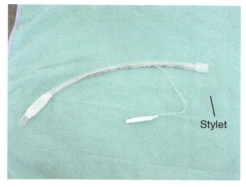

FIGURE 42–8 Endotracheal tubes.

the patient's trachea. The tube is made of heavy-duty plastic or rubber material that is slightly curved in the shape of the trachea. One end is slightly tapered to allow ease of placement into the trachea. The opposite end has an adapter that attaches to the anesthesia machine. Some tubes have a **cuff** located about ¾ of the way down the tube and is inflatable and allows the tube to be inflated with air to create a tight seal around the trachea to prevent any moisture or material from entering the lung field. The cuff is attached to the endotracheal tube by a thin tube running parallel to the endotracheal tube. When inflated, the internal and external parts of the cuff look like a balloon. The cuff is inflated using a syringe that pushes air into the cuff. Too much air can be placed into the cuff, causing the tube to overinflate, so care must be taken to not irritate or damage the trachea. The tube and the cuff should be checked prior to use to make certain no tears or leaks are present. It is important for the assistant to understand how the cuff and tube work during the intubation process, as the cuff must be deflated prior to **extubation** or endotracheal tube removal from the patient.

Actual preparation of the endotracheal tube should consist of placing a small amount of lubricant, such as KY jelly, on a gauze sponge. The end of the endotracheal tube is lubricated for ease of passage into the trachea. Some veterinarians prefer a topical anesthetic

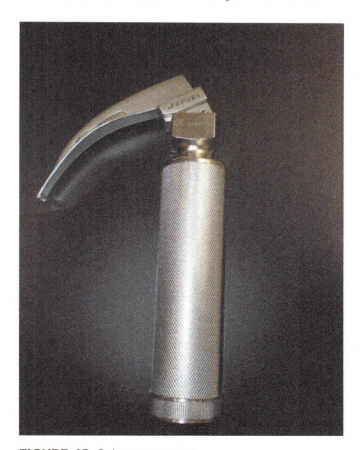

FIGURE 42–9 Laryngoscope.

the air expired from the patient or when attached to the anesthesia machine; note that the breathing is in coordination with the bag on the machine. As soon as the tube is in place, it should be secured by placing a piece of rolled gauze above the adapter lip of the endotracheal tube in a square knot, and both sides should be secured behind the patient's canine teeth and tied over the patient's nose or secured in a bow behind the patient's ears. This prevents the tube from accidental displacement. At this point, the assistant releases the patient's mouth and inflates the cuff and then rotates the patient into lateral recumbency.

The patient is then attached to the anesthesia machine by attaching the hose to the end of the tube adapter (see Figure 42–10). The patient is then placed on the machine for inhalation anesthesia, which is maintained throughout the procedure. Various vital sign monitors are then placed on the patient according to veterinarian preference and availability.

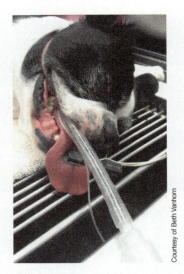

FIGURE 42–10 Endotracheal tube connection to anesthesia machine.

COMPETENCY SKILL 143

Restraint for Intubation

Objective:

To properly restrain a patient for safe insertion of an endotracheal tube

Preparation:

- Gauze sponge
- Lubricant (KY Jelly or lidocaine gel)
- Proper-sized endotracheal tube
- Gauze roll
- Laryngoscope
- 3 or 5 ml syringe
- Exam gloves

Procedure:

1. Apply gloves.
2. Place patient in sternal or lateral recumbency.
3. Extend the head and neck and open the mouth widely.
4. Apply a gauze sponge to the tongue and pull forward and down between the front teeth.
5. Follow the veterinarian's or technician's positioning needs.
6. Once the endotracheal tube is placed, close the mouth and hold the end of the tube.
7. Tie the end of the tube with a piece of rolled gauze.
8. Place the ends of the gauze roll on either side of the face and tie on top of nose or behind the ears. Tie the ends in a bow for a quick release.
9. Inflate the tube cuff using a 3 to 5 ml syringe and close the cuff.
10. Place the patient in lateral recumbency.

Patient Monitors

Patient monitors may include such equipment as a pulse oximeter, a temperature probe, a respiratory monitor, an esophageal stethoscope, a blood pressure monitor, and an electrocardiogram (ECG). These monitors are important for detailing a patient's vital signs and ensuring the animal is stable throughout the anesthesia procedure. These monitors can also detect the plane of anesthesia an animal is in and if more or less inhalant is necessary.

The **pulse oximeter** is a device used to measure a patient's vital signs by indirectly measuring the oxygen saturation within the blood and any changes in the blood volume. The tool is commonly called a *pulse ox* and is attached to the patient with a probe that may be placed on the tongue, pulse point over a paw, or a rectal probe. The screen readout always shows the oxygenation rate, and some units provide a heart rate, temperature, and respiratory rate (see Figure 42–11). The monitored signal bounces in time with the heartbeat because the arterial blood vessels expand and contract with each heartbeat. Many pulse ox devices have a printer that allows the vital signs to be printed out for each patient and the data placed in the medical record. Normal oxygenation rates are around 98 percent. This value determines how much oxygen the patient is receiving. Values below 90 percent indicate cyanosis, and the veterinarian should be notified immediately.

Pulse points are a vital source of determining pulse quality from anesthesia assessment through induction and recovery. Common pulse points in small animals include the femoral artery over the inner thigh, the pedal pulse above the paws in the front and hind limb, and sublingual or below the tongue on an anesthetized animal. Large animal pulse points include the maxillary artery on the inside of the jaw bone, the digital artery on the inside and outside of the front legs about halfway up the leg, and the femoral artery on the inside of the hind legs on the inner thigh.

An **esophageal stethoscope** is a device placed into the esophagus next to the endotracheal tube, and it emits the sound of the heartbeat throughout a procedure. This type of monitor is an inexpensive tool that consists of a tube placed through the mouth and down the esophagus to the level of the heart. This special tube can be attached to a device that amplifies the sound of the heart, which is then transmitted through the speaker on the device.

A **blood pressure monitor** or **sphygmomanometer** is used to measure an animal's blood pressure (BP) during surgery (see Figure 42–12). A cuff, similar to that used in humans, is placed on the patient's lower leg. The BP machine then automatically inflates the cuff to a pressure that stops the flow of blood through the artery in that region. The cuff will then begin to slowly deflate, allowing blood to begin flowing back into the

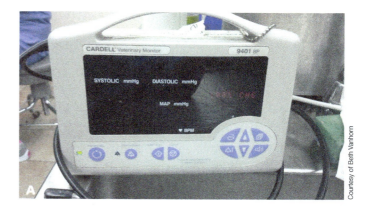

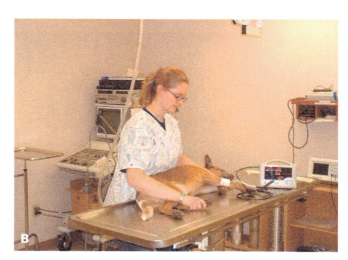

FIGURE 42–11 Pulse oximeter readings for heart rate and oxygenation.

FIGURE 42–12 (A) Blood pressure unit. (B) An assistant restrains a dog for blood pressure measurement.

844 SECTION IV Clinical Procedures

limb region, creating vibrations through the artery. Sensors in the cuff detect the vibrations, allowing the machine to determine at what pressure blood begins flowing back into the limb begins to create a reading. This will also help determine tissue perfusion as well as depth of anesthesia. If the pet is in a light surgical plane (under light anesthesia), it may have an increase in blood pressure when the surgeon starts the surgery or manipulates the tissues. The surgeon also monitors the blood loss that is occurring during the procedure, since that also affects tissue perfusion. BP cuffs come in various sizes to accommodate different species of animals. The limb being used should be kept at heart level and the patient as relaxed as possible.

An **electrocardiogram** (ECG or EKG) monitor is placed onto a patient through metal leads called **electrodes** that measure the electrical activity of the heart (see Figure 42–13). An EKG measures the electric currents generated by the heart. It is used to monitor the heart rate, its rhythm, and changes in the nerve impulses in the heart. Continuous monitoring with an EKG allows early recognition of electrical changes associated with disorders of conduction in the heart and arrhythmias that may need to be treated. Lead II is the most common lead used for monitoring surgical patients.

Surgical Preparation

All surgical patients should be prepared for surgery in a specific way that makes the patient's surgical area sterile and aseptic. This begins with making the patient urinate if possible, to prevent contamination of urine during surgery. The animal's bladder should be palpated and expressed before the preparation.

Preparation of the surgical site should be completed in the treatment area. The incision area is clipped, and a **surgical margin** should extend 2 to 4 inches beyond the anticipated incision borders (see Figure 42–14). Long-haired breeds should have any hair near the surgical margin trimmed to a length that prevents contamination of the incision area. The clipped area should be kept neat and even; smooth over all areas where hair is removed (see Figure 42–15). A surgical clipper blade of #40 or #50 should be used. The blade is held flat against the skin to avoid clipper burns or nicks in the skin. The clipper is used first with the growth of the hair coat and then against the hair growth to clip as short as possible. The surgical clip should provide a smooth area of skin with no remaining hair over the site. Some areas of certain animal species may require plucking rather than surgical clipping due to delicate tissue that may be damaged or torn by clipper blades. This is common in male tomcats and rabbits that are

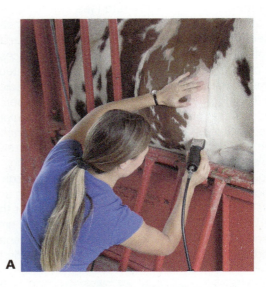

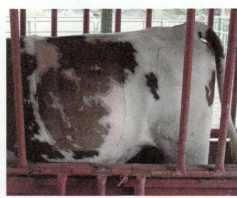

FIGURE 42–14 (A) Shaving the flank of the cow to prepare a surgical margin. (B) Properly shaved surgical margin on the flank of a cow.

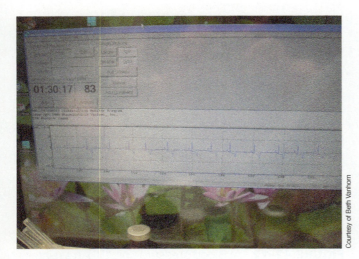

FIGURE 42–13 ECG machine screen.

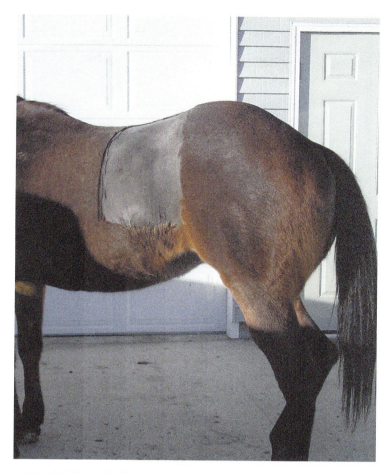

FIGURE 42-15 Flank of horse shaved in preparation for surgery.

being castrated. When preparations for orthopedic surgeries are performed on a limb, the entire circumference of the limb should be clipped free of hair. When all areas of the hair are clipped, a vacuum should be used to clean up all hair. Shop vacuums and built-in vacuum systems are common for hair collection. It is also useful to gently run the vacuum over the patient's prep site to remove any loose hair.

Skin preparation then begins with an initial scrub to rinse off any remaining hair over the surgical site. The **surgical scrub** process then begins with sterile solutions placed onto the surgical margin (see Figure 42–16). The most common surgical scrub solution is chlorhexidine, such as Nolvasan scrub. Scrub solution contains a soapy component that when mixed with water helps to break down oils, dirt, and debris on the skin to better cleanse the area. The scrub process is done multiple times by using scrub-soaked gauze sponges to cleanse the area, beginning at the center of the site and working clockwise in a circular motion to the outer edge of the hair coat. This allows the area to be cleaned without touching a previously wiped area and is considered a "rub" process to removal all dirt and debris. Always work from the incision site outward toward the remaining hair. This process is repeated as often as necessary until the gauze sponges no longer show signs of dirt or debris. An alcohol-soaked sponge may then be used in the same manner by scrubbing and then rinsing and repeating for at least three times. A surgical antiseptic such as iodine or betadine solution is then applied to the skin, and the patient may then be moved to the surgical area. A gauze sponge or several sponges can be placed over the area to prevent contamination. A second antiseptic spray may be completed when the patient is in place on the surgery table.

The patient should be placed properly onto the surgical table without contaminating the preparation area and tied onto the surgery table using rope ties. A final scrub may be completed as necessary. Any surgical monitors should then be applied to the patient and turned on for monitoring purposes.

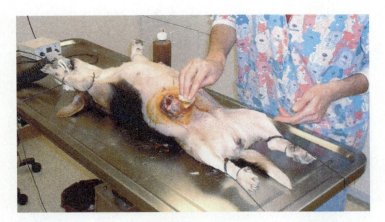

FIGURE 42-16 Surgical scrub used to prepare the surgical site.

 COMPETENCY SKILL 144

Surgical Preparation

Objective:
To properly prepare the surgical site

Preparation:
- Clippers
- #40 or #50 surgical blades
- Vacuum
- Exam gloves
- Gauze sponges
- Surgical scrub
- Water
- Alcohol
- Antiseptic (iodine or betadine)

Procedure:
1. Apply gloves.
2. Clip the patient using #40 or #50 clipper blades and clippers.
3. Clip the incision site and 2 to 4 inches around the incision site, according to the veterinarian's direction. Trim any long hair that may fall into the surgical margin.
4. Vacuum any loose hair that was removed from the patient. Vacuum patient, table, and floor areas.
5. Rinse the surgical site with a gauze sponge and water.
6. Using a gauze sponge soaked in water and surgical scrub, begin the sterile scrub at the center of the incision site and move in a circular motion outward toward the unclipped area.
7. Repeat this process until the gauze sponges show no dirt or debris from the clipped area.
8. Repeat the process using alcohol-soaked gauze sponges and repeat three times.
9. Repeat surgical scrub and water cleaning three more times.
10. Apply antiseptic spray over the incision area.
11. Apply gauze sponges over the incision area before moving the patient to the surgical suite.
12. Carefully move the patient to the surgery table and properly position and tie onto the table using the rope ties.
13. Repeat the surgical scrub and water cleaning three more times.
14. Apply another light spray of antiseptic and allow to dry on the patient's skin.

Surgical Pack Preparation

Surgical pack preparation is essential for all surgical procedures. The process must be completed with a sterile and aseptic technique. The process begins with the surgeon's items being properly opened and available for the surgery procedure. The surgeon needs the following nonsterile items:

- Hair cover or cap
- Surgical mask
- Booties or shoe covers

The surgeon requires the following items and supplies that must be sterile for each surgery:

- Scrub pack containing bristle brush
- Sterile linen hand towels
- Surgical scrub
- Surgery glove pack with correct-sized gloves
- Sterile gown pack

Surgery Packs

All packs should be sterilized with an indicator strip and tape that ensure the items have been properly sterilized. The indicator strip is located inside the pack and changes color when the heat process has reached the interior objects. The autoclave tape that seals the pack has strips that turn dark in color when heat has provided proper sterilization (see Figure 42–17). Each pack that is prepared should be closed with autoclave tape that is turned under to form a tab for easy opening. The tape should be labeled with the pack contents, date autoclaved, and initials of the person preparing the pack. The packs should be carefully opened in the surgery suite and placed on a Mayo stand. When opening the pack, the tab should be pulled, and the pack placed on the instrument stand. Then the outer pack cover should be opened with two fingers, using care not to contaminate any items inside the pack. The left and right sides are opened, and the bottom part of the pack cover pulled down to expose the surgical instruments.

Surgery packs will include the following items that are sterilized separately:

- Instrument packs (spay pack, castration pack, wound pack, orthopedic pack, etc.)
- ½ surgical drape
- ¼ surgical drape
- Towel packs
- Individual instrument envelopes
- Bowl packs

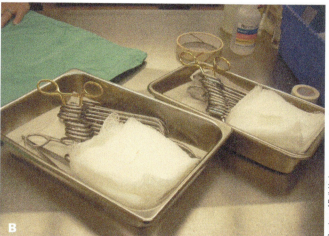

FIGURE 42–17 (A) Wrapped surgical pack. (B) Instruments in a surgical pack.

The surgical pack is put together in a specific way that may vary from facility to facility and be based on the veterinarian's recommendation or preference. Surgical packs include the instruments for each individual pack type but should also include basic supplies necessary for all surgeries (see Figure 42–18). These items are required for every type of surgery. **Surgery pack** supplies include the following:

- Gauze sponges
- Laparotomy towel
- Suture material
- Surgical blade

The surgical pack is prepared in a way that when opened should be easily accessible to the veterinarian and be kept neat and organized for surgical purposes. Typically, instruments are placed in the center of the pack on top of the laparotomy towel and gauze sponges (see Figure 42–19). The **laparotomy towel** is used to control bleeding during surgery and as a hand towel for the surgeon. The gauze sponges are placed

848 SECTION IV Clinical Procedures

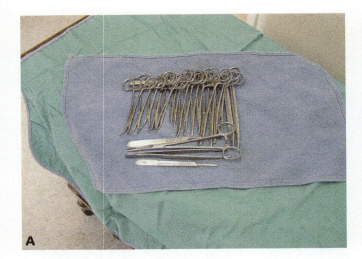

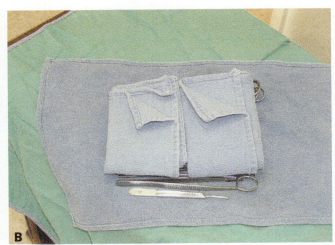

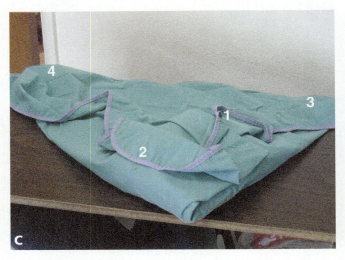

FIGURE 42-18 Contents and proper folding of surgical pack. (A) Instruments properly placed in a surgical pack. (B) Laparotomy towels properly placed in a surgical pack. (C) Proper folding directions of a surgical pack. (D) Properly completed surgical pack.

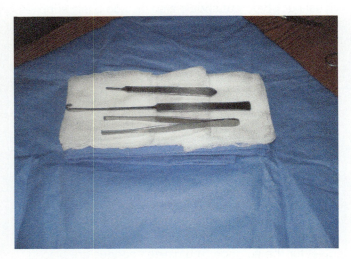

FIGURE 42-19 Proper placement of items in surgical pack.

on top of the towel and should be counted and consistent in every pack. This ensures that when the surgery is complete and packs are cleaned, every gauze sponge is accounted for and has not been accidentally left inside of the patient. The instruments are then placed on top of the gauze, and the pack is wrapped in the proper method.

COMPETENCY SKILL 145

Pack Cover Preparation

Objective:

To properly complete and fold a surgical pack for the sterilization process

Preparation:

- Surgical drape
- Scissors
- Permanent marker
- Autoclave tape

Procedure:

1. Cut a piece of surgical drape about 12 inches in length.
2. Open the drape fully.
3. Place the drape in the shape of a diamond on the work surface.
4. Place the pack item or surgical pack in the center of the pack cover.
5. Fold the bottom flap upward over the contents. Hold flap in place.
6. Fold the left flap inward and make a small dog ear at the end of the flap. Hold flaps in place.
7. Fold the right flap inward and make a small dog ear at the end of the flap. Hold flaps in place.
8. Make a small crease at the side of each side of the upper flap. This should make an envelope shape.
9. Fold the contents upward and snugly tighten the pack in place within the cover.
10. Fold again and make a dog ear folded inward on the end of the drape.
11. Apply a piece of autoclave tape with a flap tucked under the tape.
12. Write the date, pack contents, and initials of the preparer on the tape with a permanent marker.

COMPETENCY SKILL 146

Surgical Pack Preparation

Objective:

To properly prepare and fold a surgical pack

Preparation:

- Surgical drape
- Scissors
- Surgical instruments (will vary between packs)
- Indicator strip
- Permanent marker
- Autoclave tape
- Gauze sponges
- Laparotomy towel

Procedure:

1. Remove one arm's length of a piece of surgical drape or approximately 3 feet of drape.
2. Open the drape up fully.
3. Place the drape flat on the work surface.
4. One-fourth of the bottom of the drape should be folded upward twice in a fanfold and the fold be located on the back of the drape.
5. One-fourth of the top of the drape should be folded downward twice in a fanfold and the fold should be located on the back of the drape.
6. The laparotomy towel is folded lengthwise in half and then in four square fanfolds and placed in the center of the drape.
7. Apply three stacks of 4 × 4 gauze sponges, with the same number in each stack, on top of the laparotomy towel. The total number of sponges should be the same in each pack.
8. An indicator strip should be applied to the top of the sponges face up.
9. The spay hook, tissue forceps, and surgical blade handle should be placed on top of the sponges.
10. The four towel clamps should be linked together and held in one hand.
11. The curved hemostats should be placed in the hand in the opposite direction on top of the towel clamps.
12. The straight hemostats should be placed in the hand in the opposite direction on top of the curved hemostats.
13. The scissors are placed in the hand in the opposite direction on top of the straight hemostats.
14. The needle holders are placed in the hand in the opposite direction on top of the scissors.
15. The entire stack of instruments is then placed on top of the three flat instruments and held in place with one hand (see Figure 42–20).
16. While holding the instruments in place, fold the bottom half of the drape upward over the instruments. Continue to hold instruments in place.

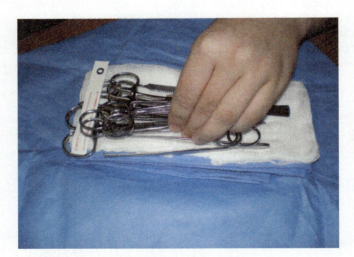

FIGURE 42–20 Hold the instruments that make up the surgical pack in one hand.

17. Fold the top half of the drape downward over the instruments.
18. Continue to hold the middle of the pack in place.
19. Fold the left side of the drape inward over the pack. Make a crease in the side of the fold.
20. Fold half of the left flap backward in a fanfold and place a dog ear fold on one corner of the drape.
21. Fold the right side of the drape inward over the pack. Make a crease in the side of the fold.
22. Fold half of the right flap backward in a fanfold and place a dog ear fold on one corner of the drape.
23. Place the surgical pack on top of a pack cover. Complete the fold of the pack cover.

COMPETENCY SKILL 147

One-Half or One-Fourth Drape Preparation

Objective:

To properly prepare surgical drapes

Preparation:

- Surgical drape
- Scissors
- Indicator strip
- Permanent marker
- Autoclave tape

Procedure:

1. Cut a piece of surgical drape that is ½ a yard or ¼ of a yard in length.
2. Open the drape fully.
3. Place the drape on the work surface.
4. Fold the drape in half lengthwise.
5. Fold the drape in half again lengthwise.
6. Fold equal amounts of the drape in fanfolds creating four equal parts.
7. Place the drape in the center of a pack cover.
8. Place an indicator strip on top of the drape.
9. Fold the pack cover.

Scrub Packs

The **scrub pack** contains the items necessary to properly prepare the surgeon and technician or assistant for surgery. This is a surgical scrub for staff members to ensure they are sterile and aseptic for the procedure. The contents in a scrub pack include the following:

- A hard-bristle brush
- Sterile hand towels
- Surgical scrub
- Surgery gloves

The bristle brush should be placed on top of a stack of folded hand towels. The towels should be folded in half lengthwise and in a fanfold. The end of the towel fold should result in a neat small square form. The number of towels should be counted and used consistently in each scrub pack. A sterile indicator strip should be placed between the hand towels and the brush. The scrub pack should be carefully opened in the same manner as the surgery pack and placed near the sterile scrub area, usually within the treatment area. The surgical scrub should be kept near the sink for hand washing and must be kept in a sterile container (see Figure 42–21).

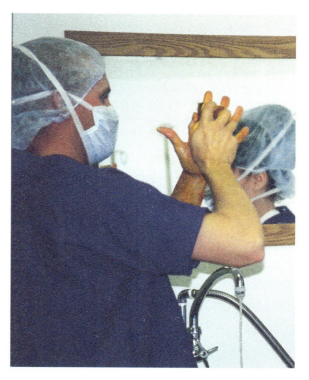

FIGURE 42–21 The surgical scrub.

Gown Pack

The **gown pack** holds the gown that is worn by the surgeon and veterinary staff assisting in the surgical area. Gowns may be disposable and meant for one-time use, or they may be reusable and require laundering after each use and then may be sterilized. The gown must be folded in the proper method to ensure that when opened and removed from the pack cover it is the proper way for applying without causing contamination (see Figure 42–22).

The **glove pack** is usually a prepackaged disposable sterile glove pack that is used one time and then discarded. The proper-sized surgery gloves should be opened by pulling down the outer plastic cover and left in the scrub area.

Gloving Techniques

The veterinary assistant should be aware of gloving techniques, as he or she may be asked to assist in surgery. Gloving begins after the sterile hand scrub is completed and the hands have been dried. Each person should know his or her glove size. The gloves should be opened and ready for use. All rings and jewelry items should be removed from the hands, and fingernails should not extend beyond the end of the finger. The inner glove pack should be opened once the person has completed the sterile hand scrub.

Open gloving is done by removing the left glove from the sterile packaging using the right hand by grasping it and the edge of the cuff. The left hand is inserted into the opening as far as possible. Then the left gloved hand is used to guide the right glove onto the right hand. This is done by holding the right glove

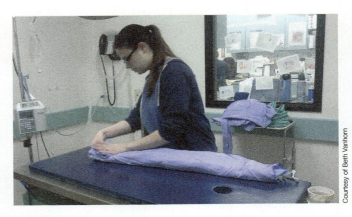

FIGURE 42–22 Folding the surgical gown.

at the folded cuff. Use the left gloved fingers to lift the right cuff upward and invert the right cuff over the gown sleeve. Pull the glove fully into place on the right hand. Once both gloves are in place, the gloved hands can be touched to better arrange the fingers within the gloves for comfort.

Closed gloving is done when the gown is placed with the hands still within the sleeves so that no skin is visible and potentially contaminating the gloves. One glove is grasped at the cuff edge through the sleeve. The other side of the cuff is grasped with the other hand still in the sleeve. The hand is then slid further down the gown sleeve and into the glove. The same procedure is repeated with the other glove. The cuffs of the gown are completely covered by the cuffs of the sterile gloves. When both hands are completely gloved, grasp them together by locking hands and adjust the gloves as necessary.

When the gloving procedure is completed, the staff member should immediately keep the elbows elevated and the hands together and not touch any nonsterile surface. Immediately move to the surgical area to assist the surgeon.

COMPETENCY SKILL 148

Gown Pack Preparation

Objective:

To properly prepare gowns in a sterile pack

Preparation:

- Surgical drape
- Surgical gown

- Scissors
- Permanent marker
- Autoclave tape

Procedure:

1. Place the surgical gown on a flat surface.
2. Place the gown front side up.
3. Place the sleeves across the front of the gown. The sleeves should slightly overlap in the center of the gown.
4. Fold the gown in half lengthwise with the center of the gown forming a fold.
5. Fold the gown in half lengthwise again.
6. Tuck the loose ends of the neck and back ties into the folds of the gown.
7. Begin at the hem of the gown and fanfold the gown in small square sections.
8. Continue folding the gown throughout its length until the neckline is reached. This should form a neat square shape.
9. Fold the top corner of the gown into a dog ear.
10. Place the gown on top of a pack cover.
11. Place a sterile indicator strip on top of the gown.
12. Fold the pack cover.

COMPETENCY SKILL 149

Opening a Pack

Objective:

To properly open a sterile pack without contaminating the contents

Preparation:

- Surgical pack

Procedure:

1. Place sterilized pack on Mayo or instrument stand in the surgical area.
2. Open the pack by pulling the tab in the autoclave tape.
3. Open the pack by pulling the top flap upward. Do not touch any part of the pack inside.
4. Open the right flap outward by pulling the dog ear flap. Do not touch any part of the pack inside.
5. Open the left flap outward by pulling the dog ear flap. Do not touch any part of the pack inside.
6. Leave the pack for the surgeon.

Sutures and Surgical Blades

The veterinary assistant should know the types and sizes of suture material and surgical blades and where the supplies are kept. These items come prepackaged individually and are already sterile (see Figure 42–23). These items are needed by the surgeon and should be placed onto the Mayo or instrument stand after opening the surgical pack and providing the surgery drape materials. Scalpel blades are numbered to indicate their size and shape. Each surgeon will have a size preference depending on the patient and procedure. The blades are wrapped in a peel-apart foil package that is opened by peeling the edges away from the blade to allow the blade to carefully drop onto the surgical pack surface. The blade should be dropped onto an open area of the surgical pack for easy visibility.

When the surgery is complete, make certain the scalpel blade is removed from the handle. Grasp the narrow end of the handle and use a pair of forceps to grasp the blade over the hooks of the handle and lift the end upward. When the blade is raised from the hooks, it can then be slid from the handle. Caution should be used when handling the blade—as it is very sharp—and all blades should be discarded after each use by placing them in the sharps container.

Sutures come in a variety of materials, sizes, and lengths. A suture is a strand of material used to close tissue or ligate blood vessels. They are dated and housed within a chemical substance to keep them sterile and soft for use. Different types of suture material are used on different tissues. There are two classes of suture in the form of absorbable and nonabsorbable materials. **Absorbable suture** is suture that is broken down within the body over a period of time and is absorbed by the body. This type of suture is used both internally and externally. **Nonabsorbable suture** is not broken down within the body and remains intact for long periods. This type of suture is commonly used on external body structures (see Figure 42–24). Nonabsorbable sutures are generally used to close skin and must be removed after wound healing is complete. There are some procedures where nonabsorbable sutures may be placed internally but cannot be removed. Some sutures have a needle already attached and are called **swaged on needles**; other suture material is on a **suture reel** and must be thread onto a needle.

Suture needles come in a wide variety of sizes and shapes. The suture needle depends on the type of tissue being sutured and the location. Needle points may have cutting, taper, or reverse-cutting sharpness. Cutting needles typically have two to three opposing cutting edges. They are often used in skin tissue that is thick and more difficult to penetrate. Taper-point needles are rounded with a single sharp tip that pierces and spreads the tissue without cutting actions. They are often used in thin tissue that is easy to penetrate. Reverse-cutting needles have a third cutting edge located on the outer curve of the needle, making them stronger and causing less damage when placed through tissue.

Monofilament suture is made of a single strand of material. This type of suture creates less drag over the tissue and has less space that may harbor disease organisms. Care must be taken with handling monofilament suture, as it is easily weakened during excessive tension and handling and is more apt to break. **Multifilament suture** consists of several strands of material twisted or braided together. This type of suture is typically more flexible than monofilament suture is.

Suture size is similar to needle size in that the smaller the number, the larger the needle size and diameter of the suture material (see Table 42–2), and the larger the number, the smaller the needle

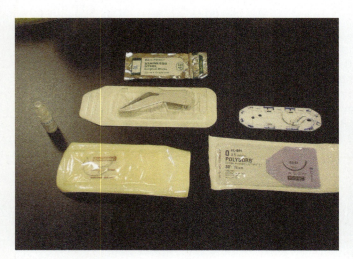

FIGURE 42–23 Suture materials.

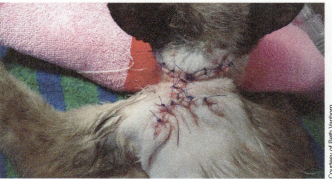

FIGURE 42–24 External sutures.

TABLE 42-2
Suture Size

EXAMPLES OF SUTURE SIZES FOR USE IN ANIMALS

10-0 to 8-0	7-0 to 5-0	4-0 to 3-0	2-0 to 0	1 to 2
Microsurgery; corneal	Ophthalmic; neurosurgery; ureters	Skin and subcutaneous; bowel; bladder	Abdominal fascia; stomach; hernia	Rib retention; cutaneous stents
5-0 to 4-0	4-0 to 0	4-0 to 2-0	4-0 to 2-0	
Oral surgery; extractions	Vessel ligation	Aural hematoma	Spay and castration	

and suture size. Larger whole numbers indicate larger sutures, so a number 2 suture is larger than 1. However, larger numbers followed by an "**ought**" (0) are smaller. A 5-0 is smaller than 2-0. The smallest sutures that are used for ophthalmic and microsurgery range down to 12-0 size.

Surgical staples are a fast, easy way to close the skin. They are useful but their application requires using a staple gun, and a specific staple remover device is required for removing them. Staples are made of metal. **Surgical glue** is useful for closure of small wounds or surgical sites and is commonly used in feline declaws. This is a sterile type of surgical glue that eventually breaks down in the body. The polymers within the substance create a strong and flexible bond in the presence of moisture. This adhesive bond usually takes less than 15 seconds to bind, but the time may be longer when excessive bleeding occurs. Common surgical glue products are Vetbond and Nexaband.

Laser Surgery

Laser stands for light amplification by stimulated emission of radiation. Laser beams are different from visual light because the light beam is focused on a specific point, thus increasing its power. The most common type of laser surgery uses laser radiation emitted by carbon dioxide. Laser surgery is used to burn, cut, or destroy tissue in a very precise way. In veterinary medicine, laser surgery is used for a variety of superficial and cutaneous procedures. Laser surgery has many benefits over traditional surgery. The most important of those include faster healing, less blood loss, sterilization of the tissue, and less postsurgical pain—all of which allow for a shortened recovery time. General anesthesia is not needed for some procedures. Another benefit of laser surgery is the possibility of operating on lesions in locations that could require a longer follow-up, such as the paws. The benefits of laser surgery are quite impressive. In fact, most patients can walk normally right after surgery with minimal or no pain. Personal protective equipment (PPE) is imperative when working around lasers. Most health concerns relate to the eyes, lungs, and skin. Laser safety goggles are required in facilities using laser treatments. Airborne vapor contaminants should be controlled via local exhaust ventilation, respiratory protection, or a combination of both.

Gowning the Surgeon and Assistants

The surgeon and surgical assistants must put on mask, cap or hair cover, and booties or shoe covers (see Figure 42–25). Next, the surgical staff should

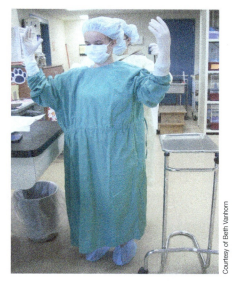

FIGURE 42-25 Surgical gown, cap, mask, and booties.

perform a surgical hand scrub preparation. Upon completing the surgical hand scrub, the hands are dried using a sterile hand towel. The gown pack should already be open and ready for the surgical staff. Then the gown is picked up by the folded dog ear at the neckline of the gown. The gown is held up and away from the body to prevent contamination. Care should be taken to not allow the gown to come in contact with the floor. The person places his or her hands through each armhole. The person then applies the sterile gloves with either an open or closed technique. This is when an assistant should stand behind the person and, without touching the external area of the gown—that is, touching only the ties—tie in the staff member. The ties should be placed in a bow. Many gowns have ties at the base of the neck and at the waist. Never touch any part of the gown other than the ties.

Patient Surgical Positioning

Patient positioning is dependent on the surgical procedure and the veterinarian's preference. Many surgeries require the animal to be placed in dorsal recumbency on the surgery table (see Figure 42–26). This is completed by keeping the patient straight and centered on the table. Each limb of the patient is then tied into the table hooks. The front limbs are pulled forward toward the head, which is centered between the front limbs. The rear limbs are pulled backward toward the tail, which is centered between the rear limbs. The ties may be placed in vicelike devices or on hooks in which figure-eights are made to hold the limbs. Other common surgical positions include sternal and lateral recumbency. The limbs should be secured as naturally as possible for the surgical procedure. Table 42–3 lists positioning for some common procedures.

The surgery table should be placed at the proper height for the procedure and the veterinarian. Some surgeons prefer to sit, and some prefer to stand—and the table should be adjusted for the necessary

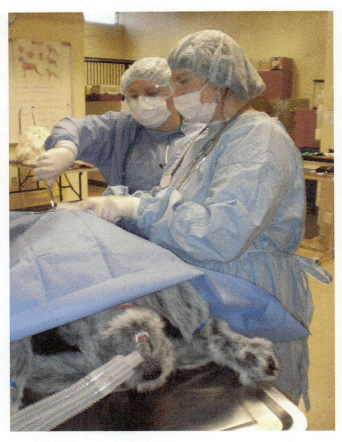

FIGURE 42–26 Patient in dorsal recumbency.

height. Some patients may require a heating pad that is placed between the patient and the table. Other patients may require a V-trough positioner to keep the animal centered on the table. If an electrocautery unit is being used, the patient should be placed on a ground plate.

Surgical lights should be positioned directly over the sight when the incision will be made. Final light positioning should be done after the animal is draped and the surgeon is in place to direct any light needs.

TABLE 42–3

Common Procedure Positions

Spay or ovariohysterectomy (OHE)	Dorsal recumbency
Dog castration	Dorsal recumbency
Cat castration	Sternal or lateral recumbency
Abdominal surgery	Dorsal recumbency
Orthopedic surgery	Lateral or dorsal recumbency
Ear crop or head surgery	Sternal recumbency
Declaw or dewclaw surgery	Lateral recumbency

Anesthesia Planes

Anesthesia planes or depths are measured by evaluating the patient's involuntary reflexes. The veterinary assistant may be asked to monitor reflexes to determine the depth of anesthesia. The **palpebral reflex** is evaluated by tapping the fingertip lightly over the surface of the eyelid. If a strong to slight blink reflex is noted, the patient is in a light plane of anesthesia, and the inhalant gas should be increased for surgical purposes. The blink reflex will decrease and eventually disappear as the plane of anesthesia deepens. The **pedal reflex** is used on the digit or the skin between digits, which is pinched to determine reflex response. A patient that moves or withdraws the foot is in a light plane of anesthesia. The reflex weakens and disappears as the plane deepens. Muscle tone is noted by the plane level and is moderate when in a light plane and lessens as the plane level deepens.

Plane I is during the induction phase when a patient becomes disoriented for a brief period and then passes into Plane II. Plane II is the excitement phase when a patient displays involuntary movements and vocalizations. This stage passes quickly and may also be visible in the recovery phase. Light sometimes affects this plane of anesthesia, and placing the patient in a lightly lit or dark area is helpful. Plane III is a light depth of anesthesia. This is the ideal plane for surgery, where vital signs remain normal, pupils are regular in size and rotate medially, and mild to no reflexes are present. Plane IV is an overdose of anesthesia and a deeper level in which the patient could potentially die. No reflexes are noted, vital signs are decreased, and pupils are centrally fixed and dilated. At this plane, cardiac arrest is possible.

The Anesthesia Machine

The veterinary assistant should have a basic knowledge and understanding of the parts of the anesthesia machine (see Figure 42–27). The most critical part of the machine is the **vaporizer**. The vaporizer converts liquid anesthesia into a gaseous form, which the patient inhales. Each type of vaporizer is meant for a specific gas inhalant. A vaporizer that is meant for isoflurane cannot be used with sevoflurane or nitrous oxide. Each vaporizer has an adjustable dial that controls the delivery rate of the gas. It is essential for all staff members to know how to turn the dial on and off, as well as adjust the dial between number settings. The

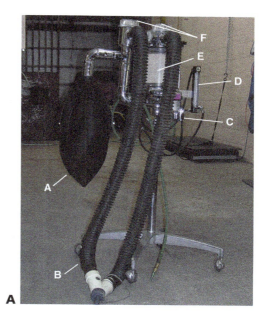

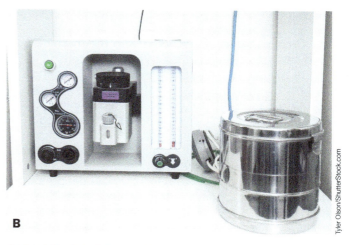

FIGURE 42–27 (A) Basic parts of an anesthesia machine. 1. Breathing bag, 2. Breathing hose, 3. Vaporizer, 4. Oxygen flow meter, 5. Absorber, 6. Inspiration/expiration valves. (B) Closer look at the vaporizer part of anesthesia machine.

delivery of gas is measured in percentage volumes. When the liquid gas form is used, it must be replenished into the vaporizer. This level must be monitored during anesthesia. The liquid level should be noted prior to sedating an animal. During the refill process, limited people should be around, as the gaseous odor may escape. All anesthetic gases are toxic, and care should be used when working around the machine. Anyone who is pregnant should not work with anesthesia. Escaped gases that are inhaled over time can cause health issues such as renal or liver damage. Short-term exposure can cause signs of lightheadedness or dizziness.

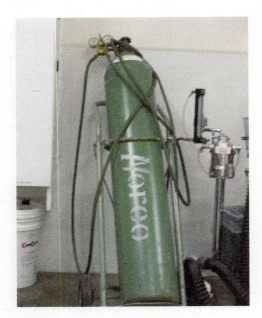

FIGURE 42–28 Oxygen tank used with the anesthesia machine.

Oxygen tanks are located under the machine or in separate cylinders next to the machine (see Figure 42–28). These **cylinders** are metal containers that hold compressed forms of oxygen or liquid oxygen to keep patients oxygenated and breathing during anesthesia. Oxygen cylinders are green in color and come in a variety of sizes. Designations for each size and use are based on a series of numbers and letters. When you look at an oxygen cylinder, you will see a number posted on the side; this number denotes the number of liters of oxygen the cylinder holds. The letters A, B, C, D, and E are used to designate the size of the cylinder. An A cylinder is the smallest of the sizes and holds only about 34 liters of compressed oxygen; an H cylinder, which is the largest and holds approximately 7107 liters of compressed air. Large cylinders are often mounted outside of the building or in low-traffic areas and deliver oxygen to the rooms where the anesthesia machines are located by hoses that descend from the wall or ceiling and attach to the machine. Often, you will see the letter M displayed on an oxygen cylinder. This indicates that this oxygen cylinder is for medical use only. This oxygen cylinder has a **pressure valve** that indicates how much oxygen is left in the tank. The valve should be checked daily; empty tanks should be removed, and a full tank should be opened and connected to the machine. It is essential that oxygen never runs out during a procedure. Tags may be placed on a tank as to its status, indicating "full," "empty," or "in use." In some facilities, nitrous oxide gas is used as an anesthetic gas and is supplied in a compressed cylinder that is blue in color.

The oxygen **flow meter** is located on the machine near the vaporizer and is adjusted for the proper rate of oxygen flow to the patient. The flow meter is calculated in liters/minute. The rate of flow is determined by patient size, method of administration, and type of breathing system used. A common rate of flow is 30 ml/kg/minute.

Example

Weight in Pounds	Weight in kg	O_2 Flow Rate
22 lb	22 lb/2.2 kg = 10 kg	30 ml × 10 kg = 300 ml
50 lb	50 lb/2.2 kg = 22.7 kg	30 ml × 22.7 kg = 681 ml

When a patient exhales during the process of breathing, carbon dioxide (CO_2) is produced and must be absorbed without entering the facility. The CO_2 is absorbed in a **canister** that contains granules called **soda lime** that collect and absorb the gas. Soda lime begins as a white color and as CO_2 is absorbed the granules change color, such as pink, blue, or purple depending on type. When 2/3 of the canister contents have changes in color, the canister should be changed.

The bag that connects to the machine and inflates and deflates as the patient breathes is called a **rebreathing bag**. The size of the bag depends on patient size and the **tidal volume** required by the animal. The average tidal volume is 10 ml/kg. The bag size is determined by multiplying the tidal volume by 6 to determine the size of bag required for the machine.

Example

A 22 pound dog weighs 10 kg

10 kg × 10 ml/kg = 100 ml

100 ml × 6 = 600 ml

Bag sizes range from 0.5 to 5 liters. In the example, 0.5 liters equals 500 ml, so a 0.5-liter bag would be too small, and the next size bag would be a 1-liter bag, which is the appropriate size for the patient. Bags are made of rubber material and slip easily onto the steel or metal tubing of the anesthesia machine. The movement of the bag is observed to evaluate the patient's rate and depth of breathing. In some cases, the patient may need assisted breathing by compressing the bag with the hands in a regular and rhythmic method. The bag should be removed and cleaned after each use. It should be dried completely before reuse.

The **pop-off valve** is used to release excess amounts of gas when the pressure becomes too high within the machine system. This valve is usually located at the top of the machine. The **scavenger hose** is attached to the pop-off valve to prevent gas from leaking into the room. The hose should be connected to an outside area of the building where the gases escape into the outside air. Veterinary team members are at greater risk of exposure to waste anesthetic gases (WAGs) that are released or leak during anesthesia procedures. Health care workers can be exposed in a variety of ways.

- Anesthetic gas may escape when filling refillable vaporizers.
- WAGs may escape during the initial hook-up and checking of the anesthesia system or the scavenging system.
- WAGs can escape from around the patient's anesthesia mask, especially if the mask is a poor fit.
- WAGs can escape from around the patient's endotracheal tube (ETT) or laryngeal mask airway (LMA) if the cuff is not properly inflated or the wrong size is used.
- Leaks in the anesthesia system can be an exposure source.
- An exposure source can be leaks in the high-pressure system between the nitrous oxide (N_2O) cylinder and yoke assembly or between the anesthetic gas column outlets and the N_2O hose.
- Exposure can occur when the system is flushed or purged at the conclusion of a medical procedure.
- Ineffective or poor ventilation of gas scavenging systems can be a source.
- Leakage from any tubing, seals, or gaskets can be a source.

Exposure to high concentrations of waste anesthetic gases, even for a short time, may cause the following health effects:

- Headache
- Irritability
- Fatigue
- Nausea
- Drowsiness
- Difficulties with judgment and coordination
- Liver or kidney disease

Proper scavenger systems with adequate ventilation, low-flow anesthesia rates, prevention of leaks, proper yearly maintenance of equipment, and air monitoring will reduce exposure of WAGs and allow for a safer working environment.

The hose that connects to the patient is made of **corrugated tubing** that is thin plastic that allows the oxygen and anesthetic gas to flow to the patient. The tubing is connected to the patient's endotracheal tube with a **Y connector**. One end of the tubing is attached to the inhalation valve, and the other end of the tube is attached to the exhalation valve. After each use, the hose should be rinsed and hung to dry. The hose must be dry prior to the next use.

Some animals may require the use of an **anesthesia chamber** for induction methods. An anesthetia chamber is a glass or plastic square or rectangular box that is clear and has two openings in which the inhalation and exhalation tubes on the anesthesia machine connect to the chamber. This allows gas and oxygen to flow into the chamber to induce anesthesia. Once the patient is in a proper plane, the patient may be intubated and placed on the machine. The chambers are large enough for cats, rodents, reptiles, and very small dogs. **Anesthesia masks** may be used for larger patients and are placed over the muzzle and face area. They must fit snugly for the animal to breathe in the gases for induction. The masks are cone shaped and made of plastic and rubber components. Once a patient is in a level plane, the patient may be intubated and placed on the machine. The mask must be cleaned after each use and properly dried.

Breathing Systems

There are two types of breathing systems used with the anesthesia machine and determined by patient size. In the **nonrebreathing system**, the canister and breathing bag are not used. This system is used for small patients such as rodents, birds, and reptiles, or small puppies and kittens that are too small to move the breathing bag with their respirations. The most common system type is called a **Bain system** and uses a single corrugated or plastic tube that connects to the patient. Another hose passes through the center of the tube that contains the gas mixture the patient inhales. The outer tube passes the exhaled gases. The pop-off valve is located at the end of the tube and is opened to allow release of pressure. The flow rate is between 100 and 300 ml based on patient size.

The **rebreathing system** allows the patient to rebreathe anesthetic gas after it has passed through the CO_2 absorbing canister. There is a closed and a semiclosed system. The closed system provides enough oxygen to meet the patient's requirements, and the pop-off valve is kept closed. This is more economical, as oxygen is not wasted. The semiclosed system has a pop-off valve open with a high oxygen flow rate.

The veterinary assistant should be able to locate the parts of the anesthesia machine and be able to adjust the knobs and valves as appropriate. It is important to know the oxygen flow and the type of systems used based on patient size. A bag of appropriate size should be selected. All manufacturer information should be reviewed to understand how each machine functions.

Postanesthetic Care

Once the animal has been removed from the anesthesia machine, it should be moved to the recovery area and placed in a warm, quiet, dark, and comfortable area to wake up. The patient should be placed in lateral recumbency with the head and neck extended. The endotracheal tube is still in place and should be untied but left intact until the animal is awake enough for safe extubation. It is important to monitor the patient's reflexes, especially the **swallow reflex**. The moment the patient begins to swallow is when the endotracheal tube should be extubated. This begins with the patient licking with the tongue or having increased jaw tone. This is a critical time to note because if the tube is extubated too soon, the patient may have difficulty breathing or aspirate materials into the lungs. If the tube is removed after the animal is waking up, the tube may be bitten or severed, causing an obstruction in the throat. When the patient is ready for extubation, the cuff should be opened and deflated, and the tube gently reversed out of the throat.

As the animal recovers, the heart rate and respiratory rate will increase. The patient will begin to move around and try to sit up or stand. As the patient wakes up, some movements may be exaggerated, and noise may cause an excited response. During this time, the patient may thrash around and should be monitored, as there is always a chance of injury. Vital signs should be monitored until the patient is mentally alert and aware of its surroundings. Some patients may require additional heat sources. Animals that are recumbent longer than 30 minutes should be rotated every 15 to 20 minutes to prevent the development of bed sores or pneumonia. Well-padded cages should be used for animals during recovery.

Postoperative Patient Care

Postoperative care is continued after the patient has recovered and continues until the patient is discharged. This is the time frame when the patient is monitored for pain, signs of infection, or hemorrhage. Once the incision is closed, the patient is disconnected from the anesthesia machine, and the incision area is cleaned with hydrogen peroxide. Do not wipe the incision itself, only areas around the site. The area should be dried with a clean gauze sponge. The veterinarian may prescribe antibiotics or pain medications that should be given to the patient. The patient should be moved to the recovery area. Some patients may require IV fluids, which should be placed on an IV pole or an infusion pump. The veterinary assistant should monitor the recovery phase to make certain the patient's vital signs are normal and the patient is comfortable.

Pain is sometimes difficult to identify and assess. Signs of pain may include the following:

- Restlessness
- Whining or vocalization
- Reluctance to move
- Chewing or biting at incision
- Anorexia
- Change in behavior

Pain control provided before and after surgery is imperative in veterinary medicine. The veterinarian may have a pain protocol for surgical procedures or may prescribe specific medications. The veterinary technician and assistant must see that the medications are administered properly and as often as necessary. Most surgical patients are sent home with pain relief medication.

Food is usually withheld from monogastric animals for a 12 hour period, and water is often given in small amounts the evening after the surgery. When food and water are begun, they are given in small amounts, as they may cause digestive upset. Large animals and ruminants are often given small amounts of food and water as soon as they are standing and alert. The surgical patient should be eating and drinking within 24 hours of surgery, unless the procedure involved the gastrointestinal (GI) tract or mouth.

When the patient is ready for discharge, the home care instruction sheet should be prepared based on the procedure and veterinarian instructions. The sheet should be reviewed with the owner at the time of discharge. Most surgical patients will require a **suture removal** appointment that may be scheduled with the technician or assistant. The suture removal may be between 7 and 14 days, depending on the surgical procedure. The incision is monitored for any signs of infection or damage, which is noted within the medical record. Any abnormal signs with the incision should be reported to the veterinarian for observation. If the incision is normal, the sutures are cut individually with **suture removal scissors** (see Figure 42–29). These scissors have one blade that has a small hook on the end to better get under the suture (see Figure 42–30). The hook area is placed under the suture, which is then cut.

CHAPTER 42 Surgical Assisting Procedures 861

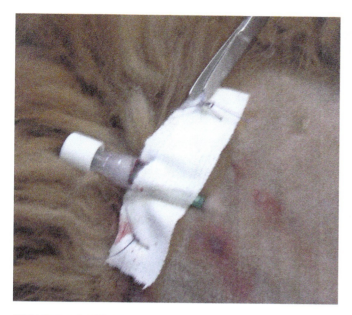

FIGURE 42–29 Suture removal with the use of suture removal scissors.

The assistant will then gently pull each suture from the skin by holding and pulling at the tie area. Staples are removed with a **staple remover**, similar to what humans use in offices. The staple remover is placed under the staple and squeezed to open up the staple, which is then gently removed from the site.

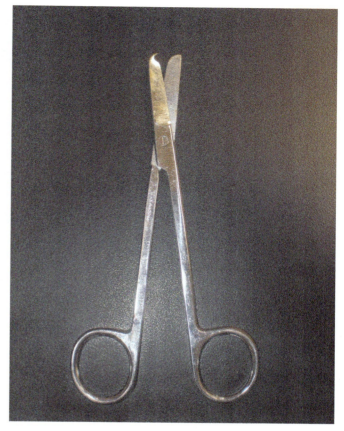

FIGURE 42–30 Suture removal scissors.

COMPETENCY SKILL 150

Suture or Staple Removal

Objective:

To properly remove sutures or staples at a surgical site

Preparation:

- Exam gloves
- Suture removal scissors
- Mosquito hemostats
- Staple remover

Procedure:

1. Apply gloves.
2. Inspect incision for signs of infection or damage.
3. Direct veterinarian to inspect incision as necessary.
4. Slip suture removal scissor blade under each knot and cut between the knot and skin.

5. Use a finger or mosquito hemostat to pull the knot through the skin and remove.
6. Slip the staple remover under the staple and squeeze the handle.
7. Remove the staple and place in sharps container.
8. Clean up and disinfect work area.
9. Disinfect all equipment.

Sterilization Techniques

The most common method for surgical sterilization of packs and instruments is the **autoclave** (see Figure 42–31). It is not used on heat-sensitive materials that are made of plastic, rubber, or other materials that may melt. Autoclaves come in a variety of shapes and sizes. **Distilled water** is heated in the chamber to allow high temperatures and pressures to produce the necessary heat to sterilize items. The distilled water is sterile water that is placed in the water reservoir to the fill line. Tap water should never be used, as it will cause mineral buildup within the equipment and clog the heat source. The higher the temperature, the quicker organisms are destroyed. Pressure forces steam into the interior packs and then destroy any living organism. Materials should not be packed into the autoclave too tightly, as this will prevent steam from entering all sites. Materials are packed onto trays that sit within the autoclave chamber. Once the packs are placed in the autoclave, a lever or button is released to fill the interior chamber, located below the trays, with the distilled water. When the area is filled, the door is closed and should be properly sealed. The heat source is turned on and the temperature set. Temperature settings and times will vary with equipment. A chart of directions should be located with the autoclave to note temperature and settings for various materials. The preheat cycle begins and is similar to an oven. Indicator lights will turn on and an audible sound will be noted as the autoclave begins to heat. The timer should be set. When items are exposed to steam for a long period of time the items become sterile. The equipment used to produce pressurized steam is called the autoclave. Gravity displacement autoclaves use sterile water that is heated inside a chamber through an electrical current and creates pressure within the chamber. The pressure allows the water to come to a boil and release steam. This is the most common type of autoclave used in veterinary medicine. The steam gradually displaces the air contained within the chamber and releases it through a vent. Proper sterilization is ensured when 250 degrees F is reached for at least 10–15 minutes. The entire cycle should be allowed to function, and observation may be required, especially if the device requires decreasing the temperature according to the type of machine. When the cycle is complete, an alarm will sound. At this time, the machine should be turned off and the pressure is allowed to decrease. When the pressure has properly decreased, the door can be partially opened to allow the steam to air out. Caution must be taken so as to not be burned by the steam or the heat of the device. Never allow direct skin contact with the autoclave. The door should be left partially ajar until the contents have cooled and dried. When the items are ready to be removed, they should be properly stored, and the autoclave door closed. Autoclaved packs and sleeves are considered sterile for 3–6 months before resterilizing is recommended. Care should be taken to properly store all sterile items in a clean, dry, low-traffic environment.

FIGURE 42–31 (A) Autoclave. (B) Autoclave tape.

COMPETENCY SKILL 151

Use of an Autoclave

Objective:

To properly fill and remove contents from an autoclave for sterilization

Preparation:

- Autoclave
- Distilled water
- Surgical packs

Procedure:

1. Fill the water-holding tank with distilled water to the fill line.
2. Close the holding tank.
3. Open the door to the autoclave.
4. Remove trays.
5. Place material on trays without stuffing items into the area.
6. Place trays into chamber.
7. Allow water chamber to fill with water.
8. Close door.
9. Set temperature and timer.
10. Allow to preheat.
11. Set temperature and timer for autoclave cycle.
12. When timer sounds, turn off autoclave and allow cooling.
13. When pressure has reduced, partially open the door to allow steam to escape.
14. Allow contents to cool and dry.
15. Unpack autoclave and store items in proper area.
16. Close door.

Surgical Instruments

Veterinary assistants should be able to identify common surgical tools, know how to clean and maintain the life of the instruments, and have an understanding of what the instruments are used for and how they work.

Cleaning Surgical Instruments

Instruments must be cleaned after each surgical use (see Figure 42–32). Surgery produces blood materials, waste materials, and other debris that may build up on instruments. Rust can occur on different types of metal instruments if not properly cleaned, dried, or maintained. It is best to use special instrument cleaners and distilled water to prevent deterioration of instruments and thus prolong their use.

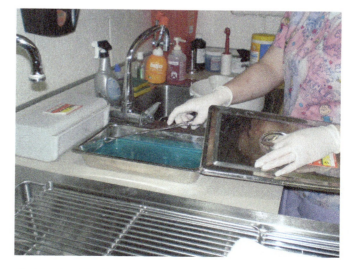

FIGURE 42–32 All surgical instruments must be properly cleaned.

Immediately after a surgery, all instruments in a pack should be removed and soaked in a disinfectant solution safe for the instruments. Organic debris should not be allowed to collect and dry on the instruments. The instruments should soak for at least 10 minutes and then be cleaned using a wire brush and manual scrubbing. Chlorhexidine solution and pink germicide are common instrument disinfectants, but manufacturer's recommendations should be consulted before using any cleaning agent. Instruments can be cleaned by hand, and particular attention should be paid to hinged areas and crevices. Mechanical cleaning is another method for cleaning instruments to aid in removing small particles of debris that are unable to be removed manually. This is done by using an **ultrasonic cleaner** (see Figure 42–33). Instruments are placed within a wire basket and the basket is seated within the cleaner, which holds a disinfectant solution. The instruments should always be placed in the basket in the open position and—ideally—spaced out. The cleaner is plugged in and turned on, then allowed to vibrate and cleanse for at least 5–10 minutes. The instruments should be fully submerged in the solution. After the cleaning cycle is complete, the instruments are rinsed with distilled water. The instruments should be thoroughly dried. They should then be dipped into a solution of **instrument milk** for 30 seconds. The instrument milk acts as a lubricant to keep the instruments protected from corrosion. Allow the instruments to drain and air dry before placing them in the surgical pack. Instruments that are not completely dry when sterilized in an autoclave are prone to rust. Any instrument that develops rust should be removed from the surgical packs and replaced with new instruments.

Scalpel Blade Handle

Scalpel blade handles are small and flat, and a blade is placed on the end to allow for making a surgical incision (see Figure 42–34). Handles come in sizes 3 and 4, and surgical blades range in size from #10 to #19 for handle size 3 and from #20 to #29 for handle size 4.

Towel Clamps

Towel clamps are used to hold the surgical drape in place over the patient during the surgery (see Figure 42–35). They attach lightly to the skin of the animal and the four corners of the drape. They are locking forceps in various sizes with sharp, curved, pointed tips (see Figure 42–36). Towel clamps are occasionally used to hold tissue. The most common type of towel clamp is the Backhaus towel clamp, which has a penetrating tip. Roeder towel clamps prevent tissue penetration due to metal balls on the tips. Four towel clamps should be placed in all surgical packs.

Needle Holders

Needle holders are used to hold a needle while placing sutures into a patient during surgery. They are hinged and lock in place to hold the needle (see

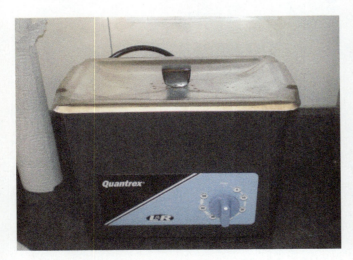

FIGURE 42–33 Ultrasonic cleaner.

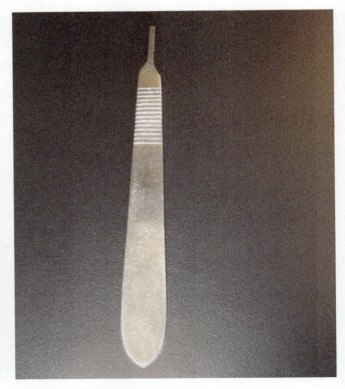

FIGURE 42–34 Scalpel blade handle.

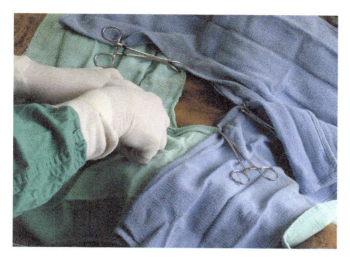

FIGURE 42–35 Use of towel clamps.

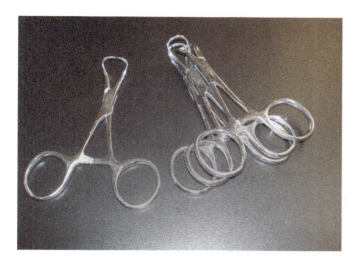

FIGURE 42–36 Towel clamps.

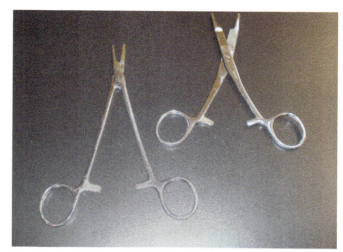

FIGURE 42–37 Needle holders.

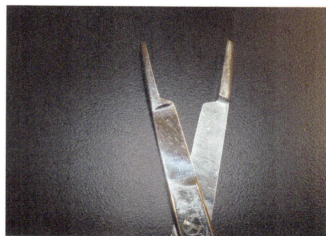

FIGURE 42–38 Head of the Olsen-Hegar needle holder.

Figure 42–37). Some types of needle holders are clamps only, and others are a combination of clamp and scissors. Mayo-Hegar needle holders are heavy instruments with mildly tapered jaws that have clamps only and no cutting blades. Olsen-Hegar needle holders include both a locking jaw and a scissor combination for holding the needle and cutting the suture (see Figure 42–38). Care must be taken when suturing to not accidentally cut the needle off from the suture material.

Tissue Forceps

Tissue forceps are used to grasp tissue and have teeth that grip the tissue (see Figure 42–39). **Dressing forceps** are used to grasp and hold tissue but have no teeth; rather, they have a smooth surface. Forceps have two handles held together at one end with a springlike action that opens and closes. **Rat tooth thumb forceps** have large teeth, similar to a rat's, that hold large, thick tissues and skin (see Figure 42–40). **Adson Brown thumb tissue forceps** have several small, delicate teeth that are used for light handling of delicate tissue (see Figure 42–41). **Debakey thumb tissue forceps** are used for the most delicate of tissues and provide atraumatic, grooved ends instead of teeth. **Allis tissue forceps** are intestinal forceps that lock in place to hold and grasp tissue (see Figure 42–42). They have small interlocking teeth used to grasp bowel tissue, intestinal tissue, and skin. When locked, Allis forceps have an opening that is wide to allow thicker tissues to be clamped. **Babcock tissue forceps** are also intestinal forceps but are less traumatic than Allis forceps. They have broad, flared ends with smooth tips. They are used to hold intestinal

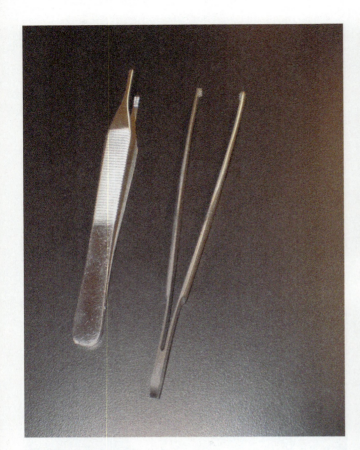

FIGURE 42–39 Tissue forceps.

FIGURE 42–40 Rat tooth forceps.

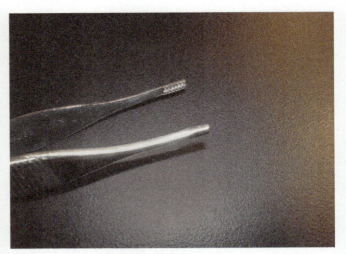

FIGURE 42–41 Adson Brown tissue forceps.

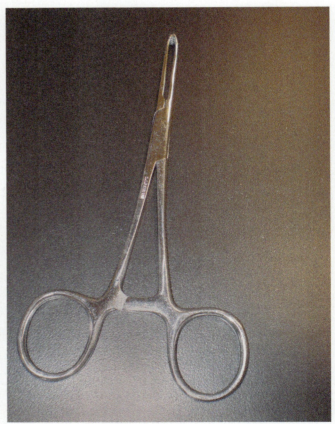

FIGURE 42–42 Allis tissue forceps.

and bladder tissue that is delicate and easily damaged. **Sponge forceps** may be straight or curved and are used to hold gauze sponges or other absorbent material to clear away bleeding and other debris (see Figure 42–43). They may have smooth or serrated edges.

CHAPTER 42 Surgical Assisting Procedures   867

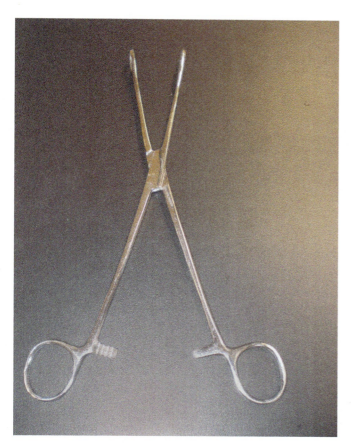

FIGURE 42–43 Sponge forceps.

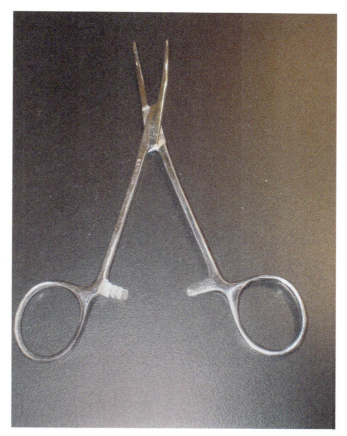

FIGURE 42–44 Hemostatic forceps.

Hemostatic Forceps

Hemostatic forceps or hemostats are hinged locking instruments that are designed to clamp and hold off blood vessels (see Figure 42–44). Hemostats may be straight or curved. **Halstead mosquito hemostats** are small in size with a fine serrated tip that is used to hold delicate small-diameter blood vessels (see Figure 42–45). **Kelly hemostatic forceps** are larger in size and are partially serrated to better hold and grasp medium-sized blood vessels (see Figure 42–46). **Crile hemostatic forceps** are similar in size and shape to Kelly forceps, except they are fully serrated. The serrations are located within the interlocking jaws and are located vertically along the blades. **Rochester-Carmalt hemostatic forceps** are much heavier and are used to hold off the ovarian stump during a spay surgery, as well as large blood vessels. The interlocking jaws are fully serrated lengthwise.

Intestinal Forceps

Intestinal forceps are used during abdominal surgeries. **Doyen intestinal forceps** or clamps are noncrushing

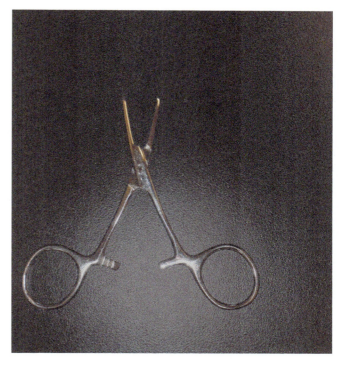

FIGURE 42–45 Mosquito hemostats.

FIGURE 42–46 Kelly hemostats.

intestinal forceps that have lengthwise serrations. They hold intestinal vessels and tissues without causing any damage. They are used for short periods of time to tie off larger vessels.

Scissors

Scissors are used to remove tissues during surgery. They may have sharp ends that are pointed or blunt ends that are rounded. They may also have straight or curved blades. **Mayo scissors** are used to cut and dissect heavy tissues and sutures (see Figure 42–47). The blades may be straight or curved. They are proportionate in size and when lying on a flat surface, the curved edges curve upward without contact at the end of the blades. **Metzenbaum scissors** are more delicate, are thinner in size, and have long handles in relation to the blade length. They are used to cut and separate delicate tissues. Suture removal scissors have a sharp tip with a section in the blade used to get under the suture material and cut it for removal. They may be straight or curved. **Sharp operating scissors** are used for cutting objects like suture or drape material rather than tissue. Blades may be straight or curved, and tips are sharp or blunt.

Retractors

Retractors may be handheld or self-retracting and are used during surgeries to hold open an animal's body cavity for exploration. Senn and Holman retractors are both handheld. **Senn retractors** have blades at each end that look like fingers on the end of a hand. They may be blunt or sharp in shape. **Holman retractors** are levers that are flat with a pointed appearance that looks similar to an arrow (see Figure 42–48). They have holes in the handle and are commonly used during orthopedic procedures. Self-retracting retractors include the Weitlaner and **Gelpi retractors**. The ends hold tissues apart for better visualization within a body cavity. The **Weitlaner retractor** has rakelike teeth on the ratchet ends and can be blunt or sharp (see Figure 42–49). The Gelpi retractor has single points on the end that are sharp.

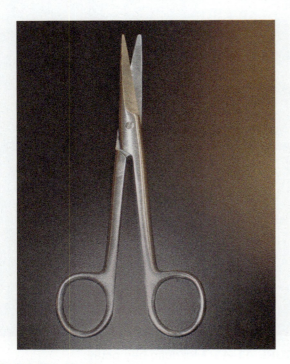

FIGURE 42–47 Mayo scissors.

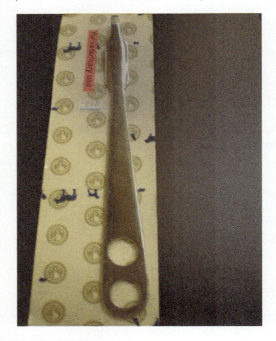

FIGURE 42–48 Holmann retractor.

CHAPTER 42 Surgical Assisting Procedures

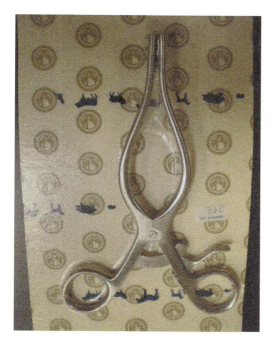

FIGURE 42–49 Weitlaner retractor.

Spay Hook

The **spay hook** is used to locate the uterus and uterine horns in small female animals. It is a long, thin metal device with a hook on the end (see Figure 42–50). The most common is the Snook spay hook.

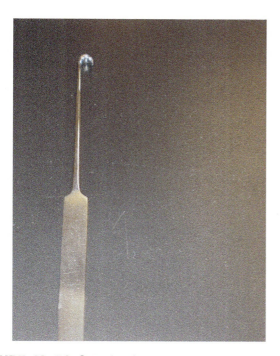

FIGURE 42–50 Spay hook.

Dental Instruments

Prophylaxis, often referred to as dental prophy, and extractions are common daily procedures in oral health medicine. The dental prophy includes scaling and polishing of the teeth under anesthesia. When extractions and other dental procedures are necessary, specialized dental instruments are often used (see Figure 42–51). A **dental mirror** is a specialized reflective face used to visualize all arcades of the teeth. Each tooth should have any pockets or depths between teeth measured using a **periodontal probe**, which is a long, thin probe used to measure pocket depths around a tooth to determine the health of the periodontium tissue. A **mouth gag** is often used to open the mouth to better visualize the teeth and allow for exam and prophy completion. The mouth gag attaches to each side of the mouth at the tooth tip to keep the mouth open (see Figure 42–52). Caution must be taken when using mouth gags for long periods of time, especially in cats, as nerve damage and blindness can result from use of improper methods. A **dental elevator** is used during a tooth extraction to elevate the area around the neck of the tooth and separate it from the gum and root. A **dental extractor** is then used to grasp the tooth to remove it from the socket. A **periosteal elevator** may be used to lift thicker soft tissue flaps during extractions. A **tartar scraper** may be used to remove calculus and tartar from the tooth surface and between teeth. Many facilities have upgraded to high-speed dental units that include instruments used in extractions and oral surgery (see Figure 42–53).

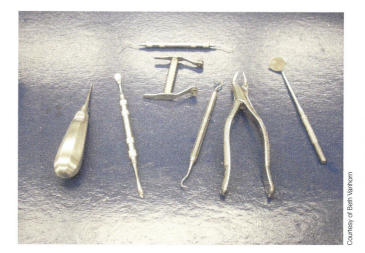

FIGURE 42–51 Dental instruments: dental elevator, periosteal elevator, periodontal probe, mouth gag, tartar scraper, dental extractor, dental mirror.

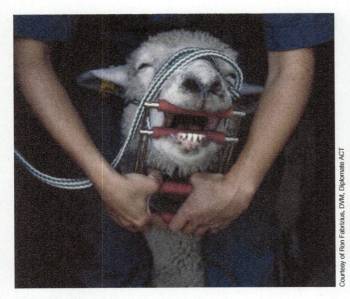

FIGURE 42–52 Mouth gag for oral exam.

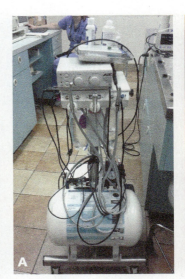 

FIGURE 42–53 (A) High-speed dental machine. (B) Use of high-speed dental scaler.

SUMMARY

Many aspects are involved in surgical assisting. Surgical preparation involves understanding care and maintenance of surgical equipment and instruments, patient preparation and care, surgeon care and assistance, a clean and aseptic surgical environment, and postoperative care of the patient. The goal of all aspects within the surgical suite is to maintain a sterile and aseptic environment that protects the patient and the veterinary health care team.

Key Terms

absorbable suture material that is broken down within the body over a period of time and is absorbed by the body

Adson Brown tissue thumb forceps instrument with several small, delicate teeth that are used for light handling of delicate tissue

Allis tissue forceps intestinal forceps that lock in place to hold and grasp tissue and when locked; have an opening that is wide to allow thicker tissues to be clamped

anesthesia chamber a glass or plastic square or rectangular box that is clear and has two openings in which the inhalation and exhalation tubes on the anesthesia machine connect to the chamber

anesthesia logbook that details the patient's status through induction, maintenance, and recovery stages of an anesthesia procedure

anesthesia mask conelike device used in larger patients; placed over the muzzle and face area; must fit snugly for the animal to breathe in the gases for induction

asepsis absence of all microorganisms or infection

aseptic technique process of obtaining a sterile and organism-free environment

autoclave device that sterilizes through extreme heat and steam

Babcock tissue forceps intestinal forceps that are less traumatic than some others and have broad, flared ends with smooth tips

Bain system   uses a single corrugated or plastic tube that connects to the patient and another hose through the center of the tube that contains the gas mixture the patient inhales

blink reflex   allows evaluation of the eye to determine which plane of anesthesia a patient is in

blood pressure monitor   device used to assess the pumping action of the heart

canister   area that absorbs carbon dioxide and contains granules to absorb it

closed gloving   process done when the gown is placed with the hands still within the sleeves so that no skin is visible and potentially contaminating the gloves

corrugated tubing   plastic and thin hose that allows the oxygen and anesthetic gas to flow to the patient by connecting from the machine to the endotracheal tube

Crile hemostatic forceps   similar in size and shape to Kelly forceps, except they are fully serrated, with serrations located within the interlocking jaws and vertically along the blades

cuff   located about ¾ of the way down the endotracheal tube; inflatable and allows the tube to be inflated with air to create a tight seal around the trachea to prevent any moisture or material from entering the lung field

cylinder   metal containers that hold compressed forms or liquid to keep patients oxygenated and breathing during anesthesia

Debakey thumb tissue forceps   used to grasp the most delicate tissue with atraumatic grooved ends instead of teeth

dental elevator   instrument used to extract a tooth by separating the tooth neck from the root

dental extractor   instrument used to remove a tooth from the socket during an extraction

dental mirror   instrument with a reflective face used to visualize all tooth surfaces

distilled water   water with impurities removed

Doyen intestinal forceps   noncrushing intestinal forceps that have lengthwise serrations and hold intestinal vessels and tissues for short time periods without causing any damage

dressing forceps   instrument used to grasp and hold tissue; have no teeth, but rather a smooth surface

electrocardiogram   monitor placed onto a patient through metal leads called electrodes that measure the electrical activity of the heart; also called an ECG or EKG

electrode   metal leads that connect a patient to the EKG unit

endotracheal tube   flexible tube that is placed into the trachea to aid in breathing

esophageal stethoscope   device placed into the esophagus next to the endotracheal tube; emits the sound of the heartbeat

extubation   process of removing the endotracheal tube from the patient

fast   no intake of food or water, or anything by mouth, prior to surgery

flow meter   located on the machine near the vaporizer; is adjusted for the proper rate of oxygen flow to the patient

Gelpi retractor   self-retracting instrument with single points on the end that are sharp

glove pack   prepackaged disposable sterile glove pack that is used one time and then discarded

gown pack   holds the sterile gown that is worn by the surgeon and veterinary staff assisting in the surgical area

Halstead mosquito hemostats   small in size with a fine tip that is used to hold delicate blood vessels

hemostatic forceps   hinged locking instruments that are designed to clamp and hold off blood vessels; also called hemostats

Holman retractor   levers that are flat with a pointed appearance; look similar to an arrow with holes in the handle; are commonly used during orthopedic procedures

induction phase time when the patient is being given anesthesia to make it sleep and remain unconscious and free of sensation during the surgery

instrument milk solution that acts as a lubricant to keep instruments protected from corrosion

intubation process of placing a tube into the trachea to establish an airway and allow the patient to continue to inhale gases that keep it under general anesthesia as long as necessary

isotonic crystalloid fluids given to an animal that is dehydrated and create a balance of elements within the body

Kelly hemostatic forceps larger in size; are partially serrated to better hold and grasp blood vessels

laparotomy towel sterile towel used to control bleeding during surgery and as a hand towel for the surgeon

laryngoscope tool made of heavy metal that has a light source on the end to help light the airway for viewing

Mayo scissors instrument used to cut heavy tissues and sutures that are proportionate in size; when lying on a flat surface, curve upward without contact at the end of the blades

Mayo stand instrument tray that is elevated to hold the surgeon's surgical instruments and supplies

Metzenbaum scissors instrument delicate and thinner in size that others; have long handles in relationship to the blade length; are used to cut and separate delicate tissues

monofilament suture single strand of material

mouth gag instrument used to hold the mouth open during a dental prophy

multifilament suture several strands of material braided together

needle holder instrument used to hold a needle while placing sutures into a patient during surgery

nonabsorbable suture material not broken down within the body; remains intact for long periods

nonrebreathing system system used for very small patients such as rodents, birds, and reptiles, or small puppies and kittens that are too small to move the breathing bag with their respirations

open gloving process done by removing the left glove using the right hand by grasping the glove at the edge of the cuff and the edge of the cuff

oxygen tank cylinder that holds oxygen

palpebral reflex evaluated by tapping the fingertip lightly over the surface of the eyelid

pedal reflex used on the digit or the skin between digits, which is pinched to determine a reflex response

periodontal probe instrument used to measure the depth of pockets around a tooth

periosteal elevator instrument used to remove thick flaps of soft skin tissue during a tooth extraction

plane level of awareness of a patient under anesthesia

pop-off valve used to release excess amounts of gas when the pressure becomes too high within the machine system

pressure valve indicates how much oxygen is left in the tank

pulse oximeter device used to measure a patient's vital signs by indirectly measuring the oxygen saturation within the blood and any changes in the blood volume

rat tooth thumb forceps instrument with large teeth, similar to a rat's, that holds large thick tissues and skin

rebreathing system bag that connects to a machine and inflates and deflates as the patient breathes; allows the patient to rebreathe anesthetic gas after it has passed through the CO_2 absorbing canister

retractor handheld or self-retracting; is used during surgeries to hold open an animal's body cavity for exploration

Rochester-Carmalt hemostats   much heavier than others; used to hold off the ovarian stump during a spay surgery; has lengthwise interlocking serrations

scavenger hose   tube attached to the pop-off valve to prevent gas from leaking into the room; pushes gases outside of the facility to escape to the outside air

scrub pack   sterile supply that contains the items necessary to properly prepare the surgeon, technician, or assistant for washing hands for surgery

Senn retractor   retractor blades located at each end that look like fingers on the end of a hand; may be blunt or sharp in shape

sharp operating scissors   used for cutting objects like suture or drape material rather than tissue

soda lime   granules within a canister that absorb carbon dioxide

spay hook   instrument used to locate the uterus and uterine horns in small female animals; has a long, thin metal device with a hook on the end

sphygmomanometer   device used to measure an animal's blood pressure during surgery; will also help determine tissue perfusion as well as depth of anesthesia; also called a blood pressure monitor

sponge forceps   straight or curved; used to hold gauze sponges or other absorbent material to clear away bleeding and other debris

staple remover   device placed under a staple and squeezed to open up the staple, which is then gently removed from the site

surgery pack   sterile instrument prepared for surgical procedure with tools and supplies required for the procedure

surgical glue   sterile substance used for closure of small wounds or surgical sites; is commonly used in feline declaws

surgical log   a record of details of all surgeries performed in the facility

surgical margin   area of the patient that should extend 2 to 4 inches beyond the anticipated surgical incision borders

surgical scrub   process then begins with sterile solutions placed onto the surgical margin to clean the patient

surgical staples   fast and easy way to close the skin with metal pieces

suture reel   container of suture material that must be cut and thread onto a needle

suture removal   occurs within 7–10 days after surgery

suture removal scissors   have one blade with a small hook on the end to better get under a suture for removal

swaged on needle   suture material that has a needle already attached

swallow reflex   the patient licking with the tongue or having increased jaw tone during extubation of the endotracheal tube

tartar scraper   instrument used to remove calculus and tartar from the tooth surface

tidal volume   amount of oxygen required by the patient

tissue forceps   instrument used to grasp tissue; has teeth that grip the tissue

towel clamp   instrument used to hold the surgical drape in place over the patient during surgery

ultrasonic cleaner   instrument is placed within a wire basket, which is then seated within the cleaner, which holds a disinfectant solution and creates vibrations to clean the item

vaporizer   converts liquid anesthesia into a gaseous form, which the patient inhales

Weitlaner retractor   self-retracting instrument with rakelike teeth on the ratchet ends; and can be blunt or sharp

Y connector   plastic device that connects the intake and outtake hoses from the anesthesia machine to the endotracheal tube of the patient

Review Questions

1. What is the importance of asepsis?
2. What is the goal of an aseptic surgical area?
3. What items are recorded in an anesthesia log?
4. Why is preanesthetic medication given to patients?
5. What is intubation?
6. How does an assistant know when to extubate a patient?
7. What vital sign monitors are found in the surgical suite?
8. What items must be sterile when working in the surgical area?
9. What items are found in a surgical pack?
10. What are signs of pain in animals?
11. What factors should veterinary assistants know about the anesthesia machine?

Clinical Situation

Sally, a veterinary assistant at Kitty Kat Klinic, is working in the surgical suite with Dr. Shamitko. Sally has been trained to open the surgical packs and prepare the surgical monitors and equipment, as well as clean and sterilize the items used in surgery. Today, Sally has noticed that the surgical pack she is beginning to open has a small tear in it. She also notices that the autoclave tape indicator is not very dark.

- What should Sally do?
- What may have caused these occurrences?
- What may prevent these situations in the future?

Glossary

abdominal system includes the stomach, intestines, and other internal organs such as the liver, pancreas, and spleen

abduction movement away from midline or the axis of the body

abductor muscle a muscle that allows a shell to open and close

abomasum last section of the ruminant stomach that acts as the true stomach and allows food to be digested

abortion a termination of pregnancy

abscess a bacterial infection that causes pus to build up in a localized area

absorbable suture material that is broken down within the body over a period of time and is absorbed by the body

absorption the movement of drug particles from the site of administration into the body system

acceptance when a person understands and accepts that a pet has passed away

account history a client's previous invoices, which can be viewed at any time to see past payments and charges

acetabulum area of the pelvis where the femur attaches to form the hip joint

acid rain polluted rainwater

acidity water pH levels below 70

ACTH adrenocorticotropic hormone; released by the adrenal glands; controls blood pressure, releases cholesterol, and regulates the body's production of steroids

actin a thin protein filament fiber found in muscle tissue

actinobacillosis a disease of the head and jaw that causes tumorlike lumps of yellow pus to build up within the jaw; also called lumpy jaw

actinomycosis a bacterial infection of the head and neck lymph nodes that causes hard lesions to form over the tongue; also called wooden tongue

active immunity developed from exposure to a pathogen through the process of vaccination or the disease itself

active transport the process in which molecules at a lower concentration move to a higher concentration

acute disease or condition occurring a short duration

ad lib giving something, such as food or water, as often as necessary or so that it is available at all times

Addison's disease occurs due to decreased production of steroids in the adrenal glands, causing the glands to fail to function; common term for hypoadrenocorticism

adduction movement toward the midline or axis of the body

adipose fat tissue

adrenal glands located cranial to the kidneys; produce and release adrenaline and other hormones

adrenaline chemical released by the nervous system in times of stress to instigate response of an animal's fight-or-flight reaction

Adson Brown tissue thumb forceps instrument with several small, delicate teeth that are used for light handling of delicate tissue

aeration oxygen or air created in the water by making bubbles

agglutination clumping

aggression behavior that makes an animal angry, difficult, and potentially unsafe to handle

aggressive refers to the behavior of an animal exhibiting an attempt to do harm, such as biting

aging allowing a water source to stand in a container for at least 5 days

agonal refers to gasps of breaths while the respiratory system shuts down at the time of death

agonistic muscles that work in pairs with each other

agouti a mixture of two or more colors

air cell an empty area located at the large end of an egg that provides oxygen

albumen egg white

albumin part of the blood that draws water into the bloodstream and provides hydration; the egg white surrounding the yolk

alevin young salmon

alfalfa hay high in protein

algae a plant-based source of food that grows spontaneously in water when left in the sunlight

alimentary canal veterinary medical terminology for the GI system

alkalinity water pH levels above 70

allergen substance that causes an allergic reaction

alligator a large reptile raised for its meat, skin, and by-products and for release into the wild

Allis tissue forceps intestinal forceps that lock in place to hold and grasp tissue; when locked, have an opening that is wide to allow thicker tissues to be clamped

alopecia hair loss

alpaca a ruminant mammal similar to but smaller than a llama; originated in South America; has a silky, fine coat and is raised for its hair

alphabetical filing system the organization of medical records by client or patient last name

alveoli tiny sacs filled with air at the end of the bronchioles; tiny structures within the udder where nutrients are converted to milk

amazon a large breed of cage bird from South or Central America that is talkative, entertaining, and very trainable

American Animal Hospital Association (AAHA) sets the standards of small animal and companion animal hospitals

American Association for Laboratory Animal Science (AALAS) an association of professionals that advances the responsible use and care of laboratory animals to benefit people and animals

American class a chicken class developed in the United States for egg and meat production

American Medical Association (AMA) the largest association of physicians and medical students in the United States; it conducts studies that benefit the advancement of human medicine and surgical procedures

American Veterinary Medical Association (AVMA) accredits college programs for veterinarian and veterinary technician state licensure and sets the standards of the veterinary medical industry

Americans with Disabilities Act (ADA) a law that prohibits discrimination against people with disabilities

amino acid simplest hormone that controls thyroid gland functions; building blocks of proteins that form chainlike structures

amphibian an animal with smooth skin that spends part of the time on land and part of the time in water

amylase enzyme produced by the pancreas that breaks down starches

anabolism a process in which smaller molecules combine into larger molecules

anal glands scent glands located on either side of the rectum

analgesic drug that reduces pain without loss of other sensations

anaphase the stage of mitosis or meiosis in which the chromosomes move toward the poles of the spindle

anaplasmosis a parasitic protozoan disease that affects dogs, cats, and ruminants

anastomosis surgical removal of a dead area of tissue along the digestive tract and re-sectioning the area back together

anatomy the study of the internal and external body structures and parts of an animal

Ancylostoma caninum hookworm species that live in the small intestine of the host animal, where they attach themselves and feed on the host's blood

anemia low red blood cell count due to the red blood cells not being replenished

anesthesia chamber restraint equipment used to sedate a cat or other small mammal under gas anesthesia that is pumped into the glass or plastic chamber

anesthesia log book that details the patient's status through the induction, maintenance, and recovery stages of an anesthesia procedure

anesthesia mask conelike device used in larger patients; placed over the muzzle and face area; must fit snugly for the animal to breathe in the gases for induction

anesthetic drug that causes loss of pain, consciousness, and all sensory sensations

anestrus time when a female animal is not actively in the heat cycle

anger a strong feeling of displeasure felt when confronting the death of a pet

Angora goat a type of goat that produces hair

animal abuse laws rules that protect against the neglect and abuse of animals; these laws are typically regulated by a humane society or dog warden

animal behavior relates to what an animal does and why it does it

animal behaviorist someone who studies animals to collect data on their behavior in captivity

animal model a live animal used for the study of behavior, diseases, procedures, treatments, and the effects of medications on the body

animal nutrition science of determining how animals use food in the body and examining all body processes that transform food into body tissues and energy for activity, as well as the processes which by animals grow, live, reproduce, and work

animal refuge a large protected area of land that houses exotic animals and is set up to resemble their own habitat

animal rights rules that govern how animals are handled and cared for, especially in research facilities

animal tissue a living part of an animal that is cultured in a lab and grown for specific needs

Animal Welfare Act (AWA) rules that regulate how animals are handled and cared for, including within research facilities

anisicoria uneven pupil size

anode rotating target that creates positively charged electrons

anogenital the area located around the far stomach between the rear legs and the base of the tail

anogenital distance space between the area of the vulva or penis and the rectum by which gender can be determined

anorexia not eating

antacid drug that reduces acidity of the stomach

antagonistic muscles that work against each other

anterior chamber area of the eye in front of the iris that holds the aqueous humor

anterior–posterior (AP) view that requires the X-ray beam to penetrate the anterior portion of the body area; the beam exits from the posterior portion of the body

anthelmintic a medication used to treat and prevent parasitic infection; also known as a dewormer

antiarrhythmic drug used to treat abnormal electrical currents of the heart

antibiotic medication used to treat infection or disease

antibody a protein that fights off disease; a substance that acts to protect the immune system

anticoagulant substance that prevents the blood from forming a clot

anticonvulsant drug used to treat seizure activity

antidiuretic hormone (ADH) hormone that promotes urine formation, water absorption, and blood pressure control; also conserves water volume in the body

antiemetic drug used to stop vomiting

antifreeze toxicity poisoning that affects the renal system and shuts it down, causing renal failure; also known as ethylene glycol toxicity

antifungal drug used to kill or treat fungus

antigen foreign materials used to create an immune response

antihistamine drug used to prevent and control an allergic reaction

anti-inflammatory drug used to treat inflammation and pain without corticosteroid components

antimicrobial drug that kills or inhibits growth of microorganisms; often called an antibiotic

antioxidants vitamins that boost the body's immune system

antiseptics effective disinfecting agents that destroy microorganisms or inhibit their growth on living tissue

antitussive drug that suppresses coughing and blocks the cough reflex within the brainstem

antiulcer drug used to treat or prevent gastric ulcers

anuria no urine production

aorta large vessel that allows blood to flow out of the heart and back into systemic circulation

apnea condition in which the patient is not breathing but then begins to breathe again

appendicular skeleton contains bones of the body that hang; the bones of the limbs or appendages

applied research a study done for a specific purpose

appointment card a written reminder for clients that includes the date and time of the appointment scheduled for their pet

aquacrop a commercially produced plant or animal living in or around a water source

aquaculture the production of living plants, animals, and other organisms that grow and live in or around water

aquarium a small rectangular vessel that holds water to contain small fish

aquarium maintenance schedule (AMS) a list of important fish tank duties that should be completed at a particular frequency

aquatic pertaining to water

aqueous humor watery fluid located in the anterior chamber of the eye

arboreal refers to an animal that lives in trees

arrhythmia change in the heart's rate or rhythm

arteries vessels that carry blood away from the heart

arterioles smaller vessels that branch off of an artery

arthritis inflammation of a joint

arthropod a crustacean with an external skeleton

artificial incubation providing a controlled temperature in which eggs hatch, such as an incubator

artificial insemination (AI) the practice of breeding female animals through the use of veterinary equipment rather than the traditional mating method

artificial vagina (AV) sterile sleeve or tube used to collect the semen of an animal during the ejaculation process

ascarid also known as roundworm; an intestinal parasite that looks like a long white worm

ascending colon first section of the large intestine

ascites a bacterial or viral infection that causes a fish to swell with fluid; also called dropsy

asepsis absence of all microorganisms or infection

aseptic free from contamination

aseptic technique process of obtaining a sterile and organism-free environment

Asiatic class a chicken class developed in Asia for show and ornamental breeds

aspiration process used with a syringe when the plunger is drawn back slightly to make certain no blood vessel has been accidently penetrated prior to administering an injection; also refers to swallowing food that enters the lung field

assessment a veterinary diagnosis of problem

associate degree a 2-year degree program

asthma condition that affects the airways in the lungs

asystole when the heart stops contracting and heart failure occurs

ataxia term used to refer to an animal that has lost control over its body, causing wobbliness and inability to control movement

atlas first cervical vertebrae located at the base of the brain; allows for up-and-down motions of the head

atopy allergy causing a skin infection

atrial fibrillation also called A-fib; condition that occurs when the SA node, the pacemaker of the heart, is not working correctly

atrophy refers to muscle that has shrunk due to disuse

auditory ossicles bones of the middle ear that transmit sound vibrations

aural pertaining to the ears

aural hematoma collection of blood within the cartilage of the ear

auscultation process of using a stethoscope to listen to the heart, chest, abdomen, and lungs for any abnormal sounds

autoclave a sealed chamber in which objects are exposed to heat and steam under pressure at extremely hot temperatures to kill all living organisms

autoimmune disease animal's red blood cells are destroyed by the immune system

autolysis the breakdown and removal of dead cells

automated inventory system a system in which all supplies and products in the facility are counted by a computer program listing how many items are in stock

automatic feeder specialized equipment that is computerized and set to release certain amounts of food at specified times throughout the day

automatic film processing equipment that develops and dries film automatically

AV valve opens and closes between the atrium and ventricle to allow blood to flow through each chamber of the heart

avian influenza a virus that affects the intestinal tract of both poultry and humans; also known as bird flu

avian pox a virus affecting poultry that is spread by mosquitoes and slowly lowers the immune system

avian system specialized digestive system of birds

axial skeleton contains bones of the body that lie perpendicular or lengthwise to the bones that are the appendages or hanging parts

axis second cervical vertebrae; allows for rotation and shaking motions of the head

axon a hairlike projection from a neuron that carries the nerve impulse

azotemia elevation of blood urea nitrogen and creatinine in the bloodstream

B lymphocytes white blood cells that develop and mature in the bone marrow

Babcock tissue forceps intestinal forceps that are less traumatic than some others and have broad, flared ends with smooth tips

babesia less common disease caused by the deer tick; causes signs of anemia, jaundice (yellow coloring of the skin and mucous membranes), fever, and vomiting; more severe stages can lead to kidney failure

bachelor of science (BS) a 2-year college degree program

back-grounding system a production system that raises calves to market size for profit

back rubber equipment that dispenses fly-control chemicals to the backs of a cow

bacon-type hog a hog raised for the large amount of meat known as bacon on its body

bacteria a living organism that invades the body, causing illness

bacterial class of disease spread through single-celled organisms that are located in all areas of an animal's environment

bactericidal kills bacteria

bacteriostatic inhibits growth of bacteria

Bain system uses a single corrugated or plastic tube that connects to the patient via a hose through the center of the tube that contains a gas mixture the patient inhales

balanced ration diet that contains all the nutrients required by an animal in correct and specific amounts

ball and socket joint that rotates in numerous directions

balling gun metal device with a long handle that has a plunger at the base; used to hold medicines to administer to large animals

band a large group

band cell immature neutrophil that indicates an infection within the body; shaped like a *U*

bandage covering applied to a wound, fracture, or area of injury to prevent movement and protect the wound

bandage scissors scissors with an angled blunt end used to carefully remove a bandage from an animal

banding method of castration in which a tight rubber band is placed over the testicle area of livestock; circulation is cut off and tissue falls away

Bang's disease *see brucellosis*

barb a device on a fish that detects food, danger, and movement; also known as a barbel

barbering the tendency of mice to chew at the fur of another mouse

bargaining stage of grief wherein the person attempts to resolve a pet's impending death by any means possible

barium enema involves a large amount of fluorescent substance passing into the colon via the rectum

barium study series of films taken as barium passes through the digestive tract

barium sulfate substance administered to the patient to show structures on an X-ray film that would otherwise be difficult to view

barrow a young castrated male pig

basic research research carried out in a lab setting with the purpose of understanding life processes and diseases

basophil white blood cell with a segmented nucleus and granules that stain very dark; aids in allergic reaction control

bathing process of cleaning the skin and hair coat of the animal or applying medicated shampoos or dips to the skin and hair coat

beak avian mouth with no teeth that forms an upper and lower bill

beak trim a procedure to shorten or round the sharp edges of a beak

beard hairs located under the wattle in turkeys

beef cattle cows specifically raised for meat

beefalo a crossbreed of bison and domesticated beef cattle that produces a beef animal with healthier and tastier meat products

benign noncancerous

benign prostatic hyperplasia (BPN) enlarged prostate gland in males

bilateral cryptorchid both testicles are retained and do not descend into the scrotum

bile yellow fluid secreted to help break down food for digestion and absorption

billy a common name for an adult male goat

biological drugs used to treat disease

biological filtration the use of bacteria and other living organisms in water to change harmful materials into safer forms

biological hazards safety concerns that pose a risk to humans and animals through contamination of living organisms through body tissues and fluids, such as blood or urine

biological oxygenation the use of plants to create oxygen in the water

biological value percentage on a food label that describes the quality of the food source

biology the study of life

bird flu an influenza virus found in avian species that may be zoonotic to humans

bird of prey a large bird that is a hunter and eats small mammals and rodents; is protected by law and not legal to hunt

bison a ruminant mammal closely related to cattle

bit equipment placed in a horse's mouth when it is being ridden

bivalve a mollusk with two shells

blackleg a fatal disease in cattle caused by bacteria; as the disease progresses, swelling causes limbs to become infected and tissue to turn black

blink reflex allows evaluation of the eye to determine which plane of anesthesia a patient is in

bloat condition that causes the abdomen to become swollen and painful due to air and gas within the intestinal tract

blood chemistry analyzer machine that runs blood samples that measure routine blood chemistries and electrolytes

blood feathers growing feathers that have a blood vessel in the center

blood glucose blood sugar within the blood

blood plasma blood sample that is frozen and then centrifuged to obtain plasma

blood pressure (BP) the heart's contractions and relaxations as blood flows through the chambers

blood pressure monitor device used to assess the pumping action of the heart

blood red liquid within the circulatory system that contains 40 percent of cells in the body and transports oxygen throughout the body

blood serum liquid portion of the whole blood sample

blood smear blood film placed on a slide that is used to look at the morphology of blood cells

blood titer measured amount of antigen within the bloodstream

blue tongue a virus spread by gnats that weakens the immune system and causes secondary infections

boar an adult male pig of reproductive age

body central part of the stomach that expands as food enters

body condition scoring (BCS) rating of how an animal appears, including weight relative to its ideal body weight

body language communication by the animal about how it feels about other animals, people, and its environment; the use of mannerisms and gestures that show how a person is feeling

bone hard, active tissue that consists mostly of calcium and forms the skeleton of the animal; gives the body support, structure, and protection

bone marrow soft substance within the medullary cavity where blood cells are produced

bone plate surgical steel plate placed around a fractured bone that holds the bone together with surgical screws

booster vaccines introduced to the immune system to build up protection and immunity over a period of time

bordetella tracheobronchitis; commonly called kennel cough

bottom-dwelling fish fish that live and feed on the bottom of a body of water

bovine a veterinary term for a cow

bovine respiratory disease complex (BRDC) a stress-related disease in young cattle, also known as shipping fever

bovine viral diarrhea (BVD) a virus in cattle with symptoms that include diarrhea and respiratory distress

Bowman's capsule cuplike sack at the beginning of the tubular component of a nephron in the kidney that performs the first step in the filtration of blood to form urine

brachycephalic a short nasal area that gives a flat-face appearance

brackish water a mixture of freshwater and saltwater

bradycardia decreased or slow heart rate

bradypnea decreased rate of breathing

brain the major organ in the nervous system that controls the body's actions

brainstem part of the brain that controls functions that maintain life

branding using extreme heat or cold to mark the skin of livestock with a number or symbol

bray a loud vocal sound that a donkey or mule makes

breeding stock the male and female animals of a species raised specifically for reproductive purposes

broiler a young chick that is between 6 and 8 weeks of age and under 4 pounds; used for meat

broiler egg production system a system in which chicks are raised for large production numbers and selected for broiler programs

broiler production a system used to produce the most poultry possible

bronchi branches at the base of each lung; connected by large passages

bronchioles smaller branches of the bronchial airways in the respiratory tract

bronchodilator drug that decreases bronchiole constriction and opens up the airway

brood a group of reptiles or amphibians

broodfish mature male and female breeding fish used for reproductive purposes

broodmare a female horse used for reproduction and breeding purposes

browse a woody plant

brucellosis a highly contagious reproductive disease for which there is no treatment; also known as Bang's disease

brumation a process in reptiles in which their body temperature decreases; similar to hibernation

brushing cleaning the hair coat using a soft brush

buccal referring to the cheek or the sides of the oral cavity

buck an adult male goat

bucking the lifting and kicking action of the hind limbs

budding yeast single-celled organisms that are in the shape of a budding flower

budgerigar commonly called a parakeet; a small bird breed that comes in a variety of bright colors with bars located over the wings and back

budgie an abbreviated name for a budgerigar or parakeet

buffy coat the layers within a hematocrit or PCV that form a tiny area of visible white composed of white blood cells and platelets

bulbourethral glands paired glands that contribute to the fluid portion of semen

bull an adult male cow of breeding age

burial the process of placing a carcass in the ground and covering it with soil to allow decomposition

burrow to dig a tunnel into an area

butting using the head to hit an object or a person

by-product a part of an animal that is not the main profit or production source, such as the hooves, antlers, hair coat, or internal organs

by-products tissue tissue that is removed from an animal when it is used for purposes, such as research or human consumption

cage a structure for holding aquatics that floats on the water

cage card identifies and helps locate a patient while in the care of the facility

calcitonin hormone that regulates the body's calcium level and is secreted when blood calcium levels are elevated

calculus mineralized plaque material that appears brown to dark yellow in color and is difficult to remove from teeth

calf a newborn cow of either gender

calicivirus a highly contagious virus that affects the upper respiratory system of cats and may cause a head tilt

California mastitis test (CMT) a test used to diagnose and measure the level of cell infection within mammary glands

caliper device or instrument with a simple rulerlike tool that slides on a fixed bar or arm that moves parallel to it and is used to measure the thickness of a body part

callus cartilage that rebuilds and thickens over a fractured bone that has healed

calories unit of measurement that specifies the amount of energy in food

calving the labor process of cows

campylobacter a bacterial infection in cattle that affects the intestinal tract or reproductive tract

canary a small breed of bird with solid bright colors that originated in the Canary Islands and is known for its singing ability

cancellous bone softer spongelike layer of bone located inside the end of bones

candling a technique of using a light to view the internal structures of an egg

canine distemper a viral infection that can be fatal to puppies; dogs who recover may have long-term neurological affects

canine teeth used to tear apart food; also known as fangs

Canis familiaris genus and species of domesticated dog

canister area that absorbs carbon dioxide and contains granules to absorb it

cannibalism the act of rodents eating their young

cannon long bones in hoofed animals located above the ankle joint

canter the three-beat running gait of English horses

canthus angle where the upper and lower eyelids meet

capillaries smallest blood vessels in the body; branch off of arterioles

capillary action the action of blood draining into a tube through gravity

capillary refill time (CRT) indicates how well an animal's blood is circulating in the body; determined by pressing on the gums and monitoring how long it takes for the color to return to the area

capon a male castrated chicken over 6 pounds and 6 months of age

caprine a veterinary term meaning of, related to, or being a goat

carapace thorax, abdomen, and limbs

carbohydrates nutrients that provide energy for body functions and allow for body structure formation; make up the largest part of an animal's dietary needs

carcass the dead body of an animal

cardia entrance of the stomach that filters food

cardiac arrest occurs when the heart is not contracting appropriately

cardiac cycle one complete contraction and relaxation of the heart

cardiopulmonary resuscitation stimulates heart to provide oxygen to the lungs; also known as CPR

carnassial tooth upper fourth premolar and lower first molar in dogs and cats that tends to become infected and abscessed

carnivore meat eating

carotene protein that promotes the health of organs

carpus joint formed by several carpal bones and arranged in two rows in the area of the wrist

carrier a boxlike object that safely holds a cat for transport; a cat that spreads a disease but shows no signs of having the disease; an animal that is infected by a disease but shows no signs

carrying capacity the number of animals living in an area that can be supported and sustained

cartilage forms at the end of bones to protect and cushion the bone

cartilage joint connects at the end of a joint where two bones meet; acts as a cushion

cashmere a fine down undercoat of goats that is warm and used in clothing

cassette tool that holds X-ray film

cast material made of a hard substance; placed over a broken bone to keep the bone in place as it heals

castration surgical removal of the testicles in male animals

casts cylinder-shaped structures in urine formed from protein secreted by the renal tubules of the kidney

cat bag restraint equipment used to hold and control a cat's body and limbs during certain procedures, such as a nail trim

Cat Fanciers' Association (CFA) a cat breed registry association that promotes the health and responsible breeding of purebred cats

cat muzzle restraint equipment placed over the face and mouth of a cat to prevent biting

catabolism a process in which large molecules are broken down into smaller molecules

cataract cloudiness or opacity of the lens in the eye

catheter thin plastic or rubber tube that is passed into the urinary bladder

catheterized refers to animal that has a long, thin rubber or plastic sterile tube inserted into the bladder through the external urinary opening to collect a urine sample

cathode area within X-ray tube that creates negatively charged electrons when heated

caudal fin tail fin that acts as a propeller

cauliflower disease another name for lymphocytosis

caustic chemicals the chemicals used to burn horn buds to prevent horn growth

cautery unit handheld equipment that heats to high temperatures to simultaneously burn through nails and clot them

cecum a sac with an opening located between the small and large intestines that helps break down roughage and protein

cell the basic unit of life and the structural basis of an animal

cell division a process in which one cell splits into two cells

cell membrane holds a cell together; also called the cell wall or plasma membrane

centimeters (cm) unit of measurement used when measuring body parts before X-rays

central nervous system (CNS) part of the body under involuntary brain control, such as the brainstem and spinal cord

central processing unit (CPU) the brain of the computer that carries out instructions asked by the software

centrifuge veterinary tool used to spin lab samples at a high rate of speed to force the separation or concentration of materials suspended in a liquid form

centrifuge tube glass or plastic tube that holds samples within a centrifuge

cephalic vein blood vessel located on the medial front limb

cephalic venipuncture a puncture to draw blood from the vein inside the front of the legs

cerclage wire surgical wire used to hold fragments of broken bone together while healing

cere the fleshy area at the top of the beak that contains the nares or nostrils

cerebellum part of the brain that controls coordination and movement

cerebrum part of the brain that controls the voluntary movements of the body and thought processes; the largest region of the brain

certified veterinary technician (CVT) a veterinary technician who has passed a state licensing exam

cervical first section of the spinal column; located over the neck area

cervix end of the vagina that opens to the uterus; the birth canal

chain shank a chain placed around the head and halter of a horse for added control

chalaza the part of an egg that anchors the yolk to the egg white

chamber box used to confine subjects during an anesthesia induction procedure.

chamois a by-product of sheep skin used as a leather product for polishing

channel a chosen route of communication

cheek pouch an area within the mouth where food is stored

chelonian a veterinary term for turtle

chem strip long, thin plastic strip separated by individual square pads containing a chemically treated paper

chemical change in the body that affects growth, sexual reproduction, and development

chemical filtration the use of special chemicals such as ozone and activated charcoal to keep water clear

chemical hazards safety concerns over products that may cause injury or over vapors that can cause injuries to the eyes, lungs, or skin

chemical restraint the use of sedatives or tranquilizers to calm and control an animal

chemoreceptor chemical receptor that allows an animal to taste, smell, and detect sounds

chemotherapy the use of chemicals to treat disease

chick a young chicken

chicken the most important type of poultry; has four different classes

chief complaint the reason an animal is being seen by the veterinarian

child labor laws laws and guidelines that regulate and protect children who work and the conditions in which they work

choana the opening to the oral cavity

chronic a long-term condition

chronic obstructive pulmonary disease (COPD) respiratory condition similar to heaves and asthma that causes difficulty breathing

ciliary body area consisting of muscles that change the shape of the lens in the eye to allow near- and far-sighted vision

circulatory system body system essential for life; includes the heart, blood, veins, arteries, and capillaries with functions that include oxygen flow, blood circulation, transport of nutrients, waste removal, and movement of hormones

clavicle bone that connects sternum to scapula; commonly called the collarbone

cleaning the process of physically removing all visible signs of dirt and organic matter

client the owner of an animal

clinical research research conducted in a lab setting that focuses on human- and veterinary-related issues

clipper guard tool that fits over the end of clippers to prevent trimming hair too short

clipping removing small amounts of hair from one or several areas

cloaca external reproductive area where egg-laying animals pass urine, feces, and eggs; also called the vent

closed gloving process done when the gown is placed with the hands still within the sleeves so that no skin is visible to potentially contaminate the gloves

Clostridium bacteria that causes tetanus; found naturally in manure and soil

clot formation of a plug in the blood to stop bleeding

cloudy eye a bacterial infection that may affect one eye or both eyes of tropical fish

cloven shaped a split in the toe of the hoof that separates the hoof into two parts

clutch a group of eggs

coat conditioner a spray placed on the hair coat to eliminate drying of skin and hair

cocci round bacteria shape

coccidia a single-celled parasite and protozoan passed in contaminated water sources

coccidiosis a protozoan disease cause by contaminated bird droppings in water and soil sources

coccygeal last section of the vertebrae that lies over the tail area

cochlea spiral-shaped passage in inner ear that receives sound vibrations and initiates an impulse to the brain for translation

cock an adult male chicken

cockatiel a small bird breed that is a combination of yellow, gray, white, and orange colors and is very easy to train

cockatoo a large bird breed that is commonly white and has a developed speaking ability

cockerel a male chicken under a year of age

coffin bone a bone located within the hoof joint that rotates when laminitis occurs

Coggins test blood test for horses that screens for equine infectious anemia

cold housing a building with no heat in which an entire herd is usually kept and where air circulates out moisture

cold sterilization the process of soaking items in a disinfectant chemical until they are needed again; not considered truly sterile for surgery procedures

cold tray a container that holds a chemical that acts as a disinfecting agent

cold water water at a temperature below 50 degrees Fahrenheit

colic condition in horses that causes severe stomach pain

collimator part of the X-ray machine that sets the size of the X-ray beam that produces radiation to take the picture

colon common term for the largest section of the large intestine

color code a method used to file medical records for visual ease of use

colostrum milk produced by a female within 24 hours of giving birth; provides antibodies that protect the immune system

colt a young male horse

columnar a tall and thin shape

coma damage to the midbrain that causes a loss of consciousness and awareness

comb a flesh-like projection on the top of a chicken's head

combing brushing the hair with a thin comb that helps remove tangles and mats

combining form a root word plus a combining vowel; often describes a body part

combining vowel a letter, usually a vowel, that is placed between the prefix and suffix; makes the word easier to pronounce

commercial refers to medium-sized rabbits that weigh between 8 and 12 pounds

commercial business a type of business activity that may distribute goods or provide services

commercial serologic test kit commonly used test kit; provides quick and accurate results related to common viruses and diseases

comminuted fracture bone breaks in several different locations; causes fragments

common laws state regulations and rules based on legal precedents

communication process the five essential steps to relaying information to others; includes the sender, the receiver, the message, the channel, and feedback

compact bone thick tissue that forms the outer layer of bone

companion animal practice a veterinary facility that treats and maintains the health of small animals, such as dogs, cats, and exotic pets

companion fish fish people keep to relieve stress or for entertainment

complete blood count (CBC) blood test

composting layering the carcass of a dead bird with waste materials and bedding to allow it to decompose or wear away

compound fracture a break in a bone that causes the bone to penetrate the skin

compression fracture when a broken bone presses into another bone

computed tomography (CT scan) procedure in which ionizing radiation, with computer assistance, passes through the patient to allow for the display of internal body structures in cross-sectional views

computer inventory system a method of monitoring supplies and products within a facility to know when items need to be reordered

computer system a computerized model that mimics animal behaviors and/or internal or external structures; also known as a simulated system

computerized appointment book a software program on the veterinary computer system used to schedule appointments and maintain medical information; allows future use of information in an organized manner

concave rounded inward

concentrates food sources high in energy and fat provided to an animal as an additional nutrient source when the primary food source is not adequate or abundant

concentration determined by the amount of food fed divided by the percentage of dry matter and how it is delivered to the animal

conception process of creating a new life that forms an embryo

conditioning process of teaching an animal an action in relationship to another action; the process of working an animal to build up its body and strengthen the animal for work

cones cells in the eye that allow animals to see colors

confidential information that is private and not to be shared by anyone outside of the facility without permission

confidentiality maintaining the privacy of medical information

confinement method a system of raising sheep in an aseptic environment for the purposes of human consumption

conformation the body's shape and form

conical refers to the end of an object that is a pointed shaped

conjunctiva thin membrane that lines the underside of the eyelids and the outer surface of the eyeball

conjunctivitis inflammation of the conjunctiva

connective tissue holds and supports body structures by connecting cells

consent form permission given by owner to treat animal

constipated unable to have a bowel movement

constipation occurrence in the digestive tract that can cause little to no bowel movement

constrict to narrow (as in the pupil in the eye)

consumable products supplies used daily that are not included on the client invoice, such as paper towels, gauze sponges, and cotton balls

contagious capable of spreading disease

contraction first stage of labor when the fetus begins to move in the uterus to prepare for birth by moving into the birth canal

contraction phase the second phase in the labor process in which wavelike motions in the uterus move the fetus into the birth canal

contrast medium material or substance administered to the patient to allow imaging of structures that would otherwise be difficult to view on an X-ray film

continuing education (CE) education provided for adults after they have left the formal education system, consisting typically of short or part-time courses in their career field.

controlled substance drug that has the potential for abuse or addiction; also called scheduled drug

controlled substance drug that has the potential for addiction or abuse

controlled substance license paper given by the DEA to allow a veterinarian to prescribe scheduled drugs

controlled substance log requires written entry for every controlled substance dispensed by the facility

conure a medium bird breed that is brightly colored; from Central America; is active and playful

cool water water at a temperature between 50 and 60 degrees Fahrenheit

cooled semen semen cooled on ice and shipped in a special container for use within 24–48 hours

cooled semen semen cooled on ice and shipped in a special container

copier machine equipment used to make exact copies of paper or other documents

Coplin jar glass container that holds chemicals for use in staining slide samples

core vaccine a vaccine considered necessary for all animals within a particular species due to the high risk for a particular disease

cornea clear outer layer of the eye; easily scratched or otherwise damaged

corpus luteum (CL) known as the yellow body; forms during the gestation period of a female to maintain pregnancy

corrugated tubing thin plastic hose that allows the oxygen and anesthetic gas to flow to the patient by connecting from the machine to the endotracheal tube

cortex outer section of the kidney structure

costal cartilage tissue located at the end of the ribs where they attach to the sternum

cotton-wool disease another name for *fish fungus*

courtesy placing another individual's needs and concerns before your own

coverslip thin piece of glass that fits over the sample on the microscope slide

cow an adult female cow of breeding age

cow kicking the way in which cows kick with their rear legs, which is off to the side

cow–calf system a production system that raises cattle to breed them once they are adults

crackles abnormal sounds within the chest that sounds like cellophane paper

cranial drawer sign in-and-out movement or motion of the knee when the cruciate ligament has been torn

cranial nerve test reflex test of the eye where a hand is moved quickly toward the eye without touching the eye or any hair around the eye

crash cart movable table that holds emergency equipment and supplies; is easily accessible to address a serious patient condition

creatinine chemical waste product filtered by the kidneys; serum creatinine levels can be analyzed to assess renal function

credit application paperwork completed by clients to allow them to obtain a line of credit to pay for veterinary services

creep feeder a small trough that allows calves to eat while preventing the adult cattle from doing so

crest the top of an animal's head

cribbing a condition in which horses chew on an object and then suck in air

Crile hemostatic forceps similar in size and shape to Kelly forceps, except they are fully serrated, with serrations located within the interlocking jaws and vertically along the blades

crop small sac that acts as a holding tank for food as it is passed from the esophagus in birds

cross ties short ropes placed on either side of a horse to make it safe for those working near the animal

crossbreeding crossing two varieties or breeds within the same species

crown the upper part of a tooth; lies above the gum line

cruciate ligament tears common knee injury to large-breed dogs; tear is in the cruciate ligament in the stifle

crumbles flakes or blocks of fish food

crustacean an aquatic animal with an outer body shell and legs

cryosurgery procedure involving freezing parts of tissue

crystals solid formations within a urine sample that form from increased components within the urinary tract or renal system

cuboidal cube shaped

cud mixture of grass sources and saliva that is chewed and regurgitated to break down food for digestion

cuff located about three-quarters of the way down the endotracheal tube; inflatable; allows the tube to be inflated with air to create a tight seal around the trachea to prevent any moisture or material from entering the lung field

culled removed from the herd

culture and sensitivity (C/S) test that determines if a bacterial issue is occurring, what type of bacteria is causing the health condition, and what antibiotic should be used to treat the problem

cursor a blinking line on a computer screen that determines where typing or some other action is occurring

Cushing's disease condition where an increase in ACTH or cortisol causes a production problem within the adrenal glands; common term for hyperadrenocorticism

cusp the center of the tooth that wears with age

cutability the quality and quantity of meat from a beef animal

cutaneous larval migrans hookworm infection in humans caused by penetration of the skin

cyanosis blue color of mouth and gums due to a lack of oxygen

cyanotic refers to bluish gray color of the mucous membranes due to a lack of oxygen

cylinder metal container that holds compressed forms of gas to keep patients oxygenated and breathing during anesthesia

cystitis term for a bladder infection; also called urinary tract infection

cystocentesis surgical puncture of the bladder using a needle to collect a urine sample

cystotomy surgical incision of the urinary bladder to remove urinary bladder stones

cytology the study of cells

cytoplasm the liquid or fluid portion of a cell that allows the internal cell structures to move

dairy a building where cows are milked

dairy cattle cows raised specifically for milk production

dairy goat a goat that produces milk

Dairy Herd Improvement (DHI) program a way to maintain records of dairy herd information

dam female parent of a horse

daughter cells one of four cells created from one cell during meiosis; each cell has half the number of chromosomes of the parent cell

Debakey thumb tissue forceps used to grasp the most delicate tissue with atraumatic grooved ends instead of teeth

debeak to remove the tip of a chick's beak to prevent pecking damage to other chicks

decant process of pouring urine out of a tube and allowing only the sediment on the bottom to remain in the tube

decapod a crustacean with five pairs of legs

deciduous baby teeth that are developed in newborn animals and eventually shed when adulthood is reached

declaw the surgical removal of the digits or claws on the feet of a cat

decongestant medicine used to reduce nasal congestion

deer a ruminant animal that grazes on pasture and has antlers

defensive aggression natural behavior that occurs when a cat is attempting to protect itself from people or other animals

defensive reaction a behavior shown by an animal to protects itself from danger

definitive host organism that provides life through the adult stages of the parasite

dehorning the process of removing horns to prevent injury to people and other animals

dehydration loss of fluids from the body

demodectic mange disease caused by the demodex mange mite on the skin and hair of dogs

dendrites multiple processes that extend from a neuron and conduct impulses received from other neurons located toward the nerve cell body

denial when a person will not accept a pet's death

dental elevator instrument used to extract a tooth by separating the tooth neck from the root

dental extractor instrument used to remove a tooth from the socket during an extraction

dental formula arrangement of teeth in the upper arcade in the numerator position and teeth in the lower arcade in the denominator position

dental mirror instrument with a reflective face used to visualize all tooth surfaces

dental prophylaxis professional cleaning that involves scaling and polishing the teeth under general anesthesia

dentifrices toothpaste used on animals

dentifrices a paste or powder for cleaning the teeth.

dentin second layer of a tooth; similar to bone

dentition arrangement of teeth within the mouth

depression a state of grief in which a person is so sad that he or she can not handle the normal functions of daily life

dermatitis inflammation of the skin

dermatophyte fungi

dermatophytosis fungal disease caused by an external parasite

descending colon third or last section of the large intestine

designer breed the cross-breeding of two breeds to form a new breed not yet recognized by the AKC

developer solution chemical that develops X-ray films

dewclaw the short inside bone that forms the first digit; acts as an opposable thumb

dewlap a flap of skin under the chin in rabbits

Dexamethasone corticosteroid administered by injection to determine pituitary problems

diabetes condition that is produced when too much or too little blood sugar is produced, and the body finds it difficult to regulate

diabetes insipidus condition that affects the water content in the bloodstream, causing the urine to become dilute

diabetes mellitus condition that causes high blood glucose levels

diagnostic imaging procedure that involve radiographs, ultrasounds, and endoscopy

diaphragm area of the chest that separates the thoracic cavity from the abdominal cavity; the chest area of a bird that must expand during the breathing process

diaphragmatic hernia condition that affects the diaphragm by causing a tear in the chest muscle; internal organs may protrude through

diaphragm mechanism part of a microscope that allows the slide sample to be moved both up and down and side to side to view the entire sample

diarrhea process of waste materials and feces becoming soft and watery

diastole relaxation phase of blood pressure; the bottom number, which is normally lower than the top number

diestrus stage of the heat cycle when pregnancy occurs; is characterized by a functional corpus luteum that releases progesterone to maintain pregnancy; if pregnancy doesn't occur, the corpus luteum breaks down, preparing the animal to cycle again

diffusion a process in which molecules move from higher to lower concentrations

digestion the process of breaking down larger particles of food into smaller particles for use by the body in order to function

digestive system the body system that contains the stomach and intestines

dilate to open, expand, or widen (as in the pupil in the eye)

dilated refers to widened or opened pupils

dilation phase the first phase of the labor process in which the birth canal opens and expands

diluent liquid mixture of a vaccine

dilution the act of diluting, or lessening in strength by adding another component, such as water

dimorphism the male and female having different body parts/appearances

dip stick long, thin plastic strip separated by individual square pads containing a chemically treated paper

dipping applying a chemical pesticide or medication to the skin and hair coat

Dipylidium caninum tapeworm species that can be up to 20 inches long; lives in the small intestine

direct contact transmission of a disease through a direct source, such as saliva or the ground

direct PLR normal pupil light response assessment

Dirofilaria immitis heartworm species that belongs to the same class as roundworms; they look similar in structure—like long, thin pieces of pasta

disaccharides double sugars

disease poor health that is a disturbance to or change in an animal's function and structure

disinfecting the process of destroying most microorganisms on nonliving things

dislocation displacement of one or more bones at a joint

displaced abomasum condition in ruminants where the stomach twists and rotates out of place

dissect to separate into pieces or break down into parts; to identify the meaning of a word

dissolved oxygen (DO) oxygen in a gas state in water

distended swollen, as in the abdomen

distilled water water with impurities removed

distribution movement of a drug into the body's tissues

diuretic drug used to increase urine production

diurnal active during the day and sleeping at night

diversity having distinct qualities

DNA deoxyribonucleic acid; a self-replicating material that is present in nearly all living organisms as the main constituent of chromosomes; the carrier of genetic information

DNA testing using blood or hair samples to identify an animal's parents

dock the tail area

doe an adult female goat

domestic a feline that is of two or more breeds and not registered

domestication taming of an animal to coexist peaceably with humans

dominance aggression behavior that refers to the pack instinct of an animal and its social status within a group

donkey a member of the equine family with large ears; also known as a burro

dorsal fin a fin on the back of a fish used to steer it through the water

dorsal recumbency restraint position with the animal placed on its back

dorsal root nerve branch of the vertebrae; contains sensory nerves

dorsal-ventral (DV) view requiring the animal to be positioned on its stomach; the X-ray beam goes through the back first and the stomach second

dosimeter X-ray badge that measures the amount of radiation a person is exposed to

dosimetry badge measures the amount of scatter radiation received during a radiographic procedure

double sugars two particles of carbohydrates that form the building blocks of nutrients; also called disaccharides

down a soft feather covering that grows under primary feathers

Doyen intestinal forceps noncrushing intestinal forceps that have lengthwise serrations and hold intestinal vessels and tissues for short time periods without causing any damage

draft capacity the amount of force an animal must exert to move or pull at a constant rate of one-tenth its body weight

draft-hitch class a contest based on the number of animals used to pull a wagon and how well the team works together

draft horse a large and muscular horse standing over 17 hands in height

drake an adult male duck

dram unit of measurement related to vials or bottles

dremel a tool used to grind nails

drenching giving medicine by mouth with a dose syringe to large numbers of animals

dressing forceps instrument used to grasp and hold tissue; have no teeth, but rather a smooth surface

dressing the amount of meat produced on one pig

drop-off appointment a set time for a client to bring the patient in for an exam, surgery, or other procedures according to the veterinarian's schedule

drop-off time a set time when an animal is expected at the veterinary facility for a later exam or surgery

dropsy another name for *ascites*

Drug Enforcement Agency (DEA) government agency that sets regulations and guidelines for ordering, storing, and dispensing controlled substances

dry-cow period a time in which milk production is stopped to give the reproductive system a rest

dry heat a method of sterilization that exposes an item to extreme heat with a flame or through incineration

dry matter the amount of nutrients in a food source without the water content

dual-purpose breed a breed that serves more than one purpose, such as production of milk and meat

duck a small water bird used for meat, feathers, and down

duckling a young duck

duodenum short first section of the small intestine

dust bath a method of grooming in which a chinchilla rolls in a dusty material to condition its skin and hair coat

dwarf small rabbit breed that weighs less than 2 pounds as an adult; *see miniature*

dynamic equilibrium a state of balance in response to rotational or angular movement

dyspnea difficulty breathing

dystocia a difficult or abnormal birth

dysuria difficult or painful urination

ear sensory organ that enables hearing

earmarks notches made on the edges of the ear flaps to identify cattle

ear mites microscopic parasites that live within the ear canal, causing severe itching and a dark, crusty discharge

ear notching the practice of making a small cut on the ears to identify the pig number and litter number

ear plucking removal of hair from the ear canal

ear tag identification tags with numbers on them fastened by piercing the ear flaps

"earthquake" technique a procedure used in training a caged bird that teaches the bird to not bite; the bird sits on a hand and when it tries to bite, the hand is shaken slightly to cause a balance check

ecdysis the process in which a snake sheds its skin

ecology the study of the environment that animals live in

ectoparasites external parasites

ectothermic the body temperature of an animal, controlled externally by the environment

edema buildup of fluid under the skin; swelling

efficacy how well a drug works

egg bound a female bird that is attempting to lay an egg that has become stuck in the cloaca

egg depositers fish that lay eggs on areas such as plants, rocks, or the bottom of a water source for protection

egg-laying fish fish that reproduce when a female releases eggs that are fertilized by the male's sperm

egg production a system used to produce a quality egg for human consumption

egg quality a standard determined by the egg's internal and external appearance

egg scatterers fish that spontaneously scatter their eggs around the tank as they swim

eggshell the outer calcium layer that protects an egg

ehrlichiosis bacterial disease caused by the lone star tick; causes fever, joint pain, depression, anemia, anorexia, weight loss, and some edema

elastrator an instrument that stretches a band over the testicles during castration

electrocardiogram monitor placed on a patient through metal leads called electrodes that measure the electrical activity of the heart; also called an ECG or EKG

electrocardiography the evaluation of the electrical currents of the heart through the use of machines

electrode metal lead that connects a patient to the EKG unit

electrolytes determine the balance of elements in the body, specifically potassium, sodium, and chloride; powdered supplements placed in food or water to encourage horses to drink more water

elimination removal of a drug from the body

ELISA enzyme-linked immunosorbent assays; simple test used to measure an antigen or antibody level within the blood

elk a ruminant animal, much like a deer, that grazes on pasture and has antlers

emaciated severely dehydrated and appearing very thin

emasculator a surgical tool used to cut the spermatic cord during castration

embryo early stage of development of a multicellular organism

emergency situation that requires immediate life-saving measures

emergency service a situation that must be seen to immediately and that may be a life or death situation

emetic drug used to induce vomiting

empathy being able to understand another person's feelings

emu a large flightless bird native to Australia that is smaller than the ostrich

enamel hardest substance in the body; covers and protects the teeth

encephalitis viral disease in horses and people; also called sleeping sickness; several strains are spread by mosquitoes and affect the nervous system

encyst enclose or become enclosed in a cyst

endangered animal species decreasing in numbers in the wild and close to becoming extinct

endocardium thin inner layer of heart muscle

endocrine system excretory system that rids the body of waste materials

endocytosis a process in which large molecules move from a higher concentration to a lower concentration

endoparasites internal parasites

endoplasmic reticulum a collection of folded membranes that attaches to the nucleus

endoscope fiberoptic instrument used to visualize structures inside the body

endoscopy procedure used to visually examine the interior of the body via an endoscope

endosteum thin connective tissue covering inner bone

endothermic the ability to regulate body temperature within the body

endotracheal tube flexible tube placed into the windpipe to create and establish an airway during anesthesia or CPR; also called E-tube

enema passes fluids into the rectum and colon to soften feces to produce a bowel movement

English class a chicken class developed in England for better meat production

enterotoxemia a bacterial disease from overeating that causes toxins to build up within the rumen; they are not absorbed by the body

enzyme chemical that changes other chemicals during metabolism

eosinophil white blood cell that fights against allergic reactions, controls inflammation, and protects the body from parasitic infection; has a large nucleus with segmented granules within the cytoplasm

epaxial located above the axis of a body part, such as above the spine

epicardium thin outer covering of the myocardium

epididymis tube that transports sperm from the seminiferous vesicles to the vas deferens

epiglottis covering or flap that prevents food from entering the airway and lungs during swallowing

epilepsy neurological disorders characterized by seizures

epinephrine short-acting chemical released during the fight-or-flight response

epithelial tissue a protective layer that covers the body's organs and lines internal and external structures

Equal Employment Opportunity (EEO) related to laws that prohibit discrimination against individuals for their race, color, gender, religion, age, sexual harassment, national origin, or pregnancy or childbirth situations and disabilities

equilibrium state of balance

equine of or related to a horse

equine encephalomyelitis a viral disease of horses found in different strains transmitted by mosquitoes; also called *sleeping sickness*

equine infectious anemia (EIA) highly infectious virus of horses; also known as swamp fever; viral disease for which there is no vaccine and no cure

Equus caballus genus and species of the horse

eructation gas buildup where belching occurs to rid the rumen of air

erythrocytes red blood cells

erythropoiesis the production of red blood cells within the bone marrow

esophageal stethoscope device placed into the esophagus next to the endotracheal tube; emits the sound of the heartbeat

esophagus tube that passes food from the mouth to the stomach

essential fatty acids nutrients that are not produced by the body; must be consumed

estimate the approximate cost of services

estrogen main female reproductive hormone; causes behavior changes in females during the estrus cycle; also allows follicle production and development to occur

estrus heat cycle that releases hormones for reproduction; the time when a female animal is receptive to breeding or mating with a male

estrus cycle the period of time when a female dog is receptive to a male dog for mating; also known as the heat cycle

ethical the act of doing what is right

ethics rules and regulations that govern proper conduct

ethology the study of animal behavior

ethylene glycol toxicity poisoning by antifreeze that causes renal failure

Eustachian tube narrow duct that leads from middle ear to nasopharynx and maintains air pressure

euthanasia process of putting an animal to sleep using humane methods

euthanasia release form signed permission from the owner to put a pet humanely to sleep, thus resulting in death

evert to turn inside out

ewe adult female sheep capable of reproduction

exam room area where patients receive routine exams

excrete to remove and rid the body of waste

exhalation act of exhaling or air leaving the lungs

exoskeleton external cartilage on the outside of the body

exotic animal species not native to the area where it is raised and rarely found in its natural habitat

exotic animal dealer someone who buys and sells exotic animals to private collectors and zoos

expectorant medicine that increases the fluid content of mucous

expelled removed

expiration to breathe out and expand the chest

expressing placing pressure on the abdomen to remove eggs or sperm

extension bending that causes a joint to open and to lengthen

external auditory canal tube that transmits sound from the pinna to the tympanic membrane

extinct animal species that is no longer found in its natural habitats

extubation process of removing an endotracheal tube from a patient

eyelids each of the upper and lower folds of skin which cover the eye when closed

eyepiece the portion of a microscope you look through

Fair Labor Standards Act (FLSA) legislation that governs the age when children can work and the duties they can perform

family-size herd a group of fewer than 100 head of cattle

farm flock method the most popular sheep production system; meat sources are the primary system and wool production the secondary system

farming demonstration use of various agricultural techniques as a learning process

farrier a professional who trims and shoes horses' hooves; also known as a *blacksmith*

farrowing the labor process of pigs

farrow to finish systems production systems in which sows farrow out and piglets are raised to market weight

fascia connective tissue that covers, supports, and separates muscles

fast no intake of food or water, or anything by mouth, prior to surgery

fats concentrated source of energy; also called lipids

fat soluble refers to vitamins that are stored in fat and released when needed within the body; vitamins A, D, E, and K

fatty acids oils that are products from fat sources; may be used as nutrients or supplements within a diet

fatty liver disease a condition in which large amounts of fat are deposited in the liver

fax machine piece of equipment used to copy and send documents via a telephone line

fear aggression defensive reaction to being harmed; an animal's instinct to protect itself

feathered edge the staggered area at the end of a slide where the veterinarian or veterinary technician reads and interprets a blood slide

feathers long hair over the lower legs of horses located just above the hoof

fecal direct smear prepared by placing a small amount of a fecal sample onto a microscope slide

fecal egg analysis a test to determine the type of parasite found in the fecal matter of an animal

fecal floatation technique done by placing feces within a liquid to determine if any parasite eggs are within the sample

fecal sample used to diagnose internal parasites and the presence of blood within the stool sample

feed materials that animals eat to obtain nutrients

feedback the return message sent by a receiver

feedlot size the ideal weight of calves that increases size and profit

feline distemper *see panleukopenia*

feline immunodeficiency virus (FIV) a fatal virus in cats that attacks the immune system

feline infectious anemia causes infection within the red blood cells of cats; also known as hemobartonella

feline infectious peritonitis (FIP) a fatal virus in cats; has a dry and wet form; affects the lungs and chest cavity

FeLV feline leukemia virus; a fatal virus in cats that affects the white blood cells and immune system

feline urologic syndrome (FLUTD) condition that commonly affects obese neutered male cats in which the urethra becomes blocked and they are unable to urinate; also known as feline lower urinary tract disease

Felis catus genus and species of cats

femur large upper bone of the rear leg; thigh bone

feral wild and not domesticated to live in an area maintained by humans

fermentation process of soaking food that allows bacteria to break down food for easier digestion

fertile an egg that has an embryo because it has been fertilized by a male

fertility the ability to reproduce

fertilization uniting egg and sperm to allow an embryo to form

fertilize the ability of a male to mate with a female to create an egg with an embryo

fetus unborn offspring of an animal that develops from an embryo

fever elevated body temperature due to excessive heat

fiber material from plant cells that is left after other nutrients are digested

fibrinogen protein that aids in blood clotting

fibrous joint fixed joint with little to no movement

fibula smaller bone that lies in the lower rear leg in the back of the limb

fight or flight natural instinct of animals faced with a threat in which they take care of themselves by either running away or protecting themselves by fighting

filly a young female horse

film hanger used to secure film for manual processing and development

filtration using a physical barrier to remove particles from the air; common in lab areas and research facilities

filtration system a mechanism to clean water and air

fin a structure that acts as arms and legs and allows movement in water

fin rot a bacterial disease that begins when fin tissue starts to erode and the tissue becomes inflamed

finch a small songbird that comes in a variety of colors and patterns and is easy to raise and breed

finger brush a small toothbrush that fits on the end of a finger; a plastic, thimblelike device that has small, soft bristles that are rubbed over the teeth

fingerling immature fish that is 2 years of age

finishing system a production system that produces a calf through its entire adult stage and then sells it as meat or for breeding

finishing system a production system where pigs are fed to market size

fish aquatic species that have scales, gills, and fins

fish fungus a fungal disease known as cotton-wool disease because of tufts of growth resembling cotton or wool that occur on the skin of the face, gills, or eyes

fixed office hours set times throughout the day when patients are seen in the order in which they enter the facility

fixer solution chemical that fixes X-ray film and helps maintain the life span of the film, allowing it to stay readable

fixing using a solution to maintain the life span of X-ray film

flagella tail of sperm that facilitates movement

flanking looking at or biting at the sides of the abdomen due to stomach pain, often in a colic situation in horses

flaxen blond hair color

flea a wingless external parasite that feeds off the blood of an animal

flea allergy dermatitis (FAD) allergic reaction to a flea bite caused by the flea's saliva that affects the animal's skin

flexion bending that causes a joint to close and shorten

flight feathers adult feathers located over the edges of the wings

float an instrument used to file down teeth; also called a rasp

floating the filing of horse teeth to remove overgrown and sharp areas

floating ribs ribs that are not connected to the sternum by cartilage

flocculent contains large amounts of particles called sediment suspended in the urine and appearing cloudy

flock a group of poultry

floss a material within a filter made of gravel, charcoal, and fiber materials

flow meter located on the anesthesia machine near the vaporizer; is adjusted for the proper rate of oxygen flow to the patient

flow rate rate at which ordered fluids are to be delivered over the course of a day

flow schedule when appointments are scheduled based on each patient being seen at a specific time by the veterinarian during the day

foal a young newborn horse of either gender

foaling the labor process of horses

focus knob the portion of a microscope that allows better visualization of the sample

follicle tiny structures within the ovary that enlarge with egg development and allow ovulation to occur

follicle-stimulating hormone (FSH) estrus hormone that allows for sperm production, regulates the female estrus cycle, forms the follicle during the breeding process, and produces estrogen

food analysis process of determining the nutrients in food and prepared mixes to ensure they serve as a balanced ration

Food and Drug Administration (FDA) agency responsible for protecting public health by ensuring the safety, efficacy, and security of human and veterinary drugs, biological products, and medical devices

food fish production raising fingerlings to market size to be sold as food product

foodstuff ingredients in animal food that determine nutrient content

foot rot a bacterial condition of the hoof that causes decay or damages the hoof tissue

forage plant-based sources of nutrition high in fiber; also called roughage

forager an animal that eats grass and pasture

force feeding placing food in a patient's mouth and forcing it to swallow

forced molting triggering feathers to be shed and regrown at specific times of the year

forcep a tool used to remove hair or parasites from the body

foreign body obstruction occurs when an animal ingests a foreign object that is not digestible and it becomes impacted within the intestinal tract

forelimb front leg of an animal

formalin chemical used to preserve tissue samples

founder the common term for laminitis

fountain a watertight structure made of plastic, fiberglass, or concrete that holds fish species

fowl cholera a bacterial infection of poultry that causes purple coloration of the head

fowl poultry

fracture break in a bone

free choice feeding method that allows animals to eat when they want; also called free access

free gas air accumulates in the dorsal rumen of a ruminant's stomach, causing the animal to choke when the esophagus becomes obstructed with food and saliva and the gas is not able to escape

freemartin a sterile adult cow that is unable to reproduce

freshening the labor process of dairy animals

freshwater fish species of fish that live in freshwater with very little salt content

freshwater water with little to no salt content

frog a V-shaped soft pad in the bottom of the hoof

frothy bloat caused by gas being trapped within small bubbles within the rumen, causing the abdomen to become swollen and painful

frozen semen process of storing and shipping sperm in liquid nitrogen which may be preserved up to 40 years

fry young newborn fish

fryer broiler

full arch a giant rabbit breed that weighs more than 14 pounds

fundus opening of the stomach

fungal class of disease spread by single-celled organisms or spores that grow on the external body and other areas of the environment, mostly in humid conditions

fungicidal kills fungus

fungus a living organism that invades the external area of the body through direct or indirect contact

FVRCP a feline vaccine combination series also known as the feline distemper series; stands for feline viral rhinotracheitis, calicivirus, and panleukopenia

gait movement or way a horse moves

gaited horse a horse that moves with a specific rhythm and motion

gall bladder organ that stores bile

gallop a fast four-beat running gait in which each foot hits the ground at a different time

game animal wildlife species hunted for food

game bird wild bird species hunted for food and sport

Game Commission state agency responsible for wildlife conservation and management

game fish aquatic species caught for food or sport

gander an adult male goose

gastric dilatation volvulus (GDV) condition in which the stomach and intestinal tract rotate after becoming swollen due to air or gas in the GI tract, causing the intestinal tract's circulation to be cut off

gastric dilation the condition known as bloat in which air or gas fills the stomach, causing the abdomen to become swollen and painful

gastrointestinal (GI) system the digestive system; contains the stomach and intestines

gauge refers to size of needle; related to its diameter

gauze a thin cotton-like material used for bandages and to make restraint muzzles

geese large water birds used for meat, eggs, feathers, and down

gelding a castrated male horse

Gelpi retractor self-retracting instrument with single points at the end that are sharp

genetic flaw an undesired trait or characteristic passed from one or both parents to the offspring

geneticist a scientist who studies genetics and uses mathematical equations to predict the outcome of breeding

genitalia reproductive organs; also called genitals

germinal disc a white spot in the egg yolk where the sperm enters the egg cell to allow fertilization

gestation length of pregnancy

giant a rabbit breed that weighs over 14 pounds

giardiasis a protozoan disease caused by water contamination

gills organs that act as lungs to allow fish to breath by filtering oxygen in the water

gilt a young female pig that has not yet been bred

gingiva soft gum tissue

gingival hyperplasia excessive amounts of gingival tissue growth

gingivitis inflammation of the gums

gizzard the digestive organ located below the crop of a bird that serves as a filter system to break down hard foods, such as seed shells or bones

glaucoma increase in pressure behind the eye

globe the eyeball; formed by connective tissues that give shape to the eye

globule a small round fat particle

globulins provide antibodies to help prevent disease

glomerulus capillaries located in a bundle at the renal pelvis

glottis opening into the larynx

glove pack prepackaged disposable sterile glove pack that is used one time and then discarded

glucagon hormone produced in the pancreas that increases blood sugar levels

glucocorticoids corticosteroids used to treat inflammation

glucose more technical term for blood sugar

goat any of various hollow-horned ruminant mammals related to sheep but of lighter build and with backwardly arching horns, a short tail, and usually straight hair; long domesticated for milk, wool, and flesh

goat pox a viral disease that affects a goat's immune system and causes flulike symptoms

Golgi apparatus a chemical processor of the cell

gonadotropin-releasing hormone (GnRH) hormone produced and released to maintain a normal estrus cycle

goose an adult female goose

gosling a young goose

gown pack holds sterile gown worn by the surgeon and veterinary staff assisting in the surgical area

Gram stain used to determine the presence of bacteria and type of bacteria in a sample

Gram-negative bacteria that stain red

Gram-positive bacteria that stain purple

grass tetany a condition in cattle caused by eating rich pasture high in nitrogen gases; causes abdominal pain

gravid pregnant, in reptiles

green foods foods such as fruits and vegetables that are green in color and have additional calcium

gregarious instinct or behavior to join a herd or group of like animals

grid series of radiolucent and radiodense material that reduces scatter radiation

grid technique plate that holds the film below the X-ray table

grief the sadness that people feel after the loss of a loved one or pet

grit a substance used to break down hard materials, such as seed shells, to allow food to digest

grooming the process of trimming and bathing in an appropriate manner

gross examination visible observation of feces

growth diet specialized food formulated to increase the size of muscles, bones organs, and body weight of young offspring

growth plate area of bone in young animals that allows the bone to grow and mature as the animal ages and the cartilage joint turns to bone

gruel a mix of equal parts of food to equal amounts of water, forming a soup-like consistency that can gradually be made into an oatmeal-like consistency

guilt when a person feels he or she is to blame for a pet's death and should have been able to save its life

guinea fowl a small wild bird domesticated for meat, eggs, and hunting

habitat an area where an animal lives

hairball a large amount of hair that collects in the digestive system of cats because of excessive grooming

half hitch tie that makes a loop around a stationary location, such as a post or fence

Halstead mosquito hemostats small in size with a fine tip that is used to hold delicate blood vessels

halter the equipment applied to the head of a horse for control and handling

hame strap a leather strap that attaches the harness and collar

hamstring muscle located on the hind limb at the mid- to distal part of the thigh

hand measurement unit for the height of a horse; equal to 4 inches

hand tanks large containers that hold chemicals and water for developing film

hardware the physical parts of a computer and its equipment

hatchery place where fish are fertilized, bred, hatched, and raised

head the part of the body containing the eyes, mouth, and gills

head exam includes examining the eyes, ears, and nose

health certificate documentation that an animal is in good health and can be transported out of state or out of the country

heart murmur occurs when an abnormal valve produces an abnormal flow of blood that creates a "swishing" noise upon auscultation

heart organ with four chambers located in the chest between the lungs

heart rate (HR) pumping action of the blood through the heart that creates a beating action

heartworm disease disease spread by mosquitoes; of most concern in dogs, cats, and occasionally ferrets; causes long white worms to build up in the heart

heat the period of time a female dog is receptive to a male dog for mating; also known as the estrus cycle

heat stroke serious condition during which the body overheats at temperatures of 105 degrees Fahrenheit or more

heaves respiratory condition in horses due to an allergic reaction, usually to dust

heifer a young female cow that has not yet been bred

hematocrit measure of the percentage of red blood cells in whole or unclotted blood; also called a PCV

hematology the study of blood

hematoma a collection of blood under the skin

hematuria blood in the urine

hemipenis the male reproductive organ in reptiles and amphibians

hemobartonella bacterial infection caused by a flea bite that affects a cat's bloodstream; infects the red blood cells and immune system

hemoglobin main component of red blood cells that allows oxygen transport; includes an iron component that allows cells to multiply

hemolysis the rupture of red blood cells; causes a pink coloration to develop in the plasma or serum

hemostatic forceps hinged locking instruments designed to and hold off blood vessels; also called hemostats

hemothorax blood within the chest cavity

hen an adult female chicken

hepatic lipidosis a condition in which large amounts of fat are deposited in the liver

hepatitis inflammation of the liver

herbivores animals that eat plant-based food sources

herd a group of animals

herd health manager someone who maintains the records and health care of a herd of cattle

herder the person in charge of a group of sheep

herpes virus a virus in bird species that causes diarrhea, weight loss, wing or leg paralysis, and death

herpetology the study of reptiles and amphibians

hibernation a natural condition in which an animal goes into a long sleep and its body systems slow down until the animal's body temperatures rises

hierarchy the place a horse has in the herd

hilus the indented center of the kidney where the renal vein and renal artery enter and exit

hinge joint that opens and closes

hinny the cross of a male horse and female donkey

hip dysplasia common genetic condition of large-breed dogs where the ball and socket joint of the femur and pelvis becomes diseased

histamine chemical released during an allergic reaction

histology the study of tissues

hobbles leg restraints that prevent a horse from kicking

hock common term for the point of the rear leg where the tibia and fibula meet

hocks the tarsal joints at the point of the rear leg

hog cholera the virus that causes brucellosis

hog snare a metal pole with a loop on one end used to restrain a pig by putting pressure over the snout

Holman retractor levers that are flat with a pointed appearance; look similar to an arrow with holes in the handle; commonly used during orthopedic procedures

homeostasis the tendency toward a relatively stable and balanced equilibrium between interdependent elements, especially as maintained by physiological processes

hooded a variety of rats that are white in body color with black or brown coloring on the head and shoulders, giving them a "hooded" look

hoof pick equipment used to remove rocks and debris from horse hooves

hoof wall the outer covering of the hoof

hookworm common intestinal parasite of dogs and occasionally cats that have either teethlike structures or cutting plates with which they attach themselves to the wall of the intestine and feed on the animal's blood

hormones chemicals in the body that regulate growth, regulate sexual production and development, and metabolize nutrients in cells

horn bud the area of horn growth on top of the head

horse pull a contest of power to see which horse can pull the heaviest load

horsepower a unit of power in an engine or motor that equals 746 watts of electricity or the amount of work done by lifting or moving 550 pounds at a distance of one mile per second

hospital administrator a person with authority over business and medical aspects of the hospital

hospital treatment board area in the facility where hospitalized patients are listed along with the procedures that are to be completed on them

host an animal that is infected by a parasite and that the parasite feeds off of

house call when a veterinarian examines a patient a the owner's home

housebreaking training a dog to eliminate outside

human–animal bond the close relationship developed between humans and their pets, where pets are often considered members of the family

humane what is considered acceptable by people in regards to an animal's physical, mental, and emotional well-being

humerus large upper bone of the forelimb

hunter a horse typically used for cross-country fox hunting and jumping smaller fences

hunting the tracking and killing of animals for sport, food, or other by-products

hurdle wood, plastic, or metal board used to direct and move pigs

hutch a special cage designed for a rabbit that has a covered area that shelters the rabbit from the weather and an open area that provides proper ventilation

hybrid the cross breeding of two breeds to form a new breed not yet recognized by the AKC

hydration status evaluated over the neck and shoulders, eye sockets, and gums to determine how much water loss has occurred

hydrolysis chemical process of breaking cells down into smaller particles

hydrophilic having a tendency to mix with, dissolve in, or be wetted by water

hydrophobic tending to repel or fail to mix with water

hyoid apparatus series of small bones at the base of the tongue that attach to muscles and aide in swallowing

hyperadrenocorticism increased production problem within the adrenal glands that results in a pituitary gland tumor when the ACTH hormone is increased or an adrenal gland tumor when cortisol hormone production is increased; also called Cushing's disease

hyperglycemia high blood glucose level

hyperkalemia increased potassium in the blood

hypersensitivity increased reaction to allergens; causes an allergic reaction

hyperthermia condition in which body temperature is above normal; causes the animal to have a fever

hyperthyroidism condition that increases thyroxine hormone production from the thyroid gland; common in cats

hyphema blood noted inside the eye

hypnotized put into a trancelike state

hypoadrenocorticism decreased production of steroids within the adrenal glands if the glands fail to function; also called Addison's disease

hypocalcemia a deficiency of calcium in the blood; also known as milk fever

hypoglycemia low blood glucose level

hyponatremia decreased sodium in the blood; sodium concentration in the blood becomes too low, and the body holds onto too much water

hypotension low blood pressure

hypothalamus organ within the brain located in front of the thalamus; controls hormone production

hypothermia condition whereby body temperature is below normal, causing the animal to become cold

hypothyroidism condition that decreases thyroxine hormone production from the thyroid gland; common in dogs

hypovolemia low blood volume

iatrogenic condition caused by humans

ich a parasitic protozoan disease caused by a one-celled organism

ideal weight breed standard based on an animal's age, species, purpose, or use and health status

ileum third, or last, section of the small intestine

ilium first section of the pelvis

IM pin intramedullary pin

immune system responsible for keeping the body healthy and protecting an animal from disease

immunity protection that begins at birth during the nursing process

immunoglobulins another term for antibodies

impacted difficult or unable to express

imprinting process of learning through attachment to an object that emits adult behaviors; can be generalized to all examples of the object

captivity housed in an enclosure or with people

incineration the burning of infectious materials or animal carcasses

incinerator a device used to burn the remains of items

incisors the front teeth located in the upper and lower jaws

incontinence uncontrolled bladder leakage

incubate to keep at a constant warm temperature

incubated eggs kept warm for hatching

incubation period the length of time until eggs hatch

incubation the process of increased temperature to allow an embryo to grow; the time in which an animal becomes infected with disease

incubator equipment used to house eggs at a controlled temperature and humidity

incus tiny ossicle bone that forms second joint within inner ear

indirect contact method via which disease spreads through ways other than touching an infected animal, such as airborne or through bedding

indirect PLR abnormal pupil light response assessment

induced ovulation occurrence of the release of an egg at the time sperm is introduced; when fertilization occurs at mating

induced ovulator when a female ovulates following mating with a male, as the release of an egg occurs at the time of mating

induction phase time when a patient is being given anesthesia to make it sleep and remain unconscious and free of sensation during surgery

infection an invasion of the body by foreign matter, causing illness and disease

infectious bovine rhinotracheitis (IBR) a respiratory virus affecting cattle; also known as red nose

infectious bronchitis a respiratory virus that affects chickens as well as egg production

infertile not able to reproduce; also known as sterile

infertility the inability to reproduce

inflammation process that causes white blood cells to build up in an area within the body; may cause pus to form, redness around the area, warm to hot skin temperatures, increased body core temperatures, edema, and pain

influenza the equine flu virus

infundibulum funnel within the uterus that catches eggs as they are released from the ovary

infusion pump equipment used to provide a constant flow of fluid at a specific rate throughout the day

inguinal rings slits in the abdominal muscles where the testicles descend into the scrotum

inhalation act of inhaling or taking air into the lungs

in-hand breeding a method of holding a mare and stallion for breeding purposes

in-house testing occurs when laboratory samples are analyzed within the veterinary facility

injection port area on an IV bag through which to remove and insert substances using a needle and syringe

inner shell membrane a thin area located outside of the albumin

inotropic drug that increases the contractile strength of the heart

insecticide spray or pour-on chemical used to control flies and other insects

insertion where a muscle ends

inspiration breathing in air; allows the chest to depress

instinctive behavior behavior that occurs naturally to an animal in reaction to a stimulus

Institutional Animal Care and Use Committee (IACUC) an organization that monitors and approves proposed research studies

instrument milk solution that acts as a lubricant to keep instruments protected from corrosion

insulin chemical produced by the pancreas that is released into the bloodstream; regulates the body's blood sugar

integumentary system includes the skin, hair, pads, horns, beak, nails, hooves, and feathers

intensifying screen located within a film cassette to help produce better film exposure

intensity biomass the number of species in a volume of water

intensive care unit (ICU) an area of the facility where critical patients needing immediate and constant care are treated and housed

intermale aggression a behavioral tendency common in adult male cats due to sexual dominance or social ranking

intermediate host organism that hosts a parasite's immature and larval stages

international health certificate documentation of an animal's health and vaccine history, stating it is approved for travel outside its home country

interneurons cells that deliver signals from one neuron to another from the brain to the spinal cord

interpersonal communication allows people to discuss and understand information with other people

interphase the interval between the end of one mitotic or meiotic division and the beginning of another

interstate health certificate documentation of an animal's health and vaccine history, stating that it is approved to travel between U.S. states; also known as certificate of veterinary inspection (CVI)

interval an amount of time set aside for a patient, such as 15 minutes

intervertebral disc soft, spongy cushioning that protects the vertebrae

intervertebral disc disease (IVDD) common injury of the back in long-backed breeds; inflammation and swelling occur when one vertebra puts pressure on another

intradermal (ID) into the layers of skin

intramedullary pin stainless steel pin or rod placed into the center of a broken bone to keep the bone in place as it heals

intramuscular (IM) refers to injection given into a muscle

intranasal (IN) refers to medication administration into the nasal passage

intraocular into the eyes

intravenous (IV) within a vein

intubation process of placing a tube into the trachea to establish an airway and allow the patient to continue to inhale gases that keep it under general anesthesia as long as necessary

intussusception condition when the stomach or intestine telescopes upon itself, cutting off circulation of the digestive tract

inventory a physical count of every medicine and supply item within the facility; allows accurate records to be kept

inventory management keeping track of the price, manufacturer, pharmaceutical company, amount, and expiration date of each item in the facility

invoice a bill given to a client that lists the costs of services and procedures

involuntary reflex occurs without the need for thinking and serves as a necessary body function, such as the heart beating

involution the process of the uterus shrinking after labor

iodine a chemical disinfectant used to destroy bacteria

Iospora canis most common coccidia species affecting animals

iris middle layer of the eye; gives the eye its color and holds the pupil

ischium second section of the pelvis

isolation ward an area that contains cages and equipment to care for patients that are contagious and may spread disease

isolation ward separate housing that groups similar patients, making it safe for all animals and staff

isotonic crystalloid fluid given to an animal that is dehydrated; creates a balance of elements within the body

isthmus shell membrane of an animal that lays eggs

IV catheter a small plastic piece of equipment that is placed within a vein to administer fluids and medications directly into a patient's bloodstream

jack a male donkey

jaundice yellow coloration of the skin or mucous membranes

jejunum second, or middle, section of the small intestine

jenny a female donkey

jird a burrowing rodent

jog the slow two-beat gait of Western horses

Johne's disease a contagious, chronic, and usually fatal infection that primarily affects the small intestine

joint ill a bacterial infection caused by contamination of the umbilical cord that results in pain and inflammation within a joint

joints where two bones meet; allows a bending motion

jugular vein blood vessel located in the neck/throat area

jugular venipuncture a puncture to draw blood from a vein in the neck

jumper a horse ridden over large fences on a set course

juvenile young stage of growth

Kelly hemostatic forceps larger in size; are partially serrated to better hold and grasp blood vessels

kennel attendant a person working in the kennel area and animal wards providing food, exercise, and clean bedding

kennel ward an area where animals are housed in cages for surgery, boarding, and care

keratin protein that allows the nail to grow and strengthen

keratoconjunctivitis sicca condition that causes decreased to no tear production in one or both eyes

ketosis a condition in dairy cattle that causes low blood sugar

keyboard buttons that have letters on them that allow typing and word formation on a computer

keyboarding the act of typing and development of skills used in computer functions and keys

kid a young newborn goat

kidding the labor process of goats

kidneys two organs located in the dorsal abdomen on either side of the spine that produce urine, are reddish brown in color, and are bean-shaped and smooth

killed vaccine vaccine manufactured from dead pathogens and placed into the animal's body in an inactive form

kilocalories amount of energy to raise 1 gram of water by 1 degree; written as a kcal

kilovoltage peak (kVp) strength of an X-ray beam

kindness being helpful and understanding

kitty burrito a restraint procedure in which a cat is wrapped in a towel

knee jerk reaction reaction to a slight tap over the knee joint that causes the knee to jerk in response; used to assess the nervous system

knee jerk reflex reflex used to assess the spinal nerves

knemidokoptes mites external parasites on the skin, beak, and feathers of birds; commonly called scaly leg mites

knob a projection on the top of a beak

knot two pieces of rope tied together so that they do not slip; tied to contain an animal

lab animal an animal raised for use in research laboratories

labial surface of the teeth located on the front area that is covered by the lips

laboratory a location in the facility where tests and samples are prepared or completed for analysis

laboratory animal technician a professional trained to care for animals used in research

laboratory veterinarian a doctor who provides medical care to animals involved in research and performs surgical procedures used in improving techniques for humans and animals

lacrimal glands tear ducts that produce tears on the surface of the eye

lactated Ringer's solution (LRS) fluid of lactic acid that is commonly used to replace fluids, as in dehydration

lactation milk production from the mammary glands

lactation diet specialized food provided to breeding females that have completed the gestation phase and are currently producing milk for their offspring

lamb a sheep less than 1 year old; also the meat from a lamb less than 1 year old

lamb dysentery a bacterial diarrhea condition in newborn lambs

lamb feeding a specialized production system of raising lambs to weaning and then selling to feed lots

lambing the labor process of sheep

laminitis inflammation of the lamina within the hoof bone and surrounding tissue

laparotomy towel sterile towel used to control bleeding during surgery and as a hand towel for the surgeon

large animal practice a veterinary facility that treats livestock animals, such as horses, cattle, and pigs

large-size herd a group of more than 100 head of cattle

larvae immature stage of a parasite

laryngoscope tool made of heavy metal that has a light source on the end to help light the airway for viewing

larynx cartilage that opens into the airway and lies within the throat area

lateral (LAT) view that requires an animal to be positioned on its side

lateral apparatus structure that produces tears and holds tear glands

lateral canthus outer corner of the eye; farther from the nose

lateral line an organ located just under the scales of fish that picks up vibrations in the water

lateral recumbency a restraint in which the animal is placed onto its side

lavaging flushing a wound using jet pressure to forcefully clean an area

laxative stool softeners or other medicine given to soften feces to produce a bowel movement

layer an adult female hen that lays eggs

lead a rope applied to a halter to handle, control, and move a horse

leading the act or motion of moving or walking a horse

learned behavior response to a stimulus that develops as a result of experience

Learning the acquisition of knowledge or skills through experience, study, or by being taught.

left atrium thin-walled chamber of the heart that is situated below the pulmonary artery on the left side of the heart

left ventricle thickest-walled chamber of the heart

lens transparent elastic structure located behind the iris; allows for focus

lens objective microscope viewer that offers a variety of different powers

leptospirosis a bacterial condition spread by urine contamination; is zoonotic to humans

lethargy inactive; tired

leukocytes white blood cells

liability to have a legal responsibility toward something

lice tiny insects that live on hair, are species specific, and are contagious

licensed veterinary technician (LVT) a veterinary technician who has passed a state licensing exam

life span the length of time an animal will live based on body size and health

life stage particular periods in one's life characterized by specific needs during that time

ligament fibrous strand of tissue that attaches bone to bone

ligated to tie off

light horse a riding horse that stands between 14 and 17 hands

limiting factor something that constrains a population's size and slows or stops it from growing

lipase enzyme produced by the pancreas; breaks down fats

lipid any of a class of organic compounds that are fatty acids or their derivatives and are insoluble in water but soluble in organic solvents

liquid manure system a costly waste-removal system used on large farms that requires above- or below-ground storage tanks with water added for ease of pumping

listening skills provide the ability to hear and understand what someone has said

litter box a tray or pan filled with a pebble-like material; used for elimination

litter trained refers to a cat that has been trained to eliminate in a litter box

litter pebble-like material used as bedding that absorb droppings

live cover a natural method of breeding a mare and stallion

live-bearing fish fish that bear or give birth to live young

liver organ behind the stomach that makes bile and produces glucose

living animal any species of animal that is alive and, in some way, responds to a stimulus

llama a ruminant animal native to South America raised for its hair and meat and used as a pack animal

log pulling a contest to see which team of draft animals can pull the heaviest load of logs

loin the lower back or lumbar area

long hair a cat's coat that is long and requires regular brushing and grooming

longe line a long lead line used to exercise a horse in circles

lope the three-beat running gait of Western horses

lordosis exhibiting a prayer position where the front end of the body is lowered to the ground with the back end held high in the air

lovebird a small songbird with bright colors that originated in Ethiopia and is a popular cage bird

lower GI contrast film of the lower digestive tract organs; the contrast medium is often given via an enema

lumbar third section of the vertebrae; lies over the lower back

lungs organs that surround the heart and inflate with air for breathing

luteinizing hormone (LH) estrus hormone that allows the production of testosterone, allows for ovulation to occur, and forms the corpus luteum (CL) during the reproductive cycle

luxation occurs when a bone or body part comes out of place

Lyme disease bacterial disease transmitted by the bite of a tick; causes joint inflammation in affected animals

lymph node small collection of tissue that produces lymph fluid

lymphocytes largest of the white blood cells; aid in immune functions that protect the body from disease; have one large single nucleus that makes up most of the cell

lymphocytosis a viral disease that causes white to gray growths to develop over the skin; also known as cauliflower disease

lysosome a structure within the cell that digests food and proteins

macaw a large bird breed that is brightly colored, originated in South and Central America, and is playful and easily trained

macrodrip line a regular IV drip that delivers 15 drops of fluid per milliliter, or 15 drops/ml

macrominerals minerals needed in large amounts, such as calcium

macrophage cell that eats and destroys organisms throughout the body

magnetic resonance imaging (MRI) procedure in which radio waves and a strong magnetic field pass through the patient to produce three-dimensional images of internal body structures

maiden a female horse that has not been bred

maintenance diet specialized diet fed to animals that may be working or competing; the goal is to keep the animal at a specific ideal weight

malignant cancerous

malleus small ossicle bone; attaches to the medial surface of the tympanic membrane

malocclusion the overgrowth of teeth; causes dental problems

malpractice the act of working below the standards of practice

mammalogy the study of animals

mandible lower jaw

mange dermatitis a condition caused by mange mites; causes itching and hair loss over the head and neck and that can lead to severe skin infection

manual developing developing X-rays by hand using chemical tanks

manure waste material high in nitrogen, phosphorus, and potassium

marbling appearance of intramuscular fat within meat

mare intact adult female horse capable of reproduction

Marek's disease a poultry disease caused by a herpes virus; also known as range paralysis, as it can cause wing or leg paralysis

marketability a measure of how well a product will sell in an area

market weight adult weight of cattle when they are sold for beef

mastectomy surgical removal of the mammary glands

mastitis inflammation of the mammary glands

mat area of hair interwoven; forms a large clump

mat splitter tool used to break down and separate matted hair

maxilla upper jaw

Mayo scissors instrument used to cut heavy tissues and sutures that are proportionate in size; when lying on a flat surface, they curve upward without contact at the end of the blades

Mayo stand instrument tray that is elevated to hold the surgeon's instruments and supplies

meal fine-ground food for young fish and aquatic specimens

mean arterial pressure (MAP) average of the systolic and diastolic blood pressures

meat goat a goat that produces meat

meat-type hog a hog raised for the large amount of meat, known as ham, on its body

mechanical filtration equipment that filters water to remove harmful particles and keep it clear

medial canthus inner corner of the eye; closer to the nose

medical record written and recorded information on the care and treatment of an animal

medical waste stored and discarded items that have been contaminated by bodily fluids and living tissues; may be infected with contagious diseases

Mediterranean class chicken class developed in the Mediterranean for large quantities of egg producers

medium gel where organism growth may occur

medium hair cat's medium-length coat; requires some brushing and grooming

medulla center section of the kidney

medulla oblongata part of the brain that controls the body's involuntary functions

medullary cavity hollow center within bone where blood cells are produced

meiosis a type of cell division that results in four daughter cells, each with half the number of chromosomes of the parent cell, as in the production of gametes and plant spores

menance response test a reflex test of the eye wherein a hand is moved quickly toward the eye without touching the eye or any hair around the eye

meniscus cartilage within the patella; forms an X shape over the cruciate ligament

mesentery connective tissue from the peritoneum; allows blood vessels and nerves to supply the small intestine

message an idea passed along through a route of communication

metabolic disease a condition that causes a chemical change within the body, usually from stress

metabolism a reaction within the body in which chemicals break down and are used by the body

metacarpal one of the bones that forms the long bones of the feet in the forelimbs

metaphase the stage of mitosis or meiosis in which the chromosomes become arranged along the equatorial plane of the spindle

metatarsal one of the bones that forms the long bones of the feet in the rear limbs

metestrus stage of the heat cycle following estrus; when ovulation occurs

methimazole medicine for cats to treat hyperthyroidism; blocks the synthesis of thyroxine

Metzenbaum scissors instrument delicate and thinner in size than others; have long handles relative to blade length; are used to cut and separate delicate tissues

microbiology the study of microscopic organisms

microchip electronic identification contained in a computer chip placed under the skin

microdrip line a small line that delivers 60 drops/ml

microfilaria tiny immature heartworm larvae that live in the bloodstream

microhematocrit measurement of the percentage of red blood cells in whole or unclotted blood; also called a PCV

microhematocrit tube small thin glass tube that holds blood within a centrifuge

microminerals minerals needed in small amounts, such as iron

microscopic very small; unable to view with the naked eye

midbrain part of the brain that controls the senses

midstream refers to urine collected shortly after urination begins and just prior to the process ending

migratory moves from location to location throughout the year

milk fever *see hypocalcemia*

milking herd a group of cattle used to produce large amounts of high-quality milk

milliamperage (mA) represents the number of X-ray beams based on time

millivolt electrical unit of measurement that equals one thousandth of a volt

milt fish sperm

mineral block minerals supplied in a block for horses to lick

minerals nutrients needed in every area of the body; found mostly in the bones and teeth; used by the body based on the animal's needs and mineral availability

miniature a small rabbit breed that weighs less than 2 pounds as an adult; also called dwarf

misrepresentation working or acting as someone you are not qualified to be

mite an insect that lives on the skin, in the hair, or in the ears

mitochondria the "powerhouse of the cell"; the part of the cell that makes energy

mitosis a type of cell division that results in two daughter cells, each with the same number and kind of chromosomes as the parent nucleus; typical in ordinary tissue growth

mixed breed an animal that is a combination of two or more breeds

mixed-animal practice a veterinary facility that treats both small and large animals

mobile services involves a vehicle equipped with veterinary equipment, tools, and supplies that travels to a client's home to examine a patient

mobile veterinary practice a veterinary facility that offers services to small animal owners, such as dogs and cats, and employs veterinarians who travel to the homes of the client to provide exams and services

modeling learning a behavior through watching other animals conduct that behavior

modem equipment for a computer that allows access to the Internet

modified live vaccine vaccine made from altered antigens created from pathogens; places small amounts of the pathogen into the animal's body

mohair long silky locks of hair from the Angora goat; used in clothing and bedding

molars a grinding teeth at the back of a mammal's mouth

mollusk an aquatic species that has a thick, hard shell

molting the process of shedding feathers to allow new feather growth

monitor computer screen

monocyte largest white blood cell; helps neutrophils rid the body of wastes and cell debris

monofilament suture single strand of material

monogastric refers to the digestive system of an animal with one simple stomach to aid in the digestion of food

monosaccharide simple sugar

moral what a person believes to be right and wrong

morphology the structure, shape, color, and appearance in numbers of cells

motility movement

motor neuron cell that delivers signals from the central nervous system (CNS) to the muscles to give a response

mouse a movable instrument attached to a computer that allows the user to move the cursor

mouth brooder fish that carry their eggs in their mouth for incubation; provides protection and allows hatching

mouth gag instrument used to hold the mouth open during a dental prophylaxis

mucosa thin connective tissue that lines the intestinal tract

mucous membranes (mm) membranes that line body cavities; the gums are an example

mule a cross between a female horse and male donkey

multifilament suture several strands of material braided together

murine veterinary term for mice or rats

murmur abnormal heart sound that presents as a swishing sound

muscle structure located throughout the animal's body that attaches to different locations and serves to protect, allow bending and movement, and aid in the strength of the animal

Mus musculus genus and species of the mouse

muscular tissue allows movement of body parts

musculoskeletal system body system consisting of bone, muscles, tendons, and ligaments

mutton meat from a sheep that is older than 1 year

muzzle a covering for the mouth of an animal to prevent biting

mycology the study of fungus

myocardium thick muscle that forms the wall of the heart

myofiber a muscle cell

myosin a thick protein filament fiber found in muscle tissue

nail bed area where the nail growth occurs; is located at the end of the digit near the hair growth of the foot

nail trim cutting the length of the nail to prevent injury and promote comfort

nanny a common name for the adult female goat

nares the nostrils

nasal septum cartilage between the nostrils

nasogastric tube tube passed through nostrils into the stomach to force-feed a patient

National Association of Veterinary Technicians in America (NAVTA) sets the standard of veterinary technology and the care of animals

National Institutes of Health (NIH) a federal agency responsible for biomedical and public health research

natural incubation the process of a female or male bird nesting on eggs to hatch them

near side the left side of the horse

neck band paper or plastic slip placed around the neck that identifies an animal

neck tag identification a tag with numbers on it worn around the neck of cows to identify them

necropsy procedure done by examining the body of a deceased animal to determine the cause of death

necrotic area of dead tissue

needle holder instrument used to hold a needle while placing sutures into a patient during surgery

needle teeth in pigs, eight long, sharp, needle-like teeth located on each side of the upper and lower jaw

negative contrast agent air, oxygen, or carbon dioxide that appears radiolucent on an image

negligence failure to do what is necessary or proper

neoplasia cancer

nephron structural and functional unit of the kidney; produces urine

nerve impulse a message sent to a nerve

nerve a whitish fiber or bundle of fibers that transmits impulses of sensation to the brain or spinal cord, and impulses from these to the muscles and organs

nervous system complex collection of nerves and cells; assessed through gait, the response of pupil reflexes, as well as other body reflexes such as those of the limbs, paws, and digits

nervous tissue contains particles that respond to a stimulus and cause a reaction in the body

neurological pertaining to the brain and spinal cord

neurons specialized cells within the nervous system that transmit nerve impulses

neurotransmitters molecules that travel across the synapse to relay information like a message transport system

neuter surgical removal of the reproductive organs of either gender

neuter certificate documentation of an animal's having been spayed or castrated and so is no longer able to reproduce

neutrophils most common type of white blood cell; destroy microorganisms within tissues; have a nucleus with segments

New World monkey a monkey species that originated in South America

Newcastle disease a respiratory virus that causes poultry to twist their necks

nictitating membrane third eyelid cartilage membrane located under the lower surface of the eye

night feces waste that a rabbit passes at night that is softer, rich in vitamins and proteins, and consumed by the rabbit

nit larva of lice

nitrogen cycle a process that converts wastes into ammonia, then into nitrites, then into nitrates

nits lice eggs

nocturnal sleeping during the day and being active at night

nonabsorbable suture material not broken down within the body; remains intact for long periods

noncore vaccine a vaccine considered only for those pets at high risk for certain diseases

nonessential fatty acids nutrients that are both produced within the body and added to the diet when necessary

nonexempt employee status that is entitled to paid overtime when working more than 40 hours per week

nongame animal species that is not hunted for food or other products

nongame bird species that is not hunted for food

nongame fish species that is too small or too large to catch and eat as a food source

nonhuman primate refers to primates in a lab setting that are being used for research

nonliving system a model that is mechanical in nature and that mimics the species of an animal under study

nonrebreathing system system used for very small patients such as rodents, birds, and reptiles, or small puppies and kittens that are too small to move the breathing bag with their respirations

nonruminant system digestive system similar to monogastric animals but with a larger, well-developed cecum for breaking down plant fibers

nonvenomous not poisonous

nonverbal communication communicating an idea without speaking

norepinephrine long-acting chemical hormone that increases heart rate, blood pressure, blood flow, blood glucose, and metabolism

normal saline solution with the same concentration level as salt

nose printing a method similar to fingerprinting in which the lines of the nose are imprinted for identification

nose ring a permanent metal ring placed through the septum of the nose that allows a cow to be led

nose tongs equipment placed in the nasal septum to apply pressure and allow a cow to be led

nosocomial infection human-caused spread of disease and animal contamination

NPO nothing by mouth, as in food or water

nuclear sclerosis drying of the lens of the eye with age

nucleic acid a complex organic substance present in living cells, especially DNA or RNA, whose molecules consist of many nucleotides linked in a long chain

nucleolus a small dense spherical structure in the nucleus of a cell during interphase; the largest structure in the nucleus of eukaryotic cells; best known as the site of ribosome biogenesis

nucleus the brain of the cell; located in the center of a cell

numerical filing system medical records organized by numbering and/or color-coding

nutrient any single class of food or group of like foods that aids in the support of life; makes it possible for animals to grow or provides energy for physiological processes

nystagmus condition where the eyes jump back and forth in rhythmic jerks due to damage in the inner ear, brainstem, or cranial nerves

obedience training a dog to obey commands

objective information measured facts (e.g., vital signs)

oblique view used on areas that need to be placed at an angle to prevent double exposure from other body organs

observation inspection of an animal at a distance to see what it looks like, how it acts, and how it moves naturally

observation skills the ability to watch and understand a person's body language and speech

occlusal top surface area of the teeth

Occupational Safety and Health Administration (OSHA) agency that governs laws regarding safety in the workplace

office manager a person who oversees the training and management of the front office staff

offspring the product of the reproductive processes of an animal

oil immersion process by which a specialized oil substance is placed over a sample to view the contents

Old World monkey a monkey that originated in Africa and Asia

olecranon bony process behind the elbow joint

olfactory sense the chemical sense that controls smell

oliguria decreased amounts of urine production

omasum third section of the ruminant stomach; absorbs nutrients and water

omentum thin lining that surrounds the organs within the abdomen

omnivore animal that eats both plant and meat sources of food

onychectomy surgical procedure that declaws a cat

oocysts single-celled eggs

open gloving process done by removing the left glove using the right hand by grasping the glove at the edge of the cuff

operculum a gill covering that divides the head from the trunk

oral barium study barium solution given by mouth to pass through the digestive system to allow X-rays to be taken over time to view internal structures of the GI tract

oral pertaining to the mouth

oral cavity includes the gums, mouth, teeth, tongue, and throat

orbits eye sockets within the skull

orchidectomy removal of the testicles; also known as castration

organ a group of tissues similar in structure and function

organelle a structure found within a cell that has a specific function

origin where a muscle begins or originates

ornamental fish popular pet kept in aquariums, small in size and varying in color and appearance

orphaned a young animal that has been rejected by or has lost its mother

orphaned lamb a young sheep that has lost or been rejected by its mother and/or must be bottle fed

Oryctolagus cuniculus genus and species of the rabbit

os penis bone within the urethra of male dogs

osmosis the movement of water or other solvent through a plasma membrane from a region of low solute concentration to a region of high solute concentration

ossification the formation of bone

ostrich large flightless bird raised for meat, feathers, oils, and other by-products; the largest and most common ratite; grows to more than 10 feet tall

otitis inflammation of the ear canal

otoscope piece of equipment used to visualize the inner ear

outer shell membrane the thin area just inside of a shell

oval window membrane that separates the middle and inner portions of the ear

ovarian cyst small masses that develop on the ovary of an intact female animal, which may cause sterility and eggs not to develop

ovariohysterectomy (OHE/OVH) term for a spay; surgical removal of the ovaries and uterus

ovary the part of the uterine horn that allows egg development and is the primary female reproductive organ in egg growth

over-the-counter (OTC) drugs medications that do not require a prescription

oviduct small tube at the end of the uterine horn that is tapered and allows the egg to pass from the ovary

ovine sheep

ovulation release of the egg during estrus in the female; allows reproduction to occur when united with sperm

ovum egg yolk

oxygenation the process of keeping air in water via dissolved oxygen

oxygen tank cylinder that holds oxygen

oxytocin hormone that causes the muscles of the uterine wall to contract and milk production to begin in the mammary glands

pacemaker system of the heart that controls heart sounds and rhythm

packed cell volume (PCV) measurement of the percentage of red blood cells in whole or unclotted blood; also called hematocrit

packing slip a paper enclosed in a package that details the items in the shipment

paddle a long-handled board that acts as an extension of the hand to move pigs

palatability how food tastes and is eaten by an animal

palpation feeling

palpebral reflex evaluated by tapping a fingertip lightly over the surface of the eyelid

pancreas organ with a dual role in endocrine and exocrine production and functions; produces insulin and regulates insulin production

panel individual blood test

panleukopenia commonly known as feline distemper; a systemic viral disease that causes a decrease in white blood cells in cats

papillae hair on the tongue that acts as taste buds

parakeet a small songbird from Australia; is brightly colored with bars located over the wings; also known as a budgie

paralysis partial or complete loss of motor skills

Parascaris equorum roundworm species in large animals; are swallowed along with contaminated hay or water; hatch in the intestinal tract

parasites living organisms found on the internal or external area of the body; feed off an animal, causing disease

parasitology the study of organisms that live off of other organisms to survive

parasympathetic system spinal cord system responsible for reactions of the sympathetic nervous system, such as decreased vital signs and regulating the body system back to normal

parathormone hormone that regulates blood calcium and is secreted when blood calcium levels become decreased

parathyroid glands glands located below the thyroid gland

parked out standing with the front and rear legs wide apart

parrot a medium-sized bird breed that comes from Central America and is bright in color, intelligent, playful, and easy to train

parturition the labor process

parvovirus a potentially fatal viral infection that causes severe bloody diarrhea

passerine a family of bird breeds that are small; have a pointed or slightly curved beak; and have four toes on each foot, three facing forward and one facing backward

passive immunity developed through antibodies acquired from another animal or source, as in colostrum, formula, or plasma

Pasteurella a naturally occurring bacteria in the respiratory tract of rabbits

pasture breeding uncontrolled mating within a flock or herd

patella kneecap; protects the front of the stifle joint

patency refers to maintaining the proper flow and purpose of IV tubing, ensuring it remains intact and usable

pathogens a bacterium, virus, or other microorganism that can cause disease

pathologist one who interprets and diagnoses changes caused by disease in tissues and body fluids

pathology the study of diseases, especially the structural and functional changes produced by them

patience behaving calmly in all situations without any negative complaints

patient history background information of past medical, surgical, nutritional, and behavioral factors that have occurred over the life of the animal

patient the animal being treated

peacock an adult male peafowl

peafowl poultry raised for their colorful, large feathers

peahen an adult female peafowl

pecking order a social organization of a group of birds where some are dominant over others

pedal reflex used on the digit or the skin between digits, which is pinched to determine a reflex response

pedigree an ancestral line

pelleted diet food fed to a bird in a size-appropriate pellet that is composed of the necessary ingredients

pellets fish meal bound into larger particles

pelvic limb rear leg of an animal

pelvis hip bone

pen a small structure placed in a body of water to confine aquacrops; connects land to water

penile area in males, where urine is passed to the outside of the body

penis male external reproductive organ; used in breeding and mating

peptides largest hormones; control proteins in the body

pericardium thin membrane that covers, protects, and maintains the beating action of the heart

perineal urethrostomy (PU) surgical creation of a new opening at the urethra and penis to eliminate the occurrence of urinary blockages

periodontal probe instrument used to measure the depth of pockets around a tooth

periosteal elevator instrument used to remove thick flaps of soft skin tissue during a tooth extraction

periosteum thin connective tissue covering outer bone

peripheral nervous system (PNS) controls the nerves, detects a stimulus, sends signals and informs the CNS, and causes a response or action to occur

peristalsis wavelike motion of the stomach that moves food through the intestines in contractions

peristalsis wavelike motion of the stomach that allows digestion of food

peritoneum clear thin lining of the abdomen

permanent adult teeth that are formed after the deciduous teeth are shed

permanent pasture grass source that regrows each year without replanting or reseeding

permit a written license granted by an authority

personal protective equipment (PPE) equipment worn for safety purposes and protection

personnel manual a handbook that describes staff job descriptions, attendance policy, dress code, and employer expectations

pesticide chemical used to treat external parasites

pet piller small device, usually made of plastic, that has a long thin handle with a plunger on the end to administer medications to small animals

pH level the quality of water on a scale of 0 to 14, with 7 being neutral

phalanges toes or digits

pharmacokinetics how drugs move into, through, and out of the body

pharmacology the study of drugs and their sources

pharmacy the area where medications are prepared and stored for veterinary patients

pharynx throat or area at the back of the mouth

pheasant a wild species of game bird raised for meat and hunting purposes

Pheromones a chemical substance produced and released into the environment by an animal, especially a mammal or an insect, affecting the behavior or physiology of others of its species

phlebitis causes the area around the IV catheter site to experience redness, swelling, pain, and inflammation

phospholipid similar to fat, with less fatty acids in its makeup

photosynthesis the process of sunlight allowing plants to grow in a water source

physical cleaning the most common method of sanitary control within a veterinary facility; involves a chemical with a cleaning object

physical examination (PE) assessment of an animal to determine its overall health status

physical hazard a situation or agent with the potential to cause physical harm to a human or animal's body

physical restraint using a person's body to control the position of an animal

physiology branch of biology that deals with the normal functions of living organisms and their parts

pigeon a small wild bird raised in captivity for meat and competition purposes

piglet a newborn pig

pigment a coloring matter in animals and plants

pill counting tray tool used to count out tablets or capsules to send home with a client

pincers a pair of structures on the head used for smelling, feeling, and protection

pinna external ear flap

pitting edema swelling around a catheter site that causes the fluid to flow outside the vein

pituitary gland "master gland" that lies at the base of the brain and controls hormone release from the endocrine glands

pivot joint that rotates around a fixed point

placenta tissue and membrane collection that connects the fetus and the mother to allow the fetus to absorb nutrients during gestation

plague disease that affects animals and humans; causes high fever, dehydration, and enlarged lymph nodes; can eventually lead to death

plan treatments or procedures provided

plane level of awareness of a patient under anesthesia

plankton tiny plants that grow in water sources

plaque soft buildup of material found over the surfaces of the teeth; composed of bacteria, food materials, and saliva

plasma formed of various proteins in the body; comprises 60 percent of the blood system; remaining portion of blood after red and white cells have been removed

plasma protein test that measures the ratio of protein within the blood; helps veterinarians determine the hydration status and inflammation occurrence in patients; also called total protein (TP)

platelet blood cell that serves to clot blood by constricting (narrowing) the vessels and allowing bleeding to stop

pleura double-membrane covering that lines the lungs

pleural cavity double-membrane covering that holds the lungs

pleural friction rub abnormal lung sound that develops with pneumonia and causes a crackling noise

plucked hair pulled out by hand

pneumonia condition of the lungs that causes inflammation; may have viral or bacterial source

pneumothorax air within the chest cavity

PO by mouth

points sharp edges or corners of teeth

policy manual information provided to all staff members as to what staff expectations are, such as dress code, policies on vacation time, personal time, and sick leave

polled no horn growth

polyarthritis a condition in young lambs that causes multiple joints to become inflamed and painful

polydactyl having many or multiple toes or digits

polydipsia increased thirst; PD

polyestrus having multiple heat or estrus cycles during a season

polysaccharide a long-chain starch made up of simple sugars

polyuria increased urine production; PU

pond a contained area of water that is not naturally occurring

pony a small equine under 14 hands in height; typically ridden by children

pool a structure made of plastic, fiberglass, or concrete that holds a large amount of water and a large number of fish

pop-off valve used to release excess amounts of gas when the pressure becomes too high within a machine system

porcine pigs, hogs, and swine

porcine somatotropin (pST) a hormone used to increase protein synthesis and food use that produces weight gain

porcine stress syndrome (PSS) a condition affecting heavily muscled pigs that can result from stress and may cause death

positive contrast agent element that absorbs more X-rays; causes white appearance on an image

posterior chamber area behind the eye that holds the nerves and is filled with vitreous humor; helps regulate pressure within the eye

posterior lobe back lobe of the pituitary gland; controls peptide hormones

postprandial refers to blood sample taken after an animal has had a meal

poult a young turkey

poultry domesticated birds; have feathers, two legs, and two wings; used as a source of eggs, meat, feathers, and by-products

poultry science the study of poultry breeding, incubation, production, and management

ppt parts per thousand; a unit used to indicate the amount of salt in water

precocial growth that occurs very quickly in bunnies that allows them to move away from danger in a short time

preconditioning building up an animal's body to prepare young animals for possible stress factors

predicted transmission ability estimation record that determines the possible traits a bull or cow will pass along to offspring

preen to clean the feathers

prefix the word part at the beginning of a term

premolars wider teeth at the back of the mouth used to grind and tear food

preprandial refers to blood sample taken after a dog or cat has fasted to reduce the amount of fats or lipids present

prepuce the skin that covers and protects the penis in male animals

prescription (Rx) type of medication, amount of medication, and directions for use of the medication prepared by the veterinarian

prescription drug medication prescribed by the veterinarian

prescription label indicates to the owner how to properly give the medication and identify the medication within a container

pressure valve indicates how much oxygen is left in a tank

primary bandage layer first bandage layer, against the surface of the animal, that acts as padding

primary response process that occurs when an antigen comes in contact with the immune system for the first time

primate an animal, such as a monkey or ape, that has movable thumbs and, in many ways, resembles a human

primed milking machine devices are applied to the udders and begin collecting milk

printer a machine that prints forms and invoices on a computer onto paper

probe method the process of passing an instrument into the cloaca to determine the sex of a reptile or amphibian

procedures manual information provided to all staff members on the workings of the veterinary facility, such as client rules, scheduling methods, payment procedures, and client visitation times

processing developing film in a darkroom that has a safe light source and no light leaks from the outside area

production cycle the time it takes for young aquatic species to reach full market size

production intensity an assurance that the size of a body of water will hold the appropriate amount of food sources and be large enough for high-production level of products

proestrus beginning stage of the heat cycle before estrus; causes follicle development and genital and behavioral changes that attract the male to the female and prepare the female for mating

profitable having gains (money) left over after paying expenses

progeny offspring

progesterone hormone produced to maintain pregnancy

proglottid a tapeworm segment that is shed as the tapeworm grows

progress notes documentation chronologically detailing the patient's history, treatments, and outcomes

prolactin hormone that controls milk production

prolapsed uterus condition wherein the uterus and vaginal tissues separate from the body and allow the uterus body to be passed outside the body

prophase a stage wherein chromatin thickens and begins to form chromosomes

prophylaxis professional cleaning of a patient's teeth

proprioception relates to the body's ability to recognize body position for posture or movement

prostate gland produces accessory secretion to add to the sperm to produce semen

protein a large molecule that works with 22 different amino acids that may join with sugars, starches, and lipids to form the strong structure of a cell; nutrients that are essential in growth and tissue repair

protozoan class of disease; protozoa are the simplest forms of life; they are single-celled organisms that are known as parasites that may live inside or outside the body

protozoastatic inhibits growth of protozoa

proventriculus acts as a monogastric stomach and begins the digestion process in birds by releasing excretions to soften food

proventriculus the tube that connects the crop to the gizzard

pseudo rabies a condition caused by a virus that has symptoms similar to rabies; spread through oral and nasal secretions

psittacine a family of bird breeds that are large parrots with a curved upper beak; have four toes on each foot, two facing forward and two backward

psittacine beak and feather disease (PBFD) a contagious and fatal viral disease that affects the beak, feathers, and immune system of birds

psittacosis a disease that has zoonotic potential, especially in young children and anyone with a poor immune system; causes severe flulike symptoms; commonly called parrot fever or chlamydia

puberty sexual maturity

pubis third section of the pelvis

pullet a young chicken raised to lay eggs

pullet production a system in which hens are raised for fertile egg production

pulmonary arteries carry oxygen-poor blood and prevent blood flow back to the heart

pulp cavity center of the tooth that holds the nerves, veins, and arteries

pulse the heart rate of an animal, or each beat created from the heart

pulse oximeter device used to measure a patient's vital signs by indirectly measuring oxygen saturation within the blood and any changes in blood volume

punch biopsy a small amount of tissue is removed for testing from a circular site

pupae immature larval stage that goes through several molting stages during growth by spinning a cocoon

pupil area in the center of the eye that dilates and constricts with light responses

pupillary light response test test in which light is shined into the animal's eyes and examiner notes constriction (shrinking) of the pupil and dilation (expanding) when the light source is removed

purebred an animal with known parentage of one breed

purebred business registered breed of cattle that has pedigrees

purebred flock method a small production system that raises purebred sheep for reproductive purposes

pushing walking toward cattle quietly and calmly to get them to move to another area

P wave first peak on an ECG that forms at the SA node (pacemaker) and reflects the flow of electricity and blood through the atrium and that a contraction has occurred

pygmy goat a small goat kept as a pet

pylorus exit passageway of the stomach

pyometra severe infection of the uterus in intact female animal that causes the uterus to fill with pus

pyothorax pus and cellular debris within the chest cavity

QRS complex series of peaks showing that electrical current and blood have flowed through the AV node and have caused a contraction in the ventricles

quadriceps muscle located on the hind limb at the cranial part of the thigh

quail a small wild bird raised for eggs; makes the whistle sound "bobwhite"

quarantined isolated

quarters sections of the mammary glands that store milk

quick blood supply within the nail bed

quills spines that project from a hedgehog's body to protect it

rabies a deadly virus that can affect any mammal, including humans; usually transmitted by the saliva of an infected animal and then is transferred into the bloodstream

rabies certificate documentation of an animal's receipt of a proper vaccination for rabies

rabies laws a state's legal standards to prevent the spread of rabies infections

rabies pole long pole used to capture a dog around the neck with a noose on the end of the pole that acts as a collar and a pole that acts as a leash; also called a snare pole

racehorse a horse bred for speed and racing purposes

raceway a long, narrow water structure with flowing water

radiation safety hazard that allows X-rays to be produced to take a radiograph; the use of ultraviolet or gamma rays to radiate and kill living organisms

radioactive iodine involved in treatment in which iodine is injected into the bloodstream to treat areas that overproduce thyroxine; damages thyroid tissue

radiology an area of a facility where X-rays are taken and developed; this area must meet specific regulations to guard against chemical and radiation exposure; study of radiation

radiolucent related to black or dark gray appearance, which may indicate soft tissue or air; may be an exposure factor related to increased exposure settings or improper patient measurement

radiopaque related to white to light gray appearance, which may indicate hard tissue such as bone; may be an exposure factor in that the machine settings were too low or the patient was not properly measured

radius larger bone on the front of the lower forelimb

rainwater a source of water from the sky that can be used if not contaminated or polluted

rale noise or crackle caused by the lungs during pneumonia

ram adult male sheep capable of reproduction

range band method raising large sheep herds on large amounts of land

range paralysis a common name for Marek's disease

raptor a large bird of prey

rare breed an animal bred and raised for a variety of purposes but not found locally

rasp an instrument used to file down horse teeth; also called a float

rat tooth thumb forceps instrument with large teeth, similar to a rat's, that holds large thick tissues and skin

ration total amount of food an animal needs within a 24-hour time frame

ratite a large flightless bird raised for poultry production purposes

Rattus norvegicus genus and species of the rat

reagent chemical used to run each individual test in an entire kit

reagent strip chemical test strip used to analyze urine or blood on a long, thin plastic strip separated by individual square pads containing a chemically treated paper

rearing rising up on the hind legs

rebreathing bag a collapsible reservoir from which gases are inhaled and into which gases may be exhaled during general anesthesia or artificial ventilation.

rebreathing system bag that connects to a machine and inflates and deflates as the patient breathes; allows the patient to rebreathe anesthetic gas after it has passed through the CO_2-absorbing canister

receiver a person who gets a message in the route of communication

reception area a location where clients arrive and check in with the receptionist

receptionist personnel greeting and communicating with patients as they come in and leave the veterinary facility

receptive accepting of

receptor found within the nerves located throughout the body; allows for detecting changes within the body and the environment

recombinant vaccine made up of a live nonpathogenic virus into which the gene for a pathogen-related antigen has been inserted

recumbency a lying position

red blood cells (RBC) most abundant cells within the body; produced in the bone marrow; transport oxygen throughout the body

redirected aggression behavior that stimulates a response from another source, such as a dog outside of the yard or a car driving down the road, in which the dog turns on the owner

reduced-calorie diet specialized food given to animals that are overweight or less active due to health status

reefer's knot single bow knot that ties in a nonslip but quick release tie

reefing using rope to place a cow on the ground for restraint

reference lab laboratory samples analyzed outside of the facility in a commercial lab

reflex units of function that produce a reaction to a stimulus

reflex hammer tool used to tap the area of the knees and other joints to elicit a response

refractometer tool used to measure the weight of a liquid and determine its specific gravity

registered veterinary technician (RVT) a veterinary technician who has passed a state licensing exam

regurgitation process of bringing food into the mouth from the stomach to break it down

renal artery located in the center of the kidney; allows blood flow out of the renal system

renal pelvis innermost section of the kidney; filters blood and urine through the veins and arteries

renal system body system that involves the kidneys and urinary production to rid the body of waste products within the bloodstream

renal vein located at the center of the kidney; allows blood flow into the renal system

replacement hens pullets that take the place of older hens that are no longer capable of egg production

repolarize reset or restart of the heart to get ready for the next electrical impulse to start

reproduction diet specialized food given to breeding animals to meet additional nutrient needs

reproductive system the organs in the body responsible for both male and female breeding and reproduction

reptile an animal with dry, scaly skin that adjusts its body temperature according to the outside temperature and environment

research studious inquiry or examination aimed at the discovery and interpretation of information

research scientist a scientist involved in research, whether on animals or humans

respiration the process of breathing through inhalation and exhalation

respiratory rate (RR) how many breaths an animal takes in a minute

respiratory system system of the body that controls oxygen circulation through the lungs and the rest of the body

restrain to hold back, check, or suppress an action and to keep something under control using safety and some means of physical, chemical, or psychological action

retained placenta reproductive condition in female animals in which the afterbirth materials have not passed 8 hours after labor

reticulum second section of the ruminant stomach; acts as a filter for food

retina inner layer of the eye; contains two types of cells that aid in seeing color and determining depth

retractable capable of projecting inward and outward, as in the nail of a cat

retractor handheld or self-retracting; is used during surgeries to hold an animal's body cavity open for exploration

rhea a moderate-sized flightless bird native to South America

rhinotracheitis a virus that affects the upper respiratory tract of cats

rib bone that attaches to individual thoracic vertebrae and protects the heart and lungs

ribosome a structure within the cell that makes protein

rickettsial class of disease spread by biting insects, such as fleas and ticks

riding horse a horse ridden for show or recreational purposes

right atrium thin-walled chamber of the heart where blood enters

right ventricle thicker-walled chamber of the heart that pumps blood to the lungs

right-to-know station an area in the facility where the OSHA and SDS binders are kept

ringworm fungal condition that affects the skin of humans and mammals and is highly contagious

rinsing using water to remove any chemicals from an X-ray film

RNA ribonucleic acid; a nucleic acid present in all living cells; principal role is to act as a messenger carrying instructions from DNA for controlling the synthesis of proteins, although in some viruses RNA—rather than DNA—carries the genetic information

roaring respiratory condition in horses that sounds like a roar due to the larynx only opening a small amount due to a trauma to the nerves of the throat area

roaster a chick over 8 weeks of age and 4 pounds that is used as meat

Rochester-Carmalt hemostatic forceps much heavier than others; used to hold off the ovarian stump during a spay surgery; have lengthwise interlocking serrations

Rocky Mountain spotted fever rickettsial disease transmitted by the American dog tick; signs include fever, joint pain (arthrodynia), depression, and anorexia

rodents mammals that have large front teeth designed for chewing or gnawing, such as rats, mice, and hamsters

rods cells that allow the eye to detect light and depth; oblong bacteria shape

rooster an adult male chicken

root part of the tooth located below the gum line; holds the tooth in place

root word the origin, or main part, of a word that gives the term its essential meaning

rotors wheels in a centrifuge that spin at a variety of speeds depending on the sample type

roughage hay or grass source fed to livestock and ruminants

round window membrane that receives sound waves through fluid passed through the cochlea

roundworm an intestinal parasite that is long and white, the most common intestinal parasite of small and large animals

routine appointment a patient visit scheduled during office hours; examples are vaccinations, yearly exams, and rechecks

rugae folds within the stomach when it is empty

rumen first section of the ruminant stomach; acts as a storage vat and softens food for fermentation

ruminant animal with a digestive system that has a stomach with four sections or compartments

rump method a restraint method in which a sheep is placed on its hind end for handling

run a group of fish

SA node the pacemaker of the heart

sac fry a yolk sac attached to a fry for one day after the hatch

sacral the fourth section of the vertebrae; lies over the pelvic area

sacrum area over the pelvis that is fused together and forms the highest point of the hip joint

safe light red light that is low in intensity and has a filter so that it doesn't damage film

Safety Data Sheet (SDS) safety information published by the manufacturer of a product that has the potential to harm humans within the facility

salinity the amount of salt in a water source

saliva fluid that helps soften and break down food for ease of swallowing and digestion

salivary glands area within the mouth that produces saliva

salmonella infection a bacterial infection that causes severe diarrhea

salmonellosis a bacterial infection that occurs from touching reptiles or amphibians

salt block salt provided in a block for horses to lick

saltwater water with a salt content of at least 165 ppt

saltwater fish species of fish that require salt in their water source

sanitation the process of keeping an area sterile to prevent the spread of disease

saphenous venipuncture a puncture to draw blood from the outer thigh area

sarcoptic mange name for the skin disease caused by infection with the *Sarcoptes scabei* mite; causes a zoonotic mite skin condition in which mites burrow into the skin and cause infection

scabies a contagious disease caused by a mite that lives on the hair and skin; known as sarcoptic mange in animals

scales a body covering that serves as skin and protects the outside of fish

scanner equipment that allows documents and images to be copied and stored electronically

scapula shoulder blade

scavenger hose tube attached to a pop-off valve to prevent gas from leaking into a room; pushes gases outside of the facility to escape to the outside air

scent gland an organ located on the head behind the horns that produces a strong hormonal odor in male goats

scheduled drug prescription drug that has the potential for abuse or addiction; also called controlled substance

scheduled feedings occur when a set amount of food is given at specific times during the day

Schirmer tear test diagnostic test that uses a paper strip to measure tear production in the eye

school a group of fish

sclera outer white layer of the eye

screen area that lines a film cassette

scrotum skin over the two testicles that protects and controls the temperature for sperm production

scrub pack sterile supply that contains the items necessary to properly prepare the surgeon, technician, or assistant for washing hands for surgery

scruff the loose skin over the base of a cat's neck that can be held to restrain the cat

scruff technique used on the base of the neck over the area where the skin is elastic and able to be grasped with a fist technique

secondary bandage layer holds the first layer in place

secondary response process that occurs when an antigen comes in contact with the immune system for the second time and provides a repeated exposure to an antigen that creates immunity to prevent disease development

sedative medication that causes drowsiness but no loss of pain sensation

sediment sandlike buildup in the urine composed of crystals and casts

seed diet a food composed of natural seeds; eaten by birds

seizure loss of voluntary control of the body with an amount of unconsciousness causing uncontrolled violent body activity

selection a natural or artificial process that results or tends to result in the survival and propagation of some individuals or organisms but not of others with the result that the inherited traits of the survivors are perpetuated

selection guidelines set of rules that state the standards of the type of cow and how it is chosen for a production program

semen fluid that transports sperm during reproduction

semen collection procedure using specialized veterinary equipment to collect semen from male animals to be used in breeding practices

semen morphology process of determining if semen and sperm is adequate for transport and breeding by analyzing the quality and quantity

semi arch a large rabbit breed that weighs between 12 and 14 pounds

semi-aquatic living part of the time in water and part of the time on land

semicircular canals sensory cells that generate nerves impulses to regulate position

seminal vesicles paired glands that contribute to the fluid portion of semen

seminiferous vesicle tubes located within the testicle that hold the sperm

sender a person relaying an idea or message in the route of communication

senior diet specialized food given to geriatric or animals older than a specific age (this is species specific) that require certain nutrient levels due to age

Senn retractor double-ended retractor blades that look like fingers on the end of a hand; may be blunt or sharp in shape

sensory neuron cell that delivers signals from the central nervous system (CNS)

sepsis a toxic condition caused by the spread of bacteria

septic refers to buildup of bacteria in the system that causes a toxic effect when it gets into the bloodstream

serpent snake

serration space on the top of a comb

serum liquid portion of blood

sharp operating scissors used for cutting objects like sutures or drape material rather than tissue

sharps sharp instruments and equipment that can injure a human or animal and cause a wound or cut that transmits a contagious disease due to contamination

sharps container an object that holds sharp items that may spread disease, such as needles, surgical blades, and glass

shaving taking away an amount of hair close to the skin

shearing a method of shaving wool from sheep

shipping fever a condition that occurs in livestock, such as horses and cattle, that are transported

shoal fish that usually live in groups of five or more breeders

shock stage of grief causing a lack of feelings caused by the sudden death of an animal or loved one; condition that occurs when an animal does not have enough blood and thus oxygen reaching the tissues

short hair a cat's coat that is short and smooth

signalment patient's name, age or date of birth, gender, color, and breed or species

silo a tall storage building that keeps food free of moisture

silver nitrate stick long slim wooden applicator with silver nitrate located on the end; is used to clot a bleeding nail by burning the quick

simple epithelial tissue single layer of cells, with every cell in direct contact with the basement membrane that separates it from the underlying connective tissue

simple fracture a break in a bone that may be complete or incomplete, but does not break through the skin; also known as a closed fracture

simple sugar single molecule of carbohydrate that forms the building blocks of nutrients; called monosaccharides

simulated system a model that mimics an animal's behavior and/or internal or external structures for the purposes of research; *see computer system*

sinus arrhythmia condition when the heart rate increases with inspiration and decreases with expiration

sinuses small open spaces of air located within the skull and nasal bone

sinus rhythm normal heart rate and rhythm

sire male parent

sitting restraint restraint used to keep an animal sitting on its rump and prevent it from standing

skin mite an insect that lives within the hair and skin; not visible to the naked eye

skin turgor assessed during the process of evaluating an animal for dehydration by lifting the skin over the base of the neck or shoulder blades

skull the bone that holds and protects the brain and head area

slander talking negatively about someone in an improper manner

slicker brush a tool used to brush the hair coat

slip knot a way of tying that has a quick release

slough tissue that has died and falls off

slurry a soft, soup-like substance fed to fry

smolt immature salmon greater than an inch in size

SNAP test a simple serologic blood test kit completed in the clinic that tests for certain disease antigens in the bloodstream

snare pole long pole with a noose on the end that acts as a leash and collar to capture a difficult dog; also called rabies pole

sneezing act of removing dust or other particles from the respiratory tract through physical force

snubbing rope a rope placed over the snout and tied to a post for the purposes of restraint

SOAP (Subjective–Objective–Assessment–Plan) subjective information–objective information–assessment–plan; a method used for documentation in the medical record

socialization an interaction with other animals and people to become used to them

soda lime granules within a canister that absorb carbon dioxide

sodium chloride NaCl; saltwater fluid

software a program or set of instructions that a computer follows

solid manure system a low-cost waste material system in which manure is collected and removed on a daily basis and stored away from cattle

solitary alone; independent behavior

somatic cell a cell that divides in mitosis; helps in the growth and repair of the body

somatotropin growth hormone that increases protein synthesis in the body, causing an increase in the animal's size

sonogram printout of an ultrasound recording

soremouth a highly contagious zoonotic viral disease that causes lesions over the mouth and lips of sheep

sorrow sadness over the loss of a pet that may result in crying and talking about the pet

sound showing no signs of lameness or injury

sow an adult female pig of reproductive age

spawn a group of fish eggs in a nest

spawning breeding or reproduction in fish

spay hook instrument used to locate the uterus and uterine horns in small female animals; has a long, thin metal device with a hook on the end

specialized payment plan a method approved by the veterinarian that allows a down payment with additional regular payments until the balance is paid

specific gravity (SG) the weight of a liquid measurement

specific-pathogen free (SPF) free of disease at birth

spectrum a wide variety of factors

speech communication passing along information verbally

spent hens adult female hens that have aged and whose egg production has decreased or stopped; used as a meat source in processed foods

sperm male reproductive cell developed in the testicle

sphygmomanometer device used to measure an animal's blood pressure during surgery; also helps determine tissue perfusion as well as depth of anesthesia; also called a blood pressure monitor

spike port site on an IV bag where the drip set tubing is placed

spinal column extends from the base of the skull to the end of the tail and allows movement; also called the spine

spinal cord strands of fibers that connect the brainstem to the PNS to send and receive signals through the body; begins at the base of the brain and continues to the lumbar vertebrae

spinal nerve reflexes evaluated for proprioceptive reflexes of the muscles and tendons

splint support applied to a broken bone with bandage and padding to allow a bone to heal

splint bone one of two small bones on the back of a horse's cannon bone that may become inflamed from large amounts of work and stress on the legs

sponge forceps straight or curved; used to hold gauze sponges or other absorbent material to clear away bleeding and other debris

spooking a startling action

spreadsheet a program used to track large quantities of items in one file

spring a natural opening in the ground that provides a clean natural water source

springing heifer a young female cow that is pregnant with the first calf

squamous flat

square knot commonly used to secure an animal; a nonslip knot that doesn't come untied or loosen easily

squeeze cage wire box with small slats that allow injections to be given to a cat, such as vaccines or sedatives

squeeze chute a cage-like structure made of metal pipes that holds a cow and prevents it from kicking during restraint; a wire cage with sliding doors that is used to control a cat for procedures such as exams and injections

stag a male pig castrated after maturity

stage part of a microscope; the flat section under the lens

stallion an intact adult male horse capable of reproduction

stanchion a head gate that holds the head of a cow in place during restraint

standing heat situation in which a female animal stands for breeding when a male is present

standing restraint restraint used to keep an animal standing and prevent it from sitting or lying down

stapes tiny third ossicle bone that forms contact with the cochlea in the inner ear

staple remover device placed under a staple and squeezed to open up the staple, which is then gently removed from the site

starches plant or grain materials that provide fiber and bulk in an animal's diet; convert to glucose or sugar during digestion

starter food food for calves as they begin eating; easily digested

state board examination an exam that veterinarians and veterinary technicians must pass to earn state licensure to practice

state board of veterinary medicine a state agency of veterinary members and members of the public who take disciplinary action in the profession as deemed necessary and also monitor the rights of the public

static equilibrium a state of rest in the body

statutory laws rules and regulations based on written government laws

steer a castrated male cow

sterile not able to reproduce

sterilizing destroying all microorganisms

sternal recumbency a restraint position in which the animal is lying on its chest and abdomen

sternum breastbone; protects the organs of the chest

steroid occurs naturally in the body; regulates chemicals, such as cholesterol, in the body that control essential life functions

stethoscope instrument used to listen to the heart, lungs, and other areas of the chest

stimulus internal or external change that causes a reaction

stirrups pieces of tape at the end of a bandage on a limb; hold the bandage in place and prevent slipping

stock horse a strong, athletic horse well suited for work

stool softeners medication given to produce a bowel movement by softening feces

stratified epithelial tissue several cell layers composed of multiple cells and many layers of thickness

stretch technique a restraint procedure in which the loose skin of a cat's neck is held with one hand and the rear limbs are pulled backward with the other hand

Sterilization the process of making something free from bacteria or other living microorganisms

strongyles common intestinal parasite in large animals; long, thin, and white in appearance

stud an intact adult male horse of breeding age

styptic powder a chemical used to stop bleeding

subcutaneous (SQ) refers to injection given under the skin

SQ fluids subcutaneous fluids placed under the skin; slowly absorbed by the body

subjective information observations of an animal's appearance and behavior made by the veterinary staff

subluxate causes bone to be partially out of the joint

substrate the material on the bottom of a cage

Submissive ready to conform to the authority or will of others; meekly obedient or passive.

suffix the word part at the end of a term

sugars type of carbohydrate that forms the simplest type of that nutrient

supernatant liquid portion of a sample

supplements additives incorporated into the diet in solid or liquid form when needed by the animal

supportive care therapy or treatment that includes such interventions as fluid replacement, medications to decrease vomiting and diarrhea, and pain medicine

surface water runoff excess water drainage from rainfall or precipitation

surface-dwelling fish fish that live and feed near the top of a body of water

surgery pack sterile instrument prepared for surgical procedure with tools and supplies required for the procedure

surgical appointment an appointment scheduled for surgery that is usually outside of routine office hours

surgical glue sterile substance used for closure of small wounds or surgical sites; is commonly used in feline declaws

surgical log a record of details of all surgeries performed in the facility

surgical margin area of the patient that should extend 2 to 4 inches beyond the anticipated surgical incision borders

surgical scrub process then begins with sterile solutions placed onto the surgical margin to clean the patient

surgical staples fast and easy way to close the skin with metal pieces

surgical suite an area where surgery and anesthesia are performed; the area must be kept sterile to prevent contamination

suture where two fibrous joints meet; fine line where little or no movement occurs

suture reel container of suture material that must be cut and thread onto a needle

suture removal occurs within 7–10 days after surgery

suture removal scissors have one blade with a small hook on the end to better get under a suture for removal

swaged on needle suture material that has a needle already attached

swallow reflex the patient licking with the tongue or having increased jaw tone during extubation of the endotracheal tube

swamp fever a common name for equine infectious anemia

swan a large water bird with a long, thin neck that is raised as ornamental poultry

swim bladder an air-filled sac that prevents fish from sinking and allows them to float and move through depths of water

swim bladder disease a bacterial disease in fish that makes it difficult to swim to the water's surface and in depths of water

swine pigs or hogs

swine dysentery a condition in pigs that causes diarrhea and spreads rapidly through contamination

sympathetic system division of the autonomic nervous system that is responsible for the fight-or-flight reaction, increased vital signs, and certain drug reactions

sympathy showing support for others at a time of sadness

synapse a junction between two nerve cells, consisting of a minute gap across which impulses pass by diffusion of neurotransmitters

synovial joint moveable joint at the area where two bones meet

synthetic seawater water made by humans

systemic circulation blood system that delivers nutrients and oxygen to all areas of the body

systole contraction phase of blood pressure, or the top number; normally higher than the bottom number

T wave last peak and the most important part of the ECG; shows that an electrical impulse has traveled through the entire heart, has completed the contraction, and has repolarized the heart

T3 most potent and active thyroid gland hormone; measured in the bloodstream to diagnosis thyroid problems

T4 thyroid hormone known as thyroxine; breaks down fats and helps control cholesterol

tabletop technique the distance differs between the X-ray tube and the top of the X-ray table surface, and the film cassette is placed on top of the table in contact with the patient

tachycardia increased or elevated heart rate

tacky slightly dry, as the gums may be

tactfulness doing and saying the appropriate things at the correct time

tadpole a young newborn frog or toad

Taenia genus of tapeworm that affects companion animals and livestock

tail the part of the body containing the caudal fin that allows movement and guidance in the water

tail docking cutting a pig's tail short to prevent chewing and injury to the tail and to keep it free of feces; a junction between two nerve cells, consisting of a minute gap across which impulses pass by diffusion of a neurotransmitter

tail flagging holding the tail high in the air during the estrus cycle

tail switch restraint a method of twisting the tail at the base to prevent a cow from kicking and moving

tail tying a method of tying the tail for the purposes of keeping it out of the way or to move a sedated horse

talon claw on a bird of prey

tank a basic container that holds water and fish

tapeworm intestinal flat worm parasite that is segmented or has individual parts that grow and shed as the parasite ages

tap water water that comes from a faucet

target tissue location where a drug travels within the body; location of its intended use

tarsus joint of the ankle of the rear limb

tartar scraper instrument used to remove calculus and tartar from the tooth surface

tattoo a mark or number placed on the skin with an electronic needle and ink

taurine a nutrient required by cats that allows for proper development

teat a nipple of the mammary glands of a female animal, from which milk is sucked by the young

technique chart list of settings on the X-ray machine based on the thickness of the area to be radiographed

telophase the final stage of mitosis and of the second division of meiosis in which the spindle disappears and the nucleus reforms around each set of chromosomes

temporary pasture grass source that does not regrow each year; needs to be replanted or reseeded

tendon fibrous strand of connective tissue that attaches muscle to bone

terrapin turtle

terrarium a plant-sourced housing for reptiles and amphibians

terrestrial living on land

territorial aggression refers to an animal's protective nature of its environment, such as an owner, offspring, or food

territory an area of land claimed by an animal

tertiary bandage layer third bandage layer that acts as a protective covering

testicle the primary reproductive organ in the male; holds sperm

testosterone main male reproductive hormone; produces male characteristics; is essential for sperm production

tetanus bacterial disease caused by a wound; commonly called lockjaw

thalamus organ within the brain; located at the top of the brainstem

thermometer a device used to measure the temperature of water

thermostat a device used to control the temperature of water

thiamine a nutrient compound of vitamin B that promotes a healthy coat

thoracic second section of vertebrae located over the chest area

thoracic cavity the chest

thrombocyte platelet

thymus a gland in young animals; has an immunologic function

thyroidectomy surgical removal of the thyroid gland

thyroid gland endocrine gland located in the neck area; produces thyroxine and calcitonin

thyroid-stimulating hormone (TSH) hormone produced in the anterior pituitary gland; controls the chemical thyroxine

thyroxine hormone that produces growth hormones and helps regulate metabolism in an animal's cells; also helps regulate body temperature

tibia larger bone in the lower rear leg; lies in front

tick wingless insect that feeds off blood of animals

tick paralysis varying degrees of paralysis caused by saliva of a female tick

tidal volume amount of oxygen required by a patient

tissue an aggregate of cells, usually of a particular kind, together with their intercellular substance that form one of the structural materials of a plant or an animal

tissue adhesive a glue-like substance used to close incisions

tissue forceps instrument used to grasp tissue; has teeth that grip the tissue

toe pinch response occurs when digits are pinched to elicit a response, which should be to pull away from the pressure; used to assess the nervous system

toner black or colored ink that allows printing with a printer, fax machine, or copier

tongue muscle within the mouth used to hold food within the mouth

tonometer equipment used to measure intraocular pressure

tooth abscess infection of the tooth root

topical the application of medication to the skin or the outside of the body

total herd large group of cows kept together and replaced when milk production decreases

total mix ration (TMR) a mixture of high-quality nutrients

total protein (TP) test that measures the ratio of protein within the blood and helps veterinarians determine the hydration status and inflammation occurrence in patients; also called plasma protein

towel clamp instrument used to hold the surgical drape in place over the patient during surgery

Toxascaris leonina roundworm species with the simplest life cycle

toxic poisonous

Toxocara canis roundworm species with a more complicated life cycle than some other roundworms; can ensure its species will be passed from generation to generation

Toxocara cati roundworm species similar to *T. canis*; may infect the lungs or abdominal tissue

Toxoplasma gondii single-celled organisms that can occur in any mammal but are shed through the feces of cats, which are the only animals that pass it through feces

toxoplasmosis a protozoan parasite passed in cat feces that may cause abortion in pregnant women

TPR temperature, pulse, and respiration

trace minerals needed in small amounts; also called microminerals

trachea the windpipe or tubelike airway within the throat

tract a major passage in the body that contains large bundles of nerve fibers and runs through the body, providing sensations and feelings

tragus opening of the ear canal structure; includes the haired area

training teaching an animal to understand specific desired habits

tranquilizer medication used to calm an animal during stress

transducer instrument used with an ultrasound; scans the body and transmits the waves back to a screen

transitional epithelium consists of multiple layers of cells with varying shapes

transverse colon second or middle section of the large intestine

trapping capturing and restraining of wild animals

travel sheet paperwork list that codes, services, and procedures that allows the staff to track fees to charge the client

treatment area a place where animals are treated or prepared for surgery

trematode flat worm lacking a body cavity; also called a fluke

triage quick general assessment of an animal to determine how urgently the animal needs a veterinarian's attention

trichobezoar a hairball

Trichuris vulpis whipworm species; infects animals when they ingest contaminated food or water

trimester period of 3 months of pregnancy in the gestation cycle

trimming removing a specific amount of hair from one or more locations

trocar plastic or metal pointed instrument placed into the rumen of a ruminant animal that has bloated; relives air pressure in the animal's stomach

tropical fish small brightly colored fish that require warm water and are naturally found in warm, tropical locations

trot the slow two-beat gait of English horses

trough a V-shaped table that holds an animal on its back for surgery and other procedures

trunk in aquatic species, part of the body that contains the body, fins, and anus

trypsin enzyme produced by the pancreas; digests proteins

tumor a mass of tissue made up of abnormal cells

turkey a poultry breed with a large body and a long, thin neck; is specifically raised for meat

twitch a chain or rope loop placed over the lip of a horse as a restraint

tympanic bulla osseous chamber at the base of the skull

tympanic membrane eardrum; tissue that separates the outer and middle portions of the ear

Tyzzer's disease a bacterial infection in rabbits; spread through spores in the air; affects the digestive system

udder a large bag-shaped organ enclosing two or more milk-producing glands, with each draining into a separate nipple on the lower surface

ulna smaller bone on the back of the lower forelimb

ultrasonic cleaner instrument placed within a wire basket, which is then seated within a cleaner that holds a disinfectant solution and creates vibrations to clean an item

ultrasound diagnostic tool that uses ultrasonic sound waves to produce images of internal organs and structures

unethical the act of knowingly doing something wrong or improper

unilateral cryptorchid condition in which one retained testicle has not descended into the scrotum

univalve a mollusk with one shell or part of a shell

upland game bird wild species that spends most of its time in the woods and does not move out of an area

upper GI contrast study of the structures of the upper digestive tract; barium is given orally

upper respiratory infection (URI) condition causing sneezing, coughing, nasal discharge, ocular discharge, wheezing, and difficulty breathing

urate the urine waste material of reptiles and amphibians

urbanization the development of cities over an area

urinary bladder muscular sac in the pelvis, just above and behind the pubic bone

urea waste product of protein metabolism by the kidney

uremia toxic condition caused by the kidneys when an increased concentration of waste products gets into the bloodstream

ureter tube made of smooth muscle that pushes and filters urine out of the kidneys through contractions

urethra the tube by which urine is passed out of the body from the bladder, and which in male vertebrates also passes semen

urinalysis breakdown and evaluation of the components of urine

urogenital system includes the kidneys, urinary bladder, external genitals, mammary glands, and rectal area

urolith bladder stone that forms within the urinary bladder

uterine horn two parts of the uterus that hold single or multiple embryos that expand as the embryos grow

uterus large, hollow reproductive organ that holds an embryo during the gestation period and is Y-shaped

vaccination antigen placed into the body to build up resistance in the immune system to disease

vaccine program injections given to protect animals from common diseases and build up the immune system

vacutainer tube tube in which blood samples are placed for future sampling

vagina internal female reproductive organ that allows mating and labor to occur

vagus nerve main nerve within the nervous system

valve a structure in mollusks that acts like gills

vaporizer converts liquid anesthesia into a gaseous form, which the patient inhales

vas deferens small tube that transports sperm from the testicles to the outside of the body

vasoconstriction diameter of a vessel decreases as blood pressure increases

vasodilatation diameter of a vessel increases as blood pressure decreases

vasodilator drug used to treat low blood pressure due to heart failure

vat a large metal or concrete structure that holds large amounts of water; commonly used for breeding and reproduction of fish

vector an insect or organism that transmits a disease

veins vessels that carry blood to the heart

velvet material shed from the antlers of a deer; used in medications and nutritional supplements

vena cava large vessel that transports blood into the heart from systemic circulation

venipuncture practice of placing a needle into a blood vessel

venison deer meat

venomous poisonous

vent the external area where eggs are laid in egg-bearing animals; also called the cloaca

ventral root nerve branch of the vertebrae that contains motor nerves

ventral-dorsal (VD) view that requires the animal to be positioned on its back with the X-ray beam going through the ventral area (stomach) first and the dorsal area (back) second

ventricular fibrillation V-fib; condition causing the ventricles to fire electrical currents rapidly; the most serious cause of cardiac arrest

verbal communication the use of speech to pass along information in the communication process

verbal restraint voice commands used to control an animal

vertebrae individual bones of the spine that surround and protect the spinal cord

vertebrate an animal with a backbone

vestibule portion of the inner ear that contains specialized receptors for balance and position

veterinarian a doctor who holds a state license to practice veterinary medicine; the vet holds the ultimate responsibility within the facility

veterinary assistant a staff member trained to assist veterinarians and technicians in various duties

veterinary hospital manager a staff member in charge of the business workings of the entire hospital, including staff scheduling, paying bills, and payroll

Veterinary Practice Act a set of rules and regulations that govern veterinary medicine in each state within the United States

veterinary referral practice a veterinary facility that offers services by veterinarians who specialize in specific areas, such as cardiology or orthopedics

veterinary technician a 2-year associate degree in specialized training to assist veterinarians

veterinary technologist a 4-year bachelor degree in specialized training to assist veterinarians

veterinary–client–patient relationship (VCPR) a relationship established between the patient, client, and veterinary staff based on trust, expertise, and duty to care for animal

viral class of disease caused by particles that are contagious and spread through the environment

virucidal kills viruses

virus something that causes a disease that is not treatable and must run it course

visceral larval migrans a roundworm infection in humans

viscous thick liquid substance

vital signs signs of life; measurements that assess the basic functions of an animal's body; include heart rate, respiratory rate, temperature, blood pressure, mucous membrane color, capillary refill time, and weight of the animal

vitamin C an essential nutrient and supplement

vitamins nutrients needed in small amounts for maintenance of life and health

vitreous humor transparent gel-like fluid located behind the lens of the eye

V-notcher a tool used to make cuts in a pig's ear

voided refers to urine sample collected as an animal is urinating

voluntary reflex occurs when an animal asks its body to perform a desired function, such as walking or running

vomiting process of bringing up partially digested or undigested food that has been in the stomach of monogastric animals

vulva external beginning of the reproductive system in the female; also where urine is excreted

waiting room an area where clients and patients wait for their appointments

walk a relaxed, slow four-beat gait

walk-in appointment when a client without an appointment arrives at the hospital wishing to see the veterinarian

want list a list of items that are needed or low in stock within the facility; items need to be reordered

warm-blooded able to control body temperature internally

warm housing a heated building with individual stalls that holds cattle in the winter

warm water water between 65 and 85 degrees Fahrenheit

water facility an area built to house aquatic animals in natural environments

water nutrient that makes up 75 percent of the body and provides several functions within the body, including controlling body temperature, maintaining body shape, transporting nutrients within the body's cells, aiding in digestion of food, breaking down food particles, and serving as a carrier for waste products

water soluble refers to vitamins that are not stored within the body, are dissolved by water, and are therefore needed in daily doses for the body to work; these include vitamins C and D

waterfowl bird that swims in water and spends a large amount of time on the water

waterless shampoo a cleansing product applied to the hair coat that does not require rinsing with water

wattle a flesh-like area located under the chin

wave schedule when a veterinarian sees a certain number of scheduled patients within a set amount of time during the day

weaning a process in which a young animal stops nursing and begins to eat solid foods

Weaned accustom to managing without something on which they have become dependent or of which they have become excessively fond.

weight tape a measurement tape used around the girth of an animal to estimate its body mass

Weitlaner retractor self-retracting instrument with rakelike teeth on the ratchet ends; can be blunt or sharp

welding gloves gloves with heavy layers of leather worn to restrain cats and other small animals

well water water that is sourced from a well

west Nile virus a mosquito-borne disease that causes inflammation or swelling of the brain and spinal cord

wet tail in hamsters, a condition caused by a bacterial disease that spreads rapidly through direct contact or bacterial spores

wether a male castrated sheep

wheelbarrow technique slightly lifting the animal's hind end off the ground and pushing the body forward in a "wheelbarrow" motion to assess the nervous system

wheezing long, high-pitched sound produced during asthma, which causes difficulty breathing

whipworm common intestinal parasite; occurs mostly in dogs; get their name from the whiplike shape of the adult parasite

white blood cells (WBCs) five types of blood cells that protect the body from infection and aid in immune system protection

white spot another name for ich

whole blood blood sample placed in a lavender-top tube to prevent clotting of the sample

wildlife any living animal that has not been domesticated

wildlife biologist a person who researches and studies data on all types of wild animals to determine how wildlife can be saved and protected

wildlife management the practice of researching the needs of wildlife, providing them the essentials of life, and monitoring their survival

wildlife rehabilitation center an area that houses and cares for sick or injured wild animals, often run by a veterinary staff

wildlife rehabilitator an individual who cares for sick, injured, or orphaned wild animals

wing trims the procedure of clipping the wings to prevent a bird from flying

winking movement of the vulva with frequent vaginal discharge and urination

withdrawal time a period of time before slaughter when pigs are given no additives

wolf teeth small teeth that may develop in front of the first molars

wool the soft fiber of sheep that acts as a coat

word-processing program software that allows typing of a document

words per minute the number of words that can be typed within 60 seconds

work diet specialized food given to animals that use a large amount of energy for work or strenuous activity

written communication writing, emailing, or texting to pass along information

X-ray common term for radiograph

X-ray log records the patient's name, client's name, date, X-ray number, X-ray position, thickness of body measurement, and area exposed

X-ray tube part of the machine that holds the radiation source

Y connector plastic device that connects the intake and outtake hoses from an anesthesia machine to the endotracheal tube of the patient

yearling a fish that is 1 year old

yolk the inner yellow cell membrane created by a hen's ovary

zoo a zoological "garden" that houses wild and exotic animals as well as plants for people to visit and observe their behaviors and beauty

zookeeper a person who manages and cares for animals in a zoo

zoology the study of animal life

zoonosis refers to disease passed from animals to humans

zoonotic a disease that is transmitted from animals to humans

zoonotic hazards safety concerns that allow contagious organisms to be spread to humans, causing infections, viruses, bacterial, fungal, and parasitic transmission

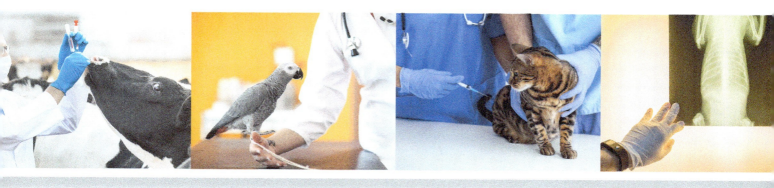

Index

A

AAFCO (Association of American Feed Control Officials), 570–571
AAHA (American Animal Hospital Association), 54, 63
AALAS (American Association for Laboratory Animal Science), 300, 307
AAVSB (American Association of Veterinary State Boards), 82
Abbreviations. See also Veterinary terminology
  common, 8
Abdominal system, 678, 680
Abduction, 440, 441, 450
Abductor muscle, 269, 276, 449
Abnormal tissues/cells, 429
Abomasum, 463–464, 472
Abortion, 323, 328, 567, 572
Abscess, 387, 390
Absorbable sutures, 854, 870
Absorption, 824, 831
Acceptance, 74, 75
Account history, 23, 30
Acetabulum, 444, 450
Acetaminophen, 423
Acidity, 264, 276
Acid rain, 264, 276
ACTH, 503, 507
Actin, 433, 434
Actinobacillosis, 388, 390
Actinomycosis, 388, 390
Active immunity, 536, 541
Active transport, 425, 434
Acute, 192, 193, 515, 516
ADA (Americans with Disabilities Act), 83, 85
Adaptive immune system vs. innate immune system, 536
Addison's disease, 506–507
Additives, 374
Adduction, 440, 441, 450
Adductor muscle, 449
ADH (antidiuretic hormone), 501–502, 507
Adipose, 432, 434
Ad lib, 159, 171, 199, 208
Adrenal glands, 503, 507
Adrenaline, 503, 507
Adson Brown tissue thumb forceps, 866, 870
Aeration, 263, 276
Age, determining, 692–694
Agglutination, 754, 780
Aggression, 602–603, 654
  dogs, 603
Aggressive behavior, 124, 148
  defensive, 159, 171

  intermale, 159, 172
  redirected, 159–160, 172
Aging, 264, 276
Agonal, 722, 741
Agonistic muscles, 449, 450
Agouti, 208, 211
AI. See Artificial insemination
Air cell, 414, 416
AKC (American Kennel Club), 119, 121
Albumen, 520, 531
Albumin, 414, 416, 477, 486
Aleutian disease, 216
Alevin, 274, 276
Alfalfa, 316, 328
Algae, 247, 254
Alimentary canal, 455, 472. See also Digestive system
Alkalinity, 264, 276
Allergens, 538, 541
Allergies, 538
Alligators, 285–286, 297
Allis tissue forceps, 865, 866, 870
Alopecia, 145, 148, 505, 507, 589, 676, 680
Alpacas, 283–284, 297
Alphabetical filing system, 26, 30
Alveoli, 326, 328, 491–492, 497
AMA (American Medical Association), 300, 307
Amazons, 181, 193
American Animal Hospital Association (AAHA), 54, 63
American Association for Laboratory Animal Science (AALAS), 300, 307
American Association of Veterinary State Boards (AAVSB), 82
American buffalo. See Bison
American class, 403, 416
American Kennel Club (AKC), 119, 121
American Medical Association (AMA), 300, 307
American paint horse, 338
American quarter horse, 338
American shetland pony, 337, 339
Americans with Disabilities Act (ADA), 83, 85
American Veterinary Medical Association (AVMA), 54, 63
Amino acids, 500, 504, 507, 564, 572
Aminoglycosides, 829
Amphibian, 242, 254
Amputation, 448
AMS (aquarium maintenance schedule), 265, 276
Amylase, 456, 461, 472
Anabolism, 425, 434
Analgesic, 829, 831
Anal glands, 148, 679, 741
  dogs, 139–140
  expression, 139–140, 737–740
  location, 738
Anaphase, 427–428, 434

Anaplasmosis, beef and dairy cattle, 325, 328
Anastomosis, 461, 471
Anatomy, 119, 148, 422, 434. See also General anatomy and disease processes
  avian, 175, 176
    digestive tract, 177
    feather, 176
    skeletal system, 177
  brain, 545
  cat, 155
  cranial nerves, 546
  crustacean, 268–269
  dog, 120
  eggs, 413–414
  electrocardiogram, 483, 484, 487
  endocrine glands in a horse, 501
  equines, 336
  external anatomy of the ear, 820
  external and internal respiration, 492
  female reproductive system, 519–520
  fish, 258
  goats, 394
  heart and blood flow, 481
  horse, 501
  layers of the heart, 481
  long bone, 439
  mollusk, 270
  monogastric digestive system, 460
  monogastric stomach, 460
  nonruminant digestive system, 464
  ornamental fish, 258
  parts of sperm, 522
  poultry, 402–403
  rabbits, 226–227
  respiratory structures of the thoracic cavity, 492
  respiratory system, 490
  ruminant stomach, 462
  sheep, 382–383
  spinal cord, 546
  structures of the ear, 557
  structures of the eye, 553–555
  structures of the renal system, 511–513
  swine, 366
  teeth, 457
Ancylostoma, 580
Ancylostoma caninum, 581, 595
Anemia, 147, 148, 478, 486, 586, 595
Anesthesia, 491
  chamber, 859, 870
  classification assessment, 839
  induction, 840
  inhalation, 840
  log book, 835–836, 870
  machine, 857–859
  mask, 859, 870

921

Anesthesia (*Continued*)
  planes, 840, 857
  post-anesthetic care, 860
  pre-anesthetic patient care, 838–839
Anesthesia chamber, 163, 171, 629, 631, 654
Anesthetic hazards, 101
Anesthetics, 829, 831
Anestrus, 522, 524, 525, 531
Angelfish, 261
Anger, 74, 75
Angora goats, 395, 399
Animal abuse laws, 84, 85
Animal behavior, 600
  animal behaviorists, 306, 307
  beef and dairy cattle, 316–317
  cats, 159–160
  chinchillas, 221
  common behavior problems
    aggression, 602-603
    barking, 126
    biting, 127
    chewing, 350
    digging, 126-127
    inappropriate elimination (house soiling), 126
  draft animals, 344, 345
  equines, 344–345
  ferrets, 214
  gerbils, 212
  goat, 396
  guinea pigs, 208
  hamsters, 204–205
  hedgehogs, 218–219
  horses, 344-345
  instinctive behaviors, 600-601
  introduction, 600
  learned, 601-602
    conditioning, 601
    imprinting, 601-602
    modeling, 601
  mice and rats, 199objectives, 600
  poultry, 407
  rabbits, 230
  rats and mice, 199
  reptiles and amphibians, 247–248
  sheep, 384–385
  swine, 369–370
Animal behaviorists, 306, 307
Animal models, 302, 307
Animal nutrition, 562–576
  beef and dairy cattle, 315–316
  body condition scoring, 568, 569, 572
  clinical situation, 575
  competency skills, 575–576
  concentrations and, 565–566
  equines, 343–344
  feeding animals, 567–568
    concentrates, 568, 572
    forages, 567
    pastures, 567
    supplements, 565, 568, 574
  feeding schedules, 571–572
  food analysis, 570, 571, 573
  goat, 396
  ideal weight, 568, 573
  introduction, 562
  needs of animals, 562
  nutrients, 562, 563–565
    carbohydrates, 563–564
    fats, 564, 573
    minerals, 565, 573
    proteins, 564–565
    vitamins, 565, 574
    water, 563
  objectives, 562
  pet food label, 571

  sheep, 383–384
  swine, 368–369
  types of diets, 566–567
    growth diets, 566
    lactation diets, 567
    maintenance diets, 566
    reduced calorie diets, 567
    reproductive diets, 566–567
    senior diets, 567
    work diets, 567
Animal refuge, 288, 297
Animal rights, 83–84, 85
Animal safety, 604
Animal terms, 10–11
Animal tissue, 302, 307
Animal Welfare Act (AWA), 83, 85, 301, 304–305, 307
Anisicoria, 671, 680
Anode, 784, 805
Anogenital, 201, 202, 223
Anogenital distance, 692, 741
Anorexia, 143, 148
Antacids, 825, 831
Antagonistic muscles, 449, 450
Anterior chamber, 553, 554, 559
Anterior posterior (AP), 788, 805
Anthelmintics, 147, 148, 325, 328, 579, 591–592, 594, 595, 830
Antiarrhythmic drugs, 826, 831
Antibacterial agents, 578
Antibacterial solution, 106
Antibiotics, 206, 223, 594, 595
Antibody, 142, 148, 423, 434, 535, 541
Anticoagulant, 756, 780
Anticonvulsants, 828, 831
Antidiarrheals, 825
Antidiuretic hormone (ADH), 501–502, 507
Antiemetics, 825, 831
Antifreeze toxicity, 515, 516
Antifungal agents, 578
Antifungals, 829, 831
Antigen, 536, 541
Antihistamine, 538, 541
Anti-inflammatory drugs, 825–826, 831
Anti-kick bars, 610, 611
Antimicrobials, 828, 831
Antioxidants, 565, 572
Antiparasitics, 830
Antiseptics, 104, 114
Antitussive drugs, 827, 831
Antiulcer, 825, 831
Anuria, 515, 516
Aorta, 482, 486
AP (anterior posterior), 788, 805
Apes, used in research, 304, 305
Appaloosa, 337, 338, 340
Apnea, 830, 831
Appendicular skeleton, 441, 443–444, 445, 450
Applied research, 302, 307
Appointment cards, 40, 47
Aquacrops, 267, 276
Aquaculture, 267, 276
Aquaculture, identification and production management, 267–275
  anatomy, 268–269
    crustacean, 268–269
    fish, 258
    mollusk, 270
  aquaculture industry, 267–268
  aquaculture production systems, 269–271
    aquatic examples, 270
  aquatic reproduction and breeding, 272–274
    broodfish and spawning, 272–273
    hatching and raising fry, 274
    producing fingerlings, 274
    readying fish for market, 274

  biology, 268–269
  breed selection, 274–275
  breeds or species, 274
  clinical situation, 281
  introduction, 256
  nutrition, 275
  objectives, 256
  veterinary terminology, 268
  water quality, 271–272
  water sources, 271
Aquarium, 262, 270–271, 276
  gerbil, 212
  hamsters, 205
  hedgehogs, 219
  reptiles and amphibians, 248–249
Aquarium maintenance schedule (AMS), 265, 276
Aquatic, 238, 254, 256, 276
Aqueous humor, 553, 554, 559
Arboreal, 248, 254
Armadillo, 301
Arrhythmias, 484–486, 675
Arterioles, 480, 486
Artery, 480, 486
Arthritis, 445–446, 450
Arthropods, 268, 276
Artificial fertilization, 272–273
Artificial incubation, 191, 193
Artificial insemination (AI), 356, 361, 528, 529, 531
  procedures, 528-529
Artificial vagina (AV), 528, 531
Ascarids (roundworms), 147, 148, 579–581, 595, 597
  host, 581
  poultry, 412
Ascending colon, 459, 472
Ascites, 267, 276
Asepsis, 112, 114, 835, 870
  surgical, 835
Aseptic, 305–306, 307
Aseptic techniques, 112–113, 114, 835, 870
Asiatic class, 403, 404, 416
Aspiration, 701, 719, 741
Assessment, 25, 30
Associate degree, 55, 63
Association abbreviations, 10
Association of American Feed Control Officials (AAFCO), 570–571
Asthma, 494, 497
Asystole, 486
Ataxia, 515, 516
Atlas, 442, 450
Atopy, 538, 541
Atrial fibrillation, 484, 486
Atrophy, 678, 680
Auditory ossicle, 556, 559
Aural, 820, 831
Aural hematoma, 558, 559, 671, 680
Aural medications, 820–821
Auscultation, 484, 486, 675–677, 680
Autoclave, 103, 114, 862, 863, 870
Autoimmune disease, 478, 486
Autolysis, 425, 434
Automated inventory system, 47
Automatic feeder, 572
Automatic film processing, 795, 799, 800–801, 805
AV (artificial vagina), 528, 531
Avian influenza, 411–412, 416
Avian pox, 411, 416
Avians. *See also* Birds
  basic health care and maintenance, 189–190
  behavior, 183–186
    biting, 185–186
    factors affecting behavior in birds, 184
    feather picking, 186
    need for a routine and stable environment, 184

socialization and stimulation, 184–185
   vocalization, 185
breed identification and production management, 174–196
   biology, 174–178
   introduction, 174
   objectives, 174
   veterinary terminology, 174
breeds, 178–182
   Amazons, 181
   canaries, 180
   cockatiels, 179
   cockatoos, 181
   conures, 180
   finches, 179
   lovebirds, 179–180
   Macaws, 181
   parakeets, 179
   parrots, 180–181
breed selection, 182
characteristics of birds by breed, 182
clinical situation, 196
common diseases, 192–193
   fatty liver disease, 192–193
   psittacine beak and feather disease, 192
   psittacosis, 192
common parasites, 193
common surgical procedures, 193
digestive system, 464–465
equipment and housing needs, 186–187
fecal exams, 191
fruits and vegetables that are safe for birds, 183
grooming, 188–189
injuries and fractures, 192
nutrition, 182–183
popularity, 174
radiology, 191
reproduction and breeding, 191–192
respiratory emergencies, 192
restraint and handling, 187–188
vaccinations, 190
vocalization, 185
Avian system, 464–465, 472
AVMA (American Veterinary Medical Association), 54, 63
AV valve, 481, 486
AWA (Animal Welfare Act), 83, 85, 301, 304–305, 307
Axial skeleton, 441–442, 450
Axis, 442, 450
Axon, 433, 434, 543, 550
Azotemia, 515, 516

B

Babcock tissue forceps, 865, 870
Babesia, 587, 595
Bachelor of Science (BS), 54, 64
Back-grounding system, 326, 328
Back rubbers, 325, 328
Bacon-type hog, 367, 378
Bacteria, 96, 114, 578
Bacterial disease, 148, 578, 595
   dogs, 143
Bactericidal, 828, 831
Bacteriostatic, 828, 831
Bain system, 859, 871
Balanced ration, 562, 570, 572
Ball and socket, 440, 450
Balling gun, 817, 831
Band, 390, 391
Bandage, 712, 713, 742
   application, 714–717
   removal, 718
   selection of material, 714–715
Bandage scissors, 716, 742
Bandaging and bandage care, 713–718
Band cell, 478, 486
Banding, 322, 328
Bang's disease, 323, 328, 376, 378, 592–593
Barbels, 268, 276
Barbering, 199, 223
Barb fish, 259
Barbs, 268, 276
Bargaining, 74, 75
Barium enema, 790, 805
Barium study, 790, 805
Barium sulfate, 790, 805
Barking, 126
Barrow, 365, 378
Bartonella, 586, 595
Basic research, 301, 302, 307
Basophil, 478, 479, 486
Basslets, 261
Bathing, 726–729, 742
   dogs, 135–136
   tools, 727
BCS (body condition scoring), 316, 328, 568, 569, 572
Beak, 174, 193, 464, 472
Beak trim, 188, 193
Beard, 402, 416
Beefalo, 284, 297
Beef and dairy cattle, breed identification and production management
   basic health care and maintenance, 321–322
      castration, 322
      dehorning, 322
   basic training, 317
   behavior, 316–317
   biology, 311–312
   body parts
      of the beef cow, 311
      of the dairy cow, 312
   breeds, 312–313
      beef cattle, 312–313
      dairy cattle, 313
   breed selection, 313–315
      beef cattle, 313–314
      culling, 314–315
      dairy cattle, 314
   clinical situation, 332
   common diseases
      bloat, 324
      bovine viral diarrhea, 323
      brucellosis, 323
      campylobacter, 323–324
      grass tetany, 324
      infectious bovine rhinotracheitis, 323
      leptospirosis, 323
      mastitis, 324
      metabolic diseases, 324–325
      retained placenta, 324
   common parasites, 325
   common surgical procedures, 325
   equipment and housing needs
      feed storage needs, 318–319
      housing, 317–318
      milking needs, 318
      pasture, 318
      waste material systems, 319
   grooming, 321
   introduction, 310
   milk production by state, 313
   nutrition, 315–316
      beef cattle, 316
      feeding dairy calves, 315–316
      feeding dry cows, 316
      lactating dairy cows, 315
   objectives, 310
   quality grades of beef, 313-314
   reproduction and breeding
      beef cattle, 325–326
      dairy cattle, 326–327
   restraint and handling, 319–321
   vaccinations, 322–323
   veterinary terminology, 310–311
Beef cattle, 311, 312–314, 316, 328
   reproduction and breeding, 325–326
Behavior. *See* Animal behavior
Belgian, 339–340
Beltsville turkey, 404
Benign, 202, 223, 429, 434
Benign prostatic hyperplasia (BPN), 530, 531
Bilateral cryptorchid, 529, 531
Bile, 457, 472
Billy, 393, 399
Biohazards, 94, 95–96
Biological, 594, 595
Biological filtration, 263, 276
Biological hazards, 95–96, 114
Biological oxygenation, 263, 276
Biological value, 564, 572
Biology, 296, 297
   beef and dairy cattle, 311–312
   equines, 335–337
   goat, 393
   sheep, 382
   swine, 366
Birds. *See also* Avians
   reproductive tract, 521
   West Nile virus, 549, 551
Birds of prey, 293, 297
Bison, 284, 297
Bit, 351, 361
Biting
   avian, 185–186
   dogs, 127
Biting flies, 588
Bivalves, 269, 276
Blackleg, 325, 328
Blacksmith, 352, 361
Blink reflex, 840, 871
Bloat, 324, 328, 398, 399, 467–468, 472
   goats, 398
Blood, 477–479, 486
   restraint for collection, 623–628, 652
   terms used in blood work evaluation, 479
Blood chemistry analyzer, 752, 757, 765, 780
Blood chemistry procedures, 755–765
Blood chemistry samples and electrolytes, 764–765
Blood collection tubes, 755, 757
Blood feathers, 189, 193
Blood flow, 480–482
Blood glucose, 503, 507
Blood plasma, 756, 780
Blood pressure (BP), 482, 487, 669
   monitor, 843, 871
Blood serum, 756, 780
Blood smear, 758–762, 780
Blood titer, 538, 541
Bloody scours. *See* Swine dysentery
Blue tongue, 388, 391
B lymphocytes, 535, 541
Boar, 365, 378
Body, 457, 472
Body condition scoring (BCS), 316, 328, 568, 569, 572
Body language, 69, 70, 75, 602–603, 654
   dogs, 602
   horse, 344-345
Boer goat, 395
Bone fractures, 446–448
Bone marrow, 438, 450

Bone plate, 447, 450
Bones, 438–439, 450
Boosters, 141, 148, 536
Booster series, 541
Booster shots, 536
Bordetella (kennel cough), 141, 494–495, 497
Borrelia burgdorferi, 587
Botflies, 584–586
Bottom-dwelling fish, 257, 276
Bovine, 310, 328
   kidney, 512
Bovine respiratory disease complex (BRDC), 323, 328, 495, 498
Bovine viral diarrhea (BVD), 323, 328
Bowman's capsule, 513, 516
BP (blood pressure), 482, 487, 669
   monitor, 843, 871
BPM (beats per minute), 482, 666
BPN (benign prostatic hyperplasia), 530, 531
Brachycephalic, 132, 149, 162
Brackish water, 271, 276
Bradycardia, 484, 485, 487, 675, 680
Bradypnea, 830, 831
Brain, 544, 550
   structures, 544–545
Brain stem, 544, 545, 550
Branding, 322, 328
Bray, 341, 361
Breathing, 491–492
Breathing systems, 859–860
Breeding stock, 283, 297
Broiler, 401
Broiler egg production system, 414–415, 416
Broiler production, 414, 416
Bronchi, 491, 497
Bronchioles, 491, 497
Bronchodilators, 494, 497, 826, 831
Brood, 238, 254
Broodfish, 272–273, 276
Broodmare, 355, 361
Brown trout, 293
Browse, 396, 399
Brucellosis, 323, 328, 376, 378, 592–593, 595
   swine, 376
Brumation, 239, 254
Brushing, 723–726, 742
   dogs, 136–137
BS (Bachelor of Science), 54, 55, 64
Buccal, 696, 742
Buck, 393, 399
Bucking, 344, 361
Budding yeast, 774, 780
Budgerigars, 179, 194
Budgie, 179, 194
Buffy coat, 762, 780
Bulbourethral glands, 521, 531
Bull, 310, 328
Burial, 410, 416
Burrow, 204, 223
Butterfly fish, 261
Butting, 385, 391
BVD (bovine viral diarrhea), 323, 328
By-products, 283, 297
By-products tissue, 301, 307

C

Cage card, 17, 22, 30, 685, 687
Cages, 271, 276
   aquaculture, 271
   bird, 186, 190
   chinchillas, 221
   cleaning, 107–109
   for drying, 727
   ferrets, 214
   gerbils, 212
   guinea pigs, 208–209
   hamsters, 205
   hospitalization, 685–686
   in lab animal management and health practices, 306
   mice and rats, 199–200
   poultry, 407–408
   rabbits, 230–231
   removing a cat, 634
Calcitonin, 502–503, 507
Calculus, 671, 673, 674, 680, 699
Calf, 310, 328
Calicivirus, 168, 171
California Mastitis Test (CMT), 324, 328
Calipers, 791, 792, 805
   measurement procedure, 792
   readings, 792
Callus, 447, 450
Calories, 564, 572
Calving, 311, 328
Campylobacter, 323–324, 328
Canaries, 180, 194
Cancellous bone, 438, 450
Cancer, 426
Candling, 414, 416
Canine, 455, 472. *See also* Dogs
Canine distemper, 143, 149
   ferrets, 216
Canis familiaris, 119, 149
Canister, 858, 871
Cannibalism, 206, 223
Cannon, 444, 450
Canter, 354, 361
Canthus, 554, 559
CAPC (Companion Animal Parasite Council), 591
Capillaria aerophilia, 580
Capillaries, 426, 480, 487
Capillary action, 762, 780
Capillary refill time (CRT), 669, 681
Capon, 401, 416
Caprine, 393, 399
Captivity, 176, 194
Carapace, 268, 276
Carbohydrates, 423, 434, 563–564, 572
Carcass, 410, 416
Cardia, 457, 472
Cardiac arrest, 484, 487
Cardiac cycle, 487
Cardiac muscle, 433
Cardinal, 294
Cardiopulmonary resuscitation (CPR), 486, 487
Careers
   in animal research, 306–307
   veterinary, 54–67
Carmalt hemostats, 867
Carnassial tooth, 456, 472, 673, 681
Carnivore, 119, 149
Carotene, 414, 416
Carpal, 443
Carpus, 443, 450
Carrier, 168, 171, 376, 378
Carrying capacity, 294, 297
Cartilage, 432, 440, 450
Cartilage joint, 440, 450
Cashmere, 395, 399
Cash register, 45
Cassette, 786, 805
Castration, 322, 328, 353, 373, 529, 531
   beef and dairy cattle, 322, 325
   dogs, 148
   draft animals, 353
   horses, 353
   swine, 373

Casts, 447, 450, 774, 780
Catabolism, 425, 434
Cataracts, 505, 507, 556, 559, 561
Cat bag, 161, 162, 171, 629, 630–631, 654
Cat carriers, 231, 642
Cat Fanciers' Association (CFA), 157, 171
Catfish, 260
Catheter, 514, 516
   IV monitoring, 709
Catheterized, 770, 780
Cathode, 784
Cat muzzles, 161, 162, 171
Cats, 301, 423
   aggression, 159-160
   barium series, 471
   basic health care and maintenance, 165–166
   basic training, 160
      inappropriate elimination, 160
   behavior, 159–160
      angry cat, 159–160
      fearful, 160
      happy cat, 159
   breed identification and production management, 154–173
      biology, 155–156
      breeds, 157
      breed selection, 157–158
      introduction, 154
      objectives, 154
      veterinary terminology, 154–155
   carrying, 635
   cephalic venipuncture, 624–625
   clinical situations, 173, 510, 561
   common diseases, 167–169
      feline immunodeficiency virus, 168
      feline infectious peritonitis, 169
      feline leukemia virus, 168
      feling calcivirus, 168
      panleukopenia, 168
      rabies, 169
      rhinotracheitis, 168
   common parasites and prevention, 169–170
   common surgical procedures, 170–171
   determining sex, 692
   diabetes mellitus, 505–506
   directional terms, 7
   dorsal recumbency, 622
   equipment and housing needs, 160
   feline urologic syndrome, 514–515
   grooming, 164–165
   head restraint, 630
   hypocalcemia, 505, 508
   instinctive behavior, 601
   jugular venipuncture, 625–626
   kidney, 512
   kitty burrito, 160, 172
   kitty taco, 633
   lateral recumbency, 622–623
   muzzles, 605–606
   nictitans, 555
   nutrition, 158–159
      feeding adult cats, 159
      feeding kittens, 158
   polydactyl, 156
   removing from a cage or carrier, 634
   removing from cage with a towel, 629
   reproduction and breeding, 166–167
   restraint and handling, 160–164
      removal from a carrier, 164
   retractable claw, 156
   saphenous venipuncture, 627–628
   scruff technique, 631–632
   sensory system, 561
   skeletal structure, 156
   standing restraint, 618–619

Index

sternal recumbency, 621
stretch technique, 632
technique for carrying an aggressive cat, 635
thyroid and parathyroid glands, 503
ultrasound, 803
upper respiratory infection, 494
use of a kitty taco, 633
vaccinations, 166
Cattle, 433. *See also* Beef and dairy cattle, breed identification and production management
   Bang's disease, 592–593
   clinical situation, 499
   handling and behaviour, 316-317, 319-321
   pneumonia, 494, 497
   reproductive system, 524–525
   ringworm, 591
   shipping fever, 495
Caudal fin, 257, 277
"Cauliflower disease," 267, 277
Caustic chemicals, 322, 328
Cautery unit, 188, 194
CBC (complete blood count), 752, 757–758, 780
CE (continuing education), 55, 64, 82
Cecum, 335, 361, 459, 472
Cell, 422, 434
   functions, 425
   parts of, 423–425
Cell division, 426–428, 434
Cell membrane, 423–424, 434
Centimeter (CM), 791–792, 805
Central nervous system (CNS), 515, 516, 543, 544, 550
Central processing unit (CPU), 43, 47
Centrifugal fecal floatation, 769
Centrifuge, 477, 487, 749–751, 780
Centrifuge tube, 750, 780
Cephalic veins, 482, 623, 654
Cephalic venipuncture, 130, 149, 163, 623–625
Cerclage wire, 447, 450
Cere, 178, 194
Cerebellum, 544, 545, 550
Cerebrum, 544, 545, 550
Certified veterinary technician (CVT), 55, 64
Cervical, 442, 450
Cervix, 387, 391, 519, 531
CFA (Cat Fanciers' Association), 157, 171
Chain shank, 347, 361, 649–650
Chalaza, 414, 416
Chammy, 383
Chamois, 383, 391
Channel, 68, 75
Cheek pouches, 204, 223
Chelonians, 238, 254
Chemical, 500, 507
Chemical filtration, 264, 277
Chemical hazards, 95, 114
Chemical inventory form, 99
Chemical restraint, 128, 149
Chemicals and cleaners, 104–107
Chemical test strips, 771–772
Chemoreceptor, 547, 550
Chemotherapy, 95, 114, 539, 541
Chem strip, 771, 780
Cherry eye, 555
Chewing, 350
Chick, 401, 413, 416
Chicken, 403–404, 416
Chief complaint, 658, 681
Child labor laws, 83, 85
Chinchillas, 220–222
   basic health care and maintenance, 222
   basic training, 221
   behavior, 221
   biology, 220
   breeds, 220, 221

breed selection, 220
common diseases, 222
common parasites, 222
equipment and housing needs, 221
genital areas, female and male, 222
grooming, 221–222
nutrition, 221
reproduction and breeding, 222
restraint and handling, 221
vaccinations, 222
veterinary terminology, 220
Choana, 178, 194
Chromatin, 426, 427
Chromosome, 428
Chronic, 192, 194, 515, 516
Chronic obstructive pulmonary disease (COPD), 495, 497
Ciliary body, 553, 554, 559
Circulatory system, 477–489, 487
   blood, 477–479
   blood flow, 480–482
   clinical situation, 489
   common diseases and conditions, 484–486
      arrhythmias, 484–486
      shock, 486
   electrocardiography, 483–484
   heart, 480
   heart sounds, 482
   introduction, 477
   objectives, 477
Circus animals, 288
CL (corpus luteum), 504, 507, 524, 531
Clavicle, 443, 450
Cleaning, 101, 114. *See also* Veterinary sanitation and aseptic technique
Client, 16, 30
   difficult, 72–73
   educating on medication use, 817–824
   information form, 18
   Clinical proceduresbasic restraint and handling procedures, 600–656
   examination procedures, 658-680
   grooming procedures, 723–745
   hospital procedures, 684–745
   laboratory procedures, 746–783
   pharmacy procedures, 809–833
   physical examination and patient history, 657–683
   radiology and imaging procedures, 784–808
   surgical assisting procedures, 834–874
   veterinary safety and aseptic technique, 90–116
Clinical research, 302, 308
   Clinical situationsanimal nutrition, 575
   aquaculture, 281
   avians, 196
   beef and dairy cattle, 332
   cats, 173, 510, 561
   cattle, 499
   circulatory system, 489
   communication and client relations, 77
   computer and keyboarding skills, 51
   digestive system, 476
   dog, 153, 476, 518, 534, 552, 575
   draft animals, 364
   endocrine system, 510
   equines, 489
   ethics and legal issues, 87grooming, 745
   horses, 364
   hospital procedures, 744
   immune system, 542
   laboratory and animal research production management, 309
   laboratory procedures, 783
   learning to dissect a veterinary term, 14
   medical records, 32

microbiology and parasitology as disease processes, 598
musculoskeletal system, 454
nervous system, 552
ornamental fish, 280
pharmacy procedures, 833
physical examination, 683
pocket pet health and production management, 224
poultry, 419
rabbits, 237
radiology procedures, 808
renal system, 518
reproductive system, 534
reptiles and amphibians, 255
respiratory system, 499
restraint and handling procedures, 656
scheduling and appointments, 51
sensory system, 561
sheep, 542
structure of living things, 437
surgical assisting procedures, 874
swine, 380
veterinary assisting procedures, 745
veterinary careers, 66
veterinary safety and aseptic technique, 116
wildlife management and rehabilitation, 299
zoo and exotic animal production management, 299
Clipper, 724, 725
   blade care and maintenance, 731–732
   electric, 730
Clipper guard, 730, 731, 742
Clipping, 730–732, 742
Cloaca, 176, 194, 239, 254, 465, 472, 520, 531
Closed gloving, 852, 871
Clostridium, 360, 361
Clot, 139, 149, 477, 487
Cloudy eye, 267, 277
Cloven-shaped, 366, 378
Clutch, 238, 254
Clydesdale, 340
CM (centimeter), 791–792, 805
CMT (California Mastitis Test), 324, 328
CNS (central nervous system), 515, 516, 543, 544, 550
Coat conditioner, 135, 149
Cocci, 774, 780
Coccidia, 253, 254, 583
Coccidiosis, 143, 149, 583, 595
   poultry, 412, 416
Coccygeal, 443, 450
Cochin hen, 404
Cochlea, 557, 558, 559
Cock, 401
Cockatiels, 179, 194
Cockatoos, 181, 194
Cockerel, 401, 416
Coffin bone, 359, 361
Coggins test, 351, 361, 588, 595
Cold housing, 317–318, 328
Cold sterilization, 102, 114
Cold tray, 102, 114
Cold water, 272, 277
Colic, 358, 361, 467, 472
   causes, 467
Collagen, 432
Collateral ligaments, 444
Collimator, 786–787, 805
Colon, 459, 472
Color code, 26, 30
Colostrum, 142, 149, 315, 328, 527, 531, 535, 541
   beef and dairy cattle, 315–316
Colt, 334, 361
Columnar, 430, 434

Coma, 544, 550
Comb, 402–403, 416
Combing, 723–726, 742
Combining form, 3, 4, 12
Combining vowel, 3, 12
Commercial, 227, 228, 229, 236
Commercial business, 326, 328
Commercial serologic test kit, 753–755, 780
Comminuted fracture, 447, 450
Common laws, 83, 85
Communication and client relations, 68–80
 appropriate communication skills, 71
 clinical situation, 77
 communication process, 68
 communication with difficult clients, 72–73
 competency skills, 77–80
 grief and communication, 74
 interacting with people, 71–72
  listening, 72
  observation, 72
  speaking, 71–72
 introduction, 68
 nonverbal communication, 69–70
 objectives, 68
 telephone skills, 73
 verbal communication, 68–69
 written communication, 70
Communication process, 68, 75
Compact bone, 438, 450
Companion Animal Parasite Council (CAPC), 591
Companion animal practice, 58, 64
Companion fish, 257, 277
Competency skills tasks
 administering an intramuscular injection, 705
 administering an intranasal injection, 706
 administering a subcutaneous injection, 704
 administering aural medication, 820–821
 administering ophthalmic medication, 823–824
 administering oral medications, 818–819
 administering SQ fluids, 708
 administering topical medications, 822
 admitting the hospitalized patient, 687
 application of a halter, 643–644
 applying a commercial leash to a dog, 638–639
 applying a commercial muzzle on dogs and cats, 605–606
 applying a gauze muzzle, 606–607
 applying a leash muzzle, 608
 applying a tape muzzle, 608
 appointment scheduling, 52
 aseptic and sterile techniques, 112
 auscultating, 676–677
 automatic film processing, 800–801
 bacterial or fungal culture procedure, 777
 bandage application, 717
 bandage removal, 718
 bathing procedure, 729
 biohazard and sharps disposal, 94
 blood smear procedure, 759–760
 blood smear stain procedure, 761–762
 brushing or combing, 725–726
 cage or kennel cleaning, 108–109
 caliper measurement procedure, 792
 carrying a cat, 635
 catheter bandage replacement, 712
 catheter patency, 711
 cephalic venipuncture restraint for dogs and cats, 624–625
 cleaning a cassette, 797
 cleaning exam rooms and the surgical suite, 111
 cleaning surfaces and supplies, 109
 cleaning surgical instruments, 111
 client communication skills, 78
 collecting a voided urine sample, 771
 competency skills, 53
 computer skills, 53
 cross tying a horse, 646–647
 dental exam, 674
 determining heart rate, 667
 determining pulse rate, 667
 determining respiratory rate, 668
 dilution of a substance or disinfectant, 106
 discharging a hospitalized patient, 688
 dorsal recumbency, 621–622
 dorsal-ventral (DV) recumbency, 639
 ear cleaning procedure, 734
 external anal gland expression procedure, 739–740
 fecal analysis procedure, 767–769
 feeding a hospitalized animal, 576, 720
 filling a syringe, 703
 food and water for the presurgical or preanesthesia patient, 575
 force feeding hospitalized patients, 721
 glove removal, 92–93
 gown pack preparation, 852–853
 gram stain procedure, 776
 grief counseling, 80
 hand hygiene, 93
 handling a fractious or scared dog, 641
 handling difficult clients, 79
 hand-washing, 107
 head restraint of a cat, 630
 how to tie a half hitch, 617
 how to tie a reefer's knot, 616
 how to tie a slip knot or quick-release knot, 615
 how to tie a square knot, 614
 infusion pump monitoring, 711
 interpersonal communication skills, 77
 jogging a horse on a lead, 645
 jugular venipuncture restraint for dogs and cats, 625–626
 laundry, 110
 laundry procedure, 740–741
 leading a cow or a horse, 644–645
 learning veterinary medical terminology and abbreviations, 15
 lifting a rabbit, 641–642
 lifting heavy objects, 100
 loading film to a cassette, 797
 longing a horse, 647
 manual film processing, 798–799
 medical records
  assembly, 33
  filing, 33–34
 mixing vaccines, 700–701
 monitoring fluid flow rate, 712
 nail trimming procedure, 737–738
 necropsy setup, 777–778
 one-half or one-fourth drape preparation, 851
 opening a pack, 853
 OSHA veterinary hospital plan, 100
 pack cover preparation, 849
 PCV procedure, 763
 performing a dental exam, 674
 personal protective equipment, 92
 pill counting tray procedure, 815
 placement of hobbles on a horse, 612
 preparation for euthanasia, 722
 preparing an IV fluid bag and IV drip set, 707
 professional appearance and dress, 66–67
 professional veterinary ethics, 88
 professional veterinary laws, 88–89
 reading a prescription, 811
 removing a bandage, 718
 removing a cat from a cage or carrier, 634
 removing a dog from a cage or kennel, 636–637
 restraint for intubation, 842
 restraint of a foal, 653
 restraint of a large animal for blood collection or an injection, 652
 returning a dog to a cage or kennel, 640–641
 saphenous venipuncture restraint in dogs and cats, 627–628
 scruff technique, 631–632
 sitting restraint, 619–620
 standing restraint of dogs and cats, 618–619
 sternal recumbency of dogs and cats, 620–621
 "stretch" technique, 632
 surgical pack preparation, 849–850
 surgical preparation, 846
 suture or staple removal, 861–862
 tail jack restraint, 651
 tail switch restraint, 652
 tail tie technique, 648
 taking rectal temperature, 665
 teeth brushing, 698
 telephone communication skills, 78–79
 total protein (TP) procedure, 764
 twitch restraint, 649
 tying a cow, 646
 tying a horse, 646
 unloading film from a cassette, 796
 urine sample procedure, 775–776
 use of a cat bag, 630–631
 use of a chain shank, 649–650
 use of a microscope, 748–749
 use of an anesthesia chamber, 631
 use of an anti-kick bar on a cow, 611
 use of an autoclave, 863
 use of an ophthalmoscope, 671
 use of a squeeze cage, 610
 use of a stanchion or head gate, 613
 use of a towel to restrain a cat, 609
 use of the "kitty taco," 633
 using an ophthalmoscope, 671
 using the otoscope, 672
 use of welding gloves, 611
 ventral-dorsal (VD) recumbency, 640
 walking a dog in the heel position, 637–638
 watering a hospitalized animal, 576
 weighing an animal using a scale, 662
 weighing an animal using a weight tape (livestock), 663
 wound cleaning, 714
Complete blood count (CBC), 752, 757–758, 780
Composting, 410, 416
Compound fracture, 447, 450
Compression fracture, 447, 450
Computed tomography (CT), 804, 805
Computer and keyboarding skills
 basic computer equipment, 43–45
 clinical situation, 51
 competency skill, 53
 components of a computer system, 44
 introduction, 35
 objectives, 35
Computer inventory system, 46, 47
Computerized appointment book, 36, 47
Computerized Veterinary Programs, 30
Computer systems, 303, 308
Concave, 252, 254
Concentrates, 326, 328, 568, 572
Concentration, 565–566, 573
Conception, 522, 531
Conditioning, 345, 361, 601
Cones, 553, 559
Confidentiality, 28, 29, 30, 81, 85
Confinement method, 390, 391
Conformation, 314, 328
Conical, 750, 780
Conjunctiva, 554, 559
Conjunctivitis, 554, 559
Connective tissue, 430, 432, 434, 440–441
 types, 430
Consent forms, 17, 21, 26–27, 31, 685

Constipated, 232, 236, 358, 361
Constipation, 467, 472
Constrict, 553, 559, 670, 681
Constriction, 432
Consumable products, 47
Contagious, 145, 149
Continental Kennel Club, 122
Continuing education (CE), 55, 64, 82
Contract, 377
Contraction phase, 387, 391, 527, 531
Contrast medium, 790, 806
Controlled substance license, 816, 831
Controlled substance log, 84, 85, 816, 831
Controlled substances, 84, 85, 816–817, 831
Conures, 180, 194
Cooled semen, 356, 361, 528, 531
Cool water, 271–272, 277
COPD (chronic obstructive pulmonary disease), 495, 497
Copier machine, 45, 47
Coplin jar, 754, 755, 780
Coprophagic, 229
Core body temperatures, 665
Core vaccine, 141, 149
Cornea, 554, 559
Corpus luteum (CL), 504, 507, 524, 531
Corrugated tubing, 859, 871
Cortex, 511, 516
Cortisone, 527
Costal cartilage, 442, 450
Cotton-wool disease, 267, 277
Courtesy, 71, 73, 75
Coverslips, 748, 780
Cow–calf system, 325–326, 329
Cow kicking, 319, 329
Coworkers, 72
Cows, 310, 328. See also Beef and dairy cattle, breed identification and production management
   bloat, 467–468
   flight zone, 321
   halter application, 643
   leading, 644–645
   preparation for surgery, 844
   prolapsed uterus, 530
   tail jack restraint, 651
   tail switch restraint, 652
   tying, 646
   use of an anti-kick bar, 611
   use of a stanchion or head gate, 613
CPR (cardiopulmonary resuscitation), 486, 487
CPU (central processing unit), 43, 47
Crackles, 494, 497, 675, 681
Cradling, 653
Cranial drawer sign, 446, 450
Cranial nerves, 546
Cranial nerve test, 547, 550
Crash cart, 690, 742
Creatinine, 511, 516
Credit application, 23, 31
Creep feeders, 316, 329
Crepitation, 448
Crepitis, 448
Crest, 239, 241, 254
Cribbing, 350, 361
Crile hemostatic forceps, 867, 871
Crocodile, 290
Crop, 176, 194, 464, 472
Crossbreeding, 284, 297
Cross ties, 346, 361, 647
Crown, 455, 472
CRT (capillary refill time), 669, 681
Cruciate ligaments, 444
Cruciate ligament tears, 446, 450
Crumbles, 275, 277

Crustaceans, 267, 277
Cryosurgery, 539, 541
Crystals, 514, 516, 774, 780
C/S (culture and sensitivity), 770, 774, 780
CT (computed tomography), 804, 805
CT scan, 804
Cuboidal, 430, 434
Cud, 461, 472
Cuff, 841, 843–844, 871
Culled, 314, 329
Culling, 314–315, 329
Culture and sensitivity (C/S), 770, 774, 780
Cursor, 43, 47
Cushing's disease, 506, 507
Cusp, 351, 361
Cutability, 311, 329
Cutaneous larval migrans, 147, 149, 581, 595
Cuterebra, 588
CVT (certified veterinary technician), 55, 64
Cyanosis, 494, 497
Cyanotic, 722, 742
Cylinder, 858, 871
Cystitis, 513, 516
Cystocentesis, 770, 772, 780
Cystotomy, 514, 516
Cytokinesis, 427, 428
Cytology, 422, 434, 777, 780
Cytoplasm, 424, 434

D

Daily energy requirements (DER), 570
Dairy, 326, 329
Dairy calves, feeding, 315–316
Dairy cattle, 311, 313, 314
   reproduction and breeding, 326–327
Dairy goats, 395, 399
Dairy Herd Improvement (DHI) program, 314, 329
Dam, 314, 329, 334, 361
Darkroom, care and maintenance, 801–802
Daughter cells, 427, 434
DEA (Drug Enforcement Agency), 84, 85, 816, 831
Debakey thumb tissue forceps, 867
De-beak, 402, 416
Decant, 772, 780
Decapod, 268, 277
Deciduous, 455, 472
Declaw, 170, 171
Decongestants, 826, 831
Deer, 285, 297
   white-tailed, 285, 290
Defensive aggression, 159, 171
Defensive reaction, 205, 223
Definitive host, 579
Degenerative joint disease, 444
Dehorning, 322, 329
Dehydration, 465–467, 472, 563, 573
   percentages, 467
Demodectic mange, 145, 149, 589, 595
Dendrites, 543, 550
Denial, 74, 75
Dental elevator, 869
Dental examination, 671–674
Dental extractor, 869
Dental formula, 456, 458, 472
Dental instruments, 869–870
Dental machine, 870
Dental mirror, 869
Dental prophylaxis, 697, 698, 699, 742
Dental scaler, 870
Dentifrices, 695, 742
Dentin, 455, 472
Dentition, 455–456, 472, 692–694, 742

   comparison between species, 694
Deoxyribonucleic acid (DNA), 434
Depression, 74, 75
DER (daily energy requirements), 570
Dermatitis, 505, 507
Dermatophyte, 578
Dermatophytosis (ringworm), 143, 151, 590–591, 595
   heifer, 591
Descending colon, 459, 472
Designer breeds, 119, 149
Developer solution, 798, 806
Dewclaw, 148, 149, 444, 451
Dewlap, 241, 252, 254
Dewormers, 579
Deworming programs, 591–592
Dexamethasone, 506, 507
DHLPPC vaccination, 141
Diabetes, 461, 472, 505–506
Diabetes insipidus, 502, 506, 507
Diabetes mellitus, 505–506, 507
Diagnostic imaging, 784, 806
Diagnostic terms, 790
Diaphragm, 187, 194, 493, 497
Diaphragmatic hernia, 495, 497
Diaphragm mechanism, 747, 780
Diaphysis, 438
Diarrhea, 465, 472, 679
Diastole, 482, 487
Diestrus, 524, 531
Diffusion, 425, 434
Digestion, 455, 472, 566, 573
Digestive system, 455–476
   clinical situations, 476
   common digestive conditions, 465–471
      bloat, 467–468
      colic, 467
      constipation, 467
      dehydration, 465–467
      foreign body obstructions, 469–471, 473
   introduction, 455
   mouth, 456, 459
   objectives, 455
   skills competency, 468
   stomach systems, 456–461
      avian system, 464–465
      monogastric systems, 457–461
      nonruminant system, 464
      ruminant system, 461–464
   teeth, 455–456
Digging, 126-127
Digital radiology, 799–800
Digital rectal thermometer, 227
Dilate, 387, 391, 553, 559, 670, 681
Dilated, 159, 171
Dilation phase, 387, 391
Diluent, 699, 742
Dilution, 104, 114
Dimorphism, 253, 254
Dipping, 726, 742
Dip stick, 771, 780
Dipylidium caninum, 580, 584, 585, 595
   life cycle of, 585
Direct contact, 96, 114, 577, 595
Directional terms, 7
   common, 7
Direct pupil light response (PLR), 671, 681
Dirofilaria immitis, 581–582, 595
Disaccharide, 564, 573
Disease, 434, 578, 595. See also Microbiology and parasitology as disease processes
   examples and sources, 96
   treatment, 594
Disinfecting, 101, 102, 114, 713
Dislocation, 446, 451

Displaced abomasums, 324, 329, 468–469, 470, 473
Dissect, 2, 12
Dissolved oxygen (DO), 264–265, 277
Distemper vaccine, 216
Distended, 468, 473
Distilled water, 862, 871
Distribution, 824, 831
Diuretics, 826, 831
Diurnal, 211, 223
Diversity, 257, 277
DNA, 521
DNA testing, 191, 194, 322, 329
   beef and dairy cattle, 322
DO (dissolved oxygen), 264–265, 277
Dock, 385, 391
Doe, 393, 399
Dog Breed Info Center, 122
Dogs, 301
   aggression, 603
   basic health care and maintenance, 137–141
      ear cleaning, 137–138
      ear plucking, 139
      expression of anal glands, 139–140
      nail trimming, 139
      teeth brushing, 140
      vaccinations, 140–141
   basic training, 125–127
      barking, 126
      biting, 127
      digging, 126–127
      inappropriate elimination, 126
   behavior, 124–125
   breed identification and production management, 118–153
      biology, 119
      breeds, 119
      breed selection, 119–122
      introduction, 118
      objectives, 118
      veterinary terminology, 118–119
   breeds prone to hypothyroidism, 505
   castration, 529
   cephalic venipuncture, 624
   clinical situations, 153, 476, 518, 534, 552, 575
   common diseases, 143–144
      bacterial, 143
      fungal, 143
      protozoan, 143
      rickettsial, 144
      viral, 143
   common parasites and prevention, 144–147
      external parasites, 144–146
      internal parasites, 146–147
   common surgical procedures, 147–148
   Demodex mange in, 590
   dentition, 458
   digging, 126-127
   directional terms, 7
   dorsal recumbency, 621–622
   edema, 538
   emaciated, 569
   equipment and housing needs, 127
   feeding adults, 123–124
   gauze muzzle, 606–607
   gingivitis, 699
   grooming, 133–137, 149
      bathing, 135–136
      brushing, 136–137
      tools, 134
      trimming or shaving, 134–135
   groups and breeds, 121
   handling a fractious or scared dog, 641
   hip dysplasia, 444–445
   hyperadrenocorticism, 506–507
   instinctive behavior, 125
   intestinal obstruction, 470
   jugular venipuncture, 625–626
   lateral recumbency, 622–623
   leash application, 638–639
   leash muzzle, 608
   life cycle of rodent tapeworm, 585
   muzzles, 605–606
   nutrition, 122–123
      feeding puppies, 122–123
      feeding requirements, 122
   patellar reflex, 547
   pneumothorax in, 496
   rabies or snare pole, 636
   red blood cells, 479
   removing from a cage or kennel, 636–637
   reproduction and breeding, 142–143
   reproductive system, 534
   reproductive tract, 520
   respiratory rate, 493, 667
   respiratory system, 490
   restraint and handling, 127–133, 603, 604, 656
      Elizabethan collar use, 133
      lifting techniques, 130–131
      muzzle use and application, 131–132, 133
      physical restraint, 128–130
      rabies or snare pole use, 132–133
      sedation use, 133
   returning to a cage or kennel, 640–641
   roundworms, 581
   saphenous venipuncture, 627–628
   seizures, 548–549
   sitting restraint, 619–620
   spay and castration, 148
   standing restraint, 618
   sternal recumbency, 620
   urine sample, 783
   walking in the heel position, 637–638
Domestication, 119, 149
Domestics, 157, 171
Dominance aggression, 125, 149, 602, 654
Donkey, 334, 341, 361
Dorsal fin, 257, 277
Dorsal recumbency, 130, 149, 618, 621–622, 654, 856
Dorsal root, 545, 550
Dorsal-ventral (D-V) recumbency, 639, 654, 788, 789, 806
Dosimeter, 785, 806
Dosimetry badge, 101, 114
Double sugar, 564, 573
Down, 404, 416
Doyen intestinal forceps, 867–868, 871
Draft animals, 333-364
Draft capacity, 342, 361
Draft-hitch classes, 342, 361
Draft horses, 334, 339–341, 342, 361
Drake, 416
Dram, 813, 831
Dremel, 139, 149, 736
Drenching, 389, 391
Dressing, 365, 378
Dressing forceps, 865, 871
Drop-off appointment, 38, 47
Drop-off time, 38, 47
Dropsy, 267, 277
Drosophila (fruit fly), 301
Drug Enforcement Agency (DEA), 84, 85, 816, 831
Drugs. See Medications
Dry cow period, 316, 326, 329
Dry cows, feeding, 316
Dryers, 728
Dry heat, 102–103, 114
Dry matter, 565, 573
Dual purpose breed, 311, 312, 329

"Dub," 482
Duckling, 416
Ducks, 404, 416
Duodenum, 457, 473
Dust bath, 211, 212, 223
DV (dorsal-ventral recumbency), 639, 654, 788, 789, 806
Dwarf, 236
   rabbits, 227
Dynamic equilibrium, 558, 559
Dysentery
   lamb, 388
   sheep, 388
   swine, 376
Dyspnea, 668, 675, 681
Dystocia, 142, 149, 387, 391, 530, 531
Dysuria, 513, 516

E

Ear cleaning, 137–138, 732–734
Earmarks, 322, 329
Ear mites, 145, 149, 588–589, 595, 820, 832
Ear notching, 373, 378
   swine, 373
Ear plucking, 139, 149
Ears, 556–557, 559
   cleaning, 137–138, 732–734
   examination, 671, 672
   external anatomy, 820
   physical examination, 671, 672
   structure, 556–557
   terms, 9
Ear tag identification, 321–322, 329
"Earthquake" technique, 185, 194
Eastern Equine Encephalitis (EEE), 593
Ecdysis, 247, 254
ECG (electrocardiogram), 483, 484, 485, 487, 843, 844, 871
Ecology, 296, 297
Ectoparasites, 586–591
Ectothermic species, 239, 254
Ectothermic temperature, 663, 681
Edema, 425–426, 538, 541
EDTA (Ethylene Diamine Tetra Acetate), 755
EEE (Eastern Equine Encephalitis), 593
Efficacy, 824, 832
Egg bound, 192, 194
Egg depositers, 266, 277
Egg-laying fish, 266, 277
Egg production system, 413–414
Eggs
   anatomy, 413–414
   quality, 414, 416
Egg scatterers, 266, 277
Eggshell, 413–414, 416
Ehrlichiosis, 587, 595
EIA (equine infectious anemia), 588, 595
EKG (electrocardiogram), 483, 484, 485, 487, 843, 844, 871
Elastrator, 322, 329
Electrocardiogram (ECG; EKG), 483, 484, 485, 487, 843, 844, 871
Electrocardiography, 483–484, 487
Electrode, 871
Electrolytes, 343, 361, 764–765, 780
Elimination, 825, 832
ELISA, 541
   testing, 538
Elk, 284–285, 297
Emaciated, 412, 416
Emasculator, 322, 329
Embryo, 522, 531

Emergency, 690, 742
Emergency services, 38, 47, 689–690
   equipment, 690
   evaluating, 689–690
Emetics, 825, 832
Empathy, 71, 75
Emus, 406, 417
Enamel, 455, 473
Encephalitis, 593, 595
Encyst, 579
Endangered animals, 291, 297
Endocardium, 480, 487
Endocrine glands, 501–504
Endocrine system, 500–510
   clinical situation, 510
   diseases and conditions, 505–507
     diabetes, 505–506
     hyperadrenocorticism, 506–507
     hyperthyroidism, 505
     hypothyroidism, 505
   estrus, 504, 508
   hormones, 504
   introduction, 500
   objectives, 500
   structures, 500–504
     adrenal glands, 503
     endocrine glands, 501–504
     gonads, 504
     pancreas, 503
     parathyroid glands, 503
     pineal gland, 504
     pituitary gland, 501–502
     thymus, 504
     thyroid gland, 502–503
Endocytosis, 425, 435
Endoparasites, 579–584
Endoplasmic reticulum, 424–425, 435
Endoscope, 803, 806
Endoscopy, 803–804, 806
Endosteum, 439, 451
Endothermic species, 119, 149
Endothermic temperature, 663, 681
Endotracheal tube, 491, 497, 691, 840–842, 871
Enema, 358, 361, 467, 473
English class, 403, 404, 417
Enterotomy, 461
Enterotoxemia, 388, 391
Enzymes, 423, 435, 504, 508
Eosinophil, 478, 479, 487
Epaxial, 704, 742
Epaxial muscle, 449
Epicardium, 480, 487
Epididymis, 521, 531
Epiglottis, 491, 497
Epilepsy, 548–549, 550
Epinephrine, 503, 508
Epiphysis, 438
Epithelial tissue, 429–432, 435
   types of, 430–432
Equal Employment Opportunity (EEO), 83, 85
Equilibrium, 558, 559
Equine encephalomyelitis, 360, 361
Equine infectious anemia (EIA), 351, 359, 361, 588, 595
Equine influenza, 360
Equines, 334, 361
   anatomy, 336
   basic health care and maintenance, 351–354
     assessment of lameness, 353–354
     castration, 353
     dental care, 351–352
     health testing, 351
     hoof care, 352–353
     lameness assessment, 353–354
   basic training, 345

   behavior, 344–345
   breed identification and production management, 333–364
     biology, 335–337
     breed selection, 342
     introduction, 333
     objectives, 333
     veterinary terminology, 333–335
   biology, 335–337
   breeds, 337–341
     donkey and mule, 339–341
     draft horses, 339–341
     gaited horses, 339
     hunters and jumpers, 339
     ponies, 339
     racehorses, 339
     riding horses, 337–338
     stock horses, 339
   breed selection, 342
   chain shank use, 649–650
   clinical situations, 489
   common diseases
     colic, 358–359
     equine encephalomyelitis, 360
     equine infectious anemia, 359
     equine influenza, 360
     laminitis, 359
     tetanus, 360
   common parasites, 360
   cribbing, 350, 361
   cross tying, 646–647
   ear positions, 344
   endocrine glands, 500–501
   equipment and housing needs, 345–346
   fetus, 526
   grooming, 350
   grooves to determine age in teeth, 694
   halter application, 643–644
   handling and behaviour, 344-345, 346–349
   jogging, 645
   leading, 644–645
   longing, 647
   nonruminant digestive system, 464
   nutrition, 343–344
   placement of hobbles, 612
   preparation for surgery, 845
   reproduction and breeding, 355–358
   reproductive tract, 522
   restraint and handling, 346–349
   roaring, 496
   shipping fever, 495
   tail tie technique, 648
   twitch restraint, 649
   tying, 646
   vaccinations, 354–355
   veterinary terminology, 333–335
   West Nile virus, 549
Equipment, 93–94
   surgical suite, 837–838
Equus caballus, 335, 361
Eructation, 464, 473
Erythrocytes, 477, 487, 762, 781
Erythropoiesis, 477–478, 487
Esophageal stethoscope, 843, 871
Esophagus, 456, 473
Essential fatty acids, 564, 573
Estimate, 23, 31
Estrogen, 504, 508, 520, 532
Estrus, 522–524, 532
Estrus cycle, 149, 504, 508, 522–524, 532
   dogs, 142
   in various species, 526
Ethical, 81, 85
Ethics, 81, 85
Ethics and legal issues, 81–89

   clinical situation, 87
   competency skills, 88–89
   federal veterinary laws, 83–84
   introduction, 81
   laws and veterinary practice acts, 82–83
   objectives, 81
   veterinary common law, 83
Ethology, 600
Ethylene Diamine Tetra Acetate (EDTA), 755
Ethylene glycol, 515
Ethylene glycol toxicity, 516
Eustachian tube, 556, 558, 559
Euthanasia, 720–723, 742
Euthanasia release forms, 27, 31
Euthanasis, 74, 75
Evert, 252, 254
Ewe, 381, 391
Examination procedures
   auscultation, 675–677
   dental examination, 671–674
   ears, 671, 672
   eye exam, 670–671
   vital signs
     blood pressure, 669
     capillary refill time, 669
     mucous membranes, 669
     pulse, 666
     respiratory rate, 666–667
     temperature assessment, 663–665
     weight assessment, 660–663
Exam room, 60, 64
   sanitation, 110–112
Excrete, 500, 508
Exercise, 718–719
Exhalation, 493, 497
Exhibition (zoo) and performance animals, 286–288
Exoskeleton, 268, 277
Exotic animal dealer, 287, 297
Exotic animals, 282, 283, 297. *See also* Zoo, exotic and wildlife identification and production management
   operating expenses of, 289
   states with regulations on, 290
Expectorants, 826, 832
Expelled, 356, 361
Expiration, 187, 194
Exploratory laparotomy, 461
Expressing, 272, 277
Extension, 440, 451
Extensor muscle, 449
External auditory canal, 556, 559
Extinct animals, 291, 297
Extracellular fluid, 425
Extubation, 841, 871
Eyelids, 554–555
Eyepiece, 747, 781
Eyes, 553–555
   examination, 670–671
   ointment, 727
   physical examination, 670–671
   structure, 554
   terms, 9

F

Face, 442
FAD (Flea allergy dermatitis), 586, 596
Fair Labor Standards Act (FLSA), 83, 85
Family-size herd, 326, 329
Farm flock method, 389, 391
Farming demonstrations, 342, 361
Farrier, 352, 361
Farrowing, 365, 378

Farrow to finish systems, 378
Fascia, 449, 451
Fast, 838, 871
Fats, 564, 573
Fat soluble, 565, 573
Fatty acids, 504, 508, 564, 573
Fatty liver disease, 192–193, 194
Fax machine, 45, 48
FDA (Food and Drug Administration), 84, 86
Fear aggression, 602–603, 654
Feathered edge, 758, 781
Feather picking, 186
Feathers, 340, 362
Fecal direct smear, 766, 781
Fecal egg analysis, 147, 149
Fecal floatation, 766, 767, 781
Fecal sample, 765–769, 781
   solutions, 766
Fecal sediment, 769
Fecal smear, 766
Feed, 567, 573
Feedback, 68, 75
Feeder pig production systems, 377
Feeding schedules, 571–572
Feedlot size, 326, 329
Feline distemper, 168, 171
Feline immunodeficiency virus (FIV), 168, 171
Feline infectious anemia (hemobartonella), 586, 596
Feline infectious peritonitis (FIP), 169, 171
Feline leukemia virus (FeLV), 168, 171
Feline lower urinary tract disease (FLUTD), 514–515, 516
Feline urologic syndrome (FUS), 514, 516
Felis catus, 155, 171
FeLV (feline leukemia virus), 168, 171
Femoral veins, 482
Femur, 444, 451
Feral, 163, 171, 288, 297
Fermentation, 461, 473
Ferrets, 213–217, 301
   basic health care and maintenance, 215–216
   basic training, 214
   behavior, 214
   biology, 213–214
   breeds, 214
   breed selection, 214
   common diseases, 216, 217
   common parasites, 216
   equipment and housing needs, 214
   grooming, 215
   nutrition, 214
   reproduction and breeding, 216, 217
   restraint and handling, 214–215
   vaccinations, 216
   veterinary terminology, 213
Fertile, 410, 417
Fertility, 356, 362
Fertilization, 410, 417, 428, 521–522, 532
   animal species ovulation types, 524
   artificial, 272–273
Fertilize, 191, 194
Fervac-D, 216
Fetlock, 444
Fetus, 524, 532
Fever, 664, 681
Fiber, 564, 573
Fibrinogen, 477, 487
Fibrous joint, 440, 451
Fibula, 444, 451
Fight or flight instinct, 545, 550, 603, 643
Filly, 334, 362
Film hanger, 798, 806
Filtration, 103–104, 114, 263, 277
Filtration system, 249, 254
Fin, 268, 277

Finches, 179, 194
Finger brush, 140, 149, 695–696, 742
Fingerling, 257, 277
   producing, 274
Finishing system, 326, 329, 378
Fin rot, 266, 277
Fins, 257, 258, 268, 277
FIP (feline infectious peritonitis), 169, 171
Fire and safety plans, 91
Fish, 256, 277. *See also* Ornamental fish
   readying for market, 274
Fish fungus, 267, 277
FIV (feline immunodeficiency virus), 168, 171
Fixed office hours, 36, 48
Fixer solution, 798, 806
Fixing, 795, 806
Flagella, 521, 532
Flanking, 467, 473
Flaxen, 339, 362
Flea allergy dermatitis (FAD), 586, 596
Flea comb, 724
Fleas, 144, 149, 586
   product comparison chart, 145, 169
   topical medications, 821
Flea tapeworm, 585
Flexion, 440, 451
Flexor muscle, 449
Flight feathers, 189, 194
Float, 352, 362
Floating, 352, 362, 674, 681
Floating ribs, 442, 451
Flocculent, 771, 781
Flock, 401, 417
Floss, 263, 277
Flow meter, 858, 871
Flow rate, 709, 742
Flow schedule, 36, 37, 48
FLSA (Fair Labor Standards Act), 83, 85
Fluid administration, 706–712
Fluid therapy, 839–840
Flukes, 584
FLUTD (feline lower urinary tract disease), 514–515, 516
Foal, 334, 362
   cradling, 653
   restraint, 653
Foaling, 334, 362
Focus knob, 747, 781
Follicle, 355, 362, 519–520, 532
Follicle stimulating hormone (FSH), 504, 508, 524, 532
Food analysis, 570, 571, 573
Food and Drug Administration (FDA), 84, 86
Food fish production, 274, 277
Foodstuff, 567, 573
Foot rot, 387, 391
Forage, 343, 567, 573
Forager, 311, 329, 362
Forced molting, 409, 417
Force feeding, 719, 721, 742
Forceps, 139, 149, 865–868
   Adson Brown, 866, 870
   Allis tissue, 865, 866, 870
   Babcock, 865, 870
   Debakey thumb tissue, 867
   Doyen, 867–868, 871
   dressing, 865, 871
   hemostatic, 867, 871
   intestinal, 867–868, 871
   rat tooth thumb, 865, 866, 872
   sponge, 866, 867, 873
   tissue, 865–867, 873
Foreign body obstruction, 469–471, 473
Forelimb, 443, 451
Formalin, 778, 781
Founder, 359, 362

Fountain, 262, 277
Fowl, 401, 417
Fowl cholera, 411, 417
Fracture, 446–447, 451
Free choice, 228, 236, 569, 571, 573
Free gas, 468, 473
Freemartin, 311, 329
Freshening, 311, 329
Freshwater, 271, 277
Freshwater fish, 258, 277
Frog, 353, 362
Frogs and toads
   anatomy, 243
   behavior, 248
   biology, 242, 243
   breeds, 245
   nutrition, 247
   reproduction and breeding, 252–253
   restraint and handling, 250
Frothy bloat, 468, 473
Frozen semen, 356, 362, 528, 532
Fruit fly (Drosophila), 301
Fry, 257, 277
   hatching and raising, 274
Fryers, 401, 417
FSH (follicle stimulating hormone), 504, 508, 524, 532
Full Arch rabbits, 227, 228, 236
Fundus, 457, 473
Fungal disease, 143, 149, 578–579, 596
Fungi, 578–579
Fungicidal, 828, 832
Fungus, 96, 114
FUS (feline urologic syndrome), 514, 516
FVRCP, 166, 172

G

Gag, 697, 698
Gait, 339, 362
Gaited horses, 339, 362
Gall bladder, 457, 464, 473
Gallop, 354
Game animals, 290–291, 297
Game birds, 293, 297
Game Commission, 292
Game fish, 292–293, 297
Gander, 417
Gas sterilization, 103
Gastric dilation, 467, 473
Gastric dilation volvulus (GDV), 467–468, 473
Gastrointestinal system (GI), 455, 473
Gastrotomy, 461
Gauge, 701–702, 742
Gauze, 131, 132, 149, 714, 715
Gauze muzzle, 606–607
GDV (gastric dilation volvulus), 467–468, 473
Geese, 404–405, 417
Gelding, 334, 362
Gelpi rectractor, 868, 871
Gender, determining, 690–692
General anatomy and disease processes. *See also* Anatomy
   animal nutrition, 562–576
   circulatory system, 477–489
   digestive system, 455–476
   endocrine system, 500–510
   immune system, 535–542
   microbiology and parasitology as disease processes, 577–598
   musculoskeletal system, 438–454
   nervous system, 543–552
   renal system, 511–518
   reproductive system, 519–534

respiratory system, 490–499
sensory system, 553–561
structure of living things, 422–437
General nursing care, 689
Genetic flaws, 313–314, 329
Geneticist, 428, 435
Genetics, 428–429
Genitalia, 519, 532, 692, 693
Genitals, 519, 532
Gerbils, 211–213
  basic training, 212
  behavior, 212
  biology, 211
  breeds, 211
  breed selection, 211
  common diseases, 213
  common parasites, 213
  equipment and housing needs, 212
  grooming, 212
  nutrition, 211
  reproduction and breeding, 213
  restraint and handling, 212
  vaccinations, 212
  veterinary terminology, 211
Germinal disc, 414, 417
Gestation cycle, 326, 329, 504, 508, 522, 524–526, 532
  length of various species, 526
GI (gastrointestinal system), 455, 473
Giant, 236
  rabbits, 227, 228
Giardia, 583
Giardiasis, 143, 149, 583
Gills, 257, 277
Gilt, 365, 378
Gingiva, 671–673, 681
Gingival hyperplasia, 674, 681
Gingivitis, 673, 681, 699
Gizzard, 176, 194, 402, 417, 464–465, 473
Glands
  adrenal, 503, 507
  endocrine, 501–504
  lacrimal, 554, 560
  parathyroid, 503, 509
  pineal, 504
  pituitary, 501–502, 509, 544, 545, 551
  thyroid, 502–503, 509
Glaucoma, 556, 560
Globe, 553, 560
Globules, 395, 399
Globulin, 477, 487
Glomerulus, 513, 516
Glottis, 491, 497
Glove pack, 852, 871
Gloves, removal, 92–93
Gloving techniques, 852
Glucagon, 500–501, 503, 508
Glucocorticoids, 826, 832
Glucose, 461, 473
Glycogen, 423
GnRH (gonadotropin), 504, 508
Goat pox, 398–399
Goats, breed identification and production management
  anatomy, 394
  basic health care and maintenance, 398
  basic training, 396–397
  behavior, 396
  biology, 393
  breeds, 395–396
    Angora, 395
    Cashmere, 395
    dairy goats, 395
    meat goats, 395
    Pygmy goats, 395–396
  breed selection, 396

  common diseases
    bloat, 398
    goat pox, 398–399
    hypocalcemia, 398
    joint ill, 398
  common parasites, 399
  equipment and housing needs, 397
  grooming, 397–398
  handling and behavior, 396, 397
  introduction, 393
  normal goat vital signs, 398
  nutrition, 396
  objectives, 393
  reproduction and breeding, 398
  restraint and handling, 397
  vaccinations, 398
  veterinary terminology, 393
Goats, 393, 399
Goldfish, 259
Golgi apparatus, 425, 435
Gonadotropin (GnRH), 504, 508, 525
Gonads, 504
Goose, 417
Gosling, 417
Gourami, 259
Gowning, 855–856
Gown pack, 852, 853, 871
Gram negative, 774, 781
Gram positive, 774, 781
Gram stain, 774, 776, 781
Grass tetany, 324, 329
Gravid, 238, 254
Green foods, 182, 194
Gregarious, 344, 362
Grid, 789, 806
Grid technique, 793, 806
Grief, 74, 75
Grit, 182, 194, 402, 417
Groomers, 57–58
Grooming, 133, 149, 685, 742. *See also* Preen
  bathing, 135–136, 726–729
  beef and dairy cattle, 321
  brushing and combing, 136–137, 723–726
  cats, 164–165
  chinchillas, 221–222
  clinical situation, 745
  clipper and clipper blade care and maintenance, 731–732
  clipping, trimming, and shaving, 730–732
  dogs, 133–137
  draft animals, 350–351
  ear cleaning, 732–734
  equines, 350
  expressing anal glands, 737–740
  ferrets, 215
  gerbils, 212
  goats, 397–398
  guinea pigs, 209–210
  hamsters, 205–206
  hedgehogs, 219
  horses, 350–351
  introduction, 723
  nail trimming, 734–738
  objectives, 723
  poultry, 409
  procedures, 723–745
  rabbits, 231–232
  rats and mice, 201
  reptiles and amphibians, 250–251
  sheep, 385
  swine, 372
  tools, 134
  trimming or shaving, 134–135
Gross examination, 765, 770–771, 781
Growling, 126

Growth diets, 566, 573
Growth plate, 440, 451
Gruel, 122, 150
Guilt, 74, 75
Guinea fowl, 405, 417
Guinea pigs, 207–211, 301, 302
  basic health care and maintenance, 210
  basic training, 208
  behavior, 208
  biology, 207–208
  breeds, 208
  breed selection, 208
  common diseases, 211
  common parasites, 211
  equipment and housing needs, 208–209
  grooming, 209–210
  nutrition, 208
  reproduction and breeding, 210–211
  restraint and handling, 209
  vaccinations, 210
  veterinary terminology, 207
Gums, 465
Guppy, 260

H

Habitat, 264, 277
Hairballs, 157, 172
Hair coat, 676, 677, 723, 724
Half hitch, 617, 654
Halstead mosquito hemostats, 867, 871
Halter, 319, 329, 346, 362, 643–644
Hame straps, 346, 362
Hamsters, 203–207, 301
  basic health care and maintenance, 206
  basic training, 205
  behavior, 204–205
  biology, 203
  breeds, 203
  breed selection, 203–204
  common diseases, 206–207
  common parasites, 207
  equipment and housing needs, 205
  grooming, 205
  nutrition, 204
  reproduction and breeding, 206
  restraint and handling, 205
  vaccinations, 206
Hamstring, 704, 742
Hands, 334, 362
  cleansing, 95
  hygiene, 93, 104–107
Hand tanks, 795, 806
Hardware, 43, 48
Hatchery, 272, 277
HBC (hit by car), 495
Head, 268, 278
  exam, 670, 681
Head gate, 612, 613
Head restraint, 630
Health certificate, 27, 29, 31
Hearing, 557–558
Heart, 487
  internal structure and blood flow, 481
  layers, 481
Heart murmur, 484, 487
Heart rate (HR), 480–481, 487, 545, 666, 667, 675, 681
  normal, 482, 666
Heart sounds, 482
Heartworm disease, 150, 581–582, 596
  dogs, 146
  life cycle, 582
  product comparison chart, 146, 170

Heartworms, 581–582
  life cycle, 582
Heat, 150
  dogs, 142
Heat stroke, 664, 681
Heaves, 495, 497
Hedgehogs, 217–220
  basic health care and maintenance, 219
  basic training, 219
  behavior, 218–219
  biology, 218
  breeds, 218
  breed selection, 218
  common diseases, 220
  common parasites, 220
  equipment and housing needs, 219
  grooming, 219
  nutrition, 218
  reproduction and breeding, 219–220
  restraint and handling, 219
  vaccinations, 219
  veterinary terminology, 217
Heifer, 310, 329, 494
  ringworm, 591
Hematocrit, 762, 763, 781
Hematocrit scale, 763
Hematology, 477, 487, 765, 781
Hematology sample collection, 755
Hematoma, 235, 236, 820, 832
Hematuria, 376, 378, 513, 516
Hemipenis, 251, 252, 254
Hemobartonella (feline infectious anemia), 586, 596
Hemoglobin, 478, 487
Hemolysis, 755, 781
Hemostatic forceps, 867, 871
Hemostats
  Crile, 867, 871
  Halstead mosquito, 867, 871
  Kelly, 867, 868, 872
  Rochester-Carmalt, 867, 873
Hemothorax, 496, 497
Hen, 417
Hepatic lipidosis, 192, 194
Hepatitis, 143, 150
Herbivores, 176, 194, 455, 473
Herd, 311, 329, 393, 399
Herders, 390, 391
Herd health manager, 321, 329
Herpes virus, 411, 417
Herpetology, 238, 254
Hibernation, 204, 223
Hierarchy, 344, 362
Hilus, 511, 516
Hinge, 440, 451
Hinny, 334, 362
Hip dysplasia, 444, 451
  dogs, 444–445
Histamine, 478, 487, 538, 541
Histology, 429, 435, 777, 781
Hit by car (HBC), 495
Hobbles, 347, 362, 610, 612
Hocks, 232, 236, 444, 451
Hog cholera, 376, 378
Hog snare, 372, 378
Holman retractor, 868, 871
Homeostasis, 425, 435, 500, 508
Hooded, 198, 223
Hoof pick, 353, 362
Hoof tester, 448
Hoof wall, 353, 362
Hookworm, 147, 150, 580, 581, 596
Hormones, 500, 504, 508, 522, 532
Horn bud, 322, 329
Horner's syndrome, 433
Horsepower, 342, 362

Horse pull, 342, 362
Horses. See Equines
Hospital administrator, 57, 64
Hospital procedures, 684–745
  administering injections, 701–706
    filling a syringe, 701–702
    intramuscular injections, 704–705
    intranasal injections, 706
    IV catheter monitoring, 709
    IV fluid monitoring, 709–710
    labeling a syringe, 702
    subcutaneous injections, 702–704
  bandaging and bandage care, 713–718
  clinical situations, 744, 745
  common, 690–694
  dental care, 694–699
    daily dental care and brushing, 695–698
    prophylaxis, 699
  determining animal gender, 690–692
  determining approximate animal age, 692–694
  euthanasia, 720–723
  evaluating emergency situations, 689–690
  feeding and watering hospitalized patients, 719–720, 721
  fluid administration, 706–708
    subcutaneous fluid therapy, 706–708
  general nursing care, 689
  hospitalized patient, 685–689
    admitting, 685–686, 687
    discharging, 686–688
    recording observations, 688
  identification information, 685
  introduction, 684–685
  laundering materials, 740–741
  objectives, 684
  pain assessment, 689
  socialization and exercise of patients, 718–720
  vaccine preparation, 699–701
  wound cleaning and care, 713, 714
Hospital treatment board, 686, 702, 742
Host, 579, 596
Housebreaking, 150
  dogs, 125
House calls, 38–39, 48
Housekeeping, 107–110
HR (heart rate), 480–481, 487, 545, 666, 667, 675, 681
Humane, 74, 75
Humerus, 443, 451
Hunters, 339, 362
Hunting, 292, 297
Hurdle, 371, 378
Hutches, 230, 233, 236
Hybrid, 119, 150
Hydration status, 676, 681
Hydrolysis, 563, 573
Hydrophilic, 423, 435
Hydrophobic, 423, 435
Hygiene, hands, 93, 104–107
Hyoid apparatus, 491, 497
Hyperadrenocorticism, 506–507, 508, 827
  dogs, 506
Hyperglycemia, 505, 508
Hyperkalemia, 507, 508
Hyperpnea, 668
Hypersensitivity, 538, 541
Hyperthermia, 664, 681
Hyperthyroidism, 505
Hyperventilation, 668
Hyphema, 671, 681
Hypnotize, 230, 231, 236
Hypoadrenocorticism, 506–507, 508, 828
Hypocalcemia, 325, 329, 399, 400, 433, 505, 508
  cats, 505
Hypoglycemia, 505, 508

Hyponatremia, 506–507, 508
Hypotension, 826, 832
Hypothalamus, 501–502, 508, 544, 550
Hypothermia, 663, 681
Hypothyroid drugs, 827
Hypothyroidism, 505, 508
Hypovolemia, 826, 832

I

IACUC (Institutional Animal Care and Use Committee), 304, 305, 308
Iatrogenic, 506, 508
Ich, 267, 278
ICU (intensive care unit), 60, 64
ID (intradermal), 537, 538, 541
Ideal weight, 568, 573
Ileum, 457, 473
Ilium, 444, 451
IM (intramuscular), 141, 150, 537, 541, 701, 704, 705, 742
  injections, 704–705
Immobilization, 448
Immune system, 535–542
  allergies, 538
  clinical situation, 542
  ELISA testing, 538
  infection, 537–538
  introduction, 535
  objectives, 535
  vaccination routes, 537
  vaccines, 536, 537
Immunity, 535, 541
Immunization, 536
Immunoglobulins, 535, 541
Impacted, 139, 150
IM (intramedullary) pin, 447, 451
Imprinting, 601–602
IN (intranasal), 537, 541, 701, 706, 742
  injections, 706
Incineration, 103, 114
Incinerator, 102–103, 114
Incisors, 456, 473
Incontinence, 513, 516
Incubate, 191, 194, 251, 254
Incubation, 143, 150, 524–527, 532
Incubation period, 410, 417
Incubator, 191, 194, 417
Incus, 556, 560
Indirect contact, 96, 114, 577, 596
Indirect PLR, 671, 681
Induced ovulation, 523, 532
Induced ovulators, 167, 172, 523
Induction phase, 840, 872
Infection, 434, 435, 537–538, 541
Infectious bovine rhinotracheitis (IBR), 323, 329
Infectious bronchitis, 411, 417
Infertility, 191, 194, 323, 330
Inflammation, 434, 435, 537–538, 541
Influenza, 360, 362, 578
Infundibulum, 519, 520, 532
Infusion pump, 709, 710, 711, 742
Inguinal rings, 520, 532
Inhalation, 492–493, 497
In-hand breeding, 356, 362
In-house testing, 747, 781
Injection port, 707, 742
Injections
  administering, 701–706
    IV catheter monitoring, 709
    IV fluid monitoring, 709–710
  routes, 701
Injury, 434

Index

Innate immune system *vs.* adaptive immune system, 536
Inner shell membrane, 414, 417
Inotropic drugs, 826, 832
Insecticides, 325, 330
Insertion, 449, 451
Inspiration, 187, 194
Instinctive behaviors, 600–601
Institutional Animal Care and Use Committee (IACUC), 304, 305, 308
Instrument milk, 864, 872
Instruments, 93–94
Insulin, 461, 473, 500–501, 503, 508, 828
Integumentary system, 676–677, 681
Intensifying screen, 793, 794, 796, 806
Intensity biomass, 267, 278
Intensive care unit (ICU), 60, 64
Intercostal muscle, 449
Intermale aggression, 159, 172
Intermediate host, 579
International health certificate, 27, 31
Interneurons, 543, 550
Interpersonal communication, 71, 75
Interphase, 426, 427, 435
Interstate health certificate, 27, 31
Intervals, 36, 48
Intervertebral disc, 442, 451
Intervertebral disc disease (IVDD), 442, 451, 548, 550
Intestinal forceps, 867–868, 871
Intradermal (ID), 537, 538, 541
Intramedullary (IM) pin, 447, 451
Intramuscular (IM), 141, 150, 537, 541, 701, 704, 705, 742
   injections, 704–705
Intranasal (IN), 537, 541, 701, 706, 742
   injections, 706
Intraocular, 537, 541
Intravenous (IV), 466, 473
Intubation, 840, 872
Intubation procedure, 840–842, 872
Intussusception, 461, 467, 469, 473
Inventory, 46, 48
Inventory management, 46, 48
Invoice, 20–24, 31
Involuntary reflex, 545, 550
Involution, 527, 532
Iodine, 357, 362
Iospora canis, 583, 596
Iris, 553, 554, 560
Ischium, 444, 451
Isolation ward, 62, 64, 113, 114
Isotonic crystalloid, 839, 872
Isthmus, 520, 532
IV (catheter), 709, 742
   monitoring, 709
IV (intravenous), 466, 473
IVDD (intervertebral disc disease), 442, 451, 548, 550
IV fluid
   administration, 706–708
   comparison of drop size in macrodrip and microdrip infusions, 710
   fluid rate calculation examples, 710
   monitoring, 709–710
Ixodes spp., 587

J

Jack, 334, 362
Jaundice, 670, 681
Jejunum, 457, 473
Jellyfish, 301
Jenny, 334, 362
Jird, 211, 223
Jog, 354, 362, 645
Johne's disease, 388, 391
Joint capsule, 440
Joint ill, 398, 400
Joints, 440, 451
   types and locations, 440
Jugular veins, 482, 623, 654
Jugular venipuncture, 130, 150, 163, 625–626
Jumpers, 339, 362
Juvenile, 242, 254

K

Kangaroo, 301
KBr (potassium bromide), 549
Kcal (kilocalories), 564, 573
Kelly hemostats, 867, 868, 872
Kennel attendant, 57, 64
Kennel cough (bordetella), 141, 494–495, 497
Kennel ward, 62, 63, 64
Keratin, 735, 742
Keratoconjunctivitis sicca, 555, 560
Ketosis, 325, 330
   beef and dairy cattle, 325
Keyboard, 43, 45–46, 48
Keyboarding, 45–46, 48
Kid, 393, 400
Kidding, 393, 400
Kidneys, 511–513, 517
Killed vaccine, 536, 541
Kilocalories (kcal), 564, 573
Kilovoltage peak (kVp), 787, 806
Kindness, 71, 75
Kitty burrito, 160, 172
Kitty taco, 633
Kneecap, 444
Knee jerk reaction, 677, 681
Knee jerk reflex, 547, 550
Knemidokoptes mites, 193, 195
Knob, 402, 417
Knots, 612–617, 654
   half hitch, 617
   quick release, 615
   reefer's, 615, 616, 655
   slip, 615
   square, 613–614, 655
Koi, 259
KVp (kilovoltage peak), 787, 806

L

Lab animals, 301, 307
Labial, 696, 743
Laboratory, 60, 61, 64
   procedures. *See* Laboratory procedures
   terms and abbreviations, 9
Laboratory and animal research production management, 300–309
   animal models and laboratory medicine, 302–303
   animal research, 300–301
   animal use research facts, 301
   careers, 306–307
   clinical situation, 309
   common species of animals used in research, 303–304
   introduction, 300
   management and health practices, 305–306
   objectives, 300
   research regulations and guidelines, 304–305
   types of research, 301–302
Laboratory animal technician, 306, 308
Laboratory procedures, 746–783
   blood chemistry procedures, 755–765
      blood smear, 758–762
      complete blood count, 752, 757–758
      packed cell volume, 762–763
      plasma protein, 764
   blood chemistry samples and electrolytes, 764–765
   clinical situation, 783
   cytology and histology, 777–779
   elements in urinary sediment, 774–778
   fecal sample, 765–769
   hematology sample collection, 755
   introduction, 746–747
   microscopic exam, 772–774
   necropsy procedure, 778–779
   objectives, 746
   recording laboratory results, 755
   serologic testing kits, 753–754
   urine sample, 769–772
      chemical test strips, 771–772
      culture and sensitivity, 774
      gram stain, 774, 776
      gross examination of urine, 770–771
      urine collection, 769–770
      urine specific gravity, 771
   veterinary laboratory equipment, 747–755
      blood chemistry analyzers, 752, 757, 765, 780
      centrifuge, 749–751
      microscope, 747–749
      refractometer, 751–752
      serological test kits, 753–755
Laboratory veterinarian, 306, 308
Labored breathing, 668
Labor process, 527
Labyrinth fish, 259
Lacrimal glands, 554, 560
Lactated Ringer's Solution (LRS), 466, 473, 706, 743, 839
Lactating dairy cows, 315
Lactation, 314, 330, 527, 532
Lactation diets, 567, 573
Lamb, 381, 391
Lamb dysentery, 388, 391
Lamb feeding, 390, 391
Lambing, 381, 391
Laminitis, 359, 363
Landrace pig, 367
Laparotomy towel, 847, 872
Large animal practice, 58–59, 64
Large roundworms, 412
Large-size herd, 326, 330
Larvae, 579, 581, 582, 584, 596
Laryngoscope, 491, 691, 841, 872
Larynx, 491, 497
Laser, 855
Laser surgery, 855
LAT (lateral), 788, 806
Lateral (LAT), 788, 806
Lateral apparatus, 554, 560
Lateral canthus, 554, 560
Lateral line, 257, 278
Lateral recumbency, 128–129, 150, 618, 622–623, 654
Laundry, 110
   laundering materials, 740–741
Lavaging, 713, 743
Laxatives, 232, 236, 467, 473, 825, 832
Layer, 401, 417
Lead, 319, 330, 345, 363
Leading, 346, 363, 644–645
Learned behaviors, 601–602
Learning, 601

Leash muzzle, 608
Left atrium, 482, 487
Left ventricle, 482, 487
Legal issues. *See* Ethics and legal issues
Legislation. *See also* Animal abuse laws; Child labor laws
   Americans with Disabilities Act, 83, 85
   Animal Welfare Act, 83, 85, 301, 304–305, 307
   Fair Labor Standards Act, 83, 85
   federal veterinary laws, 83–84
   laws and veterinary practice acts, 82–83
   Veterinary Practice Act, 82–83, 86
Leicester sheep, 384
Lens, 553–554, 556, 560
Lens objective, 747, 781
Leptospirosis, 143, 150
   beef and dairy cattle, 323, 330
   swine, 376, 379
Lethargy, 143, 150
Leukocytes, 478, 487
LH (luteinizing hormone), 504, 508, 524, 532
Liability, 83, 85
Lice, 146, 150, 588
Licensed veterinary technician (LVT), 55, 64
Life cycle, 579
   parasite, 144
Life span, 119, 150
   dogs, 119
   goats, 393
Life stage, 122, 150
Lifting techniques, dogs, 130–131
Ligaments, 432, 435, 440–441, 451
Ligated, 148, 150
Light horses, 334, 363
Limiting factor, 294, 297
Linea alba, 449
Lingual route, 537
Lipase, 461, 473
Lipids, 423, 435, 564, 573
Liquid manure system, 319, 330
Listening, 72
Listening skills, 72, 75
Litter, 407, 417
Litter box, 160, 172
Litter trained, 160, 172
Live-bearing fish, 266, 278
Live cover, 356, 363
Liver, 457, 473
Living animals, 302, 308
Lizards
   behavior, 248
   biology, 241
   breeds, 243–245
   nutrition, 246
   reproduction and breeding, 252
   restraint and handling, 249–250
Llama, 283, 297
Log pulling, 342, 363
Loin, 442, 451
Longe line, 345, 363
Long hair, 172
   cats, 157, 164, 172
Longing, 647
Lope, 354, 363
Lordosis, 167, 172, 524, 532
Lovebirds, 179–180, 195
Lower GI, 790, 806
LRS (Lactated Ringer's Solution), 466, 473, 706, 743, 839
"Lub," 482
Lumbar, 442, 451, 544
Lumpy jaw. *See* Actinobacillosis
Lungs, 492, 493, 497
Lungworm, 580
Luteinizing hormone (LH), 504, 508, 524, 532

Luxation, 446, 451
LVT (licensed veterinary technician), 55, 64
Lyme disease, 141, 143, 150, 587, 596
Lymph node, 535, 536, 541
Lymphocytes, 478, 479, 487
Lymphocytosis, 267, 278
Lysosome, 425, 435

M

MA (milliamperage), 787, 794, 806
Macaws, 181, 195
Macrodrip line, 710, 743
Macrominerals, 565, 573
Macrophage, 478, 487
Maggots, 589–590
Magnetic resonance imaging (MRI), 804, 805, 806
Maiden, 334, 363
Maintenance diet, 150, 566, 573
   dogs, 124
Malignant, 202, 223, 429, 435
Malleus, 556, 560
Malocclusion, 235, 236
Malpractice, 83, 85
Mammalogy, 295, 297
Mammary glands, 530
Mandible, 441, 451
Mange dermatitis, 376, 378
Manual developing, 795, 798, 806
Manure, 319, 330
MAP (mean arterial pressure), 669
Marbling, 311, 330
Mare, 333, 363
Marek's disease, 411, 417
Marketability, 289, 297
Market weight, 326, 330
Mastectomy, 530, 532
Master problem list, 17, 18
Mastitis, 324, 330, 387–388, 391
Mat, 136, 150
Material safety data sheets (MSDS), 97
Maternal instinct, 600–601
Mating behavior, 601
Mats, 724, 743
Mat splitter, 724, 743
Maxilla, 451
Mayo scissors, 868, 872
Mayo stand, 837, 872
Meal, 275, 278
Mean arterial pressure (MAP), 669
Meat goats, 395, 400
Meat-type hog, 367, 378
Mechanical filtration, 263, 278
Medial canthus, 554, 560
Medical records, 16–34
   clinical situations, 32
   competency skill, 33–34
   consent forms and certificates, 26–28
   creating, 17–23
   filing, 26
   introduction, 16
   invoicing, 23–24
   as legal documents, 28
   recording information, 24–25
   veterinary, 16–17
Medical terminology and abbreviations, 2–15. *See also* Veterinary terminology
   learning to dissect a veterinary term, 2–4
      clinical situation, 14
      common directional terms, 7
      common terms and abbreviations used in veterinary practice, 8–11
      competency skills task, 15

      objectives, 2
      overview, 2
      spelling, 11
Medical waste, 84, 95–96, 114
Medications
   amount to dispense, 814–815
   aural, 820–821
   containers, 813
   controlled substances, 816–817
   dispensing, 813
   educating client on use, 817–824
   ethics and legal issues, 84
   ophthalmic, 822–824
   oral, 817–819
   pill counting tray, 815
   scheduled drug classes, 816
   schedules, 84
   storage, 824
   topical, 821–822
   types, 816–817, 825–829
Medicinal administration, routes, 10
Mediterranean class, 403, 404, 417
Medium, 774, 781
Medium hair, 157, 172
Medulla, 511, 516
Medulla oblongata, 544, 545, 550
Medullary cavity, 438, 451
Meiosis, 427, 435
   phases, 427–428
Menace response test, 555, 560
Meniscus, 441, 446, 451
MER (metabolic energy requirements), 570
Mercury rectal thermometer, 664
Mesentery, 460, 474
Message, 68, 75
Metabolic diseases, beef and dairy cattle, 324–325
Metabolic energy requirements (MER), 570
Metabolism, 425, 435, 825, 832
Metacarpal, 443–444, 452
Metaphase, 427–428, 435
Metatarsal, 444, 452
Metestrus, 522–523, 532
Methimazole, 505, 508
Metzenbaum scissors, 868, 872
Mice, 197–203, 301, 303
   basic health care maintenance, 201
   basic training methods, 199
   behavior, 199
   biology, 198
   breeds, 198
   breed selection, 198
   common diseases, 201
   common parasites, 201–202
   common surgical procedures, 202–203
   equipment and housing needs, 199–200
   grooming, 201
   nutrition, 199
   reproduction and breeding, 201
   restraint and handling, 200–201
   used in research, 303
   vaccinations, 201
   veterinary terminology, 198
Microbiology, 577
Microbiology and parasitology as disease processes, 577–598
   clinical situations, 598
   deworming programs, 591–592
   disease treatment, 594
   health maintenance, 593–594
   introduction, 577
   objectives, 577
   parasitology, 579–591
      external, 586–591
      internal, 579–584
   types of microorganisms, 578–579

Index

bacterial disease, 578
   fungal disease, 578–579
   protozoal disease, 579
   rickettsial disease, 578
   viruses, 578
zoonosis, 592–593
   brucellosis, 592–593
   encephalitis, 593
   toxoplasmosis, 592
Microchips, 322, 330
Microdrip line, 710, 743
Microfilaria, 146, 150, 582, 596
Microhematocrit, 762, 781
Microhematocrit tube, 750, 781
Microminerals, 565, 573
Microscope, 747–749
   magnification powers, 748
Microscopic, 144–145, 150
Midbrain, 544, 550
Midstream, 769, 770, 781
Migratory birds, 293, 294, 297
Milk fever. *See* Hypocalcemia
Milking herd, 326, 330
Milk production, by state, 313
Milliamperage (mA), 787, 794, 806
Millivolt, 545, 550
Milt, 272, 278
Mineral block, 343, 363
Minerals, 565, 573
Miniature, 236
   rabbits, 227
Misrepresentation, 81, 85
Mites, 144–145, 150, 588–589
   poultry, 412
Mitochondria, 425, 435
Mitosis, 426, 435
   phases of, 426–428
Mixed-animal practice, 59, 64
Mixed breed, 150
   dog, 119
MM (mucous membranes), 465, 474, 587, 596
Mobile services, 38, 48
Mobile veterinary practice, 59, 64
Modeling, 601
Modem, 43, 48
Modified live vaccine, 536, 541
Mohair, 395, 400
Molars, 456, 474
Mollusk, 269, 278
Molly, 260
Molting, 189, 195, 409–410, 417
Monitor, 43, 48
Monkeys, 301
   characteristics of monkey types, 304
   used in research, 304, 305
Monocyte, 478, 479, 487
Monofilament suture, 854, 872
Monogastric, 119, 150, 155–156
Monogastric system, 457–461
Monosaccharide, 423, 435, 564, 573
Mopping, 101, 102
Moral, 81, 85
Morgan horse, 339
Morphology, 356, 363, 758, 781
Mosquitoes, 588
Mosquito hemostats, 867, 871
Motility, 356, 363
Motor neuron, 543, 550
Mouse (animal). *See* Mice
Mouse (computer), 43, 48
Mouth, 456, 459
Mouth brooders, 272, 278, 521–522, 532
Mouth gag, 869, 870
MRI (magnetic resonance imaging), 804, 805, 806

MSDS (material safety data sheets), 97
Mucosa, 456, 474
Mucous membranes (MM), 465, 474, 587, 596, 669
Mule, 334, 341, 363
Multifilament suture, 854, 872
Murine, 198, 223
Murmur, 675–676, 681
Muscles, 440–441, 452
Muscle tissue, 432–433
Muscular system, 449
Muscular tissue, 430–431, 435
Musculoskeletal system, 438–454, 681
   anatomy of a long bone, 439
   appendicular skeleton, 443–444
   axial skeleton, 441–442
   bones, 438–439
   clinical situation, 454
   common musculoskeletal diseases and conditions, 444–448
      arthritis, 445–446
      bone fractures, 446–448
      cruciate ligament tears, 446, 450
      hip dysplasia, 444
      intervertebral disc disease, 442
   introduction, 438
   joints, 440
   muscles and connective tissues, 440–441
   objectives, 438
   physical examination, 677–678
Mus musculus, 303, 308
Mutton, 381, 391
Muzzles, 131, 150
   cats, 161, 162, 171, 604–606
   dogs, 604–606
Mycology, 578
Myelogram, 548
Mylo, 449
Myocardium, 480, 487
Myofiber, 433, 435
Myopathy, 449
Myoplasty, 449
Myosin, 433, 435
Myositis, 449
Myotomy, 449

N

NaCl (sodium chloride), 466, 475
Nail bed, 735, 743
Nail trimming, 139, 150, 734–738
   dogs, 139
Nanny, 393
Nares, 491, 497
Nasal septum, 319, 330
Nasogastric tube, 719, 743
National Association of Veterinary Technicians in America (NAVTA), 55, 64
National Institutes of Health (NIH), 306, 308
National Research Council (NRC), 570
Natural incubation, 191, 195
Navicular bone, 444
Navicular syndrome, 448
NAVTA (National Association of Veterinary Technicians in America), 55, 64
Near side, 346, 363
Neckband, 685, 743
Neck tag identification, 322, 330
Necropsy, 778, 781
   procedure, 778–779
Necrotic tissue, 471, 474
Needle holders, 864–865, 872
Needle teeth, 374, 378
Negative contrast agent, 790, 806

Negligence, 83, 85
Neoplasia (cancer), 539–540, 541
   common cancers in animals, 540
Neoplasm, 429
Nephron, 511–513, 517
Nerve, 433, 435
Nerve impulse, 433, 435
Nerve testing, 547
Nerve tissue, 433
Nervous system, 543–552, 681
   clinical situation, 552
   common diseases and conditions, 547–549
      epilepsy, 548–549
      intervertebral disc disease, 548, 550
      rabies, 549
      West Nile virus, 549, 551
   coordinated function of the central nervous system and peripheral nervous system, 545
   divisions, 547
   introduction, 543
   objectives, 543
   physical examination, 677, 678
   receptors, 547
   structures, 543–545
      central nervous system, 543–544
      peripheral nervous system, 545
   testing, 547
Nervous tissue, 430, 433, 435
Neurological, 143, 150
Neurons, 433, 435, 543, 550
Neurotransmitters, 543, 550
Neuter, 119, 147–148, 150
Neuter certificate, 27, 31
Neutering, 529
Neutrophils, 478, 479, 488
Newcastle disease, 411, 417
New World monkeys, 304, 308
Nictitans, 555
Nictitating membrane, 555, 560
Night feces, 229, 236
NIH (National Institutes of Health), 306, 308
Nitrogen cycle, 264, 278
Nits, 146, 150, 588, 596
Nocturnal, 198, 223
Non-absorbable suture, 854, 872
Noncore vaccine, 141, 150
Nonessential fatty acids, 564, 573
Nonexempt, 83, 86
Nongame animals, 292, 297
Nongame birds, 293, 297
Nongame fish, 293, 297
Nonhuman primates, 304, 308
Nonliving systems, 302, 308
Nonproprietary, 809
Non-rebreathing system, 859, 872
Nonruminants, 226, 236, 335, 363
Nonruminant system, 464, 474
Nonsteroidal antiinflammatory drugs (NSAIDs), 825–826
Nonvenomous, 239, 254
Nonverbal communication, 69–70, 75
Norepinephrine, 503, 508
Normal heart rates, 666
Normal saline, 466, 474
Nose printing, 322, 330
Nose ring, 319, 330
Nose tongs, 319, 330
Nosocomial infection, 112, 114
NPO, 466, 474, 719
NRC (National Research Council), 570
NSAIDs (nonsteroidal antiinflammatory drugs), 825–826
Nubian goat, 395
Nuclear sclerosis, 556, 560
Nucleic acid, 423, 435

Nucleolus, 424, 436
Nucleus, 424, 436
Numerical filing system, 26, 31
Nutrients, 562, 563–565, 573
Nutrition. *See* Animal nutrition
Nystagmus, 555, 560, 671, 682

O

Obedience, 150
  dogs, 126
Obesity, hedgehogs, 220
Objective information, 25, 31
Oblique, 788, 806
Oblique muscles, 449
Observation, 72, 660, 682, 688, 743
  recording, 688
Observation skills, 72, 75
Occlusal, 696, 743
Occupational Safety and Health Administration (OSHA). *See* OSHA (Occupational Safety and Health Administration)
Odors, 670, 671, 679
OFA (Orthopedic Foundation of Animals), 445
Office manager, 57, 64
Offspring, 206, 223, 314, 330
OHE/OVH (ovariohysterectomy), 147, 148, 151, 170, 529, 532
  rabbits, 235
Oil immersion, 747, 781
Old World monkeys, 304, 308
Olecranon, 443, 452
Olfactory sense, 558, 560
Oliguria, 515, 517
Olsen-Hegar needle holder, 865
Omasum, 463, 474
Omentum, 460, 474
Omnivores, 176, 195, 239, 254, 257, 278, 455, 474
Onychectomy, cats, 170–171, 172
Oocysts, 583, 596
Open gloving, 852, 872
Operculum, 268, 278
Ophthalmic medications, 822–824
Ophthalmoscope, 670, 671
Oral, 144, 150
Oral barium study, 469, 474
Oral cavity, 669, 673, 682
Oral medications, 817–819
Oral route, 537
Orbits, 452
Orchidectomy, 151, 172
  cats, 170, 172
  dogs, 148, 151
Organ, 433, 436
Organelle, 423, 436
Organ system, 433
Origin, 449, 452
Ornamental fish, 256, 278
  aquaculture industry, 267–268
  basic health care and maintenance, 265
  breeding and reproduction, 266
  breeds, 258–261
    angelfish, 261
    barb fish, 259
    basslets, 261
    butterfly fish, 261
    catfish, 260
    goldfish, 259
    gourami, 259
    guppy, 260
    koi, 259
    labyrinth Fish, 259
    molly, 260
    platy, 260
    swordtail, 260
    tetras, 260
  catching and handling, 265
  clinical situation, 280
  equipment needs, 261–264
  fish disease and health, 266–267
    ascites, 267
    cloudy eye, 267
    fin rot, 266
    fish fungus, 267
    ich, 267
    lymphocytosis, 267
    swim bladder disease, 267
  identification and production management, 256–281
    anatomy, 258
    biology, 257–258
    breed selection, 261
    introduction, 256
    objectives, 256
    veterinary terminology, 257
  nutrition, 261
  water sources and quality, 264–265
Orphaned, 151
  lamb, 387, 391
  puppies, 123
Orthopedic Foundation of Animals (OFA), 445
Oryctolagus cuniculus, 303, 308
OSHA (Occupational Safety and Health Administration), 83, 86, 97, 114, 785
  anesthetic hazards, 101
  controlled substances, 84
  guidelines and regulations, 97–101
  labeling secondary containers, 99–100
  lifting heavy objects, 100
  radiation hazards, 101
  radiation safety, 785–787
  safety data sheets, 97–99
  veterinary hospital plan, 100
Osmosis, 425–426, 436, 513, 517
Os penis, 513, 517
Ossification, 438, 452
Ostectomy, 448
Osteitis, 446
Osteoarthritis, 446
Osteoblasts, 439
Osteochondrosis, 446
Osteoclasts, 439–440
Osteocytes, 439
Osteomalacia, 446
Osteopexy, 448
Osteoplasty, 448
Osteosarcoma, 539
Osteosclerosis, 446
Ostrich, 285, 297, 406, 417
OTC (over-the-counter drugs), 816, 832
Otitis, 557, 558, 560, 732, 743
Otodectes mite, 589
Otoscope, 557, 558, 560, 670, 672, 734
"Ought," 855
Outer shell membrane, 414, 417
Oval window, 556, 560
Ovarian cyst, 530, 532
Ovariohysterectomy (OHE/OVH), 147, 148, 151, 170, 529, 532
  rabbits, 235
Ovary, 519–520, 532
Over-the-counter drugs (OTC), 816, 832
Oviduct, 519–520, 532
Ovine, 381, 391
Ovulation, 356, 363, 504, 508, 520, 523–524, 532
  animal species types, 524
Ovum, 520, 532
Owner information, 17

Oxen, 342
Oxygenation, 262, 278
Oxygen tank, 858, 872
Oxytocin, 501, 509

P

Pacemaker, 482, 488
Packed cell volume (PCV), 762–763, 781
Packing slip, 47, 48
PACS (picture archiving and communications system), 800
Paddle, 371, 378
Pain, 678
Pain assessment, 689
Palatability, 315, 330, 564, 566, 574
Palpation, 355, 363
Palpebral reflex, 857, 872
Pancreas, 460, 474, 503, 509
Panels, 765, 781
Panleukopenia, 168, 172
Papillae, 456, 461, 474
Parakeet, 179, 195
Paralysis, 143, 151, 548, 550
Parascaris equorum, 580, 596
Parasites, 96, 114, 144, 151
  beef and dairy cattle, 325
  birds, 193
  cats, 169–170
  chinchillas, 222
  equines, 360
  external, 144–146
  ferrets, 216
  gerbils, 213
  goat, 399
  guinea pigs, 211
  hamsters, 207
  hedgehogs, 220
  internal, 146–147
  poultry, 412
  rabbits, 235
  rats and mice, 201–202
  reptiles and amphibians, 253
  sheep, 389
  swine, 376–377
Parasitology, 579–591, 596. *See also* Microbiology and parasitology as disease processes
  external, 586–591
  internal, 579–584
Parasympathetic system, 545, 550
Parathormone, 503, 509
Parathyroid glands, 503, 509
Parked-out, 359, 363
"Parrot fever," 192
Parrots, 180–181, 195
Parturition, 356, 363, 501, 509, 525, 527, 532
Parvovirus, 143, 151
Passerine, 178, 195
Passive immunity, 536–537, 541
Pasteurella, 234, 236
Pasture breeding, 356, 363
Pastures, 567
Patella, 444, 452
Patency, 709, 743
Pathogen, 372, 535, 536, 542, 577, 596
Pathologist, 429, 436
Pathology, 422, 436
Patience, 71, 75
Patient, 16, 31
  exercise, 718–719
  feeding and watering hospitalized patients, 719–720, 721
  hospitalized, 685–689

identification information, 685
information, 17
information form, 18
measurement, 791–792
physical examination. See Physical examination
socialization, 718–720
Patient history, 657–658, 682
   examples of questions, 658
   terms, 8
Patient monitors, 843–844
PBFD (psittacine beak and feather disease), 192, 195
PCV (packed cell volume), 762–763, 781
PD (polydipsia), 502, 505, 509
PE. See Physical examination
Peacock, 405, 417
Peafowl, 405, 417
Peahen, 417
Pecking order, 184, 195
Pectoral muscle, 449
Pedal reflex, 857, 872
Pedigree, 313, 330
Pelleted diet, 182, 195
Pellets, 275, 278
Pelvic limb, 444, 452
Pelvis, 443, 452
Pen, 271, 278
Penile area, 201, 223
Penis, 513, 517, 520, 532
Peptides, 504, 509
Percheron, 340
Performance animals, 286–288
Pericardium, 480, 488
Perineal urethrostomy (PU), 514–515, 517
Periodontal probe, 869
Periosteal elevator, 869
Periosteum, 439, 452
Peripheral nervous system (PNS), 545, 550
Peristalsis, 456, 474, 545, 551
Peritoneum, 460, 474
Permanent, 455, 474
Permanent pasture, 567, 574
Permit, 286, 298
Personal protective equipment (PPE), 91–93, 114, 855
Personnel manual, 56, 64
Pesticide, 594, 596
Pet food label, 571
Pet piller, 817, 818, 832
Phalanges, 444, 452
Phalanx, 444
Pharmacokinetics, 824–825, 832
Pharmacology, 809, 832
Pharmacy, 60, 61, 64, 809, 832
Pharmacy procedures, 809–833
   anthelmintics and antiparasitics, 830
   clinical situations, 833
   dispensing medications, 813
      amount to dispense, 814–815
      drug identification, 814
      medication containers, 813
      pill counting tray, 815
   educating client on medication use, 817–824
      administering aural medications, 820–821
      administering ophthalmic medications, 822–824
      administering oral medications, 817–819
      administering topical medications, 821–822
   introduction, 809
   labeling a prescription, 811–813
   medication storage, 824
   objectives, 809
   pharmacokinetics, 824–825
   preanesthetics and anesthetics, 829–830
   reading a prescription, 811
   terms and abbreviations, 9
   types of medications, 816–817, 825–829
      controlled substances, 816–817
      scheduled drug classes, 816
Pharynx, 456, 474, 491, 497
Pheasant, 405, 406, 418
Phenobarbital, 549
Pheromones, 601
Phlebitis, 709, 743
pH level, 264, 278
Phospholipid, 423, 436
Photons, 784
Photosynthesis, 263, 278
Physical cleaning, 101–102, 114
Physical examination (PE), 151, 658–683, 682
   abdominal system, 678
   blood pressure, 669
   capillary refill time, 669
   clinical situations, 683
   dental examination, 671–674
   dogs, 137, 151
   ear exam, 671
   eye exam, 670–671
   head exam, 670
   heart rate, 666, 667
   introduction, 657
   mucous membranes, 669
   musculoskeletal system, 677–678
   nervous system, 677, 678
   objectives, 657
   observation, 660
   oral cavity, 669, 673
   pulse, 666
   recording, 679–680
   respiratory rate, 666–667
   skin and integumentary system, 676–677
   surgery patients, 838–839
   temperature assessment, 663–665
   terms, 8–9
   thoracic cavity, 675
   tools, 659–660
   urogenital system, 678–679
   vital signs, 663
   weight assessment, 660–663
Physical hazards, 95, 114
Physical restraint, 128–130, 151
Physiology, 119, 151, 304, 308, 422, 436
Picture archiving and communications system (PACS), 800
Pigeons, 405, 418
Piglets, 365, 378
Pigment, 257, 278
Pigs, 301
   alopecia, 589
   sarcoptic mange, 589
Pill counting tray, 815, 832
Pincers, 268, 278
Pineal gland, 504
Pinna, 556, 560, 732, 743, 820, 832
Pinworms, 584–586
Pitting edema, 425–426, 709, 743
Pituitary gland, 501–502, 509, 544, 545, 551
Pivot, 440, 452
Placenta, 357, 363, 524, 526, 527, 533
Plague, 586, 596
Plan, 25, 31
Plane, 840, 872
Plankton, 263, 275, 278
Plaque, 671, 673, 682
Plasma, 432, 477, 478, 488, 537
Plasma protein, 764, 781
Platelet, 478–479, 488
Platy, 260
Pleura, 492, 497
Pleural cavity, 492, 497
Pleural friction rub, 494, 497
PLR (pupillary light response), 671
Plucked, 233, 236
Plymouth Rock, 403
Pneumonia, 494, 497
   cats, 169, 172
   swine, 376, 379
Pneumothorax, 496, 497
PNS (peripheral nervous system), 545, 550
PO, 466, 474
Pocket pets, health and production management, 197–224
   chinchillas, 220–222
   clinical situation, 224
   ferrets, 213–217
   gerbils, 211–213
   guinea pigs, 207–211
   hamsters, 203–207
   hedgehogs, 217–220
   introduction, 197
   mice and rats, 197–203
   objectives, 197
   popularity, 197
Points, of teeth, 351, 363
Policy manual, 43, 48
Polled, 311, 330
Polyarthritis, 388, 391
Polydactyl, 156, 172
Polydipsia (PD), 502, 505, 509
Polyestrous, 524, 533
Polyestrus, 166, 172
Polysaccharide, 423, 436, 564, 574
Polyuria (PU), 502, 505, 509
Pond, 269, 278
Ponies, 334, 339, 363
Pool, 262, 278
Pop-off valve, 858, 859, 872
Porcine, 365, 378
Porcine somatotropin (pST), 374, 378
Porcine stress syndrome (PSS), 376, 378
Positive contrast agent, 790, 806
Post-anesthetic care, 860
Posterior chamber, 553, 560
Posterior lobe, 501, 509
Postoperative patient care, 860–861
Postprandial, 755, 781
Potassium bromide (KBr), 549
Poult, 418
Poultry, breed identification and production management, 401–418
   anatomy, 402–403
   basic health care and maintenance, 409–410
   basic training, 407
   behavior, 407
   biology, 402–403
   clinical situation, 419
   common diseases, 411–412
      avian influenza, 411–412
      avian pox, 411
      fowl cholera, 411
      infectious bronchitis, 411
      Marek's disease, 411
      Newcastle disease, 411
   common parasites, 412
      coccidiosis, 412
      large roundworms, 412
      mites, 412
      tapeworms, 412
   equipment and housing needs, 407–408
   grooming, 409
   introduction, 401
   nutrition, 407
   objectives, 401
   poultry production, 412

Poultry, breed identification and production management (*Continued*)
   poultry production systems, 412–415
      broiler egg production system, 414–415
      broiler production, 413
      egg production system, 413–414
      pullet production, 413
   reproduction and breeding, 410–411
   restraint and handling, 408
   selection of species or breed, 407
   species and classes, 403–407
   vaccinations, 410
   veterinary terminology, 401–402
Poultry science, 412, 418
PPE (personal protective equipment), 91–93, 114, 855
PPT, 271, 278
Pre-anesthetic patient care, 838–839
Preanesthetics and anesthetics, 829–830
Precocial, 233, 236
Preconditioning, 594, 596
Predicted transmission ability, 314, 330
Preen, 188, 195. *See also* Grooming
Pre-exposure vaccine, 143
Prefixes, 2, 12
   common, 4, 5
Premolars, 456, 474
Preprandial, 755, 781
Prepuce, 513, 517, 520, 521, 533
Prescription (Rx), 832
   labelling, 811–813
   reading, 811
Prescription drugs, 816, 832
Prescription label, 811, 832
Pressure valve, 858, 872
Primary bandage layer, 714, 743
Primary response, 536, 542
Primates, 304, 308
   used in research, 301, 304, 305
Primed, 326, 330
Printer, 43, 48
Probe method, 251, 254
Procedures manual, 43, 48, 56, 64
Processing, 795, 806
Production cycle, 272, 278
Production intensity, 267, 278
Proestrus, 522, 524, 525, 533
Profitable business, 288, 298
Progeny, 314, 330, 658
Progesterone, 504, 509, 520, 524, 533
Proglotids, 584, 596
Proglottid, 147, 151
Progress notes, 17, 20, 31
Prolactin, 502, 509
Prolapsed uterus, 529–530, 533
Prophase, 426–428, 436
Prophylaxis, 673, 682, 697, 698, 699
   dental care, 699
Proprietary, 809
Proprioception, 547, 551
Prostaglandins, 500
Prostate gland, 521, 533
Proteins, 423, 436, 564–565, 574
Protozoa, 579
Protozoal disease, 579, 596
Protozoan disease, 143, 151
Protozoans, 96, 115, 412, 418
Protozoastatic, 828, 832
Proventriculus, 176, 195, 464, 474
Pseudo rabies, 376, 378
Psittacine, 178, 195
Psittacine beak and feather disease (PBFD), 192, 195
Psittacosis, 192, 195
PU (perineal urethrostomy), 514–515, 517
PU (polyuria), 502, 505, 509

Puberty, 142, 151
Pubis, 444, 452
Pullet, 401, 418
Pullet production, 413
Pulmonary artery, 481–482, 488
Pulp cavity, 455, 474
Pulse, 480, 488, 666
Pulse oximeter, 840, 843, 872
Pulse rate
   determining, 667
   normal, 482
Punch biopsy, 777, 782
Pupae, 585, 596
Pupil, 553, 554, 555, 560
Pupillary light response (PLR), 671
Pupillary light response test, 555, 560
Purebred, 119, 151
   dog, 121
Purebred business, 326, 330
Purebred flock method, 389, 391
Pushing, 319, 330
P wave, 483, 488
Pygmy goats, 395–396, 400
Pylorus, 457, 474
Pyometra, 529, 530, 533
Pyothorax, 496, 497

Q

QRS complex, 483, 488
Quadriceps, 704, 743
Quail, 405, 418
Quarantined, 251, 254
Quarter horse, 337, 338
Quarters, 326, 330
Queening, 167
Quick, 139, 151, 735, 743
Quick-release knot, 615
Quills, 218, 223

R

Rabbits, 301
   basic health care and maintenance, 232–233
   basic training, 230
   behavior, 230
   clinical situation, 237
   common diseases, 234–235
      Pasteurella, 234
      respiratory, 234
      Tyzzer's disease, 234
   common parasites, 235
   common surgical procedures, 235
   equipment and housing needs, 230–231
   grooming, 231–232
   identification and production management, 225–237
      anatomy, 226–227
      biology, 226
      breeds, 226–227
      breed selection, 227–228
      introduction, 225
      objectives, 225
      popularity, 225
      veterinary terminology, 225
   lifting, 641–642
   nutrition, 228–230
   reproduction and breeding, 233–234
   restraint and handling, 231
   used in research, 304
   vaccinations, 233

Rabies, 96–97, 151, 537, 549, 551
   cats, 169
   certificate, 27, 28, 31
   dogs, 143
   pole use, 132–133
   snare pole use, 132
   swine, 376, 379
   vaccine, 549
Rabies laws, 143, 151
Rabies pole, 132, 133, 151, 636, 654
Rabies suspect handling and sample submission, 779
Raccoon, 292
Racehorses, 339, 363
Raceways, 269–270, 278
Radiation, 103, 115, 784, 806
Radiation hazards, 101
Radiation safety, 785–787
Radioactive iodine, 505, 509
Radiograph artifacts and errors, 802
Radiographic film, 794
Radiology, 60, 61, 64, 191, 784, 806
Radiology and imaging procedures, 784–808
   clinical situations, 808
   darkroom care and maintenance, 801–802
   developing film, 795–800
      automatic film processing, 799, 800–801
      digital radiology, 799–800
      film cassettes, 796–797
      film hangers, 798
      manual developing, 795, 798
      pros and cons of development methods, 795
   endoscopy, 803–804
   filing film, 800–801
   film identification, 794–795
   introduction, 784
   machine setting procedures, 793–794
   objectives, 784
   patient measurement, 791–792
   radiation safety, 785–787
   radiographic detail, 789–790
   radiographic film, 794
   radiology log, 791
   radiology terminology, 787–789
      diagnostic terms, 790
      x-ray machine terminology, 787–788
      x-ray positioning terms, 788–789
      x-ray terms, 787
   technique chart, 792–793
   ultrasound diagnostics, 802–803
   veterinary animal restraint, 790–791
Radiology generation, 784–785
Radiology log, 791
Radiolucent, 787, 788, 789, 806
Radiopaque, 469, 471, 474, 787, 788, 806
Radius, 443, 452
Rainwater, 264, 278
Rales, 494, 497, 675, 682
Ram, 381, 391
Range band method, 389–390, 391
Range paralysis, 411, 418
Raptors, 293, 298
Rare breeds, 286, 298
Rasp, 352, 363
Ration, 124, 151, 562, 574
Ratite, 406, 418
Rats, 197–203, 301
   basic health care maintenance, 201
   basic training methods, 199
   behavior, 199
   biology, 198
   breeds, 198
   breed selection, 198
   common diseases, 201
   common parasites, 201–202

common surgical procedures, 202–203
equipment and housing needs, 199–200
grooming, 201
nutrition, 199
reproduction and breeding, 201
restraint and handling, 200–201
used in research, 303
vaccinations, 201
veterinary terminology, 198
Rat tooth thumb forceps, 865, 866, 872
Rattus norvegicus, 303, 308
RBC (red blood cells), 477–478, 479, 488
Reagent, 753, 754, 782
Reagent strip, 771, 772, 782
Rearing, 344, 363
Rebreathing bag, 858
Rebreathing system, 859, 872
Receiver, 68, 75
Reception area, 59, 64
Receptionist, 57, 64
Receptive, 142, 151
Receptors, 547, 551
Recombinant vaccine, 536, 542
Rectal temperature, 664, 665, 679
Rectal thermometer, 664
Rectus muscles, 449
Recumbency, 128, 151, 618, 655
Red blood cells (RBC), 477–478, 479, 488
Redirected aggression, 125, 151, 159–160, 172, 602, 603, 655
Red nose. *See* Infectious bovine rhinotracheitis (IBR)
Reduced calorie diets, 151, 567, 574
  dogs, 124
Reefer's knot, 615, 616, 655
Reefing, 319, 330
Reference lab, 747, 782
Reflexes, 545, 551
  palpebral, 857
  pedal, 857
  swallow, 860
Reflex hammer, 547, 551
Reflex testing, 555–556
Refractometer, 751–752, 782
Registered veterinary technician (RVT), 55, 64
Regurgitation, 283, 461, 464, 474
Renal artery, 511, 517
Renal failure, 515
Renal pelvis, 511, 517
Renal system, 511–518
  clinical situation, 518
  common diseases and conditions, 513–515
    feline urologic syndrome, 514–515
    renal failure, 515
    toxicity, 515
    urinary blockage, 514
    urinary incontinence, 513
  introduction, 511
  objectives, 511
  structures, 511–513
    kidneys, 511–513
    ureters, 513
    urethra, 513
    urinary bladder, 513
Renal vein, 511, 517
Replacement hens, 413, 418
Re-polarize, 484, 488
Reproduction diets, 566–567, 574
Reproductive system, 519–534
  artificial insemination, 528, 529
  clinical situation, 534
  estrus, 522–524
  female, 519–520
  fertilization, 521–522
  gestation and incubation, 524–527

introduction, 519
labor process, 527
male, 520–521
mammary gland problems, 530
neutering, 529
objectives, 519
reproductive problems, 529–530
semen collection, 528
Reptile, 239, 254
Reptiles and amphibians
  basic health care and maintenance, 251
  basic training, 248
  behavior, 247–248
  breed identification and production management, 238–255
    biology, 239–242
    breeds, 243–246
    introduction, 238
    objectives, 238
    popularity, 239
    veterinary terminology, 238–239
  breed selection, 246
  clinical situation, 255
  common diseases, 253
  common parasites, 253
  common surgical procedures, 253
  equipment and housing needs, 248–249
  frogs and toads
    anatomy, 243
    behavior, 248
    biology, 242, 243
    breeds, 245
    nutrition, 247
    reproduction and breeding, 252–253
    restraint and handling, 250
  grooming, 250–251
  lizards
    behavior, 248
    biology, 241
    breeds, 243–245
    nutrition, 246
    reproduction and breeding, 252
    restraint and handling, 249–250
  nutrition, 246–247
  reproduction and breeding, 251–253
  restraint and handling, 249–250
  salamanders
    anatomy, 244
    behavior, 248
    biology, 242–243
    breeds, 245, 246
    nutrition, 247
    restraint and handling, 250
  snakes
    behavior, 247–248
    biology, 240
    breeds, 243
    nutrition, 246
    reproduction and breeding, 251–252
    restraint and handling, 249
  turtles
    behavior, 248
    biology, 241–242
    breeds, 245
    nutrition, 247
    reproduction and breeding, 252–253
    restraint and handling, 250
RER (resting energy rates), 570
Research, 300, 308
  use of mice, 303
  use of monkeys, 304, 305
  use of rabbits, 304
  use of rats, 303
Research scientists, 307, 308
Respiration, 492–494, 497

Respiratory rate (RR), 493–494, 544, 545, 666–667, 675, 682
  normal, 668
Respiratory system, 490–499, 579
  clinical situations, 499
  diseases and conditions, 494–496
    asthma, 494
    diaphragmatic hernia, 495
    heaves, 495
    hemothorax, 496
    kennel cough (bordetella), 494–495
    pneumonia, 494
    pneumothorax, 496
    pyothorax, 496
    roaring, 496
    shipping fever, 495
    upper respiratory infection, 494
  external and internal respiration, 491–492
  introduction, 490
  mechanics of breathing, 493
  objectives, 490
  respiration, 492–494
  respiratory rates, 493–494
  respiratory structures of the thoracic cavity, 492–494
  structures, 491–492
Resting energy rates (RER), 570
Restrain, 603, 655
Restraint and handling procedures, 600–656
  animal safety, 604
  avian, 187–188
  beef and dairy cattle, 319–321
  cats, 160–164
  chinchillas, 221
  clinical situation, 656
  competency skills, 605–653
  dogs, 127–133, 603, 604, 656
  equines, 346–349
  ferrets, 214–215
  frogs and toads, 250
  gerbils, 212
  goat, 397
  guinea pigs, 209
  hamsters, 205
  hedgehogs, 219
  interpretation of an animal's body language and behavior, 602–603
    angry animal, 602–603
    happy animal, 602
    scared animal, 602
  introduction, 600
  lifting techniques, 130–131
  lizards, 249–250
  mice, 200–201
  muzzle use and application, 604–608
  objectives, 600
  physical restraint, 128–130, 151
  planning the restraint procedure, 612
  poultry, 408
  procedures and techniques, 628–653
    large animal restraint, 643–653
    small animal restraint, 628–642
  rabbits, 231
  rabies or snare pole use, 132–133
  radiology, 790–791
  rats, 200–201
  reptiles and amphibians, 249–250
  restraint considerations, 603–604
  restraint equipment, 604–612
    anti-kick bars, 610, 611
    hobbles, 610, 612
    leather welding gloves, 610, 611
    squeeze cages, 609, 610
    stanchions or head gates, 612, 613
    towels, 609

Restraint and handling procedures, 600–656
   animal safety (Continued)
     restraint for blood collection, 623–628
     restraint knots, 612–617
       half hitch, 617
       reefer's knot, 615, 616
       square knot, 613–614
     restraint positions, 618, 619–623
     salamanders, 250
     sedation use, 133
     sheep, 385
     snakes, 249
     swine, 371–372
     tail switch, 652
     turtles, 250
     verbal, 128, 152
     wildlife, 295
Retained placenta, 324, 330, 387, 391
Reticulum, 463, 474
Retina, 553, 554, 560
Retractable claw, 156, 172
Retractors, 868–869, 872
   Gelpi, 868, 871
   Holman, 868, 871
   Senn, 868, 873
   Weitlaner, 868, 869, 873
Rhea, 418
Rhinotracheitis, 168, 172
Rhode Island Red, 403
Rib, 442, 452
Ribonucleic acid (RNA), 436
Ribosomes, 424, 436
Rickettsial disease, 96, 115, 144, 151, 578, 596
Riding horses, 337–338, 363
Right atrium, 481, 488
Right-to-know station, 97, 115
Right ventricle, 481, 488
Ringworm (dermatophytosis), 143, 151, 399, 400, 590–591, 595, 596
   heifer, 591
Rinsing, 795, 806
Roaring, 496, 498
Roaster, 401, 418
Rochester-Carmalt hemostatic forceps, 867, 873
Rocky Mountain Spotted Fever, 144, 151, 587, 597
Rodents, 197, 223
Rods, 553, 560, 774, 782
Rooster, 401, 418
Root, 455, 474
Root words, 2, 12
   common, 3–4
Rotation, 440
Rotator muscle, 449
Rotors, 749, 750, 782
Roughage, 316, 331
Rough endoplasmic reticulum, 425
Round window, 556, 560
Roundworms (ascarids), 147, 151, 579–581, 597
   host, 581
   poultry, 412
Routine appointment, 38, 48
RR (respiratory rate), 493–494, 544, 545, 666–667, 675, 682
   normal rates, 493
Rugae, 457, 474
Rumen, 461–464, 474
Ruminant, 283
Ruminant system, 461–464
Rumination, 464
Rump method, 385, 391
Run, 257, 278
RVT (registered veterinary technician), 55, 64
Rx. See Prescription

S

Sac fry, 274, 278
Sacral vertebrae, 442, 452
Sacrum, 443, 452
Safe light, 795, 806
Safety, radiation, 785–787. See also OSHA (Occupational Safety and Health Administration); Veterinary safety
Safety data sheets (SDS), 97–99, 115
Salamanders
   anatomy, 244
   behavior, 248
   biology, 242–243
   breeds, 245, 246
   nutrition, 247
   restraint and handling, 250
Salinity, 265, 278
Saliva, 456, 474
Salivary glands, 456, 459, 475
Salmonella infection, 251, 253, 254, 255
Salmonellosis, 253, 255
Salt block, 343, 363
Saltwater, 271, 278
Saltwater fish, 258, 279
Sanitation, 101, 115
   exam room, 110–112
   filtration, 103–104
   methods, 101–104
   physical cleaning, 101–102
   sterilization, 101, 102–103
SA node, 482, 488
Saphenous veins, 482, 623
Saphenous venipuncture, 130, 131, 151, 627–628
Sarcoptes scabei, 589
Sarcoptic mange, 145, 151, 589, 597
Scabies, 145, 151, 589, 597
Scales, 257, 279, 660–662
   hematocrit, 763
Scalpel blade handle, 864
Scanner, 43, 48
Scapula, 443, 452
Scavenger hose, 859, 873
Scent glands, 139, 151, 393, 400
Scheduled drugs, 84, 86, 816–817, 832
Scheduled feeding, 571–572, 574
Scheduling and appointments, 35–53
   appointment dos and don'ts, 42
   blocking times, 40, 41
   cancellations, 40
   clinical situation, 51
   competency skill, 52
   introduction, 35
   for multiple veterinarians, 40–42
   objectives, 35
   organization of the appointment book, 39
   policies and procedures, 43
   scheduling mishaps, 41, 42, 43
   types of appointments, 36–37
   types of appointment schedules, 35–37
   veterinary appointment book, 35–36
Schirmer tear test, 555, 560
School, 257, 279
Scissors, 868
   Mayo, 868, 872
   Metzenbaum, 868, 872
   suture removal, 860, 861, 873
Sclera, 554, 560
Screen, 786, 806
Scrotum, 201, 223, 520, 533
Scrub pack, 851–852, 873
Scruff, 162, 172
Scruff technique, 629, 631–632, 655
SDS (safety data sheets), 97–99, 115

Secondary bandage layer, 715, 743
Secondary containers, labeling, 99–100
Secondary response, 536, 542
Sedatives, 128, 151, 604, 655, 829, 832
   dogs, 128
Sediment, 514–515, 517, 749, 782
Seed diet, 182, 195
Seizure, 548–549, 551, 552
Selection, 428, 436
Selection guidelines, 313, 331
Semen, 521, 528, 533
Semen collection, 528, 533
Semen morphology, 521, 533
Semi Arch, 236
   rabbits, 227, 228
Semicircular canals, 556–557, 560
Seminal vesicles, 521, 533
Seminiferous vesicle, 521, 533
Sender, 68, 75
Senior diet, 124, 151, 567, 574
Senn retractor, 868, 873
Sensory neuron, 543, 551
Sensory system, 553–561
   clinical situation, 561
   ear, 556–557
   equilibrium, 558
   eye, 553–555
   hearing, 557–558
   introduction, 553
   objectives, 553
   reflex testing, 555–556
   smell, 558–559
   taste, 559
Sepsis, 324, 331
Septic, 529, 533
Serologic test kit, 753–755
   types, 754
Serpents, 238, 255
Serration, 402, 418
Serum, 477, 488
Sesamoid bones, 444
SG (specific gravity), 265, 279
Sharp operating scissors, 868, 873
Sharps, 93–94, 115
Sharps container, 94, 115
Shaving, 730–732, 743
Shearing, 385, 391
Sheep, breed identification and production management, 381–390
   anatomy, 382–383
   basic health care and maintenance, 385–386
   behavior, 384–385
   biology, 382
   breeding and reproduction, 386–387
   breeds, 382
   breed selection, 382–383
   classified by wool types, 382
   clinical situations, 542
   common diseases
     actinobacillosis, 388
     actinomycosis, 388
     blue tongue, 388
     enterotoxemia, 388
     foot rot, 387
     Johne's disease, 388
     lamb dysentery, 388
     mastitis, 387–388
     polyarthritis, 388
     soremouth, 388
   common parasites, 389
   equipment and housing needs, 385
   grooming, 385
   introduction, 381
   nutrition, 383–384
   objectives, 381

restraint and handling, 385
sheep production, 389–390
    confinement method, 390
    farm flock, 389
    lamb feeding, 390
    purebred flock, 389
    range bed, 389–390
vaccinations, 386
veterinary terminology, 381
Shipping fever. See Bovine respiratory disease complex (BRDC)
Shire, 341
Shoal, 266, 279
Shock, 74, 75, 486, 488
Short hair, 172
  cats, 157, 172
Shots, 536
Signalment, 658, 682
Silage, 563, 567
Silo, 319, 331
Silver nitrate stick, 139, 151, 735, 736, 743
Simple columnar epithelium, 431
Simple cuboidal epithelium, 431
Simple epithelial tissue, 430, 436
Simple fracture, 446–447, 452
Simple squamous epithelium, 431
Simple sugar, 564, 574
Simulated systems, 303, 308
Sinus arrhythmia, 484, 488
Sinuses, 452
Sinus rhythm, 483, 488
Sire, 314, 331, 334, 363
Sitting restraint, 618, 619–620, 655
Skeletal muscle, 432–433
Skills competency, converting pounds to kilograms, 468
Skin, 676–677
Skin lesions, 713
Skin mite, 145, 151
Skin turgor, 465, 475
Skull, 441, 452
Slander, 81, 86
Slicker brush, 135, 137, 152
Slip knot, 346, 363, 615
Slough, 322, 331
Slurry, 274, 279
Smell, 558–559
Smolt, 274, 279
Smooth endoplasmic reticulum, 425
Smooth muscle, 432
Snakes
  behavior, 247–248
  biology, 240
  breeds, 243
  nutrition, 246
  reproduction and breeding, 251–252
  restraint and handling, 249
SNAP test, 165, 168, 172, 538, 539, 542, 754, 782
Snare pole, 132, 152, 636, 655
Sneezing, 494, 498
Snubbing rope, 372, 378
SOAP (Subjective, Objective, Assessment, and Plan), 25, 31
Socialization, 152, 718–720
  birds, 184–185
  dogs, 126, 143
Soda lime, 858, 873
Sodium chloride (NaCl), 466, 475
Software, 43, 48
Solid manure system, 319, 331
Solitary, 218, 223
Somatic cells, 426–427, 436
Somatotropin, 502, 509
Sonogram, 803, 807
Soremouth, 388, 391

Sorrow, 74, 75
Sound, 314, 331
Sow, 365, 378
Spawn, 272, 279
Spawning, 272–273, 279
Spay
  cats, 170
  dogs, 148
Spay hook, 869, 873
Speaking, 71–72
Specialized payment plan, 23, 31
Species, terms, 8
Specific gravity (SG), 265, 279, 782
Specific-pathogen free (SPF), 372, 378
Spectrum, 104, 115
Speech communication, 71, 75
Spent hen, 401, 418
Sperm, 520–521, 533
SPF (specific-pathogen free), 372, 378
Sphincter muscles, 449
Sphygmomanometer, 843, 873
Spike port, 707, 743
Spinal column, 441, 452
Spinal cord, 441, 452, 544, 546, 551
Spinal nerve reflexes, 547, 551
Spinous process, 441
Spleen, 538
Splint, 447, 452
Splint bone, 444, 452
Spondylitis, 443
Spondylosis, 443
Sponge forceps, 866, 867, 873
Spooking, 344, 363
Spreadsheet, 46, 48
Springing heifer, 310, 331
Spring water, 272, 279
SQ. See Subcutaneous
SQ fluids, 706–707, 708, 743
Squamous, 430, 436
Square knot, 319, 331, 613–614, 655
Squeeze cage, 609, 610, 629, 655
Squeeze chute, 163, 172, 319, 331
Stag, 365, 378
Stage, 747, 782
Stallion, 333–334, 363
  reproductive tract, 522
Stanchion, 319, 331, 612, 613
Standing heat, 524, 533
Standing restraint, 618–619, 655
Stapes, 556, 560
Staple remover, 861, 873
Starches, 563, 574
Starter food, 316, 331
State Board examination, 55, 64
State board of veterinary medicine, 81, 86
Static equilibrium, 558, 560
Statutory laws, 83, 86
Steer, 310, 331
Sterile, 530, 533
Sterile techniques, 112
Sterilization, 101, 102–103, 115, 862
  in surgery, 862–863
Sterilizing, 101, 115
Sternal recumbency, 129–130, 152, 618, 620–621, 655
Sternum, 442, 452
Steroids, 500, 504, 509
Stethoscope, 482, 483, 488, 676
  esophageal, 843, 871
Stifle, 444
Stimulus, 600
Stirrups, 716, 743
Stock horse, 339, 363
Stomach systems, 456–461
Stool softeners, 467, 475
Stratified epithelial tissue, 430, 436

Stratified squamous epithelium, 431
Stretch technique, 162, 163, 172, 629, 632, 655
Strongyles, 582–583, 597
Structure of living things, 422–437
  cells, 422–428
    cell division, 426–428
    functions, 425
    phases of meiosis, 427–428
    structure, 423–425
  clinical situations, 437
  disease and injury, 434
  introduction, 422
  objectives, 422
  organs, 433
  tissues, 429–433
Stud, 334, 363
Styptic powder, 139, 152, 735, 743
Subcutaneous (SQ), 141, 152, 466, 475, 537, 542, 701, 702–703, 704, 743
  fluid therapy, 706–708
  injections, 702–704
Subjective, Objective, Assessment, and Plan (SOAP), 25, 31
Subjective information, 25, 31
Subluxate, 444, 452
Submissive, 602
Substrate, 239, 255
Suffixes, 3, 12
  common, 4, 6–7
  meaning pertaining to, 6
Suffolk, 341, 384
Sugar, 564, 574
Sulfonamides, 411
Supernatant, 749, 782
Supplements, 565, 568, 574
Supportive care, 578, 597
Surface-dwelling fish, 257, 279
Surface water runoff, 272, 279
Surgery
  birds, 193
  cats, 170–171
  cleaning surgical instruments, 111
  dogs, 147–148
  mice, 202–203
  rabbits, 235
  rats, 202–203
  reptiles and amphibians, 253
  surgical assisting procedures, 834–874
Surgery pack, 847–851, 873
Surgical appointment, 38, 48
Surgical assisting procedures, 834–874
  anesthesia induction, 840
  anesthesia log book, 835–836
  anesthesia machine, 857–859
  anesthesia planes, 857
  breathing systems, 859–860
  clinical situation, 874
  fluid therapy, 839–840
  gowning the surgeon and assistants, 855–856
  introduction, 835
  intubation procedure, 840–842
  objectives, 834
  patient monitors, 843–844
  patient surgical positioning, 856
  post-anesthetic care, 860
  postoperative patient care, 860–861
  pre-anesthetic patient care, 838–839
  sterilization techniques, 862–863
  surgical asepsis, 835
  surgical instruments, 863–870
    cleaning, 863–864
    hemostatic forceps, 867, 871
    intestinal forceps, 867–868, 871
    needle holders, 864–865
    retractors, 868–869, 872

Surgical assisting procedures (*Continued*)
    scalpel blade handle, 864
    scissors, 868
    spay hook, 869, 873
    tissue forceps, 865–867
    towel clamps, 864, 865
  surgical log book, 835–836
  surgical pack preparation, 847–855
  surgical preparation, 844–846
  surgical suite maintenance, 837–838
    ceiling sanitation, 837
    counter and shelf sanitation, 837
    equipment, 837–838
    floor sanitation, 837
    wall sanitation, 837
Surgical glue, 855, 873
Surgical instruments, 863–870
Surgical log book, 835–836, 873
Surgical margin, 844, 873
Surgical positions, 856
Surgical preparation, 844–846
Surgical scrub, 845, 846, 873
Surgical staples, 855, 873
Surgical suite, 60, 62, 64, 113
Suture reel, 854, 873
Suture removal, 860–862, 873
Suture removal scissors, 860, 861, 873
Sutures, 440, 452
Sutures and surgical blades, 854–855
  absorbable, 854, 870
  monofilament, 854, 872
  multifilament, 854, 872
  non-absorbable, 854, 872
  removal, 860–862
  sizes, 855
Swaged on needles, 854, 873
Swallow reflex, 860, 873
Swamp fever, 359, 363
Swan, 405, 418
Swelling, 678–679
Swim bladder, 258, 279
Swim bladder disease, 267, 279
Swine, 365, 378
Swine, breed identification and production management, 365–379
  anatomy, 366
  basic health care and maintenance, 372–374
    additives, 374
    castration, 373
    ear notching, 373
    tail docking, 373
    trimming needle teeth, 374
    vaccinations, 374–375
  basic training, 370
  behavior, 369–370
  biology, 366
  breeds, 366–367
  breed selection, 367–368
  clinical situation, 380
  common diseases, 375–376
    brucellosis, 376
    leptospirosis, 376
    pneumonia, 376
    porcine stress syndrome, 376
    pseudo rabies, 376
    swine dysentery, 376
  common parasites, 376–377
  equipment and housing needs, 370–371
  grooming, 372
  introduction, 365
  nutrition, 368–369
  objectives, 365
  production industry, 377
  production methods and systems, 377–378
  reproduction and breeding, 375
  restraint and handling, 371–372
  veterinary terminology, 365
Swine dysentery, 376, 378
Swordtails, 260
Sympathetic system, 545, 551
Sympathy, 71, 75
Synapse, 543, 544, 551
Synovial fluid, 440
Synovial joint, 440, 452
Synovial membrane, 440
Synthetic seawater, 264, 279
Syringe, 701–702
  filling, 701–702
  labeling, 702
Systemic circulation, 480–481, 488
Systole, 482, 488

T

T_3, 503, 509
T_4, 503, 509
Tabletop technique, 793, 807
Tachycardia, 484, 488, 675, 682
Tachypnea, 668
Tacky, 465, 475
Tactfulness, 71, 75
Tadpoles, 253, 255
Taenia pisiformis, life cycle of, 585
Taenia spp., 584, 597
Tail, 268, 279
Tail docking, 148, 152
  dogs, 148, 152
  sheep, 386,
  swine, 373, 378
Tail flagging, 524, 533
Tail jack restraint, 651
Tail switch restraint, 319, 331, 652
Tail tie technique, 648
Tail tying, 347, 364
Talons, 293, 298
Tamarin, 301
Tank, 262, 270–271, 279
Tape, 715
Tape muzzle, 608
Tapeworm, 147, 152, 580, 583–584, 597
  life cycle, 585
  poultry, 412
Tap water, 264, 279
Target tissue, 824, 832
Tarsal, 444
Tarsus, 444, 452
Tartar scraper, 869
Taste, 559
Tattoos, 322, 331
Taurine, 158, 172
Teats, 314, 331
Technique chart, 792–793, 807
Teeth, 455–456
  anatomy, 457
  dental care, 694–699
    daily care and brushing, 695–698
    dental examination, 671, 673–674
    dental tools, 695, 696, 697
  dentition of adult domestic animals, 457
  dogs, 140
  horse, 694
  identification, 692, 693
Telephone skills, 73
Telophase, 427–428, 436
Temperature, 545
  assessment, 663–665
  fever, 664, 681
  normal body, 537
  normal core body, 665
  rectal, 664, 665, 679
Temperature, pulse, and respiration (TPR), 663, 682
Temperature assessment, 663–665
Temporary pasture, 567, 574
Tendons, 432, 436, 441, 452
Tennessee walking horse, 337, 340
Terrapins, 238, 255
Terrariums, 239, 255
Terrestrial, 238, 255
Territorial aggression, 603, 655
Territory, 295, 298
Tertiary bandage layer, 715, 743
Testicles, 201, 223, 520–521, 529, 533
Testosterone, 504, 509, 521, 533
Tetanus, 143, 152, 360, 364
Tetras, 260
Thalamus, 544, 545, 551
Thermometers, 264, 279
  digital, 664
  rectal, 664
Thermostat, 264, 279
Thiamine, 158, 172
Thoracic, 442, 453
Thoracic cavity, 492–494, 498, 675
Thrombocyte, 478, 488
Thymus, 504, 509
Thyroidectomy, 505, 509
Thyroid gland, 502–503, 509
Thyroid stimulating hormone (TSH), 502–503, 509
Thyroxine, 502–503, 509
Tibia, 444, 453
Tick paralysis, 587
Ticks, 144, 152, 587
  beef and dairy cattle, 325
  species and disease, 587
  topical medications, 821
Tidal volume, 858, 873
Tissue, 429–433, 436
Tissue adhesive, 170, 172
Tissue forceps, 865–867, 873
Toe pinch response, 677, 678, 682
Toner, 44, 48
Tongue, 456, 459, 475
Tonometer, 556, 560
Tools, 659–660
Tooth abscess, 673, 682
Tooth float, 697, 698
Topical, 144, 152, 207, 223, 266, 279, 821, 832
Topical medications, 821–822
Total herd, 326, 331
Total mix ration (TMR), 315, 331
Total protein (TP), 764, 782
Towel clamps, 864, 865, 873
Towels, 609
  for grooming, 728
  laparotomy, 847, 872
Toxascaris leonina, 579, 580, 597
Toxic, 264, 279
Toxicity, 515
Toxocara canis, 579, 580, 581, 597
Toxocara cati, 579, 597
Toxoplasma gondii, 592, 597
Toxoplasmosis, 170, 172, 592
  life cycle, 593
Toys, 695
TP (total protein procedure), 764, 782
TPR (temperature, pulse, and respiration), 663, 682
Trace minerals, 565, 574
Trachea, 491, 498
Tract, 544, 551
Tragus, 820, 832
Training, 152
  cats, 160
  dogs, 125

Index

Tranquilizers, 128, 151, 295, 298, 604, 655, 829, 832
Transducer, 802, 807
Transitional epithelium, 430, 432, 436
Transmission, 145
Transverse colon, 459–460, 475
Transverse muscles, 449
Transverse process, 441–442
Trapping, 291, 298
Travel sheet, 20–21, 24, 31
Treatment area, 60, 64
Treatment board, 11
Trematodes, 584
Triage, 659, 682
Trichobezoar, 222, 223
Trichuris vulpis, 580, 581, 597
Trimester, 566–567, 574
Trimming, 730–732, 743
Trocar, 468, 470, 475
Tropical fish, 257, 279
Trot, 354, 364
Trough, 372, 378
Trunk, 268, 279
Trypsin, 461, 475
TSH (thyroid stimulating hormone), 502–503, 509
Tumor, 426, 429, 436
Turkeys, 404, 418
Turtles
   behavior, 248
   biology, 241–242
   breeds, 245
   nutrition, 247
   reproduction and breeding, 252–253
   restraint and handling, 250
T wave, 483, 488
Twitch, 346–347, 364
Twitch restraint, 649
Tympanic bulla, 556, 560
Tympanic membrane, 556, 557, 560, 733, 743
Tyzzer's disease, 234, 236

U

Udders, 314, 331
Ulna, 443, 453
Ultrasonic cleaner, 864, 873
Ultrasound, 103, 115, 355, 364, 807
   diagnostics, 802–803
Unethical, 81, 86
Unilateral cryptorchid, 529, 533
Univalve, 269, 279
University of Pennsylvania PennHIP program, 445
Upland game birds, 293, 298
Upper GI, 790, 807
Upper respiratory infection (URI), 494
Urates, 251, 255
Urbanization, 292, 298
Urea, 511, 517
Uremia, 515, 517
Ureters, 513, 517
Urethra, 201, 223, 513, 519, 533
URI (upper respiratory infection), 494
Urinalysis, 513, 514, 517, 769, 782
Urinary bladder, 513, 514
Urinary blockage, 514
Urinary catheter maintenance, 713
Urinary incontinence, 513
Urinary tract infections (UTI), 513
Urine
   collection, 769–770
   crystals, cells, and casts, 773
   gross examination, 770–771
   sample, 679, 769–772
   specific gravity, 771

Urogenital system, 678–679, 682
Urolith, 514, 517
USDA (U.S. Department of Agriculture), 301, 305, 306
U.S. Department of Agriculture (USDA), 301, 305, 306
U.S. Fish and Wildlife Department, 292
Uterine horns, 519–520, 533
Uterus, 519–520, 533
UTI (urinary tract infections), 513

V

Vaccination, 536, 542
Vaccination routes, 537
Vaccine program, 152, 699
   beef and dairy cattle, 322–323
   birds, 190
   cats, 166
   chinchillas, 222
   dogs, 140–141
   equines, 354–355
   ferrets, 216
   gerbils, 212
   goat, 398
   guinea pigs, 210
   hamsters, 206
   hedgehogs, 219
   for kennel cough (bordetella), 494–495
   for maintenance of good animal health, 593–594
   mice and rats, 201
   mixing vaccines, 700–701
   poultry, 410
   preparation, 699–701
   rabbits, 233
   rabies, 549
   sheep, 386
   storage, 699
   swine, 374–375
Vaccines, 536, 537
   injection routes, 594
   storage and use, 537
Vacutainer tube, 755, 756, 782
Vagina, 519, 533
Vagus nerve, 545, 551
Valve, 269, 279
Vaporizer, 857, 873
Vas deferens, 521, 533
Vasoconstriction, 482, 488
Vasodilatation, 482, 488
Vasodilators, 826, 832
Vat, 262, 270–271, 279
VCPR (veterinary-client-patient relationship), 16, 31
VD (ventral-dorsal recumbency), 640, 655, 788, 789, 790, 807
Vector, 146, 152, 579, 597
VEE (Venezuelan Equine Encephalitis), 593
Vein, 480, 488
Velvet, 284, 298
Vena cava, 481, 488
Venezuelan Equine Encephalitis (VEE), 593
Venipuncture, 623, 655
Venison, 285, 298
Venomous, 239, 255
Vent, 176, 195, 239, 255, 465, 475, 520, 533
Ventral-dorsal (V-D) recumbency, 640, 655, 788, 789, 790, 807
Ventral root, 545, 551
Ventricular fibrillation, 486, 488
Verbal communication, 68–69, 75
Verbal restraint, 128, 152
Vertebrae, 441, 453, 544
Vertebral foramen, 442

Vertebrates, 268, 279
Vestibule, 556, 560
Veterinarian, 54–55, 64
Veterinary animal production
   aquaculture, identification and production management, 267–275
   avian breed identification and production management, 174–196
   beef and dairy cattle, breed identification and production management, 310–332
   cat breed identification and production management, 154–173
   dog breed identification and production management, 118–153
   equine breed identification and production management. see Equines
   goat, breed identification and production management, 393–400
   laboratory and animal research production management, 300–309
   ornamental fish identification and production management, 256–281
   pocket pet health and production management, 197–224
   poultry breed identification and production management, 401–418
   rabbit identification and production management, 225–237
   reptile and amphibian breed identification and production management, 238–255
   sheep, breed identification and production management, 381–390
   swine, breed identification and production management, 365–379
   wildlife management and rehabilitation, 282–299
   zoo, exotic and wildlife identification and production management, 282–299
Veterinary assistant, 55–56, 58, 64
Veterinary assisting procedures, 684–745
   administering injections, 701–706
   common hospital procedure, 690–694
   clinical situations, 745
   dental care, 694–699
   evaluating emergency situations, 689–690
   euthanasia, 720–723
   fluid administration, 706–708
   general nursing care, 689
   grooming skills in veterinary facility, 723–741
   hospitalized patient, 685–689
   introduction, 684–685
   objectives, 684
   pain assessment, 689
   socialization and exercise of patients, 718–720
Veterinary careers, 54–67
   clinical situation, 66
   competency skills, 66–67
   introduction, 54
   objectives, 54
   professional dress and appearance, 58–59
   veterinary facility, 59–63
   veterinary team, 54–58
Veterinary-client-patient relationship (VCPR), 16, 31
Veterinary hospital manager, 56–57, 65
Veterinary Practice Act, 55, 65, 82–83, 86
Veterinary practice management
   common procedures and time estimates, 39
   communication and client relations, 68–80
   ethics and legal issues, 81–89
   laws and veterinary practice acts, 82–83
   medical records, 16–34
   medical terminology and abbreviations, 2–15
   scheduling and appointments, 35–53
Veterinary referral practice, 59, 65
Veterinary safety and aseptic technique, 90–116

944 Index

aseptic techniques, 112–113
    surgical suite, 113
chemicals and cleaners, 104–107
clinical situation, 116
exam room sanitation, 110–112
housekeeping and general cleaning, 107–110
introduction, 90
isolation ward, 113
methods of sanitation, 101–104
    filtration, 103–104
    gas sterilization, 103
    physical cleaning, 101–102
    sterilization, 102–103
objectives, 90
OSHA guidelines and regulations, 97–101
safety equipment, 99
in the veterinary facility, 90–97
    biological hazards, 95–96
    chemical hazards, 95
    fire and safety plans, 91
    instruments and equipment, 93–94
    personal protective equipment, 91–93
    physical hazards, 95
    rabies suspects, 96–97
    zoonotic hazards, 96
Veterinary technician, 55, 65
    specialties, 56
Veterinary technologist, 55, 65
Veterinary terminology. *See also* Abbreviations; Medical terminology and abbreviations
    aquatic-specific, 268
    avian-specific, 174
    beef and dairy cattle, 310–311
    canine-specific, 119
    chinchillas, 220
    diagnostic terms, 790
    equines, 333–335
    feline-specific, 155
    ferrets, 213
    gerbils, 211
    goat, 393
    guinea pigs, 207
    hedgehogs, 217
    mice and rats, 198
    ornamental fish, 257
    pharmacy terms and abbreviations, 9
    poultry-specific, 401–402
    rabbits, 225
    reptiles and amphibians, 238–239
    sheep, 381
    swine, 365
    terms used in blood work evaluation, 479
    x-ray machine terminology, 787–788
    x-ray positioning terms, 788–789
    x-ray terms, 787
Vietnamese Potbellied Pig, 367
Viral disease, 143, 152, 578, 597
Virucidal, 828, 832
Viruses, 96, 115, 578
Visceral larval migrans, 147, 152
Viscous, 702, 743
Vital signs, 545, 551, 663, 682
Vitamin A, 565
Vitamin B12, 565
Vitamin C, 208, 223, 565
Vitamin D, 565
Vitamin E, 565
Vitamin K, 565
Vitamins, 565, 574
Vitreous humor, 553, 554, 560
V-notcher, 373, 378
Vocalization, avian, 185

Voided, 769, 782
Voluntary reflex, 545, 551
Vomiting, 465, 475
Vulva, 513, 517, 519, 533

W

Waiting room, 59, 65
Walk, 354, 364
Walk-in appointment, 38, 48
Want list, 46, 48
Warm-blooded animals, 337, 364
Warm housing, 317–318, 331
Warm water, 272, 279
Water, 563, 574
    distilled, 862, 871
Water facility, 269, 279
Waterfowl, 293, 298
Waterless shampoo, 164, 165, 172
Water soluble, 565, 574
Wattle, 402, 418
Wave schedule, 36, 48
WBC (white blood cells), 478, 488
Weaned, 122, 152
WEE (Western Equine Encephalitis), 593
Weight
    assessment, 660–663
    ideal, 568, 573
    management, 568–569
Weight tape, 661, 663, 682
Weitlaner retractor, 868, 869, 873
Welding gloves, 164–165, 172, 610, 611
Well water, 264, 279
Western Equine Encephalitis (WEE), 593
West Nile virus, 549, 551, 578
Wether, 381, 391
Wet tail, 206, 223
Wheel barrow technique, 682
Wheezing, 494, 498
Whipworm, 147, 152, 580, 581, 597
White blood cells (WBC), 478, 488
White spot, 267, 279
Whole blood, 756–757, 782
Wildlife, 282, 298
Wildlife biologist, 295, 298
Wildlife management, 282, 298
Wildlife management and rehabilitation, 282–299
    capture and restraint of wildlife, 295
    classes of wildlife, 292–294
    control of wildlife, 292
    importance, 290–292
    role of veterinary medicine in wildlife management, 292
    wildlife as a food source, 290–291
    wildlife as a resource, 291
    wildlife as entertainment, 291
    wildlife health and management practices, 294–295
    wildlife rehabilitation, 295–296
Wildlife rehabilitation center, 295, 298
Wildlife rehabilitator, 296, 298
Wing trim, 189, 195
Winking, 524, 533
Withdrawal time, 374, 378
Wolf teeth, 351, 364
Wooden tongue. *See* Actinomycosis
Wool, 383, 391
Word-processing program, 44, 48
Words per minute, 46, 48
Work diets, 567, 574

Wounds, cleaning and care, 713, 714
Written communication, 70, 75

X

X-ray, 784, 807
X-ray film
    automatic processing, 799, 800–801
    cassettes, 796–797
    developing film, 795–800
    digital radiology, 799–800
    filing, 800–801
    hangers, 798
    identification, 794–795
    manual developing, 795, 798
X-ray log, 786, 787, 807
X-ray machine
    setting procedures, 793–794
    terminology, 787–788
X-ray positioning terms, 788–789
X-ray terms, 787
X-ray tube, 784, 785, 793, 807

Y

Y connector, 859, 873
Yearlings, 257, 279
Yolk, 414, 418
Yorkshire pig, 367

Z

Zoo, exotic and wildlife identification and production management, 282–299
    businesses, 286
    capture and restraint, 295
    care and management, 289–290
    classes of wildlife, 292–294
    control of wildlife, 292
    health and management practices, 294–295
    importance of wildlife
        entertainment, 291
        food source, 290–291
        most endangered in the United States, 291–292
        resource, 291
    performance animals, 286–288
    purpose and uses
        alligators, 285–286
        alpaca, 283–284
        bison, 284
        deer, 285
        dehorning, 322
        elk, 284–285
        llama, 283
        ostrich, 285
        rare breeds, 286
    rehabilitation, 295–296
    veterinary medicine, 292
Zookeeper, 287, 298
Zoological garden, 286
Zoology, 287, 298
Zoonosis, 592–593, 597
Zoonotic, 143, 152
Zoonotic hazards, 96, 115
Zoos, 286–288, 298